# Lecture Notes in Electrical Engineering

Volume 528

***Lecture Notes in Electrical Engineering (LNEE)*** is a book series which reports the latest research and developments in Electrical Engineering, namely:

- Communication, Networks, and Information Theory
- Computer Engineering
- Signal, Image, Speech and Information Processing
- Circuits and Systems
- Bioengineering
- Engineering

The audience for the books in LNEE consists of advanced level students, researchers, and industry professionals working at the forefront of their fields. Much like Springer's other Lecture Notes series, LNEE will be distributed through Springer's print and electronic publishing channels.

More information about this series at http://www.springer.com/series/7818

Yingmin Jia · Junping Du · Weicun Zhang
Editors

# Proceedings of 2018 Chinese Intelligent Systems Conference

Volume I

Springer

*Editors*
Yingmin Jia
Beihang University
Beijing, China

Junping Du
Beijing University of Posts
and Telecommunications
Beijing, China

Weicun Zhang
University of Science
and Technology Beijing
Beijing, China

ISSN 1876-1100 ISSN 1876-1119 (electronic)
Lecture Notes in Electrical Engineering
ISBN 978-981-13-4758-0 ISBN 978-981-13-2288-4 (eBook)
https://doi.org/10.1007/978-981-13-2288-4



This Springer imprint is published by the registered company Springer Nature Singapore Pte Ltd.
The registered company address is: 152 Beach Road, #21-01/04 Gateway East, Singapore 189721, Singapore

# Contents

# Research on Cosine-Type Non-uniform Air Gap Structure Based on Finite Element Analysis

**Yuanyuan Yang, Bulai Wang, Xiutao Ji, Xiangsheng Liu and Qiangqiang Xu**

**Abstract** That the air gap magnetic field of PMSM contains a large number of harmonics, aiming at the problem, this paper presents a cosine-type non-uniform air gap motor structure. Taking a 22 kW 3000 rpm permanent magnet synchronous motor as the researching object, a motor model was established by finite element analysis software to calculate the air gap flux density, no-load back EMF and output torque distribution curve respectively. The comparing results prove that the cosine type non-uniform air gap structure can effectively reduce the no-load back EMF harmonic components, reduce the torque ripple and improve the motor efficiency.

**Keywords** Cosine-type non-uniform air gap · MagneForce
No-load back EMF · Torque ripple

## 1 Introduction

The requirements of PMSM for automotive applications such as electric vehicles, electric power ships and other industries include high efficiency, high power density, low vibration and acoustic noise. However, PMSMs have the problem with a large amount of harmonic components. Because its permanent magnetic flux density approximates the trapezoidal wave distribution, so that the air gap magnetic field contains a large amount of harmonic components, which increases the harmonic current of the motor and the additional harmonic loss, and the harmonic magnetic field produces torque fluctuations, causing vibration and noise. The quality of permanent magnet synchronous motor is affected seriously.

To this end, a non-uniform magnetic steel slot motor structure is proposed in [1], which reducing the air-gap magnetic density harmonic content, but the back-EMF Wave amplitude is decreased due to non-uniform magnetic steel tank internal air

Y. Yang (✉) · B. Wang · X. Ji · X. Liu · Q. Xu
School of Electrical and Electronic Engineering,
Shanghai Institute of Technology, Shanghai 201418, China
e-mail: yangyuanyuanfly@163.com

 
Y. Jia et al. (eds.), *Proceedings of 2018 Chinese Intelligent Systems Conference*,
Lecture Notes in Electrical Engineering 528,
https://doi.org/10.1007/978-981-13-2288-4_1

consumption part of the magnetomotive force. An eccentric air gap motor structure is proposed in [2], which also reducing the air-gap magnetic density harmonic content and getting the air gap magnetic flux density harmonic content with eccentricity curve. An unequal thick magnet design method is proposed in [3], it is high in cost while optimizing the empty air gap magnetic density wave shape. A cosine-type non-uniform air-gap motor structure is designed in this paper from the view of improving the air-tight flux density sinusoid and reducing the torque ripple. Several performances of the motor are evaluated based on air-Airgap Cosine Amplitude from obtaining the expression of the ideal sinusoidal [4].

## 2 Analysis of Cosine Non-uniform Air Gap Structure

Assuming that the total permanent magnetomotive force F provided by each permanent magnet is of constant value, the magnetomotive force consumed on the air gap between the fixed rotors is F1, the magnetomotive force consumed in the stator core is F2, then

$$F = F_1 + F_2$$

Taking into account the stator core material permeability is far greater than the air permeability, to facilitate the analysis, ignoring the core of the magnetomotive force consumption, that

$$F_1 = F \tag{1}$$

Figure 1 shows the cosine type non-uniform air-gap rotor structure, setting the rotor pole center line where the arc is 0, the gap length minimum $\delta_{min}$, air gap flux density maximum $B_{\delta max}$. The length of the air gap at the arc x position from the magnetic pole center is δ(x), the air gap is $B_\delta$(x), and the length of the air gap is $\delta_{max}$ at arc ± π/2 from the center of the pole. The air gap flux density is Minimum $B_{\delta min}$.

The proposed air gap length is:

$$\delta(x) = 0.5 + A - A \cdot \cos(2x) \tag{2}$$

In the formula, A is the cosine of the air gap amplitude in mm, assuming the air gap in the permeability of μ0:

$$\begin{cases} B_\delta(x) = \frac{\mu_0 F}{k_\delta \cdot \delta(x)} \\ B_{\delta\max} = \frac{\mu_0 F}{k_\delta \cdot \delta_{\min}} \\ B_{\delta\min} = \frac{\mu_0 F}{k_\delta \cdot \delta_{\max}} \end{cases} \tag{3}$$

$k_\delta$ is the air gap coefficient. The formula (2) into Eq. (3) available:

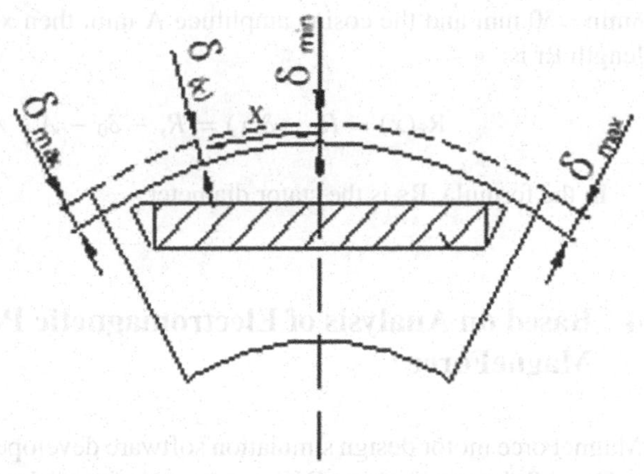


**Fig. 1** Cosine-type non-uniform air gap rotor structure

$$
\begin{aligned}
B_\delta(x) &= \frac{\mu_0 F}{k_\delta \cdot \delta(x)} \\
&= \frac{\mu_0 F}{k_\delta \cdot (0.5 + A)} \cdot \frac{1}{1 - \frac{A}{A+0.5}\cos(2x)}
\end{aligned}
\tag{4}
$$

Taylor expansion of the function $\frac{1}{1+z}(|z| \leq 1)$, available from mathematical analysis

$$
\frac{1}{1+z} = 1 - z + z^2 + \cdots + (-1)^n z^n + \cdots, (|z| \leq 1)
$$

Analysis (4), the second factor to meet Taylor expansion conditions. Due to the coefficient $\frac{A}{A+0.5}$ (Due to the size of the rotor itself structure size, A range of 0–0.4 mm) of the second and above the smaller value of the second-order, to facilitate the calculation, will be ignored, then Eq. (4) can be further reduced to:

$$
B_\delta(x) = \frac{\mu_0 F}{k_\delta \cdot (0.5 + A)} \cdot \left(1 + \frac{A}{A + 0.5}\cos(2x)\right) \tag{5}
$$

Analysis (5) shows that, the air gap flux density could get a good sine when the air gap length according to the type (2) law changes.

## 3 Cosine Non-uniform Air Gap Rotor Size Determination

Taking a 22 kW 3000 rpm permanent magnet synchronous motor as the research object, the motor has a uniform air gap and a length of δ0 mm. To ensure the electrical performance of the motor, the minimum cosine-type non-uniform air gap length δmin = δ0 mm and the cosine amplitude A mm, then x arc at the outer rotor radial length Rr is:

$$R_r(x) = R_s - \delta(x) = R_s - \delta_0 - A + A \cdot \cos(2x) \tag{6}$$

In the formula, Rs is the stator diameter.

## 4 Based on Analysis of Electromagnetic Properties MagneForce

MagneForce motor design simulation software developed by the United States MagneForce Software Systems INC company, is focused on the rapid design of various types of high-performance motor products, professional software. MagneForce software uses finite element (FEA) electromagnetic field simulation technology based on time-stepping to design and simulate the electromagnetic parameters of the motor [5]. At the same time MagneForce internally couples the spice-based driving circuit system, which can accurately measure the electromagnetic field and Control system performance analysis and evaluation.

The air gap structure is improved based on a 22 kW 3000 rpm PMSM with average air gap in this paper, compared with the original motor by using finite element analysis. Table 1 shows that 22 kW 3000 rpm permanent magnet synchronous motor main design data.

MagneForce realizes the rapid establishment of the finite element electric motor model by using the method of parameterized model library. The model library contains rich stator and rotor models. Meanwhile, the software also supports the customers to import dxf format model files to realize the modeling of the punch slice. The dxf format is used to import the MagneForce motor model In view of the special structure of the rotor punching sheet, as shown in Fig. 2.

**Table 1** 22 kW 3000 rpm permanent magnet synchronous motor main design data

| Project | Parameters | Project | Parameters |
|---|---|---|---|
| Rated power (kW) | 22 | Rated voltage (V) | 380 |
| Outer diameter of the stator (mm) | 260 | Inner diameter of the stator (mm) | 152 |
| Rated current (A) | 35 | Stator chute (slot) | 1 |
| Rated speed (rpm) | 3000 | Winding | Double winding |
| Magnet size (mm) | 63 * 6 | Wire gauge/turns | 3–0.85/33 |
| Magnetic steel grades | N38UH | Unilateral air gap length (mm) | 0.7 |
| Iron core length (mm) | 90 | Silicon steel grades | B50A470 |

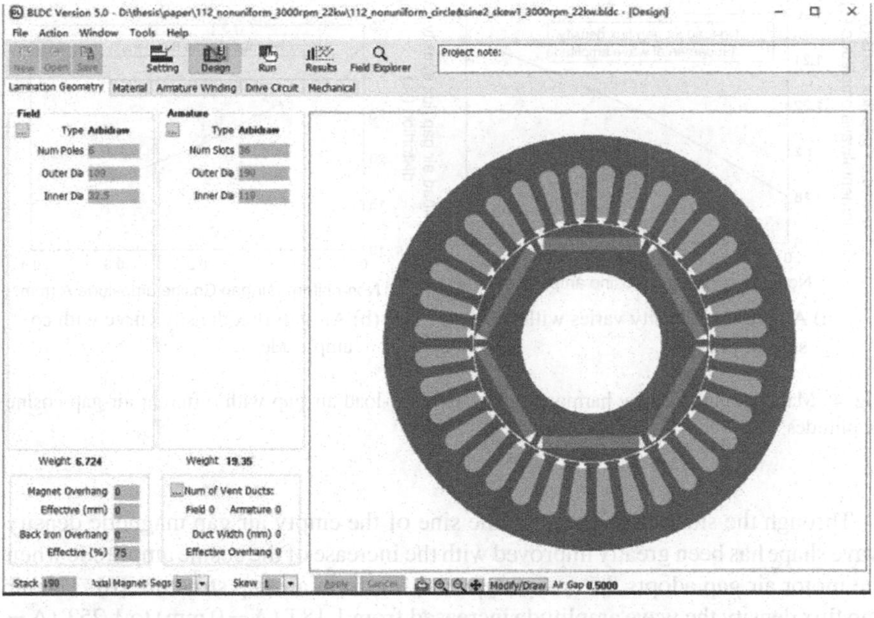


**Fig. 2** 22 kW 3000 pm MagneForce model permanent magnet synchronous motor

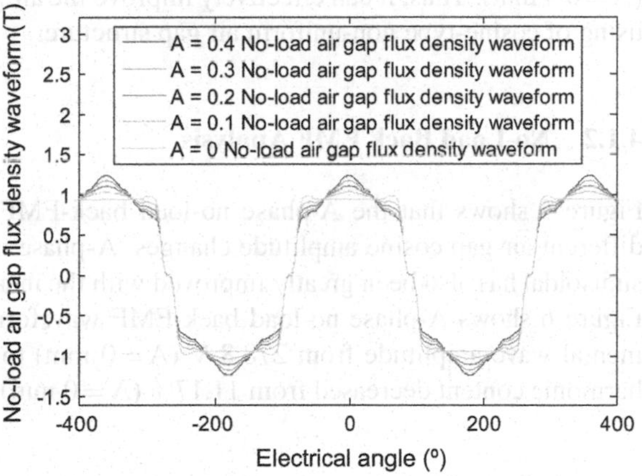


**Fig. 3** Curves of air gap flux density versus of electric angle under different cosine amplitudes 1.5 analysis of experimental results

## *4.1 No-Load Analysis*

### 4.1.1 No-Load Air Gap Flux Density Analysis

MagneForce software using a non-uniform air gap in the motor air gap and different uniform amplitude cosine of empty air gap magnetic density analysis, the results shown in Figs. 3 and 4.

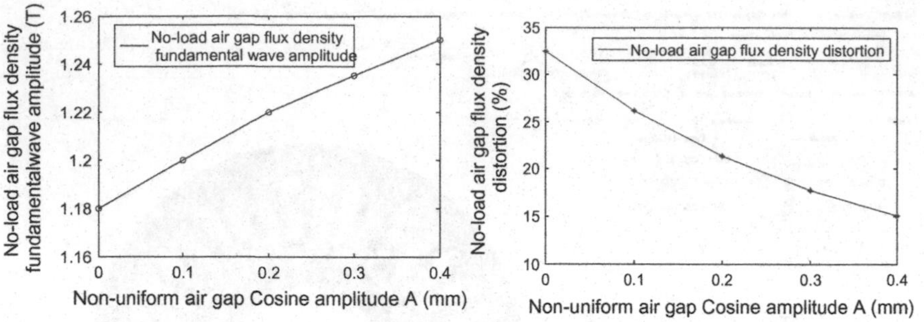

(a) Airgap flux density varies with cosinet sine amplitude

(b) Air gap flux density curve with co amplitude

**Fig. 4** Magnetic flux density harmonic analysis of no-load air gap with different air-gap cosine amplitudes

Through the simulation analysis, the sine of the empty air gap magnetic density wave shape has been greatly improved with the increase of the cosine amplitude when the motor air gap adopts the cosine-type non-uniform air gap structure [6]. The air gap flux density the wave amplitude increased from 1.18T (A = 0 mm) to 1.25T (A = 0.4 mm), while the harmonic content decreased from 32.48% (A = 0 mm) to 15.04% (A = 0.4 mm). Thus, it can effectively improve the air gap flux density waveform by using of cosine-type non-uniform air gap structure.

### 4.1.2 No-Load Back EMF Analysis

Figure 5 shows that the A-phase no-load back-EMF waveform change law with different air-gap cosine amplitude changes. A-phase no-load back-EMF waveform sinusoidal has also been greatly improved with the increase of the cosine amplitude. Figure 6 shows A-phase no-load back-EMF waveform and distortion rate, fundamental wave amplitude from 278.8 V (A = 0 mm) to 282.9 V (A = 0.4 mm), The harmonic content decreased from 11.17% (A = 0 mm) to 6.81% (A = 0.4 mm).

### 4.1.3 Cogging Torque Analysis

The permanent magnet synchronous motor cogging torque expression is:

$$T_{cog} = \frac{\pi z l}{4\mu_0} \cdot (R_2^2 - R_1^2) \sum_{n=1}^{\infty} n G_n B_{r\frac{nz}{2p}} \sin nz\alpha \tag{7}$$

In the formula, l is armature core axial length, R1 and R2 are the armature outer radius and the stator yoke inner radius respectively, and n is an integer such that nz/2p is a positive integer.

**Fig. 5** A-phase back-EMF with air gap cosine amplitude curve

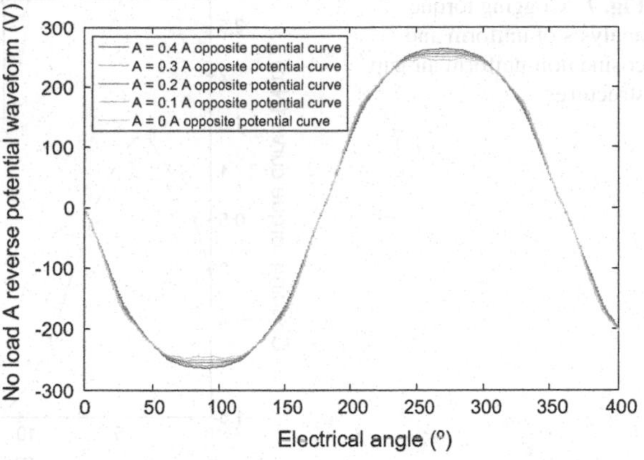


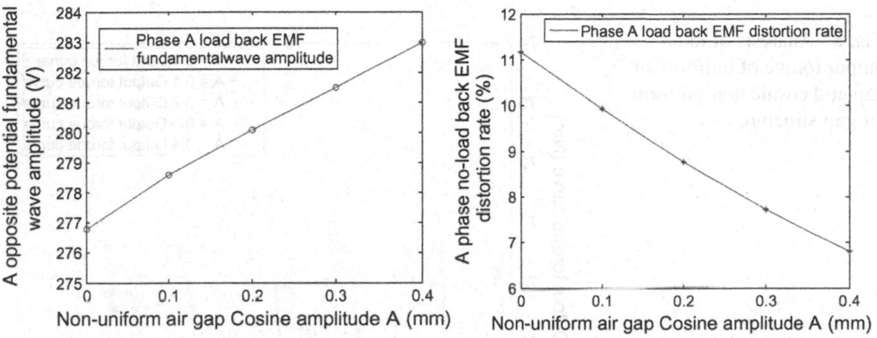


(a) A opposite potential fundamental wave amplitude with cosine amplitude curve

(b) A reverse-polarity potential distortion rate curve with the cosine amplitude

**Fig. 6** A-phase back-EMF harmonic analysis with different air-gap cosine amplitudes

From Eq. (7), when R1, R2, l and other parameters are constant, the cogging torque amplitude is related to the size of the $B_r \frac{nz}{2p}$, nz/2p subharmonic of the no-load air gap flux density can affect the motor Cogging torque Tcog amplitude, so reducing the no-load air gap magnetic flux density n/2p harmonic content is needed only, it can weaken cogging torque [7]. In this paper, a 6 pole 36 slot built-in permanent magnet synchronous motor is used. Thus, the amplitude of the motor Tcog could be affected by load harmonic component of 6 N the magnetic flux density.

From Fig. 7, it can be analyzed that when the motor adopts the cosine-type non-uniform air gap structure, the amplitude of the motor cogging torque is significantly reduced because the harmonics of air gap flux density Substantial decline due to components. It can be seen that the cosine-type non-uniform air gap structure helps to reduce the cogging torque amplitude, thereby increasing the stability of the motor output torque and improving the motor performance [8].

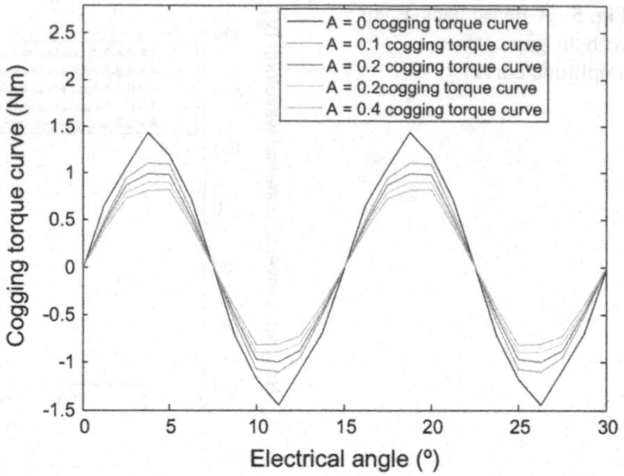


**Fig. 7** Cogging torque analysis of uniform and cosine non-uniform air gap structures

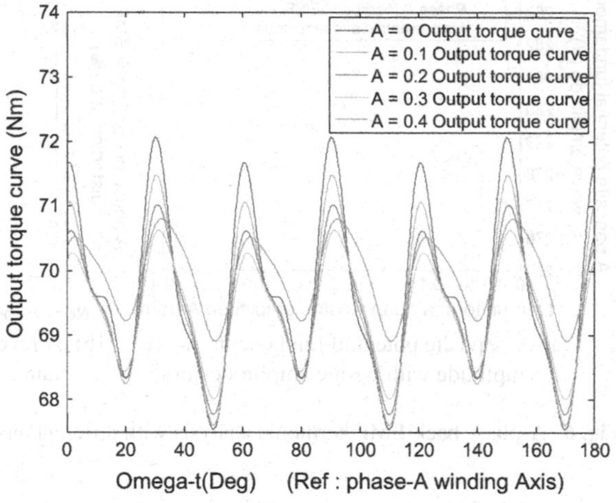


**Fig. 8** Analysis of rated output torque of uniform air gap and cosine non-uniform air gap structure

### 4.1.4 Output Torque Analysis

Set the motor full load condition in MagneForce software as shown in Table 2.

Motor access ideal three-phase AC voltage source, the frequency of 150 Hz, line voltage 380 V RMS, set the motor output torque of 70 N m, full load conditions set the stator temperature are 75 °C, the use of uniform air gap and cosine type When the uniform air gap structure of the motor output torque waveform shown in Fig. 8.

**Table 2** 22 kW 3000 rpm permanent magnet synchronous motor full load condition parameter settings

| Hz | Rpm | Vll_rms | Torque | T Arm | T RotPm |
|---|---|---|---|---|---|
| 150 | 3000 | 380 V | 70 N m | 75 °C | 75 °C |

As can be seen from Fig. 8, when the motor adopts a cosine-type non-uniform air gap structure, the output torque ripple is obviously lower than that of a motor with a uniform air gap structure [9]. The motor operates more smoothly. The motor vibration and noise are reduced effectively. In particular, the power angle of the motor with uniform air-gap structure and the motor with cosine-type non-uniform air gap structure are respectively 44° electrical angle and 43° electrical angle, 42.5° electrical angle and 42° electrical angle when rated condition simulation. 41.5° electrical angle. It can be seen that the non-uniform air-gap structure changes the electrical performance of the motor, causing the change of the electrical angle characteristics of the motor.

## 5 Conclusions

In this paper, a cosine-type non-uniform air-gap motor structure is proposed. Compared with the average air-gap structure, the sinusoidal properties of air-gap flux density and back EMF waveforms have been greatly improved in no-load operation. The harmonic content is obviously reduced; the fluctuation of torque during full load operation obviously drops, which increases the stability of the motor during operation. Thus, the cosine-type non-uniform air gap structure can significantly improve the quality of permanent magnet synchronous motor.

## References


1. B. Wang, Z. Wu, Y. Wang et al., Low harmonic high efficiency permanent magnet synchronous machines based on non-uniform magnet slot, in *2011 International Conference on Electrical and Control Engineering, ICECE 2011—Proceedings*, pp. 233–236
2. T.M. Jahns, W.L. Soong, Torque ripple reduction in interior permanent magnet synchronous machine using the principle of mutual harmonics exclusion, in *Industry Applications Conference, 2007. 42nd IAS Annual Meeting. Conference Record of the 2007*, 23–27 Sept 2007 (IEEE, 2007), pp. 558–565
3. J. Youhua, W. Hongwei, Optimization of air-gap flux density of permanent magnet synchronous motor based on Ansoft. Micro-motor **46**(12), 84–87 (2013)
4. E. Carraro, N. Bianchi, Design and comparison of interior permanent magnet synchronous motors with non-uniform air gap and conventional rotor for electric vehicle applications. IET Electr. Power Appl. **8**(6), 240–249 (2014)
5. L. Jia, Z. Lin, W. Baocheng et al., Design and finite element analysis of a new rotor structure of permanent magnet synchronous motor. Microelectronics **46**(1):21–23 (2013)
6. X.Y. Lei, L.Q. Zhan, W. Tao, Permanent magnet synchronous motor no-load air gap flux density waveform optimization. J. Southwest Jiaotong Univ. **44**(4), 513–516 (2009)
7. C. Dang, W. Zhou, L. Yin et al., Analysis and reducing methods of cogging torque on permanent magnet AC servo motor, in *17th International Conference on Electrical Machines and Systems (ICEMS)* (IEEE, 2014), pp. 2136–2140

8. N. Bianchi, S. Bolognani, D. Bon et al., Rotor flux-barrier design for torque ripple reduction in synchronous reluctance and PM-assisted synchronous reluctance motors. IEEE Trans. Ind. Appl. **45**(3), 921–928 (2009)
9. X. Tang, W. Xiuhe, S. Shumin et al., Analysis and analysis of cogging torque of asynchronous starting permanent magnet synchronous motor. Proc. CSEE **36**(5), 1395–1403 (2016)

# Quasi-Interval Bipartite Consensus Problems on Discrete-Time Signed Networks

**Jianqiang Liang and Deyuan Meng**

**Abstract** In this paper, discrete-time signed networks with cooperative and antagonistic interactions are considered in the presence of time-varying topologies. A separation approach is proposed such that the cooperations and antagonisms can be clearly distinguished. It is shown that given the repeated joint strong connectivity, signed networks can achieve bipartite consensus (respectively, stability) if and only if the repeated joint structural balance (respectively, unbalance) are ensured. Furthermore, when only joint spanning tree condition is satisfied, quasi-interval bipartite consensus (respectively, stability) holds if and only if the repeated joint structural balance (respectively, unbalance) can be guaranteed for the signed digraphs formed by only these joint root nodes. The simulation tests are included to verify the effectiveness of our obtained results.

**Keywords** Discrete-timedynamics · Signed networks · Time-varying topology
Repeated joint structural balance

## 1 Introduction

Signed networks have drawn impressive attention from various fields in current years, such as social media (see [1]), opinion dynamics (see, e.g., [2, 3]) and multi-agent works (see, e.g., [4–6]). The interactions can be not only cooperative but also antagonistic. If antagonisms are absent, signed networks will collapse into traditional (or

---

J. Liang · D. Meng (✉)
The Seventh Research Division, Beihang University (BUAA), Beijing, China
e-mail: dymeng@buaa.edu.cn

J. Liang · D. Meng
School of Automation Science and Electrical Engineering, Beihang University (BUAA), Beijing, China
e-mail: liangjianqiang@buaa.edu.cn



Y. Jia et al. (eds.), *Proceedings of 2018 Chinese Intelligent Systems Conference*,
Lecture Notes in Electrical Engineering 528,
https://doi.org/10.1007/978-981-13-2288-4_2

called conventional) ones, in which consensus problems are widely studied (see, e.g., [7–9]). Nevertheless, antagonisms are unavoidable in real networks, since disagreement usually exists. The counterpart problems in signed networks are called bipartite consensus problems (see, e.g., [5, 6, 10]).

Our primary objective is to discover novel ways to build up the relation between signed and traditional networks, then to analyze discrete-time dynamics in signed networks. In [11], signed graphs are transformed to conventional ones by gauge transformation. By such approach, bipartite consensus problems with continuous-time models have been discussed widespreadly (see, e.g., [5, 6]). Extended problems are also involved, such as modulus consensus (see, e.g., [3]) and interval bipartite consensus (see [12]). For discrete-time models, there are also relevant results, such as [13] for modulus consensus and [14] for exponential bipartite consensus. Besides, [15] via lifting approach and [10] for opinion evolution discuss polarization in continuous- and discrete-time systems. Novel approaches, such as lifting approach (see [15]) and M-matrix approach (see [16]) are also proposed. However, all of these methods demand some assumptions on networks, e.g., digon sign-symmetry, spanning tree or structural balance.

In this paper, discrete-time dynamics with switching signed networks are studied. A separation approach to linking signed and trivial networks is proposed. Accordingly, by applying technical theories in [14, 17, 18], bipartite and quasi-interval bipartite consensus are accomplished. Dynamics will achieve bipartite consensus (or stability) if and only if signed digraphs (or directed graphs) are repeatedly jointly structurally balanced (or unbalanced), when the digraphs are repeatedly jointly strongly connected. Moreover, when repeated joint spanning tree condition is given, quasi-interval bipartite consensus (or stability) is reached if and only if the rooted subgraph is repeatedly jointly structurally balanced (or unbalanced). For space limitation, all proofs are omitted.

We organise our paper as follows. In Sect. 2, preliminaries and problem statement are introduced. Next, necessary and sufficient conditions are proposed to guarantee bipartite and quasi-interval bipartite consensus respectively based on the separation approach presented in Sect. 3. Finally, illustrative simulations are carried out in Sect. 4 and conclusion is shown in Sect. 5.

*Notations*: Denote $\mathscr{I}_n = \{1, 2, \ldots, n\}$, $\mathbb{Z} = \{1, 2, \ldots\}$, $\mathbb{Z}_+ = \{0, 1, 2, \ldots\}$ and $\mathscr{D}_n = \{D = \text{diag}\{d_1, d_2, \ldots, d_n\} : d_i \in \{\pm 1\}, \forall i \in \mathscr{I}_n\}$. For any $a \in \mathbb{R}$, $a^+ = a$ holds if $a \geq 0$ and $a^+ = 0$ otherwise. Let sign$(a)$ represent the sign function of $a \in \mathbb{R}$. $A = [a_{ij}] \in \mathbb{R}^{m \times n}$ is called a nonnegative matrix, denoted by $A \geq 0$, if $a_{ij} \geq 0, \forall i \in \mathscr{I}_n, \forall j \in \mathscr{I}_m$. $A \geq B$ indicates $A - B \geq 0$ for $A, B \in \mathbb{R}^{m \times n}$. Three extra matrices are specified next: $|A| = [|a_{ij}|]$; $A^+ = [a_{ij}^+]$ and $A^- = -(-A)^+$. Denote $1_n = [1, 1, \ldots, 1]^{\mathrm{T}} \in \mathbb{R}^n$. For any $0 \leq A \in \mathbb{R}^{n \times n}$, it is called a stochastic (respectively, substochastic) matrix if $A1_n = 1_n$ (respectively, $A1_n \leq 1_n$).

## 2 Preliminaries and Problem Statement

### 2.1 Preliminaries of Signed Digraphs

Let $\mathscr{G}(t) = (\mathscr{V}, \mathscr{E}(t), \mathscr{A}(t))$ represent a signed digraph, where $\mathscr{V} = \{v_1, v_2, \ldots, v_n\}$ denotes the node set, $\mathscr{E}(t) \subseteq \mathscr{V} \times \mathscr{V} = \{(v_i, v_j) : v_i, v_j \in \mathscr{V}\}$ represents the edge set, in which $(v_i, v_j)$ is a directed edge from $v_i$ to $v_j$, and $\mathscr{A}(t) = [a_{ij}(t)] \in \mathbb{R}^{n\times n}$ is an adjacency weight matrix. For limitation of space, here we omit the definitions of strong connectivity, spanning trees and structural balance, which can be found in [11] for more details. The Laplacian matrix of $\mathscr{G}(t)$ is formulated as $L(t) = \left[l_{ij}(t)\right] \in \mathbb{R}^{n\times n}$ with

$$l_{ij}(t) = \begin{cases} \sum\limits_{k\in\mathscr{N}_i(t)} |a_{ik}(t)|, & j = i \\ -a_{ij}(t), & j \neq i. \end{cases} \tag{1}$$

Consider switching topologies and let $\mathscr{G}(t)$ switch among a sequence of digraphs including $M \in \mathbb{Z}$ digraphs, i.e., $\mathscr{G}(t) \in \{\mathscr{G}_p = (\mathscr{V}, \mathscr{E}_p, \mathscr{A}_p) : p \in \mathscr{I}_M\}$. Denote the union of $\mathscr{G}_1, \mathscr{G}_2, \ldots, \mathscr{G}_j (j \in \mathbb{Z}_+)$ as $\bigcup_{p=1}^{j} \mathscr{G}_p$. By following [14], joint and repeated joint strong connectivity, spanning trees, structural balance are defined similarly, which are omitted for simplicity.

### 2.2 Discrete-Time Dynamics

Let $x(t) = [x_1(t), x_2(t), \ldots, x_n(t)]^{\mathrm{T}}$ $(t \geq t_0, t \in \mathbb{Z}_+)$ represent system states to describe the dynamics in [12], which are given by

$$x_i(t+1) = x_i(t) + r u_i(t), i \in \mathscr{I}_n, t \geq t_0 \tag{2}$$

where $t_0$ is the initial time and $r > 0$ is the step-size such that

$$r\left(\max_{i\in\mathscr{I}_n} \sum_{j\in\mathscr{N}_i(t)} \left|a_{ij}(t)\right|\right) < 1, \forall t \geq t_0, t \in \mathbb{Z}_+. \tag{3}$$

Inspired by [12], the protocol below is applied to system (2):

$$u_i(t) = \sum_{j\in\mathscr{N}_i(t)} a_{ij}(t)\left[x_j(t) - \mathrm{sign}(a_{ij}(t))x_i(t)\right], \tag{4}$$

so that the dynamics (2) can be rewritten as the following form:

$$x(t+1) = [I - rL(t)]x(t) \triangleq F(t)x(t) = F(t)F(t-1), \ldots, F(t_0)x(t_0). \tag{5}$$

Thus the state transition matrix of system (2) is

$$\Phi_{F(t)}(t, t_0) = F(t-1)F(t-2), \ldots, F(t_0). \tag{6}$$

# 3 Main Results

In this section, a separation approach to linking signed and classical networks are proposed. Moreover, necessary and sufficient conditions to guarantee bipartite and quasi-interval bipartite consensus are given, respectively.

## 3.1 Technical Method: A Separation Approach

Denote $\Phi(t, t_0) = \Phi_{F(t)}(t, t_0)$ and $x_0 = x(t_0)$. We focus on how to separate cooperations and antagonisms. By the approach applying to linear continuous plants in [19], we extend it to discrete-time system (2) successfully.

By means of (1), it is deduced that $F(t)$ can be restated as

$$F(t) = I - rL(t) = \underbrace{[I - r(\mathscr{L}_{\mathscr{A}^+(t)} + \Delta_{|\mathscr{A}^-(t)|})]}_{a(t)} - \underbrace{r\left|\mathscr{A}^-(t)\right|}_{b(t)}. \tag{7}$$

From (3) and the decomposition in (7), $a(t)$ and $b(t)$ are both substochastic matrices for any $t \geq t_0, t \in \mathbb{Z}$. Further, (6) can be rewritten as:

$$\Phi(t, t_0) = \prod_{k=t_0}^{t-1} [a(k) - b(k)] = (-1)^{\omega} \sum c(t-1)c(t-2), \ldots, c(t_0), \tag{8}$$

where $c(k) \in \{a(k), b(k)\}, \forall k \in \{t_0, t_0+1, \ldots, t-1\}$ and $\omega$ represents the frequency of $c(k) = b(k)$. The theorem below is presented to decompose $\Phi(t, t_0)$.

**Theorem 1** *For any initial time $t_0 \geq 0$ and $t \geq t_0, t \in \mathbb{Z}_+$, the state transition matrix $\Phi(t, t_0)$ has the decomposition which can be stated as:*

$$\Phi(t, t_0) = \Phi_{pos}(t, t_0) - \Phi_{neg}(t, t_0) \tag{9}$$

*where $\Phi_{pos}(t, t_0)$ and $\Phi_{neg}(t, t_0)$ are given by*

$$\Phi_{pos}(t, t_0) = \sum_{\omega \text{ is even}} c(t-1)c(t-2), \ldots, c(t_0) \tag{10}$$

$$\Phi_{neg}(t, t_0) = \sum_{\omega \text{ is odd}} c(t-1)c(t-2), \ldots, c(t_0) \tag{11}$$

With (10) and (11), $\Phi_{\text{pos}}(t, t_0)$ and $\Phi_{\text{neg}}(t, t_0)$ are the nonnegative and nonpositive terms of $\Phi(t, t_0)$, respectively. Inspired by [19], it is considerable what $\Phi_{\text{pos}}(t, t_0) + \Phi_{\text{neg}}(t, t_0)$ implies. We notice $\Phi_{I-r\mathcal{L}_{|\mathcal{A}(t)|}}(t, t_0)$ is a stochastic matrix apparently whatever properties $\mathcal{G}(t)$ keeps. The theorem below is proposed to build up the relation between $\Phi(t, t_0)$ and $\Phi_{I-r\mathcal{L}_{|\mathcal{A}(t)|}}(t, t_0)$.

**Theorem 2** *The state transition matrix $\Phi_{I-r\mathcal{L}_{|\mathcal{A}(t)|}}(t, t_0)$, $\forall t \geq t_0, t \in \mathbb{Z}_+$ is nonnegative and can be decomposed as the sum of two nonnegative matrices:*

$$\Phi_{I-r\mathcal{L}_{|\mathcal{A}(t)|}}(t, t_0) = \Phi_{pos}(t, t_0) + \Phi_{neg}(t, t_0). \tag{12}$$

*What's more, $\Phi(t, t_0)$ and $\Phi_{I-r\mathcal{L}_{|\mathcal{A}(t)|}}(t, t_0)$ hold the bounded relation as*

$$|\Phi(t, t_0)| \leq \Phi_{I-r\mathcal{L}_{|\mathcal{A}(t)|}}(t, t_0), \forall t \geq t_0, t \in \mathbb{Z}_+. \tag{13}$$

Some straightforward results can be derived by Theorems 1 and 2.

**Corollary 1** *$\Phi_{pos}(t, t_0)$ and $\Phi_{neg}(t, t_0)$ are both substochastic matrices for any $t \geq t_0, t \in \mathbb{Z}_+$, and $\Phi(t, t_0)$ is bounded with*

$$\|\Phi(t, t_0)\|_\infty \leq 1, \forall t \geq t_0, t \in \mathbb{Z}_+. \tag{14}$$

*Remark 1* By means of Theorems 1 and 2, the cooperations and antagonisms are separated successfully, and the relation between $\Phi(t, t_0)$ and $\Phi_{I-r\mathcal{L}_{|\mathcal{A}(t)|}}(t, t_0)$ has been established that the difference and sum of $\Phi_{\text{pos}}(t, t_0)$ and $\Phi_{\text{neg}}(t, t_0)$ are $\Phi(t, t_0)$ and $\Phi_{I-r\mathcal{L}_{|\mathcal{A}(t)|}}(t, t_0)$ respectively. Through Corollary 1, if $\mathcal{A}^-(t) \equiv 0$, $\Phi_{\text{pos}}(t, t_0)$, $\Phi_{\text{neg}}(t, t_0)$ and (14) collapse into a stochastic matrix, 0 and an equation, respectively. Owing to (14), $\|x(t)\|_\infty \leq \|x_0\|_\infty$ holds for any $t \geq t_0, t \in \mathbb{Z}_+$, which implies that ultimate states are bounded with $\|x_0\|_\infty$. This in fact provides us possibility on convergence analysis of system (2) with protocol (4).

## 3.2 Bipartite Consensus and Stability

The convergence conditions of dynamics (2) are of great interests. It is time to analyze discrete-time dynamics (2) by the separation approach. Before convergence analysis, a fundamental assumption should be prepared in advance.

**Assumption 1** $\mathcal{G}(t)$ switches among $\{\mathcal{G}_1, \mathcal{G}_2, \ldots, \mathcal{G}_M\}$ repeatedly, so that for $\forall p \in \mathcal{I}_M$ and $\tau \geq t_0$, there exists some $t' > \tau$ such that $\mathcal{G}(t') = \mathcal{G}_p$.

From Assumption 1, each $\mathcal{G}_p, \forall p \in \mathcal{I}_M$ works no matter how large $t$ is. It is said bipartite consensus is reached if there exist some $0 < h \in \mathbb{R}$ and $D \in \mathcal{D}_n$ subject to

$\lim_{t\to\infty}x(t) = h(D1_n)$. If $\lim_{t\to\infty}x(t) = 0$, then it is called system (2) stabilizes. Based on Assumption 1, we utilize Theorems 1 and 2 to explore the convergence of $\Phi(t, t_0)$ under some topology conditions of $\mathscr{G}(t)$. The results are shown as follows.

**Lemma 1** *If Assumption 1 holds and $\mathscr{G}(t)$ is repeatedly jointly strongly connected, then for any initial time $t_0 \geq 0$, it follows that $\lim_{t\to\infty}\Phi_{I-r\mathscr{L}_{|\mathscr{A}(t)|}}(t, t_0) = 1_n v^{\mathrm{T}}(t_0)$, where $v(t_0)$ is a nonnegative vector such that $v^{\mathrm{T}}(t_0)1_n = 1$. Furthermore, the limit is reached exponentially fast.*

Lemma 1 induces us to detect what behavior $\Phi(t, t_0)$ keeps. By linking $\Phi(t, t_0)$ with r.j.s.b. property of $\mathscr{G}(t)$, the results are concluded in the following theorem.

**Theorem 3** *Suppose that $\mathscr{G}(t)$ is repeatedly jointly strongly connected and Assumption 1 holds. For any initial time $t_0 \geq 0$, the following results hold.*

1. *$\Phi(t, t_0)$ converges exponentially fast and follows*

$$\lim_{t\to\infty}\Phi(t, t_0) = D1_n v^{\mathrm{T}}(t_0)D, \tag{15}$$

*where $v(t_0)$ is the same as that in Lemma 1, if and only if $\mathscr{G}(t)$ is r.j.s.b..*

2. *$\Phi(t, t_0)$ converges exponentially fast and follows*

$$\lim_{t\to\infty}\Phi(t, t_0) = 0 \tag{16}$$

*if and only if $\mathscr{G}(t)$ is r.j.s.ub..*

The relation between $\Phi(t, t_0)$ and $\Phi_{I-r\mathscr{L}_{|\mathscr{A}(t)|}}(t, t_0)$ has been built up by similarity transformation under both r.j.s.b. and r.j.s.ub. circumstances. By Theorem 3, some extra properties of $\Phi_{\mathrm{pos}}(t, t_0)$ and $\Phi_{\mathrm{neg}}(t, t_0)$ can be derived.

**Corollary 2** *Suppose that repeated joint strong connectivity of $\mathscr{G}(t)$ and Assumption 1 hold. If $\mathscr{G}(t)$ is r.j.s.b., then for any $t_0 \geq 0$, $\Phi_{pos}(t, t_0)$ and $\Phi_{neg}(t, t_0)$ both converge exponentially fast and follow*

$$\lim_{t\to\infty}\Phi_{pos}(t, t_0) = \frac{1}{2}[1_n v^{\mathrm{T}}(t_0) + D1_n v^{\mathrm{T}}(t_0)D] \tag{17}$$

$$\lim_{t\to\infty}\Phi_{neg}(t, t_0) = \frac{1}{2}[1_n v^{\mathrm{T}}(t_0) - D1_n v^{\mathrm{T}}(t_0)D]. \tag{18}$$

Owing to Theorem 3, when Assumption 1 holds and $\mathscr{G}(t)$ repeatedly jointly strongly connected, bipartite consensus (or stability) is achieved with r.j.s.b. (or r.j.s.ub.) $\mathscr{G}(t)$, i.e., $\lim_{t\to\infty}x(t) = [v^{\mathrm{T}}(t_0)Dx_0]D1_n$ (or $\lim_{t\to\infty}x(t) = 0$). Then a question emerges naturally: what behaviors will system (2) keep if the connectivity condition is relaxed to repeated joint spanning trees?

**Lemma 2** *Suppose Assumption 1 holds. For $\forall t_0 \geq 0$, $\lim_{t\to\infty}\Phi_{I-r\mathscr{L}_{|\mathscr{A}(t)|}}(t, t_0) = 1_n v^{\mathrm{T}}(t_0)$, where $v(t_0)$ is defined the same as that in Lemma 2, holds with an exponentially fast speed if and only if $\mathscr{G}(t)$ has a repeated joint spanning tree.*

Lemma 2 makes it possible to discuss convergence results of $\Phi(t, t_0)$ under the spanning tree condition. An equivalent condition, which ensures bipartite consensus, is summarized in the theorem below.

**Theorem 4** *Suppose that $\mathcal{G}(t)$ is r.j.s.b. and Assumption 1 holds. For any $t_0 \geq 0$, (15) holds and converges with an exponentially fast speed, and system (2) reaches bipartite consensus if and only if $\mathcal{G}(t)$ has a repeated joint spanning tree.*

Bipartite consensus problems are both solved successfully with joint strong connectivity and spanning trees. The results coincide with existing ones.

### *3.3 Quasi-interval Bipartite Consensus*

Bipartite consensus of $\mathcal{G}(t)$ with spanning trees links with r.j.s.b. property, which inspires us to explore what behaviors dynamics (2) keeps with r.j.s.ub. $\mathcal{G}(t)$ instead. Assume $\bigcup_{p=1}^{M} \mathcal{G}_p$ has $m \in \mathcal{I}_n$ roots, denoted as $v_1, v_2, \ldots, v_m$, forming a joint root set $\mathcal{V}_r$. The remaining nodes make up a non-root set $\mathcal{V}_{nr}$. Then rooted subgraph and non-rooted subgraph are induced and represented as $\mathcal{G}_r(t) = (\mathcal{V}_r, \mathcal{E}_r(t), \mathcal{A}_r(t))$ and $\mathcal{G}_{nr}(t) = (\mathcal{V}_{nr}, \mathcal{E}_{nr}(t), \mathcal{A}_{nr}(t))$, respectively, where $\mathcal{E}_r(t)$ is the rest of $\mathcal{E}(t)$ by removing edges connected $\mathcal{V}_{nr}$ and $\mathcal{A}_r(t) = [a_{ij,r}(t)] \in \mathbb{R}^{m \times m}$ satisfies $a_{ij,r}(t) = a_{ij}(t), i, j \in \mathcal{I}_m$. $\mathcal{E}_{nr}(t)$ and $\mathcal{A}_{nr}(t)$ are defined similarly.

Next, by reviewing (1) and (6), $\mathcal{A}(t)$, $L(t)$ and $\Phi(t, t_0)$ can be rewritten as the form of lower triangular partitioned matrices, i.e., for any $t \geq t_0, t \in \mathbb{Z}_+$,

$$\mathcal{A}(t) = \begin{bmatrix} \mathcal{A}_r(t) & 0 \\ \mathcal{A}_{rnr}(t) & \mathcal{A}_{nr}(t) \end{bmatrix}, \tag{19}$$

and correspondingly, $L(t)$ has the expression as

$$L(t) = \begin{bmatrix} L_r(t) & 0 \\ -\mathcal{A}_{rnr}(t) & L_{nr}(t) + \Delta_{|\mathcal{A}_{rnr}(t)|} \end{bmatrix}, \tag{20}$$

where $L_r(t)$ and $L_{nr}(t)$ are respectively the Laplacian matrices of $\mathcal{A}_r(t)$ and $\mathcal{A}_{nr}(t)$. Besides, $\Phi(t, t_0)$ also has the form as

$$\Phi(t, t_0) = \begin{bmatrix} \Phi_r(t, t_0) & 0 \\ \Phi_{rnr}(t, t_0) & \Phi_{nr}(t, t_0) \end{bmatrix}, \tag{21}$$

where

$$\Phi_r(t,t_0)=\prod_{k=t_0}^{t-1}[I-rL_r(t)]\quad \Phi_{nr}(t,t_0)=\prod_{k=t_0}^{t-1}\left[I-r(L_{nr}(t)+\Delta_{|\mathscr{A}_{rnr}(t)|})\right]$$

$$\Phi_{rnr}(t,t_0)=r\sum_{k=t_0}^{t-1}[\Phi_{nr}(t,k)\mathscr{A}_{rnr}(k)\Phi_r(k,t_0)]. \tag{22}$$

Obviously $\mathscr{G}_r(t)$ is repeatedly jointly strongly connected. Following Theorem 3, bipartite consensus holds if and only if $\mathscr{G}_r(t)$ is r.j.s.b.. However, $\Phi_{nr}(t,t_0)$ may not converge but oscillate persistently. We are encouraged to probe conditions required to guarantee quasi-interval bipartite consensus, i.e., $\lim_{t\to\infty}x_i(t)=\pm h, \forall i\in\mathscr{I}_m$ and $\limsup_{t\to\infty}x_j(t)\in[-h,h], \forall j\in\mathscr{I}_n\backslash\mathscr{I}_m$ hold for some $h>0$ associated with $x_{r0}=x_r(t_0)$. Equivalent conditions are proposed to admit quasi-interval bipartite consensus in the theorem below.

**Theorem 5** *Suppose that $\mathscr{G}(t)$ has a repeated joint spanning tree and Assumption 1 holds. For any initial time $t_0\geq 0$, system (2) reaches quasi-interval bipartite consensus (respectively, stability) if and only if the rooted digraph $\mathscr{G}_r(t)$ is r.j.s.b. (respectively, r.j.s.ub.).*

*Remark 5* Compared with connectivity restriction, the condition is relaxed into repeated joint spanning trees. Based on Theorem 5, it holds quasi-interval bipartite consensus (respectively, stability) of discrete-time dynamics (2) depends on r.j.s.b. (respectively, r.j.s.ub.) property of $\mathscr{G}_r(t)$, instead of $\mathscr{G}(t)$. Actually, the spanning tree condition cannot guarantee bipartite consensus, instead only joint root consensus is ensured and others oscillate consistently.

## 4 Simulation Results

In this section, simulations are provided to verify the results in Sect. 3. Consider $\mathscr{G}(t)$ switches once per second among the digraphs shown in Fig. 1 repeatedly when $\mathscr{G}_r(t)$ is r.j.s.b. or r.j.s.ub., respectively. In both two cases, set $t_0=0$ and $x_0=[4,-3,2,-1,4]^{\mathrm{T}}$, and only spanning trees are ensured. Notice $v_1$, $v_2$ and $v_3$ are joint roots and $v_4$ and $v_5$ are non-roots in both two circumstances.

From Fig. 2a, it is shown that $x_1$, $x_2$ and $x_3$ achieve bipartite consensus, while $x_4$ and $x_5$ do not converge, instead their "amplitudes" are less than others, leading to quasi-interval bipartite consensus. Besides, Fig. 2b tells stability is reached. These results doubtlessly proof the validity of Theorems 3 and 5.

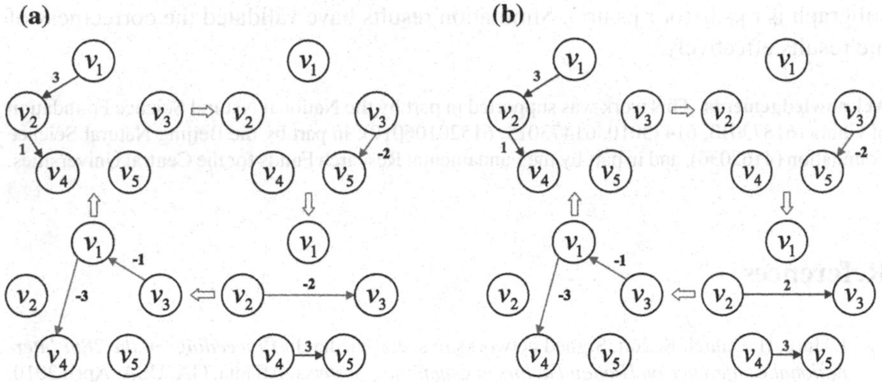
Fig. 1 Switching patterns of $\mathscr{G}(t)$ with repeated joint spanning trees: **a** $\mathscr{G}_r(t)$ is r.j.s.b.; **b** $\mathscr{G}_r(t)$ is r.j.s.ub.

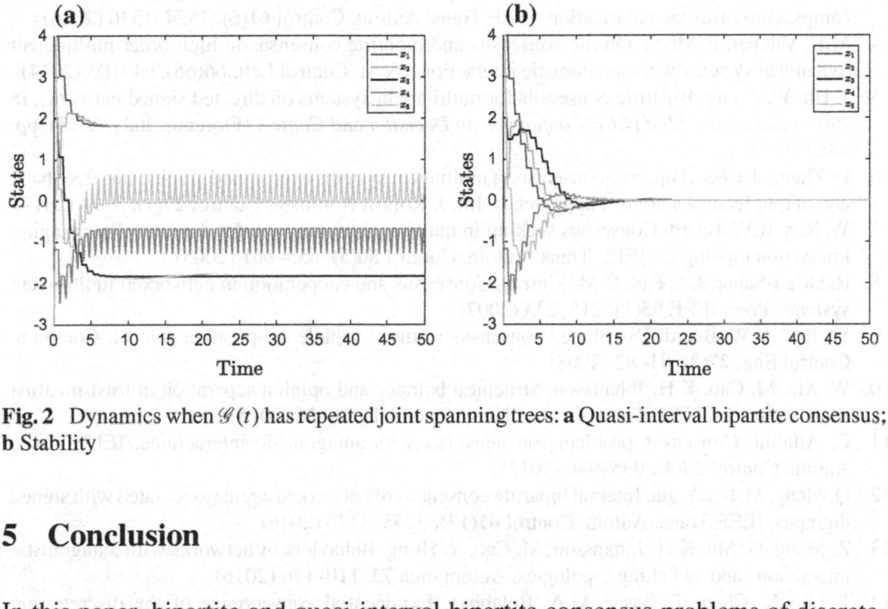

**Fig. 2** Dynamics when $\mathscr{G}(t)$ has repeated joint spanning trees: **a** Quasi-interval bipartite consensus; **b** Stability

# 5 Conclusion

In this paper, bipartite and quasi-interval bipartite consensus problems of discrete-time signed networks with both cooperative and antagonistic interactions have been dealt with. To tell apart the cooperations and antagonisms, a separation approach based on state transition matrices has been proposed, so that the relation between signed and conventional networks has been disclosed. Besides, this approach does not require digraphs to meet any properties, such as connectivity, digon sign-symmetry or structural balance. It is shown that given repeatedly jointly strongly connected signed digraphs, bipartite consensus and stability of discrete-time dynamics are achieved if and only if the switching signed digraph is r.j.s.b. and r.j.s.ub., respectively. Furthermore, when the signed digraph is relaxed to have a repeated joint spanning tree, quasi-interval bipartite consensus (or stability) is achieved if and only if the rooted subgraph is r.j.s.b. (or r.j.s.ub.). Simulation results have validated the correctness of the results effectively.

**Acknowledgements** This work was supported in part by the National Natural Science Foundation of China (61873013, 61473010, 61473015, 61520106010), in part by the Beijing Natural Science Foundation (4162036), and in part by the Fundamental Research Funds for the Central Universities.

## References

1. L. Jure, H. Daniel, K. Jon, Signed networks in social media, in *Proceedings of the 28th International Conference on Human Factors in Computing Systems*, Atlanta, GA, USA, April 2010, pp. 1361–1370
2. A. Daron, O. Asuman, Opinion dynamics and learning in social networks. Dyn. Games Appl. **1**(1), 3–49 (2011)
3. A.V. Proskurnikov, A. Matveev, M. Cao, Opinion dynamics in social networks with hostile camps: consensus vs. polarization. IEEE Trans. Autom. Control **61**(6), 1524–1536 (2016)
4. M.E. Valcher, P. Misra, On the consensus and bipartite consensus in high-order multi-agent dynamical systems with antagonistic interactions. Syst. Control Lett. **66**(66), 94–103 (2014)
5. J. Hu, W. Zheng, Bipartite consensus for multi-agent systems on directed signed networks, in *Proceeding of the 52nd IEEE Conference on Decision and Control* (Florence, Italy, 2013), pp. 3452–3456
6. H. Zhang, J. Chen, Bipartite consensus of multiagent systems over signed graphs: state feedback and output feedback control approaches. Int. J. Robust Nonlinear Control **27**(1), 3–14 (2017)
7. W. Ren, R.W. Beard, Consensus seeking in multi-agent systems under dynamically changing interaction topologies. IEEE Trans. Autom. Control **50**(5), 655–661 (2005)
8. R. Olfati-Saber, J.A. Fax, R.M. Murray, Consensus and cooperation in networked multi-agent systems. Proc. IEEE **95**(1), 215–233 (2007)
9. W. Ren, R.W. Beard, Distributed consensus in multi-vehicle cooperative control. Commun. Control Eng. **27**(2), 71–82 (2008)
10. W. Xia, M. Cao, K.H. Johansson, Structural balance and opinion separation in trust-mistrust social networks. IEEE Trans. Control Netw. Syst. **3**(1), 46–56 (2015)
11. C. Altafini, Consensus problems on networks with antagonistic interactions. IEEE Trans. Autom. Control **58**(4), 935–946 (2013)
12. D. Meng, M. Du, Y. Jia, Interval bipartite consensus of networked agents associated with signed digraphs. IEEE Trans. Autom. Control **61**(12), 3755–3770 (2016)
13. Z. Meng, G. Shi, K.H. Johansson, M. Cao, Y. Hong, Behaviors of networks with antagonistic interactions and switching topologies. Automatica **73**, 110–116 (2016)
14. J. Liu, X. Chen, T. Basar, M.A. Belabbas, Exponential convergence of the discrete and continuous-time Altafini's models. IEEE Trans. Autom. Control **62**(12), 6168–6182 (2017)
15. J.M. Hendrickx, A lifting approach to models of opinion dynamics with antagonisms, in *Proceedings of the IEEE Conference on Decision and Control* (Los Angeles, CA, USA, 2014), pp. 2118–2123
16. D. Meng, Convergence analysis of directed signed networks via an M-matrix approach. Int. J. Control, https://doi.org/10.1080/00207179.2017.1294263
17. D. Cartwright, F. Harary, Structural balance: a generalization of Heider's theory. Psychol. Rev. **63**(5), 9–25 (1977)
18. R.A. Horn, C.R. Johnson, *Matrix Analysis* (Cambridge University Press, 1985)
19. D. Meng, Z. Meng, Y. Hong, A state transition matrix-based approach to separation of cooperations and antagonisms in opinion dynamics. arXiv:1705.04430

# The Research on Force-Magnetic Effect of Wheelset of High-Speed Train Based on Metal Magnetic Memory Method

**Zhenfa Bi and Le Kong**

**Abstract** High-speed railway in the world is developing towards the direction of high-speed and heavy load. The safety of train is an important part of the research of high-speed train. The online detection based metal magnetic memory method is proposed to meet the safety requirement. In order to detect the characteristics of the magnetic memory signal of the wheelset or axle in real-time, the research on the force-magnetic effect of the wheelset or axle material is necessary. This paper takes 25CrMo4 as an example to analyze the relationship between the force and the magnetic. The experimental results show that the magnetic memory signal has a tendency to become smaller with the increase of the load on the experimental specimens.

**Keywords** Force-magnetic effect · 25CrMo4 · Quasi-static tensile
Metal magnetic memory method

## 1 Introduction

### *1.1 The Development of CRH*

China's high-speed rail construction is known as the miracle of China's economic development. It has taken 10 years from the introduction of technology to lead the world and won the world's attention. High-speed rail has become an indispensable part of the life. By the end of 2015, high-speed railway network in China has basically formed. The high-speed train fleet has amounted to 1300 vehicles and the total mileage of operation has reached 14,000 km. One billion and 400 million passengers

---

Z. Bi (✉) · L. Kong
School of Railway Transportation, Shanghai Institute of Technology, Shanghai 201418, China
e-mail: bizhenfa@sit.edu.cn

L. Kong
e-mail: 1924880680@qq.com



Y. Jia et al. (eds.), *Proceedings of 2018 Chinese Intelligent Systems Conference*,
Lecture Notes in Electrical Engineering 528,
https://doi.org/10.1007/978-981-13-2288-4_3

have been transported and China's rail has transited into the era of high-speed rail. High-speed rail not only has narrowed the distance between the cities and has become a strong power to promote the regional economic development, but also has narrowed the distance between people's hearts and has narrowed the distance between each people and the dream.

## *1.2 Safety Requirement*

With the development of the rail to the direction of High-speed and heavy load, the safety of the high-speed rail is a common concern, especially the wheelset which is the rotating parts of the train. The wheelset bears static load, assembly stress, dynamic load, thermal stress, centrifugal force result in curved sections, so there would be stress concentration and fatigue crack initiation (both called incipient fault, is the abnormal symptom of early small) among the wheelset tread, rim, spoke and plate hole and this would give rise to deterioration of the dynamic performance [1, 2]. When the vibration and noise are repeated to the wheelset, it accelerates stress concentration and fatigue initiation to the development of the macro defects. Crack expansion may cause the wheel to rupture, causing a major accident. For example, in 1998 due to the tiny fault of wheelset can't be diagnosed as early as possible, which results in ICE catastrophic accident in Germany. Therefore, it is of great practical significance to detect the incipient fault on the wheelset of the high-speed train.

To achieve the goal of real-time dynamic prediction of early fault, the real-time detecting system is proposed based on metal magnetic memory method [3].

Metal magnetic memory method is a new nondestructive testing method in 21 Century, first proposed by the professor Doubov from Russia. It is based on the residual magnetic field (RMF) of a component or equipment which allows the localization of defects (such as cracks, inclusions, blowholes) or stress concentration zones [4–8].

For the real-time detecting the incipient fault, the metal magnetic memory signal should be measured. The total metal magnetic memory signal is mainly include three parts, such as background magnetic memory signal, the magnetic memory signal caused by incipient fault and the magnetic memory signal caused by online load [9–11]. The total metal magnetic memory signal and the background magnetic memory signal can be measured by sensors. For getting the magnetic memory signal caused by incipient fault, it will be of great use to carry out the research on the force-magnetic effect of the material of wheelset. Through the research of force-magnetic effect, the magnetic memory signal caused by the online load can be calculated.

China now has four types of high-speed rail, in which the CRH3 high-speed rail wheelset material used is 25CrMo4. In this paper, 25CrMo4 as example is used to study the force-magnetic effect.

**Table 1** Sample raw material is annealed 25CrMo4/plate material

| Materials | Tensile strength σ/(MPa) | Yield strength $\sigma_y$/(MPa) | Elastic modulus $E$/(MPa) | Poisson's ratio |
|---|---|---|---|---|
| 25CrMo4 | 708 | 554 | 214 | 0.32 |
| 34CrNiMo6 | 1035 | 967 | 210 | 0.29 |

### *1.3 Introduction of 25CrMo4*

The 25CrMo4 is a high quality alloy structural steel, it is widely used in engineering construction and manufacture of mechanical parts.

A series of experiments on the static performance of 25CrMo4, analysis of the effect on the magnetic memory signal of different static conditions, a magnetic force model, and analyze the recrystallization grain size and grain uniformity dynamically through metallographic observation, hardness uniformity index for forging to provide directional guidance (Table 1).

## 2 The Manufacture of Specimen, Test, Step, Device

### *2.1 The Selection of Specimen*

The choice of metal magnetic memory specimen size is based on the following reasons:

(1) For the reason that fixture materials and most components of the tensile testing machine are metal, the measurement results of metal magnetic memory signal in the static state are greatly influenced.
(2) The overall projected size of the sensors and the support frame is larger than 35 mm × 140 mm. In order to measure the magnetic memory signal under the static force, the test specimen should be selected as the non-standard size.
(3) The metal magnetic memory signal is very weak. For obtaining a relatively accurate magnetic memory signal, the thickness of the specimen is thicker than that of the calibration specimen.

Based on the above reasons, the size of the specimen in the study of the force-magnetic effect is selected as the non-standard size. The size of the specimen is shown as Fig. 1.

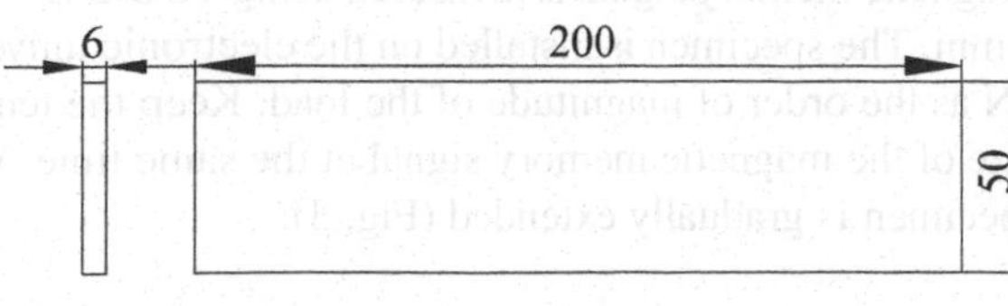

**Fig. 1** Experimental template size

## 2.2 Experimental Procedure and Equipment Used

(1) 25CrMo4 plate processing with electric spark cutting machine;
(2) Demagnetization at 1000 °C with SXL-1200 high temperature heating furnace;
(3) Grinding with grinding machine;
(4) Static stretching with OMT 5305 electronic universal testing machine;
(5) Metal magnetic memory signal was measuring with TSC-2M-8 stress concentration magnetic detector.

## 2.3 Metal Magnetic Memory Method

Metal magnetic memory method is anew nondestructive testing method in 21 Century, first proposed by the professor Doubov from Russia. It is based on the residual magnetic field (RMF) of a component or equipment which allows the localization of defects (such as cracks, inclusions, blowholes) or stress concentration zones. As shown in Fig. 2, the maximum value of leakage magnetic field *Hp* exits in the concentrated area of stress and deformation. It is said that the tangential component of magnetic leakage field *Hp*(*x*) has the maximum value and the normal component *Hp*(*y*) changes sign and has zero points. The irreversible changes in the magnetic state will continue to be retained after the elimination of the working load. It can deter mine the areas of high stress density (fatigue crack initiation), and realize the early diagnosis of the micro fault by the measurement of the normal component of the scattered magnetic leakage field.

Compared with the traditional nondestructive testing method, metal magnetic memory method can find the microscopic faults such as stress concentration, which can be applied to the prediction of failure [12–14]. The traditional nondestructive testing method can only be applied to the detection of the incipient fault, which belongs to the post detection. On the other hand, the traditional nondestructive testing method requires a medium in the detection process, such as water or powder, is not conducive to real-time detection, and metal magnetic memory method don't need medium, can realize the real-time detection. On the basis of the above comparison, the dynamic detection of the incipient fault on wheel-set of high-speed train is shown in Fig. 2.

## 2.4 Magnetic Memory Signal Test Under Static Force

Magnetic memory signal is measured using TSC-2M-8 with collecting a signal every 2 mm. The specimen is installed on the electronic universal testing machine with 50 kN as the order of magnitude of the load. Keep the tension constant, measuring the size of the magnetic memory signal at the same time. With the increase of time, the specimen is gradually extended (Fig. 3).

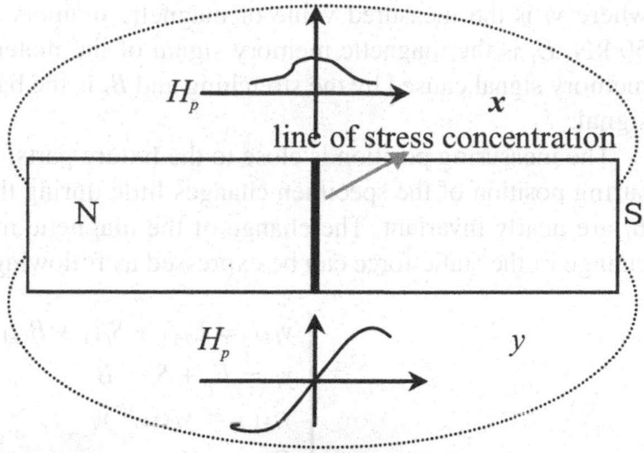


**Fig. 2** Principle diagram of magnetic memory method

**Fig. 3** Tensile on a universal testing machine

# 3 Data Analysis: Algorithm, Actual Result

## *3.1 Acquisition of Magnetic Memory Signal Caused by Stretching*

Magnetic memory signal of 25CrMo4 under stress state is measured with magnetic memory sensor. In the whole tensile process of 25CrMo4, the sensor is in ferromagnetic environment. The value of the magnetic memory sensor includes the magnetic memory signal of the material itself, the magnetic memory signal caused by the stretching, and the background magnetic memory signal. The composition of the signal of the magnetic memory sensor is shown as following:

$$y_i = E_i + S_i + B_i \tag{1}$$

where $y_i$ is the measured value of magnetic memory sensor, $i$ is the multiples of 50 kN, $E_i$ is the magnetic memory signal of the material itself, $S_i$ is the magnetic memory signal caused by the stretching and $B_i$ is the background magnetic memory signal.

The measuring position is close to the fixture parts. Because the size of the measuring position of the specimen changes little during the stretching process, $E_i$ and $B_i$ are nearly invariant. The change of the magnetic memory signal caused by the change of the static force can be expressed as following:

$$\begin{cases} y_{i+1} = E_{i+1} + S_{i+1} + B_{i+1} \\ y_i = E_i + S_i + B_i \\ y_{i+1,i} = y_{i+1} - y_i \\ E_{i+1} = E_i \\ B_{i+1} = B_i \end{cases} \tag{2}$$

In order to show the distribution of metal magnetic memory signals across the board being measured, the data measured on the 4 channel need to be expanded. Data is expanded using a squeezed cubic spline interpolation method.

Using the first derivatives of the known two end points $S'(x_0)$, $S'(x_N)$, to write the equation:

$$m_0 = \frac{3}{h_0}\big(d_0 - S'(x_0)\big) - \frac{m_1}{2} \tag{3}$$

$$m_N = \frac{3}{h_{N-1}}\big(S'(x_N) - d_{N-1}\big) - \frac{m_{N-1}}{2} \tag{4}$$

where the values $m_1, m_2, \ldots, m_{N-1}$ in the equation are determined by the following equations.

$$\begin{cases} \left(\frac{3}{2}h_0 + 2h_1\right)m_1 + h_1 m_2 = u_1 - 3\big(d_0 - S'(x_0)\big), \\ h_{k-1}m_{k-1} + 2(h_{k-1} + h_k)m_k + h_k m_{k+1} = u_k, \quad k = 2, 3, \ldots, N-2, \\ h_{N-2}m_{N-2} + \left(2h_{N-2} + \frac{3}{2}h_{N-1}\right)m_{N-1} = u_{N-1} - 3\big(S'(x_N) - d_{N-1}\big). \end{cases} \tag{5}$$

### 3.2 *Data Analysis*

When the tensile force is F = 50 kN, F = 100 kN, F = 150 kN, the test piece in the middle of 44 mm range of leakage magnetic field density, and save all the measured data (Fig. 4).

This experiment uses the TSC-2M-8 magnetic memory detector, which has 8 signal channels, respectively, to detect the corresponding four path tangential and normal leakage magnetic field density component. As the fixture and the base of the

**Table 2** Tangential component of magnetic field density (*Hp*(*x*): A/m)

| Load (kN) | Max | Min | Mean |
|---|---|---|---|
| 50 | −167 | −297 | −214.9 |
| 100 | −147 | −291 | −193.4 |
| 150 | −120 | 242 | −158.5 |

**Table 3** Normal component of magnetic field density (*Hp*(*y*): A/m)

| Load (kN) | Max | Min | Mean |
|---|---|---|---|
| 50 | 3.3409 | −8 | −3.2892 |
| 100 | 2.1212 | −8.2086 | −3.8444 |
| 150 | 0.3958 | −8.9875 | −5.1694 |

experimental equipment are ferromagnetic components, the specimen signal acquisition caused a certain impact, in the collected data appeared in the local value is too large. The simple features of all the data are extracted to obtain the peak and mean of the magnetic memory signal as shown in the Tables 2 and 3, which can directly reflect the influence of the tensile stress on the magnetic memory signal.

In order to filter out the environmental noise generated by the geomagnetic field and the experimental equipment, the difference between the magnetic memory signals is obtained. Get the image shown in Fig. 5.

From the table and the image, we can see that the tangential component of the leakage magnetic field on the surface of the specimen increases with the increase of

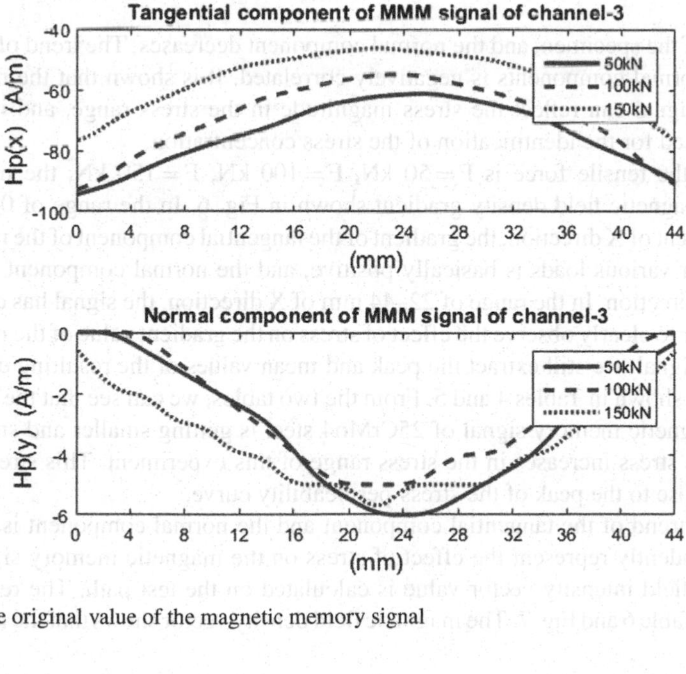


**Fig. 4** The original value of the magnetic memory signal

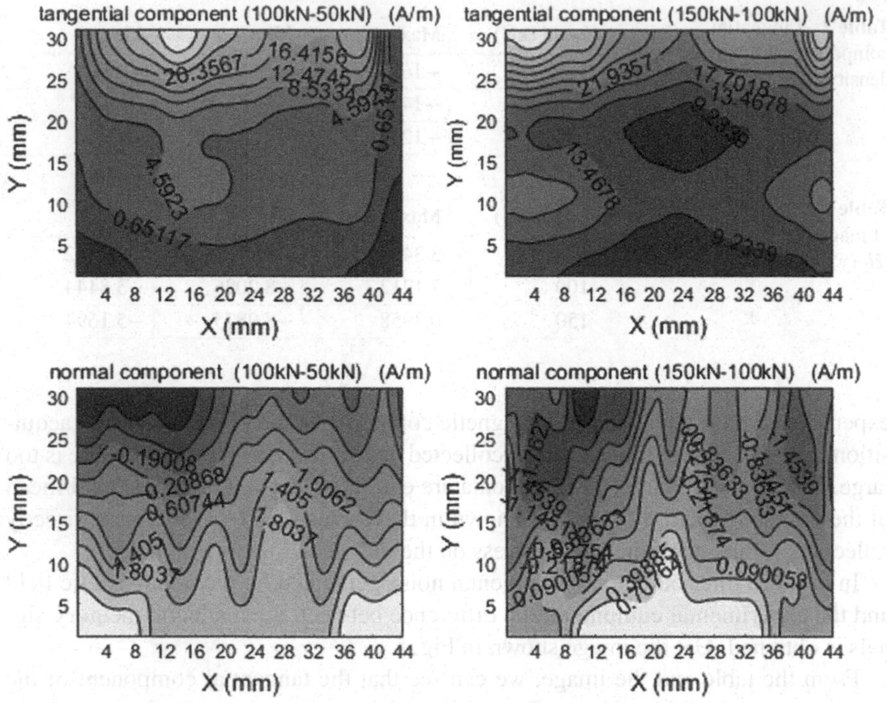


**Fig. 5** Magnetic memory signal difference and its rate of change

the load of the specimen, and the normal component decreases. The trend of tangential and normal components is negatively correlated. It is shown that the magnetic memory signal can reflect the stress magnitude in the stress range, and it can be directly used for the identification of the stress concentration.

When the tensile force is F = 50 kN, F = 100 kN, F = 150 kN, the measured leakage magnetic field density gradient shown in Fig. 6. In the range of 0–22 mm displacement of X direction, the gradient of the tangential component of the magnetic field under various loads is basically positive, and the normal component is in the opposite direction. In the range of 22–44 mm of X direction, the signal has changed.

In order to clearly observe the effect of stress on the gradient value of the magnetic memory signal, we still extract the peak and mean values of the resulting data. The results are shown in Tables 4 and 5. From the two tables, we can see that the gradient of the magnetic memory signal of 25CrMo4 steel is getting smaller and smaller as the tensile stress increases in the stress range of this experiment. This stress range may be close to the peak of the stress-permeability curve.

As the trend of the tangential component and the normal component is difficult to independently represent the effect of stress on the magnetic memory signal, the magnetic field intensity vector value is calculated on the test path. The results are shown in Table 6 and Fig. 7. The magnetic field density vector shows that the magnetic

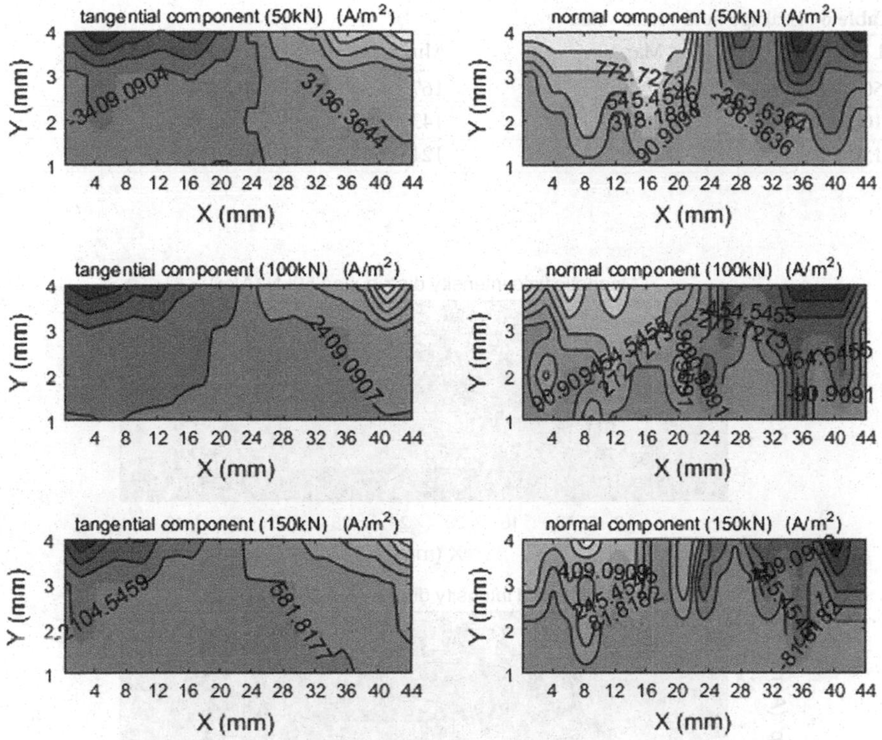


**Fig. 6** The gradient of the magnetic memory signal

**Table 4** Tangential gradient of leakage magnetic field

| Load (kN) | Max | Min | Mean |
|---|---|---|---|
| 50 | 16,500 | −19,500 | 780 |
| 100 | 16,000 | −24,500 | 339 |
| 150 | 12,850 | −16,699 | 45 |

**Table 5** Normal gradient of leakage magnetic field

| Load (kN) | Max | Min | Mean |
|---|---|---|---|
| 50 | 1500 | −1000 | −58 |
| 100 | 1000 | −1000 | −19 |
| 150 | 900 | −900 | 0 |

field density of the specimen decreases with the increase of the load, which is basically consistent with the energy least theory.

The magnetic field density vector value is again subtracted to remove the ambient noise, and the image shown in Fig. 8 is obtained. This image illustrates the effect of stress on the magnetic memory signal of 25CrMo4 steel. During stress loading from 50 to 100 kN, stress anomalies occurred at 12 and 38 mm of X direction, and the stress anomaly was intensified during the loading of 150 kN.

**Table 6** Leakage field vector value

| Load (kN) | Max | Min | Mean |
|---|---|---|---|
| 50 | 297 | 167 | 215 |
| 100 | 291 | 143 | 193 |
| 150 | 242 | 121 | 158 |

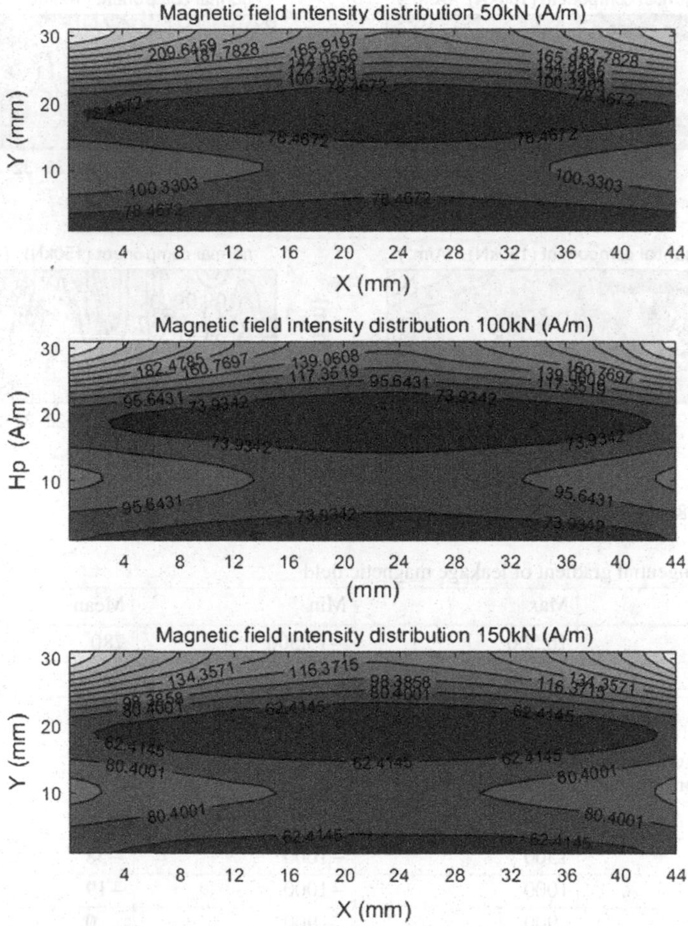


**Fig. 7** The vector value of the magnetic memory signal

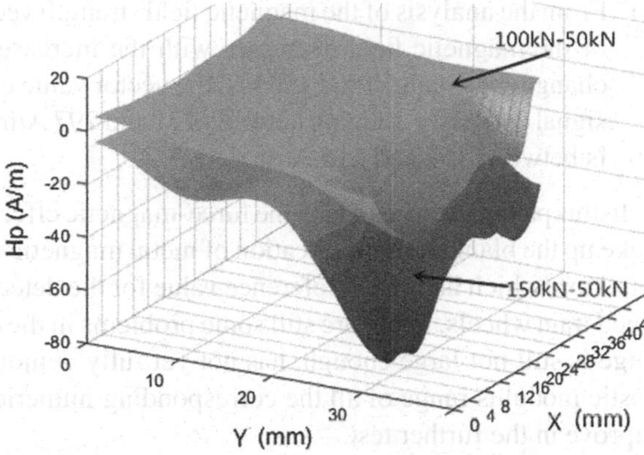


**Fig. 8** The vector difference of magnetic memory signal

# 4 Conclusions

In order to study the application of magnetic memory test in the safety of high-speed trains, this paper has carried on the static drawing of the 25CrMo4 steel plate specimen of the "Harmony" high-speed train. The magnetic memory signal values of the 25CrMo4 steel were measured with different tensile stress. The change of the magnetic memory signal with the stress was obtained.

(1) Through the comparative analysis of the leakage magnetic field density, the tangential component of the leakage magnetic field on the sample surface increases as the load increases. When the load changes between 50 and 150 kN, the magnetic memory signal change at −297 to −120 A/m, and the mean value of the corresponding magnetic memory signal ranges from −215 to −158 A/m. The normal component of the magnetic memory signal is reduced and the variation trend of the two components is negatively correlated. When the load change is between 50 and 150 kN, the magnetic memory signal fluctuates from −9 to 3.5 A/m, the mean value of the corresponding magnetic memory signal ranges from −5.2 to −3.2 A/m.

(2) In the analysis of the gradient of the magnetic memory signal, it is found that the tangential gradient of the magnetic memory signal decreases and the normal gradient increases gradually with the increase of the tensile stress. The degree of dispersion is getting smaller and smaller. When the load value is between 50 and 150 kN, the tangential component gradient of the magnetic memory signal is reduced, and the scope of change is between −24,500 and 16,500 A/m$^2$. The corresponding gradient is in the range of 45–780 A/m$^2$ between. The normalized component gradient increases when the load varies between 50 and 150 kN, the magnetic storage signal changes from −1000 to 1500 A/m$^2$ and the magnetic storage signal varies from −58 to 0 A/m$^2$.

(3) From the analysis of the magnetic field strength vector, it shows that the strength of the magnetic field decreases with the increase of the load. When the load changes between 50 and 150 kN, the vector value of the metal magnetic memory signal decreases, ranging between 121 and 297 A/m. The corresponding gradient is between 158 and 215 A/m.

In this paper, the research on the force-magnetic effect of high-speed train wheelset make up the blank of the application of metal magnetic memory test in train wheelset detection, which has a high reference value for the detection of micro-faults on High-speed train wheels. There are still some problems in the experiment, such as the stress range is still not large enough, has not yet fully demonstrated 25CrMo4 steel in its elastic modulus range of all the corresponding numerical situation, will continue to improve in the further test.

**Acknowledgements** The research work has been financially supported by the National Natural Science Foundation of China (51405303), the City University Youth Teacher Training Fund (ZZyy15110) and the Shanghai University of Applied Technology Talent Fund (YJ2014-17).

## References

1. C.P. Zhao, *Study on Electromagnetic Ultrasonic Launch System for On-line Flaw Detection of Train Wheelset* (Harbin Institute of Technology, Harbin, 2012)
2. J. Qi, W. Zhang, Y.J. Chu, S.B. Hu, Introduction of metal magnetic memory detection technology. Chem. Equip. Pipeline (5), 60–61 (2008)
3. J. Zhang, Damage memory element research, in *International Conference on Robotics and Automation* (Nanjing University of Aeronautics and Astronautics, Nanjing, 2004)
4. H.Z. Yan, L.J. Gong, Constitutive model of 20CrMo material and its finite element simulation. J. Cent. South Univ. (Nat. Sci. Ed.) **11**, 4268–4273 (2012)
5. X.X. Li, H.X. Wang, P.Z. Ying, Shearer drum material selection. Coal Mine Mach. **9**, 134–135 (2015)
6. L.Q. Luo, W. Ju, R.Q. Yu, S.J. Wang, Feasibility study of metal magnetic memory technology for stress relief annealing. Heat Treat. Technol. Equip. (5), 17–19 (2011)
7. A.A. Doubov, Screening of weld quality using the metal magnetic memory, in *Welding in the World* (1998), pp. 196–199
8. Q.M. Yang, G.Z. Li, D.S. Wang, Development and preliminary application of metal magnetic memory detector for railway. China Railway Sci. **1**, 138–141 (2005)
9. Y. Liu, *Rectangular Steel Tube Axial Stress Electromagnetic Test Research and Sensor Miniaturization*, 5–8 March 2001 (Northeastern University, Shenyang, USA, 2001), pp. 1–8
10. J.L. Ren, J.M. Liu et al., *Metal Magnetic Memory Detection Technology* (China Electric Power Press, Beijing), I7-5083-0504-3
11. Q. Pan, *Magnetic Memory Testing Technology Force-Magnetic Effect of the Experimental Study* (Nanchang Aviation University, Nanchang, 2010)
12. D.H. Xiao, D.Y. Luo, Study on the influence of crack width on magnetic memory signal. J. Changsha Aeronaut. Vocat. Tech. Coll. **1**, 42–44 (2007)
13. J. Wang, *Metal Magnetic Memory Detection Signal Analysis* (Nanchang Aviation University, Nanchang, 2010)
14. J.Y. Yao, Based on the force/magnetic coupling of remanufacturing components magnetic memory detection

# Application of Single Neuron LADRC in Trajectory Tracking Control of Parafoil System

**Hongchen Jia, Qinglin Sun and Zengqiang Chen**

**Abstract** In order to further reduce the nonlinearity of the parafoil system and the effect of environmental disturbance on its trajectory tracking control. On the basis of linear active disturbance rejection control (LADRC), using the self-learning ability of neural network, a single neuron is used to construct adaptive parameters, so that parameters can be adjusted accordingly based on the change of system errors, so as to achieve on-line self-tuning of parameters. The simulation results of track tracking by parafoil show that the effect of external interference can be effectively overcome and high precision tracking control can be realized. Compared with the traditional LADRC, the anti-interference ability and robustness are obviously improved.

**Keywords** Parafoil system · Trajectory tracking
Liner active disturbance rejection control · Single neuron · Parameter self-tuning

## 1 Introduction

Parafoil system has been widely used in aerospace, military and civil fields due to its high lift drag ratio, aerodynamic performance, excellent gliding ability, good handling and stability.

In the trajectory tracking control of parafoil system, Xiong [1] analyzed and studied traditional proportional integral differential control (PID), fuzzy PID control and hybrid PID control. Li et al. [2] proposed a trajectory tracking control method of parafoil system based on fuzzy control and predictive control switching. The

H. Jia · Q. Sun (✉) · Z. Chen
College of Computer and Control Engineering, Nankai University, Tianjin 300350, China
e-mail: sunql@nankai.edu.cn

H. Jia
e-mail: jhcthink@163.com

Z. Chen
e-mail: chenzq@nankai.edu.cn

 
Y. Jia et al. (eds.), *Proceedings of 2018 Chinese Intelligent Systems Conference*,
Lecture Notes in Electrical Engineering 528,
https://doi.org/10.1007/978-981-13-2288-4_4

adaptive control method put forward by Benjamin [3] is applied in large parafoil homing control. The above studies have achieved better control effect.

However, the above research, the intelligent control methods are required to obtain accurate dynamic model. Because parafoil system always has parameter uncertainty and modeling error and external environment interference, it inevitably leads to errors between mathematical models and actual systems, which makes it difficult to guarantee the accuracy of trajectory tracking control. In order to overcome the influence of the nonlinear characteristics of the parafoil system and the disturbance of external environment on trajectory tracking control, Tao [4] designed a trajectory tracking controller based on LADRC. And through the method of bandwidth parameterization by Gao [5], Cleveland State University, the ADRC is simplified to LADRC. It greatly reduces the number of parameters to adjust, and is easy to debug and engineering. However, there is still a problem that the controller has more adjustable parameters and is not easy to adjust.

In order to overcome the influence of the uncertainty of the parafoil system itself and the environment on the trajectory tracking control, simplify the parameter adjustment and realize the self-tuning of the parameters, The single neuron method proposed in [6] is used in this paper to adjust the parameters of the error feedback loop of the LADRC. The feasibility of the improved controller to the tracking control of the parafoil trajectory is verified by the simulation experiment.

## 2 The Mathematical Model of the Parafoil System

Considering the relative yaw angle between payload and a parafoil, an eight degree of freedom dynamic model is established in this paper. As shown in Fig. 1, the eight degree of freedom model of the parafoil is used in three coordinate systems [7] and they are the geodetic coordinate system $O_I x_I y_I z_I$, the parafoil coordinate system $O_s x_s y_s z_s$ and the payload coordinate system $O_p x_p y_p z_p$.

$V_s = [u_s\ v_s\ w_s]^T$ and $W_s = [p_s\ q_s\ r_s]^T$ are used to represent the velocity and angular velocity of the parafoil, respectively. $V_p = [u_p\ v_p\ w_p]^T$ and $W_p = [p_p\ q_p\ r_p]^T$ represent the speed and angular velocity of the payload, respectively. The parafoil and the payload are connected by a rope, and the motion state of the two is constrained by the following constraints:

$$V_p + W_p \times L_{pc} = V_s + W_s \times L_{sc} \tag{1}$$

$$W_p = W_s + \tau_s + \kappa_p \tag{2}$$

where $\tau_s = [0\ \ 0\ \ \varphi_r]^{\mathrm{T}}$ and $\kappa_s = [0\ \ \theta_r\ \ 0]^{\mathrm{T}}$ are the relative yaw angle and the relative pitching angle of the parafoil and the payload, respectively. $L_{pc}$ and $L_{sc}$ are the distance between the payload centroid and the centroid of the parafoil to the c point, and the c point is located at the center point of the two suspension lines on the payload.

**Fig. 1** Structure diagram of parafoil system

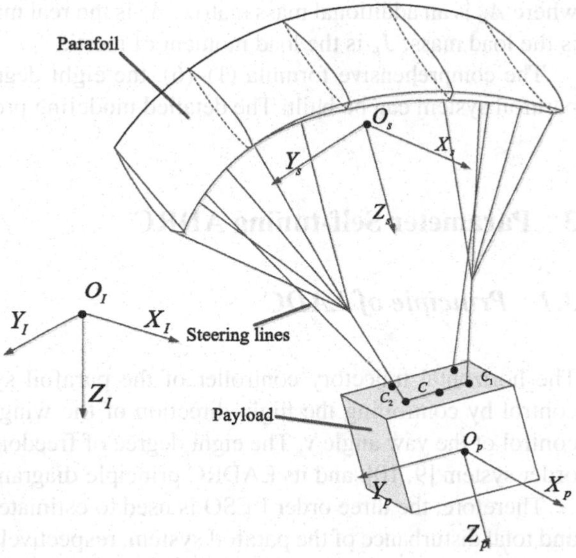


By using the momentum theorem and the moment of momentum theorem, the force of the payload and the parafoil are analyzed, which can be expressed as:

$$\frac{\partial P_p}{\partial t} + W_p \times P_p = F_p^{aero} + F_p^{te} + F_p^{g} + F_p^{th} \tag{3}$$

$$\frac{\partial H_p}{\partial t} + W_p \times H_p = M_p^{aero} + M_p^{f} + M_p^{te} \tag{4}$$

$$\frac{\partial P_s}{\partial t} + W_s \times P_s = F_s^{aero} + F_s^{g} + F_s^{te} \tag{5}$$

$$\frac{\partial H_s}{\partial t} + W_p \times H_p + V_s \times P_s = M_s^{aero} + M_s^{f} + M_s^{g} + M_s^{te} \tag{6}$$

where superscripts $aero, f$, $te$, $th$ and $g$ represent aerodynamic force, friction force, rope tension, thrust and gravity respectively. The aerodynamic force is a variable related to the velocity of the movement of the air relative to the parafoil system. $F$ and $M$ are the corresponding forces and moments, respectively.

The momentum and momentum moment of the load and parafoil are expressed as:

$$\begin{cases} P_p = m_p V_p \\ H_p = J_p W_p \end{cases} \tag{7}$$

$$\begin{bmatrix} P_s \\ H_s \end{bmatrix} = [A_a + A_r] \begin{bmatrix} V_s \\ W_s \end{bmatrix} \tag{8}$$

where $A_a$ is an additional mass matrix; $A_r$ is the real mass matrix of the parafoil; $m_p$ is the load mass; $J_p$ is the load moment of inertia.

The comprehensive formula (1)–(6), the eight degree of freedom model of the parafoil system can be built. The detailed modeling process is described in [7, 8].

## 3 Parameter Self-tuning ADRC

### 3.1 Principle of LADC

The horizontal trajectory controller of the parafoil system realizes the trajectory control by controlling the flight direction of the wing parafoil system, that is, the control of the yaw angle $y$. The eight degree of freedom parafoil system is a second order system [9, 10], and its LADRC principle diagram is shown in Fig. 2.

Therefore, the three order LESO is used to estimate the yaw angle, yaw velocity and total disturbance of the parafoil system, respectively. The expression of the third order LESO can be expressed as:

$$\left.\begin{aligned} e &= z_1 - y \\ \dot{z}_1 &= z_2 - \beta_1 e \\ \dot{z}_2 &= z_3 + bu - \beta_2 e \\ \dot{z}_3 &= -\beta_3 e \end{aligned}\right\} \tag{9}$$

where $z_1$ is the observation value of yaw angle $y$, and $u$ is the output of the controller, that is, the trailing edge of the parafoil. The $e$ is the observation error of the yaw angle, and the $b$ is the input gain. $r$ and $\dot{r}$ denote the input of the parafoil system and its derivative.

The error feedback control law of the horizontal trajectory control channel LSEF is controlled by PD [11], and that can be expressed as:

$$u_0 = K_P e + K_D \dot{e} \tag{10}$$

Therefore $z_3 \approx f$, then

Fig. 2 The schematic diagram of LADRC controller

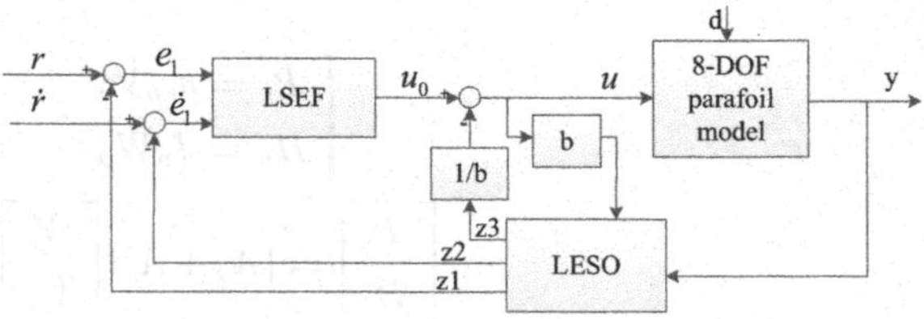

$$u = \frac{u_0 - z_3}{b} \tag{11}$$

where $u_0$ is the control amount of linear PD output.

## 3.2 Single Neuron LADRC

The design of the single neuron auto disturbance rejection controller is inspired by the single neuron PID controller. To increase the stability of the neuron structure, add an input [12]. The basic structure diagram is shown in Fig. 3.

Neurons can generate control signals by association search, which can be expressed as:

$$u_0(k) = u_0(k-1) + K\sum_{i=1}^{3} w_i(k)x_i(k) \tag{12}$$

where $x_i$ is the state variable required for single neuron learning, $w_i$ is the corresponding weight coefficient, and $K$ is the ratio coefficient of the neuron, $K>0$. In this way, the neuron is self-tuning and self-organizing by adjusting the input weight coefficient $w_i$. And the adjustment of the weighting coefficients uses supervised Hebb learning rules. It is related to the input, output, and system deviation of the neuron. The aforementioned learning algorithm is normalized as following:

$$u_0(k+1) = u_0(k) + K\sum_{i=1}^{3} w'(k)x_i(k) \tag{13}$$

$$w'(k) = w_i(k)/\sum_{i=1}^{3}|w_i(k)| \tag{14}$$

$$\begin{cases} w_1(k+1) = w_1(k) + \eta_P z(k+1)u_0(k+1)x_1(k+1) \\ w_2(k+1) = w_2(k) + \eta_D z(k+1)u_0(k+1)x_2(k+1) \\ w_3(k+1) = w_3(k) + \eta_I z(k+1)u_0(k+1)x_3(k+1) \end{cases} \tag{15}$$

The single neuron is added to the parafoil control system, and the schematic diagram is shown in Fig. 4. The transmission state is respectively $x_1(k) = e_1(k)$,

Fig. 3 Single neuron structure diagram

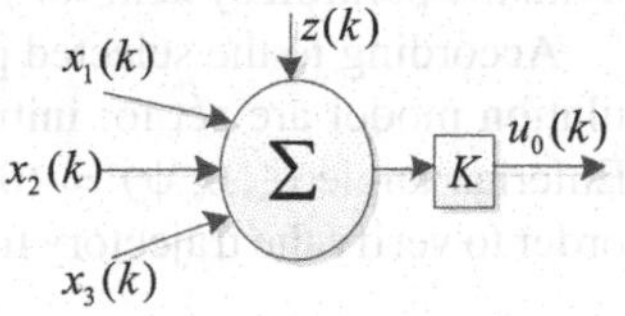

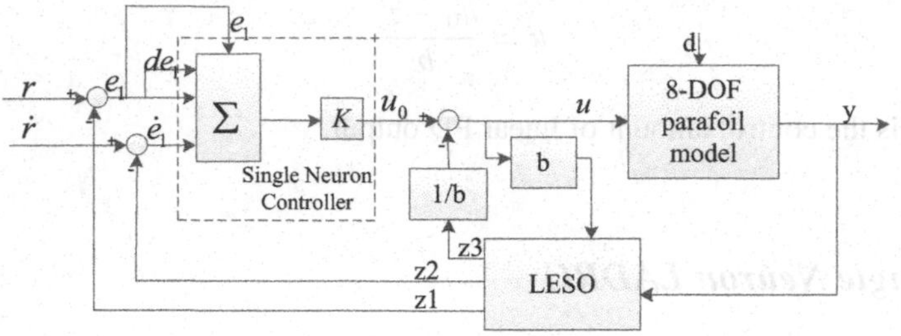


**Fig. 4** The schematic diagram of single neuron LADRC controller

**Table 1** The parameters of parafoil

| | | | |
|---|---|---|---|
| Span | 10.62 m | Length of lines | 6.8 m |
| Chord | 3.17 m | Installation angle | 10° |
| Aspect ratio | 3.35 m | Mass of canopy | 9.68 kg |
| Area of canopy | 34 $m^2$ | Mass of payload | 100 kg |

$x_2(k) = \dot{e}_1(k)$, $x_3(k) = de_1(k)$. Among them, $de_1(k) = e_1(k) - e_1(k-1)$. $\eta$ is the learning rate, $z(k) = e_1(k)$.

A lot of simulation showed that the system could tend to be stable, when $\eta_P$, $\eta_D$, $\eta_I$ gave (0, 1) a random positive number, and the parameter $K$ was selected properly [6]. The weight coefficient in LSEF could be self-adjusted, that was, parameters $w_1$, $w_2$ and $w_3$ could be automatically adjusted by single neuron algorithm according to the error. So the system could accomplish the self-tuning of parameters, and the tunable parameter in LSEF is only one $K$. The bigger the K, the faster the better, but it is prone to overshoot or even make the system unstable. If $K$ is too small, the speediness of the system will become worse, and there may be error in the system response.

## 4 Simulation Results and Analysis

In order to illustrate the improvement of single neuron LADRC compared with LADRC, the LADRC algorithm is added in the simulation results. Before, the LADRC control algorithm used by the project group has achieved a better control effect in the trajectory tracking control of the parafoil. The parameters of the dynamic parafoil system for modeling and simulation are shown in Table 1.

According to the selected parafoil type, the initial motion parameters of the simulation model are set to: initial velocity (u, v, w) = (14.9 m/s, 0, 2.1 m/s); initial Eulerian angle (ξ, θ, ψ) = (0, 0, 0); initial angular velocity (p, q, r) = (0, 0, 0). In order to verify the trajectory tracking control performance, the reference trajectory is

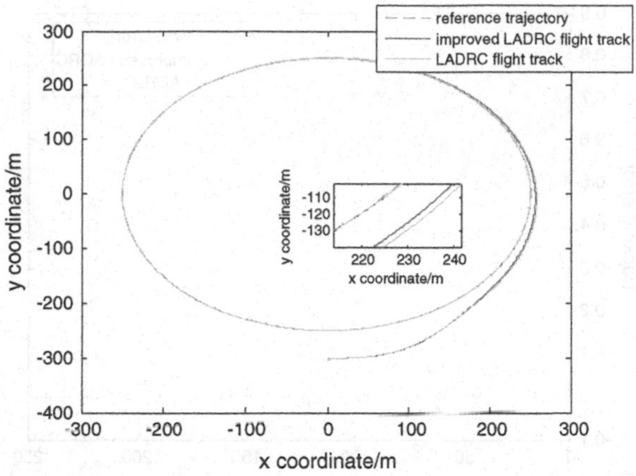


**Fig. 5** The flight tracking trajectory

set to a circle with a center of (0, 0) and a radius of 250 m. The initial plane position of the parafoil system is (0, −300 m), the altitude of the airdrop is 2000 m.

The simulation conditions are set as follows (1)–(2), and the simulation time is set to 250 s.

(1) No disturbance.
(2) Atmospheric disturbance.

The controller parameters are selected as follows: the parameters of LADRC are: $w_0 = 2.7$, $K_p = 0.08$, $K_d = 1.8$, $b = 0.064$; The single neuron improved LADRC parameters are set as: $w_0 = 2.7$, $\eta_P = 0.0003$, $\eta_D = 0.4$, $\eta_I = 0.08$, $K = 2.1$. The simulation results are shown in the following diagram.

Figure 5 is a tracking trajectory. The track is round in the horizontal direction and is represented by the red dotted line. The green line is the flight trajectory is controlled by LADRC controller. The blue line is the flight trajectory controlled by a single neuron improved LADRC controller. In general, they have the same tracking effect, but it can be clearly seen from the local large map that the blue line can follow the track quickly. Figure 6 is the tracking error curve, and the green line for the tracking error curve of LADRC; the blue line is the error curve of the single neuron improved LADRC. It can be seen from the graph that all two curves can reduce the error quickly and do not produce overshoot. The tracking error of LADRC is less than 0.001 in 93 s, however, the tracking error of single neuron LADRC is less than 0.001 at 79 s. In comparison, the improved LADRC is faster and shorter.

After 50 s, the average wind of 3 m/s along the positive direction of the Y axis of the geodetic coordinate system is added. The simulation results are shown in Figs. 7 and 8.

In the case of atmospheric turbulence, the single neuron LADRC and LADRC can also track the reference trajectory very well. However, the single neuron LADRC is better than LADRC in terms of speed or anti-disturbance effect. After the perturbation

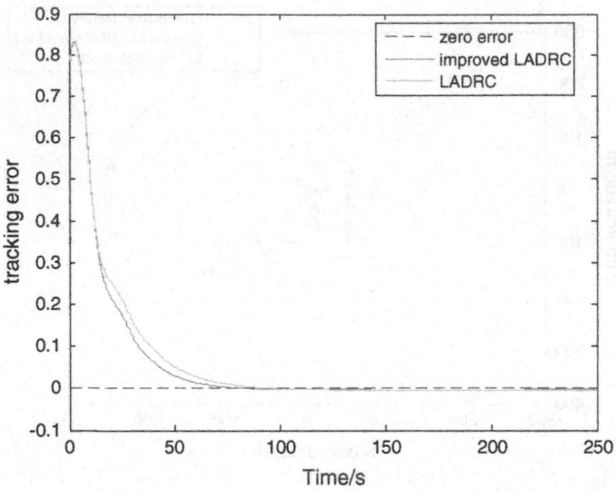


**Fig. 6** The diagram of tracking error

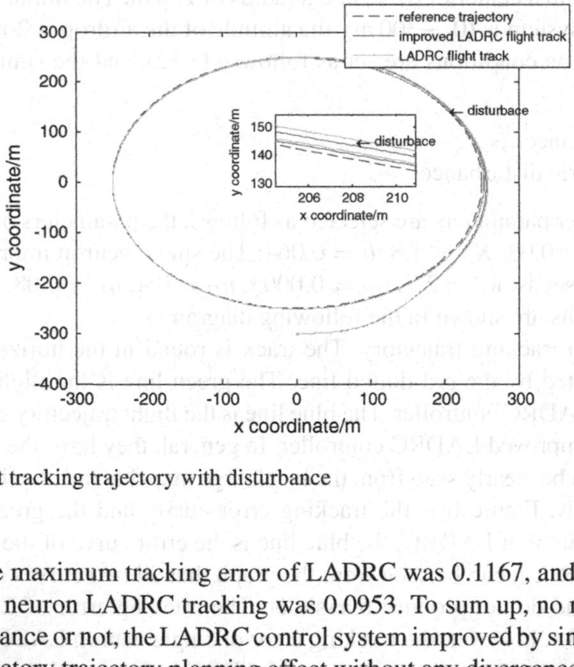


**Fig. 7** The flight tracking trajectory with disturbance

was added, the maximum tracking error of LADRC was 0.1167, and the maximum error of single neuron LADRC tracking was 0.0953. To sum up, no matter whether there is disturbance or not, the LADRC control system improved by single neuron can achieve satisfactory trajectory planning effect without any divergence phenomenon. Compared with the control system built by LADRC, the single neuron improved LADRC control system has better anti-interference performance and robustness.

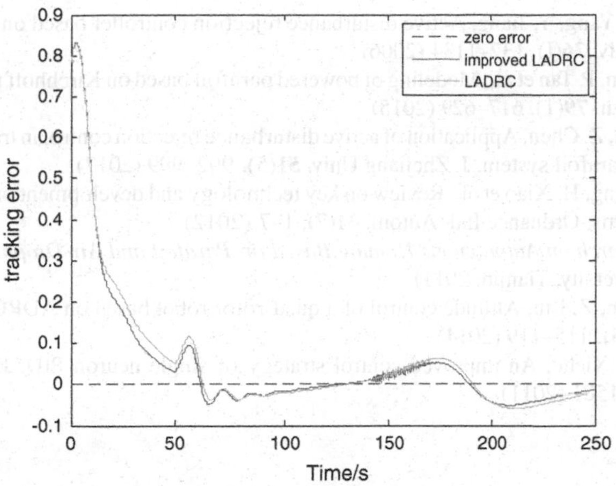


**Fig. 8** The diagram of tracking error with disturbance

# 5 Conclusion

On the basis of the application of LADRC to track trajectory of parafoil and good control effect of disturbance rejection, the LADRC is improved. In order to realize the self-tuning of the adjustable parameters, the parameter tuning method of single neuron is introduced to realize the parameter adaptation of LADRC. Compared with the LADRC trajectory tracking method, the LADRC trajectory tracking control method modified by single neuron has better tracking effect and stronger disturbance rejection ability.

**Acknowledgements** This work is supported by National Natural Science Foundation of China under Grant (61273138, 61573197), National Key Technology R&D Program under Grant (2015BAK06B04).

# References

1. J. Xiong, *Research on the Dynamics and Homing Project of Parafoil System* (National University of Defense Technology, Changsha, 2005)
2. Y. Li, Z. Chen, Q. Sun, Flight path tracking of a parafoil system based on the switching between fuzzy control and predictive control. CAAI Trans. Intell. Syst. **7**(6), 481–488 (2012)
3. S.C. Benjamin, Adaptive control of a 10 K parafoil system, in *23rd AIAA Aerodynamic Decelerator Systems Technology Conference*, Daytona Beach, FL (2015), pp. 1–27
4. J. Tao, LADRC based trajectory tracking control for a parafoil system. J. Harbin Eng. Univ. **38**(3), 1–7 (2018)
5. Z. Gao, Scaling and bandwidth-parameterization based controller tuning, in *Proceedings of the American Control Conference*, Denver, Colorado (2003), pp. 4989–4996

6. Z. Zhang, H. Yang, Y. Jiang, Active disturbance rejection controller based on mono neuron. J. Southeast Univ. **36**(I), 132–1134 (2006)
7. E. Zhu, Q. Sun, P. Tan et al., Modeling of powered parafoil based on Kirchhoff motion equation. Nonlinear Dyn. **79**(1), 617–629 (2015)
8. P. Tan, Q. Sun, Z. Chen, Application of active disturbance rejection control in trajectory tracking of powered parafoil system. J. Zhejiang Univ. **51**(5), 992–999 (2017)
9. Y. Han, C. Yang, H. Xiao et al., Review on key technology and development of parafoil precise airdrop systems. Ordnance Ind. Autom. **31**(7), 1–7 (2012)
10. L. Jiao, *Research on Autonomous Homing Based on Parafoil and Air-Dropped Robot System* (Nankai University, Tianjin, 2011)
11. Y. Li, Z. Chen, Z. Liu, Attitude control of a quad-rotor robot based on ADRC. J. Harbin Inst. Technol. **46**(3), 115–119 (2014)
12. X. Wang, H. Xiehe, An improved control strategy of single neuron PID. J. Zhejiang Univ. **45**(8), 1498–1501 (2011)

# A Design Scheme of Adaptive Switching Neural Control with Uncertain Nonlinearity and External Disturbance

**Lei Yu, Junyi Hou, Jun Huang, Yongju Zhang and Wei Zhang**

**Abstract** In this paper, we concern with the robust adaptive tracking control problem using neural networks for switched nonlinear systems with uncertain nonlinearity and external disturbance. The hypothesis condition that the sign of control gain is known has been relaxed by the proposed control strategy. RBF neural networks (NNs) are utilized to model the unknown nonlinear functions and a robust adaptive neural tracking control method is recommended to enhance the switching the system robustness. Based on switched multiple Lyapunov function strategy, we have derived the adaptive updated control law and the appropriate switching law. It is shown that the technique proposed is able to guarantee that the resulting closed-loop system is asymptotically stable in the Lyapunov sense such that the system output tracking error performance can be well obtained. The effectiveness of the presented control method is demonstrated by the simulation results.

**Keywords** Tracking control · Switched nonlinear systems · RBF neural networks

L. Yu (✉) · J. Hou · J. Huang
School of Mechanical and Electric Engineering, Soochow University, Suzhou, China
e-mail: slender2008@163.com

L. Yu
Collaborative Innovation Center of Industrial Energy-Saving and Power Quality Control, Anhui University, Hefei, China

L. Yu
Key Laboratory of Intelligent Perception and Systems for High-Dimensional Information of Ministryof Education, Nanjing University of Science and Technology, Nanjing, China

Y. Zhang · W. Zhang
Special Equipment Safety Supervision Inspection Institute of Jiangsu Province, Wujiang Branch, Wujiang, China



Y. Jia et al. (eds.), *Proceedings of 2018 Chinese Intelligent Systems Conference*, Lecture Notes in Electrical Engineering 528,
https://doi.org/10.1007/978-981-13-2288-4_5

## 1 Introduction

Switched control systems in control theory are found to be useful in many practical control systems such as robotic manipulator systems, automotive industry, electrical systems, mechanical systems, industrial control system, railroad control, power systems, network and many other practical control areas, and have been intensively studied [1–3]. Recently, a great deal of significant results about the switching control analysis and synthesis have been researched on this subject [1–9].

However, some variables or parameters of switched nonlinear systems is not exactly known in a priori. It is not easy to set up an admissible mathematical model for the controller design of uncertain switched nonlinear systems. Also, in many practical control systems, it often exists nonlinearities and uncertainties. As a common knowledge, stability, robustness, and control performance properties of the switched nonlinear systems are seriously affected in the presence of uncertainties which are inherent in the practical control systems.

For the control of complex uncertainties and highly nonlinear dynamics, due to the approximation properties of RBF NNs which is a powerful approach, it's found that robust adaptive neural control algorithm which has attracted tremendous interest in both control theory and practical applications is particularly powerful [6, 9–16]. However, the robust adaptive neural tracking control problem for switched nonlinear systems with uncertain nonlinearity and the external disturbances is of recurrent interest. At present, only few results have been developed and reported on this research subject. How to handle this control problem is also an important issue.

In this paper, in order to deal with the uncertain nonlinearity and external disturbance of switched systems, a robust adaptive neural switching controller is developed to obtain the satisfactory tracking performance. RBF NNs as an approximator is employed to approximate the uncertain nonlinear terms and a robust adaptive neural tracking switching controller is designed to enhance the system robustness [9, 13, 16]. With the presented switching control scheme, adaptive neural network update law, together with the admissible switching law can guarantee that the resulting closed-loop switched system is asymptotically stable while the tracking error performance can be well achieved. The advantage of this switching control method in this paper are that not only the system output tracking performance can be assumed but also that appropriate neural network parameters are chosen to obtain favorable approximation.

## 2 Problem Formulation and Preliminaries

This paper focuses on the design of the adaptive tracking NN control for a class of switched nonlinear systems with uncertain nonlinearity and external disturbance which can be described by:

$$y^{(n)} = f_{\sigma(t)}(y, \dot{y}, \ldots, y^{(n-1)}, u, \dot{u}, \ldots, u^{(m-1)}) + \frac{1}{\theta} \cdot u^{(m)} + \Delta f_{\sigma(t)}(y, \dot{y}, \ldots, y^{(n-1)}, u, \dot{u}, \ldots, u^{(m-1)}) + d_{\sigma(t)}(t) \quad (1)$$

where, u, y are control input and control output, respectively. $d$ is the external disturbance. $y^{(i)}$ and $u^{(i)}$ are the $i$th derivative of $y$ and $u$, respectively. $\theta = [\theta_1, \theta_2, \ldots \theta_m]^T$ are the unknown constant parameters. But the sign of $\theta$ is known. In general, we assume $\theta > 0$. In this paper, $\sigma(t)$ is a piecewise constant function called switching law. We assumed this law as a piecewise continuous (from the right) function of time and the switching values are taken in the compact set $\Xi$. Let $x_1 = y, x_2 = y^{(1)}, \ldots, x_n = y^{(n-1)},\ z_1 = u, z_2 = u^{(1)}, \ldots, z_m = u^{(m-1)}$.

Then Eq. (1) could be written as:

$$\begin{cases} \dot{x}_i = x_{i+1} \quad (i = 1, 2, \ldots, n-1) \\ \dot{x}_n = f_{\sigma(t)}(x, z) + \frac{\upsilon}{\theta} + \Delta f_{\sigma(t)}(x, z, \omega) + d_{\sigma(t)}(t) \\ \dot{z}_i = z_{i+1} \quad (i = 1, 2, \ldots, m-1) \\ \dot{z}_m = \upsilon \end{cases} \quad (2)$$

where, $x = (x_1, x_2, \ldots, x_n)^T \in R^n, z = (z_1, z_2, \ldots, z_m)^T \in R^m$. System control input $\upsilon = u^{(m)}$ is for the augmented system. The $\omega \in R^l$ in $\Delta f(x, z, \omega)$ is the unmodeled dynamics described by:

$$\dot{\omega} = q(\omega, x, z) \quad (3)$$

In this paper, our goal is to develop a robust adaptive neural switching control scheme with switched multiple Lyapunov function scheme for a class of switched nonlinear systems with uncertain nonlinearity and external disturbance such that the resulting closed-loop switched system is robustly stable. To achieve the presented control goal, we get the following assumptions.

## 3 Design of Robust Switching Adaptive Neural Controller

From (2), the output tracking error dynamic equation can be expressed as in the following form [10]:

$$\begin{cases} \dot{e} = A \cdot e + B \cdot \left[ f_{\sigma(t)}(e + Y_d, z) + \Delta f_{\sigma(t)}(e + Y_d, z, \omega) + \frac{1}{\theta} \cdot \upsilon - y_d^{(n)} + d_{\sigma(t)}(t) \right] \\ \dot{z} = A \cdot z + B \cdot \upsilon \end{cases} \quad (4)$$

Define $A_c = A - BK$, where $A_c$ is Hurwitz. Then the derivative of $e$ is as follows:

$$\overset{\bullet}{e} = A_c \cdot e + B \cdot \left[ K \cdot e + f_{\sigma(t)}(e + Y_d, z) + \Delta f_{\sigma(t)}(e + Y_d, z, \omega) + \frac{1}{\theta} \cdot \upsilon - y_d^{(n)} + d_{\sigma(t)}(t) \right] \tag{5}$$

As we know, $f_i(e + Y_d, z)$ and $\Delta f_i(e + Y_d, z, \omega)$ are both unknown nonlinear functions. So combining with them together as uncertain nonlinear terms gives as:

$$F_i(e + Y_d, z) = f_i(e + Y_d, z) + \Delta f_i(e + Y_d, z, \omega) \tag{6}$$

Substituting (6) into (4) and, we get:

$$\overset{\bullet}{e} = A_c \cdot e + B \cdot \left[ K \cdot e + F_i + \frac{1}{\theta} \cdot \upsilon - y_d^{(n)} + d_i(t) \right] \tag{7}$$

The RBF NNs are adopted to approximate the nonlinear function and can be expressed as [4, 6, 9, 13–16]:

$$F_i(e + Y_d, z) = W_f^T \cdot S_{fi}(e + Y_d, z) + \delta_f(e + Y_d, z) \tag{8}$$

where $i \in \Xi$, $\forall x \in \Omega$ for some compact set $\Omega \in R^n$, $W_f$ are the adaptive NN weights, defining as:

$$W_f = \left[ W_{f1}, W_{f2}, \ldots W_{fn} \right] \tag{9}$$

$S_{fi}(x) : x \to R^n$ indicate vectors of Gaussian basis function as:

$$S_{fi}(e + Y_d, z) = \left[ S_{fi1}(e + Y_d, z), S_{fi2}(e + Y_d, z), \ldots, S_{fin}(e + Y_d, z) \right] \tag{10}$$

Then throughout the paper, the evaluate values of the unknown nonlinear functions $F_i(e + Y_d, z)$ is updated by:

$$\widehat{F_i}(e + Y_d, z) = \widehat{W_f}^T \cdot S_{fi}(e + Y_d, z) \tag{11}$$

where $\widehat{W_f}$ are the evaluate values of weights vector $W_f$. Then the NN weights vector error has been developed as the forms:

$$\widetilde{W_f} = W_f - \widehat{W_f} \tag{12}$$

$$W_f^* := \arg \min_{W_f \in \Omega_f} \left\{ \min_{i \in \Xi} \left\{ \sup_{x \in R^n} \left| F_i(e + Y_d, z) - \widehat{F_i}(e + Y_d, z) \right| \right\} \right\} \tag{13}$$

$$\begin{aligned} \delta_i &= F_i(e + Y_d, z) - \widehat{F_i}(e + Y_d, z, W_f^*) \\ &= \widetilde{W_f}^T \cdot S_{fi}(e + Y_d, z) + \delta_{fi}(e + Y_d, z) \end{aligned} \tag{14}$$

**Lemma 1** [6, 10] *Choosing the system output tracking error dynamic Eq. (14), given a positive constant $\gamma > 0$, if it exists the symmetric matrices $Q_i = Q_i^T > 0$, the necessary and sufficient conditions have satisfied two conditions such that the resulting switching closed-loop system is stable in the following:*

*(i) matrices $P_i$ are the positive definite solutions of Lyapunov equality, having the following properties:*

$$A_c^T P_i + P_i A_c = -Q_i \tag{15}$$

*(ii) there exists an adaptive switching NN control law to handle the given system tracking control problem of uncertain switched nonlinear systems with uncertain nonlinearity and external disturbance:*

$$\upsilon = \hat{\theta}\left[-K \cdot e - \widehat{W_f}^T \cdot S_{fi} + y_d^{(n)} - \varepsilon_i\right] \tag{16}$$

*Meanwhile, for $i, j \in \Xi$, the following matrix inequality is characterized by [6, 14]:*

$$\begin{bmatrix} -P_i & (\Pi_{i,j} + I)^T \cdot P_j \\ P_j \cdot (\Pi_{i,j} + I) & -P_j \end{bmatrix} < 0 \tag{17}$$

*In the following, the adaptive controller and the switching law $\sigma(t)$ which are both deduced by the switched multiple Lyapunov function candidate can be designed as:*

$$\begin{cases} \dot{\widehat{W}}_f = \Gamma_f e^T P_i B S_{fi} \\ \dot{\hat{\theta}} = -e^T P_i B(-K \cdot e - \widehat{W_f}^T \cdot S_{fi} + y_d^{(n)} - \varepsilon_i) \end{cases} \tag{18}$$

$$\sigma(t) = \arg\ min\left\{D_i < \sup_{i \in \Xi} \frac{\|E\|_{min}^2}{\|Z\|^2}\right\} \tag{19}$$

*where,*

$$\begin{cases} \|E\|_{min} = min\left\{\int_0^t e^T Q_i e\, dt\right\} \\ \|Z\| = max\left\{\int_0^t 2e^T P_m B\, dt\right\} \end{cases} \tag{20}$$

*Based on the above analysis, for the switched nonlinear system (1) with uncertain nonlinearity and external disturbance, we can get the theorem in the following.*

**Theorem 1** *Consider the system given by (1), the presented switching NN controller (16), adaptive neural update laws (18), together with switching law (19) can prove that the resulting closed-loop systems are asymptotically Lyapunov stable and uniformly ultimately bounded such that the tracking error performance can be well obtained.*

*Proof* Choose the switched multiple Lyapunov function candidate as:

$$V = \sum_{i=1}^{n} \vartheta_i(t) \cdot e^T P_i e + \frac{1}{\theta} \cdot \tilde{\theta}^2 + \Gamma_f^{-1} \cdot \widetilde{W_f}^T \cdot \widetilde{W_f} \tag{21}$$

For $m \in \Xi$, $t \in (t_{k-1}, t_k] \in \Omega_m$, and $t \in (t_k, t_{k+1}] \in \Omega_{m+1}$, according to (17) and (21), we have:

$$\begin{aligned} \Delta V(t) &= V(t_{k+1}) - V(t_k) \\ &= e^T(t_{k+1}) P_{m+1} e(t_{k+1}) - e^T(t_k) P_m e(t_k) < 0 \end{aligned} \tag{22}$$

By the switching law (19), the tracking error dynamic Eq. (4) and Lemma 1, the time derivative of V is derived in the following:

$$\begin{aligned} \dot{V} &= \dot{e}^T P_m e + e^T P_m \dot{e} + \frac{2}{\theta} \cdot \tilde{\theta} \cdot \dot{\tilde{\theta}} + 2\Gamma_f^{-1} \cdot \widetilde{W_f}^T \cdot \dot{\widetilde{W_f}} \\ &= e^T (P_m A_c + A_c^T P_m) e - \frac{2}{\theta} \cdot \tilde{\theta} \cdot \dot{\hat{\theta}} - 2\Gamma_f^{-1} \cdot \widetilde{W_f}^T \cdot \dot{\widehat{W_f}} \\ &+ 2e^T P_m B \left\{ K \cdot e + F_m + \frac{(\theta - \tilde{\theta})}{\theta} \left[ -K \cdot e - \widehat{W_f}^T \cdot S_{fm} + y_d^{(n)} - \varepsilon_i \right] - y_d^{(n)} + d_m(t) \right\} \end{aligned} \tag{23}$$

Then from (16), (23) and Lemma 1, we obtain:

$$\begin{aligned} \dot{V} &= -e^T Q_m e + 2e^T P_m B (W_f^T \cdot S_{fm} - \widehat{W_f}^T \cdot S_{fm} - \varepsilon_i) \\ &\quad - 2\Gamma_f^{-1} \cdot \widetilde{W_f}^T \cdot \dot{\widehat{W_f}} + 2e^T P_m B d_m(t) \\ &= -e^T Q_m e + 2\widetilde{W_f}^T (e^T P_m B S_{fm} - \Gamma_f^{-1} \cdot \dot{\widehat{W_f}}) + 2e^T P_m B d_m(t) \\ &= 2e^T P_m B d_m(t) - e^T Q_m e \\ &\le 2e^T P_m B D_m - e^T Q_m e \end{aligned} \tag{24}$$

It then follows (19) and (24) that $\dot{V}(t) < 0$, i.e., also, the above inequality is integrated as:

$$\int_0^t \dot{V}\, dt \le \int_0^t 2e^T P_m B D_m\, dt - \int_0^t e^T Q_m e\, dt \Rightarrow V(t) - V(0) \le D_m \|Z\| - \|E\|_{\min} \tag{25}$$

By Lyapunov stability theorem, from (19), (20) and (25), we can have: $\lim_{t\to\infty} e(t) = 0$. Therefore, the proof of Theorem 1 has been finished.

## 4 Design Example

Now, we provide a simulation example to show the effectiveness of our results, the following switched system (1) with uncertain nonlinearity and external disturbance is considered:

$$\sum\nolimits_1 : y^{(2)} = f_1(y, \dot{y}, u) + \frac{1}{2} \cdot \dot{u} + \Delta f_1(y, \dot{y}, u) + 0.5 \cdot \cos(t) \tag{26}$$

$$\sum\nolimits_2 : y^{(2)} = f_2(y, \dot{y}, u) + \frac{1}{3} \cdot \dot{u} + \Delta f_2(y, \dot{y}, u) + 0.5 \cdot \sin(t) \tag{27}$$

where, $\dot{\omega} = -\omega + y^2 + \dot{y}^2 + 0.5, f_1(y, \dot{y}, u) = y + 2u + y\dot{y}, f_2(y, \dot{y}, u) = 2y + u + 3y\dot{y}$, $\Delta f_2(y, \dot{y}, u) = 3\omega$, $\Delta f_1(y, \dot{y}, u) = 2\omega$. The design of control objective is the actual output $y = x_1$ follows the desired output signal $y_d = 0.6 \sin \mathrm{t} - 0.4 \cos \mathrm{t}$. Due to the switching controller design procedures in Sect. 3, the parameters are selected as: $\Gamma_f = 0.5$, $k = (k_1, k_2) = (1,\ 2)$, $A = \begin{pmatrix} 0 & 1 \\ 3 & 5 \end{pmatrix}$, $B = \begin{pmatrix} 0 \\ 1 \end{pmatrix}$, $\Pi_{1,2} = \begin{pmatrix} 1 & 0 \\ 1 & 1 \end{pmatrix}$, $\Pi_{2,1} = \begin{pmatrix} 1 & 1 \\ 0 & 1 \end{pmatrix}$, and the initial value $x(0) = [-0.4\ \ 0.6]^T$. Selecting that the control parameter matrices $Q_1$ and $Q_2$ are chosen as diagonal matrices with diagonal elements 4 and 8, respectively, i.e., $Q_1 = \begin{pmatrix} 4 & 0 \\ 0 & 4 \end{pmatrix}$, $Q_2 = \begin{pmatrix} 8 & 0 \\ 0 & 8 \end{pmatrix}$, from (15) and (18), we have: $P_1 = \begin{pmatrix} -\frac{11}{3} & 1 \\ 1 & \frac{1}{3} \end{pmatrix}$, $P_2 = \begin{pmatrix} -\frac{22}{3} & 2 \\ 2 & \frac{2}{3} \end{pmatrix}$. The initial weights values of RBF neural networks are selected randomly between 0 and 1. The number of RBF neural networks hidden units is chosen as 35. The switching law is provided in (17).

For the experimental results, Fig. 1 presents the output tracking performance, it can be seen that the actual system output y in (1) can follow the any given bounded trajectory signal $y_d$. From Fig. 2, we can observe that the output error of the system tracking performance converges to 0 nearby. Therefore, the tracking objective is well obtained. In addition, the control input and the closed-loop phase plane are respectively shown in Figs. 3 and 4. With the suggested switching control method, the robust adaptive NN switching controller can achieve the expected satisfactory tracking performance by choosing proper control parameters. Therefore, we can conclude that our controller design has obtained the expected results.

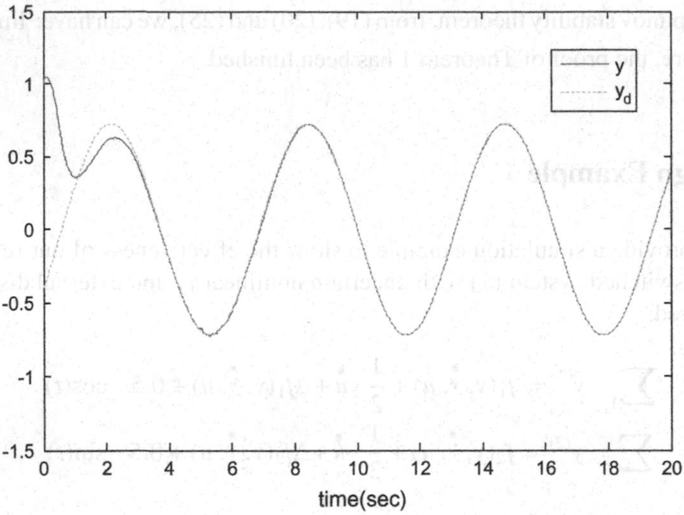


**Fig. 1** Tracking performance

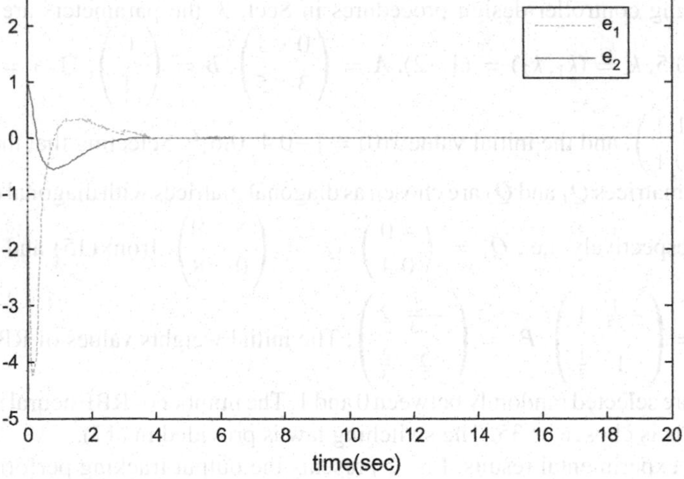


**Fig. 2** Output tracking error performance

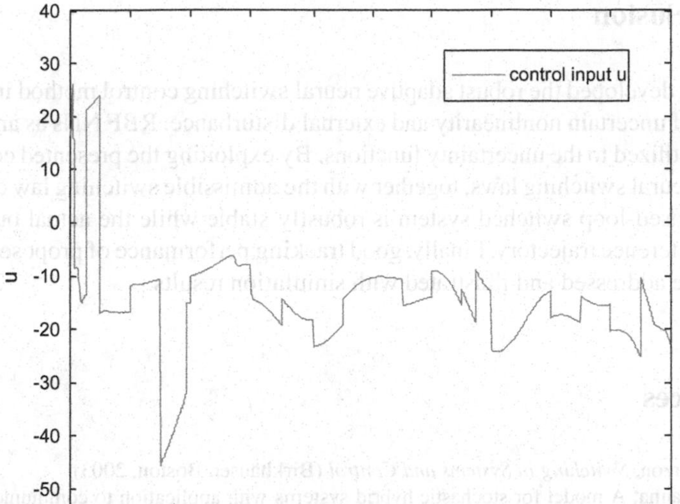


**Fig. 3** Control input (u)

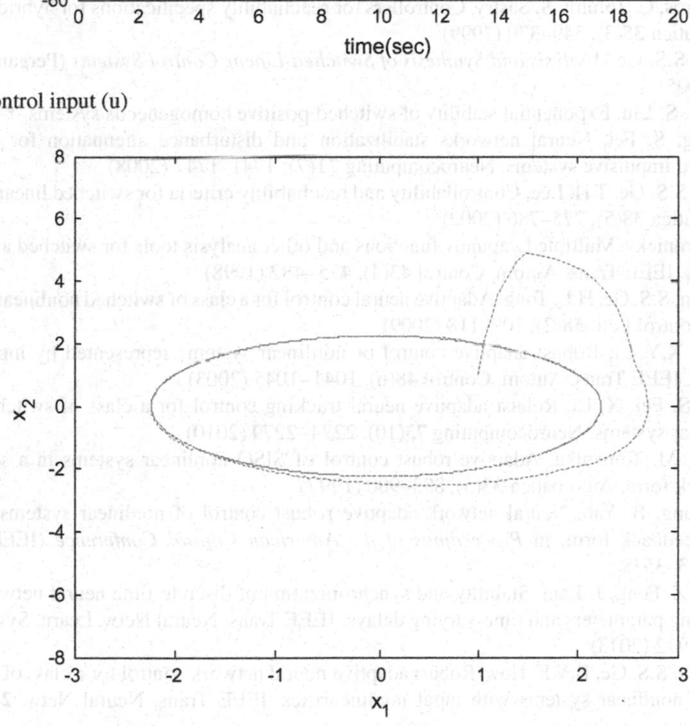


**Fig. 4** Curve of closed-loop phase plane

## 5 Conclusion

This paper developed the robust adaptive neural switching control method in the presentence of uncertain nonlinearity and external disturbance. RBF NNs as an approximator is utilized to the uncertainty functions. By exploiting the presented controller, adaptive neural switching laws, together with the admissible switching law can prove that the closed-loop switched system is robustly stable while the actual output follows the reference trajectory. Finally, good tracking performance of proposed control scheme are addressed and illustrated with simulation results.

## References

1. D. Liberzon, *Switching in Systems and Control* (Birkhäuser, Boston, 2003)
2. J. Hespanha, A model for stochastic hybrid systems with application to communication networks. Nonlinear Anal. **62**(8), 1353–1383 (2005)
3. J. Lygeros, C. Tomlin, S. Sastry, Controllers for reachability specifications for hybrid systems. Automatica **35**(3), 349–370 (1999)
4. Z. Sun, S.S. Ge, *Analysis and Synthesis of Switched Linear Control Systems* (Pergamon Press Inc., 2005)
5. D. Tian, S. Liu, Exponential stability of switched positive homogeneous systems. 1–8 (2017)
6. F. Long, S. Fei, Neural networks stabilization and disturbance attenuation for nonlinear switched impulsive systems. Neurocomputing **71**(7), 1741–1747 (2008)
7. Z. Sun, S.S. Ge, T.H. Lee, Controllability and reachability criteria for switched linear systems. Automatica **38**(5), 775–786 (2002)
8. M.S. Branicky, Multiple Lyapunov functions and other analysis tools for switched and hybrid systems. IEEE Trans. Autom. Control **43**(4), 475–482 (1998)
9. T.T. Han, S.S. Ge, H.L. Tong, Adaptive neural control for a class of switched nonlinear systems. Syst. Control Lett. **58**(2), 109–118 (2009)
10. Y. Liu, X.Y. Li, Robust adaptive control of nonlinear systems represented by input-output models. IEEE Trans. Autom. Control **48**(6), 1041–1045 (2003)
11. L. Yu, S. Fei, X. Li, Robust adaptive neural tracking control for a class of switched affine nonlinear systems. Neurocomputing **73**(10), 2274–2279 (2010)
12. B. Yao, M. Tomizuka, Adaptive robust control of SISO nonlinear systems in a semi-strict feedback form. Automatica **33**(5), 893–900 (1997)
13. J.Q. Gong, B. Yao, Neural network adaptive robust control of nonlinear systems in semi-strict feedback form, in *Proceedings of the American Control Conference* (IEEE, 2001), pp. 3533–3538
14. L. Wu, Z. Feng, J. Lam, Stability and synchronization of discrete-time neural networks with switching parameters and time-varying delays. IEEE Trans. Neural Netw. Learn. Syst. **24**(12), 1957–1972 (2013)
15. M. Chen, S.S. Ge, B.V.E. How, Robust adaptive neural network control for a class of uncertain MIMO nonlinear systems with input nonlinearities. IEEE Trans. Neural Netw. **21**(5), 796 (2010)
16. H. Bouzaouache, Calculus of variations and nonlinear optimization based algorithm for optimal control of hybrid systems with controlled switching. Complexity 1–11 (2017)

# Trajectory Tracking of Intelligent Vehicles Based on Decoupling Performance

**Jianrui Wu and Yingmin Jia**

**Abstract** This paper presents a novel method to achieve trajectory tracking of intelligent vehicles. Both kinetic control and kinematical control are considered in the design procedure, which are called execution layer and decision layer respectively. In the execution layer, input-output decoupling is applied to velocity-varying 4WS vehicles kinetic model. As a result, 3-DOF movements of vehicles can be tracked independently. In the decision layer, the trajectory tracking error model is used to design the desired movements of intelligent vehicles based on sliding mode control. Finally, the two layers can be combined into a system. The simulation results show the system can make the intelligent vehicles track a certain trajectory accurately.

**Keywords** Intelligent vehicles · Trajectory tracking · Decoupling control
Sliding mode control

## 1 Introduction

With the development of new technology, intelligent vehicles are becoming more and more popular nowadays. It's common sense that safety and comfort are two important factors when driving [1]. Intelligent vehicles move automatically, which reduce accidents caused by human factor and improve safety obviously [2]. Trajectory tracking means vehicles move along the preinstalled trajectory automatically. It is an essential technology in automatic driving and now receives more and more attention.

In order to achieve trajectory tracking, both kinetic control and kinematical control should be considered. However, some researches only considered kinematical control

J. Wu · Y. Jia (✉)
The Seventh Research Division and the Center for Information and Control, School of Automation Science and Electrical Engineering, Beihang University (BUAA), Beijing 100191, China
e-mail: ymjia@buaa.edu.cn

J. Wu
e-mail: 1603120@buaa.edu.cn



Y. Jia et al. (eds.), *Proceedings of 2018 Chinese Intelligent Systems Conference*,
Lecture Notes in Electrical Engineering 528,
https://doi.org/10.1007/978-981-13-2288-4_6

[3] or just use geometric control [4], which does not accord with the actual situation. In kinetic control, decoupling has been proved to be effective and practical to make vehicles easy to control [5]. However, velocity-varying motion should be considered into decoupling control but many researches ignored [6]. Meanwhile, the decoupling control method in [6] provides a simple way to connect kinetics with kinematics of vehicles. As for generalized kinematical control of vehicles, building error models to describe the tracking problem is the premise. Different kinematical models have been presented [7–9]. Longitudinal velocity and yaw rate are used to describe vehicle motion in [9]. Sliding mode control [10] is proved to be useful to solve this problem.

In this paper, a new trajectory tracking system is built. Both kinetic control and kinematical control are considered, which are called execution layer and decision layer respectively. Decision layer is used to calculate the desired movements and execution layer can accomplish the movements. In the execution layer, kinetic model of velocity-varying 4WS vehicles is briefly introduced. Input-output decoupling is used to decouple the model into three two-order subsystems. As a result, 3-DOF movements of vehicles can be tracked independently in the subsystems. In the decision layer, trajectory tracking error model is introduced. Sliding mode control is used to calculate the desired vehicle movements, which makes the tracking error very close to zero. Finally, execution layer and decision layer are combined to build the trajectory tracking system.

The rest of paper is organized as follows. In Sect. 2, decoupling control and velocity tracking are finished based on kinetic model. In Sect. 3, sliding mode control is derived for the kinematical error model. In Sect. 4, the system structure is built. In Sect. 5, simulation results are presented. Finally, conclude this paper in Sect. 6.

## 2 Execution Layer

Vehicles' 3-DOF movements are used to describe vehicles' motion. It refers to longitudinal motion, lateral motion and yaw motion. The function of execution layer is to control the 3-DOF movements of vehicles and make longitudinal velocity $v_x$, lateral velocity $v_y$ and yaw rate $r$ track their desired value. This is not an easy task because kinetic models of vehicles are nonlinear and highly coupled. However, the work of [6] provides a way to control the 3-DOF movements of vehicles independently. Tracking task can be easily solved based on it.

### *2.1 Kinetic Model*

The velocity-varying 4WS vehicles kinetic model in [6] can be simply described by (1)

$$\dot{x} = A(x) + Bu$$
$$z = Hx \tag{1}$$

where

$$x = \begin{bmatrix} x_1 \; x_2 \; x_3 \; x_4 \; x_5 \; x_6 \end{bmatrix}^T = \begin{bmatrix} v_x \; v_y \; r \; \delta_f \; \delta_r \; F_{le} \end{bmatrix}^T$$
$$u = \begin{bmatrix} u_1 \; u_2 \; u_3 \end{bmatrix}^T$$
$$B = \begin{bmatrix} 0 \\ I_{3\times3} \end{bmatrix}, \; H = \begin{bmatrix} I_{3\times3} \; 0 \end{bmatrix}$$

The kinetic model has six states, which are longitudinal velocity $v_x$, lateral velocity $v_y$, yaw rate $r$, front/rear tire steering angle $\delta_f/\delta_r$ and acceleration/braking force $F_{le}$. Output $z$ are $v_x$, $v_y$ and $r$. $A(x)$ is a nonlinear expression that makes the longitudinal velocity $v_x$, lateral velocity $v_y$ and yaw rate $r$ couple with each other. It's clear that the kinetic model is a system with three inputs and three outputs but we can't control three outputs independently. That makes the trajectory tracking very difficult to achieve because we want three outputs to track their desired value independently in trajectory tracking.

## 2.2 Input-Output Decoupling and Tracking

Differentiate the output until input appears

$$\dot{z} = HA(x)$$
$$\ddot{z} = H\frac{\partial A(x)}{\partial x}(A(x) + Bu) = H^*(x) + D^*(x)u \tag{2}$$

The relative degree is 2, let

$$u = F^*(x) + G^*(x)\eta, \quad \eta = \begin{bmatrix} \eta_1 \; \eta_2 \; \eta_3 \end{bmatrix}^T \tag{3}$$

where

$$F^*(x) = -D^{*-1}(x)\left[H^*(x) + M^*(x)\right]$$
$$G^* = D^{*-1}(x)$$
$$M^*(x) = \begin{bmatrix} a_{11}x_1 + a_{12}\dot{x}_1 \\ a_{21}x_2 + a_{22}\dot{x}_2 \\ a_{31}x_3 + a_{32}\dot{x}_3 \end{bmatrix} \quad (a_{ij} > 0)$$

$M^*(x)$ is the desired linear form of decoupled system. Nonlinear terms of system have been accurately counteracted by design of input $u$. Decoupling can be applied when $D^*(x)$ is invertible. This condition always holds which has been proved by [6]. Then we get three two-order subsystems, which are linear time invariant systems controlled by virtual control input:

$$\ddot{z} = -M^*(x) + \eta$$

or

$$\begin{bmatrix} \dot{x}_i \\ \ddot{x}_i \end{bmatrix} = \begin{bmatrix} 0 & 1 \\ -a_{i1} & -a_{i2} \end{bmatrix} \begin{bmatrix} x_i \\ \dot{x}_i \end{bmatrix} + \begin{bmatrix} 0 \\ 1 \end{bmatrix} \eta_i$$

$$z_i = x_i, i = 1, 2, 3 \tag{4}$$

Assume the desired value to be tracked is $x_{id}$, then tracking error is $e_i = x_i - x_{id}$. According to (4), we can get second order differential value of $e_i$ and it satisfies:

$$\begin{bmatrix} \dot{e}_i \\ \ddot{e}_i \end{bmatrix} = \begin{bmatrix} 0 & 1 \\ -a_{i1} & -a_{i2} \end{bmatrix} \begin{bmatrix} e_i \\ \dot{e}_i \end{bmatrix} + \begin{bmatrix} 0 \\ \eta_i - \ddot{x}_{id} - a_{i2}\dot{x}_{id} - a_{i1}x_{id} \end{bmatrix}, i = 1, 2, 3 \tag{5}$$

Obviously, if:

$$\eta_i = \ddot{x}_{id} + a_{i2}\dot{x}_{id} + a_{i1}x_{id}, \ i = 1, 2, 3 \tag{6}$$

The tracking error system becomes:

$$\begin{bmatrix} \dot{e}_i \\ \ddot{e}_i \end{bmatrix} = \begin{bmatrix} 0 & 1 \\ -a_{i1} & -a_{i2} \end{bmatrix} \begin{bmatrix} e_i \\ \dot{e}_i \end{bmatrix}, \ i = 1, 2, 3 \tag{7}$$

If we choose $a_{ij}$ appropriately according to the design principle of second order system, the above system is stable and a satisfactory convergence speed can be achieved. As a result, vehicle can get the desired value of longitudinal velocity $v_x$, lateral velocity $v_y$ and yaw rate $r$ independently through controlling three subsystems.

## 3 Decision Layer

When intelligent vehicles move along the desired trajectory set by computer automatically, they must be told how they move. The function of decision layer is to calculate the desired longitudinal velocity $v_{xd}$, lateral velocity $v_{yd}$ and yaw rate $r_d$ and command the vehicles to move in the desired motion. Because lateral velocity will make people uncomfortable and get sick sometimes, the desired lateral velocity

$v_{yd}$ just sets zero. It can be realized by setting $\eta_2$ zero based on decoupling performance. In a word, the task is to design $v_{xd}$ and $r_d$ to make trajectory tracking error model converge to zero.

## 3.1 Kinematical Model

Define a desired trajectory in world coordinate system. The reference position of the vehicle set by computer is $(x_r, y_r, \theta_r)$ and current position is $(x, y, \theta)$. The error between them is $(x_e, y_e, \theta_e)$ (Fig. 1).

Trajectory tracking error model [9]:

$$\begin{cases} \dot{x}_e = v_{xr}\cos\theta_e - v_x + ry_e \\ \dot{y}_e = v_{xr}\sin\theta_e - rx_e \\ \dot{\theta}_e = r_r - r \end{cases} \tag{8}$$

where $v_{xr}$ and $r_r$ is the reference longitudinal velocity and yaw rate belonging to the reference position of the vehicle. Now we want to design the desired value of $v_x$ and $r$ to make the error model converge. As a result, vehicle can move to the reference position from current position and track the reference motion all the time.

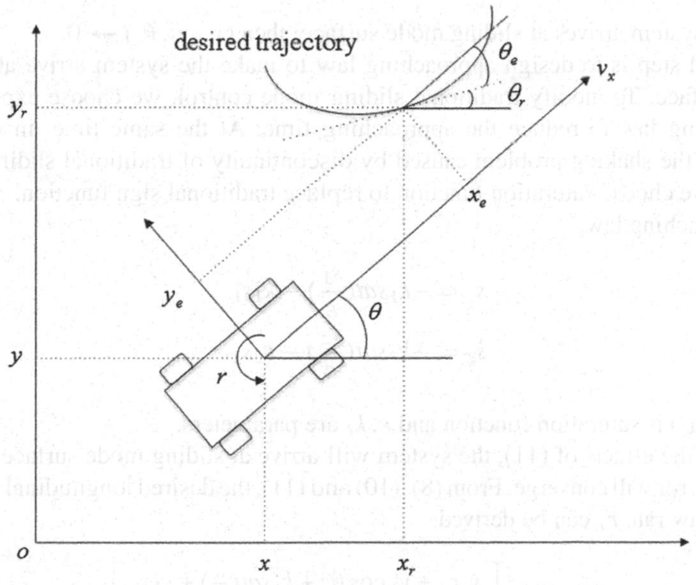


**Fig. 1** Kinematical model of trajectory tracking

## 3.2 Sliding Mode Control

According to sliding mode control theory, sliding mode surface is first to be designed [10]. Obviously, if:

$$\begin{aligned} &x_e \to 0 \\ &\theta_e \to -\arctan(v_r y_e) \end{aligned} \tag{9}$$

Let Lyapunov function

$$V = \frac{1}{2} y_e^2$$

Then

$$\dot{V} = y_e \dot{y}_e = -v_{xr} y_e \sin(\arctan(v_{xr} y_e)) \leq 0$$

If $v_{xr} \neq 0$, according to Lyapunov stability theory, $y_e \to 0$.

As a result, $\theta_e \to -\arctan(v_r y_e) \to 0$.

So sliding mode surface can be chose by (10):

$$s = \begin{bmatrix} s_1 \\ s_2 \end{bmatrix} = \begin{bmatrix} x_e \\ \theta_e + \arctan(v_r y_e) \end{bmatrix} \tag{10}$$

If the system arrives at sliding mode surface, then $(x_e, y_e, \theta_e) \to 0$.

Second step is to design approaching law to make the system arrive at sliding mode surface. To modify traditional sliding mode control, we choose exponential approaching law to reduce the approaching time. At the same time, in order to eliminate the shaking problem caused by discontinuity of traditional sliding mode control, we choose saturation function to replace traditional sign function.

Approaching law:

$$\begin{aligned} \dot{s}_1 &= -k_1 sat(\frac{s_1}{\varepsilon}) - k_1 s_1 \\ \dot{s}_2 &= -k_2 sat(\frac{s_2}{\varepsilon}) - k_2 s_2 \end{aligned} \tag{11}$$

where $sat(\cdot)$ is saturation function and $\varepsilon$, $k_i$ are parameters.

Under the effects of (11), the system will arrive at sliding mode surface and the tracking error will converge. From (8), (10) and (11), the desired longitudinal velocity $v_{xd}$ and yaw rate $r_d$ can be derived:

$$\begin{bmatrix} v_{xd} \\ r_d \end{bmatrix} = \begin{bmatrix} y_e r_d + v_r \cos\theta_e + k_1 sat(\frac{s_1}{\varepsilon}) + k_1 s_1 \\ \dfrac{r_r + \frac{y_e \dot{v}_r}{1+(v_r y_e)^2} + \frac{v_r^2 \sin\theta_e}{1+(v_r y_e)^2} + k_2 sat(\frac{s_2}{\varepsilon}) + k_2 s_2}{1 + \frac{v_r x_e}{1+(v_r y_e)^2}} \end{bmatrix} \tag{12}$$

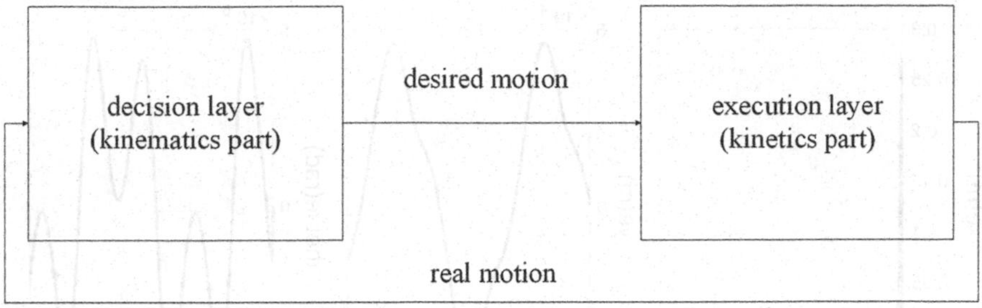


**Fig. 2** System structure

## 4 System Structure

Through analyzing the two layers, it is clear that the input of the execution layer is desired longitudinal velocity $v_{xd}$ and yaw rate $r_d$, which is exactly the output of the decision layer. The output of the execution layer is real longitudinal velocity $v_x$ and yaw rate $r$, which is exactly the input of the decision layer. So two layers can be directly connected. The system works as the following process.

First, the decision layer receives the real motion from the execution layer and gets new current position. Then the decision layer calculates the desired motion according to the error between reference position and current position. After getting the desired motion, it is transferred to execution layer. The execution layer tracks the desired motion and outputs real motion to the decision layer. After a period of time, the system will get into a stable state. The current position will catch the reference position. That is to say, trajectory tracking is accomplished (Fig. 2).

## 5 Simulation

In our simulation, all the vehicle related parameters are obtained from [6]. The vehicle tracks the trajectory denoted by:

$$\begin{aligned} x_r &= 20t \\ y_r &= 100\sin(0.1t) \end{aligned} \tag{13}$$

Equation (13) contains the reference information of $\theta_r$, $v_r$ and $r_r$. The initial tracking error is zero, Figs. 3, 4, 5, 6 and 7 show the results.

Simulation results show that the tracking error is very close to zero, which can be ignored in trajectory tracking. The actual trajectory completely coincides with the desired trajectory. The lateral velocity is very small, so it will not influence the trajectory tracking. The reference and real longitudinal velocity (yaw rate) finally

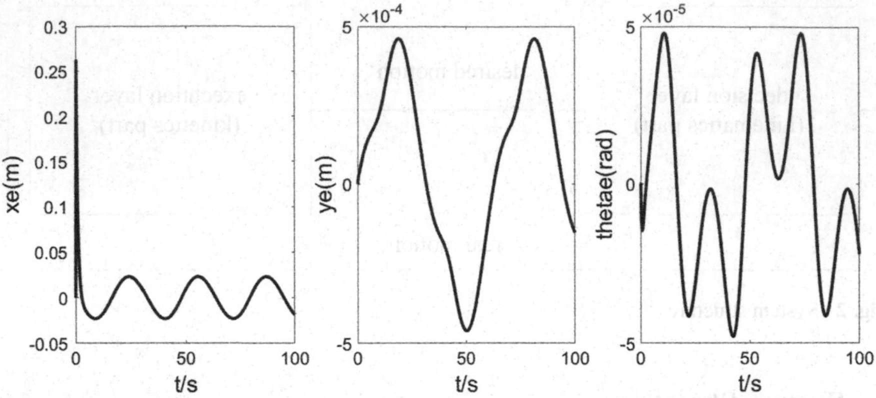


**Fig. 3** Tracking error

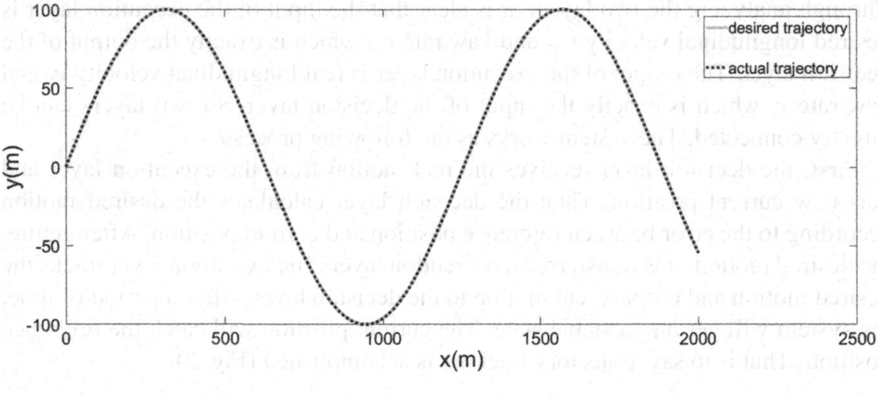


**Fig. 4** Trajectory comparison

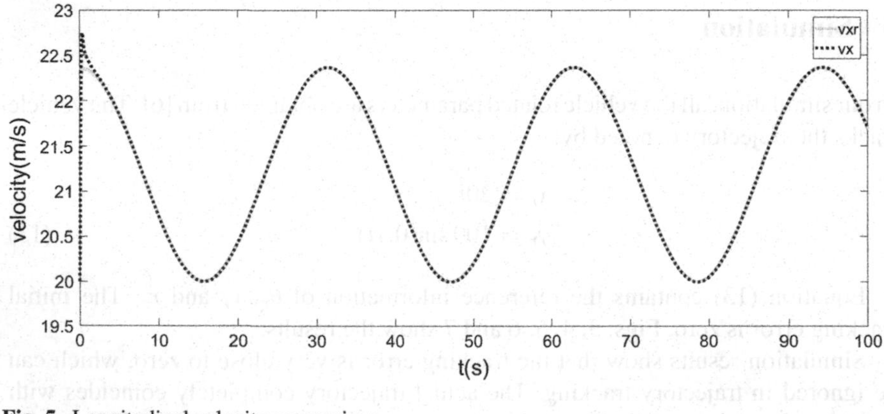


**Fig. 5** Longitudinal velocity comparison

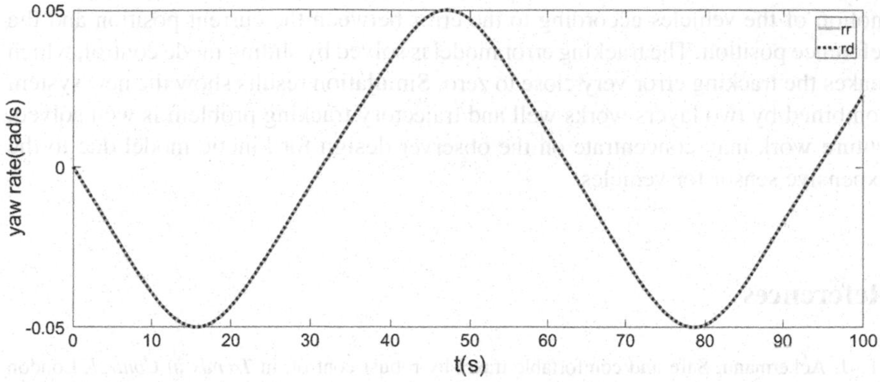


**Fig. 6** Yaw rate comparison

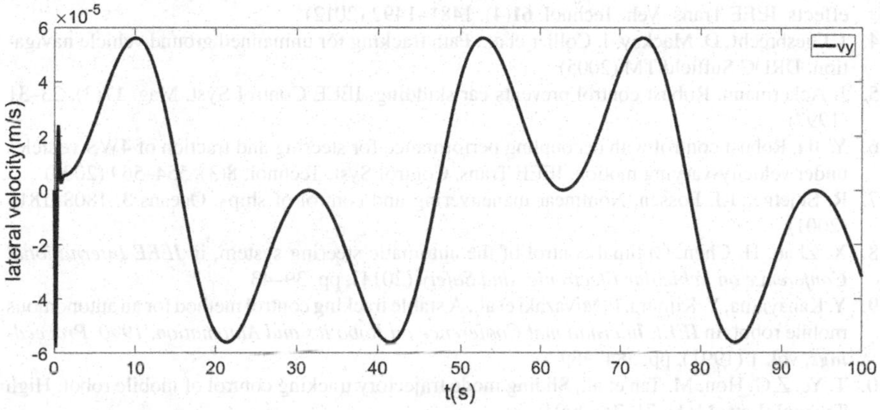


**Fig. 7** Lateral velocity

tend to the same value. All these results show that the system connected by two layers works very well and the trajectory tracking is well achieved.

## 6 Conclusion

In this paper, a novel method is presented that solves the problem of trajectory tracking of intelligent vehicles. Two layers (kinetic control and kinematical control) of the tracking system are designed independently. The execution layer decouples the 3-DOF movements of vehicles into three independent motions and each motion is controlled by a two order subsystem. Based on decoupling performance, longitudinal velocity and yaw rate are designed to track their desired value independently in the subsystem and lateral velocity sets zero. The decision layer calculates the desired motion of the vehicles according to the error between the current position and the reference position. The tracking error model is solved by sliding mode control, which makes the tracking error very close to zero. Simulation results show the new system combined by two layers works well and trajectory tracking problem is well solved. Future work may concentrate on the observer design for kinetic model due to the expensive sensor for vehicles.

## References

1. J. Ackermann, Safe and comfortable travel by robust control, in *Trends in Control*, London (1995), pp. 1–16
2. R. Rajkumar, A cyber-physical future. Proc. IEEE **100**, 1309–1312 (2012)
3. S.A. Arogeti, N. Berman, Path following of autonomous vehicles in the presence of sliding effects. IEEE Trans. Veh. Technol. **61**(4), 1481–1492 (2012)
4. J. Giesbrecht, D. Mackay, J. Collier et al., Path tracking for unmanned ground vehicle navigation. DRDC Suffield TM (2005)
5. J. Ackermann, Robust control prevents car skidding. IEEE Control Syst. Mag. **17**(3), 23–31 (1997)
6. Y. Jia, Robust control with decoupling performance for steering and traction of 4WS vehicles under velocity-varying motion. IEEE Trans. Control Syst. Technol. **8**(3), 554–569 (2000)
7. R. Skjetne, T.I. Fossen, Nonlinear maneuvering and control of ships. Oceans **3**, 1808–1815 (2001)
8. X. Zhao, H. Chen, Optimal control of the automatic steering system, in *IEEE International Conference on Vehicular Electronics and Safety* (2011), pp. 39–43
9. Y. Kanayama, Y. Kimura, F. Miyazaki et al., A stable tracking control method for an autonomous mobile robot, in *IEEE International Conference on Robotics and Automation, 1990. Proceedings*, vol. 1 (1991), pp. 384–389
10. T. Ye, Z.G. Hou, M. Tan et al., Sliding mode trajectory tracking control of mobile robot. High Technol. Lett. **11**(1), 71–74 (2004)

# Smart Home System Based on Open Source Hardware Development Platform

**Lei Yu, Changdi Li, Haina Ji, Tianyuan Miao, Yongju Zhang and Wei Zhang**

**Abstract** Smart home products have quickly integrated into our lives. The smart home comprehensively utilizes advanced various hybrid smart technologies to organically combine home electrical devices related to the home life field and regards each electrical device as a subsystem. This article selected an open source hardware platform to design a smart home system. It combines long-distance communication and short-distance communication, and wired communication and wireless communication combine to enable mobile phones to control home appliances anytime and anywhere. This set of smart home control management system has the characteristics of low price, superior performance, and has a very large market size and potential.

**Keywords** Smart home · Open source hardware platform · Computer technology
Home appliances

## 1 Introduction

Smart home products have been quickly integrated into our lives. For example, in the morning you can wake up with soothing music, such as smart lights that can be controlled at your heart, central controls, lights to the touch, lights to go off, smart

L. Yu (✉) · C. Li · H. Ji · T. Miao
School of Mechanical and Electric Engineering, Soochow University, Suzhou, China
e-mail: slender2008@163.com

L. Yu
Key Laboratory of Intelligent Perception and Systems for High-Dimensional Information of Ministry of Education, Nanjing University of Science and Technology, Nanjing, China

L. Yu
Collaborative Innovation Center of Industrial Energy-Saving and Power Quality Control, Anhui University, Hefei, China

Y. Zhang · W. Zhang
Special Equipment Safety Supervision Inspection Institute of Jiangsu Province, Wujiang Branch, Wujiang, China



Y. Jia et al. (eds.), *Proceedings of 2018 Chinese Intelligent Systems Conference*,
Lecture Notes in Electrical Engineering 528,
https://doi.org/10.1007/978-981-13-2288-4_7

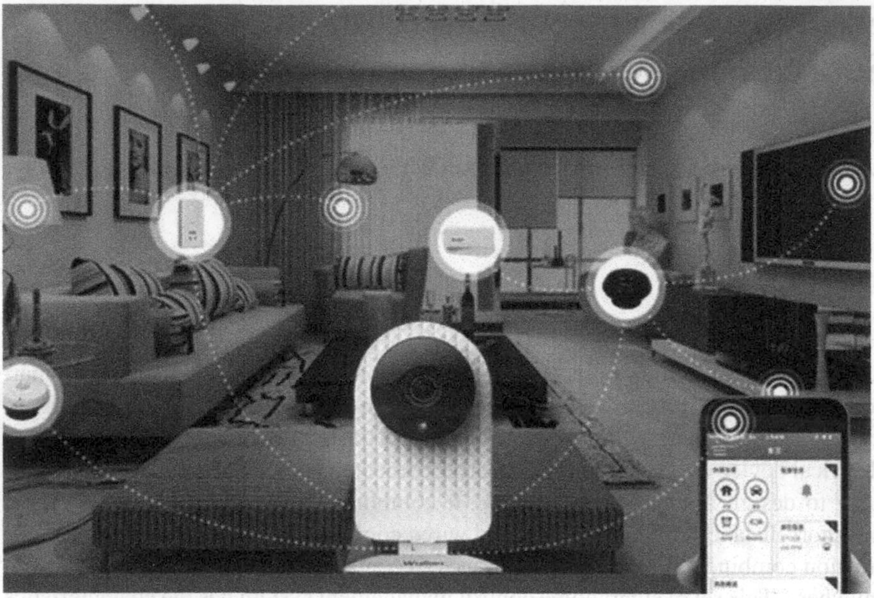

Fig. 1 Smart home products

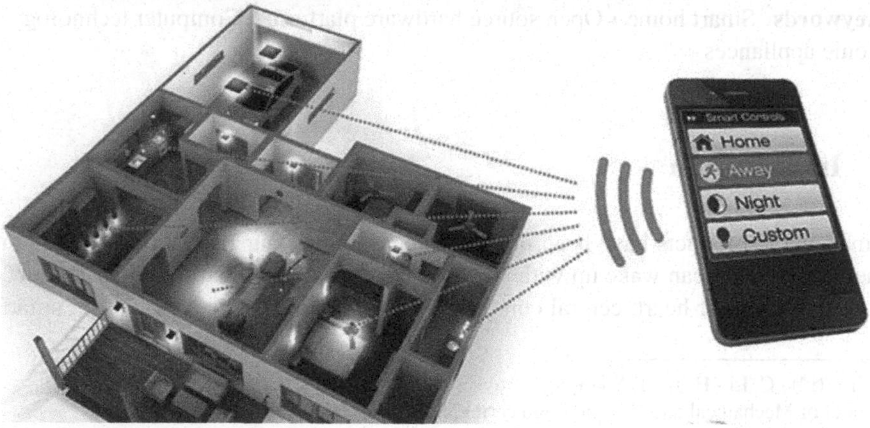


Fig. 2 The smart home system diagram

door locks, and visuals in Fig. 1. Intercom and so on, these all play an important role in our lives. Smart homes usually can be set up as a comprehensive management system in Fig. 2, featuring intelligence, information, energy saving, and humanity [1–5].

Today's more famous smart home systems include Xiaomi's Mijia system (centered on Xiaomi's smart hardware and mobile phone APP), Apple's HOMEKIT (authorized third-party manufacturers to produce compatible smart hardware) and so on. Xiaomi, Apple and other large companies have powerful technical advantages, and the introduction of smart home products has initially formed a certain market size. In recent years, there have been a large number of open source programs for the control of home appliances in the open source hardware area of the Internet. Using the built-in class libraries, a simple LED lamp controlled mobile phone can be developed or a mobile phone remote control air conditioner can be realized [4, 6–10]. There are many smart home reference cases on smart home community forums. However, the means of implementation adopted by them vary widely and are relatively fragmented. Just for communication, there are Bluetooth communication, WiFi communication, zigbee communication, serial communication and so on [4, 9, 11–13]. This article selects an open source platform for network hardware, thereby reducing the difficulty of program preparation and quickly implementing an intelligent smart home system.

This article will focus on building a home appliance control platform, using an open source hardware platform to achieve control of indoor lighting, temperature, humidity and other parameters through the mobile phone appliance. Smart home is currently a new industry with development potential. In the future, with the further implementation of the promotion and popularization of smart home products, consumer demand will become increasingly diversified and intelligent. Therefore, the consumption potential of the smart home market is bound to be huge, and the industrial outlook is bright.

## 2 Smart Home Control System Based on Hardware Open Source Platform

The smart home system integrates various conventional technologies (such as intelligent control technology, linear system technology, cabling technology, network technology, audio intelligent technology, etc.) to implement organic integration and continuously improves the convenience of smart home system [5, 10, 13–15]. The main technical solution of the smart home control system based on the hardware open source platform designed in this paper is to firstly complete the network open source hardware platform construction, and simultaneously compile open source control program and mobile terminal control software. The hardware of the system consists of: control board, sensor module, home appliance, relay/wireless switch/smart socket module (for controlling non-infrared remote-control home appliances), infrared signal transmission module (for controlling infrared remote-control home appliances), Bluetooth module (for building a Bluetooth intranet, sensor module (collecting room information), WiFi module (connecting to an external network).

The overall hardware framework is shown in Fig. 3. The working principle of the entire product is that the sensor collects indoor information and transmits it to the main board. It interacts with the mobile phone through the external network, returns command information, and transmits information to the infrared information transmission module and relay/wireless switch/smart socket module via the Bluetooth

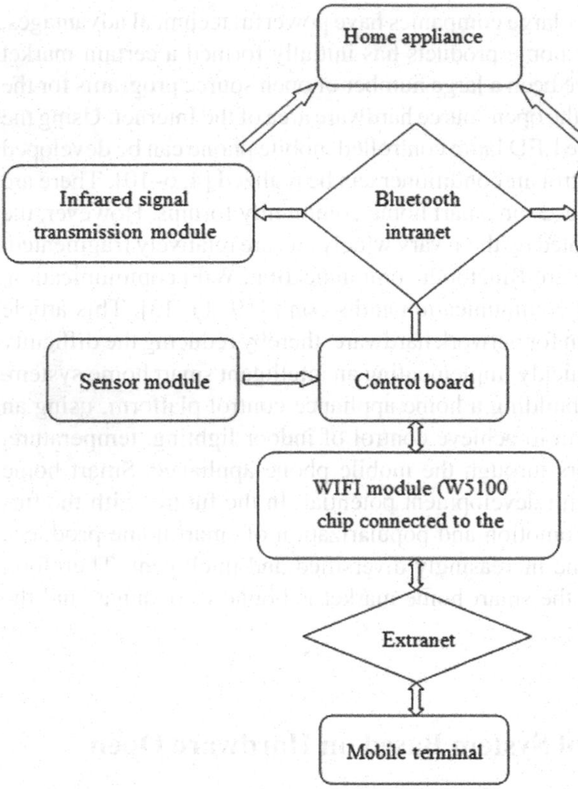


**Fig. 3** Diagram of hardware design of the overall framework

intranet to control home appliances. The Bluetooth internal network consists of a main board (upper computer), a lower computer, and a HC-05 Bluetooth module. The open source hardware platform and the Bluetooth module are connected via a serial port, and the Bluetooth modules communicate wirelessly. The software function design framework is shown in Fig. 4. The arrow that the program sends points to the function that it wants to realize. For different functions, use the open source libraries on the network rationally, and study the user-friendly interface design program carefully while developing the mobile terminal to make the software used as easy as possible, control the hardware and software costs better, achieve a low price, high performance smart home system easily.

The smart home system based on the open source hardware platform designed by this article has many advantages and characteristics, as follows: (1) Using an open source hardware development platform for the network results in lower cost, shorter development cycle, easier program modification and debugging, and more convenient development of new functions. It is less difficult to maintain software in later stages. For example, if the development hardware open source board of the product

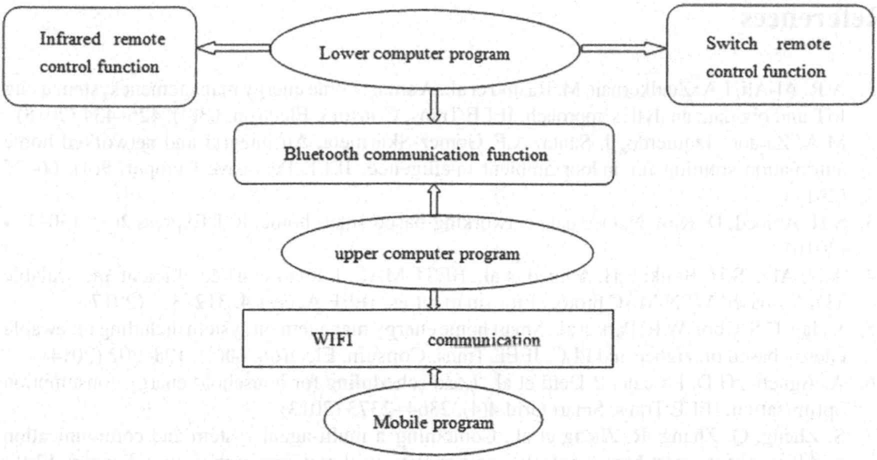


**Fig. 4** Diagram of software function design of the framework

fails, users follow the prompts to purchase a new board and follow the instructions to burn the program onto the board, even without any electronic technology foundation. (2) The combination of long-distance communication and short-distance communication, wired communication and wireless communication, can flexibly adjust the number of lower computers and upper computers and improve the indoor network structure. Avoid the security problems caused by a single communication method. (3) There is no need to purchase expensive and matched smart home appliances, and the adaptability is relatively strong. Not only can it be controlled on the mobile phone, but it can also be accessed via a computer. It is more convenient than the HOMEKIT system.

# 3 Conclusion

This article designed a smart home system based on open source hardware platform, achieving a new home life experience through a comprehensive network of integrated intelligent control and management. According to the new-generation artificial intelligence development plan issued by the State Council, by the end of 2020, China's major industrial value in the core areas of artificial intelligence (AI) will reach 1 trillion Yuan, bringing the scale of relevant industries to more than 10 trillion Yuan. Among these industries, home is the one most closely connected with the daily life of the general public, and undoubtedly a key point in the development of the artificial intelligence industry. Smart home products have now become an important trend in the development of the home industry. With the continuous improvement of the level of smart home product technology and the continuous accumulation of installation technologies, more users can see the high quality performance of smart products.

## References


1. A.R. Al-Ali, I.A. Zualkernan, M. Rashid et al., A smart home energy management system using IoT and big data analytics approach. IEEE Trans. Consum. Electron. **63**(4), 426–434 (2018)
2. M.A. Zamora-Izquierdo, J. Santa, A.F. Gomez-Skarmeta, An integral and networked home automation solution for indoor ambient intelligence. IEEE Pervasive Comput. **9**(4), 66–77 (2011)
3. S.H. Ahmed, D. Kim, Named data networking-based smart home. ICT Express **2**(3), 130–134 (2016)
4. A.N. Alvi, S.H. Bouk, S.H. Ahmed et al., BEST-MAC: bitmap-assisted efficient and scalable TDMA-based WSN MAC protocol for smart cities. IEEE Access **4**, 312–322 (2017)
5. J. Han, C.S. Choi, W.K. Park et al., Smart home energy management system including renewable energy based on zigbee and PLC. IEEE Trans. Consum. Electron. **60**(2), 198–202 (2014)
6. A. Agnetis, G.D. Pascale, P. Detti et al., Load scheduling for household energy consumption optimization. IEEE Trans. Smart Grid **4**(4), 2364–2373 (2013)
7. S. Zheng, Q. Zhang, R. Zheng et al., Combining a multi-agent system and communication middleware for smart home control: a universal control platform architecture. Sensors **17**(9), 2135 (2017)
8. J.Y. Son, J.H. Park, K.D. Moon et al., Resource-aware smart home management system by constructing resource relation graph. IEEE Trans. Consum. Electron. **57**(3), 1112–1119 (2011)
9. D.M. Han, J.H. Lim, Smart home energy management system using IEEE 802.15.4 and zigbee. IEEE Trans. Consum. Electron. **56**(3), 1403–1410 (2010)
10. M. Pipattanasomporn, M. Kuzlu, S. Rahman, An algorithm for intelligent home energy management and demand response analysis. IEEE Trans. Smart Grid **3**(4), 2166–2173 (2012)
11. P. Vlacheas, R. Giaffreda, V. Stavroulaki et al., Enabling smart cities through a cognitive management framework for the internet of things. IEEE Commun. Mag. **51**(6), 102–111 (2013)
12. L. Liu, Y. Liu, L. Wang et al., Economical and balanced energy usage in the smart home infrastructure: a tutorial and new results. IEEE Trans. Emerg. Top. Comput. **3**(4), 556–570 (2015)
13. S. Rani, R. Talwar, J. Malhotra et al., A novel scheme for an energy efficient Internet of Things based on wireless sensor networks. Sensors **15**(11), 28603–28626 (2015)
14. S. Rani, S.H. Ahmed, *Multi-hop Reliability and Network Operation Routing. Multi-hop Routing in Wireless Sensor Networks* (Springer, Singapore, 2016), pp. 29–44
15. S.Y. Shin, H.S. Park, S. Choi et al., Packet error rate analysis of ZigBee under WLAN and Bluetooth interferences. IEEE Trans. Wirel. Commun. **6**(8), 2825–2830 (2007)

# Exponential Stability of Neural Network with General Noise

**Xin Zhang, Yiyuan Zheng, Yiming Gan, Wuneng Zhou, Yuqing Sun and Lifei Yang**

**Abstract** The problem of exponential stability of neural network (NN) with general noise is considered in this article. The noise in our neural network model which can be a mixture of white and non-white noise is more suitable for real nervous systems than white noise. By utilizing the random analysis method and Lyapunov functional method techniques, we obtain the conditions of the exponential stability for neural network with general noise. Unlike the NN with white noise in the existing papers, which are modeled as stochastic differential equations, our model with general noise is based on the random differential equations. Finally, an illustrative example is presented to demonstrate the effectiveness and usefulness of the proposed results.

**Keywords** Exponential stability · Neural networks · General noise

## 1 Introduction

It is well known that neural network have played an significant role in many areas for their practical applications including automatic control, image processing, optimization, pattern recognition (see [1–6]). Since the applications depend heavily on the stability of the equilibrium point of neural network, there have been growing research

---

X. Zhang · Y. Gan · W. Zhou (✉) · Y. Sun
School of Information Science and Technology, Donghua University,
Shanghai 201620, China
e-mail: zhouwuneng@163.com

Y. Zheng
Wenlan School of Business, Zhongnan University of Economics and Law,
Wuhan 430073, China

Y. Gan
Bros Eastern Stock Co., Ltd., Ningbo 315040, China

L. Yang
Glorious Sun School of Business and Management, Donghua University,
Shanghai 201620, China



Y. Jia et al. (eds.), *Proceedings of 2018 Chinese Intelligent Systems Conference*,
Lecture Notes in Electrical Engineering 528,
https://doi.org/10.1007/978-981-13-2288-4_8

interests on the stability analysis for neural network, and many results about the stability of neural network have been given. For example, Arik derived several sufficient conditions for the asymptotic stability of delayed neural network in [7], Blythe and Mao discussed almost sure exponential stability for a stochastic delay neural network in [8], Li and Sun proposed the sufficient conditions to guarantee the mean square exponential stability of an equilibrium solution in [9], etc.

On the other hand, noise perturbation which is usually considered as white noise affects the stability of neural network largely. Stochastic differential equation theory such as Itô-type equation is the basic tool to study the stochastic neural network. Several stability results of different stochastic neural network with white noise perturbations have been deeply investigated in recent years, such as robust stability analysis of delayed recurrent network (see [10–17]), exponential stability of recurrent neural network with time delays [18], etc. For the framework of the stochastic integration theory, the stability of neural network with white noise was investigated and some interesting results had been obtained. But the white noise is not suit very well for many human brain neural network. Many noises under the specific environment are non-white. For example, it is more reasonable to describe the final effects of stochastic disturbances to other electric elements by stationary processes than white noise in a circuit system [19]. Therefore, the study about the neural network with general noise is important. But the study about this kind of model was blank so far. As stochastic differential equations are the basic tools to study the neural network with white noise, random differential equations are the basic tools for the neural network with general noise. In [19–22] the asymptotic stability of nonlinear system is proved by Lyapunov function. Wu [23] relaxed the stability conditions of random nonlinear systems by Lyapunov second method. Based on this method, this paper studied the exponential stability of neural networks with general noise. We not only enriched the NN models but also extended the research methods of NN.

The papers main work and contributions are as follows. In this paper, the problem of exponential stability has been investigated for NN with general noise. Using Lyapunov functional and random analysis method, some sufficient conditions have been proposed for neural network with general noise, which is more general and appropriate. And a numerical simulation example has been presented to demonstrate the theoretical analysis. The organization of the rest of this article is as follows. Section 2 gives the problem and the preliminaries. In Sect. 3, the criterions of exponential stability in mean square and $p$th moment have been set up by using Lyapunov stability theory. Section 4 presents a numerical example to demonstrate the effectiveness of the main results. Finally, Sect. 5 is conclusion.

Notations: Throughout this paper, $|\cdot|$ designates the Euclidean norm or 2-norm of a matrix X; $X^T$ represents the transpose of matrix X; $\lambda_{max}(\cdot)$ and $\lambda_{min}(\cdot)$ stand for the maximal and minimal eigenvalues of a matrix, respectively; diag$\{\rho_1, \ldots, \rho_n\}$ is a diagonal matrix with the elements $\rho_1, \ldots, \rho_n$ on the diagonal line; $E(\cdot)$ denotes the expectation operator; $\mathbb{R}_+$ is the set of all nonnegative real numbers.

## 2 Problem Formulation and Preliminaries

In this paper, the following neural network with general noise is considered:

$$\begin{aligned} dx(t) &= -Ax(t) + Bf(x(t)) + g(x(t), t)\sigma(t), \\ x(t_0) &= x_0. \end{aligned} \tag{1}$$

where $x(t) = [x_1(t), x_2(t), \ldots, x_n(t)]^T \in \mathbb{R}^n$ is the state vector associated with n neurons. In addition, $A$ =diag$\{a_1, a_2, \ldots, a_n\}$ is a diagonal matrix with $a_j > 0$, and $B = (b_{jk})_{n\times n}$ is the connection weight matrix. $f(x(t)) = [f_1(x_1(t)), f_2(x_2(t)), \ldots, f_n(x_n(t))]^T$ denotes the neuron activation function, which satisfy

$$f(0) = 0, \rho_i^- \leq \frac{f_i(x_i(t))}{x_i(t)} \leq \rho_i^+, i = 1, 2, \ldots, n \tag{2}$$

$g(x(t), t) : \mathbb{R}^n \times \mathbb{R}_+ \to \mathbb{R}^{n\times n}$ is the noise intensity matrix. $\sigma(t) = [\sigma_1(t), \sigma_2(t), \ldots, \sigma_n(t)]^T$ is the disturbance defined on a complete probability space $(\Omega, \mathcal{F}, P)$ with a natural filtration $\{\mathcal{F}_t\}_{t\geq 0}$ generated by $\{\sigma(s) : 0 \leq s \leq t\}$.

**Definition 1** The solution of the system (1) is said to be exponentially stable in $p$th moment if there exist parameters $k_1, k_2 > 0$ such that $\forall t \geq t_0, x_0 \in \mathbb{R}^n \setminus \{0\}$.

$$E|x(t)|^m < k_1|x_0|^m e^{-k_2(t-t_0)} \tag{3}$$

Stochastic process $\sigma(t) \in \mathbb{R}^l$ satisfies the following assumption:

**Assumption 1** Process $\sigma(t)$ is $\mathcal{F}_t$-adapted and piecewise continuous, and there exists a constant $K > 0$ such that

$$\sup_{t\leq t_0} E|\sigma(t)|^2 < K. \tag{4}$$

The following two preliminary assumptions ensured the existence and uniqueness of global solution of system (1).

**Assumption 2** There exists a constant $L_M > 0$, $L_0 > 0$, such that $\forall x_1, x_2 \in \mathbb{R}^n$, $x_1 \neq x_2$

$$|g(x_1(t), t) - g(x_2(t), t)| \leq L_M|x_2(t) - x_1(t)|, \tag{5}$$

$$|g(0, t)| < L_0. \tag{6}$$

The following assumption is necessary in the theory of exponential stability of random system:

**Assumption 3** For stationary process $\sigma(t)$, there exists a time function $\delta(\cdot)$ such that $\forall \varepsilon > 0$, $t_1 \neq t_0$,

$$Ee^{\varepsilon \int_{t_0}^{t_1} |\sigma(s)|ds} \leq e^{\delta(\varepsilon)(t_1 - t_0)}. \tag{7}$$

In the follows, we present a preliminary lemma which play an important role in the proof of the main theorems. We present a result about the exponential stability to the random systems with general noise.

**Lemma 1** *Under Assumptions 1–3, assume there exist a function $V \in C^1$ and constants $a_1, a_2, c_1, c_2 > 0$, such that*

$$a_1|x|^p \leq V(x) \leq a_2|x|^p, \tag{8}$$

$$\frac{\partial V}{\partial x}[-A(x(t)) + Bf(x(t))] \leq -c_1 V(x), \tag{9}$$

$$|\frac{\partial V}{\partial x} g(x,t)| \leq c_2 V(x). \tag{10}$$

*If*

$$c_1 > \delta(c_2), \tag{11}$$

*Then there exists a unique solution to system (1) and the system is exponentially stable in pth moment.*

*Proof* From (9) and (10), we have

$$\begin{aligned} \dot{V}(x(t)) =& \frac{\partial V}{\partial x}((-Ax(t) + Bf(x(t))) + \frac{\partial V}{\partial x} g(x(t),t)\xi(t) \\ \leq& (-c_1 + c_2|\xi(t)|)V(x(t)). \end{aligned} \tag{12}$$

According to Lemma 1 in [2], we can immediately deduce

$$V(x(t \wedge \tau_k)) \leq V(x_0)e^{\int_{t_0}^{t}(-c_1 + c_2|\xi(t)|)ds} = V(x_0)e^{-c_1(t-t_0)}d^{c_2 \int_{t_0}^{t} |\xi(s)|ds} < \infty. \tag{13}$$

Denote $L_3 := max\{|\rho_1^-|, \ldots, |\rho_n^-|, \rho_1^+, \ldots, \rho_n^+\}$; $L_4 := \|A\|$. We have the following inequalities

$$|-A * 0 + Bf(0)| + |g(0,t)| < L_1, \tag{14}$$

$$\begin{aligned} &|(-Ax_1 + Bf(x_1)) - (-Ax_2 + Bf(x_2))| + |g(x_1,t) - g(x_2,t)| \\ &\leq (L_2 + L_3 + L_4)|x_2 - x_1|. \end{aligned} \tag{15}$$

From Assumption 1, it is obvious that $|\sigma(t)| < \infty$. Then based on the Lemma 1 in [1], we can obtain that there exists a unique solution to system (1) on $[t_0, \infty)$.

According to Lemma 1, it comes from (8) and (11) that

$$a_1|x(t \wedge \tau_k)|^p \leq V(x(t \wedge \tau_k)) \leq a_2|x_0|^p e^{-c_1(t-t_0)} e^{c_2 \int_{t_0}^{t} |\xi(s)|ds}. \tag{16}$$

Let $k \to \infty$, we have

$$E|x(t)|^p \le \frac{a_2}{a_1}|x_0|^p e^{-c_1(t-t_0)} E e^{c_2 \int_{t_0}^t |\sigma(s)|ds}. \tag{17}$$

Using Assumption 1, one can obtain that

$$E e^{c_2 \int_{t_0}^t |\sigma(s)|ds} \le e^{\delta(c_2)(t-t_0)}. \tag{18}$$

Substituting (18) into (17), yields

$$E|x(t)|^p \le \frac{a_2}{a_1}|x_0|^p e^{-(c_1-\delta(c_2))(t-t_0)}. \tag{19}$$

This completes the proof.

*Remark 1* In this section, a general result which can be applied widely has been established. By the method of random analysis, a criterion of exponential stable in $p$th moment for system (1) is given.

## 3 Exponentially Stable in $p$th Moment of Neural Network with General Noise

In this section, we give a criterion of exponential stable in $p$th moment for system (1). Here we denote $K_M$ =diag$\{\rho_1^+,\ldots,\rho_n^+\}$, $K_m$=diag$\{\rho_1^-,\ldots,\rho_n^-\}$, $\Gamma_1$ =diag$\{\gamma_{11},\ldots,\gamma_{1n}\}$, $\Gamma_2$ =diag$\{\gamma_{21},\ldots,\gamma_{2n}\}$, $M_1 = K_m k_m$, $M_2 = \frac{K_M+k_m}{2}$, $\alpha = \lambda_{max}P$, $\beta = \lambda_{max}(-A)$, $\eta = \lambda_{max}B$, $\vartheta = \lambda_{max}U$, $\mu = \lambda_{max}(M_1U)$, $\nu = \lambda_{max}(M_2U)$, $\iota = \lambda_{max}(K_M^T\Gamma_2 - K_m^T\Gamma_1)$, $\kappa = \lambda_{max}(\Gamma_1+\Gamma_2)(K_M-K_m)$, $\psi = \lambda_{max}(\Gamma_1-\Gamma_2)$ respectively.

**Theorem 1** *Let Assumptions 1–3 hold, and $p > 2$. If there exist an $n \times n$ positive definite symmetric matrix $P$, $n \times n$ positive definite diagonal matrix $U$, such that*

$$Q_1 + Q_2 + Q_3 + Q_4 \le 0, \tag{20}$$

*where*

$$Q_1 = (p\alpha^{\frac{p}{2}}\beta - \mu + c_1(\alpha+\kappa)^{\frac{p}{2}} + p\mu^{\frac{p-2}{2}}\iota\beta),$$

$$Q_2 = \frac{1+L_M^2}{2}(p\alpha^{\frac{p}{2}}\eta - \nu + p\kappa^{\frac{p}{2}}(\iota\eta)),$$

$$Q_3 = \frac{1 + L_M^2}{2}(p\kappa^{\frac{p}{2}}\psi\beta + \nu),$$

$$Q_4 = L_M(p\kappa^{\frac{p}{2}}\psi\eta - \vartheta),$$

*then there exists a unique global solution to system (1), and its equilibrium is exponentially stable in pth moment.*

*Proof* For each $i \in \mathbb{R}_+$, choose the following Lyapunov functional

$$\begin{aligned} V =&(x^T(t)Px(t))^{\frac{p}{2}} + 2(\sum_{i=1}^{m}\gamma_{1i}\int_0^{x_i(t)}(f_i(s) - \rho_i^- s)ds)^{\frac{p}{2}} \\ &+ 2(\sum_{i=1}^{m}\gamma_{2i}\int_0^{x_i(t)}(\rho_i^+ s - f_i(s))ds)^{\frac{p}{2}}, \end{aligned} \tag{21}$$

There exist $q_1 \in [0, 1]$ such that the following inequalities hold from differential intermediate value theorem.

$$\begin{aligned} 0 \le 2\sum_{i=1}^{m}\gamma_{1i}\int_0^{x_i(t)}(f_i(s) - \rho_i^- s)ds &= 2\sum_{i=1}^{m}\gamma_{1i}x_i(t)(f_i(q_1x_i(t)) - \rho_i^- q_1x_i(t)) \\ &\le 2\sum_{i=1}^{m}\gamma_{1i}(\rho_i^+ - \rho_i^-)x_i^2(t), \end{aligned} \tag{22}$$

and there exist $q_2 \in [0, 1]$ such that

$$0 \le 2\sum_{i=1}^{m}\gamma_{2i}\int_0^{x_i(t)}(f_i(s) - \rho_i^- s)ds \le 2\sum_{j=1}^{m}\gamma_{2i}(\rho_i^+ - \rho_i^-)x_i^2(t). \tag{23}$$

So

$$\begin{aligned} (\lambda_{min}(P))^{\frac{p}{2}}|x(t)|^p \le V(x) \le&((\lambda_{max}(P))^{\frac{p}{2}} + (\lambda_{max}\Gamma_1(K_M - K_m))^{\frac{p}{2}} \\ &+ (\lambda_{max}\Gamma_2(K_M - K_m))^{\frac{p}{2}})|x(t)|^p \\ \le&(\lambda_{max}(P) + \lambda_{max}(\Gamma_1 + \Gamma_2)(K_M - K_m))^{\frac{p}{2}}|x(t)|^p. \end{aligned} \tag{24}$$

Then (8) holds, where $a_1 = (\lambda_{min}P)^{\frac{p}{2}}$, $a_2 = (\lambda_{max}P + \lambda_{max}(\Gamma_1 + \Gamma_2)(K_M - K_m))^{\frac{p}{2}}$. Furthermore,

$$
\begin{aligned}
\frac{\partial V}{\partial x}((-Ax(t)+Bf(x(t))) \leq\ & p(x^T(t)Px(t))^{\frac{p-2}{2}}x^T(t)P((-Ax(t)+Bf(x(t)) \\
&+p(\sum_{i=1}^{m}\gamma_{1i}\int_0^{x_i(t)}(f_i(s)-\rho_i^- sds)^{\frac{p-2}{2}}(f(X(t)) \\
&-K_m x(t))^T\Gamma_1((-Ax(t)+Bf(x(t)) \\
&+p(\sum_{j=1}^{m}\gamma_{2i}\int_0^{x_i(t)}(\rho_i^+ s - f_i(s)ds)^{\frac{p-2}{2}}(K_M x(t). \\
&-f(X(t)))^T\Gamma_2((-Ax(t)+Bf(x(t)). \\
\leq & p(\lambda_{max}P)^{\frac{p-2}{2}}|x(t)|^{p-2}x^T(t)P(Ax(t)+Bf(x(t))) \\
&+p(\lambda_{max}(\Gamma_1+\Gamma_2)(K_M-K_m))^{\frac{p-2}{2}}|x(t)|^{p-2}((f(x(t)) \\
&-K_m x(t))^T\Gamma_1+(K_M x(t) \\
&-f(x(t)))^T\Gamma_2)(Ax(t)+Bf(x(t)))
\end{aligned}
\tag{25}
$$

From (2), there exists a diagonal matrix $U$ such that

$$
\begin{aligned}
&-|x(t)|^{p-2}(x^T(t)M_1Ux(t)-f^T(x(t))M_2Ux(t) \\
&-x^T(t)M_2Uf(x(t))+f^T(x(t))Uf(x(t)))\geq 0.
\end{aligned}
\tag{26}
$$

So

$$
\begin{aligned}
\Delta_1 :=& \frac{\partial V}{\partial x}((-Ax(t)+Bf(x(t)))+c_1V \\
\leq&(p\alpha^{\frac{p}{2}}\beta-\mu+c_1(\alpha+\kappa)^{\frac{p}{2}}+p\mu^{\frac{p-2}{2}}\iota\beta)|x(t)|^p \\
&+(p\alpha^{\frac{p}{2}}\eta-\nu+p\kappa^{\frac{p}{2}}(\iota\eta))|x(t)|^{p-2}(\frac{x^T(t)x(t)}{2}+\frac{f^T(x(t))f(x(t))}{2}) \\
&+(p\kappa^{\frac{p}{2}}\psi\beta+\nu)|x(t)|^{p-2}(\frac{x^T(t)x(t)}{2}+\frac{f^T(x(t))f(x(t))}{2}) \\
&+(p\kappa^{\frac{p}{2}}\psi\eta-\vartheta)|x(t)|^{p-2}f^T(x(t))f(x(t)) \\
=&(Q_1+Q_2+Q_3+Q_4)|x(t)|^p.
\end{aligned}
\tag{27}
$$

where

$$
c_1=\delta(2p(L_M^2+h)+\varepsilon)+\varepsilon,\ c_2=2p(L_M^2+h)+\varepsilon, \tag{28}
$$

$h \leq 0$ is a constant, $\varepsilon$ is an arbitrarily small positive constant.

$$\begin{aligned}\Delta_2 :=& |\frac{\partial V}{\partial x} g(x(t), t)| - c_2 V \\ \leq& p(L_M^2 + 1)(\alpha^{\frac{p}{2}} + \kappa^{\frac{p}{2}}) - c_2(\alpha + \kappa)^{\frac{p}{2}} \\ =& (2p(L_M^2 + 1) - c_2)(\alpha + \kappa)^{\frac{p}{2}} \\ \leq& 0.\end{aligned} \tag{29}$$

Therefore, by Lemma 1, the system (1) is exponential stability in $p$th moment. This completes the proof.

From Theorem 1, a corollary about the exponential stable in mean square for system (1) can be easily obtained. Here we denote $\alpha = \lambda_{max} P$, $\beta = \lambda_{max}(-A)$, $\eta = \lambda_{max} B$, $\vartheta = \lambda_{max} U$, $\mu = \lambda_{max}(M_1 U)$, $\nu = \lambda_{max}(M_2 U)$, $\gamma_1 = \lambda_{max} \Gamma_1$, $\gamma_2 = \lambda_{max} \Gamma_2$, $\rho^- = \lambda_{max} K_m$, $\rho^+ = \lambda_{max} K_M$, respectively.

**Corollary 1** *Let Assumptions 1–3 hold, and $p = 2$, if there exists an $n \times n$ positive definite symmetric matrix $P$, $n \times n$ positive definite diagonal matrix $U$, such that*

$$Q_1 + Q_2 + Q_3 + Q_4 \leq 0, \tag{30}$$

where

$$Q_1 = 2\alpha\beta - 2\rho^- \gamma_1 \beta + 2\rho^+ \gamma_2 \beta + c_1(\alpha + 2\kappa) - \mu,$$

$$Q_2 = (2\alpha\eta - 2\rho^- \gamma_1 \eta + 2\rho^+ \gamma_2 \eta - \upsilon)\frac{1 + L_M^2}{2},$$

$$Q_3 = (2\gamma_1 \beta - 2\gamma_2 \beta - \upsilon)\frac{1 + L_M^2}{2},$$

$$Q_4 = (2\gamma_1 \eta - 2\gamma_2 \eta + \vartheta) L_M^2,$$

then there exists a unique global solution to system (1), and its equilibrium is exponentially stable in $p$th moment.

*Proof* For each $i \in \mathbb{R}_+$, choose the following Lyapunov functional

$$\begin{aligned}V =& x^T(t) P x(t) + 2\sum_{i=1}^{m} \gamma_{1i} \int_0^{x_i(t)} (f_i(s) - \rho_i^- s) ds \\ &+ 2\sum_{i=1}^{m} \gamma_{2i} \int_0^{x_i(t)} \rho_i^+ s - f_i(s) ds,\end{aligned} \tag{31}$$

The proof next is similar in spirit to Theorem 1.

*Remark 2* In this section, some conditions of exponential stable in $p$th moment for system (1) are given by the method of random analysis and Lyapunov functional.

According to the general result given in Lemma 1, it is easy to derive the criterions in this section. And we divided the proof into two cases according to the value of $p$, corresponding to Theorem 1 and Corollary 1 respectively.

# 4 Numerical Example

In this section, a numerical example will be presented to illustrate our main results.

*Example 1* Consider a two-neuron neural network (1) with initial condition $x(t_0) = 0$, and

$$A_1 = \begin{bmatrix} 14.9 & 0 \\ 0 & 5.3 \end{bmatrix}, \; B_1 = \begin{bmatrix} -1.8 & -4.1 \\ -1.2 & -1.4 \end{bmatrix}.$$

The neuron activation function is $f(x) = \tanh(x)$, the noise intensity function $g(\cdot)$ is $g(\cdot) = \sin(x)$ and the noise function $\sigma(\cdot)$ satisfies $\sigma(t) = a\sin(\lambda t + U)$, where $a = 6$, $\lambda = 0.05$, $U$ is a random variable distributed on the interval $[0, \pi]$, which is shown in Fig. 1. We set

$$Q_1 = \begin{bmatrix} 0.4212 & -0.2209 \\ -0.2209 & 1.1878 \end{bmatrix} \times 10^4, \; U_1 = \begin{bmatrix} 1.7574 & 0 \\ 0 & 0.9474 \end{bmatrix},$$

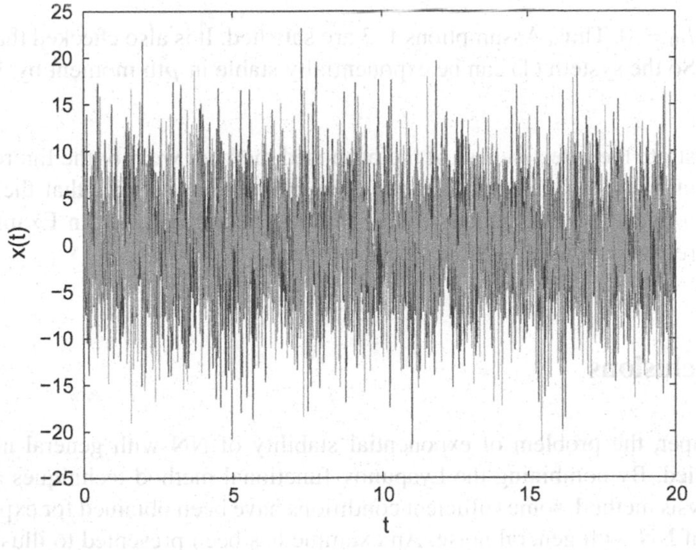


**Fig. 1** The noise of the NN

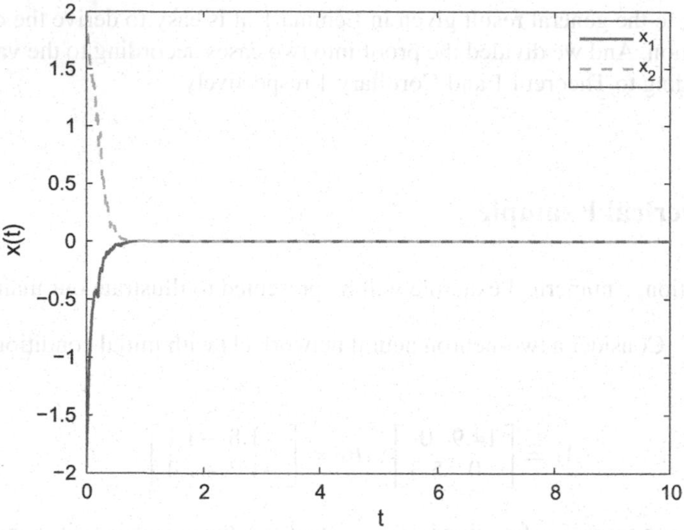


**Fig. 2** The curve of system state

$$K_m = \begin{bmatrix} 0.1 & 0 \\ 0 & 0.1 \end{bmatrix}, \; K_M = \begin{bmatrix} 2.6325 & 0 \\ 0 & 2.6325 \end{bmatrix},$$

$$\Gamma_1 = \begin{bmatrix} 0.0035 & 0 \\ 0 & 0.9474 \end{bmatrix}, \; \Gamma_2 = \begin{bmatrix} 0.0018 & 0 \\ 0 & 0.0115 \end{bmatrix},$$

$L_M = 1, L_0 = 0$. Thus, Assumptions 1–3 are satisfied. It is also checked that (20) is satisfied. So the system (1) can be exponentially stable in $p$th moment by Theorem 1.

To illustrate the effectiveness of the proposed method, we plot the figures of the time evolution of the state variable $x(t)$. We can see from Fig. 2 that the state of system (1) tends to zero, which verifies that the neural network in Example 1 is almost surely exponentially stable in $p$th moment.

## 5 Conclusions

In this paper, the problem of exponential stability of NN with general noise has been studied. By combining the Lyapunov functional method techniques and random analysis method, some sufficient conditions have been obtained for exponential stability of NN with general noise. An example has been presented to illustrate the effectiveness and potential of the proposed results.

**Acknowledgements** This work was partially supported by the Natural Science Foundation of China (grant no. 61573095).

## References

1. A. Cichocki, R. Unbehauen, *Neural Networks for Optimization and Signal Processing* (Wiley, Hoboken, NJ, 1993)
2. Z. Wu, P. Shi, H. Su, J. Chu, Passivity analysis for discrete-time stochastic Markovian jump neural networks with mixed time delays. IEEE Trans. Neural Netw. **22**(10), 1566–1575 (2011)
3. W. Zhou, Q. Zhu, P. Shi, H. Su, J. Fang, L. Zhou, Adaptive synchronization for neutral-type neural networks with stochastic perturbation and Markovian switching parameters. IEEE Trans. Cybern. **44**(12), 2848–2860 (2014)
4. J. Zhou, X. Ding, L. Zhou, W. Zhou, J. Yang, D. Tong, Almost sure adaptive asymptotically synchronization for neutral-type multi-slave neural networks with Markovian jumping parameters and stochastic perturbation. Neurocomputing **214**(19), 44–52 (2016)
5. L.O. Chua, L. Yang, Cellular Neural Networks: applications. IEEE Trans. Circuits Syst. **35**(1), 1273–1290 (1988)
6. Z.G. Wu, P. Shi, H. Su, Stochastic synchronization of Markovian jump neural networks with time-varying delay using sampled data. IEEE Trans. Cybern. **43**(6), 1796–1806 (2013)
7. S. Arik, Stability analysis of delayed neural networks. IEEE Trans. Circuits Syst. **47**(7), 1089–1092 (2000)
8. S. Blythe, X. Mao, X. Liao, Stability of stochastic delay neural networks. J. Frankl. Inst. **338**(4), 481–495 (2001)
9. W. Li, J. Sun, Mean square exponential stability of stochastic delayed Hopfield neural networks. Phys. Lett. A **343**(4), 306–318 (2005)
10. X. Mao, *Stochastic differential equations and applications* 2nd edn. (Horwood Publishing Chichester, Chichester, 2008)
11. S. Xu, J. Lam, A survey of linear matrix inequality techniques in stability analysis of delay systems. Int. J. Syst. Sci. **39**(12), 1095–1113 (2008)
12. W. Zhou, D. Tong, Y. Gao, C. Ji, H. Su, Mode and delay-dependent adaptive exponential synchronization in pth moment for stochastic delayed neural networks with markovian switching [J]. IEEE Trans. Neural Netw. Learn. Syst. **23**(4), 662–668 (2012)
13. C. Li, X. Liao, Robust stability and robust periodicity of delayed recurrent neural networks with noise disturbance. IEEE Trans. Circuits Syst. Video Technol. **53**(10), 2265–2273 (2006)
14. P. Shi, Y. Xia, G.-P. Liu, D. Rees, On designing of sliding-mode control for stochastic jump system. IEEE Trans. Autom. Control **51**(1), 97–103 (2006)
15. P. Yan, H. *ö*zbay, Stability analysis of switched time delay systems. SIAM J. Control Optim. **2**(47), 936–949 (2008)
16. X.M. Sun, J. Zhao, D. Hill, Stability and $L_2$-gain analysis for switched delay systems: a delay-dependent method. Automatica **10**(42), 1769–1774 (2006)
17. X. Mao, C. Yuan, *Stochastic differential equations with Markovian switching* (Imperial College Press, London, 2006)
18. Y. Shen, J. Wang, Robustness analysis of global exponential stability of recurrent neural networks in the presence of time delays and random disturbances. IEEE Trans. Neural Netw. **23**(1), 87–96 (2012)
19. J. Bertram, P. Sarachik, Stability of circuits with randomly timevarying parameters. IRE Trans. Inf. Theory **5**(5), 260–270 (1959)
20. T.T. Soong, Random differential equations in science and engineering. J. Appl. Mech. **103**(5), 372–402 (1973)
21. Y. Jia, Robust control with decoupling performance for steering and traction of 4WS vehicles under velocity-varying motion. IEEE Trans. Control Syst. Technol. **8**(3), 554–569 (2000)

22. Y. Jia, Alternative proofs for improved LMI representations for the analysis and the design of continuous-time systems with polytopic type uncertainty: a predictive approach. IEEE Trans. Autom. Control **48**(8), 1413–1416 (2003)
23. Z. Wu, Stability criteria of random nonlinear systems and their applications. IEEE Trans. Autom. Control **60**(4), 1038–1049 (2015)

# Fault Detection Method Based on the Monitoring State Synchronization for Industrial Process System

**Hao Ren, Yi Chai, Jian-feng Qu, Ke Zhang and Qiu Tang**

**Abstract** Currently, with the requirements of quality, safety and service in many modern industrial process systems, fault detection and diagnosis become a significant issue to ensure the high control performances. Under these circumstances, this paper presents a fault detection method based on the monitoring state synchronization to perceive and evaluate the system operation situation. This model firstly considers the system states measured by some sensors as the nodes and defines the distributed monitoring state network with changing dynamically over time, all of which have been given their definition, calculation method and actual physical meaning, and finally its synchronization state of distributed network can be employed to achieve fault detection. Furthermore, this method can be used to provide a novel and feasible research method for the global assessment of the operation states and for monitoring the local operation fault of modern industrial process systems. The application of this method on a simple multivariate process system example shown that it can not only track the operation state of the whole system well to detect the fault, and not only monitor the situation of each distributed network node in real time to achieve fault diagnosis, but also make use of the correlation between the network node states to effectively locate the system operation fault.

**Keywords** Industrial process system · Monitoring State Synchronization · Fault Detection

---

H. Ren · Y. Chai (✉) · J. Qu · K. Zhang · Q. Tang
Key Laboratory of Complex System Safety and Control, Ministry of Education and the School of Automation, Chongqing University, Chongqing 400044, People's Republic of China
e-mail: chaiyi@cqu.edu.cn

H. Ren
e-mail: renhao@cqu.edu.cn

J. Qu
e-mail: qujianfeng@cqu.edu.cn

K. Zhang
e-mail: zhangke@cqu.edu.cn

Q. Tang
e-mail: tangqiu@cqu.edu.cn



Y. Jia et al. (eds.), *Proceedings of 2018 Chinese Intelligent Systems Conference*, Lecture Notes in Electrical Engineering 528,
https://doi.org/10.1007/978-981-13-2288-4_9

## 1 Introduction

With the development of modern computers, sensors and communications technologies, the scale of modern industrial process systems becomes more and more complex; the degree of automation becomes more and more higher; the requirements of operation fault analysis and assessment become higher and higher. In modern industrial process systems, a large number of facilities and subsystems are coupled together, which have been bound to bring a series of challenges [1–3]. Generally, fault can be considered as the state that the system performs its function under the conditions of acceptable minimum accident loss, viz fault can be defined as at least one system characteristic or the variable appears one deviation that is not allowed. Therefore, it is extremely difficult to establish an accurate and complete mathematical model to achieve fault detection for distribution industrial process system [4–12].

In the past decades, there exists in the literature one main method namely process monitoring and evaluation methods [11, 13] to achieve fault detection for complex industrial process systems. Fault analysis and assessment based on process monitoring can be considered as the effective monitoring method, which can be employed to track the health state of the system in time by using the measured process and condition parameters [2, 13–18].

With the reason that the amazing size of data obtained from modern industrial process systems, it has accelerated the progresses and breakthroughs in related fields. Currently, process data-sets are applied to improve the safety of modern industrial process system, and the research on information science could be further benefited from the process monitoring-based techniques to employ considerable amount of data-sets more efficiently [19–21].

Generally, the process monitoring-based methods can be considered as the reveal in early period via using multivariate statistic analysis methods for operation fault detection [19]. The method via the monitoring data to achieve real-time monitoring of system states can often be employed in practical engineering [22].

It should be noted that the process system fault can be reflected by some indicators, which are called operation state indicators and the information of these indicators can be obtained by the process condition monitoring. In the actual engineering system, the states of the dynamic system changes constantly during the operation, which make the index reflecting the state of modern industrial process system changed [6–9, 20, 21].

There are obvious some highlights in applying monitoring state synchronization to achieve fault detection for industrial process monitoring. Its main advantages can be considered as the following aspects [13]:

**(i)** Suitability to describe the virtual and discrete system. Generally, industrial process system can be considered as that is similar to the complex system in physical structures, all of which have a large amount of vertexes and strong relationships between vertexes and their neighboring vertexes [13].

**(ii)** Global assessment and local monitoring. Complex network can be considered as an intersection between graph theory and statistical mechanics, which can been employed to solve the problem of traditional pattern recognition [13].

The remainder of this paper can be organized as follows. In Sect. 2, basic theory of dynamical network synchronization has been elaborated, and the comprehensive and novel method based on monitoring state synchronization has been proposed to achieve fault detection for distributed industrial process system. Section 3 presents a case study that can be used to illustrate the effectiveness of the proposed methods. Conclusions and further works are also given in Sect. 4.

## 2 Basic Theory of Fault Detection Method

Consider a dynamical network consisting of $N$ nonidentical nodes with linearly diffusive couplings and each node can be considered as an $n$-dimensional dynamic system, and it can be described as follows.

$$\dot{x}_i(t) = f_i(t, x_i(t)) + \sigma \sum_{j=1}^{N} \delta_{ij} \Gamma g_i(t, x_i(t)), i = 1, 2, \ldots, N. \quad (1)$$

where $x_i(t) = x_{i1}(t), x_{i2}(t), \ldots, x_{in}(t)^T \in \Re^n$ donates the state variable of the $i-th$ node, $f_i, g_i : \Re^+ \times \Re^n \rightarrow \Re^n$ is the vector-valued continuous function, the positive constant $\sigma$ donates the coupling strength, the inner-coupling matrix $\Gamma \in \Re^{nn}$ represents positive definite, and $\Im = (\delta_{ij})_{N \times N}$ represents the coupling configuration matrix, which meets the diffusive condition donated by $\delta_{ij} = -\sum_{j \neq i} \delta_{ij}$.

**Definition 1** Suppose that the synchronization manifold of dynamical network can be donated by $\Re^{n \times N}$, and its linear sub-space donated by $M = \{x = (x_1, x_2, \ldots, x_N) : x_i = x_j \in \Re^n, \forall i, j = 1, 2, \ldots, N\}$. If $t \rightarrow \infty$, and the solution donated by $x = (x_1, x_2, \ldots, x_N)$ of Eq. (1) convergence into $M$, then the dynamical network described by Eq. (1) can be considered as the synchronization state, viz the following equation can be established under any initial conditions for all the nodes of the dynamical network ($\varepsilon > 0$).

$$\|x_i(t) - x_j(t)\| \leq Me^{-\varepsilon t}, t \rightarrow \infty, i, j = 1, 2, \ldots, N. \quad (2)$$

**Definition 2** Definite the synchronization state, and it can be described as follows, viz isolated node solution $s(t)$.

$$\dot{s}(t) = f(s(t)). \quad (3)$$

**Definition 3** If $\delta_{ij} \geq 0, \delta_{ij} = -\sum_{j=1, i \neq j}^{m} \delta_{ij}, j = 1, 2, \ldots, m$, and $rank(\Re) = N - 1$, then the left eigenvalue $\xi = (\xi_1, \xi_2, \ldots, \xi_m)$ corresponding to minimal eigenvalue of matrix $\Im$ are non-negative, viz $\xi_i \geq 0$, $1 \leq i \leq N$. Suppose $\sum_{j=1}^{N} \xi_i = 1$, let $x_i$ weight with the components of the left eigenvalue of $\Re$, and it can be described as follows.

$$\bar{x}(t) = \sum_{j=1}^{N} \xi_j x_j(t). \tag{4}$$

where $x_i(t)$ is the solution of the dynamical network described by Eq. (1), $\bar{x}(t)$ donates the weighted average state.

**Theorem 1** *If the dynamical network described by Eq. (1) can be considered as the synchronization state in the sense of Definition 1, then its necessary and sufficient conditions can be described as follows.*

$$\lim_{t \to \infty} \|x_i(t) - \bar{x}(t)\| = 0, \forall i, j = 1, 2, \ldots, N. \tag{5}$$

**Theorem 2** *Assume that $f_i(\cdot)$ is linear homogeneous one. If the network is in synchronization state, then synchronization state $\bar{x}(t) = \sum_{j=1}^{N} \xi_j x_j(t)$ can be considered as the isolated node solution, viz $\bar{x}(t)$ meets the following equation.*

$$\dot{\bar{x}}(t) = f(\bar{x}(t)). \tag{6}$$

**Theorem 3** *Assume that $f_i(\cdot)$ meets the Lipschitz condition, viz there exists the constant $L > 0$, which can be used to hold $\|f(x) - d(y)\| \leq L\|x - y\|$ for $\forall x, y \in \Re^n$. If the network is in synchronization state, then $\bar{x}(t) = \sum_{j=1}^{N} \xi_j x_j(t)$ can hold the following equation.*

$$\lim_{t \to \infty} \|\dot{x}(t) - f(\bar{x}(t))\| = 0. \tag{7}$$

*Remark 1* It is clearly received from Theorems 1–3 (relevant proofs can be referred to Refs. [8, 12]) that if the coupling network is in synchronization state, then this synchronization state can be defined as Eq. (4), and the synchronization state $\bar{x}(t)$ can be considered as the isolated solution of isolated node equation $\dot{s}(t) = f(s(t))$ [8, 12].

*Remark 2* There are many systems that meet the Lipschitz condition, including all the linear system and Lure system, thus, it can not be considered as a problem [8, 12].

In order to improve the effectiveness and the safety in complex industrial process system, a comprehensive and novel method based on monitoring synchronization state is proposed to investigate the characteristics of operation faults interacted with each other to affect the safety of the entire system [2, 23].

Suppose a sample state vector $x_k \in \Re^m$ consists of $m$ sensor measurements at sampling time $k$. With the effect of dynamic processes, measurement samples at different time $k$ are not independent, which can be both auto-correlated. The core advantage of adaptive threshold method lies at the effectiveness under varied signal disturbance or uncontrolled effects in time sequence [2]. The threshold should be derived from a segment of the measurement signal based on statistical principle [2].

Generally, the measurement signal approximately meets the normal distribution, the mean and variance of $n$ samples in the segment of the measurement signal can be calculated as the follows [2].

$$\begin{cases} m_t(k) = \frac{1}{n} \sum\limits_{t=k-n}^{k} x_i(t), \\ v_t(k) = \frac{1}{n-1} \sum\limits_{t=k-n}^{k} (x_i(t) - m_t(k))^2. \end{cases} \tag{8}$$

where $0 < n < k$, $x_i(t)$ donates the measurement signal. Under the assumption of the statistical model of the measurement signal, a threshold can be calculated by using $m_i(k)$, $v_i(k)$ according to the following equation.

$$T_i(k) = \pm t_\beta v_i(k) + m_i(k). \tag{9}$$

where $T_i(k)$ donates the adaptive threshold, and $t_\beta$ is the quantile of $t$-distribution with probability $\beta$. It should be noticed that the size of the time window $n$ should be chosen properly. The threshold would almost become a constant with the large enough $n$ [23]. And in contrast, the threshold would be very sensitive to any changes in the measurement signal with the small enough $n$ [23].

From the threshold system and monitoring state estimated model, as shown in Fig. 1, the dynamic monitoring threshold state estimation can be calculated as follows [2].

$$\begin{cases} 0, x_i(t) \le T_{fixedi}(t), \\ \frac{x_i(t) - T_{fixedi}(t)}{T_{adaptivei}(t) - T_{fixedi}(t)}, x_i(t) > T_{fixedi}(t). \end{cases} \tag{10}$$

where $r_i(t)$ donates the dynamic monitoring threshold state estimation, $T_{adaptivei}(t)$ donates the adaptive threshold, which can be obtained by Eq. (9), and correspondingly, $T_{fixedi}(t)$ is the fixed threshold, which can be calculated as follows [11].

**Fig. 1** Threshold system and monitoring state estimated model

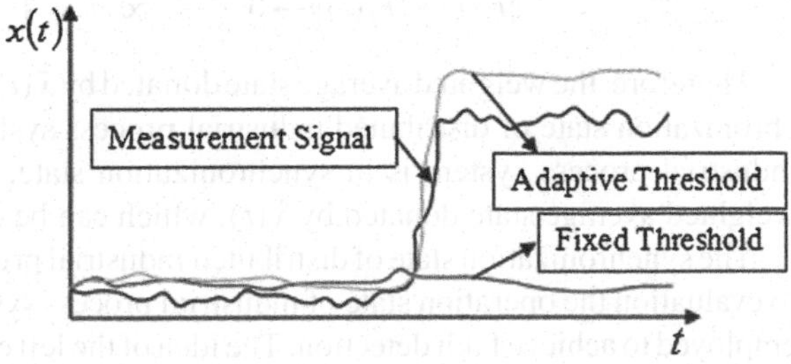

$$T_{fixedi}(t) = \begin{cases} T_{fixedi}(t_1), t_1 \in t_{normal}, \\ \lambda T_{fixedi}(t_2), t_2 \neq t_{normal}. \end{cases} \quad (11)$$

where $T_{fixedi}(t_2)$ donates the threshold state of the system, $T_{fixedi}(t_1)$ means the normal threshold operation state, which can be calculated from the normal adaptive threshold, and the following method can be considered as follows.

$$T_{fixedi}(t_1) = \frac{1}{nn} \sum_{t=kk-nn}^{kk} T_{adaptivei}(t_{normal}). \quad (12)$$

where $nn$ donates the reference length of the normal adaptive state.

As mentioned before, the actual industrial process system can be considered as a complex and scalable large-scale network system, and its synchronization state does not rely on the initial conditions, viz the nodes of the distributed industrial process system will have corresponding synchronization state solutions at any state at any time, which can be described by the complex network synchronization state to monitor the topology and node dynamics of industrial process system in real time [23].

Assume that $r_i(k)$ donates the monitoring state at sampling time $k$ of $i$-th node or subsystem. Defining the coupling matrix $\Im = (\delta_{ij})_{N \times N}$ of distributed industrial process network system, which can be described as follows.

$$\begin{cases} \delta_{ij} = \frac{1}{1+(r_i - r_j)^2} > 0, i \neq j, \\ \delta_{ij} = -\sum_{j=1, i=j}^{N} \delta_{ij}, i, j = 1, 2, \ldots, N. \end{cases} \quad (13)$$

It can be easily known that the rank of the coupling matrix of industrial process network system is $rank(\Im) = N - 1$, the synchronization manifold of dynamical network can be donated by $\Im^{n \times N}$, and its linear sub-space donated by $M = \{r = (r_1, r_2, \ldots, r_N) : r_i = r_j \in \Im^n, \forall i, j = 1, 2, \ldots, N\}$. If $t \to \infty$, and the solution donated by $r - (r_1, r_2, \ldots, r_N)$ convergence into $M$, then the distributed industrial process system can be considered as in synchronization state, viz for all nodes of distributed industrial process network system under any initial conditions, it can be considered to meet the following equation.

$$\|r_i(t) - r_j(t)\| \to 0, t \to \infty, i = 1, 2, \ldots, N. \quad (14)$$

Therefore, the weighted average state donated by $\bar{x}(t)$ can be considered as the synchronization state of distributed industrial process system, viz when the distributed industrial process system is in synchronization state, each node can converges to weighted average state donated by $\bar{x}(t)$, which can be calculated by Eq. (4).

The synchronization state of distributed industrial process system can be employed to evaluation the operation state of industrial process system in general, which can be employed to achieve fault detection. The idea of the left eigenvalue $\xi = \xi_1, \xi_2, \ldots, \xi_m$

corresponding to minimal eigenvalue of matrix $\Im$ can be employed to reflect the local details of industrial process network system, and this can be used to diagnosis the fault. The fault location can be achieved by integrate all of the knowledge information, including historical data knowledge information, expert knowledge.

## 3 A Simple Multivariate Process System Example.

This section is employed to demonstrate and verify the effectiveness of the proposed fault detection and diagnosis approach based monitoring state synchronization by further testing on a simple multivariate process system example, whose multiple state variables have been monitored to achieve fault detection [23].

Due to a industrial process system can be considered as a typical nonlinear system, following dynamic and non-linear process can be simulated and employed to have a comparison between our proposed method and other methods, which is modified version of the process system referred in Ref. [13]. Assume that the industrial process system can be described as follows [13].

$$\begin{cases} \begin{bmatrix} z_1(k) \\ z_2(k) \\ z_3(k) \end{bmatrix} = \begin{bmatrix} 0.118 & -0.191 & 0.287 \\ 0.847 & 0.264 & 0.943 \\ -0.333 & 0.514 & -0.217 \end{bmatrix} \begin{bmatrix} z_1(k-1) \\ z_2(k-1) \\ z_3(k-1) \end{bmatrix} + \begin{bmatrix} 1 & 2 \\ 3 & -4 \\ -2 & 1 \end{bmatrix} \begin{bmatrix} u_1^2(k-1) \\ u_2^2(k-1) \end{bmatrix} \\ \begin{bmatrix} y_1(k) \\ y_2(k) \\ y_3(k) \end{bmatrix} = \begin{bmatrix} z_1(k) \\ z_2(k) \\ z_3(k) \end{bmatrix} + \begin{bmatrix} v_1(k) \\ v_2(k) \\ v_3(k) \end{bmatrix} \end{cases} \quad (15)$$

where $z = [z_1(k), z_2(k), z_3(k)]^T$, and $u = [u_1(k), u_2(k)]^T$ donates the correlated inputs. The input $w = [w_1(k), w_2(k), w_3(k)]^T$ represents the random vector there each element is uniformly distributed across the interval $(-0.5, 0.5)$. The output $y = [y_1(k), y_2(k), y_3(k)]^T$ can be considered as equal to $z = [z_1(k), z_2(k), z_3(k)]^T$ with a random noise vector $v = [v_1(k), v_2(k), v_3(k)]^T$, and each element has a zero mean and a variance of 0.1.

Then, the data vector for process monitoring can be considered as the input and output measurements donated by $x = [u_1, u_2, y_1, y_2, y_3]^T$. About 1000 samples with four states data have been used to simulate the healthy, control, actuator fault and sensor fault state and another three about 1000 samples with abnormal data have been employed for online monitoring and simulating operation faults, as shown in Fig. 2.

As shown in Fig. 2, from top to bottom, there are the input and output measurements donated by $x(k) = [y_1, y_2, y_3, u_1, u_2]^T$. And from left to right, there are four operation states including healthy, control, actuator fault and sensor fault. In the example of normal control operation states, a step change of a random noise vector $v$ with a mean of 2.0 and a variance of 0.1. In the sensor fault operation states, a gain fault exists in the output signal marked as $y_1$. In the actuator operation states, there are some changes in output measurements $y_1$, $y_2$, $y_3$, the faults are mainly manifested

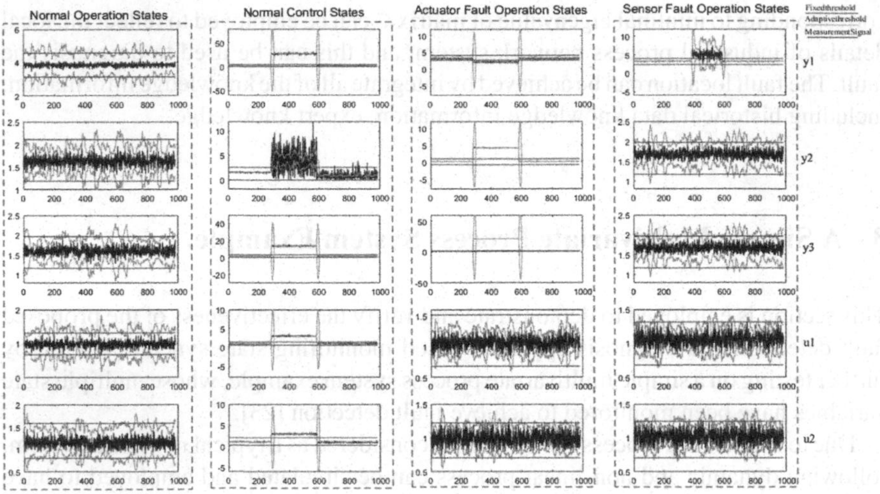


**Fig. 2** The online monitoring state of healthy, control, actuator fault and sensor fault

as follows.

$$\begin{bmatrix} 0.118 & -0.191 & 0.287 \\ 0.847 & 0.264 & 0.943 \\ -0.333 & 0.514 & \underline{-0.217} \end{bmatrix} \Rightarrow \begin{bmatrix} 0.118 & -0.191 & 0.287 \\ 0.847 & 0.264 & 0.943 \\ -0.333 & 0.514 & \underline{4.000} \end{bmatrix} \tag{16}$$

These four simulated operation states basically represent the fault conditions of this simple multivariate process system example, and their monitoring state synchronization and their left eigenvalues corresponding to minimal eigenvalue can be seen in Fig. 3.

As shown in Fig. 3a, the red curve can be considered as a very calm line, which means that the industrial process system is in a smooth operation state without any faults or any control commands, and these have been viewed in the monitoring threshold state estimation marked as $y_1, y_2, y_3, y_4, y_5$. Correspondingly, the system output would respond according to the control command with the system input, as shown in Fig. 3b. From Fig. 3b, the synchronization state (red curve) has first changed violently in about sample 300, and second changed violently in about sample 600, which means that the process system has some changes.

However, it is not fault information with the reason that the mean of the synchronization state tends to 0. Furthermore, the left eigenvalue $\xi$ corresponding to minimal eigenvalue of matrix $\Im$ can be employed to illustrate this, and this is also explained in the monitoring threshold state estimation.

In contrast, Fig. 3c shown that the synchronization state contains some fault information with the mean of it tends to 0.4 and then landed to 0.1, which means that the process system has some faults, which has been illustrated in the monitoring threshold state estimation with the fault occurred in about sample 300, and this has been

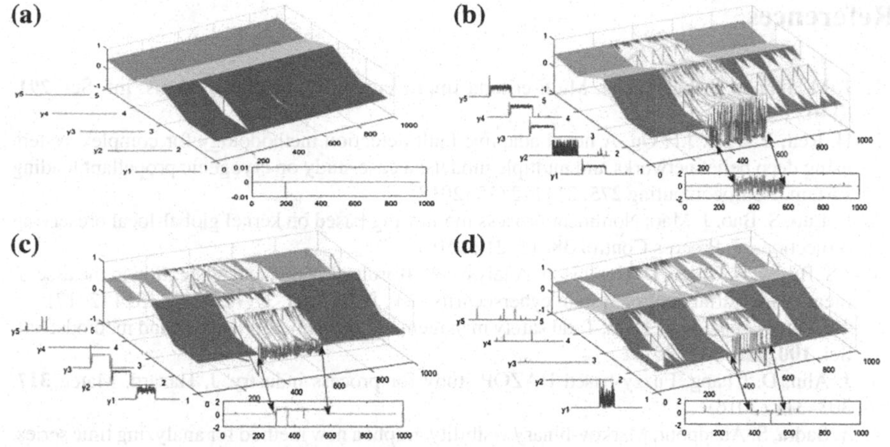

**Fig. 3** Process monitoring result of our proposed fault detection and diagnosis for healthy (**a**), control (**b**), actuator fault (**c**) and sensor fault (**d**) (Color figure online)

illustrated. The sensor fault can be viewed in Fig. 3d, whose mean of synchronization state tends to 0, while its variance is very smaller.

## 4 Conclusions and Further Works

In this paper, a novel fault detection method based on the monitoring state synchronization has been proposed to perceive and evaluate the system operation situation. Firstly, the synchronization state of distributed network has been proved and employed to achieve fault detection. Simultaneously, the left eigenvalue corresponding to minimal eigenvalue of matrix can be employed to achieve fault diagnosis. Finally, from the case study, it can be summarized that this method can be used to provide a novel and feasible research method for the global assessment of the operation states and for monitoring the local operation fault of modern industrial process systems.

However, this method also has its limitations, which mainly reflects in the effective setting of parameters, and this will be our further study. Besides, this method only gives the features of data globally and locally, which needs to further identify the fault from the perspective of pattern recognition, and this is also the direction for further study.

**Acknowledgements** This work was supported by the National Natural Science Foundation of China Under Grant 61633005, 61673076 and 61773080, and it is also funded by the Natural Science Foundation of Chongqing City, China (cstc2016jcyjA0504).

## References

1. J. Montmain, C. Labreuche, Multi-criteria Improvement of complex systems. Inf. Sci. **291**, 61–68 (2015)
2. H. Ren, Y. Chai, J.F. Qu, A novel adaptive fault detection methodology for complex system using deep belief networks and multiple models: a case study on cryogenic propellant loading system. Neurocomputing **275**, 2111–2125 (2018)
3. L. Luo, S. Bao, J. Mao, Nonlinear process monitoring based on kernel global-local preserving projections. J. Process Control **38**, 11–21 (2016)
4. J.S. Busby, B. Green, D. Hutchison, Analysis of affordance, time, and adaptation in the assessment of industrial control system cybersecurity risk. Risk Anal. **37**(7), 1298–1314 (2017)
5. T. Kontogiannis, M.C. Leva, Total safety management: principles, processes and methods. Saf. Sci. **100**, 128–142 (2017)
6. J. Ahn, D. Chang, Fuzzy-based HAZOP study for process industry. J. Hazard. Mater. **317**, 303–311 (2016)
7. Y. Sadra, S. Ahadpour, Markov-binary visibility graph: a new method for analyzing time series. Inf. Sci. **274**, 286–302 (2014)
8. Y. Tang, F. Qian, Synchronization in complex networks and its application-a survey of recent advances and challenges. Annu. Rev. Control **38**, 184–198 (2014)
9. Y. Kim, S.J. Lee, T. Park, Robust leak detection and its localization using interval estimation for water distribution network. Comput. Chem. Eng. **92**, 1–17 (2016)
10. Q. Zhang, S. Geng, Dynamic uncertain causality graph applied to dynamic fault diagnoses of large and complex systems. IEEE Trans. Reliab. **64**(3), 910–927 (2015)
11. H. Ren, J. Qu, Y. Chai, et al., Deep learning for fault diagnosis: the state of the art and challenge. Control Decis. **32**(8) 1345–1358 (2017)
12. H. Yu, F. Khan, V. Garaniya, Self-organizing map based fault diagnosis technique for non-gaussian processes. Ind. Eng. Chem. Res. **53**(21), 8831–8843 (2014)
13. E. Cai, D. Liu, L. Liang, Monitoring of chemical industrial processes using integrated complex network theory with PCA. Chemom. Intell. Lab. Syst. **140**, 22–35 (2015)
14. L. Bakule, Decentralized control: status and outlook. Annu. Rev. Control **38**(1), 71–80 (2014)
15. F. Khan, S.J. Hashemi, N. Paltrinieri, Dynamic risk management: a contemporary approach to process safety management. Curr. Opin. Chem. Eng. **14**, 9–17 (2016)
16. A. Primadianto, C.N. Lu, A review on distribution system state estimation. IEEE Trans. Power Syst. **32**(5), 3875–3883 (2017)
17. Z.E. Bhatti, P.S. Roop, R. Sinha, Unified functional safety assessment of industrial automation systems. IEEE Trans. Ind. Inf. **13**(1), 17–26 (2017)
18. G. Li, S.J. Qin, D. Zhou, A new method of dynamic latent-vatiable modeling for process monitoring. IEEE Trans. Ind. Electron. **61**(11), 6438–6445 (2014)
19. S. Yin, X. Li, H. Gao, Data-based techniques focused on modern industry: an overview. IEEE Trans. Ind. Electron. **62**(1), 657–667 (2015)
20. Y. Jia, Robust control with decoupling performance for steering and traction of 4WS vehicles under velocity-varying motion. IEEE Trans. Control Syst. Technol. **8**(3) 554–569 (2000)
21. Y. Jia, Alternative proofs for improved LMI representations for the analysis and the design of continuous-time systems with polytopic type uncertainty: a predictive approach. IEEE Trans. Autom. Control **48**(8), 1413–1416 (2003)
22. N. Alileche, V. Cozzani, Thresholds for domino effects and safety distances in the process industry: a review of approaches and regulations. Reliab. Eng. Syst. Saf. **143**, 74–84 (2015)
23. F. Baghernezhad, K. Khorasani, Computationally intelligent strategies for robust fault detection, isolation, and identification of mobile robots. Neurocomputing **171**, 335–346 (2016)

# Circulating Current Minimization in MC-WPT System with Multiple Inverter Modules Operate in Parallel

**Zhou Xu, Zhihui Wang and Jianhua Wu**

**Abstract** The magnetic coupled wireless power transfer (MCWPT) system with multiple inverter modules operated in parallel is investigated in this paper. The circulating current among the inverter modules is analyzed first. A novel topology with a control scheme is proposed aiming to minimize the circulating current among the inverter modules based on the active and reactive current decomposition. In the proposed circulating minimization method, there is only one loop which is proved to be a phase control loop. It is shown that the circulating current minimization can be achieved if the phase of all the modules output voltages are the same. Performance is verified with both simulations and experiments on a prototype MCWPT system where two resonant inverter modules are operated in parallel. Finally, a conclusion is given.

**Keywords** Magnetic coupled wirelcss power transfer system
Inverter modules operated in parallel · Current decomposition
Circulating current minimization

## 1 Introduction

Magnetic coupled wireless power transfer (MCWPT) technology uses modern electronics technology to convert direct current energy into high-frequency alternating current within high-frequency inverter modules, and then the power is transferred from the primary coil to the secondary coil without contact [1–3]. With the development of electric vehicles (EVs), MCWPT technology is more and more applied

Z. Xu (✉) · Z. Wang · J. Wu
College of Automation, Chongqing University, Chongqingg 400044, China
e-mail: XZ_muyang@163.com

Z. Wang
e-mail: wangzhihui@cqu.edu.cn

J. Wu
e-mail: 429059800@qq.com

 
Y. Jia et al. (eds.), *Proceedings of 2018 Chinese Intelligent Systems Conference*,
Lecture Notes in Electrical Engineering 528,
https://doi.org/10.1007/978-981-13-2288-4_10

to wireless charging, because of the superiority of MCWPT technology. With the increase of charging power, MCWPT system with a single inverter module generally can't meet the demand of power capacity. Inverter modules operated in parallel is the commonly used method nowadays. The circulating current among the inverter modules is one of the problems that need to be solved first.

In order to suppression the circulating current among the inverter modules in high-frequency systems, many approaches have been achieved. In a master–slave control [4, 5], the current control reference is generated by a master module for all other modules. The method has the advantage of simple implementation. However, this method suffers from the poor dynamic current sharing [6, 7]. One feasible approach is to make sure that all of the module are working in a power sharing mode, and then control the module voltages to be the same as the load voltage which is choose to be the reference amount [8]. Another feasible approach is to decompose the module currents into reactive component and active component [9], hence, with the minimization of the reactive component and sharing of the active component, the circulating current among the inverter modules is suppressed [10]. when analyzing these methods, There are two loops: (1) the current sharing loop and (2) the minimization of reactive current loop. So, the complex and the computational burden increases.

In this paper, a novel topology with a phase control approach is proposed. There is only a single control loop, which decreases the computational burden and the complexity of the control algorithm. Without losing the generality, the circulating current in a MCWPT system with two inverter modules operated in parallel is analyzed in Sect. 2. And then a novel topology with a phase control method is proposed based on the analysis above in Sect. 3. By this method, the minimization of the circulating current is guaranteed. In Sect. 4, the effectiveness of the proposed approach is verified with both simulation and experiment. And finally, the conclusion is given in Sect. 5

## 2 Circulating Current Analysis

The Power architecture of the MCWPT system with two inverter modules operated in parallel is shown in Fig. 1, which consists of two series resonant tank with the same resonant frequency. For the $i$th series resonant tank ($L_1$, $L_2$, $C_1$, $C_2$), the following equation can be written:

$$\omega_0^2 L_i C_i = \omega_0^2 L_p C_p = \omega_0^2 L_s C_s = 1 \ (i = 1, 2) \tag{1}$$

Without seeing the inner structure of the inverter modules, the simplified architecture of Fig. 1 is shown in Fig. 2, which references transformer T-equivalent circuit.

As we can see from Fig. 2, the output current of inverter modules can be described as the combination of the $u$, $e$, and $Z$, so, the following circuit equations can be written down for the inverter modules.

$$\dot{I}_1 = \frac{\dot{U}_1 - \dot{E}_1}{Z_1} \tag{2}$$

$$\dot{I}_2 = \frac{\dot{U}_2 - \dot{E}_2}{Z_2} \tag{3}$$

Assuming that all the high-frequency transformers are designed to be the same parameters, so the equation $Z_m = L_m + R_m$ is a known amount. And then, the following equations can be written down for the transformers.

$$\dot{E}_1 = (\dot{I}_1 + \dot{I}_o)Z_m \tag{4}$$

$$\dot{E}_2 = (\dot{I}_2 + \dot{I}_o)Z_m \tag{5}$$

$$\dot{I}_o = \frac{\dot{E}_1 + \dot{E}_2}{Z_{eq}} \tag{6}$$

From Eqs. (2) to (6), the output current of the inverter modules can be written down as follows:

$$\dot{I}_1 = \frac{(\dot{U}_1 + \dot{U}_2)Z_m^2 + (2Z_2 + Z_{eq})Z_m\dot{U}_1 + Z_2Z_{eq}\dot{U}_1}{(Z_1 + Z_2 + Z_{eq})Z_m^2 + (Z_1Z_{eq} + Z_2Z_{eq} + Z_1Z_2)Z_m + Z_1Z_2Z_{eq}} \tag{7}$$

$$\dot{I}_2 = \frac{(\dot{U}_1 + \dot{U}_2)Z_m^2 + (2Z_1 + Z_{eq})Z_m\dot{U}_2 + Z_1Z_{eq}\dot{U}_2}{(Z_1 + Z_2 + Z_{eq})Z_m^2 + (Z_1Z_{eq} + Z_2Z_{eq} + Z_1Z_2)Z_m + Z_1Z_2Z_{eq}} \tag{8}$$

Define that the circulating current is the half of the current difference between the two inverter modules, so the circulating current can be written down as follow:

$$\dot{I}_H = \frac{\dot{I}_1 - \dot{I}_2}{2} = \frac{1}{2}\frac{(2Z_2 + Z_{eq})Z_m\dot{U}_1 + Z_2Z_{eq}\dot{U}_1 - (2Z_1 + Z_{eq})Z_m\dot{U}_2 - Z_1Z_{eq}\dot{U}_2}{(Z_1 + Z_2 + Z_{eq})Z_m^2 + (Z_1Z_{eq} + Z_2Z_{eq} + Z_1Z_2)Z_m + Z_1Z_2Z_{eq}} \tag{9}$$

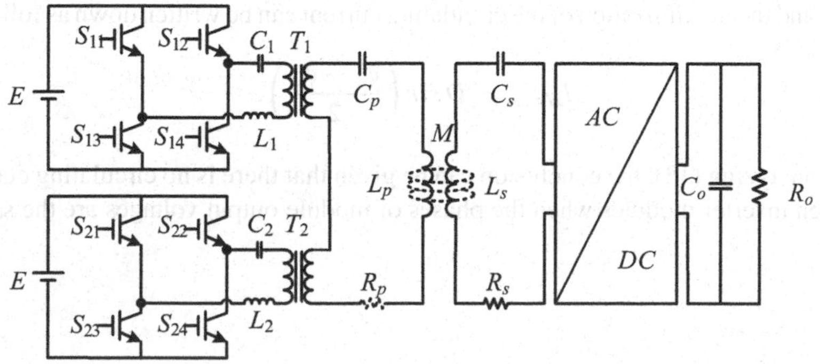


**Fig. 1** Power architecture of MCWPT system based on two inverter modules operated in parallel

Fig. 2 Simplified architecture of Fig. 1

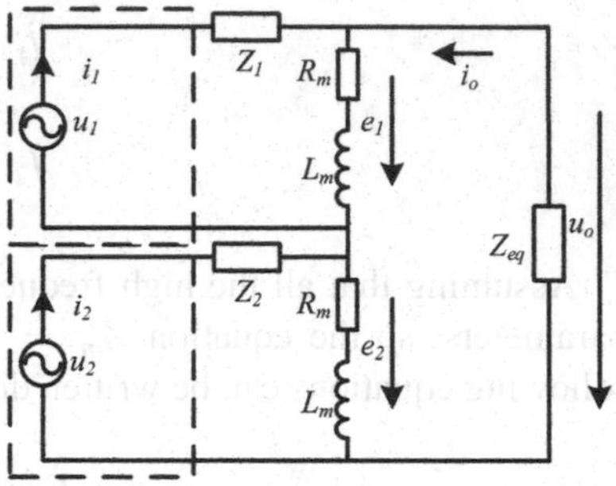


Assuming that the high-frequency transformers are designed to satisfy the following equation:

$$Z_m \gg Z_{eq} \gg Z_1, Z_2 \tag{10}$$

So the Eq. (9) can be rewritten down as follow:

$$\dot{I}_H = \frac{\dot{I}_1 - \dot{I}_2}{2} = \frac{Z_{eq}(\dot{U}_1 - \dot{U}_2)}{2(Z_1 + Z_2 + Z_{eq})Z_m} \tag{11}$$

Assuming that the all the resonant tanks are working in resonance, and then the resistance equivalent load ca be achieved. The circulating current in a two-inverter-module MCWPT system is written with time domain expression as follows:

$$\frac{2\dot{I}_H}{D} = cos(\omega t + \phi_1) - cos(\omega t + \phi_2) = -2sin\left(\omega t + \frac{\phi_1 + \phi_2}{2}\right)sin\left(\frac{\phi_1 - \phi_2}{2}\right) \tag{12}$$

where, $D = Z_{eq}/(Z_1 + Z_2 + Z_{eq})Z_m$ is a coefficient. As we can see from Eq. (12), the circulating current between the two parallel-connected inverter modules is a sine curve, and the maximization of the circulating current can be written down as follows:

$$I_{Hmax} = D\,sin\left(\frac{\phi_1 - \phi_2}{2}\right) \tag{13}$$

By analyzing (13), the conclusion can be given that there is no circulating current between inverter modules when the phases of module output voltages are the same.

## 3 The Control Strategy

As the analysis above shows, if the phases of module output voltages are the same, there will be no circulating current between inverter modules. So, the most important thing is to control the phases of module output voltages to be the same. In this Section, an approach is proposed to control the voltage phases of inverter modules without phase-locked loop (PLL) and signal communication between the two parallel-connected inverter modules, which decreases the computational burden and complexity of the circuit.

If the output voltage is chosen to be the reference, which means that the phase of output voltage is regarded to be zero. And then, the following equations can be written down for a time-domain expression:

$$u_o = U_{om}\, cos(\omega t) \tag{14}$$

Assuming that the phase shift between the signal $u_o$ and signal $u_{oi}$ is $\phi_i$. Hence, the following equation can be written down:

$$u_i = u_o u_{oi} = U_{om} U_{oim} \cos(\omega t)\cos(\omega t + \phi_i) = \frac{U_{om} U_{oim}[\cos(2\omega t + \phi_i) + \cos(\phi_i)]}{2} \tag{15}$$

As we can see from the Eq. (15), the signal $u_i$ include a DC component and an oscillating component. And by placing a low pass filter, the DC component can be written down:

$$u_{i\_dc} = \frac{U_{oim} U_{oim} cos(\phi_i)}{2} \tag{16}$$

So, the phase shift between signal $u_{oi}$ and $u_o$ can be written as follow:

$$cos(\phi_i) = \frac{2u_{i\_dc}}{U_{oim} U_{om}} \tag{17}$$

Hence, the steady-state control objective is to minimize the phase shift of each inverter module, which is to control the phase shift is lower than a constant $\varepsilon$ that is small enough, as is given in Eq. (18).

$$\{\phi_i \le \varepsilon\} \tag{18}$$

With the $\phi_i$ approaches zero, the $cos\ \phi_i$ infinity approaches 1, so the Eq. (18) can be deformed to be the following form:

$$\{1 - cos(\phi_i) \le \varepsilon\} \tag{19}$$

**Table 1** Parameters of simulation and experiment

| Parameters | Values | Units |
|---|---|---|
| Filter inductance, $L_1, L_2$ | 350.2, 345.3 | μH |
| Filter capacitor, $C_1, C_2$ | 100, 100 | pF |
| Primary and secondary coil $L_p, L_s$ | 258.0, 256.4 | μH |
| Primary and secondary capacitor, $C_p$ $C_s$ | 13.6, 13.3 | nF |
| Output filter capacitor, $C_o$ | 1000 | μF |
| Load, $R_o$ | 38 | Ω |

According to Eqs. (14) and (17), the decomposition schematic of the phase shift between signal $u_{oi}$ and signal $u_o$ is shown in Fig. 3, and the number of the modules should be as much as the inverter modules.

The control block diagram of proposed method is shown in Fig. 4. As is shown in Fig. 4, *DC.i* is the *i*th phase decomposition circuit, $K_i(s)$ is phase shift minimization controller and all the controllers are PI controller, *PG.i* is the *i*th pulse generator. The $\cos\phi_i$ is calculated according to the module output voltage $U_{oi}$ and the reference voltage $U_o$, and then the $\cos\phi_i$ is compared to constant 1, so the *i*th error signal is given. The error signal is used to control the phase shift of the inverter modules. The output voltages of inverter modules are regulated by the pulse which is generated by the pulse generator. And then, the *i*th module voltage $U_{oi}$ and reference voltage are detected by the voltage sensors respectively, In this way, the circulating current minimization is achieved without PLL which is proved to decrease the computational burden and the complexity of control circuit.

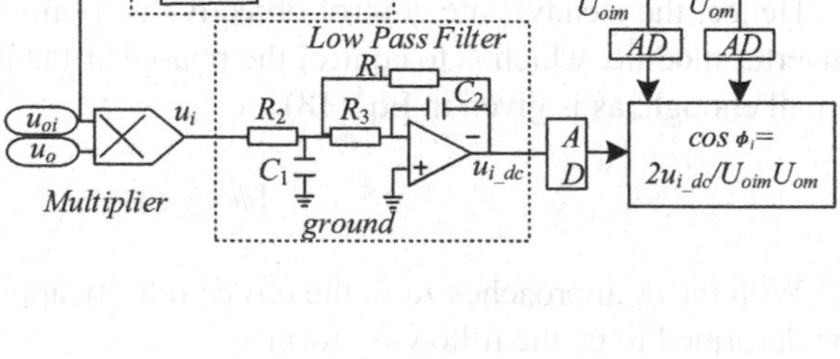


**Fig. 3** Diagram for the phase decomposition

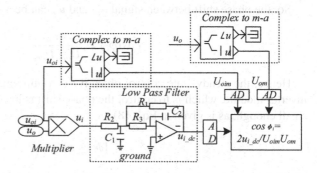

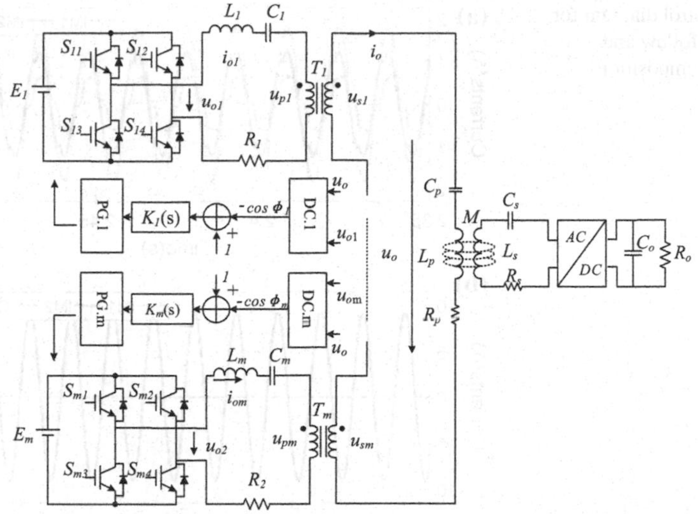


**Fig. 4** Control diagram for the phase control

## 4 Prototype and Test Results

In order to verify the effectiveness of the proposed method, both simulation and experiment are provided. The output power of the prototype is 1 KW. The circulating current is suppressed by regulating the phase of the modules output voltages. The circuit parameters of the prototype are provided in Table 1. In order to compare the experimental and simulation results, all the devices are measured to get their actual values. The STP47N60C3 is employed to be the power switch of proposed prototype, the STM32F407VGT6 microcontroller is used to signal detection and data processing, the EP2C5T144C8N field programmable gate array (FPGA) is employed to generate the PWMs. The low pass filters are butter-worth low pass filter. The PW6001 power analyzer is used to measure the power and the efficiency which appear in this paper. The experimental waveforms are measured and displayed by TPS2024B scope.

As we can see from Fig. 5a, there is a periodic circulating current between the two inverter modules, and a distortion point exists in the curve IM2 which is caused by the circulating current. In Fig. 5b, the circulating current between the two inverter modules is controlled to be lower than 0.5 A.

The experimental diagrams are shown in Fig. 6. Without the phase control approach employed into the MCWPT prototype, the circulating current between the inverter modules is large and there is a distortion in the module current curve. The efficiency of the experiment is 91.65%. But when the phase control approach is employed into the MCWPT prototype, the circulating current between the inverter modules is suppressed to a much lower level (lower than 0.5 A), which is shown in

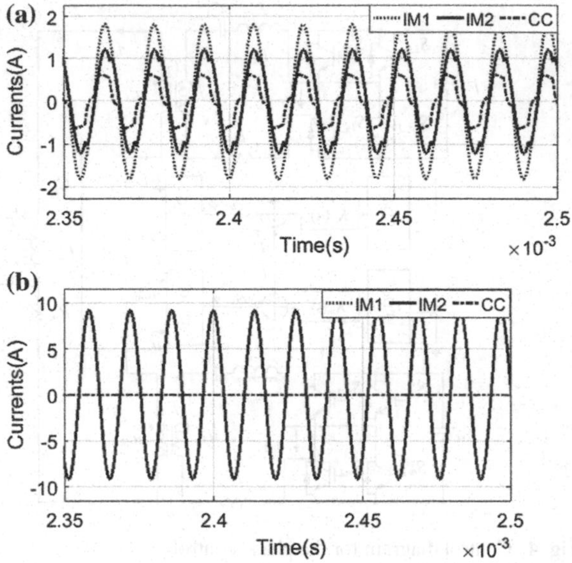


**Fig. 5** Control diagram for the current follow and voltage decomposition

Fig. 6b. There is no distortion in any of the module current. The efficiency of the experiment is 92.50%.

## 5 Analysis of Experimental Results

A MCWPT system connected with inverter modules is described in this paper to upgrade the power capability of MCWPT system with low-power-capability semiconductor switches. It is point out that the circulating current between inverter modules is caused by a slight unbalance of the series-parallel parameters, the output module voltages and other non-linear factors. With the proposed topology and the proposed phase control approach, the circulating current minimization is achieved without PLL, which is proved that the computational burden decreases and the complexity of control algorithm is simplified. Finally, in order to verify the effectiveness of the proposed topology and the control approach, a 1 KW MCWPT prototype with two inverter modules operated in parallel is provided. The experimental results show that the circulating current minimization is achieved and the efficiency of experiment with control approach employed into the prototype is 92.50%, which is higher than 91.65%, the one without control approach.

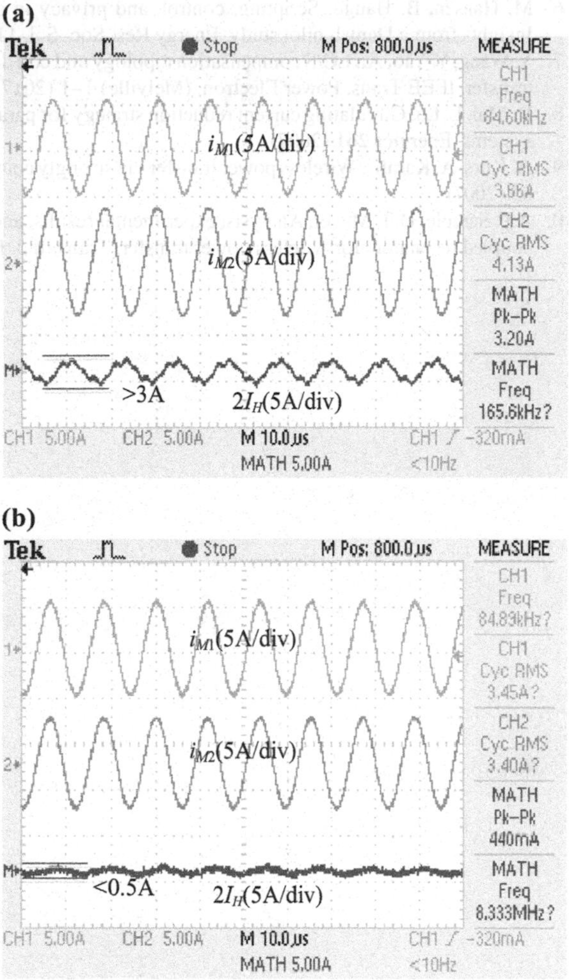


**Fig. 6** Experimental diagram for phase control approach

# References


1. Y. Sun, Z.H. Wang, Study of frequency stability of contactless power transmission system. Trans. China Electrotech. 56–59 (2005)
2. Y.G. Su, C.S. Tang, Load adaptive technology of contactless power transfer system. Trans. China Electrotech. Soc. 153–157 (2009)
3. M. Liu, S. Liu, A high-efficiency/output power and low-noise megahertz wireless power transfer system over a wide range of mutual inductance. IEEE Trans. Microw. Theory Tech. 1–9 (2017)
4. Z. Ye, P.K. Jain, Circulating current minimization in high-frequency AC power distribution architecture with multiple inverter modules operated in parallel. IEEE Trans. Power Electron. 2673–2687 (2007)
5. M. Borage, S. Tiwari, Analysis and design of an LCL-T resonant converter as a constant-current power supply. IEEE Trans. Ind. Electron. 1547–1554 (2000)

6. M. Hansen, B. Hauge, Scripting, control, and privacy in domestic smart grid technologies: Insights from a Danish pilot study. Energy Res. Soc. Sci. 112–123 (2017)
7. Y. Wang, Y. Yao, An LC/S compensation topology and coil design technique for wireless power transfer. IEEE Trans. Power Electron. (Melville) 1–1 (2017)
8. R. Mai, L. Lu, Circulating current reduction strategy for parallel-connected inverters based IPT systems. Energies **261** (2007)
9. A. Kurs, A. Karalis, Wireless power transfer via strongly coupled magnetic resonances. Science **83** (2007)
10. A.P. Sample, D.T. Meyer, Analysis, experimental results, and range adaptation of magnetically coupled resonators for wireless power transfer. Trans. Ind. Electron. 544–554 (2011)

# An Optimized Scheme for Monitoring Data Transmission of Complex Engineering Systems

**Ke Zhang, Zhuo Liu and Yi Chai**

**Abstract** In operation process of complex engineering systems, some problems of low transmission performance have severely burst out because of high concurrency and large monitoring data size. Three aspects such as operating mode, pre-compression pretreatment and data compression are considered to optimize the transmission process of system monitor data in this paper. Improvement of coding output process in LZSS algorithm is proposed as the compression processing method before data transmission. Finally, an optimization scheme is applied to an air launch site in mass monitoring data transmission process. The experimental results demonstrate that the proposed method significantly improves the transmission process. With the addition of the optimization scheme, transmission time is shortened nearly 75%.

**Keywords** Complicated engineering system · High concurrency
Transmission performance · LZSS algorithm

## 1 Introduction

With the promotion of communication and sensing technology, global industry is facing the fourth industrial revolution. Meanwhile, it attracts more and more attention that large amount of monitoring data is generated in complex engineering systems. Complex engineering system is a large-scale special system based on user-demand, which has some characteristics of one-time, multidisciplinary and complex system integration [1]. Complex engineering systems have widespread occurrence in many domains such as national economy, national defense construction, natural science and social sciences, especially in aerospace engineering systems, environmental engineering systems, military engineering systems and traffic engineering systems, etc.

---

K. Zhang (✉) · Z. Liu · Y. Chai
State Key Laboratory of Power Transmission Equipment and System Security and New Technology, School of Automation, Chongqing University, Chongqing 400030, China
e-mail: smeta@163.com



Y. Jia et al. (eds.), *Proceedings of 2018 Chinese Intelligent Systems Conference*,
Lecture Notes in Electrical Engineering 528,
https://doi.org/10.1007/978-981-13-2288-4_11

Complex engineering system monitoring data are generated during the operation process and provide support for the system status monitoring. With the high-speed operation of engineering system, the amount of data is also great. The division of the complex engineering system is detailed. Different systems contain a large amount of equipment, different equipment contain multiple parameters. Due to high operation speed of the system, the amount of data generated, collected and processed by device is much greater than the data generated by enterprise computers and artificially. Meanwhile, the requirement of real-time data processing is much higher than that of artificial data generated by common enterprises [2]. Therefore, problems faced in monitoring data processing are more complicated in certain cases [3].

The characteristics of monitoring data of complex engineering systems are mainly reflected in the large volume of data, low data density, relatively fixed data type and high real-time acquisition requirements [4]. Therefore, due to high collection frequency, a large amount of monitoring data will be generated in the operation process of system. During the peak period of monitoring data request, data transmission will be blocked and delayed because of the large amount of data and high concurrency and reduce overall transmission efficiency of the system. It may also cause the system's overall paralysis to affect the operation because of high server load. Although it can improve its overall performance by using load balancing or increasing the bandwidth of public network, it is difficult to be practical because of its high cost. For small and medium-sized systems, it is necessary to adopt a more suitable scheme. Therefore, in the complex engineering monitoring system with massive data, transmission is often a hot and difficult point of attention.

To address this problem, it is feasible to optimize transport protocols (such as optimizing TCP transport protocol by sliding Windows), increase network bandwidth, and reduce the amount of data by compressing transmitted data. On the one hand, the researchers proposed the TCP proxy technology [5] due to the long delay transmission characteristics of WAN. TCP proxy are deployed instead of the receiving and sending end respectively. The delay between sending end and proxy end is greatly reduced and the delay between the sending end and the receiving end of the proxy is also greatly reduced. In case of two proxy devices, data is quickly transmitted through some unique transmission technologies, which ultimately speeds up the transmission and reception at both ends. Another example of optimizing TCP transport protocol is to add the TCP congestion control algorithm-FAST TCP [6]. When data is transmitted at the transport layer, sending window is used to determine the amount of data to be sent each time. The sending window is determined by receiving window of receiving end, sliding window of the sending end and the congestion window abstracted according to network conditions. For the optimization of congestion control in TCP transport, it is mainly to solve the problem of slow growth of TCP window but drop when dropping packet, which makes TCP transport more stable. However, it is difficult to further optimize protocol due to the interior of TCP transport protocol optimization system has been very mature. On the other hand, the cost is too high to speed up the access speed by increasing network bandwidth of the server. For the monitoring system with a large increase in data volume in a specific time, the network bandwidth will be wasted in the case of non-access peak or small amount of monitoring node data.

It is objective to reduce the amount of data transfer in network by using data reduction methods, indirectly increase the transfer rate. The high-compression efficiency is more realistic. However, the existing series of mainstream data compression methods are not common used. The results of algorithm for compression of different data types are sometimes significantly different. Reason for the question above is that convergence speed of the lossless data compression algorithm is also related to the intrinsic properties of the internal structure [7]. The compression efficiency of the same compression algorithm is different for different types of files, such as text, image, audio, etc. It is necessary to select an appropriate compression method for the characteristics of compressed data.

The rest of this paper is organized as follows. In Sect. 2, we briefly review some background materials on space launch site. For transmission situation of the specific monitoring system studied, this paper optimizes the response mechanism of traditional service-client to reduce the bottleneck of transmission data in high concurrent situations in Sect. 3. Section 4 reports performances of the proposed methods on the transmission of monitoring data. With the addition of the optimization scheme, transmission time is shortened nearly 75%. Finally, conclusion and future directions of this paper are arranged in Sect. 5.

## 2 Background and Related Work

### *2.1 Background*

A space launch site is considered as the research object in this paper. The complex engineering system consists of 12 subsystems including power subsystem, air conditioning subsystem, elevator subsystem, fire control subsystem, gas alarm subsystem and crane subsystem. In each subsystem, a large number of devices are distributed. Data is transmitted through UDP protocol group, and collected data is processed and displayed in real time through the monitoring data integration platform. The monitoring data of this complex engineering system has the characteristics of short data collection period, large data collection, low correlation of data sources and fixed data type [8–10]. Characteristics of monitoring data in this application background are explained in the following three aspects:

(1) Large amount of data

The complex engineering system has 12 subsystems. Each subsystem contains a lot of devices with a lot of parameters. The amount of data generated in this process is enormous under long running time of devices.

(2) Fixed data type

Monitoring data include three types: time labels, parameter values and parameter states. All kinds of data has different functions. Time label represents the time when the parameters are collected; Parameter value represents the real value of parameters in devices; Parameter state represents the operation of devices, generally 0 or 1, which respectively represents closed or disconnected state.

(3) High real-time requirement

Different subsystems require a different time interval for the received data. It needs to transmit the recorded data about 20,000 under the extreme cases. It is a higher requirement for the system's data processing and real-time transmission.

## 2.2 The Improved Data Compression Scheme

With the large size of engineering systems, the equipment in various subsystems of engineering systems is increasing. With the consequent increase in the amount of historical data required to be recorded and processed. While the biggest problem encountered during project monitoring is the IO bottleneck caused by transmission. During monitoring process, load of server is increased rapidly, causing a blockage of transmission channel to affect monitoring quality and even directly cause the server to become disabled. It is necessary to realize efficient lossless compression of monitoring data to reduce transmission load.

Lossless data compression technology reduces the coding redundancy of data through certain models and coding methods. To shorten the length of the unified data encoding technology, its core principle is given shorter coding for high frequency information and given longer coding for low frequency information, and it forms a certain encoding rules at the same time. Combining with the data characteristics of researched object and application environment, this paper adopts an optimized LZSS compression algorithm to reduce the pressure of transmission channel to solve the problem about the monitoring data transmission.

LZSS replaces the triples with binary groups, and specifies the minimum matching length. Only when the length of the longest matched string found in sliding window is greater than the minimum matching length, matching is successful [11]. At the same time, a binary group (*off*, *len*) is outputted and the sliding window is moved backward to *len*. If length of maximum matched string is less than the minimum length of string, match fails, then the first character in coded area will be exported. This case guarantees the result of zip code will not exceed the amount of data, thus reduce data redundancy. In addition, in order to distinguish the "match failure to output first character" and "binary group" in decompression, it is necessary to add a switch bit before output result, usually 0 means "match failure to output the first character", and 1 means the binary group of output.

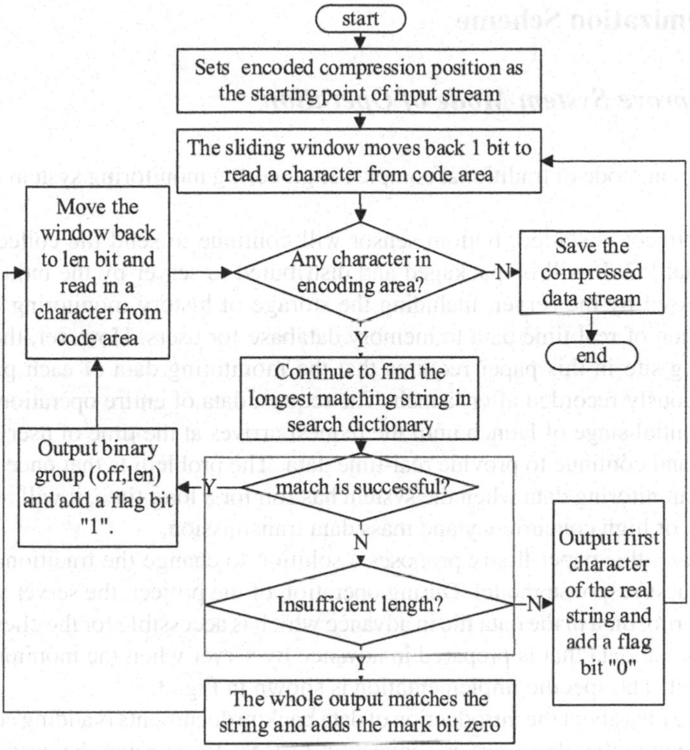


**Fig. 1** Improved algorithm flow chart of compression

In the process of string matching, if the maximum string length matched is less than the specified length, then the first character of the matched string is exported and the sliding window is moved backward by one bit. If the length of the match is *n* (*n* is less than the minimum matching length N), then it is necessary to have a window slide of *n* times to output a string of length *n*. So this paper will make improvements to the output above rules. If the length of the matching result is less than the minimum matching length in the process of searching, then the first character of the match is replaced with the string matched by the integer output. Therefore, it is necessary to judge the current matching result before the result output to do further processing, and the flow chart of the improved compression algorithm is shown in Fig. 1.

The improved algorithm reduces the time complexity in the process of encoding output and reduces unnecessary matching operations to make a larger data dictionary possible.

# 3 Optimization Scheme

## *3.1 Improve System Mode of Operation*

The operation mode of traditional complex engineering monitoring system is shown in Fig. 2.

Once project is started, bottom sensor will continue to send the collected data to industrial PC. It will be packaged and distributed to server by the industrial PC and processed by the server, including the storage of history monitoring data and the provision of real time data to memory database for users. However, the task of a launching site in this paper requires that the monitoring data of each parameter be continuously recorded after launch. The request data of entire operation process from the initial stage of launch until the request arrives at the time of user's visit to the client and continue to provide real-time data. The problem is that once multiple users ask monitoring data when the system has run for a long time, it will result in a bottleneck of high concurrency and mass data transmission.

Therefore, this paper firstly proposes a solution to change the traditional client-server request/response model. During operation of the project, the server will keep the monitoring data in the data file in advance which is accessible for the client. Users will obtain the data that is prepared in advance by server when the monitoring data is requested. The specific implementation is shown in Fig. 3.

The core idea about the introduction of data backup documents is adding additional space to reduce the client-server connection process to improve the access speed. The subsequent compression technology provides the possibility to reduce the impact of additional space. The main feature of the improved work pattern is that reduces

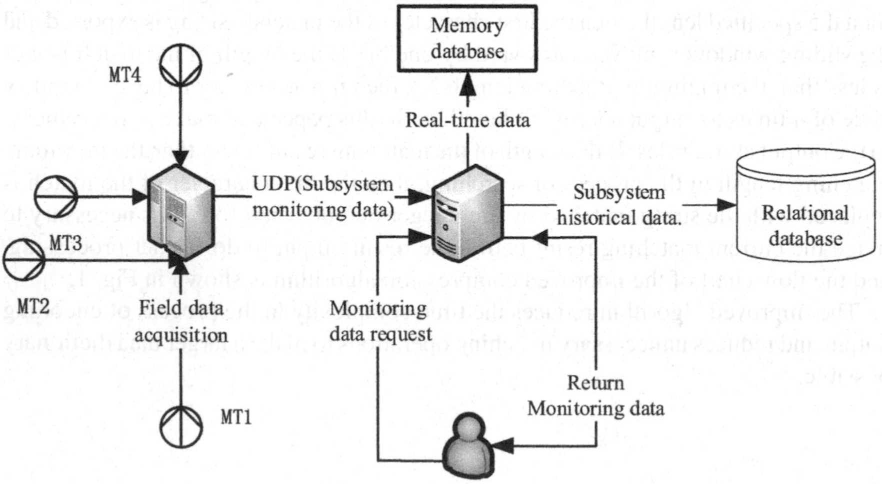


**Fig. 2** Traditional operation mode

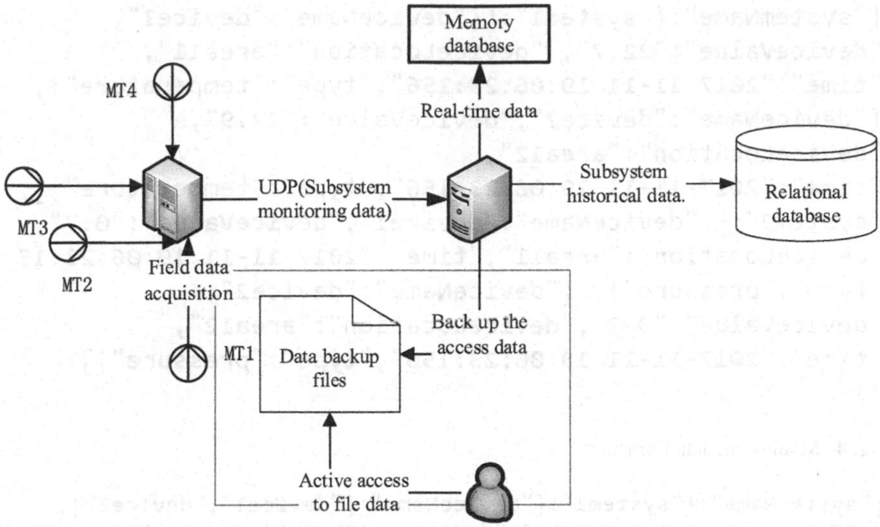


**Fig. 3** Improved operation mode

the time. Thus establish a connection between the server and the client. Data which is accessed frequently is backed up in the data file directly. When the user requests the monitoring data, it will actively request the data in the backup document. Connection time is the main cause of transmission congestion during high concurrent access in a given period of time. After the improvement of operation mode, when the customer requests all the current monitoring data, the response mode is changed to the client to actively obtain the files to reduce the pressure of server. This paper takes it as the first step of transmission optimization.

## 3.2 *Pretreatment Before Compression*

### 3.2.1 Lightweight Serialization Processing

Some monitoring data formats of the researched object subsystem are shown in Fig. 4.

It can be seen that there are many key values in monitoring data such as "deviceName", "deviceValue", "time", "type" and "deviceLoation", etc. So the duplicate key value can be extracted as the public key value. It can make lightweight processing of the original data structure to reduce the redundancy of data. The data format after processing is shown in Fig. 5.

From the data format after the light weight processing, we can obviously see that the overall space of the data is smaller. For the researched object of this paper, the empirical evidence of data information after the light weight processing is reduced by 14.3%. Although the effect is not very ideal, it is reasonable as the light weight treatment before compression.

```
{"systemName":{"system1":[{"deviceName":"device1",
"deviceValue":"22.7", "deviceLocation":"area11",
"time":"2017-11-11 19:06:23:156","type":"temperature"},
{"deviceName":"device2","deviceValue":"22.9",
"deviceLocation":"area12",
"time":"2017-11-11 19:06:23:156","type":"temperature"}],
"system2":[{"deviceName":"device1","deviceValue":"0.7",
"deviceLocation":"area11","time":"2017-11-11 19:06:23:15
"type":"pressure"}, {"deviceName":"device2",
"deviceValue":"0.7","deviceLocation":"area12",
"time":"2017-11-11 19:06:23:156","type":"pressure"}]
}}
```


**Fig. 4** Monitoring data formats


```
{"systemName":{"system1":{"deviceName":["device1","device2"],
  "deviceValue":[22.7,22.9],"deviceLoation":["area11","area12"],
  "type":["tempreature","pressure"]},
  "systme2":{"deviceName":["device1","device2"],
    "deviceValue":[0.7,0.7],"deviceLoation":["area21","area22"],
    "type":["tempreature","pressure"]}}}
```


**Fig. 5** Improved monitoring data format

### 3.2.2 Create a Custom Map

The data key value characters of the objects studied in this paper are long. They can be replaced with shorter characters in the premise of ensuring uniqueness, such as currentdevice1-curdev1 to reduce the data placeholder. Only establish custom map dictionary and restored dictionary respectively on the server and client.

After above two steps, overall proportion of monitoring data can be reduced from macro level. It is important to note that when the client receives the monitoring data. The corresponding data reduction processing method is needed to guarantee the data integrity.

In summary, for the system objects studied in this paper, the overall data transmission optimization scheme flow chart is shown in Fig. 6.

# 4 Results Analysis

This paper selects the historical monitoring data generated during the task of a launch site as an example to carry out the verification analysis of transmission optimization scheme. This verification is the peak of simulation request. At the same time, 20 computers simultaneously launched 2000 concurrent requests, and selected the three most important subsystems of power supply subsystem, fire protection subsystem and gas alarm to perform optimization analysis of 10 sets of monitoring data. The results of the experiment are shown in Fig. 7 (select a set of monitoring data of the largest fire control subsystem), request sent is the time that takes the client to send the request to the server. Waiting (TTFB) is the time that takes to initiate a request from the client until the server returns the first byte. Content Download is the time that client downloads the received resources.

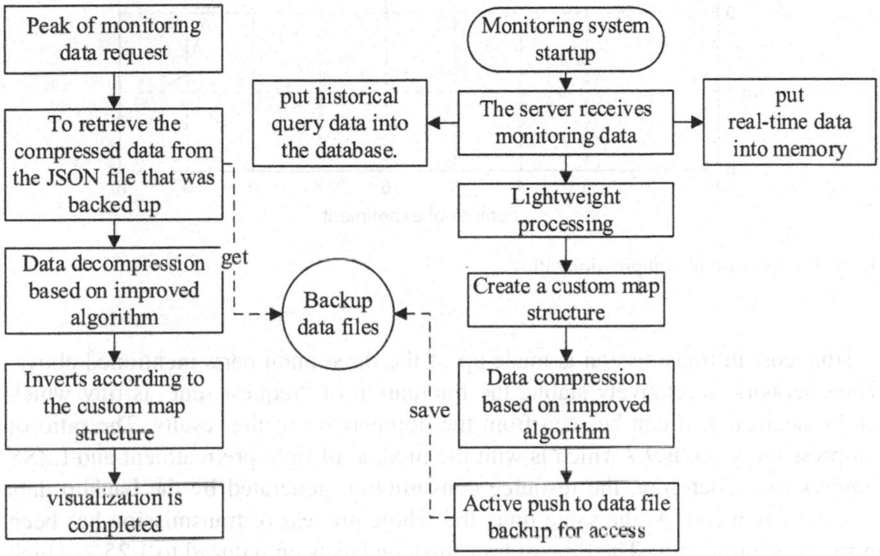


**Fig. 6** System overall transmission optimization flow chart

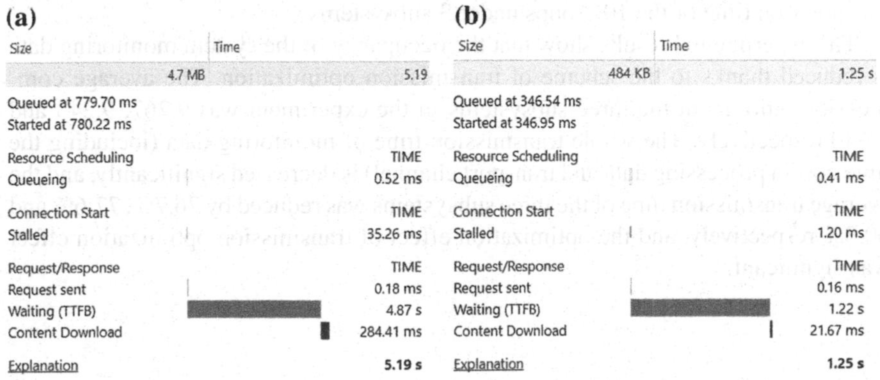


**Fig. 7** **a** Performance before optimization **b** performance after optimization

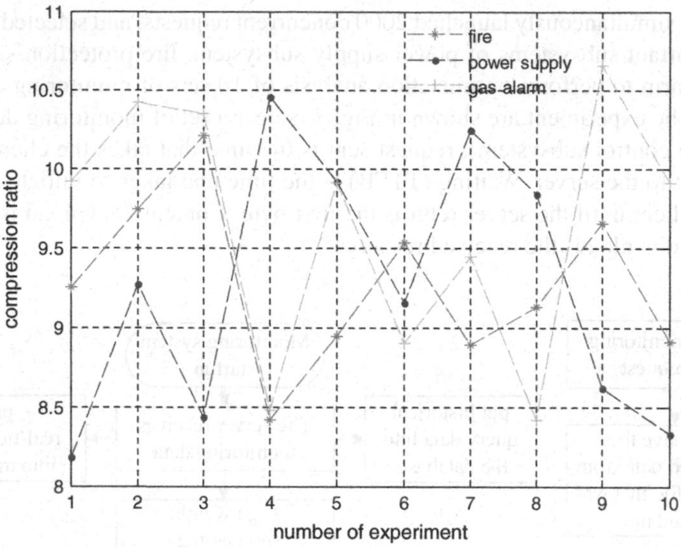


**Fig. 8** Comparison of compression ratio

Time cost in transmission is made up of the three main parts mentioned above. When network is relatively stable, the fluctuation of "request sent" is tiny which can be neglected. It can be seen from the comparison of the results. The ratio of compression is about 9.7 which is with the method of light pretreatment and LZSS compression. Therefore, the resource consumption generated by the backup data file can be reduced. At the same time, the whole process of transmission has been improved significantly. The time of transmission has been reduced to 1.25 s, which has basically been able to satisfy the monitoring request of the mass data in peak. Figures 8 and 9 are the results of the data compression ratio and the reduction of transmission time of the 10 groups under 3 subsystems.

The experimental results show that the occupancy of the system monitoring data is reduced thanks to the scheme of transmission optimization. The average compression ratio about the three subsystems in the experiment was 9.267, 9.243 and 9.640 respectively. The whole transmission time of monitoring data (including the time cost in processing data and transport channel) is decreased significantly, and the average transmission time of the three subsystems was reduced by 76.7%, 77.6% and 80.3% respectively, and the optimization effect of transmission optimization effect was significant.

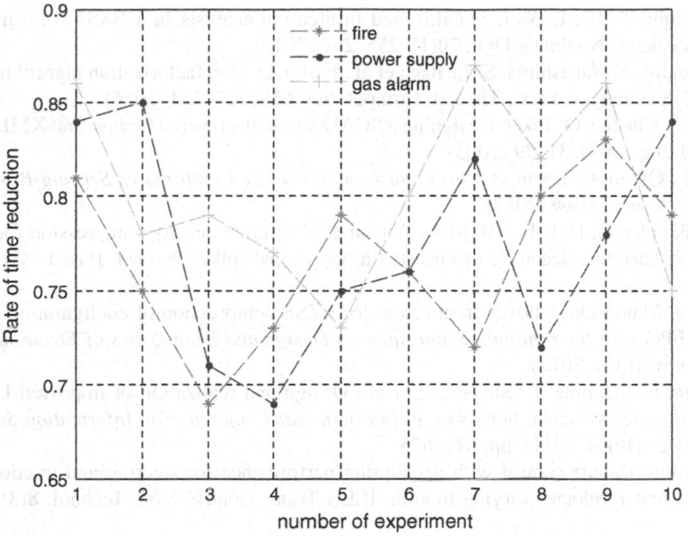


**Fig. 9** Comparison of transmission time

# 5 Conclusion

Aiming at the transmission process of complex engineering systems, this paper proposes an improved operation mode combined with optimized LZSS compression algorithm to optimize the transmission of monitoring data. It is better to adapt the data transmission demand of complex monitoring systems. But there is room for further improvement of LZSS algorithm [12–14]. The higher degree of compression, the smaller amount of data transmission, and the shorter time of transmission. On the other hand, it causes the increase of the compression time. Therefore, the feedback mechanism of adjustable parameters should be used to find the optimal transmission efficiency of the system.

# References


1. F. Ren, T. Zhao, J. Jiao et al., Resilience optimization for complex engineered systems based on the multi-dimensional resilience concept. IEEE Access (99), 1–1 (2017)
2. L. Boyuan, H. Shuangxi, F. Wenhui et al., Data driven uncertainty evaluation for complex engineered system design. Chin. J. Mech. Eng. **29**(5), 889–900 (2016)
3. G.V. Kalabin, V.I. Gorny, S.G. Kritsuk, Environmental appraisal of the area of Kachkanar mining-and-processing plant by satellite monitoring data. J. Min. Sci. **52**(2), 394–400 (2016)
4. C. Fang, F. Marle, E. Zio et al., Network theory-based analysis of risk interactions in large engineering projects. Reliab. Eng. Syst. Saf. **106**(2), 1–10 (2012)
5. Y. Moon, D. Kim, Y. Go et al., Cedos: a network architecture and programming abstraction for delay-tolerant mobile apps. IEEE/ACM Trans. Networking **99**, 1–16 (2017)

6. Z. Zhenqiu, Z. Jie, L. Wei, Stability and bifurcation analysis in a FAST TCP model with feedback delay. Nonlinear Dyn. **70**(1), 255–267 (2012)
7. I. Tomohiro, Y. Nakashima, S. Inenaga et al., Faster Lyndon factorization algorithms for SLP and LZ78 compressed text. Theoret. Comput. Sci. **656**, 215–224 (2016)
8. R. Fan, L. Cheded, O. Toker, Designing a SCADA system powered by Java and XML. Comput. Control Eng. **16**(5), 31–39 (2005)
9. W. Ji, L. Qilian, C. Kwan, *A Novel and Comprehensive Compressive Sensing-Based System for Data Compression* (2012)
10. D. Sodkomkham, D. Ciliberti, M.A. Wilson et al., Kernel density compression for real-time Bayesian encoding/decoding of unsorted hippocampal spikes. Knowl. Based Syst. **94**, 1–12 (2016)
11. R. Iša, J. Matoušek, A novel architecture for LZSS compression of configuration bitstreams within FPGA, in *International Symposium on Design and Diagnostics of Electronic Circuits & Systems* (IEEE, 2017)
12. Z. Huan, F. Xiaoping, L. Shaoqiang et al., Design and realization of improved LZW algorithm for wireless sensor networks, in *International Conference on Information Science and Technology* (IEEE, 2011), pp. 671–675
13. J. Yingmin, Robust control with decoupling performance for steering and traction of 4WS vehicles under velocity-varying motion. IEEE Trans. Control Syst. Technol. **8**(3), 554–569 (2000)
14. J. Yingmin, Alternative proofs for improved LMI representations for the analysis and the design of continuous-time systems with polytopic type uncertainty: a predictive approach. IEEE Trans. Autom. Control **48**(8), 1413–1416 (2003)

# Auxiliary Rotor Slot Optimization Design for Improving Back-EMF Waveform of PMSM Based on MagneForce

**Yuanyuan Yang, Bulai Wang, Xiutao Ji, Xiangsheng Liu and Panyuan Ren**

**Abstract** As lots of harmonics are in the back-electromotive force (EMF) of Permanent Magnet Synchronous Motor (PMSM), an auxiliary rotor slot is presented to reduce the Total Harmonic Distortion (THD), to improve the waveform of back-EMF, and to make the waveform of it close to sinusoid. A 22 kW 3000 rpm PMSM is simulated by MagneForce software. The simulation results show that the THD is decreased by 48.4%, while the fundamental voltage amplitude is increased by 4.4%, and the waveform is closer to sinusoid after the auxiliary rotor slot is opened.

**Keywords** PMSM · Back-EMF · MagneForce · An auxiliary rotor slot

## 1 Introduction

Permanent magnet synchronous motor is increasingly concerned by scholars and experts in related fields such as electric vehicles, aerospace, and industrial applications because of its high power density, high efficiency, and good dynamic response capability. Compared with surface-mounted permanent magnet synchronous motors, the IPM machine have the advantages of large torque density, wide range of field weakening, and high mechanical strength. it is necessary to weaken the harmonic content of the motor back EMF and improve the back EMF waveform,with its magnetic circuit structure, the back EMF contains a large number of harmonics, which will reduce the control accuracy and increase the harmonic loss and torque ripple [1, 2].

Optimizing the magnetic pole width to weaken the harmonic content in the back EMF is proposed for the surface-mount magnetic pole structure in [3]. One method through reducing the back EMF distortion rate by using a non-uniform magnetic steel groove, with its easy processing, verified by a prototype is proposed in [4]. A method

Y. Yang (✉) · B. Wang · X. Ji · X. Liu · P. Ren
School of Electrical and Electronic Engineering, Shanghai Institute of Technology, Shanghai 201418, China
e-mail: yangyuanyuanfly@163.com



Y. Jia et al. (eds.), *Proceedings of 2018 Chinese Intelligent Systems Conference*,
Lecture Notes in Electrical Engineering 528,
https://doi.org/10.1007/978-981-13-2288-4_12

is designed for improving no-load back EMF waveform based on the optimization of air gap magnetic density, for using the technique of unequal thickness magnetic steel in [5]. Method in [6] uses uneven air gap and stator chute, increasing machining accuracy and high process requirements for magnetic steel and rotor core, in spite of reducing the harmonic content in the back EMF waveform.

Aims at an IPM machine, the novel method weakens the harmonic components and increases the fundamental current amplitude, using ordinary magnetic steel and auxiliary slots on the rotor. The method has more simple structure, low cost. The processing is convenient with the auxiliary groove of the rotor stamped.

## 2 Analysis of No-Load Back Electromotive Force of Original Motor

The IPM machine with a uniform air gap, the no-load back electromotive force approximates a trapezoidal wave distribution for the air gap flux density approximates a trapezoidal wave distribution. The harmonic content is large and the waveform distortion rate is high. It will cause motor vibration and noise, result in harmonic loss, seriously affect the quality of permanent magnet synchronous motor. Optimizing the back-EMF waveform by opening the auxiliary slot on the rotor is proposed, which reducing the harmonic content and improving the sinusoidal nature of the no-load back EMF. The position and size of the auxiliary slot is determined with optimization analysis.

In this paper, a 22 kW 3000 rpm permanent magnet synchronous motor is taken as an object to study the influence of auxiliary slots on the performance of the rotor. The main structural parameters of the motor are shown in Table 1.

The finite element model of the motor is established quickly by using MagneForce software with the parametric model library. Lots of stator and rotor models are

**Table 1** The main structural parameters of the motor

| Project | Parameter | Project | Parameter |
|---|---|---|---|
| Number of poles | 6 poles | Number of parallel paths | 3 |
| Number of slots | 36 slots | Unilateral air gap length | 0.5 mm |
| Outer diameter of stator | 190 mm | Silicon steel grade | DW315-50 |
| Stator inner diameter | 110 mm | Magnet size | 45 * 5 mm |
| Winding type | Double layer winding around Y | Magnetic steel grade | N38UH |
| Wire gauge/number of turns | 3–1.12/12 | Iron core length | 190 mm |

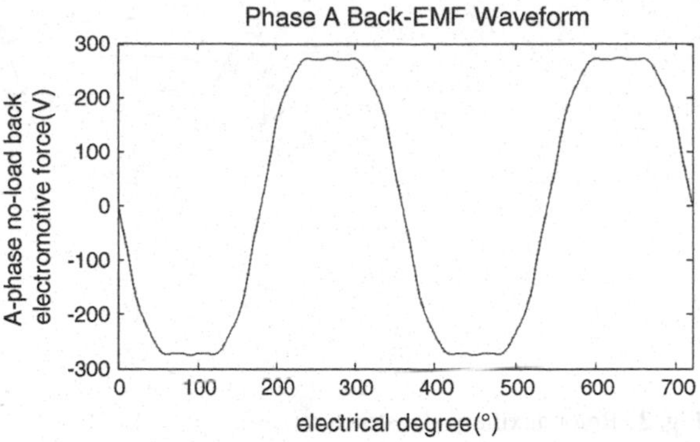


**Fig. 1** A-phase no-load back EMF waveform

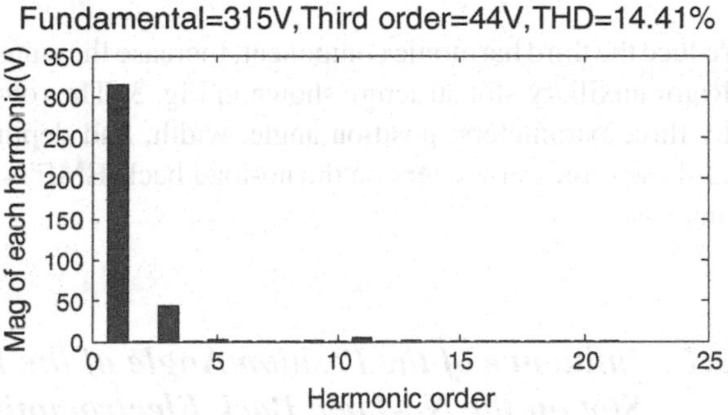


**Fig. 2** Analysis of harmonic content of A-phase no-load back EMF

contained the model library. The software also supports that importing dxf format model files to set the slice model. The MagneForce motor model is built by using the dxf format for the special structure of the rotor punching plate. Its simulation is performed.

Figure 1 shows the back-EMF waveform diagram of the no-load motor of the original model. It can be seen from the figure that the waveform has an approximate trapezoidal wave distribution, and the harmonic content is relatively large, resulting in harmonic loss and severely affecting the performance of the motor; Fig. 2 shows the phase A space. The harmonic content of the back-EMF analysis chart shows that the fundamental amplitude is 315 V, the third harmonic in the harmonic content accounts for the largest, and the amplitude is 44 V. The remaining harmonics account for a relatively small percentage, and the harmonic distortion rate is 14.41%.

## 3 Rotor Auxiliary Slot Design

Aiming at the problem of large harmonic content (especially the third harmonic component) of no-load back EMF, this paper proposes a method of opening the auxiliary slot on the rotor to reduce the air-gap magnetic-amplitude amplitude and

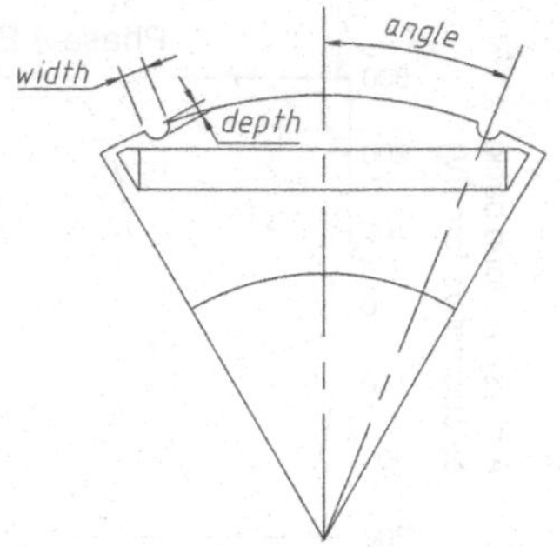


**Fig. 3** Rotor auxiliary slot structure

reduce the third harmonic component. Increase the sinusoidal nature of the waveform. Rotor auxiliary slot structure shown in Fig. 3. The rotor auxiliary slot is determined by three parameters: position angle, width, and depth. In this paper, the influence of these three parameters on the no-load back-EMF is studied through optimization analysis.

### 3.1 Influence of the Position Angle of the Rotor Auxiliary Slot on the No-Load Back Electromotive Force

Figure 4 shows the variation curve of the back-EMF waveform with the angle of the rotor auxiliary slot angle. From the figure, it can be analyzed: the amplitude of the fundamental wave changes inversely to the V-shape of the angle and reaches its maximum value at around 24°; third harmonic the amplitude and the distortion rate are V-shaped for the angle and reach a minimum around 22°. It can be seen that the auxiliary slot position angle could have a great impact on the no-load back EMF. From the perspective of reducing the third harmonic content and increasing the sinusoidality of the back EMF, this paper finally selected angle = 22° as the auxiliary slot position of the rotor and continued to study it based on this.

### 3.2 The Effect of Width and Depth of Rotor Auxiliary Slot Width on No-Load Back EMF

Figure 5 is the variation curve of the no-load back-EMF waveform with the width and depth of the rotor auxiliary slot width, from which the following conclusion can be drawn: When the width is constant, the amplitude of the fundamental wave [7] increases with depth, and the third harmonic amplitude. The value and the waveform distortion rate decrease with the increase of the depth; when the depth is fixed, the

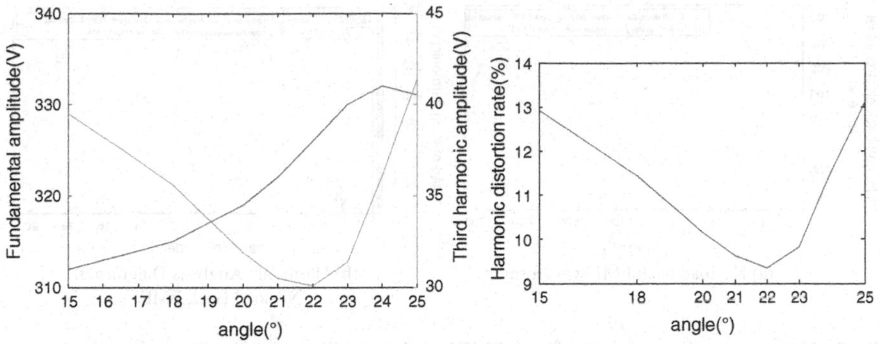


(a) The effect of angle on fundamental and third harmonic amplitudes

(b) The effect of angle on the rate of distortion

**Fig. 4** Effect of rotor auxiliary slot position angle on back-EMF waveform (width = 2 mm, depth = 2 mm)

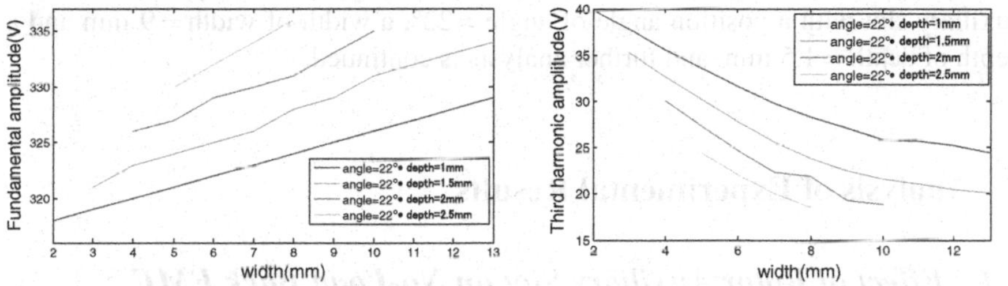


(a) The amplitude of the fundamental amplitude varies with the width of the auxiliary slot width

(b) Variation of the third harmonic amplitude the width of the auxiliary slot width

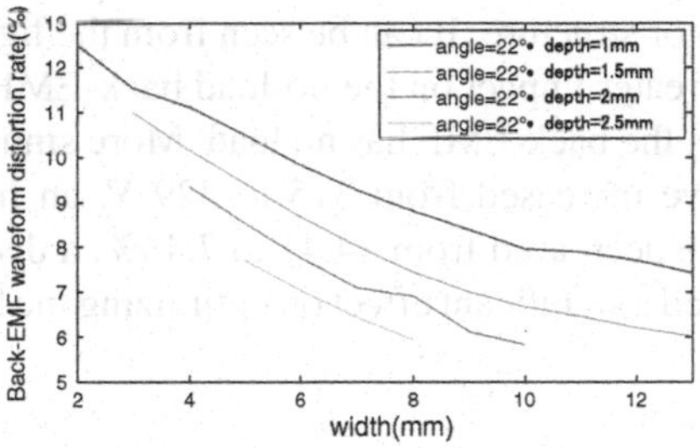


(c) Variation in back-EMF waveform distortion rate with auxiliary slot width

**Fig. 5** Effect of width and depth of rotor auxiliary slot width on no-load back-EMF waveform (angle = 22°)

amplitude of the fundamental wave increases with the increase of the width, and the third harmonic amplitude and the waveform distortion rate increase with the width. decline. In view of the limitation of the structure size of the rotor itself, the width and the depth may vary in some cases.

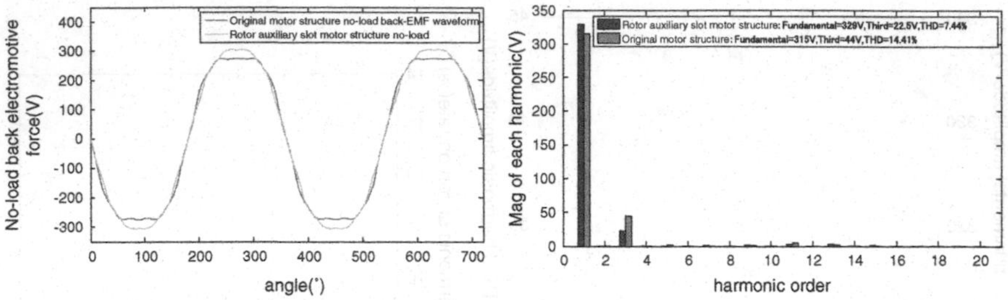


(a) No-load back EMF waveform

(b) Harmonic Analysis Diagram of No-load Back EMF

**Fig. 6** Harmonic analysis of no-load back EMF based on auxiliary rotor groove and original motor structure

Based on the above analysis, taking into account the mechanical strength of the rotor and the electrical performance of the motor, this paper finally selects a rotor auxiliary slot with a position angle of angle = 22°, a width of width = 9 mm and a depth of depth = 1.5 mm, and further analysis is continued.

# 4 Analysis of Experimental Results

## *4.1 Effect of Rotor Auxiliary Slot on No-Load Back EMF*

Figure 6 shows the comparison of the back-EMF waveforms of the rotor auxiliary slot and the original motor structure. It can be seen from the figure that the auxiliary slot on the rotor has a greater impact on the no-load back-EMF. Compared with the original motor structure, the back-EMF has no load. More sinusoidal, the amplitude of the fundamental wave increased from 315 to 329 V, an increase of 4.4%, the waveform distortion rate decreased from 14.41 to 7.44%, a decrease of 48.4%, and the rotor auxiliary slot had a significant effect on optimizing the back-EMF waveform of no-load.

## *4.2 Effect of Rotor Auxiliary Slot on Rated Output Torque*

Figure 7 shows the effect of the rotor on the rated output torque of the motor after the auxiliary slot is opened. It can be seen from the figure that the output torque ripple peak value is reduced from 3.9 to 2.3 N m, and the torque is more stable than the original motor model. This is because the rotor reduces the back-EMF harmonic content after opening the auxiliary slot, which in turn reduces torque ripple [8].

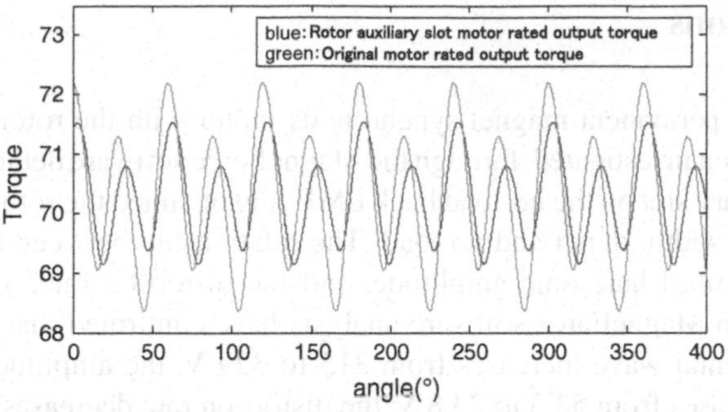


**Fig. 7** Effect of rotor auxiliary slot on rated output torque

**Table 2** Effect of rotor auxiliary slot on iron loss efficiency and $L_d$, $L_q$

| | Stator iron consumption (W) | Rotor iron consumption (W) | Total loss (W) | Output power (W) | Effectiveness (%) | $L_d$ (mH) | $L_q$ (mH) |
|---|---|---|---|---|---|---|---|
| Original motor structure | 524 | 28.8 | 797.1 | 22,000 | 96.5 | 2.4 | 6.5 |
| Rotor auxiliary slot motor structure | 478.5 | 22.8 | 750.7 | 21,990 | 96.7 | 2.2 | 5.6 |

## *4.3 Effect of Rotor Auxiliary Slot on Iron Consumption, Efficiency and $L_d$, $L_q$*

Table 2 reflects the effect of the rotor auxiliary slot on iron loss, efficiency, and Ld, Lq. It can be seen from the table, after the auxiliary slot is opened on the rotor, the stator iron loss is reduced by 8%, iron consumption decreases by 20.8%, and motor efficiency increases 96.7% from 96.5%. Because it reduces the harmonic content. The rotor auxiliary slot could cause Ld and Lq to decrease in different degrees, especially the influence of the Lq is more significant, which makes the salient pole rate decreased. Because the auxiliary slot is close to the q-axis position of the rotor. Compared with the original motor structure, it increases the length of the air gap near the q-axis and makes the Lq drop more obviously.

## 5 Conclusions

In this paper, a permanent magnet synchronous motor with the rotor auxiliary slot structure has been investigated, through the MagneForce software detailed analysis of the rotor auxiliary slot on the no-load back EMF, and obtained the rotor auxiliary slot position angle, width, depth and no load. The relationship between the back-EMF amplitude, the third harmonic amplitude, and the distortion rate. The simulation results based on MagneForce software analysis have confirmed that the amplitude of the fundamental wave increases from 315 to 329 V, the amplitude of the third harmonic decreases from 53.3 to 23.8 V, the distortion rate decreases from 14.41 to 7.44%, and the output torque pulsation decreases and the efficiency increases.

In summary, the back-EMF waveform can be improved significantly for the proposed rotor auxiliary slot structure. It is feasible in the manufacturing process.

## References


1. J.A. Guemes, A.M. Iraolagoitia, J.I. Del Hoyo, Torque analysis in permanent-magnet synchronous motors: a comparative study. IEEE Trans. Energy Convers. **26**(1), 55–63 (2011)
2. H. Jing, F. Hongyu, W. Chen, Optimization design of built-in permanent magnet synchronous motor based on Ansoft. J. Sichuan Univ. Sci. Technol. **29**(4), 35–38 (2016)
3. L. Quanwu, D. Manfeng, L. Zhaojie, Optimization of pole width optimization method for back-EMF waveform of permanent magnet synchronous motor. Micro Motors **40**(9), 6–8 (2012)
4. B. Wang, Z. Wu, Y. Wang, Z. Hou, Low harmonic high efficiency permanent magnet synchronous machines based on non-uniform magnet slot, in *2011 International Conference on Electrical and Control Engineering, ICECE 2011—Proceedings*, pp. 233–236
5. J. Youhua, W. Hongwei, Optimization of air gap flux density of permanent magnet synchronous motor based on Ansoft. Micromotor **46**(12), 84–87 (2013)
6. Z. Lei, L. Guang, Analysis of the back-EMF waveform distortion rate of permanent magnet synchronous generator based on Ansoft. Explos. Proof Motor **47**(4), 28–31 (2012)
7. A. Kioumarsi, M. Moallem, B. Fahimi, Mitigation of torque ripple in interior permanent magnet motors by optimal shape design. IEEE Trans. Magn. **42**(11), 3706–3711 (2006)
8. L. Dosiek, P. Pillay, Cogging torque reduction in permanent magnet machines. IEEE Trans. Ind. Appl. **43**(6), 1565–1571 (2007)

# Research on Three-Phase Grid-Connected Inverter Model Predictive Control

**Lin Cao, Shuaiyi Wang and Jing Bai**

**Abstract** This paper presents a method of three-phase grid-connected inverter model predictive control (MPC) for the shortcomings of slow dynamic response and poor robustness of three-phase grid-connected inverters. Through MPC prediction, the optimal control quantity is obtained based on the optimization function. The establishment and implementation process of the MPC algorithm are introduced. The corresponding simulation model is established in Matlab/Simulink environment. The simulation results show that the MPC has the characteristics of fast dynamic response and strong robustness. It has good steady-state performance and is suitable for high-performance control of three-phase grid-connected inverters.

**Keywords** MPC · Three-phase grid-connected inverter · Optimization function

## 1 Introduction

With the rapid development of new energy, the system needs to convert direct current into alternating current, and the inverter has played a huge role. Optimize the system's control strategy will get high-quality electrical energy. There are many problems in the traditional control: the steady-state dynamic error of the PI control is slow, the adjustment of the PR control parameter is difficult, and the large harmonics and filter settings in the hysteresis control are difficult. The disadvantages of slow dynamic response and poor robustness based on three-phase grid-connected inverters [1], a model predictive control algorithm is proposed. Model predictive control (MPC) is a model-based optimization control technique. It is easy to model and has good robustness. It is an effective way to control complex systems. Since Richalet et al. proposed the model predictive heuristic control (MPHC) in 1978, the model predictive control has been greatly developed. MPC is considered to be the most promising advanced control algorithm in the industrial process. Since its

L. Cao · S. Wang · J. Bai (✉)
School of Electrical & Information Engineering, Beihua University, Jilin 132021, China
e-mail: jlbyj@163.com

 
Y. Jia et al. (eds.), *Proceedings of 2018 Chinese Intelligent Systems Conference*,
Lecture Notes in Electrical Engineering 528,
https://doi.org/10.1007/978-981-13-2288-4_13

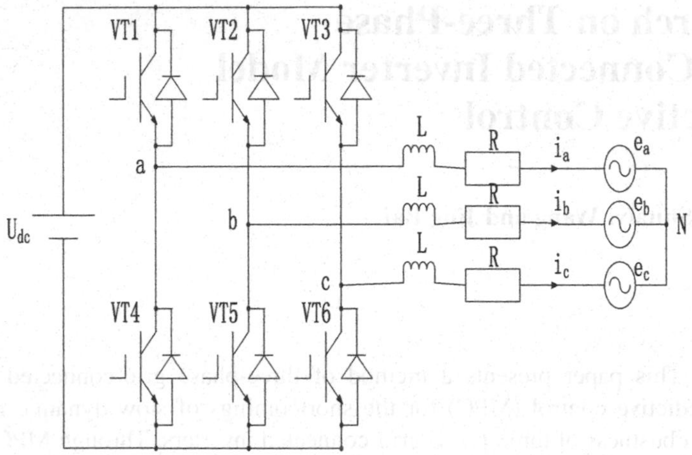


**Fig. 1** Topology of three-phase grid-connected inverter

birth, MPC has attracted the attention of many researchers and gradually attracted the attention of industrial control and theoretical circles. So far, model predictive control has been widely used in power generation, oil refining, metallurgy, chemical industry, automotive, aerospace and other fields. MPC is a kind of computer control algorithm, which mainly includes steps of model prediction, rolling optimization, feedback correction, etc. It has a very good control effect, and the method is simple and easy to implement. Model prediction of direct current control [2–4] (MPDCC) and model prediction direct power control [5–7] (MPDPC) as an extension of model predictive control has gradually become a research hotspot. Direct current control for model prediction. The main focus is on the study of the stationary coordinates of grid-connected currents and the current components in rotating coordinates. The model predictive current control strategy can be summarized as the following steps: Define the cost function [8]. Establish the converter model and its possible switching states. Establish a load forecasting model.

In recent years, the model predictive control has been used in many applications in the inverter. Photovoltaic grid-connected inverter model prediction direct power control research [9] has good stability. However, the difference in grid connection; photovoltaic grid-connected inverter finite-state model predictive current control [10], has a good current tracking effect, but the dynamic response is poor; three-phase boost grid-connected inverter discrete-time predictive control [11] has good grid-connected performance. However, the dynamics are relatively poor.

## 2 Mathematical Model of Three-Phase Grid-Connected Inverter

The topology of three-phase grid-connected inverters is shown in Fig. 1. In Fig. 1, $U_{dc}$ is the DC side voltage, $L$ is the filter inductance, $R$ is the load resistance, $i_a$, $i_b$, $i_c$ are respectively $a$, $b$, $c$ three-phase grid-connected current $e_a$, $e_b$, $e_c$ is $a$, $b$, $c$ three-phase grid voltage. The switching state of the inverter is controlled by $S_a$, $S_b$, $S_c$. A-phase output voltage value,

$$u_{aN} = S_a U_{dc} \tag{1}$$

Output voltage vector

$$U = \frac{2}{3}\left(U_{aN} + aU_{bN} + a^2 U_{cN}\right) \tag{2}$$

Among them: $U_a$, $U_b$, $U_c$ inverter output voltage.

The states of $U_a$, $U_b$, $U_c$ may have 8 voltage vectors, among them $U_0 = U_7$. The voltage vector in the output complex plane of the inverter is shown in Fig. 2.

A-phase load current

$$U_{aN} = L\frac{di_a}{dt} + Ri_a + e_a + U_{nN} \tag{3}$$

Load current and space back electromotive force vector

$$i = \frac{2}{3}\left(i_a + ai_b + a^2 i_c\right) \tag{4}$$

$$e = \frac{2}{3}\left(e_a + ae_b + a^2 e_c\right) \tag{5}$$

Fig. 2 Inverter output in-plane voltage vector

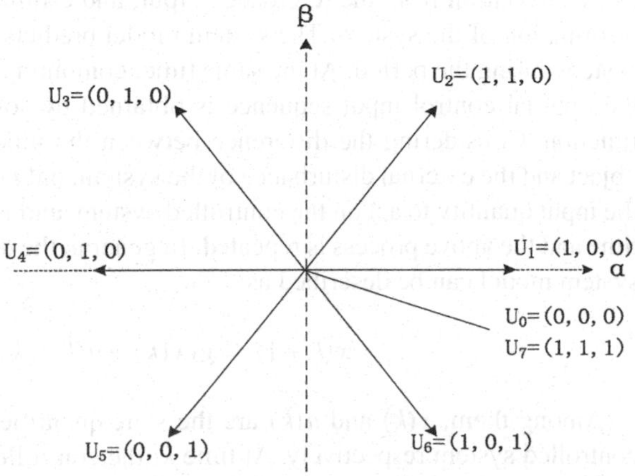

Load voltage dynamic vector equation:

$$U = Ri + L\frac{di}{dt} + e \tag{6}$$

Discretization of the load current equation to obtain a suitable discrete-time model for calculation prediction:

$$\frac{di}{dt} = \frac{i(k+1) - i(k)}{T_S} \tag{7}$$

where: $T_s$ is the sampling time.

The output current dynamic vector is discretized:

$$i^p(k+1) = \left(1 - \frac{RT_S}{L}\right)i(k) + \frac{T_S}{L}(U(K) - e(k)) \tag{8}$$

Each voltage vector generates a predicted current, which gives the value of the cost function. It can be observed that selecting the vector yields the lowest value of the cost function. Then, the voltage vector is selected and applied to the inverter.

## 3 The Principle of Model Predictive Control

The model predictive control obtains the state quantity of the current system through the state estimation algorithm, uses the system nominal model to predict, and uses the dynamic optimization algorithm to calculate the control sequence online while satisfying the performance index and the constraint condition. The system nominal model, dynamic optimization algorithm, performance index and system constraints constitute the three major elements of the controller.

At a certain sample time, the controller measures the output current of the system and its deviation from the reference output, and estimates the current state quantity information of the system. The system model predicts the dynamic behavior of the system during the period. At the same time, combining the given reference quantity, the optimal control input sequence is obtained by solving the performance index function. Considering the difference between the model and the actual controlled object and the external disturbance of the system, only the first element is selected as the input quantity to act on the controlled system, and is held until the next sampling time and the above process is repeated. In general, the discrete form of the controlled system model can be described as:

$$x(k+1) = g(x(k) + u(k)), k \geq 0 \tag{9}$$

Among them, $x(k)$ and $u(k)$ are the state quantities and input quantities of the controlled system respectively. At time $k$, the controlled system state quantity $x(k)$

is measurable, $N$ is the number of steps corresponding to the predicted time domain, and the control optimization problem is expressed as

$$J(x(k), U) = \sum_{i=0}^{N-1} L(x(k+i/k), u(k+i/k)) + V(x(k+N/k)) \quad (10)$$

Satisfy:

$$x(k+i+1/k) = g(x(i/k), u(i/k)), i = 0, 1 \cdots\cdots \mathrm{N}-1 \quad (11)$$

$x(k+i/k)$ denotes the predicted value of the controller state quantity for the future $k+i$ moment at time $k$. Assuming that there is an optimal control sequence $U^*$ and take its first element into the controlled system, the control law of the MPC controller can be expressed as:

$$u_{MPC}(k) = U_0^* \quad (12)$$

## 4 Three-Phase the Model Predictive Control of Three-Phase Grid-Connected Inverter

Model predictive control (MPC) is a kind of non-linear optimization control algorithm that predicts the current control action by predicting the state of the system in the finite time domain. The main features of this algorithm are good control effect and strong robustness. Create a cost function based on the desired goal and then select the optimal input based on the minimum value of the cost function. The model predictive control is actually to find the optimal value of the cost function at a fixed sampling time, and then repeat the cycle in each sampling time. The goal of current control is to minimize the error between the current and the reference, the error between the reference and the predicted current:

$$g = /i_\alpha^*(k+1) - i_\alpha^p(k+1)/ + /i_\beta^*(k+1) - i_\beta^p(k+1)/ \quad (13)$$

Assume that this reference current does not change sufficiently within one sampling interval, consider $i^*(k+1) = i^*(k)$. Model predictive current control algorithm block diagram shown in Fig. 3.

According to the system model, all existing states of the $k+1$ sampling time can be predicted, and the smallest cost function is selected at the next sampling time. At the $k$ sampling time, the state $S_2$ minimizes the cost function, then the state $S_2$ is selected at the $k$ sampling time; At the $k+1$ sampling time, $S_3$ minimizes the cost function, and $S_3$ is selected at the $k+1$ sampling time, and the same process is performed for each sampling time. The model predictive control principle is shown in Fig. 4.

**Fig. 3** The algorithm block diagram of model predictive current control

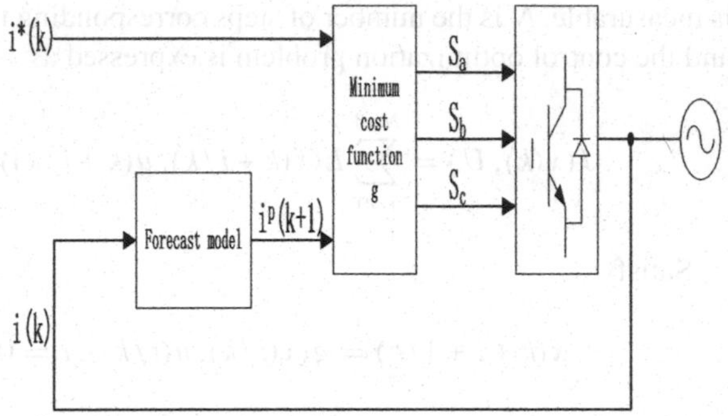


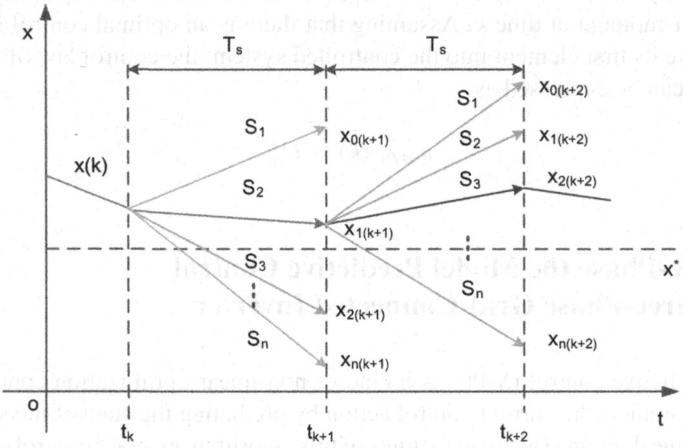


**Fig. 4** Model predictive control schematic

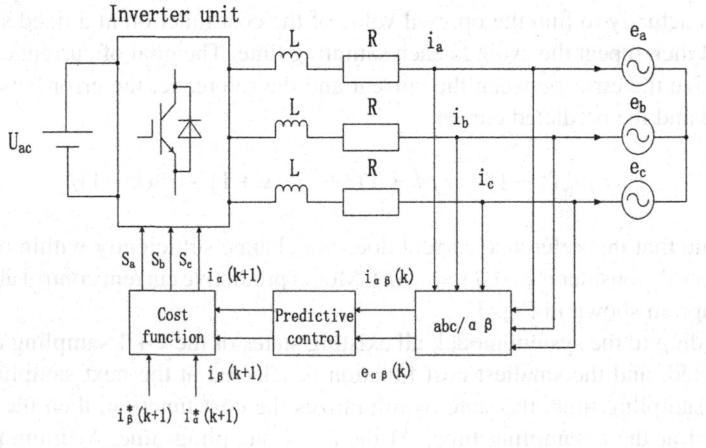


**Fig. 5** Three-phase grid-connected inverter predictive control structure

At the $k$ sampling time, the load current $i_a$, $i_b$, $i_c$ the grid voltage $e_a$, $e_b$, $e_c$ is measured and transformed into $i_{\alpha\beta}(k)$, $e_{\alpha\beta}(k)$; $i_{\alpha\beta}(k)$, $e_{\alpha\beta}(k)$ can be obtained through predictive control $i_\alpha(k+1)$, $i_\beta(k+1)$; Finally, the reference value $i_\alpha^*(k+1)$, $i_\beta^*(k+1)$ of the $k+1$ sampling time is selected by the cost function to select the optimal switching state $S_a$, $S_b$, $S_c$ acting on the switch. The three-phase grid-connected inverter predictive control structure is shown in Fig. 5.

## 5 Simulation Analysis

The simulation model was built by Matlab/Simulink and simulation analysis was performed. The simulation results show that the model predicts fast dynamic response, strong robustness, and meets grid connection.

Figure 6 shows that the amplitude of three-phase reference current is 40 A.

Figure 7 shows that the amplitude three-phase grid-connected current is 40 A. Figures 6 and 7, the model predictive control has a good current tracking.

Figure 8 shows the three-phase grid voltage with an amplitude of 311 V.

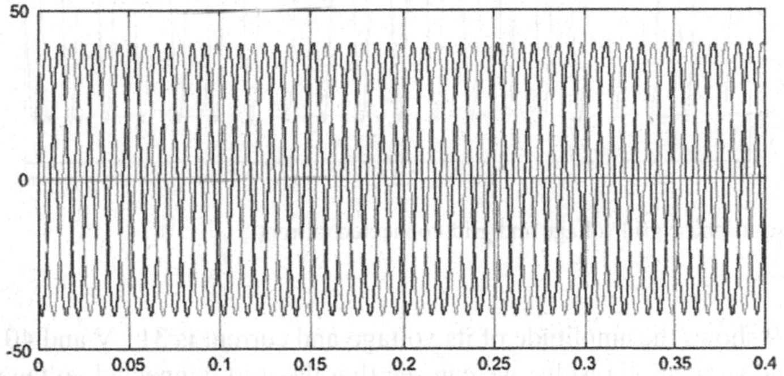


**Fig. 6** Three-phase reference current

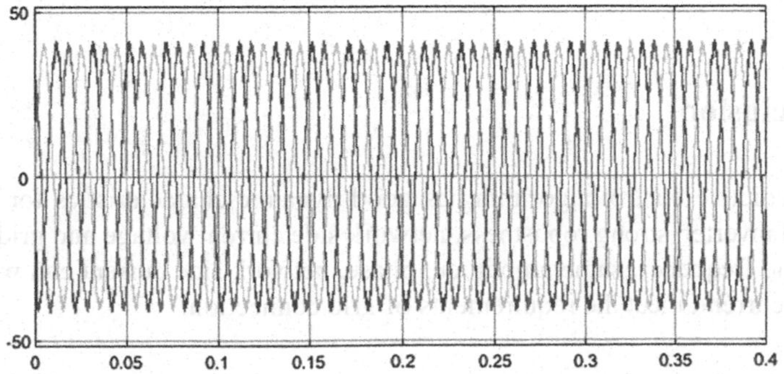


**Fig. 7** Three-phase grid-connected current

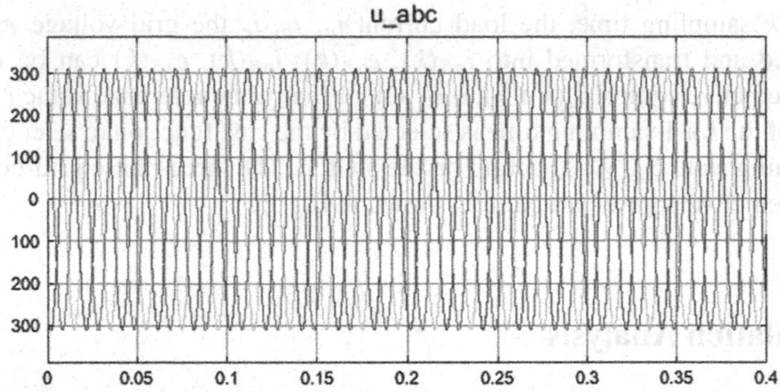


**Fig. 8** Three-phase grid voltage

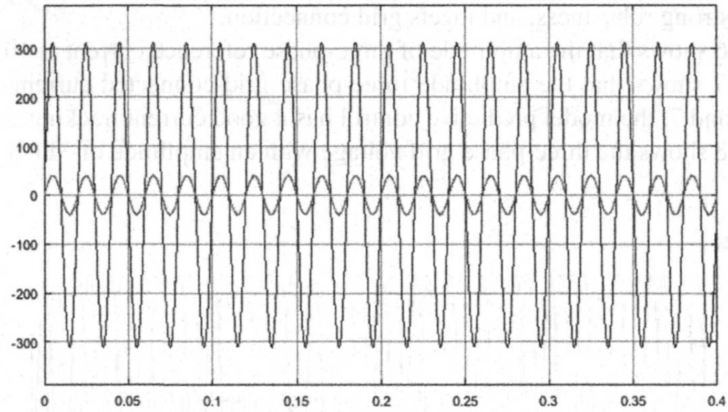


**Fig. 9** Phase-to-phase grid voltage and grid-connected current

Figure 9 shows the amplitude of its voltage and current is 311 V and 40 A.

From the simulation results, we can see that the grid-connected voltage and the grid-connected current have the same frequency, the same amplitude, and the same phase, meet the grid connection requirements, and complete the grid connection.

## 6 Conclusion

Model predictive control algorithm has good dynamic characteristics for PV grid connected inverter, strong robustness, network side current voltage and grid voltage of the same frequency, same amplitude, phase, network side current can well track reference current, meet the requirements of grid connection.

## References

1. X. Chen, M. Zhang, Z.C. Wang, Research on the dynamic interaction between photovoltaic grid-connected inverter and grid based on impedance analysis method. Chin. J. Electr. Eng. **34**(27), 4559–4567 (2014)
2. P. Karamanakos, T. Geyer, S. Manias, Direct model predictive current control strategy of dc-dc Boost converters. IEEE J. Emerg. Sel. Top. Power Electron. **1**(4), 337–346 (2013)
3. B.S. Riar, T. Geyer, U.K. Madawala, Model predictive direct current control of modular multi-level converters: modeling, analysis, and experimental evaluation. IEEE Trans. Power Electron. **30**(1), 431–439 (2015)
4. S. Kun, Z. Hao, W. Ling et al., Three-phase voltage-type inverter model predictive control. Chin. J. Electrotech. **28**(12), 283–289 (2013)
5. J. Hu, J. Zhu, D.G. Dorrell, Model predictive direct power control of doubly fed induction generators under unbalanced grid voltage conditions in wind energy applications. IET Renew. Power Gener. **8**(6), 687–695 (2014)
6. V. Yaramasu, B. Wu, J. Chen, Model-predictive control of grid-tied four-level diode-clamped inverters for high-power wind energy conversion systems. IEEE Trans. Power Electron. **29**(6), 2861–2873 (2014)
7. Z. Xiaojie, Y. Yi, W. Fei, Single-phase grid-connected converter predicts direct power control strategy research. China J. Electr. Eng. **34**(30), 5269–5276 (2014)
8. J. Hu, J. Zhu, D.G. Dorrell, Model predictive control of grid-connected inverters for PV systems with flexible power regulation and switching frequency reduction. IEEE Trans. Ind. Appl. **51**(1), 587–594 (2015)
9. J. Nan, D. Xuanxuan, C. Guangzhao, Photovoltaic grid-connected inverter model prediction direct power control method research. Chin. J. Electr. Eng. **11**(6), 13–18 (2016)
10. H. Jin Nan, C.G. Shiyang, Photovoltaic grid-connected inverter finite state model predictive current control. Chin. J. Electr. Eng. **35**(S1), 190–196 (2015)
11. L. Yuling, W. Kerou, L. Huipin, Three-phase boost grid-connected inverter discrete-time predictive control. Chin. J. Electr. Eng. **31**(15), 190–196 (2011)

# Semi-supervised Learning Based on Coupled Graph Laplacian Regularization

**Xuejuan Zhao, Di Wang, Xiaoqin Zhang, Nannan Gu and Xiuzi Ye**

**Abstract** This paper aims at constructing coupled graph to discover the intrinsic sample structures under the Semi-Supervised Learning (SSL). Specifically, we first select some anchors by a clustering method such as K-means, and build the weight matrix by local reconstruction coefficients that represent each sample as a linear combination of its neighboring anchors. Then the graph Laplacian matrices over anchors and samples are respectively constructed by the weight matrix. On one hand, the anchor graph gives the coarse data structure and reduces the influences of the noise of training samples and outliers. On the other hand, the sample graph gives the detailed description for the fine structures of samples. We integrate the two graphs into a unified optimization framework, and propose the coupled graph Laplacian regularized semi-supervised learning approach. Experiments on several publicly datasets show that the proposed approach achieves the superior classification performances, while the computational costs are comparable to state-of-the-art methods.

**Keywords** Semi-supervised learning · Coupled graph · Anchor graph

## 1 Introduction

In the real-world classification applications, we can easily collected a massive amount of data from the Internet surfing and daily social communication. However, most of the data is unlabeled, and the labeling process is quite costly in human labor. Therefore, Semi-Supervised Learning (SSL) methods which utilize both labeled data and unlabeled data for classification are attracted numerous attention in recent years.

---

X. Zhao · D. Wang (✉) · X. Zhang · X. Ye
Department of Computer Science, Wenzhou University, Wenzhou 325035, China
e-mail: wangdi@amss.ac.cn

N. Gu
School of Statistics, Capital University of Economics and Business, Beijing 100070, China

 
Y. Jia et al. (eds.), *Proceedings of 2018 Chinese Intelligent Systems Conference*, Lecture Notes in Electrical Engineering 528,
https://doi.org/10.1007/978-981-13-2288-4_14

The emphasis of SSL is to fully explore the intrinsic structure of samples for inferring the unknown labels. Graph has proved to be successful in characterizing pairwise data relationship and manifold exploration, and a large number of approaches utilize graph as a tool for designing SSL algorithms. Graph-based SSL approaches are built based on the assumption that nearby samples (or samples of the same cluster or data manifold) are likely to have the same label. They can be divided into two main groups: the transductive learning approach [1–12] and the inductive learning approach [13–23]. The transductive algorithms take unlabeled samples as the test samples and predict the labels of unlabeled data points without an explicit decision function. The inductive algorithms predict the labels for unlabeled samples try to induce decision functions with low expected risk.

However, traditional graph-based SSL approaches focus on the graph construction of samples and ignore the scale of the graph which is related to the number of training samples. They will suffer from the high computational complexity and large storage requirement when datasets are large-scale. To overcome these limitations, recent works [24–30] concentrate on making SSL practical on large scale data collections by skillfully constructing large graphs over all data. Zhang et al. [24] introduce the prototype vectors to approximate the graph-based regularizer and model representation, and the problem size is drastically reduced. Liu et al. [25] construct a novel graph over samples via a small number of anchor points which can approximately cover the entire data, and propose the Anchor Graph Regularization (AGR) model for SSL. Wang et al. [31] introduce a novel fast local anchor embedding method, and propose the Efficient Anchor Graph Regularization (EAGR) model which designs the normalized graph over anchors by considering the commonly linked data samples. Wu et al. [30] present a landmark-based label propagation method which first locally approximates the soft label of each sample by a linear combination of labels on its nearby landmarks, and then carried out within a Bayesian inference framework. The above anchor graph based models have achieved satisfactory performance in scalable classification tasks [32, 33]. However, the graph Laplacian over anchors puts more emphasis on the manifold relationship between anchors and ignores the fine structure of samples, and the graph Laplacian over samples could be biased when the dataset includes noises and outliers.

In this paper, we continue the fruitful line of the anchor graph based methods, and propose a novel model termed as Coupled Graph Laplacian Regularization (CGLR) for SSL. On one hand, the anchors are learned from training samples by a clustering method, and they can inherit the manifold structure of training samples. More importantly, they are more stable than a single original training sample. Hence, graph Laplacian over anchors gives the coarse data structure and can somewhat overcome the problem caused by the noise data and outliers. On the other hand, graph Laplacian over samples gives the detailed description for the fine structures of samples, which guarantees that the nearby samples tend to have similar labels. The main contributions of this paper are summarized below:

- The proposed model makes the prediction of the label via coupled graphs, which fully explores the intrinsic geometry of the samples.

- A convex regularization framework is developed, so that the obtained solution is globally optimal.
- Experimental evaluations on several publicly available datasets show that our algorithm is effective in classification tasks.

The rest of this paper is organized as follows. The motivation is illustrated in Sect. 2. In Sect. 3, we propose the coupled graph laplacian regularized semi-supervised learning model and give a closed-form solution for handing large scale SSL tasks. In Sect. 4, experiments are conducted on several benchmark datasets to illustrate the validity and effectiveness of the proposed model. Conclusion is given in Sect. 5.

## 2 Motivation

In this section, we present a synthetic example in $R^2$ (see Fig. 1) to explain the basic idea of the proposed method. As shown in Fig. 1, the red stars and the green crosses represent positive and negative samples, respectively. The black dot denoted by $x$ is an unlabeled sample. The red circles and the green circles are selected anchors by a clustering method.

Recall that the label for $x$ is generally estimated as a weighted average of the labels of its nearest samples in classical graph based SSL methods, where the weights are related to the similarities between $x$ and its nearest samples. In Fig. 1a, we select 4 nearest samples of sample $x$ denoted by $x_1$, $x_2$, $x_3$ and $x_4$ to construct the local adjacency graph, where $x_3$ is a negative outlier surrounded by positive samples. It's easy to see that the unlabeled sample $x$ should belong to positive class from Fig. 1a. However, $x$ will probably be misclassified as "negative sample" since $x$ is much closer to the outlier $x_3$ than it would be from the other adjacency samples $x_1$, $x_2$ and $x_4$, and it has great similarity to $x_3$. In anchor graph based models, the label of sample $x$ is predicted as a weighted average of the labels of its nearest anchors denoted by $u_1$, $u_2$, $u_3$ and $u_4$ as shown in Fig. 1a. Then $x$ will be correctly classified

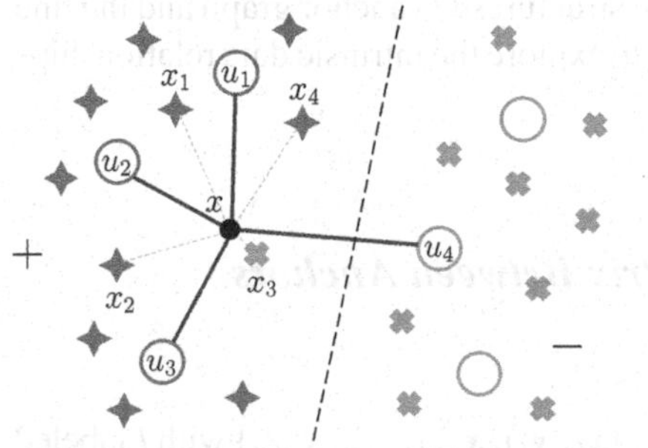

(a) The role of anchor graph.

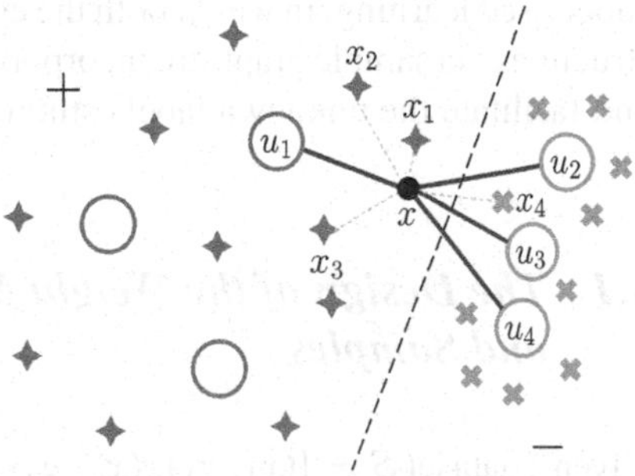

(b) The role of sample graph.

**Fig. 1** 2-D toy example for graph constructions

because the three positive anchors $u_1$, $u_2$ and $u_3$ are closer to $x$ than the negative anchor $u_4$. From the toy example, it is concluded that the anchor graph can somewhat overcome the problem caused by the noise data and outliers, because the anchors learned from training samples by a clustering method are more stable than a single original training sample.

Another toy example is illustrated in Fig. 1b, where the sample distribution of the negative class is denser than that of the positive class. We can see that $x$ should be positive with high probability. But if we use labels of the anchors $u_1, u_2, u_3$ and $u_4$ to predict the label of sample $x$, $x$ will be misclassified as "negative" samples because the distances from $x$ to the three negative anchors $u_2$, $u_3$ and $u_4$ are almost same as the distance from $x$ to the positive anchor $u_1$. In this case, anchor graph is not reliable in predicting unknown labels. In fact, if we use the sample graph to predict the unknown labels, i.e., the labels of sample $x_1, x_2, x_3$ and $x_4$ are utilized to estimate the label of $x$, then $x$ will be correctly classified as positive sample. This is because the sample graph gives fine structures of local relationships and guarantees that the nearby samples tend to have similar labels.

From above two toy examples, we can draw the following conclusions. Firstly, the anchor graph gives the coarse data structure and is robust to the noise data and outliers. Nevertheless, it can not characterize the detailed relationships between samples, especially when there are a small number of anchors. Secondly, sample graph emphasizes the local relationships of samples and can describe the fine structures of the data, but it is susceptible to outliers and noise data. Hence, it is desirable to integrate the anchor graph and the sample graph to discover the intrinsic sample structures under the semi-supervised learning setting.

# 3 Coupled Graph Laplacian Regularized Semi-supervised Learning

In this section, we formulate anchor graph and sample graph into a single unified optimization framework—termed as coupled graph Laplacian regularized semi-supervised learning, in which both the coarse structures via anchor graph and the fine structures via sample graph are incorporated to explore the intrinsic data relationships and facilitate the unknown label estimation.

## *3.1 The Design of the Weight Matrix Between Anchors and Samples*

Given a dataset $\mathcal{S} = \{(\boldsymbol{x}_1, \boldsymbol{y}_1), (\boldsymbol{x}_2, \boldsymbol{y}_2), \ldots, (\boldsymbol{x}_l, \boldsymbol{y}_l), \boldsymbol{x}_{l+1}, \ldots, \boldsymbol{x}_{l+u}\}$ with $l$ labeled samples and $u$ unlabeled samples in $d$ dimensions, a small number of anchor points which can approximately cover the entire data can be selected by a clustering method such as K-means, and denoted as $\mathcal{U} = \{\boldsymbol{u}_1, \boldsymbol{u}_2, \ldots, \boldsymbol{u}_m\} \in R^{d\times m}$. Based on the assumption that the way the anchors represent the sample should resemble the way the labels of anchors represent the label of the sample, once the weight vector $z_i = [z_{i1}, z_{i2}, \ldots, z_{im}]^\top$ that measures the potential relationships between the sample $\boldsymbol{x}_i$ and the anchors obtained, the label of sample $\boldsymbol{x}_i$ can be estimated as a weighted average of the labels of the anchors, i.e.,

$$\hat{y}_i = \sum_{k=1}^{m} z_{ik}\boldsymbol{a}_k = Az_i, \tag{1}$$

where $\boldsymbol{a}_k$ represents the prediction label vector of the anchor $\boldsymbol{u}_k$ and $A = [\boldsymbol{a}_1, \boldsymbol{a}_2, \ldots, \boldsymbol{a}_m]$.

To enhance the coding efficiency and make the weights between anchors and samples more robust, we adopt FLAE [31] like objective function with a rational geometric characteristics:

$$\begin{cases} \min\limits_{c_{\langle i\rangle}} \|x_i - U_{\langle i\rangle}c_{\langle i\rangle}\|^2 + \lambda\|d_{\langle i\rangle} \odot c_{\langle i\rangle}\|^2 \\ \text{s.t. } \mathbf{1}^\top|c_{\langle i\rangle}| = 1. \end{cases} \tag{2}$$

Here, $\langle i\rangle$ is the index set of $s$ closest anchors of sample $\boldsymbol{x}_i$, $U_{\langle i\rangle} \in R^{d\times s}$ is the related matrix composed of $s$ closest anchors of $\boldsymbol{x}_i$, $\odot$ denotes the Hadamard product, and $\boldsymbol{d}_{\langle i\rangle}$ is the locality adaptor whose elements are

$$\boldsymbol{d}_{\langle i\rangle}(k) = \exp\left(\frac{\|x_i - U_{\langle i\rangle}(:, k)\|^2}{\sigma}\right), k = 1, \ldots, s. \tag{3}$$

The parameter $\sigma$ is used to adjust the weight decay speed for the locality adaptor. To follow the advantage of the shift-invariant in the geometric reconstruction problem, weight vector $z_i$ between sample $\boldsymbol{x}_i$ and the anchors is defined by $z_i(\langle i\rangle) = |c_{\langle i\rangle}|$ and $z_i(\langle \bar{i}\rangle) = \mathbf{0}$, where $\langle \bar{i}\rangle$ is the complement of $\langle i\rangle$. Accordingly, the weight matrix that measures the potential relationships between samples and anchors is denoted as $Z = [z_1, z_2, \ldots, z_{l+u}]$.

### 3.2 Graph Laplacian over Anchors and Samples

If we define the $j$-th row of matrix $Z$ as the profile of the anchor $\boldsymbol{u}_j$ and denote it as $\hat{z}^j$, the basic relationships between samples and anchors can be approximately written as follows:

$$X \approx \boldsymbol{u}_1 \cdot \hat{z}^1 + \cdots + \boldsymbol{u}_i \cdot \hat{z}^i + \cdots + \boldsymbol{u}_j \cdot \hat{z}^j + \cdots + \boldsymbol{u}_m \cdot \hat{z}^m, \tag{4}$$

where $X = [\boldsymbol{x}_1, \boldsymbol{x}_2, \ldots, \boldsymbol{x}_{l+u}]$ is the matrix of training samples. It can be seen that there are one-to-one correspondences between the profiles and the atoms. The profiles can be used to measure the contributions of anchors to reconstruct the training samples. Moreover, the similar profiles can encourage the corresponding anchors to be similar, and the similar anchors tend to have similar labels. Therefore, we can use the profiles to construct the adjacency matrix between anchors as follows:

$$W = ZZ^\top. \tag{5}$$

Accordingly, the normalized graph-based regularization term over anchors can be formulated as

$$\frac{1}{2}\sum_{i,j=1}^{m} W_{ij}\left\|\frac{\boldsymbol{a}_i}{\sqrt{D_{ii}}} - \frac{\boldsymbol{a}_j}{\sqrt{D_{jj}}}\right\|^2 = \mathrm{tr}(A\bar{L}_1A^\top), \tag{6}$$

where $D$ is the diagonal matrix with element $D_{ss} = \sum_{t=1}^{m} W_{st}$, $\bar{L}_1$ is the normalized Laplacian matrix $\bar{L}_1 = I - D^{-1/2}WD^{-1/2}$, and $\mathrm{tr}(\cdot)$ is the trace of matrix.

The anchors inherit the structure information of the training samples and are more stable than a single original training sample, thus the anchor graph gives the coarse data structure and reduces the influence of the noise of samples and outliers. However, it can not guarantee that the nearby samples tend to have the similar labels as shown in Fig. 1b. Therefore, we also need to construct a graph over samples to characterize the fine structures of the data. Since $z_i$ is the coding vector of sample $\boldsymbol{x}_i$ with respect to the anchor set $\mathcal{U}$, $z_i$ can be considered as another kind of features for sample $\boldsymbol{x}_i$. Similar codings can encourage the corresponding samples to be similar, and the similar samples also tend to have similar labels. Hence, we adopt the normalized adjacency matrix which satisfies the one-order transition probability [25]

$$G = Z^\top \Lambda^{-1} Z, \tag{7}$$

where the diagonal matrix $\Lambda \in R^{m\times m}$ is defined as $\Lambda_{kk} = \sum_{i=1}^{l+u} Z_{ki}$. Then the normalized graph-based regularization term over samples can be written as

$$\frac{1}{2}\sum_{i,j=1}^{l+u} G_{ij}\|\boldsymbol{f}_i - \boldsymbol{f}_j\|^2 = \mathrm{tr}(FL_2F^\top) = \mathrm{tr}(AZL_2Z^\top A^\top) = \mathrm{tr}(A\bar{L}_2A^\top). \tag{8}$$

Here, $\boldsymbol{f}_i = Az_i$ is the label prediction for sample $\boldsymbol{x}_i$, and $F = AZ$. $L_2 = I - Z^\top\Lambda^{-1}Z$ is the normalized Laplacian matrix, and $\bar{L}_2 = ZZ^\top - ZZ^\top\Lambda^{-1}ZZ^\top$ is the reduced Laplacian matrix.

### 3.3 Building the Coupled Graph Laplacian Framework

We integrate the anchor graph and the sample graph into a single unified optimization framework as follows:

$$\arg\min_{A} \|AZ_l - Y_l\|_F^2 + \theta_1 \mathrm{tr}(A\bar{L}_1 A^\top) + \theta_2 \mathrm{tr}(A\bar{L}_2 A^\top). \tag{9}$$

In (9), $Y_l = [y_1, y_2, \ldots, y_l] \in R^{c\times l}$ denotes a class indicator matrix on labeled data points with element $Y_{ij} = 1$ if $x_i$ belongs to class $j$ and $Y_{ij} = 0$ otherwise. $\theta_1, \theta_2 > 0$ are the trade-off parameters, and $\mathrm{tr}(A\bar{L}_1 A^\top)$ and $\mathrm{tr}(A\bar{L}_2 A^\top)$ are respectively the Laplacian regularization terms over anchors and samples. On setting the derivative of the objective function in (9) to zeros, we can easily obtain the globally optimal solution:

$$A = Y_l Z_l^\top \left(Z_l Z_l^\top + \theta_1 \bar{L}_1 + \theta_2 \bar{L}_2\right)^{-1}. \tag{10}$$

Because the scale of the matrix $Z_l Z_l^\top + \theta_1 \bar{L}_1 + \theta_2 \bar{L}_2 \in R^{m\times m}$ is related to the number of anchors and independent of the size of training set, it yields a close-form solution for dealing with large scale SSL. Subsequently, the hard label for any unlabeled sample can be obtained by utilizing the soft labels associated with anchors as follows [25, 31]:

$$\hat{y}_i = \arg\max_{j\in\{1,\ldots,c\}} \frac{(Az_i)_j}{a^j Z\mathbf{1}}, \quad i = l+1, \ldots, n, \tag{11}$$

where $(Az_i)_j$ is the $j$th element of the vector $Az_i$, $a^j$ is the $j$th row of $A$ and $\mathbf{1}$ is a $c$-dimensional vector with elements equal to 1.

### 3.4 Computational Complexity and Storage Cost

We first review the symbols used in this part: $l$ and $u$ are the numbers of labeled and unlabeled samples respectively, $n = l + u$ and $m$ are the numbers of samples and anchors respectively, $s$ is the number of the closest anchors selected for each sample, $d$ is the dimension of samples, and $\kappa$ is the number of iterations in the K-means algorithm. The computational complexity and storage cost are analyzed as follows.

- **Computational complexity**: (i) In the training process, selecting the anchors via K-means takes $O(mnd\kappa)$ operations. Computing the weight matrix $Z$ takes $O(mns + s^2dn)$ operations. The complexity of the evaluations of the graph Laplacians $\bar{L}_1$ and $\bar{L}_2$ is $O(m^2n)$. Solving the label matrix of anchors takes $O(clm + lm^2 + m^3)$ operations. Hence, the computational complexity of the training process is $O(m^2n + s^2dn)$ since $s \ll m \ll n$, $c < m$ and $l < n$. This guarantees that our method can deal with large scale data since the complexity is exactly liner with the data size. (ii) In the classification stage, predicting the hard label of an unlabeled sample via (11) takes $O(cm)$ operations.

- **Storage cost**: The storage costs for the training samples and the anchors are respectively $O(nd)$ and $O(md)$. The weight matrix $Z$ requires $O(mn)$ storages. The graph Laplacians $\bar{L}_1$ and $\bar{L}_2$ occupy $O(m^2)$ storages. The storage space for label matrix $A$ is $O(cm)$. Hence, the maximal space complexity of our method is $O(mn + dn)$.

# 4 Experiments

In this section, we conduct experiments on six benchmark datasets. An overall description of the datasets is presented in Table 1. "#" stands for the number of Data, Class, anchor and labeled samples, respectively. To evaluate the performance of the coupled graph regularization, we define two versions of our proposed CGLR as L-CGLR and F-CGLR, which first employ LAE [25] and FLAE [31] for calculating weight $Z$ respectively, and then use the unified framework (9) for estimating the label matrix $A$. We compare our methods with three baseline methods including 1NN, ELM [34] and GGMC [2], and two state-of-the-art anchor based SSL approaches including AGR [25] and EAGR [31]. The parameters of all methods are obtained by using cross validation. For each dataset, we randomly select a fixed number of samples as labeled samples in each class, and the rest are left for unlabeld samples. Following the common evaluation procedure, we repeat the experiments 20 times with different random spits of the datasets to report the average classification accuracy together with standard deviation, and the best classification results are in boldface. We present the classification performances of the seven approaches in Table 2, from which we can obtain the following observations. Firstly, the accuracies of the anchor graph based methods including AGR, EAGR, L-CGLR, and F-CGLR are higher than those of the baseline methods. This is because the former utilize the potential relationships between samples and anchors to explore the underlying data structure, and the anchors are more stable than a single original training samples. While the latter emphasize the local structures of samples to estimate the unknown labels. Sec-

**Table 1** Overall description of the datasets

| Datasets | Dimension | Data # | Class # | Anchor # | Labeled # |
|---|---|---|---|---|---|
| Ionosphere | 34 | 351 | 2 | 50 | 10 |
| Pima | 8 | 768 | 2 | 300 | 10 |
| German | 24 | 1000 | 2 | 300 | 30 |
| Spam | 57 | 4601 | 2 | 100 | 10 |
| USPS | 256 | 11,000 | 10 | 500 | 10 |
| MNIST | 780 | 60,000 | 10 | 500 | 10 |

**Table 2** The comparison on classification accuracies (Mean ± std)

| Datasets | 1NN | ELM | GGMC | AGR | EAGR | L-CGLR | F-CGLR |
|---|---|---|---|---|---|---|---|
| Ionosphere | 77.89 (5.33) | 73.27 (2.93) | 59.34 (7.49) | 77.69 (2.04) | 78.62 (1.53) | 79.00 (3.53) | **79.94** (2.73) |
| Pima | 62.51 (2.80) | 58.17 (8.11) | 56.50 (0.71) | 65.40 (9.28) | 65.66 (2.41) | 65.81 (1.74) | **66.35** (0.51) |
| German | 64.36 (3.98) | 59.32 (2.09) | 64.67 (4.81) | 56.38 (3.98) | 52.19 (4.40) | 64.55 (3.69) | **67.28** (4.33) |
| Spam | 71.85 (6.18) | 67.71 (3.10) | 67.78 (14.65) | 76.30 (3.50) | 76.39 (3.36) | 76.93 (3.21) | **77.16** (2.43) |
| USPS | 70.15 (1.97) | 48.95 (0.94) | 82.62 (6.03) | 86.53 (1.34) | 87.21 (1.33) | 87.17 (1.02) | **88.64** (1.15) |
| MNIST | 71.79 (1.59) | 74.39 (1.19) | 84.98 (0.81) | 86.66 (1.14) | 88.45 (0.94) | 89.26 (1.05) | **89.42** (1.14) |

ondly, EAGR and F-CGLR respectively performs better than AGR and L-CGLR in the most time. The reason is that FLAE follows the shift-invariant requirement in geometric reconstruction and better measures the local weights between samples and their neighboring anchors than LAE. Thirdly, the coupled graph regularized methods L-CGLR and F-CGLR achieve better classification performances than AGR and EAGR. The reason for this is that the coupled graph regularized methods simultancously characterize the coarse structures of anchors and the fine structures of samples, while AGR and EAGR consider only one data structure.

Table 3 shows the time costs of the seven different approaches. The time costs for 1NN and ELM are not high, because they belong to supervised method and the training time is related to the size of the labeled sample set, which is not large in our experiments. The traditional graph based method GGMC is quite slow on large-scale datasets due to the high complexity arisen from the kNN strategy for graph construction and the inverse calculation of the Laplacian matrix for optimization. Anchor graph based approaches reduces the computational cost by utilizing the relationships between samples and selected anchors, and both time costs and memory grow linearly with the data size. Hence, they are very efficient especially for the large

**Table 3** The comparison on training time costs (seconds)

| Datasets | 1NN | ELM | GGMC | AGR | EAGR | L-CGLR | F-CGLR |
|---|---|---|---|---|---|---|---|
| Ionosphere | 0.09 | 0.08 | 0.50 | 0.85 | 0.17 | 0.86 | 0.17 |
| Pima | 0.15 | 0.10 | 1.21 | 1.70 | 0.20 | 1.71 | 0.20 |
| German | 0.52 | 0.12 | 2.07 | 2.26 | 0.32 | 2.28 | 0.34 |
| Spam | 0.88 | 0.36 | 104.92 | 8.92 | 0.52 | 8.99 | 0.71 |
| USPS | 21.89 | 1.91 | 377.39 | 24.95 | 4.87 | 25.01 | 4.94 |
| MNIST | 436.71 | 23.52 | 2024.24 | 338.15 | 114.76 | 390.58 | 117.01 |

datasets. The time costs for L-CGLR and F-CGLR are respectively similar with AGR and EAGR, which verify the computational complexities analyzed in Sect. 3.4 and [31]. F-CGLR and EAGR are faster than L-CGLR and AGR for the reason that FLAE is more efficient than LAE for calculating the weight matrix between samples and anchors. In a word, the proposed approaches achieve the superior classification performances, while the computational cost are comparable to state-of-the-art anchor based methods.

## 5 Conclusion

In this paper, we propose a novel anchor graph based semi-supervised learning approach called as Coupled Graph Laplacian Regularization (CGLR), which simultaneously characterizes the coarse structures of anchors and the fine structures of samples by combining the anchor graph regularization and sample graph regularization in a unified optimization framework. The anchor graph reduces the influences of the noise data and outliers, and the sample graph via anchors gives the detailed local relationships between samples. Experiments on publicly available datasets have validated the efficiency for our proposed method.

**Acknowledgements** This work is supported in part by the National Natural Science Foundation of China [grant nos. 61772374, 61503263, 61472285], in part by the Zhejiang Provincial Natural Science Foundation [grant nos. LY17F030004, LR17F030001, LY16F020023], in part by the project of science and technology plans of Zhejiang Province [grants no. 2015C31168], in part by the Key Innovative Team Support and Project of science and technology plans of Wenzhou City [grant nos. G20160002, C20170008, G20150017, ZG2017016].

## References

1. T. Joachims, *Learning to Classify Text Using Support Vector Machines: Methods, Theory and Algorithms* (Kluwer Academic Publishers, Dordrecht, The Netherlands, 2002)
2. X. Zhu, Z. Ghahramani, J. Lafferty, Semi-supervised learning using Gaussian fields and harmonic functions, in *Proceeding of the Twentieth International Conference on Machine Learning* (2003), pp. 912–919
3. D. Zhou, O. Bousquet, T.N. Lal, J. Weston, B. Schlkopf, Learning with local and global consistency, in *Neural Information Processing Systems* (2004), pp. 321–328
4. O. Chapelle, A. Zien, Semi-supervised classification by low density separation, in *Proceeding of the International Conference on Artificial Intelligence and Statistics* (vol. 1) (2005), pp. 57–64
5. B.B. Liu, Z.M. Lu, Image colourisation using graph-based semi-supervised learning. Image Process. **3**, 115–120 (2009)
6. B.B. Ni, S.C. Yan, A.A. Kassim, Learning a propagable graph for semisupervised learning: classification and regression. IEEE Trans. Knowl. Data Eng. **24**(1), 114–126 (2012)
7. M. Sokol, K. Avrachenkov, P. Goncalves, A. Mishenin, Generalized optimization framework for graph-based semi-supervised learning, in *Siam International Conference on Data Mining* (2012), pp. 966–974

8. T. Zhang, R. Ji, W. Liu, D. Tao, G. Hua, Semi-supervised learning with manifold fitted graphs, in *Proceeding of the International Joint Conference on Artificial Intelligence* (2013), pp. 1896–1902
9. C. Gonga, K. Fu, Q. Wu, E. Tu, J. Yang, Semi-supervised classification with pairwise constraints. Neurocomputing **139**, 130–137 (2014)
10. Z. Lu, L. Wang, Noise-robust semi-supervised learning via fast sparse coding. Pattern Recognit. **48**(2), 605–612 (2015)
11. D. Wang, X. Zhang, M. Fan, X. Ye, Semi-supervised dictionary learning via structural sparse preserving, in *Proceeding of the Thirtieth American Association for Artificial Intelligence* (2016), pp. 2137–2144
12. L.S. Zhuang, Z.H. Zhou, S.H. Gao, J.W. Yin, Z.C. Lin, Y. Ma, *Label information guided graph construction for semi-supervised learning* (IEEE Trans, Image Process, 2017)
13. X.J. Zhu, J. Lafferty, Harmonic mixtures: combining mixture models and graph-based methods for inductive and scalable semi-supervised learning, in *Proceeding of the International Conference on Machine Learning* (2005)
14. M. Belkin, P. Niyogi, V. Sindhwani, Mandifold regularization: a geometric framework for learning from labeled and unlabeled examples. J. Mach. Learn. Res. **7**(11), 2399–2434 (2006)
15. P.K. Mallapragada, R. Jin, A.K. Jain, Y. Liu, Semiboost: boosting for semi-supervised learning. IEEE Trans. Pattern Anal. Mach. Intell. **31**(11), 2000–2014 (2009)
16. M.Y. Fan, N.N. Gu, H. Qiao, B. Zhang, Sparse regularization for semi-supervised classification. Pattern Recognit. 1777–1784 (2011)
17. L. Chen, I.W. Tsang, D. Xu, Laplacian embedded regression for scalable manifold regularization. IEEE Trans. Neural Netw. **23**(6), 902–915 (2012)
18. R.G. Soares, H. Chen, X. Yao, Semisupervised classification with cluster regularization. IEEE Trans. Neural Netw. **23**(11), 1779–1792 (2012)
19. Z. Zhang, T.W.S. Chow, M.B. Zhao, Trace ratio optimization-based semi-supervised nonlinear dimensionality reduction for marginal manifold visualization. IEEE Trans. Knowl. Data Eng. **25**(5), 1148–1161 (2013)
20. M. Zhang, Z. Zhou, Exploiting unlabeled data to enhance ensemble diversity. Int. Conf. Data Min. **26**(1), 98–129 (2013)
21. G. Huang, S. Song, J.N. Gupta, C. Wu, Semi-supervised and unsupervised extreme learning machines. IEEE Trans. Syst. Man Cybern. **44**(12), 2405–2417 (2014)
22. D. Wang, X.Q. Zhang, M.Y. Fan, X.Z. Ye, An efficient semi-supervised classifier based on block-polynomial mapping. IEEE Trans. Signal Process. **22**(10), 1776–1780 (2015)
23. Z. Zhang, M.B. Zhao, T.W.S. Chow, Graph based constrained semi-supervised learning framework via propagation over adaptive neighborhood. IEEE Trans. Knowl. Data Eng. **27**(9), 2362–2374 (2015)
24. K. Zhang, J.T. Kwok, B. Parvin, Prototype vector machine for large scale semi-supervised learning, in *Proceeding of the International Conference on Machine Learning* (2009), pp. 1233–1240
25. W. Liu, J.F. He, S.F. Chang, Large graph construction for scalable semi-supervised learning, in *Proceeding of the International Conference on Machine Learning* (2010)
26. W. Liu, J. Wang, S.F. Chang, Robust and scalable graph-based semisupervised learning. Proc. IEEE **100**(9), 2624–2638 (2012)
27. C. Deng, R. Ji, W. Liu, D. Tao, X. Gao, Visual reranking through weakly supervised multi-graph learning, in *Proceeding of the International Conference on Computer Vision* (Dec. 2013), pp. 2600–2607
28. S. Kim, S. Choi, Multi-view anchor graph hashing, in *in Proceeding of the International Conference on Acoustics, Speech, and Signal Processing* (May 2013), pp. 3123–3127
29. Y. Xiong, W. Liu, D. Zhao, X. Tang, Face recognition via archetype hull ranking, in *Proceeding of the International Conference on Computer Vision* (Dec. 2013), pp. 585–592
30. Y. Wu, M. Pei, M. Yang, J. Yuan, Y. Jia, Robust discriminative tracking via landmark-based label propagation. IEEE Trans. Image Process. **24**(5), 1510–1523 (2015)

31. M. Wang, W.J. Fu, S.J. Hao, D.C. Tao, X.D. Wu, Scalable semi-supervised learning by efficient anchor graph regularization. IEEE Trans. Knowl. Data Eng. **28**(7), 1864–1875 (2016)
32. D. Cai, X. Chen, Large scale spectral clustering with landmark-based representation, in *Proceeding of the National Conference on Artificial Intelligence* (vol. 45, no. 8) (2015), pp. 1669–1680
33. Z. Yang, E. Oja, Clustering by low-rank doubly stochastic matrix decomposition, in *Proceeding of the International Conference on Machine Learning* (2012), pp. 831–838
34. G.-B. Huang, Q.-Y. Zhu, C.-K. Siew, Extreme learning machine: theory and applications. Neurocomputing **70**(1), 489–501 (2006)

# Title Fault Diagnosis of Transmission Network Based on Fusion of Time Sequence and Hierarchical Transitional WFPN

**Shasha Zhao, Xiaoxiao Lin, Fantao Meng and Xuezhen Cheng**

**Abstract** In order to reduce the complexity of the model and improve the accuracy of the model, and make full use of the timing information of the alarm signal. A transmission network fault diagnosis method based on time sequence and hierarchical transitional WFPN was proposed. Firstly, the existing model was improved to reduce the complexity of the model, then the time correlation characteristics of component, protection and circuit breaker were constructed, and the protection and circuit breaker that did not conform to the correlation characteristics were found by time sequence reasoning. Finally, the fault diagnosis of power network was carried out by fuzzy inference. Through the analysis of typical examples, it is found that this method improves the accuracy and fault tolerance of fault diagnosis.

**Keywords** Transmission network · Diagnosis · Time sequence · Petri

## 1 Introduction

With the development of science and technology, electricity plays an increasingly important role in today's society. It is especially important to find the faulty component in time and to quickly restore the power supply when the grid fails.

At present, the main fault diagnosis methods of power system include: expert system [1], artificial neural network [2], Petri net [3, 4], etc. In recent years, many scholars apply fuzzy theory to Petri net, which fully reflects the advantages of fuzzy Petri net in fault diagnosis of power network. However, these methods do not take the time sequence of the alarm information and the time correlation characteristics between the information into account.

S. Zhao · X. Lin · F. Meng · X. Cheng (✉)
College of Electrical Engineering and Automation, Shandong University of Science and Technology, Qingdao 266590, China
e-mail: zhenxc6411@163.com

S. Zhao
e-mail: 121230793@qq.com



Y. Jia et al. (eds.), *Proceedings of 2018 Chinese Intelligent Systems Conference*, Lecture Notes in Electrical Engineering 528,
https://doi.org/10.1007/978-981-13-2288-4_15

In order to make full use of the time attribute of alarm information. Literature [5] used the timing Bayesian method to express the timing constraint relationship between events, and proposes a timing constraint consistency checking method. Literature [6] applied timing information of protection and circuit breaker to power grid fault diagnosis based on Petri nets. Literature [7] defined the univariate and binary time constraints between component fault and protection action, and breaker tripping. Probability information for protection and circuit breaker place that do not meet timing constraints is corrected.

Based on literature [4], this paper established a hierarchical transition WFPN model, and then established time correlation characteristics between fault component and protection, circuit breakers. Finally, improved the checking method of the time correlation characteristics in [8]. After those steps, disoperation, maloperation and time scale error for protection and circuit breakers can be effectively judged.

## 2 Hierarchical Transitional WFPN (Weighted Fuzzy Petri Net)

### 2.1 *Definition 1*

Hierarchical transitional WFPN can be defined as a 7 tuples: $U = \{N, L, P, T, I, O, M\}$

(1) $N$ is the number of layer of hierarchical transition, and each layer contains several transitions.
(2) $L$ is the number of the layer where the place locates, in this paper, the sub-models of line, transformer, and bus all have two layers. The number of transitions in the first layer is 2, while the second layer is 1.
(3) $P$ is a place collection, $P = \{p_1, p_2, \ldots, p_m\}$.
(4) $T$ is a transition collection, $T = \{t_1, t_2, \ldots, t_n\}$.
(5) $I$ is an input function, $I = W \cdot I_1$, where $W = diag(w_1, w_2, \ldots, w_m)$ is the weight matrix for the place to the transition, $w_m \in (0, 1]$. $I_1$ is an input correlation matrix of $m \times n$, when there is a directed arc of $p_i$ to $t_j$, $I_{1ij} = 1$, otherwise, $I_{1ij} = 0$. $i = 1, 2, 3, \ldots, m, j = 1, 2, 3, \ldots, n$.
(6) $O$ is an output function, $O = O_0 \cdot U$, where $U = diag(u_1, u_2, \ldots, u_n)$ is the weight matrix for the transition to the place, $u_n \in (0, 1]$. $O_0$ is an output correlation matrix of $m \times n$, when there is a directed arc of $t_j$ to $p_i$, $O_{0ji} = 1$, otherwise, $O_{0ji} = 0$. $i = 1, 2, 3, \ldots, m, j = 1, 2, 3, \ldots, n$.
(7) $M = [\beta(p_1), \beta(p_2), \ldots, \beta(p_m)]^T$ is the confidence vector of the place.

### 2.2 Definition 2

(1) The moment of an event $t$ can be certain or uncertain [9]. When $t$ is a certain moment, it can be represented by a point on the time axis. When $t$ is an uncertain moment, it can be expressed as a closed interval. $[t^-, t^+]$ is the closed interval of the uncertain moment.
(2) Assuming that the moment of occurrence of events $e_i$ and $e_j$ are $t_i$ and $t_j$ respectively. When the moment of their occurrence is uncertain, the time interval in which event $e_i$ occurs is $[t_i^-, t_i^+]$, while event $e_j$ occurs is $[t_j^-, t_j^+]$. If the event $e_i$ is preceded by the event $e_j$, the time distance between events $e_i$ and even $e_j$ can be expressed as $\Delta t = [\Delta t^-, \Delta t^+] = [t_j^- - t_i^-, t_j^+ - t_i^+]$.

## 3 Establishment of Time Sequence WFPN Model

(1) There are three layers model according to the literature [3, 7] in the lines, the sending end (receiving end) of the first layer consists of the main protection, the near back-up protection and the remote back-up protection. The second layer is a comprehensive diagnostic subnet based on above protections. The sending end and the receiving end of the third layer is the comprehensive diagnostic model. In this paper, the line is modeled in two layers. The first layer is the sub model of the sending end and receiving end and has two layer transitions. The second layer is a comprehensive diagnosis model and has one layer transition. The number of input place is determined according to the number of remote back-up protection in the first level model. When the topological structure of the power grid changes, it is necessary to increase or decrease the number of remote back-up protection in the first layer and the number of the input place in the second layer, so there are small changes in the model.

Literature [3] improved algorithm, the probability value was greater than 1 in the inference calculation. Literature [4] used the Gaussian function to correct the output probability of the place, but did not satisfy the fuzzy production rule when calculating the probability of ending place. Therefore, based on the hierarchical transition, this paper changes the "or" rule of the first-level model ending place to the "and" rule containing only one transition. Therefore, it can avoid the occurrence of extreme values when reasoning and it is also consistent with the reasoning process of Petri nets. Besides, the calculating process is the same as literature [4].
(2) A typical power grid structure is shown in Fig. 1.
Generally, the lines and transformers in the power grid are equipped with main protection, near back-up protection and remote back-up protection.

When the fault occurs, the acting order of the protection and circuit breakers is main protection action, circuit breaker action corresponding to main protection, near back-up protection action, circuit breaker action corresponding to near back-up pro-

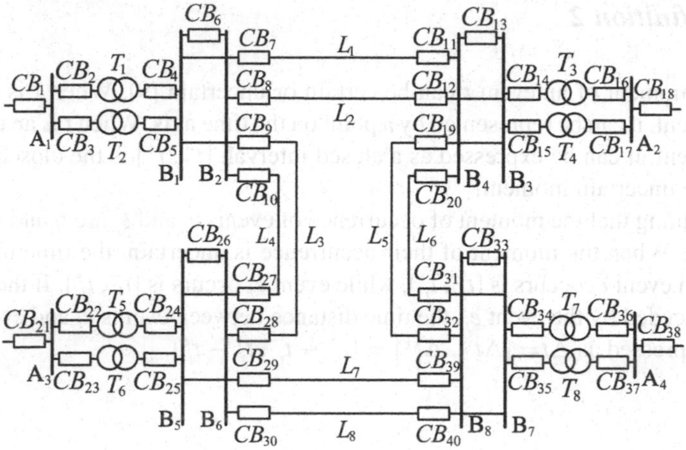


**Fig. 1** Typical power grids structure

tection, remote back-up protection action, and circuit breaker action corresponding to remote back-up protection.

(3) The numbers in Figs. 2 and 3 represent the serial number of the protection, circuit breaker or component. *S* represents sending end (*R* represents receiving end). *m*, *p*, *s* are expressed as main protection, near back-up protection, and far back-up protection respectively. For instance, $L_{7Sm}$ is behalf of the main protection of the line $L_7$ at the sending end, and $B_{6m}$ indicates the main protection of the bus $B_6$. The sending end and receiving end subnets of line $L_7$ are established respectively in Fig. 2. Then establishing a comprehensive diagnostic subnet for line $L_7$ based on the number of far back-up of the sending end and the receiving end. The dashed box indicates all the far back-up protections of the sending (receiving) end (in this paper, there are a total of 10 far back-up protections). $S_N$ indicates the remote back-up protection for each outlet directions of $L_7$ , and $CB_S$ indicates the circuit breaker corresponding each remote back-up protection. As show in Fig. 3, the modeling method for $B_6$ is the same as $L_7$.

# 4 Petri Net Fault Diagnosis Based on Time Sequence

## *4.1 Timing Constraints in Fault Diagnosis of Power Grids*

According to [7], the timing constraints between the component action and the corresponding main protection action, near back-up protection action, and remote back-up protection action are [10, 20], [485, 545], [960, 1070], respectively. The timing constraints between main protection, near back-up protection, remote back-up protection

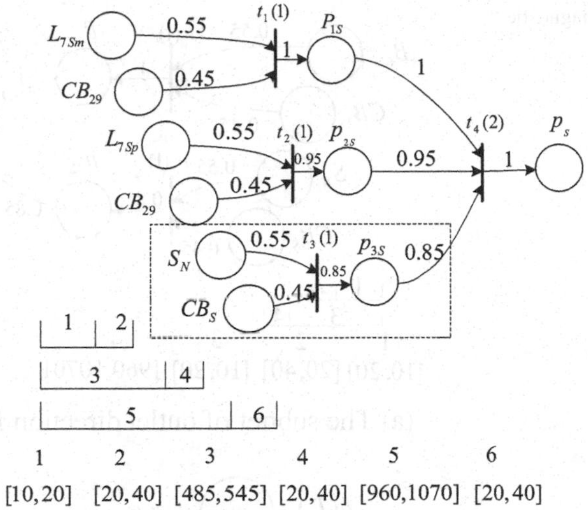


(a) The subnet of outlet direction for line $L_7$

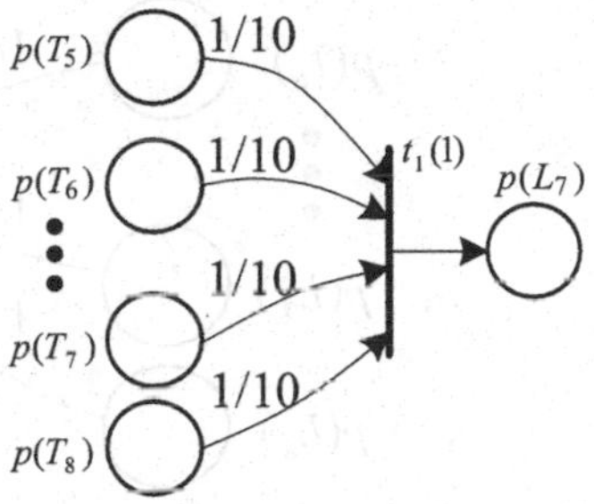


(b) Comprehensive diagnostic subnet for line $L_7$

**Fig. 2** Layered diagnostic model of line $L_7$

and corresponding circuit breaker tripping are [20, 40], [20, 40], [20, 40], respectively. As shown in Fig. 2 and Fig. 3, 1, 2, 3, 4, 5, 6 represent time intervals from component fault to protection action and protection action to circuit breaker tripping respectively.

When components failed, timing constraints between components and protections, and circuit breakers are shown in Table 1.

According to Definition 2, the timing constraints between main protection and near back-up protection, near back-up protection and remote back-up protection, main protection and remote back-up protection are [475, 525], [475, 525] and [950, 1050] respectively.

**Fig. 3** Layered diagnostic model of bus $B_6$

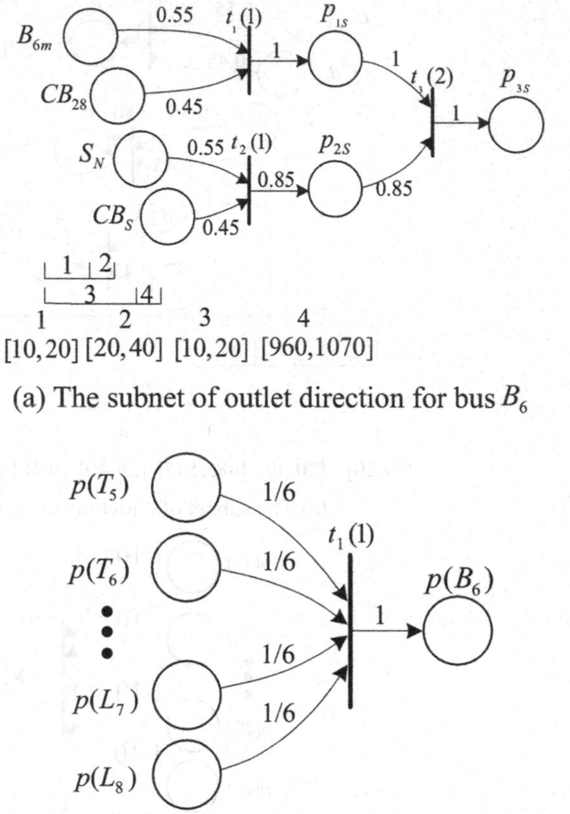


(a) The subnet of outlet direction for bus $B_6$

(b) Comprehensive diagnostic subnet for bus $B_6$

**Table 1** The timing constraints between elements and protections, protections and circuit breakers

| | Component (e) | Circuit breaker (cb) |
|---|---|---|
| Main protection (mp) | $\left\{t_{mp}^-, t_{mp}^+\right\} = \left\{t_e^- + 10, t_e^+ + 20\right\}$ | $\left\{t_{cb}^-, t_{cb}^+\right\} = \left\{t_{mp}^- + 20, t_{mp}^+ + 40\right\}$ |
| Near back-up protection (bp) | $\left\{t_{bp}^-, t_{bp}^+\right\} = \left\{t_e^- + 485, t_e^+ + 545\right\}$ | $\left\{t_{cb}^-, t_{cb}^+\right\} = \left\{t_{bp}^- + 20, t_{bp}^+ + 40\right\}$ |
| Remote back-up protection (rp) | $\left\{t_{rp}^-, t_{rp}^+\right\} = \left\{t_e^- + 960, t_e^+ + 1070\right\}$ | $\left\{t_{cb}^-, t_{cb}^+\right\} = \left\{t_{rp}^- + 20, t_{rp}^+ + 40\right\}$ |

### 4.2 Fault Diagnosis Method Based on Time Sequence

According to the principle of protection, element failure will inevitably trigger a protection acting, which in turn will cause the circuit breaker to trip. And both of them are indispensable.

When a fault occurs in the system, the information of protection actions and circuit breaker tripping are collected, in the same time they are put into set $\Omega$. Checking whether the components in $\Omega$ conform to the time sequence requirements. If the components do not conform to time sequence, the initial probabilities would be corrected. Before the system fault is detected, it is necessary to analyze the timing of each component in the system in order to find out the components with inconsistent timing and correct their initial probability.

Firstly, according to the collected component information, the components that may fail are identified. Then, time sequence hierarchical transitional WFPN is established. And put the elements of the set $\Omega$ into the corresponding component fault information set $N$.

For the fault information set $N_i$ corresponding to each component, find out the main protection, near back-up protection and far back-up protection contained in the set and express them by $x_i$.

If the protection $x_i$ and its corresponding circuit breaker meet the time sequence, then it can be judged whether meet time sequence constraints with other protections in set $N_i$. For the protection that meets the timing constraints, judging whether they meet the timing constraints with the corresponding circuit breaker. The elements meet the above timing constraints are placed in the set $A(x_i)$, and the elements don't meet the timing constraints are placed in the set $M(x_i)$. Select the set $A(x_i)$ containing the most elements as the final time-related consistent set. Because the protection and the circuit breaker can only act on one component at a time, when the time-related consistent set of different component contains several same elements, the same element (protection or circuit breaker) is considered to be caused by the action of component containing many elements. Removing the same elements from the set containing fewer elements, the actions of these elements are not caused by the fault of this component.

## 5 Prototype and Test Results

Case 1: Obtaining alarm information: $B_{1m}$ (50 ms), $CB_4$ (85 ms), $CB_5$ (80 ms), $CB_6$ (82 ms), $CB_7$ (82 ms), $CB_9$ (88 ms) [6].

The suspect component is found to be: $B_1 \cdot N_1 = \{B_{1m}, CB_4, CB_5, CB_6, CB_7, CB_9\}$. According to 2.2, a set of components that meet timing constraints is $A(B_1) = \{B_{1m}, CB_4, CB_5, CB_6, CB_7, CB_9\}$. A set of components that not meet timing constraints is an empty set $M(B_1)$.

After temporal time sequence reasoning, only the moment of tripping of the circuit breakers $CB_4$, $CB_5$ meet the time sequence constraints of the acting moment of the main protection of the bus $B_1$. As a result, the probability of fault of the circuit breakers $CB_4$, $CB_5$ for the remote back-up protection of the transformer $T_1$, $T_2$ is changed from 0.2, 0.2 to 0.75, 0.75 (The reliability of protection and circuit breaker Ref. [10]).

Therefore, the fault probability of the bus $B_1$ is 0.957 when hierarchical transitional WFPN is applied. The fault probability of the bus $B_1$ is 0.995 when the time sequence hierarchical transition WFPN is used. According to 2.1, the time interval of bus $B_1$ action is [30, 40].

Case 2: Obtaining alarm information: $L_{2Sm}$(35 ms), $T_{3S}$(1034 ms), $T_{4S}$ (1035 ms), $L_{5Rs}$(1036 ms), $CB_8$ (66 ms), $CB_{13}$ (1020 ms), $CB_{14}$ (1064 ms), $CB_{15}$ (1065 ms), $CB_{32}$ (1066 ms) [6]. The suspicious elements are found: $B_3$, $L_2$, $L_5$ .

Dubious components of $B_3$ are $N_1 = \{T_{3S}, T_{4S}, L_{5Rs}, CB_{13}, CB_8, CB_{14}, CB_{15}, CB_{32}\}$. According to 2.2, a set of components that meet timing constraints is $A(B_3) = \{T_{3S}, T_{4S}, L_{5Rs}, CB_{14}, CB_{15}, CB_{32}\}$. A set of components that not meet timing constraints are $M(B_3) = \{CB_8, CB_{13}\}$.

Dubious components of $L_2$ are $N_2 = \{L_{2Sm}, T_{3s}, T_{4s}, L_{5Rs}, CB_8, CB_{13}, CB_{14}, CB_{15}, CB_{32}\}$. According to 2.2, a set of components that meet timing constraints is $A(L_2) = \{L_{2Sm}, CB_8, T_{3s}, T_{4s}, L_{5Rs}, CB_{14}, CB_{15}, CB_{32}\}$. A set of components that not meet timing constraints is $M(B_3) = \{CB_{13}\}$.

Dubious components of $L_5$ are $N_3 = \{T_{3s}, T_{4s}, CB_{14}, CB_{15}, CB_{13}, CB_8\}$. According to 2.2, a set of components that meet timing constraints is $A(L_5) = \{T_{3S}, T_{4S}, CB_{14}, CB_{15}\}$. Otherwise, $M(L_5) = \{CB_{13}, CB_8\}$.

Because of $T_{3s}$, $T_{4s}$, $CB_{14}$, $CB_{15}$ belong to the set $A(B_3)$, set $A(L_2)$, set $A(L_5)$ and the elements of the set $A(L_2)$ are more than the set $A(B_3)$ and set $A(L_5)$, the action probabilities of $T_{3s}$, $T_{4s}$, $CB_{14}$, $CB_{15}$ in set $A(B_3)$, $A(L_5)$ are corrected from 0.7, 0.7, 0.75, 0.75 to 0.4, 0.4, 0.2, 0.2, respectively. $L_{5Rs}$, $CB_{32}$ belong to the set $A(B_3)$, set $A(L_2)$ and the elements of the set $A(L_2)$ are more than the set $A(B_3)$, the action probabilities of $L_{5Rs}$, $CB_{32}$ in set $A(B_3)$ are corrected from 0.7, 0.75 to 0.2, 0.2, respectively. Therefore, the fault probability of bus $B_3$, line $L_2$, and line $L_5$ are 0.630, 0.547, 0.859 respectively, when hierarchical transitional WFPN is used. The fault probability of the bus $B_3$, line $L_2$, and line $L_5$ are 0.232, 0.824, 0.286 when time sequence hierarchical transition WFPN is used. The probability of fault of bus $B_3$ decreases from 0.630 to 0.232. The probability of fault of the line $L_2$ changes from 0.547 to 0.824. The probability of fault of line $L_5$ decreases from 0.859 to 0.286. So the faulty component is $L_2$ [3]. According to 2.1, the time interval of line $L_2$ action is [15, 25]. $L_{2Rm}$, $L_{2Rp}$, $CB_{12}$ refuse to act.

## 6 Conclusion

In this paper, based on the hierarchical transition WFPN, a fault diagnosis method about time sequence was proposed. Considering the action time of element, protection and circuit breaker, a consistent set of time association was established firstly, and then the probability of protection and circuit breakers that do not satisfy the timing was corrected. The results of example analysis show that this method can not only increase the accuracy and fault tolerance of fault diagnosis, but also calculate the time interval of component fault occurrence.

## References


1. H.J. Cho, J.K. Park, An expert system for fault section diagnosis of power systems using fuzzy relations. IEEE Trans. Power Syst. **12**(1), 342–347 (1997)
2. W.A. Dos Santos Fonseca, U.H. Bezerra, M.V.A. Nunes et al., Simultaneous fault section estimation and protective device failure detection using percentage values of the protective devices alarms[J]. IEEE transactions on Power Systems, 2012, 8(1): 170-180
3. H. Xie, X. Tong, A method of synthetical fault diagnosis for power system based on fuzzy hierarchical Petri net. Power Syst. Technol. **36**(1), 246–252 (2012)
4. C. Xuezhen, C. Qiang, Y. Yongjin, et al., The analytical method of power grid fault based on hierarchical transition weighted fuzzy Petri net. Trans. China Electrotech. Soc. **31**(15), 125–135 (2016)
5. M. Sun, X. Tong, X. Liu et al., A power system fault diagnosis method using temporal Bayesian knowledge based. Power Syst. Technol. **38**(3), 715–722 (2012)
6. J. Yang, Z. He, Power system fault diagnosis approach based on time sequence fuzzy Petri net. Autom. Electr. Power Syst. **35**(15), 46–51 (2011)
7. X. Tong, H. Xie, M. Sun, Power system fault diagnosis model based on layered fuzzy Petri net considering temporal constraint checking[J]. Autom. Electr. Power Syst. **37**(2), 63–68 (2013)
8. X. Cheng, X. Lin, C. Zhu et al., power system fault analysis based on hierarchical fuzzy petri net considering time association character. Trans. China Electrotech. Soc. **32**(14), 229–237 (2017)
9. Z. Yan, *Enhanced Models and Methods for Power System Fault Diagnosis Utilizing Temporal Information of Alarm Messages* (Zhejiang, Zhejiang University)
10. J. Chen, *Study of Fault Diagnosis in Uncertain Power System on Information Fusion* (Southwest Jiaotong University, Chengdu)

# Hopfield Neural Network Identification for Prandtl-Ishlinskii Hysteresis Nonlinear System

**Xuehui Gao, Shubo Wang, Ruiguo Liu and Bo Sun**

**Abstract** A new Hopfield Neural Network (HNN) identification approach is proposed for a Prandtl-Ishlinskii (P-I) hysteresis nonlinear system. Firstly, The P-I hysteresis nonlinear system is transformed into canonical form by linear state transformation with $B^{\perp}$ to suit the identification design. Then, we define a energy function $E$ which is constituted by the transformed canonical state space system coefficients. Another suitable energy function $E_n$ is proposed with HNN to identify the hysteresis system. Finally, simulation results have verified the performance of the proposed identification.

**Keywords** HNN · Identification · Hysteresis · P-I model

## 1 Introduction

Many researchers have paid attention to the task of the identification and control of nonlinear systems for many years [1, 2]. For hysteresis nonlinear system, the presence of hysteresis can degrade the system performance due to the strict nonlinearity, thus, to deal with the hysteresis nonlinearity becomes an important issue. In order to improve the system performance, system identification is the most important section. With the development of the technology, the more precision system identifications are desiderated to suit the new technology. Nowadays, some results of the hysteresis identification have obtained, but it needs more in-depth researches and appropriate methods to approximate the unknown parameters.

X. Gao (✉) · R. Liu · B. Sun
Department of Mechanical and Electrical Engineering, Shandong University of Science and Technology, Tai'an 271019, China
e-mail: xhgao@163.com

S. Wang
College of Automation and Electrical Engineering, Qingdao University, Qingdao 266071, People's Republic of China



Y. Jia et al. (eds.), *Proceedings of 2018 Chinese Intelligent Systems Conference*, Lecture Notes in Electrical Engineering 528,
https://doi.org/10.1007/978-981-13-2288-4_16

In general, the popular hysteresis models include Jiles-Atherton model [3], Preisach model [4], Prandtl-Ishlinskii (P-I) model [5] and Bouc-Wen model [6], Jiles-Atherton model belongs to physical model which parameters are associated with the physical materials. Most Jiles-Atherton model are utilized magnetic systems. Chen and Szewczyk [7] modified the Jiles-Atherton model of magnetic hysteresis. The modified Jiles-Atherton model improved the differential equation and which was verified on the base of magnetic hysteresis loops of non grain oriented electrical steel. Pop and Caltun et. al. [8] adopted a calculating algorithm to identify the Jiles-Atherton model of the magnetic hysteresis curves. The root mean square deviation was utilized to simulated the regression curve and obtained the fitting parameters.

Different from Jiles-Atherton model, other models are phenomenological models. In these phenomenological models, The P-I model is important and popular using a lot of systems. Gao et al. [5] proposed a new Extended State Observer (ESO) and adaptive neural network controller for P-I hysteresis nonlinearity system. The ESO and adaptive control based the P-I model were designed and guaranteed by Lyapunov theory. Zou and Gu [9] applied the P-I model to describe the viscoelastic hysteresis nonlinearity of Dielectric elastomer actuators (DEAs) which can be utilized for electric-driven artificial muscle in soft robotics. A modified rate-dependent P-I model (MRPIM) characterized the asymmetric curve and tested by seven different frequencies in the experiments. Al Janaideh and Aljanaideh [10] utilized a rate-dependent P-I model with dead-band function to characterize the magnetostrictive actuators with asymmetry and saturation under relatively higher frequencies or large magnitude inputs. An inverse model compensated the asymmetric hysteresis nonlinearities as a feed-forward compensator. But to the authors' knowledge, using HNN to identify the P-I model has not been founded in the literatures.

HNN is a class of feedback neural network and it can be identify the unknown parameters of the hysteresis nonlinearity systems. Atencia et .al. [11] proposed a new ability of Hopfield neural networks optimisation algorithm for estimating time-varying parameters of dynamical systems. It relaxed some usual statistical hypothesis conditions and advanced the robustness analysis with deterministic disturbances. Wang and Hung [12] developed a high-order Hopfield-based neural network (HOHNN) with functional link net (FLN) for dynamic system identification. The inputs of each neuron was added such that had faster convergence rate and less computational load. Xuehui and Bo [6] investigated a HNN identification for Buc-Wen hysteresis nonlinear system to approximate the unknown coefficients.

In this paper, we investigate the issue of P-I hysteresis nonlinearity identification. Firstly, the P-I hysteresis nonlinear system will be transformed into canonical form with $B^{\perp}$. Then, an energy function will be adopted and a HNN will be designed. Another energy function will defined by the HNN to estimate the unknown coefficients. Finally, simulations demonstrate the effectiveness of the identification.

## 2 Problem Formulation

Considering a class of hysteresis system with P-I model are defined as follows:

$$\begin{cases} \dot{x} & = Ax + B\Psi(u) \\ y & = Cx \end{cases}, \tag{1}$$

where $A \in \mathbb{R}^{n\times n}$, $B \in \mathbb{R}^{n}$ are unknown coefficients of state vector, $\Psi(u)$ represents hysteresis nonlinearity, which is defined by P-I model in this paper. Then, the hysteresis P-I model is expressed as follows:

$$\Psi(u) = v_0 u(t) + \int_0^R v(r)\Omega_r[u](t)dr, \tag{2}$$

where $v_0$ denotes the initial value of P-I model, such $v_0 > 0$ is a positive constant, the integrable density function $v(r)$ satisfies $\int_0^R rv(r)dr < \infty$ for $v(r) \geqslant 0$, $\Omega_r[u](t)$ represents play or stop operator. In this paper, we chose the play operator defining $\Omega_r[u](t)$, the play operator can be expressed as follows [13]:

$$\begin{array}{l} \varpi_0(0) = \Omega_r[u](0) = g_r(u(0), \varpi_{m-1}(0)) \\ \varpi_m(t) = \Omega_r[u](t) = g_r(u(t), \varpi_m(t_i)) \end{array}, \tag{3}$$

where $g_r(u, \varpi_m) = max\{(u - r), min\{(u + r, \varpi_m)\}\}$, $r \geqslant 0$, $t_i \leqslant t \leqslant t_{i+1}$ and $\varpi_{m-1}$ is given as initial condition.

In the system (1), Assumption 1 are given as follows:

**Assumption 1** (A1). The system (1) is controllable and observanle.
(A2). The transfer function $G(s) = C(SI - A)^{-1}B$ has full rank.
(A3). The gain $K_g = CB$ is known.

To identify the unknown parameters, the system (1) will be transformed into a canonical state space form based on the Assumption 1. It is known that the following holds through the nonsingular linear state transformation [14]:

$$\begin{cases} \begin{bmatrix} w \\ z \end{bmatrix} = \begin{bmatrix} B^{\perp} \\ C \end{bmatrix} x \\ B^{\perp}B = 0 \end{cases}. \tag{4}$$

Thus, the P-I hysteresis system (1) is deduced to the canonical state space form as follows:

$$\begin{cases} \dot{w} = A_{11}w + A_{12}z \\ \dot{z} = A_{21}w + A_{22}z + K_g\Psi(u) \end{cases}, \tag{5}$$

where $w \in \mathbb{R}^{n-m}$ and $z \in \mathbb{R}^{m}$.

Without loss of generality, we chose the dimension of $z$ as $m = 1$ to simplify the control design. Therefore, the canonical form (5) can be rewritten as:

$$\begin{aligned}\begin{bmatrix}\dot{w}\\ \dot{z}\end{bmatrix} &= \begin{bmatrix}A_{11} & A_{12}\\ A_{21} & A_{22}\end{bmatrix}\begin{bmatrix}w\\ z\end{bmatrix} + \begin{bmatrix}0\\ K_g\end{bmatrix}\Psi(u)\\ &= \bar{A}\begin{bmatrix}w\\ z\end{bmatrix} + \bar{B}\Psi(u).\end{aligned} \tag{6}$$

# 3 Identification Design

In this section, a HNN will be designed to approximate the unknown parameters $\bar{A}$, which is transformed by $A$. Firstly, we introduce the HNN and rewrite it as matrix notation. Then, an error is defined and an energy function $E$ formulated by this error and the unknown $\bar{A}$ is designed for identification. According to the HNN, another energy function $E_n$ is defined which equal to the energy $E$ such that can be approximated the unknown $\bar{A}$.

## 3.1 Hopfield Neural Network

The HNN is defined in this section. The $i$ neuron HNN can be described as follows:

$$\frac{du_i}{dt} = -\frac{1}{D_i}\left(\frac{1}{S_i}u_i(t) + \sum_j W_{ij} f_i(u_j(t)) + L_i\right), \tag{7}$$

where $u_i$ is the input, $D_i$, $S_i$, $W_{ij}$, $L_i$ represent adjust parameters, the weigh, the external input,respectively, and $f_i$ denotes a strictly increasing, bounded, nonlinear continuous function. In this paper, it is assumed that $R_i = \infty$ and $C_i = 1$.

Considering that the HNN has $M$ neurons, it is expressed as:

$$\frac{du_i}{dt} = -\left(\sum_{j=1}^{M} W_{ij} Q_j(t) + L_i\right), \tag{8}$$

where $Q_j$ denotes the output, and it has

$$Q_j(t) = \rho tanh\left(\frac{u_i(t)}{\eta}\right), \tag{9}$$

where $\rho, \eta > 0$.

To design the identification, the HNN is rewritten as matrix notation:

$$\frac{du}{dt} = -(W(t)Q(t) + L(t))$$
$$Q(t) = \rho tanh\left(\frac{u(t)}{\eta}\right) \tag{10}$$

## 3.2 Identification Approach

In order to identify the unknown parameters $\bar{A}$, the Eq. (6) is expressed as:

$$\dot{\bar{x}} = \bar{A}\bar{x} + \bar{B}\Psi(u). \tag{11}$$

We assume a estimated system as follows:

$$\dot{\hat{x}} = A_p\bar{x} + \bar{B}\Psi(u). \tag{12}$$

Then, the identification error is defined as:

$$e_x = \bar{x} - \hat{x}. \tag{13}$$

The following theorem holds:

**Theorem 1** *The hysteresis nonlinear system with P-I model is described as* (1),(2) *and* (3), *it's canonical state space form is defined as* (11), *the HNN is selected as* (10), *then, the designed HNN can be approximated the hysteresis system* (11) *with* (12).

*Proof* Considering the identification error (13), the derivative of the error can be deduced from (11) and (12):

$$\begin{aligned}\dot{e}_x &= \dot{\bar{x}} - \dot{\hat{x}} = \dot{\bar{x}} - (A_P\bar{x} + \bar{B}\Psi(u)) \\ &= \dot{\bar{x}} - (A_p\bar{x} + \bar{B}(v_0u(t) + \int_0^R v(r)\Omega_r[u](t)dr)).\end{aligned} \tag{14}$$

The energy function is defined as follows:

$$E = \frac{1}{2}\dot{e}_x^T\dot{e}_x. \tag{15}$$

Considering (14), we have

$$
\begin{aligned}
E &= \frac{1}{2}(\dot{\bar{x}} - A_p\bar{x} - \bar{B}(v_0 u(t) + \int_0^R v(r)\Omega_r[u](t)dr))^T \\
&(\dot{\bar{x}} - A_p\bar{x} - \bar{B}(v_0 u(t) + \int_0^R v(r)\Omega_r[u](t)dr)) \\
&= \frac{1}{2}(\dot{\bar{x}}^T\dot{\bar{x}} - \dot{\bar{x}}^T A_p\bar{x} - v_0 u(t)\dot{\bar{x}}^T\bar{B} - \dot{\bar{x}}^T\bar{B}\int_0^R v(r)\Omega_r[u](t)dr \\
&- \bar{x}^T A_p^T\dot{\bar{x}} + \bar{x}^T A_p^T A_p\bar{x} + v_0 u(t)\bar{x}^T A_p^T\bar{B} + \bar{x}^T A_p^T\bar{B}\int_0^R v(r)\Omega_r[u](t)dr \\
&- v_0 u(t)\bar{B}^T\dot{\bar{x}} + v_0 u(t)\bar{B}^T A_p\bar{x} + v_0^2 u^2(t)\bar{B}^T\bar{B} + v_0 u(t)\bar{B}^T\bar{B}\int_0^R v(r)\Omega_r[u](t)dr \\
&- \bar{B}^T\dot{\bar{x}}\int_0^R v(r)\Omega_r[u](t)dr + \bar{B}^T A_p\bar{x}\int_0^R v(r)\Omega_r[u](t)dr \\
&+ v_0 u(t)\bar{B}^T\bar{B}\int_0^R v(r)\Omega_r[u](t)dr + \bar{B}^T\bar{B}(\int_0^R v(r)\Omega_r[u](t)dr)^2).
\end{aligned} \tag{16}
$$

To estimate the unknown parameters $\bar{A}$, we define a vector as:

$$
V = [\bar{A}, \bar{B}] = [a_{11}, a_{12}, \ldots, a_{nn}, 0, \ldots, 1]. \tag{17}
$$

Thus, we define another energy function based on HNN as:

$$
E_n = \frac{1}{2}WV - LV \tag{18}
$$

Let $E = E_n$, then, the following solutions can be deduced as:

$$
\begin{aligned}
WV &= \dot{\bar{x}}^T\dot{\bar{x}} + \bar{x}^T A_p^T A_p\bar{x} + v_0 u(t)\bar{x}^T A_p^T\bar{B} + \bar{x}^T A_p^T\bar{B}\int_0^R v(r)\Omega_r[u](t)dr \\
&+ v_0 u(t)\bar{B}^T A_p\bar{x} + v_0^2 u^2(t)\bar{B}^T\bar{B} + v_0 u(t)\bar{B}^T\bar{B}\int_0^R v(r)\Omega_r[u](t)dr \\
&+ \bar{B}^T A_p\bar{x}\int_0^R v(r)G_r[u](t)dr + v_0 u(t)\bar{B}^T\bar{B}\int_0^R v(r)\Omega_r[u](t)dr \\
&+ \bar{B}^T\bar{B}(\int_0^R v(r)\Omega_r[u](t)dr)^2,
\end{aligned} \tag{19}
$$

$$
\begin{aligned}
LV &= \dot{\bar{x}}A_p\bar{x} + v_0 u(t)\dot{\bar{x}}^T\bar{B} + \dot{\bar{x}}^T\bar{B}\int_0^R v(r)\Omega_r[u](t)dr \\
&+ \bar{x}^T A_p^T\dot{\bar{x}} + v_0 u(t)\bar{B}^T\dot{\bar{x}} + \bar{B}^T\dot{\bar{x}}\int_0^R v(r)\Omega_r[u](t)dr.
\end{aligned}
$$

Calculating the Eq. (19), the unknown parameters $V$ are acquired.
The proof completed. □

## 4 Simulations

The proposed HNN identification approach will be verified in this section. Firstly, Considering a hysteresis nonlinear system is described as (11), where

$$\bar{A} = \begin{bmatrix} 0 & 0.3 \\ -3.2 & -2 \end{bmatrix}, \ \bar{B} = \begin{bmatrix} 0 \\ 1 \end{bmatrix}, \tag{20}$$

and $\Psi(u)$ is described as (2), where $v_0 = 2.5$, $v(r) = 0.053e^{-0.062(r-1)^2}$, $R = 50$.

Considering the identification model (12), the estimated parameters $A_p$ can be defined as:

$$A_p = \begin{bmatrix} a_{11} & a_{12} \\ a_{21} & a_{22} \end{bmatrix}, \tag{21}$$

and $V$ is defined as $V = [a_{11}, a_{12}, a_{21}, a_{22}, 0, 1]^T$.

According to (19) and $E = E_n$, the weight $W$ and $I$ of HNN is deduced as

$$W = \begin{bmatrix} \bar{x}_1^2 & \bar{x}_1\bar{x}_2 & 0 & 0 & \bar{x}_1 u & 0 \\ \bar{x}_2\bar{x}_1 & \bar{x}_2^2 & 0 & 0 & \bar{x}_2 u & 0 \\ 0 & 0 & \bar{x}_1^2 & \bar{x}_1\bar{x}_2 & 0 & \bar{x}_1 u \\ 0 & 0 & \bar{x}_2\bar{x}_1 & \bar{x}_2^2 & 0 & \bar{x}_2 u \\ u\bar{x}_1 & u\bar{x}_2 & 0 & 0 & u^2 & 0 \\ 0 & 0 & u\bar{x}_1 & u\bar{x}_2 & 0 & u^2 \end{bmatrix}, \ I = \begin{bmatrix} \bar{x}_1\dot{\bar{x}}_1 \\ \bar{x}_2\dot{\bar{x}}_1 \\ \bar{x}_1\dot{\bar{x}}_2 \\ \bar{x}_2\dot{\bar{x}}_2 \\ u\dot{\bar{x}}_1 \\ u\dot{\bar{x}}_2 \end{bmatrix}. \tag{22}$$

Then, a simulation with (20) by (22) is performed to verify the effectiveness of the proposed HNN identification approach. The identification results and the identification errors are illustrated in Figs. 1 and 2. It is clearly illustrated that the proposed HNN can precisely estimate the unknown parameters $\bar{A}$. From Fig. 2, the errors are also converged to a small neighbourhood of zero. Furthermore, the mean absolute error (MAE) of $a_{11}, a_{12}, a_{21}, a_{22}$ are illustrated in Table 1.

From the Table 1, it is shown that the HNN identification has small enough errors and it has a short identified time for this hysteresis nonlinear system which the hysteresis is described by P-I model. Therefore, the simulation results demonstrate that the proposed HNN identification is effectiveness for the hysteresis nonlinear system.

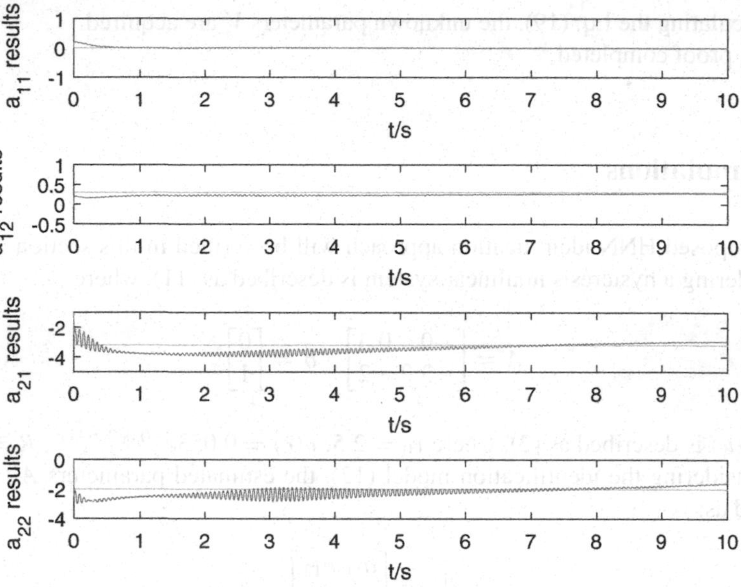


**Fig. 1** The identification results of $\bar{A}$

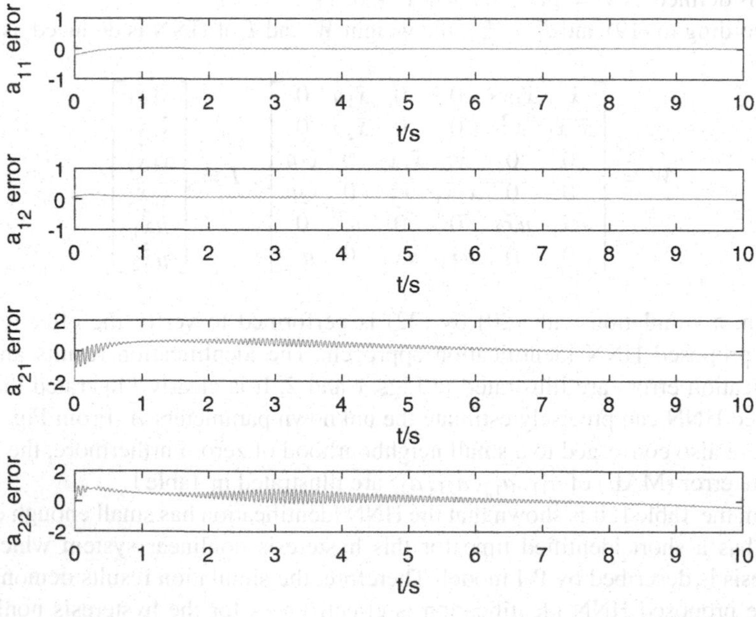


**Fig. 2** The identification error of $\bar{A}$

**Table 1** The MAE of the identification results

| $a_{11}$ | $a_{12}$ | $a_{21}$ | $a_{22}$ |
|---|---|---|---|
| 0.0076 | 0.471 | 0.0613 | 0.5011 |

## 5 Conclusion

This paper designed a HNN to estimate the unknown parameters for hysteresis nonlinear system where the hysteresis nonlinearity was described by P-I model. The system was first transformed into regular a state space form to simplify the identification design. Then, two energy functions were adopted for the HNN and that could calculate the unknown parameters. Finally, simulations were employed to verify the proposed identification and the results demonstrated the precision of the identification approaches.

**Acknowledgements** This work is Supported by the National Natural Science Foundation of China (61433003), Shandong Natural Science Foundation of China(ZR2017MF048), Shandong Key Research and Development Programme (2016GGX105013), Shandong Science and technology program of higher education (J17KA214), Scientific Research Foundation of Shandong University of Science and Technology for Recruited Talents (2016RCJJ035), Tai'an Science and Technology development program (2017GX0017).

## References

1. J. Na, A.S. Chen, G. Herrmann, R. Burke, C. Brace, Vehicle engine torque estimation via unknown input observer and adaptive parameter estimation. IEEE Transactions on Vehicular Technology **67**(1), 409–422 (2018). Jan
2. S. Wang, J.Na, X. Ren, Rise-based asymptotic prescribed performance tracking control of nonlinear servo mechanisms. IEEE Trans. Syst. Man Cybern. Syst. (99), 1–12 (2017)
3. D. Zhang, M. Jia, Y. Liu, Z. Ren, C.S. Koh, Comprehensive improvement of temperature-dependent jiles-atherton model utilizing variable model parameters. IEEE Trans. Mag. **54**(3), 1–4 (2018). March
4. X. Gao, X. Ren, C. Zhu, C. Zhang, Identification and control for hammerstein systems with hysteresis non-linearity. IET Control Theory Appl. **9**(13), 1935–1947 (2015)
5. X. Gao, R. Liu, B. Sun, D. Shen, *Neural Network Adaptive Control for Hysteresis Hammerstein System*, vol. 459 (China, Mudanjiang, 2018), pp. 259–269
6. G. Xuehui, S. Bo, Identification for Bouc-Wen hysteresis system with hopfield neural network, in *2017 9th International Conference on Modelling, Identification and Control (ICMIC)*, July 2017, pp. 248–253
7. P. Cheng, R. Szewczyk, *Modified Description of Magnetic Hysteresis in Jiles-Atherton Model*, vol. 743 (Warsaw, Poland, 2018), pp. 648–654
8. N. Pop, O. Caltun, Jiles-atherton magnetic hysteresis parameters identification. Acta Phys. Polon. A **120**(3), 491–496 (2011)
9. J. Zou, G. Gu, Modeling the viscoelastic hysteresis of dielectric elastomer actuators with a modified rate-dependent Prandtl-Ishlinskii model. Polymers **10**(5) (2018)
10. M. Al Janaideh, O. Aljanaideh, Further results on open-loop compensation of rate-dependent hysteresis in a magnetostrictive actuator with the Prandtl-Ishlinskii model. Mech. Syst. Signal Process. **104**, 835–850 (2018)
11. M. Atencia, G. Joya, F. Sandoval, Identification of noisy dynamical systems with parameter estimation based on hopfield neural networks. Neurocomputing **121**, 14–24 (2013)
12. C.-H. Wang, K.-N. Hung, Dynamic system identification using high-order hopfield-based neural network (HOHNN). Asian J. Control **14**(6), 1553–1566 (2012)
13. S. Liu, C.-Y. Su, Inverse error analysis and adaptive output feedback control of uncertain systems preceded with hysteresis actuators. Control Theory Appl. IET **8**(17), 1824–1832 (2014)
14. P.V.N.M. Vidal, E.V.L. Nunes, L. Hsu, Output-feedback multivariable global variable gain super-twisting algorithm. IEEE Trans. Autom. Control **62**(6), 2999–3005 (2017). June

# Weighted Tensor Schatten $p$-norm Minimization for Image Denoising

**Yufang Yan, Xiaoqin Zhang, Jingjing Zheng and Li Zhao**

**Abstract** In the traditional non-local similar patches based denoising algorithms, the image patches are firstly flatted into a vector, which ignores the spatial layout information within the image patches that can be used for improving the denoising performance. To deal with this issue, we propose a weighted tensor Schatten $p$-norm minimization (WTSN) algorithm for image denoising and use alternating direction method (ADM) to solve it. In WTSN, the image patches are treated as matrix instead of vectorizing them, and thus make full use of information within the structure of the image patches. Furthermore, the employed Schatten $p$-norm requires much weaker incoherence conditions and can find sparser solutions than the nuclear norm, and thus is more robust against noise and outliers. Experimental results show that the proposed WTSN algorithm outperforms many state-of-the-art denoising algorithms in terms of both quantitative measure and visual perception quality.

**Keywords** WTSN · Alternating direction method · Nonlocal self-similarity
Image denoising

## 1 Introduction

Image denoising is a fundamental problem in image processing. It has attracted much research interest and numerous denoising algorithms have been proposed over the past several decades. Image denoising methods are roughly categorized into two categories: local methods based on the intrinsic sparsity of natural images and nonlocal methods based on the nonlocal self-similarity. Local methods [1, 2] usually use a sort of kernel to do convolution computation with the whole image by utilizing the spatial connection between pixels to realize denoising. However, local methods are likely to make the denoised image vague and lose detail since they ignore the global structure information of image. Among non-local denoising methods [3–5], a

Y. Yan · X. Zhang (✉) · J. Zheng · L. Zhao
Department of Computer Science, Wenzhou University, Wenzhou 325035, China
e-mail: xqzhang@wzu.edu.cn



Y. Jia et al. (eds.), *Proceedings of 2018 Chinese Intelligent Systems Conference*,
Lecture Notes in Electrical Engineering 528,
https://doi.org/10.1007/978-981-13-2288-4_17

pioneering work applying the non-local strategy to image denoising is the Nonlocal Means (NLM) algorithm. NLM makes full use of redundant information in the image to realize denoising for current pixel through weighted averaging of a set of similar pixels. This method can keep the image details at removing noise. As soon as NLM algorithm was put forward, it has aroused a lot of attention from scholars. Then a series of non-local image denoising methods also have been proposed.

Out of non-local denoising methods, Low rank matrix approximation (LRMA) has attracted significant research interest. LRMA aims to recover the underlying low rank matrix from its degraded observation. Low rank matrix approximation can be generally obtained by the low rank matrix factorization (LRMF) methods [6, 7] and the nuclear norm minimization (NNM) methods [8, 9]. In this paper, we focus on the latter category. The nuclear norm of a matrix $X \in R^{m \times n}$ is defined as the sum of its singular values, i.e., $\|X\|_* = \Sigma_{i=1}^{\min(m,n)} |\sigma_i(X)|_1$. According to [4], the nuclear norm is strictly convex relaxation of the original rank minimization problem. Given a matrix $Y$, NNM aims to approximate $Y$ by $X$, which satisfies the following objective function:

$$\hat{X} = \arg\min_X \|Y - X\|_F^2 + \lambda \|X\|_* \tag{1}$$

where $\lambda$ is a tradeoff parameter to balance the data fidelity and regularization. Candes and Recht [10] have showed that the low rank matrix can be perfectly recovered by NNM under certain conditions, and Cai et al. [11] proposed a soft-thresholding operation to solve (1) efficiently and obtained $\hat{X} = U S_\lambda(\Sigma) V^T$, where $Y = U \Sigma V^T$ is the SVD of $Y$ and $S_\lambda(\Sigma)$ is the soft-thresholding function: $S_\lambda(\Sigma)_{ii} = \max(\Sigma_{ii} - \lambda, 0)$.

Despite the convexity of the NNM model and with theoretical guarantee, it still has limitations. Chartrand [12] has showed that NNM, in certain case, (such as image denoising) the solution can seriously deviate from the original solution of rank minimization problem. Therefore, it has been proposed in [13, 14] to enhance low rank regularization by using the Schatten $p$-norm, which is defined as the $l_p$ norm of the singular values, i.e., $(\Sigma_{i=1}^{\min(m,n)} \sigma_i^p)^{\frac{1}{p}}$, where $0 < p < 1$. Theoretically, Schatten $p$-norm will guarantee a more accurate recovery of signal [15]. Since the importance of singular values are different, both the nuclear norm and the Schatten $p$-norm have a common drawback that treats each singular value equally, resulting to be not enough flexible in dealing with many practical problems.

In addition, although low rank approximation methods have achieved great success in image denoising, they entirely ignore the spatial layout information which may contain vital cues for improving the denoising performance. Motivated by the above discussions, we propose a weighted tensor Schatten $p$-norm minimization (WTSN) algorithm for image denoising. The features of the proposed algorithm are two-fold: (1) WTSN treats the image patches as matrix to preserve the structure information within the image patches. (2) WTSN adopts Schatten $p$-norm instead of nuclear norm, which guarantees a more accurate low-rank recovery.

The rest of this paper is organized as follows. In Sect. 2, we describe the proposed WTSN model in detail and analyze the optimization scheme to solve it. In Sect.

3, WTSN is applied to image denoising. The experimental results are presented in Sect. 4, and Section 5 concludes the paper.

# 2 Weighted Tensor Schatten p-norm Minimization

## 2.1 Model

Traditional non-local similar patches based denoising algorithms make all the patches as the column vector and then stack them as a matrix. In fact, these algorithms ignore the spatial layout information within the image patches. In this paper, we treat image patches as matrix instead of vectorizing them and define the weighted Schatten $p$-norm of tensor $\mathcal{X}$ as follows:

$$\|\mathcal{X}\|_{\mathbf{w},S_p}^p = \sum_{i=1}^{N} \lambda_i \|X_{(i)}\|_{\mathbf{w}_i,S_p}^p \tag{2}$$

where $\lambda_i$ is positive constant satisfying $\Sigma_{i=1}^N \lambda_i = 1$; $X_{(i)}$ is unfolding matrix of tensor $\mathcal{X}$ on the $i$th mode; $\mathbf{w} = [\mathbf{w}_1^T, \ldots, \mathbf{w}_N^T]$, $\mathbf{w}_i$ is the weight vector of each weighted Schatten $p$-norm; $\|X_{(i)}\|_{\mathbf{w}_i,S_p}^p = \Sigma_{j=1}^r (\mathbf{w}_i)_j \sigma_j^p (X_{(i)})$, where $r$ is the rank of matrix $X_{(i)}$. We adopt the weighted tensor Schatten $p$-norm (2) as the regularization, and define the weighted tensor Schatten $p$-norm minimization model:

$$\min_{\mathcal{X}} \frac{1}{2} \|\mathcal{Y} - \mathcal{X}\|_F^2 + \|\mathcal{X}\|_{\mathbf{w},S_p}^p \tag{3}$$

where $\frac{1}{2}$ is for ease of calculation.

## 2.2 Optimization Algorithm

To remove these the interdependence among the unfolding matrices $X_{(i)}$ and optimize these terms independently, we introduce a set of auxiliary matrices $M_i$ to replace $X_{(i)}$, and the optimization problem changes to:

$$\begin{cases} \min_{\mathcal{X},M_i} \frac{1}{2} \|\mathcal{Y} - \mathcal{X}\|_F^2 + \sum_{i=1}^{N} \lambda_i \|M_i\|_{\mathbf{w}_i,S_p}^p \\ \text{s.t. } M_i = X_{(i)}, i = 1, 2, \ldots, N \end{cases} \tag{4}$$

We apply the Augmented Lagrange Multiplier (ALM) [16] to the problem (4) for relaxing the equality constraints and obtain the following function:

$$f_\mu(M_i, \mathcal{X}, Q_i) = \frac{1}{2}\|\mathcal{Y} - \mathcal{X}\|_F^2 + \sum_{i=1}^{N} \lambda_i \|M_i\|_{\mathbf{w}_i, S_p}^p + \sum_{i=1}^{N} (\langle Q_i, X_{(i)} - M_i \rangle + \frac{\mu_i}{2}\|X_{(i)} - M_i\|_F^2) \tag{5}$$

where $Q_i$ is Lagrange multiplier matrix, $\langle \cdot, \cdot \rangle$ is the inner product of matrix and $\mu_i$ is a positive scalar. In this part, we use alternating direction method (ADM) [16, 17] to optimize the following problem (6):

$$\begin{cases} M_i^{k+1} = \arg\min_{M_i} f_\mu(M_i, \mathcal{X}^k, Q_i^k) \\ \mathcal{X}^{k+1} = \arg\min_{\mathcal{X}} f_\mu(M_i^{k+1}, \mathcal{X}, Q_i^k) \\ Q_i^{k+1} = Q_i^k + \mu_i (X_{(i)}^{k+1} - M_i^{k+1}) \end{cases} \tag{6}$$

For term $M_i^{k+1}$:

$$\begin{aligned} M_i^{k+1} &= \arg\min_{M_i} \lambda_i \|M_i\|_{\mathbf{w}_i, S_p}^p + \langle Q_i^k, X_{(i)}^k - M_i \rangle + \frac{\mu_i}{2}\|X_{(i)}^k - M_i\|_F^2 \\ &= \arg\min_{M_i} \frac{\lambda_i}{\mu_i}\|M_i\|_{\mathbf{w}_i, S_p}^p + \frac{1}{2}\|X_{(i)}^k + \frac{1}{\mu_i}Q_i^k - M_i\|_F^2 \end{aligned} \tag{7}$$

Ordering $X_{(i)}^k + \frac{1}{\mu_i}Q_i^k = U\Sigma V^T$, $\Sigma = \mathrm{diag}(\sigma_1, \sigma_2, \ldots, \sigma_r)$. according to [18], $\delta_j = GST(\sigma_j, \frac{\lambda_i}{\mu_i}\mathbf{w}_{i(j)}, p)$, $j = 1, 2, \ldots, r$. $\Delta = \mathrm{diag}(\delta_1, \delta_2, \ldots, \delta_r)$, the solution of $M_i^{k+1}$ can be obtained by $M_i^{k+1} = U\Delta V^T$.

For term $\mathcal{X}^{k+1}$:

$$\begin{aligned} \mathcal{X}^{k+1} &= \arg\min_{\mathcal{X}} \frac{1}{2}\|\mathcal{Y} - \mathcal{X}\|_F^2 + \sum_{i=1}^{N} (\langle Q_i^k, X_{(i)} - M_i^{k+1} \rangle + \frac{\mu_i}{2}\|X_{(i)} - M_i^{k+1}\|_F^2) \\ &= \arg\min_{\mathcal{X}} \frac{1}{2}\|\mathcal{Y} - \mathcal{X}\|_F^2 + \sum_{i=1}^{N} \frac{\mu_i}{2}\|X_{(i)} + \frac{1}{\mu_i}Q_i^k - M_i^{k+1}\|_F^2 \\ &= \arg\min_{\mathcal{X}} \frac{1}{2}\|\mathcal{Y} - \mathcal{X}\|_F^2 + \sum_{i=1}^{N} \frac{\mu_i}{2}\|\mathcal{X} - \mathrm{refold}_i(M_i^{k+1} - \frac{1}{\mu_i}Q_i^k)\|_F^2 \end{aligned} \tag{8}$$

where $\mathrm{refold}_i(\cdot)$ is the inverse operation of the mode-$i$ unfolding, which restores the original tensor from the mode-$i$ unfolding matrix. Calculate the partial derivative of (8) with respect to $\mathcal{X}$, and set it to zero.

$$-\mathcal{Y} + \mathcal{X} + \sum_{i=1}^{N} \mu_i \mathcal{X} - \sum_{i=1}^{N} \mu_i \mathrm{refold}_i(M_i^{k+1} - \frac{1}{\mu_i}Q_i^k) = 0 \tag{9}$$

After rearranging the term with $\mathcal{X}$, we can obtain:

$$\mathcal{X} = \frac{\mathcal{Y} + \Sigma_{i=1}^{N} \mu_i \mathrm{refold}_i(M_i^{k+1} - \frac{1}{\mu_i}Q_i^k)}{1 + \Sigma_{i=1}^{N} \mu_i} \tag{10}$$

## 3 WTSN for Image Denoising

Image denoising aims to reconstruct the original image $\mathbf{x}$ from its noisy observation $\mathbf{y} = \mathbf{x} + \mathbf{n}$, where $\mathbf{n}$ is assumed to be additive Gaussian white noise with zero mean and variance $\sigma_n^2$. In this section, we will apply the proposed WTSN algorithm to image denoising.

For a local patch $\mathbf{y}_s$ in image $\mathbf{y}$, we can search for its $Z$ nonlocal similar patches across the image by block matching [19]. Then, we stack those similar patches into a 3rd-order tensor $\mathcal{Y}_s$. Thus, the obtained tensor satisfies $\mathcal{Y}_s = \mathcal{X}_s + \mathcal{N}_s$, where $\mathcal{X}_s$ and $\mathcal{N}_s$ are the patch tensor of original image and noise, respectively. Finally, we have the following optimization problem:

$$\hat{\mathcal{X}}_s = \min_{\mathcal{X}_s} \frac{1}{2} \|\mathcal{Y}_s - \mathcal{X}_s\|_F^2 + \|\mathcal{X}_s\|_{\mathbf{w},S_p}^p \tag{11}$$

where the first term of (11) represents the $F$-norm data fidelity term, and the second term represents the regularization term. We adopt the weighting technique proposed in [20] for our WTSN. The weight vector $\mathbf{w}_i$ is determined as

$$(\mathbf{w}_i)_j = \sigma_n^2 c\sqrt{Z}/(\delta_j^{\frac{1}{p}}(\hat{X}_{s,(i)}) + \varepsilon) \tag{12}$$

where $\varepsilon = 10^{-16}$ is added to avoid dividing by zero, $c$ is a positive constant, $\sigma_n^2$ is the variance of noise and $Z$ is the number of similar patches. Since $\delta_j(\hat{X}_{s,(i)})$ is unavailable before $\hat{\mathcal{X}}_s$ is estimated, it can be initialized by

$$\delta_j(\hat{X}_{s,(i)}) = \sqrt{\max\{\sigma_j^2(Y_{s,(i)} + \frac{1}{\mu_i} Q_i) - Z\sigma_n^2, 0\}} \tag{13}$$

We adopt the iterative regularization scheme in [7]

$$\mathbf{y}^t = \hat{\mathbf{x}}^{t-1} + \alpha(\mathbf{y} - \hat{\mathbf{x}}^{t-1}) \tag{14}$$

where $t$ is the iteration number and $\alpha$ is a relaxation. Finally, by aggregating all the denoised patches together, the image $\mathbf{x}$ can be reconstructed. The whole denoising algorithm is summarized in Algorithm 1.

**Algorithm 1** Image Denoising by **WTSN**

**Input:** Noisy image $\mathbf{y}$
**Output:** Denoised image $\hat{\mathbf{x}}^T$
Initialization: $\hat{\mathbf{x}}_0 = \mathbf{y}, \hat{\mathbf{y}}_0 = \mathbf{y}$
**for** $t = 1 : T$ **do**
  Iterative regularization $\mathbf{y}^t = \hat{\mathbf{x}}^{t-1} + \alpha(\mathbf{y} - \hat{\mathbf{x}}^{t-1})$
  **for** each patch $\mathbf{y}_s$ in $\mathbf{y}^t$ **do**
    Find similar patches to form tensor $\mathcal{Y}_s$ from $\mathbf{y}^t$
    Determine weight vector $\mathbf{w}_i$ via Eq. (12)
    Calculate $\Delta$ by using GST
    Estimate tensor $\hat{\mathcal{X}}_s$ via Eq. (10)
  **end**
  Aggregate $\hat{\mathcal{X}}_s$ to form the denoised image $\hat{\mathbf{x}}^t$
**end**
**Return** The final denoised image $\hat{\mathbf{x}}^T$

## 4 Experimental Results and Analysis

To validate the effectiveness of the proposed algorithm, we compare it with recently proposed state-of-the-art denoising methods, including block-matching 3D filtering [19] (BM3D), patch-based near-optimal image denoising [21] (PBNO), expected patch log likelihood for image denoising [22] (EPLL), global image denoising [23] (GID) and spatially adaptive iterative singular-value thresholding [3] (SAIST).

The parameter setting of proposed approach is as follows: the searching window $L \times L$ for similar patches is set to be $30 \times 30$. Iteration number and patch size are set based on noise level. For higher noise level, we need to choose bigger patches and run more times. The size of each patch $\sqrt{d} \times \sqrt{d}$ is set to be $6 \times 6$, $7 \times 7$, $8 \times 8$, $9 \times 9$ and $p = \{0.95, 0.8, 0.7, 0.5\}$ for $\sigma_n \leq 20$, $\sigma_n \leq 30$, $\sigma_n \leq 60$ and $\sigma_n \leq 100$, respectively. $T$ is set to 8, 8, 12, and 14 respectively, on these noise levels. We evaluate the proposed algorithm and the competing algorithms on 12 images (see Fig. 1). Zero mean additive white Gaussian noise with variance $\sigma_n^2$ are added to those test images to generate the noisy observations. In this test, six noise levels $\sigma_n = \{20, 30, 50, 60, 75, 100\}$ are used.

Tables 1 and 2 show the peak-signal-to-noise-ratio (PSNR) results. Due to space limit, we replace the full name of denoised images, including House, Monarch, Airplane, Barbara, Boat, Bridge, Couple, F.print, Hill, Lena, Man and Straw, with the numbers from one to twelve respectively. Similarly, "A." stands for average and "Im" denotes Image. The highest PSNR result for each image and on each noise level is highlighted in bold. We can see that our method outperforms other state-of-the-art denoising methods on average, and we have the following observations. The method of WTSN achieves 0.2–0.3 dB improvement over the SAIST method on average and outperforms the benchmark BM3D method by 0.2–0.4 dB on average consistently

**Fig. 1** The 12 test images

**Table 1** Denoising results (PSNR) by different methods

| Im | BM3D | PBNO | EPLL | GID | SAIST | WTSN | BM3D | PBNO | EPLL | GID | SAIST | WTSN |
|---|---|---|---|---|---|---|---|---|---|---|---|---|
| $\sigma_n = 20$ | | | | | | | $\sigma_n = 30$ | | | | | |
| 1 | 33.77 | 33.58 | 32.98 | 32.81 | 33.75 | **33.78** | 32.08 | 31.92 | 31.22 | 30.35 | 31.39 | **32.14** |
| 2 | 30.35 | 29.55 | 30.48 | 29.65 | 30.76 | **30.90** | 28.36 | 27.85 | 28.35 | 27.60 | 28.03 | **28.46** |
| 3 | 29.55 | 32.06 | 29.67 | 31.48 | 29.65 | **32.63** | 27.56 | 30.21 | 30.41 | 29.47 | 29.35 | **30.55** |
| 4 | 31.77 | 31.06 | 29.76 | 30.21 | **32.10** | 31.90 | **29.81** | 29.50 | 27.56 | 27.95 | **30.04** | 29.68 |
| 5 | **30.88** | 30.39 | 30.66 | 29.53 | **30.84** | 30.82 | **29.11** | **28.81** | **28.89** | 27.66 | **28.83** | 28.77 |
| 6 | 27.27 | 26.70 | 27.49 | 26.49 | 27.31 | 27.42 | 25.46 | 25.22 | **25.68** | 24.78 | 25.43 | 25.47 |
| 7 | **30.76** | 30.22 | 30.54 | 29.28 | **30.66** | 30.56 | **28.86** | **28.58** | **28.61** | 27.15 | **28.58** | 28.46 |
| 8 | 28.80 | 27.76 | 28.28 | 27.95 | 28.99 | **29.12** | 26.82 | 26.35 | 26.18 | 26.00 | 26.82 | **26.97** |
| 9 | **30.72** | 30.32 | 30.49 | 29.59 | 30.58 | 30.67 | **29.15** | **28.95** | **28.90** | 27.75 | **28.94** | 28.86 |
| 10 | **33.05** | 32.75 | 32.61 | 31.74 | **33.08** | 32.92 | **31.26** | **31.16** | 30.78 | 29.83 | 30.77 | 31.06 |
| 11 | 30.59 | 30.15 | 30.63 | 29.59 | 30.54 | **30.65** | **28.86** | **28.65** | **28.82** | 27.82 | **28.68** | 28.61 |
| 12 | 26.98 | 25.86 | 26.80 | 26.63 | 27.23 | **27.62** | 24.84 | 24.70 | 24.74 | 24.59 | 24.74 | **25.17** |
| **A.** | 30.37 | 30.03 | 30.03 | 29.58 | 30.46 | **30.75** | 28.51 | 28.49 | 28.35 | 27.58 | 28.47 | **28.68** |
| $\sigma_n = 50$ | | | | | | | $\sigma_n = 60$ | | | | | |
| 1 | 29.69 | 29.44 | 28.76 | 27.62 | **29.99** | 29.77 | 28.73 | 28.62 | 27.84 | 26.66 | 28.88 | **29.08** |
| 2 | 25.81 | 25.53 | 25.77 | 24.97 | **26.09** | 26.07 | 24.97 | 24.64 | 24.85 | 24.15 | 24.94 | **25.20** |
| 3 | 25.10 | 27.77 | 27.88 | 26.91 | 28.25 | **28.26** | 27.32 | 26.98 | 26.97 | 25.82 | 26.64 | **27.35** |
| 4 | 27.22 | 26.95 | 24.82 | 25.17 | 27.49 | **27.50** | 26.28 | 26.08 | 23.87 | 24.19 | 26.40 | **26.40** |
| 5 | **26.78** | 26.67 | 26.65 | 25.59 | 26.63 | 26.72 | **26.02** | **25.94** | **25.84** | 24.68 | 25.52 | 25.71 |
| 6 | 23.57 | 23.49 | **23.69** | 22.88 | 23.49 | 23.60 | **23.02** | **22.90** | **23.08** | 22.19 | 22.85 | 22.83 |
| 7 | **26.46** | **26.30** | 26.23 | 24.64 | 26.29 | 26.29 | **25.66** | **25.43** | 25.40 | 24.01 | 24.98 | 25.41 |
| 8 | 24.52 | 24.29 | 23.59 | 23.09 | 24.54 | **24.72** | 23.75 | 23.57 | 22.65 | 21.90 | 23.71 | **23.86** |
| 9 | **27.19** | 27.02 | 26.95 | 25.93 | 27.04 | 27.08 | **26.52** | 26.27 | 26.27 | 25.32 | **26.39** | 26.30 |
| 10 | **29.05** | 28.81 | 28.42 | 27.69 | **29.01** | 28.98 | **28.27** | 27.92 | 27.59 | 26.91 | 28.00 | 28.10 |
| 11 | **26.80** | **26.72** | **26.72** | 25.83 | 26.67 | 26.70 | **26.13** | **26.00** | **26.00** | 25.14 | 25.78 | 25.92 |
| 12 | 22.29 | 22.81 | 22.00 | 21.98 | 22.65 | **22.93** | 21.63 | 22.01 | 21.06 | 20.93 | **22.13** | 22.12 |
| **A.** | 26.21 | 26.32 | 25.96 | 25.19 | 26.51 | **26.55** | **25.69** | 25.53 | 25.12 | 24.33 | 25.52 | **25.69** |

**Table 2** Denoising results (PSNR) by different methods

| $\sigma_n = 75$ | | | | | | | $\sigma_n = 100$ | | | | | |
|---|---|---|---|---|---|---|---|---|---|---|---|---|
| Im | BM3D | PBNO | EPLL | GID | SAIST | WTSN | BM3D | PBNO | EPLL | GID | SAIST | WTSN |
| 1 | 27.50 | 27.15 | 26.68 | 25.16 | **27.90** | 27.85 | 25.87 | 25.42 | 25.19 | 23.59 | **26.45** | 26.34 |
| 2 | 23.90 | 23.62 | 23.71 | 22.77 | 23.95 | **24.01** | **22.51** | 22.19 | 22.23 | 20.83 | **22.63** | 22.49 |
| 3 | **23.47** | 25.83 | 25.83 | 24.69 | 25.82 | 26.28 | 22.11 | 24.31 | 24.35 | 23.28 | 24.55 | **24.90** |
| 4 | 25.12 | 24.94 | 22.94 | 23.06 | 25.35 | **25.37** | 23.62 | 23.42 | 22.14 | 21.76 | 23.98 | **24.03** |
| 5 | **25.14** | 24.85 | 24.88 | 23.81 | 24.80 | 24.98 | **23.97** | 23.62 | 23.71 | 22.74 | 23.67 | 23.79 |
| 6 | **22.40** | 22.26 | **22.39** | 21.52 | 22.07 | 22.26 | **21.60** | 21.42 | **21.58** | 20.74 | 21.21 | 21.47 |
| 7 | **24.70** | **24.51** | **24.44** | 23.27 | 24.17 | 24.43 | 23.51 | 23.28 | 23.32 | 22.38 | 23.01 | **23.79** |
| 8 | 22.83 | 22.67 | 21.46 | 20.43 | 22.72 | **22.94** | 21.61 | 21.50 | 19.84 | 18.74 | 21.51 | **21.74** |
| 9 | **25.67** | 25.45 | 25.45 | 24.62 | 25.50 | 25.52 | **24.58** | 24.33 | 24.42 | 23.79 | 24.29 | 24.53 |
| 10 | **27.25** | 27.00 | 26.57 | 25.96 | 26.97 | 27.17 | **25.95** | 25.60 | 25.30 | 24.64 | 25.81 | 25.84 |
| 11 | **25.31** | **25.11** | **25.14** | 24.38 | 25.06 | 25.10 | **24.22** | 23.98 | 24.07 | 23.33 | 23.98 | 24.06 |
| 12 | 20.56 | 21.04 | 20.07 | 19.55 | 21.08 | **21.13** | 19.43 | 19.86 | 19.01 | 18.41 | 19.54 | **20.08** |
| **A.** | 24.50 | 24.54 | 24.13 | 23.27 | 24.62 | **24.75** | 23.25 | 23.24 | 22.93 | 22.02 | 23.39 | **23.59** |

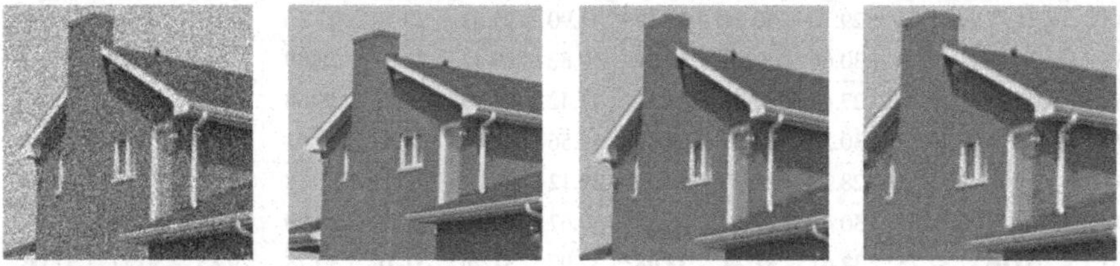

(a) Noisy (b) Original (c) Proposed (d) SAIST

**Fig. 2** Denoised image of "house", $\sigma_n = 30$

on all the six noise levels. To sum up, our proposed WTSN outperforms averagely all the other competing methods at all noise levels.

In terms of visual quality, we pick up the "house" image in Fig. 2 and the "straw" image in Fig. 3 to show the denoised images. At this point, we only compare WTSN with SAIST because compared with other algorithms, SAIST denoising effect is the best. One can observe that the denoised image by the proposed method have much less noise than those by the SAIST. It is obvious that WTSN generates much less artifacts and preserves much better the image edge structures than other competing methods. In summary, WTSN presents strong denoising capability, producing promising visual quality while keeping higher PSNR.

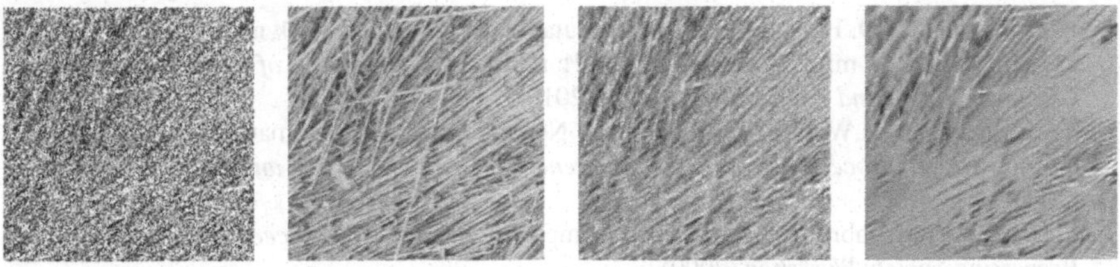

(a) Noisy (b) Original (c) Proposed (d) SAIST

**Fig. 3** Denoised image of "straw", $\sigma_n = 100$

## 5 Conclusions

In this paper, we have proposed a new image denoising model: the weighted tensor Schatten $p$-norm minimization model (WTSN). WTSN treats the image patches as matrix instead of vectorizing these image patches. Therefore, WTSN make full use of information within the structure of the image patches, which is very helpful to enhance the denoising effect of the image. Moreover, WTSN adopts Schatten $p$-norm in its model which guarantees a more accurate low-rank recovery. The proposed WTSN was then applied to image denoising to show its effectiveness. The experimental results showed that WTSN is better than state-of-the-art methods. In the future, we expect WTSN has more success in the field of other computer vision applications.

**Acknowledgements** This work is supported in part by the National Natural Science Foundation of China [grant nos. 61772374, 61503263, 61472285], in part by the Zhejiang Provincial Natural Science Foundation [grant nos. LY17F030004, LR17F030001, LY16F020023], in part by the project of science and technology plans of Zhejiang Province (Grants no. 2015C31168), in part by the Key Innovative Team Support and Project of science and technology plans of Wenzhou City [grant nos. C20170008, G20160002, G20150017, ZG2017016].

## References

1. J. Portilla, V. Strela, M.J. Wainwright, E.P. Simoncelli, Image denoising using scale mixtures of gaussians in the wavelet domain. IEEE Trans. Image Process. **12**(11), 1338–1351 (2003)
2. A.L.D. Cunha, J. Zhou, M.N. Do, The nonsubsampled contourlet transform: theory, design, and applications. IEEE Trans. Image Process. **15**(10), 3089–3101 (2006)
3. W. Dong, G. Shi, X. Li, Nonlocal image restoration with bilateral variance estimation: a low-rank approach. IEEE Trans. Image Process. **22**(2), 700–11 (2013)
4. S. Gu, L. Zhang, W. Zuo, and X. Feng. Weighted nuclear norm minimization with application to image denoising, in *Proceedings of IEEE Conference on Computer Vision and Pattern Recognition* (2014)
5. A. Buades, B. Coll, J.M. Morel, A non-local algorithm for image denoising, in *Proceedings of IEEE Conference on Computer Vision and Pattern Recognition* (2005)

6. A. Eriksson, A.V.D. Hengel, Efficient computation of robust low-rank matrix approximations in the presence of missing data using the l1 norm, in *Proceedings of IEEE Conference on Computer Vision and Pattern Recognition* (2010)
7. A.M. Buchanan, A.W. Fitzgibbon, Damped Newton algorithms for matrix factorization with missing data, in *Proceedings of IEEE Conference on Computer Vision and Pattern Recognition* (2005)
8. M. Partridge, M. Jabri, Robust principal component analysis, in *Proceedings of IEEE Signal Processing Society Workshop* (2000)
9. M. Fazel, H. Hindi, S.P. Boyd, A rank minimization heuristic with application to minimum order system approximation, in *Proceedings of American Control Conference* (2001)
10. E.J. Candes, B. Recht, Exact low-rank matrix completion via convex optimization, in *Proceedings of Allerton Conference on Communication, Control, and Computing* (2008)
11. J.F. Cai, E.J. Cands, Z. Shen, A singular value thresholding algorithm for matrix completion. SIAM J. Optim. **20**(4), 1956–1982 (2008)
12. R. Chartrand, Exact reconstruction of sparse signals via nonconvex minimization. IEEE Signal Process. Lett. **14**(10), 707–710 (2007)
13. F. Nie, H. Huang, C. Ding, Low-rank matrix recovery via efficient Schatten $p$-norm minimization, in *Proceedings of AAAI Conference on Artificial Intelligence* (2012)
14. K. Mohan, M. Fazel, Iterative reweighted algorithms for matrix rank minimization. J. Mach. Learn. Res. **13**(1), 3441–3473 (2012)
15. L. Liu, W. Huang, D. Chen, Exact minimum rank approximation via schatten $p$-norm minimization. J. Comput. Appl. Math. **267**(6), 218–227 (2014)
16. X. Zhang, D. Wang, Z. Zhou, Y. Ma, Simultaneous rectification and alignment via robust recovery of low-rank tensors, in *Proceedings of Advances in Neural Information Processing Systems* (2013)
17. X. Zhang, Z. Zhou, D. Wang, Y. Ma, Hybrid singular value thresholding for tensor completion, in *Proceedings of AAAI Conference on Artificial Intelligence* (2014)
18. W. Zuo, D. Meng, L. Zhang, X. Feng, A generalized iterated shrinkage algorithm for non-convex sparse coding, in *Proceedings of IEEE International Conference on Computer Vision* (2013)
19. K. Dabov, A. Foi, V. Katkovnik, K. Egiazarian, Image denoising by sparse 3-d transform-domain collaborative filtering. IEEE Trans. Image Process. **16**(8), 2080–2095 (2007)
20. Y. Xie, S. Gu, Y. Liu, W. Zuo, W. Zhang, L. Zhang, Weighted schatten $p$-norm minimization for image denoising and background subtraction. IEEE Trans. Image Process. **25**(10), 4842–4857 (2015)
21. P. Chatterjee, P. Milanfar, Patch-based near-optimal image denoising. IEEE Trans. Image Process. **21**(4), 1635–1649 (2012)
22. D. Zoran, Y. Weiss, From learning models of natural image patches to whole image restoration, in *Proceedings of IEEE International Conference on Computer Vision* (2011)
23. H. Talebi, P. Milanfar, Global image denoising. IEEE Trans. Image Process. **23**(2), 755–768 (2014)

# Vibration Signal EMD Filter Detection Method for Blast Furnace Opening Machine

**Zhen Guo, Xiaobin Li, Tianyang Yu and Xiaoyu Fang**

**Abstract** The vibration signals generated during the operation of the blast furnace opener contain various noises, which are disturbing and superimposed, and it is difficult to identify the operating status of the open machine, put forward a kind of based on Empirical Mode Decomposition (EMD) filter method. From the physical structure of the opening machine, the complexity of the vibration signal is qualitatively analyzed. The EMD technology is used to adaptively decompose the vibration signal into a single intrinsic mode function (IMF) with different frequency components. the high frequency noise components is filtered in the IMF component, the remaining IMF components are reconstructed to form a new vibration signal and compared with the results of the wavelet threshold denoising way. The consequences show that the EMD filtering method can overcome the disadvantages of glitches and signal superposition after wavelet denoising, and can fully preserve the nonlinear characteristics of the vibration signal. It is an effective method for filtering and denoising detection of mechanical vibration signals of blast furnace opening machines.

**Keywords** Blast furnace opening machine · Vibration signal
Empirical mode decomposition (EMD) · Filtering

## 1 Introduction

The opening machine is an important part of the operation equipment of the blast furnace, and its safe operation determines the stability of the blast furnace production efficiency and production operation. For the blast furnace opening machine, the vibration signal is the carrier of the reaction equipment operation status, real-time monitoring and signal collection, and the use of effective signal processing methods

Z. Guo · X. Li (✉) · T. Yu · X. Fang
School of Electrical and Electronic Engineering, Shanghai Institute of Technology, Shanghai 201400, China
e-mail: lixiaobinauto@163.com; 1486126523@qq.com



Y. Jia et al. (eds.), *Proceedings of 2018 Chinese Intelligent Systems Conference*, Lecture Notes in Electrical Engineering 528,
https://doi.org/10.1007/978-981-13-2288-4_18

to analyze and process the signal has always been the focus of research on safe and stable operation of the equipment in front of the blast furnace.

The vibration waves generated during the operation of the opening machine not only contain the signals of the running status of the equipment, including many aliasing signals of noise. Therefore, the vibration signal must first be filtered before it can be further analyzed to obtain the operation status of the equipment. The existing vibration signal filtering methods include fast Fourier transform [1], wavelet transform [2] and Hlibert–Huang transform [3]. Among them, the fast Fourier transform is only suitable for analyzing the linear stable signal, don't have very good filtering effect to other complicated signals; Wavelet analysis can deal with nonlinear and unstable signals, but when the wavelet basis is difficult to select and the noise is too large, the filtering effect will be greatly interfered, and it does not get rid of the constraints caused by the traditional FFT transform. Hlibert–Huang transform is currently applied in engineering more algorithms, mainly used for the analysis of the complex signal aliasing, it not only avoids the problem of selecting wavelet base, also after decomposition of the signal has the adaptability, retain the inherent characteristics of the original signal.

For the denoising problem of EMD, many studies have been done at home and abroad. Wu and Huang [4] verified that the EMD method has the characteristics of a filter through a large number of experiments; Flandrin [5] proposed the idea of reconstructing the filter based on the intrinsic modal function IMF, and selected the corresponding order of IMF components to reconstruct, equivalent to a signal passing through an adaptive low-pass, high-pass, or band-reject filter; Boudraa and Cexus [6] denoised each IMF component by different thresholds; Yu et al. [7] used Hlibert–Huang transform to carry out extensive research on mechanical faults; Yang et al. [8] compared the application of Hlibert–Huang transform and wavelet transform in rotating machinery and verified its superiority; Yang [9] introduced the EMD filtering method into the processing of vibration signals of rotating machinery; Li and Meng [10] used the EMD method to extract the harmonic components of a signal under noise and vibration signal interference and achieved outstanding results. It can be seen that the EMD filtering detection method has been applied in various fields and has achieved certain results.

The EMD method overcomes the traditional Fourier transform in dealing with the problem of vibration signal, moreover, this method is based on the signal decomposition, avoids the problem of selecting the basis function, and does not need to use the prior knowledge of the signal, so it has good adaptability. The decomposition result of EMD is to obtain several IMF components, they are distributed from high to low in frequency and each IMF component contains different components, and is not equal bandwidth, and changes with the signal itself. Therefore, the signal after recombining the various signals is not distorted, and the characteristics of the original signal are retained. The results are compared with the signal after wavelet threshold denoising, verifying the superiority and flexibility of EMD filtering method and the feasibility of applying it to this field, which creates the conditions for the determination of opening time of blast furnace.

# 2 Empirical Mode Decomposition (EMD)

## 2.1 The Principle of EMD [11]

At present, the vibration signal of the opening machine of ironmaking blast furnaces is generated by superimposed vibrations that periodically act on muddy mortars with different amplitudes and frequency impact forces, and is coupled with other signals that are not related to the opening machine, and is complex and hard to explain.

For the problem of noise reduction filtering of complex signals, Huang et al. [12] and others have conducted extensive research and proposed a signal processing method called EMD decomposition method, which decomposes a nonlinear, non-stationary aliased signal into a series of sums of IMF components and margins. Among them, the IMF needs to meet two conditions. In general, complex signals do not satisfy these two conditions and often require decomposition.

EMD decomposition steps:

(1) Find the maximum value and minimum point of the signal, and connect the extremum points with three interpolation methods, respectively constituting the upper and lower envelope.
(2) The mean of the upper and lower envelope is $m(t)$, The original signal is recorded as $x(t)$, use the original signal $x(t)$, minus $m(t)$ the difference obtained is recorded as, If $h(t)$ meet IMF conditions, will be $x(t)$ the first IMF component, otherwise it will $h(t)$ as the original signal, repeat steps (1), (2) until the conditions are met.
(3) Record the first IMF component as $h_1$, will $h_1$ from the original signal $x(t)$ separate from, marked as $c_1 = x(t) - h_1$.
(4) Again $c_1$ as a new original signal, repeat the previous steps, in order $h_2, h_3, \ldots, h_n$, until $c_n$ Stop when it's monotonous. $c_n$ is the residual component.

The original signal is decomposed into:

$$x(t) = \sum_{i=1}^{n} h_i + c_n \tag{1}$$

## 2.2 EMD Filtering Method

It can be seen from the above EMD algorithm that as the decomposition order increases, the frequency of the IMF component gradually decreases. The IMF component that is first decomposed represents the highest frequency part of the original signal [13]. If the first few high-frequency components of the IMF component are removed and the remaining components are combined, it can be considered that the original signal is adaptively low-pass filtered:

$$x_L(t) = \sum_{i=k}^{n} h_i(t) + C_n(t) \tag{2}$$

If the IMF component is removed from the first few high-frequency components and the last few low-frequency components, the remaining components can be combined as a band-pass filter that adapts the original signal:

$$x_B(t) = \sum_{i=b}^{k} h_i(t) \tag{3}$$

If the lower several components of the IMF component are removed and the remaining components are combined, the original signal can be treated as an adaptive high-pass filter:

$$x_H(t) = \sum_{i=1}^{k} h_i(t) \tag{4}$$

This filter based on IMF components is implemented in the time domain. Due to the adaptive and hierarchical nature of the filtering process, the original signal's nonlinearity and inherent characteristics are fully preserved. Select different IMF component recombination to achieve different filtering effects. For the characteristics of blast furnace opener, each component of IMF that is decomposed will indicate the operating status of a device, therefore, by selecting the IMF component containing the device state and removing the useless component, the filtering process of the vibration signal can be completed.

## 3 Vibration Signal Experimental Study

Blast furnace opening machine is a kind of rotating machinery equipment. When opening the iron mouth, it will produce vibration signal. This signal is made up of vibration and noise, and it has certain periodicity. According to the opening machine impact frequency is 20–50 Hz, the physical structure of the opening machine has a certain uniqueness, and its vibration signal includes sine signal, amplitude modulation signal and interference signal. Construct a sinusoidal cosine and amplitude modulated oscillating mixed signal with a frequency of 20–50 Hz. According to the site environment, two kinds of interference signals are added to be used as the interference generated by the interior of the opening machine and the interference caused by the external environment. Through EMD filtering of this signal, the result is compared with the wavelet threshold noise reduction to verify the superiority of EMD filtering and the feasibility of applying to the vibration signal processing of blast furnace opening machine.

According to the characteristics of blast furnace opening machine tested in the field, the structural experiment function is as shown in formula (5):

$$x = 2 * \sin(2\pi * 25t) + (1 + \sin(2\pi * 5t)) \sin(2\pi * 40t + \cos(2\pi * 15t)) \tag{5}$$

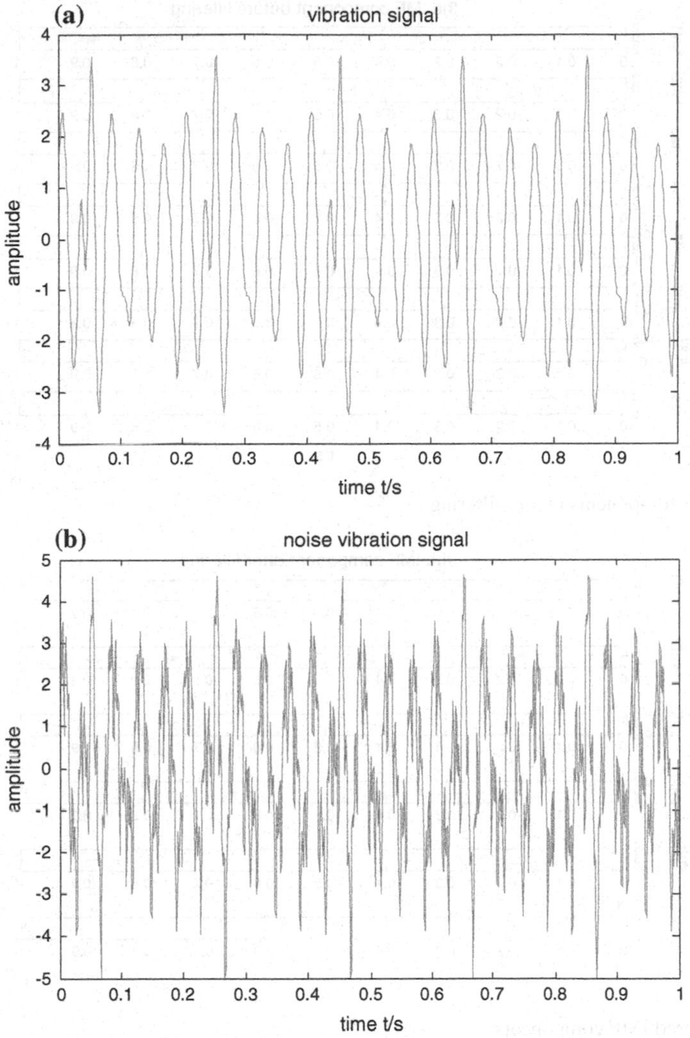


**Fig. 1** **a** Time domain diagram of vibration signal. **b** Time domain diagram of noise signal added with noise

It can be seen from Eq. (5) that the signal consists of a baseband signal with a frequency of 40 Hz, an amplitude modulated oscillation signal with a modulation frequency of 15 Hz and a sinusoidal signal with a frequency of 25 Hz. Figure 1a is a time domain diagram of vibration signal with a sampling frequency of 1,000 Hz and a sampling point of 1,000. Figure 1b is a time domain diagram of a vibration signal with added noise.

The vibration signal after aliasing is decomposed by EMD. Figure 2 shows the filtered IMF components. From top to bottom are $h1$, $h2$, …, $h7$ and $c_n$, $h1$, $h2$ is the high frequency component, contains a lot of noise, and the remaining components gradually decrease in frequency from top to bottom.

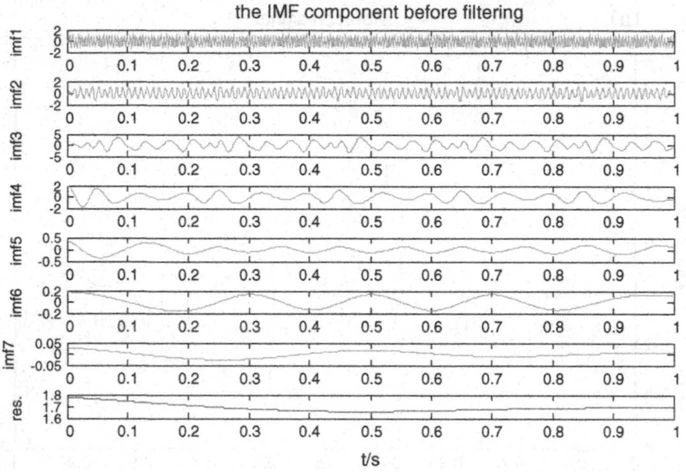


**Fig. 2** IMF components before filtering

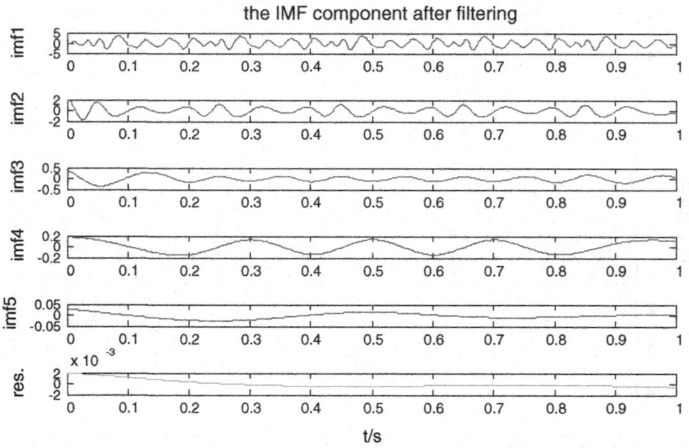


**Fig. 3** Filtered IMF components

For high-frequency noise components, an adaptive low-pass filter is used. The remaining components are the processed IMF components. As shown in Fig. 3, the first IMF function corresponds to the sine part of the experimental function of 40 Hz, the amplitude and spacing of the second IMF function vary, corresponding to the modulation part of the experimental function, and the third IMF component corresponds to the cosine part of the experimental signal. The residual components should not appear in terms of analytical expressions, but because EMD uses a cubic spline interpolation algorithm, the first three IMF components are not true original signal waveforms, so that the remaining tiny components are formed because of the amplitudes. The value is too small and has no effect on subsequent analysis.

The filtered IMF components are combined and recombined. Figure 4a shows the reconstructed vibration signal. In order to verify the effectiveness of EMD filtering, this paper introduces wavelet noise reduction and its comparison. The wavelet basis

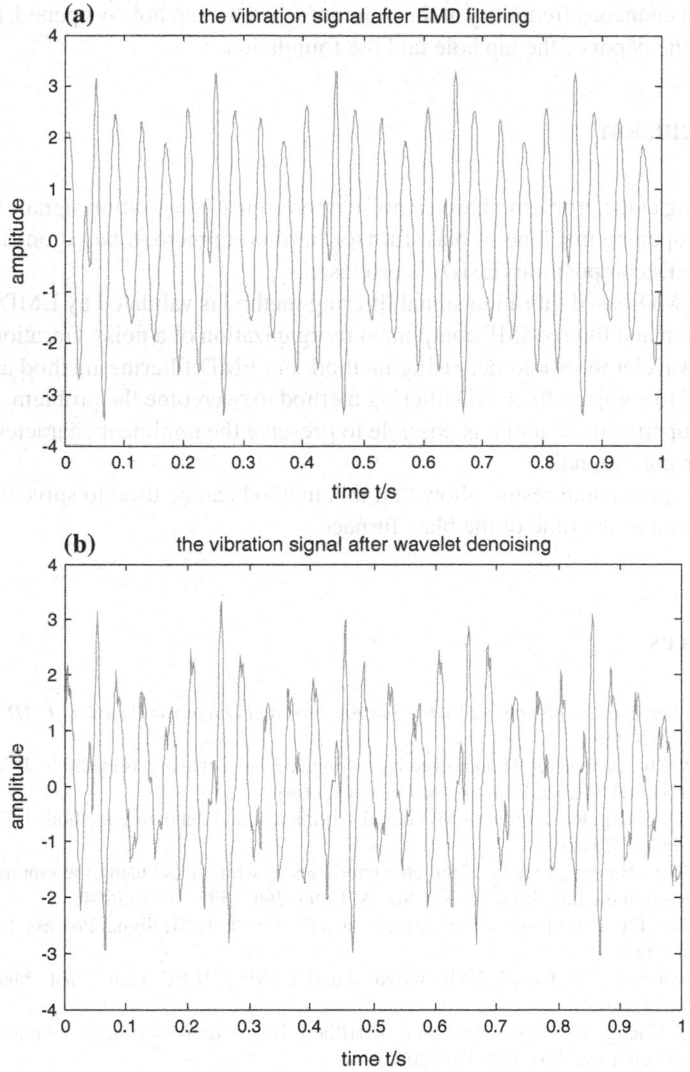


**Fig. 4** **a** Reconstructed vibration signal. **b** Vibration signal after wavelet denoising

db1 is used to decompose the experimental signal into 4 parts, and the coefficients are extracted. Then the filter is filtered by the default soft threshold. Figure 4b is the vibration signal after wavelet threshold denoising.

From analysis of Fig. 4a, b, EMD filtering and wavelet threshold denoising can all play a role in filtering the original signal. However, the vibration signal after wavelet denoising still contains the phenomenon of high-frequency glitches and signal superposition, and it is not very stable throughout the sampling interval. The EMD filtering is an adaptive filtering method that overcomes these problems and preserves the characteristics of the original signal as much as possible, so as to find out the instantaneous frequency at the moment when the tap hole is opened, and then calculate, the depth of the tap hole laid the foundation.

## 4 Conclusion

(1) Aiming at the problem of mixed noise interference of vibration signals from the front opening machine of blast furnace, a filtering method based on Empirical Mode Decomposition (EMD) is proposed.
(2) The EMD-based vibration signal filtering method is validated by EMD decomposition and filtered IMF component reorganization of a noisy vibration signal.
(3) The wavelet threshold denoising method and EMD filtering method are introduced to compare the EMD filtering method to overcome the problems of burrs and superposition, and it is possible to preserve the nonlinear characteristics of the original signal.
(4) The experimental results show that this method can be used to solve the determination of the time of the blast furnace.

## References

1. C. Cao, *Research on Mechanical Vibration Analysis and Diagnosis Based on EMD* (Zhejiang University, 2009)
2. D. Li, W. Zhang, Structural nondestructive testing and wavelet analysis methods. J. Anhui Inst. Archit. (Natural Science Edition) **16**(03), 1–7 (2008)
3. H. Ma, D. Zhang, Research on EMD based vibration signal denoising method. J. Vib. Shock **35**(22), 38–40 (2016)
4. Z. Wu, N.E. Huang, A study of the characteristics of white noise using the empirical mode decomposition method. Proc. R. Soc. Ser. A, Lond. **460**, 1597–1611 (2004)
5. P. Flandrin, Empirical mode decomposition as a filter bank. IEEE Signal Process. Lett. **11**(2), 112–114 (2003)
6. A.O. Boudraa, J.-C. Cexus, EMD-based signal filtering. IEEE Trans. Inst. Meas. **56**(6), 2196–2202 (2007)
7. D. Yu, J. Cheng, Y. Yang, Application of Hilbert–Huang transform in gear fault diagnosis. Chin. J. Mech. Eng. (06), 102–107 (2005)
8. S. Yang, J. Hong, Rotating machinery vibration signal based on EMD Hilbert transform and wavelet transform time-frequency analysis comparison. Proc. CSEE **23**(06), 102–107 (2003)
9. J. Yang, M. Jia, Vibration signal denoising processing of rotating machinery based on empirical mode decomposition. Eng. Sci. (08), 66–69 (2005)
10. H.G. Li, G. Meng, Detection of harmonic signals from chaotic interference by empirical mode decomposition. Chaos, Solitons Fractals **30**(4), 930–935 (2006)
11. J. Hu, S. Yang, Research on filtering technology of rotating machinery vibration signal based on empirical mode decomposition. J. Vib. Meas. Diagn. (02), 20–22 (2003)
12. N.E. Huang, Z. Shen, S.R. Long, et al., The empirical mode decomposition and the Hilbert spectrum for nonlinear and nonstation time series analysis. Proc. R. Soc. A **454**, 903–995 (1998)
13. T. Wang, *EMD Algorithm Research and Its Application in Signal Denoising* (Harbin Engineering University, 2003)

# Distribution Network Fault Location Based on Improved Binary Particle Swarm Optimization

**Fantao Meng, Shasha Zhao, Zhen Li and Shiguang Li**

**Abstract** Due to the local optimum and the inaccurate fault location of distribution network when DG (distributed generation) access using the traditional BPSO (binary particle swarm optimization), an IBPSO (improved binary particle swarm optimization) to locate the fault places is proposed. Firstly, the locating model of distribution network fault is established, which mainly includes the improved coding mode, the improved switching function and the improved fitness function. Then, the BPSO is improved, in which the inertial weight in the algorithm has the adaptive ability, so that the particle can maintain better. At last, this algorithm is used to simulate and locate the fault of distribution network with DG. The results prove that the algorithm and the improved function can accurately locate fault places when the single point fault and multi point fault in the distribution network with DG.

**Keywords** Distribution network · Fault location
Binary particle swarm optimization · Adaptive weight

## 1 Introduction

As different types of DG are connected to the distribution network, the traditional distribution network becomes a multi-power supply network [1], which increases load variation and makes the wiring more complicated, so it is very easy to have failures in the distribution network. According to the data of relevant departments, about 95% of power failure accidents occur in power distribution systems, so it is a very important task for the distribution automation system to locate the fault positions as soon as possible when the distribution network failure occurs [2].

F. Meng (✉) · S. Zhao · S. Li
College of Electrical Engineering and Automation, Shandong University of Science and Technology, Qingdao 266590, China
e-mail: 1491573205@qq.com

Z. Li
Qingdao Civil-Military Integration College, Qingdao 266400, China

 
Y. Jia et al. (eds.), *Proceedings of 2018 Chinese Intelligent Systems Conference*, Lecture Notes in Electrical Engineering 528,
https://doi.org/10.1007/978-981-13-2288-4_19

At present, the fault location algorithms of distribution network mainly include matrix algorithm, genetic algorithm, binary particle swarm algorithm, neural network algorithm and so on [3–6]. Literature [3] uses matrix algorithm to construct the network matrix, and then some modifications are made to the network through the fault current information collected by the FTU. According to the modified matrix and related fault criterion, the fault location function is completed, but the algorithm runs with a large number of matrices, and the fault tolerance is poor. In literature [4], GA (genetic algorithm) is used to locate distribution network fault location. GA has a strong fault tolerance ability, allowing some fault information to distort, but the direction of GA search is not clear, and it's easy to fail evolution when genetic manipulation promotes population evolution. Literature [5] uses BPSO to locate the fault in the distribution network. Comparing to the more complex procedures such as the non-variant operation of the GA algorithm, the computational complexity is reduced, but the BPSO algorithm has a slow optimization iteration and is easy to fall into the local optimality. Literature [6] uses neural network training fault data to locate, with strong robustness, memory ability and strong self-learning ability, but the neural network operation requires a large amount of data support, and its data quality is unstable and uninterpretable.

In this paper, BPSO algorithm, coding method, switch function and fitness function are all improved, and then the location experiment of single point fault and multi-point fault of the distribution network with DG are carried out. Simulation results verify its effectiveness.

## 2 Fault Location Model of Distribution Network

The fault location of the distribution network is to judge that when the fault current detected by FTU in the sectional switch is greater than the presetting threshold, the fault current is considered to have the fault current, and then the fault information is uploaded to the SCADA (Supervisory Control and Data Acquisition), and the fault location is located by the effective method. The fault information should be coded properly, and the switching function is used to express the relationship between the over current information and the switch state information of the line, and then the fitness function which conforms to the characteristics of the DG distribution network is defined, and the fault location is accurately located by the effective algorithm [7]. Since the power flow direction of distribution network has changed to two way after joining DG, we need to improve the coding mode, the switching function and the fitness function.

(a) In the study of fault location, 0 and 1 encoding is often used, that is, 0 indicates that there is no fault current flowing through the line segment, and 1 represents the fault current flowing through the section [8].

However, this coding method is mainly used for fault location of distribution network with unidirectional flow without DG access, and it cannot accurately locate

**Fig. 1** Distribution network fault topology

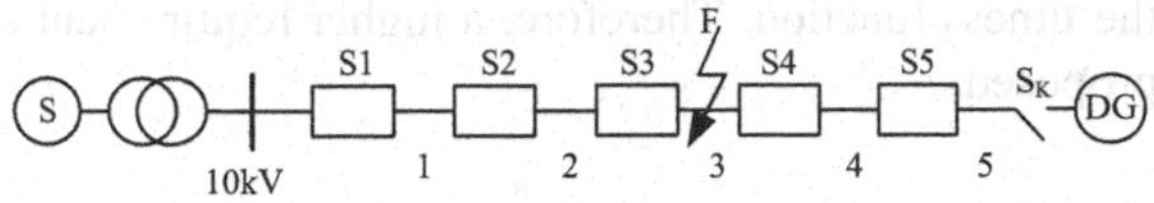


the fault of distribution network with DG access due to the two-way flow of the tidal current. In this paper, the positive direction of the system is the relative direction of the current flowing out of the source power supply node, and the 1, 0 and 1 encoding method is adopted, namely 1 represents the fault current is consistent with the direction, while the opposite is −1, 0 means that no current flows through.

(b) The switch function is usually expressed in formula (1).

$$I^*(s_j) = \prod_{i}^{n} s_i \tag{1}$$

In the formula, $I^*(s_j)$ represents the function value of switch $j$, $\prod$ indicates the operation 'or' operation of each section, and $s_i$ represents the current state of the segment $i$ in the network, $j = i$.

However, this method cannot accurately locate fault location when accessing DG. As shown in Fig. 1, when $S_k$ is disconnected from DG and the position F is in failure, the value of each switch function is $I^*(s_1) = 0 + 0 + 1 + 0 + 0 = 1$, $I^*(s_2) = 0 + 1 + 0 + 0 = 1$, $I^*(s_3) = 1 + 0 + 0 = 1$, $I^*(s_4) = 0 + 0 = 0$, $I^*(s_5) = 0$, The switch function value set is [1, 1, 1, 0, 0], which can judge the fault location is 3. But when $S_k$ is closed, that means the DG is connected to the distribution network. The switch function value set is [1, 1, 1, 1, 1] when the place F is in failure, in this way, it is unable to determine the fault location.

Therefore, it is necessary to improve the switching function in order to accurately locate the fault location when the fault occurs. The improved switch formula is shown below.

$$I^*(s_j) = K_{DG} * \left(0 - \prod_{s}^{m} s_{js}\right) + \prod_{d}^{m} s_{jd} \tag{2}$$

In the formula, $K_{DG}$ is used to determine whether there is a DG access. When DG is connected, its value is 1, otherwise it is 0; $\prod_s^m s_{js}$ and $\prod_d^m s_{jd}$ respectively represent the "or" operation of the fault state value in the upstream and downstream sections of the switch, and 1 indicates the failure of the section, and 0 means no failure. After improvement, when the position F fails, when the switch $S_k$ is disconnected from the DG, the set of the switch function value is [1, 1, 0, 0]; When the switch $S_k$ is connected to DG, the set of values is [1, 1, −1, −1]. Both results indicate that the fault occurs at location 3 in Fig. 1.

(c) The algorithm mainly uses the fault current information, uses the constructed switch function, and finally finds the location of the fault through the iteration of the fitness function. Therefore, a higher requirement on fitness function should be proposed.

$$f(x) = \sum_{j=1}^{n} \left| I^*(s_j) - I(s_j) \right| + \partial \sum_{j=1}^{M} |S_B(j)| \tag{3}$$

In the formula: $n$ represents the number of FTU, $I(s_j)$ represents the actual current of the switch, and $I^*(s_j)$ represents the value of the switch function. $\partial \sum_{j=1}^{M} |S_B(j)|$ is the correction term, which can avoid misjudgment or omission in fault location, and $M$ is the sum of network segments. $\partial$ is the factor between (0, 1), which takes 0.5 in this chapter.

## 3 The Introduction of IBPSO Algorithm

The basic idea of PSO algorithm is that the population is equal to the bird population, and the optimal value of the target function is equivalent to the food. In addition, each bird in the group is considered to be the best solution to the function itself. The process of finding food is regarded as the process of the particle moving towards the optimal solution and finally reaching the optimal solution [9].

PSO is mainly used to solve the optimal solution of the continuous function, but solving the fault location of such discrete problem, usually adopts the BPSO algorithm. The position of each particle in the algorithm can only be 0 or 1, and the particle swarm is a collection of 1 or 0. When the particle is approaching the optimal position, the velocity of its approximation is assumed to be the probability that the value of the particle is 0 or 1. Its speed and position update formula is as follows [10]:

$$v_{id}^{t+1} = c_1 r_1 (p_{\text{best}}^t - x_{id}^t) + c_2 r_2 (g_{\text{best}}^t - x_{id}^t) + \omega v_{id}^t \tag{4}$$

$$\begin{cases} x_{id}^{t+1} = 1 \ \text{randn}() \le \text{sigmoid}(v_{id}^{t+1}) \\ x_{id}^{t+1} = 0 \ \text{randn}() \ge \text{sigmoid}(v_{id}^{t+1}) \end{cases} \tag{5}$$

Among them, sigmoid function is defined as follows:

$$\text{sigmoid}(v_{id}^{t+1}) = \begin{cases} 0.98 & v_{id}^{t+1} > 4 \\ \frac{1}{1+e^{-v_{id}^{t+1}}} & -4 \le v_{id}^{t+1} \le 4 \\ -0.98 & v_{id}^{t+1} < -4 \end{cases}$$

In the formula, $v_{id}^t$ and $x_{id}^t$ are respectively the velocity and position of the particle $i$ in $d$ dimension at the moment $t$, $c_1$ and $c_2$ are the acceleration coefficients affected by the individual and the accelerated factor affected by the group, respectively. $r_1$ and $r_2$ are the random coefficients of independent uniform distribution in the interval [0, 1]. $w$ is the inertial weight; $p_{\text{best}}$ and $g_{\text{best}}$ are respectively the best in individual history and the best in group history. In order to balance the global search ability and local improved ability of the algorithm, here with nonlinear dynamic adaptive inertia weight coefficient formula, expression is as follows:

$$w = w_{\min} - \begin{cases} w_{\min} - \frac{(w_{\max} - w_{\min}) * (f - f_{\min})}{(f_{avg} - f_{\min})}, f \le f_{avg} \\ w_{\max}, \quad f > f_{avg} \end{cases} \tag{6}$$

In the formula, $w_{\max}$ and $w_{\min}$ represent the maximum and minimum values of $w$ respectively, and $f$ represents the current objective function value of the particle, that is, the value of Eq. (3). $f_{avg}$ and $f_{\min}$ respectively represent the average target value and the minimum target value of all the particles at present.

When the target value of each particle consistent or tending to local optimum, will make the inertia weight increase, and the particles in the target is dispersed, will reduce the inertia weight, and the objective function value is superior to the average particles, its corresponding inertia weights were smaller, thus protecting the particles. However, the corresponding inertia weight of the particle with the target function value is worse than the average target value, which is close to the better search area.

On the basis of the improved coding method, switch function formula (2) and fitness function formula (3), the steps of IBPSO algorithm for fault location of distribution network are as follows:

(1) Determine the particle swarm size $m$, the maximum number of iterations $T\max$, acceleration constant $c_1$, $c_2$, inertial weight $w_{\max}$, $w_{\min}$ and particle swarm dimension. Input the data information of each feeder section and switch state in the topology structure of the distribution network to form the switch function and fitness function needed in the network.
(2) Random generation of m feasible solutions, each of which is a set of 0, 1, −1, and random initialization velocity $v_i \in (-4, 4)$.
(3) By formula (3), the fitness of each particle is calculated, and the position and adaptive value of the current particles are stored in $p_{\text{best}}$ of each particle, and the position and adaptive value of the optimal individual in all $p_{\text{best}}$ is stored in $g_{\text{best}}$.
(4) According to formulas (4) and (5), the velocity and displacement of the particle are updated, and the inertia weight is updated according to formula (6).
(5) For each particle, the fitness value is compared with the best position that has been experienced, and if it is good, it is the best place to be in the current position, comparing all current $p_{\text{best}}$ and $g_{\text{best}}$ and updating $g_{\text{best}}$.
(6) If the number of iterations reaches the maximum number of iterations $T\max$, the algorithm runs stop, and the formula (2) outputs the optimal switch function

value set. Formula (3) obtains the global best position of the particle group, which is the actual fault state of each feeder section of the distribution network. Otherwise return step (4) to continue the search.

## 4 Case Analysis

Through the simulation analysis of the single point of failure and multipoint fault in the distribution network with DG topology (Fig. 2), to verify the accuracy running the improved BPSO algorithm for fault location of distribution network with DG. In addition, by comparing the performance of common BPSO and IBPSO for fault location, which further verifies the advantages of improved algorithm for fault location of distribution network with DG.

Figure 2 is a multi-power supply network with DG1, DG2 and DG3, where V is the main power supply, and there are 20 switches in the system, which can be represented by s1–s20, and all of them can upload the value information of the switch function. In addition, there are 15 sections with the number 1–15 respectively. The algorithm parameters are: particle swarm size $m=60$, maximum number of iterations $T\max=50$, particle dimension 15, $c_1 = c_2 = 2$, inertia weight $w_{\max} = 0.9$, $w_{\min} = 0.4$.

When running the algorithm simulation, the network load is not working, and the improved coding mode produces the switching current information. The switch is flowing through the fault current at 1, the reverse flow is −1, and the no-fault current flows to 0. Through FTU, the function collection information of the switch function is continuously uploaded. Finally, the fitness function set information is obtained, that is, the output shows the fault location information. The output of the section failure is 1, normal is 0.

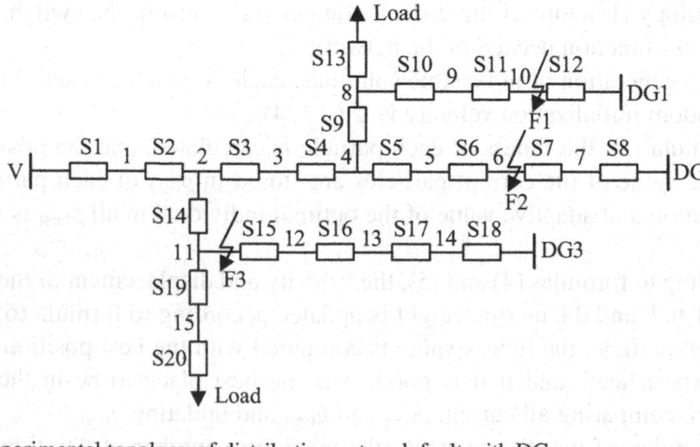


**Fig. 2** Experimental topology of distribution network fault with DG

**Table 1** Partial experimental simulation results

| Information uploaded by FTU | Fault condition | Distorted information | The output | Fault section |
|---|---|---|---|---|
| [1 1 −1 −1 −1 −1 −1 −1 −1 −1 −1 −1 0 −1 −1 −1 −1 −1 0 0] | Single point failure | No | [0 1 0 0 0 0 0 0 0 0 0 0 0 0 0] | 2 |
| [1 1 0 1 −1 −1 −1 −1 0 −1 −1 −1 0 −1 −1 −1 −1 −1 0 0] | Single point failure | 3, 9 | [0 0 0 1 0 0 0 0 0 0 0 0 0 0 0] | 4 |
| [1 1 −1 −1 −1 −1 −1 −1 1 1 −1 −1 0 −1 −1 −1 −1 −1 0 0] | Two point failure | No | [0 1 0 0 0 0 0 0 1 0 0 0 0 0 0] | 2, 9 |
| [1 1 −1 −1 −1 0 −1 −1 1 −1 −1 −1 0 −1 −1 −1 −1 −1 0 0] | Two point failure | 6 | [0 1 0 0 0 0 0 1 0 0 0 0 0 0 0] | 2, 8 |
| [1 1 1 0 1 −1 −1 −1 1 1 1 −1 0 1 −1 −1 0 0 1 0] | Multi-point failure | 4, 17, 18 | [0 0 0 0 1 0 0 0 0 1 0 0 0 0 1] | 5, 10, 15 |

(1) Single point failure experimental analysis. Suppose that the position F1 in the topology 2 has a failure, the information uploaded by FTU is [1, 1, 1, 1, 1, 1, 1, 1, 1, 1, 1, −1, 0, 1, 1, 1, 1, 1, 0, 0]. The output of the algorithm is [0, 0, 0, 0, 0, 0, 0, 0, 0, 1, 0, 0, 0, 0, 0], which can directly locate the fault section at segment 10. The BPSO and the IBPSO run respectively 80 times, with the former falling into the local optimal number five times, and the accuracy rate was 93.8%, while the latter was in the local optimum two times and the accuracy was 97.5%.

(2) Multi-point failure experiment analysis. Suppose that the positions F2 and F3 in the topology 2 have failures, the information uploaded by FTU is [1, 1, 1, 1, 1, 1, −1, −1, 1, 1, 1, 1, 0, 1, −1, −1, −1, −1, 0, 0]. The output of the algorithm is [0, 0, 0, 0, 0, 1, 0, 0, 0, 0, 1, 0, 0, 0, 0], which can directly locate the fault sections at segment 6 and segment 11. The BPSO and the IBPSO run respectively 80 times, with the former falling into the local optimal number nine times, and the accuracy rate was 88.75%, while the latter was in the local optimum five times and the accuracy was 93.8%.

In order to show the effectiveness of this algorithm in the distribution network fault location more intuitively, 80 simulation experiments were conducted on random faults. Due to the space problem, only some typical simulation results were selected, as shown in Table 1.

Table 1 shows that, even though the switch fault information uploaded by FTU has been distorted, but as a result of improved fitness function has the ability to prevent

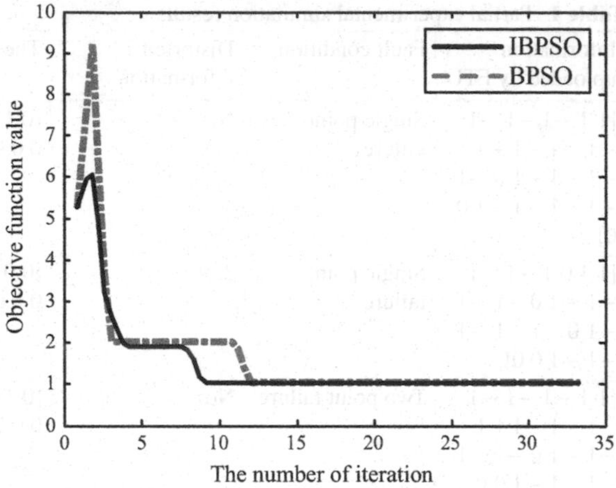


**Fig. 3** Simulation iteration curve diagram

distortion, so at the end of the improved BPSO algorithm running can still generate accurate results.

Figure 3 shows the iterative curves of BPSO and IBPSO for single point failure of distribution network respectively, and the comparison results show that both of them can accurately locate the fault location of the distribution network (the target function value is 1). The traditional BPSO algorithm has an iteration number of about 13 generations for the single point fault location of the distribution network, while the IBPSO algorithm runs a convergence algebra of about 9 generations, and the algorithm is faster than the traditional BPSO algorithm. In addition, the IBPSO algorithm is more stable than the BPSO algorithm, and the disadvantages of local convergence are improved.

## 5 Conclusion

In view of the fault of distribution network with DG, this paper puts forward the improved BPSO algorithm used in distribution network fault location, and improves the operation efficiency of the common BPSO algorithm, improves the global search ability of the optimal solution, to avoid the algorithm converges to local optimal solution which leads the failure of distribution network fault location. The improvement of coding mode and switch function makes the algorithm not only suitable for fault location of traditional distribution network without DG, but also can accurately locate the distribution network fault containing DG. Even though the set information of the switch function has been distorted, the improvement of fitness function still makes the algorithm run output correctly and locate the fault places accurately.

The simulation results show that the improved algorithm and its improvement of related distribution network fault location function, which is also very effective for multi-point fault location of multiple DG distribution network and makes up for the shortage of the previous algorithm and function only applicable to the fault location of single power distribution network. In addition, the research on fault location of distribution network is greatly enriched.

## References


1. J. Sun, R. Chen, S. Cai, et al. A new fault location scheme for distribution system with distributed generations. Power Syst. Technol. **37**(6), 1645–1650 (2013)
2. J. Liu, X. Dong, X. Chen, et al., *Fault Location and Restoration for Distribution Systems* (China Electric Power Press, Beijing, 2012), pp. 1–8
3. F. Hu, S. Sun, Fault location of distribution network by applying matrix algorithm based on graph theory. Electr. Power **49**(3), 94–98 (2016)
4. S. Chen, Y. Ma, J. Fang, Application and analysis of genetic algorithm in fault location of intelligent distribution network. China Energy Environ. Prot. (12), 219–222 (2017)
5. C. Li, Z. He, H. Zhang, et al., Fault location for radicalized distribution networks based on BPSO algorithm. Power Syst. Prot. Control **37**(7), 35–39 (2009)
6. X. Ruan, X. Zhang, Fault location of multi-sources power distribution network based on neural network. Coal Mine Mach. **35**(2), 239–240 (2014)
7. Y. Wang, *Application Study of Fault Location for Distribution Network Containing Distributed Generation* (Taiyuan University of Technology, Taiyuan, 2017)
8. J. Zhao, Z. Tu, Y. Xie, Application of improved binary particle swarm optimization for fault location in distribution network with distributed generation. J. Heilongjiang Univ. Sci. Technol. **24**(3), 277–281 (2014)
9. J. Kennedy, R.C. Eberhart, A discrete binary version of the particle swarm algorithm, in *Proceeding of IEEE International Conference on Systems, Man, and Cybernetics*, Orlando, USA (1997)
10. Y. Yao, Z. Wang, K. Guo, et al., Distribution network service restoration using a multi-objective binary particle swarm optimization based on e-dominance. Power Syst. Prot. Control (23), 76–81 (2014)

# A Fault Detection Method for Non-Gaussian Industrial Processes via Joint KICA and FVS

**Li Zhang, Xinying Zhong, Yi Chai and Ke Zhang**

**Abstract** The data in industrial processes have the features of non-linear and non-Gaussian. In order to enhance the accuracy of fault detection for industrial processes, and to reduce the calculation time consumption, a method is proposed to combine the joint kernel independent component analysis and the feature vector selection (FVS) to achieve fault detection in this paper. Firstly, the joint kernel function of Gaussian radial basis kernel function and polynomial kernel function is used to improve the learning and generalization ability of kernel independent component analysis (KICA) algorithm, and this can be employed to improve the accuracy of fault detection. Secondly, FVS is given to reduce the computational complexity of Joint KICA, especially in the case of large sample size. Finally, the simulation results of Tennessee Eastman (TE) process can be used to verify the effectiveness of this proposed method.

**Keywords** Kernel independent component analysis · Feature vector selection
TE process · Fault detection

## 1 Introduction

The main characteristics of the development of modern industrial processes systems can be considered as its increasing large-scale and complexity, which result in many new necessities on maintaining a high level of safety and reliability. Due to the large number of available data collected during the operation, it should be made full use with depth information to ensure reliable and safe operation [1–3]. Recently, the multivariate statistical process monitoring (MSPM) had been studied and used

L. Zhang (✉) · X. Zhong · Y. Chai · K. Zhang
School of Automation, Chongqing University, Chongqing 400044, China
e-mail: zl15823051535@163.com

L. Zhang · Y. Chai · K. Zhang
Key Laboratory of Complex System Safety and Control, Ministry of Education, Chongqing University, Chongqing 400044, China



Y. Jia et al. (eds.), *Proceedings of 2018 Chinese Intelligent Systems Conference*,
Lecture Notes in Electrical Engineering 528,
https://doi.org/10.1007/978-981-13-2288-4_20

to describe the correlation between these variables, and this becomes one of most research hot topics in the field of process control [4]. The typical MSPM methods, such as principal component analysis (PCA), partial least square (PLS), etc., had been widely applied in fault detection for industrial processes [5–8]. However, all of these methods had been studied at the assumption that the monitoring data are subject to normal distribution, and furthermore, these data often have the features of non-linear and non-Gaussian.

To cope with this problem, Lee et al. [9], can be considered as the first time to introduced independent component analysis (ICA) to the fault detection field, and this method can be employed to separate non-Gaussian independent components from linear data [10]. In order to solve nonlinear problem, the kernel method was applied to renovate ICA [11]. Bach and Jordan [12], proposed a novel KICA method aiming at non-Gaussian data in nonlinear system, which can perform better on feature extraction and fault detection. However, there still exists some shortcomings in KICA, which attracts many scholars' attentions. Firstly, the parameters of kernel function are always selected with experience, and Jia et al. [13], proposed the genetic algorithm to obtain the correct types and parameters of kernel function. Secondly, Zeyu and Peiliang [14], proposed an improved KICA algorithm to reduce the false alarm rate and the missed alarm rate.

In terms of the time consumption increased by the kernel function. FVS scheme can be generally used to reduce the size of kernel matrix and the calculation time consumption when the number of samples becomes larger [15]. Peiling et al. [16], proposed KPCA algorithms, applying the FVS to kernel principal component (KPCA), to achieve fault detection with low time consumption.

Therefore, this paper aims to solve the problems on fault detection for industrial process systems, and it seeks to improve KICA for fault detection in two ways to further enhance the accuracy of fault detection. The joint kernel function is given to improve the learning and generalization ability of KICA algorithm, which combines the Gaussian radial basis kernel function and the new polynomial kernel function. FVS scheme can be used to select fewer sampling vectors and effectively reduce the computational complexity of KICA.

The rest of this paper is organized as follow. In Sect. 2, Joint KICA is briefly introduced. The principles and implementation steps of FVS are presented in Sect. 3. In Sect. 4, the proposed fault detection method are described in detail. In Sect. 5, fault detection results of two scheme by TE process simulations are reported. Finally, conclusions are also outlined.

## 2 Brief Introduction of Joint KICA

The KICA algorithm was first proposed by Lee et al. [17]. The main ideas can be considered as to map the sample data into the high dimensional feature space by KPCA, and which then can be extracted independent components by ICA. Joint KICA improves the kernel function by combining Gaussian radial basis kernel function and polynomial kernel function. The brief introduction of Joint KICA algorithm can be described as follows.

## 2.1 Whitening Data in the Feature Space

The purpose of whitening observed data is to remove the correlation between the observed variables. In this paper, the improved KPCA algorithm is used to whiten processing to simplify ICA, which is particularly important for processing high-dimensional data.

Suppose that the original observed data is a distribution with its $N$ data length, where $x_k \in R^m, k = 1, \ldots, N$, and it can be mapped into a higher dimensional feature space $F$ by nonlinear mapping $\Phi : R^m \to F$. The covariance matrix in the feature space can be calculated as follows.

$$C^F = \frac{1}{N}\sum_{k=1}^{N} \Phi(x_k)\Phi(x_k)^T \tag{1}$$

where $\Phi(x_k), k = 1, \ldots, N$ with 0 mean and 1 variance.

Let $\Theta = [\Phi(x_1), \ldots, \Phi(x_N)]$, Eq. (1) can be simplified as follows, mathematically.

$$C^F = \frac{1}{N}\Theta\Theta^T \tag{2}$$

Then, the whitening matrix $z$ can be obtained by KPCA algorithm.

$$\begin{aligned} z &= \sqrt{N}\Lambda^{-1}H^T\Theta^T\Phi(x) \\ &= \sqrt{N}\Lambda^{-1}H^T[\Phi(x_1), \ldots, \Phi(x_N)]^T\Phi(x) \\ &= \sqrt{N}\Lambda^{-1}H^T\left[\tilde{k}_{scl}(x_1, x), \ldots, \tilde{k}_{scl}(x_N, x)\right]^T \end{aligned} \tag{3}$$

In Eq. (3), $\tilde{K}_{scl}$ is the centralization of the kernel matrix $K$.

$$\begin{aligned} \tilde{K} &= K - I_N K - K I_N + I_N K I_N \\ \tilde{K}_{scl} &= \frac{\tilde{K}}{\operatorname{trace}(\tilde{K})/N} \end{aligned} \tag{4}$$

where $I_N = \frac{1}{N}\begin{bmatrix} 1 & \cdots & 1 \\ \vdots & \ddots & \vdots \\ 1 & \cdots & 1 \end{bmatrix} \in R^{N\times N}$, $K_{ij} = \langle \Phi(x_i), \Phi(x_j) \rangle = \mathrm{k}(x_i, x_j)$.

The kernel function $k(x_i, x_j)$ is the joint kernel function that satisfies the Mercer condition [18]. If $k_1, k_2$ are kernel function, like $k = ak_1 + bk_2$ also is a kernel function. Therefore, we combine the local kernel function with strong learning ability and global kernel function with strong generalization ability as a new kernel function, so that the joint kernel function has both advantages and achieves better in the whitening process.

In this paper, a more flexible kernel function is constructed by combining the polynomial kernel function and Gaussian radial basis kernel function in a convex way. The joint kernel function can be described as follow:

$$\begin{aligned} k(x, y) &= \alpha k_1 + (1-\alpha)k_2 \\ &= \alpha\left[\langle x, y\rangle + 1\right]^\eta + (1-\alpha)\exp\left(-\|x-y\|^2/\sigma\right) \end{aligned} \tag{5}$$

where $\alpha$ is weighted coefficient of the joint kernel function, $\alpha \in [0,1]$. $\eta$ is the order of polynomial kernel function, and $\eta \in [1, 3]$ is an integer, $\eta \in [1, 3]$. The parameter $\sigma$ represents the kernel parameters of the Gauss radial basis function. $H$ is the orthonormal eigenvectors corresponding to the $d$ largest eigenvalues of $\widetilde{K}_{scl}$, the $d$ largest eigenvalues of $C^F$ are given by $\lambda_1/N, \lambda_2/N, \ldots, \lambda_d/N$.

$$\Lambda = diag(\lambda_1/N, \lambda_2/N, \ldots, \lambda_d/N) \tag{6}$$

## 2.2 Implementation of ICA Algorithm

The task of the ICA algorithm is to extract the independent components $s$ in the KPCA feature space. That is to say, to find the orthogonal matrix $C_n$ makes the irrelevant whitening matrices $z$ become independent.

Currently, there exists many ICA algorithms, such as the maximum likelihood gradient method proposed by Amari, the self-organizing neural network method proposed by Bell, and the fixed point method proposed by Hyvarinen and Oja [19]. This paper draws on the FastICA algorithm proposed by Hyvarinen and Oja to extract the independent components $s$ [20].

Hyvarinen and Oja introduced an approximation of the neg-entropy as a measure of non-Gaussian.

$$J(y) \approx k[E\{G(y)\} - E\{G(v)\}]^2 \tag{7}$$

where $y$ is a random variable of zero mean and unit variance, and $v$ is a Gaussian variable of zero mean and unit variance. $G$ is non-quadratic function, such as:

$$G_1(\mu) = \frac{1}{a_1}\log\cosh(a_1\mu) \tag{8}$$

**Table 1** Steps for FastICA

| |
|---|
| **Input**: the whitening matrix $z$ and the number $q$ of independent components |
| **Output**: the orthogonal matrix $C_n$ |
| 1. Select the estimated number of independent components, set counter $i \leftarrow 1$; |
| 2. Replace the initial vector $C_{n,i}$ into i-th row of the matrix $C_n^T$; |
| 3. Maximize the approximated negentropy: $c_{n,i} \leftarrow E\left\{zg(c_{n,i}^T z)\right\} - E\left\{g'(c_{n,i}^T z)\right\}c_{n,i}$, where $g$ and $g'$ are the first derivative and second derivative of $G$; |
| 4. Orthogonalize: $c_{n,i} \leftarrow c_{n,i} - \sum_{j=1}^{i-1}(c_{n,i}^T c_{n,j} c_{n,j})$; |
| 5. Normalize: $c_{n,i} \leftarrow \frac{c_{n,i}}{\lVert c_{n,i} \rVert}$; |
| 6. If $c_{n,i}$ has not converged, return to step 3, else output vector $c_{n,i}$. If $i \leq p$, then set $i \leftarrow i + 1$ and return to step 2. |

$$G_2(\mu) = \exp\left(-\frac{a_2\mu^2}{2}\right) \tag{9}$$

$$G_3 = \mu^4 \tag{10}$$

Here, $1 \leq a_1 \leq 2, a_2 \approx 1$. In this paper, we select Eq. (8) as the optimization objective function of the ICA algorithm. The following Table 1 is the descriptions of the whole process.

The independent components of the observed data can be extracted from the KPCA feature space by the orthogonal matrix $C_n$ and the diagonal matrix $D$. $D$ represents the diagonal matrix of eigenvalues, which are retained when the independent components are extracted, $D = diag(\lambda_1, \lambda_2, \ldots, \lambda_p)$, $p \leq d$.

$$s = D^{1/2} C_n^T z \tag{11}$$

## 2.3 Calculation of Statistics for Fault Detection

Similar to the KPCA algorithm, $T^2$ and squared prediction error (SPE) statistics can be conducted as the control statistics for fault detection [21, 22], and described by

$$T^2 = s^T D^{-1} s \tag{12}$$

$$SPE = e^T e = \left(x - \hat{x}\right)^T \left(x - \hat{x}\right) \tag{13}$$

$T^2$ indicates the degree of each sample deviating from the model in variation tendency and amplitude. SPE expresses the degree of the measured value $x_i$ deviating from the kernel independent component model at the $i$ moment. For Eq. (13), when we map the observed data $x$ to the high-dimensional feature space, it will lead to the difficulty of obtaining the estimated value $\widehat{x}$. So it can make $z$ instead of $x$.

$$SPE = e^T e = \left(z - \widehat{z}\right)^T \left(z - \widehat{z}\right) = z^T (I - C_n C_n^T) z \tag{14}$$

were $e = z - \widehat{z}$, $z = \sqrt{N}\Lambda^{-1} H^T \widetilde{K}_{scl}$ and $\widehat{z} = C_n D^{-1/2} s = C_n C_n^T z$. Since the independent component can't obey the Gaussian distribution, the control limits of the $T^2$ and SPE can be solved by kernel density estimation (KDE), which can be refer to Ref. [23] for more details.

## 3 Feature Vector Selection

Generally, the computation of the joint kernel function is too complexity with the large size of process data. This paper combines the FVS method with the joint KICA method to avoid this situation. The FVS algorithm aims at reducing the calculation complexity and saving the calculation time with preserving the geometric structure of all the data [24–26]. The main idea of FVS is to find a subset of samples as small as possible, and the mapping of this subset in the feature space can be used to represent the mapping of all sample data.

Suppose that $S = \left\{x_{S_1}, x_{S_2}, \ldots, x_{S_{L_S}}\right\}$ is the selected subset of sample set $X = \{x_1, x_2, \ldots, x_N\}$, $L_s$ is the length of the selected vectors. The estimation of any sample data $x_i$ mapping in the feature space can be expressed as a linear combination of $S$.

$$\widehat{\phi}_i = \phi_S \cdot \tau_i \tag{15}$$

where $\phi_S = \left(\phi_{S_1}, \phi_{S_2}, \ldots, \phi_{S_{L_S}}\right)$ is the mapping of sample subset in the feature space, $\tau_i = \left(\tau_i^1, \tau_i^2, \ldots, \tau_i^{S_{L_S}}\right)$ is the coefficient vector.

Then, the following minimization problem should be solved to find coefficient vector $\tau_i$, which makes the estimated value $\widehat{\phi}_i$ as closer as to the real value $\phi_i$.

$$\min(\delta_i) = \frac{\left\|\phi_i - \widehat{\phi}(x_i)\right\|^2}{\|\phi_i\|^2} \tag{16}$$

Let $\partial\delta_i / \partial\tau_i = 0$, and rewritten it:

$$\min(\delta_i) = 1 - \frac{k_{si}^T k_{ss}^{-1} k_{si}}{k_{ii}} \tag{17}$$

**Table 2** Feature vector selection procedure

**Input**: the sample vector set $X=\{x_1,x_2,\ldots,x_N\}$; kernel function and its parameter; the required vector number $L_S$ $(0<L_S<N)$.

**Output**: the selected subset of the sample $S$

**Parameters initialization**: $S=\varnothing$ and $L=0$

**First iteration:** select a sample that maximizes $J_s$ and add it to $S$, let $L=1$.

**Do iteration:**

Combining with previous L samples, select a sample from the remaining samples.

If $k_{SS}$ is invertible and $L<L_S$

Select the sample vector that maximizes $J_s$, append it to $S$ and let $L=L+1$.

Else

Stop iteration.

End if

End iteration

where $k_{ii}=k(x_i,x_i)$, $k_{ss}=\left(k\left(x_{s_p},x_{s_q}\right)\right)_{1\le p\le L_s,1\le q\le L_s}$ is the product of the selected vector, $k_{si}=\left(k\left(x_{s_p},x_i\right)\right)_{1\le p\le L_S}$ is the inner product vector between point $x_i$ and sample subset $S$.

Let Eq. (17) satisfies all the samples, and this is the ultimate goal.

$$\max_{s} J_s=\frac{1}{N}\sum_{x_i}\frac{k_{si}^T k_{ss}^{-1} k_{si}}{k_{ii}} \tag{18}$$

This problem can be solved through an iterative process. The iterative terminating condition is $k_{SS}$ no longer invertible or $J_s$ reaches a given boundary value, or a predetermined number of selected vectors is reached. The FVS algorithm specific is outlined in Table 2.

## 4 Fault Detection Based on Joint KICA and FVS

As mentioned before, a novel fault detection method for non-Gaussian industrial processes via joint KICA and FVS is proposed in this section. The fault detection procedure mainly includes offline modeling and online detection. Figure 1 is the flow chart of fault detection based on Joint KICA and FVS. The specific steps are as following two phases:

*Offline modeling phase*:

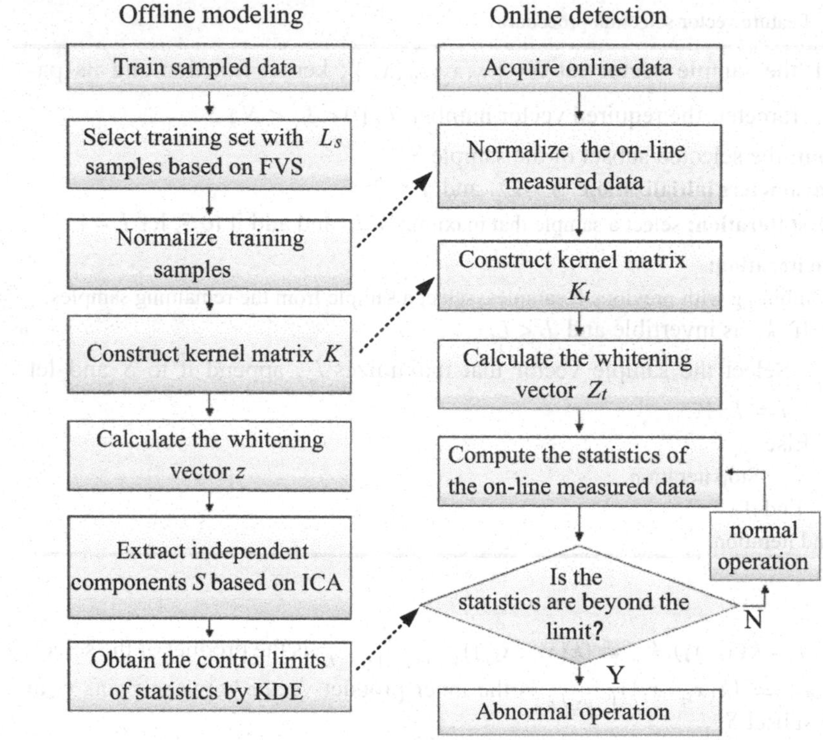


**Fig. 1** The flow chart of fault detection based on Joint KICA and FVS

1. Obtain the original training set $X$ in normal operation and select training set with $L_S$ samples based on FVS;
2. Normalize $L_S$ training samples with zero mean and unit variance;
3. Construct kernel matrix $K$ with combining kernel function, and normalize the kernel matrix in the feature space;
4. Calculate the whitening vector $z$ according to Eq. (4), and apply the ICA in the feature space to extract independent components $s$;
5. Obtain the $T^2$ and SPE statistics based on Eqs. (12) and (14);
6. Complete the control limits of the $T^2$ and SPE statistics by KDE.

*Online detection phase*:

1. Standardize the newly acquired online data $x_t, x_t \in R^m, t = 1, \ldots, N_t$ with the mean and variance of the offline KICA;
2. Calculate the kernel matrix $K_t \in R^{N_t \times N}$ by the following formula:

$$[K]_{tj} = \langle \Phi(x_t), \Phi(x_j) \rangle = k(x_t, x_j)$$

where $x_j$ is offline training data, $x_j \in R^m$, $j = 1, \ldots, N$;

3. Normalize the kernel matrix in the feature space using $\widetilde{K}_t = K_t - E_{N_t}K - KE_N + E_{N_t}KE_N$ and $\widetilde{K}_{tscl} = \frac{\widetilde{K}_t}{trace(\widetilde{K})/N}$;
4. Obtain the whitened data vector $z_t$ by $z_t = \sqrt{N}\Lambda^{-1}H^T\widetilde{K}_{tscl}$, and extract independent components $s$ with ICA In the feature space;
5. Calculate the $T^2$ and SPE statistics of $x_t$, if this statistics exceed the control limits, this process is abnormal.

## 5 Simulation and Result

### *5.1 TE Process Description*

TE process is a chemical process testing platform based on real cases. It is mainly composed of five reaction unit, including a reactor, a product condenser, a recycle compressor, a vapor-liquid separator and a product stripper. The whole industrial processes consist of 41 measured variables and 12 manipulated variables. It can simulate the process of normal operation and 21 fault conditions. The details can be referred in Ref. [27].

In this simulation, the TE data-sets are composed of one training data-set under normal operation and 21 testing data-sets corresponding to the 21 classes of faults. The sampling time is 3 min, and each data-set has 52 observed variables. The training data set is obtained under the 24 h normal operation simulation, and the number of samples is 480. The testing data sets are obtained through 48 h operation simulation, the total number of testing data is 960. The fault are introduced from sample 160, thus the first 160 samples are normal data, and the latter 800 samples are fault data. The simulation data in this paper can be obtained from http://brahms.scs.uiuc.edu.

### *5.2 Analysis of Fault Detection Results*

In this paper, fault 1, fault 4, fault 5, fault 8, fault 15 and fault 20 are detected based on ICA, KICA and Joint KICA. Through comparing the fault detection rate with this three methods, the validity of the proposed method is verified. The kernel function of KICA algorithm select Gaussian radial basis kernel function, and the kernel parameter $\sigma$ is 3000. Letting kernel parameter $\sigma = 2500$, $\lambda = 0.1$, $d = 2$ of Joint KICA. The simulation results are shown in Table 3.

As show in Table 3, the Joint KICA algorithm can be used to achieve better detection accuracy for most faults, but for some minor fault, such as fault 5 and fault 15, the fault detection rate is very low. In addition, the fault detection rate of KICA is higher than ICA. The reason is that the KICA algorithm updates the kernel function and solves the nonlinear problem of the ICA algorithm well. Compared to KICA, the

**Table 3** Fault detection rate (%) comparison

| Case | ICA | | KICA | | Joint KICA | |
|---|---|---|---|---|---|---|
| | $T^2$ | SPE | $T^2$ | SPE | $T^2$ | SPE |
| 1 | 71.23 | 78.34 | 80.28 | 90.08 | 87.62 | 94.37 |
| 4 | 10.29 | 67.50 | 81.12 | 84.25 | 88.63 | 90.75 |
| 5 | 23.25 | 36.14 | 34.15 | 40.52 | 32.13 | 39.27 |
| 8 | 89.72 | 95.53 | 91.51 | 92.65 | 93.61 | 95.68 |
| 15 | 14.21 | 16.67 | 20.88 | 21.29 | 19.25 | 18.63 |
| 20 | 48.17 | 52.19 | 79.50 | 77.13 | 83.12 | 79.75 |

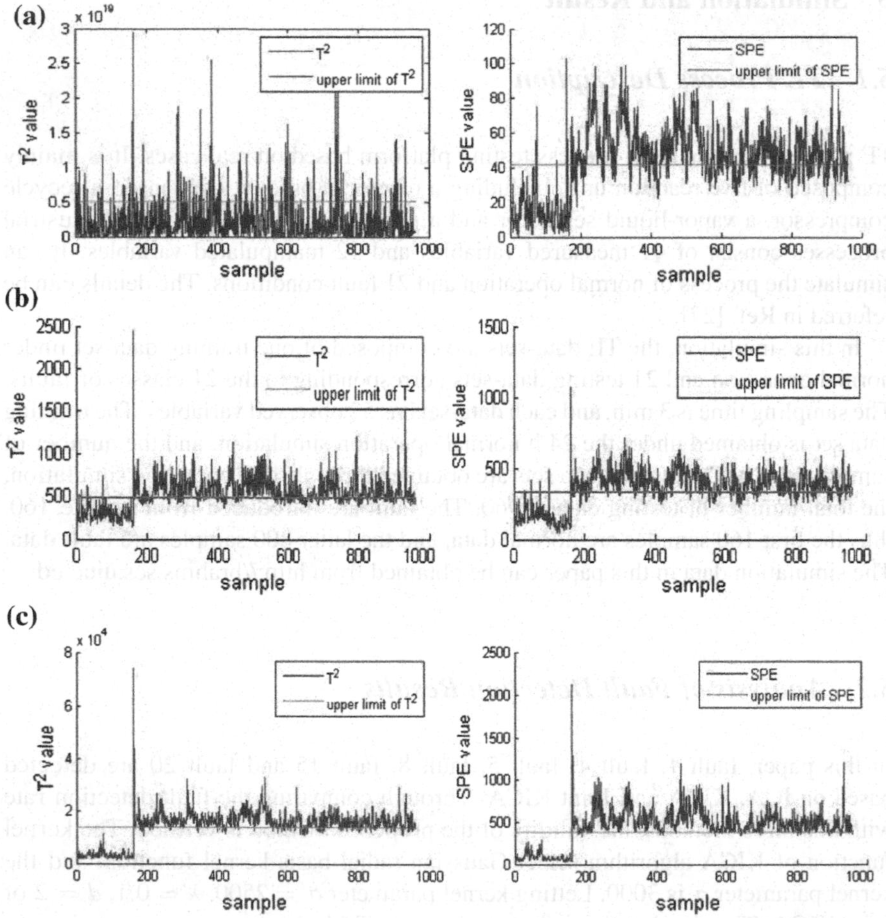


**Fig. 2** Fault detection results of TE process in the case of fault 4: **a** ICA, **b** KICA and **c** Joint KICA

fault detection rate of Joint KICA is increased in small increments with Joint KICA improve the original kernel function.

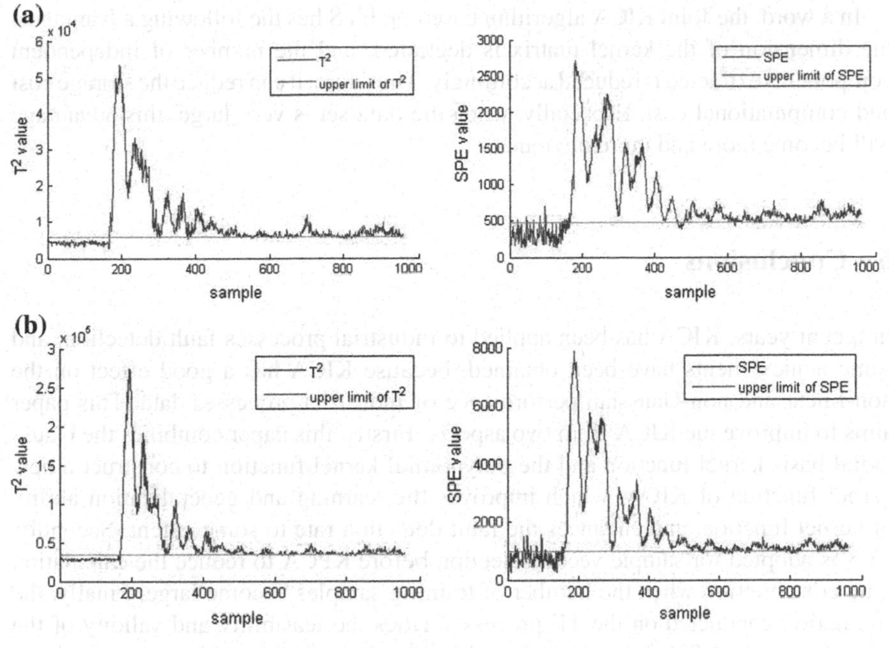


**Fig. 3** Fault detection results of TE process in the case of fault 1: **a** KICA and **b** Joint KICA based on FVS

The fault 4 involves a step change of the inlet temperature in the reactor cooling water. Figure 2 shows the $T^2$ and SPE statistics charts for fault detection of TE process in the case of fault 4 by ICA, KICA and Joint KICA.

In Fig. 2a, the fault detection based on ICA, the $T^2$ statistic can hardly detect fault 4, the SPE statistic can timely detect fault 4 in 161 sampling time. But there still exists in a lot of omissions. In Fig. 2b, the KICA method can detect the fault in 161 sampling time, and the miss detection ratio is much lower than ICA. Compared with the previous two methods, the Joint KICA algorithm has less fluctuation in the statistics of $T^2$ and SPE charts, and the miss detection ratio is lower. So the Joint KICA algorithm is better than the original KICA algorithm.

On the other hand, FVS method is introduced at the initial stage of the Joint KICA algorithm to reduce computational complexity. Through verification, the selected training sample number $L_S = 60$ based on FVS. In terms of independent components extraction, the number of independent components detected for Joint KICA based on FVS is 21, while for KICA is 32. The detection result of fault 1 is analyzed below.

According to Fig. 3, the fault detection rate of the $T^2$ and SPE statistics are 87.62% and 94.37% respectively for fault 1 using Joint KICA algorithm, while the fault detection rate are 83.13% and 92.37% by Joint KICA algorithm based on FVS. Therefore, it can be seen that the Joint KICA algorithm based on FVS have not a negative impact on fault detection rate. It only has a small fluctuation.

In a word, the Joint KICA algorithm based on FVS has the following advantages: the dimension of the kernel matrix is degraded, and the number of independent components extracted is reduced accordingly. Therefore, it can reduce the storage cost and computational cost. Especially, when the data-set is very large, this advantage will become more and more obvious.

## 6 Conclusions

In recent years, KICA has been applied to industrial processes fault detection, and some achievements have been obtained, because KICA has a good effect on the non-linear and non-Gaussian performance of industrial processes data. This paper aims to improve the KICA from two aspects. Firstly, this paper combines the Gauss radial basis kernel function and the polynomial kernel function to construct a new kernel function of KICA, which improves the learning and generalization ability of kernel function, and enhances the fault detection rate to some extent. Secondly, FVS is adopted for sample vector selection before KPCA to reduce the calculation time consumption when the number of training samples become large. Finally, the simulation conducted on the TE process verifies the feasibility and validity of the proposed method. The Joint KICA based on FVS have not a negative impact on fault detection rate. Moreover, it can reduce the computational complexity significantly. Hence, this method is superior to the KICA algorithm in fault detection.

## References


1. V. Venkatasubramanian, Prognostic and diagnostic monitoring of complex systems for product lifecycle management: challenges and opportunities. Comput. Chem. Eng. **29**(6), 1253–1263 (2005)
2. Z. Ke, Z. Donghua, C. Yi, Review of multiple fault diagnosis methods. Control Theory Appl. **32**(9), 1143–1157 (2015)
3. W. Bing, Y. Chunmei, L. Qiang, Fault diagnosis for process industry (Science Press, 2012)
4. L. Han, X. Deyun, Survey on data driven fault diagnosis methods. Control Dec. **26**(1), 1–9 (2011)
5. D. Rongxing, Z. Huilin, A new fault diagnosis method based on fault tree and Bayesian networks. Energy Proc. **17**(17), 1376–1382 (2012)
6. M. Tipping, Sparse kernel principal component analysis. Adv. Neural Inf. Process. Syst. 633–639 (2001)
7. G. Jingyu, H. Jianbing, L. Yuan, X. Jinxue, Fault diagnosis of complex chemical process based on local fisher discriminant analysis. Appl. Res. Comput. **35**(4) (2018). http://kns.cnki.net/kcms/detail/51.1196.tp.20170401.1738.038.html
8. L. Qiang, Q.S. Joe, C. Tianyou, Decentralized fault diagnosis of continuous annealing processes based on multilevel PCA. IEEE Trans. Autom. Sci. Eng. **10**(3), 687–698 (2013)
9. J.M. Lee, C.K. Yoo, I.B. Lee, Statistical process monitoring with independent component analysis. Process Control **14**, 467–485 (2004)
10. A. George Stefatos, B. Hamza, Dynamic independent component analysis approach for fault detection and diagnosis. Expert Syst. Appl. **37**(12), 8606–8617 (2010)

11. S. Saitoh, Theory of reproducing kernels. Trans. Am. Math. Soc. **68**(3), 337–404 (2015)
12. F.R. Bach, M.I. Jordan, Kernel independent component analysis. J. Mach. Learn. Res. **3**, 1–48 (2003)
13. J. Mingxing, X. Hengyuan, L. Xiaofei et al., The optimization of the kind and parameters of kernel function in KPCA for process monitoring. Comput. Chem. Eng. **46**(15), 94–104 (2011)
14. Y. Zeyu, W. Peiliang, Fault detection method for non-linear systems based on kernel independent component analysis and support vector data description. Inf. Control **46**(2), 153–158 (2017)
15. W. Hongyan, H. Daoping, Kernel principal component analysis based on feature vector selection. Comput. Sci. **36**(7), 185–255 (2009)
16. C. Peiling, L. Junhong, W. Guizeng, Improved kernel principal component analysis for fault detection. Expert Syst. Appl. **2**(34), 1210–1219 (2008)
17. J.-M. Lee, S.J. Qin, I.B. Lee, Fault detection of non-linear processes using kernel independent component analysis. Canad. J. Chem. Eng. **85**(4), 526–536 (2007)
18. S. Saitoh, Theory of reproducing kernels. Trans. Am. Math. Soc. **68**(3), 337–404 (2003)
19. A. Hyvärinen, E. Oja, Independent component analysis: algorithm and applications. Neural Netw. **13**(4–5), 411–430 (2000)
20. A. Hyvärinen, Fast and robust fixed-point algorithms for independent component analysis. IEEE Trans. **10**, 1129–1159 (1999)
21. J.-M. Lee, C.K. Yoo, S.W. Choi et al., Nonlinear process monitoring using kernel principal component analysis. Chem. Eng. Sci. **59**(1), 223–234 (2004)
22. J.Yingmin, Alternative proofs for improved LMI representations for the analysis and the design of continuous-time systems with polytopic type uncertainty: a predictive approach. IEEE Trans. Autom. Control **48**(8), 1413–1416 (2003)
23. Z.I. Botev, J.F. Grotowski, D.P. Kroese, Kernel density estimation via diffusion. Ann. Stat. **38**(5), 2916–2957 (2010)
24. G. Baudat, F. Anouar, Kernel-based methods and function approximation, in *Proceedings of International Conference on Neural Networks* (Washington DC, 2001), pp. 1244–1249
25. G. Baudat, F. Anouar, Feature vector selection and projection using kernels. Neurocomputing **55**, 21–38 (2003)
26. J. Yingmin, Robust control with decoupling performance for steering and traction of 4WS vehicles under velocity-varying motion. IEEE Trans. Control Syst. Technol. **8**(3), 554–569 (2000)
27. A. Zidan, M. Khairalla, A.M. Abdrabou et al., Fault detection, isolation, and service restoration in distribution systems: state-of-the-art and future trends. IEEE Trans. Smart Grid **24**(5), 1–16 (2016)

# Improved Deep Deterministic Policy Gradient Algorithm Based on Prioritized Sampling

**HaoYu Zhang, Kai Xiong and Jie Bai**

**Abstract** Deep reinforcement learning tends to have low sampling efficiency, and prioritized sampling algorithm can improve the sampling efficiency to a certain extent. The prioritized sampling algorithm can be used in deep deterministic policy gradient algorithm, and a small sample sorting method is proposed to solve the problem of high complexity of the common prioritized sampling algorithm. Simulation experiments prove that the improved deep deterministic policy gradient algorithm improves the sampling efficiency and the training performance is better.

**Keywords** Deep reinforcement learning · Deep deterministic policy gradient
Prioritized sampling

## 1 Introduction

In recent years, deep reinforcement learning (DRL) has achieved great success in such fields as go [1, 2] and video games [3, 4]. Deep Q network (DQN) has solved the problem that using neural network to fit the value function usually leads to the non-convergence of training results by introducing experience replay technology [3]. Lillicrap combined experience replay technology with deterministic policy gradient (DPG) [5] and put forward the deep deterministic policy gradient (DDPG) [6], which successfully extended the problem with low dimensional discrete action space that

H. Zhang · K. Xiong (✉)
Science and Technology on Space Intelligent Control Laboratory, Beijing Institute of Control Engineering, Beijing 100190, China
e-mail: 17600517255@163.com

H. Zhang
e-mail: Haoy_Zhang@163.com

J. Bai
Beijing Key Laboratory of Intelligent Space Robotic Systems Technology and Applications, Beijing Institute of Spacecraft System Engineering, Beijing, China



Y. Jia et al. (eds.), *Proceedings of 2018 Chinese Intelligent Systems Conference*, Lecture Notes in Electrical Engineering 528,
https://doi.org/10.1007/978-981-13-2288-4_21

can be solved through general reinforcement learning algorithm to high dimension continuous action space.

In experience replay technology, a large number of training samples are collected and stored in a sample set. A batch of training samples are selected randomly from the sample set during training, in which case can break the correlation between the training data, and this method is proved to be effective. However, the disadvantage of this method is that the sampling efficiency is rather low, and it is difficult to select effective training data for training, so the convergence rate of the algorithm is slow and it is even difficult to converge. In order to solve this problem, Schual put forward the prioritized experience replay technology [7] and applied it to the DQN, the importance of each sample is determined according to the temporal difference (TD) error in this method, and the sample with higher priority is more likely to be picked up for training, so as to speed up the training process.

In this paper, we focus on the deterministic policy gradient algorithm, study the feasibility of applying the prioritized experience replay method to the deep deterministic policy gradient algorithm. Because the priority for all samples need to be computed and sorted in prioritized experience replay algorithm, which results in high computation complexity, so in this paper, we propose a method in which only a few samples are sorted, and use the two training models CartPole and Pendulum to verify the effectiveness of the algorithm.

## 2 Related Work

### 2.1 Deep Deterministic Policy Gradient Algorithm

The reinforcement learning (RL) usually can be described as markov decision process [8] (MDP), the environment E can be described through an array $(S, A, P, R)$, where $S$ represents for the state set, $A$ represents for the action set, $P$ represents for the state transition probability and $R$ represents for the reward function. In reinforcement learning, the policy is denoted by $\pi$, which represents the mapping of state to action, that is $S \rightarrow A$.

The return from a state at $t$ is defined as the sum of discounted future reward [9],

$$R_t = \sum_{i=t}^{T} \gamma^{(i-t)} r(s_i, a_i) \tag{1}$$

where $\gamma$ is the discounted factor, $r(s_i, a_i)$ represents for the reward received at state $s_i$ after select the action $a_i$, $T$ is the final time step. Suppose that the initial state is $s_1$ and the applied policy is $\pi$, the state distribution obeys $\rho^{\pi}$, then the task of reinforcement learning is to learn a policy which can maximize the expectation of initial state return. The objective of reinforcement learning is defined as follows:

$$J = \mathbb{E}_{s\sim\rho,a\sim\pi}[R_1] \tag{2}$$

The state action value function $Q^\pi(s, a)$ represents the sum of reward that apply the policy $\pi$ after the execution of action $a$,

$$Q^\pi(s_t, a_t) = \mathbb{E}_{s\sim\rho,a\sim\pi}[R_t|s_t, a_t] \tag{3}$$

In temporal difference method [10], the current state action value function is estimated through the state action value function of the next moment,

$$Q^\pi(s_t, a_t) = \mathbb{E}_{r,s_{t+1}\sim E}\left[r(s_t, a_t) + \gamma\mathbb{E}_{a_{t+1}\sim\pi}\left[Q^\pi(s_{t+1}, a_{t+1})\right]\right] \tag{4}$$

where the distributions of $r(s_t, a_t)$, $s_{t+1}$ are determined by the environment $E$, the distribution of $a_{t+1}$ is determined by the policy $\pi$. As a convenience, the reward $r(s_t, a_t)$ is usually replaced by $r_t$.

The TD error is defined as

$$\delta_t = r_t + \gamma Q^\pi(s_{t+1}, a_{t+1}) - Q^\pi(s_t, a_t) \tag{5}$$

In the case that we don't need to specify a particular policy, $Q^\pi$ can often be replaced by $Q$ directly.

When use neural network to approximate state action value function $Q(s_t, a_t)$, set the parameters of the network as $\boldsymbol{\theta}^Q$, the loss function of the network can be defined as

$$L(\boldsymbol{\theta}^Q) = \mathbb{E}_{s\sim\rho,a\sim\pi}\left[\left(y_t - Q(s_t, a_t|\boldsymbol{\theta}^Q)\right)^2\right] \tag{6}$$

where

$$y_t = r_t + \gamma Q(s_{t+1}, a_{t+1}|\boldsymbol{\theta}^Q) \tag{7}$$

In deterministic policy gradient, for a particular state, the action won't be taken according to a probability, instead of it, the action will be chosen in a deterministic way. Compared to the stochastic policy gradient, it is more effective to apply the deterministic policy gradient because there is no need to integral in the action space. In deterministic policy gradient, the policy is expressed in terms of $\mu$, and its parameters is $\boldsymbol{\theta}$. The target of reinforcement learning can be expressed as a function of the policy, and the gradient descent method can be used to optimize the parameters. It has been proved that the gradient of the target function $J(\mu_\theta)$ to the parameter $\boldsymbol{\theta}$ is [5]

$$\nabla_\theta J(\mu_\theta) = \mathbb{E}_{s\sim\rho^\mu}\left[\nabla_\theta\mu_\theta(s)\nabla_a Q^\mu(s, a)|_{a=\mu_\theta(s)}\right] \tag{8}$$

In the Actor-Critic framework [11], the Critic network is used to evaluate actions and guide the updating of network parameters, the parameters is $\theta^{Q}$, meanwhile the Actor network is used to choose action, the parameter is $\theta^{\mu}$. According to the training data selected, the TD error can be obtained.

$$\delta_t = r_t + \gamma Q^{\theta^Q}(s_{t+1}, \mu_{\theta^\mu}(s_{t+1})) - Q^{\theta^Q}(s_t, a_t) \tag{9}$$

The gradient descent method is used to update the network parameters, and the learning rate is alpha $\alpha$,

$$\theta_{t+1}^{Q} = \theta_t^{Q} + \alpha_Q \delta_t \nabla_{\theta^Q} Q^{\theta^Q}(s_t, a_t) \tag{10}$$

$$\theta_{t+1}^{\mu} = \theta_t^{\mu} + \alpha_\mu \nabla_{\theta^\mu} \mu_{\theta^\mu}(s_t) \nabla_a Q^{\theta^Q}(s_t, a_t)|_{a=\mu_\theta(s)} \tag{11}$$

The idea of DQN is adopted in the deep deterministic policy gradient, and the independent target networks are setup in the Actor network and the Critic network, whose parameters are updated more slowly, which can increase the training stability [12].

## 2.2 *Experience Replay*

Prior to DQN, fitting value function by means of neural network typically results in the problem of divergence. The reason is that the labeled data adopted to train neural network is assumed to be independent and identically distributed, but there exists some relation between data collected through reinforcement learning, and thus if the data is used for neural work training directly, such data may result in instability. In DQN, a critical method which is called experience replay, is proposed to break the connection between the training data. To be specific, the method firstly establishes an experience set D to store historical experience $e_t = (s_t, a_t, r_t, s_{t+1})$, and then randomly sample a batch of historical experience from the experience set to estimate the expected value function in Q-learning algorithm, following which executes gradient descent once. The relation between training data can be effectively broken by applying this method, which making convergent final training.

However, the training data is randomly sampled from the experience set when the ordinary experience replay technology is applied, so the differences between samples are tended to be neglected, which will result in quite low sampling efficiency. As a result , it typically takes quite a long time for them to achieve convergence, and even worse, they sometimes fail to achieve the convergence in the end. Targeting at the training process of the DQN, Schaul regard the magnitude of the TD error as an index measuring the importance of samples [7] and proposes the method of prioritized experience replay, which can largely improve the sampling efficiency and increase the convergence speed. However, the method just mentioned requires computing TD errors of all the samples in the experience set and properly arranging corresponding sequence, which makes the algorithm far more complex.

## 3 Improved Deep Deterministic Policy Gradient Algorithm

The prioritized experience replay technique has been successfully applied in DQN. In this paper, targeting at the problem of inefficient training in DDPG, we apply the experience replay technique in DDPG Algorithm, and propose a small sample sorting method instead of the complex algorithm adopted in the nature prioritized experience replay technique.

The deep deterministic policy gradient is trained by Actor-Critical framework (see Fig. 1 [10]), the deterministic policy gradient method is used to select the action in the Actor network while the temporal difference method is applied to estimate the value function in the Critic network. In addition, the TD error can be used to evaluate the action that selected by the Actor network. In this framework, the TD error signal generated by the Critic network is also used to guide the Critic network and Actor network to update, so it is reasonable to use TD error to determine the priority of training samples.

**Fig. 1** Actor-critic framework

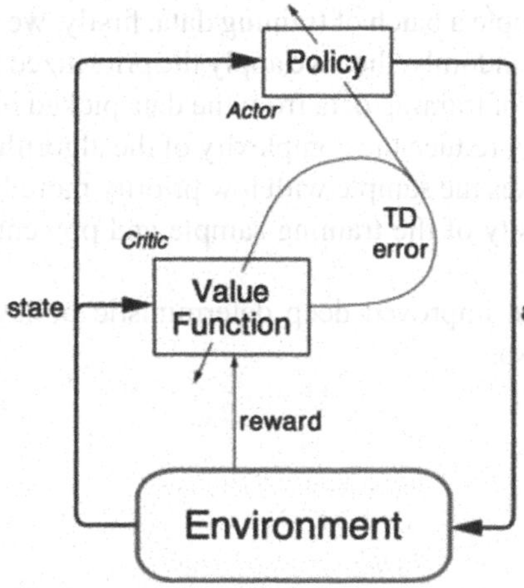

At any time $t$, the priority of each sample is expressed as

$$p_t = |\delta_t| = |r_t + \gamma Q(s_{t+1}, a_{t+1}) - Q(s_t, a_t)| \tag{12}$$

According to the priority of each sample, the probability distribution of selecting the $i$th sample can be determined as

$$P(i) = \frac{p_i}{\sum_k p_k} \tag{13}$$

According to the probability distribution of sample selection mentioned above, we will more likely to select the samples with higher priority for training, but in the early training process, the TD error may be not accurate. Take the above factor into consideration, we need to decrease the priority weights when sampling the training data early in the training process. Meanwhile, in order to ensure the sample with an extremely low priority can even be selected, a bias is added to the priority.

The probability distribution of selecting the sample can be changed to

$$P(i) = a \times \frac{p_i^\alpha}{\sum_k p_k^\alpha} + b \tag{14}$$

where $\alpha$ is the priority weight, when $\alpha$ is 0, the priority of each sample is the same, and the priority sampling is not used at all. $b$ is a small constant almost equal to zero, it enables the sample with a priority of almost 0 is also likely to be selected. $a$ is the normalized ratio, which makes the sum of probabilities of the whole samples still equal to 1.

The TD errors for all samples need to be computed and ordered in the ordinary prioritized experience replay method, which increases the cost of computation. In this paper, we propose a small sample sorting method. To be specific, when we need to sample a batch of training data, firstly, we sample several times batches of training data randomly, then we apply the prioritized experience replay technique and select a batch of training data from the data picked in the first step. With this method, we can greatly reduce the complexity of the algorithm. At the same time, to a certain extent, it makes the sample with low priority more likely to be selected, which increases the diversity of the training sample and prevents the over estimation of value function [13].

The improved deep deterministic policy gradient algorithm is summarized as follows.

**Algorithm : Improved DDPG**

Randomly initialize actor network $\mu$ and critic network $Q$ with weights $\theta^{\mu}$ and $\theta^{Q}$
Initialize target network $\mu'$ and $Q'$ with weights $\theta^{\mu'} \leftarrow \theta^{\mu}$ and $\theta^{Q'} \leftarrow \theta^{Q}$
Initialize replay buffer $R$
Set the parameters $a, b, \alpha, \tau$
For episode =1,$M$ do
  Initialize a random process $\mathcal{N}$ for action exploration
  Receive initial observation state $S_1$
  For $t = 1, T$ do
    Select action $a_t = \mu(s_t|\theta^{\mu}) + \mathcal{N}$ according to the current policy and exploration noise
    Execute action $a_t$ and observe reward $r_t$ and observe new state $s_{t+1}$
    Store transition $(s_t, a_t, r_t, s_{t+1})$ in $R$
    Randomly sample several times batches of transitions from $R$
    Compute the priority of each transition using $P(i) = a \times \frac{p_i^{\alpha}}{\sum_k p_k^{\alpha}} + b$
    Sample a batch of $N$ transitions according to the priority
    Set $y_i = r_i + \gamma Q'(s_{i+1}, \mu'(s_{i+1}|\theta^{\mu'})|\theta^{Q'})$
    Update the critic network by minimize the loss $\frac{1}{N}\sum_i (y_i - Q(s_i, a_i|\theta^{Q}))^2$
    Update the actor network using the sampled policy gradient

$$\nabla_{\theta^{\mu}} J \approx \frac{1}{N}\sum_i \nabla_{\theta^{\mu}} \mu(s|\theta^{\mu})|_{s_i} \nabla_a Q(s, a|\theta^{Q})|_{a=\mu_\theta(s)}|_{s=s_i, a=\mu(s_i)}$$

    Update the target networks

$$\theta^{Q'} \leftarrow \tau\theta^{Q} + (1-\tau)\theta^{Q'}$$
$$\theta^{\mu'} \leftarrow \tau\theta^{\mu} + (1-\tau)\theta^{\mu'}$$

  End for
End for

## 4 Results

In this paper, the improved DDPG algorithm is applied to such classical control problems as CartPole and Pendulum to verify its effectiveness, and it is also compared with the nature DDPG algorithm for analysis. The action space of CartPole is discrete, and we have the straight pole maintained perpendicular through adjusting the position of the sliding block connected beneath it. The action space of Pendulum is continuous, and we adjust the state of the pendulum from being vertically downward to being vertically upward and have it maintained at the adjusted state through manipulating the output torque of the motor at the bottom of the pendulum.

In reinforcement learning, the reward returned from environment is typically regarded as the evaluation index for training effects, and in this paper, we adopt average award to evaluate relevant algorithms and have the average award normalized to compare the convergence rates of such algorithms. To make our experiment results more convincing, the simulation data is recorded 10 times through the two algorithms and the average values are acquired, meanwhile the same random seed is used in all the experiments. Figure 2 shows the average reward using the two algorithms under CartPole environment. It can be seen that at the 60th round, the average reward is constantly improve when use the improved DDPG algorithm, meanwhile the average reward in nature DDPG algorithm begin to raise at the 170th round. As shown in Fig. 3, the improved DDPG algorithm has a faster convergence rate. Figure 4 shows the reward changes for each turn, which can reflect the training process more directly.

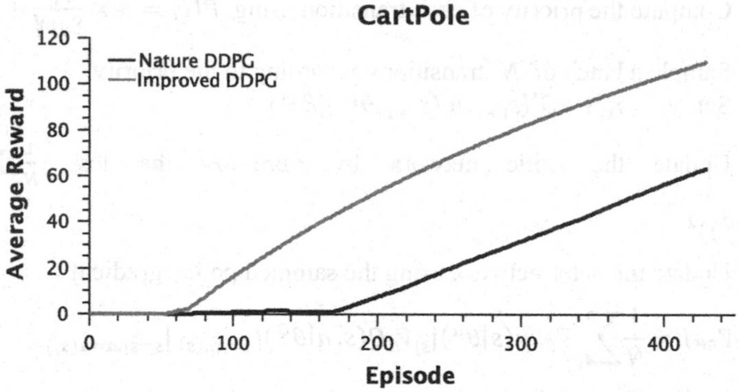


**Fig. 2** CartPole average reward

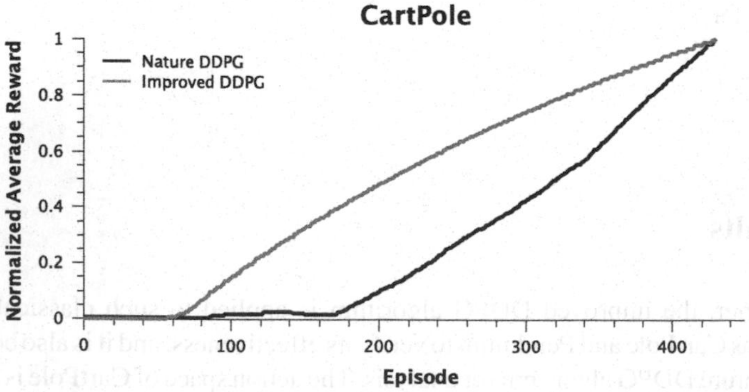


**Fig. 3** CartPole normalized average reward

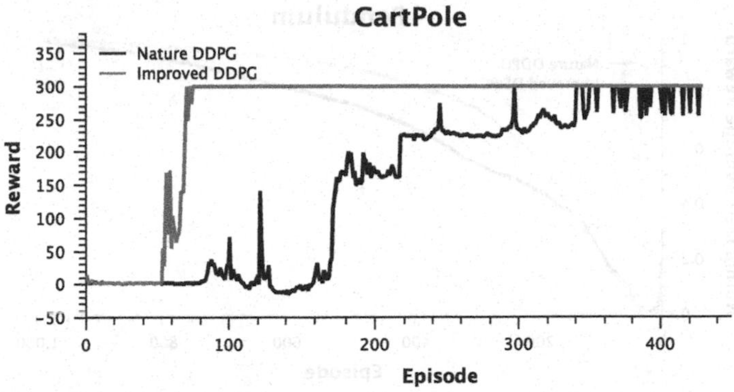


**Fig. 4** CartPole each episode reward

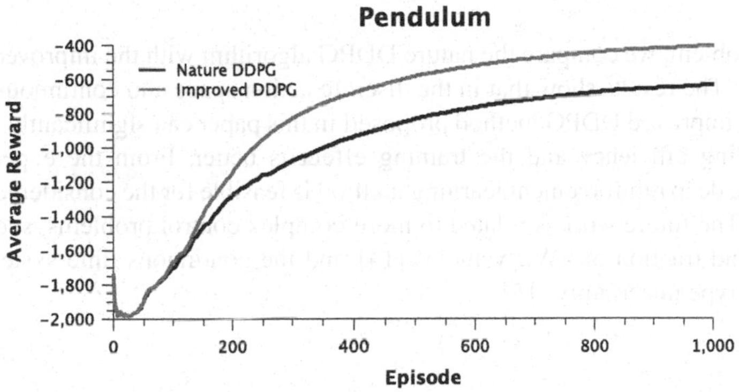


**Fig. 5** Pendulum average reward

Figure 5 shows the average reward curve obtained by using two algorithms in environment Pendulum. It can be seen that the improved DDPG algorithm improves the average reward faster. As shown in Fig. 6, the convergence rate of the improved DDPG algorithm is relatively fast.

The simulation results show that compared with the nature DDPG algorithm, the improved DDPG algorithm makes learning more efficient.

## 5 Conclusion

In this paper, we focus on the deep deterministic policy gradient algorithm, and study the feasibility of applying the prioritized experience replay method to speed up training process of the deep deterministic policy gradient algorithm. Meanwhile, we propose a small sample sorting method to solve the problem of high complexity that the prioritized experience replay algorithm may cause. Based on the classical

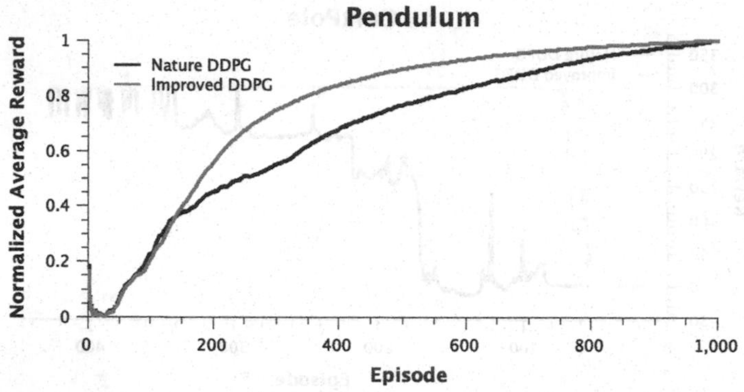


**Fig. 6** Pendulum normalized average reward

control problem, we compare the nature DDPG algorithm with the improved DDPG algorithm. The results show that in the discrete action space and continuous motion space, the improved DDPG method proposed in this paper can significantly improve the sampling efficiency and the training effect is better. From the experimental results, the deep reinforcement learning method is feasible for the considered control problem. The future work is related to more complex control problems, such as the steering and traction of 4WS vehicles [14] and the continuous-time systems with polytopic type uncertainty [15].

## 6 Acknowledgements

The study is supported in part by Beijing Nature Science Foundation (4162070) and Naturel Science Foundation of China (61573059)

## References

1. D. Silver, A. Huang, C.J. Maddison et al., Mastering the game of Go with deep neural networks and tree search. Nature **529**(7587), 484 (2016)
2. D. Silver, J. Schrittwieser, K. Simonyan et al., Mastering the game of Go without human knowledge. Nature **550**(7676), 354–359 (2017)
3. V. Mnih, K. Kavukcuoglu, D. Silver et al., Playing Atari with deep reinforcement learning. Comput. Sci. (2013)
4. V. Mnih, K. Kavukcuoglu, D. Silver et al., Human-level control through deep reinforcement learning. Nature **518**(7540), 529 (2015)
5. D. Silver, G. Lever, N. Heess, T. Degris, D. Wierstra, M. Riedmiller, Deterministic policy gradient algorithms, in *The International Conference on Machine Learning (ICML)* (2014)
6. T.P. Lillicrap, J.J. Hunt, A. Pritzel et al., Continuous control with deep reinforcement learning. Comput. Sci. **8**(6), A187 (2015)

7. T. Schaul, J. Quan, I. Antonoglou et al., Prioritized experience replay. Comput. Sci. (2015)
8. Zhou, *Machine Learning* (Tsinghua University Press, Beijing, 2016), pp. 377–382
9. J. Schulman, P. Moritz, S. Levine et al., High-dimensional continuous control using generalized advantage estimation. Comput. Sci. (2015)
10. R.S. Sutton, A.G. Barto et al., Introduction to reinforcement learning. Mach. Learn. **16**(1), 285–286 (2005)
11. V. Konda, Actor-critic algorithms. SIAM J. Control Optim. **42**(4), 1143–1166 (2006)
12. H.V. Van, A. Guez, D. Silver, Deep reinforcement learning with double q-learning, in *Proceedings of the AAAI Conference on Artificial Intelligence*. Phoenix, USA (2016), pp. 2094–2100
13. S. Thrun, A. Schwartz, Issues in using function approximation for reinforcement learning, in *Proceedings of the 1993 Connectionist Models Summer School*, Hillsdale, NJ, ed. by M. Mozer, P. Smolensky, D. Touretzky, J. Elman, A. Weigend (1993)
14. Y. Jia, Robust control with decoupling performance for steering and traction of 4WS vehicles under velocity-varying motion. IEEE Trans. Control Syst. Technol. **8**(3), 554–569 (2000)
15. Y. Jia, Alternative proofs for improved LMI representations for the analysis and the design of continuous-time systems with polytopic type uncertainty: a predictive approach. IEEE Trans. Autom. Control **48**(8), 1413–1416 (2003)

# A Target Information Conversion Method of Antiaircraft Weapon System

**Shujun Yang, Jianqiang Zheng, Yiming Liang, Qinghua Ma, Shuaiwei Wang and Haipeng Deng**

**Abstract** A detailed conversion method was provided to create good conditions for the search and tracking of phased array radar, and improve the target acquisition probability of the system. By using attitude compensating and coordinate transformation we eliminated the system's transmission error, eventually provided the target information that launcher truck required. Finally, the flight experiment results are presented to validate it.

**Keywords** Target information conversion · Antiaircraft weapon system

## 1 Introduction

In Antiaircraft Weapon System, it's very important to carry on an effective conversion design aiming to the characteristics of distributed target information [1, 2]. There are mutual information between the launcher truck and surveillance radar [3, 4]. By using attitude compensating and coordinate transformation we eliminated the system's transmission error, eventually provided the target information that launcher truck required. This method can create good conditions for the search and tracking of phased array radar, and improve the target acquisition probability of the system.

## 2 I/O

Input information includes vehicle attitude angles provided by launcher truck attitude sensor (launcher truck pitch angle, launcher truck tilt angle), and the information by surveillance radar (distance of the target, azimuth angle of the target, elevation angle of the target). Output information includes target Information based on launcher

S. Yang · J. Zheng · Y. Liang (✉) · Q. Ma · S. Wang · H. Deng
Xian Institution of Modern Control Technology, Xian 201848, China
e-mail: 372614045@qq.com

 
Y. Jia et al. (eds.), *Proceedings of 2018 Chinese Intelligent Systems Conference*,
Lecture Notes in Electrical Engineering 528,
https://doi.org/10.1007/978-981-13-2288-4_22

**Table 1** Input information

| Information name | Unit | Mode | Period (ms) | Sender | Remark |
|---|---|---|---|---|---|
| Pitch angle $\vartheta_V$ | Radian | RS422 | 100 | Launcher truck attitude sensor | Based on launcher truck coordinates |
| Truck tilt angle $\gamma_V$ | | | | | |
| Distance of the target rT_HK | Meter | RS485 | 40 | Surveillance radar | Based on launcher truck horizontal coordinates |
| Azimuth angle of the target btT_HK | Radian | | | | |
| Alevation angle of the target epT_HK | Radian | | | | |

**Table 2** Output information

| Information name | Unit | Mode | Period (ms) | Receiver | Remark |
|---|---|---|---|---|---|
| Distance of the target r_TarHK | Meter | Net | 20 | Launch-control device and guidance computer | Based on launcher truck coordinates |
| Azimuth angle of the target bt_TarHK | Radian | | | | |
| Alevation angle of the target ep_TarHK | Radian | | | | |

truck coordinates (distance of the target, azimuth angle of the target, elevation angle of the target). The communication mode and period of Input and Output was showed in Tables 1 and 2 respectively.

# 3 Call Conditions

The design project's computing period is 20 ms, outputting period is 20 ms. When a "Target guide" signal or "system reset" signal is received, the design project will be contemplated begins or ends.

## 4 Target Information Conversion Method

(1) The information provided by the surveillance radar is based on WGS-84 North east down. The surveillance radar convert the location and orientation information (the space position, north angle) to launcher truck horizontal coordinate, and send the transformational information to the information interchange unit of the launcher truck in the same time. The target information is converted from polar coordinates to rectangular coordinates and the location information of the target is obtained.

$$X_T^{vh} = \mathrm{rT\_HK} \times \cos(\mathrm{epT\_HK}) \times \cos(\mathrm{btT\_HK});$$
$$Y_T^{vh} = \mathrm{rT\_HK} \times \sin(\mathrm{epT\_HK});$$
$$Z_T^{vh} = \mathrm{rT\_HK} \times \cos(\mathrm{epT\_HK}) \times \sin(\mathrm{btT\_HK})$$

(2) Combining the pitch angle and tilt angle from the launcher truck attitude sensor, convert the target information from launcher truck horizontal coordinates to launcher truck coordinates.

$$\begin{aligned} X_T^V &= \cos\vartheta_v \cdot X_T^{vh} + \sin\vartheta_v \cdot Y_T^{vh} \\ Y_T^V &= -\sin\vartheta_v \cdot \cos\gamma_V \cdot X_T^{vh} + \cos\vartheta_v \cdot \cos\gamma_V \cdot Y_T^{vh} + \sin\gamma_V \cdot Z_T^{vh} \\ Z_T^V &= \sin\vartheta_v \cdot \sin\gamma_V \cdot X_T^{vh} - \cos\vartheta_v \cdot \sin\gamma_V \cdot Y_T^{vh} + \cos\gamma_V \cdot Z_T^{vh} \end{aligned}$$

(3) Convert the information from rectangular coordinates to polar coordinates, specify the angles according to the definition of the polar in launcher truck system simultaneous.

$$r\_TarHK = \sqrt{(X_T^V)^2 + (Y_T^V)^2 + (Z_T^V)^2}$$

$$ep\_TarHK = \begin{cases} \arctan\left(Y_T^V / \sqrt{(X_T^V)^2 + (Z_T^V)^2}\right) & Y_T^V \geq 0 \\ \arctan\left(Y_T^V / \sqrt{(X_T^V)^2 + (Z_T^V)^2}\right) + 2\pi & Y_T^V < 0 \end{cases}$$

$$bt\_TarHK = \begin{cases} \arctan(Z_T^V / X_T^V) + \pi & X_T^V < 0 \\ \arctan(Z_T^V / X_T^V) & X_T^V > 0\ Z_T^V \geq 0 \\ \arctan(Z_T^V / X_T^V) + 2\pi & X_T^V > 0\ Z_T^V < 0 \\ \pi/2 & X_T^V = 0\ Z_T^V > 0 \\ 3\pi/2 & X_T^V = 0\ Z_T^V < 0 \end{cases}$$

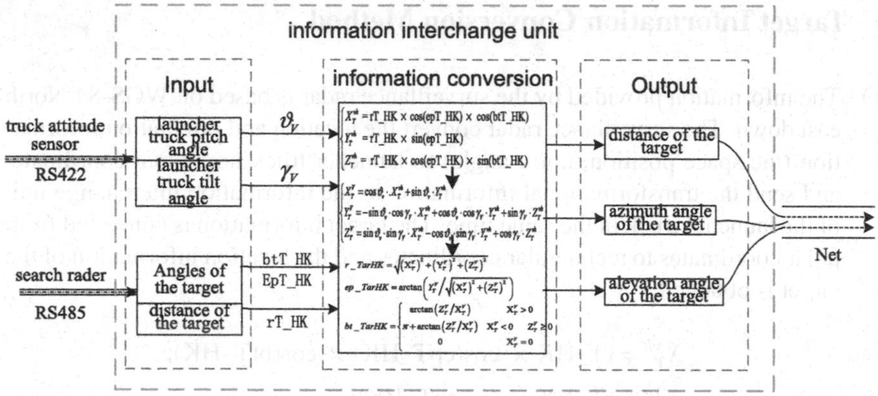


**Fig. 1** Conversion process of the target information

**Table 3** Test plans

| Number | Input | | | | | Output | | |
|---|---|---|---|---|---|---|---|---|
| | $\vartheta_V$ | $\gamma_V$ | rT_HK | btT_HK | epT_HK | r_TarHK | bt_TarHK | ep_TarHK |
| 1 | −0.1 | 0.5 | 30,000 | 5 | 5 | Correctness is based on back calculation results | | |
| 2 | −0.08 | 0.12 | 20,000 | 105 | 21 | | | |
| 3 | −0.1 | 0.08 | 10,000 | 195 | 49 | | | |
| 4 | −0.1 | 0.15 | 8000 | 325 | 66 | | | |
| 5 | 0.02 | −0.05 | 6000 | 5 | 355 | | | |
| 6 | 0.1 | −0.12 | 5000 | 105 | 339 | | | |
| 7 | 0.05 | −0.15 | 4000 | 195 | 311 | | | |
| 8 | 0.15 | −0.01 | 3000 | 325 | 291 | | | |
| 9 | 0 | 0 | 2000 | 90 | 45 | | | |
| 10 | 0 | 0 | 1000 | 270 | 325 | | | |
| 11 | S1 | S2 | S3 | S4 | S5 | | | |

Annotation
S1–S5 Definition:
S1: Amplitude is ±6°, frequency is 1 Hz sine curve;
S2: Amplitude is ±6°, frequency is 1 Hz sine curve;
S3: Line, initial is 5000 m, slop is −5 m/s (when S3 <= 0, keep 0);
S4: Amplitude is 0–360°, frequency is 1 Hz sine curve;
S5: Amplitude is 0–360°, frequency is 1 Hz sine curve.

# 5 Design Assurance

According to the process of information conversion (Fig. 1), the space relationship between there surveillance radar and missile launcher truck' attitude angles. we designed 11 test plans (Table 3) to test the information conversion scheme thoroughly. For each input, there will be an output to ensure the effective implementation of the design project.

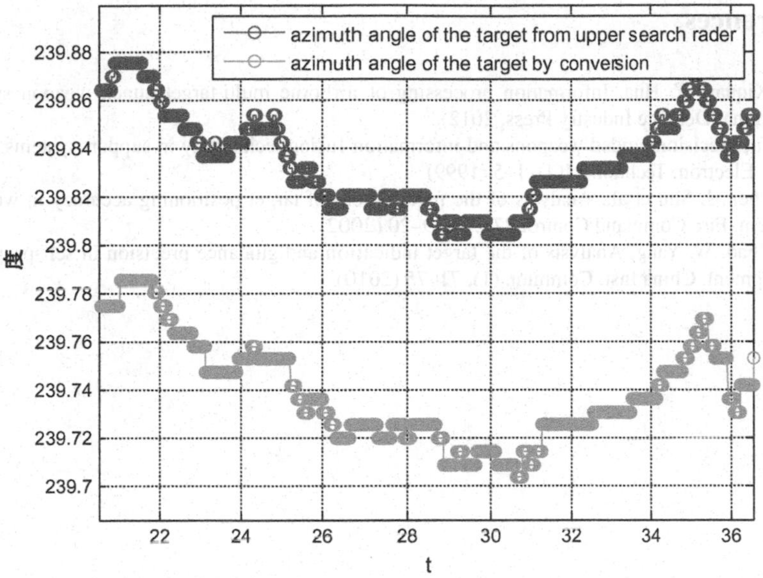


**Fig. 2** Target information conversion

## 6 Verification and Analysis of the Experimental Results

In a flight experiment, target information conversion from the surveillance radar to launcher truck is precise achieved. Track radar completed the target acquisition successfully, and the result was showed in Fig. 2.

The test results show that this method can improve the probability of radar interception by 67% and ensure the reliable implementation of target tracking.

## 7 Conclusion

This method is commonly used in cooperative/independent combat in antiaircraft weapon system successfully now. The target indication of weapon system operation and the reliable implementation of target tracking are guaranteed. This method has obtained good application effect in the development of some type of anti-aircraft missile projects and has a broad prospect of military application.

## References

1. X. Xiquan, Y. Hua, Information processing of airborne multi-target attack weapon system (National Defence Industry Press, 2012)
2. L. Yu, Precision guided weapons and information fusion technology in support systems. Syst. Eng. Electron. Technol. **21**(4), 1–5 (1999)
3. Y. Chen, J. Shu et al., Analysis of the requirement of target positioning accuracy in weapon system. Fire Command Control **27**(z1), 19–20 (2002)
4. K. Zhao, W. Yang, Analysis of the target indication and guidance precision in self-propelled equipment. China Inst. Commun. (1), 71–75 (2010)

# Prediction of Remaining Life of Rolling Bearing Based on Optimized EEMD

**Tong Wu, Caixia Gao and Ziyi Fu**

**Abstract** Aiming at the problem that the early vibration signal has a weak decay characteristic in the prediction of the remaining life of the rolling bearing, a method for optimizing the bearing residual life prediction based on the optimized ensemble empirical mode decomposition (EEMD) is proposed. First, the eigenmode decomposition of the vibration signal is performed. The effect depends on two important parameters: the average number of times and the size of the added noise. Therefore, white noise criteria are added to the set of empirical mode decomposition. Then, the decomposed intrinsic mode function (IMF) is filtered with the gray correlation degree of the envelope spectrum to filter out IMF components with decay characteristics and reconstruct signals. Finally, Multi-feature parameter vector of the reconstructed signal, its redundancy is removed by principal component analysis (PCA), and then input neural network to predict bearing residual life. Experiments show that the proposed method has higher prediction accuracy and stability.

**Keywords** Life prediction · Ensemble empirical mode decomposition · Nuclear principal component analysis · Rolling bearing

## 1 Introduction

In rotating machinery, rolling bearings are the most widely used and important components, and their health may directly affect the operation of the entire system. Of the various failures in rotating machinery, 30% are caused by rolling bearing failures. In 2002, the aircraft engine of a Chinese-Korea passenger aircraft, because of a bearing failure, caused the decision-making of the control system to be mistaken and the

---

T. Wu · C. Gao (✉) · Z. Fu
School of Electrical Engineering and Automation, Henan Polytechnic University, Jiaozuo 454000, China
e-mail: gcx81@126.com

T. Wu
e-mail: gegewutong12345@sina.com



Y. Jia et al. (eds.), *Proceedings of 2018 Chinese Intelligent Systems Conference*, Lecture Notes in Electrical Engineering 528,
https://doi.org/10.1007/978-981-13-2288-4_23

plane crashed. Of course, there was a reason for the error of the control system [1, 2], but the cause was the bearing damage; therefore, the status monitoring of the bearings and the prediction of the remaining life became extremely important. It could find the bearing damage as early as possible. Active maintenance or replacement to minimize production loss and casualties [3, 4].

The EEMD decomposition algorithm is to overcome the problem of mode aliasing caused by pulse interference and other phenomena in the EMD decomposition process. The algorithm adds white noise uniformly distributed throughout the signal, performs EMD calculations multiple times, and utilizes the statistical property of uniform distribution of Gaussian white noise frequencies to eliminate discontinuities in the signal and effectively suppress modal aliasing problems. However, the effect depends on the two important parameters of the size of the white noise added and the overall average number of times. However, in the existing EEMD method, these two parameters need to be artificially set in accordance with experience, lacking reliability for different signals, and at the same time destroying the adaptability of the EEMD method. Therefore, the study of the criteria for adding white noise in the EEMD method is of great significance for the analysis of nonlinear and nonstationary signals [5].

After decomposing the signal, since the characteristics of the rolling bearing recession or fault features are mainly concentrated in the high-frequency part, the adaptive decomposition algorithm can be used to select the first few high-frequency components in the prediction and discard the remaining components. Liu et al. [6] used adaptive decomposition of bearing vibration signals to select the first few high-frequency components to obtain eigenvalues; however, direct selection of the first few components may cause partial loss characteristics to be lost early in the bearing. In this regard, Lei [7] used correlation analysis to obtain the components of the decomposed component and the original signal's sensitive factor, and successfully screened the component sensitive to the fault feature. However, the correlation analysis is to analyze the linear relationship between the original signal and the component. It is not enough for the selection of the decaying feature component, and when the original signal contains large noise, the non-sensitive component may be erroneously selected. Zhang et al. [8] used grey correlation and mutual information to filter fault-sensitive components, which is as broad as the former, but insufficiently targeted.

In order to solve the difficulty of selecting the IMF components caused by the weak decay characteristics of early vibration signals, an IMF component selection method that is sensitive to the decline characteristics is needed. In this paper, the combination of envelope spectrum and grey correlation method is used to screen and reconstruct IMF components. The screening method is divided into two steps. The first step is to calculate the envelope spectrum of the original signal and the decomposition component. In the second step, the gray correlation between their envelope spectra is calculated, and the IMF components are selected for reorganization using the magnitude of their envelope correlation. This method first uses the characteristics of the envelope spectrum sensitive to shocks, captures the weak impact generated by the performance degradation of the early rolling bearing vibration signal, and displays it in a simple and clear manner in the envelope spectrum. Then using the grey relational degree to reflect the characteristics of the two groups of variables' changing situation, the change trend between the original signal and the decomposition component is analyzed. Finally, reorganization of the fading feature-sensitive IMF components is performed in one step, which not only reduces the dimensions of the subsequent feature vectors, but also reduces the noise impact caused by unwanted IMF components. After obtaining the reconstructed signal, taking into account the selection of a single feature parameter for calculation is not sufficient to comprehensively characterize the declining characteristics of the bearing over the entire period. Therefore, this paper selects several features such as kurtosis and rms from the time and frequency domain for calculation. Taking into account the correlation between multiple features, this paper uses PCA to de-redundancy, and finally input into Probabilistic Neural Network (PNN) to predict the bearing residual life.

## 2 Guidelines for Adding White Noise to the EEMD Method

In the EEMD algorithm, the amplitude of the white noise added to the original signal is too large and a spurious modal component is generated during the decomposition process. If the amplitude of the added noise is too small, it may not be sufficient to cause the local extreme point of the original signal to change, and the modal confusion problem cannot be solved [9]. For any discontinuous signal, satisfying the two conditions that need to be satisfied for adding white noise, the key is how to effectively determine the two parameters of adding white noise in the EEMD decomposition method, and add the white noise criterion.

Since the RMS value reflects the magnitude of the vibration energy, it is not suitable for the slow wear damage caused by the bearing over time. Therefore, the rms standard deviation method is proposed in this paper:

$$0 < \alpha < \beta/2 \tag{1}$$

In the above formula: $\alpha = \sqrt{RMS_n}/\sqrt{RMS_o}$, $\beta = \sqrt{RMS_h}/\sqrt{RMS_o}$.

Among them, $RMS_n$ is the root mean square deviation of white noise, $RMS_o$ is the root mean square deviation of the original signal, $RMS_h$ is the root mean square deviation of the high frequency component of the signal, and $\alpha$ and $\beta$ are the ratio coefficients.

From (1) available:

$$0 < RMS_n < RMS_h/4 \tag{2}$$

Generally taking A can effectively avoid modal aliasing.

In the EEMD method, Qiu et al. [10] studied the relationship between the ratio coefficient of adding white noise and the choice of the overall average number of times M:

$$e = \alpha/\sqrt{M} \tag{3}$$

In the above equation, $e$ is an error, so it can be seen that if $\alpha$ is smaller, the error $e$ is smaller, which contributes to the improvement of the resolution accuracy. However, when the value of $\alpha$ is too small, it may not be sufficient to cause changes in the local extreme points of the signal, and it is not possible to change the local time span of the signal, failing to exert the advantages of EEMD. If M is larger, $e$ will also decrease, but it also increases the computational burden. Therefore, it is not ideal to artificially determine the value of the overall average number of times M. Using the proposed white noise criterion, $\alpha$ is calculated first from the original signal, then the expected error $e$ is set (generally 1%), and the value of the overall average M can be obtained by using formula (3).

## 3 IMF Selection and Weighted Reconstruction Based on Gray Correlation of Envelope Spectrum

During the operation of the rolling bearing, whether the inner ring, the outer ring or the rolling element fails, the vibration generated by the rolling bearing is often accompanied by the impact component. Before the failure has occurred, the weak component of the bearing or the early performance degradation period, the impact component is often more subtle. Therefore, the characteristics of the envelope are particularly sensitive to the impact of the envelope to capture the original signal and the IMF component of the subtle impact component, and then use the gray correlation to compare the original signal envelope spectrum and IMF component envelope spectrum changes. Compared with the use of correlation analysis to compare the linear relationship between the two variables is more macro, to avoid the interference of chance factors on the selection of sensitive IMF components.

Methods as below:

(1) Decompose the original signal with EEMD to obtain m IMF components $IMF_k = \{IMF_k(j)|k = 1, 2, 3, \ldots, m; j = 1, 2, 3, \ldots, n\}$.
(2) Calculate the envelope spectral data of the original signal $X'_o = \{x'_o(j)|j = 1, 2, 3, \ldots, n\}$ and the IMF component envelope data $IMF'_k = \{imf'_k(j)|k = 1, 2, 3, \ldots, m; j = 1, 2, 3, \ldots, n\}$.
(3) Calculate the gray correlation between the envelope spectrum of A and the original signal envelope B. The formula for calculating the gray correlation is:

$$\frac{\min_k \min_j \left|x'_o(j) - imf'_k(j)\right| - \lambda \max_k \max_j \left|x'_o(j) - imf'_k(j)\right|}{\left|x'_o(j) - imf'_k(j)\right| - \lambda \max_k \max_j \left|x'_o(j) - imf'_k(j)\right|} \tag{4}$$

Among them, λ is the resolution coefficient. In order to satisfy the maximum information resolution, 0.6 is calculated in this paper, and then the gray correlation degree is obtained by averaging the correlation coefficients.

(4) The correlation degree sequence $\{y_m\}$ is obtained by arranging the gray correlation degrees from the largest to the smallest, and the first s corresponding IMF components are taken so that the sum of the correlation degrees is greater than 0.9.

(5) Sum the grey correlation of the first s IMF components:

$$R_z = y_1 + y_2 + \cdots + y_s \tag{5}$$

The weight coefficient is:

$$P_x(x = 1, 2, \ldots, s) = \frac{y_x}{R_z} \tag{6}$$

Weighting the IMF component reconstruct signal with weight coefficients:

$$X_c = P_1 \cdot IMF_1 + P_2 \cdot IMF_2 + \cdots + P_s \cdot IMF_s \tag{7}$$

## 4 Prediction of Remaining Life of Rolling Bearing Based on Optimized EEMD

The overall bearing life data used in this paper is from the Intelligent Maintenance System Center of the University of Cincinnati, USA [10]. The bearing test bed is equipped with four experimental bearings on a drive shaft. The data in this paper is the first channel data in the second test. The bearing model is Rexnord ZA-2115 double row cylindrical roller bearing. The bearing speed is, and the data sampling frequency is, the experiment is to collect the signal once every 10 min, and the time span is 164 h until the failure of the bearing outer ring fails. In this experiment, 6 sets of data were taken every hour for 60 h. A total of 360 sets of data were used for experiments. The remaining life prediction method for rolling bearings is shown in Fig. 1.

The forecasting steps are as follows:

(1) Extract 5 randomly selected groups of training samples per hour for a total of 300 groups of samples. One group of each remaining hour is a prediction sample, and a total of 60 groups of prediction samples.
(2) EEMD decomposition is performed on the prediction sample and the training sample, and the sensitive IMF component is selected by the gray correlation degree of the envelope spectrum for weighted reconstruction.
(3) Perform PCA analysis on the reconstructed signal to obtain the first several principal elements, and enter PNN for prediction.
(4) Make a comparison chart between actual remaining life and predicted remaining life.

The number of principal elements is selected based on the cumulative contribution rate, and some of the principal element contributions are shown in Table 1.

As can be seen from the above table, when the first two principal elements were selected, the cumulative contribution rate reached 86.3%, which is equivalent to the first two principal elements covering most of the information of the pre-PCA feature set. Therefore, we choose the first two principal elements to input PNN for prediction. The prediction results are as follows:

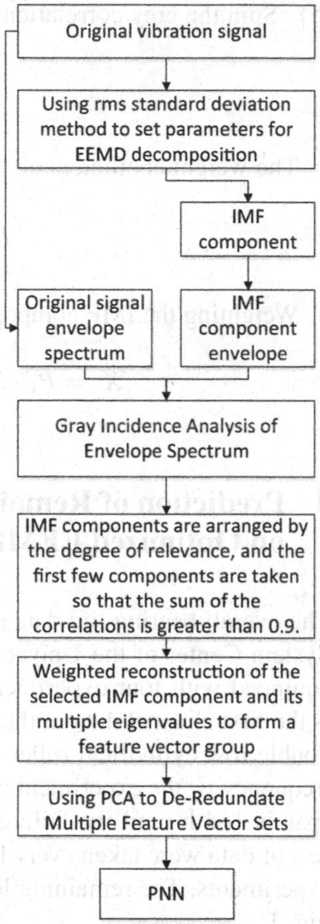


**Fig. 1** Rolling bearing residual life prediction method

From the Fig. 2, it can be seen that the predicted remaining life will fluctuate from the actual remaining life, but it will not fluctuate. The overall trend of the predicted remaining life is consistent with the actual remaining life. Then, in the rest of the steps unchanged, this paper only uses the correlation analysis method to filter sensitive IMF components, and takes the IMF component with correlation coefficient greater than

**Table 1** Partial principal contribution table

| Main ingredient number | Contribution rate (%) | Cumulative contribution rate (%) |
|---|---|---|
| 1 | 56.5 | 56.5 |
| 2 | 29.8 | 86.3 |
| 3 | 3.2 | 89.5 |
| 4 | 0.2 | 89.7 |
| 5 | 0.1 | 89.8 |

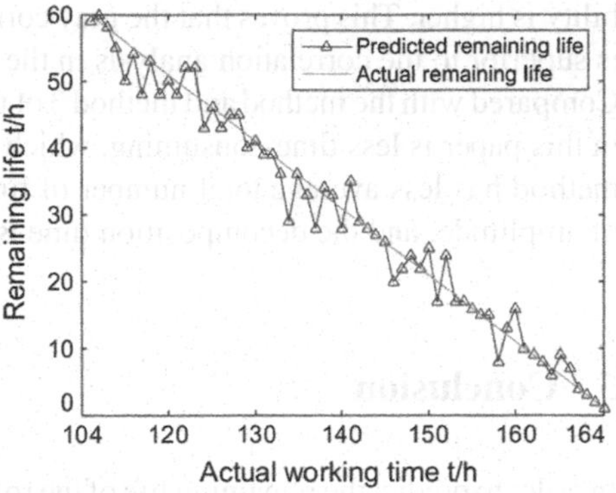


**Fig. 2** Comparison of remaining life prediction and actual life

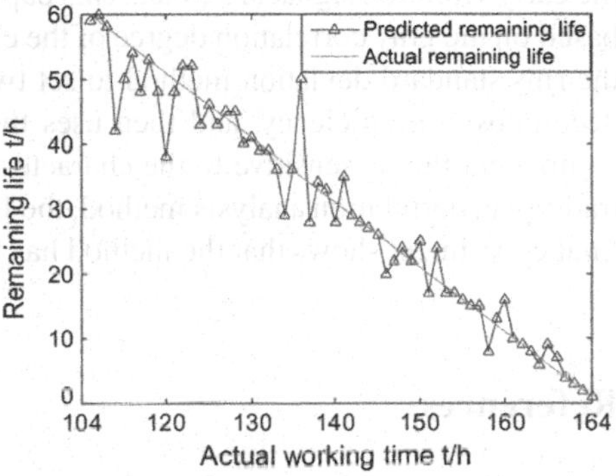


**Fig. 3** Comparison of remaining life and actual life of method two

**Table 2** Comparison of the errors in the remaining life prediction of the method and method II of this paper

| Method | Standard error | Time/s |
|---|---|---|
| Method 2 | 5.53 | 55.6 |
| Method 3 | 3.01 | 78.3 |
| Method of this article | 2.78 | 52.3 |

0.3 to reconstruct, and seeks multiple features for PCA analysis and input into PNN. This method is named Method 2, and the forecast results obtained by Method 2 are shown below.

As can be seen from the Fig. 3, in the first half of the figure, that is, the prediction of the remaining life of the bearing, there is an obvious phenomenon of excessive error. This is due to the weak decay characteristics of the early vibration signal and the incorrect selection of non-sensitive IMF components. Then, under the condition that the rest of the method of the method is not changed, the parameter setting is performed using the traditional method of standard deviation of the amplitude, which is named method 3. The performance of the three methods is shown in Table 2.

From Table 2, it can be seen that the standard deviation of this method is smaller than that of method 2, which means that the predicted volatility is smaller and the stability is higher. This proves that the gray correlation degree of the envelope spectrum is superior to the correlation analysis in the selection of sensitive IMF components. Compared with the method and method 3 of this paper, we found that the method used in this paper is less time-consuming, which means that the RMS standard deviation method has less average total number of times than the standard deviation method of amplitude, and the decomposition time is shortened.

## 5 Conclusion

In order to predict the remaining life of the rolling bearing, the decay characteristics of the early vibration signal are weak. This paper proposes an improved EEMD method based on the gray correlation degree of the envelope spectrum. The method first uses the rms standard deviation method to set two parameters of EEMD to improve the decomposition efficiency, and then uses the envelope spectrum to select the IMF component that is sensitive to the characteristics of the impact. Compared with the traditional correlation analysis method, the IMF component is more targeted, and the final experiment shows that the method has higher prediction accuracy and stability.

## References


1. Y. Jia, Robust control with decoupling performance for steering and traction of 4WS vehicles under velocity-varying motion. IEEE Trans. Control Syst. Technol. **8**(3), 554–569 (2000)
2. Y. Jia, Alternative proofs for improved LMI representations for the analysis and the design of continuous-time systems with polytopic type uncertainty: a predictive approach. IEEE Trans. Autom. Control **48**(8), 1413–1416 (2003)
3. F. Wang, X. Chen, C. Liu et al., Reliability assessment and life prediction of rolling bearings based on KPCA and WPHM. Vib. Test. Diagn. **37**(3), 476–483 (2017)
4. Z. Shen, X. Chen, Z. He et al., Residual life prediction of rolling bearing based on relative features and multivariable support vector machines. Chin. J. Mech. Eng. **49**(2), 183–189 (2013)
5. C. Lue, G. Tang, S. Yanyang et al., Application of adaptive EEMD method in ECG signal processing. Data Acquis. Process. **26**(13), 361–368 (2011)
6. J. Liu, J. Cheng, Y. Liu, Life prediction of rolling bearing based on LCD and GMM. Mod. Manuf. Eng. **7**(7), 120–124 (2016)
7. Y. Lei, Mechanical fault diagnosis based on improved Hilbert-Huang transform. Chin. J. Mech. Eng. **47**(5), 71–77 (2011)
8. Z. Zhang, X. Shi, Q. Shi et al., Fault feature extraction of rolling bearing based on improved EMD and spectral kurtosis. J. Vib. Meas. Diagn. **33**(3), 478–482 (2013)
9. Wang Y, Jiang Yicheng, S. Kang et al., Roller bearing fault location and performance degradation degree diagnosis method based on optimized set EMD. J. Instrum. Instrum. **1**(7), 1834–1840 (2013)
10. H. Qiu, J. Lee, J.Lin, Wavelet filter-based weak signature detection method and its application on roller bearing prognostics. J. Sound Vib. 1066–1090 (2006)

# Research on Load Balancing Method of Object Storage System Based on Data Heat Prediction and Migration

**Hao Li, Yi Chai, Ke Zhang and Qiulin Dan**

**Abstract** Since most existing object storage systems balance the load by scheduling task requests, there is no concern about the uneven load generated by random access to the data, resulting in a large number of requests to be concentrated on a small number of servers. Based on Weighted Least-Connection (WLC) and Consistent Hashing, a load balancing algorithm for Heat Prediction and Migration (HPM) is proposed. By refining thc number of connections to the object layer and comprehensively considers the number of connections, the difference of heterogeneous nodes and objects as the object heat, and predicts the heat to guide the scheduling of objects. Finally, the effectiveness of the algorithm is verified by simulation.

**Keywords** Object storage · Load balancing · Heat prediction and migration

## 1 Introduction

With the vigorous development of cloud computing, the demand for data storage is also increasing, and object storage is attracting attention due to its high availability, good scalability. Using data object as basic unit, through the network and service clusters, to provide users with online data storage and access capabilities. However, as the cluster size increases, the storage nodes will experience unbalanced load and result in a decrease in overall performance. Therefore, balancing the load of the storage cluster can improve resource utilization and system stability [1–3].

The consistent hash algorithm divides storage physical nodes into multiple virtual nodes according to their performance, and evenly maps them to the ring. By calculating the hash value of the object, and mapping it to the ring to find the corresponding virtual node, then the physical node where the object is located can be confirmed. Consistent hashing algorithm can guarantee to change the mapping rela-

H. Li (✉) · Y. Chai · K. Zhang · Q. Dan
State Key Laboratory of Power Transmission Equipment and System Security and New Technology, College of Automation, Chongqing University, Chongqing 400044, China
e-mail: haoli@cqu.edu.cn



Y. Jia et al. (eds.), *Proceedings of 2018 Chinese Intelligent Systems Conference*,
Lecture Notes in Electrical Engineering 528,
https://doi.org/10.1007/978-981-13-2288-4_24

tionship between the object and the node as little as possible when changing the cluster size, and a good hash algorithm can make the object distributed roughly in each virtual node to balance the load, but the Consistent hashing algorithm does not take into account the load tilt problem caused by the centralized access of users to the data, and may cause system failure due to a large number of centralized access [4]. The traditional load balancing algorithm based on request redistribution is simple and fast, such as Weighted Least-Connection algorithm. By recording the current request connection number of each server, the request is assigned to the server with smallest ratio of the connection number to the server performance weight. To a certain extent, the impact of centralized access to hot data by users on the load is considered. However, due to the static storage characteristics of objects, there are still a large number of requests for different objects concentrated on a small number of nodes. Therefore, the intelligent optimization algorithm to optimize the heat data scheduling to balance the node load is a hot topic in recent years [5].

Gongye et al. [6] proposes that load balance is carried out through load prediction and migration, and the temperature ratio is used as the load of the object storage device. By pairing the high temperature node with the low temperature node, the heat is transmitted with the data migration, but it does not indicate the data to be migrated, and the heat may be still high after migration, and need to be migrated again which is a waste of resources. Yan et al. [3] proposes to balance the load by duplicating the object so that the access request is shared by multiple replica nodes. However, a large number of replicas can also cause a waste of storage resources. Zhao [7] and Wang et al. [8] proposes to group different categories of virtual nodes, and through splitting and merging between nodes to balance the load. However, splitting and merging need to migrate large amounts of data to increase the burden of the system.

This paper proposes a load balancing algorithm for object heat prediction and migration on the basis of Consistent hashing and Weighted Least-Connection algorithm. The record of the request connection number is refined to the object in the node. The heat value of the object is calculated according to the number of object requests, the difference of the object attributes and the node attributes, and predict the heat of the next serval cycles. The load of the node is characterized by the sum of all predicted heat values of the stored objects, and the migration of the objects in the high heat nodes is performed to balance the loads among the nodes. The prediction of heat can avoid the migration failure due to the change of heat. The global optimization scheduling of the heat object is a good solution to the limitations of the static storage of the traditional load balancing algorithm based on the request redistribution.

The rest of this paper is organized as follows. Section 2 gives the method of heat analysis and prediction. Section 3 abstracts the migration scheduling of objects to a linear programming problem. Then, Simulation results are presented to illustrate the effectiveness of the proposed method in Sect. 4. Finally, followed by some concluding remarks in Sect. 5.

## 2 Heat Analysis

By calculating the sum of the heat of each object as the load of the node, then how to select the appropriate parameters to calculate the heat of each object relates to the determination of the node load. For an object storage system, each storage node has independent computing and processing capabilities [1]. The main parameters response to the load include: CPU utilization, memory occupancy rate, request response time, connection number, network usage rate, etc. [9–11]. The acquisition of CPU utilization rate, memory occupancy rate and network usage rate needs nodes to collect and transmit to the load balancing server regularly. However, because of the load of the network or the node itself, the load balancing server cannot obtain information in a timely manner, which leads to the problem of negative load balancing control of the nodes. Therefore, this paper uses the number of requests that can be counted by the proxy server, records the number of request access of each object, takes the difference of the performance of nodes in the heterogeneous cluster and the different access requests into account, then combines the attribute of the object, the attribute of the node and the number of connections, finally calculates the heat value of the object $H_{ij}(t)$ and the load $L_i(t)$ of the node $i$

$$H_{ij}(t) = \frac{C_{ij}(t) * M_{ij}}{\sigma_i} \tag{1}$$

$$L_i(t) = \sum_{j=1}^{n} H_{ij} \tag{2}$$

$C_{ij}, M_{ij}$ respectively represent the connection number and size of the object $j$ in the node $i$ at time $t$. $\sigma_i$ indicates the performance of node $i$, in order to avoid the addition and deletion nodes to change the $\sigma_i$, resulting in large-scale updates of object heat $H_{ij}(t)$, the $\sigma_i$ is calculated by the ratio of the three performance parameters of the CPU, Memory, and Network bandwidth of the node to the standard value and the corresponding weights.

$$\sigma_i = \frac{1}{2}\left(\alpha \frac{Cpu_i}{Cpu_s} + \beta \frac{Memory_i}{Memory_s} + \varphi \frac{Band_i}{Band_s}\right) \tag{3}$$

where $\alpha$, $\beta$, $\varphi$ represent the weight coefficients of the CPU, Memory, and Network bandwidth, and the sum is 1, that is, when $\sigma_i = 0.5$, the node $i$ has the same performance as the standard node. $Cpu_i, Cpu_s, Memory_i, Memory_s, Band_i, Band_s$ represents 3 performance indicators of node $i$ and 3 performance indicators of standard respectively.

However, load balancing of nodes only through real-time object heat may cause inefficient migration due to changes in access and load, which will increase the system burden. Therefore, the time series prediction model is established to predict the object heat values of the next $k$ periods, and the mean of object heat and the node load are calculated to guide the specific object of the migration. In the prediction algorithm of time series, Recurrent Neural Network (RNN) bases on the traditional neuron structure, the input of the hidden layer includes not only the output of the layer above it but also the output of itself at the previous moment, which makes the state of the last time affect the present calculation, but it has the problem of gradient disappearance. As a special type of RNN, Long Short-Term Memory (LSTM) adds valve on the basis of RNN to make information selectively participate in the calculation of this moment, so that the problem of gradient disappearance of RNN is well overcome [12–14], and can be well applied to time series prediction. By predicting the heat of the next $k$ periods of each object, the predicted heat mean $\overline{H}_{ij}(t, t+kT)$ and the node load mean $\overline{L}_i(t, t+kT)$ are calculated.

## 3 Load Scheduling

In the previous section, the storage object's heat and node load are analyzed, and the LSTM is used to predict the object's heat and get the load of the node. Through the migration scheduling of hot data, the problem of unbalanced load between high load nodes and idle nodes is solved.

Assuming that there are $n$ nodes in the object storage cluster $N = \{N_1, N_2, \ldots, N_n\}$, the corresponding load is $\overline{L}(t, t+kT) = \{\overline{L}_1(t, t+kT), \overline{L}_2(t, t+kT), \ldots, \overline{L}_n(t, t+kT)\}$. When the cluster load is relatively balanced, in order to reduce the extra burden caused by unnecessary migration, this paper adds a threshold $\Delta$ to the cluster load mean $\overline{\overline{L}}(t, t+kT)$ to form an equilibrium interval $[\overline{\overline{L}}(t, t+kT) - \Delta, \overline{\overline{L}}(t, t+kT) + \Delta]$, When the node load in cluster falls within the equilibrium range, it is considered that the cluster load is more balanced and load balancing is not required.

$$\overline{\overline{L}}(t, t+kT) = \frac{1}{n}\sum_{i=1}^{n}\overline{L}_i(t, t+kT) \tag{4}$$

When the cluster nodes are not all in the equalization interval, the heat objects need to be migrated to balance the load. Because of the consistent hash characteristics of the object storage, the number of data objects stored in each node is roughly the same. Assume that there are $m$ objects in node $i$, the corresponding heat value construct into a diagonal matrix $A_i$

$$A_i = diag(H_{ij}(t, t+kT)), j = 1, 2, \ldots, m \tag{5}$$

In order to maintain the consistency of the consistent hash, the heat in the high load node is transferred to the low load node through the exchange of elements between the matrices, then the migrated matrix $\dot{A}_i$

$$\dot{A}_i = P_{i1}A_1 + P_{i2}A_2 + \ldots + P_{ii}A_i + \ldots + P_{in}A_n \tag{6}$$

$P_{ij}(j = 1, 2, \ldots, i-1, i+1, \ldots, n)$ is the migration coefficient matrix of node $i$, which represents the migration of the node $j$ to the node $i$. $P_{ii}$ is the migration coefficient matrix of node $i$, indicates that the objects of node $i$ migrate to other nodes.

In order to make the node load within the equilibrium interval after migration, the following objective functions are included.

$$f(\dot{L}_i(t, t+kT)) - \begin{cases} \dot{L}_i(t, t+kT) \leq \overline{\overline{L}}(t, t+kT) + \Delta \\ \dot{L}_i(t, t+kT) \geq \overline{\overline{L}}(t, t+kT) - \Delta \end{cases}, i = (1, 2, \ldots, n) \tag{7}$$

$\dot{L}_i(t, t + kT)$ is the load of the node $i$ after the migration and can be calculated from the sum of the data heat after the migration. Note that due to the change of the weight of the node after the migration, the heat of the migrated data will also change accordingly.

In order to maintain the amount of node data, the amount of data to be migrated and moved in should be the same, that is, the rank of the outgoing coefficient matrix is equal to the sum of the ranks of all the immigration coefficient matrices, the constraint conditions are

$$rank(P_{ii}) = \sum_{\substack{j=1 \\ j \neq i}}^{n} rank(P_{ij}), i = (1, 2, \ldots, n) \tag{8}$$

In order to avoid large-scale migration and bring burden to the cluster, the migration volume should be as small as possible, that is, the sum of the ranks of all the node's migration coefficient matrix should be as small as possible, that is, the optimal solution minimizes $g(P_{ii})$

$$\min g(P_{ii}) = \sum_{i=1}^{n} rank(P_{ii}) \tag{9}$$

## 4 Analysis of Experimental Results

Through Matlab simulation, the effectiveness of HPM load balancing algorithm is verified, and the effect of WLC and HPM on the balance of system load is compared. Assume that the number of nodes in the system is $n = 50$, which contains four performance indicators $\sigma = \{0.3, 0.5, 0.6, 0.85\}$, the number of objects stored by each node is 100, the object size obeys an exponential distribution of 4, and the access request of the object obeys the Poisson distribution, the arrival rate is $\lambda = 100$. Predict the heat value of the last 15 periods of a certain object. The predicted and actual values are shown in Fig. 1.

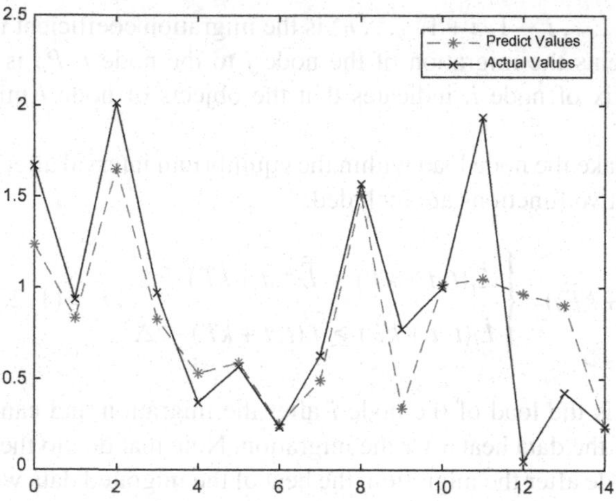


**Fig. 1** Object heat prediction value and actual value

**Fig. 2** Comparison of load balancing effect between WLC and HPM

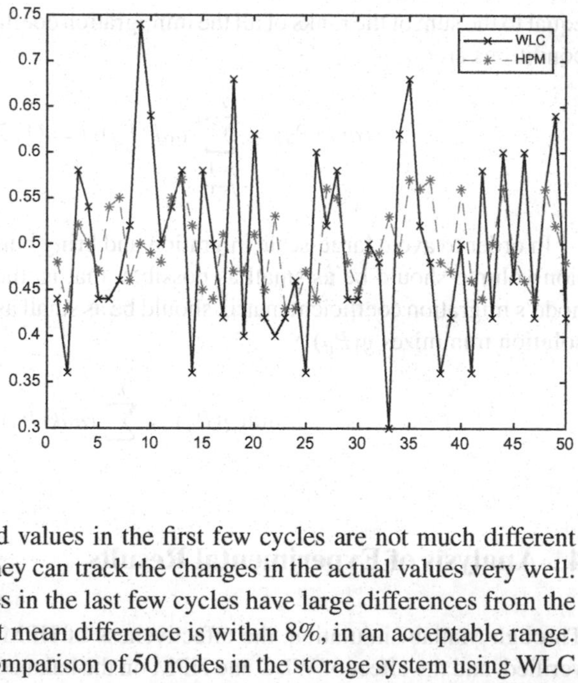


From Fig. 1, the predicted values in the first few cycles are not much different from the actual values, and they can track the changes in the actual values very well. Although the predicted values in the last few cycles have large differences from the actual values, the overall heat mean difference is within 8%, in an acceptable range.

Figure 2 shows the load comparison of 50 nodes in the storage system using WLC and HPM respectively for load balancing, most nodes load of the WLC algorithm is more balance, but there are still a small number of high-load nodes and low-load nodes, and the overall load of the HPM algorithm is within the control range, which effectively avoids the single-point overload phenomenon and proves the effectiveness of the method.

## 5 Conclusions

This paper comprehensively considers the number of object request connections, the heterogeneity of heterogeneous nodes and objects, calculates the heat value of the object, and establishes a time series forecasting model to predict the heat value, then migrates the heat object from a global perspective. The balance of the cluster load effectively eliminates the single-point overload and overcomes the limitations of the traditional load balancing algorithm. Predictive heat-based migration scheduling also avoids invalid migration due to expired heat.

## References


1. M. Mesnier, G.R. Ganger, E. Riedel, Object-based storage. Commun. Mag. IEEE **41**(8), 84–90 (2003)
2. M. Mesnier, G. Ganger, E. Riedel, Object-based storage: pushing more functionality into storage. Potent. IEEE **24**(2), 31–34 (2005)
3. L. Yan, D. Feng, L. Zeng et al., Dynamic load balancing algorithm based on object storage System. Comput. Sci. **33**(5), 88–91 (2006)
4. H. Yang, B. Lin, Research on consistent hash algorithm in distributed storage system. Comput. Knowl. Technol. **07**(22), 5295–5296 (2011)
5. Y. Luo, X. Li, R. Sun, A review of load balancing algorithms. Library Informat. Sci. **18**(23), 134–136 (2008)
6. Z. Gongye, W. Lei, J. Chen, Hotspot data balancing strategy based on object storage system. J. Huazhong Univ. Sci. Technol. (Nature Sci.), 2007, 35(12): 28–31
7. D. Zhao, *Research and Implementation of Read Load Balancing in Swift* (Southeast University, 2016), 8
8. Z. Wang, H. Chen, Y. Fu et al., Workload balancing and adaptive resource management for the swift storage system on cloud. Fut. Generat. Comput. Syst. **51**, 120–131 (2015)
9. X. Min, M. Li, J. Zheng et al., Swift load balancing algorithm based on OpenStack. J. Comput. Syst. **1**, 127–131 (2018)
10. R. Lee, B. Jeng, Load-balancing tactics in cloud, in *International Conference on Cyber-Enabled Distributed Computing and Knowledge Discovery* (IEEE, 2011), pp. 447–454
11. Z. Zhang, X. Zhang, A load balancing mechanism based on ant colony and complex network theory in open cloud computing federation, in *The, International Conference on Industrial Mechatronics and Automation* (2010) 240–243
12. A. Graves, *Long short-term memory supervised sequence labelling with recurrent neural networks* (Springer, Berlin Heidelberg, 2012), pp. 1735–1780
13. J. Yingmin, Robust control with decoupling performance for steering and traction of 4WS vehicles under velocity-varying motion. IEEE Trans. Control Syst. Technol. **8**(3), 554–569 (2000)
14. Y. Jia, Alternative proofs for improved LMI representations for the analysis and the design of continuous-time systems with polytopic type uncertainty: a predictive approach, IEEE Trans. Automat. Control **48**(2003), 8, 1413–1416

# Intelligent Sensor Detection Technology in Lighting Design and Application

**Yongsheng Xie**

**Abstract** Sensors are an important part of intelligent lighting. They can sense the sound and brightness of the environment, the movement and presence of the human body, and sense the temperature, humidity and air quality of the environment. After sensor detection, it can further control the movement of lighting products, switch light and shade, color temperature, switch and scene changes, achieve the lighting effects of functional lighting, scene lighting, and environmental protection requirements for energy saving and emission reduction. The application of sensors and light sensors in intelligent lighting control has increased the level of intelligence in lighting control and saved a lot of energy.

**Keywords** Sensor · Lighting · Smart · Smart lighting

## 1 Introduction

With the progress of social civilization, sustained and healthy economic development, and the gradual depletion of non-renewable energy resources, we are now facing serious resource shortages in our country and even the world. Energy has always been the guarantee for us human beings to survive, and the exhaustion of resources has caused us widespread. The importance attached. For a long time, protecting the environment and saving energy has become a common concern for all of us. In view of the contradiction between the ever-increasing energy demand and energy consumption, various fields are seeking effective solutions [1]. Electrical energy is closely related to people's lives. In recent decades, human material civilization has made great progress. At the same time, spiritual civilization has also been developing rapidly. People's demands for living, office, and learning environment quality and comfort have been continuously improved. The indoor environment of a building is

Y. Xie (✉)
College of Mathematics and Computer, Guangxi Science & Technology Normal University, Laibin 546199, Guangxi, People's Republic of China
e-mail: 551907834@qq.com



Y. Jia et al. (eds.), *Proceedings of 2018 Chinese Intelligent Systems Conference*, Lecture Notes in Electrical Engineering 528,
https://doi.org/10.1007/978-981-13-2288-4_25

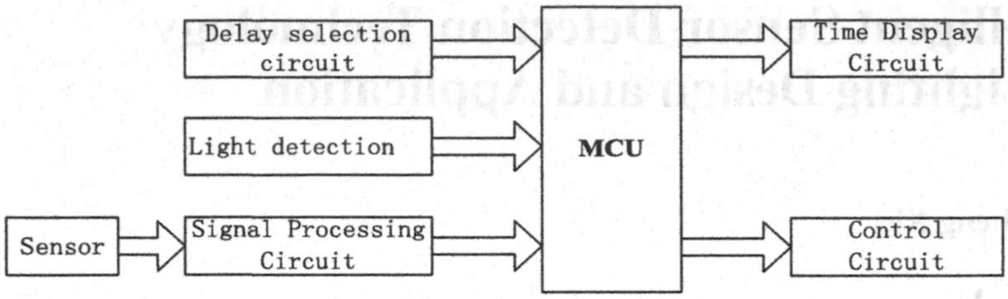


**Fig. 1** Intelligent lighting system hardware structure

the most frequent and intimate environment for people to contact and therefore brings about huge energy consumption. People's lives continue to be extravagant, indoor lighting decoration more and more beautiful, more and more lighting, lighting the building also brought a lot of pressure, if you can save 10% of the lighting power, in the increasingly scarce energy This is a considerable figure. Therefore, the application of smart sensor detection technology in lighting is of great significance in today's society [2].

## 2 Introduction to Lighting Sensors

Today's intelligent wave sweeps across the globe, and the concept and development of the Internet of Things is rapid, and the intelligence of lighting systems is also increasingly valued by people. As an important component of intelligent lighting, the sensor is the input device of the entire system, can perceive the surrounding environmental changes, communicate with users, and automatically control the light changes to achieve optimal lighting and energy-saving effects. Therefore, how to better design and apply sensors is worthy of our consideration and exploration.

The sensing products required for different application fields in lighting are different. In the outdoor road lighting, the street light senses the ambient light intensity to automatically turn on the lights in the night and turn off the lights automatically during the day; the road is measured through the detection of traffic flow. The traffic flow pressure; in the park on the tree-lined road while the lights are automatically adjusted to the brightest, illuminate the walking path, and people walk away to lower the brightness and save energy. Indoors, the lighting fixtures are always kept brightest when people are in office meetings, and the lights are automatically turned off when people leave [3]. At times, the light is moved along with the malls, aisles and other places to ensure that the pedestrians requirements for the ambient lightness are achieved at the same time. Optimal energy-saving effect. Different types of sensor products are combined with each other to suit different occasions, giving them more flexible lighting effects, creating more user-friendly and smarter application scenarios, and achieving optimal intelligent lighting control and energy-saving and emission-reducing effects (Fig. 1).

## 2.1 LED Sensor Lighting Circuit

Fengfeng LED street lamp is a kind of highly intelligent and unattended road lighting lamp. It uses wind power and sunlight to generate electricity and uses batteries to store energy. Therefore, the automatic management of energy is very important. The photo sensor is an ideal electronic sensor that can control the circuit to automatically switch due to changes in the illumination intensity during daylight and daylight (sunrise and sunset).

Photosensitive sensors can automatically control the opening and closing of LED lighting fixtures in shopping malls according to the weather, time period and region. Reduce the power consumption by reducing the output power in bright daylight hours. Compared with the use of fluorescent lamps, a convenience store with a store area of 200 $m^2$ can reduce power consumption by up to 53%. The longevity is also about 5–10 million hours. Under normal circumstances, the life of LED lighting is about 40,000 h [4]; the color of light can also be RGB (red, green, blue) colorful change, so that the mall lighting more colorful, more active atmosphere; and supporting the use of yellow Compared with the original blue LED of the phosphor, the purple LED supporting the use of the three-color phosphors of red, green and blue has higher color rendering [5].

## 2.2 Infrared Sensor Lighting Application

Infrared sensors work by detecting infrared radiation emitted by the human body. The main principle is that the infrared ray emitted by the human body is enhanced by the filter lens and then is concentrated on the pyroelectric element PIR (passive infrared) detector. When people move, the emission position of infrared radiation will change, and the element will change. The charge balance will be lost, and the pyroelectric effect will be released to release the charge. The infrared sensor will convert the infrared radiation energy of the Fresnel filter lens into an electrical signal, i.e. thermoelectric conversion. When there is no body movement in the detection area of the PIR detector, the infrared sensor senses only the background temperature. When the human body enters the detection area, it is sensed by the Fresnel lens [6]. The pyroelectric infrared sensor is The difference between the human body temperature and the background temperature, the signal is collected and compared with the existing probe data in the system to determine whether someone really waits for the infrared source to enter the detection area. Passive infrared sensors have three key components: Fresnel filter lenses, pyroelectric infrared sensors, and matched low-noise amplifiers [7]. The Fresnel lens has two functions: one is focusing, which is to refract the pyro-infrared signal on the PIR; the second is to divide the detection area into several bright and dark areas so that the moving object/person entering the detection area can Changes in the temperature of the pyroelectric infrared signal generated on the PIR. Low-noise amplifiers will generally be matched. When

Fig. 2 Infrared lighting sensor outline

the ambient temperature on the detector rises, especially near normal human body temperature, the sensitivity of the sensor decreases, gain is compensated through it, and its sensitivity is increased. The output signal can be used to drive an electronic switch to achieve switching control of the LED lighting circuit [8] (Fig. 2).

### 2.3 The Principle of Intelligent Lighting System

Simply take the following figure as an example to illustrate the special status of sensing in intelligent lighting systems. The device is composed of various types of signals (optical signals and human body signals), acquisition devices, control devices, interface circuits, etc., which is an intelligent lighting device system. Under the lighting conditions of the building corridor, the intelligent lighting system will use appropriate sensors to detect the presence of the human body, so as to match the intensity of the appropriate light in the circuit detection environment [9]. These human presence signals and ambient light intensity signals are identified and intelligently judged and the information is packaged in a single-chip microcomputer. The micro controller then performs switching operations on the lighting device through the control circuit based on these information. If no human body is detected, the light is automatically adjusted to be off. Instead, the system will automatically turn on the lighting device. This will not only achieve lighting control, but also achieve the purpose of energy saving (Fig. 3).

### 2.4 The Application of Sensors

The independent mode of operation is the most basic application, using a sensor to control one or more lamps. Simple to realize, it can be integrated in the lighting products and can also be independent externally. There are a large number of lamps with induction, bulb lamps, ceiling lamps and other products are widely used. The application scenarios are mostly independent rooms or corridors, corridors and other small areas where the flow of people is not large. When there are no people, the brightness of the lights can be reduced to save energy, and when people pass by, the brightness of the lights can be increased to provide better illumination effect. The illuminance sensor can also be used as an independent control work, such as in small offices and conference rooms, using ambient light sensors to measure the ambient brightness, and by controlling the lightness and darkness of the illuminance to control the light and darkness of the light to achieve a constant illuminance of the entire environment, as well as saving the overall power consumption. Another independent mode of operation is voice control, which is used to control the lighting switch effect by detecting whether the walking sound perception is passing through, and is suitable for use in corridors and corridors. In addition to using a separate sensor module, it is also possible to achieve more intelligent control through a combination of multiple sensor modules, such as the combination of light sensing and motion sensing to turn off the lights during the day and light people at night. Turn off the light function; can also use microwave and sensor induction to achieve the function of approximate presence of induction, even if office lights stay in office for a long time will remain lit. No matter what kind of application, the introduction of sensors to the lighting products is equipped with intelligent wings, which can simply realize interaction with people.

# 3 Application of Smart Lighting Sensor Technology in Smart Buildings

Today's world economy is gradually becoming globalized, and regional economies are also showing an integrated model. With the continuous pursuit of material spiritual life, people's requirements for work and living environment are getting higher and higher, especially in terms of flexibility, efficiency and comfort. In this sense, the traditional work and residential residences are no longer adapted, intelligent buildings, intelligent communities, intelligent homes, etc. Based on this, various

**Fig. 3** Infrared sensor circuit diagram

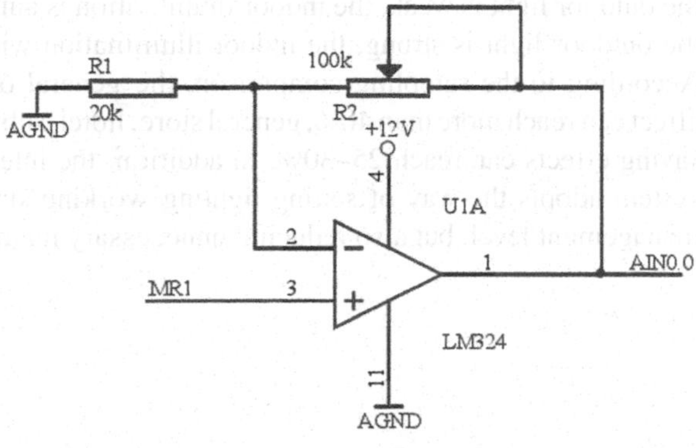

automation systems such as building automation, communication automation, office automation, and home automation are constantly emerging. The building intelligence of new sensor application has become the mainstream technology of today's building development. Its service range covers a wide range, from air-conditioning systems, fire-fighting systems to security systems, and perfect computer networks and communication systems. In the architecture, the superiority of the intelligent lighting system is more and more obvious, and its application in the intelligent building is more and more extensive. Its application effect is as follows.

### 3.1 Intelligent Lighting Control

Intelligent lighting architecture uses an intelligent lighting control system to make the lighting system work into a completely automatic state, and the system is working along a number of pre-set states and instructions. The status of these jobs will automatically adapt to the needs of the surrounding environment within the specified time. As the weather gets late, the light can no longer meet the needs of passing passengers, the system will automatically enter the night working state, and automatically adjust the brightness of each area. Moreover, the lights in the unmanned area will be automatically shut down under the command of the system's detection function, and the lights in the pedestrian area will be adjusted to the most suitable brightness. In the face of special needs, we can also change the illuminance of each area by programming in order to meet the scene requirements from various occasions.

### 3.2 The Obvious Effect of Energy Saving

The intelligent lighting control system with sensors as its main component uses advanced power electronics technology to enable the most comfortable intelligent dimming. For many lamps and lanterns such as incandescent lamps, fluorescent lamps, with special ballast sodium lamps, mercury lamps, neon lights, etc. When the outdoor light is weak, the indoor illumination is automatically brightened; when the outdoor light is strong, the indoor illumination will be automatically dimmed. According to the sampling comparison, the general office building energy-saving effect can reach more than 40%, general store, hotel, subway station and other energy-saving effects can reach 25–30%. In addition, the intelligent lighting management system adopts the way of setting lighting working status, not only improving the management level, but also reducing unnecessary maintenance costs.

### 3.3 Intelligent Lighting Application of Sensor Technology in Buildings

The advanced application as an independent model is the linkage of sensors. Linkage means that sensors can communicate with each other to achieve more complex functions and scenarios. Wired and wireless transmissions can be used. Wired mode can use lighting communication to transfer data between sensors. In office or commercial lighting places, lighting devices in different places are divided into multiple areas. Each area is controlled by one or more sensors, and lighting communication is used between sensors. When sensors in the corridor sense that someone passes, they will tell other sensors to turn on the lights in the office or conference room in advance. When they reach the office or conference room, they will automatically turn off the lights in the hallway after no one has entered. The lighting effect of the entire scene linkage.

### 3.4 Smart Sensor Wireless Communication

Wireless communication can be used, including passive infrared or radio frequency, to achieve a dynamic effect of light changes with the human. This type of scenario is more widely used in offices, shopping malls and other places where there are a large number of luminaries. Each luminaire is equipped with an independent sensor, and the sensors communicate with each other through infrared reflection or radio frequency, and when one of the sensors detects Some people will automatically send instructions to the surrounding sensors after moving, according to a certain algorithm to control the brightness of the light to produce a hierarchical brightness effect, and with the movement of the person, the nearest fixture will always remain the brightest state, the farther the lamp is kept the most Dark state. Each sensor sends commands and receives commands to maintain a dynamic working state to show a kind of lighting effect that light moves with people. It ensures better energy saving effect under the premise that the human eye adapts to light and darkness. System integration due to the diversity and complexity of the scene, integrating the sensors into the control system can better achieve different work requirements. The sensor acts as a child node in the entire system and no longer actively executes control commands. All work is managed centrally by the controller. The centralized control sends query instructions to obtain various parameters detected on the sensor. Control the lighting, curtains, air conditioning and other electrical appliances to work together to achieve a truly intelligent office and life.

### *3.5 Application of Smart Sensor Lighting Technology in People's Life*

The application of sensors in lighting is diversified, and independent work mode, linkage work mode or integrated work mode can be combined with each other to achieve lighting requirements in different places. For example, in parks, gardens and other outdoor places, lighting can give people a sense of security, the effect of linkage in the pedestrian walking on the road to meet the illuminance requirements of the road, in some scenic spots rest area using an independent illuminance sensor. When the night comes, the light is automatically turned on to the brightest, so that visitors can understand the specific location of the rest area to play a guiding role, but also to meet the lighting needs. Demand gives people a sense of security while also saving unnecessary electricity. Another example is installing the illuminance sensor and traffic flow sensor at the entrance of the tunnel and coordinating with the centralized control system to reach the illuminance requirements of different areas of the tunnel. At the same time, when the weather is dark, the lights in the entrance and exit will be automatically dimmed. If the weather is fine, the lights will be brightened. The entrance and exit lights reduce the safety hazards caused by white holes and black hole effects. It can be said that as long as there are lights in the place where the sensors are used, the sensor equipment makes the lighting products become more intelligent and humanized.

## 4 Conclusion

Compared with traditional sensors, smart sensors have advantages in accuracy, reliability, signal-to-noise ratio resolution, and adaptability. It will have a profound influence on the future of mankind. It changed the design concept and application mode of the original traditional sensor and represents the development trend of sensor technology in the future. It introduces micro-processing technology into the sensor, so that the sensor has a certain degree of intelligence. It provides a brand-new method in terms of information acquisition, processing, etc., increases the types and quantities of detection parameters, improves the information processing technology, and obtains a comprehensive and comprehensive test data, which is especially suitable for workers who are not suitable for direct Measurement monitoring area. Since smart sensors include almost all functions of instrumentation, with the continuous development and application of new technologies such as control technology, microelectromechanical and nanometer, and smart materials development, their functions will be further improved and performance will be further improved. It will also promote the continuous development of measurement and control technology. Nowadays, there are various kinds of sensor products that are used in lighting. There are applications that meet basic needs, such as illumination, motion sensing, and voice control. There are also more complex applications such as linkage and system integration, regardless of functional lighting. Scene lighting has brought more possibilities. The future of the lighting sensor should be more intelligent, there will be more and more new types of sensor devices. It is now possible to control lighting and home appliances through voice recognition. In the near future, users can perceive what they are thinking and change different lighting scenarios without speaking about exports. When customers cannot find a place to go in the mall, the lights will automatically guide the correct road. There is no longer a wall switch at home or in the office. Everything changes with the people's thoughts. Future sensors will increasingly understand people's thoughts and understand the emotions expressed by people's facial expressions and body language.

**Acknowledgements** The Key Disciplines for Operational Research and Cybernetics of the Education Department of Guangxi Province & Project of improving the basic ability of young and middle-aged teachers in Colleges and universities in Guangxi in 2018 (the application of wireless sensor network in agriculture) (2018KY0698).

## References

1. R.F. Hughes et al., Substantial energy savings through adaptive lighting. IEEE Electr. Power Energy Conf. **9**(12), 215–218 (2008)
2. S. Onaygil, O. Guler et al., Determination of the energy saving by daylight responsive lighting control systems with an example from Istanbul, Elsevier Sci. Ltd **1, 6**(12), 215–219 (2003)
3. S.F.S. Fadzil et al., Improved illumination levels and energy savings by uplamping technology for office Buildings. IEEE Conf. Publ. **5**(9), 599–603 (2009)
4. A. Guillemin et al., An innovative lighting controller integrated in a self-adaptive building control system. Elsevier Sci. B.V **6**(10), 477–487 (2001)
5. X. Suiru, Z. Yan, L. Guojun et al., Illumination automatic control system based on PID control. Comput. Digit. Eng. **5**(38), 70–73 (2010)
6. L. Huai, C. Yifei et al., Research on indoor illumination control system based on fuzzy neural network. Light. Design **1**(10), 27–30 (2008)
7. G. Xiaojing et al., Indoor lighting level prediction in smart lighting systems. Low Volt. Apparat. **2**(6), 12–14 (2004)
8. X. Xiangdong, Y. Lihua et al., Natural lighting and indoor artificial lighting (PSALI) control. J. Light. Eng. **3**(16), 23–26 (2005)
9. W. Jinguang, X. Hui et al., Intelligent integration strategy of natural lighting and artificial lighting. Artif. Intell. Recogn. Technol. **1**(21), 829–832 (2007)

# Intelligent Monitoring System of Cremation Equipment Based on Internet of Things

**Lin Tian, Fengguang Huang, Lingyu Fang and Yu Bai**

**Abstract** The cremation of the remains and the burning of relics and sacrifices are the core and key to the funeral and funeral services. With the development of computer and numerical computation, the research of cremation process is becoming more and more important. Changing the traditional combustion method is of great significance for efficient operation of equipment and energy saving and emission reduction. In this paper, we transmit the combustion data collected by the smart sensor to the remote server terminal in real time through GPRS data transmission technology. Then we set up a database for data storage. Logistic regression, random forest, XGBoost algorithm three data analysis models were used to establish a multi-input and multi-output simulation model of cremation equipment. And the actual working conditions in the process of cremation equipment were simulated to provide guidance. An intelligent monitoring system for cremation equipment is established, which integrates computer technology, sensing technology, automatic control technology, network technology and communication technology. This is of great significance for promoting the scientific development of modern funeral business.

**Keywords** Cremation equipment · Database · Logistic regression · Random forest XGBoost

## 1 Introduction

Cremation is the core and key equipment of funeral services. The level of technology reflects the service capacity of the funeral service agencies and the development level of the entire funeral industry, and it also concerns whether science and technology

L. Tian · F. Huang
The 101 Research Institute of Ministry of Civil Affairs, Beijing 100071, China

L. Fang (✉) · Y. Bai
School of Automation and Electrical Engineering, University of Science and Technology Beijing, Beijing 100083, China
e-mail: fang_ly1993@163.com

 
Y. Jia et al. (eds.), *Proceedings of 2018 Chinese Intelligent Systems Conference*, Lecture Notes in Electrical Engineering 528,
https://doi.org/10.1007/978-981-13-2288-4_26

can play a supporting role in building civilized funerals and green funerals. Because of the special nature of the funeral industry, there are many limitations to the experimental research that can be done more often [1]. The experimental method can't be widely used in the research and development of the working process of the cremator. Simultaneously, the complicated and changeable characteristics of the cremation process raises a great difficulty for the collection of related thermal parameters. With the development of computer and numerical calculation, the research of cremation process is more and more important, which plays an important guiding role in analyzing the optimal combustion conditions. It is of great significance to change the traditional test method, use the data analysis model to predict the data, assist the field test, realize the high efficient operation of the cremation equipment, and realize the energy saving and emission reduction.

In this paper, we first receive data collected by smart sensors that are transmitted to remote server terminals. Then, we build data storage data through SQL Server 2008 to facilitate data viewing. And the simulation model of multi input and multi output cremation equipment is established by using three data analysis models of logical regression, random forest and the XGBoost algorithm. The actual working conditions in the process of the cremation equipment are simulated, and the model is close to the ideal working state, and the guiding opinions of the operation and improvement of the cremation equipment are put forward.

## 2 Data Storage

In an information-based society, the full and effective management and use of various types of information resources is a prerequisite for conducting scientific research and decision-making management. A database is a unit or a common data processing system in an application field. It stores a collection of related data belonging to companies and business units, groups, and individuals. After receiving the burning data on the remote server, we chose to build a database to store the data for easy viewing and sharing.

### *2.1 Microsoft SQL Server 2008*

Microsoft SQL Server is a relational database management system introduced by Microsoft Corporation. It is easy to use, scalable, and highly integrated with related software, and can be used across multiple platforms. SQL Server is comprehensive and efficient. It can be tightly integrated with the Windows operating system, whether it is the application development speed or the system transaction processing speed, can be greatly improved. For various enterprise-level information management systems developed on the Windows platform, SQL Server is a good choice whether it is a C/S (client/server) architecture or a B/S (browser/server) architecture.

Microsoft SQL Server 2008 is a major product version. It introduced many new features and improvements, making it the most powerful and comprehensive version of Microsoft SQL Server to date [2]. SQL Server 2008 appears on the Microsoft Data Platform vision because it allows companies to run their most mission-critical applications while reducing the cost of managing data infrastructure and sending observations and information to all users [3].

This platform has the following features:

(1) Trusted—allows companies to run their most mission-critical applications with high security, reliability, and scalability.
(2) Efficient—allows companies to reduce the time and cost of developing and managing their data infrastructure.
(3) Smart—provides a comprehensive platform to send observations and information to your users when they need them.

Therefore, our data storage module chose Microsoft SQL Server 2008 as our main experimental platform to build a database to facilitate data storage and viewing.

## 2.2 *Database Building*

Figure 1 is the way to store data in our database. There are 12 sensors.

When importing data into the database, we chose to save the 10-day data as a table. This facilitates data management and viewing. We can also use SQL Server's query function to query historical data of a certain day through SQL statements. Figure 2 is the result of querying the First day combustion data.

Because of database creation, query, add and delete, all these require the support of SQL statement. It is not convenient for managers who are not familiar with database operation. In addition, when you query data in a database, you need to input SQL statements every time. The efficiency of data management and query process is not high. Therefore, we chose to design the interface in Visual Studio 2015 and connect it to the database through the interface. Through the design button, one key query is achieved, which greatly improves the efficiency of data storage and query.

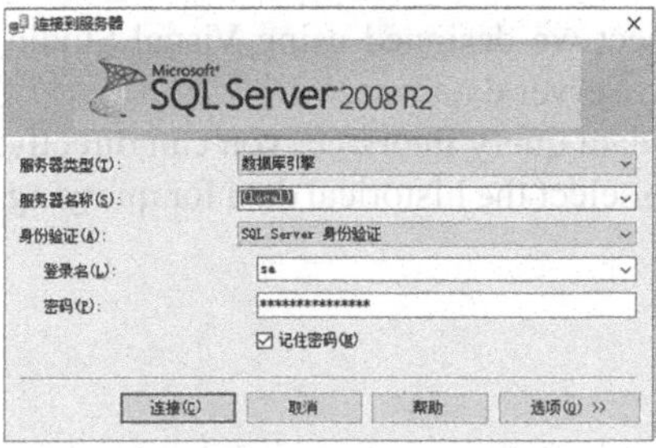


| | Date | Temperature1 | Temperature2 | Temperature3 | Temperature4 | Temperature5 | Temperature6 | Pressure1 |
|---|---|---|---|---|---|---|---|---|
| 1 | Day-0 | 63 | 814 | 984 | 548 | 696 | 679 | -43 |
| 2 | Day-0 | 61 | 811 | 971 | 543 | 698 | 679 | -27 |
| 3 | Day-0 | 57 | 811 | 960 | 547 | 701 | 681 | -21 |
| 4 | Day-0 | 51 | 810 | 945 | 545 | 702 | 681 | -20 |
| 5 | Day-0 | 765 | 804 | 923 | 543 | 695 | 678 | -20 |
| 6 | Day-0 | 761 | 802 | 919 | 546 | 698 | 676 | -19 |
| 7 | Day-0 | 758 | 802 | 902 | 540 | 697 | 681 | -19 |
| 8 | Day-0 | 758 | 804 | 889 | 544 | 699 | 678 | -19 |
| 9 | Day-0 | 754 | 801 | 879 | 547 | 697 | 672 | -19 |
| 10 | Day-0 | 755 | 798 | 859 | 541 | 693 | 675 | -18 |
| 11 | Day-0 | 755 | 791 | 857 | 546 | 691 | 677 | -18 |
| 12 | Day-0 | 751 | 796 | 839 | 546 | 693 | 672 | -18 |
| 13 | Day-0 | 748 | 792 | 819 | 542 | 696 | 670 | -18 |
| 14 | Day-0 | 744 | 787 | 808 | 547 | 693 | 674 | -20 |
| 15 | Day-0 | 739 | 63 | 797 | 542 | 691 | 673 | -19 |
| 16 | Day-0 | 736 | 63 | 62 | 542 | 688 | 665 | -18 |
| 17 | Day-0 | 735 | 786 | 57 | 547 | 690 | 666 | -19 |
| 18 | Day-0 | 733 | 785 | 758 | 540 | 690 | 669 | -19 |

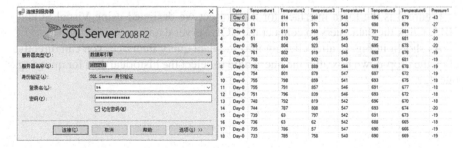

**Fig. 1** Database login interface and data storage form

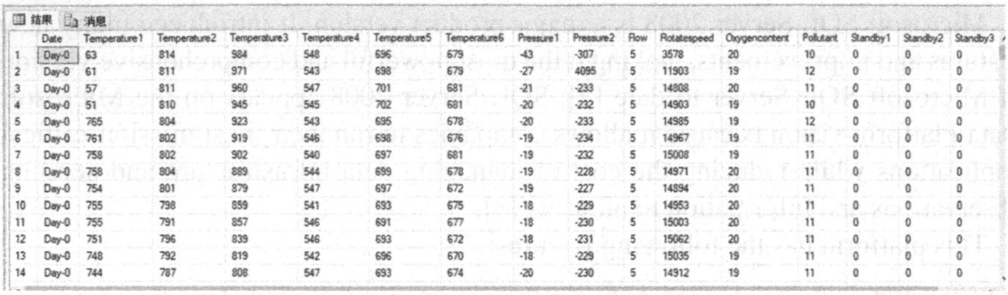

| | Date | Temperature1 | Temperature2 | Temperature3 | Temperature4 | Temperature5 | Temperature6 | Pressure1 | Pressure2 | Flow | Rotatespeed | Oxygencontent | Pollutant | Standby1 | Standby2 | Standby3 |
|---|---|---|---|---|---|---|---|---|---|---|---|---|---|---|---|---|
| 1 | Day-0 | 63 | 814 | 984 | 548 | 696 | 679 | -43 | -307 | 5 | 3578 | 20 | 10 | 0 | 0 | 0 |
| 2 | Day-0 | 61 | 811 | 971 | 543 | 698 | 679 | -27 | 4095 | 5 | 11903 | 19 | 12 | 0 | 0 | 0 |
| 3 | Day-0 | 57 | 811 | 960 | 547 | 701 | 681 | -21 | -233 | 5 | 14808 | 20 | 11 | 0 | 0 | 0 |
| 4 | Day-0 | 51 | 810 | 945 | 545 | 702 | 681 | -20 | -232 | 5 | 14903 | 19 | 10 | 0 | 0 | 0 |
| 5 | Day-0 | 765 | 804 | 923 | 543 | 695 | 678 | -20 | -233 | 5 | 14985 | 19 | 12 | 0 | 0 | 0 |
| 6 | Day-0 | 761 | 802 | 919 | 546 | 698 | 676 | -19 | -234 | 5 | 14967 | 19 | 11 | 0 | 0 | 0 |
| 7 | Day-0 | 758 | 802 | 902 | 540 | 697 | 681 | -19 | -232 | 5 | 15008 | 19 | 11 | 0 | 0 | 0 |
| 8 | Day-0 | 758 | 804 | 889 | 544 | 699 | 678 | -19 | -228 | 5 | 14971 | 20 | 11 | 0 | 0 | 0 |
| 9 | Day-0 | 754 | 801 | 879 | 547 | 697 | 672 | -19 | -227 | 5 | 14894 | 20 | 11 | 0 | 0 | 0 |
| 10 | Day-0 | 755 | 798 | 859 | 541 | 693 | 675 | -18 | -229 | 5 | 14953 | 20 | 11 | 0 | 0 | 0 |
| 11 | Day-0 | 755 | 791 | 857 | 546 | 691 | 677 | -18 | -230 | 5 | 15003 | 20 | 11 | 0 | 0 | 0 |
| 12 | Day-0 | 751 | 796 | 839 | 546 | 693 | 672 | -18 | -231 | 5 | 15062 | 20 | 11 | 0 | 0 | 0 |
| 13 | Day-0 | 748 | 792 | 819 | 542 | 696 | 670 | -18 | -229 | 5 | 15035 | 19 | 11 | 0 | 0 | 0 |
| 14 | Day-0 | 744 | 787 | 808 | 547 | 693 | 674 | -20 | -230 | 5 | 14912 | 19 | 11 | 0 | 0 | 0 |

**Fig. 2** The query result of a day's combustion of data

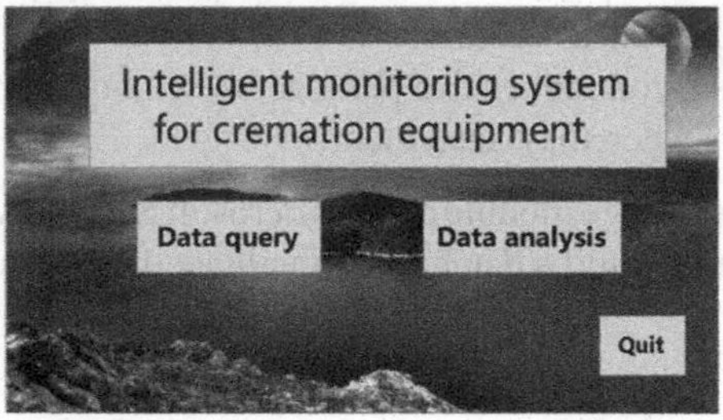


**Fig. 3** The login interface of the project in vs2015

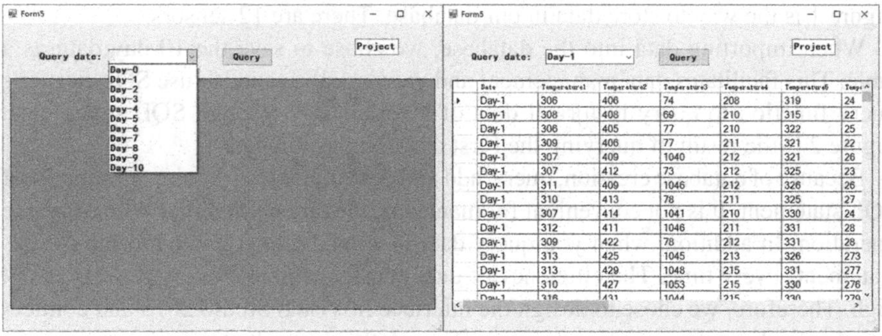


**Fig. 4** The process and results of data query

Figure 3 is the login interface for the project we designed using Visual Studio 2015. Click the data query to connect the SQL Server database.

As shown in Fig. 4, after clicking into the data query interface, you can directly click the drop-down box after the query date to select the historical data for querying one day.

# 3 Data Analysis

The main purpose of data storage is to facilitate data viewing and data analysis. In modeling, we predict oxygen content by burning data of 11 sensors, such as temperature, pressure and speed, and further predict the combustion status of the furnace. Three methods of logistic regression, random forest and the XGBoost algorithm are used to model the data, and the results of the three algorithms are compared in the end.

## 3.1 Logistic Regression

Logistic regression, also known as logistic regression analysis, is a generalized linear regression analysis model. It is often used in fields such as data mining, automatic disease diagnosis, and economic forecasting [4, 5]. Logistic regression is mainly evolved from linear regression. The main formula for linear regression is as follows:

$$z = \theta_0 + \theta_1 x_1 + \theta_2 x_2 + \theta_3 x_3 + \cdots + \theta_n x_n = 0^T x \quad (1)$$

For Logistic regression, the idea is based on linear regression. The formula is as follows:

$$h_\theta(x) = \frac{1}{1 + e^{-z}} = \frac{1}{1 + e^{-\theta^T x}} \quad (2)$$

where y is called the sigmoid function, the Logistic regression algorithm maps the result of the linear function to the sigmoid function.

$$y = \frac{1}{1 + e^{-x}} \quad (3)$$

The function output of sigmoid is between (0, 1) and the middle value is 0.5, so the meaning of the previous formula $h_\theta(x)$ is well understood because the output of $h_\theta(x)$ is between (0, 1). Between the two, it also shows the probability that the data belongs to a category, such as:

$h_\theta(x) < 0.5$ indicates that the current data belongs to Class A;
$h_\theta(x) > 0.5$ indicates that the current data belongs to Class B.

So, we can think of the sigmoid function as the probability density function of the sample data. After that, the parameter $\theta_0$ can be estimated.

First, the value of the $\theta$ function has a special meaning. It represents the probability that the result of $h_\theta(x)$ is 1. Therefore, the probability that the input $x$ classification result is category 1 and category 0 is:

$$P(y = 1|x; \theta) = h_\theta(x) \quad (4)$$

$$P(y = 0|x;\theta) = 1 - h_\theta(x) \tag{5}$$

According to the above formula, next we can use the method of maximum likelihood estimation in probability theory to solve the loss function. First, we get the probability function:

$$P(y|x;\theta) = (h_\theta(x))^y * (1 - h_\theta(x))^{1-y} \tag{6}$$

Because the sample data are independent, their joint distribution can be expressed as the product of the marginal distributions. The likelihood function is:

$$L(\theta) = \prod_{i=1}^{m} P(y^{(i)}|x^{(i)};\theta) \tag{7}$$

$$L(\theta) = \prod_{i=1}^{m} (h_\theta(x^{(i)}))^{y^{(i)}} * \left(1 - h_\theta(x^{(i)})\right)^{1-y^{(i)}} \tag{8}$$

Logarithm likelihood function:

$$l(\theta) = \log(L(\theta)) = \sum_{i=1}^{m} \log((h_\theta(x^{(i)}))^{y^{(i)}}) + \log\left(\left(1 - h_\theta(x^{(i)})\right)^{1-y^{(i)}}\right) \tag{9}$$

$$l(\theta) = \log(L(\theta)) = \sum_{i=1}^{m} y^{(i)}\log(h_\theta(x^{(i)})) + (1 - y^{(i)})\log(1 - h_\theta(x^{(i)})) \tag{10}$$

The maximum likelihood estimation is the $\theta$ required to make $l(\theta)$ take the maximum value, which can be solved using the gradient ascent method. Let's change it slightly:

$$J(\theta) = -\frac{1}{m} l(\theta) \tag{11}$$

Then use the gradient descent method to solve the parameters.

### 3.2 Random Forest

Random Forest (RF) refers to a classifier that uses multiple trees to train and predict samples [6]. The basic idea of a random forest is to randomly generate a number of irrelevant decision trees. Each decision tree can independently use training samples for training. After the decision tree is generated and each new sample enters, each decision tree classifies it and reports the classification results to the random forest. Each decision tree has its own voting rights. The random forest uses the category with the highest number of votes as the final classification result of the sample [7, 8].

Random forest belongs to the bagging algorithm in Ensemble Learning. Bagging algorithm is used to generate training samples, and decision attributes can be obtained by random subspace algorithm.

The Bagging algorithm was proposed by Breiman in 1996. It essentially uses Bootstrap sampling to perform sampling on the sample and continuously extracts samples until multiple sets of different training samples are constructed [9]. Each set of samples is assigned to a weak classifier for training. Finally, all the weak classifiers summarize the classification results and report them to the strong classifier in the form of voting. The strong classifier selects the class with the most votes as the final classification result.

The random forest algorithm is Breiman's improvement to the previously proposed Bagging algorithm. It combines the back-in extraction of training samples in the Bagging algorithm and the non-return extraction of attributes in the stochastic subspace algorithm. It can simultaneously have the advantages of the two algorithms, so that it can perform global search better and achieve higher results [10]. The algorithm flow is as follows:

(1) Determine the parameters necessary to train the random forest—the training sample set, the number of decision tree L, the number of random attributes m, and the pruning threshold of the training decision tree.
(2) Sampling is performed on the training sample set until a sample set equal to the number of training sample sets is extracted as a training sample for a decision tree.
(3) The non-return sampling of the attribute set is performed. After extracting the m attributes, only the data corresponding to the m attribute are retained as training samples.
(4) Use the training samples generated in Step 2 and Step 3 to train a decision tree.
(5) Pruning thresholds are used to prune the trained decision tree.
(6) If the number of trained decision trees is less than L, return to Step 2 to continue execution. Otherwise, all L decision trees are cascaded through voting strategies to form a random forest.

### 3.3 XGBoost Algorithm

The XGBoost algorithm is a lifting algorithm. Prior to this, the AdaBoost algorithm and the GBDT algorithm and other lifting algorithms have achieved good modeling results in the industry, and the XGBoost algorithm is based on these two algorithms. In general, the accuracy of the model depends on the optimization of the objective function. The closer the predicted value is to the true value, the better the model's generalization ability. Minimize loss functions and increase model complexity of punishment terms can achieve both goals [11]. We optimize the objective function to achieve an optimal combination of error and complexity. Give the objective function $Obj$:

$$Obj(\theta) = L(\theta) + O(\theta) \tag{12}$$

The objective function $Obj(\theta)$ consists of two parts: the error function $L(\theta)$ and the complexity function $O(\theta)$. The two ultimate goal of machine learning is to make the deviation of loss smaller and the complexity of function lower. When the complexity is high, it means that the learning of the model has reached the limit and may lead to the risk of overfitting. Therefore, our final focus is on finding a balance between deviation and variance.

The principle of the XGBoost algorithm is to split the original data set into multiple sub-data sets, and randomly allocate each sub-data set to the base classifier for prediction. Then the weakly classified results are calculated according to certain weights to predict the result [12].

Base classifiers work together only to achieve good results, but traditional SGD models do not. The purpose of the conventional SGD algorithm is to find a function that minimizes the squared loss function. The prediction model of the results of such an additive base classifier is generally called an additive model. The formula is as follows:

$$F = \sum_{i=1}^{m-1} f + f_m \tag{13}$$

The additive model's predictive fitting is the spline smoothing of the backwards fitting algorithm for each base classifier's prediction result. It makes the fitting error reach an equilibrium state.

## 3.4 Comparison of the Results of the Three Algorithms

Converts the collected hexadecimal data to decimal data. According to the decimal data, it can be found that some data such as temperature and rotation speed occasionally jump during the acquisition process. In order to avoid the impact of jump data on the final data analysis, we first preprocessed the data and smoothed it. Figure 5 is a comparison of the results before and after preprocessing.

The processed data was regressed using logistic regression, random forest, and the XGBoost algorithm in machine learning, and the results were compared by three evaluation indicators. The main evaluation indicators are as follows:

MSE: Mean-Square Error, a measure that reflects the degree of difference between an estimate and an estimate.
MAD: Mean Absolute Deviation, the average of the absolute values of the deviations of all individual observations from the arithmetic mean.
RMSLE: Root Mean Squared Logarithmic Error. It is also called the standard error, which is the square root of the ratio of the square of the observed value to the true deviation and the ratio of the observed times n.

**(a)**

| | | | | | | | | | |
|---|---|---|---|---|---|---|---|---|---|
| 565 | 617 | 550 | 409 | 598 | 532 | -18 | -226 | 5 | 14844 |
| 566 | 619 | 548 | 411 | 598 | 540 | -18 | -226 | 5 | 14872 |
| 567 | 619 | 544 | 405 | 597 | 538 | -17 | -228 | 5 | 14804 |
| 568 | 618 | 534 | 409 | 596 | 532 | -18 | -226 | 5 | 933 |
| 566 | 616 | 44 | 410 | 594 | 533 | -18 | -228 | 5 | 14921 |
| 564 | 620 | 42 | 406 | 594 | 536 | -17 | -223 | 5 | 943 |
| 562 | 620 | 36 | 411 | 593 | 534 | -18 | -224 | 5 | 14962 |
| 561 | 618 | 506 | 409 | 594 | 529 | -18 | -225 | 5 | 14908 |
| 562 | 616 | 511 | 407 | 596 | 534 | -18 | -222 | 5 | 943 |
| 563 | 614 | 502 | 411 | 596 | 536 | -17 | -225 | 5 | 14840 |
| 564 | 614 | 497 | 408 | 596 | 533 | -18 | -223 | 5 | 14844 |

**(b)**

| | | | | | | | | | |
|---|---|---|---|---|---|---|---|---|---|
| 565 | 617 | 550 | 409 | 598 | 532 | -18 | -226 | 5 | 14844 |
| 566 | 619 | 548 | 411 | 598 | 540 | -18 | -226 | 5 | 14872 |
| 567 | 619 | 544 | 405 | 597 | 538 | -17 | -228 | 5 | 14804 |
| 568 | 618 | 534 | 409 | 596 | 532 | -18 | -226 | 5 | 14736 |
| 566 | 616 | 524 | 410 | 594 | 533 | -18 | -228 | 5 | 14921 |
| 564 | 620 | 520 | 406 | 594 | 536 | -17 | -223 | 5 | 15106 |
| 562 | 620 | 518 | 411 | 593 | 534 | -18 | -224 | 5 | 14962 |
| 561 | 618 | 506 | 409 | 594 | 529 | -18 | -225 | 5 | 14908 |
| 562 | 616 | 511 | 407 | 596 | 534 | -18 | -222 | 5 | 14854 |
| 563 | 614 | 502 | 411 | 596 | 536 | -17 | -225 | 5 | 14840 |
| 564 | 614 | 497 | 408 | 596 | 533 | -18 | -223 | 5 | 14844 |

**(c)**

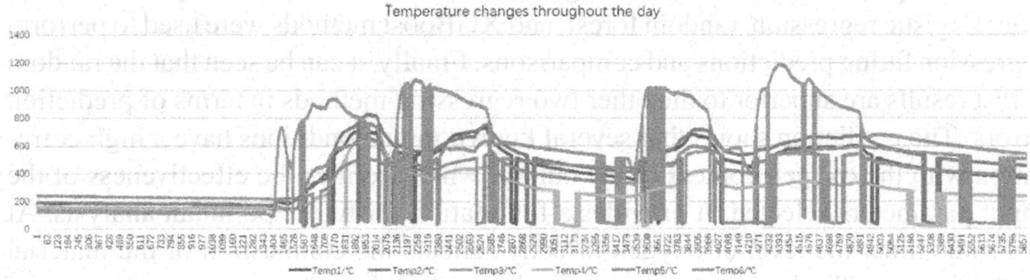


**(d)**

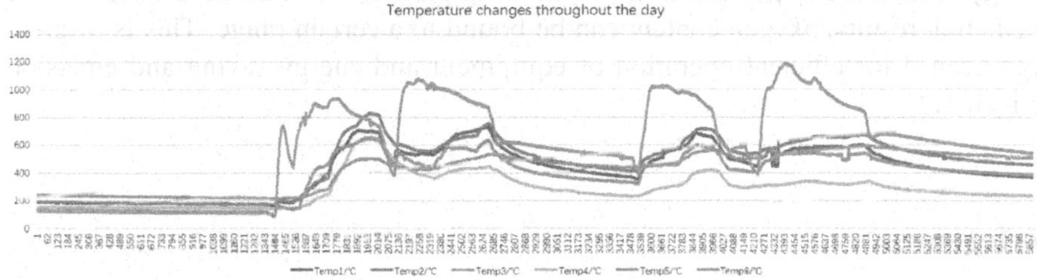

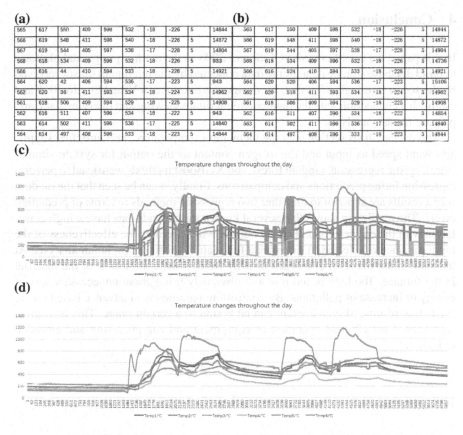

**Fig. 5** Comparison of results before and after data preprocessing. **a**, **c** are the pre processing data, and **b**, **d** are the processed data

**Table 1** Comparison of the prediction results of the three algorithms

| | Logistic regression | Random forest | XGBoost |
|---|---|---|---|
| MSE | 27.304 | 0.670 | 2.400 |
| MAD | 2.817 | 0.494 | 0.998 |
| RMSLE | 0.594 | 0.094 | 0.232 |

The comparison results are shown in Table 1.

From the results of the operation, it can be seen that the random forest prediction is the best, followed by XGBoost, and the less effective is the logistic regression. Among them, random forests are more suitable for dealing with high-dimensional data. Random forests have good generalization ability, and the prediction accuracy is higher. Logistic regression is easy to under-fit and the accuracy is not high enough, so the accuracy of regression prediction in these three algorithms is also worse.

## 4 Conclusion

In this paper, we use Microsoft SQL Server 2008 as the main experimental platform to build database storage data for the data collected by the intelligent sensor received by the server terminal. After that, we design the interface in Visual Studio 2015 and establish a connection with the database to facilitate viewing and querying data, and prepare for data analysis.

The data analysis section uses the incinerator temperature, pressure, fuel supply, and wind speed as input and the oxygen content as the output for system simulation. Logistic regression, random forest, and XGBoost methods were used to perform regression fitting predictions and comparisons. Finally, it can be seen that the random forest results are superior to the other two regression methods in terms of prediction errors. The prediction shows that several known input conditions have a high correlation with the oxygen content of the output, which verifies the effectiveness of the input parameters selected. It provides a foundation for future pollutant analysis. At the same time, the level of oxygen directly reflects the combustion of the material in the furnace. Too high or too low will obviously bring about unnecessary loss of energy or increase of pollutants. By controlling parameters in advance based on the prediction results, oxygen content can be bound to a certain range. This is of great significance for efficient operation of equipment and energy saving and emission reduction.

## References

1. L. Renqing, Green burials the way of the future (China Daily, 06 April 2011)
2. D. Yun Wang, Advanced studying on microsoft SQL server 2008 data mining. Appl. Mech. Mat. **893**(20) (2010)
3. Anonymous, Review: microsoft SQL server 2008 R2. Network World (2010)
4. L. Saro, J. Seong Woo, O. Kwan-Young, L. Moung-Jin, The spatial prediction of landslide susceptibility applying artificial neural network and logistic regression models: a case study of Inje, Korea. Open Geosci. **8**(1) (2016)
5. G.M. Fitzmaurice, N.M. Laird, Binary response models and logistic regression (Elsevier Inc., 06 April 2011)
6. C. Su, S. Ju, Y. Liu, Z. Yu, Improving random forest and rotation forest for highly imbalanced datasets. Intell. Data Anal. **19**(6) (2015)
7. A. Sim, D. Tsagkrasoulis, G. Montana, Random forests on distance matrices for imaging genetics studies. Stat. Appl. Genetics Mol. Biol. **12**(6) (2013)
8. P. Cichosz, Ł. Pawełczak, Imitation learning of car driving skills with decision trees and random forests. Int. J. Appl. Math. Comput. Sci. **24**(3) (2014)
9. J. Błaszczyński, J. Stefanowski, Neighbourhood sampling in bagging for imbalanced data. Neurocomputing **150** (2015)
10. A.M. Prasad, L.R. Iverson, A. Liaw, Newer classification and regression tree techniques: bagging and random forests for ecological prediction. Ecosystems **9**(2), 181–199 (2006)
11. B. Pan, Application of XGBoost algorithm in hourly PM2.5 concentration prediction. IOP Conf. Series Earth Environ. Sci. **113**(1) (2018)
12. A. Kadiyala, A. Kumar, Applications of python to evaluate the performance of decision tree-based boosting algorithms. Environ. Progress Sustain. Energy **37**(2) (2018)

# Design and Implementation of Remote Monitoring System for Working Conditions of Cremation Equipment

**Fengguang Huang, Lin Tian, Yu Bai and Wei Wang**

**Abstract** With the embedded technology and the 4G TD-LTE wireless data transfer technology, a remote acquisition system for the working data of cremation equipment has been researched in this paper. The acquisition system realizes acquiring data and uploading data to internet, centralizing storage and monitoring the data timely. The data will support engineers in realizing remote cremation equipment maintenance, fault diagnosis, operation monitoring and upgrade of cremation equipment.

**Keywords** STM32 · 4G TD-LTE · Cremation equipment · Remote monitoring

## 1 Introduction

With the rapid development and wide application of Internet technology in the national information industry, the information industry and the Internet of Things technology are rapidly integrating into various fields [1] Its application brings more and more significant technical, economic and social benefits [2].

The domestic funeral industry is a unique industry. Conventional funeral cremation device has the problems such as low fuel combustion efficiency, lack of equipment operation technology, inconvenience of equipment trouble-shooting, and

F. Huang · L. Tian · W. Wang
101 Institute of the Ministry of Civil, Beijing 100070, China

F. Huang · W. Wang
Key Laboratory of Pollution Control of the Ministry of Civil Affairs, Beijing 100070, China

L. Tian
Key Laboratory of Cremation Equipment of the Ministry of Civil Affairs, Beijing 100070, China

Y. Bai (✉)
School of Automation and Electrical Engineering, University of Science and Technology Beijing, Beijing 100082, China
e-mail: 18811348078@163.com



Y. Jia et al. (eds.), *Proceedings of 2018 Chinese Intelligent Systems Conference*, Lecture Notes in Electrical Engineering 528,
https://doi.org/10.1007/978-981-13-2288-4_27

excessive discharge of pollutants. Funeral parlors are usually distributed in remote areas in various provinces and cities. So, these problems cannot be solved in time because of geographical factors. Meantime, for geographical reasons of the funeral parlors also makes cremation equipment manufacturers and technical personnel of cremation equipment scene investigation more difficult. The shortage of on-site operating condition data for cremation equipment also makes equipment upgrades slower. Therefore, to introduce information technology for the funeral industry and to realize the remote data collection, data storage, and data analysis of the cremation equipment are imperative.

The rapid progress of informationization has made it possible for the funeral industry to achieve nationwide networked data collection. Currently, wireless communication technologies commonly used for data acquisition include Bluetooth, WIFI, and GPRS, 3G, and 4G of mobile networks. Data collection using Bluetooth requires establishing a Bluetooth serial connection between the remote monitoring terminal and the cremation device. However, the Bluetooth anti-interference ability is weak, the transmission distance is short, and the information security cannot be guaranteed. Even though the WIFI network construction has high rate, the WIFI cannot be connected to each device without affecting the scene. Because of the generation 4G network, GPRS and 3G have been basically eliminated. 3G access network air interface is not uniform, which is easy to cause confusion communications. 4G technology is more mature now. It has the advantages of wide network coverage, low cost, short network delay, high security performance, and no geographical restrictions [3]. 4G has been applied to data acquisition in multiple fields [4–6]. In this paper, using 4G technology can effectively achieve the collection of on-site operating condition data of the cremation equipment, which has high R&D significance and practical value [7].

This article uses STM32, 4G TD-LTE wireless data transmission technology and database to establish a data collection and monitoring system. This system realizes the functions of remotely collecting mortal cremation equipment terminal data, uploading data, and real-time online monitoring. The system fills the gap in the field of remote data collection for domestic cremation equipment and provides powerful support for further monitoring and management of equipment for burial and cremation as well as equipment renewal.

## 2 System Design

This design uses 4G network to achieve data communication between on-site conditions and servers to obtain field data.

The overall structure of the system is shown in Fig. 1. The system consists of four parts: Data Collection Terminal, Data Processing Terminal, Data Upload Terminal, and Data Monitoring Terminal. Among them, Data Processing Terminal is one of the key points of this design implementation. The main controller is STM32F103ZET6. It is responsible for processing analysis data and distributing it to SD card, LED screen

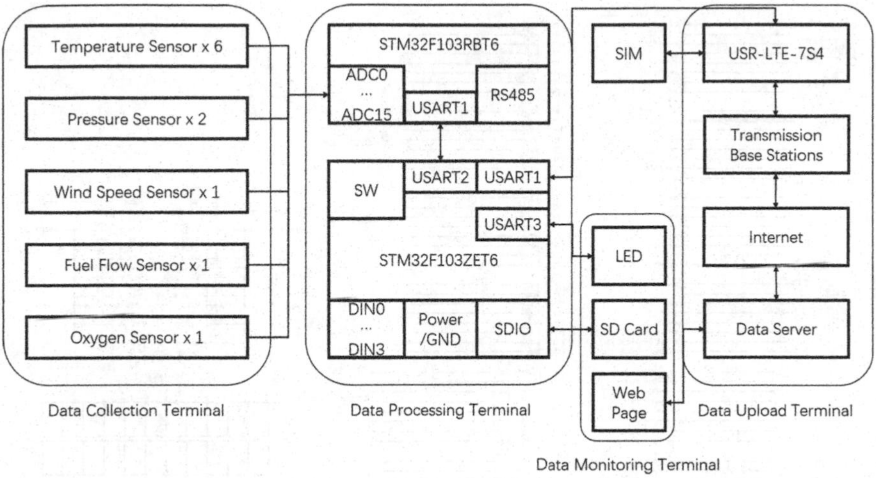


**Fig. 1** System structure

and 4G module. The auxiliary controller is STM32F103RBT6, whose function is to get the current signal sent by the sensor and send the processing package to the main chip. Data Collection Terminal is a type-selected sensor after field investigation, after determining the range, use environment and other factors. Data Upload Terminal shows the process of transferring data through 4G to the database in the server. The Data Monitoring Terminal is the part that the staff can view. Get the data from the Data Collection Terminal in this section. On the server, we also built a database and designed a web interface to facilitate the funeral industry staff to view the data anytime, anywhere.

# 3 Hardware Circuit Design Method

The hardware of the system designed in this paper is mainly composed of four parts: main control module, data acquisition module, data transceiver module and power supply module.

## *3.1 Main Control Module*

The core chip used in this paper design uses STM32F103ZET6-144 as the main controller (MCU) [8–12]. It is a 32-bit Coretex-M3 core processor operating at 72 MHz and has built-in high-speed memory (512 KB of flash, 64 KB of SRAM).

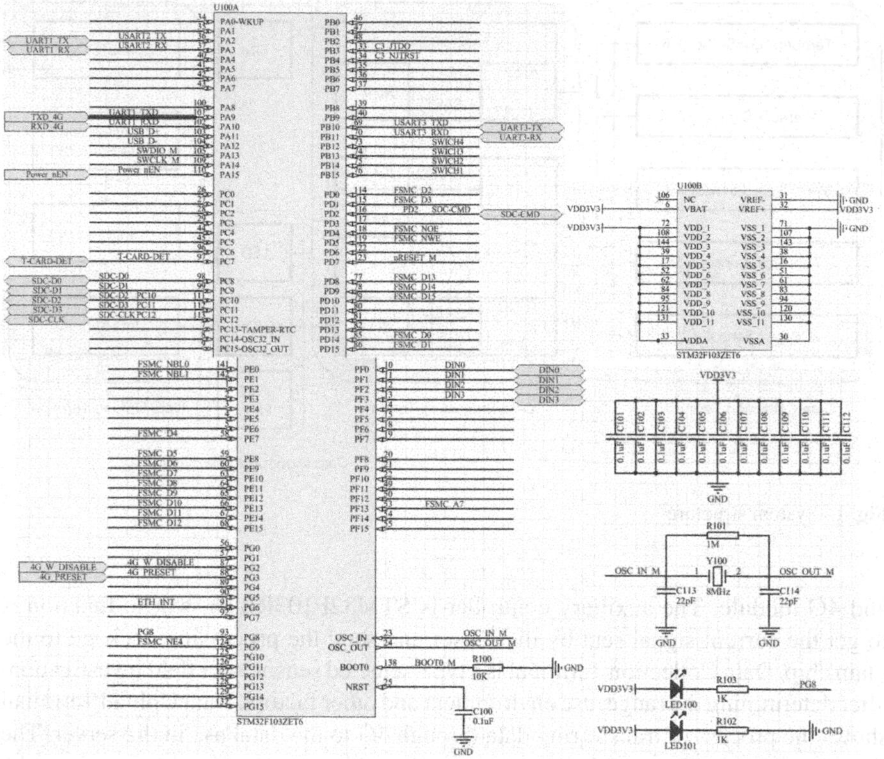


**Fig. 2** STM32F103ZET6 and its minimum system design

This chip can provide multiple pins that can be used as a serial port, thus enabling high-speed data transfer between peripherals and memory.

In the design, USART1 using STM32F103ZET6 is interconnected with 4G module USR-LTE-7S4. USART2 is interconnected with USART1 of STM32F103RBT6, and USART3 is interconnected with LED module DIN, DOUT. PC8, 9, 10, 11, 12, PD16 is connected to SD card according to SDIO protocol. PF0-3 connects DIN0-3 and receives digital signals. PA13, 14 as the chip download port. The rest of the pins are used to connect other functions such as network port chips. STM32F103ZET6 circuit diagram is shown in Fig. 2.

## 3.2 *Data Acquisition Module*

The data acquisition module includes an auxiliary chip STM32F103RBT6-64 and its peripheral circuits, including 16 analog signal receiving circuits. Another RS485 communication module connected to this chip and four digital signal receiving cir-

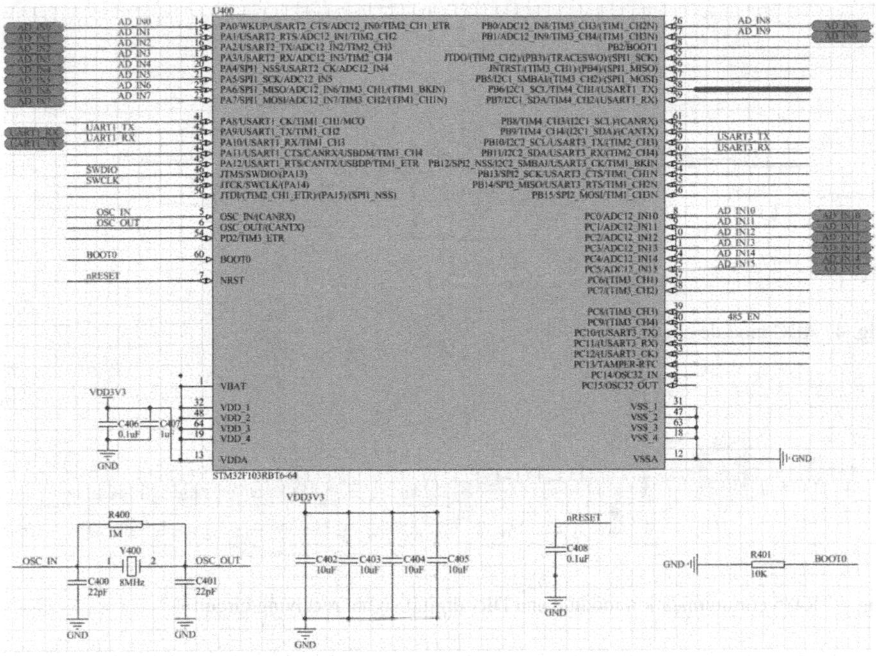


**Fig. 3** STM32F103RBT6 and its minimum system design

cuits connected to the main chip are left as scalable functions, which were not used in this experiment. STM32F103RBT6 circuit diagram is shown in Fig. 3.

The STM32F103RBT6 uses a high-performance ARMCortex-M3™ 32-bit RISC core operating at 72 MHz, high-speed embedded memory (flash and SRAM up to 128 bytes to 20 bytes) and enhanced I/O peripherals connected to two The APB bus is extensive. This design uses a 64-LQFP packaged device to provide two 12-bit ADCs to meet the 16-sensor analog signal.

For ADC module circuit is shown in Fig. 4, R506 and R508 are not simultaneously welding. They correspond to two different wiring methods. When only welding R506, it means that the circuit board does not supply power to the sensor. Corresponding to the three-wire connection mode, the signal current is directly connected to the circuit board. When only welding R508, it means that the circuit board provides 24 V voltage. Corresponding to the two-wire connection mode, the power supply current is the signal current.

RS485 and 4 digital signal receiving modules were not used in this design. So these two parts are retained as extensible features. This part of the circuit diagram is shown in Fig. 5.

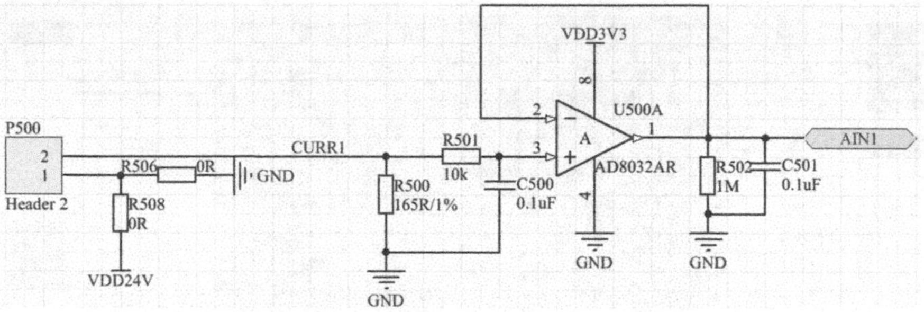


**Fig. 4** ADC module circuit

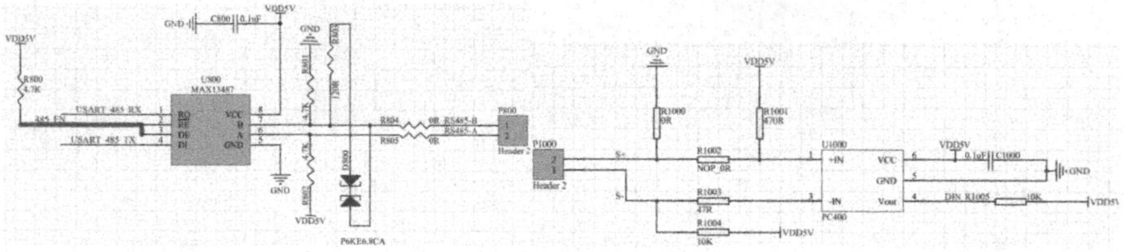

**Fig. 5** R485 communication module and DIN digital signal receiving circuit

## 3.3 *Data Transceiver Module*

The data transceiver module consists of four parts: 4G module, network port communication module, local LED display module and local SD card storage module. The network port module is not used in this experiment and is reserved as an extensible function.

4G module: USR-LTE-7S4 is a compact, feature-rich M2M product for mobile, China Unicom, Telecom 4G and mobile, Unicom 3G and 2G network standards. And support custom registration package, heartbeat package function, support 4-way socket connection. It has high speed, low latency, and supports the FTP HE upgrade protocol and FTP self-upgrade protocol. Based on the above advantages, the system uses USR-LTE-7S4 4G wireless terminal products. The module can realize bidirectional data transparent transmission between the serial device and the network server through the carrier network. The USR-LTE-7S4 4G wireless terminal product is shown in Fig. 6.

As shown in Fig. 6, three switches are designed to correspond to RESET, RELOAD, and POWER_KEY, respectively. According to the pin description manual, all are active low. Pins UART1_RX, UART1_TX are used for data transmission and reception signal transmission.

Local LED display module: For the convenience of local burial staff to view real-time data, this design incorporates an LED display module. As shown in Fig. 7, a DGUS screen with a 7.0-in., 65-K color LED backlight model

DMT80480C070_04 W is selected for this design [13]. DGUS multimedia screen is a kind of configuration screen that can be developed by users from DWIN.

As shown in Fig. 8, the DGUS screen takes in 8 pins. According to the TTL standard, connect DIN and DOUT to send and receive signals. Both VDD and GND are also provided on the board.

Local SD card storage module: To store data more conveniently and prevent data loss caused by disconnection, this design incorporates the SD card local storage function. Drive circuit design SD memory circuit as shown in Fig. 9. The SD card communicates with the microcontroller through the SDIO communication protocol [14]. The SD card is mainly divided into four parts: external pins, internal registers, interface controllers, and internal storage media. The same as the SD bus, there are several signals as follows:

(1) CLK signal: Clock signal for HOST to DEVICE.
(2) CMD signal: A bidirectional signal used to transmit commands and responses.
(3) DATA0–DATA3 signal: Four data lines for transmission.
(4) VDD signal: power signal.
(5) VSS: Power ground signal.

In the SDIO bus definition, the DAT1 signal line is multiplexed as an interrupt line. In 1BIT mode of SDIO, DAT0 is used to transmit data, and DAT1 is used as an interrupt line. In the 4BIT mode of SDIO, DAT0–DAT3 is used to transmit data, where DAT1 is used as the interrupt line.

Network port communication module: The network port function is not used, and the circuit design is a universal design, which will not be described here.

## 3.4 Power Supply Module

Figure 10 shows the power module circuit diagram. The ACDC module converts 220 V AC–24 V DC. And through the regulator chip and other components to generate 5, 3.3 V voltages for the LED screen, the chip to provide power. In another part, components such as the LM2596-ADJ convert 24–3.9 V and provide it to the 16th pin of the 4G module.

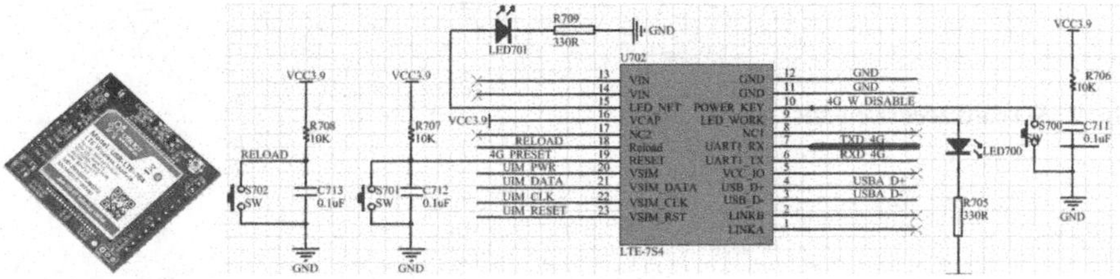


**Fig. 6** USR-LTE-7S4 and 4G transceiver circuit

**Fig. 7** DMT80480C070_04 W from DWIN

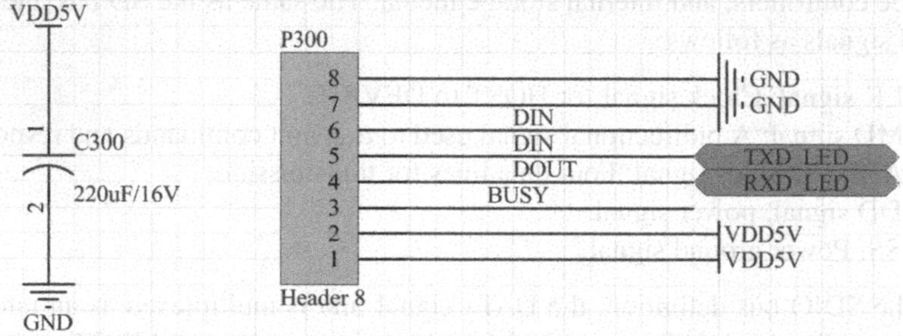


**Fig. 8** LED access circuit

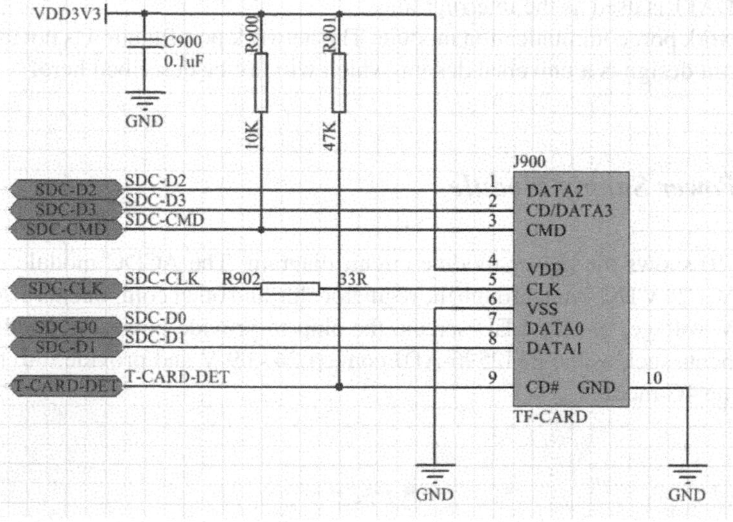


**Fig. 9** SD card access circuit

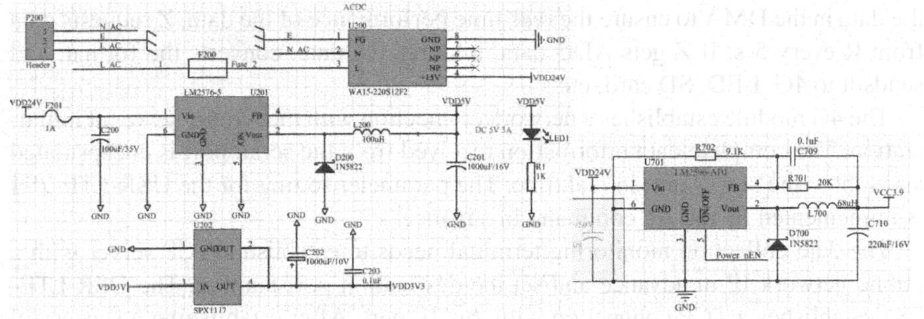


Fig. 10 Power module circuit diagram

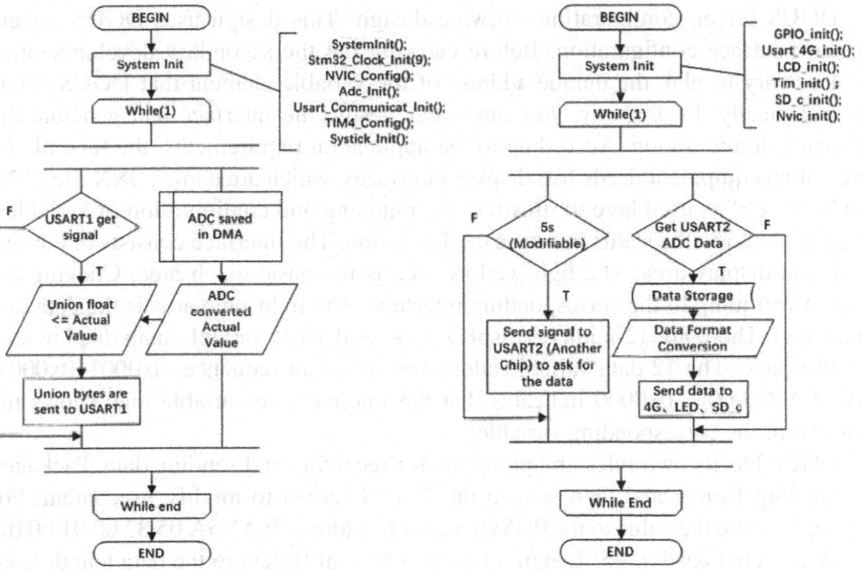


Fig. 11 Flow chart of data acquisition software and data transceiver software

# 4 Software Design Method

The system software design consists of four parts: data acquisition software, data transceiver software, DGUS screen configuration software design, and server-side software design. They correspond to the auxiliary chip, the main control chip, the DGUS screen, and the server. Among them, the flow chart of data acquisition software and data transceiver software is shown in Fig. 11.

As shown in the flow chart, after the system is powered on, the two chips (Z = main chip, R = auxiliary chip) are initialized, including the detection settings of the module and SD card status. After completing hardware initialization such as interrupts and serial ports, R enters the wait data phase. Every time R gets the ADC data, it stores the data in the DMA to ensure the real-time performance of the data. Z requests data from R every 5 s. If Z gets ADC data, it stores the data, converts the format, and sends it to 4G, LED, SD card, etc.

The 4G module establishes a network connection with the remote server at initialization. The complete data information received from the serial port is encapsulated and sent to the server port in real time. The parameter setting for the USR-LTE-7S4 is implemented by the AT command in advance.

The data collection monitoring terminal needs to establish a TCP server with a public network IP in advance and set the destination port number. The USR-LTE-7S4 establishes a TCP connection with this IP port. After establishing a successful network connection, the 4G module will send data according to the transparent transmission protocol [15].

DGUS screen configuration software design: This design uses DGUS system visual interface configuration. Before carrying out the secondary development, it is necessary to plan the unique address of the variable element that DGUS wants to realistically. In this way, you can better design the interface and generate the interface configuration. According to the application requirements, the funeral and cremation equipment needs five display interfaces, which are named 0XX files. The "0 Status FF" is used here to illustrate the planning and configuration of variables. Figure 12 shows the status interface configuration. The interface consists of buttons and data display area. The light-yellow area is the basic touch area. Clicking the button will jump to the corresponding interface. The light-blue area is the data display area. There are 12 numeric display areas and 1 0/1 variable map display area on this page. The 12 data variable addresses are set in sequence: 0x0001–0x0009, 0x000A–0x000C. 0x0000 indicates that the intermediate variable map shows the address of the corresponding variable.

DGUS has its own rules and protocols for receiving and sending data. Packaged in the chip format and then sent to the DGUS screen to modify its content. For example, write the value to the 0x0001 variable address 2: A5 5A 05 82 00 01 00 02. A5 5A: Frame header; 05: Length of data to be sent (refers to the data length from the beginning of the instruction to the last data, where 5 bytes are transmitted from the 82 instruction); 82: Write data storage area instruction; 00 01: Variable address (two bytes); 00 02: Number 2 (two bytes).

Server-side software design: To ensure the smooth development and application of the server in the remote monitoring system of the cremation equipment, it is necessary to first select an effective development platform. Currently in the development of the system platform, the main Spring3, Struts2, MyBatis framework, data communication using Apache Mina, database using MySQL. The development language uses Java. Software deployment uses Tomcat 7.0. The server software includes four functions: login and permissions, device management, data query, and data export. The system architecture includes three parts: data acquisition monitoring port, data processing, and data persistence.

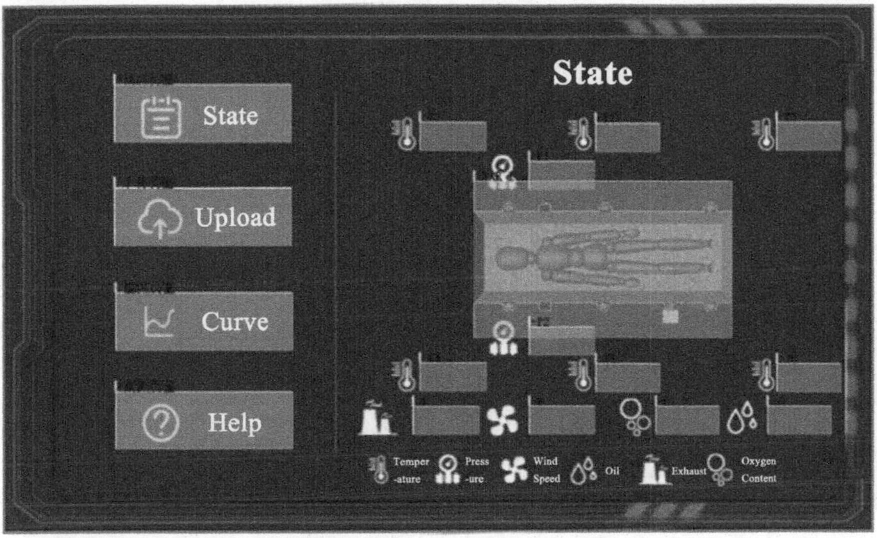


**Fig. 12** DGUS visual interface design

# 5 Prototype and Test Results

Figure 13 shows the last version of the PCB design and final product. To verify the data collection device for the remote working conditions of the cremation equipment developed in this paper, we will arrange the sensors and the equipment we designed and developed for the data acquisition in the Boxing funeral parlor in Shandong.

The field test is shown in Fig. 14. After the on-site sensors are connected, the resulting values can be viewed on the web page. Figure 15 shows the web page login and query interface.

# 6 Analysis of Experimental Results

This design combines the STM32 and TD-LTE (4G) wireless transmission technologies to successfully implement a remote acquisition and monitoring system for working condition data of burial and cremation equipment. The system fills the gap in on-line data collection for burial and cremation equipment.

Experiments show that the remote monitoring system has the characteristics of high real-time performance, accurate data transmission, reliable results, and convenient export analysis. Through this system, the monitoring personnel can easily view the specific working conditions of the cremation equipment randomly connected to the system and whether there are any abnormal conditions on the equipment that can access the external network and improve the security and maintainability of the cre-

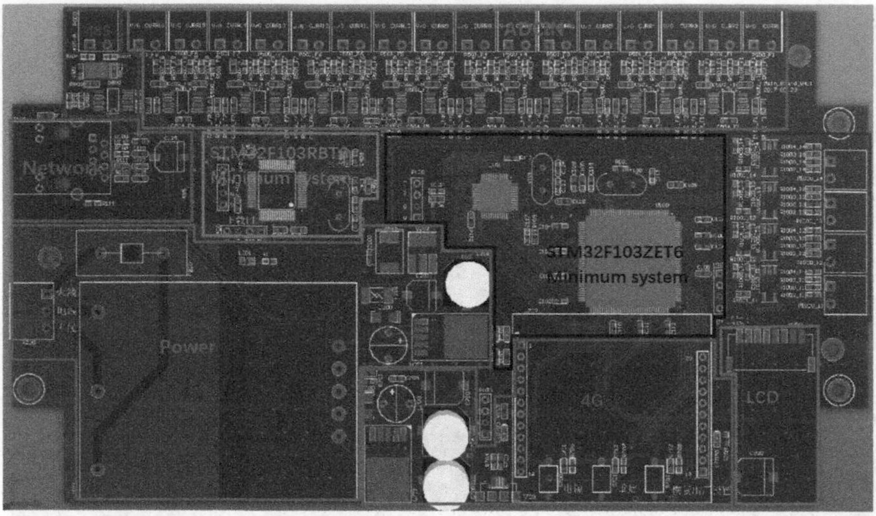


**Fig. 13** The last version of the PCB design and final product

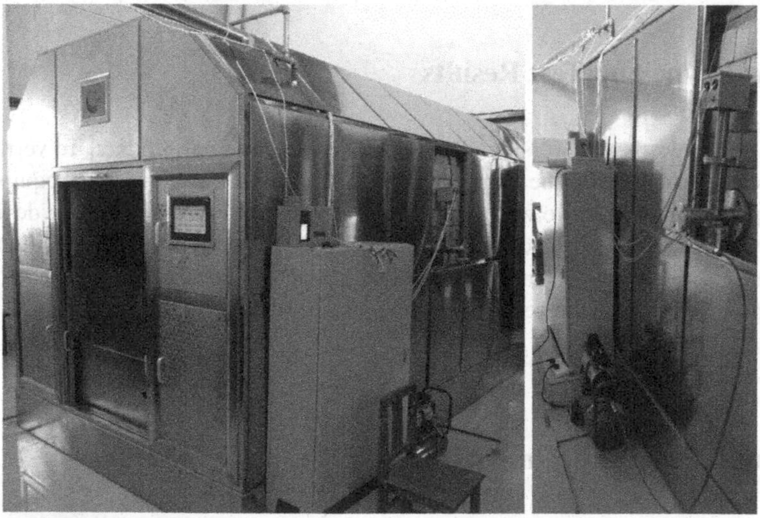

**Fig. 14** Field test show

mation site. And for the future field control, equipment maintenance, fault diagnosis, pollutant monitoring and other applied research provides reliable data support.

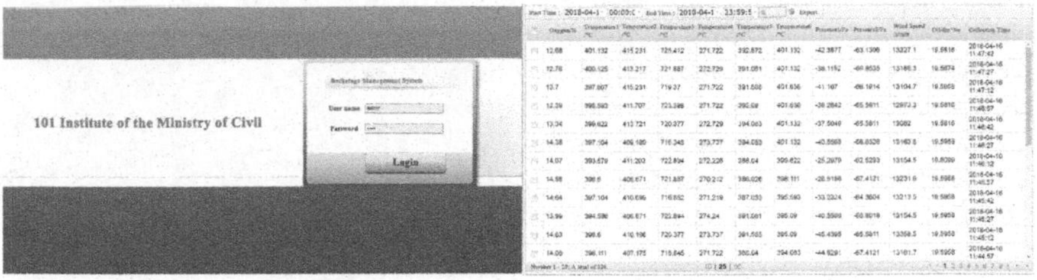


**Fig. 15** The web page login and query interface

# References


1. E.F. Nakamura, A.A.F. Loureiro, Information fusion in wireless sensor networks. ACM Comput. Surv. **39**(3), 381–69075 (2008)
2. Z. Lv, Y. Cheng, A system design of AMR based on ARM and GPRS. Comput. Knowl. Technol. **4**(36), 2687–2688 (2008)
3. Z. Liang, Analysis of the development on wireless communication technology. China New Telecommun. (13), 6 (2014)
4. W.U. Ke, L. Cheng, Y. Huang, Design and implementation of an automotive performance data acquisition system based on 4G. Comput. Digit. Eng. **45**(02), 397–402 (2017)
5. H.E. Maohui, Design of multi-terminal mobile data acquisition system utilizing 4G network for architectural engineering field. Modern Electr. Tech. **39**(15), 25–27 (2016)
6. W. Zhu, W. Zhou, An intelligent agricultural microclimate monitoring system based on 4G technology. Exp. Technol. Manage. **33**(04), 82–85 (2016)
7. C.H.E.N. Jilong, T.E.N.G. Lin, W.U. Yingchun et al., Study on remote diagnosis system based on 4G communication. Eng. Test **55**(02), 104–106 (2015)
8. B. Qiu, Q. Zhang, J. Li, Research and development of chlorine leak detector based on ARM, in *International Conference on Robots & Intelligent System*. IEEE Computer Society, pp. 242–245 (2017)
9. X. Zheng, S. Hou, Design of fault concentrator based on STM32 and μC/OS-II, in *IEEE International Conference on Computer and Communications* (IEEE, 2017)
10. S. Chen, J. Wu, J. Bao et al., An acoustic emission system for detecting failures of check valves based on STM32 and high-speed USB bulk transfer, pp. 6205–6209 (2015)
11. C. Dong, Y. Zhang, J. Li et al., The integrated design for micro—environment monitoring system of showcase in museum, 05009 (2017)
12. X.F. Shi, T. Feng, X.M. Zhang, Remote data acquisition system based on embedded RT-thread. Appl. Mech. Mat. **241–244**, 2238–2241 (2013)
13. Y. Weng, Application of DGUS multimedia screen in underground drill rig of coal mine. Modern Electr. Tech. **40**(06), 96–99 (2017)
14. D. Boswarthick, M2M communications: a systems approach (Wiley, 2010), pp. 54–56
15. Y.U. Yue, Investigation and evaluation to cremation equipment works condition on shenyang. Environ. Monitor. China **22**(4), 84–86 (2006)

# Stabilizing Quadrotor Helicopter with Uncertainties Based on Controlled Lagrangians and Disturbance Observer

**Zhonglin Li and Wei Huo**

**Abstract** How to apply the Controlled Lagrangian method to stabilization controller design for the quadrotor helicopter with uncertainties is investigated in this paper. The dynamical model of the uncertain quadrotor is transformed to a linear model without uncertainties and an uncertain term to facilitate controller design. First, a stabilization controller for linearized model is design based on the Controlled Lagrangian method. For the under-actuated quatrotor, its uncertainties and control inputs are mismatched, the mismatched uncertainties are replaced with equivalent matched uncertainties by utilizing the equivalent disturbance method, then a disturbance observer is constructed to estimate the matched uncertainties, and added to the controller for linearized model to compensate the effect of uncertainties. It is proved that states of the controlled quadrotor are uniformly ultimately bounded and converge to a small neighborhood of the desired equilibrium point. Simulation results verify effectiveness of the proposed controller with the observer.

**Keywords** Quadrotor helicopter · Controlled Lagrangians
Disturbance observer · Stabilization

## 1 Introduction

In recent years, due to the characteristics of vertical takeoff-and-landing and hovering flight, quadrotor helicopter has attracted more and more attention, and has been widely used in aerial photography, exploration, rescue and so on.

The quadrotor is an under-actuated nonlinear system, i.e. it has six motion degrees of freedom, but there are only four control inputs. Numerous researchers have

---

Z. Li · W. Huo (✉)
The Seventh Research Division, School of Automation Science and Electrical Engineering, Beihang University, Beijing 100191, People's Republic of China
e-mail: weihuo@buaa.edu.cn

Z. Li
e-mail: lizhonglin@buaa.edu.cn

 
Y. Jia et al. (eds.), *Proceedings of 2018 Chinese Intelligent Systems Conference*,
Lecture Notes in Electrical Engineering 528,
https://doi.org/10.1007/978-981-13-2288-4_28

designed many control methods for the quadrotor helicopter, for example, the classical PID and LQ algorithm [1], backstepping method [2], trajectory linearization control [3], neural network approach [4], robust control technique [5] and so on. However, most of control methods divide the system into inner/outer loops structure. The outer loop is used to control the position and the inner loop is used to control the attitude. Accordingly, controller parameters must satisfy the time-scale separation principle, i.e. the convergence rate of inner loop must be much faster than that of outer loop. This makes the design of the controller more complex and the stability analysis of closed loop system more difficult. In contrast to above methods, Controlled Lagrangians (CL) method does not divide the system into inner/outer loops structure, which provides a new approach to design controller of under-actuated nonlinear mechanical systems.

The CL method was first proposed in 1997 [6], after that, it is constantly developed [7, 8]. The CL method is a way to design the mechanical system controller based on the idea of energy reconstruction, which is especially suitable to design controller for under-actuated systems. As a novel control method, it has been applied to many mechanical systems, for example, pendubot [9], Furuta pendulum [10], flexible structure [11], quadrotor helicopter [12, 13], etc. But most applications are based on precise mathematical models and do not consider modeling uncertainties, such as uncertain parameters, unmodeled dynamics and external disturbances. In this paper, taking the quadrotor helicopter with uncertainties as an example, a controller with disturbance observer is designed by using CL method, such that the controlled quadrotor can be stabilized to a small neighborhood of the desired equilibrium point. It provides an novel approach of applying the CL method to uncertain under-actuated mechanical systems.

The contents of this paper are arranged as follows. In Sect. 2, the problem statement is illustrated. In Sect. 3, the controller of linearized model and the disturbance observer are designed. Stability of the closed loop system is proved in Sect. 4. Simulation results for the proposed controller are shown and compared in Sect. 5. Some conclusions are drown in Sect. 6.

## 2 Problem Statement

### 2.1 *Dynamical Model of the Quadrotor Helicopter*

As shown in Fig. 1, the model of quadrotor helicopter is illustrated by two reference frames, namely, the inertial frame $\mathcal{I} = \{O_e x_e y_e z_e\}$ and the body frame $\mathcal{B} = \{Oxyz\}$. The inertial frame $\mathcal{I} = \{O_e x_e y_e z_e\}$ is fixed to the earth, its origin $O_e$ is located at a point on the earth surface, the $x_e$ axis points to the east, the $z_e$ axis points upright vertically and the $y_e$ axis is confirmed by the right-hand rule, pointing to the north. The body frame $\mathcal{B} = \{Oxyz\}$ is fixed to the fuselage of the quadrotor helicopter, its origin $O$ is located at the center of gravity of the fuselage, and directions of $x$, $y$ and $z$

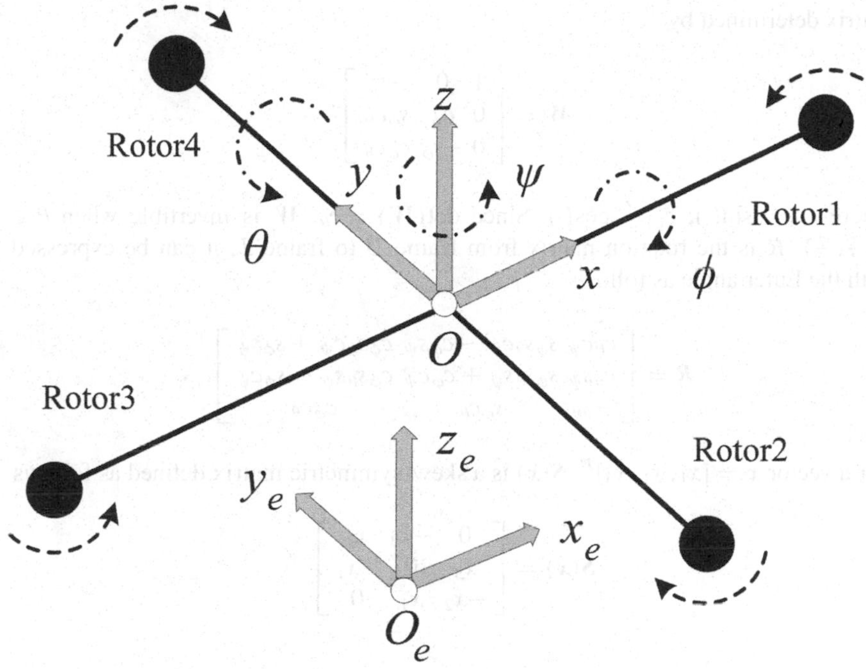


**Fig. 1** Quadrotor helicopter model

axis are illustrated in Fig. 1. In this paper, the quadrotor helicopter is regarded as a six-degree-of-freedom rigid body by neglecting the elastic deformation, its mechanical structure is symmetrical and the gyroscopic effect caused by rotation of the rotors is ignored. From the Newton-Euler equations, mathematical model of the quadrotor helicopter can be derived as follows [14]

$$\dot{\boldsymbol{p}} = \boldsymbol{v} \tag{1}$$

$$\boldsymbol{\omega} = \boldsymbol{W}\dot{\boldsymbol{\gamma}} \tag{2}$$

$$m\dot{\boldsymbol{v}} = -mg\boldsymbol{e}_3 + f\boldsymbol{R}\boldsymbol{e}_3 + \boldsymbol{d}_f \tag{3}$$

$$\boldsymbol{J}\dot{\boldsymbol{\omega}} = -\boldsymbol{S}(\boldsymbol{\omega})\boldsymbol{J}\boldsymbol{\omega} + \boldsymbol{\tau} + \boldsymbol{d}_\tau \tag{4}$$

where $\boldsymbol{p} \triangleq [x, y, z]^T$ and $\boldsymbol{v} \triangleq [v_x, v_y, v_z]^T$ denote position and velocity of the center of mass in frame $\mathcal{I}$, respectively, $\boldsymbol{\gamma} \triangleq [\phi, \theta, \psi]^T$ denotes the Euler angle, $\boldsymbol{\omega} \triangleq [\omega_x, \omega_y, \omega_z]^T$ represents angular velocity of the quadrotor in frame $\mathcal{B}$, $m$ is mass of the quadrotor, $g$ is the gravity acceleration, $\boldsymbol{e}_3 \triangleq [0, 0, 1]^T$, $f$ denotes total thrust in frame $\mathcal{B}$, $\boldsymbol{J}$ =diag$\{J_x, J_y, J_z\}$ denotes inertial matrix of the quadrotor in frame $\mathcal{B}$, $\boldsymbol{\tau} = [\tau_1, \tau_2, \tau_3]^T$ denotes total torque in frame $\mathcal{B}$, $\boldsymbol{d}_f$ and $\boldsymbol{d}_\tau$ denote uncertainties such as unmodeled dynamics and external disturbances, $\boldsymbol{W}$ is attitude transformation matrix determined by

$$W = \begin{bmatrix} 1 & 0 & -s_\theta \\ 0 & c_\phi & s_\phi c_\theta \\ 0 & -s_\phi & c_\phi c_\theta \end{bmatrix}$$

where $s_{(\cdot)} \triangleq \sin(\cdot)$, $c_{(\cdot)} \triangleq \cos(\cdot)$. Since $\det(W) = c_\theta$, $W$ is invertible when $\theta \in (-\frac{\pi}{2}, \frac{\pi}{2})$. $R$ is the rotation matrix from frame $\mathcal{B}$ to frame $\mathcal{I}$, it can be expressed with the Euler angle as follows

$$R = \begin{bmatrix} c_\theta c_\psi & s_\phi s_\theta c_\psi - c_\phi s_\psi & c_\phi s_\theta c_\psi + s_\phi s_\psi \\ c_\theta s_\psi & s_\phi s_\theta s_\psi + c_\phi c_\psi & c_\phi s_\theta s_\psi - s_\phi c_\psi \\ -s_\theta & s_\phi c_\theta & c_\phi c_\theta \end{bmatrix}$$

For a vector $x = [x_1, x_2, x_3]^T$, $S(x)$ is a skew-symmetric matrix defined as follows

$$S(x) = \begin{bmatrix} 0 & -x_3 & x_2 \\ x_3 & 0 & -x_1 \\ -x_2 & x_1 & 0 \end{bmatrix}$$

## 2.2 Model Transformation

Combining (1) and (3) gives

$$m\ddot{p} = -mge_3 + fRe_3 + d_f \tag{5}$$

Combining (2) and (4) leads to

$$JW\ddot{\gamma} = -J\dot{W}\dot{\gamma} - S(W\dot{\gamma})JW\dot{\gamma} + \tau + d_\tau \tag{6}$$

**Assumption 1** In movement process of the quadrotor, $\phi, \theta \in (-\frac{\pi}{2}, \frac{\pi}{2})$.

In this paper, our control objective is to design controller $f$ and $\tau$ such that states $[x, y, z, \phi, \theta, \psi, \dot{x}, \dot{y}, \dot{z}, \dot{\phi}, \dot{\theta}, \dot{\psi}]^T$ of the uncertain quadrotor model (5) and (6) are uniformly ultimately bounded and converge to a small neighborhood of the desired equilibrium point $[x_d, y_d, z_d, 0, 0, \psi_d, 0, 0, 0, 0, 0, 0]^T$.

Taking invertible input transformations

$$f = \frac{m(\widetilde{f} + g)}{c_\phi c_\theta} \tag{7}$$

$$\tau = J\dot{W}\dot{\gamma} + S(W\dot{\gamma})JW\dot{\gamma} + JW\widetilde{\tau} \tag{8}$$

and substituting (7) and (8) into (5) and (6) yield

$$\ddot{\boldsymbol{p}} = \boldsymbol{k}g + \boldsymbol{h}\widetilde{f} + \frac{\boldsymbol{d}_f}{m} \tag{9}$$

$$\ddot{\boldsymbol{\gamma}} = \widetilde{\boldsymbol{\tau}} + \boldsymbol{W}^{-1}\boldsymbol{J}^{-1}\boldsymbol{d}_\tau \tag{10}$$

where $\boldsymbol{k} = [t_\theta c_\psi + t_\phi \frac{s_\psi}{c_\theta}, t_\theta s_\psi - t_\phi \frac{c_\psi}{c_\theta}, 0]^T$, $\boldsymbol{h} = [t_\theta c_\psi + t_\phi \frac{s_\psi}{c_\theta}, t_\theta s_\psi - t_\phi \frac{c_\psi}{c_\theta}, 1]^T$, $t_{(\cdot)} \triangleq \tan(\cdot), \widetilde{\boldsymbol{\tau}} = [\widetilde{\tau}_1, \widetilde{\tau}_2, \widetilde{\tau}_3]^T$. The Taylor series expansions of $\boldsymbol{k}$ and $\boldsymbol{h}$ at the desired equilibrium point $[x_d, y_d, z_d, 0, 0, \psi_d, 0, 0, 0, 0, 0, 0]^T$ are

$$\boldsymbol{k} = \boldsymbol{l} + \boldsymbol{O}_1, \quad \boldsymbol{h} = \boldsymbol{e}_3 + \boldsymbol{O}_2 \tag{11}$$

where $\boldsymbol{l} = [\theta c_{\psi_d} + \phi s_{\psi_d}, \theta s_{\psi_d} - \phi c_{\psi_d}, 0]^T$, $\boldsymbol{O}_1$ and $\boldsymbol{O}_2$ are the high-order nonlinear terms. Substituting (11) to (9), (9) and (10) can be rewritten as

$$\ddot{\boldsymbol{p}} = \boldsymbol{l}g + \boldsymbol{e}_3\widetilde{f} + g\boldsymbol{O}_1 + \widetilde{f}\boldsymbol{O}_2 + \frac{\boldsymbol{d}_f}{m} \triangleq \boldsymbol{l}g + \boldsymbol{e}_3\widetilde{f} + \boldsymbol{d}_1 \tag{12}$$

$$\ddot{\boldsymbol{\gamma}} = \widetilde{\boldsymbol{\tau}} + \boldsymbol{W}^{-1}\boldsymbol{J}^{-1}\boldsymbol{d}_\tau \triangleq \widetilde{\boldsymbol{\tau}} + \boldsymbol{d}_2 \tag{13}$$

Take invertible state transformation

$$\begin{bmatrix} \widetilde{x} \\ \widetilde{y} \\ \widetilde{z} \\ \widetilde{\psi} \end{bmatrix} = \begin{bmatrix} c_{\psi_d} & s_{\psi_d} & 0 & 0 \\ s_{\psi_d} & -c_{\psi_d} & 0 & 0 \\ 0 & 0 & 1 & 0 \\ 0 & 0 & 0 & 1 \end{bmatrix} \begin{bmatrix} x - x_d \\ y - y_d \\ z - z_d \\ \psi - \psi_d \end{bmatrix} \tag{14}$$

and define $\boldsymbol{\eta}_1 = [\widetilde{x}, \theta, \dot{\widetilde{x}}, \dot{\theta}]^T$, $\boldsymbol{\eta}_2 = [\widetilde{y}, \phi, \dot{\widetilde{y}}, \dot{\phi}]^T$, $\boldsymbol{\eta}_3 = [\widetilde{z}, \dot{\widetilde{z}}]^T$, $\boldsymbol{\eta}_4 = [\widetilde{\psi}, \dot{\widetilde{\psi}}]^T$, $\boldsymbol{\eta} = [\boldsymbol{\eta}_1^T, \boldsymbol{\eta}_2^T, \boldsymbol{\eta}_3^T, \boldsymbol{\eta}_4^T]^T$, $\boldsymbol{u} = [\widetilde{\tau}_2, \widetilde{\tau}_1, \widetilde{f}, \widetilde{\tau}_3]^T$, $\boldsymbol{d} = [\boldsymbol{d}_1^T, \boldsymbol{d}_2^T]^T$, then the uncertain model (12) and (13) can be converted to the form of state equation as

$$\dot{\boldsymbol{\eta}} = \boldsymbol{A}\boldsymbol{\eta} + \boldsymbol{B}\boldsymbol{u} + \boldsymbol{B}_d\boldsymbol{d} \tag{15}$$

where $\boldsymbol{A} = \mathrm{diag}\{\boldsymbol{A}_1, \boldsymbol{A}_2, \boldsymbol{A}_3, \boldsymbol{A}_4\}$, $\boldsymbol{A}_1 = \boldsymbol{A}_2 = \begin{bmatrix} 0&0&1&0 \\ 0&0&0&1 \\ 0&g&0&0 \\ 0&0&0&0 \end{bmatrix}$, $\boldsymbol{A}_3 = \boldsymbol{A}_4 = \begin{bmatrix} 0&1 \\ 0&0 \end{bmatrix}$,

$\boldsymbol{B} = \mathrm{diag}\{\boldsymbol{B}_1, \boldsymbol{B}_2, \boldsymbol{B}_3, \boldsymbol{B}_4\}$, $\boldsymbol{B}_1 = \boldsymbol{B}_2 = \begin{bmatrix} 0 \\ 0 \\ 0 \\ 1 \end{bmatrix}$, $\boldsymbol{B}_3 = \boldsymbol{B}_4 = \begin{bmatrix} 0 \\ 1 \end{bmatrix}$, $\boldsymbol{B}_d = \begin{bmatrix} \boldsymbol{B}_{d11} & \boldsymbol{B}_{d12} \\ \boldsymbol{B}_{d21} & \boldsymbol{B}_{d22} \\ \boldsymbol{B}_{d31} & \boldsymbol{0} \\ \boldsymbol{0} & \boldsymbol{B}_{d42} \end{bmatrix}$,

$\boldsymbol{B}_{d11} = \begin{bmatrix} 0&0&0 \\ 0&0&0 \\ c_{\psi_d}&s_{\psi_d}&0 \\ 0&0&0 \end{bmatrix}$, $\boldsymbol{B}_{d12} = \begin{bmatrix} 0&0&0 \\ 0&0&0 \\ 0&0&0 \\ 0&1&0 \end{bmatrix}$, $\boldsymbol{B}_{d21} = \begin{bmatrix} 0&0&0 \\ 0&0&0 \\ s_{\psi_d}&-c_{\psi_d}&0 \\ 0&0&0 \end{bmatrix}$, $\boldsymbol{B}_{d22} = \begin{bmatrix} 0&0&0 \\ 0&0&0 \\ 0&0&0 \\ 1&0&0 \end{bmatrix}$,

$\boldsymbol{B}_{d31} = \begin{bmatrix} 0&0&0 \\ 0&0&1 \end{bmatrix}$, $\boldsymbol{B}_{d42} = \begin{bmatrix} 0&0&0 \\ 0&0&1 \end{bmatrix}$.

With above model transformation, our control objective is equivalent to design $\boldsymbol{u} = [\widetilde{\tau}_2, \widetilde{\tau}_1, \widetilde{f}, \widetilde{\tau}_3]^T$ such that the state $\eta$ of uncertain quadrotor model (15) can be uniformly ultimately bounded and converge to a small neighborhood of the origin.

## 3 Controller Design

The main idea of the CL method is to design a desired controlled Lagrangian for closed loop system, and to make the Lagrangian of the original system equivalent to the controlled Lagrangian, so that the matching conditions can be derived. By solving the matching conditions, the controller structure can be obtained and the controller parameters can be determined by stability theory. But for a mechanical system with uncertainties, since the uncertainties are unknown, it is impossible to solve the matching conditions. This leads to the failure of applying the CL method. This paper provides a method to solve this problem. The method is divided into two steps. The first step is to design a controller for the linearized nominal model without uncertainties by using the CL method. The second step is to design a disturbance observer to estimate the uncertainties. Finally, the estimation of uncertainties is added to the controller designed in the first step, such that the uncertainties can be compensated.

### *3.1 Controller Design of Linearized Model Based on CL Method*

If the uncertainty item $\boldsymbol{B}_d\boldsymbol{d}$ in (15) is removed, the linearized nominal model of (15) can be obtained as follows

$$\dot{\eta} = \boldsymbol{A}\eta + \boldsymbol{B}\boldsymbol{u} \tag{16}$$

Because $\boldsymbol{A}$ and $\boldsymbol{B}$ are both diagonal matrices, the linearized model (16) without uncertainties can be decoupled into four linearized subsystems:

$$\ddot{\widetilde{x}} = g\theta, \ddot{\theta} = \widetilde{\tau}_2 \tag{17}$$

$$\ddot{\widetilde{y}} = g\phi, \ddot{\phi} = \widetilde{\tau}_1 \tag{18}$$

$$\ddot{\widetilde{z}} = \widetilde{f} \tag{19}$$

$$\ddot{\widetilde{\psi}} = \widetilde{\tau}_3 \tag{20}$$

In the literature [12], the CL method is applied to design a stabilization controller for the linearized subsystem (17) and (18). Firstly, design the controller $\widetilde{\tau}_2$ to asymptotically stabilize linearized subsystem (17). Choose $\boldsymbol{q} = [\widetilde{x}, \theta]^T$ as generalized coordinates, construct $\overline{E}_k(\dot{\boldsymbol{q}}) = \frac{1}{2}\dot{\boldsymbol{q}}^T\overline{\boldsymbol{M}}\dot{\boldsymbol{q}}$ and $\overline{E}_p(\boldsymbol{q}) = \frac{1}{2}\boldsymbol{q}^T\overline{\boldsymbol{P}}\boldsymbol{q}$ as the desired kinetic energy and potential energy of the controlled system, respectively, where $\overline{\boldsymbol{M}} = \begin{bmatrix} m_1 & m_2 \\ m_2 & m_3 \end{bmatrix} = \overline{\boldsymbol{M}}^T > 0$, i.e. $m_1 > 0, m_1 m_3 > m_2^2$, $\overline{\boldsymbol{P}} = \begin{bmatrix} p_1 & p_2 \\ p_2 & p_3 \end{bmatrix} = \overline{\boldsymbol{P}}^T > 0$, and the controlled Lagrangian is defined as

$$\overline{L}(\boldsymbol{q}, \dot{\boldsymbol{q}}) = \overline{E}_k(\dot{\boldsymbol{q}}) - \overline{E}_p(\boldsymbol{q}) = \frac{1}{2}\dot{\boldsymbol{q}}^T\overline{\boldsymbol{M}}\dot{\boldsymbol{q}} - \frac{1}{2}\boldsymbol{q}^T\overline{\boldsymbol{P}}\boldsymbol{q} \tag{21}$$

The controlled Euler-Lagrange equation of the closed loop system is as follows

$$\frac{d}{dt}\frac{\partial \overline{L}}{\partial \dot{\boldsymbol{q}}} - \frac{\partial \overline{L}}{\partial \boldsymbol{q}} = \overline{\boldsymbol{u}} \tag{22}$$

where $\overline{\boldsymbol{u}}$ is the desired generalized force to be designed. Combining (21) and (22) gives the controlled closed loop system equation

$$\overline{\boldsymbol{M}}\ddot{\boldsymbol{q}} + \overline{\boldsymbol{P}}\boldsymbol{q} = \overline{\boldsymbol{u}} \tag{23}$$

Assign total energy $\overline{E}$ of the controlled system as a Lyapunov function

$$\overline{E} = \overline{E}_k(\dot{\boldsymbol{q}}) + \overline{E}_p(\boldsymbol{q}) = \frac{1}{2}\dot{\boldsymbol{q}}^T\overline{\boldsymbol{M}}\dot{\boldsymbol{q}} + \frac{1}{2}\boldsymbol{q}^T\overline{\boldsymbol{P}}\boldsymbol{q} \tag{24}$$

and its time derivative along the system (23) satisfies

$$\dot{\overline{E}} = \dot{\boldsymbol{q}}^T\overline{\boldsymbol{M}}\ddot{\boldsymbol{q}} + \dot{\boldsymbol{q}}^T\overline{\boldsymbol{P}}\boldsymbol{q} = \dot{\boldsymbol{q}}^T\overline{\boldsymbol{u}} \tag{25}$$

Design the desired generalized force as

$$\overline{\boldsymbol{u}} = -\overline{\boldsymbol{D}}\dot{\boldsymbol{q}} \tag{26}$$

where $\overline{\boldsymbol{D}} = \begin{bmatrix} d_1 & d_2 \\ d_2 & d_3 \end{bmatrix} = \overline{\boldsymbol{D}}^T \geq 0$. Substituting (26) into (25) yields

$$\dot{\overline{E}} = -\dot{\boldsymbol{q}}^T\overline{\boldsymbol{D}}\dot{\boldsymbol{q}} \leq 0 \tag{27}$$

According to the Lyapunov stability theory, the controlled closed loop system (23) is stable.

From (23) and (26), we can rewrite the controlled closed loop system equation as

$$\ddot{\boldsymbol{q}} = \overline{\boldsymbol{M}}^{-1}(-\overline{\boldsymbol{D}}\dot{\boldsymbol{q}} - \overline{\boldsymbol{P}}\boldsymbol{q}) \tag{28}$$

In order to make the controlled closed loop system (28) equivalent to the original linearized subsystem (17), following condition should be satisfied

$$\overline{\boldsymbol{M}}^{-1}(-\overline{\boldsymbol{D}}\dot{\boldsymbol{q}}-\overline{\boldsymbol{P}}\boldsymbol{q})=\begin{bmatrix} g\theta \\ \widetilde{\tau}_2 \end{bmatrix} \tag{29}$$

From the first row of (29), a matching condition is obtained as

$$-\frac{1}{m_1m_3-m_2^2}[(m_3d_1-m_2d_2)\dot{\widetilde{x}}+(m_3d_2-m_2d_3)\dot{\theta} \\ +(m_3p_1-m_2p_2)\widetilde{x}+(m_3p_2-m_2p_3)\theta]=g\theta \tag{30}$$

Considering the independence of $\dot{\boldsymbol{q}}=[\dot{\widetilde{x}},\dot{\theta}]^T$ and $\boldsymbol{q}=[\widetilde{x},\theta]^T$, the matching condition (30) can be decomposed into

$$m_3d_1-m_2d_2=0,\quad m_3d_2-m_2d_3=0 \tag{31}$$

$$m_3p_1-m_2p_2=0,\quad -(m_3p_2-m_2p_3)=(m_1m_3-m_2^2)g \tag{32}$$

Solving (31) and (32) gives, separately

$$\overline{\boldsymbol{D}}=d_1\begin{bmatrix} 1 & \frac{m_3}{m_2} \\ \frac{m_3}{m_2} & (\frac{m_3}{m_2})^2 \end{bmatrix} \tag{33}$$

$$\overline{\boldsymbol{P}}=\begin{bmatrix} p_1 & \frac{m_3}{m_2}p_1 \\ \frac{m_3}{m_2}p_1 & \frac{m_1m_3-m_2^2}{m_2}g+(\frac{m_3}{m_2})^2p_1 \end{bmatrix} \tag{34}$$

Select $d_1>0$ so that $\overline{\boldsymbol{D}}$ is a positive semidefinite matrix. Choose $p_1>0, m_2>0$ so that $\overline{\boldsymbol{P}}$ is a positive definite matrix.

From the second row of (29), the controller to stabilize linearized subsystem (17) is obtained as

$$\widetilde{\tau}_2=-\frac{p_1}{m_2}\widetilde{x}-\frac{m_3p_1+m_1m_2g}{m_2^2}\theta-\frac{d_1}{m_2}\dot{\widetilde{x}}-\frac{m_3d_1}{m_2^2}\dot{\theta}\triangleq\boldsymbol{k}_1\boldsymbol{\eta}_1\triangleq\widetilde{\tau}_{20} \tag{35}$$

Characteristic equation of the matrix $\boldsymbol{A}_1+\boldsymbol{B}_1\boldsymbol{k}_1$ is as follows

$$\lambda^4+\frac{m_3d_1}{m_2^2}\lambda^3+\frac{m_3p_1+m_1m_2g}{m_2^2}\lambda^2+\frac{d_1g}{m_2}\lambda+\frac{p_1g}{m_2}=0 \tag{36}$$

Considering controller parameters $m_1>0, m_2>0, m_3>\frac{m_2^2}{m_1}, d_1>0, p_1>0$, and using Hurwitz criterion we can verify that all roots of the characteristic equation (36) have negative real parts. Therefore linearized subsystem (17) is asymptotically stable with the controller $\widetilde{\tau}_{20}$.

By using the same method as above, the controller to asymptotically stabilize linearized subsystem (18) can be designed as

$$\tilde{\tau}_1 = -\frac{p_4}{m_5}\tilde{y} - \frac{m_6 p_4 + m_4 m_5 g}{m_5^2}\phi - \frac{d_4}{m_5}\tilde{\dot{y}} - \frac{m_6 d_4}{m_5^2}\dot{\phi} \triangleq k_2\eta_2 \triangleq \tilde{\tau}_{10} \tag{37}$$

where the controller parameters $m_4 > 0, m_5 > 0, m_6 > \frac{m_5^2}{m_4}, d_4 > 0, p_4 > 0$.

With the PD method, the controller to asymptotically stabilize linearized subsystem (19) and (20) can be designed as

$$\tilde{f} = -a_1\tilde{z} - a_2\tilde{\dot{z}} \triangleq k_3\eta_3 \triangleq \tilde{f}_0 \tag{38}$$

$$\tilde{\tau}_3 = -b_1\tilde{\psi} - b_2\tilde{\dot{\psi}} \triangleq k_4\eta_4 \triangleq \tilde{\tau}_{30} \tag{39}$$

where the controller parameters $a_1 > 0, a_2 > 0, b_1 > 0, b_2 > 0$.

## 3.2 Uncertainty Matching

For the under-actuated quatrotor, its uncertainties and control inputs are mismatched, i.e. the dimension of the uncertainty terms is different from that of the control inputs, which makes it difficult to suppress the uncertainty. In this subsection, the mismatched uncertainties are replaced with equivalent matched uncertainties by utilizing the equivalent disturbance method.

**Lemma** [15] For a linear time-invariant system with disturbance $\boldsymbol{d}$ as follows

$$\begin{cases} \dot{\boldsymbol{x}} = \boldsymbol{A}\boldsymbol{x} + \boldsymbol{B}\boldsymbol{u} + \boldsymbol{B}_d\boldsymbol{d} \\ \boldsymbol{y} = \boldsymbol{C}\boldsymbol{x} \end{cases} \tag{40}$$

where $\boldsymbol{x}$ is $n$-dimension state vector, $\boldsymbol{u}$ is $m$-dimension input vector, $\boldsymbol{d}$ is $r$-dimension disturbance vector, $\boldsymbol{y}$ is $p$-dimension output vector. If the system satisfies: (1) $(\boldsymbol{A}, \boldsymbol{B})$ is controllable, $(\boldsymbol{C}, \boldsymbol{A})$ is observable; (2) $(\boldsymbol{A}, \boldsymbol{B}, \boldsymbol{C})$ doesn't have zeros on the imaginary axis, then there must exist a equivalent disturbance $\boldsymbol{d}_e$ corresponding to the disturbance $\boldsymbol{d}$, such that following linear system

$$\begin{cases} \dot{\boldsymbol{x}} = \boldsymbol{A}\boldsymbol{x} + \boldsymbol{B}\boldsymbol{u} + \boldsymbol{B}\boldsymbol{d}_e \\ \boldsymbol{y} = \boldsymbol{C}\boldsymbol{x} \end{cases} \tag{41}$$

satisfies that the output of (41) is identically equal to the output of (40).

From the expressions of $\boldsymbol{B}$ and $\boldsymbol{B}_d$ in uncertain quadrotor model (15), it can be seen that the uncertainty $\boldsymbol{d}$ is not matched with the control $\boldsymbol{u}$. Above **Lemma** is used to match the uncertainty and the control input. Select $\boldsymbol{C} = \boldsymbol{I}_{12\times 12}$, it is easy to verify the system (15) satisfying conditions of the **Lemma**, so the state equation (15) can be equivalently written as

$$\dot{\boldsymbol{\eta}} = \boldsymbol{A\eta} + \boldsymbol{Bu} + \boldsymbol{Bd_e} \tag{42}$$

where $\boldsymbol{d_e}$ is the equivalent disturbance corresponding to the disturbance $\boldsymbol{d}$. In this way, the uncertainty and the control can be matched, which provides convenience for suppressing the uncertainty. But it is should be noted that the **Lemma** only shows existence of the equivalent disturbance $\boldsymbol{d_e}$, but do not provide an algorithm to calculate the equivalent disturbance. Therefore, we propose a equivalent disturbance observer to estimate the disturbance $\boldsymbol{d_e}$ in the next subsection.

## 3.3 Equivalent Disturbance Observer Design

In order to estimate and compensate the equivalent disturbance $\boldsymbol{d_e}$ in system (42), a disturbance observer is constructed by [16]

$$\begin{cases} \dot{\boldsymbol{\xi}} = -\alpha_2(\boldsymbol{\xi} + \boldsymbol{H\eta}) - \alpha_1\widehat{\boldsymbol{d}}_e - \boldsymbol{H}(\boldsymbol{A\eta} + \boldsymbol{Bu}) \\ \dot{\widehat{\boldsymbol{d}}}_e = \boldsymbol{\xi} + \boldsymbol{H\eta} \end{cases} \tag{43}$$

where $\boldsymbol{\xi}$ and $\widehat{\boldsymbol{d}}_e$ are the observer states, $\widehat{\boldsymbol{d}}_e$ is the estimate of $\boldsymbol{d_e}$, $\boldsymbol{H} = \alpha_1(\boldsymbol{B}^T\boldsymbol{B})^{-1}\boldsymbol{B}^T$, $\alpha_1$ and $\alpha_2$ are two positive constants. From (42) and (43), we know

$$\begin{aligned} \ddot{\widehat{\boldsymbol{d}}}_e &= \dot{\boldsymbol{\xi}} + \boldsymbol{H}\dot{\boldsymbol{\eta}} \\ &= -\alpha_2(\boldsymbol{\xi} + \boldsymbol{H\eta}) - \alpha_1\widehat{\boldsymbol{d}}_e - \boldsymbol{H}(\boldsymbol{A\eta} + \boldsymbol{Bu}) + \boldsymbol{H}(\boldsymbol{A\eta} + \boldsymbol{Bu} + \boldsymbol{Bd_e}) \\ &= -\alpha_2\dot{\widehat{\boldsymbol{d}}}_e - \alpha_1\widehat{\boldsymbol{d}}_e + \alpha_1\boldsymbol{d_e} \end{aligned} \tag{44}$$

Denote $\boldsymbol{d_e} = [d_{e1}, \ldots, d_{e4}]^T$, $\widehat{\boldsymbol{d}}_e = [\widehat{d}_{e1}, \ldots, \widehat{d}_{e4}]^T$, the transfer function from $d_{ei}$ to $\widehat{d}_{ei} (i = 1, \ldots, 4)$ can be obtained as

$$\frac{\widehat{d}_{ei}(s)}{d_{ei}(s)} = \frac{\alpha_1}{s^2 + \alpha_2 s + \alpha_1} \tag{45}$$

So if $\alpha_1$ is sufficiently large, $\widehat{\boldsymbol{d}}_e \approx \boldsymbol{d_e}$, it implies that $\widetilde{\boldsymbol{d}}_e = \boldsymbol{d_e} - \widehat{\boldsymbol{d}}_e$ is bounded. It is should be noted that design of the observer (43) is independent of controller $\boldsymbol{u}$, i.e. design of controller $\boldsymbol{u}$ will not affect convergence of the observer.

## 3.4 Total Controller Design

The total controller consists of two parts: the controller of linearized model and the equivalent disturbance observer; i.e. the controller for the uncertain system (15) can be expressed by

$$\boldsymbol{u} = \begin{bmatrix} \widetilde{\tau}_2 \\ \widetilde{\tau}_1 \\ \widetilde{f} \\ \widetilde{\tau}_3 \end{bmatrix} = \begin{bmatrix} \widetilde{\tau}_{20} \\ \widetilde{\tau}_{10} \\ \widetilde{f}_0 \\ \widetilde{\tau}_{30} \end{bmatrix} - \widehat{\boldsymbol{d}}_e = \boldsymbol{K}\eta - \widehat{\boldsymbol{d}}_e \tag{46}$$

where $\boldsymbol{K}\eta$ is controller for the linearized model (16) and $\boldsymbol{K}$ =diag$\{\boldsymbol{k}_1, \boldsymbol{k}_2, \boldsymbol{k}_3, \boldsymbol{k}_4\}$ can be calculated by (35), (37), (38) and (39); $\widehat{\boldsymbol{d}}_e$ is generated via equivalent disturbance observer (43). Finally, the original controller $f$ and $\boldsymbol{\tau}$ for the uncertain quadrotor (5)–(6) can be obtained by (7) and (8), where $\widetilde{f}$ and $\widetilde{\boldsymbol{\tau}} = [\widetilde{\tau}_1, \widetilde{\tau}_2, \widetilde{\tau}_3]^T$ are determined by (46).

## 4 Stability Analysis

**Theorem** If the uncertain nonlinear quadrotor model (5) and (6) satisfies **Assumption** 1, the controller (7), (8) and (46) with disturbance observer (43) can guarantee that the states of the quadrotor are uniformly ultimately bounded and converge to the small neighborhood of the desired equilibrium point $[x_d, y_d, z_d, 0, 0, \psi_d, 0, 0, 0, 0, 0, 0]^T$.

*Proof* Substituting (46) into (42) yields

$$\dot{\eta} = (\boldsymbol{A} + \boldsymbol{B}\boldsymbol{K})\eta + \boldsymbol{B}\widetilde{\boldsymbol{d}}_e \triangleq \boldsymbol{A}_K\eta + \boldsymbol{B}\widetilde{\boldsymbol{d}}_e \tag{47}$$

where $\boldsymbol{A}_K$ =diag$\{\boldsymbol{A}_1 + \boldsymbol{B}_1\boldsymbol{k}_1, \boldsymbol{A}_2 + \boldsymbol{B}_2\boldsymbol{k}_2, \boldsymbol{A}_3 + \boldsymbol{B}_3\boldsymbol{k}_3, \boldsymbol{A}_4 + \boldsymbol{B}_4\boldsymbol{k}_4\}$. Since all the characteristic roots of $\boldsymbol{A}_K$ have negative real parts, for any positive definite matrix $\boldsymbol{Q}$, the Lyapunov equation

$$\boldsymbol{A}_K{}^T\boldsymbol{P} + \boldsymbol{P}\boldsymbol{A}_K = -\boldsymbol{Q}$$

always has a positive definite matrix solution $\boldsymbol{P}$. Assign

$$V(\eta) = \eta^T\boldsymbol{P}\eta$$

as a Lyapunov function, and its time derivative along system (47) satisfies

$$\dot{V} = -\eta^T\boldsymbol{Q}\eta + 2\eta^T\boldsymbol{P}\boldsymbol{B}\widetilde{\boldsymbol{d}}_e \le -\lambda_m(\boldsymbol{Q})\|\eta\|^2 + 2\|\eta\|\|\boldsymbol{P}\boldsymbol{B}\|\|\widetilde{\boldsymbol{d}}_e\|$$

where $\lambda_m(\boldsymbol{Q})$ is the minimum eigenvalue of $\boldsymbol{Q}$, $\|\cdot\|$ represents Euclidean norm. As shown in Sect. 3.3, if $\alpha_1$ is sufficiently large, $\widetilde{\boldsymbol{d}}_e$ is bounded, namely, $\|\widetilde{\boldsymbol{d}}_e\| < \rho$. So

$$\dot{V} < 0, \text{ when } \|\eta\| > \frac{2\|\boldsymbol{P}\boldsymbol{B}\|\rho}{\lambda_m(\boldsymbol{Q})} \triangleq \varepsilon$$

It follows that $\eta$ is uniformly ultimately bounded and will converge to a small neighborhood $\|\eta\| \le \varepsilon$ of the origin. That means the states of the uncertain quadrotor are uniformly ultimately bounded and will converge to a small neighborhood of the desired equilibrium point. According to the definition of $\varepsilon$, it can be arbitrarily small via increasing $\alpha_1$.

## 5 Simulations

In order to verify effectiveness of the proposed controller and compare performance of the controllers with or without disturbance observer, some simulations are performed for the quadrotor helicopter nonlinear model (5) and (6) in the MATLAB environment. According to [4], the model parameters are chosen as $m = 0.9\,\mathrm{kg}$, $g = 9.81\,\mathrm{m/s^2}$, $\boldsymbol{J} = \mathrm{diag}\{0.32, 0.42, 0.63\}\,\mathrm{kgm^2}$, the model uncertainties are chosen as

$$\boldsymbol{d}_f = \begin{bmatrix} c_1 + c_2|v_x| & 0 & 0 \\ 0 & c_3 + c_4|v_y| & 0 \\ 0 & 0 & c_5 + c_6|v_z| \end{bmatrix} \begin{bmatrix} v_x \\ v_y \\ v_z \end{bmatrix} \mathrm{N}$$

$$\boldsymbol{d}_\tau = \begin{bmatrix} c_7 + c_8|\omega_x| & 0 & 0 \\ 0 & c_9 + c_{10}|\omega_y| & 0 \\ 0 & 0 & c_{11} + c_{12}|\omega_z| \end{bmatrix} \begin{bmatrix} \omega_x \\ \omega_y \\ \omega_z \end{bmatrix} \mathrm{Nm}$$

where $c_1, \ldots, c_{12}$ are damping coefficients, given by $\{c_1, \ldots, c_{12}\} = \{0.06, 0.1, 0.06, 0.1, 0.06, 0.1, 0.1, 0.15, 0.1, 0.15, 0.1, 0.15\}$. The controller and disturbance observer parameters are chosen as $m_1 = m_4 = 3, m_2 = m_5 = 3, m_3 = m_6 = 5, d_1 = d_4 = 2, p_1 = p_4 = 1, a_1 = b_1 = 1, a_2 = b_2 = 2, \alpha_1 = 5, \alpha_2 = 1$.

The desired position and attitude are given by $[x_d, y_d, z_d]^T = [1, 1, 5]^T$ m, $[\phi_d, \theta_d, \psi_d]^T = [0, 0, \pi/6]^T$ rad. The initial states are assigned as $[x(0), y(0), z(0)]^T = [-5, -5, 1]^T$ m, $[\dot{x}(0), \dot{y}(0), \dot{z}(0)]^T = [0, 0, 0]^T$ m/s, $[\phi(0), \theta(0), \psi(0)]^T = [0, 0, -0.2]^T$ rad, $[\dot{\phi}(0), \dot{\theta}(0), \dot{\psi}(0)]^T = [0, 0, 0]^T$ rad/s.

Figure 2 illustrates the simulation results for the proposed controller with observer. From the results, we can see that under the action of the controller, the quadrotor helicopter moves from the initial state to the desired equilibrium point, which validates the effectiveness of the controller designed in this paper.

If the parameters are still selected as above, but the estimated value of the disturbance observer is not used in the controller, the simulation results are shown as Fig. 3. It is shown from the results that the position variables, attitude variables and control inputs of the quadrotor helicopter will become very large after a period of flight. The reason is that the attitude angle $\phi$ and $\theta$ reached the singular point $\pi/2$, since the uncertainties can not be efficiently compensated, the quadrotor helicopter has overturned.

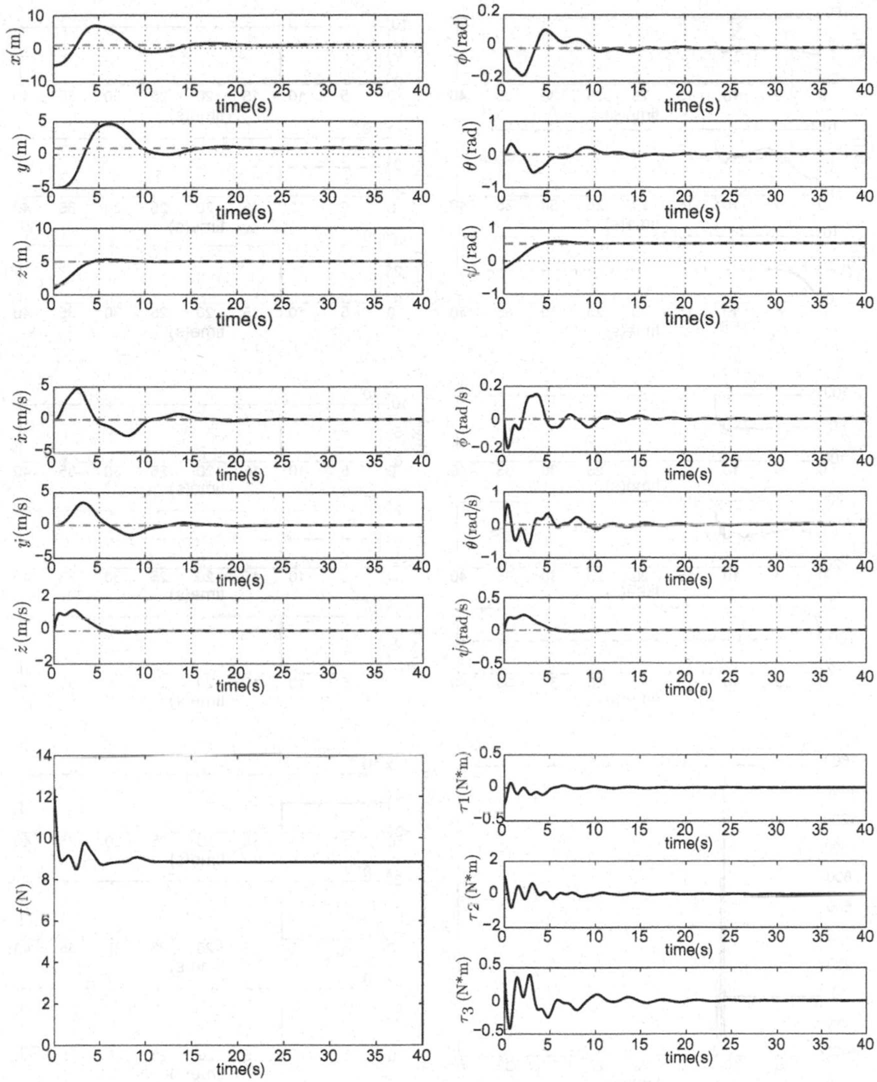


**Fig. 2** Simulation results with disturbance observer

By comparing the two simulation results, it can be seen that the controller with disturbance observer designed in this paper has better robustness than the controller without disturbance observer. In practical engineering, the quadrotor helicopter model contains unmodeled dynamics and external disturbances, so the controller designed in this paper has more practical value.

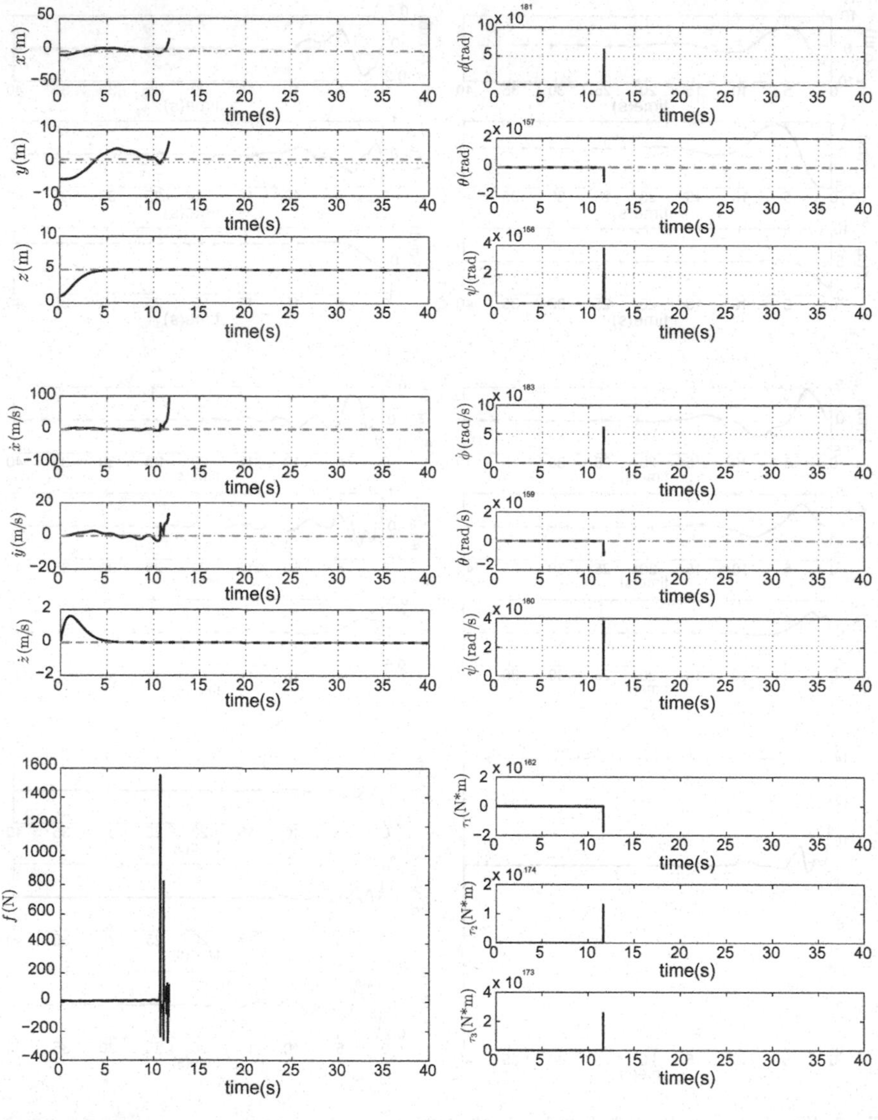

**Fig. 3** Simulation results without disturbance observer

# 6 Conclusions

In this paper, a stabilization controller with disturbance observer is designed for the nonlinear model of quadrotor helicopter with uncertainties. It provides an novel approach of applying the CL method to controller design of uncertain under-actuated mechanical systems. The simulation results verify the effectiveness and robustness of the proposed controller.

**Acknowledgements** This work is supported by National Natural Science Foundation (NNSF) of China under Grant No. 61673043.

## References

1. S. Bouabdallah, A. Noth, R. Siegwart, PID vs LQ control techniques applied to an indoor micro quadrotor, in *IEEE/RSJ International Conference on Intelligent Robots and Systems* (Sendai, 2004), pp. 2451–2456
2. T. Madani, A. Benallegue, Backstepping control for a quadrotor helicopter, in *IEEE/RSJ International Conference on Intelligent Robots and Systems* (Beijing, 2006), pp. 3255–3260
3. B. Zhu, W. Huo, Trajectory linearization control for a quadrotor helicopter, in *IEEE International Conference on Control and Automation* (Xiamen, 2010), pp. 34–39
4. T. Dierks, S. Jagannathan, Output feedback control of a quadrotor UAV using neural networks. IEEE Trans. Neural Netw. **21**(1), 50–66 (2010)
5. H. Liu, Y. Bai, L. Geng, Z. Shi, Y. Zhong, Robust tracking control of a quadrotor helicopter. J. Intell. Robot. Syst. **75**(3-4), 595–608 (2014)
6. A.M. Bloch, N. Ehrich Leonard, J.E. Marsden, Stabilization of mechanical systems using controlled Lagrangians, in *IEEE Conference on Decision and Control* (San Diego, 1997), pp. 2356–2361
7. M.A. Bloch, N. Ehrich Leonard, J.E. Marsden, Controlled Lagrangians and the stabilization of mechanical systems. I. The first matching theorem. IEEE Trans. Auto. Control **45**(12), 2253–2270 (2000)
8. M.A. Bloch, D.E. Chang, N. Ehrich Leonard, J.E. Marsden, Controlled Lagrangians and the stabilization of mechanical systems. II. Potential shaping. IEEE Trans. Auto. Control **46**(10), 1556–1571 (2001)
9. M.-Q. Li, W. Huo, Controller design for mechanical systems with underactuation degree one based on controlled Lagrangians method. Int. J. Control **82**(9), 1747–1761 (2009)
10. K. Machleidt, J. Kroneis, S. Liu, Stabilization of the Furuta Pendulum using a nonlinear control law based on the method of controlled Lagrangians, in *IEEE International Symposium on Industrial Electronics* (Vigo, 2007), pp. 2129–2134
11. Z. Liao, W. Huo, Tracking control of underactuated mechanical systems based on controlled Lagrangians, in *IEEE International Conference on Industrial Technology* (Athens, 2012), pp. 278–283
12. B. Zhang, W. Huo, Stabilizing quadrotor helicopter based on controlled Lagrangians, in *Chinese Intelligent Systems Conference* (Mudanjiang, 2017), pp. 685–695
13. B. Zhang, W. Huo, Stabilization of quadrotor with air drag based on controlled Lagrangians method, in *Chinese Automation Congress* (Jinan, 2017), pp. 5798–5801
14. Z. Zuo, C. Wang, Adaptive trajectory tracking control of output constrained multi-rotors systems. IET Control Theory Appl. **8**(13), 1163–1174 (2014)
15. J.-H. She, M. Fang, Y. Ohyama, H. Hashimoto, W. Min, Improving disturbance-rejection performance based on an equivalent-input-disturbance approach. IEEE Trans. Industr. Electr. **55**(1), 380–389 (2008)
16. W.-H. Chen, Disturbance observer based control for nonlinear systems. IEEE/ASME Trans. Mechatr. **9**(4), 706–710 (2004)

# A Fault Diagnosis Approach Based on Deep Belief Network and Its Application in Bearing Fault

**Qiulin Dan, Xuyu Liu, Yi Chai, Ke Zhang and Hao Li**

**Abstract** With the development of Industry 4.0, not only the equipment but also the operational conditions in industrial manufacturing are becoming more and more complex. It is necessary to diagnose failures, whose probability is now increasing violently. As a typical deep learning model, the Deep Belief Network (DBN) can be employed to extract features from the original data directly. Compared with traditional fault diagnosis methods, the DBN can get rid of the dependence on signal processing technology and diagnosis experience. In this paper, the fault diagnosis approach based on DBN is studied to identify the bearing failure. First of all, the basic principles of DBN and the steps of fault diagnosis are described. Then some key parameters of DBN which affect the fault identification performance are analyzed and determined according to the simulation experiments. The practicability of this method is verified by comparing with Support Vector Machine (SVM) and Back Propagation Neural Network (BPNN) at last.

**Keywords** Signal processing · Fault diagnosis · Deep Belief Network · Feature extraction

## 1 Introduction

In general, the fault is defined as an event that causes the function of some components in system to be ineffective which causes deterioration of entire system. The reason for fault diagnosis is to estimate the state of the internal components of equipment by analyzing the data of related information of the equipment, so as to determine whether to repair or protect it [1].

Q. Dan (✉) · X. Liu · Y. Chai · K. Zhang · H. Li
School of Automation, Chongqing University, Chongqing City 400044, China
e-mail: 20161302019t@cqu.edu.cn

Q. Dan · Y. Chai · K. Zhang · H. Li
Key Laboratory of Complex System Safety and Control, Ministry of Education, Chongqing University, Chongqing City 400044, China



Y. Jia et al. (eds.), *Proceedings of 2018 Chinese Intelligent Systems Conference*, Lecture Notes in Electrical Engineering 528,
https://doi.org/10.1007/978-981-13-2288-4_29

Traditional methods of fault diagnosis are mainly based on analytic model, knowledge and signal processing. But it's difficult to establish clear and accurate mathematical model for some complex systems. Moreover, there are some technical difficulties in how to obtain expert knowledge efficiently. In addition, the fault diagnosis method which is based on signal processing (such as wavelet analysis, support vector machine, etc.) has strong dependence on signal processing technology which may lead to dimensionality disaster for multidimensional data.

The DBN was proposed by Geoffrey Hinton in 2006, has powerful automatic feature extraction capability. It can obtain high-level feature representations by layer-by-layer greedy learning from original signals, thus can avoid extracting the fault feature artificial, and enhance the intelligence of the recognition process [2, 3].

The early DBN is mainly applied to speech recognition [4], image recognition and other fields, and has achieved good research results. Due to the advantage of DBN in feature extraction, more and more people have begun to apply it to fault diagnosis in recent years. For example, Tamilselvan used DBN to carry out health testing of aircraft engines and health diagnosis of power transformers firstly, which has got fine research findings [5]. In addition, Liao Ning also utilized Empirical Mode Decomposition (EMD) to extract feature and then combined with DBN to realize the fault classification and recognition of rolling bearings [6]. The DBN has been used in the fault diagnosis of some mechanical equipment, such as compressor [7], bearing [8] and so on.

However, most of the related researches only use DBN as a classifier for fault recognition, still need some traditional methods to extract feature, such as EMD, et al. Therefore, the strong feature extraction capabilities of DBN have not been fully utilized. Besides, there is no effective method to determine the structural parameters of DBN. Thus, this paper studied an approach of fault diagnosis based on DBN and applied it to bearing fault diagnosis. Then analyzed the key parameters of DBN by simulation experiments, such as length of the input sample, number of iterative learning and number of hidden layer nodes. Finally, the key parameters were determined according to the analysis of simulation results, and the effectiveness and practicality of this method was verified by comparing with traditional fault diagnosis methods.

## 2 Basic Principle of the DBN

The DBN is a deep learning method, which is composed of probabilistic statistics, machine learning and neural networks. Its component is the Restricted Boltzmann Machines (RBM). Optimizing the connection weight of DBN with a greedy learning algorithm is the core of DBN. That is, using positive unsupervised layer-by-layer training to extract the features of equipment fault effectively at first, then optimizing the fault identification capabilities of DBN by reverse supervised fine-tuning [9].

## 2.1 Restricted Boltzmann Machines

The RBM is a randomly generated neural network that can learn probability distribution by input data sets. Each RBM contains a visible layer and a hidden layer, in which the visible layer and the hidden layer are bidirectionally connected, but there is no connection between units in the same layer [10]. The RBM can be regarded as an undirected graph model with bisection structure, as shown in Fig. 1.

Where the bottom layer is the visible layer unit $v = \{v_1, v_2, v_3, \ldots, v_n\} \in \{0, 1\}$, it's used to represent the observed data, and $n$ is the number of visible layer neurons. The top layer is the hidden layer unit $h = \{h_1, h_2, h_3, \ldots, h_m\} \in \{0, 1\}$, which is used to extract abstract features from input data, and $m$ is the number of hidden layer neurons. Besides, $w$ represents the weight matrix between hidden layer and visible layer, $b$ and $a$ respectively represent the bias of visible layer element and hidden layer element.

Multiple RBM stacks form a DBN, you can see it in Fig. 2. The DBN model can extract feature from original data layer by layer, and finally get a higher level of expression. The weight and the bias in RBM are continuously updated before the maximum number of iterations is reached.

The training process of DBN includes two aspects: the unsupervised forward stacking RBM learning process and the supervised backward fine-tuning learning process [11].

## 2.2 Forward Stacking RBM Learning

The essence of forward stacking RBM learning is to use a greedy algorithm to perform unsupervised training on RBM layer by layer [12]. The output of the last layer of RBM training is used as the input of next layer, and to obtain the initial parameters of each RBM through layer-by-layer training.

According to the basic model of RBM described in Sect. 2.1, we can define $\theta = \{w, a, b\}$, so the energy function can be expressed as

$$E(v, h; \theta) = -\sum_{i=1}^{n} b_i v_i - \sum_{j=1}^{m} a_j h_j - \sum_{i=1}^{n}\sum_{j=1}^{m} v_i w_{ij} h_j \tag{1}$$

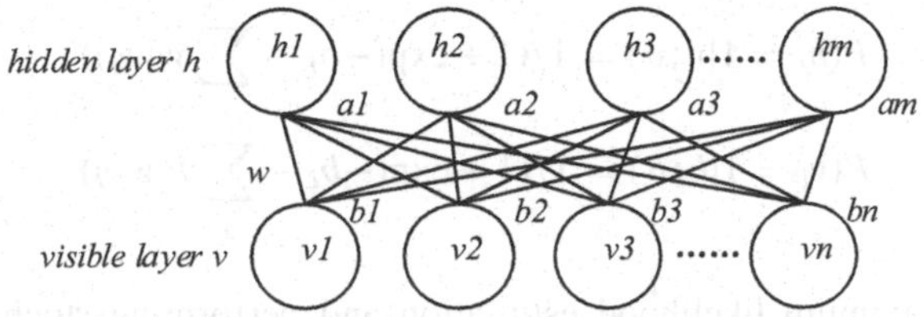


**Fig. 1** RBM basic model

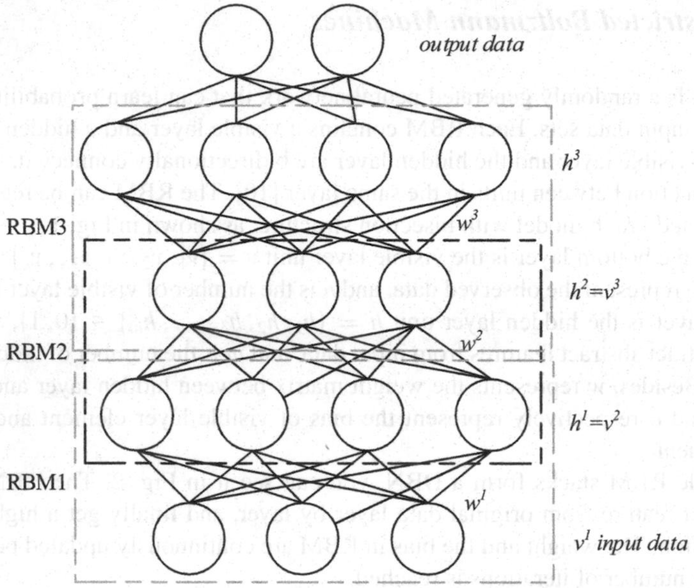


**Fig. 2** DBN structure

Then defining the joint probability of hidden layer and visible layer

$$P(v, h; \theta) = \frac{1}{Z(\theta)} \exp(-E(v, h; \theta)) \tag{2}$$

where $Z(\theta) = \sum \exp(-E(v, h; \theta))$, the conditional probability of hidden layer and visible layer can be obtained

$$P(h|v; \theta) = P(v, h; \theta)/P(v; \theta) = \prod_j P(h_j|v; \theta) \tag{3}$$

$$P(v|h; \theta) = P(v, h; \theta)/P(h; \theta) = \prod_i P(v_i|h; \theta) \tag{4}$$

Since there is no connection inside the visible layer and the hidden layer, the activation function is represented as follows

$$P(h_j = 1|v; \theta) = 1/(1 + \exp(-a_j - \sum_i v_i w_{ij})) \tag{5}$$

$$P(v_i = 1|h; \theta) = 1/(1 + \exp(-b_i - \sum_j h_j w_{ij})) \tag{6}$$

By using maximum likelihood estimation and performing stochastic gradient descent [13], the parameter of RBM model $\theta = \{b_i, a_j, w_{ij}\}$ is obtained from training samples

$$\frac{\partial \ln P(v)}{\partial \theta} = \sum_h p(h/v)[-\frac{\partial E(v,h)}{\partial \theta}] - \sum_v \sum_h p(v,h)[-\frac{\partial E(v,h)}{\partial \theta}] \tag{7}$$

The subtracted number of Formula (7) can be understood as the expected value of the energy partial derivative function of input sample under the distribution of the sample itself. The subtraction is the expected value of the energy partial derivative function under RBM model representation distribution, it's difficult to obtain its value [14].

Using the Contrast Divergence (CD) algorithm proposed by Professor Hinton to solve appeal problem, the parameters are updated as follows

$$\Delta w_{ij} = \varepsilon(< v_i h_j >_{p(h/v)} - < v_i h_j >_{recon}) \tag{8}$$

$$\Delta b_i = \varepsilon(< v_i >_{p(h/v)} - < v_i >_{recon}) \tag{9}$$

$$\Delta a_j = \varepsilon(< h_j >_{p(h/v)} - < h_j >_{recon}) \tag{10}$$

where $\varepsilon$ is learning rate, $< \ldots >_{p(h/v)}$ is the expectation of partial derivative function in the distribution of $p(h/v)$, and $< \ldots >_{recon}$ is the partial derivative function expectation under reconfiguration model distribution.

In general, the vector generated by the first RBM visible layer is passed to the hidden layer through the RBM network. On the contrary, the hidden layer is used to reconstruct the visible layer, and the weight is updated according to the difference between the reconstructed layer and the visible layer, until reaching the maximum number of iterations.

## 2.3 Backward Fine-Tuning Learning

Backward fine-tuning learning adds a small amount of label data to the top layer of DBN after the forward stacking RBM learning finished. It is supervised and trained by using Back Propagation (BP) to fine tune relevant parameters.

The fine-tuning process utilizes the BP algorithm to distribute errors to each layer of RBM, thus realizing supervised learning of the entire network. Perform optimization operations on all hidden layers at the same time, and the weight update criteria for each layer are descried as follows

$$\Delta w^{(t)} = m_b^{(t)} \Delta w^{(t-1)} + \eta_b^{(t)} w^{(t-1)} \tag{11}$$

$$w^{(t)} = w^{(t-1)} - \Delta w^{(t)} \tag{12}$$

where $w^{(t)}$ represents the weight coefficient between each layer in DBN, $m_b^{(t)}$ and $\eta_b^{(t)}$ respectively represent the momentum factor and the learning rate between two layers in the DBN structure.

The supervised training of DBN can reduce the training errors and improve the accuracy of DBN classification model.

# 3 Process of Fault Diagnosis Based on DBN

The process of fault diagnosis mainly extracts the features of fault sample data through DBN, then classifies the faults combing the Softmax regression model. The main steps are descried as follows:

(1) Defines the problem of fault diagnosis and the fault types, so as to determine the number of nodes in the output layer of DBN.
(2) Since the DBN requires input data in the range of [0, 1], the fault data needs to be normalized. Because we will diagnose the bearing faults next, but there are positive and negative values in the vibration signal, so we choose the linear normalization method

$$\tilde{x}_i = (x_i - x_{\min})/(x_{\max} - x_{\min}) \tag{13}$$

(3) Divides the normalized data into training set and test set for subsequent DBN learning.
(4) Initializes the relevant parameters of DBN, such as input sample length, number of nodes per layer, number of iterations, momentum factor, learning rate, weight matrix and so on.
(5) Uses the training set as input data of the DBN to conduct deep learning, which including forward stacking learning and backward fine-tuning learning.
(6) Input the test set into the DBN model which is trained, and the output vectors of each hidden layer are recorded.

As shown in Fig. 3, the DBN is a comprehensive model that combines feature extraction and fault classification. The core of this network is to extract features through layer-by-layer learning from original fault data and classify the faults through the Softmax model.

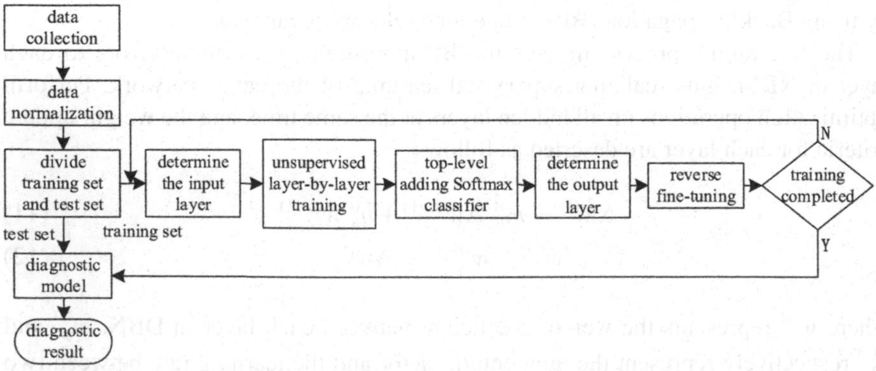


**Fig. 3** Fault diagnosis framework based on DBN

# 4 Simulation Experiments of Bearing Fault Diagnosis

Bearing as an indispensable part of rotating machinery, whose health condition has great influence on the stability of the machine [15]. According to the statistics, more than 30% of mechanical fault is caused by the bearing fault [16], so the fault diagnosis of bearing has a measure of engineering significance.

This paper selects the bearing data published by Case Western Reserve University which simulated the normal state and three kinds of fault state (ball fault, inner race fault, and outer race fault) of the bearing. There are three kinds of fault diameters (0.007, 0.014 and 0.021 in.) of each fault state, which can be understood as the degree of fault (mild, moderate and severe fault). Besides, the data set was acquired at a sampling frequency of 12 kHz at 4 different loads (0–3 HP) [9]. This study mainly focused on the fault data with a motor speed of 1797 rpm and a load of 0, and the continuous sampling time is about 5 s. The experimental environment is described as follows: the computer processor is Intel(R) Core(TM) i5-7500 3.40 GHz, the memory is 8 GB, the operating system is 64-bit, and the simulation tool is MATLAB R2016a.

## 4.1 Effect of Sample Length on Fault Diagnosis Results

Sample length is one of the initial parameters of DBN, and its selection has some randomness. Besides, it's necessary to set different numbers of hidden layer nodes due to sample length, thus affecting the fault recognition results. There is no doubt that we need to study and analyze the selection principle of sample length.

The number of data collected by sensor in one cycle is $12{,}000 \times 60/1797 \approx 400$. For research the effect of different sample length on fault recognition results, 100 data points will be taken as step length to construct 20 sets of the fault data with different sample length (the minimum is 100 and the maximum is 2000).

70% of fault samples are selected as training samples and the remaining 30% are selected as test samples. The number of layers of the DBN model is 5; the number of hidden layer nodes is set lower than the upper layer nodes number, the length of the sample is equal to the number of nodes in the input layer, and the number of output layer nodes is 10, which equal to the number of fault types. The rest of the parameters settings for DBN training process are shown in the following Table 1.

**Table 1** The initialization parameters of the DBN learning process

| Momentum factor | Learning rate $\varepsilon$ | Learning rate $\eta$ | Number of iterations | Weight matrix | Bias b | Bias a |
|---|---|---|---|---|---|---|
| 0.9 | 0.1 | 0.01 | 100 | Normal random distribution | 0 | 0 |

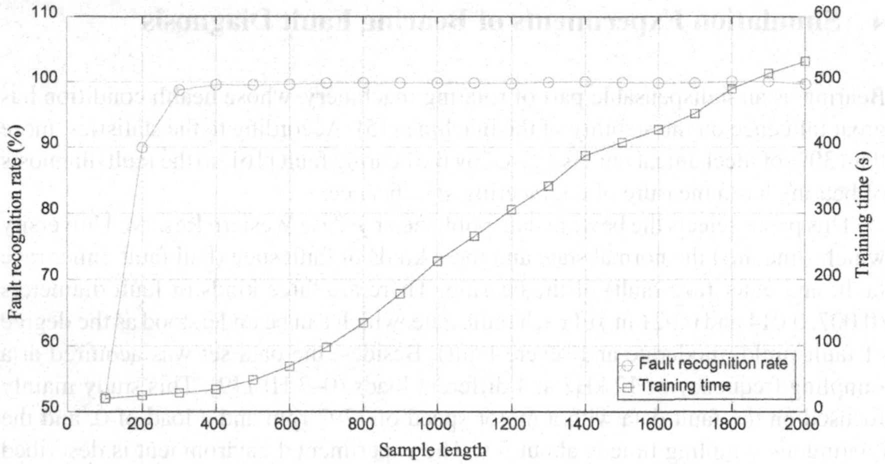


**Fig. 4** Fault identification results of different sample lengths

In order to eliminate the randomness of the algorithm, the experiment repeated 10 times to get the average value. Figure 4 shows the bearing fault recognition results for different sample lengths.

From the results, the fault recognition rate increases with the increase of sample length. Among them, when the sample length is less than 400, the fault recognition rate increases rapidly as the sample length increases. But when it is greater than 400, the fault recognition rate is basically stable to a certain extent.

The analysis shows that when the sample length is less than the number of bearing data collected by sensor in a cycle, the fault recognition rate is low. When it is greater than the number of data in a cycle, the fault recognition rate has reached a high level, but the training time of DBN also increases rapidly. Considering the fault recognition rate and the time cost, we always select the number of data sampled in one cycle of the bearing as sample length.

### *4.2 Effect of Number of Iterations on Fault Diagnosis Results*

Before reaching the maximum number of iterations, the weight and the bias in RBM are continuously updated, so the number of iterations determines when the training end. In order to study the influence of the number of iterations on fault recognition results, 20 sets of different iterations of simulation experiments will be conducted, and other parameters of DBN are consistent with those in Sect. 4.1. Similarly, in order to enhance the reliability of the classification results, the average value of 10 experimental results [17] was taken, and Fig. 5 shows the fault recognition results for different number of iterations.

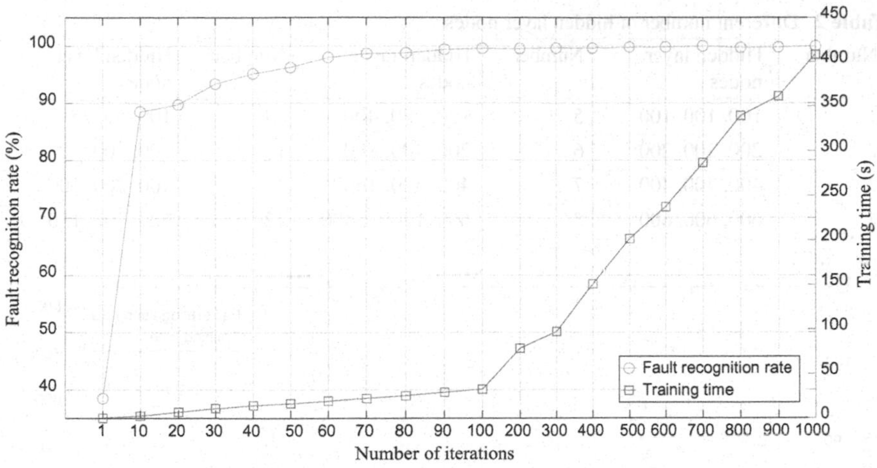


**Fig. 5** Fault identification results of different number of iterations

We can get from the fault identification results: when the number of iterations is less than 100, the rate of fault recognition increases with the number of iterations. When it is greater than 100, the fault recognition rate changes little with the increasing of iterations, but the training time of DBN increases quickly.

In general, the increase of the number of iterations is beneficial to improve the fault recognition rate, but when it is large enough, the rate of fault recognition increases little with high time cost. Therefore, this study setting the number of iterations to 100 is a good choice, which can not only guarantee the recognition rate of the fault, but also reduce the training time.

## 4.3 *Effect of Number of Hidden Layer Nodes on Fault Diagnosis Results*

At present, most of the number of hidden layer nodes are set according to some empirical formula, which is subjective. Therefore, for the purpose of studying the influence of the number of hidden layer nodes on fault identification results, 12 sets of simulation experiments with different hidden layer nodes will be taken, as shown in Table 2, and the rest parameters of the training process are the same as before.

Similarly, each experiment was repeated 10 times, and the average of the results was taken. Figure 6 shows the fault recognition results for different number of hidden layer nodes.

From the fault identification results, it is obvious to conclude that when the number of hidden layer nodes is set to a combination of 400-200-100, we can get a good fault recognition rate with short training time. When the number of hidden layer nodes in DBN is too large, too many nodes may analyze the same information and interfere with each other. On the other hand, if it is far less than the number of input layer nodes, some details of the initial data may not be expressed, thus reducing the fault recognition rate.

**Table 2** Different number of hidden layer nodes

| Number | Hidden layer nodes | Number | Hidden layer nodes | Number | Hidden layer nodes |
|---|---|---|---|---|---|
| 1 | 100, 100, 100 | 5 | 100, 200, 400 | 9 | 100, 50, 25 |
| 2 | 200, 200, 200 | 6 | 200, 400, 800 | 10 | 200, 100, 50 |
| 3 | 400, 400, 400 | 7 | 400, 800, 1600 | 11 | 400, 200, 100 |
| 4 | 600, 600, 600 | 8 | 600, 1200, 2400 | 12 | 600, 300, 150 |

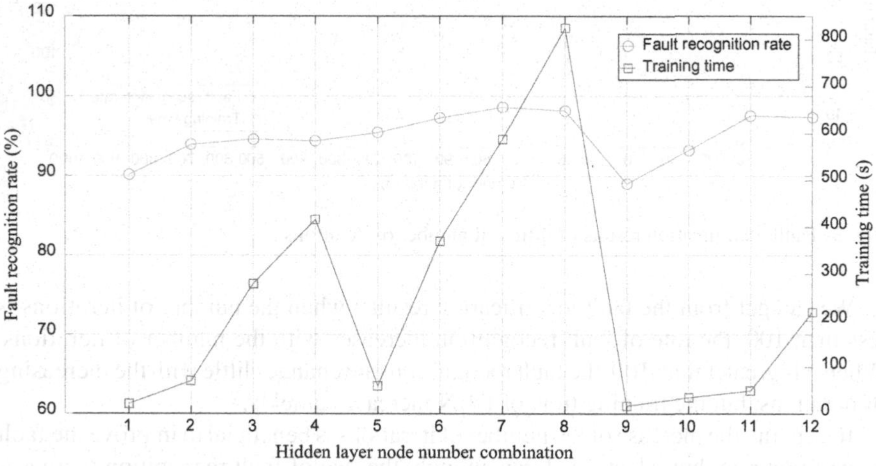


**Fig. 6** Fault identification results of different number of hidden layer nodes

The analysis shows that the more nodes of the hidden layer, the longer training time required, but the fault recognition rate will not necessarily increase. When the hidden layer nodes are closer to the input layer nodes, the higher fault recognition will be got. If we want to reduce the time cost on the premise of ensuring the fault recognition rate, the number of hidden layer nodes can be set slightly lower than the number of visible layer nodes.

## 4.4 Experimental Results and Analysis

In order to verify the fault diagnosis performance of DBN under the parameter structure set up in this paper, two traditional fault diagnosis methods (Support Vector Machine (SVM) and Back Propagation Neural Network (BPNN)) were compared with this method. Among them, the parameters of DBN are consistent with previous research results, SVM uses a radial basis function, the penalty and nuclear parameters are 2.62 and 1.83, and the network structure of BPNN is set to 400-200-10. The following Fig. 7 shows the fault identification results of different fault diagnosis methods.

As can be seen from the experimental data that the fault diagnosis results of different methods are basically stable. Among them, the bearing fault recognition rate based on DBN is greater than 98%. The fault recognition rate based on SVM method can reach 93.3%. And BPNN method has the lowest fault recognition rate, which is near 90%.

The analysis shows that DBN can form more abstract high-level representations by combining low-level features to discover the internal characteristics of data [11], which shows good performance in the fault diagnosis of bearing. SVM and BPNN are shallow learning fault diagnosis methods, the expression ability of complex functions is limited [9], and the effect of bearing fault diagnosis is less than DBN. In addition, when the training sample is large and feature dimension is large, the SVM may lead to the dimension disaster, thus the fault recognition rate was reduced. In the training process of BPNN, it may fall into the local optimal, thus the higher fault recognition rate [18–20] can't be obtained. Generally speaking, the approach of fault diagnosis based on DBN can extract the characteristics from original data, which is more conducive to the fault diagnosis of bearing than other traditional fault diagnosis methods.

## 5 Conclusions

In this paper, the fault diagnosis method based on DBN is studied. First of all, the basic principle of DBN and the basic steps to realize fault diagnosis are described in detail. And the DBN is introduced to address the problem of fault diagnosis of bearing. Meanwhile, some key parameters (such as sample length, number of iterations and so on) which affect the fault diagnosis result are studied by simulation experiments. Then the parameters of DBN are set according to the experimental results and used for bearing fault diagnosis. Finally, by compared with traditional fault diagnosis

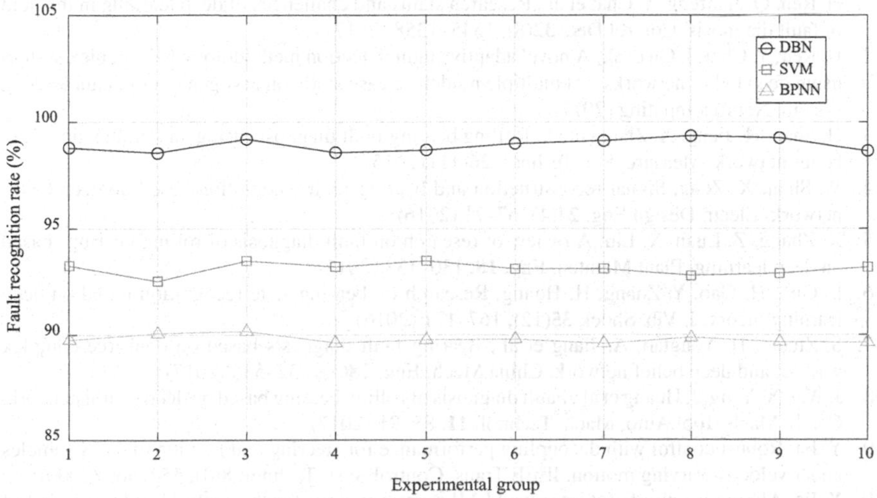


**Fig. 7** Comparison of fault diagnosis results of 10 experiments in different methods

methods, we can sum up that the fault diagnosis method base on DBN has a high fault recognition rate than others, and it has good practicality and effectiveness in fault diagnosis of bearing.

## References


1. J. Xiong, Q. Zhang, Z. Li, J. Xiong, Q. Zhang, G. Sun et al., An information fusion fault diagnosis method based on dimensionless indicators with static discounting factor and KNN. IEEE Sens. J. **16**(7), 2060–2069 (2016)
2. Q. Zhang, L.T. Yang, Z. Chen, Deep computation model for unsupervised feature learning on big data. IEEE Trans. Serv. Comput. **9**(1), 161–171 (2016)
3. D. Yu, L. Deng, Deep learning and its applications to signal and information processing [exploratory DSP]. IEEE Signal Process. Mag. **28**(1), 145–154 (2010)
4. G. Hinton, L. Deng, D. Yu et al., Deep neural networks for acoustic modeling in speech recognition: The shared views of four research groups. IEEE Signal Process. Mag. **29**(6), 82–97 (2012)
5. P. Tamilselvan, P.F. Wang, Failure diagnosis using deep belief learning based health state classification. Reliab. Eng. Syst. Safety **115**(7), 124–135 (2013)
6. N. Liao, J. Tao, D. Yang, Fault diagnosis of double row conical rolling bearing based on deep learning and empirical mode decomposition. J. Hunan Univ. Sci. Technol. (Natural Science) **32**(2), 70–77 (2017)
7. A. Tranvt, A. Ball, An approach to fault diagnosis of reciprocating compress or valves using Teager-Kaiser energy operator and deep belief networks. Expert Syst. Appl. **41**(9), 4113–4122 (2014)
8. W. Li, W. Shan, X.Q. Zeng, Bearing fault classification and recognition based on DBN. J. Vibr. Eng. **29**(2), 340–347 (2016)
9. G. Zhao, Q. Ge, X. Liu et al., Research on fault feature extraction and diagnosis method based on DBN. Chinese J. Sci. Instr. **37**(9), 1946–1953 (2016)
10. J. Liu, Y. Liu, X. Luo, Research progress of Boltzmann machine. J. Comput. Res. Develop. **51**(1), 000001–16 (2014)
11. H. Ren, Q. Jianfeng, Y. Chai et al., Research status and challenges of deep learning in the field of fault diagnosis. Control Dec. **32**(8), 1345–1358 (2017)
12. H. Ren, Y. Chai, J. Qu et al., A novel adaptive fault detection methodology for complex system using deep belief networks and multiple models: a case study on cryogenic propellant loading system. Neurocomputing (2017)
13. H. Shao, H. Jiang, X. Zhang et al., Rolling bearing fault diagnosis using an optimization deep belief network. Measure. Sci. Technol. **26**(11) (2015)
14. W. Shan, X. Zeng, Signal reconstruction and bearing fault recognition based on deep belief network. Electr. Design Eng. **24**(4), 67–71 (2016)
15. X. Zhang, Z. Luan, X. Liu, A review of research on fault diagnosis of rolling bearings based on deep learning. Plant Mainten. Eng. **18**, 130–133 (2017)
16. L. Guo, H. Gao, Y. Zhang, H. Huang, Research on bearing state recognition based on deep learning theory. J. Vib. Shock **35**(12), 167–171 (2016)
17. S. Zhang, H. Yongtao, A. Jiang et al., Bearing fault diagnosis based on dual tree complex wavelet and deep belief network. China Mech. Eng. **28**(5), 532–536 (2017)
18. J. Wei, X. Yang, J. Huang et al., Fault diagnosis of rolling bearing based on deep neural network. Comb. Mach. Tool Auto. Mach. Technol. **11**, 88–91 (2017)
19. Y. Jia, Robust control with decoupling performance for steering and traction of 4WS vehicles under velocity-varying motion. IEEE Trans. Control Syst. Technol. **8**(3), 554–569 (2000)
20. Y. Jia, Alternative proofs for improved LMI representations for the analysis and the design of continuous-time systems with polytopic type uncertainty: a predictive approach. IEEE Trans. Auto. Control **48**(8), 1413–1416 (2003)

# Reconstructed Multi-innovation Gradient Algorithm for the Identification of Sandwich Systems

**Linwei Li, Xuemei Ren and Yongfeng Lv**

**Abstract** Inspired by multi-innovation stochastic gradient identification algorithm, a reconstructed multi-innovation stochastic gradient identification algorithm (RMISG) is presented to estimate the parameters of sandwich systems in this paper. Compared with the traditional multi-innovation stochastic gradient identification algorithm, the RMISG is constructed by using the multistep update principle which solves the multi-innovation length problem and improves the performance of the identification algorithm. To decrease the calculation burden of the RMISG, the key-term separation principle is introduced to deal with the identification model of sandwich systems. Finally, simulation example is given to validate the availability of the proposed estimator.

**Keywords** Parameter estimation · Sandwich systems
Multi-innovation gradient algorithm · Key-term separation principle

## 1 Introduction

Block-oriented nonlinear systems provide an excellent modeling approach for physical systems in engineering applications [1]. To describe the more complex nonlinear physical systems, sandwich systems are proposed by using the combination of the Wiener system and Hammerstein system. The sandwich systems are also called Wiener-Hammerstein systems which consist of two linear systems and a nonlinear submodel, as shown in Fig. 1, where $L_1$ and $L_2$ are the linear subsystems, $f(\cdot)$ is nonlinear model. Such sandwich nonlinear systems can represent many actual systems such as, suspension system [2], power amplifier system [4], vehicle system [5] and magnetorheological fluid dampers [3], etc. It is of practical significance to study the parameter identification of sandwich systems.

L. Li · X. Ren (✉) · Y. Lv
Beijing Institute of Technology, Beijing 100081, China
e-mail: xmren@bit.edu.cn



Y. Jia et al. (eds.), *Proceedings of 2018 Chinese Intelligent Systems Conference*,
Lecture Notes in Electrical Engineering 528,
https://doi.org/10.1007/978-981-13-2288-4_30

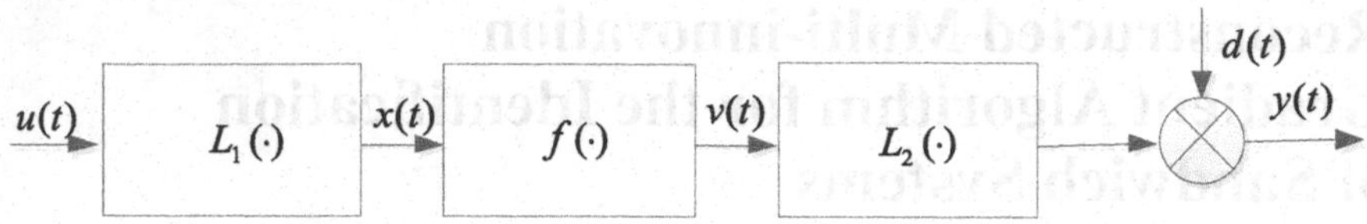


**Fig. 1** Diagrammatic sketch of sandwich systems

The parameter identification of sandwich systems have been attracted large research works in the past [6–9]. Vörös solved the parameter estimation of sandwich systems with backlash nonlinearity [10] based on the decomposition technique, and estimated the parameters of considered systems by using an iterative identification method. Tan et al. [12] then developed a recursive general identification approach to identify the parameters of the X-Y moving positioning based on the sandwich systems with dead-zone nonlinearity, and used Lyapunov stability theorem to prove the convergence of the proposed identification algorithm. In [11], Schoukens et al. applied the best linear approximation to obtain the initial values of sandwich systems with polynomial nonlinearity, and proposed a Levenberg-Marquardt algorithm to optimize the final parameter values by using the initial results of the first stage.

In recent decades, to improve the estimation accuracy and convergence speed of the identification algorithm, many novel estimation algorithms have been developed such as, the auxiliary model ideas [13], the coupled identification ideas [14] and the multi-innovation identification idea [15], etc. Inspired by multi-innovation identification algorithm, in this paper, a reconstructed identification method (RMISG) based on multi-innovation stochastic gradient algorithm is developed for the sandwich systems with backlash nonlinearity. The contributions of this paper are listed as follows:

(1) A reconstructed multi-innovation stochastic gradient algorithm is designed for the purpose of solving the multi-innovation length problem which restricts the accuracy of the parameter estimation. The modified algorithm improves further the accuracy of parameter estimation.
(2) The key-term separation principle is applied to transform the considered systems into a linear regression model by choosing the key-term, which reduces the computational burden of the proposed estimator.

The remaining sections are summarized as follows. Section 2 introduces the sandwich systems with backlash nonlinearity and the identification model. The RMISG approach is described in Sect. 3. Section 4 provides simulation example. Finally, some conclusions are given in Sect. 5.

## 2 System Description

As shown in Fig. 1, the backlash nonlinearity is chosen as nonlinear submodel. In Fig. 1, $u(t)$ and $y(t)$ are the input and output sequences, $x(t)$ and $v(t)$ are unknown middle variables. $d(t)$ is white noise. $L_1$ and $L_2$ can be described as

$$L_1 : B(z^{-1})x(t) = A(z^{-1})u(t) \tag{1}$$

$$L_2 : D(z^{-1})y(t) = C(z^{-1})v(t) + d(t) \tag{2}$$

where $A(z^{-1}),\ B(z^{-1}),\ C(z^{-1})$ and $D(z^{-1})$ are polynomials in the shift operator $z^{-1}$ $[z^{-1}y(t) = y(t-1)]$, which are expressed as follows: $A(z^{-1}) = a_1z^{-1} + a_2z^{-2} + \cdots + a_{n_a}z^{-n_a}$, $B(z^{-1}) = 1 + b_1z^{-1} + b_2z^{-2} + \cdots + b_{n_b}z^{-n_b}$, $C(z^{-1}) = c_1z^{-1} + c_2z^{-2} + \cdots + c_{n_c}z^{-n_c}$, $D(z^{-1}) = 1 + d_1z^{-1} + d_2z^{-2} + \cdots + d_{n_d}z^{-n_d}$.

**Assumption 1** (1) $L_1$ and $L_2$ are stable; (2) $n_a$, $n_b$, $n_c$ and $n_d$ are equal and unknown, and the coefficients $a_i, b_j, c_i$ and $d_j$ are unknown; (3) $u(t) = 0, x(t) = 0, v(t) = 0$ and $y(t) = 0$ for $t \leq 0$.

Backlash is multivalued mapping function with memory [16], the structure of the backlash is depicted in Fig. 1,

$$v(t) = \begin{cases} m_L(x(t) + c_L) & if\ x(t) \leq z_L \\ v(t-1) & if\ z_L < x(t) < z_R \\ m_R(x(t) - c_R) & if\ x(t) \geq z_R \end{cases} \tag{3}$$

where $x(t)$ is the input of backlash nonlinearity, and $v(t)$ are the output of the nonlinearity, $m_L$ and $m_R$ are two slopes, $c_L$ and $c_R$ are two endpoints, $m_L,\ m_R,\ c_L$ and $c_R$ are constant and $z_L = v(t)/m_L - c_L$, $z_R = v(t)/m_R + c_R$.

Then, the switching function is introduced to linearize the backlash. Define the switching function,

$$h(s) = \begin{cases} 1 & if\ s \leq 0 \\ 0 & if\ s > 0 \end{cases} \tag{4}$$

Based on the $h(s)$, two intermediate functions are described as follows:

$$f_1(t) = h[x(t) - z_L]$$
$$f_2(t) = h[z_R - x(t)]$$

The backlash (3) can be written as

$$\begin{aligned} v(t) &= m_Lx(t)f_1(t) + m_Lc_Lf_1(t) + m_Rx(t)f_2(t) - m_Rc_Rf_2(t) \\ &\quad + v(t-1) - f_1(t)v(t-1) - f_2(t)v(t-1) \\ &\quad + v(t-1)f_1(t)f_2(t) \end{aligned} \tag{5}$$

From (5), it can be seen that the backlash (3) can be described by an analytical equation and the relations of the input-output are identical on (5) and (3).

The models of the sandwich systems with backlash are listed as follows

$$x(t) = \sum_{i=1}^{n_a} a_i u(t-i) - \sum_{j=1}^{n_b} b_j x(t-j) \tag{6}$$

$$\begin{aligned} v(t) = {} & m_L x(t) f_1(t) + m_L c_L f_1(t) + m_R x(t) f_2(t) - m_R c_R f_2(t) + v(t-1) \\ & - f_1(t) v(t-1) - f_2(t) v(t-1) + v(t-1) f_1(t) f_2(t) \end{aligned} \tag{7}$$

$$y(t) = \sum_{m=1}^{n_c} c_m v(t-m) - \sum_{n=1}^{n_d} d_n y(t-n) + d(t) \tag{8}$$

When we substitute (6), (7) into (8), we can obtain the identification model, it leads to a complex identification model. Then, to simplify the identification model, the first terms $x(t)$ in (7) and $v(t)$ in (8) are regarded as the key-terms [10], which can solve the above problem.

By substituting (6) into (7), (7) can be expressed as

$$\begin{aligned} v(t) = {} & m_L f_1(t) \sum_{i=1}^{n_a} a_i u(t-i) - m_L f_1(t) \sum_{j=1}^{n_b} b_j x(t-j) + m_L c_L f_1(t) + m_R x(t) f_2(t) \\ & - m_R c_R f_2(t) + v(t-1) - f_1(t) v(t-1) - f_2(t) v(t-1) + v(t-1) f_1(t) f_2(t) \end{aligned} \tag{9}$$

By substituting (9) into (8), $y(t)$ yields

$$\begin{aligned} y(t) = {} & m_L c_1 f_1(t-1) \sum_{i=1}^{n_a} a_i u(t-i-1) - m_L c_1 f_1(t-1) \sum_{j=1}^{n_b} b_j x(t-j-1) \\ & + m_L c_L c_1 f_1(t-1) + m_R c_1 x(t-1) f_2(t-1) - m_R c_R c_1 f_2(t-1) \\ & + c_1 v(t-2) - c_1 f_1(t-1) v(t-2) - c_1 f_2(t-1) v(t-2) \\ & + c_1 v(t-2) f_1(t-1) f_2(t-1) + \sum_{m=2}^{n_c} c_m v(t-m) - \sum_{n=1}^{n_d} d_n y(t-n) + d(t) \end{aligned} \tag{10}$$

Define the vector of parameter

$$\begin{aligned} \theta = {} & [m_L c_1 a_1, m_L c_1 a_2, \ldots, m_L c_1 a_{n_a}, m_L c_1 b_1, \ldots, m_L c_1 b_{n_b}, m_L c_1 c_L, m_R c_1, \\ & m_R c_R c_1, c_1, c_2, \ldots, c_{n_c}, d_1, d_2, \ldots, d_{n_d}]^T \end{aligned}$$

where $m_L = m_L c_1 a_1$, $a_2 = m_L c_1 a_2 / m_L, \ldots, a_{n_a} = m_L c_1 a_{n_a} / m_L$, $b_1 = m_L c_1 b_1 / m_L, \ldots, b_{n_b} = m_L c_1 b_{n_b} / m_L$, $m_R = m_R c_1$, $c_L = m_L c_L c_1 / m_L$, $c_R = m_R c_R c_1 / m_R$.

Define the information vector

$$\begin{aligned}\varphi(t) = [&f_1(t-1)u(t-2), f_1(t-1)u(t-3), \ldots, f_1(t-1)u(t-n_a-1),\\ &-f_1(t-1)x(t-2), \ldots, -f_1(t-1)x(t-n_b-1), f_1(t-1), f_2(t-1)x(t-1),\\ &-f_2(t-1), v(t-2)[1-f_1(t-1)][1-f_2(t-1)], v(t-2), v(t-3), \ldots, v(t-n_c),\\ &-y(t-1), -y(t-2), \ldots, -y(t-n_d)]^T.\end{aligned}$$

Then, the output Eq. (10) can be described as follows

$$y(t) = \varphi^T(t)\theta + d(t) \tag{11}$$

Equation (11) is the estimation model of the sandwich systems with backlash.

## 3 Reconstructed Multi-innovation Stochastic Gradient Algorithm

In this section, we show that the Reconstructed multi-innovation stochastic gradient algorithm is derived.

Defining the criterion function, $J(\theta) = [y(t) - \varphi^T(t)\theta]^2$ and minimizing $J(\theta)$ by applying the gradient search theory, then we can obtain the stochastic gradient algorithm as follows:

$$\hat{\theta}(t) = \hat{\theta}(t-1) + \frac{1}{r(t)}\varphi(t)[y(t) - \varphi^T(t)\hat{\theta}(t-1)] \tag{12}$$

$$e(t) = y(t) - \varphi^T(t)\hat{\theta}(t-1) \tag{13}$$

$$r(t) = r(t-1) + \|\varphi(t)\|^2, r(0) = 1 \tag{14}$$

where $\hat{\theta}(t)$ denotes the estimation of $\theta(t)$, $e(t)$ be the scalar innovation [13, 17], and $\frac{1}{r(t)}$ represents the step-size.

Expanding the scalar innovation $e(t)$ to the innovation vector (multi-innovation) $\mathbf{E}(p, t)$ by using the newest $p$ data [17]:

$$\mathbf{Y}(p,t) = \begin{bmatrix} y(t) \\ y(t-1) \\ \vdots \\ y(t-p+1) \end{bmatrix}, \quad \mathbf{E}(p,t) = \begin{bmatrix} e(t) \\ e(t-1) \\ \vdots \\ e(t-p+1) \end{bmatrix},$$

$$\boldsymbol{\phi}(p,t) = [\varphi(t), \varphi(t-1), \ldots, \varphi(t-p+1)]$$

where $p$ represents the innovation length ($p \geq 1$).

Then, the multi-innovation gradient algorithm (MISG) can be rewritten as

$$\hat{\theta}(t) = \hat{\theta}(t-1) + \frac{1}{r(t)}\boldsymbol{\phi}(p,t)\mathbf{E}(p,t) \tag{15}$$

$$\mathbf{E}(p,t) = \mathbf{Y}(p,t) - \boldsymbol{\phi}^T(p,t)\hat{\theta}(t-1) \tag{16}$$

$$r(t) = r(t-1) + \|\boldsymbol{\phi}(p,t)\|^2, r(0) = 1 \tag{17}$$

If $p = 1$, the MISG identification algorithm for sandwich systems with backlash reduces to stochastic gradient algorithm. In MISG algorithm, to fully excite system characteristic, the multi-innovation length must be large enough to satisfy the accuracy of the parameter estimation. Thus, the multi-innovation length restricts the accuracy of the estimation. How to improve the accuracy of the parameter identification becomes a difficult problem. To deal with the above problem, the multi-innovation length $p$ is decomposed into $p$ sub-innovation update steps.

Define $y(t,i) = y(t-p+i)$, $\hat{\varphi}(t,i) = \hat{\varphi}(t-p+i)$ $(i = 1, 2, \ldots, p)$. i.e., $y(t,i) \in \{y(t-p+1), y(t-p+2), \ldots, y(t)\}$, $\hat{\varphi}(t,i) \in \{\hat{\varphi}(t-p+1), \hat{\varphi}(t-p+2), \ldots, \hat{\varphi}(t)\}$. Then, RMISG algorithm is summarized as follows:

$$\hat{\theta}_i(t) = \hat{\theta}_{i-1}(t) + \frac{1}{r_i(t)}\hat{\varphi}(t,i)[y(t,i) - \hat{\varphi}^T(t,i)\hat{\theta}_{i-1}(t)] \tag{18}$$

$$e(t,i) = y(t,i) - \hat{\varphi}^T(t,i)\hat{\theta}_{i-1}(t) \tag{19}$$

$$r_i(t) = r_{i-1}(t) + \|\hat{\varphi}(t,i)\|^2, r_0(0) = 1 \tag{20}$$

$$\hat{\theta}_0(t) = \hat{\theta}_p(t-1), r_0(t) = r_p(t-1), \ i = 1, 2, \ldots, p \tag{21}$$

where $i$ denotes the step of the sub-innovation, and $t$ denotes the above MISG step. If $i = p$, then the next multi-innovation updating will be activated.

Noting that the (18)–(21) involve the unknown variables $\varphi(t)$, $x(t-i)$ and $v(t-i)$, the solutions are to reconstruct the auxiliary model through the usage of the auxiliary model method [13]. Replacing $x(t-i)$ and $v(t-i)$ in $\varphi(t)$ with the outputs $\hat{x}(t-i)$ and $\hat{v}(t-i)$ of the auxiliary models. $\hat{\varphi}(t,i)$ is obtained by using the auxiliary model scheme.

## 4 Example

In this section, the proposed algorithm (RMISG) is verified on the illustrative examples of the sandwich systems with backlash. To validate the performance of the proposed algorithm, we provide a comparison of MISG algorithm.

Consider the following sandwich systems with backlash. The linear subsystems are given by:

$$L_1 : x(t) = u(t-1) + 0.2u(t-2) - 0.5x(t-1) - 0.35x(t-2)$$
$$L_2 : y(t) = v(t-1) + 0.45v(t-2) - 0.55y(t-1) - 0.3y(t-2) + d(t)$$

The backlash can be described by using the following four parameters, $m_L = m_R = 1.2$, $c_L = c_R = 0.1$. The inpout signal $u(t)$ is a stochastic signal. The $d(t)$ is a white noise signal with the variance $\sigma^2 = 0.1^2$. The corresponding SNR is SNR=15.20. The proposed estimator is conducted based the parameters: the length of the sample N=3000, $\hat{\theta}(0) = \mathbf{1}_{11}/10^3$, $p = 3$. Applying the RMISG to estimate the parameters of considered systems. The curves of parameter estimates are plotted in Fig. 2a–c. The actual output with solid blue line and the predicted output with dash red line is depicted in Fig. 2d.

To validate the advantage of the proposed algorithm, the estimation errors of identification algorithms are shown in Fig. 3. It can be seen that the proposed estimator provides faster convergence speed and higher accuracy in comparison with MISG

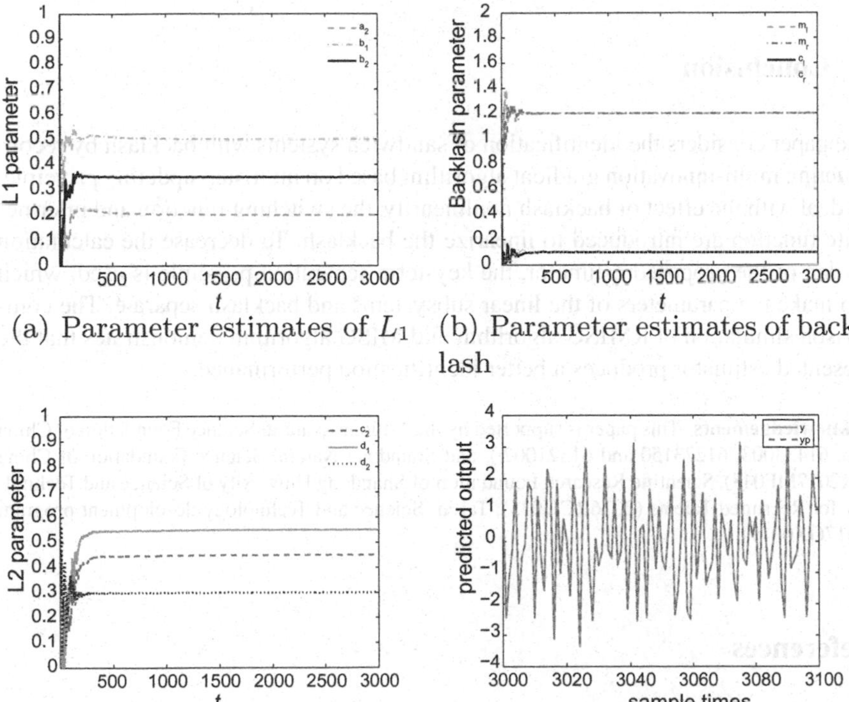


(a) Parameter estimates of $L_1$ (b) Parameter estimates of backlash

(c) Parameter estimates of $L_2$ (d) model validation for RMISG

**Fig. 2** The estimation of the parameters by RMISG ($\sigma^2 = 0.1^2$)

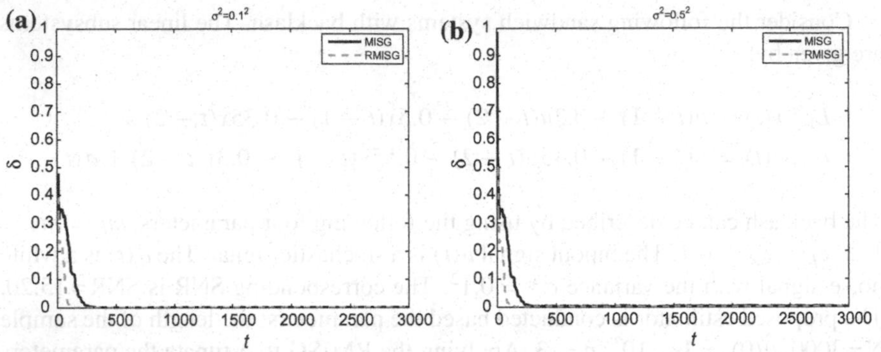


**Fig. 3** Estimation errors of RMISG and MISG algorithms with noise variances $\sigma^2$. **a** Variance is $\sigma^2 = 0.1^2$. **b** Variance is $\sigma^2 = 0.5^2$

algorithm, which illustrates the advantage of the proposed method. Comparing with the parameter estimation errors with variance $\sigma^2 = 0.1^2$, the parameter estimation errors with variance $\sigma^2 = 0.5^2$ has a large estimation error.

## 5 Conclusion

The paper considers the identification of sandwich systems with backlash by reconstructing multi-innovation gradient algorithm based on multistep updating principle. To deal with the effect of backlash nonlinearity, the switching function and intermediate function are introduced to linearize the backlash. To decrease the calculation burden of the proposed estimator, the key-term separation principle is used, which can make the parameters of the linear subsystems and backlash separate. The comparison simulation of RMISG algorithm and MISG algorithm demonstrates that the presented estimator produces a better identification performance.

**Acknowledgements** This paper is supported by the National Natural Science Foundation of China (No. 61433003, 61273150 and 61321002.), and Shandong Natural Science Foundation of China (ZR2017MF048), Scientific Research Foundation of Shandong University of Science and Technology for Recruited Talents (2016RCJJ035), Tai'an Science and Technology development program (2017GX0017).

## References

1. F. Giri, E.W. Bai, Block-oriented nonlinear system identification, vol. 1 (Springer, 2010)
2. Y. Rochdi, F. Giri, J.B. Gning, F.Z. Chaoui, Identification of block-oriented systems in the presence of nonparametric input nonlinearities of switch and backlash types. Automatica **46**(5), 864–877 (2010)

3. J. Vörös, Parametric identification of the Dahl model for large scale MR dampers. Struct. Control Health Monitor. **19**(3), 332–347 (2012)
4. J. Vörös, Modeling and identification of nonlinear cascade and sandwich systems with general backlash. Struct. Control Health Monitor. **4**(4), 282–290 (2010)
5. Y.M. Jia, Robust control with decoupling performance for steering and traction of 4WS vehicles under velocity-varying motion. IEEE Trans. Control Syst. Technol. **8**(3), 554–569 (2000)
6. L.W. Li, X.M. Ren, Decomposition-based recursive least-squares parameter estimation algorithm for Wiener-Hammerstein systems with dead-zone nonlinearity. Int. J. Syst. Sci. **48**(11), 2405–2414 (2017)
7. J. Vörös, Identification of cascade systems with backlash. Int. J. Control **83**(6), 1117–1124 (2010)
8. Z.P. Zhou, Y.H. Tan, Y.Q. Xie, R.L. Dong, State estimation of a compound non-smooth sandwich system with backlash and dead zone. Mech. Syst. Signal Process. **83**, 439–449 (2017)
9. Y.M. Jia, Alternative proofs for improved LMI representations for the analysis and the design of continuous-time systems with polytopic type uncertainty: a predictive approach. IEEE Trans. Auto. Control **48**(8), 1413–1416 (2003)
10. J. Vörös, Modeling and identification of systems with backlash. Automatica **46**(2), 369–374 (2010)
11. M. Schoukens, R. Pintelon, Y. Rolain, Identification of Wiener-Hammerstein systems by a nonparametric separation of the best linear approximation. Automatica **50**(2), 628–634 (2014)
12. Y. Tan, R. Dong, R. Li, Recursive identification of sandwich systems with dead zone and application. IEEE Trans. Control Syst. Technol. **17**(4), 945–951 (2009)
13. Y. Wang, F. Ding, The auxiliary model based hierarchical gradient algorithms and convergence analysis using the filtering technique. Signal Process. **128**, 212–221 (2016)
14. F. Ding, Coupled-least-squares identification for multivariable systems. IET Control Theory Appl. **7**(1), 68–79 (2013)
15. Q. Jin, Z. Wang, X. Liu, Auxiliary model-based interval-varying multi-innovation least squares identification for multivariable OE-like systems with scarce measurements. J. Process Control **35**, 154–168 (2015)
16. V. Cerone, D. Regruto, Bounding the parameters of linear systems with input backlash. IEEE Trans. Auto. Control **52**(3), 531–536 (2007)
17. Z.Y. Wang, Y. Wang, Z.C. Ji, Filtering based multi-innovation extended stochastic gradient algorithm for Hammerstein nonlinear system modeling. Appl. Math. Model. **39**(18), 5724–5732 (2015)

# Using Big Data to Enhance the Capability of the Situational Awareness of Battlefield

**Hongpeng Wen and Jing Song**

**Abstract** This paper studies the basic methods of using big data to enhance the capability of the battlefield situational awareness. In the paper, the big data structure of the battlefield situational awareness is built based on the idea of confirming representation dimension of the big data of the battlefield situational awareness, defining their memory granularity, distinguishing their update frequency and establishing their association graphs. The big data sources of the battlefield situational awareness are expanded through pre-war basic data preparation, dynamic data acquisition, implicit data mining. The paths of effective use of the big data of the battlefield situational awareness are put forward such as the general-purpose Operations View via broadcast, Sub-domain publishing local situation, direct pushing priori data, and customizing personalized data.

**Keywords** Big data · Enhance · Battlefield · Situational awareness

## 1 Introduction

Integrated joint operations require accurate, efficient and shared battlefield situational awareness. All kinds of goals, relationships and confrontations between both sides will generate diverse, complex and massive data on the information battlefield. These data can effectively enhance the situational awareness of the battlefield, with clear characteristics of big data. Relying on the network information system to sense the battlefield situation will inevitably require that the capability of the battlefield situational awareness based on big data should be formed as soon as possible.

H. Wen (✉) · J. Song
Information and Communications College, National University of Defense Technology, Wuhan 430010, Hubei, China
e-mail: wenhp2002@163.com



Y. Jia et al. (eds.), *Proceedings of 2018 Chinese Intelligent Systems Conference*, Lecture Notes in Electrical Engineering 528,
https://doi.org/10.1007/978-981-13-2288-4_31

## 2 Building the Big Data Structures that Can Support the Battlefield Situational Awareness

To use big data to enhance the capability of situational awareness of battlefield, we must first define big data that can be used to support the battlefield situational awareness. Then we should standardize and format them. Only after such processing, these big data, which can support the battlefield situational awareness, can be used to support real-time battlefield situational awareness, operation planning, operation regulation and post-war assessment and analysis. These data can be enabled to adapt to the capabilities of various sensing terminals and information communication networks and to meet the needs of joint operations and command [1].

### *2.1 Confirming the Representation Dimension of the Big Data of the Battlefield Situational Awareness*

The representation dimension is the reflection of different levels and characteristics of entities or events. In general, the larger the representation dimension, the more comprehensive and precise the representation of the entity or event is. On the information battlefield, the marking of military targets, the description of combat operations, or the delineation of the situation of both sides are difficult to accomplish by using data of a single dimension. They can be represented fully and accurately only through using data of multiple dimensions. Timing relationships, spatial relationships and information interaction relationships among different battlefield goals and operations need also be represented by data of different dimensions. At the same time, sufficient representation dimensions facilitate the analysis and mining of data. Therefore, when building the big data structures of the battlefield situational, the characteristics of battlefield targets and the requirements of operations should be took into account to make a unified specification and determination of the representation dimension of all big data of the battlefield situational awareness.

### *2.2 Defining the Memory Granularity of the Big Data of the Battlefield Situational Awareness*

Data granularity is the refinement or comprehensiveness of entity or event characterization. The smaller the granularity is, the higher the degree of refinement of the big data of the battlefield situational awareness is, the more detailed the descriptions of entities or events are. Joint operation commanders, command organization, arms and services have totally different requirements on data granularity of the same target. When using big data to characterize the battlefield situation, the higher the command level is, the more macroscopic the command activity is, the lower requirement for the refinement of the big data of the battlefield situational awareness is, and the larger the data granularity is. Conversely, the higher requirement for the refinement is, the smaller the data granularity is. Therefore, when building the big data structures, the data standard granularity of all command seats or battle positions should be defining appropriately based on different requirements and standardized various situation data according to the determined standard granularity.

## *2.3 Distinguishing the Update Frequency of the Big Data of the Battlefield Situational Awareness*

Update frequency is a representation of the dynamic change of an entity or event. The higher the frequency of update is, the more timely the feedback on the battlefield situation changes is, the more comprehensive the dynamic situation becomes. It is necessary to continuously update the big data of the battlefield situational awareness to ensure that commanders and command organization at all levels understand the battlefield situation accurately and timely, know better the battlefield dynamics and development trends and grasp the fleeting opportunity for operations. The update of the big data of the battlefield situational awareness is not required to achieve overall iterations and it is difficult to achieve overall iterations. Instead, local or individual association iterations should be conducted in accordance with reality. In general, main battle weapon platforms have a great impact on combat deployment and progress. The battlefield situation of high-speed maneuvering targets changes rapidly and time sensitivity of information objectives are high. These must have higher requirements on the frequency of updating the situation data. Therefore, when building the big data structures, the frequency of updating the big data of the battlefield situational awareness should be distinguished based on the operational requirements of the commanders and command organization at all levels of the joint operations and the different nature of the various battlefield objectives [2].

## *2.4 Establishing the Association Graphs of the Big Data of the Battlefield Situational Awareness*

The association graphs reflect the relationship between entity or event representations. The more detailed the association graphs are, the more clearly the correlations and relationships among multiple targets or operations on the battlefield are, and the more accurate the forecast of the battlefield trend changes is. The association graph is a relationship network formed by linking different types of situational data, which facilitates the understanding and control of the battlefield situation as a whole by commanders and command organization at all levels of the joint operations. In general, the big data of the battlefield situational awareness are collected and organized by taking a human-machine approach. Dominant association graphs are established based on the basic rules of command activities and operations. The special associations hidden in the big data of the battlefield situational awareness are searched and mined through the methods of extraction, clustering, reasoning, analysis, etc. to establish implicit association graphs. Therefore, when building the big data structures, an initial association graph of situation data should be established based on the existing situation data and association rules. Then, according to the new findings of data processing and trend judgments, new relationships should be established, link relationships should be changed, and correlation coefficients should be adjusted.

## 3 Expanding the Big Data Sources of the Battlefield Situational Awareness

To achieve the situational awareness of battlefield, we must have the comprehensive, accurate and effective big data of the situational awareness of battlefield. The main sources of the big data include pre-war preparation, wartime collection, and in-deep data mining of implicit data.

### *3.1 Pre-war Preparation of the Basic Data of the Battlefield Situation*

The basic data of the battlefield situation, mainly refers to the basic data acquired by using various existing intelligence and data resources in the operation preparation phase, including geo-information data, electromagnetic environment data, organization and equipment data of both sides, various operational model data, pre-war situation data, etc. The first is the data processing of various intelligence information. Intelligence information in the form of texts, charts, graphics, images, videos, etc. is processed into data to construct battlefield situational initial database. The second is the comparison and organization of the basic big data of the battlefield situational awareness. Redundant and wrong data are eliminated, outdated and changed data are updated, and missing data are supplemented. The third is processing and analysis of the basic big data. Representation dimension, granularity, update frequency of all the situational data are defining detail by detail, and the initial association graphs are established based on existing analysis results. The fourth is the initial deployment of the data of the situational awareness of battlefield. The relevant situational data are deployed to different types of command seats or battle positions to form a basic operational view [3].

### 3.2 Acquisition of the Dynamic Data of the Battlefield Situation in Real Time

The dynamic data of the battlefield situation, mainly refers to all the situational data collected by various terminals of the battlefield situational awareness in the process of planning, organizing and implementing operations, and various situational data acquired by superiors, friends or other forces. The first is the collection of data. Battlefield situation awareness is formed through "perception network" composed of personnel, weapon platform and sensors. Various targets and actions in the battlefield are reconnoitered and monitored to get real-time or near-real-time entity or event characterization data. The second is the fusion of data. Acquired dynamic data are verified through distributed calculation processing, and compared with basic data to finish data fusion and form a battlefield common combat view. The third is the interconnection of data. Situational awareness capabilities at the same level are integrated. Then situational data requirements are proposed to superior or friends, or imported into superior or friends to fulfill the interconnection of the situational data of this level with superior or friends. The fourth is the interoperability of data. The corresponding authority of data processing for each command seat according to operational needs is defined. Verified and approved data are updated in real time to realize the interoperability of the situational data of this level with superior or friends [4, 5].

### 3.3 The Deep Mining of Implicit Data of the Battlefield Situation

The implicit data of the battlefield situation, mainly refers to the new situation data or correlation acquired through in-depth analysis and in-depth mining of various explicit data in the process of the battlefield situational awareness. The implicit data does not appear in a visualized way. The contents it represents are mainly reflected in the history of formation, degree of association and development trends etc. The first is the statistical analysis of data. Statistics on situation data of the same target or a certain action are made to find out the quantitative changes, qualitative changes and development trends in the battlefield situation. The second is the association analysis of data. Classify, cluster and correlate multidimensional and massive situation data. Through the relationship and variation laws of situation data, we study the changing base point and conduction path of battlefield situation. The third is the evaluation analysis of data. After all situation data are traced back, the mechanism and impact of relevant data on operational processes and outcomes will be evaluated and analyzed. The fourth is the predictive analysis of situation changes. Any trend is seen at the micro. Predictive analysis of possible trends in a certain target or action are made based on small changes in one or more data to further clarify its impact on the battlefield situation [6, 7].

# 4 Optimization of the Big Data Utilization of the Shared Battlefield Situational Awareness

Using big data to enhance the capability of situational awareness of battlefield is bound to be achieved through the scientific and rational big data distribution and utilization. We must taking into account the large capacity, multiple structure, flow aging requirement of the big data of the battlefield situational awareness, and the limited bandwidth of the battlefield network information system, the user's personalized needs, and other complex conditions. When distributing and using battlefield situational data, it is necessary to pay attention to the specific contents of the big data of the battlefield situational awareness, but also the distribution and use of the big data.

## *4.1 In Order to Meet the Basic Situation Awareness Requirements, a General-Purpose Operational View Is Released Through Broadcast*

According to the common needs of all users of the battlefield for situation awareness, the battlefield network information system is used by joint operations command organizations to disseminate battlefield situation data that needs to be shared across the entire network to all command seats and battle positions in a broadcast way. Battlefield Common Operations View will be shared. All command seats and battle positions in the network can receive the same situation data via one broadcast. However, the broadcast battlefield situation data are not pertinent, and they occupy more channel resources in the course of broadcasting.

## *4.2 In Accordance with the Effective Scope of Mission Command, Local Situational Data Are Published in the Sub-domain Way*

Mission command is a very important command method in informationalized operations. Commanders in the corresponding levels organize limited forces to complete a specific task. According to the importance, situational needs and levels of authority of operational missions, and according to the corresponding command hierarchy, operational area, operational mission or collaborative relationship, the local situational data related to operational missions are issued to command seats or battle positions. Command seats or battle positions in the same area or performing the same combat mission can receive the same local situational data. However, it is necessary to accurately know the content and scope of the local situational data, and to make corresponding dynamic adjustments based on the changes in combat processes and tasks.

### 4.3 According to Functions and Business of Different Positions, the Priori Local Situational Data Are Pushed Directly

Different command seats or battle positions have different responsibilities and functions. Different commanders and staff officers also have different methods and habits of data collection, processing and analysis. According to the professional characteristics of operational missions and forces, and statistical analysis of different users' records of application, inquiry, browsing or utilization of professional situational data, situational data directly related to or potentially relevant to the completion of missions will be pushed directly to specific command seats or battle positions for reference. The priori data pushed will be updated according to the latest users' records of data utilization. The priori data pushed are more professional and have a better applicability. They can publish situational data that command seats or battle positions don't know but may want to know. However, there may be invalid local situational data.

### 4.4 Based on the Needs of Special Tasks, Customized Personalized Data Are Customized

When performing a particular task, or when a certain command seat or battle position is carrying on the operational mission, there may be lack or deficiency of current situational data. At this point, we can apply to situation awareness networks or superiors and friends for personalized situational data requirements. When situation awareness networks or superiors and friends acquire the relevant situational data, these data can be transmitted to designated command seat or battle position according to authorization or corresponding procedures. Customized personalized situational data should not be too much. When the required personalized situational data are enough to meet operational needs, or when the mission is over, customization should be terminated in time.

## 5 Conclusion

Using big data to enhance the capability of the battlefield situational awareness refers to real-time collection of the multi-structured big data of battlefield situation that can support the commander's perception and understanding, relying on the terminal network distributed in the all dimensions of the battlefield. And it refers to interaction of multidimensional big data of battlefield situation via the battlefield information communication network, realizing standardization, formatting and effective integration of the big data of battlefield situation by using the Command Information System, building mechanisms of generation, circulation, processing and distribution of the big data of the battlefield situation after discovering hidden correlations and potential values of the big data of battlefield situation through deep data mining, and generating accurate, real-time, easy-to-understand, calculated and easily distributed the digitization map of battlefield situation in accordance with the requirement of command and operations, to provide material basis and necessary conditions for the realization of data-based operational command and accelerated formation of the command capabilities of joint operations.

## References

1. X. Zhang, The U.S. DOD and the services seek a cross-domain distributed common ground system. China Nat. Def. Sci. Technol. Inf. Netw. (2017)
2. A.N. Dewan (Translator), Development situation and enlightenment of US military COP. Know-far Institute for Strategic & Defence Studies (2017)
3. J. Wan, The U.S. DOD conducted the radical transformation construction through joint information environment. CICC Collection (2016)
4. X. Zhan, Seizing the leading data of future battlefield. Lib. Army Daily (2017)
5. H. Li, The U.S. army improves the situational awareness and mission execution capabilities of drone system. Mod. Mil. (2017)
6. E Security, Real-time analysis of big data enables drones to realize situational awareness. E Security News (2016)
7. E Security. How does the U.S. military use situational awareness and situational understanding. E Security News (2017)

# Optimal Controls for Dual-Driven Load System with Synchronously Approximate Dynamic Programming Method

**Yongfeng Lv, Xuemei Ren, Linwei Li and Jing Na**

**Abstract** This paper applies a synchronously approximate dynamic programming (ADP) scheme to solve the Nash controls of the dual-driven load system (DDLS) with different motor properties based on game theory. First, a neural network (NN) is applied to approximate the dual-driven servo unknown system model. Because the properties of two motors are different, they have different performance indexes. Another NN is used to approximate performance index function of each motor. In order to minimize the performance index, the Hamilton function is constructed to solve the approximate optimal controls of the load system. Based on parameter error information, an adaptive law is designed to estimate NN weights. Finally, the practical DDLS is simulated to demonstrate that the optimal control inputs can be studied by ADP algorithm.

**Keywords** Servo system · Approximate dynamic programming
Nash equilibrium · Multi-input system · Neural networks

## 1 Introduction

Considering that the single motor driven capacity is limited, the high power motor is expensive and the technical conditions are limited, the original single motor driven servo system cannot meet the needs of the present applications [1]. To resolve the above problems, dual-motor driven servo system is developed and has been widely used in industrial and military fields [2]. Zhao et al. realized the synchronization and tracking control for the dual-driven servo system by using a novel switching controller incorporated with backlash and friction compensation [2], and then solved

Y. Lv · X. Ren (✉) · L. Li
School of Automation, Beijing Institute of Technology, Beijing 100081, China
e-mail: xmren@bit.edu.cn

J. Na
Faculty of Mechanical & Electrical Engineering, Kunming University of Science & Technology, Kunming 650500, China



Y. Jia et al. (eds.), *Proceedings of 2018 Chinese Intelligent Systems Conference*,
Lecture Notes in Electrical Engineering 528,
https://doi.org/10.1007/978-981-13-2288-4_32

the tracking problem by designing an adaptive sliding mode control [3], where a novel estimation algorithm was developed. And many other controls are designed for the dual-motor servo system, for example adaptive control [4, 5], back-stepping control [6].

Although the above control methods can ensure the stability of the system, the optimality is not considered. The traditional optimal method for the linear system is to construct a quadratic performance and solve the problem with the Riccati method. For the nonlinear multi-driven servo system, the optimal tracking control was designed by combing the back-stepping algorithm and Riccati equation in [7], but it divided the optimal controller into the optimal tracking controller and the optimal synchronization controller, that may lead to a suboptimal problem. Moreover, all the above control methods require that all motors have the same power, which restricts the application of multiple motor driven system.

Aiming at the above problems, this paper developed an approximate dynamic programming (ADP) algorithm as presented in [8–10] to solve the unknown dual-driven load system (DDLS) by using the nonzero-sum (NZS) game theory [11, 12]. The unknown DDLS is first identified by NN structure. The ADP NN [13] is used to approximate the value functions of two motor with different power. According to the above NNs, the optimal inputs can be calculated, where an adaptive law as in [14, 15] is used to estimate the NN weights.

## 2 Dual-Motor Driven Servo System Description

DDLS is that two motor drive the same load, and the properties of each motor are different. The dynamic system of DDLS is constructed as

$$J_l\ddot{\theta}_l(t) + b_l\dot{\theta}_l(t) + T_d(t) = \sum_{i=1}^{2} g_i(t)\tau_i(t) \tag{1}$$

where $\theta_l$ is the load angular position, $\dot{\theta}_l$ is the angular velocity and $\ddot{\theta}_l$ denotes the acceleration. $J_l$ denotes the load inertia, and $b_l$ is the viscous friction coefficient of the load. $g_i(t)$ is the $i$-th motor system dynamic and $\tau_i(t)$ is the $i$-th motor input. $T_d(t)$ is the unknown disturbance.

Regarding the load rotation and speed as state variables, the control gain will be a time-varying function, which will increase the design burden and computational complexity of the controller. To avoid those problems, before designing the controller we first define the new system state variables as

$$\begin{cases} x_1 = \theta_l(t) \\ x_2 = \dot{\theta}_l(t) \\ u_i = \tau_i(t) \end{cases} \tag{2}$$

Then, the dynamic equation of the DDLS can be expressed as

$$\begin{cases} \dot{x}_1 = x_2 \\ \dot{x}_2 = -\frac{b_l}{J_l} x_2 + \frac{1}{J_l} \sum_{i=1}^{2} g_i u_i - T_d \end{cases} \tag{3}$$

The purpose of this paper is to design the optimal controller to tune the speed to be a given value, and ensure the system uniformly ultimate boundedness (UUB). We denote $f(x) = [x_2, -b_l x_2 / J_l - T_d]^T$, $\sum_{i=1}^{2} g_i u_i = [0; \frac{1}{J_l} \sum_{i=1}^{2} g_i \tau_i]$. The load system is represented as

$$\dot{x} = f(x) + \sum_{j=1}^{2} g_j(x) u_j \tag{4}$$

where $x$ is the state of the dual motor load system: $x = [x_1, x_2]^{\mathrm{T}}$.

## 3 NN Observer Design

We assumed that the load system is unknown, then a NN is used to approximate the system as

$$\dot{x} = W_1^T \varphi_1(x, u, t) + \xi_1 \tag{5}$$

where $W_1 = [W_f, W_{g1}, W_{g2}]^T \in \mathbb{R}^{4\times 2}$ is the connection weight matrix, $\varphi_1$ is the output activation function, $\xi_1$ is the approximation error. To update the unknown NN weight $W_1$, we design the following estimation method. First, the state and excitation of the dual motor load system are established as in [16], in which the filtering variable of $x$ is $x_f$, and the filtering variable of $\varphi_1$ is $\varphi_{1f}$.

$$\begin{cases} k\dot{x}_f + x_f = x \\ k\dot{\varphi}_{1f} + \varphi_{1f} = \varphi_1 \end{cases} \tag{6}$$

$\dot{x}_f$ is the derivative of $x_f$, $\dot{\varphi}_{1f}$ is the derivative of $\varphi_{1f}$, and $k$ is the preset filter parameter. Then establishment of auxiliary matrix $P_1$ and $Q_1$ is given by

$$\begin{cases} \dot{P}_1 = -\ell P_1 + \varphi_{1f} \varphi_{1f}^T, \\ \dot{Q}_1 = -\ell Q_1 + \varphi_{1f} \left[\frac{x - x_f}{k}\right]^T \end{cases} \tag{7}$$

Define $M_1$ as

$$M_1 = P_1 \hat{W}_1 - Q_1 \tag{8}$$

where $\hat{W}_1 = [\hat{W}_f, \ \hat{W}_{g1}, \ \hat{W}_{g2}]^T$ is the $W_1$'s estimation. Finally, the adaption is given as

$$\dot{\hat{W}}_1 = -\Gamma_1 M_1 \tag{9}$$

where $\Gamma_1$ is an empirical value of adaptive law.

**Theorem 1** *For the unknown DDLS (3) and (4), if $\varphi_1$ in the neural network (5) is continuously excited, the adaptive law (9) is applied to identify the approximate weights $\hat{W}_1$, they will converge to the true values, and the unknown load system can be accurately identified by the neural network (5).*

Because the limited space, the similar proof is referred to [14].

From Theorem 1, the approximation DDLS can be given as

$$\dot{\hat{x}} = \hat{W}_1^T \varphi_1(x, u, t) \tag{10}$$

where $\hat{x}$ is the approximation state of the load position and speed.

## 4 Optimal Controls with ADP Method

This section will construct the Hamilton-Jacobi-Bellman (HJB) equation to solve the optimal problem of DDLS model, then use an approximate dynamic programming method to learn the cost function of each motor.

Because the properties of dual motors are different, the value function of the $i$-th motor of the load is first given as

$$V_i^*(x) = \min_{u_i} \int_t^{\infty} \left( x^T Q_i x + \sum_{j=1}^{2} u_j^T R_{ij} u_j \right) d\tau, i = 1, 2 \tag{11}$$

$Q_i$ and $R_{ij}$ is the symmetric positive definite matrix with matching dimension. To simplify the optimal control calculating process, critic NNs based on the ADP methods is proposed to learn the optimal value functions of each motor as:

$$V_i(x) = W_{ci}^T \varphi_{ci}(x) + \xi_{ci}, i = 1, 2 \tag{12}$$

with the weight $W_{ci}$, regressor vector $\varphi_{ci}$ and NN error $\xi_{ci}$. Then the approximation of the value functions are depicted as

$$\hat{V}_i(x) = \hat{W}_{ci}^T \varphi_{ci}(x), i = 1, 2 \tag{13}$$

with the approximations $\hat{V}$ and $\hat{W}_{ci}$ of the optimal value function (12). Then the coupled Hamilton-Jacobi-Behrman (HJB) equations are given by based on the NN observer (10):

$$H_i(x, \nabla V_i, u_1, u_2) = r_i(x, u_1, u_2) + \left(\nabla\varphi_{ci}^T(x)\hat{W}_{ci}\right)^T \left(\hat{W}_f\varphi_f(x) + \sum_{j=1}^{2}\left(\hat{W}_{gj}^T\varphi_{gj}(x)\right)u_j\right) = \varepsilon_{HJi}, i = 1, 2 \tag{14}$$

where $\nabla V_i = \nabla\varphi_{ci}^T(x)\hat{W}_{ci}$ is the derivative of the approximate value function (13), $r_i(x, u_1, u_2) = x^T Q_i x + \sum_{j=1}^{2} u_j^T R_{ij} u_j$ is the cost function, $\nabla\varphi_{ci}(x)$ is the derivative of $\varphi_{ci}(x)$. According to $\partial H_i(x, \nabla V_i, u_1, u_2)/\partial u_i = 0$, one can derive the optimal input torque of the $i$-th motor in DDLS as

$$\hat{u}_i = -\frac{1}{2}R_{ii}^{-1}\left(\hat{W}_{gi}^T\varphi_{gi}\right)^T\left(\nabla\varphi_{ci}^T\hat{W}_{ci}\right), i = 1, 2 \tag{15}$$

$\hat{W}_{gi}$ is the estimated value of the $W_{gi}$ obtained by the estimated value $\hat{W}_1$.

Denote $\varepsilon_{HJi} = -\left(\nabla\varphi_{ci}^T(x)\tilde{W}_{ci} + \nabla\xi_{ci}\right)^T\left(\hat{W}_f\varphi_f(x) + \sum_{j=1}^{2}\left(\hat{W}_{gj}^T\varphi_{gj}(x)\right)u_j\right)$ with $\tilde{W}_{ci} = W_{ci} - \hat{W}_{ci}$ and the $\xi_{ci}$'s derivative $\nabla\xi_{ci}$, $\Theta_i = r_i(x, u_1, u_2)$ and $\Xi_i = \nabla\varphi_{ci}^T(x)\left(\hat{W}_f\varphi_f(x) + \sum_{j=1}^{2}\left(\hat{W}_{gj}^T\varphi_{gj}(x)\right)u_j\right)$.

Then the auxiliary matrixes $P_{2i}$ and $Q_{2i}$ for the $i$-th performance index of DDLS is defined as

$$\begin{cases} \dot{P}_{2i} = -\ell P_{2i} + \Xi_i\Xi_i^T, & P_{2i}(0) = 0 \\ \dot{Q}_{2i} = -\ell Q_{2i} + \Xi_i\Theta_i, & Q_{2i}(0) = 0 \end{cases} \tag{16}$$

$M_2$ is designed as

$$M_{2i} = P_{2i}\hat{W}_{ci} + Q_{2i}, i = 1, 2 \tag{17}$$

The adaptive law for the parameter $\hat{W}_{ci}$ is designed as

$$\dot{\hat{W}}_{ci} = -\Gamma_{2i}M_{2i}, i = 1, 2 \tag{18}$$

$\Gamma_{2i}$ is the adaptive learning gain with parameter experience value.

**Theorem 2** *For the unknown dual-motor load system (1), if the neural network input vector $\varphi_1$ and $\varphi_{ci}$ are continuously excited, the adaptive law (9) and (18) are applied to calculate the optimal control inputs (15), the unknown dual-motor load system is controlled, and the performance index (10) is approximately optimal.*

*Proof* It is omitted here because of the space limitation, the similarity proof can be found in [9].

*Remark 1* Because dual-motor may have different properties, for example different power, the performance indexes of each motor are different. Thus, the game theory can be used to obtain the optimal control problem of DDLS with ADP algorithm.

## 5 Simulation Results

To illustrate the proposed methods, we denote that $J_l = 4$, $b_l = -3$, $g_1 = 4$ and $g_2 = 8$, $T_d = \sin(x) + \tanh(x)$ in the dual-motor load system (1). Then the dynamic system (4) can be represented as

$$\dot{x} = -\frac{3}{4}x - \sin(x) - \tanh(x) + u_1 + 2u_2 \tag{19}$$

First an identifier NN is used to reconstruct the unknown dual-driven servo system with the regressor vector $\varphi_1 = \begin{bmatrix} x \ \sin(x) + \tanh(x) \ u_1 \ u_2 \end{bmatrix}^T$. We set the simulating parameters as $x(0) = 1$, $W_1(0) = [0.5\ 0.5\ 0.5]^{\mathrm{T}}$, $W_{c1}(0) = 0$ and $W_{c2}(0) = [0.5\ 0\ 0]^T$. In the identifier, the simulation parameters are set as $k = 0.001$, $\ell_1 = 1$, $\Gamma_1 = 5500I$. Figure 2 presents that the identifier NN weights

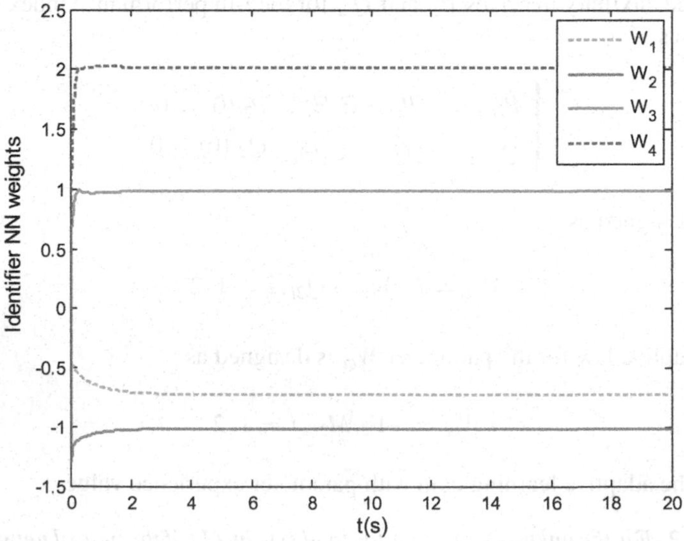


**Fig. 1** Observer NN weights

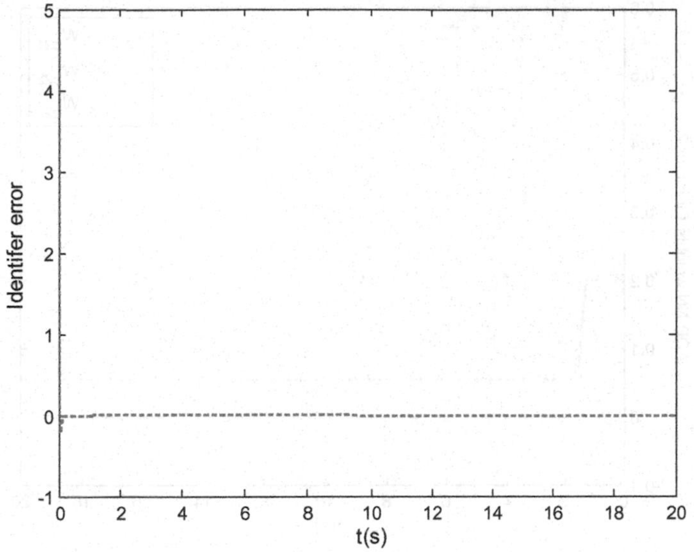


**Fig. 2** Load system observer error

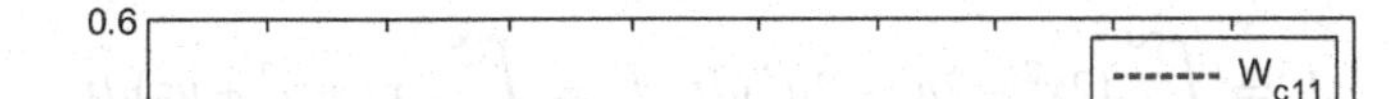

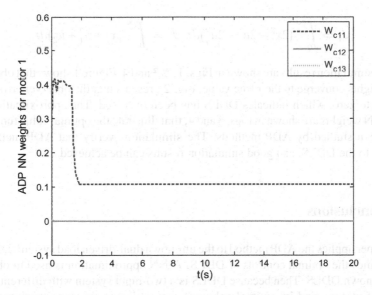


**Fig. 3** ADP NN weights for load input 1

converge to their truth values $[-3/4, -1, 1, 2]^T$. And the critic NN parameters are set as $\ell_{21} = \ell_{22} = 2$, $\Gamma_{21} = \Gamma_{22} = 30diag([1, \ 1, \ 1])$.

The performance indexes for each input are defined as

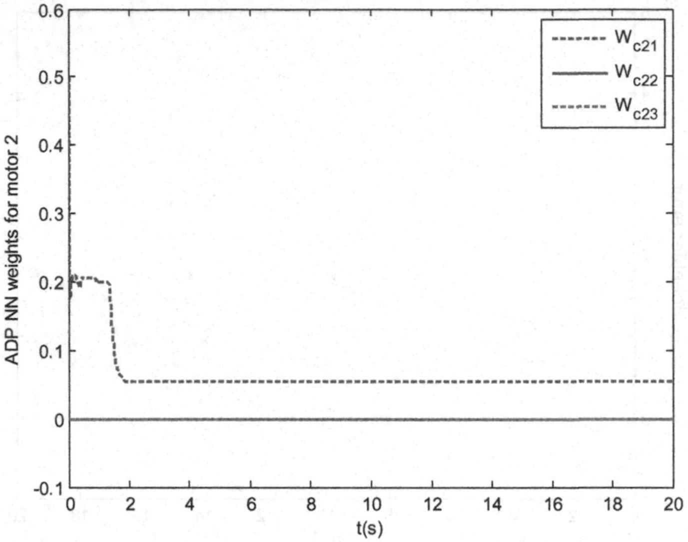


**Fig. 4** ADP NN weights for load input 2

$$J_1 = \int_0^\infty (2x^2 + 2u_1^2 + 2u_2^2)dt,\ J_2 = \int_0^\infty (x^2 + u_1^2 + u_2^2)dt. \tag{20}$$

The simulation results are shown in Figs. 1, 2, 3 and 4. Figure 1 shows that observer NN weights converge to their true value, Fig. 2 presents that the observer error goes closely to zero, which indicates DDLS has been observed. The approximations of ADP NN weights are shown in Figs. 3 and 4, that illustrate the optimal value functions have been studied by ADP methods. The simulations verify that ADP method is applied to the DDLS, and good simulation results can be achieved.

# 6 Conclusions

This paper applies the ADP method to the unknown dual-driven load system (DDLS), and studies the optimal controls of DDLS. A NN approximation is used to observe the unknown DDLS. Then because DDLS is a two-input system with different properties of motors, and its optimal solutions are similar to the nonzero-sum (NZS) game, we developed the NZS game theory to solve DDLS optimal control problem. Moreover, the approximate dynamic programming (ADP) method is applied to learn the optimal value functions, which can be used to calculate the optimal controls. And the simulating results verify that ADP method can be applied to the DDLS, and good results can be achieved.

**Acknowledgments** The work was supported by National Natural Science Foundation of China (No. 61433003 and No. 61273150).

## References

1. S. Wang, J. Na, X. Ren, RISE-based asymptotic prescribed performance tracking control of nonlinear servo mechanisms. IEEE Trans. Syst. Man Cybernet. Syst (2017)
2. W. Zhao, X. Ren, X. Gao, Synchronization and tracking control for multi-motor driving servo systems with backlash and friction. Int. J. Robust Nonlinear Control **26**(13), 2745–2766 (2016)
3. W. Zhao, X. Ren, S. Wang, Parameter estimation-based time-varying sliding mode control for multimotor driving servo systems. IEEE/ASME Trans. Mechatron. **22**(5), 2330–2341 (2017)
4. Y. Jia, Robust control with decoupling performance for steering and traction of 4WS vehicles under velocity-varying motion. IEEE Trans. Control Syst. Technol. **8**(3), 554–569 (2000)
5. Y. Jia, Alternative proofs for improved LMI representations for the analysis and the design of continuous-time systems with polytopic type uncertainty: a predictive approach. Automat. Control IEEE Trans. **48**(8), 1413–1416 (2003)
6. M. Wang, X. Ren, Q. Chen, S. Wang, X. Gao, Modified dynamic surface approach with bias torque for multi-motor servomechanism. Control Eng. Pract. **50**, 57–68 (2016)
7. M. Wang, X. Ren, Q. Chen, Cascade optimal control for tracking and synchronization of a multimotor driving system. IEEE Trans. Control Syst. Technol. (2018)
8. Y. Lv, J. Na, Q. Yang, X. Wu, Y. Guo, Online adaptive optimal control for continuous-time nonlinear systems with completely unknown dynamics. Int. J. Control **89**(1), 99–112 (2016)
9. Y. Lv, X. Ren, J. Na, Online optimal solutions for multi-player nonzero-sum game with completely unknown dynamics. Neurocomputing (2017)
10. A. Al-Tamimi, F.L. Lewis, M. Abu-Khalaf, Discrete-time nonlinear HJB solution using approximate dynamic programming: convergence proof. IEEE Trans. Syst. Man Cybernet. Part B (Cybernetics) **38**(4), 943–949 (2008)
11. J. Nash, Non cooperative games. Ann. Math. 286–295 (1951)
12. D. Liu, H. Li, D. Wang, Online synchronous approximate optimal learning algorithm for multi-player non-zero-sum games with unknown dynamics. IEEE Trans. Syst. Man Cybernet. Syst. **44**(8), 1015–1027 (2014)
13. D. Zhao, Q. Zhang, D. Wang, Y. Zhu, Experience replay for optimal control of nonzero-sum game systems with unknown dynamics. IEEE Trans. Cybernet. **46**(3), 854–865 (2016)
14. J. Na, M.N. Mahyuddin, G. Herrmann, X. Ren, P. Barber, Robust adaptive finite-time parameter estimation and control for robotic systems. Int. J. Robust Nonlinear Control **25**(16), 3045–3071 (2015)
15. J. Na, G. Herrmann, Online adaptive approximate optimal tracking control with simplified dual approximation structure for continuous-time unknown nonlinear systems. IEEE/CAA J. Automat. Sinica **1**(4), 412–422 (2014)
16. Y. Lv, J. Na, X. Ren, Online H$\infty$ control for completely unknown nonlinear systems via an identifier–critic-based ADP structure. Int. J. Control, 1–12 (2017)

# Path Planning for Unmanned Campus Sightseeing Vehicle with Linear Temporal Logic

**Mengtian Jiao and Yunzhong Song**

**Abstract** In order to solve the global path planning problem of unmanned campus sightseeing vehicles, this paper proposes a path optimization method based on linear temporal logic (LTL). First, the plan avoids the cumbersome and huge modeling for the actual road environment, and all the stops are modeled as a weighted finite-state transition system. Second, use LTL language to describe the tasks that the unmanned sightseeing vehicle needs to perform in actual operations. Next, construct a Product automaton that contains the environment model and task requirements. Finally, use a path search method based on Dijkstra algorithm to search for the optimal route on the Product automaton, and the optimal route is mapped back to the stops transition system in the actual environment, so that the route which the vehicle needs to perform during actual operation is obtained. Simulation results show that this method can completely solve the problem of patrolling between multiple stops, and can guarantee the optimality of the operating route.

**Keywords** Unmanned campus sightseeing vehicle · Linear temporal logic
Path planning · Multi-point patrol

## 1 Introduction

With the development of higher education in China, colleges and universities have increased enrollment, the campus has continued to expand, and functional areas such as teaching buildings, dormitories, dining area, supermarkets, stadium area have become more dispersed. The need for student travel has led to the emergence of campus sightseeing vehicles. However, in view of the fact that many universities only

M. Jiao · Y. Song (✉)
School of Electrical Engineering and Automation, Henan Polytechnic University,
Jiaozuo 454000, China
e-mail: songhpu@126.com

M. Jiao
e-mail: jiaomengtianxaz@163.com



Y. Jia et al. (eds.), *Proceedings of 2018 Chinese Intelligent Systems Conference*,
Lecture Notes in Electrical Engineering 528,
https://doi.org/10.1007/978-981-13-2288-4_33

rely on manual route planning for the operation of sightseeing vehicles, they have the disadvantages of wasting fleet personnel and vehicle resources. In order to better improve the service quality, reduce operating costs and reduce traffic time, people began to attach importance to the research on unmanned campus sightseeing vehicle. As the unmanned sightseeing vehicle is a controlled application under fixed lines and fixed scenes, the global path planning problem is a core part of the realization of autonomous vehicle navigation systems. Meanwhile, as the traditional methods such as $A^*$ algorithm [1], genetic algorithm [2], and ant colony algorithm [3], are mainly focused on satisfying simple task instructions, and the optimal path cannot be well planned for complex tasks contain time sequence and cyclicity, so the research based on the theory of temporal logic [4] is undoubtedly a future direction of development.

In recent years, a lot of attention has been devoted to patrol surveillance for path planning research, and many scholars have proposed different methods. In [5, 6], a controller synthesis framework inspired by model predictive control is proposed, where the rewards are locally optimized at each time-step over a finite horizon, and the immediate optimal control is applied. In [7], the authors proposed an embedded control software method that combines LTL specifications to implement path planning with fault tolerance. However, none of the methods proposed in the above literature can directly solve the problem of multipoint monitoring with complex requirements. Although the literature [8, 9] solve the multipoint recurring monitoring problem by building a transition system and Büchi automaton [10]. However, the method is affected by the order of task nodes, so that the searched path is not a global optimal path.

The main issues studied in this paper are: fixed-point operations during peak periods, point-to-point driving, that is, when an unmanned sightseeing vehicle is patrolling on multiple stops, it is necessary to find out the optimal route that meets task requirements. Therefore, this paper proposes an optimal patrol path planning method based on LTL in the light of the actual application requirements of unmanned campus sightseeing vehicles. According to the environmental characteristics of functional areas and the task requirements for the sightseeing vehicles, this method plans a global optimal patrol route that meets the needs of complex tasks.

## 2 Path Optimization Method for Unmanned Sightseeing Vehicles via LTL

First, according to the geographic location and environmental characteristics of functional areas in the university, the connection status between all the stops is modeled as a weighted transition system. The transition relationship between state nodes is based on the time required for the unmanned sightseeing vehicles to travel from one stop to another. Second, use LTL language to describe the tasks that unmanned sightseeing vehicles need to perform in the practical application. Next, construct a Product automaton that contains the environment model and task requirements. Finally, use

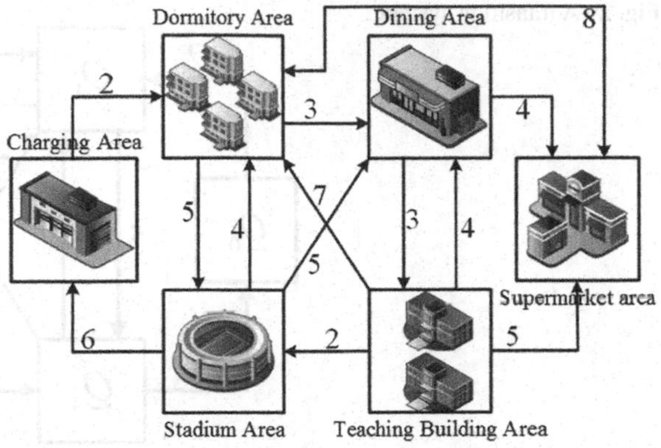


**Fig. 1** Functional area transition relation

a path search method based on Dijkstra algorithm to search for the optimal path on the Product automaton, and the optimal path is mapped back to the functional area transition system in the actual environment, so as to obtain the route that needs to be run in the environment.

## *2.1 Problem Description*

### 2.1.1 Environmental Modeling

First, according to the actual geographical environmental characteristics of functional areas in a university, all fixed stops need to be modeled as a finite state weighted transition system.

**Definition 1 (*transition system*)** A transition system is represented as a tuple $\Gamma = (Q, Q_{init}, \delta, AP, \omega)$, where Q is a finite set of states representing the set of nodes in the weighted transition system; $S_{init} \in S$ is the initial state, which represents the initial position of the campus sightseeing vehicles; $\delta \subseteq Q \times Q$ is the transition function between nodes; AP is expressed as a set of finite atomic propositions, namely the set of tasks that the vehicles need to complete; $\omega$ represents a weight function that is always positive, namely the time the unmanned sightseeing vehicles need to run from one region to another in real situations.

For simplicity, a simplified diagram sampled from the actual environment of a college (the transition system shown in Fig. 1) can be used to describe the connectivity of the various functional areas in the university. The weights indicate the time required for the unmanned sightseeing vehicles to run between stops.

To facilitate the description of the areas represented by each functional area, we use the states $Q_0$ to $Q_5$ which indicate the location of each area in Fig. 1 to describe the stops in the five functional regions (charging area, dormitories area, dining area,

**Fig. 2** A transition system

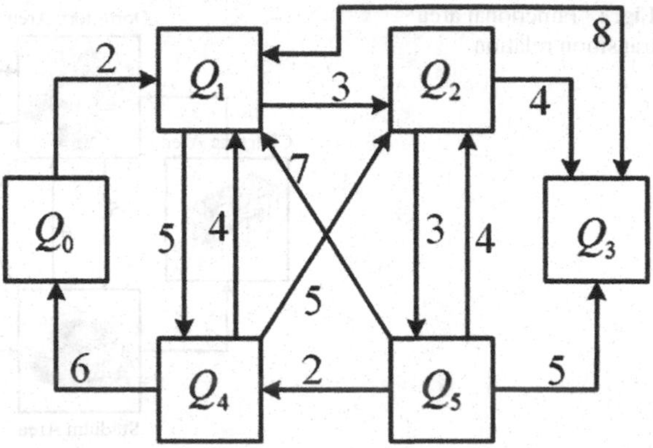


**Table 1** Algorithm 1

| **Algorithm 1** Construct weighted transition system |
|---|
| (1) define the number of nodes N in the transition system according to the total number of transition system vertices in Fig. 2 |
| (2) define the initial position node of the state transition system according to the starting position of the unmanned sightseeing vehicle in the actual environment |
| (3) define observed state T.obs based on task area |
| (4) construct an initial matrix, all assigned INF, and supplement the adjacency matrix based on the connectivity between the stop areas and the time between any two areas |
| (5) define the nodes that need to be monitored according to the task requirements of students taking a sightseeing vehicle |

supermarket area, stadium area, and teaching building area), where the node $Q_0$ serves as the initial position of the unmanned campus sightseeing vehicle. Here, we can convert the transition system $\Gamma$ of Fig. 1 into the form of Fig. 2.

In the simulation, the weighted transition system $\Gamma$ is constructed by executing the Algorithm 1 according to the transition system connection diagram of Fig. 2 (Table 1).

#### 2.1.2 Task Description

When colleges start normal routines, unmanned campus sightseeing vehicles need to select the optimal route to carry students according to actual needs. As a mission specification language, we use LTL to describe the task requirements, for its resemblance to natural language, and expressive power.

**Definition 2 (*linear temporal logic formula*)** A linear temporal logic formula is the combination of the atomic propositions of the transition system, the Boolean connectives [¬(negation), ∨(disjunction), ∧ (conjunction)] and the temporal operators [G(Always), F(Eventually), X(Next), U(Until)]. For example, XQ indicates that the stop point corresponding to state Q is the next target position that the unmanned campus sightseeing vehicle needs to reach. $Q_1 U Q_2$ indicates that the unmanned campus sightseeing vehicle must pass the stop corresponding to state $Q_2$ to reach the stop corresponding to $Q_1$.

In this article, aiming at the peak of classes in universities, we make a common situation where a large number of students need to go from the dormitory to the teaching building and from the teaching building back to the dormitory as the task of an unmanned campus sightseeing vehicle. That means it is necessary to use a sightseeing vehicle to travel from point to point between the dormitory area and the teaching building area, and to reciprocate. According to the characteristics of the linear temporal logic formula describing the mission requirements, we can describe the mission of the unmanned campus sightseeing vehicle between the dormitory area ($Q_1$) and the teaching building ($Q_5$) as

$$\phi = \mathrm{GF}Q_1 \,\&\, \mathrm{GF}Q_5 \tag{1}$$

## 2.2 Method Presention

The main tool here is automaton, so it is better for us to say something about that.

**Definition 3 (*automaton*)** *An automaton is a tuple $A = (S, S_{init}, \Sigma, \delta, T)$, where $S$ is a finite set of states; $S_{init} \in S$ is an initial state; $\Sigma$ indicates an input alphabet; $\delta \subseteq S \times \Sigma \times S$ is a transition function; $F \subseteq S$ is a set of accepting conditions. Any automaton $(S, S_{init}, \Sigma, \delta, F)$ can be viewed as a graph $(V, E)$ with the vertexes $V = S$ and the edges $E$ given by $\delta$ in the expected way.*

In order to obtain an optimal path that satisfies the weighted transition system and the linear temporal logic task formula, it is necessary to construct a network topology that integrates environment model information and task requirement information. The construction method requires the following two steps:

(1) Construct a Büchi automaton

Linear temporal logic task formulas and automata have corresponding relations. After getting the task requirements as shown in formula (1), it is necessary to convert the linear temporal logic expression into text form through a tool (LTL2BA), and then convert the text into a matrix by the string processing and finally convert the adjacency matrix to its corresponding chart form (Büchi automaton).

We can obtain the corresponding Büchi automaton as shown in Fig. 3 by the transition system $\Gamma$ and the task formula (1), while T in the Fig. 3 represents all the states in the transition system.

(2) Construct a Product automaton

We can obtain a feasible network topology map (Product automaton) as shown in Fig. 4 via the Cartesian product of the finite state weighted transition system

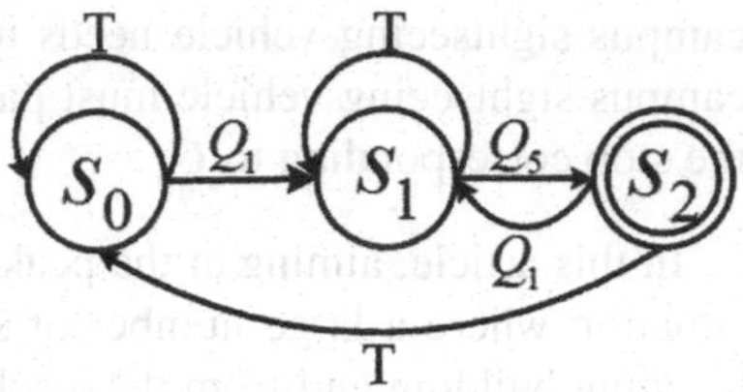


**Fig. 3** Büchi automaton

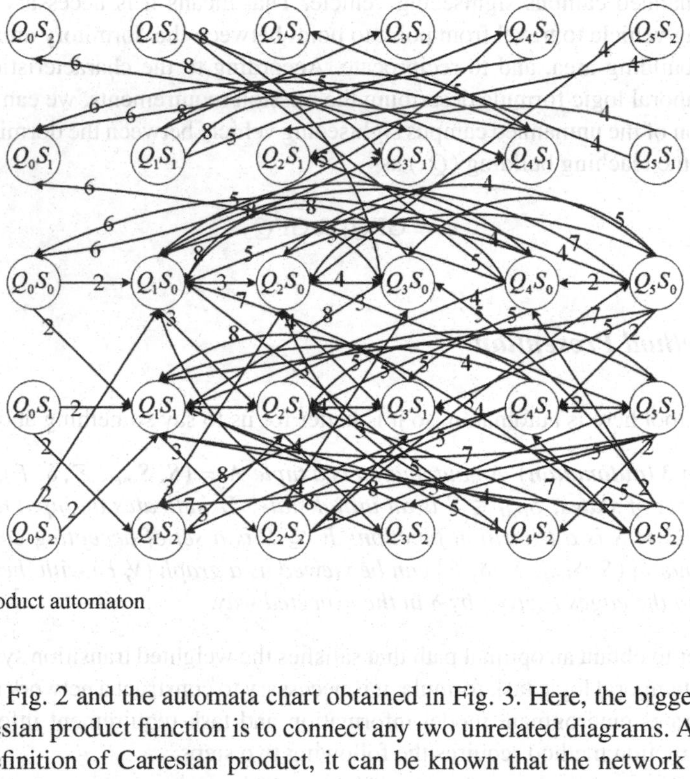


**Fig. 4** Product automaton

model in Fig. 2 and the automata chart obtained in Fig. 3. Here, the biggest role of the Cartesian product function is to connect any two unrelated diagrams. According to the definition of Cartesian product, it can be known that the network topology of Product automata not only satisfies the actual environment characteristics of the stops, but also satisfies the various carrying task requirements.

*Remark 1* The Product automaton combines actual environmental model with complex task requirements. The transition relationships between the 18 states ($Q_0S_0$, $Q_0S_1$, $Q_0S_2$, $Q_1S_0$, $Q_1S_1$, $Q_1S_2$, $Q_2S_0$, $Q_2S_1$, $Q_2S_2$, $Q_3S_0$, $Q_3S_1$, $Q_3S_2$, $Q_4S_0$, $Q_4S_1$, $Q_4S_2$, $Q_5S_0$, $Q_5S_1$, $Q_5S_2$) on the network topology are determined by the transition system $\Gamma$ and the Büchi automaton.

## 2.3 Optimal Path Optimization Method

Since the above-mentioned task requirement formula for the unmanned campus sightseeing vehicle is $\phi = \mathrm{GF}Q_1\&\mathrm{GF}Q_5$, in order to find the optimal patrol path, the first step is to determine whether a closed loop including states $Q_1$ and states $Q_5$ could be found in the task feasible network topology of the Product automaton, then find out the optimal patrol route with the shortest running time by optimal path optimization method based on Dijkstra algorithm. That is to say, we firstly need to determine the whether there is a closed loop in all path combinations of initial state $Q_1$ and end state $Q_5$. At the same time, calculate the time required for the optimal route that meets the task requirements and obtain the specific vertex order.

The optimal path search algorithm for the unmanned campus sightseeing vehicle is as follows: (Table 2)

By executing Algorithm 2, the optimal path that meets the task requirements can be searched on the task feasible network topology. The result is shown in Fig. 5. Since the time and weight of running this path are the minimum and the number of path nodes is the minimum, it can be judged that the patrol path is the optimal patrol path.

*Remark 2* The network topology merges transition system and Büchi automaton information, and we can search the optimal path that satisfies both the environmental information and the specified task on the network topology.

After obtaining the optimal patrol path ($Q_1S_0 \rightarrow Q_2S_1 \rightarrow Q_5S_2 \rightarrow Q_4S_0 \rightarrow Q_1S_0$) shown in Fig. 5, and the path node sequence in the actual environment can be obtained, as shown in Fig. 6. According to Fig. 6, it can be learned that when the

**Table 2** Algorithm 2

| |
|---|
| **Algorithm 2** Optimal path search algorithm based on Dijkstra |
| **Input**: two patrol stops (s, e) for the unmanned campus sightseeing vehicle, weighted transition system T, task formula $\phi$<br>**Output**: the minimum time required from the start point to the end point and the specific route that the unmanned campus sightseeing vehicle needs to perform (the vertices are sorted in order)<br>(1) call the transition system $\Gamma$<br>(2) build a Büchi automaton: convert the task formula into an automaton chart using the LTL2BA tool<br>(3) build a product automaton: make Cartesian product of the transition system $\Gamma$ and the automaton to get the product automaton network topology<br>(4) write an input function to enter the task patrol stops for the unmanned sightseeing vehicles<br>(5) execute the command [time, path] = dijkstra(A,s,e) to call Dijkstra algorithm, where time represents the sum (time) of the weights of the shortest path to be run, and path indicates the vertices sequence of the shortest path.<br>(6) define screen output statements which relate to the total time of the shortest path and the vertex sequencing of the shortest path<br>(7) run the program segment and input the patrol stops to return the cost and the vertices sequence of the corresponding optimal patrol path, and get the optimal route |

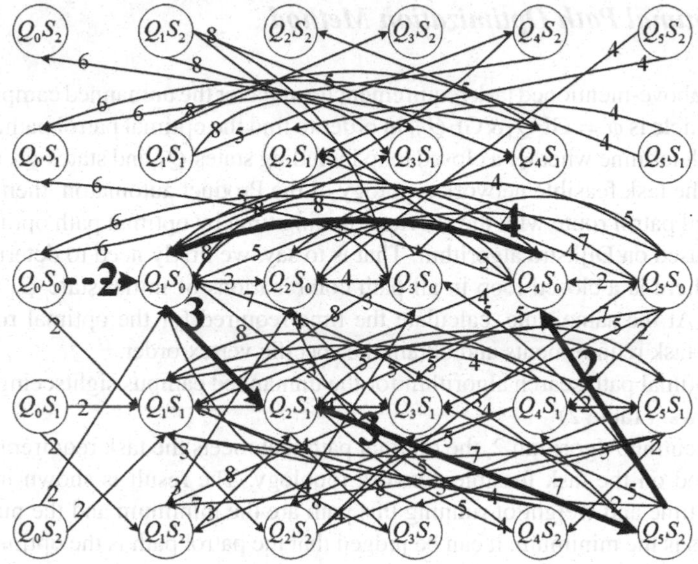


**Fig. 5** The optimal path on product automaton

**Fig. 6** The optimal path in the actual environment

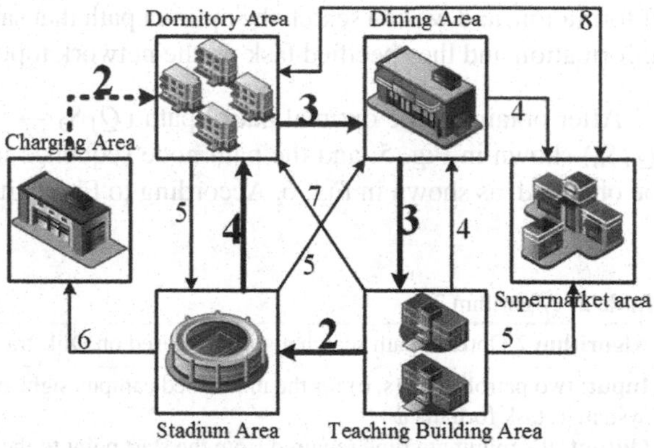


campus sightseeing vehicle executes a command that circulates between dormitory area and the teaching building area during the peak period of class, the sightseeing vehicle will select an optimal path by the path planning method based on LTL. The optimal patrol route, namely (charging area → dormitory area → dining area → teaching building area → stadium area → dormitory area →…).

*Remark 3* In the weighted transition system $\Gamma$ of the actual environment, there exists a path corresponding to an arbitrary path that meets the task requirements searched out on the feasible network topology, and the path satisfies the task requirement while ensuring the path optimality.

# 3 Conclusion

This paper proposes a global optimal path planning system for unmanned campus sightseeing vehicles based on LTL. With a more efficient and accurate model to carry students, it can effectively increase the utilization rate of sightseeing vehicles, reduce operating costs, reduce travel time. However, the problem of the exponential blow-up caused by the construction of the automaton accepting the language satisfying a LTL formula is well known. Therefore, for this research, the direction of future efforts can be devoted to reducing the complexity of the algorithm. The scheme proposed here is promising. It can be extended to the much more complex systems, such as 4WS Vehicles [11]. And after some advancement of our method, it can also be employed to the uncertain environments, such as [12].

**Acknowledgments** This work is supported by Natural Science Fund of Henan Provice, China (182300410112).

## References

1. F. Duchoň, A. Babinec, M. Kajan et al., Path planning with modified a star algorithm for a mobile robot. Proc. Eng. **96**(96), 59–69 (2014)
2. E. Onieva, J.E. Naranjo, V. Milanés et al., Automatic lateral control for unmanned vehicles via genetic algorithms. Appl. Soft Comput. **11**(1), 1303–1309 (2011)
3. K. Ghoseiri, B. Nadjari, An ant colony optimization algorithm for the bi-objective shortest path problem. Appl. Soft Comput. **10**(4), 1237–1246 (2010)
4. H. Lin, W. Zhang, Model checking: theories, techniques and application. Acta Electron. Sinica **S1**, 1907–1912 (2002). (in Chinese)
5. X. Ding, M. Lazar, C. Belta, Receding horizon temporal logic control for finite deterministic systems. Automatica, 2012, **50**(2), 399–408
6. A. Ulusoy, C. Belta, Receding horizon temporal logic control in dynamic environments. Int. J. Robot. Res. **33**(12), 1593–1607 (2014)
7. T. Wongpiromsarn, U. Topcu, R.M. Murray, Receding horizon temporal logic planning. IEEE Trans. Autom. Control **57**(11), 2817–2830 (2012)
8. S.L. Smith, J. Tůmová, C. Belta et al., Optimal path planning for surveillance with temporal-logic constraints. Int. J. Robot. Res. **30**(14), 1695–1708 (2011)
9. S.L. Smith, C. Belta, D. Rus, Optimal path planning under temporal logic contstraints, in International Conference on Intelligent Robots and Systems. IEEE, 3288–3293 (2010)
10. J. Guo, M. Bian, J. Han, Translation from LTL formula into Büchi automata. Comput. Sci. **35**(7), 241–243 (2008). (in Chinese)
11. Y. Jia, Robust control with decoupling performance for steering and traction of 4WS vehicles under velocity-varying motion. IEEE Trans. Control Syst. Technol. **8**(3), 554–569 (2000)
12. Y. Jia, Alternative proofs for improved LMI representations for the analysis and the design of continuous-time systems with polytopic type uncertainty: a predictive approach. IEEE Trans. Autom. Control **48**(8), 1413–1416 (2003)

# Sampled-Data Based Mean Square Bipartite Consensus of Double-Integrator Multi-Agent Systems with Measurement Noises

**Yifa Liu and Long Cheng**

**Abstract** A distributed sampled-data based bipartite consensus protocol is proposed for double-integrator multi-agent systems with measurement noises under signed digraph. A time-varying consensus gain and the agents' states feedback are adopted to counteract the noise effect and achieve bipartite consensus. By determining the state transition matrix of the multi-agent system, we describe the dynamic behaviour of the system. Under the proposed protocol, the states of some agents converge in mean square to one random vector while the rest of agents' states are convergent to another random vector. It is noted that these two vector are at the same amplitude, however their signs are different. It is proved that sufficient conditions for achieving the mean square bipartite consensus are: (1) the topology graph is weighted balanced, structurally balanced and has a spanning tree; and (2) the time-varying consensus gain satisfies the stochastic approximation conditions. We verify the validity of the proposed protocol by numerical simulations.

**Keywords** Bipartite Consensus · Multi-agent Systems (MASs)
Measurement Noises · Double-integrator · Sampled-data

## 1 Introduction

As the most important and key problem in distributed coordination of multi-agent systems (MASs), consensus plays a fundamental role of control protocol design and distributed optimization. Most researchers assume that MASs work in an ideal communication environment, however, measurement noises always exist in reality, which means agents cannot get accurate information of each other.

To reduce the effect of noise, there has been lots of available publications. A stochastic approximation type consensus protocol for discrete-time first-order inte-

---

Y. Liu · L. Cheng (✉)
State Key Laboratory of Management and Control for Complex Systems,
Institute of Automation, Chinese Academy of Sciences, Beijing 100190, China
e-mail: long.cheng@ia.ac.cn

 
Y. Jia et al. (eds.), *Proceedings of 2018 Chinese Intelligent Systems Conference*,
Lecture Notes in Electrical Engineering 528,
https://doi.org/10.1007/978-981-13-2288-4_34

gral MASs with communication noises was proposed and the concepts of mean square and almost sure consensus were introduced in [1]. The necessary and sufficient conditions for ensuring the mean square and almost sure consensus were proved and it was shown that the stochastic approximation type gain was necessary in [2, 3]. Both communication noises and delays were considered for the mean square consensus of first-order integral MASs in [4]. Leader-following consensus of first-order integral MASs with measurement noises was studied in [5]. It is found that for the continuous-time mean square leader-following consensus problem, the necessary and sufficient conditions of time-varying consensus gains can be relaxed [6]. Most publications concerning communication noises focus on the first-order integral MASs, more realistic, many control systems can be modeled by the second-order integral dynamics, such as the multi-robot system. Necessary and sufficient conditions were given for the mean square consensus of continuous-time second-order integral MASs under fixed topology in [7]. And a sampled-data based consensus protocol was proposed for second-order integral MASs in [8]. For the generic linear MASs, protocols for solving the mean square consensus with measurement noises were proposed under the fixed topology [9] and under the switching topology [10], respectively. And Cheng modified the consensus protocol and gave each agent its own noise-attenuation gain [11].

It is noted that those publications mentioned above are about cooperative multi-agent systems, however competition and opposition always exist in our world, for example military and politics [12]. For the conventional consensus problem, agents achieve average consensus through collaboration. However in a more realistic situation, some agents may compete and therefore the states of all agents converge to the same value except their signs and achieve bipartite consensus instead of the conventional average consensus [13].

Attempts were made to modify existing protocols for conventional consensus to solve the bipartite consensus problem. There were also many control strategies adopted. State feedback and output feedback control laws were designed to stabilize generic linear time-invariant (LTI) MASs and achieve bipartite consensus in [14]. Switching topologies were taken into consideration in [15] and a input saturation strategy for bipartite consensus on generic linear MASs was put forward in [16]. For bipartite consensus most studies did not take noises into consideration. There are very few available papers considering measurement noises, necessary and sufficient conditions were given for the mean square bipartite consensus of first-order integral MASs [17] and the mean square bipartite consensus of second-order integral MASs [18]. The proof of sufficient conditions for the high-order MASs was given in [19].

However because of the wide application of digital equipment, only sampled data can be used in the real control engineering. In this paper, a sampled-data based bipartite consensus protocol is proposed for double-integrator MASs with measurement noises under signed digraph, the bipartite consensus can be achieved by using a time-varying consensus gain and agents' states feedback to counteract the noise effect. The sufficient conditions for achieving the mean square bipartite consensus are given in this paper.

## 2 Problem Formulation

### 2.1 Preliminaries of Signed Graph

In the literature of bipartite consensus of multi-agent systems, the network is usually modeled by a weighted signed digraph $\mathcal{G} = (\mathcal{V}, \mathcal{E}, \mathcal{A})$, where $\mathcal{V} = (v_1, v_2, \ldots, v_N)$ denotes the set of nodes, $v_i$ represents the $i$th agent; $\mathcal{E} \subseteq \mathcal{V} \times \mathcal{V}$ denotes the set of edges, $e_{ij} = (v_i, v_j) \in \mathcal{E}$ if and only if there is the information flow from agent $j$ to agent $i$; $\mathcal{A} = [a_{ij}] \in \mathbb{R}^{N\times N}$ denotes the weighted adjacency matrix, $a_{ij} \neq 0 \Leftrightarrow e_{ij} \in \mathcal{E}$ and $a_{ij} = 0 \Leftrightarrow e_{ij} \notin \mathcal{E}$. $a_{ij} > 0$ means cooperation and $a_{ij} < 0$ means competition between agent $i$ and agent $j$. In this paper, we always assume that $a_{ii} = 0$ and $a_{ij}a_{ji} \geq 0, i, j = 1, \ldots, N$. The neighborhood of node $v_i$ is denoted by $\mathcal{N}_i = \{v_j \in \mathcal{V} | e_{ij} \in \mathcal{E}\}$. $\deg_{in}(i) = \sum_{j=1}^{N} |a_{ij}|$ is called the in-degree of node $v_i$ and $\deg_{out}(i) = \sum_{j=1}^{N} |a_{ji}|$ is called the out-degree of node $v_i$. A weighted signed digraph $\mathcal{G}$ is said to be weighted balanced if $\deg_{in}(i) = \deg_{out}(i)$. The Laplacian matrix $\mathcal{L}$ of $\mathcal{G}$ is $\mathcal{L} = \mathcal{D} - \mathcal{A}$, where $\mathcal{D} = \operatorname{diag}\left(\sum_{j=1}^{N} |a_{1j}|, \ldots, \sum_{j=1}^{N} |a_{Nj}|\right)$.

**Lemma 1** ([20]) *A weighted balanced digraph is connected if and only if it has a spanning tree.*

A weighted signed digraph $\mathcal{G} = (\mathcal{V}, \mathcal{E}, \mathcal{A})$ is said structurally balanced if there exist two sets of nodes $\mathcal{V}_1$ and $\mathcal{V}_2$, $\mathcal{V}_1 \cup \mathcal{V}_2 = \mathcal{V}$, $\mathcal{V}_1 \cap \mathcal{V}_2 = \varnothing$, such that $a_{ij} \geq 0, \forall v_i, v_j \in \mathcal{V}_l (l \in \{1, 2\})$ and $a_{ij} \leq 0, \forall v_i \in \mathcal{V}_l, v_j \in \mathcal{V}_q, l \neq q (l, q \in \{1, 2\})$ [17].

**Lemma 2** ([21]) *If a weighted signed digraph $\mathcal{G} = (\mathcal{V}, \mathcal{E}, \mathcal{A})$ is structurally balanced, the Laplacian matrix of $\mathcal{G}$ has at least one zero eigenvalue and all other eigenvalues have positive real parts. Particularly, $\mathcal{G}$ has exactly one zero eigenvalue if and only if $\mathcal{G}$ has a spanning tree.*

### 2.2 Consensus Protocol

Consider a MAS with $N$ agents, where the $i$th agent is described by the following sampled-data double-integrator dynamics

$$z_i[k+1] = Az_i[k] + Bu_i[k], \tag{1}$$

where $z_i[k] = (x_i[k], v_i[k])^T (k \in \mathbb{N}, i = 1, \ldots, N)$, $A = \begin{bmatrix} 1 & T \\ 0 & 1 \end{bmatrix}$, $B = \begin{bmatrix} \frac{T^2}{2} \\ T \end{bmatrix}$, $T$ is the sampling period, $x_i[k] \in \mathbb{R}$, $v_i[k] \in \mathbb{R}$ and $u_i[k] \in \mathbb{R}$ are the position, velocity and control input of the $i$th agent at the $k$th sampling point respectively.

Because of measurement noises, the $i$th agent receives the neighbor agent's information $y_{ij}[k] = (y_{xij}[k], y_{vij}[k])^T = (x_j[k] + n_{xij}[k], v_j[k] + n_{vij}[k])^T$, $j \in \mathcal{N}_i$, where $y_{xij}[k]$ and $y_{vij}[k]$ denote the measurements of the $j$th agent's position and velocity. $\{n_{xij}[k], n_{vij}[k], k \in \mathbb{N}, i, j = 1, \ldots, N\}$ is the independent random noise sequence with zero mean, and

$$\sup_{k\in\mathbb{N}} \max_{i,j=1,\ldots,N} E(n_{xij}^2[k]) < \sigma_{max}, \quad \sup_{k\in\mathbb{N}} \max_{i,j=1,\ldots,N} E(n_{vij}^2[k]) < \sigma_{max}.$$

The bipartite consensus problem under measurement noises is to design a distributed protocol for system (1) such that the states of some agents converge in mean square to one random vector while the rest of agents' states are convergent to another random vector with the same amplitude and opposite sign.

Define the transformation matrix $T_z = \begin{bmatrix} \frac{T^2}{2} & T^2 \\ -3T & 0 \end{bmatrix}$. Let $\bar{z}_i[k] = T_z^{-1} z_i[k] = (\bar{z}_{xi}[k], \bar{z}_{vi}[k])^T$, then system (1) can be transformed into

$$\bar{z}_i[k+1] = \bar{A}\bar{z}_i[k] + \bar{B}u_i[k], \tag{2}$$

$$\bar{A} = T_z^{-1} A T_z = \begin{bmatrix} 1 & 0 \\ -3 & 1 \end{bmatrix}, \quad \bar{B} = T_z^{-1} B = \begin{bmatrix} -\frac{1}{3} \\ \frac{2}{3} \end{bmatrix}.$$

In this paper, we propose the following distributed bipartite consensus protocol

$$u_i[k] = K_1 \bar{z}_i[k] + a[k] K_2 \sum_{j\in\mathcal{N}_i} |a_{ij}| (\mathrm{sgn}(a_{ij}) \bar{y}_{ij}[k] - \bar{z}_i[k]), \tag{3}$$

where $K_1 = \begin{bmatrix} 4.5 & 0 \end{bmatrix}$, $K_2 = \begin{bmatrix} 1.5 & 1.5 \end{bmatrix}$, $\bar{y}_{ij}[k] = T_z^{-1} y_{ij}[k]$, and $a[k] > 0$ is the non-increasing consensus gain sequence.

# 3 Main Result

Applying (3) into system (1) leads to the following closed-loop system

$$\begin{aligned} \bar{Z}[k+1] = {} & [I_N \otimes (\bar{A} + \bar{B}K_1) - a[k]\mathcal{L} \otimes (\bar{B}K_2)]\bar{Z}[k] \\ & + a[k][\Sigma \otimes (\bar{B}K_2 T_z^{-1})]N[k], \end{aligned} \tag{4}$$

where $\mathcal{L}$ is the Laplacian matrix of $\mathcal{G}$, $\bar{Z}[k] = (\bar{z}_1^T[k], \ldots, \bar{z}_N^T[k])^T \in \mathbb{R}^{2N}$, $\Sigma = \mathrm{diag}(\tilde{a}_1, \ldots, \tilde{a}_N) \in \mathbb{R}^{N\times N^2}$, $\tilde{a}_i = (a_{i1}, \ldots, a_{iN}) \in \mathbb{R}^{1\times N}$, $N[k] = (n_{x11}[k], n_{v11}[k], \ldots, n_{x1N}[k], n_{v1N}[k], \ldots, n_{xNN}[k], n_{vNN}[k])^T \in \mathbb{R}^{2N^2}$.

For system (1) we adopt the following assumptions

(A1) $\mathcal{G}$ is weighted balanced and has a spanning tree.

(A2) $\mathcal{G}$ is structurally balanced.

(A3) $\sum_{k=0}^{\infty} a[k] = \infty.$ (A4) $\sum_{k=0}^{\infty} a^2[k] < \infty.$

If assumption (A2) holds, then $\mathcal{V}$ can be divided into two subsets $\mathcal{V}_1$ and $\mathcal{V}_2$, for simplification, we assume $\mathcal{V}_1 = \{v_1, v_2, \ldots, v_{K_0}\}$, $\mathcal{V}_2 = \{v_{K_0+1}, \ldots, v_N\}$, and if assumption (A1) holds, $\mathcal{L}$ has the following eigenvector $\omega$ of eigenvalue 0 [13] $\omega = [\underbrace{1, \ldots, 1}_{K_0}, -1, \ldots, -1]^T$.

Consider the homogeneous equation

$$\bar{Z}[k+1] = [I_N \otimes (\bar{A} + \bar{B}K_1) - a[k]\mathcal{L} \otimes (\bar{B}K_2)]\bar{Z}[k]. \tag{5}$$

Define $\Phi(k, k_0) = \prod_{i=k_0}^{k-1} [I_N \otimes (\bar{A} + \bar{B}K_1) - a[k]\mathcal{L} \otimes (\bar{B}K_2)]$ $(\Phi(k_0, k_0) = I_{2N})$, which represents the state transition matrix of (5) and $Q_N = \frac{1}{N}\omega\omega^T \otimes \begin{bmatrix} 0 & 0 \\ 0 & 1 \end{bmatrix}$, and define $F[k] = I_N \otimes (\bar{A} + \bar{B}K_1) - a[k]\mathcal{L} \otimes (\bar{B}K_2) - Q_N$.

Then we can find it easily that $Q_N \times Q_N = Q_N \times Q_N^T = Q_N^T \times Q_N = \Phi(k, k_0) \times Q_N = Q_N \times \Phi(k, k_0) = Q_N$, it can be calculated that

$$\begin{aligned} \|F[k]\|_2^2 = {} & \|[I_N \otimes (\bar{A} + \bar{B}K_1) - a[k]\mathcal{L} \otimes (\bar{B}K_2) - Q_N][I_N \otimes (\bar{A} + \bar{B}K_1) \\ & - a[k]\mathcal{L} \otimes (\bar{B}K_2) - Q_N]^T\|_2 \\ \le {} & \|I_N \otimes (\bar{A} + \bar{B}K_1)(\bar{A} + \bar{B}K_1)^T - a[k]\mathcal{L}^T \otimes (\bar{A} + \bar{B}K_1)(\bar{B}K_2)^T \\ & - a[k]\mathcal{L} \otimes (\bar{B}K_2)(\bar{A} + \bar{B}K_1)^T - Q_N\|_2 + a^2[k]\|\mathcal{L}\mathcal{L}^T \otimes (\bar{B}K_2)(\bar{B}K_2)^T\|_2. \end{aligned}$$

Because $(\bar{A} + \bar{B}K_1)(\bar{B}K_2)^T = (\bar{B}K_2)(\bar{A} + \bar{B}K_1)^T = \begin{bmatrix} \frac{1}{4} & -\frac{1}{2} \\ -\frac{1}{2} & 1 \end{bmatrix} \triangleq C$, then $\|F[k]\|_2^2 \le \|I_N \otimes \Gamma_A - a[k](\mathcal{L} + \mathcal{L}^T) \otimes C - Q_N\|_2 + a^2[k]\|\mathcal{L}\mathcal{L}^T \otimes (\bar{B}K_2)(\bar{B}K_2)^T\|_2$, where $\Gamma_A = (\bar{A} + \bar{B}K_1)(\bar{A} + \bar{B}K_1)^T$.

If assumptions (A1) and (A2) hold, according to Lemma 1, there exists a matrix $T \in \mathbb{C}^{N \times N}$, whose first column is $\frac{1}{\sqrt{N}}\omega$ such that $T^{-1}(\mathcal{L} + \mathcal{L}^T)T = \mathrm{diag}(0, \lambda_2, \ldots, \lambda_N)$, where $0 < \lambda_2 \le \cdots \le \lambda_N$ is the eigenvalue of $(\mathcal{L} + \mathcal{L}^T)$. Then $F[k]$ can be calculated by (6).

$$\begin{aligned} & I_N \otimes \Gamma_A - a[k](\mathcal{L} + \mathcal{L}^T) \otimes C - Q_N = (T^{-1} \otimes I_2) \times \\ & \mathrm{diag}\,(\Gamma_A - \Gamma_1, \Gamma_A - a[k]\lambda_2 C, \ldots, \Gamma_A - a[k]\lambda_N C) \times (T \otimes I_2), \end{aligned} \tag{6}$$

where $\Gamma_1 = \begin{bmatrix} 0 & 0 \\ 0 & 1 \end{bmatrix}$, $\|\Gamma_A - a[k]\lambda_i C\|_2 = \left\| \begin{bmatrix} \frac{1}{4} - a[k]\lambda_i & \frac{1}{2}a[k]\lambda_i \\ \frac{1}{2}a[k]\lambda_i & 1 - a[k]\lambda_i \end{bmatrix} \right\|_2$ $= \sqrt{\frac{1}{4}(a[k]\lambda_i)^2 + \frac{17}{32}(1 - a[k]\lambda_i)^2 + \frac{5}{32}|1 - a[k]\lambda_i|\sqrt{(3 - 3a[k]\lambda_i)^2 + (4a[k]\lambda_i)^2}}$.

If assumptions (A3) and (A4) hold, there exists $D_1$, such that for $\forall k > D_1$, $0 < a[k]\lambda_i < 1$, $i = 2, \ldots, N$

$$\|(\bar{A}+\bar{B}K_1)(\bar{A}+\bar{B}K_1)^T - a[k]\lambda_i C\|_2 \le \sqrt{\frac{1}{4}(a[k]\lambda_i)^2 + (1-a[k]\lambda_i)^2} \le 1 - a[k]\lambda_i.$$

Then $\|F[k]\|_2^2 \le 1 - a[k]\lambda_2 + a^2[k]\|\mathcal{L}\mathcal{L}^T \otimes (\bar{B}K_2)(\bar{B}K_2)^T\|_2$.

There exists $D_2 > D_1$, such that for $\forall k > D_2$, $\frac{1}{2}a[k]\lambda_2 > a^2[k]\|\mathcal{L}\mathcal{L}^T \otimes (\bar{B}K_2)(\bar{B}K_2)^T\|_2$, it can be obtained that

$$\|F[k]\|_2 \le \sqrt{1 - a[k]\lambda_2 + a^2[k]\|\mathcal{L}\mathcal{L}^T \otimes (\bar{B}K_2)(\bar{B}K_2)^T\|_2}$$
$$\le \sqrt{1 - \frac{1}{2}a[k]\lambda_2} \le 1 - \frac{1}{4}a[k]\lambda_2.$$
$$\|\prod_{i=k_0}^{k} F[i]\|_2 \le \prod_{i=k_0}^{k}\|F[i]\|_2 \le \prod_{i=k_0}^{D_2-1}\|F[i]\|_2 \prod_{i=D_2}^{k}(1 - \frac{1}{4}a[k]\lambda_2)$$
$$\le \prod_{i=k_0}^{D_2-1}\|F[i]\|_2 e^{(-\frac{1}{4}\lambda_2 \sum_{i=D_2}^{k} a[i])}. \tag{7}$$

If assumption (A3) holds,

$$\lim_{k\to\infty}\|\prod_{i=k_0}^{k} F[i]\|_2 = 0, \qquad \lim_{k\to\infty}\Phi(k,k_0) = Q_N. \tag{8}$$

The stochastic difference equation (4) can be solved as follows

$$\bar{Z}[k] = \Phi(k,0)\bar{Z}[0] + \sum_{i=0}^{k-1} a[i]\Phi(k,i+1)\Sigma \otimes (\bar{B}K_2T_z^{-1})N[i]. \tag{9}$$

If assumption (A4) holds, there exists $N_1$, such that for $\forall \varepsilon > N_1, \sum_{k=N_1}^{\infty} a^2[k] < \varepsilon$.

Define $Q[k] = \sum_{i=0}^{k-1} a[i]\Phi(k,i+1)\Sigma \otimes (\bar{B}K_2T_z^{-1})N[i]$.

According to (7) and (8), there exists $N_2 > N_1$, such that for $\forall m > n > N_2$, $\|\Phi(m,s) - \Phi(n,s)\|_F < \varepsilon$ $(s = 0, \ldots, N_1)$, and there exists $\Phi_{max} > 0$, such that for $\forall n_1, n_2 \in \mathbb{N}$, $\|\Phi(n_1,n_2)\|_F < \Phi_{max}$, by using the similar method in [7].

$$\|E(Q[m]-Q[n])E(Q[m]-Q[n])^T\|_F \le 2\|(\sum_{i=n}^{m-1} a[i]\Phi(m,i+1)\Sigma \otimes (\bar{B}K_2T_z^{-1})N[i])$$
$$(\sum_{i=n}^{m-1} a[i]\Phi(m,i+1)\Sigma \otimes (\bar{B}K_2T_z^{-1})N[i])^T + 2(\sum_{i=0}^{n-1} a[i][\Phi(m,i+1) - \Phi(n,i+1)]$$
$$\Sigma \otimes (\bar{B}K_2T_z^{-1})N[i])(\sum_{i=0}^{n-1} a[i][\Phi(m,i+1) - \Phi(n,i+1)]\Sigma \otimes (\bar{B}K_2T_z^{-1})N[i])^T\|_F$$
$$= 2\|\sum_{i=n}^{m-1} a^2[i]\Phi(m,i+1)\Sigma \otimes (\bar{B}K_2T_z^{-1})N[i]N^T[i]\Sigma^T \otimes (\bar{B}K_2T_z^{-1})^T\Phi(m,i+1)^T\|_F$$

$$+2\|\sum_{i=0}^{n-1} a^2[i]\Phi_{cmn}\Sigma \otimes (\bar{B}K_2T_z^{-1})N[i]N[i]^T\Sigma^T \otimes (\bar{B}K_2T_z^{-1})^T\Phi_{cmn}^T\|_F$$

$$\leq 2\varepsilon\alpha_{max}^2 N^4\sigma_{max}\|\bar{B}K_2T_z^{-1}\|_F\ (1+4\Phi_{max}^2+\varepsilon\sum_{i=0}^{N_1-1} a^2[i]),$$

where $\alpha_{max} = \max_{i,j=1,\ldots,N,k\in\mathbb{N}}\{|a_{ij}[k]|\}$, $\Phi_{cmn} = \Phi(m,i+1) - \Phi(n,i+1)$.

Then $Q[k]$ converges in mean square to a random vector $Q^*$, such that

$$E(Q^*Q^{*T}) = \lim_{k\to\infty} E(Q[k]Q^T[k]) = \lim_{k\to\infty}\sum_{i=0}^{k-1} a^2[k]\Phi(k,i+1)\Sigma\otimes(\bar{B}K_2T_z^{-1})\Xi$$

$$\Sigma^T\otimes(\bar{B}K_2T_z^{-1})^T\Phi^T(k,i+1).$$

If assumption (A4) holds, there exists $N_3$, such that for $\forall\varepsilon > 0, k > N_3$

$$\sum_{i=N_3}^{k-1} a^2[i]\|\Phi(k,i+1)\Sigma\otimes(\bar{B}K_2T_z^{-1})\Xi\ \Sigma^T\otimes(\bar{B}K_2T_z^{-1})^T\Phi^T(k,i+1)\|_F \leq \varepsilon,$$

$$\sum_{i=N_3}^{\infty} a^2[i]\|Q_N\Sigma\otimes(\bar{B}K_2T_z^{-1})\Xi\ \Sigma^T\otimes(\bar{B}K_2T_z^{-1})^T Q_N^T\|_F \leq \varepsilon,$$

and there exists $N_4 > N_3$, such that for $\forall k > N_4$

$$\sum_{i=0}^{N_3-1} a^2[i]\|\Phi(k,i+1)\Sigma\otimes(\bar{B}K_2T_z^{-1})\Xi\ \Sigma^T\otimes(\bar{B}K_2T_z^{-1})^T\Phi^T(k,i+1)$$

$$-Q_N\Sigma\otimes(\bar{B}K_2T_z^{-1})\Xi\ \Sigma^T\otimes(\bar{B}K_2T_z^{-1})^T Q_N^T\|_F \leq \varepsilon.$$

Hence it can be found that

$$\lim_{k\to\infty}\sum_{i=0}^{k-1} a^2[k]\Phi(k,i+1)\Sigma\otimes(\bar{B}K_2T_z^{-1})\Xi\ \Sigma^T\otimes(\bar{B}K_2T_z^{-1})^T\Phi^T(k,i+1)$$

$$=\sum_{i=0}^{\infty} a^2[i]Q_N\Sigma\otimes(\bar{B}K_2T_z^{-1})\Xi\ \Sigma^T\otimes(\bar{B}K_2T_z^{-1})^T Q_N^T = \omega\omega^T\otimes\begin{bmatrix}0 & 0\\ 0 & \Omega\end{bmatrix},$$

where $\Omega = \frac{1}{N^2}\sum_{k=0}^{\infty} a^2[k]\left(\sum_{i=0}^{N}\sum_{j\in\mathcal{N}_i} a_{ij}^2[k]\left(\frac{E(n_{xij}^2[k])}{T^4} + \frac{E(n_{vij}^2[k])}{36T^2}\right)\right)$.

By using the similar method in [7], it can be proved $D(\bar{z}_{xi}^*) = 0$, $D(\bar{z}_{vi}^*) = \Omega$, $Cov(\bar{z}_{xi}^*, \bar{z}_{vi}^*) = 0$. For $\forall v_i, v_j \in \mathcal{V}_l\ (l\in\{1,2\})$, $Cov(\bar{z}_{vi}^*, \bar{z}_{vj}^*) = \Omega$, $\rho(\bar{z}_{vi}^*, \bar{z}_{vj}^*) = 1$. It can be calculated that

$$\lim_{k\to\infty} E(\bar{z}_{vi}[k]-\bar{z}_{vj}^*)^2 = \lim_{k\to\infty} E(\bar{z}_{vi}[k]-\bar{z}_{vi}^*+\bar{z}_{vi}^*-\bar{z}_{vj}^*)^2$$

$$\leq \lim_{k\to\infty} E(\bar{z}_{vi}[k]-\bar{z}_{vi}^*)^2 + E(\bar{z}_{vi}^*-\bar{z}_{vj}^*)^2 + \lim_{k\to\infty} 2(E(\bar{z}_{vi}[k]-\bar{z}_{vi}^*)^2)^{\frac{1}{2}}(E(\bar{z}_{vi}^*-\bar{z}_{vj}^*)^2)^{\frac{1}{2}}$$

$$= E(\bar{z}_{vi}^*)^2 + E(\bar{z}_{vj}^*)^2 - 2E(\bar{z}_{vi}^*\bar{z}_{vj}^*) = \Omega+\Omega-2\Omega = 0.$$

By the same process, $\lim_{k\to\infty} E(\bar{z}_{xi}[k] - \bar{z}^*_{xj})^2 = 0$.

Similarly for $\forall v_i \in \mathcal{V}_l,\ v_j \in \mathcal{V}_q,\ l \neq q\ (l, q \in \{1, 2\}),\ Cov(\bar{z}^*_{vi}, \bar{z}^*_{vj}) = -\Omega$, $\rho(\bar{z}^*_{vi}, \bar{z}^*_{vj}) = -1$,

$$\lim_{k\to\infty} E(\bar{z}_{vi}[k] + \bar{z}^*_{vj})^2 = 0, \quad \lim_{k\to\infty} E(\bar{z}_{xi}[k] + \bar{z}^*_{xj})^2 = 0.$$

Therefore for $\forall v_i \in \mathcal{V}_1$, $\bar{z}_{xi}[k]$ and $\bar{z}_{vi}[k]$ converge in mean square to the random vector $\bar{z}^*_x$ and $\bar{z}^*_v$, while for $\forall v_i \in \mathcal{V}_2$, $\bar{z}_{xi}[k]$ and $\bar{z}_{vi}[k]$ converge in mean square to the random vector $-\bar{z}^*_x$ and $-\bar{z}^*_v$.

$$\begin{aligned}
E(z^*_x) &= E(z^*_x)^2 = 0,\\
E(z^*_v) &= \frac{1}{N}\left(\sum_{i=1}^{K_0}(\frac{x_i[0]}{T^2} + \frac{v_i[0]}{6T}) - \sum_{i=K_0}^{N}(\frac{x_i[0]}{T^2} + \frac{v_i[0]}{6T})\right),\\
E(z^*_v)^2 &= \Omega + \frac{1}{N^2}\left(\sum_{i=1}^{K_0}(\frac{x_i[0]}{T^2} + \frac{v_i[0]}{6T}) - \sum_{i=K_0}^{N}(\frac{x_i[0]}{T^2} + \frac{v_i[0]}{6T})\right)^2.
\end{aligned}$$

$$\begin{aligned}
&\lim_{k\to\infty} Z[k] = (I_N \otimes T_z)Q_N(I_N \otimes T_z^{-1})Z[0] = \frac{1}{N}\omega\omega^T \otimes \begin{bmatrix} 1 & \frac{T}{6} \\ 0 & 0 \end{bmatrix} Z[0],\\
&\lim_{k\to\infty} E(x_i[k] - x^*)^2 = 0,\ \lim_{k\to\infty} E(v_i[k] - v^*)^2 = 0,\ v_i \in \mathcal{V}_1,\\
&\lim_{k\to\infty} E(x_i[k] + x^*)^2 = 0,\ \lim_{k\to\infty} E(v_i[k] + v^*)^2 = 0,\ v_i \in \mathcal{V}_2.
\end{aligned}$$

$$E(x^*) = \frac{1}{N}\left(\sum_{i=1}^{K_0}(x_i[0] + \frac{T}{6}v_i[0]) - \sum_{i=K_0}^{N}(x_i[0] + \frac{T}{6}v_i[0])\right), \quad E(v^*) = 0.$$

The above analysis gives the states of agents at the sampling instant. Using the same analysis method proposed in [8], it can be proved

$$\begin{aligned}
&\lim_{t\to\infty} E(x_i(t) - x^*)^2 = 0, \quad \lim_{t\to\infty} E(v_i(t) - v^*)^2 = 0,\ v_i \in \mathcal{V}_1,\\
&\lim_{t\to\infty} E(x_i(t) + x^*)^2 = 0, \quad \lim_{t\to\infty} E(v_i(t) + v^*)^2 = 0,\ v_i \in \mathcal{V}_2.
\end{aligned}$$

Hence the protocol (3) can solve the mean square bipartite consensus problem of double-integrator multi-agent systems (1) with measurement noises.

## 4 Simulation Examples

*Example 1* Consider a double-integrator MAS with six agents. The communication topology $\mathcal{G}$ is shown in Fig. 1. Obviously, $\mathcal{G}$ is weighted balanced, structurally bal-

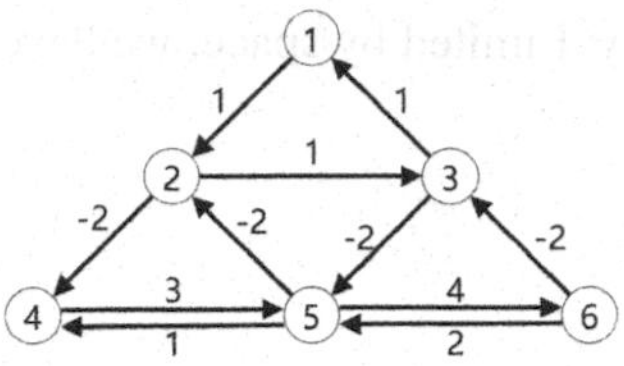


| agent | 1 | 2 | 3 | 4 | 5 | 6 |
|---|---|---|---|---|---|---|
| $x_i[0]$ | 9.439 | 7.241 | 2.004 | 1.888 | -0.158 | -6.227 |
| $v_i[0]$ | 5.235 | -1.571 | 5.834 | -3.289 | -3.204 | 0.133 |

**Fig. 1** Topology and initial states of multi-agent system in Example 1

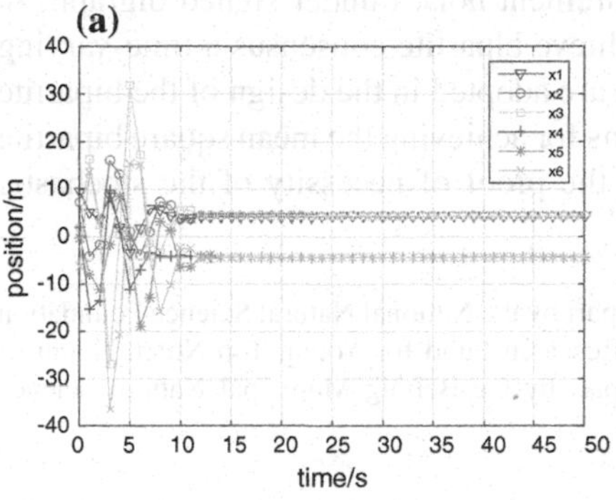

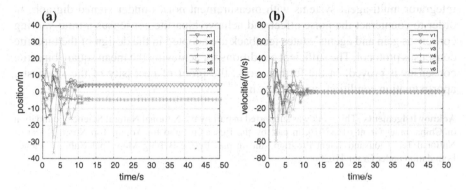


**Fig. 2** Evolutions of agents in Example 1. **a** Positions; **b** velocities

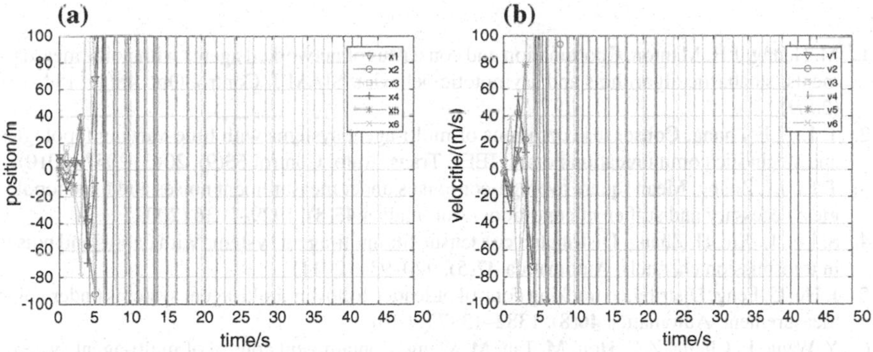


**Fig. 3** Evolutions of agents in Example 2. **a** Positions; **b** velocities

anced and has a spanning tree. The sampling period is set to 1. With the effect of measurement noises, we take the time-varying consensus gain $a[k] = 1/(k+1),\ k \in \mathbb{N}$ in protocol (3) and $a[k]$ satisfies the stochastic approximation conditions. $E(x^*) = 4.304,\ \ E(x^*)^2 < \infty, \quad E(v^*) = E|v^*|^2 = 0$ (Fig. 2).

*Example 2* If the stochastic approximation conditions are not satisfied, it should be verified whether agents can achieve bipartite consensus. Therefore we take a constant gain $a[k] = 1$, The evolutions of positions/velocities of agents are shown in Fig. 3. It is shown that assumptions (A3) and (A4) are necessary. Limited by space, we have not given proof of necessity.

## 5 Conclusion

In this paper, a mean square bipartite consensus protocol is proposed for double-integrator multi-agent systems with measurement noises under signed digraph. In order to counteract the noise effect and achieve bipartite consensus a time-varying consensus gain and agents' states feedback are adopted in the design of the bipartite consensus protocol. The sufficient conditions for achieving the mean square bipartite consensus is proved. In the further work, the proof of necessity of the stochastic approximation conditions will be given.

**Acknowledgements** This work was supported in part by the National Natural Science Foundation of China under Grant 61633016, in part by the Research Fund for Young Top-Notch Talent of National Ten Thousand Talent Program, and in part by the Beijing Municipal Natural Science Foundation under Grant 4162066.

## References

1. M. Huang, J.H. Manton, Coordination and consensus of networked agents with noisy measurements: stochastic algorithms and asymptotic behavior. SIAM J. Control Opt. **48**(1), 134–161 (2009)
2. T. Li, J.F. Zhang, Consensus conditions of multi-agent systems with time-varying topologies and stochastic communication noises. IEEE Trans. Auto. Control **55**(9), 2043–2057 (2010)
3. T. Li, J.F. Zhang, Mean square average-consensus under measurement noises and fixed topologies: necessary and sufficient conditions. Automatica **45**(8), 1929–1936 (2009)
4. S. Liu, L. Xie, H. Zhang, Distributed consensus for multi-agent systems with delays and noises in transmission channels. Automatica **47**(5), 920–934 (2011)
5. J. Hu, G. Feng, Distributed tracking control of leader-follower multi-agent systems under noisy measurement. Automatica **46**(8), 1382–1387 (2010)
6. Y. Wang, L. Cheng, Z.G. Hou, M. Tan, M. Wang, Containment control of multi-agent systems in a noisy communication environment. Automatica **50**(7), 1922–1928 (2014)
7. L. Cheng, Z.G. Hou, M. Tan, X. Wang, Necessary and sufficient conditions for consensus of double-integrator multi-agent systems with measurement noises. IEEE Trans. Auto. Control **56**(8), 1958–1963 (2011)
8. L. Cheng, Y. Wang, Z.G. Hou, M. Tan, Z. Cao, Sampled-data based average consensus of second-order integral multi-agent systems: switching topologies and communication noises. Automatica **49**(5), 1458–1464 (2013)
9. L. Cheng, Z.G. Hou, M. Tan, A mean square consensus protocol for linear multi-agent systems with communication noises and fixed topologies. IEEE Trans. Auto. Control **59**(1), 261–267 (2014)
10. Y. Wang, L. Cheng, W. Ren, Z.G. Hou, M. Tan, Seeking consensus in networks of linear agents: communication noises and markovian switching topologies. IEEE Trans. Auto. Control **60**(5), 1374–1379 (2015)

11. L. Cheng, Y. Wang, W. Ren, Z.G. Hou, M. Tan, On convergence rate of leader-following consensus of linear multi-agent systems with communication noises. IEEE Trans. Auto. Control **61**(11), 3586–3592 (2016)
12. H. Noel, The dynamics of two-party politics: party structures and the management of competition by alan ware. Polit. Sci. Quart. **125**(3), 515–517 (2010)
13. C. Altafini, Consensus problems on networks with antagonistic interactions. IEEE Trans. Auto. Control **58**(4), 935–946 (2013)
14. H. Zhang, J. Chen, Bipartite consensus of multi-agent systems over signed graphs: state feedback and output feedback control approaches. Int. J. Robust Nonlinear Control **27**(1), 3–14 (2017)
15. J. Li, W. Dong, H. Xiao, Signed consensus problems on networks of agents with fixed and switching topologies. Int. J. Control **90**(2), 148–160 (2017)
16. J. Qin, W. Fu, W.X. Zheng, H. Gao, On the bipartite consensus for generic linear multiagent systems with input saturation. IEEE Trans. Cybern. **47**(8), 1948–1958 (2017)
17. C. Ma, Z. Qin, Bipartite consensus on networks of agents with antagonistic interactions and measurement noises, in *IET Control Theory & Applications*, vol. 10 (2016), pp. 2306–2313
18. C. Ma, W. Zhao, Y. Zhao, Bipartite linear $\chi$-consensus of double-integrator multi-agent systems with measurement noise. Asian J. Control **20**(1), 577–584 (2018)
19. J. Hu, Y. Wu, Y. Zhao, Consensus control of high-order multi-agent systems with antagonistic interactions and measurement noises. IFAC **50**(1), 2482–2487 (2017)
20. J. Cortes, Distributed algorithms for reaching consensus on general functions. Automatica **44**(3), 726–737 (2008)
21. D. Meng, M. Du, Y. Jia, Interval bipartite consensus of networked agents associated with signed digraphs. IEEE Trans. Auto. Control **61**(12), 3755–3770 (2016)

# Image Recognition of Engine Ignition Experiment Based on Convolutional Neural Network

**Shangkun Huang, Fengshun Lu, Yufei Pang and Sumei Xiao**

**Abstract** In the engine ignition experiment, the specific instant of the ignition is usually obtained from a large quantity of high-resolution pictures taken with high-speed cameras, which puts forward an urgent request for the rapid image recognition. To address this issue, a picture recognition method based on convolutional neural network (CNN) is described. First, a training data set for the CNN model is made based on the original experimental images. Second, the constructed CNN model is trained to obtain the classification result. Finally, the CNN model is evaluated and optimized for the image recognition of engine ignition. The experimental results show that the method can quickly and accurately recognize the engine ignition.

**Keywords** Convolution neural network · Data set · Image recognition
Engine ignition

## 1 Introduction

In order to investigate the state of engine ignition, a large number of engine ignition experiments are required. The engine ignition experiment is a continuous and dynamic process, during which the engine states change dramatically between the before- and after-ignition, namely from static to dynamic. It is very important to observe and study the fuel combustion state at the ignition moment and the impact of state transition on various components of the engine. In order to understand and accurately capture the engine ignition, the entire experimental process is photographed

S. Huang (✉) · S. Xiao
China Aerodynamics Research
and Development Center, Computational Aerodynamics Institute, Mianyang 621000, China
e-mail: caesarhskanne@163.com

F. Lu · Y. Pang
School of Manufacturing Science & Engineering, Southwest University
of Science & Technology, Mianyang 621010, China
e-mail: lufengshun@nudt.edu.cn

 
Y. Jia et al. (eds.), *Proceedings of 2018 Chinese Intelligent Systems Conference*,
Lecture Notes in Electrical Engineering 528,
https://doi.org/10.1007/978-981-13-2288-4_35

with a high-speed camera and a large number of images are generated from which the engine ignition could be recognized. Traditionally, the image recognition is usually performed manually and it becomes the bottleneck to the in-time data feedback required for the experiment. After the experiment, the identification of the engine ignition test image is usually performed manually by searching a large number of image data, and then the corresponding engine status is validated according to the display time.

Artificial intelligence [1–4] is an external extension to the human consciousness, whose working process is thought to be intelligent, efficient and rational. The artificial intelligence represented by deep learning has achieved remarkable achievements in the past decade and outperformed human intelligence in specific areas, such as the image classification [5], the game of Go [6, 7] and autopilot [8–11]. The deep learning technology uses massive training data and has demonstrated a very powerful induction learning ability. Combined with a deep network structure, it can perform the multi-level feature extraction in an automatic, efficient and accurate manner, without the need for prior knowledge and known features. Nowadays the artificial intelligence based on deep learning has been widely utilized in the field of picture recognition and understanding.

Specially, the image recognition model based on the convolutional neural network (CNN) [12–16] has inherited advantages in the image recognition. It can obtain the features for distinguishing the images after training through an enormous data set, establish an abstract concept for image classification, and then perform automatic image recognition for the engine ignition.

The rest of the paper is organized as follows. Section 1 introduces the convolutional neural network. In the Sect. 2, the image recognition method based on CNN is described. Section 3 demonstrates the details for the training and evaluating of the CNN model. The experimental results are presented and analyzed in Sect. 4. The conclusion and future work is given in Sect. 5.

## 1.1 Convolutional Neural Network

Deep learning is a machine learning method that made artificial intelligence algorithms more intelligent. Combined with the multi-layer structure, it allows the machine to extract characteristics from the data set, and gains the ability of abstract concepts obtained through repeated training [17]. In order to learn complex functions that can represent high-level abstract features, deep structures including multi-layer nonlinear operators are needed to increase the accuracy of classification or prediction, such as the convolutional neural network (CNN) with multiple hidden nodes [18, 19].

CNN is a kind of deep neural network with convolutional structure, which can reduce the amount of memory occupied by the deep network and alleviate the overfitting problem. As an important direction of artificial intelligence technology, CNN learns principles from the training dataset and then use them to predict new data. The

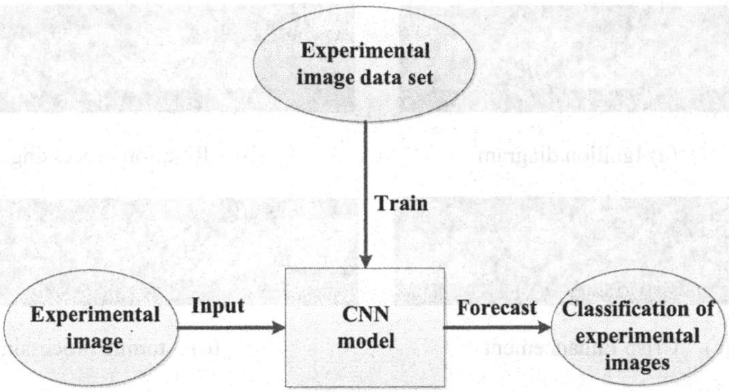


**Fig. 1** Block diagram of experimental picture classification and recognition based on convolutional neural network

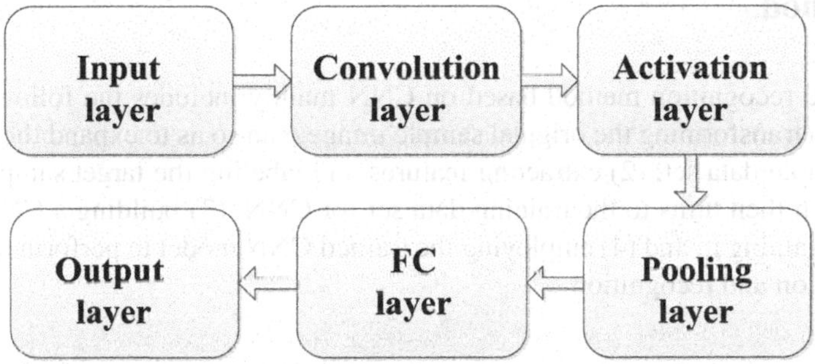


**Fig. 2** Convolutional neural network model structure

classifier based on CNN is essentially a mapping $c' : X_S \rightarrow Y_S, c'(x)$ is an estimate of the unknown real mapping $c(x)$. The sample format that are used to train the classifier is $(x_i^S, y_i^S)$. The purpose of CNN is to construct a function c′ that approximates c as closely as possible and then to predict the quality attribute information as accurately as possible.

Given the known image sample $x_i^S \in X_s$ and its corresponding classification tag $y_i^S \in Y$ as its inputs, the classifier performs a series of training and gains the ability to identify images. Then, the classifier can perform the classification and recognition processes for new experimental images. The basic ideas for the image recognition of engine ignition based on CNN is depicted in Fig. 1.

A convolutional neural network is a type of deep learning. It is a multi-layer sensor specially designed to identify two-dimensional shapes. As shown in Fig. 2, the initial layers are usually convolutional layer, activation layer and pooling layer. The last layers close to the output layer are fully connected layers. The purpose of training CNN is to learn network parameters such as convolution kernel parameters and inter-layer connection weights, while the prediction process mainly calculates the category label based on the input images and learned network parameters.

(a) Ignition diagram

(b) Rotation processing

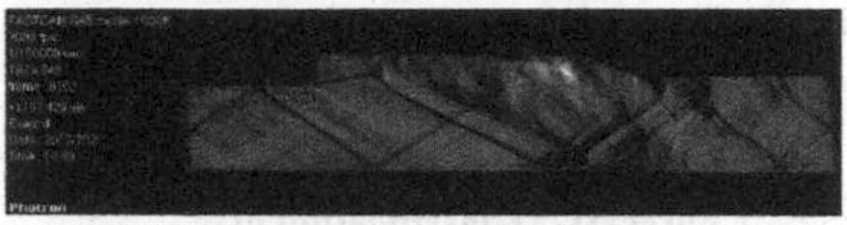

(c) Crisp Enhancement

(d) toning processing

**Fig. 3** Ignition image processing

# 2 Method

The image recognition method based on CNN mainly includes the following processes: (1) transforming the original sample image data so as to expand the capacity of the sample data set; (2) extracting features and labeling the target sample image data, which then turns to the training data set for CNN; (3) building a CNN model and then training it; and (4) employing the trained CNN model to perform the image classification and recognition.

## *2.1 Data Processing*

During the engine ignition experiment, a high-speed camera recorded 2161 images, including 132 images before ignition, one ignition image and 2028 images after ignition. The time interval between two adjacent images was 143 ms. The training of the CNN model requires a huge amount of images, Increasing the training set data volume will allow the CNN to achieve better training results. Compared with the manual data collection, many image transformation methods can efficiently enlarge the dataset volume, such as grayscale changes, horizontal flipping, vertical flipping, random grouping, color value jump, sharpening, translation, rotation and so on. The original images are processed with three techniques as shown in Fig. 3, namely rotation, sharpeness and color mixing.

## *2.2 Image Data Set Design*

Data sets are key to deep learning. It is the release of large public data sets that makes CNN develop rapidly. Therefore, the production of data sets is particularly

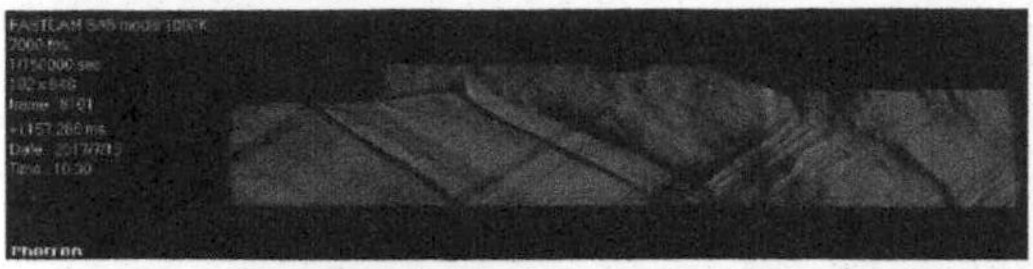

(a) Pre ignition picture

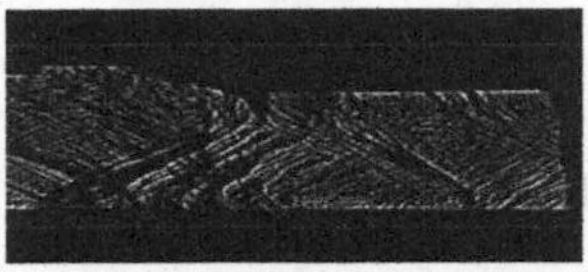

(b) Pre ignition image processing

(c) Ignition picture

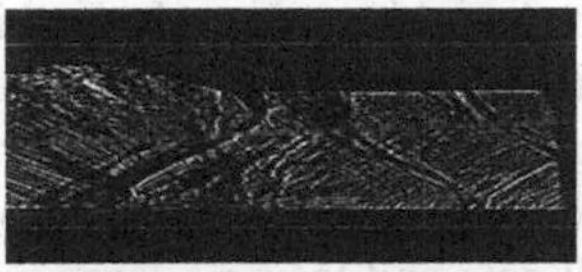

(d) Ignition picture processing

(e) Post ignition picture

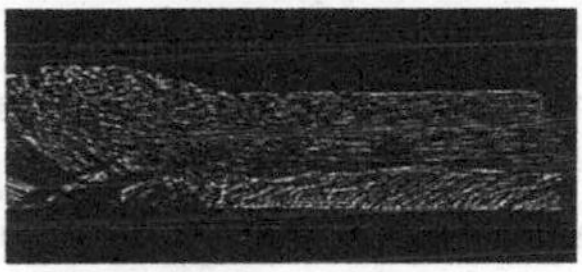

(f) Post ignition picture processing

**Fig. 4** Comparison of experimental pictures before and after processing

important for the training quality and efficiency of CNN [20–22]. For the engine ignition experiment, there three types of images, namely the before-ignition, after-ignition and ignition ones. Since the number of images relevant to each type is relatively small, it is necessary to expand the data sample size and perform data enhancement to meet the training requirements. It is possible to expand the sample capacity without destroying the quality of the experimental picture feature itself. For instance, the main feature information of the images are maintained and the other parts are eliminated so as to make the CNN easily identify and capture the key features from input images. It can be done as follows: for the images that have been processed with the techniques in Sect. 2.1, the ImageMagick image processing tool is utilized to reduce the redundant information and to extract boundary feature, which makes the images more streamlined and the flow field characteristics more obvious. Figure 4 compares the images before and after the picture processing. The simplification of experimental images can reduce the computational cost and improve the efficiency of CNN to extract features from the same type of images. Consequently, the training time for image recognition is reduced.

For the engine ignition images, the image data set after processing and feature extraction can be expressed as $D_S = \{(x_1^S, y_1^S), (x_2^S, y_2^S), \ldots, (x_{N_S}^S, y_{N_S}^S)\}$, where $N_S$ denotes the size of data set, $x_i^S \in X_S$ represents the eigenvector of the ith sample of the data set, and $Y_j^S \in Y_S$ corresponds to the category label.

## 2.3 *CNN Model Design*

### 2.3.1 Convolution Layer

Convolutional layer is the core part of CNN. Compared with the traditional methods, CNN takes advantages of the local connections and weight sharing to improve the network architecture and to achieve better performance. During the design of convolutional layers, the most important factor is the choice of filters, because they can directly affect the effect of feature extraction. For a smaller filter, it has the advantage of extracting more detailed information; however, it also has the risk of obtaining blank data and losing a large number of features. Besides, smaller filters can also result in the increment in computational cost, which in turn affects the efficiency of CNN. Based on practical needs and prior testing, the size of the filter is set to 5 * 5. After the introduction of the filter into CNN, the matrix is transferred between the nodes by the filter and is transformed into a unit node matrix for the next layer. Consequently, we can effectively reduce the parameters that need to be trained and suppress the overfitting problem.

In this research, CNN takes images as its input. In the convolutional layer, the characteristic images generated by the previous layer is convoluted with its convolution kernel. The relevant result is mapped by the activation function and becomes the characteristic image of the next layer. The process can be expressed by the formula (1).

$$G(i, j) = \sum_{m} \sum_{n} X(m, n) * W(i - m, j - n) \quad (1)$$

where $X$ represents the input, which is generally a two-dimensional image, $W$ represents a convolution kernel, $m$ and $n$ are the sizes of the convolution kernel, and the resulting feature map is $G(i, j)$.

### 2.3.2 Activation Layer

In our CNN model, the activation function of the activation layer is used to add nonlinear factors, since the linear model does not have enough ability to express the nonlinearity in the experimental images. The feature of the experimental image is preserved and mapped by the activation function, and the redundant information has been eliminated. We consider the following activation functions:

$$\text{sigmoid function: } f(x) = \frac{1}{1 + e^{-x}} \quad (2)$$

$$\text{tanh function: } f(x) = tanh(x) \quad (3)$$

$$\text{ReLU function: } f(x) = max(x, 0) \quad (4)$$

$$\text{softplus function: } f(x) = log(1 + exp(x)) \quad (5)$$

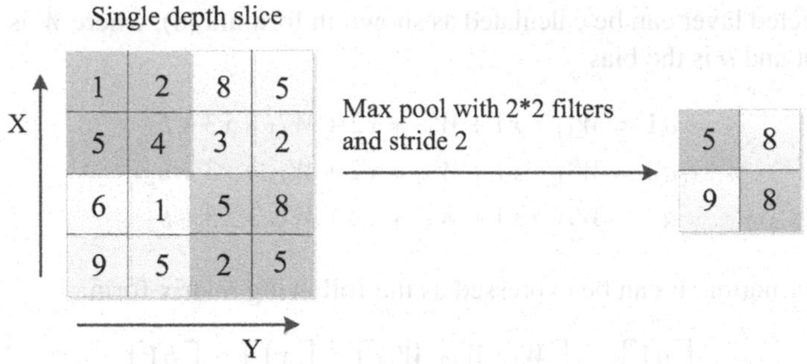


**Fig. 5** Max pooling schematic

The tanh function will be very effective when the eigen values are quite different from each other, and it can amplify the eigen value. The sigmoid function performs better when the feature difference is not large enough and requires more subtle classification. The ReLU function does not need normalized input to prevent saturation, and it can build sparse matrices to reduce the redundancy in the data. Most CNNs currently use the ReLU function as their activation function. Consequently, we also use the ReLU as the activation function.

#### 2.3.3 Pooling Layer

Usually, the output of convolutional layer has a huge volume. To address this issue, a pooling layer is often added between two convolution layers to effectively reduce the complexity. In our CNN model, a pooling layer is periodically inserted between two adjacent convolution layers. The pooling unit calculates the value of a local block in the characteristic images. The adjacent pooling unit reads the data row by row, which can decrease the dimension of the characteristic image, maintain the translation invariance of the data and reduce the number of parameters. We utilize the Max pooling scheme in our CNN model, which is shown as in Fig. 5.

#### 2.3.4 Fully Connected Layer

The fully connected layer works as a "classifier" in the design of CNN. The operations of the convolutional layer, pooling layer and activation layer map the original data into the hidden feature space and then save the mapping parameters. The full connectivity layer maps the learned "distributed feature representation" to the sample tag space.

Each node of the fully connected layer is connected to all the nodes in the previous layer, so as to integrate all the features extracted from the previous layer. Due to its intrinsic nature, the fully connected layer usually has the largest amount of parameters compared with other layers. In the forward calculation process, each output of the full connected layer can be calculated as shown in formula (6), where $W$ is a weight coefficient and $b$ is the bias.

$$\begin{aligned} a1 &= W_{11} * x1 + W_{12} * x2 + W_{13} * x3 + b_1 \\ a2 &= W_{21} * x1 + W_{22} * x2 + W_{23} * x3 + b_2 \\ a3 &= W_{31} * x1 + W_{32} * x2 + W_{33} * x3 + b_3 \end{aligned} \quad (6)$$

After deformation, it can be expressed as the following matrix form:

$$\begin{bmatrix} a1 \\ a2 \\ a3 \end{bmatrix} = \begin{bmatrix} W_{11} & W_{12} & W_{13} \\ W_{21} & W_{22} & W_{23} \\ W_{31} & W_{32} & W_{33} \end{bmatrix} * \begin{bmatrix} x1 \\ x2 \\ x3 \end{bmatrix} + \begin{bmatrix} b1 \\ b2 \\ b3 \end{bmatrix} \quad (7)$$

where $x1$, $x2$, and $x3$ are the inputs of the fully connected layer, and $a1$, $a2$, and $a3$ are the outputs. In our CNN model, the fully connected layer has 64 * 10 * 10 input nodes and 1024 output nodes; consequently, There are totally 6,553,600 weight parameters and 1024 bias parameters.

# 3 Experiments

The integrated development environment utilized is the open source deep learning framework Tensorflow, and the GPU is employed to accelerate the training process of the CNN model. The CNN training includes the following key steps: (1) Dividing the enlarged image data set into two parts, namely the training set and the test set. The former is utilized to train the CNN model. (2) Constructing a CNN model is and producing various CNNs based on the different combinations of training parameters. (3) Training the convolutional neural network model, loading model training features and weights in the fully connected layer, and finally using Softmax to obtain the classification recognition results.

## *3.1 Training Data Set*

The CNN presented in Sect. 3.2 is trained with the designed data set aforementioned. As listed in Table 1, the training set and the test set both contain three types of pictures. The training data set contains 63,816 samples, and the test set contains 20,832 samples. The images are all stored in the *png* format. The training set and the test set are independent from each other and they share none common images.

**Table 1** Experimental picture data set

| Data set | Pre ignition | Ignition | Post ignition |
|---|---|---|---|
| Training set | 14,766 | 24,876 | 24,174 |
| Test set | 4796 | 6042 | 9994 |

**Table 2** Convolutional neural network model training parameters

| Training parameters | Training settings |
|---|---|
| Activation function | tanh, ReLU, sigmoid |
| Pooling method | Max pooling, mean pooling |
| Loss function | Cross-entropy |
| Optimization function | Gradient descent, αk = 0.01 |
| Droupt | 0.5 |

## 3.2 Training CNN Model

We employ the customed image data set in Sect. 2.1 to train the CNN and the training parameters are listed in Table 2. The Activation function is *tanh*, *ReLU* and *sigmoid*; the Pooling method is *Max pooling*, *Mean pooling*; the Loss function is *cross-entropy*, Optimization function is *Gradient Descent* whose αk = 0.01.

After the parameters for each CNN layer are set, the CNN is then iteratively trained. In each iteration, a hundred training samples are loaded; the weight relevant to every layer is gradually revised, and consequently the accuracy of image classification is gradually improved.

## 3.3 Evaluating CNN Model

After the CNN model has been trained, the test set are employed to evaluate the accuracy of the CNN model in recognizing the input images. We take advantage of the *Test Accuracy* metric to perform the evaluation process, which is defined as:

$$TesAcc = \frac{TestImagesCurrently}{TestImages} \tag{8}$$

where *TestImagesCurrently* indicates the number of images that have been correctly tested, and *TestImages* denotes the total number images in the test data set. The greater the test accuracy value is, the better performance the CNN model can achieve in the image recognition.

**Table 3** Test accuracy of the CNN with different parameters

| CNN parameter | tanh | ReLU | Sigmoid |
|---|---|---|---|
| Max pooling | 94.47 ± 0.3 | 96.55 ± 0.5 | 92.37 ± 0.6 |
| Mean pooling | 91.21 ± 0.2 | 93.32 ± 0.8 | 86.68 ± 0.5 |

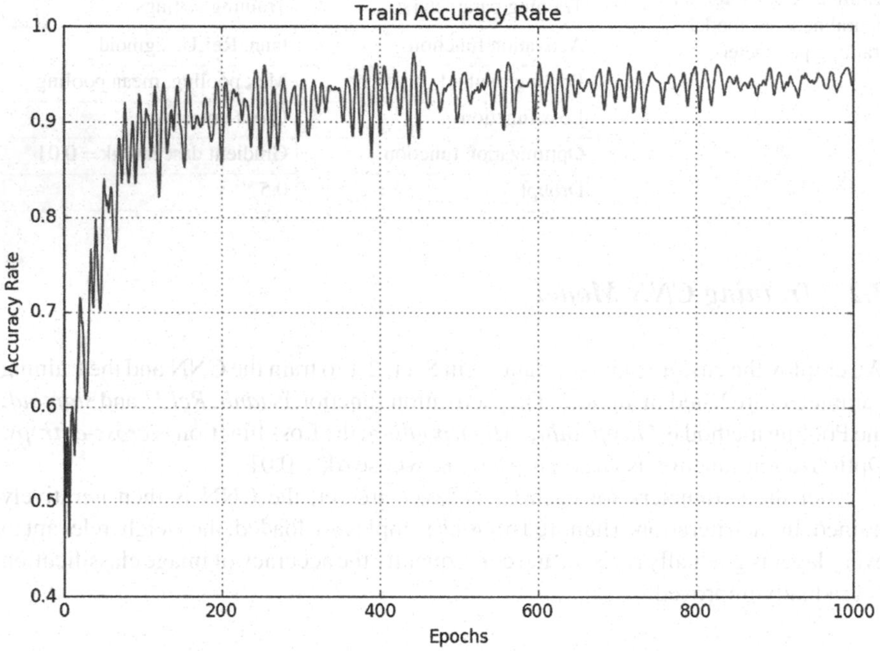


**Fig. 6** Convergence process for the recognition accuracy of the optimal CNN

## 4 Experimental Results and Analysis

As shown in Table 3, there are three activation functions and two pooling methods. Therefore, the CNN are trained with 6 different parameters and the relevant results are gathered after 1000 iterations for each parameter.

We obtain four observations from the experimental results. (1) Different CNN models perform variously from each other given the same engine ignition image set. (2) The parameter combination sigmoid and Mean pooling resulted in the lowest test accuracy. (3) The CNN model under the parameter combination tanh and Max pooling has a moderate performance. (4) The parameter combination ReLU and Max pooling indicates the best CNN model (ReLUMax), and the convergence process for the recognition accuracy is shown in Fig. 6.

In Fig. 6, *Epochs* represents the number of trainings, and *Accuracy Rate* represents the accuracy of image recognition. It can be seen that the optimal CNN model *ReLUMax* converges quickly during the training process. The test accuracy of

*ReLUMax* has reached to 90% with only 100 trainings. In the next 500 trainings, its performance fluctuated around 93% in a relatively small amplitude. Afterwards, the convergence tends to be stable. Finally, the test accuracy converged to about 96%. The image recognition models without the CNN are often prone to overfitting, which results in a local optimum and a final recognition accuracy of only about 90%.

Compared with the traditional image recognition method, the image recognition method based on the CNN can extract enough high-level features and experience in the training process. We believe that our method is superior to the traditional counterpart, which can avoids the overfitting problem in the training process and achieves a better performance in the image recognition.

## 5 Summary

In this paper, we have designed CNN model to perform the image recognition for engine ignition experiment. Based on the original images taken with the high-speed cameras, we expanded the sample data set by the image transformation technique and enhanced the features of the included images. We designed several CNN models that had different parameters to identify the engine ignition image. Extensive experiments were performed and the results showed that the proposed CNN model had a better performance in the classification problem than the traditional method. Concretely speaking, the CNN model had higher recognition rate, lower mean square error and faster convergence speed. It could simplify the tedious steps involved in the traditional way.

**Acknowledgements** This work was supported by the National Key Research and Development Plan of China under Grant No. 2017YFB0202101. We also express our gratitude to Lanying Ge (ROMTEC) for his technical supports.

## References

1. Z. Kaifei, Applications of artificial intelligence and its future prospects. J. Lvliang High. College **26**(4), 79–81 (2010)
2. R.K. Lindsay, E.A. Feigenbaum, B.G. Buchanan et al., Applications of artificial intelligence for chemical inference: the dendral project. Electr. Power Syst. Res. **41**(96), 117–131 (1980)
3. B. Fuqing, Image recognition and classification based on artificial neural network (Chengdu University of Technology, 2010)
4. W. Yongqing, Principle and method of artificial intelligence (Xi'an Jiao Tong University press, 1998)
5. Y. Wei, X. Wei, M. Lin et al., HCP: a flexible cnn framework for multi-label image classification. IEEE Trans. Pattern Anal. Mach. Intell. **38**(9), 1901–1907 (2016)
6. D. Silver, A. Huang, C.J. Maddison et al., Mastering the game of go with deep neural networks and tree search. Nature **529**(7587), 484 (2016)
7. D. Silver, J. Schrittwieser, K. Simonyan et al., Mastering the game of go without human knowledge. Nature **550**(7676), 354–359 (2017)

8. Hu Kui, Zhang Dongping, Yang Li, Multi-scale pedestrian detection based on convolutional neural networks. J. China Univ. Metrol. **4**, 472–477 (2017)
9. B. Chenjia, Research on autonomous driving method based on computer vision and deep learning (Harbin Institute of Technology, 2017)
10. S. Guibin, Research on road scene object detection based on depth convolutional neural network (Northern Polytechnical University, 2017)
11. S. Ai, L. Jia, C. Zhuang et al., A registration method for 3D point clouds with convolutional neural network (2017)
12. A. Krizhevsky, I. Sutskever, G.E. Hinton, ImageNet classification with deep convolutional neural networks, in *International Conference on Neural Information Processing Systems* (Curran Associates Inc. 2012), pp. 1097–1105
13. X. Ke, Study of convolutional neural network application on image recognition (Zhejiang University, 2012)
14. W. Zhen, G. Maoting, Design and implementation of image recognition algorithm based on convolution neural network, in *Modern Computer*, vol. 7 (Professional Edition, 2015), pp. 61–66
15. J. Fu, H. Zheng, T. Mei, Look closer to see better: recurrent attention convolutional neural network for fine-grained image recognition, in *IEEE Conference on Computer Vision and Pattern Recognition* (IEEE Computer Society, 2017), pp. 4476–4484
16. J. Bai, Z. Chen, B. Feng et al., Image character recognition using deep convolutional neural network learned from different languages, in *IEEE International Conference on Image Processing* (IEEE, 2015), pp. 2560–2564
17. G.E. Hinton, R.R. Salakhutdinov, Reducing the dimensionality of data with neural networks. Science **313**(5786), 504–507 (2006)
18. Y. Lecun, B. Boser, J.S. Denker et al., Backpropagation applied to handwritten zip code recognition. Neural Comput. **1**(4), 541–551 (2014)
19. Y. Lecun, L. Bottou, Y. Bengio et al., Gradient-based learning applied to document recognition. Proc. IEEE **86**(11), 2278–2324 (1998)
20. T.C. Wang, J.Y. Zhu, E. Hiroaki et al., A 4D light-field dataset and CNN architectures for material recognition, in *Computer Vision—ECCV 2016* (Springer International Publishing, 2016), pp. 121–138
21. Z. Xiaohua, S.S. Guang, C. Bo et al., CAS2PEAL: a large2scale Chinese face database and some primary evaluations. J Comput-Aid Design Comput. Graph. **17**(1), 9–17 (2005)
22. J. Guo, S. Gould, Deep CNN ensemble with data augmentation for object detection (Computer Science, 2015)

# Evolutionary Characteristics and the Adaptability Improvement of an Innovation Ecosystem Based on an Extension NK Model

**Nanping Feng, Furong Ruan and Fenfen Wei**

**Abstract** Based on the influence of members' roles and their relationship strengths on the adaptability of an innovation ecosystem, this paper extends the basic NK model by introducing weight parameter and relationship strength parameter. According to the extended model, three evolutionary characteristics are obtained. Firstly, for an innovation ecosystem, compared with the traditional view of "strengthening the weakness", "strengthening the strengths" will achieve a higher efficiency for its adaptability enhancement. Secondly, a suitable relationship strength but not a highest one will help to enhance its adaptability. Finally, a global vision will help to achieve the optimal evolution path. Moreover, on the basis of the evolutionary characteristics, we put forward three strategies to improve the system adaptability.

**Keywords** Innovation ecosystem · System evolution · System adaptability NK model

## 1 Introduction

Evolution is a fundamental question of an innovation ecosystem [1, 2], which can bring substantial enlightenment to academics, business managers and policy makers [3]. Many scholars have stressed the need to deepen the understanding of the evolution of innovation ecosystems [4, 5], and a number of related studies have emerged in recent years. For example, Chen and Xie studied the evolution path of China's photovoltaic industry ecosystems based on the Lotka-Volterra model [6]. Still constructed an evolution framework of an enterprise innovation ecosystem by using basic relationship and network-centric method [7]. Rabelo studied evolution stages of an innovation ecosystem by adapting the concept of enterprise architecture framework [8]. Sun constructed an evolution model of an innovation ecosystem based on the

N. Feng · F. Ruan (✉) · F. Wei
School of Management, Institute of Computer Network System,
Hefei University of Technology, Hefei 230009, China
e-mail: lotus087@163.com

 
Y. Jia et al. (eds.), *Proceedings of 2018 Chinese Intelligent Systems Conference*,
Lecture Notes in Electrical Engineering 528,
https://doi.org/10.1007/978-981-13-2288-4_36

multi-level perspectives framework (MLP) [9]. Yao and Zhou established a symbiotic evolution model of Mobile Internet Industrial Platform Innovation Ecosystem (MIPIE) based on the logistic model [10]. Although these studies have analyzed the evolution related issues of an innovation ecosystem from different perspectives, little attention has been paid to the optimization behavior of an innovation ecosystem during its evolution. Moreover, an innovation ecosystem is a typical complex adaptive system, and optimization is regarded as an important feature during its process of evolution.

NK model is an universal and applicable tool for the study of complex systems [11]. Compared with other analysis tools, NK model focuses on the optimization process and the evolution path of a system, and embodies the characteristics of mutual influence and mutually beneficial symbiosis of the members [12]. Based on the NK model, Luo [13] examined how the inter-firm network structure of an innovation ecosystem condition its system evolution. However, he did not consider the influence of different roles of members and their different relationship strengths on the evolution. In fact, the influences of those differences on the evolution of the entire system cannot be ignored [14, 15].

Consequently, in view of the above issues, this paper attempts to extend the basic NK model by introducing weight parameter and relationship strength parameter to consider the different roles of members and the different relationship strengths among members. Then, on the basis of this extended model, we analyze the optimization of evolution path for innovation ecosystems. Finally, some suggestions on the adaptability promotion of an innovation ecosystem are given.

## 2 The Adaptability and the Fitness Landscape of an Innovation Ecosystem

Adaptability refers to the coordination degree that the members of a natural ecosystem can reach during the interaction with the environment by proactively adjusting their own behaviors. In the process of pursuing the improvement of adaptability, the members could achieve the goals of self-survival and self-development, and thus promote the evolution of the natural ecosystem [16]. Similarly, as a collaborative mechanism [17], the innovation ability and the development of an innovation ecosystem are also promoted through its members' mutual interactions and complementary advantages. At the same time, the members also improve their own innovation abilities and achieve self-developments. Therefore, the adaptability of an innovation ecosystem refers to the abilities of innovation and development.

The fitness landscape, originally used to describe the evolution of biological organisms, is a rugged landscape of similar peaks shaped by the adaptability values of different individuals in space. High adaptability values form peaks while low adaptability values form valleys, as shown in Fig. 1. The evolution of biology is represented by the individual's wandering and climbing on the fitness landscape,

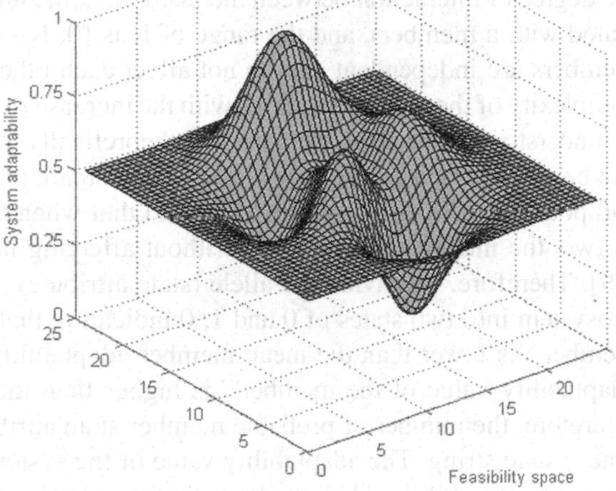


**Fig. 1** The concept map of fitness landscape

and finally reaching the optimal state being the highest peak in the landscape [18]. Similarly, the fitness landscape of an innovation ecosystem is a landscape formed by the search process of the global highest peak (i.e. the optimal state). Different members of an innovation ecosystem interact and consciously coordinate with each other in order to improve the adaptability of the whole system and form the peaks and valleys.

# 3 An Extended NK Model and Evolutionary Characteristics of an Innovation Ecosystem

## *3.1 The Basic NK Model*

The NK model is a method for studying the evolution of biological organism, which was originally proposed by Kauffman in Wright's adaptive landscape theory. N represents the total number of genes that an individual has, and K represents the relationship between genes which can be understood as the degree of interaction. The different state of each gene is the allele, and the fitness landscape of the system is the feasibility space formed by different combinations of states. Because of the complexity of the interaction between genes, it is difficult to quantify the degree of adaptability of genes in the system. Therefore, Kauffman defined that when a gene mutates itself or the one having a relationship with it mutates, a random number can be selected from the uniform distribution of (0, 1) as the gene's adaptability value [12].

The NK model can be applied to an innovation ecosystem, where N represents the total number of members, $i$ represents the i-th member ($i = 1, \ldots$ N), K indicates the degree of interaction between members (i.e., the number of other members associated with a member), and the range of K is [0, N − 1]. When K = 0, the system members are independent and do not affect each other. When K ∈ (0, N − 1], the complexity of the system increases with the increase of K. The allele of member $i$ can be understood as its attribute state, and theoretically, the value of this attribute state can be varied (such as two states of large and small, or three states of good, medium and poor). Kauffman's research has found that when the attribute state was defined as two, the model was simplified without affecting its general qualitative features [19]. Therefore, we divide the allele (state attribute) of member $i$ in an innovation ecosystem into two states of 0 and 1, 0 indicating that the adaptability value of the member $i$ is lower than the mean member adaptability, while 1 indicating that the adaptability value of the member $i$ is higher than the mean member adaptability. Therefore, the number of probable member state attributes is 2 N, represented by a binary state string. The adaptability value of the system is the mean value of all the member adaptabilities [12], as shown in formula (1):

$$F = \frac{1}{N}\sum_{i=1}^{N} f_i \tag{1}$$

In the above formula, adaptability $f_i$ indicates the contribution of member $i$ to the system's fitness landscape. $f_i$ is not only determined by i-th member itself, but also related to the other K members. The definition of adaptability embodies the interaction and co-evolution of the members in an innovation ecosystem, and confirms the applicability and rationality of the introduction of NK model to discuss its evolution.

The formula (1) implies two basic hypotheses. First, there is no difference in the importance of the members. Second, there is no difference in the relationship strength between the members. However, in reality, the contribution of different members to an innovation ecosystem is different. Non-core members generally engaged in the innovation activities around core members [20–23]. Thus core members always play more important roles than non-core members in the evolution [9, 14]. In addition, the relationship strengths between each two members are also different, influencing the innovation effect [24], the information exchange among members [25, 26], the system evolution, etc. [21]. Therefore, the difference in the relationship strength is also significant to the evolution [15]. Based on the above consideration, we introduce the weight parameter and the relationship strength parameter to extend the NK basic model.

### 3.2 *The Extended NK Model and the Evolutionary Characteristics of Innovation Ecosystems*

This paper divides the extension of the basic NK model into two steps. Firstly, we only introduce the weight parameter to analyze the effect of different member roles on the evolution of an innovation ecosystem. Secondly, the relationship strength parameter is further introduced, to analyze the influence of different relationship strength on the evolution.

(1) **The extended NK model with weight parameter introduced and the evolutionary characteristics of the "strengthening the strengths" effect**

Attribute importance is a kind of main basis for determining its weight [27–29]. The adaptability value can be understood as the contribution of an individual to the innovation ecosystem [30–32], reflecting the importance degree of the member to promote system evolution. Therefore, this paper chooses the original member adaptability value as the degree index, and based on this, the weight is introduced to recalculate the system adaptability.

The adaptability recalculation process is shown in Fig. 2. First, the state combinations of an innovation ecosystem is determined. These state combinations are the peaks with different heights in the fitness landscape, thus the feasibility space S can be obtained. According to Kauffman, if the state attribute of any member changes, a random value from the uniformly distributed interval (0, 1) can be selected as its adaptability value. We name this adaptability value as the original adaptability one. Then the original adaptability value is normalized and the weight parameter of the member is obtained. Finally, the adaptability $F$ of the member can be recalculated by considering the weight parameters and the original adaptability values, as shown in the formula (2):

$$F' = \sum_{i=1}^{N} \omega_i f_i \tag{2}$$

Although there are many members of an innovation ecosystem, most studies have integrated them into 3–4 categories. For example, an innovation ecosystem can be understood to consider three major groups naming research, development and application [33], or consider three types of members of core actors, supportive actors, and individual actors [14]; or consider four levels of core enterprises, technology R&D, product application and innovation platform [9].

Therefore, This paper assumes that the number of system members is four (N = 4), in order to enhance the understanding of the simulation process. In addition, considering the interactions among members [21–23], and the emphasis is not the influence of the relationship numbers between members on the system evolution,

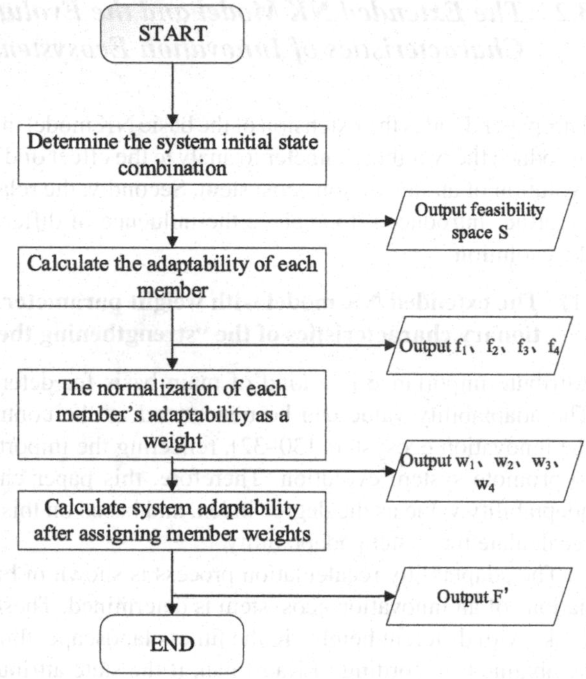


**Fig. 2** The calculation process of adaptability when considering weights

This paper thus assigns K the default N − 1, meaning K = 3. Each member has two attribution states of 0 and 1. Thus the possible state combinations of feasibility space (S) are 24. The set is represented as S = {0000, 0001, 0010, 0100, 1000, 0011, 0110, 1100, 0101, 1010, 1001, 1110, 1101, 1011, 0111, 1111}. Normalizing and giving weights according to Fig. 2. This paper calculates the system adaptability $F'$, as shown in Table 1. In order to compare the difference between the adaptability with weights and the original one. This paper lists both the adaptability value $F$ calculated according to the NK basic model and $F'$ calculated according to the extended model, as shown in Table 1.

Comparing $F$ and $F'$ in Table 1, two results can be found. (1) For all state combinations, $F < F'$, meaning that the system adaptabilities would improve when considering the weights. Moreover, the larger the adaptability gap among members, the more significant the improvement. (2) Core members (members who contribute more to the system adaptability) play key roles in improving the overall adaptabilities. The less other members contribute to the system, the more the core members contribute. As shown in Table 1, for the combination of 0011 and 0101, the system adaptability values significantly increased after considering weights. The weights of core members who contribute more to the system adaptability are larger, so the system's overall adaptability necessarily improves with the effect of larger weights and higher adaptability. The influence of the core members on the adaptability is more significant when the adaptability of the members varies greatly. The model with

**Table 1** The comparison of innovation ecosystem's adaptabilities based on two methods

| | $S$ | $f_1$ | $f_2$ | $f_3$ | $f_4$ | $F$ | $F'$ |
|---|---|---|---|---|---|---|---|
| 1 | 0000 | 0.51 | 0.57 | 0.63 | 0.32 | 0.51 | 0.53 |
| 2 | 0001 | 0.66 | 0.34 | 0.64 | 0.71 | 0.59 | 0.62 |
| 3 | 0010 | 0.11 | 0.96 | 0.85 | 0.72 | 0.66 | 0.83 |
| 4 | 0100 | 0.99 | 0.36 | 0.26 | 0.45 | 0.52 | 0.67 |
| 5 | 1000 | 0.35 | 0.22 | 0.89 | 0.65 | 0.53 | 0.66 |
| 6 | 0011 | 0.90 | 0.41 | 0.06 | 0.04 | 0.35 | 0.70 |
| 7 | 0110 | 0.44 | 0.70 | 0.12 | 0.57 | 0.46 | 0.56 |
| 8 | 1100 | 0.16 | 0.22 | 0.79 | 0.29 | 0.37 | 0.54 |
| 9 | 0101 | 0.47 | 0.05 | 0.07 | 0.68 | 0.32 | 0.54 |
| 10 | 1010 | 0.35 | 0.06 | 0.89 | 0.58 | 0.47 | 0.67 |
| 11 | 1001 | 0.93 | 0.74 | 0.86 | 0.67 | 0.80 | 0.81 |
| 12 | 1110 | 0.86 | 0.23 | 0.29 | 0.84 | 0.56 | 0.72 |
| 13 | 1101 | 0.23 | 0.20 | 0.31 | 0.57 | 0.33 | 0.39 |
| 14 | 1011 | 0.26 | 0.50 | 0.90 | 0.49 | 0.54 | 0.64 |
| 15 | 0111 | 0.45 | 0.54 | 0.06 | 0.55 | 0.40 | 0.50 |
| 16 | 1111 | 0.99 | 0.49 | 0.79 | 0.97 | 0.81 | 0.86 |

the weights embodies the importance of the core members for the fitness landscape formation, thus is more realistic than the basic model.

The above findings show that, in an innovation ecosystem, compared with other non-core members, the improvement of the core member's adaptability has a more significant effect on the improvement of the system adaptability. Therefore, different from the "strengthening the weakness" principle in "the cask theory", "strengthening the strengths" can achieve better results in an innovation ecosystem.

(2) **The extended NK model with weights and relationship strength introduced and the evolution characteristics of suitable relationship strength**

This paper refers to "contact frequency" produced by Gao [31] to measure relationship strength between two members of the system. $\theta_{ij}$ is used to represent relationship strength among $i$ and $j$ ($\theta_{ij} \in [0, 1]$). $\theta_i$ is total relationship strength of $i$ equaling to the mean of relationship strengths between $i$ and any other member. The formula is as below:

$$\theta_i = \frac{1}{N-1}\sum_{j \neq i} \theta_{ij} \tag{3}$$

Thus, the formula of system relationship strength is:

$$\theta = \frac{1}{N}\sum_{i=1}^{N} \theta_i \tag{4}$$

Existing research concludes that only in a suitable range can the relationship strength play a positive role to promote an innovation ecosystem's evolution. When the relationship strength is beyond the critical point, the system adaptability will decrease [31]. Thus the adaptability formula of member $i$ with the relationship strength $f_i^*$ is as the following:

$$f_i^* = \begin{cases} (1+\theta_i)f_i, 0 < \theta_i < \alpha \\ \left(1+\theta_i - \frac{3}{4}\theta_i^2\right)f_i, \alpha \le \theta_i < 1 \end{cases} \quad (5)$$

In the formula, α is the demarcation point.

Based on the aforementioned methods, in which the weights and relationship strength are introduced, the new formula of adaptability is as below:

$$F^* = \sum_{i=1}^{N} \omega_i' f_i^* \quad (6)$$

(*$\omega_i'$ and $\omega_i$ (in Formula 2) with different values but the same meaning).

We defined [0, 1/3), [1/3, 2/3) and [2/3, 1] as weak, strong and over strong contact separately [31]. Here we chose 0.2, 0.6 and 0.9 for θ to discuss. The curve below shows adaptability trends when considering both weights and relationship strength.

In Fig. 3, N = 4, K = 3. The three curve lines represent system adaptability changes for different value as well as considering weights. We can see that when the relationship strength increases to 0.9 from 0.6, the system adaptability decreases. The explanation is that, as members contact with each other at an appropriate frequency, their heterogeneity can help to innovate, thus improve the system adaptability, while too much contact (for example relationship strength over 0.9 in this paper) between members could increase homogeneity and thus lower innovative vitality.

The above findings show that the relationship strength among members of innovation ecosystems should remain within a certain range. A suitable relationship strength can promote the improvement of the innovation ecosystem, whereas an immoderate intensity can hinder its further development.

(3) **The fitness landscape based on the extended NK model and the evolutionary characteristics of inconsistent optimal points**

The Boolean hypercube is used to describe fitness landscape (as shown in Fig. 4) with both weights and the relationship strength (θ = 0.6) introduced.

In Fig. 4, the upper numbers in the boxes show different status of an innovation ecosystem, and the adaptability values in parenthesis represent peak or valley in landscape by higher or lower value respectively. The arrows stand for possible evolution directions under various combination states. "0" means a lower fitness value than the mean value for member $i$ and "1" higher than the mean value. The evolution requires the adaptability improvement of a certain member. For example, the adaptability of the fourth member should be improved when the system evolves from "0000" to "0001".

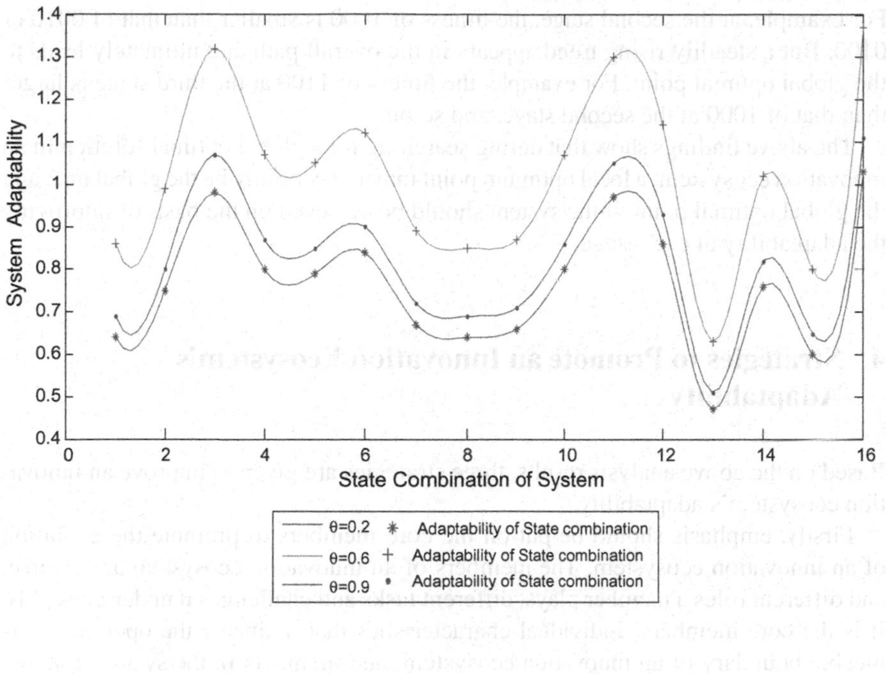


**Fig. 3** The adaptability curve when considering weights and relationship strength

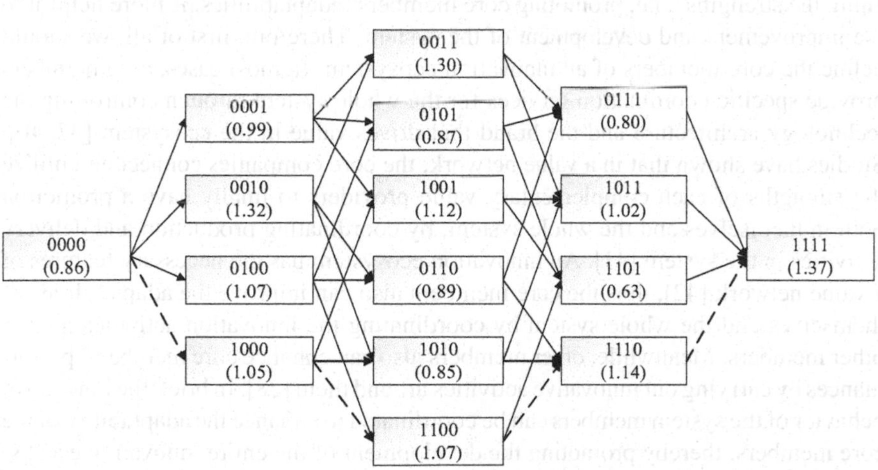


**Fig. 4** The fitness landscape of an innovation ecosystem considering both weights and relationship strength

According to Fig. 4, there is only one path leading to the global optimal point, which is the optimal evolution path (the path in dashed, 1000—1100—1110—1111). For the optimal path, the evolution state at every stage is not the local optimum one.

For example, at the second stage, the fitness of 1000 is smaller than that of 0010 or 0100. But a steadily rising trend appears in the overall path and ultimately leads to the global optimal point. For example, the fitness of 1100 at the third stage is larger than that of 1000 at the second stage, and so on.

The above findings show that during searching for a global optimal solution in an innovation ecosystem, a local optimum point may not certainly be the global one, and the global optimal point of the system should be achieved on the basis of improving the adaptability at each stage.

## 4 Strategies to Promote an Innovation Ecosystem's Adaptability

Based on the above analysis results, three strategies are given to improve an innovation ecosystem's adaptability.

Firstly, emphasis should be put on the core members to promote the evolution of an innovation ecosystem. The members of an innovation ecosystem are diverse, and different roles a member plays, different tasks and challenges it undertakes [34]. It is the core members' individual characteristics that delineate the open and permeable boundary of an innovation ecosystem, and members in the system conduct innovative activities within the boundary [35, 36]. The research shows that "strengthening the strengths", i.e. promoting core members' adaptabilities, is more helpful to the improvement and development of the system. Therefore, first of all, we should define the core members of an innovation ecosystem. In most cases, core members provide specific coordination services for the whole system through controlling the technology architecture and the brand that drives value in the ecosystem [37–40]. Studies have shown that in a value network, the core companies connect and utilize the strengths of each complementary value providers to finally have a promotion both in themselves and the whole system, by coordinating production and delivery activities in the system [41]. An innovation ecosystem has the necessary features of a value network [42], thus the core members also can improve the adaptabilities of themselves and the whole system by coordinating the innovation activities among other members. Meanwhile, other members also can enhance core members' performances by carrying out innovative activities around them [28]. In brief, the innovative behavior of the system members can be coordinated to enhance the adaptability of the core members, thereby promoting the development of the entire innovation ecosystems and ultimately benefiting all members of the system. This also embodies the symbiotic evolution of an innovation ecosystem.

Secondly, an appropriate relationship strength between members should be maintained in the process of symbiotic evolution. The study shows that the relationship strength between members is not the stronger the better, only in a suitable range can it maximally promote the whole system's follow-up evolution. Since the non-core members mostly carry out innovative activities around the core members [26, 27], the core members can control the relationship strength at a suitable degree through coordinating the innovation activities based on the platform [4]. Such a coordinating approach has a key role in the health and stability of an innovation ecosystem [43]. This activity is also one of the way to exert the core members' platform leadership [37]. At the same time, with a suitable relationship strength, the members can learn from each other, simultaneously maintaining their heterogeneity in the system.

Thirdly, the members should choose their own development paths under an overall vision. Each member of an innovation ecosystem ultimately bears the common fate and risk with the entire system [39]. From a perspective of value creation, the only source of competitive advantage is to conceive a whole value creation system and make it come into play [44]. The members will obtain more benefits from this system than working on its own. Therefore, the members of an innovation ecosystem generally are willing to coordinate their own behaviors initiatively in order to enhance the core members' adaptability [45]. The research shows that the local optimal point in an innovation ecosystem is not necessarily the global one. Therefore, when choosing their own development path, the members should not blindly pursue their own maximal adaptability, but take into account other members' development and seek the maximal adaptability of the whole system, because each participating member influences and shares the success or failure of the network [46].

## 5 Conclusion

Considering the important influence of different members and the relationship strength among members on the evolution of an innovation ecosystem, this paper extends the NK basic model and analyzes the evolutionary characteristics based on the extended model. According to the fundamental connotation of weight and the definition of adaptability, we measure members' weights with their original adaptabilities to analyze their contributions to system evolution. We construct relationship strength function based on contact frequency to analyze the influence of relationship strength between members on the system evolution.

The results indicate that, in an innovation ecosystem,: (1) "strengthening the strengths" will be more helpful to enhance the system adaptabilities, that is, compared to other non-core members, the further improvement of the core member's adaptability can promote the adaptability of the whole system better; (2) the relationship strength between the members is not the higher the better, that is, the relationship between the members is not necessarily as close as possible, and an appropriate intensity of relationship can promote the development of an innovation ecosystem while too close relationship will be detrimental to the further development of the system; (3) the search process for optimization is complicated and difficult, and without an overall vision, the system is liable to stop evolution at a local optimal position.

The research can be further expanded in the future. In this paper, we use the random value from 0 to 1 in the basic NK model to calculate the system adaptability value, and in the further study, more value sampling methods can be used. The influence critical point of the relationship strength is not determined in our study, and we can try to explore it through more comprehensive simulation in the future.

## References


1. A. Gawer, M.A. Cusumano, Industry platforms and ecosystem innovation. J. Prod. Innov. Manag. **31**(3), 417–433 (2014)
2. A. Ginsberg, M. Horwitch, S. Mahapatra et al., Ecosystem strategies for complex technological innovation: the case of smart grid development, in *Technology Management for Global Economic Growth* (IEEE, 2010), pp. 1–8
3. O. Dedehayir, S.J. Mäkinen, J.R. Ortt, Roles during innovation ecosystem genesis: a literature review. Technol. Forecast. Soc. Change (2016)
4. E. Autio, L.D.W. Thomas, Innovation ecosystems: implications for innovation management, in *The Oxford Handbook of Innovation Management* (2014), pp. 204–228
5. A. Gawer, Bridging differing perspectives on technological platforms: toward an integrative framework. Res. Policy **43**(7), 1239–1249 (2014)
6. Y. Chen, F.J. Xie, Bionics study on innovation ecosystems' evolution path of photovoltaic industry based on Lotka-Voterra model. Res. Develop. Manage. **24**(3), 74–84 (2012) (In Chinese)
7. K. Still, J. Huhtamäki, M.G. Russell et al., Insights for orchestrating innovation ecosystems: the case of EIT ICT Labs and data-driven network visualizations. Int. J. Technol. Manage. **66**(2/3), 243 (2014)
8. R.J. Rabelo, P.A. Bernus, Holistic model of building innovation ecosystems, in *Ifac Symposium on Information Control Problems in Manufacturing* (2015), pp. 2250–2257
9. B. Sun, X.F. Xu, H.T. Yao, Research on the evolution of innovation ecosystems based on the MLP framework. Stud. Sci. Sci. **34**(8), 1244–1254 (2016). (In Chinese)
10. Y. Yao, H. Zhou, The dynamic equilibrium and simulation of mobile internet platform innovation ecosystem: a symbiotic evolution model. Kybernetes **45**(9), 1406–1420 (2016)
11. D. Xu, Research on the innovation complexity of business model (Economic Management Press, 2005) (In Chinese)
12. S.A. Kauffman, The origins of order: self-organization and selection in evolution (New York & Oxford University Press, 1993)
13. J. Luo, Network structure and evolvability of innovation ecosystems. Acad. Manage. Ann. Meet. Proc. **2014**(1), 11518–11518 (2014)
14. K. Mohannak, J. Matthews, O. Dedehayir et al., The birth and development of innovation ecosystems: a literature review. Radiat. Meas. **44**(5), 571–575 (2009)
15. B.H. Hou, H. Liu, Analysis on NK model of enterprise network adaptability. China Industr. Econ. **4**, 94–104 (2009). (In Chinese)
16. J. Holland, Emergence: from chaos to order (Century Publishing Group, Shanghai Science and Technology Press, 2006) (In Chinese)
17. R. Adner, Match your innovation strategy to your innovation ecosystem. Harvard Bus. Rev. **84**(4), 98–107; 148 (2006)
18. S. Wright, The roles of mutation, inbreeding, crossbreeding and selection in evolution, in *Proceedings of the Sixth International Congress on Genetics* (1932), pp. 356–366
19. S.A. Kauffman, Adaptation on rugged fintess landscapes, in *Lectures in the Science of Complexity*, ed. by D. Stein (Addison-Wesley, Longman, 1989), pp. 527–618
20. E. Autio, L.D.W. Thomas, Innovation ecosystems: implications for innovation management (The Oxford Handbook of Innovation Management, 2014), pp. 204–228
21. E. Almirall, R. Casadesus-Masanell, Open versus Closed innovation: a model of discovery and divergence. Acad. Manag. Rev. **35**(1), 27–47 (2010)
22. M. Iansiti, R. Levien, Strategy as ecology. Harvard Bus. Rev. **82**(3), 68–78, 126 (2004)

23. J.F. Moore, The death of competition: leadership and strategy in the age of business ecosystems. Ecosystems (1996)
24. S.T. Pan, Y.L. Zheng, Network relationship strength and enterprise technology innovation performance—based on the empirical research of exploratory learning and exploitative learning. Stud. Sci. Sci. **29**(11), 1736–1743 (2011) (In Chinese)
25. M.S. Granovetter, The strength of weak ties (1973)
26. D. Krackhardt, The strength of strong ties: the importance of philos in organizations. Netw. Org. 216–239 (1992)
27. T.L. Saaty, What is the analytic hierarchy process? Mathematical models for decision support (Springer, New York, Inc. 1988), pp. 109–121
28. J. E. J. The Delphi Method: Techniques and Applications, ed. by H.A. Linstone, M. Turoff. J. Market. Res. **18**(3), 363–364 (1975)
29. Y.M. Wang, Using maximization of deviations for multiple index decisions and sorting. Syst. Eng. Electr. **7**, 24–26 (1988). (In Chinese)
30. G. Zhang, Q.S. Bo, The theory of modular organization structure based on NK model. Soft Sci. **23**(6), 24–27 (2009). (In Chinese)
31. C.Y. Gao, X.Y. He, Research on the Improvement of Knowledge Innovation in HTVIC based on NK model. Stud. Sci. Sci. **32**(11) (2014) (In Chinese)
32. J. Ma, H. Zhang, Y.M. Xi, Simulation Research on Maneger's Recognition based on NK model[J]. Operations Research and Management Science **17**(1), 000137–000143 (2008). (In Chinese)
33. Judy estrin. American innovation is in decline? [M]. China Machine Press,2010
34. R. Adner, R. Kapoor, Value Creation in Innovation Ecosystems: How the Structure of Technological Interdependence Affects Firm Performance in New Technology Generations[J]. Strateg. Manag. J. **31**(3), 306–333 (2010)
35. M. Iansiti, R. Levien, Strategy as ecology [J]. Harvard Bus. Rev. **82**(3), 68 (2004)
36. R. Gulati, P. Puranam, M. Tushman, Meta-organization design: rethinking design in interorganizational and community contexts. Strateg. Manag. J. **33**(6), 571–586 (2012)
37. M.A. Cusumano, A. Gawer, The elements of platform leadership. IEEE Eng. Manage. Rev. **31**(1), 8–8 (2003)
38. M. Iansiti, R. Levien, The keystone advantage: what the new dynamics of business ecosystems mean for strategy, innovation, and sustainability. Future Surv. **20**(2), 88–90 (2004)
39. Y.R. Li, The technological roadmap of Cisco's business ecosystem. Technovation **29**(5), 379–386 (2009)
40. L. Pierce, Big losses in ecosystem niches: how core firm decisions drive complementary product shakeouts. Strateg. Manag. J. **30**(3), 323–347 (2009)
41. M. Weiner, N. Nohria, A. Hickeman et al., Value networks—the future of the U.S. electric utility industry. Sloan Manage. Rev. **38**(4), 21–34 (1997)
42. J.F. Moore, Predators and prey: a new ecology of competition. Harvard Bus. Rev. **71**(3), 75 (1993)
43. D.S. Evans, A. Hagiu, R. Schmalensee, Invisible engines: how software platforms drive innovation and transform industries. Mit Press Books **1**(4), 410–411 (2007)
44. R. Normann, R. Ramírez, J. Champy et al., From value chain to value constellation (1996)
45. K. Möller, S. Svahn, Role of knowledge in value creation in business nets. J. Manage. Stud. **43**(5), 985–1007 (2010)
46. M. Pagani, C.H. Fine, Value network dynamics in 3G-4G wireless communications: a systems thinking approach to strategic value assessment. J. Bus. Res. **61**(11), 1102–1112 (2008)

# Adaptive Compensation for MIMO Nonlinear Systems Against Actuator Failures

**Hao Ren, Gang Zhu and Yan Lin**

**Abstract** In this note, an adaptive backstepping scheme is proposed for a class of multi-input multi-output (MIMO) nonlinear minimum phase systems aiming at accommodating three types of common actuator failures: stuck failure, bias failure and loss of effectiveness failure. These failures are uncertain in failure type, value and time. The scheme ensures all signals boundedness and output asymptotic tracking. An aircraft application illustrates the effectiveness of the given scheme.

**Keywords** Adaptive backstepping · Actuator failure · MIMO nonlinear systems Output tracking

## 1 Introduction

In modern control systems, there are various failures in sensors, actuators and the system components. These failures will have serious impacts on the system performance inevitably, even leading to instability of the system. To improve the safety and reliability of the system, many prominent fault-tolerant methods and techniques have been developed by researchers over the past decades, such as multiple-model designs [1, 2], fault diagnosis and fault tolerant control designs [3, 4], direct adaptive compensation designs [5–10]. Direct adaptive compensation methods can accommodate the expected failures automatically without any knowledge of the failure, so that they can handle large uncertainties both in controlled plant and failure with the

H. Ren · Y. Lin (✉)
School of Automation, Beihang University, Beijing 100191, China
e-mail: linyan@buaa.edu.cn

H. Ren
e-mail: 1075109703@qq.com

G. Zhu
Institute of Electronic Engineering, China Academy of Engineering Physics, Mianyang 621900, Sichuan, China
e-mail: 690163352@qq.com



Y. Jia et al. (eds.), *Proceedings of 2018 Chinese Intelligent Systems Conference*, Lecture Notes in Electrical Engineering 528,
https://doi.org/10.1007/978-981-13-2288-4_37

simpler controller structure, while it is difficult to give the strict proof due to various uncertainties. Tang et al. [5] gives the adaptive compensation scheme for MISO nonlinear systems considering one type of actuator failure. Based on [5], Wang and Wen [7] presents the adaptive scheme for MISO nonlinear systems to deal with two types of actuator failures. Tang et al. [6] extends the method to MIMO nonlinear systems considering actuator stuck failure. Ouyang and Lin [9, 10] proposed a switching scheme, some healthy actuators serve as backups, the designed monitoring functions monitor the behavior of the system in real time so that the actuator once fails, the switching to the healthy actuator will take place. In practical engineering, there are three types of common failures on actuators which may occur, including stuck failure, bias failure and loss of effectiveness failure. In this paper, our goal is to develop a direct adaptive compensation method for a class of MIMO nonlinear minimum phase systems considering three types of actuator failures simultaneously, which is significantly important.

## 2 Problem Formulation

In this paper, a class of MIMO nonlinear minimum phase systems is taken into consideration:

$$\begin{aligned} \dot{x} &= f_0(x) + \sum_{j=1}^{l} \theta_j f_j(x) + \sum_{k=1}^{m} g_k(x) u_k, \\ y &= h(x), \end{aligned} \tag{1}$$

where $x \in R^n$ is the state vector of the plant, $y \in R^q$ is the output vector of the plant, and $u_k \in R$, $k = 1, 2, \ldots, m$ is the output of the $k$th actuator. $f_0(x) \in R^n$, $f_j(x) \in R^n$, $j = 1, 2, \ldots, l$, $g_k(x) \in R^n$, $k = 1, \ldots, m$. $\theta_j$, $j = 1, 2, \ldots, l$, are parameters which are unknown.

During system operation, three important types of failures on actuators may occur:

- stuck failure

$$u_k(t) = \overline{u}_k \tag{2}$$

- loss of effectiveness failure

$$u_k(t) = \lambda_k v_k(t) \tag{3}$$

- bias failure

$$u_k(t) = v_k(t) + a_k \tag{4}$$

where failure type, failure actuator $k$, failure value $\overline{u}_k$, $\lambda_k \in [0,1)$, $a_k$ and failure time are all unknown. With expressions (2)–(4), the control inputs $u = [u_1 \ldots u_k \ldots u_m]^{\mathrm{T}}$ are rewritten as:

$$u(t) = \lambda v(t) + \sigma_1(\overline{u} - \lambda v(t)) + \sigma_2 a \tag{5}$$

where $v = [v_1, \ldots, v_m]^{\mathrm{T}} \in R^m$ are the desired inputs, which will be designed from the feedback.

where

$$\overline{u} = [\overline{u}_1, \ldots, \overline{u}_m]^T \tag{6}$$

$$a = [a_1, \ldots, a_m]^T \tag{7}$$

$$\sigma_1 = diag\{\sigma_{11}, \sigma_{12}, \ldots, \sigma_{1m}\} \tag{8}$$

$$\sigma_{1k} = \begin{cases} 1 \text{ if the kth actuator undergoes stuck failure} \\ 0 \quad \text{otherwise} \end{cases} \tag{9}$$

$$\rho = diag\{\rho_1, \rho_2, \ldots, \rho_m\} \tag{10}$$

$$\rho_k = \begin{cases} \rho_k \text{ if the kth actuator undergoes loss of effectiveness failure} \\ 0 \quad \text{otherwise} \end{cases} \tag{11}$$

$$\sigma_2 = diag\{\sigma_{21}, \sigma_{22}, \ldots, \sigma_{2m}\} \tag{12}$$

$$\sigma_{2k} = \begin{cases} 1 \text{ if the kth actuator undergoes bias failure} \\ 0 \quad \text{otherwise} \end{cases} \tag{13}$$

Considering the expression (5), we rewrite the plant (1) as the following form:

$$\begin{aligned} \dot{x} &= f_0(x) + F(x)\theta + g(x)\sigma_1\overline{u} + g(x)\sigma_2 a + g(x)\lambda(I - \sigma_1)v \\ y &= h(x) \end{aligned} \tag{14}$$

where $F(x) = [f_1, \ldots, f_j, \ldots, f_l] \in R^{n \times l}$, $\theta = [\theta_1, \ldots, \theta_j, \ldots, \theta_l]^{\mathrm{T}}$, $g(x) = [g_1, \ldots, g_k, \ldots, g_m] \in R^{n \times m}$.

The control objective is that the designed controller can accommodate three kinds of possible actuator failures (2)–(4) so that the all signals are not only bounded and the outputs $y(t) = [y_1(t), y_2(t), \ldots, y_q(t)]^{\mathrm{T}}$ track the reference signals $y_r(t) = [y_{r1}(t), y_{r2}(t), \ldots, y_{rq}(t)]^{\mathrm{T}}$ asymptotically. To guarantee $q$ outputs tracking of the $q$ arbitrary reference signals, it is necessary to have at least $q$ actuators in operation (not stuck), which can be understood intuitively, for example, to achieve the tracking of two outputs, one control input is not enough obviously. Hence there are at most $m$-$q$ stuck type of actuators all the time, while all actuators are permitted to undergo bias failure or loss of effectiveness failure simultaneously.

## 3 Direct Adaptive Compensation Design

To achieve $q$ outputs tracking of $q$ reference signals, each output must be directly connected to some inputs, thus the control characteristic indices for system (1) $\{\rho_1, \ldots \rho_q\}$ are assumed to exist. Besides, to handle the redundancy of actuators, we group the $m$ actuators into $q$ classes with the proportional actuation scheme in the same class, the grouping results are shown as

$$\Omega : \{\{C_{1,1}, C_{1,2}, \ldots, C_{1,c1}\}, \ldots, \{C_{i,1}, C_{i,2}, \ldots, C_{i,ci}\}, \ldots, \{C_{q,1}, C_{q,2}, \ldots, C_{q,cq}\}\}$$

and $v_i = b_i w_j, I = C_{i,1}, C_{i,2}, \ldots, C_{i,ci}\, j = 1, 2, \ldots, q$, $w_j$ is the basic control law in the $j$th class. From the above expression, we can obviously see that $Ci$ actuators are grouped into one group and their actuation are proportional to each other. Then the desired control input vector $v = [v_1, v_2, \ldots, v_m]^{\mathrm{T}}$ can be expressed as

$$v = Bw \tag{15}$$

where $B = \begin{bmatrix} b_{11} & \ldots & b_{1q} \\ \vdots & \ddots & \vdots \\ b_{m1} & \cdots & b_{mq} \end{bmatrix}$, $\mathrm{b_{ij}} = \begin{cases} \mathrm{b_j}, & \text{if } \mathrm{v_i} \text{ belongs to jth gruop}, \\ 0, & \text{otherwise.} \end{cases}$

$$w = \left[w_1, w_2, \ldots w_q\right]^T.$$

To use the backstepping technique, the following assumption is indispensable.

**Assumption 1** The functions $f_k(x)$ satisfy the following triangular condition:

$$d(L_{f_k} L_{f_0}^{r_i - 1} h_i) \in span\{d(h_1), \ldots, d(L_{f_0}^{\rho_1 - \rho_i + r_i - 1} h_1), d(h_2), \ldots, d(L_{f_0}^{\rho_2 - \rho_i + r_i - 1} h_2), \ldots$$
$$, \ldots d(L_{f_0}^{\rho_j - \rho_i + r_i - 1} h_j), \ldots, d(h_q), \ldots, d(L_{f_0}^{\rho_q - \rho_i + r_i - 1} h_q)$$

for $I = 1, 2, \ldots, q$, $r_i = 1, 2, \ldots, \rho_i$, $k = 1, \ldots, l$.

Then system (14) can be transformed into the following triangular form via the diffeomorphism $[\xi\ \eta]^{\mathrm{T}} = T(x) = [T_c(x)\ T_z(x)]^{\mathrm{T}}$

$$\begin{aligned} \dot{\xi}_{i,1} &= \xi_{i,2} + \varphi_{i,1}^T \theta \\ \dot{\xi}_{i,2} &= \xi_{i,3} + \varphi_{i,2}^T \theta \\ &\vdots \\ \dot{\xi}_{i,r_i} &= \xi_{i,r_{i+1}} + \varphi_{i,r_i}^T \theta \\ \dot{\xi}_{i,\rho_i} &= \phi_i + \varphi_{i,\rho_i}(\xi, \eta)\theta + \beta_i \sigma_1 \overline{u} + \beta_i \sigma_2 a + \beta_i (I - \sigma_1) \lambda B w \\ \dot{\eta} &= \psi(\xi, \eta) + \phi(\xi, \eta)\theta + \Delta_{\sigma_1}(\xi, \eta)\overline{u} + \Delta_{\sigma_2}(\xi, \eta) a \end{aligned} \tag{16}$$

for $i=1, 2, \ldots, q$, $r_i = 1, 2, \ldots, \rho_i$, where $\xi = [\xi_{11}, \xi_{12}, \ldots, \xi_{1\rho 1}, \ldots, \xi_{2\rho 2}, \ldots, \xi_{q\rho q}]^T \in R^{\rho 1+\rho 2+\cdots+\rho q}$, $\eta \in R^{n-(\rho 1+\rho 2+\cdots+\rho q)}$, $T_c(x) = [h_1, L_{f0}h_1, \ldots, L_{f0}^{\rho_1-1}h_1, h_2, \ldots, L_{f0}^{\rho_2-1}h_2 L_{f0}^{\rho_q-1}h_q]$, $\phi_i = L_{f0}^{\rho_i}h_i$, $\beta_i = [L_{g1}L_{f0}^{\rho_i-1}h_i(x), \ldots, L_{gm}L_{f0}^{\rho_i-1}h_i(x)]$, $\dot{\eta} = \psi(\xi, \eta) + \phi(\xi, \eta)\theta + \Delta_{\sigma_1}(\xi, \eta)\overline{u} + \Delta_{\sigma_2}(\xi, \eta)a$ is the zero dynamics of the system. The backstepping procedure is shown below.

The error variables are:

$$z_{i,1} = y_i - y_{ri} \tag{17}$$

$$z_{i,r_i} = \xi_{i,r_i} - \alpha_{i,r_i-1} \tag{18}$$

for $i=1, 2, \ldots, q$ and $r_i = 2, \ldots, \rho_i$.

The stabilizing functions $\alpha_{i,ri}$ are:

$$\begin{aligned}\alpha_{i,r_i} &= -z_{i,r_i-1} - c_{i,r_i}z_{i,r_i} - \varphi_{i,r_i}^T\hat{\theta} + \sum_{j=1}^{q}\sum_{k=1}^{\rho_j-\rho_i+r_i-1}\frac{\partial\alpha_{i,r_i-1}}{\partial\xi_{j,k}}(\xi_{j,k+1} + \varphi_{j,k}^T\hat{\theta})\\ &+ \frac{\partial\alpha_{i,r_i-1}}{\partial\hat{\theta}}\tau_{i,r_i-1} + \Big(\sum_{j=1}^{q}\sum_{k=1}^{\rho_j-\rho_i+r_i-1} z_{j,k+1}\frac{\partial\alpha_{j,k}}{\partial\hat{\theta}}\Big)\Gamma_\theta\\ &\times \Big(\varphi_{i,r_i} - \sum_{j=1}^{q}\sum_{k=1}^{\rho_j-\rho_i+r_i-1}\frac{\partial\alpha_{i,r_i-1}}{\partial\xi_{j,k}}\varphi_{j,k}\Big) + \sum_{k=1}^{r_i}\frac{\partial\alpha_{i,r_i-1}}{\partial y_{ri}^{(k-1)}}y_{ri}^{(k)}\end{aligned} \tag{19}$$

for $i=1, 2, \ldots, q$ and $r_i = 1, 2, \ldots, \rho_i$.

The tuning functions:

$$\tau_{i,r_i} = \sum_{j=1}^{q}\tau_{j,\rho_j-\rho_i+r_i-1} + z_{i,r_i}\Gamma_\theta \times \Big(\varphi_{i,r_i} - \sum_{j=1}^{q}\sum_{k=2}^{\rho_j-\rho_i+r_i-1}\frac{\partial\alpha_{i,r_i-1}}{\partial\xi_{j,k}}\varphi_{j,k}\Big) \tag{20}$$

for $i=1, 2, \ldots, q$ and $r_i = 1, 2, \ldots, \rho_i$.

Then we make $K_1 = (I - \sigma_1)\lambda$, $K_2 = \sigma_1\overline{u} + \sigma_2 a$, and $M = \begin{bmatrix}\beta_1(I-\sigma_1)\lambda B\\ \beta_2(I-\sigma_1)\lambda B\\ \vdots\\ \beta_q(I-\sigma_1)\lambda B\end{bmatrix}$.

To ensure the existence of the control law of the system, the following assumption is given:

**Assumption 2** The actuation matrix $M$ is nonsingular.

If all knowledge about actuator failure is known, the control law can be designed as:

$$w = \begin{bmatrix} w_1 \\ w_2 \\ \vdots \\ w_q \end{bmatrix} = \begin{bmatrix} \beta_1 K_1 B \\ \beta_2 K_1 B \\ \vdots \\ \beta_q K_1 B \end{bmatrix}^{-1} \begin{bmatrix} \alpha_{1,\rho_1} - \phi_1 - \beta_1 K_2 \\ \alpha_{2,\rho_2} - \phi_2 - \beta_2 K_2 \\ \vdots \\ \alpha_{q,\rho_q} - \phi_q - \beta_q K_2 \end{bmatrix}, \quad (21)$$

However, we cannot obtain any knowledge about actuator failure in the process of system operation, thus the adaptive control law is given as:

$$w = \begin{bmatrix} \beta_1 \widehat{K}_1 B \\ \beta_2 \widehat{K}_1 B \\ \vdots \\ \beta_q \widehat{K}_1 B \end{bmatrix}^{-1} \begin{bmatrix} \alpha_{1,\rho_1} - \phi_1 - \beta_1 \widehat{K}_2 \\ \alpha_{2,\rho_2} - \phi_2 - \beta_2 \widehat{K}_2 \\ \vdots \\ \alpha_{q,\rho_q} - \phi_q - \beta_q \widehat{K}_2 \end{bmatrix}, \quad (22)$$

with updating law

$$\dot{\widehat{K}}_1 = \Gamma_1 B w \sum_{i=1}^{q} z_{i,\rho_i} \beta_i, \quad (23)$$

$$\dot{\widehat{K}}_2 = \Gamma_2 (\sum_{i=1}^{q} z_{i,\rho_i} \beta_i)^T, \quad (24)$$

$$\dot{\widehat{\theta}} = \sum_{i=1}^{q} \tau_{i,\rho_i}, \quad (25)$$

where $\widehat{K}_1$ and $\widehat{K}_2$ are the estimates of the $K_1$ and $K_2$, respectively.

## 3.1 Stability Analysis

Consider the Lyapunov function candidate during one time interval $[t_{i-1}, t_i]$:

$$V(t) = \frac{1}{2} z^T z + \frac{1}{2} \widetilde{\theta}^T \Gamma_\theta^{-1} \widetilde{\theta} + \frac{1}{2} \widetilde{K}_1^T \Gamma_1^{-1} \widetilde{K}_1 + \frac{1}{2} \widetilde{K}_2^T \Gamma_2^{-1} \widetilde{K}_2 \quad (26)$$

where $z = [z_{11}, \ldots, z_{\rho 1}, \ldots, z_{2\rho 2}, \ldots, z_{q\rho q}]^{\mathrm{T}}, \widetilde{\theta} = \widehat{\theta} - \theta, \widetilde{K}_1 = \widehat{K}_1 - K_1, \widetilde{K}_2 = \widehat{K}_2 - K_2$. By computing the derivative of $V$, we get $\dot{V}(t) = -\sum_{i=1}^{q} \sum_{j=1}^{\rho_i} z_{ij}^2 \le 0$ we see $V$ decreases with time in each interval during which the knowledge about failure is unchanged. When new failure occurs, there will be finite jumping value in V because of the finite change in $\widetilde{K}_1$ and $\widetilde{K}_2$. Thus $V$ is always bounded only if the initial value is bounded, so $z$, $\widetilde{K}_1$, $\widetilde{K}_2$, $\widetilde{\theta}$, $\xi$ are bounded and $\eta$ is also bounded with respect to $\xi$, $\overline{u}$, $a$ as the input because of the ISS of the zero dynamics. Hence all signals are bounded and the closed-loop is stable. Considering $\dot{z} \in L^{\infty}$ and $z \in L^2$, according to barbalat lemma, we can get the conclusion that $\lim_{t\to\infty} z(t) = 0$ and so $\lim_{t\to\infty}(y_i(t) - y_{ri}(t)) = 0$.

## 4 Simulation Results

In this section, a classical aircraft nonlinear model is used for our simulation study:

$$\dot{V} = \frac{F_x \cos(\alpha) + F_z \sin(\alpha)}{m},$$
$$\dot{\alpha} = q + \frac{-F_x \sin(\alpha) + F_z \cos(\alpha)}{mV},$$
$$\dot{\theta} = q, \dot{q} = \frac{M}{I_y}, \tag{27}$$

where $V, \alpha, m, \theta, q, I_y$ are velocity, attack angle, the mass, the pitch angle, the pitch rate, the moment of inertia, respectively.

$$F_x = \overline{q}S(C_{x1}\alpha + C_{x2}\alpha^2 + C_{x3} + C_{x4}(d_1\delta_{e1} + d_2\delta_{e2})) + T_1 \cos\gamma_1 + T_2 \cos\gamma_2 - mg\ \sin(\theta)$$
$$F_z = \overline{q}S(C_{z1}\alpha + C_{z2}\alpha^2 + C_{z3} + C_{z4}(d_1\delta_{e1} + d_2\delta_{e2}) + C_{z5}q_1) + T_1 \sin\gamma_1 + T_2 \sin\gamma_2 + mg\ \sin(\theta)$$
$$M = \overline{q}cS(C_{m1}\alpha + C_{m2}\alpha^2 + C_{m3} + C_{m4}(d_1\delta_{e1} + d_2\delta_{e2}) + C_{m5}q) \tag{28}$$

where $\overline{q} = \frac{1}{2}\rho V^2$ is the dynamic pressure, $S$ is the wing area, $\rho$ is the air density, $c$ is the mean chord, $T_1$ and $T_2$ are the thrusts, and $\delta_{e1}, \delta_{e2}$ are the output angles of two elevators. We choose $x=[x_1, x_2, x_3, x_4]^T=[V, \alpha, \theta, q]^T$ as the state variables, $u=[u_1, u_2, u_3, u_4]^T=[\delta_{e1}, \delta_{e2}, T_1, T_2]^T$ as the control inputs, and $y=[x_3, x_1]^T$ as the outputs. The signals to be tracked are the outputs of the systems 1/(s$^2$ +5s+6) and 1/(s+1), with 0.01sin(0.05t) and 60+0.01sin(0.05t) as the inputs, respectively. The uncertain parameters are obtained as: $m$=4600 kg, $I_y$=31,027 kg m$^2$, $S$=39.02 m$^2$, $c$=1.98 m, $\rho$=0.7377 kg/m$^3$, $C_{x1}=0.39$, $C_{x2}=2.9099$, $C_{x3}=-0.0758$, $C_{x4}=0.0961$, $d_1=0.6$, $d_2=0.4$, $C_{z1}=-7.0186$, $C_{z2}=4.1109$, $C_{z3}=-0.3112$, $C_{z4}=-0.2340$, $C_{z5}=-0.1023$, $C_{m1}=-0.8789$, $C_{m2}=-3.8520$, $C_{m3}=-0.0108$, $C_{m4}=-1.8987$, $C_{m5}=-0.6266$, $\gamma_1=\arctan(53/1216)$, $\gamma_2=\arctan(2/45)$. All state variables and four control inputs are listed in Figs. 1 and 2, as well as two output signals, two reference signals and two tracking errors. From the figures, we can see all signals are bounded and two tracking errors both converge to the neighbor of zero, so asymptotic tracking are guaranteed although there are some failures on elevators in the process of the operation. In the simulation study, the failure mode is that actuator 1 is stuck at $u_1=0.04$ rad at $t$=50 s, then actuator 2 suffers from bias of 1.2 at $t$=100 s, then actuator 3 loses 40% of effectiveness at $t$=300 s, and actuator 4 is stuck at $u_4=800$ N at $t$=500 s finally.

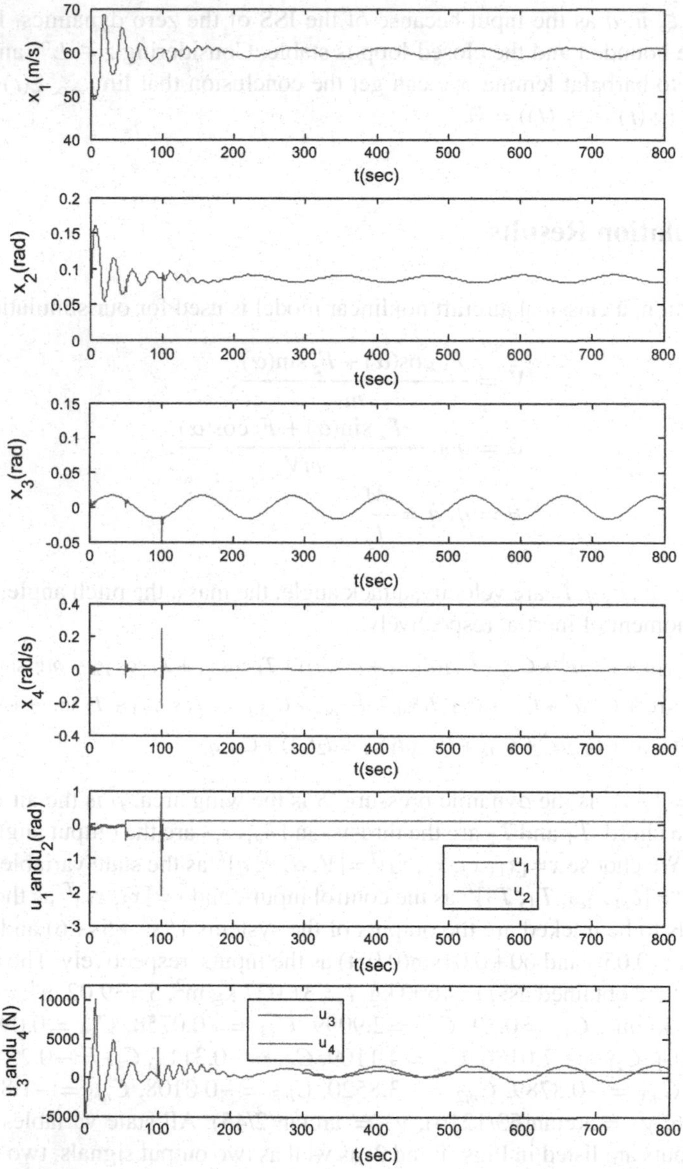


**Fig. 1** State variables and four control inputs

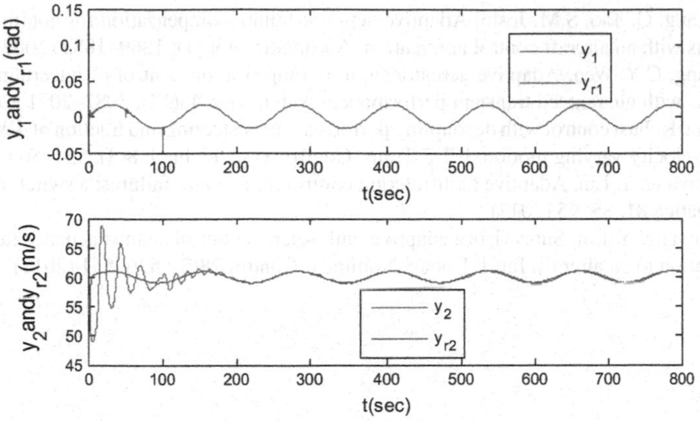


**Fig. 2** Two outputs and two reference signals

# 5 Conclusion

In this paper, we first give the actuator failures model considering three types of common actuator failures simultaneously. These failures including stuck failure, bias failure and loss of effectiveness failure are all unknown in failure type, value and time. Then we give the direct adaptive compensation scheme for a class of MIMO nonlinear minimum systems to accommodate these failures via the grouping and proportional actuation techniques. The scheme ensures all signals bounded and output asymptotic tracking. The simulation about the elevator failure on the aircraft demonstrates the scheme is effective. Adaptive compensation control for nonlinear nonminimum phase systems is still a challenging problem. Besides, further study about direct adaptive compensation for nonlinear systems considering other failures such as sensor failure and the controlled plant failure are also needed.

# References


1. J.D. Boskovic, J.A. Jackson, R.K. Mehra, N.T. Nguyen, Multiple-model adaptive fault-tolerant control of a planetary lander. J. Guid. Control Dyn. **32**(6), 1812–1826 (2009)
2. J.D. Boskovic, R.K. Mehra, N.T. Nguyen, A decentralized fault tolerant control system for accommodation of failures in higher oder flight control actuators. IEEE Trans. Control Syst. Technol. **18**(5), 1103–1115 (2010)
3. X.D. Zhang, T. Parisini, M.M. Polycarpou, Adaptive fault-tolerant control of nonlinear uncertain systems: an information-based diagnostic approach. IEEE Trans. Autom. Control **49**(8), 1259–1274 (2004)
4. X.D. Zhang, M.M. Polycarpou, T. Parisini, Adaptive fault diagnosis and fault-tolerant control of MIMO nonlinear uncertain systems. Int. J. Control **83**(5), 1054–1080 (2004)
5. X.D. Tang, G. Tao, S.M. Joshi, Adaptive actuator failure compensation for parametric strict feedback systems and an aircraft application. Automatica **39**(11), 1975–1982 (2003)

6. X.D. Tang, G. Tao, S.M. Joshi, Adaptive actuator failure compensation for nonlinear MIMO systems with an aircraft control application. Automatica **43**(11), 1869–1883 (2007)
7. W. Wang, C.Y. Wen, Adaptive actuator failure compensation control of uncertain nonlinear systems with guaranteed transient performance. Automatica **46**(12), 2082–2091 (2010)
8. Y.M. Jia, Robust control with decoupling performance for steering and traction of 4WS vehicles under velocity-varying motion. IEEE Trans. Control Syst. Technol. **8**(3), 554–569 (2000)
9. H.P. Ouyang, Y. Lin, Adaptive fault-tolerant control for actuator failures: a switching strategy. Automatica **81**, 85–95 (2017)
10. H.P. Ouyang, Y. Lin, Supervisory adaptive fault tolerant control against actuator failures with application to an aircraft. Int. J. Robust Nonlinear Control **28**(2), 536–551 (2017)

# A Mixed Approach for Fetal QRS Complex Detection

**Lijuan Liao, Wei Zhong, Xuemei Guo and Guoli Wang**

**Abstract** Non-invasive fetal electrocardiogram (NI-FECG) plays an important role in detecting and diagnosing fetal diseases. Fetal electrocardiogram (FECG) is used to know the information of the fetal health. In this paper, we propose a mixed approach for extracting FECG from maternal abdominal ECG (AECG) recording. The proposed method is based on a combination of the wavelet transform and Support Vector Machines (SVM). As a first tier, the wavelet transform is used to detect maternal QRS complex from abdominal ECG recording. Then, a coherent averaging method was using to construct MECG and remove MECG from AECG recording. After removing MECG, SVM is used to locate fetal QRA complex from residual signal. The accuracy (84.53%) and Positive predictive value (PPV) (89.6%) in this study are much higher than other method.

**Keywords** NI-FECG · Wavelet transform · SVM · Signal quality assessment · PPV

## 1 Introduction

The statistics show that 30% of cerebral palsy and serious mental retardation is due to fetal hypoxia [1]. Therefore, an effective technique is required to monitor the fetal condition during the pregnancy and at delivery.

L. Liao
School of Electronics and Information Technology, Sun Yat-Sen University, Guangzhou 510006, China

W. Zhong · X. Guo · G. Wang (✉)
School of Data and Computer Science, Sun Yat-Sen University, Guangzhou 510006, People's Republic of China
e-mail: isswgl@mail.sysu.edu.cn

X. Guo · G. Wang
Key Laboratory of Machine Intelligence and Advanced Computing, Ministry of Education, Beijing, People's Republic of China



Y. Jia et al. (eds.), *Proceedings of 2018 Chinese Intelligent Systems Conference*, Lecture Notes in Electrical Engineering 528,
https://doi.org/10.1007/978-981-13-2288-4_38

Electronic fetal monitoring (EFM) techniques can be divided into two kinds of invasive and non-invasive methods. Invasive method that is including scalp ECG (SECG), can provide accurate fetal heart rate (FHR) time series. However, this method can only be performed at delivery. Non-invasive methods include fetal phonocardiography (PCG), Doppler ultrasound, Cardiotocography (CTG), NI-FECG.

PCG obtains acoustic recording from maternal abdomen. These recordings have the lowest SNR of all methods, however, it is high prone to noise and can be only used late in pregnancy.

Doppler ultrasound is routinely used to monitor FHR during pregnancy and at delivery. However, despite being a non-invasive method, it has not been proved that ultrasound radio frequency is completely safe for the fetus. Moreover, while using Doppler ultrasound, the maternal HR can be detected instead of FHR.

CTG, which is the most widespread method, uses ultrasound transducer and uterine contraction pressure-sensitive transducer. It is prone to maternal and fetal HR confusion signs and can only be used in clinical practice.

Among all the methods, the Non-invasive fetal electrocardiogram (NI-FECG) is recommended for continuous monitoring which can combine the advantages of both the CTG and SECG. That is it can not only provide an accurate FHR, but also the additional information on the electrical activity of the heart. Several electrodes are placed on maternal abdomen to obtain recordings. So, in this subject, the NI-FECG is a considerable research.

In the field of NI-FECG signal processing, it is often difficult to detect the fetal QRS (FQRS) complexes and other fetal ECG features in the abdomen recordings due to their low SNR. The main reason of the low SNR is the relatively low SNR of FECG compared to the maternal ECG (MECG) [2]. MECG, which is the obvious interference source, often has greater amplitude than FECG. Moreover, the frequency spectrum of MECG partially overlaps that of FECG. Other interference sources include baseline drift 50 or 60 Hz power-line interference, respiration interference, maternal electromyogram (EMG), electrode contact noise and so on [3].

Despite all of the difficulties, many approaches have been attempted to extract the FECG. Among those techniques are adaptive filtering methods (AM), or other filter methods, blind source separation (BSS) [4], template subtraction methods (TS), and artificial neural network (ANN) or fusion of multiple source separation.

AM make use of one or more maternal reference channel(s), in order to remove the MECG on each abdominal signal [5]. Some of the methods have been proposed in the field, such as the least mean square ($AM_{lms}$) [6], the recursive least square algorithm ($AM_{rls}$), and the echo state neural network ($AM_{esn}$) [7]. Although this method is very simple, it depends on the sign SNR and its stability.

Other filter methods, such as the Wavelet transform [8], Short time Fourier Transform, and matching pursuits [9], are also developed and evaluated for locating fetal QRS complexes.

BSS techniques usually assume that the signal mixture is stationary (or short-term stationary). These methods attempt to decompose the multichannel abdominal mixture into different components without a priori knowledge about the signal itself. Independent Component Analysis (ICA) is one of the most widely used and a suitable technique for detaching the FECG "source" from the rest [10]. Other techniques of BSS includes principal component analysis (PCA) [11], singular value decomposition (SVD) [12] and periodic component analysis [$\pi$CA] (semi-BSS) [13].

TS depend on building an average MECG signal (the so-called 'template') by means of coherent averaging several maternal beats. This approach mainly relies on accurate maternal QRS detection. After being constructed MECG, abdomen recordings subtracted template, leaving residual FECG and noise [14, 15].

In this study, we propose a novel extraction system by using of a mixed approach. Firstly, we used a wavelet transform to reconstruct maternal ECG. Secondly, we used TS to obtain FECG and noise signal. Thirdly, Support Vector Machine (SVM) was used as classifier to discriminate whether it was the location of a R wave.

# 2 Methods

## 2.1 Dataset

In this work, we consider the database that provided by the PhysioNet/Computing in Cardiology Challenge Database (PCDB). This database represents the largest publicly available dataset to date. All records from this database were formatted to have a 1 kHz sampling frequency, 1 min duration, and four channels of non-invasive abdominal maternal ECG leads. It had three data sets, but only in set A, both records and expert annotations were made public. In Set A, there were seventy-five recordings (a01–a75). We selected a subset of 69 records from the learning set by excluding partially/badly annotated records (a33, a38, a52, a54, a71, a74) [16].

The classical adult acceptance interval is 150 ms [7]. However, to account for the higher FHR, if a detected fetal QRS is with 50 ms from the reference annotation, it is considered a true positive. So the databases are divided into 100 ms per segment. At first the data equally divided into 600 segments. Then, if the reference location of fetal QRS was within the segment, the NI-FECG signal was considered a fetal QRS complex, otherwise considered not fetal QRS complex. To consider the balance of databases, another segment would be taken in the databases, in which the fetal QRS complex is in the center. Each NI-FECG signal is annotated into one of two classes according to the reference annotation. Taking a10 as an example, after re-building the databases, there are 340 segments, which fetal QRS complexes are located in, and 425 segments, which fetal QRS complexes are not in. In 69 recordings, a total of 19,164 segments considered fetal QRS complexes, and 31,818 segments considered not fetal QRS complexes.

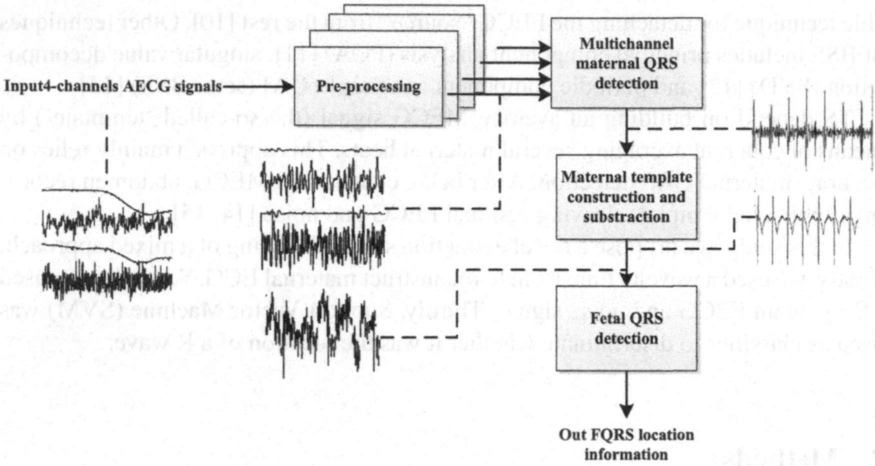


**Fig. 1** Block diagram of the proposed mixed method

## 2.2 *Method Description*

Our software consisted in a sequence of steps to remove most of the undesired 'noisy' components before the FQRS detection. The block diagram of the algorithm is shown in Fig. 1. In the following sections, its different parts will be analyzed in detail.

### 2.2.1 Pre-processing

Before extracting MECG, some fundamental noises should be removed. Baseline wander and power-line interference appear in many ACG recordings. In this study, a low-pass filter based on the wavelet transform method [17] is used to remove the baseline wander which is below 1 Hz. And then, high frequency noise is removed by wavelet soft-threshold de-noising. Power-line interference consists in a sinusoidal component at a frequency around 50 Hz and its harmonics. Whether this component existe, the power spectral density is estimated by comparing the peak of the power density in a narrow interval around 50 Hz with the average power density in such nearby frequencies. If the power-line interference is explored, it will be removed by notch filters (forward-backward, zero phase,1 Hz bandwidth).

Among the databases we considered, there are four channels. Not every channel signal quality is good enough to detect FQRS. It is very common for the AECG recordings to have other undesirable physiological signals and noises. Therefore, we considered the sample entropy (SampEn) for noise identification [18]. The SampEn is defined as:

| | SampEn | State |
|---|---|---|
| Channel1 | 1.9998 | excluded |
| Channel2 | 1.5073 | reserved |
| Channel3 | 2.201 | excluded |
| Channel4 | 1.3727 | reserved |

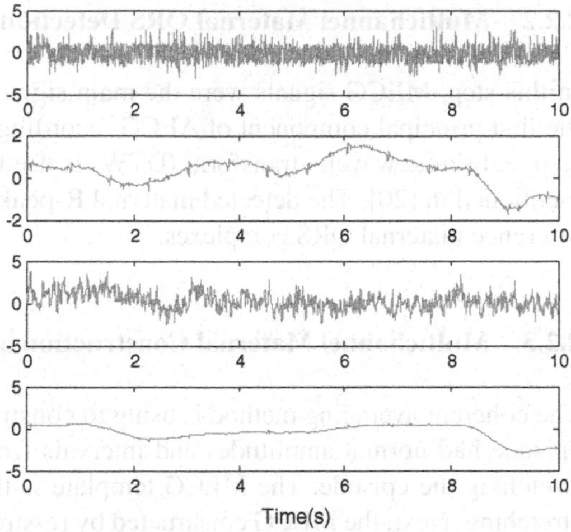

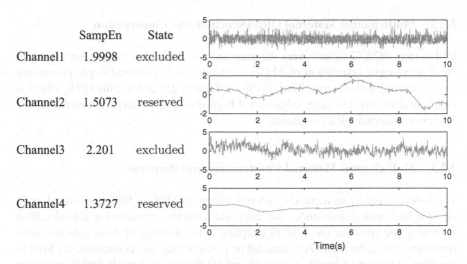

**Fig. 2** An example of signal quality assessment for a signal episode from recording a53. There are only one channel SampEn value less than $r(1.5)$, channel 2 (SampEn value $= 1.5073$) will also be reserved

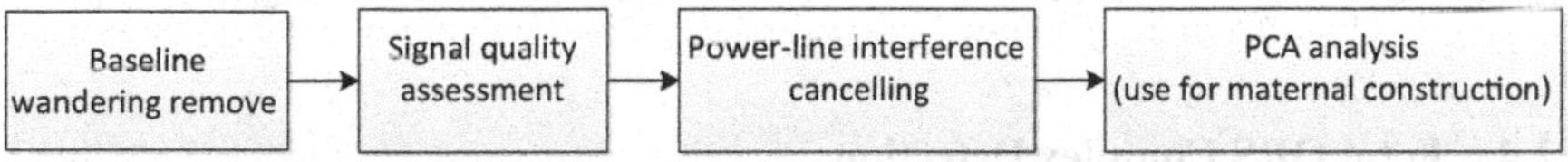


**Fig. 3** Block diagram of the pre-processing stage

$$SampEn(m, r, N) = -\ln\left(\sum_{i=1}^{N-m} A_i^m(r) / \sum_{i=1}^{N-m} B_i^m(r)\right) \quad (1)$$

where $m$ is embedded dimension, $r$ to threshold, $N$ to data length. More information about SampEn, the article from [19] is recommended. In this study, $m$ is set at 2, while $r$ is set at 1.5 and $N$ is set at 500. If SampEn values are less than $r$, they would be regarded as good quality. However, if there are less than two channels of good quality, the channels with the smallest and penultimate will be used for next study. Figure 2 shows an example of signal quality assessment on 10 s signal episode from record a53.

After signal quality assessment, the maternal R-peaks should be detected, if we want to construct MECG. Therefore, the PCA method is used to obtain the first principal component of AECG recordings before constructed MECG. A synthetic block scheme of the pre-processing steps is shown in Fig. 3.

### 2.2.2 Multichannel Maternal QRS Detection and Construction

In this step, MECG signals were the main signals, and other signals are noise. As the first principal component of AECG recordings were obtained in pre-processing step, a discrete wavelet transform (DTW) is used to detect maternal QRS, which is mentioned in [20]. The detected maternal R-peaks (or Q-peaks) are regarded as the reference maternal QRS complexes.

### 2.2.3 Multichannel Maternal Construction and Removal

The coherent averaging method is using to construct MECG. Slicing of signal, each episode had normal amplitudes and intervals from two consecutive R-peaks, then stretching the episode. The MECG template is the average of these episodes after stretching. Next, the MECG constructed by re-stretching and contrasting the MECG template to the actual length of each period from the first R-peak, and then joining them segment by segment.

When MECG template reconstructed, the residual signals, which are only containing FQRS signals and other noise, will be obtained by subtracting the template.

### 2.2.4 Fetal QRS Complex Detection

In this step, fetal heart rate is what this paper care about. Just located the R-peaks of fetal can calculate the fetal heart rate. Then SVM is used to detect the location of fetal R-peaks. A SVM classifies data by finding the best hyperplane that separates all data points of one class from the other class. It could describe as following:

$$\begin{aligned} &\min \tfrac{1}{2}\sum_{i}^{N}\sum_{j}^{N}\alpha_i\alpha_j y_i y_j K(x_i, x_j) - \sum_{i}^{N}\alpha_i \\ &\text{s.t} \ \sum_{i}^{N}\alpha_i y_i = 0 \\ &\qquad \alpha_i > 0, i = 1, 2, \ldots, N \end{aligned} \tag{2}$$

where N corresponds to data length, $\alpha_i$ to a sample $(x_i, y_j)$ and $K(x_i, x_j)$ to a Mercer Kernel.

In this study, the Radial Basis Function (RBF) kernel or Gaussian Kernel is used to learn nonlinear features. The kernel formula is:

$$K(x_i, x_j) = \exp(-\|x_i - x_j\|^2)/2\sigma^2 \tag{3}$$

This kernel has a $\sigma$ parameter to control the performance of the model.

# 3 Experimental Result and Analysis

## 3.1 Statistical Assessment

Sensitivity (SE), Positive predictive value (PPV) and $F_1$ accuracy measure ($F_1$) are typically used for assessing the performances of FQRS detection. SE, PPV and $F_1$ are defined as:

$$SE = \frac{TP}{TP + FN} \tag{4}$$

$$PPV = \frac{TP}{TP + FP} \tag{5}$$

$$F_1 = 2 \cdot \frac{PPV \cdot SE}{PPV + SE} = \frac{2 \cdot TP}{2 \cdot TP + FN + FP} \tag{6}$$

where *TP* are the number of true positive (correctly detected FQRS complexes), *FP* to the number of false positive (wrongly detected FQRS complexes), *FN* to false negative (missed detected FQRS complexes). Notice that *FN* and *FP* are equally affect the $F_1$ accuracy measure. SE tells how good an algorithm is at finding the true FQRS. PPV tells how good an algorithm is at identifying true FQRS out of all the detections it makes. $F_1$ is a harmonic mean, which can be used as an accuracy measure when training an algorithm and to summarize the final over performance of it in accurately detecting the R-peak locations. The performance of the classification algorithm is computed by accuracy:

$$Accuracy = \frac{TP + TN}{TP + TN + FP + FN} \tag{7}$$

## 3.2 Contrast Experimental Results

To confirm the effectiveness of the approach in this study, we compared with other state-of–the-art classifiers, that were KNN, NB and the algorithm mentioned in [21, 22]. On the other hand, to test the effectiveness of the signal quality assessment, experiments with (/without) signal quality assessment are employed. As $F_1$ could be calculated by SE and PPV, the results of SE and PPV are given.

Table 1 gives a comparison of the performance of two processes (with or without signal quality assessment). It illustrates that better accuracies are obtained by implementing the SampEn based signal quality assessment procedure. So it is necessary to handle signals with signal quality assessment.

Table 2 shows results of several other classifiers and new method on physionet dataset. This table illustrates the efficiency of proposed method. Here, we can note that PPV and Accuracy are higher than other methods.

**Table 1** Accuracies results with (without) signal quality assessment

| | With signal quality assessment | Without signal quality assessment |
|---|---|---|
| Accuracies (%) | **84.53** | 80.27 |

**Table 2** Classification results

| | SE (%) | PPV (%) | Accuracy (%) |
|---|---|---|---|
| Proposed method | 68.47 | **89.6** | **84.53** |
| KNN | 30.26 | 87.47 | 70.98 |
| NB | 7.67 | 40.35 | 59.38 |
| Method from [21] | 71.43 | 71.48 | 82.26 |
| Method from [22] | **80.54** | 75.33 | 77.38 |

## 4 Discussion and Conclusion

In order to achieve satisfactory fetal QRS performance, this work presents a new model for detecting the fetal QRS complexes. In this model, we combined signal processing and machine learning to detecting FQRS. The accuracy of classification processed and two evaluation metrics including Sensitivity and Positive predictive value are used to compare the performance proposed method to KNN, NB and method from [21, 22]. The experiments have proved that the proposed method can improve the classification accuracy.

It is clearly that SE is lower than the method from [22]. It is because that the number of false negative is too much, as noted in the Eq. (1). It is worth to improve SE by reducing the number of false negative in the future study. On the other hand, although the accuracy and the PPV are much higher, the processing of this method is little complexity, so the real-time performance is little poor. That is another important issue in the future.

**Acknowledgements** This work was supported by the National Natural Science Foundation of P. R. China under Grant No. 61375080, and the Key Program of Natural Science Foundation of Guangdong, China under Grant No. 2015A030311049. The Guangzhou science and technology project under Grant Nos. 201510010017, 201604010101.

## References

1. H.-B. Li, S.-Y. Fang, Development of internet-based home telemonitoring system for fetus. Chin. Med. Equip. J. **2**, 17–19 (2006)
2. N. Ivanushkina, K. Ivanko E. Lysenko, et al., *Fetal Electrocardiogram Extraction from Maternal Abdominal Signals* (Kyiv, 2014) pp. 334–338
3. S.B. Barnett, D. Maulik, Guidelines and recommendations for safe use of doppler ultrasound in perinatal applications. J. Matern. Fetal Med. **10**(2), 75–84 (2001)

4. M. Sato, Y. Kimura, S. Chida et al., A novel extraction method of fetal electrocardiogram from the composite abdominal signal. IEEE Trans. Biomed. Eng. **1**, 49–58 (2007)
5. F. Andreotti, J. Behar, S. Zaunseder, J. Oster, G.D. Clifford, An open-source framework for stress-testing non-invasive foetal ecg extraction algorithms. Physiol. Meas. **37**(5), 627 (2016)
6. B. Widrow, J.R. Glover, J. McCool, J. Kaunitz, C. Williams, R. Hearn, J. Zeidler, J. Eugene Dong, R. Goodlin, Adaptive noise cancelling: principles and applications. Proc. IEEE **63**, 1692–1696 (1975)
7. J. Behar, A. Johnson, G.D. Clifford, J. Oster, A comparison of single channel foetal ECG extraction methods Ann. Biomed. Eng. **42**, 1340–1353 (2014)
8. R. Bhoker, J.P Gawande, Fetal ECG extraction using wavelet transform. ITSI Trans. Electr. Electron. Eng. (1), 19–22 (2013)
9. M. Akay, E. Mulder, Examining fetal heart-rate variability using matching pursuits. IEEE Eng. Med. Biol. **15**, 64–72 (1996)
10. C.J. James, C.W. Hesse, Independent component analysis for biomedical signals. Physiol. Meas. **26**, 15–39 (2005)
11. C. Di Maria, W.F. Duan, M. Bojarnejad, F. Pan, S. King, D.C. Zheng, A. Murray, P. Langley, An algorithm for the analysis of foetal ECGs from 4-channel non-invasive abdominal recordings. Proc. Comput. Cardiol **4**, 305–308 (2013)
12. P.P. Kanjilal, S. Palit, G. Saha, Fetal ECG extraction from single-channel maternal ECG using singular value decomposition. IEEE Trans. Biomed 44 51–3 (1997)
13. R. Sameni, Extraction of fetal cardiac signals from an array of maternal abdominal recordings. Ph.D. Thesis, Sharif University of Technology—Institute National Polytechnique deGrenoble, 2008, www.sameni.info/Publications/Thesis/PhDThesis.pdf
14. J. Behar, J. Oster, G.D. Clifford, Combining and comparing benchmarking methods of foetal ECG extraction without maternal or scalp electrode data. Physiol. Meas. **35**, 1569–89 (2014)
15. P. Podziemski, J. Gierałtowski, Fetal heart rate discovery: algorithm for detection of fetal heart rate from noisy. Comput. Cardiol. **40**, 333–336 (2013)
16. J. Behar, J. Oster, G.D. Clifford, Non-invasive FECG extraction from a set of abdominal sensors. Comput. Cardiol. 297–300 (2013)
17. P. Quan, D. Zhang, D. Guanzhong, Z. Hongcai, Two denoising methods by wavelet transform. IEEE Trans. Signal Proces. 47, 3401–6 (1999)
18. C. Liu, P. Li, C. Di Maria, L. Zhao, H. Zhang, Z. Chen, A multi-step method with signal quality assessment and fine-tuning procedure to locate maternal and fetal QRS complexes from abdominal ECG recordings. Physiol. Meas. **35**, 1665–1683 (2014)
19. C.Y. Liu, P. Li, L.N. Zhao, F.F. Liu, R.X. Wang, Real-time signal quality assessment for ECGs collected using mobile phones. Proc. Comput. Cardiol. **38** 357–60 (2011)
20. S. Banerjee, R. Gupta, M. Mitra, Delineation of ECG characteristic features using multiresolution wavelet analysis method. Measurement **45**, 474–487 (2012)
21. R. Kahankova, R. Martinek.et al. Fetal ECG Extraction from Abdominal ECG Using RLS based Adaptive Algorithms, in *2017 18th International Carpathian Control Conference (ICCC)* (IEEE Conferences, 2017)
22. W. Zhong, L. Liao, X. Guo, G. Wang, A deep learning approach for fetal QRS complex detection. Physiol. Meas. (9), 045004 (2018)

# A Multi-sensor Characteristic Parameter Fusion Analysis Based Electrical Fire Detection Model

**Xuewu Yang, Ke Zhang, Yi Chai and Yuan Li**

**Abstract** Electrical fires are mostly caused by the release of thermal energy from electrical equipment. It is more difficult to detect the cause of fire than normal fires. A fire detection model that combines characteristics of many types of electrical fire sensors is proposed. These fire detection systems are difficult to accurately monitor the cause of electrical fires. The proposed model uses smoke, CO concentration, temperature, and electrical line residual current as characteristic parameters of electrical fire. It analyzes a three-tier structure including information layer, feature layer, and decision layer. Fire-risk-factor and warning duration are defined as decision factor. When the model is working, it firstly collects the residual current signal that characterizes the fault of the electrical equipment through multiple types of sensors. It conducts multi-parameter real-time monitoring of the main characteristic signals of the early stage of the electrical fire. Then it completes the fusion of the detected characteristics of electrical fire to achieve accurate identification of electrical fires. The proposed model is simulated according to national standard fire test dataset. The simulation result shows that it can quickly and accurately forccast the electrical fire and effectively reduce false alarm rate in of electrical fire detection process.

**Keywords** Electrical fire · Characteristic parameters · Multi-sensor fusion · Data fusion · Early fire

## 1 Introduction

Fire is one of major disasters faced by human life. Electrical fire is caused by the release of thermal energy from electrical equipment [1]. It is one of main causes of serious and large fires and is also the main disaster factor that affects the fire safety of the society. Because electrical equipment releases thermal energy and has both

X. Yang (✉) · K. Zhang · Y. Chai · Y. Li
State Key Laboratory of Power Transmission Equipment and System Security and New Technology, College of Automation, Chongqing University, Chongqing 400044, China
e-mail: 20161302003t@cqu.edu.cn



Y. Jia et al. (eds.), *Proceedings of 2018 Chinese Intelligent Systems Conference*, Lecture Notes in Electrical Engineering 528,
https://doi.org/10.1007/978-981-13-2288-4_39

fault and non-failure characteristics, such as leakage current, short circuit, overload, contact resistance, static electricity and other reasons [2]. It is possible to rapidly accumulate temperature locally. These fire factors are often not easily found by traditional fire monitoring methods. Therefore, the monitoring and prevention of electrical fires has always been the focus of fire safety [3].

From the research point of view, the traditional detection of electrical fires has a high degree of directionality, such as the use of certain types of sensors for specific fire factors to monitor [4]. Such monitoring methods only focus on specific characteristic parameters and it is difficult to obtain monitoring targets from the overall situation. Fire and related information of the surrounding environment; and single sensor anti-interference ability is weak [5]. It is difficult to ensure long-term monitoring accuracy, but also leads to higher perception of information deviation. With the further improvement of the response sensitivity and anti-interference capability of fire detection, multi-sensor composite fire detection technology (especially electrical fire) has become the foremost issue in the current fire detection technology research [6, 7].

Multi-sensor fire detection technology captures and analyzes data from different types of sensors to obtain a more complete and detailed description of the target. Compared with a single type of sensor, higher detection reliability and accuracy can be obtained. At present, although the research on this area has developed rapidly and has achieved certain research results. There are still some problems that can be summarized in two aspects: First, the combination of fire characteristic parameters can hardly fully reflect all the characteristics of the fire. Second, there are certain problems in the data fusion algorithm used. Traditional data fusion algorithms mainly include threshold methods, relational methods, and fuzzy logic inference algorithms [8–11]. Threshold method uses logic and sum of logic criteria for various types of sensor data to determine the fire and non-fire source algorithm. The relational method is proposed based on the threshold method. The algorithm has better performance than the threshold method. The fuzzy logic reasoning algorithm uses fuzzy inference rules and language control rules to judge the fire, which is in line with human thinking.

However, the use of multiple types of sensors to comprehensively detect and evaluate electrical fire research is currently developing slowly. There are two main reasons: First, it is difficult to make an effective and direct link between an ordinary electrical circuit fault and an electrical fire; Second, electrical fires have the fault concealment caused by the cumulative effect of time. The current related researches mostly include collecting fault arcs with special parameters of electrical fires [12], and combining the traditional fire characteristic parameters (such as CO concentration and temperature) for fusion recognition in order to obtain a complete electrical fire signal. However, the occurrence of a faulty arc on an electrical line indicates that the probability of an electrical fire is extremely high. It is difficult to detect possible fires earlier in the existence of hidden dangers. Moreover, the detection method of arc faults is relatively complicated, which takes a long time and has the more fatal flaws on real-time nature of detection.

According to the analysis above, this paper focus on the technical requirements for detection of electrical fires through multi-feature parametric fusion analysis.

On the basis of existing studies, the knowledge that can significantly characterize electrical fire characteristics is analyzed. Smoke, CO concentration, and temperature are analyzed as the characteristic parameters of electrical fire. Taking into account the characteristics of the excess current generated prior to the generation of a fault arc [13], the residual current is used as a feature of electrical fire to replace the fault arc. In accordance with the requirements of data fusion, an electrical fire detection method architecture including information layer, feature layer, and decision layer, and fire risk factor and warning duration were defined as decision factors. Through the detection of the residual current in the electrical system of the building, combined with the data fusion algorithm, the fire parameters are comprehensively identified by other characteristic parameters of the electrical fire. The system output is obtained through fuzzy inference in order to obtain accurate electrical fire detection results.

## 2 Electrical Fire Detection Characteristic Parameter Analysis

According to statistics, the proportion of electrical fires in fires exceeds 20%, and the severity of casualties and property losses caused by them has an absolute status in fires. In electrical fires, electrical circuit fires accounted for more than 60% [14]. Unlike traditional fires, electrical fires (especially electrical circuit fires) are caused by the release of faulty heat energy after long-time operation of electrical circuits. They are highly concealed and not easily detectable.

The direct line faults that cause electrical fires include leakage, electrical short-circuit, overload, and poor contact of electrical components [15]. Leakage (residual current) is both a fault and a major factor in causing other line faults. The detection of leakage is usually performed by detecting in real time whether the residual current in the electrical line exceeds a set value. The timely and accurate detection of residual current can provide a basis for preventing early fire hazards in electrical circuits. Therefore, this paper proposes an electrical fire detection model based on residual current detection, which combines detection and early warning by combining several characteristic parameters in the early stages of electrical fires.

Since the detection of electrical fire focuses on early warning, the key to whether the system can accurately identify the fire is the selection of the type and number of fire parameters. In addition to the residual current, which is a unique fire parameter for electrical fires, the fire parameters that are necessary for general fires are also important characteristic parameters. In feature selection, the requirements of different parameter types can be complemented with each other to improve the ability of the entire system to resist various environmental factors and to accurately identify the nature of the fire source.

As shown in Fig. 1, there is a corresponding relationship between the fire detector and the fire development process.

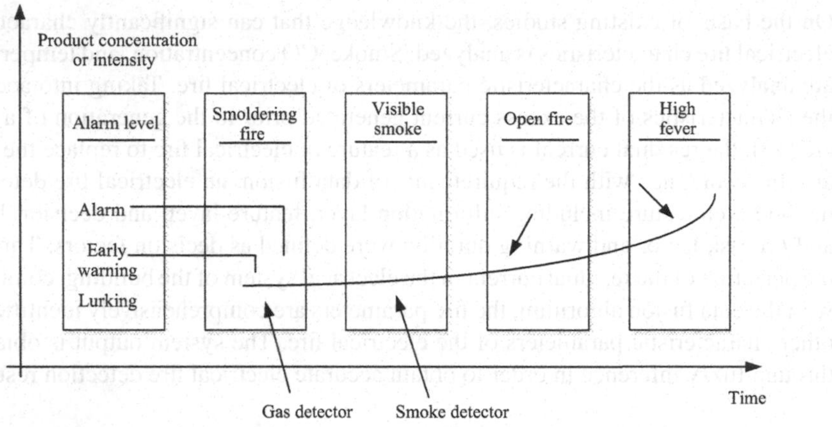


**Fig. 1** Correspondence between fire detectors and fire development process in most situations

With reference to existing research, this model analyzes the main characteristic parameters of electrical fires:

(1) Smoke

Smoke is a clear physical phenomenon that can be clearly observed in most fires. It is also an important feature to characterize fire [16]. However, due to its versatility, it cannot identify a specific type of fire alone.

(2) CO concentration

A large amount of gas will be produced during the fire. The components mainly include CO, $CO_2$, and $O_2$. Before and after the fire, the concentration of CO will change drastically. Therefore, it is an important method to determine whether the fire has occurred by detecting the concentration of CO. Compared with smoke, CO concentration can not identify the type of fire, but it can be used as the main reference for fire development.

(3) Temperature

In traditional fires, early temperature changes were not significant, and only significant changes occurred after deflagration. A major feature of electrical fires is the release of thermal energy from electrical equipment and the accumulation of heat. Especially the rapid increase in local temperature due to aging damage (such as showing power series growth) has become an important monitoring parameter to distinguish fire types. Therefore, this model needs to focus on the temperature change under a specific trend.

(4) Electric line residual current

Residual current is a unique feature of electrical fires. According to the definition, the residual current refers to the vector sum of the instantaneous value of the three-phase current flowing through the electrical line [17]. There is inevitably a slight amount of normal residual current in the actual electrical circuit. But if the residual current exceeds the alarm value that the electrical line has a leakage fault, it can easily cause fire. This model requires that the electrical fire monitoring system continuously monitor the changes of the residual current of the line in real time for a long period of time. When the residual current is too large, an alarm will be given in time and an alarm location will be indicated.

In addition, the parameters such as gas concentration, voltage fluctuation, acousto-optic and fuse-body resistance generated by scorching of insulation skin after over-heating of the electrical circuit can also provide a basis for electrical fire detection. However, they all have defects that are difficult to detect or have unclear identification characteristics. Therefore, in this model, the above four indicators, such as smoke, CO concentration, temperature and electric line residual current are taken as the characteristic parameters of the fusion analysis.

## 3 Electrical Fire Detection Model

According to the foregoing, the four main characteristic parameters that identify the electrical fire show different physical and chemical characteristics. The characteristics of the electrical fire signal are combined to ensure the timelines of the fire alarm, reduce the false alarm rate of the alarm system and improve the reliability of the alarm system. As well as the requirements of the forecasting system, we believe that the residual current should be different from the other three features in the way and effect of monitoring. Smoke, CO concentration and temperature are the basic characteristics of a fire. They can be triggered by sensors such as smoke, gas, and temperature. Only when one of the parameters has an abnormality exceeds the limit, the fusion model is triggered; The current exceeds the limit before the fire signal is generated. Therefore, the continuous online detection of the residual current alone provides a reference for fire detection while also realizing the rapid warning of the fire; in particular, when the residual current exceeds the limit The launched fusion system fuses the fire characteristic signals collected by the sensors to obtain the corresponding fire probability, and finally obtains the final fire information through the decision-making reasoning.

Therefore, this model has designed a data fusion system with three layers of information, feature and decision-making layers. As shown in Fig. 2, combined with signal preprocessing, artificial neural networks, fuzzy inference and other technologies, the model will be used for multiple types of sensors. The data was integrated with fire detection analysis.

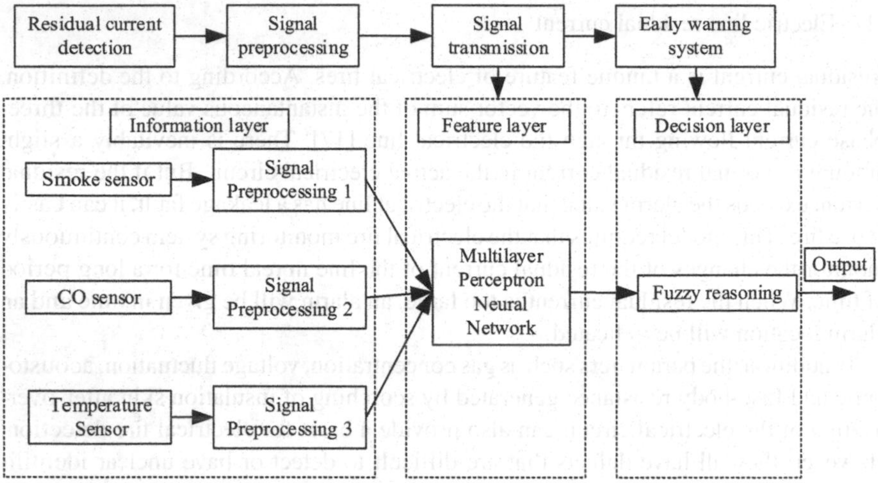


**Fig. 2** Electrical fire detection algorithm fusion model structure

## *3.1 Information Layer*

The information layer is responsible for collecting various parameters of the site and preprocessing the information. If the collected information shows signs of non-stationary, that is, the rate of change of each parameter exceeds the corresponding threshold, the feature layer is prompted to extract information.

According to the form of expression, electrical fires can be roughly divided into two types: smoldering fire and open fire. Under the smolder fire condition, the residual current far exceeds the set value, the CO concentration and the smoke concentration increase gradually, and the temperature is basically guaranteed to be stable. Under the open flame condition, the electrical line is damaged due to the flame combustion. The residual current is transformed into short-circuit arc. The smoke density and temperature increase sharply, and the CO concentration slowly increases. In addition, due to the high sampling frequency of the sensor, in order to ensure the signal processing speed, only when one or more sensors reach the preset alarm limit, they are sent to the feature layer for feature recognition. In this way, the early identification of fires can be maximized, and the processing of information that does not have fire characteristics can be avoided, the workload is reduced, and the real-time performance is higher.

Due to the high sampling frequency of the sensor, in order to ensure the signal processing speed, this model adopts a trigger mechanism. That is, when at least one sensor reaches the preset alarm limit, the measured values of all types of sensors are sent to the characteristic layer for feature recognition. This will not only maximize the early identification of fires, but also avoid the handling of information that does not have fire characteristics.

When an electrical fire occurs, the residual current in the electrical circuit increases rapidly, and other parameters such as smoke and temperature in the surrounding environment also gradually increase. Therefore, the change rate detection method is used for the processing of a single signal. That is, the rate of change of each characteristic signal is detected to determine whether it continuously exceeds the set threshold, that is:

Let the detected signal (that is, the residual current signal) be discretized and sampled. The original sampling sequence of the signal is $X(n)$.

$$S_n = \sum_{n=1}^{N} (X_n - X_{n-1}) \tag{1}$$

If $S_n > S_{ft}$, $S_{ft}$ is the signal of fixed threshold, $a_i = 0$, where $i = 1$, 2, and 3 represent raw data for smoke, temperature and CO concentration respectively. After processing the smoke, temperature and CO concentration data according to the above method, respectively:

$$A = a_1 \cup a_2 \cup a_3 \tag{2}$$

If $A = 1$, then one or more data in the collected data in this system has undergone non-stationary changes. At this time, the group of information needs to be sent to the feature layer for fusion recognition.

## 3.2 Feature Layer

The feature layer integrates various feature information sent from the information layer, uses pattern recognition methods to fuse, realizes the association of multiple feature vectors, and completes the recognition of target object features. This model uses smoldering fire and open flame as the recognition target objects of the feature layer. The recognition algorithm uses a multi-layer perceptron neural network and uses two independent single-hidden multilayer perceptron artificial neural networks to ensure two different type of fire identification.

In terms of structure, the multi-layer network structure for different types of fires is basically the same. The difference lies in the difference in the weights of the four types of characteristic parameters of the smoldering fire and open flame. As shown in Fig. 3, the input layer has three input parameters $IN_1$, $IN_2$ and $IN_3$, which represent CO concentration, smoke concentration, and temperature respectively. The output layer output two parameters $OT_1$ and $OT_2$, which are the smolder fire probability and the open flame probability. The hidden layer number is 1, and the node number is 4. They are $M_1$, $M_2$, $M_3$ and $M_4$. The weight between $IN_i$ and $M_j$ is $W_{ij}$, and the weight between $M_j$ and $OT_k$ is $V_{jk}$. When the input is $IN_i$, the sum of the hidden layer inputs is

$$NET_1(j) = \sum_{i=1}^{3} (N_i \cdot W_{ij}) \quad (3)$$

$NET_1(j)$ is converted to 0 to 1 using the Sigmoid function, which means

$$M_j = \frac{1}{1 + \exp[-NET_1(j)]} \quad (4)$$

Similarly, the sum of the inputs to the output layer is

$$NET_2(k) = \sum_{j=1}^{4} (M_j \cdot V_{jk}) \quad (5)$$

When $NET_2(k)$ is converted to 0 to 1, it means that

$$OT_k = \frac{1}{1 + \exp[-NET_2(k)]} \quad (6)$$

In order to achieve the fastest convergence speed, the neural network training algorithm uses trainlm (Levenberg-Marquardt optimization algorithm) algorithm in the process of fire detection. The weight value learning function adopts the learngdm algorithm, which provides the network with faster convergence speed.

Neural network training samples were obtained by analyzing the national standard smolder fire and open flame data. Since the smoldering fire is quite different from the open flame, the smoldering fire is different from the open fire in defining the relationship between the input and output of the sample point. The important characteristic of smoldering fire is CO concentration, so its training sample is mainly characterized

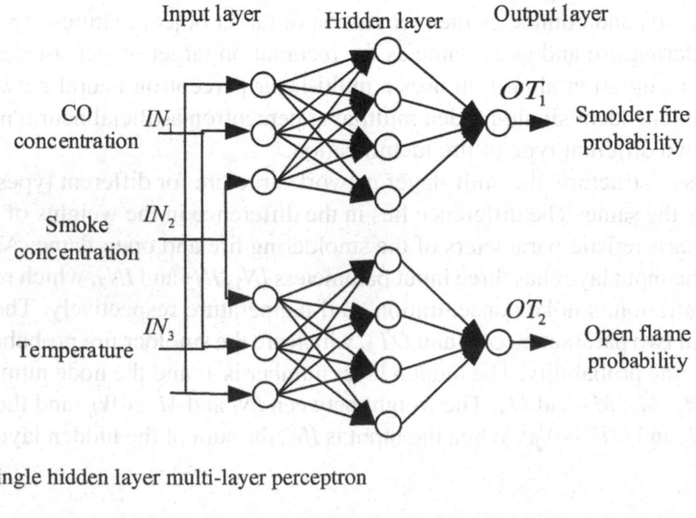


**Fig. 3** Single hidden layer multi-layer perceptron

by CO concentration, smoke and temperature are auxiliary characteristic parameters; the main characteristic parameters of open flame are temperature, CO concentration and smoke as auxiliary characteristic parameters.

## 3.3 *Decision Layer*

The decision-making layer obtains the final decision output of the system by combining the fusion results obtained by the feature layer (open flame probability and smoldering fire probability) with other information [16]. In order to achieve an accurate prediction of the fire situation in the scene, the model introduces a fire risk factor as an element that reflects the site information to supplement the integrity of the fusion analysis results obtained through the feature layer. At the same time, the residual current continues to exceed the limit time to characterize the duration of possible failure of the electrical equipment, and can provide corrections to reduce the impact of fire site disturbance factors. Both of these factors are used as the decision factor for the determination of a fire. As shown in Fig. 4, the model uses fuzzy inference algorithms to make decisions.

(1) Fire risk factor

The fire risk factor represents the probability of a fire in the detection area. The greater the fire risk factor, the more likely it is that the area is prone to fire. Conversely, the less likely it is that a fire will occur [17–19]. The necessary condition for a fire in the detection area is that the material in the field has a combustion condition. Fire-prone field elements are called ignition factors, such as high temperature, overcurrent, electric spark, etc.; fire burning medium is called combustion factor, such as gasoline, paper, cotton articles, etc. Therefore, the fire risk factor can be divided into ignition risk factor and combustion performance factor.

$$R_i = \frac{M_i N_i}{C} \tag{7}$$

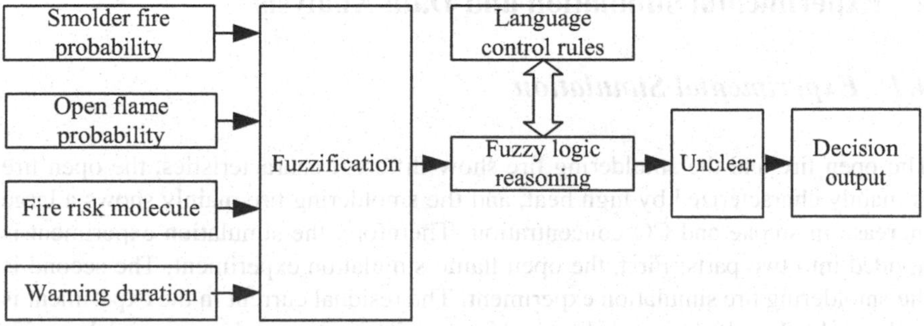


**Fig. 4** Fuzzy reasoning decision model

In the formula, $R_i$ is the fire risk factor corresponding to the $i$ region, $M_i$ is the combustion performance factor corresponding to the $i$ region, $N_i$ is the ignition risk factor corresponding to the $i$ region, $C$ is the fire risk assessment factor, and the value should satisfy $C \geq \max(M_i, N_i)$.

(2) Warning duration

In the early stages of an electrical fire, the residual current in the electrical line continued to increase and gradually exceeded the set threshold. At this time, the time was started and the early warning system was triggered to issue an alarm. The longer the alarm time, the greater the possibility of an electrical fire and the higher the risk of fire in the corresponding area.

(3) Fuzzy Reasoning Decision Module

This module uses fuzzy inference rules to make fuzzy inference at the input of the decision-making layer to obtain the final system decision. Inputs to this module include the smolder fire probability ($OT_1$), the open flame probability ($OT_2$), the fire risk factor ($R_i$), and the residual current warning duration ($T$). These parameters are blurred to 3 levels, the possibility of a small fire (PL), the possibility of fire (PM), the possibility of fire (PH). The decision-making output ($O$) is blurred into 4 levels, namely, no fire (NH), warning (ZL), alarm (ZM), and severe fire (ZH).

Language control rules and fuzzy logic inference are the core of the fuzzy inference module. The inference rule adopted by the system is the Mamdani method, and the language control rules are determined according to the relationship between the actual signal characteristics. For example, in areas of fire risk, the probability of smoldering fire is large, the probability of open flame is small, and the duration of warning is medium, and the output of the system will be an alarm.

The electrical fire detection model uses the three-layer structure of information layer, feature layer and decision layer to fuse and identify the data of each detection parameter of the electrical fire, which not only preserves the integrity of the data, but also accurately identifies the characteristics of the fire in the data. It satisfies the reliability and real-time nature of fire detection.

# 4 Experimental Simulation and Data Analysis

## 4.1 Experimental Simulation

The open fire and the smoldering fire show different characteristics, the open fire is mainly characterized by high heat, and the smoldering fire mainly shows a large increase in smoke and CO concentration. Therefore, the simulation experiment is divided into two parts: First, the open flame simulation experiment. The second is the smoldering fire simulation experiment. The residual current in the experiment is collected by the self-designed electrical fire monitoring host and can control the residual current overrun time. The experimental simulation platform adopts MATLAB.

**Table 1** Experimental data of standard fire

| | | | Open fire | | Smolder fire | | |
|---|---|---|---|---|---|---|---|
| Smoke | CO | Temperature | Expected | Actual | Expected | Actual | Decision layer output |
| 0.24 | 0.53 | 0.85 | 0.78 | 0.79 | 0.12 | 0.11 | ZH |
| 0.41 | 0.46 | 0.82 | 0.82 | 0.65 | 0.11 | 0.09 | ZM |
| 0.31 | 0.72 | 0.96 | 0.85 | 0.83 | 0.13 | 0.14 | ZH |
| 0.15 | 0.65 | 0.78 | 0.75 | 0.75 | 0.09 | 0.10 | ZH |
| 0.23 | 0.53 | 0.93 | 0.80 | 0.78 | 0.10 | 0.09 | ZH |
| 0.11 | 0.48 | 0.82 | 0.72 | 0.71 | 0.15 | 0.13 | ZH |
| 0.26 | 0.42 | 0.86 | 0.72 | 0.73 | 0.16 | 0.12 | ZH |
| 0.33 | 0.45 | 0.90 | 0.86 | 0.85 | 0.12 | 0.14 | ZH |
| 0.44 | 0.56 | 0.81 | 0.64 | 0.84 | 0.10 | 0.12 | ZM |
| 0.35 | 0.54 | 0.86 | 0.67 | 0.85 | 0.13 | 0.15 | ZM |

The fire data used in the simulation experiment comes from the national standard open fire SH4 and the standard negative fire SH1 data set.

(1) Standard open fire simulation experiment and data analysis

From the standard open flame SH4, randomly select 10 sets of fire data, normalized and sent to the neural network fusion device to obtain the smolder fire probability $OT_1$ and the open flame probability $OT_2$, and compared with the corresponding expected probability. The output of the decision layer is basically the same. The data is shown in Table 1. It can be seen that the effect of the open flame experiment is very satisfactory.

(2) Standard smolder fire simulation experiment and data analysis

From the standard smoldering fire SH1, randomly select 10 sets of data and perform the same simulation as in Experiment 1. Similarly, the smolder fire probability and the open flame probability are obtained and compared with the corresponding expected probability. The experimental data is shown in Table 2. The output of the decision-making layer of the third and eighth groups of data is no fire, and the output of the other decision-making layers of other groups of data has forecasted the ignition fire to varying degrees. It can be seen that the system can also be smoldering in the vast majority of cases. Fire can also be well recognized.

From the above-mentioned open flame and smoldering fire simulation experiments, the system characteristic layer can effectively identify open fire type fire characteristics, and output a warning (ZM) or a serious alarm (ZH) at the decision layer even if there is strong interference in the environment. The signal does not produce false positives. The fire characteristics of most types of smoldering fires can also be effectively identified, and the corresponding fire decision information is output at the decision-making level. It can be seen that the electrical fire detection

**Table 2** Experimental data of standard smoldering fire

| | | | Open fire | | Smolder fire | | |
|---|---|---|---|---|---|---|---|
| Smoke | CO | Temperature | Expected | Actual | Expected | Actual | Decision layer output |
| 0.58 | 0.53 | 0.20 | 0.13 | 0.20 | 0.62 | 0.55 | ZL |
| 0.62 | 0.60 | 0.16 | 0.20 | 0.18 | 0.72 | 0.71 | ZM |
| 0.54 | 0.72 | 0.23 | 0.16 | 0.17 | 0.68 | 0.72 | ZM |
| 0.53 | 0.50 | 0.12 | 0.10 | 0.06 | 0.45 | 0.35 | NH |
| 0.73 | 0.70 | 0.18 | 0.14 | 0.12 | 0.80 | 0.81 | ZM |
| 0.65 | 0.62 | 0.16 | 0.23 | 0.20 | 0.79 | 0.82 | ZL |
| 0.59 | 0.75 | 0.25 | 0.15 | 0.16 | 0.78 | 0.79 | ZM |
| 0.63 | 0.60 | 0.24 | 0.13 | 0.12 | 0.81 | 0.80 | NH |
| 0.82 | 0.65 | 0.18 | 0.14 | 0.15 | 0.85 | 0.84 | ZM |
| 0.76 | 0.68 | 0.26 | 0.20 | 0.19 | 0.83 | 0.85 | ZM |

**Table 3** Time and accuracy of each model

| | Waste time (s) | Accuracy (%) |
|---|---|---|
| Traditional model | 0.016 | 75.6 |
| Residual current based detection model | 0.053 | 95.7 |
| Fault arc based detection model | 0.476 | 94.8 |

model can greatly improve the accuracy of fire detection, effectively avoid the false alarm rate of fire alarms. Therefore, the design of the detection model is reliable and can accurately determine the electrical fire.

## *4.2 Comparison Between Different Models*

The traditional electrical fire detection model is a general fire detection model which is judged only by the integration of smoke concentration, CO concentration, and temperature. The complexity is low, but the accuracy is greatly reduced. The electrical fire detection model based on fault arc detection is based on the traditional fire detection model, which detects fault arcs to achieve accurate monitoring of electrical fires. However, the detection of fault arcs is complex and time-consuming, and does not apply to emphasize real-time fire monitoring environment. Comparing the simulation results of this model with the other two models, the simulation data is shown in Table 3, which shows that the residual current based detection model has lower complexity and higher accuracy.

## 5 Conclusion

This paper proposes an improved electrical fire detection model based on data fusion, aiming at the characteristics of electrical fire and detection principle. By detecting whether the residual current in the electrical circuit exceeds the limit, it is possible to quickly determine whether an electrical fire may occur. At the same time, fire characteristic parameters such as smoke concentration, CO concentration, and temperature are collected at the information layer and sent to the feature layer for fusion recognition. Smoldering fire and the probability of open flame, and finally the introduction of on-site fire risk factors and the duration of the residual current limit alarm as decision factors at the decision-making level, integrate more field information, and help the system to detect and prevent fires more specifically. Improve the ability of the system to identify electrical fires. Compared with the detection system that judges electrical fire by collecting fault arcs. The detection model determines the possibility of electrical fire by detecting whether the residual current exceeds the limit, with lower complexity and higher sensitivity. It is very suitable for the accurate detection of electrical fires.

## References


1. X. Liu, Application of data fusion technology in highway tunnel fire detection. Transp. Constr. Manag. **8**, 222–224 (2015)
2. F. Li, Application of multi-sensor information fusion technology in fire detection. Sci. Mosaic **8**, 88–90 (2013)
3. S.G. Kong, D. Jin, S. Li, H. Kim, Fast fire flame detection in surveillance video using logistic regression and temporal smoothing. Fire Saf. J. **79** (2016)
4. G.S. Walia, A. Gupta, R. Kapoor, Intelligent fire-detection model using statistical color models data fusion with Dezert Smarandache method. Int. J. Image Data Fusion **4**(4), 324–341 (2013)
5. T.H. Chen, P.H. Wu, Y.C. Chou, An early fire-detection method based on image processing, in *International Conference on Image Processing*, vol. 3 (IEEE, 2005), pp. 1707–1710
6. Z.G. Liu, Y. Yang, X.H. Ji, Flame detection algorithm based on a saliency detection technique and the uniform local binary pattern in the YCbCr color space. SIViP **10**(2), 277–284 (2016)
7. S. Verstockt, S. Hoecke, P. Potter et al., Multi-modal time-of-flight based fire detection. Multimed. Tools Appl. **69**(2), 313–338 (2014)
8. S. Verstockt, P.D. Potter, S.V. Hoecke, et al., Multi-sensor fire detection using visual and time-of-flight imaging, in *IEEE International Conference on Multimedia and Expo* (IEEE, 2011) pp. 1–6
9. Q.F. Yu, *Electric Fire Forecasting System Based on Wavelet Analysis and Data Fusion and its Application* (Yanshan University, 2013)
10. Z. Zhang, X. Kaili, Z. Li, Research on multi-sensor fire detection technology based on PNN algorithm. J. Sci. Technol. Fire Prot. **36**(10), 1404–1406 (2017)
11. L. Xie, *Research on Signal Processing of Residual Current in Electrical Fire* (Wuhan University of Technology 2013)
12. S. Tian, Design of intelligent electric fire alarm system based on neural network. Fire Sci. Technol. **9**, 1201–1204 (2015)
13. P. Xie, B. Liu, Neural network data fusion algorithm for fire detection. J. Yanshan Univ. **25**(1), 84–87 (2001)

14. C. Xiong, W. Ma, W. Liao et al., Research on intelligent electric fire alarm system. Commun. Tech. **45**(2), 19–21 (2012)
15. S. Liu, J. Lv, S. Wang, Fire detection based on data fusion and neural network. Appl. Sci. Technol. **38**(5), 9–12 (2011)
16. T. Xiong, Analysis of big data processing and application of electrical fire forecast system. Contemp. Educ. Pract. Teach. Res. E-news (7) (2017)
17. W. Wang, D. Ma, Q. Jiang et al., Research on electric arc fire detection methods. Autom. Instrum. **34**(6), 16–19 (2013)
18. Y. Jia, Robust control with decoupling performance for steering and traction of 4WS vehicles under velocity-varying motion. IEEE Trans. Control Syst. Technol. **8**(3), 554–569 (2000)
19. Y. Jia, Alternative proofs for improved LMI representations for the analysis and the design of continuous-time systems with polytopic type uncertainty: a predictive approach. IEEE Trans. Autom. Control **48**(8), 1413–1416 (2003)

# A Novel Multi-exposure Image Fusion Approach Based on Parameter Dynamic Selection

**Yuanyuan Li, Mingyao Zheng, Hexu Hu, Huan Wang and Zhiqin Zhu**

**Abstract** This paper propose a parameter dynamic selection approach for multi-exposure image fusion (MEF) that based on image cartoon-texture and structural patch decomposition. The image texture component is obtained by using texture-cartoon decomposition from the input image. The dynamic parameter is achieved by calculating the image texture entropy. The image patch is divided into three conceptually independent components by using structural patch decomposition. Respectively processing and fusing these three components, a fusion patch and aggregate fused patches are reconstruct into a fused image. This novel MEF method achieves dynamic parameter selection by utilizing texture-cartoon decomposition to obtain fusion images with more details.

**Keywords** Multi-exposure image fusion · High dynamic range imaging
Parameter dynamic selection · Structural patch decomposition

## 1 Introduction

Natural scenes contain luminance levels that span a very high dynamic range (HDR). Its visual information is difficult to be captured by a normal camera. Multi-exposure image fusion (MEF) solved this problem by taking multiple images of the same scene under different exposure levels and synthesizes an low dynamic range (LDR) image from them that is expected to be more informative.

Most existing MEF algorithms are pixel-wise methods, which can be divided into two main domains: transform domain and spatial domain. Transform domain-based fusion methods decompose source images into transform domain and select the corresponding fusion coefficients from source images to generate a set of new fusion coefficients. The fused image is obtained by inverse transformation. The most

Y. Li · M. Zheng (✉) · H. Hu · H. Wang · Z. Zhu
School of Automation, Chongqing University of Posts and Telecommunications, Chongqing 400065, China
e-mail: ZMYzhengmingyao@126.com



Y. Jia et al. (eds.), *Proceedings of 2018 Chinese Intelligent Systems Conference*, Lecture Notes in Electrical Engineering 528,
https://doi.org/10.1007/978-981-13-2288-4_40

MEF methods are the multi-scale and sparse representation-based methods. In which, Mertens et al. [1] proposed a multi-resolution based on Laplacian pyramid to fuse a exposure sequence into a HDR image. The weighted value determined by contrast, saturation and well-exposedness take a weighted average to obtain pyramid coefficient and recontract it to get the final image. Although eliminating the gap phenomenon, it will introduce halo phenomenon at the edge. Wang et al. [2] proposed a novel shift-invariant and rotation-invariant steerable pyramid-based exposure fusion algorithm to reduce the influence induced by the misalignment. This method improved the fusion effect of the severely underexposed part of the image. Wang et al. [3] used sparse representation method to encode multi-exposure images by approximate K-SVD dictionary. Shen et al. [4] proposed a exposure fusion approach based on the hybrid exposure weight and the novel boosting Laplacian pyramid (BLP) to consider the gradient vectors between different exposure source images. This method used the improved Laplacian pyramid to decompose input signal into detail and base layers. Then we can obtain a fusion image with rich color and detail information.

Spatial domain-based MEF extract original information in different exposure source images to obtain fusion image. The most spatial domain-based MEF methods are the image patch and image pixel-based fusion methods. Shen et al. [5] proposed a probabilistic model for MEF. Based on probabilistic model, Shen et al. [6] proposed another MEF method based on perceptual quality measure. This method uses contrast and color information to model the probability of human visual system, and derives the optimal fusion weighs by Hierarchical multivariate Gaussian conditional Random Field. This method can improve the performance of MEF, and provide a better visual experience for the audience. Hara et al. [7] determined the global and local weights via gradient-based contrast maximization and image saliency detection. Oh et al. [8] proposed a MEF image algorithm that based on the rank-1 structure of LDR images. This algorithm formulates HDR generation as a rank minimization problem which simultaneously estimates a set of geometric transformations to align LDR images and detects both moving objects and under/over-exposed regions. In patch-based fusion algorithm, Song et al. [9] proposed a fusion method to suppress gradient inversion by integrating local adaptive scene details. Bertalmio and Levine [10] proposed a variational method combining with color matching and gradient direction information. They introduced an energy functional consisting of two terms, the first one measuring the difference in edge information with the short-exposure image and another measuring the local color difference with a warped version of the long-exposure image. This method is able to handle camera and subject motion as well as noise. Combining the principle of the structural similarity (SSIM) index [11], Ma et al. [12] proposed a structural patch decomposition (SPD) based MEF (SPD-MEF) for MEF. For each patch, the SPD-MEF method first decomposition it into three conceptually independent components: signal strength, signal structure and mean intensity. Then, each component is fused according to patch structure, exposedness and structural consistency measures. Comparing with the multi-exposure image fusion algorithm based on transform domain, the MEF based on spatial domain directly fuses multiple images with different exposures. This spatial domain based method contain less detail information, such as image brightness, color, contrast, texture, etc that can improve the human visual performance.

In this paper, based on image cartoon-texture decomposition and structural patch decomposition [12], we propose a novel MEF method which named parameter dynamic selection based MEF (PDS-MEF). Moreover, we apply texture-cartoon decomposition for input images to obtain image texture components. We decompose an image patch into three conceptually independent components by using structural patch decomposition (SPD) [12] method. This novel MEF method achieves dynamic parameter selection by utilizing texture-cartoon decomposition to obtain fusion images with more details.

The rest of the paper is organized as follows. Section 2 presents in detail the PDS-MEF algorithm. Section 3 is experimental part and we conclude the paper in Sect. 4.

# 2 Parameter Dynamically Selection for MEF

In this section, the detail of the proposed parameter dynamically selection (PDS) framework for MEF is exhibited. The schematic diagram of the proposed fusion framework is shown in Fig. 1. Firstly, the image texture component is obtained by texture-cartoon decomposition, then parameter dynamic selection is achieved by calculating the image texture entropy. Secondly, patches are extracted from the source sequence by using a moving window with a dynamic stride D, and each patch is decomposed into three conceptually independent components by using structural patch decomposition method. Next, respectively processing and fusing these three components, a fusion patch is reconstructed. Finally, the image fusion is achieved by aggregating fused patches.

## *2.1 Image Cartoon-Texture Decomposition*

Cartoon-texture decomposition can decompose images into texture and cartoon components, which mainly describe the detailed information and structure information, respectively [13–15]. First, input images are decomposed into cartoon and texture components in proposed fusion framework. In this work, VeseCOsher (VO) model is implemented for image decomposition [13]. The VO model is shown in Eq. (1):

$$\inf_{u,\mathbf{g}} \left\{ VO_P(u, \mathbf{g}) = |u|_{TV} + \lambda \, \|f - u div(\mathbf{g})\|_{L^2}^2 + \mu \|\mathbf{g}\|_{L^P} \right\} \tag{1}$$

where $\mathbf{g} = (g_1, g_2)$ is a vector in G space to represent digital images. $\lambda$ and $\mu$ are regularization parameters. $u$ represents the cartoon component of image. $f$ is the input image. $\|\mathbf{g}\|_{L^P}$ represents $L_P$ norm of $\mathbf{g}$, which can be calculated by Eq. (2):

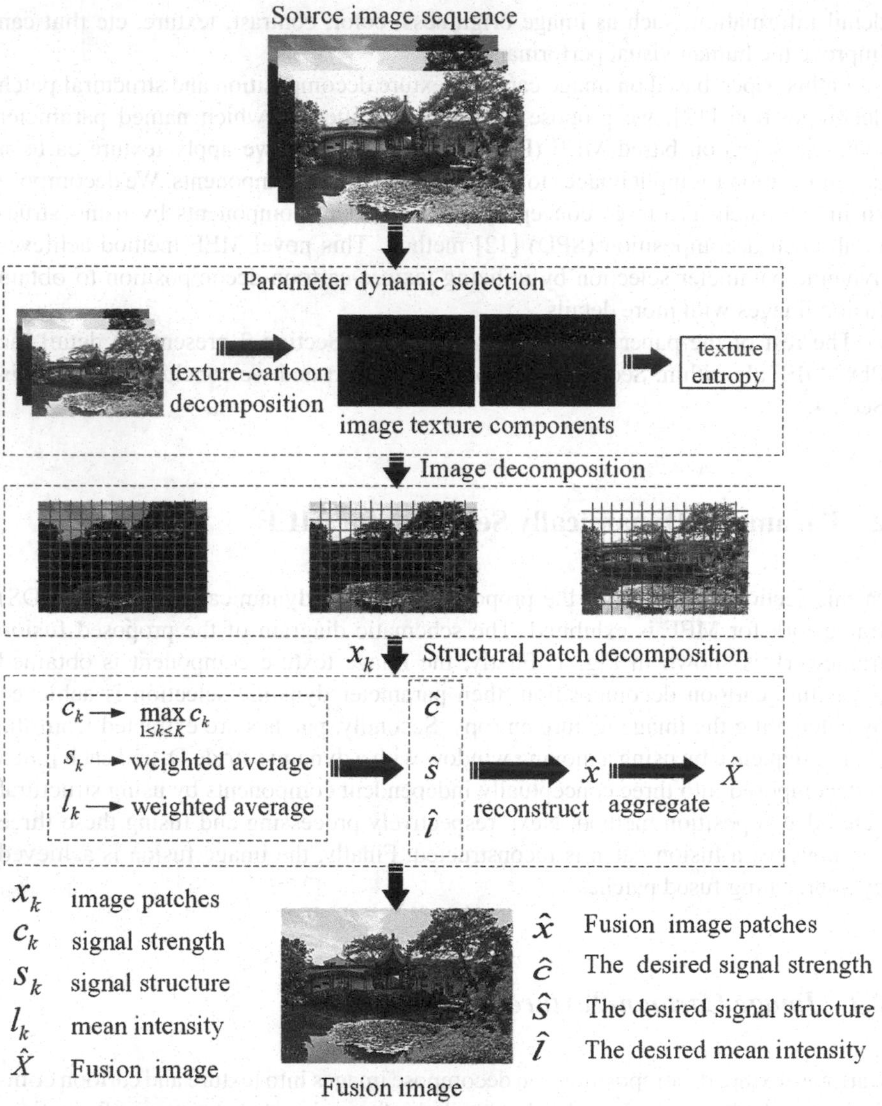


**Fig. 1** The schematic diagram of the proposed fusion framework

$$\|\mathbf{g}\|_{L^P} = \left[\int \left(\sqrt{g_1^2 + g_2^2}\right)^P dxdy\right]^{\frac{1}{P}} \tag{2}$$

This model can fast calculate cartoon component $u$ of image $f$ by setting p between 1 to 10. When the cartoon component $u$ is calculated, the texture component $v$ can be simply calculated by $v = I - u$.

## 2.2 Parameter Dynamically Selection

In this method, the image texture component is converted into a grayscale image. The grayscale difference statistical method is used to obtain the entropy value of the image texture feature to achieve the dynamic parameter selection.
**Step I**: $(x, y)$ represents a point in the image. $(x + \Delta x, y + \Delta y)$ means a point with a quite small distance from $(x, y)$, of which grayscale value can be represented as Eq. (3):

$$g_\Delta(x, y) = g(x, y) - g(x + \Delta x, y + \Delta y) \tag{3}$$

where $g_\Delta$ represents gray-value differential.
**Step II**: Assumed that all possible values of grayscale difference have $m$ levels. Let $(x, y)$ move through the entire image and count the number of times each value of $g_\Delta$. From this, a histogram of $g_\Delta$ is obtained. The probability value of each gray-level difference obtained from histogram statistics is $p(i)$, the entropy of the image texture can use Eq. (4) to indicate:

$$ent = -\sum_{i=0}^{m} p(i)log_2[p(i)] \tag{4}$$

**Step III**: This paper iterates the above process for all input images to obtain all image texture feature entropies $\{ent_1, ent_2, \ldots, ent_n\}$. where n denotes the number of input multi-exposure images.

**Step IV**: According to the image texture feature entropy value, it can achieve dynamic parameters selection, shown as Eq. (5):

$$wSize = \begin{cases} wSize/2 - 3 \times \frac{1}{n}\sum_{i=1}^{n} ent_i, & if \frac{1}{n}\sum_{i=1}^{n} ent_i \le 3 \\ 2 \times \frac{1}{n}\sum_{i=1}^{n} ent_i, & if 3 < \frac{1}{n}\sum_{i=1}^{n} ent_i < 5 \\ wSize + \frac{1}{n}\sum_{i=1}^{n} ent_i, & if 5 < \frac{1}{n}\sum_{i=1}^{n} ent_i < 7 \\ wSize + 3 \times \frac{1}{n}\sum_{i=1}^{n} ent_i, & if \frac{1}{n}\sum_{i=1}^{n} ent_i \ge 7 \end{cases} \tag{5}$$

where $wSize = 16$ denotes the patch size.

## 2.3 Structural Patch Decomposition

**Step I**: Patches from the source sequence are extracted by using a moving window with a dynamic stride D.

**Step II**: An image patch is decomposed into three conceptually independent components by using structural patch decomposition: signal strength $c_i$, signal structure $s_i$, and mean intensity $l_i$.
**Step III**: As for the component of signal strength, the highest signal strength that is selected of all source image patches determine the desired signal strength of the fused image patch. Shown as Eq. (6).

$$\hat{c} = \max_{1 \leq i \leq n} c_n = \max_{1 \leq i \leq n} \|\tilde{x}_i\| \tag{6}$$

As for signal structure, the desired structure of the fused image patch is expected to best represent structures of all source image patches. Hence, weighted average processing is performed on the signal structure $s_i$.

$$\hat{s} = \frac{\bar{s}}{\|\bar{s}\|} \tag{7}$$

where the definition of $\bar{s}$ is shown as Eq. (8):

$$\bar{s} = \frac{\sum_i^n S(\bar{x}_i) s_i}{\sum_i^n S(\bar{x}_i)} \tag{8}$$

where $S(\cdot)$ is a weighting function that determines the contribution of each source image patch in the structure of the fused image patch. This power weighting function is given by Eq. (9):

$$S(\bar{x}_i) = \|\bar{x}_i\|^4 \tag{9}$$

With regard to the mean intensity of the local patch, we take a similar form of Eq. (8)

$$\hat{l} = \frac{\sum_{i=1}^n L(\mu_i, l_i) l_i}{\sum_{i=1}^n L(\mu_i, l_i)} \tag{10}$$

where $L(\cdot)$ is a weighting function that given by Eq. (11)

$$L(\mu_i, l_i) = \exp(-\frac{(\mu_i - 0.5)^2}{2\sigma_g^2} - \frac{(l_i - 0.5)^2}{2\sigma_l^2}) \tag{11}$$

where $\sigma_g = 0.2$ and $\sigma_l = 0.5$ respectively control the spreads of the profile along $\mu_i$ and $l_i$ dimensions.
**Step IV**: Once $\hat{c}$, $\hat{s}$ and $\hat{l}$ are computed, they uniquely define a new vector.

$$\hat{x} = \hat{c} \cdot \hat{s} + \hat{l} \tag{12}$$

**Step V**: This paper iterates the above II–IV process for all source image sequence to obtain all the fused patches. Fused patches are aggregated into fused image.

## 3 Experimental Results

In our experiments, 25 pairs of multi-exposure source sequences are applied for fusion performance testing. Multi-exposure source sequences for testing are collected by Ma et al. [16] and can be downloaded at www.kedema.org. All the experiments are programmed in Matlab 2016a on an Intel (R) Core (TM) i7-7700k CPU @ 4.20 GHz Desktop with 16.00 GB RAM. Experimental results on the multi-exposure images are shown as Figs. 2 and 3.

Figures 2 and 3 show two examples of PDS-MEF fusion result. In Fig. 2a–c are under-exposure, normal- and over-exposure source images respectively, Fig. 2d is the fusion image. Comparing with source image (a), the fusion image preserves the color information appearance on the sky, and enhances the global intensity. Comparing with source image (b), the fusion image decreases the global intensity, and enhances the local contrast. Comparing with source image (c), the fusion image with clearer texture detail information appearance on the water surface. Therefore, the PDS-MEF produces more natural and vivid color appearance on the sky and the road surface. Moreover, it does a better job on structure preservation around the over-exposure area. Meanwhile the overall appearance of the fused image is more appealing.

In Fig. 3a–c are under-exposure, normal- and over-exposure source images respectively, Fig. 3d is the fusion image. Comparing with source image (a), the fusion image enhances the global intensity, and with more detail information. Comparing with source image (b), the fusion image preserves the texture information appearance on the wall, and enhances the local contrast. Comparing with source image

**Fig. 2** Results of proposed exposure fusion method

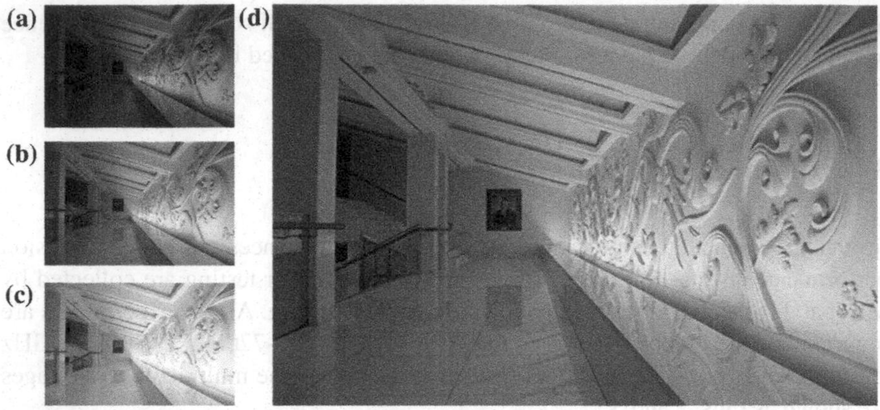


**Fig. 3** Results of proposed exposure fusion method

(c), the fusion image decreases the global intensity, and with clearer texture detail information appearance on the photo frame and the wall. Therefore, the PDS-MEF can preserve more texture and color detail information, and produce natural and vivid color appearance on the fusion image. Moreover, it does a better job on texture preservation around the over-exposure area. Meanwhile the overall appearance of the fused image is more appealing.

## 4 Conclusions

This paper proposes a novel MEF method to achieve parameter dynamic selection. In the method, the dynamic parameter selection is achieved by calculating the image texture entropy, the structural patch decomposition method is used to decompose an image patch. The details of the proposed fusion method based on parameter dynamic selection are first analyzed theoretically. Then the advantage of the proposed fusion method experimentally verified. In our experiments, 25 pairs of multi-exposure source sequences are applied for fusion performance testing. Experimental results demonstrate our proposed MEF method can preserve the texture and color detail information, and produce MEF images with sharp details, vivid color appearance.

In the future, we will improve computational efficiency and enhance global intensity of the fusion image. Particularly, the PDS-MEF method will be continuously optimized for HDR images fusion, and we will explore its potential use in HDR reconstruction to generate high quality HDR images with little ghosting artifacts.

## References


1. T. Mertens, J. Kautz, F. Van Reeth, Exposure fusion: a simple and practical alternative to high dynamic range photography. Comput. Gr. Forum 161–171 (2009)
2. J. Wang, D. Xu, B. Li, Exposure fusion based on steerable pyramid for displaying high dynamic range scenes. Spie Opt. Eng. **48**(11) 117003–1170010 (2009)
3. J. Wang, H. Liu, N. He, Exposure fusion based on sparse representation using approximate K-SVD. Neurocomputing **135**(135), 145–154 (2014)
4. J. Shen, Y. Zhao, S. Yan, X. Li, Exposure fusion using boosting Laplacian pyramid. IEEE Transactions on Cybernetics **44**(9), 1579–1590 (2014)
5. R. Shen, I. Cheng, J. Shi, A. Basu, Generalized random walks for fusion of multi-exposure images. IEEE Trans. Image Process. Publ. IEEE Signal Process. Soc. **20**(12), 3634–3646 (2011)
6. R. Shen, I. Cheng, A. Basu, QoE-based multi-exposure fusion in hierarchical multivariate Gaussian CRF. IEEE Trans. Image Process. Publ. IEEE Signal Process. Soc. **22**(6), 2469–2478 (2012)
7. K. Hara, K. Inoue, K. Urahama, A differentiable approximation approach to contrast-aware image fusion. IEEE Signal Process. Lett. **21**(6), 742–745 (2014)
8. T.H. Oh, J.Y. Lee, T.Y. Wing, I.S. Kweon, Robust high dynamic range imaging by rank minimization. IEEE Trans. Pattern Anal. Mach. Intell. **73**(6), 1219–1232 (2015)
9. M. Song, D. Tao, C. Chen, J. Bu, J. Luo, C. Zhang, Probabilistic exposure fusion. IEEE Trans. Image Process. Publ. IEEE Signal Process. Soc. **21**(1), 341–357 (2012)
10. M. Bertalmio, S. Levine, Variational Approach for the Fusion of Exposure Bracketed Pairs. IEEE Trans. Image Process. Publ. IEEE Signal Process. Soc. **22**(2), 721–723 (2013)
11. Z. Wang, A.C. Bovik, H.R. Sheikh, E.P. Simoncelli, Image quality assessment: from error visibility to structural similarity. IEEE Trans. Image Process. **13**(4), 600–612 (2004)
12. K. Ma, H. Li, H. Yong, Z. Wang, D. Meng, L. Zhang, Robust Multi-exposure image fusion: a structural patch decomposition approach. IEEE Trans. Image Process. **26**(5), 2519–2532 (2017)
13. Z.Q. Zhu, H. Yin, Y. Chai, Y. Li, G. Qi, A novel multi-modality image fusion method based on image decomposition and sparse representation. Inf. Sci. **432**, 516–529 (2017)
14. Z. Zhu, Y. Chai, H. Yin, Y. Li, Z. Liu, A novel dictionary learning approach for multi-modality medical image fusion. Neurocomputing **214**, 471–482 (2016)
15. K. Wang, G. Qi, Z. Zhu, Y. Chai, A novel geometric dictionary construction approach for sparse representation based image fusion. Entropy **19**(7), 306 (2017)
16. K. Ma, Z. Duanmu, H. Yeganeh, Z. Wang, Multi-exposure image fusion by optimizing a structural similarity index. IEEE Trans. Comput. Imaging **4**(1), 60–72 (2017)

# Train Velocity Tracking Control with Considering Wheel-Rail Adhesion

**Zhechen Wang and Yingmin Jia**

**Abstract** Velocity tracking plays a key role in train safe operation. Furthermore, the complex relation between wheels and rail affects velocity tracking. How to obtain a precise tracking algorithm with considering wheel-rail adhesion is a hard problem. This paper designs controllers in two cases respectively. The first algorithm is obtained based on the known wheel-rail adhesion model, while the second one is given with considering the unknown parameters of adhesion model. The simulation results comfirm the availability of the proposed controllers.

**Keywords** Velocity tracking · Wheel-rail adhesion · Uncertainties
Sliding mode control

## 1 Introduction

Build road before you build wealth, that is to say the importance of transportation in the development of economics. Furthermore, the rail transport, which is able to offer safe operation, high loading capacities, cost-efficient travelling and accurate arrival time, plays a key role in development of national transportation. As one of the significant theme, velocity speeding-up raises higher requirements for train control which increasingly become the focused point of vital concernment for researchers. One of the significant problem is cruise control which aims to achieve high-precise velocity tracking [1].

In the last few decades, numerous research focused on train cruise control or speed tracking control. The proposed methods were established on two different motion

Z. Wang · Y. Jia (✉)
The Seventh Research Division and the Center for Information and Control,
School of Automation Science and Electrical Engineering,
Beihang University (BUAA), Beijing 100191, China
e-mail: ymjia@buaa.edu.cn

Z. Wang
e-mail: yipin_yipin@163.com



Y. Jia et al. (eds.), *Proceedings of 2018 Chinese Intelligent Systems Conference*,
Lecture Notes in Electrical Engineering 528,
https://doi.org/10.1007/978-981-13-2288-4_41

models, single-point mass model and multi-body model respectively. Single-point model was a more simplified model which considered the train as a single mass point, while multi-body model considered the interaction between rolling stocks. From the control methods prospective, papers could be classified into three main categories which were called proportional integral derivative (PID) control methods, intelligent control methods and adaptive control methods respectively [2]. PID control was widely applied in traditional papers. Although PID methods was easy to use in practice, the determination of appropriate PID parameters and the adaptive ability improving of control methods were challenging problems. In follow-up studies, intelligent control methods were applied to deal with these problems. In [3], a fuzzy controller was designed for single-mass model (SM) and unit-displacement multi-particle model (UDMP) respectively. In [4], expert system and reinforcement learning were applied to the control system which, in the result, developed the performance of train operation. In addition, adaptive iterative learning control methods were used in [5] to find the control scheme which achieved speed tracking under unknown speed delays and input saturations.

Compared with above methods which were not depend on high-accuracy models, adaptive methods were more effective on realize speed tracking control in complicated system model, especially in the high speed railway with multi-body model. In recent years, different adaptive methods were proposed for train speed control [1, 6–12]. Zhu presented a nonlinear consensus control methods for locomotive control in [8]. Dong used the backstepping design and dynamic surface control methods to deal with the cooperative control of multiple trains in [9]. In [10, 11], decentralized control were proposed for multiple train control, where the input constraints and communication delays were taken into consideration in [10] and event-triggered strategy were designed in [11].

Most of existing research models are dedicated to designing a control law which can be directly used to drag the train forward. However, the engine of trains provide driving torque to the wheels, and the traction force is actually the adhesion force which is produced by complicated interactions between rail and rotating wheels [12]. Therefore, it is critical to calculate the proper driving torque value, instead of traction force, on the purpose of achieving train speed tracking.

The interaction between wheel and rail is nonlinear, which is affected by many factors such as temperature, humidity, surface roughness, surface cleanliness, etc [13]. Considering the control problems with adhesion, most of scholars have been interested in the re-adhesion control or the optimal slip ratio control. Xu et al. [14] proposed an optimal slip ratio control scheme by using robust sliding mode control. In [12], a super-twisting sliding mode algorithm was presented to complete the optimal slip ratio tracking in four different cases. In [15], a comparison between these two methods was made. Different from above research, Cai et al. [13] took advantage of the Barrier Lyapunov Function method to achieve the optimal slip ratio tracking, which makes the slip ratio convert to the bounded region near the target value with the precondition of the train operation stability. While the limitation of this method is that the target value can only be obtained by experience. Due to the difficulty of obtaining some states and parameters values in wheel-rail system, large amount of papers were dedicated to overcoming this problem by designing observers. A disturbance observer and a closed-loop force estimator were presented in [16, 17] respectively. In addition, Zhang proposed a cascading sliding mode observer to estimate the adhesion coefficient [19]. The objective of the re-adhesion control is to ensure the slip ratio converts to the optimal value in order to maximal the traction or braking force. However, the desired velocity of train can be different from the maximum available velocity of train. As a consequence of various limitations during the train cruise, velocity tracking is required for us to ensure the safe operation.

In this paper, a backstepping adaptive controller is proposed to deal with thc train control problem, and we use sliding mode method to realize the uncertainty disturbance attenuation. In contrast with earlier researches, the method in this work can achieve the velocity tracking with considering the rail-wheel nonlinear relationship. As a result, we are able to control the wheel torque directly to stable the train operation.

## 2 Problem Formulation

Generally, a train motion model can be established by the Newton's law of motion, which is expressed as

$$M\dot{v} = F - F_r \tag{1}$$

where $M$ is the mass of train, $v$ is train velocity, $F$ is traction force and $F_r$ is the basic resistance including aero-dynamic force and rail resistance, which is described, in this paper, as the Davis formula [18]

$$F_r = a + bv + cv^2 \tag{2}$$

where $a$, $b$, $c$ are the unknown constant, which can be determined approximately in practice. In the actual train driving, the train motor driving the wheels of the locomotive for freight train or motor cars for high speed trains. When the wheels rotate, the adhesion force is produced and drives the car forward. Thus, the wheel driving model and the wheel-rail model should be showed thereinafter.

The wheel driving model is described as

$$J\dot{\omega}_i = T_i - F_i \cdot R \tag{3}$$

where $J$ is the inertia of each wheel, $R$ is the wheel radius, $\omega_i$ is the angular velocity of the $i$th wheel, $T_i$ is the driving torque of wheels and $F_i$ is the adhesion force between the $i$th wheel and the rail. It is necessary to notice that the adhesion force $F$ in (1) is the total adhesion force between wheels and rail, that is to say $F = \sum_{i=1}^{n} F_i$. Without loss of generality, $n = 1$ is assumpted in the following of this paper.

The adhesion force is shown as

**Fig. 1** Burckhardt model

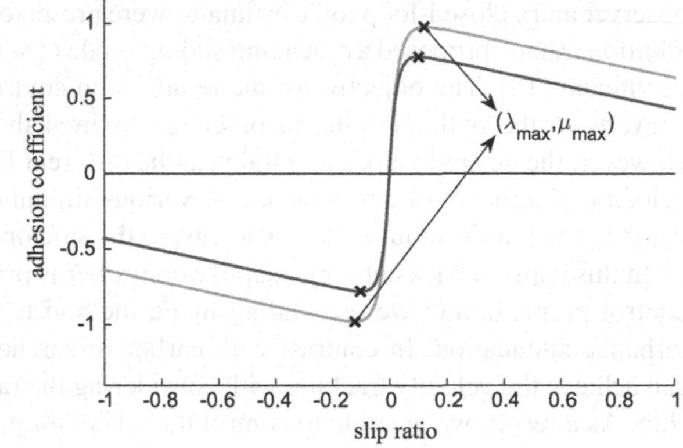


$$F = \mu(\lambda) \cdot M \cdot g, \tag{4}$$

where $g$ is the gravity constant and $\mu(\lambda)$ is the adhesion coefficient, which is a function of slip ratio $\lambda$

$$\lambda = \frac{\omega R - v}{v}. \tag{5}$$

The relationship between adhesion coefficient and ratio slip is described by Burckhardt model [20]

$$\mu(\lambda) = \begin{cases} \alpha_1 \left(1 - e^{-\alpha_2 \lambda}\right) - \alpha_3 \lambda, & \lambda \geq 0, \\ -\alpha_1 \left(1 - e^{\alpha_2 \lambda}\right) - \alpha_3 \lambda, & \lambda < 0, \end{cases} \tag{6}$$

where $\alpha_1, \alpha_2, \alpha_3$ are variational coefficients in different conditions. It is obvious that, $\mu(\lambda)$ is an unimodel function in the first quadrant and the third quadrant respectively. As is shown in Fig. 1, when $|\lambda| \leq \lambda_{max}$ which is also called stable region, the wheel-rail system is stable. In contrast, $|\lambda| > \lambda_{max}$ means the wheel skid happens. Thus, the wheel-rail system keeping in the stable region is desired.

The objective of this paper is to find a control scheme to realize the velocity tracking and the wheel-rail system run in stable region simultaneously.

## 3 Controller Design and Stability Analysis

From (1), and (4), we obtain

$$\dot{v} = \mu(\lambda) g - \frac{F_r}{M} \tag{7}$$

and from (3) and (5), we find

$$\begin{aligned} \dot{\mu}(\lambda) &= \frac{d\mu}{d\lambda} \cdot \dot{\lambda} \\ &= \frac{d\mu}{d\lambda} \cdot \left[ \frac{R}{Jv} \cdot T - \frac{MR^2 + J(\lambda+1)}{Jv} \cdot g\mu(\lambda) + \frac{1}{Mv} \cdot F_r \right] \end{aligned} \tag{8}$$

Define the velocity tracking error as

$$e_v = v - v_d \tag{9}$$

and the adhesion coefficient tracking error as

$$e_\mu = \mu - \mu_d, \tag{10}$$

where $v_d$ is the desired velocity and $\mu_d$ is a virtual control law respectively.

In order to ensure the tracking error (9) converges to zero asymptotically, we differential (9) and obtain

$$\begin{aligned} \dot{e}_v &= \dot{v} - \dot{v}_d \\ &= \mu(\lambda) g - \frac{F_r}{M} - \dot{v}_d + \mu_d g - \mu_d g \\ &= g e_\mu + \left( g\mu_d - \frac{F_r}{M} - \dot{v}_d \right). \end{aligned} \tag{11}$$

The virtual control law $\mu_d$ can be defined as

$$\mu_d = \frac{1}{g} \left( \frac{F_r}{M} + \dot{v}_d - k_v e_v \right). \tag{12}$$

Plug (12) into (11), we have

$$\dot{e}_v = g e_\mu - k_v e_v. \tag{13}$$

It is indispensable to notice that the virtual control law $\mu_d$ is actually the desired adhesion coefficient which ensures the velocity tracking error (9) converges to zero asymptotically when $e_\mu = 0$. In fact, as is mentioned above, adhesion coefficient $\mu$ must satisfy the condition $|\mu| \leq \mu_{\max}$. Therefore, the virtual control law should satisfy the limitations

$$k_l \leq k_v \leq k_u \tag{14}$$

where

$$k_l = \max \left\{ \frac{F_r/M + \dot{v}_d - \mu_{\max} g}{\gamma}, -\frac{F_r/M + \dot{v}_d + \mu_{\max} g}{\gamma} \right\}$$

$$k_u = \min \left\{ -\frac{F_r/M + \dot{v}_d - \mu_{\max} g}{\gamma}, \frac{F_r/M + \dot{v}_d + \mu_{\max} g}{\gamma} \right\}$$

and $|e_v| \leq \gamma$.

Considering the tracking error (10). Differential both side of the equation, we have

$$\begin{aligned}\dot{e}_\mu &= \dot{\mu}(\lambda) - \dot{\mu}_d \\ &= \frac{d\mu}{d\lambda} \cdot \dot{\lambda} - \frac{1}{g}\left(\frac{\dot{F}_r}{M} + \ddot{v}_d - k_v g e_\mu + k_v^2 e_v\right)\end{aligned} \tag{15}$$

To ensure the asymptotical stability of $e_\mu$, the driving torque $T$ is designed as

$$T = \frac{Jv}{R}\left[\frac{MgR^2 + Jg(\lambda+1)}{Jv} \cdot \mu - \frac{F_r}{Mv} + \frac{\dot{\mu}_d - k_\mu e_\mu}{d\mu/d\lambda}\right] \tag{16}$$

where $k_\mu$ is the designed parameter.

**Theorem 1** *Suppose that the velocity tracking error is initially bounded by $|e_v(0)| \leq \gamma$, and the adhesion coefficient is initially bounded by $|\lambda(0)| \leq \lambda_{max}$. If the controller (16) with the limitation conditions (14) and (17) is applied to the error dynamic system (11) and (7), then $e_v$ converges to zero asymptotically (i.e., $e_v \to 0$ as $t \to \infty$) and $|\lambda| \leq \lambda_{max}$ is always guaranteed.*

$$k_v, k_\mu > \frac{1}{2}g \tag{17}$$

*Proof* Choosing a Lyapunov candidate as

$$V_1 = \frac{1}{2}e_v^2 + \frac{1}{2}e_\mu^2 \tag{18}$$

Obviously, $V_1 \geq 0$, and $V_1 = 0$ if and only if $e_v = e_\mu = 0$. Taking the time derivative of $V_1$ and substituting (11), (16) and (17) into the result lead to

$$\begin{aligned}\dot{V}_1 &= e_v\dot{e}_v + e_\mu\dot{e}_\mu \\ &= e_v(ge_\mu - k_v e_v) + e_\mu\left(\frac{d\mu}{d\lambda} \cdot \dot{\lambda} - \dot{\mu}_d\right) \\ &= -k_v e_v^2 + g e_v e_\mu \\ &\quad + e_\mu\left(\frac{d\mu}{d\lambda} \cdot \left(\frac{R}{Jv} \cdot T - \frac{MR^2 + J(\lambda+1)}{Jv} \cdot g\mu(\lambda) + \frac{F_r}{Mv}\right) - \dot{\mu}_d\right) \\ &= -k_v e_v^2 + g e_v e_\mu - k_\mu e_\mu^2 \\ &\leq -\left(k_v - \frac{1}{2}g\right)e_v^2 - \left(k_\mu - \frac{1}{2}g\right)e_\mu^2 \leq 0\end{aligned} \tag{19}$$

which implies that the velocity tracking error and slip ratio error both converges to zero asymptotically. Notice that $|e_v(0)| \leq \gamma$ ensures $|e_v| \leq \gamma$ which further guarantees the satisfication of limitations (14) and (17). As is shown in Fig. 1, $|\mu_d| \leq \mu_{max}$

is always satisfied under the limitation conditions, in other words, $|\lambda| \leq \lambda_{max}$ is guaranteed if $|\lambda(0)| \leq \lambda_{max}$.

# 4 Controller Design Considering Parameters Uncertainty

Considering the uncertainties in the Davis formular (2), the parameters $a$, $b$, and $c$ are variant in different environment. We make hypothesis that the Davis formular can be expressed as

$$F_r = (a^* + \Delta a) + (b^* + \Delta b)\, v + (c^* + \Delta c)\, v^2 \tag{20}$$

where $a^*, b^*, c^*$ are the determined constant, $\Delta a, \Delta b, \Delta c$ are the variant uncertainties which satisfy the bounded condition $|\Delta a| \leq \bar{a}, |\Delta b| \leq \bar{b}, |\Delta c| \leq \bar{c}$. The existence of uncertainty inevitably bring negative effect on the stability of control performance. To overcome this problem, a sliding mode control scheme is presented in this paper. Define the sliding mode surface $s$ as

$$s := \dot{e}_v + k e_v \tag{21}$$

where $k > 0$. The control law T is redesigned as

$$T = \frac{Jv}{gR \cdot d\mu/d\lambda}\left[T_1 + \ddot{v}_d + k\dot{v}_d + \frac{\dot{F}_r^*(v)}{M} - \left(\frac{d\mu}{d\lambda} \cdot \frac{g}{Mv} - \frac{k}{M}\right) F_r^*(v) + T_n\right] \tag{22}$$

where $T_1 = \left(\frac{d\mu}{d\lambda} \cdot \frac{MgR^2 + Jg(\lambda+1)}{Jv} - k\right)\mu g$, $F_r^* = a^* + b^* v + c^* v^2$, $T_n = -\beta sign\,(s)$, $\left|\frac{\dot{F}_r^* - \dot{F}_r}{M} + \left(\frac{k}{M} - \frac{g\frac{d\mu}{d\lambda}}{Mv}\right)\left(F_r^* - F_r\right)\right| \leq \beta_0 < \beta$ and $sign\,(\cdot)$ is the signal function.

**Theorem 2** *Suppose that the velocity tracking error is initially bounded by* $|e_v(0)| \leq \gamma$, *and the adhesion coefficient is initially bounded by* $|\lambda(0)| \leq \lambda_{max}$. *If the controller (22) with the limitation (14) is applied to the error dynamic system (11) and (7) where the dynamic resistance is described as (20), then* $e_v$ *converges to zero asymptotically (i.e.,* $e_v \to 0$ *as* $t \to \infty$*) and* $|\lambda| \leq \lambda_{max}$ *is always guaranteed.*

*Proof* Choosing a Lyapunov candidate as

$$V_2 = \frac{1}{2}s^2 \tag{23}$$

Obviously, $V_2 \geq 0$, and $V_2 = 0$ if and only if $s = 0$. Taking the time derivative of $V_2$ and substituting (11), (22) and (17) into the result lead to

$$\begin{aligned}
\dot{V}_2 &= s\dot{s} \\
&= s\left[g\frac{d\mu}{d\lambda}\left(\frac{R}{Jv}T - \frac{MR^2 + J(\lambda+1)}{Jv}g\mu + \frac{F_r}{Mv}\right)\right. \\
&\quad \left. -\ddot{v}_d - k\dot{v}_d + k\mu g - \frac{\dot{F}_r}{M} - \frac{kF_r}{M}\right] \\
&= s\left[\frac{\dot{F}_r^* - \dot{F}_r}{M} + \left(\frac{k}{M} - \frac{g\frac{d\mu}{d\lambda}}{Mv}\right)\left(F_r^* - F_r\right) - \beta sign\,(s)\right] \\
&\leq -(\beta - \beta_0)\,|s| \leq 0
\end{aligned} \tag{24}$$

which implies that the sliding mode surface $s$ converges to 0 asymptotically. And then, we obtain $\dot{e}_v + ke_v = 0$, which ensures that $e_v$ convergers to zero asymptotically. Notice that $\dot{s} = \dot{\mu}g - \frac{\dot{F}_r}{M} - \ddot{v}_d + k\dot{e}_v = g\dot{e}_\mu$ if $k = k_v$. As is confirmed in Theorem 1, $|e_v(0)| \leq \gamma$ ensures the limitation (14) which further guarantees the satisfication of $|\lambda| \leq \lambda_{max}$.

## 5 Simulation Results

To verify the effectiveness of the proposed controller, simulation results are shown in this section. The parameters of system model are choosen from Chen's paper [12] which are given as: the mass of train $M = 10000$ (kg), the inertia of each wheel $J = 36\ (\text{N s}^2)$, the radius of each wheel $R = 0.31$ (m), the nominal coefficient of David formular $a = 1600$ (N), $b = 32\ (\text{N s m}^{-1})$, $c = 0.63\ (\text{N s}^2\ \text{m}^{-2})$, the gravity constant $g = 9.8\ (\text{m s}^{-2})$, the coefficients of Burckhardt model $\alpha_1 = 0.3302$, $\alpha_2 = 40$, $\alpha_3 = 0.2419$. To express the uncetainty of the David formula's coefficient in (20), the oscillation excitation force from the rail is added to $F_r$. Therefore the basic

**Fig. 2** Velocity tracking

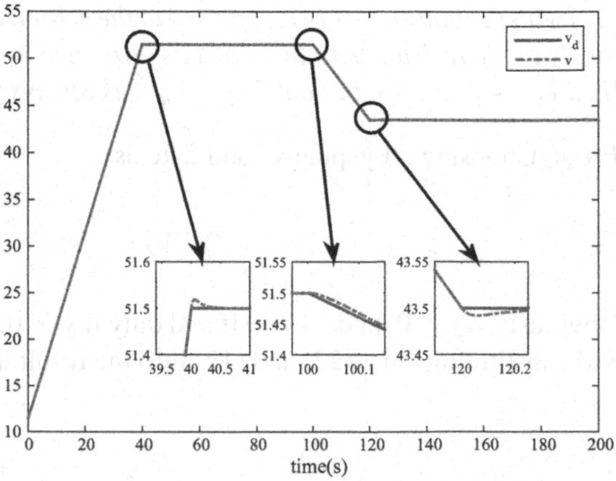

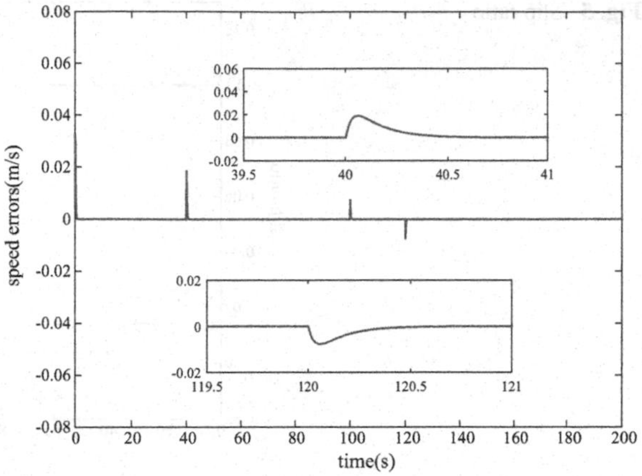


**Fig. 3** Velocity tracking error

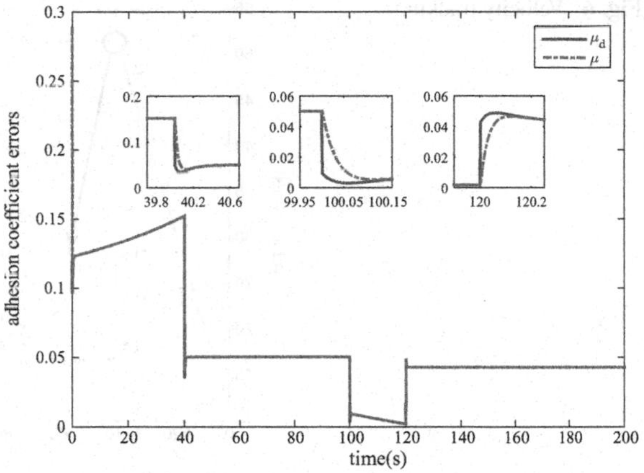


**Fig. 4** Adhesion coefficient error

resistance is given as

$$F_r = a + bv + cv^2 + 1500\sin(0.02vt). \tag{25}$$

The initial states of the train system are set as: $\omega(0) = 40\,(\mathrm{rad/s})$, $v(0) = 11.5$ (m/s). The reference velocity and acceleration are set as: $v_{ref}(0) = 11.5\,(\mathrm{m/s})$, $\dot{v}_{ref} = 1\left(\mathrm{m/s^2}\right)$ if $0 \le t \le 40\,(\mathrm{s})$, $\dot{v}_{ref} = 0\left(\mathrm{m/s^2}\right)$ if $40 < t \le 100\,(\mathrm{s})$, $\dot{v}_{ref} = -0.8\left(\mathrm{m/s^2}\right)$ if $100 < t \le 120\,(\mathrm{s})$, $\dot{v}_{ref} = 0\left(\mathrm{m/s^2}\right)$ if $120 < t \le 200\,(\mathrm{s})$. The control coefficients are set as: $k = 15$, $\beta = 10$ for controller (22) and $k_v = 8$, $k_\mu = 30$ for controller (16).

The performance of controller (16) without considering the uncertainty of train system is shown as Figs. 2, 4 and 5. Figure 2 indicates that the velocity tracking performs well at each mode of train moving. The velocity tracking error can be observed from Fig. 3, which indicates that the maximum of tracking error is lower than 0.05 m/s in each mode of train moving. As showned in Figs. 2, 3, 4 and 5, there

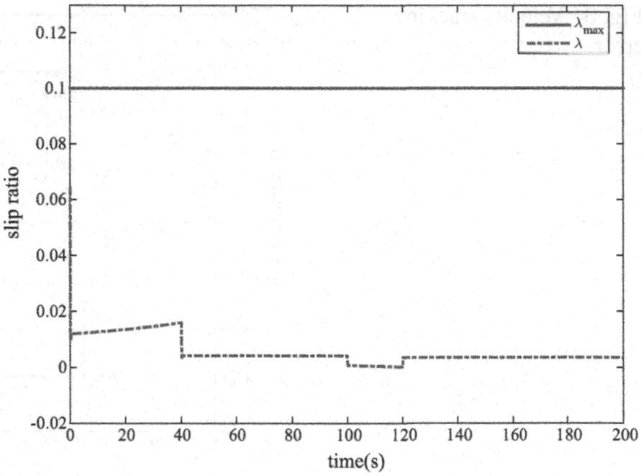


**Fig. 5** Slip ratio

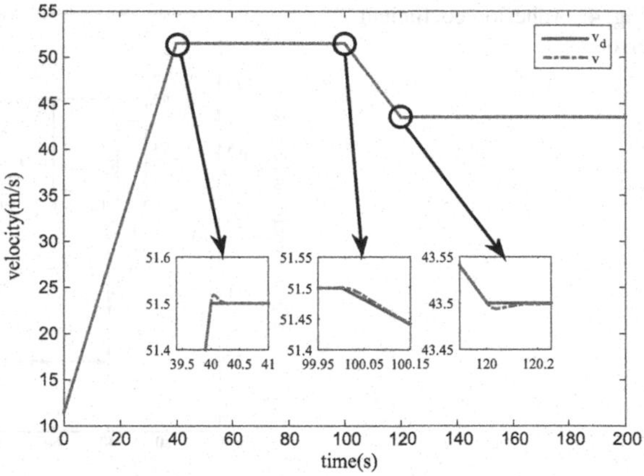


**Fig. 6** Velocity tracking

exist a delay after the transition from one mode to another mode of train moving. This phenomenon is due to a sudden jump of reference acceleration which can not be catched in the essentially discrete simulation. Figure 5 shows that the slip ratio is always in the stable area $|\lambda| \leq \lambda_{max}$.

The performance of controller (22) with considering the uncertainty of train system is shown as Figs. 6, 7 and 8. Different from the controller (16), the velocity tracking error and slip ratio profile jitter are obviously finded from simulation results. The reasons are summarized as follows. Firstly, the oscillation from the model itself leads to the jitter. Secondly, the switching control laws designed by the sliding mode controller causes the flaw of velocity tracking. As is showned in Fig. 8, the slip ratio always stay in stable area $|\lambda| \leq \lambda_{max}$.

In conclusion, the simulation results verifies that: (1) the velocity tracking can be realized in every modes and the transition between one mode to another mode of train moving, and the convergence is able to be ensured simultaneously; (2) the

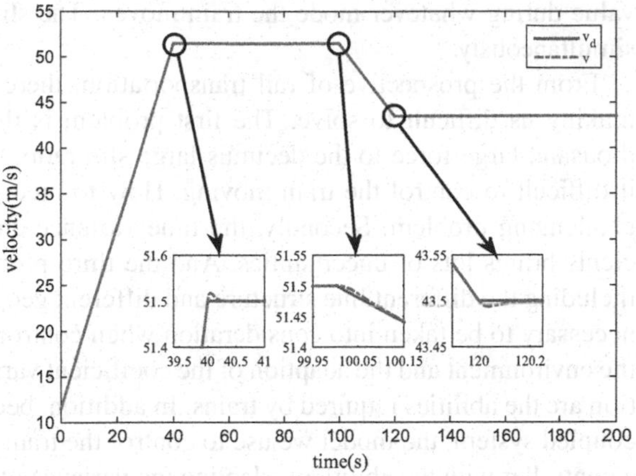

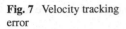

**Fig. 7** Velocity tracking error

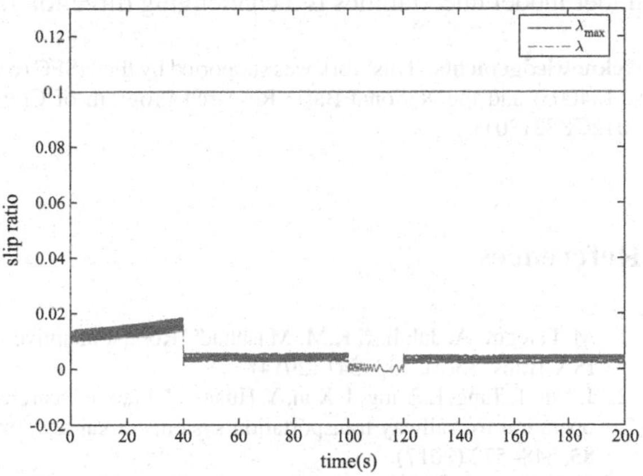


**Fig. 8** Slip ratio

velocity tracking can be realized with or without considering the model uncertainties simultaneously; (3) the velocity tracking error remains low during the control process.

# 6 Conclusions

This paper proposed a backstepping method for the system without considering uncertainties and a sliding mode method for system with uncertainties respectively. These two methods firstly design the velocity tracking controller considering the adhesion relationship between wheels and rail. Compared with traditional methods, controller in this research gives the control law which drives the wheels, rather than the track force on the train body which actually is not practical to realize. The simulation results implicate that the velocity tracking errors remain under a low value during whatever mode the train moves. The slip ratio stay in the stable area simultaneously.

From the prospective of rail transportation, there always exist three problems making us difficult to solve. The first problem is the scale difference. From the thousand-large force to the decimus-large slip ratio, the big scale difference make it difficult to control the train moving. How to precisely give the control law is a challenging problem. Secondly, the time variance characteristic of system coefficients brings lots of uncertainties. And the third problem is the spacial invariance including the different line structure and different geographical conditions, which is necessary to be taken into consideration when controlling trains. The perception of the environment and the adaption of the coefficient variance according to the perception are the abilities required by trains. In addition, because of the complexity of the coupled system, the model we use to control the train is not precise. How to design a controller with the ability of adapting the variance of environment and a controller under model uncertainties is a challenging direction in future work.

**Acknowledgements** This work was supported by the NSFC (61327807, 61521091, 61520106010, 61134005) and the National Basic Research Program of China (973 Program: 2012CB821200, 2012CB821201).

## References

1. M. Faieghi, A. Jalali, S.K.M. Mashhadi, Robust adaptive cruise control of high speed trains. ISA Trans. **53**(2), 533–541 (2014)
2. J. Yin, T. Tang, L. Yang, J. Xun, Y. Huang, Z. Gao, Research and development of automatic train operation for railway transportation systems: a survey. Transp. Res. Part C Emerg. Technol. **85**, 548–572 (2017)
3. H.-R. Gao, S.-G. Dong, B. Ning, L, Li, Extended fuzzy logic controller for high speed train. Neural Comput. Appl. **22**(2), 321–328 (2013)
4. J. Yin, D. Chen, L. Li, Intelligent train operation algorithms for subway by expert system and reinforcement learning. IEEE Trans. Intell. Transp. Syst. **15**(6), 2561–2571 (2014)
5. H. Ji, Z. Hou, R. Zhang, Adaptive iterative learning control for high-speed trains with unknown speed delays and input saturations. IEEE Trans. Autom. Sci. Eng. **12**(1), 260–273 (2016)
6. Q. Song, Y.d. Song, T. Tang, B. Ning, Computationally inexpensive tracking control of high-speed trains with traction/braking saturation. IEEE Trans. Intell. Transp. Syst. **12**(4), 1116–1125 (2011)
7. Z. Mao, G. Tao, B. Jiang, X.G. Yan, Adaptive compensation of traction system actuator failures for high-speed trains. IEEE Trans. Intell. Transp. Syst. **18**(11), 2950–2963 (2017)
8. B. Zhu, X. Xia, Nonlinear trajectory tracking control for heavy-haul trains. IFAC-PapersOnLine **48**(11), 41–46 (2015)
9. H. Dong, S. Gao, B. Ning, Cooperative control synthesis and stability analysis of multiple trains under moving signaling systems. IEEE Trans. Intell. Transp. Syst. **17**(10), 2730–2738 (2016)
10. K. Gao, Z. Huang, J. Wang, J. Peng, W. Liu, Decentralized control of heavy-haul trains with input constraints and communication delays. Control Eng. Pract. **21**(4), 420–427 (2013)
11. F. Zhou, Z. Huang, K. Gao, L. Li, H. Liao, J. Peng, Distributed cooperative tracking control for heavy haul trains with event-triggered strategy, in *2016 American Control Conference (ACC)* (2016), pp. 770–775

12. Y. Chen, H. Dong, J.L, X. Sun, L. Guo, A super-twisting-like algorithm and its application to train operation control with optimal utilization of adhesion force. IEEE Trans. Intell. Transp. Syst. **17**(11), 3035–3044 (2016)
13. W.C. Cai, D.Y. Li, Y.D. Song, A novel approach for active adhesion control of high-speed trains under antiskid constraints. IEEE Trans. Intell. Transp. Syst. **16**(6), 3213–3222 (2015)
14. K. Xu, G.-Q. Xu, C.-H. Zheng, Analysis of torque transmitting behavior and wheel slip prevention control during regenerative braking for high speed EMU trains. Acta Mechanica Sinica **32**(2), 244–251 (2016)
15. C. Uyulan, M. Gokasan, S. Bogosyan, Comparison of the re-adhesion control strategies in high-speed train. Proc. Inst. Mech. Eng. Part I J. Syst. Control Eng. **232**(1), 92–105 (2018)
16. L. Diao, L. Zhao, Z. Jin, L. Wang, S.M. Sharkh, Taking traction control to task: high-adhesion-point tracking based on a disturbance observer in railway vehicles. IEEE Ind. Electron. Mag. **11**(1), 51–62 (2017)
17. J. Liu, Q. Peng, Z. Huang, H. Li, D. Wang, Y. Chen, F. LinZhou, A novel estimator for adhesion force of railway vehicles braking systems and reference speed calculation. in *2017 29th Chinese Control and Decision Conference (CCDC)* (2017), pp. 7606–7611
18. J.W. Davis Jr., The tractive resistance of electric locomotives and cars. Gen. Electr. Rev. **29**, 2–24 (1926)
19. C.F. Zhang, J. Sun, J. He, L.F. Liu, Online estimation of the adhesion coefficient and its derivative based on the cascading SMC observer. J. Sens (2017)
20. U. Kiencke, L. Nielsen, *Automotive Control Systems* (Springer, Berlin, Germany, 2000)

# Subspace Clustering Based Association Analysis Between Multiple Process-Variable-Parameters and Faults

**Yuyang Zhong, Ke Zhang and Yi Chai**

**Abstract** Aiming at the problem of large amount of data and low utilization rate in complex industrial systems and processes, an association analysis method of process variables and faults is proposed. Because of the characteristic that large number of process variables and large data volume consist in complex industrial system, a subspace clustering based quantitative association rule mining method is proposed to the association analysis between multiple process-variables and faults. The validity and efficiency of the method is verified by using the fault datasets of TE process.

**Keywords** Multiple process-variable-parameter · Association analysis
Quantitative association rules · Subspace clustering · Data mining · Fault diagnosis

## 1 Introduction

In modern complex industrial systems, most of the equipment will generate a large amount of operating data, which contains knowledge related to the system running status and quality. Generally, these data will be used as the basis on data mining of fault-tolerance control and fault diagnosis. This kind of data is focused by people, and the relevant information will form a dataset of key parameters reflecting overall operating status through certain technical methods. However, these data features are not intuitively obvious. Under the condition that system mechanism model is not clear enough, it is difficult to find effective information contained in the data by qualitative empirical knowledge. Therefore, people are very concerned about effective use of historical process data, and hope to extract knowledge or rules from data mining to discover the causes and laws of faults, thus providing support for improving system security and reliability [1].

Y. Zhong (✉) · K. Zhang · Y. Chai
State Key Laboratory of Power Transmission Equipment and System Security and New Technology, College of Automation, Chongqing University, Chongqing 400044, China
e-mail: yuyang_hotaru@cqu.edu.cn

 
Y. Jia et al. (eds.), *Proceedings of 2018 Chinese Intelligent Systems Conference*, Lecture Notes in Electrical Engineering 528,
https://doi.org/10.1007/978-981-13-2288-4_42

As an important part of data mining, association analysis can objectively mine potential fault rules in large-scale datasets, which can guide the detection and location of faults and predict potential faults [2, 3]. It has become a hot research direction in fault-tolerance control in recent years. For example, Dong proposed a data mining system frame of ship fault diagnosis based on association rules, and discussed the mining method about association rules of ship equipment system's single-layer and multi-layer faults [4]. Li studied the quantitative association rules mining method for liquid rocket engine fault, and applied this method to the measured steady data of a large-scale liquid propellant rocket engine and successfully mined the fault rules [5]. Nie introduced the association rule mining method into the power network fault diagnosis, and results of calculation examples showed that this method was correct and effective [6].

However, in existing research, more cases are focused on the fault rule mining in a single device or a specific object (such as power systems). In addition, there is relatively little association analysis between multiple process-variable-parameters and faults (multiple process-variable-parameters are more universal in complex systems). In view of the problem that the operating process of complex industrial systems will generate a large amount of operating data but utilization rate is low, people are eager to find association rules between process-variable-parameters and faults through association analysis. This paper takes this as the breakthrough point, aiming at the problem of large number of process-variable-parameters and large data volume. After analyzing the common quantitative association rule mining methods, a subspace clustering based quantitative association rule mining method is proposed. It is applied to mining incidence relation between multiple process-variable-parameters and faults. It can provide decision support for fault diagnosis and verify that the method can effectively mine the incidence relation between multiple process-variables and faults by using the fault datasets of TE process.

## 2 Analysis of Quantitative Association Rule Problem of System Operation Process Parameters

During the operation of a complex industrial system, the parameters of its monitoring variables, including key parameters of each equipment and overall operating status of the system, exceed tens or even more. Moreover, after years of operation, the accumulated monitoring data is extremely large. This dimension superposition situation makes general association rule mining methods not able to quickly and effectively find the effective knowledge. Especially in the event of a system fault, the direct sources that can be referenced are these data which are massive and lacking of effective point. Therefore, these original massive data need to be further processed to obtain key information.

As a kind of mature data mining technology, association analysis can mine valuable knowledge describing the incidence relation of variables from a large amount of data. Let the process variable parameter set $P = \{P_1, P_2, \ldots, P_n\}$ and the fault set $F = \{F_1, F_2, \ldots, F_m\}$. In reality, the occurrence of fault is related to some characterization variables of fault, that is, there is such a mapping relation $P = \{P_1, P_2, \ldots, P_n\} \Rightarrow F_j$, where $1 \leq i \leq n$, $1 \leq j \leq m$. If association analysis method is used to mine the mapping relation, that is, association rules, between process-variable-parameters and faults, it can provide assistant decision-making for finding the cause of fault [7].

In the common association analysis, according to the types of processing variables, association rules can be divided into Boolean association rules for discrete and Boolean data, and quantitative association rules for continuous numerical and categorical data. In actual operation process of complex industrial systems, a large number of process monitoring variables are continuous, such as pressure value, liquid level value, etc. However, the more mature ones are Boolean association rule mining methods at present, such as Apriori and FP-growth [8, 9]. Therefore, to conduct association analysis between multiple process-variable-parameters and faults, we must consider quantitative association rule mining problems.

However, massive system operating data raises three requirements for association analysis [10]:

(1) How to mine the quantitative association rules of continuous multiple process-variable-parameters;
(2) How to efficiently mine association rules of large datasets;
(3) How to find connection between fault events and forecast derivative faults during the operation of systems.

The above three requirements also have limitations on the applicable quantitative association rules. At present, mature quantitative association rule mining methods include equal depth division method [11], mining association among quantitative attributes (MAQA) [12], and competitive agglomeration based mining quantitative association rules (CA-MQAR) [13]. Among them:

(1) Equal depth division method. This method divides the N values of a quantitative attribute into M disjoint intervals, each of which contains approximately the same number of data, and then considers merging of adjacent intervals during the generation of association rules. This method can solve the problem of mini-confidence and mini-support to a certain degree. It can quickly obtain the segmentation interval for massive datasets. However, it is difficult to reflect the characteristics of attribute value distribution, so it is not suitable for association analysis of multiple process-variables and faults.
(2) MAQA. This method uses inductive partitioning or clustering analysis for different data types to divide intervals, and uses Apriori to mine frequent item sets after attribute discretization, and then uses frequent item sets to generate interesting association rules. The advantage of this method is that it can reflect the distribution of attribute values. But, because the way of merging adjacent cells is not unique, different classification results may lead to different output association rules, which is not conducive to application of mining results to provide decision support for fault diagnosis.

(3) CA-MQAR. This method uses an optimized competitive agglomeration algorithm to divide numerical attributes into intervals, and then uses Apriori to mine association rules. It has the advantages of partition clustering and hierarchical clustering. It can continuously eliminate the less competitive class according to the competitiveness of class during the iterative process of the method, thus dividing numerical attributes into several optimized intervals. However, for a large data volume fault dataset, the method mining efficiency is not high.

In summary, the existing mature quantitative association rule mining methods have the disadvantages of varying degrees for multiple process-variable-parameters and faults association analysis. Therefore, it is necessary to find a new method to carry out association analysis for multiple process-variable-parameters and faults.

# 3 Association Analysis Between Multiple Process-Variable-Parameters and Faults

From the point of view of existing general association analysis methods, the basic idea of quantitative association rule mining is to transform the quantification association rule problem into Boolean association rule problem, and then use the existing Boolean association rule mining method to obtain interesting rules. The process of this idea is shown in Fig. 1. For the quantitative association rule mining problem of multiple process-variable-parameters and faults, the most important thing is to find an interval division method of multi-dimensional continuous variables, which has a decisive influence on the quality of mined fault rules, and thus affects the subsequent decision of fault diagnosis.

In the process shown in Fig. 1, quantitative attribute discretization is the key to solving the quantitative association rule mining problem. The quantitative attribute discretization is to divide the attribute range into multiple subintervals. Different dividing methods will result in different subranges, which determines the quality of quantitative association rule mining. If the interval is improperly divided, there will be the problem of mini-confidence, mini-support and interval combination explosion. At the same time, there are many monitoring variables in the process of complex industrial systems. Anomalies in one or several process variables will lead to the occurrence of faults, that is, the occurrence of faults is often associated with one or more process variable parameters.

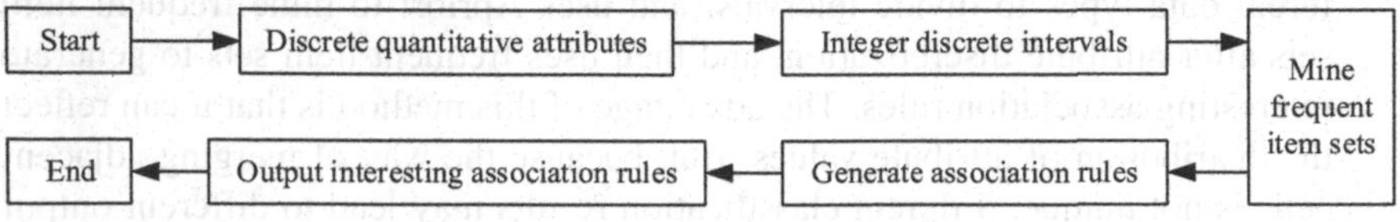


**Fig. 1** The flowchart of quantitative association rules mining

According to general incidence relation between system operating process parameters and faults, the following quantitative association rules are defined:

**Definition 1** (*quantitative association rules*). Set the process variable parameter set $P = \{P_1, P_2, \ldots, P_n\}$, the fault set $F = \{F_1, F_2, \ldots, F_m\}$, the variable $P_i$ value set is $I_i$, then a record $d$ in the fault dataset $D$ is $\{p_1, p_2, \ldots, p_n, f\}$, where $p_i \in I_i,\ 1 \le i \le n, f \in F$.

The value interval $[l_i, u_i]$ of the variable $P_i$ is discretized to divide it into $r$ subintervals $[l_{i1}, u_{i1}],\ [l_{i2}, u_{i2}], \ldots, [l_{ir}, u_{ir}]$, and the *j-th* subinterval $\left[l_{ij}, u_{ij}\right]$ of the variable $P_i$ is denoted as $P_{ij}(1 \le i \le n,\ 1 \le j \le r)$, then the record $d$ in the fault dataset $D$ may be rewritten as $\{P_{1a}, P_{2b}, \ldots, P_{nx}, f\}$, where $P_{1a}$ represents the value $p_1$ of the variable $P_1$ is in the range of its *a-th* subinterval $[l_{1a}, u_{1a}]$.

Then, the quantitative association rules for process-variable-parameters and faults are expressed as $P_{1j} \cap P_{2j} \cap \cdots \cap P_{ij} \Rightarrow F_k(1 \le i \le n,\ 1 \le k \le m)$, which represents the value of the process variable $P_i$ is in the interval $\left[l_{ij}, u_{ij}\right]$, the fault $F_k$ may occur.

**Definition 2** (*support and confidence*). Support $s\%$ of quantitative association rule $P_{ij} \Rightarrow F_k$ is defined as $support\left(P_{ij} \Rightarrow F_k\right) = P\left(P_{ij} \cup F_k\right)$, indicating that $s$ percent of the records in the fault dataset $D$ contain $P_{ij} \cup F_k$, and confidence $c\%$ is defined as $confidence\left(P_{ij} \Rightarrow F_k\right) = P\left(F_k | P_{ij}\right)$, indicating that $c$ percent of the records in the fault dataset $D$ containing $P_{ij}$ also contain $F_k$.

**Definition 3** (*quantitative association rule problem*). The quantitative association rule problem is the process of generating all quantitative association rules that satisfy the minimum support and minimum confidence in a given fault dataset $D$.

From operation process of complex systems, although its structure is complex, its functions are significantly different. Different functional areas are mapped to relatively independent subsystem structures, and certain dataset boundaries are also defined. In this case, in order to facilitate local feature selection of the dataset, the method of subspace clustering [14] is considered in the interval division of multiple process-variable-parameters.

For a bounded and ordered process variable parameter set $P = \{P_1, P_2, \ldots, P_n\}$, then $S = P_1 \times P_2 \times \cdots \times P_n$ is an n-dimensional numerical space, where $P_1, P_2, \ldots, P_n$ is the $n$ dimensions of $S$, and $P_{t_1} \times P_{t_2} \times \cdots \times P_{t_k}\left(k \le n \,\&\, t_i \le t_j\right)$ is a subspace unit of $S$. Assuming that the dimensions of $S$ are divided into $r$ intervals on average, then $S$ is divided into $n^r$ mutually disjoint grid cells, denoted by $v = \left\{ v_i^j \middle| i \in (1, 2, \ldots, n), j \in (1, 2, \ldots, r) \right\}$, which indicates that the *i-th* dimension of the grid cell $v$ is the *j-th* interval on $P_i$.

When the value $p_i$ of each dimension in a fault data record $\{p_1, p_2, \ldots, p_n\}$ is satisfied in the *j-th* interval of the *i-th* dimension, the record point is said to be within the grid cell $v$. Then for a fault data set $D$ containing $n$ process variable parameters, $n^r$ grid cells $v = \left\{ v_i^j \middle| i \in (1, 2, \ldots, n),\ j \in (1, 2, \ldots, r) \right\}$ are associated with fault set $F = \{F_1, F_2, \ldots, F_m\}$. Let the relation $g(\cdot)$ be able to express this incidence

**Table 1** Steps of subspace clustering based quantitative association rule mining method

| |
|---|
| **Input**: fault data set *D*, density threshold *density*, minimum support *min_sup*, minimum confidence *min_conf*, the number of division intervals in each dimension *interval* |
| **Output**: quantitative association rules |
| 1. Scan the fault dataset *D* and construct a high density interval list; |
| 2. Create DGFP-tree, compress the information in fault dataset D and store it to DGFP-tree; |
| 3. Use a "bottom-up" method to search for high-density subspace unit set on the created DGFP-tree; |
| 4. Linking adjacent high-density subspace units to form subspace clustering; |
| 5. Adjust the clustering result to the form of quantitative association rules; |
| 6. Output interesting quantitative association rules |

relation, that is, $F_k = \left\{ v_1^j, v_2^j, \ldots, v_i^j \middle| i \in (1, 2, \ldots, n), j \in (1, 2, \ldots, r) \right\}$. If a certain method is used to solve the representation of $g(\cdot)$, the fault $F_k$ can be mapped to a specific grid cell $v$, that is, the value interval $\left[l_{ij}, u_{ij}\right]$ of the process variable $P_i$ that generates the fault $F_k$ can be found.

Therefore, a subspace clustering based quantitative association rule mining method (it is named SC-MQAR) is proposed, which combines the classic subspace clustering algorithm CLIQUE [15] with the efficient frequent itemset mining algorithm FP-growth [16]. A tree structure DGFP-tree is designed to compress and store information in dataset, and low-dimensional subspace with clustering is discovered by searching path in this tree, thereby transforming the quantitative association rule mining problem into the process of creating a DGFP-tree and using this tree to search for high-density units to form clusters [17]. The main steps of this method are described in Table 1.

In this method, time overhead is mainly used to create DGFP-tree. However, since DGFP-tree fully compresses the dataset, there is no need to scan dataset multiple times, which reduces the overall time complexity of the method. At the same time, in the process of searching for high-density subspace units, there is no need to generate a large number of candidate units, which reduces the space complexity.

# 4 Analysis of Experimental Results

The TE process is a realistic industrial process created by Tennessee Eastman Company. The TE process totally has 52 controllable process variables, including 41 measure variables and 11 operating variables. There are 21 preset faults, including 5 unknown faults and 16 known faults [18–20].

In this paper, SC-MQAR is simulated by using the known fault dataset of the TE process. Experiments show that SC-MQAR can accurately mine quantitative association rules between multiple process-variable-parameters and faults in the TE process. After comparing with the mining efficiency of CA-MQAR, it is found that the mining efficiency of SC-MQAR is greatly improved.

**Table 2** Partial fault data for TE process

| 01 | 02 | 03 | 04 | 05 |
|---|---|---|---|---|
| 0.61110 | 3654.5 | 4515.8 | 9.2627 | 26.776 |
| 0.23364 | 3718.5 | 4678.7 | 9.5413 | 26.750 |
| 0.26154 | 3692.7 | 4509.5 | 9.3630 | 27.256 |
| 0.23848 | 3696.8 | 4595.7 | 9.3569 | 26.804 |
| 0.34527 | 3662.3 | 4266.1 | 9.4362 | 26.923 |
| 0.24571 | 3583.1 | 4486.8 | 9.2782 | 26.855 |
| 0.26228 | 3624.9 | 4527.3 | 9.3483 | 27.051 |

**Table 3** Mining results of SC-MQAR

| Rules | Confidence (%) |
|---|---|
| 29[20.9040, 27.9100] → F | 99.85 |
| 51[43.91, 46.31] → D | 99.53 |
| 02[3747, 3805] → C | 98.66 |
| 52[21.84, 24.72] → E | 97.73 |
| 01[0.6476, 0.8601] ∩ 04[8.1554, 9.1414] → A | 97.23 |
| 09[120.1, 120.3] → K | 93.67 |
| 24[9.2265, 10.0500] → B | 92.68 |
| 30[14.1320, 15.2530] → B | 91.53 |
| 51[41.84, 42.61] → I | 91.13 |
| 21[94.19, 94.39] → O | 90.43 |

The simulation experiment selected 15 datasets with known faults and a total of 7200 fault data. 52 process variables number 01-52 are output as previous items of the rule, and 15 known faults number A-O are output as subsequent items of the rule. Table 2 shows some of the fault data.

Let $min\_sup = 0.1$, $min\_conf = 0.9$, $density = 4$, $interval = 8$, and the finally mined association rules are shown in Table 3.

The first rule in the above table indicates that if the value of process variable 29 is in the range of 20.9040–27.9100, the fault F will occur. The confidence of this rule is 99.85%. If the value of process variable 29 is within this range in a new data record, the fault F may occur at this time.

Through analysis of the TE process model and comparison with fault diagnosis results with other methods of TE process, the above association rule mining results are consistent with it. Therefore, SC-MQAR can effectively mine the incidence relation for process variables and faults in complex industrial systems. These mining association rules can form a fault rule base to provide decision support for future fault diagnosis.

In order to verify the mining efficiency of the proposed method, it is compared with the CA-MQAR. Set $density = 4$, $interval = 8$, and $min\_sup$ is 0.1, 0.2, 0.3, 0.4,

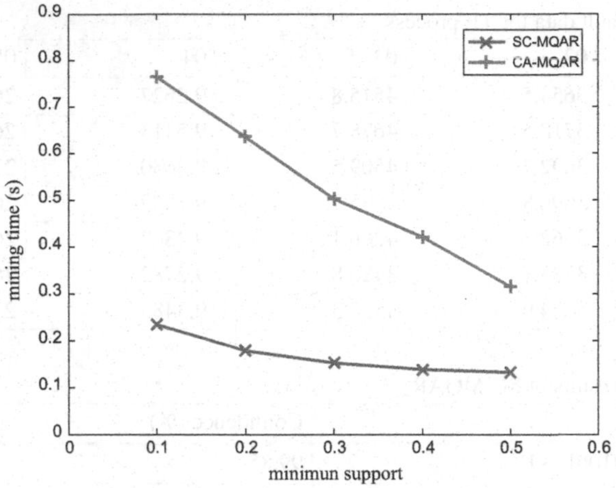


**Fig. 2** Mining time comparison chart of SC-MQAR and CA-MQAR

and 0.5 respectively. The curve of the mining time of SC-MQAR and CA-MQAR is shown in Fig. 2, which shows that the mining efficiency of SC-MQAR is higher.

## 5 Conclusions

Association analysis enables people to use process variable parameter data no longer to stay in simple query and analysis processing. It provides a new method of mining knowledge in data and guiding decision support with knowledge, and also solves the problem of information explosion but lack of knowledge in a large extent. The association analysis between multiple process-variable-parameters and faults can mine the association rules between multiple process-variable-parameters and faults, locate the causes of faults, and prevent the occurrence of faults in the subsequent system operation.

This paper discusses the feasibility and advantages of association analysis between multiple process-variable-parameters and faults, as well as the problems of multiple process variable parameters, large-scale fault data and continuous variables. A subspace clustering based quantitative association rule mining method is proposed to address this problem. Through the simulation experiments on TE process fault datasets, the effectiveness and high efficiency of the proposed method (SC-MQAR) are verified. The mining fault rules can provide decision support for fault diagnosis of complex industrial systems' operation process.

## References


1. Z. Jianming, R. Gang, Fault diagnosis method and research based on association rules. Control Instrum. Chem. Ind. **30**(5), 11–14 (2003)
2. J. Han, M. Kamber, J. Pei et al., *Data Mining: Concepts and Techniques* (China Machine Press, 2012)
3. H. Chunli, S. Jinyan, Application on the fault diagnosis of rotary machinery for the digging methods of related regulations. Huadian Technol. **29**(1), 57–59 (2007)
4. D. Doudou, L. Dengfeng, C. Yuwen, Data mining system frame of ship fault diagnosis based on association rules. Ship Eng. (4):61–64 (2001)
5. L. Li Jinghao, H. Xiaoping, H. Quandong, Mining quantitative association rules of liquid propellant rocket engine. J. Rocket Propuls. **33**(2), 7–11 (2007)
6. N. Qianwen, G. Wei, A power network fault diagnosis method based on data mining association rules. Power Syst. Prot. Control **37**(9), 8–14 (2009)
7. C. Yuting, Z. Baowen, H. Dequan, Research progress on association analysis. Inf. Secur. Commun. Priv. **2**, 84–87 (2010)
8. R. Agrawal, Mining association rule between sets of items in large databese, in *ACM SIGMOD Conference on Management of Data* (1993)
9. Y. He, Q. Ding, Application research of association analysis in business intelligence, in *Asia-Pacific Conference on Information Processing* (2009), pp. 304–307
10. Z. Huanyin, L. Jinsheng, L. Ming et al., Data mining technologies based on fault diagnosis. Microcomput. Inf. **24**(1), 157–159 (2008)
11. C. Zhang, W. Xu, W. Zhang, Analysis on quantitative association rules mining and several typical algorithms. Comput. Program. Skills Maint. **4**, 20–21 (2012)
12. Z. Zhang, Y. Lu, B. Zhang, An algorithm for mining quantitative association rules. J. Softw. **9**(11), 801–805 (1998)
13. Z. Wang, *Data Mining and Condition Recognition on Characteristic Data of Quayside Container Crane* (Shanghai Jiao Tong University, 2008)
14. L. Parsons, E. Haque, H. Liu, Subspace clustering for high dimensional data: a review. ACM SIGKDD Explor. Newsl **6**(1), 90–105 (2004)
15. R. Agrawal, J. Gehrke, D. Gunopulos, P. Raghavan, Automatic subspace clustering of high dimensional data. Data Min. Knowl. Disc. **11**(1), 5–33 (2005)
16. N. Kavitha, S. Karthikeyan, An efficient algorithm for mining frequent k-item sets for association rule mining in large databases. Data Min. Knowl. Eng. (2011)
17. F. Zhang, *The Research on The Algorithm of Mining Quantitative Association Rules* (Xi'an University of Science and Technology, 2010)
18. Z. Wu, *Study of Fault Detection Based on Tennessee-eastman Process* (East China Jiaotong University, 2016)
19. Y. Jia, Robust control with decoupling performance for steering and traction of 4WS vehicles under velocity-varying motion. IEEE Trans. Control Syst. Technol. **8**(3), 554–569 (2000)
20. Y. Jia, Alternative proofs for improved LMI representations for the analysis and the design of continuous-time systems with polytopic type uncertainty: a predictive approach. IEEE Trans. Autom. Control **48**(8), 1413–1416 (2003)

# Simulation Research Based on Asynchronous Motor Vector Control Technology

**Panyuan Ren, Bulai Wang, Xiutao Ji, Xiangsheng Liu and Yuanyuan Yang**

**Abstract** In order to improve the dynamic performance of the three-phase asynchronous motor open-loop control system, a current closed-loop vector control strategy is adopted. In the rotor flux-oriented control system, the closed-loop PI control of the decoupled exciting current and torque current is used to obtain the stator voltage in the dynamic coordinate system; finally, the asynchronous motor is controlled by the SVPWM and the asynchronous motor is realized the control of the flux and speed of the asynchronous motor. The simulation results of the vector control of the asynchronous motor show that the current closed-loop control strategy can control the stable operation of the motor.

**Keywords** Vector control · Closed loop current · SVPWM · PI
Open loop control

## 1 Introduction

Asynchronous motors are modeled in a synchronous rotating coordinate system after the Clarke transformation, Park transformation and rotor flux orientation. The stator current is decomposed into an excitation component and a torque component, and the rotor flux is determined only by the excitation component of the stator current. The electromagnetic torque is determined by the rotor flux linkage and the stator current torque component. The decoupling of the two components of the stator current is achieved. In this paper, the principle of the closed-loop control of the two components of the stator current is used to realize the control of the flux linkage and the rotation speed of the asynchronous motor [1]. Compared with the open-loop control, the current closed-loop vector control method makes the waveform of the asynchronous motor more ideal.

P. Ren (✉) · B. Wang · X. Ji · X. Liu · Y. Yang
School of Electrical and Electronic Engineering,
Shanghai Institute of Technology, Shanghai 201418, China
e-mail: 1721154916@qq.com



Y. Jia et al. (eds.), *Proceedings of 2018 Chinese Intelligent Systems Conference*,
Lecture Notes in Electrical Engineering 528,
https://doi.org/10.1007/978-981-13-2288-4_43

## 2 Mathematical Model

After the orientation of the rotor flux, the equation of state in MT coordinates is obtained [2].

$$\begin{aligned}
\frac{d\omega}{dt} &= \frac{n_p^2 L_m}{J L_r} i_{st}\psi_r - \frac{n_p}{J} T_L \\
\frac{d\psi_r}{dt} &= -\frac{1}{T_r}\psi_r + \frac{L_m}{T_r} i_{sm} \\
\frac{di_{sm}}{dt} &= \frac{L_m}{\sigma L_s L_r T_r}\psi_r - \frac{R_s L_r^2 + R_r L_m^2}{\sigma L_s L_r^2} i_{sm} + \omega_1 i_{st} + \frac{u_{sm}}{\sigma L_s} \\
\frac{di_{st}}{dt} &= -\frac{L_m}{\sigma L_s L_r}\omega\psi_r - \frac{R_s L_r^2 + R_r L_m^2}{\sigma L_s L_r^2} i_{st} + \omega_1 i_{sm} + \frac{u_{st}}{\sigma L_s}
\end{aligned} \tag{1}$$

Slip angle and sync angle frequency in the MT coordinate system:

$$\omega_s = \frac{L_m}{T_r \psi_r} i_{st} \quad \omega_1 = \omega_s + \omega n_p \tag{2}$$

Electromagnetic torque expression:

$$T_e = \frac{n_p L_m}{L_r} i_{st}\psi_r \tag{3}$$

From the above formula, the mathematical model of rotor flux orientation is equivalent to a DC motor model [1].

## 3 Control Principle

The orientation of the rotor flux linkage achieves the decoupling of the stator current excitation component and the torque component setpoint, but there is still coupling in the voltage equation. With the closed-loop control of the current PI, the voltage equation can be decoupled. In this way, the dynamic and static performance of the asynchronous motor speed control system can be equivalent to the DC speed control system [2].

## 3.1 Slip Frequency Control

This paper first gives the control principle of the open-loop control system. Based on the open-loop control system, the closed-loop control is applied and the two are compared.

Slip frequency control is the use of mathematical models of asynchronous motors, in turn, open-loop control of the motor. Control schematic is shown in Fig. 1.

In the open-loop control, the $u_{st}$ voltage of the mathematical model of the asynchronous motor is not only controlled by $i_{st}$, but also controlled by $i_{sm}$. In the same way, the $u_{sm}$ voltage is also controlled by $i_{st}$ and $i_{sm}$. Because the voltages controlled by the currents $i_{st}$ which contain differential terms, the differentials are particularly sensitive to disturbances, which can easily cause the interference voltage of the motor and cause the output waveform of the motor to be unstable [3].

## 3.2 Current Closed Loop Control

The detected three-phase current is subjected to 3/2 transformation and rotation transformation to obtain the $i_{st}$ and $i_{sm}$ in the MT coordinate system. The PI control is used to obtain the $u_{st}$ and $u_{sm}$. After reverse rotation transformation, the sum is obtained. After SVPWM, the three-phase motor is controlled. The control schematic is shown in Fig. 2.

The current PI closed-loop control decouples the stator current excitation component and the torque component from the corresponding voltage component. After decoupling, the destabilization of the voltage component given by the coupling of the current component can be reduced, resulting in a more stable control system. [3–5].

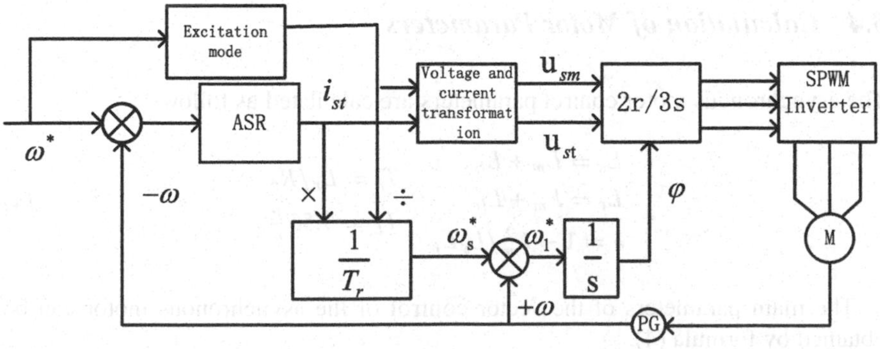


**Fig. 1** Schematic diagram of slip frequency control

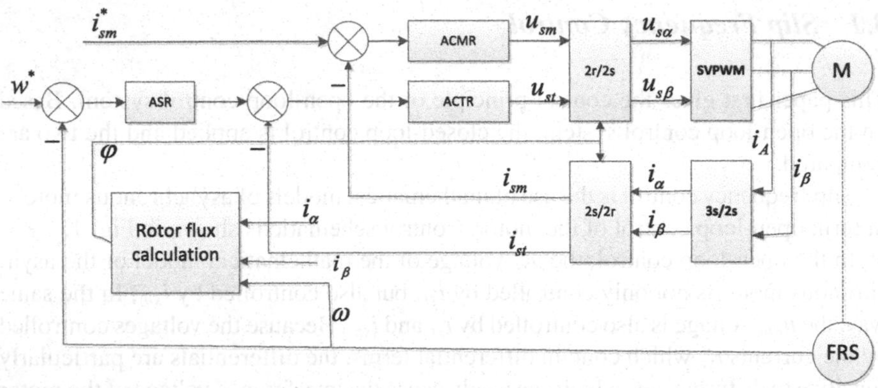


**Fig. 2** Current closed-loop control schematic

## 3.3 SVPWM Principle

The main purpose of SPWM control is to make the output voltage of the inverter as close to a sine wave as possible, and not to pay attention to the waveform of the output current. The ultimate goal of the vector control is to obtain a circular rotating magnetic field, which requires the inverter output current waveform to be close to the sine wave. Locked to get the goal of a circular magnetic field, SVPWM control technology uses different combinations of switching control signals of each bridge arm of the inverter, so that the output trajectory of the output voltage space vector of the inverter is as close as possible to a circle. SVPWM is based on the motor's point of view, focusing on making the motor obtain a circular magnetic field with a constant amplitude [1].

## 3.4 Calculation of Motor Parameters

The asynchronous motor control parameters are calculated as follows:

$$\begin{matrix} L_s = L_m + L_{ls} \\ L_r = L_m + L_{lr} \\ \delta = 1 - L_m^2 / L_s L_r \end{matrix} \quad \begin{matrix} T_r = L_r / R_r \\ T_L = 9.55 \frac{P}{n} \end{matrix} \tag{4}$$

The main parameters of the vector control of the asynchronous motor can be obtained by formula (4).

**Table 1** Motor parameters

| Motor parameters | Parameter value |
|---|---|
| $R_s$ | 1.405 Ω |
| $R_r$ | 1.395 Ω |
| $L_s = L_r$ | $5.839 \times 10^{-3}$ H |
| $L_m$ | 0.1722 H |
| $J$ | $1.31 \times 10^{-2}$ kg m$^2$ |
| $n_p$ | 2 |

# 4 Simulation System Build

Through the above analysis, the MATLAB-based simulation control diagram of the asynchronous motor can be constructed based on the schematic diagram. Use a motor with a rated power of 4 kW. The basic parameters are shown in the Table 1.

## 4.1 *Slip Frequency Control Simulation*

In the case of slip frequency control simulation, the PWM wave generation module is used in the simulation. From the dynamic coordinate system to the coordinate conversion module of the three-phase coordinate system, the given $i_{sm}$ value is calculated by formula (4), and the value $i_{st}$ obtained by slipping the closed loop. Its value can be obtained using the function module, then the voltage function module is established according to the voltage equation, and the voltage of the output is changed by matching with the PWM module's comparison voltage, and then the output voltage amplitude is sent to the PWM module, and then 6 trigger pulses are provided to control the motor [6].

## 4.2 *Current Closed-Loop Control Simulation*

In closed-loop control of current PI, the closed-loop stator current excitation component and closed-loop torque component are added to the simulation compared to the open-loop control. The phase calculation of the rotor flux linkage is added and realized through the coordinate transformation module. The decoupled voltage component is obtained by the closed-loop control of the stator current component and sent to the SVPWM module. The six-phase trigger pulse drives the three-phase bridge inverter to control the motor.

The SVPWM module is mainly divided into four modules, a sector calculation module, a time calculation module, a time cooperation module, and a trigger pulse generation module. In the simulation model, the PWM cycle is set to 0.0002 s, and

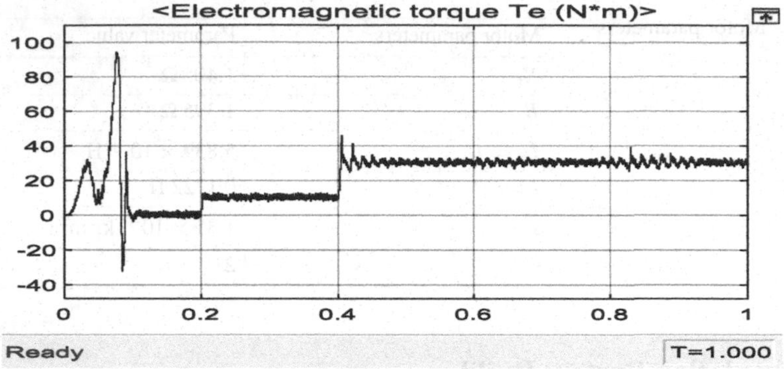


**Fig. 3** Open-loop control torque waveform

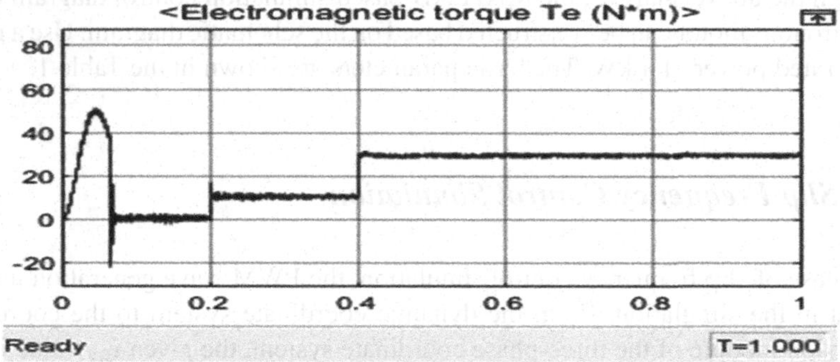


**Fig. 4** Current closed-loop control torque waveform

the DC side voltage is set to 566 V, where the current pulse hysteresis comparator is used in the trigger pulse generation module. Simulations have found that the current hysteresis comparator is more stable than the average torque waveform obtained by the comparator.

## *4.3 Current Closed-Loop Control Simulation*

True simulation system inverter DC voltage 566 V, rated speed 1500 r/min, excitation given value 5.2 A, rated load 25 N m, simulation time is 1 s [1, 2, 7]. A load of 10 N m is added at 0.2 s, and a load of 30 N m is added at 0.4 s.

If the electromagnetic torque waveforms of Figs. 3 and 4 are added to the load of 10 N m at 0.2 s, both control overshoots are smaller and the adjustment time is smaller. When the load is increased by 30 N m at 0.4 s, Larger than the rated load 25 N m, the open-loop control overshoot is larger and the adjustment time is longer. Current closed-loop control response speed faster and more stable.

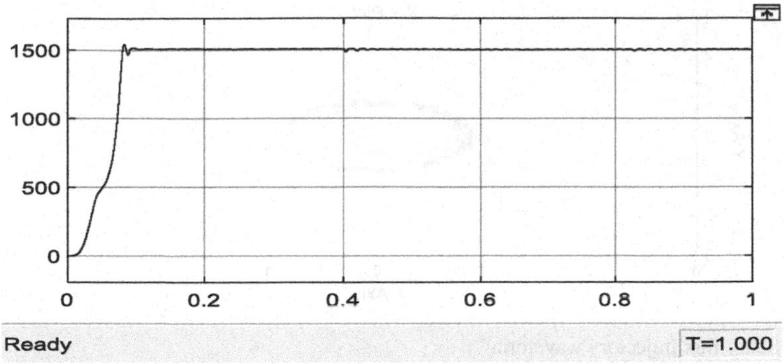


**Fig. 5** Open loop control speed waveform

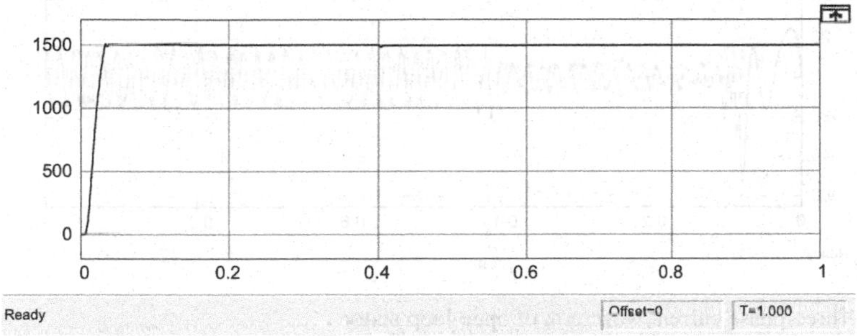


**Fig. 6** Current closed loop control speed waveform

As shown in the speed waveforms of Figs. 5 and 6, the open-loop control reaches the rated speed at 0.08 s, but the adjustment time is very long and the overshoot is large, and the closed-loop control reaches the rated speed at 0.05 s. After the load is added, the closed-loop control of the speed drop is not obvious and returns to the rated speed within a short period of time. After stabilization, the amplitude of the up-and-down fluctuation is at 0.1 rpm; the open-loop control speed drops greatly, and the amplitude of the up-and-down fluctuation after stabilization is 3 turns.

Figure 7 shows the stator flux waveform. The change of the stator flux chain is not regular at first, but the flux linkage becomes a regular circle after pushing the time. When the motor starts, there is a process of establishing the magnetic field. During this process, the flux linkage changes irregularly. The change of the magnetic field affects the torque and the torque fluctuates greatly in this process. When the flux linkage becomes a regular circle, The torque is also almost stable [8, 9]. Figures 8 and 9 show the stator current waveforms. The simulation waveform changes correspond to the torque.

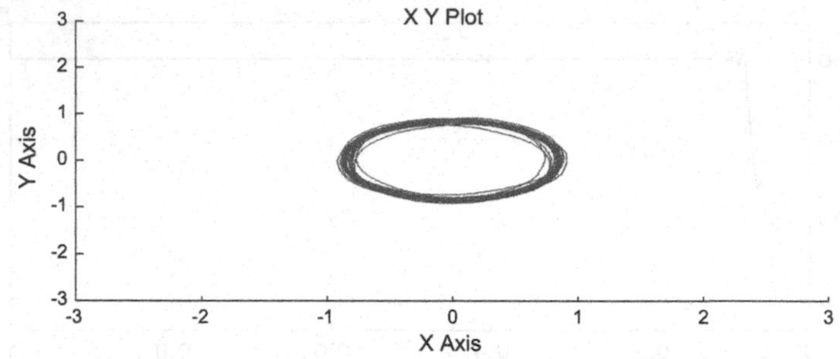


**Fig. 7** Stator flux trajectory waveform

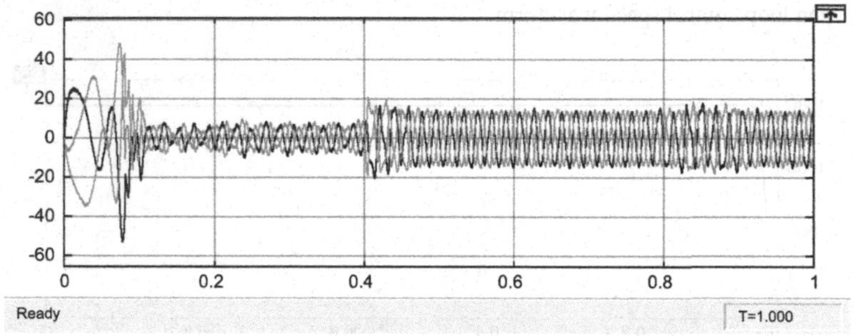


**Fig. 8** Three-phase current waveform of open-loop stator

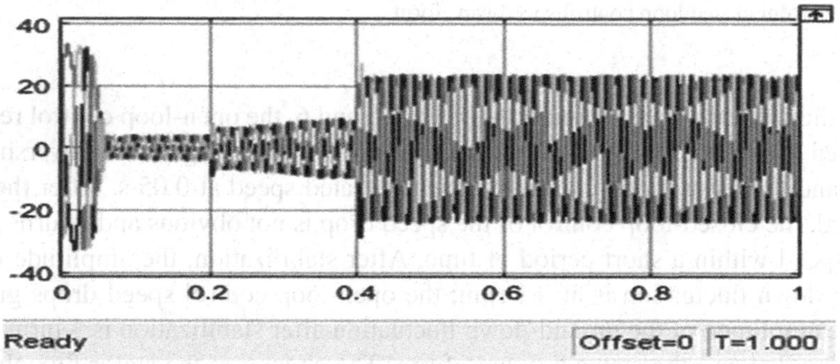


**Fig. 9** Closed-loop stator three-phase current waveform

Compared with the simulation waveforms of the slip frequency control and the current closed loop control, in the slip frequency control, the amplitude of the speed waveform and the torque waveform fluctuate greatly after the load is added, and the adjustment time is very long. The current closed-loop control, the mediation time is short, the waveform is very stable. The reason is that the closed-loop control accurately calculates the position of the stator flux through the formula, but the open-loop control orientation is not accurate. The current closed-loop PI control indirectly regulates the flux linkage and the torque so that the actual value can quickly follow the given value, thus ensuring The magnetic chain is a stable circle. The torque is the product of the flux linkage and the current torque component. When the load is added, the torque is quickly recovered by adjusting the torque component of the current.

## 5 Conclusions

This article uses the current closed-loop control strategy to control the three-phase asynchronous motor speed, torque program. On the basis of the slip frequency control, the stator current excitation component and the torque component are closed-loop controlled to obtain a given voltage component. The SVPWM principle is used for vector control to ensure the stability of the motor speed and flux linkage. Simulation results show that the closed-loop control strategy achieves stable operation of the asynchronous motor. It lays a theoretical foundation for studying the dynamic performance simulation and testing of speed control system under different control strategies.

## References

1. R. Yi, C. Boshi, *Electric Drag Automatic Control System—Motion Control System*, 4th edn. (Machinery Industry Press, Beijing, 2009)
2. W. Zhaoan, L. Jinjun, *Power Electronics Technology*, 5th edn. (Machinery Industry Press, Beijing, 2009)
3. H. Naigang, *Power Electronics, Motor Control System Modeling and Simulation* (Mechanical Industry Press, Beijing, 2012)
4. W. Tao, *Three-phase Asynchronous Motor Vector Control Design and Implementation* (Hubei University of Technology, Hubei, 2016)
5. H. Huishan, C. Long, C. De, Induction motor vector control system design and simulation. Comput. Simul. **29**(2), 400–403 (2012)
6. Z. Jun, Hercules et al., Slip frequency vector control of the motor speed control system design and research. Power Technol. **20**(331), 171–173 (2010)
7. W. Chengyuan, X. Jiakuan, S. Yibiao, *Modern Motor Control*, 2nd edn. (Machinery Industry Press, Beijing, 2014)
8. L. Hongmei, L. Zhongji, D. Shijun, Simulation of dynamic performance of asynchronous motor powered by SVPWM inverter. J. Electr. Eng. Control **5**(3), 146–148 (2001)
9. L. Hanqiang, L. Yujuan, G. Chengwei, Voltage feed forward decoupled vector control system with current error compensation. J. Wuhan Univ. Technol. **27**(1), 29–31 (2003)

# Designed of Wind Power Generation Control System Based on Matrix Converter

**Xiuli Liu, Huida Duan and Xin Peng**

**Abstract** In this paper, the matrix converter (MC) as the full power converter of the direct-drive permanent magnet synchronous wind power generation system (DD-PMSG) is analyzed. The space vector modulation for the MC is deduced and calculated, respectively. And the field oriented control is used as the control strategy for the DD-PMSG. On the one hand, the simulation model of the MC fed DD-PMSG is built by MATLAB/Simulink, and the simulation on the RL load is carried out, which verifies the correctness of the theoretical analysis and the modulation method. On the other hand, the simulation results show that the MC has the advantages of good input and output waveforms, controllable amplitude and frequency, high input power factor and so on. It also shows that the MC is suitable for DD-PMSG, which provides a theoretical basis for further research.

**Keywords** Direct-drive wind power generation system · Matrix converter
Space vector modulation

## 1 Introduction

The direct-drive permanent magnet synchronous wind power generation system is composed of wind turbine, permanent magnet synchronous generator (PMSG), full power converter and other components. Among them, due to the disappearance of the variable speed gearbox, PMSG and wind turbine are directly coupled, so that the maintenance of the whole machine, the volume and cost of the system are greatly reduced. At present, the full power converter is mainly divided into AC-DC-AC converter and AC converter.Among the AC-DC-AC converter contains large inductance or large capacitance as DC-link energy storage element, which meets the system requirements, but the volatilization of electrolyte of the inductance or capacitor will seriously affect the life of the converter; In addition, most of the AC-DC-AC con-

X. Liu · H. Duan (✉) · X. Peng
School of Electrical & Information Engineering, Beihua University, Jilin 132021, China
e-mail: huida_duan@163.com



Y. Jia et al. (eds.), *Proceedings of 2018 Chinese Intelligent Systems Conference*,
Lecture Notes in Electrical Engineering 528,
https://doi.org/10.1007/978-981-13-2288-4_44

verter circuit is used in the phase control mode, which will cause the dead current , current discontinuous and other consequences, leading to a larger harmonic content of the output current. Although to use reactive compensation and active filtering, but it is difficult to fundamentally solve the harmonic pollution problem of the power grid. The fundamental measure is to develop power electronic converters topologies withhigh input power factor and low harmonic pollution. Therefore, the matrix converter (MC) topology is proposed [1, 2]. The MC is a kind of direct AC-AC converter. It has the advantages of high input power factor, bidirectional flow of energy, compact structure and high efficiency without the intermediate DC-link energy storage components. Therefore, the matrix converter can make up for the shortcomings of traditional AC-DC-AC converters [3]. In the paper, the matrix converter is used in DD-PMSG system by the method of combining the theoretical research and simulation, then in-depth study of the control method for MC and its application in the DD-PMSG to achieve a better input and output performances of the entire system.

## 2 Matrix Converter and Space Vector Modulation

As shown in Fig. 1, the three-phase to three-phase MC consists of 3 × 3 array, it has total 9 bidirectional switches $S_{i,j}$ (i = A, B, C; j = a, b, c) and each bidirectional switch can accomplish two-way conduction and bidirectional turn off [4]. Space vector modulation technique [5–7] is applied for the equivalent circuit of the matrix converter, which consists of a current source rectifier and a voltage source inverter. The input phase current space vector modulation on the rectifier stage is carried out, and the output line voltage space vector modulation for the inverter stage is carried out. Then the intermediate DC-link is eliminated by synthesize the switch states of both the rectifier stage and the inverter stage. Finally, the SVPWM method for the MC is implemented.

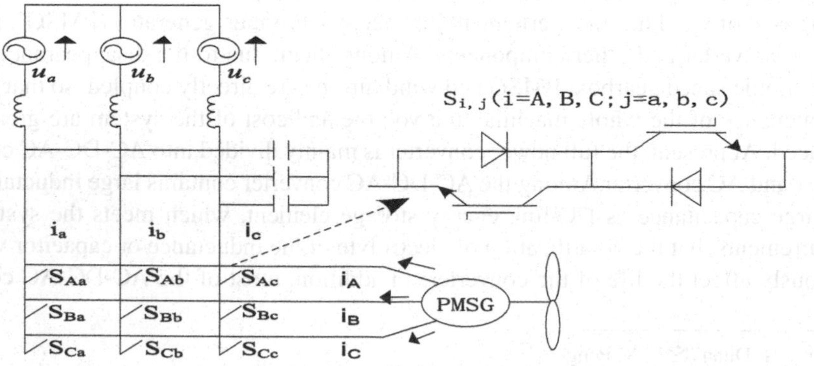


**Fig. 1** Topology of three-phase to three-phase matrix converter

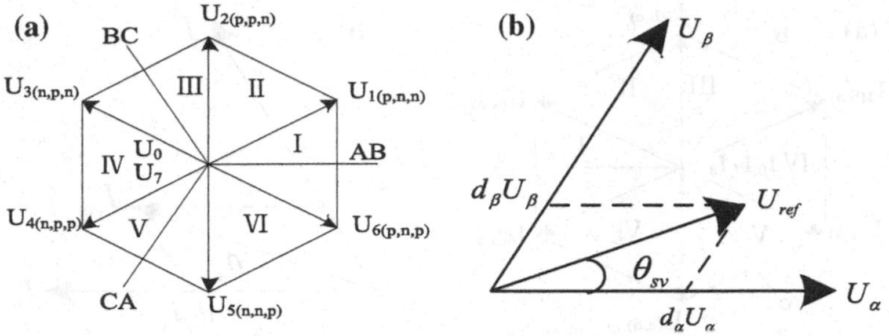


**Fig. 2** Voltage space vector sector map and vector composition diagram of inverter stage

There are 8 kinds of output line voltage SVPWM which are formed different combinations of the 6 switches in the inverter SVPWM, namely 6 kinds of active vector $U_1$–$U_6$, and the corresponding switch states are expressed as (p, n, n), (p, p, n), (n, p, n), (n, p, p), (n, n, p) and (p, n, p), 2 kinds of zero vectors of $U_0$ and $U_7$, and the corresponding switch states are expressed as (n, n, n) and (p, p, p), respectively. The spatial position of each vector and relations is shown in Fig. 2. Through the different arrangements of 8 fundamental voltage space vectors, a new voltage space vector can be generated to form a set of voltage space vectors with equal amplitude and different phases.

The reference voltage vector $U_{ref}$ of the output line voltage is synthesized by two adjacent active vectors $U_\alpha$, $U_\beta$ and one zero vector $U_0$ in the sector. The duty cycle of each vector is calculated according to the SVPWM principle, corresponding expressions are as follows:

$$U_{ref} = d_\partial U_\partial + d_\partial U_\partial + d_{ov} U_O \tag{1}$$

$$\begin{cases} d_\alpha = T_\alpha / T_s = m_v \sin(60^\circ - \theta_{sv}) \\ d_\beta = T_\beta / T_s = m_v \sin(\theta_{sv}) \\ d_{0v} = T_{0v} / T_s = 1 - d_\alpha - d_\beta \end{cases} \tag{2}$$

where $m_v = \sqrt{3} U_{om} / U_{pn}$ is the voltage modulation index; $U_{om}$ is the amplitude of output phase voltage; $T_s$ is the period of a switching cycle; $d_\alpha$, $d_\beta$ and $d_{ov}$ are the duty cycle of voltage vector $U_\alpha$, $U_\beta$ and $U_0$; $T_\alpha$, $T_\beta$ and $T_{ov}$ is the corresponding turn-on time, respectively; $\theta_{sv}$ is the phase angle of the reference voltage vector, which is in the range of 0–60°.

According to the corresponding relationships between the voltage vectors of inverter stage and switching states, so the corresponding relation between the corresponding synthesized reference voltage vector and the switching state can be obtained. Then the relationship between the local average of the output line voltage and the virtual DC-link voltage can be obtained. For example, when the reference voltage vector $U_{ref}$ is located in the sector I, the output line voltages of the inverter

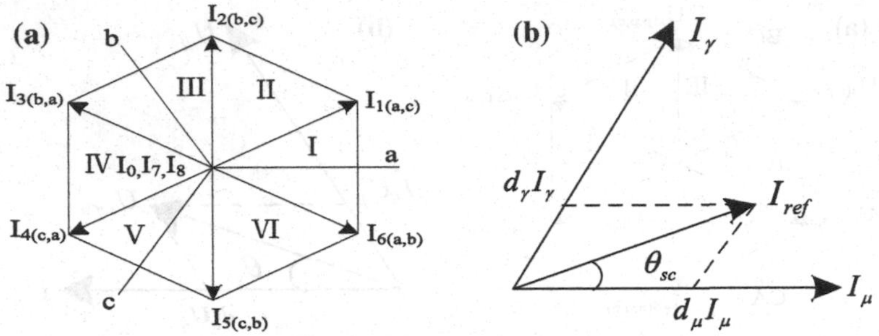


**Fig. 3** The input current space vector diagram and vector diagram of sector

stage $U_6$ and $U_1$ correspond to the switching state and the local average of the output line voltage:

$$u_{ol} = \begin{bmatrix} u_{AB} \\ u_{BC} \\ u_{CA} \end{bmatrix} = \begin{bmatrix} d_\alpha + d_\beta \\ -d_\alpha \\ -d_\beta \end{bmatrix} \cdot \mathrm{U_{pn}} = \mathrm{T_{VSI}} \cdot \mathrm{U_{pn}} \tag{3}$$

where, $T_{VSI}$ is the modulation matrix of the inverter stage.

Similarly, the local average value of the DC-link input current of the inverter stage can be obtained:

$$\mathrm{I_{pn}} = d_\alpha(i_A + i_C) + d_\beta i_A = \mathrm{T_{VSI}^T} \cdot i_{ol} \tag{4}$$

The modulation of the virtual rectifier stage is similar to the space vector modulation of the virtual inverter stage. The reference space vector $I_{ref}$ of the input phase current is synthesized by two active vectors $I_\mu$, $I_\gamma$ and one zero vector $I_0$ in the sector at one time. The duty cycle of each vector is calculated according to the SVPWM principle, and corresponding expressions are as follows (Fig. 3):

$$I_{ref} = d_\mu I_\mu + d_\gamma I_\gamma + d_{0c} I_0 \tag{5}$$

$$\begin{cases} d_\mu = T_\mu / T_s = m_c \sin(60° - \theta_{sc}) \\ d_\gamma = T_\gamma / T_s = m_c \sin(\theta_{sc}) \\ d_{0c} = T_{0c} / T_s = 1 - d_\mu - d_\gamma \end{cases} \tag{6}$$

where, $m_c = I_{im}/I_{pn}$, and $m_c$ is the current modulation index; $I_{im}$ is the amplitude of output phase current; $T_\mu$, $T_\gamma$ and $T_{oc}$ are the turn-on time of current vector $I_\mu$, $I_\gamma$ and $I_0$, respectively; $d_\mu$, $d_\gamma$ and $d_{oc}$ are the corresponding duty cycle; $\theta_{sc}$ is the phase angle of the reference current vector, which is in the range of 0–60°.

The relationship between the local average of the input phase current and the virtual DC-link current can be obtained. For example, when the reference current vector is located in the sector I, so the vector $I_6$, $I_1$ and $I_0$ are synthesized to obtain the local average of the input phase current at the rectifier stage:

$$i_i = \begin{bmatrix} i_a \\ i_b \\ i_c \end{bmatrix} = \begin{bmatrix} d_u + d_r \\ -d_u \\ -d_r \end{bmatrix} \cdot I_{pn} = T_{CSK} \cdot I_{pn} \quad (7)$$

where, $T_{CSR}$ is the modulation matrix of the rectifier stage.

And the local average value of the DC-link input voltage of the rectifier stage can be obtained:

$$U_{pn} = d_\mu u_{ab} + d_r u_{ac} = T^{T}_{VSR} \cdot u_i \quad (8)$$

Because there is no intermediate DC-link in the actual matrix converter, the input DC-link voltage and current of the inverter stage are equal to the output DC-link voltage and current of the rectifier stage. Therefore, the voltage and current are obtained by the formula (3), (8) and (4), (7).

$$u_{ol} = T_{VSI} \cdot T^{T}_{VSR} \cdot u_{ip} = T \cdot u_{ip} \quad (9)$$

$$i_l = T_{VSR} \cdot T^{T}_{VSI} \cdot i_{ol} = T^{T} \cdot i_{ol} \quad (10)$$

The T is called the entire transformation matrix. The input current and the output voltage are all located in the sector I as an example.

$$\begin{aligned} d_{\mu\alpha} &= m \cdot \sin(60^\circ - \theta_{sc}) \cdot \sin(60^\circ - \theta_{sv}) \\ d_{\mu\beta} &= m \cdot \sin(60^\circ - \theta_{sc}) \cdot \sin(\theta_{sv}) \\ d_{\gamma\alpha} &= m \cdot \sin(\theta_{sc}) \cdot \sin(60^\circ - \theta_{sv}) \\ d_{\gamma\beta} &= m \cdot \sin(\theta_{sc}) \cdot \sin(\theta_{sv}) \end{aligned} \quad (12)$$

where, $d_{\mu\alpha}$ and $d_{\mu\beta}$ and $d_{\gamma\alpha}$ and $d_{\gamma\beta}$ is the duty cycle of vector combination $I_6$–$U_6$ and $I_6$–$U_1$ and $I_1$–$U_6$ and $I_1$–$sU_1$, respectively.

The modulation index of the entire MC is

$$m = m_v \cdot m_c = \frac{2}{\sqrt{3}} \cdot \frac{U_{om}}{U_{im}} \cdot \frac{1}{\cos\varphi_i} \quad (13)$$

where, $\varphi_i$ is the input phase displacement angle.

And the duty cycle of zero vector is:

$$d_0 = 1 - d_{\alpha\mu} - d_{\beta\mu} - d_{\alpha\gamma} - d_{\beta\gamma} \tag{14}$$

## 3 Control Strategy of Direct-Drive Wind Power Generation System

In order to verify the feasibility of MC used in DD-PMSG and the effectiveness of vector control in permanent magnet synchronous generator, the vector control strategy for permanent magnet synchronous generator is controlled by MC,which is shown in Fig. 4.

The system active power P* and reactive power Q* are determined by the optimal speed of the wind turbine and the actual demand of the power grid, respectively. As shown in Fig. 4, the two coordinate transformations are employed through the angular position signal from the generator's shaft encoder. The rotor currents are transformed by Park transformation, thus, the d-axis and q-axis components of the rotor currents are obtained. The decoupling voltage is compensated, and the final decoupling voltage $u_{sd}$ and $u_{sq}$ are obtained. The voltage of the three-phase coordinate system is obtained by the anti-Park coordinate transformation of the decoupling voltage. Finally, the SVPWM modulation voltage is sent to the MC to modulate the three-phase AC voltage with the adjustable amplitude, frequency and phase to meet the demand of the grid [8–10].

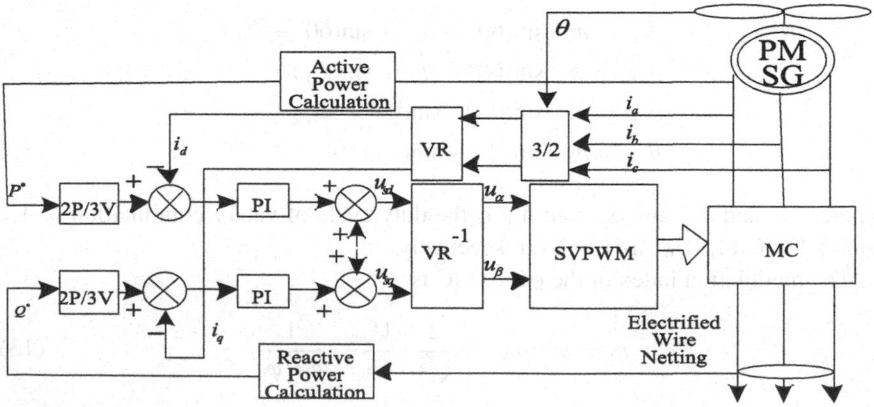


**Fig. 4** Block diagram of vector control for direct-drive wind power generation system

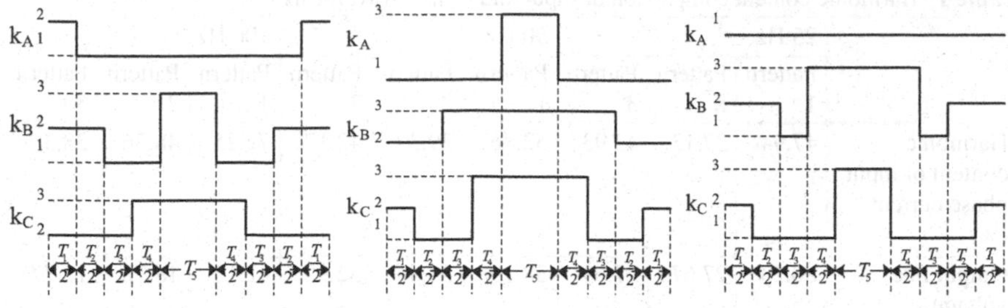

Fig. 5 The output phase pulse waveform of mode I-I sector output phase one or two and three

## 4 Simulation Implementation of SVPWM Control

In each SVPWM period T, the switch duty ratio of the 5 vectors involved in the synthesis is calculated, and the control of the matrix converter can be realized according to the order control switch of some vector action sequence. The sequence of vector action in each modulation period and the position of zero vector in the vector sequence have a great influence on the switching loss and the quality of the input and output waveforms of the matrix converter. According to the different position of zero vector in the vector sequence, 3 typical vector sequences are given in this section to study the influence of the position of zero vector on the quality of the input and output waveforms of MC. The first vector sequence, in each modulation period, the vector sequence begins with zero vector and ends with a zero vector (called a pulse output mode); the second vector sequence, the middle of the vector sequence is inserted into the zero vector, and the zero vector is centred on the symmetric distribution (which is called the pulse output mode two) and the third vector sequences. The zero vectors are inserted into the first half and the latter half of the vector sequence, respectively, which is called pulse output mode three. This section gives three pulse output modes of MC SVPWM, and compares the three modes according to the input and output waveform spectra obtained by simulation experiments (Fig. 5).

## 5 Simulation Analysis

To illustrate the effect of the position of zero vector in the vector sequence on the quality of the input and output waveforms of a matrix converter, the simulation of the three output modes of the above three kinds of pulse is carried out by using the simulation tool.

**Table 1** Harmonic content comparison of input and output waveforms

| | 25 Hz | | | 50 Hz | | | 100 Hz | | |
|---|---|---|---|---|---|---|---|---|---|
| | Pattern 1 | Pattern 2 | Pattern 3 | Pattern 1 | Pattern 2 | Pattern 3 | Pattern 1 | Pattern 2 | Pattern 3 |
| Harmonic content of input phase current (%) | 47.94 | 37.12 | 43.93 | 52.86 | 39.39 | 42.17 | 72.25 | 48.36 | 58.38 |
| Output line voltage harmonic content (%) | 80.23 | 27.67 | 27.69 | 38.19 | 31.86 | 32.92 | 42.12 | 18.70 | 18.77 |
| Harmonic content of output phase current (%) | 79.86 | 27.58 | 35.76 | 10.10 | 10.46 | 15.26 | 44.50 | 20.30 | 21.17 |

The parameters in the simulation are set as follows: Three phase symmetric input voltage 220 V and 50 Hz, The initial phase is 0. Permanent magnet synchronous generator, rated power 2.2 kW, rated voltage 380 V and 50 Hz, Stator resistance $R_s = 0.435\ \Omega$, stator inductance $L_s = 0.0713$ H, stator flux 0.7 Wb, moment of inertia $J = 0.089\ \text{kg m}^2$, resistive load $R = 5\ \Omega$, $L = 5$ ml (Fig. 6).

Because of the limited space, the output waveform and the spectrum of the output mode one or two and three when the output frequency is 50 Hz are given. Table 1 lists the harmonic content of the simulation waveforms when the output frequency 25, 50 and 100 Hz are output (Figs. 7, 8 and 9).

It can be seen from Table 1 that the input and output waveform harmonic content of the MC is the lowest when the output frequency 25, 50 and 100 Hz are output, and the harmonic content of the input and output waveforms at low frequency 25 Hz and

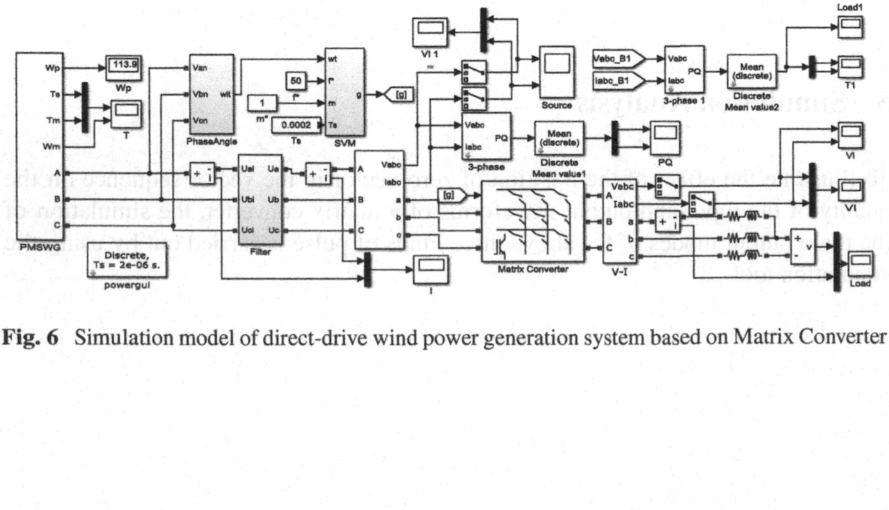


**Fig. 6** Simulation model of direct-drive wind power generation system based on Matrix Converter

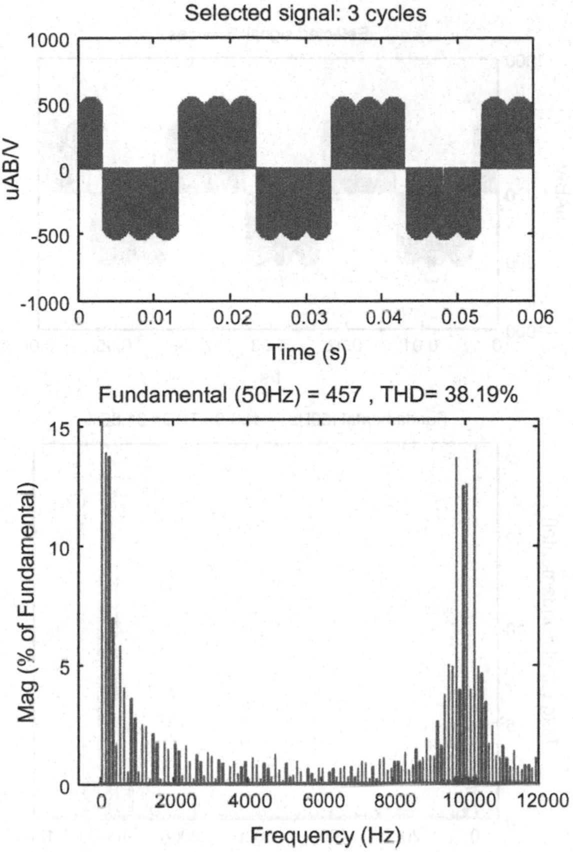


**Fig. 7** Output line voltage and spectrum of mode

high frequency 100 Hz is close to that of 50 Hz, and the quality of the waveform is better. The input and output waveform of the MC of pulse output mode three is larger than that of the pulse output mode two, and the waveform quality is slightly worse. When the output frequency is 50 Hz, the harmonic content of the input and output waveform of the MC is close to the pulse output mode two, but the harmonic content is higher in the low frequency 25 Hz and the high frequency 100 Hz, and the quality of the input and output waveform is poor. When the output frequency is 50 Hz, the harmonic content of the input and output waveform of the MC is close to the pulse output mode two, but the harmonic content is higher in the low frequency 25 Hz and the high frequency 100 Hz, and the quality of the input and output waveform is poor.

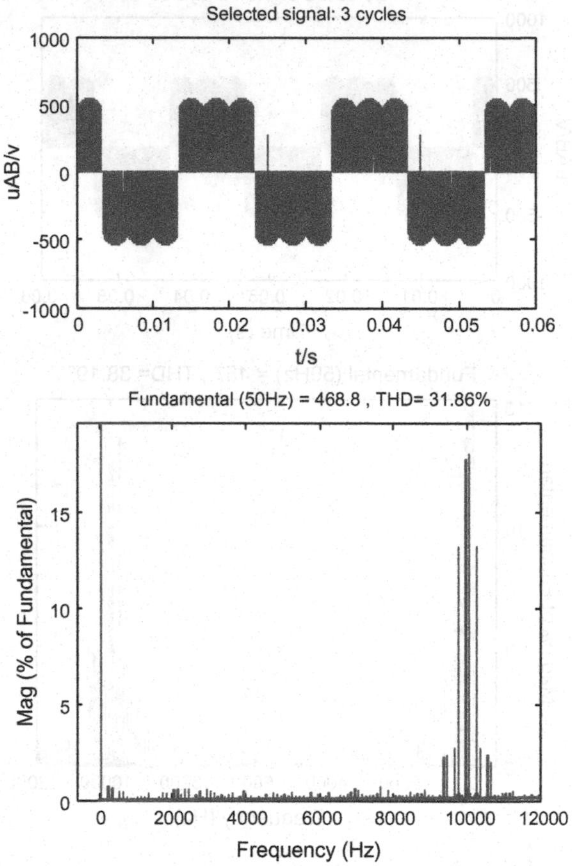


**Fig. 8** Output line voltage and spectrum of mode

# 6 Analysis of Experimental Results

This paper first analyzes the advantages of the DD-PMSG and the requirements for the full power converter. On this basis, the MC is used as the full power converter of the DD-PMSG. The advantages of the MC make the performance advantage of the DD-PMSG more advantageous. Add protruding. In the SVPWM method of MC, the insertion of zero vector in the middle of the vector sequence is the best pulse output mode in the 3 typical pulse output modes in a modulation period, and the quality of the input and output waveforms is better. Finally, the modulation strategy of the rectifier and inverter level is simulated and verified by MATLAB, and the stable output of the MC, the good power quality and the controll ability of the input power factor are verified. It also provides a theoretical basis for future research.

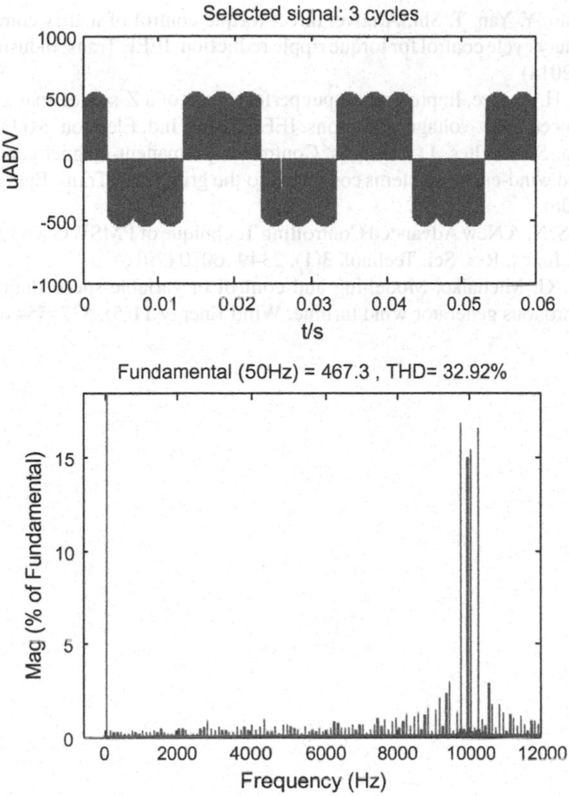


**Fig. 9** Output line voltage and spectrum of modc

**Acknowledgements** This work was financially supported by the Fund Project of Jilin Education Department (2015162, 2015148, 2016056).

# References

1. H. Hojabri, H. Mokhtari, L. Chang, Reactive power control of permanent-magnet synchronous wind generator with matrix converter. IEEE Trans. Power Deliv. **28**(2), 575–584 (2013)
2. P.W. Wheeler, J. Rodríguez, J.C. Clare, L. Empringham, A. Weinstein, Matrix converters: a technology review. IEEE Trans. Ind. Electron. **49**(2), 276–288 (2002)
3. K. Park, K.-B. Lee, Improving output performance of a Z-source sparse matrix converter under unbalanced input-voltage conditions. IEEE Trans. Power Electron. **27**(4), 2043–2054 (2012)
4. K. Tawfig, A.F. Abdou, E.E. EL-Kholy, Application of matrix converter connected to wind energy system, in *Eighteenth International Middle East Power Systems Conference* (2016)
5. K. Park, K.-B. Lee, Improving output performance of a Z-source sparse matrix converter under unbalanced input-voltage conditions. IEEE Trans. Power Electron. **27**(4), 2043–2054 (2012)

6. C. Xia, J. Zhao, Y. Yan, T. Shi, A novel direct torque control of matrix converter-fed PMSM drives using duty cycle control for torque ripple reduction. IEEE Trans. Industr. Electron. **61**(6), 2700–2713 (2014)
7. T.D. Nguyen, H.-H. Lee, Improving output performance of a Z-source sparse matrix converter under unbalanced input-voltage conditions. IEEE Trans. Ind. Electron. **51**(1), 129–140 (2012)
8. M. Chinchilla, S. Amaltes, J.C. Burgos, Control of permanent-magnet generators applied to variable-speed wind-energy systems connected to the grid. IEEE Trans. Energy Convers **21**(1), 130–135 (2006)
9. D.P. Dilip, H.S. N, A New Advanced Controlling Technique of PMSWG with Matrix Convertor. IJIRST-Int. J. Innov. Res. Sci. Technol. **3**(1), 2349–6010 (2016)
10. A.D. Hansen, G. Michalke, Modelling and control of variable-speed multi-pole permanent magnet synchronous generator wind turbine. Wind Energy **11**(5), 537–554 (2008)

# Prescribed Performance Control for Robotic Systems with Unknown Dynamics

**Chao Zhang, Jing Na, Bin Wang and Guanbin Gao**

**Abstract** In this paper, a prescribed performance controller for robotic systems with guaranteed transient and steady-state performance is proposed. A performance function that characterizes the convergence rate, maximum overshoot and steady-state error is employed to construct a new coordinate system. Then, the tracking error of the original system can be retained within a prescribed bound by stabilizing the transformed system. The unknown dynamics are accurately estimated by an estimator. The merit of the estimator is that the structure is simple and only one parameter needs to be tuned. The stability of the control system including the prescribed performance control and the unknown dynamics estimator is proved via Lyapunov theory. Simulation results are carried out to validate the effectiveness of the proposed control scheme.

**Keywords** Prescribed performance control · Nonlinear control · Robotic system

## 1 Introduction

In the past few decades, robots have been widely used in various industries. With the rapid development of robot technology and the changes of market demand, the new generation of robots have gradually replaced human in monotonous, repetitive and dangerous works. However, the works of the robots may be filled with uncertainties, such as the sudden change from the external environment or the payload. Therefore, one of the key problems in robotic control is to ensure the transient and steady-state performance of the robotic systems.

Many control schemes have been proposed to improve the aforementioned robotic system control goals. In the classical framework, the adaptive control algorithms were driven by the control error, which can guarantee the convergence of the tracking error.

C. Zhang · J. Na (✉) · B. Wang · G. Gao
Faculty of Mechanical and Electrical Engineering, Kunming University of Science and Technology, Kunming 650500, China
e-mail: najing25@163.com

 
Y. Jia et al. (eds.), *Proceedings of 2018 Chinese Intelligent Systems Conference*, Lecture Notes in Electrical Engineering 528,
https://doi.org/10.1007/978-981-13-2288-4_45

Although adaptive control has been widely used to achieve tracking control of robotic systems with unknown parameters, it is difficult to justify control schemes based on local linearization techniques, or slowly time varying assumptions [1]. In [2], sliding mode control (SMC) was presented which provides inherent order reduction, direct incorporation of robustness against system uncertainties and disturbances. In order to reduce the effects of the system parameter variations on control performance, robust control scheme was proposed [3, 4].

In fact, the unknown dynamics are usually unavoidable in practical robotic systems, which deteriorate the control performance significantly. To address these unknown uncertainties, neural networks (NNs) were presented to approximate the robot model uncertainties [1, 5]. However, the explosion of computational burden caused by NNs has not been fully resolved. In the past decades, extended state observer (ESO) [6], uncertainty and disturbance estimator (UDE) [7] and disturbance observer (DOB) [8, 9] were proposed to address the unknown dynamics. However, the parameter tuning of these methods is not a trivial task. In our previous work, a direct estimation scheme [10] was proposed by introducing filter operations. The estimator has a simple structure and only one parameter needs to be adjusted.

On the other hand, although the closed-loop system's steady-state tracking error can be guaranteed by using the aforementioned methods, transient tracking performance (e.g., overshoot, undershoot and the rate of convergence) cannot be quantitatively studied. Recently, a prescribed performance control algorithm was presented in [11, 12], where the maximum overshoot, the convergence rate and the steady-state error are all addressed.

Motivated by these observations, this paper will propose a prescribed performance control for robotic systems with unknown dynamics. This work was inspired by [13, 14]. To accommodate unknown time-varying dynamics, an estimator is employed. The estimated dynamics are then integrated into the prescribed performance controller, such that the transient and steady-state performance of tracking error can be guaranteed. Moreover, a suitable prescribed performance function (PPF) is suggested to derive a transformed error system. In the new framework, the tracking error of the original system can be retained within a prescribed bound when the transformed error system is stable. Simulations based on a Selective Compliance Assembly Robot Arm (SCARA) model are provided to verify the effectiveness of the proposed control method. The results show that the guaranteed satisfactory transient and steady-state performance can be obtained.

This paper is organized as follows. Problem formulation is provided in Sect. 2. Section 3 proposes the controller and the closed-loop stability analysis of the proposed control scheme. Simulation results demonstrating the effectiveness of the proposed control are provided in Sect. 4. Some conclusions are given in Sect. 5.

# 2 Problem Formulation

## 2.1 System Description

In this paper, an $n$-degree of freedom (DOF) robot arm can be described as:

$$M(q)\ddot{q} + C(q,\dot{q})\dot{q} + G(q) = \tau \tag{1}$$

where $q, \dot{q}, \ddot{q} \in R^n$ are the joint position, velocity and acceleration, respectively. $n$ is the number of the DOF, $\tau \in R^n$ is the control input torque, $M(q) \in R^{n \times n}$ is the inertia matrix, $C(q,\dot{q}) \in R^{n \times n}$ represents the Coriolis/centripetal matrix, and $G(q) \in R^n$ denotes the gravity vector.

According to [1], the robotic system (1) has the following properties:

**Property 1** *The inertia matrix $M(q) \in R^{n \times n}$ is symmetric positive-definite.*

**Property 2** *The matrices $M(q)$, $C(q,\dot{q})$ and $G(q)$ are all bounded.*

## 2.2 System Transformation

We define the system state as $x = [x_1, x_2]^T = [q, \dot{q}]^T$, $u = \tau$, then the model of robotic systems (1) can be described as

$$\begin{cases} \dot{x}_1 = x_2 \\ \dot{x}_2 = -M^{-1}(Cx_2 + G) + M^{-1}u \end{cases} \tag{2}$$

Then $\dot{x}_2$ can be reformulated as

$$\dot{x}_2 = F(x) + M^{-1}u \tag{3}$$

where $F(x) = -M^{-1}(Cx_2 + G)$ is the unknown dynamics. Hence, robotic system can be transformed to the following form:

$$\dot{x}(t) = f(x) + Bu(t) \tag{4}$$

where $f(x) = [x_2\ F(x)]^T$ is assumed to be a smooth function with $f(0) = 0$, $B = [0\ M^{-1}]^T$ is the known part of the robotic system, and $u(t) \in \mathbb{R}^n$ denotes the input of the investigated system. The system output is considered as $y = x_1$.

**Assumption 1** In (3), $F(x)$ is the unknown dynamics of the robotic system with $F(0) = 0$ and the derivative of $F(x)$ is bounded by $\sup_{t \le 0}\|\dot{F}(x)\| \le z_\xi$ for a constant $z_\xi > 0$.

# 3 Controller Design and Stability Analysis

## 3.1 Online Estimation of the Unknown Robotic Dynamics

To address the unknown robotic dynamics, we introduce a simple estimator to obtain $F(x)$. To this end, we first define the filtered variables $\dot{x}_{2f}$, $u_f$ of $\dot{x}_2$, $u$ as follows

$$\begin{cases} k\dot{x}_{2f} + x_{2f} = x_2, & x_{2f}(0) = 0 \\ k\dot{u}_f + u_f = u, & u_f(0) = 0 \end{cases} \tag{5}$$

where $k > 0$ is a scalar filter parameter. Then the estimator can be given as

$$\hat{F}(x) = \frac{x_2 - x_{2f}}{k} - M^{-1}u_f \tag{6}$$

**Lemma 1** *For system (2) with the estimator (6), the estimation error* $z = F(x) - \hat{F}(x)$ *is bounded by*

$$\|z(t)\| \le \sqrt{z^2(0)e^{-\frac{t}{k}} + \frac{3}{2}k^2(\gamma^2 + \xi^2 + z_\xi^2)} \tag{7}$$

*and* $\hat{F}(x) \to F(x)$ *for any* $k \to 0$.

*Proof* We apply a stable filter operation $1/(ks + 1)$ on both sides of (3) as

$$\frac{1}{ks+1}[\dot{x}_2] = \frac{1}{ks+1}[F(x)] + \frac{1}{ks+1}[M^{-1}u] \tag{8}$$

Define the filtered variables $F(x)_f$ of $F(x)$, $k\dot{F}_f(x) + F_f(x) = F(x)$, $F_f(0) = 0$. Consider the equation of (5) and the Swapping Lemma [15] for the term $\frac{1}{ks+1}[M^{-1}u]$, we rewrite (8) as

$$\dot{x}_{2f} = F_f(x) + M^{-1}u_f + \zeta \tag{9}$$

where residual term $\zeta = -\frac{k}{ks+1}\left[\dot{M^{-1}}u_f\right]$. The fact implies that $\zeta$ is bounded, for any $k > 0$, (i.e., $\|\zeta\| \le \gamma$ for a positive constant $\gamma$). And $\sup_{t\le 0}\|\dot{\zeta}\| \le \xi$ for a constant $\xi > 0$. From (5), we have $\dot{x}_{2f} = (x_2 - x_{2f})/k$, substituting it into (9), one can obtain $\hat{F}(x) = F_f(x) + \zeta$. Now, we have $\dot{F}_f(x) = \frac{1}{k}\left(F(x) - F_f(x)\right) = \frac{1}{k}z + \frac{1}{k}\zeta$.

Then we get the estimation error as

$$z = F(x) - \hat{F}(x) = F(x) - \left(F_f(x) + \zeta\right) \tag{10}$$

the derivative of the estimation error (10) can be rewritten as

$$\dot{z} = \dot{F}(x) - \dot{\hat{F}}(x) = -\frac{1}{k}z - \frac{1}{k}\zeta - \dot{\zeta} + \dot{F}(x). \tag{11}$$

Select a Lyapunov function as $V = \frac{1}{2}z^T z$, then its derivative can be derived as

$$\begin{aligned}\dot{V} &= z^T \dot{z} \\ &= -\frac{1}{k}z^T z - \frac{1}{k}z^T \zeta - \frac{1}{k}z^T \dot{\zeta} + z^T \dot{F}(x) \\ &\leq -\frac{1}{k}V + \frac{3k}{2}\gamma^2 + \frac{3k}{2}\xi^2 + \frac{3k}{2}z_\xi^{\ 2}\end{aligned} \tag{12}$$

From (12), we have $V(t) \leq e^{-\frac{t}{k}}V(0) + \frac{3}{2}k^2(\gamma^2 + \xi^2 + z_\xi^2)$, and then (7) can be obtained. In this case, $z(t)$ can exponentially converge to zero as $k \to 0$.

## 3.2 Prescribed Performance Function

To obtain the satisfactory prescribed performance, the tracking error of the original system will be reconstructed by using a PPF.

We first define the tracking error as

$$e(t) = q - q_d \tag{13}$$

where $q_d$ is the reference signal.

Then, we introduce a positive decreasing smooth function $\mu(t) : R^+ \to R^+$ as the PPF [11, 12]:

$$\mu(t) = (\mu_0 - \mu_\infty)e^{-\alpha t} + \mu_\infty \tag{14}$$

where $\mu_0 > \mu_\infty > 0$ and $\alpha > 0$ are the parameters selected by the designer.

Then, the tracking error $e(t)$ can be retained within a prescribed bound by the following condition

$$-\underline{\delta}\mu(t) < e(t) < \overline{\delta}\mu(t) \quad \forall t > 0 \tag{15}$$

where $\underline{\delta}$ and $\overline{\delta}$ are positive parameters selected by the designers.

In (14) and (15), we can get that $-\underline{\delta}\mu(t)$ and $\overline{\delta}\mu(t)$ represent the lower bound of the undershoot and the upper bound of the maximum overshoot. $\alpha$ denotes the convergence speed and $\mu_\infty$ defines the steady-state tracking error. Hence, the transient and steady-state performances can be guaranteed by tuning the parameters $\underline{\delta}$, $\overline{\delta}$, $\alpha$, $\mu_0$ and $\mu_\infty$.

To design the prescribed performance controller, an output error transform will be introduced by transforming condition (15) into an equivalent "unconstrained" one [12]. According to [13], a smooth and strictly increasing function $S(z_1)$ of the transformed error $z_1 \in$ R is defined such that:

(1) $-\underline{\delta} < S(z_1) < \overline{\delta}, \forall z_1 \in L_\infty$
(2) $\lim\limits_{z_1 \to +\infty} S(z_1) = \overline{\delta}$, and $\lim\limits_{z_1 \to -\infty} S(z_1) = -\underline{\delta}$

From the properties of $S(z_1)$, condition (15) can be rewritten as

$$e(t) = \mu(t) S(z_1) \tag{16}$$

Since $S(z_1)$ is strictly monotonic increasing and $\mu(t) \geq \mu_\infty > 0$, the transformed error $z_1$ can be deduced from the inverse function of $S(z_1)$ as

$$z_1 = S^{-1}\left[\frac{e(t)}{\mu(t)}\right] \tag{17}$$

Note that the parameters $\underline{\delta}$, $\overline{\delta}$, $\alpha$, $\mu(0)$, $\mu_\infty$ of PPF (15) and $S(z_1)$ can be selected by the designer. For any initial condition $e(0)$, if $\underline{\delta}$, $\overline{\delta}$, and $\mu(0)$ are chosen to satisfy $-\underline{\delta}\mu(0) < e(0) < \overline{\delta}\mu(0)$, then $z_1$ is bounded, and the condition $-\underline{\delta} < S(z_1) < \overline{\delta}$ holds. Thus, the condition (15) is guaranteed, and the tracking control of system (2) is transformed to stabilize the transformed system (17).

In this paper, we select the function $S(z_1)$ as

$$S(z_1) = \frac{\overline{\delta} e^{z_1} - \underline{\delta} e^{-z_1}}{e^{z_1} + e^{-z_1}} \tag{18}$$

So that the transformed error $z_1$ can be calculated as

$$z_1 = S^{-1}\left[\frac{e(t)}{\mu(t)}\right] = \frac{1}{2}\ln\frac{\lambda(t) + \underline{\delta}}{\overline{\delta} - \lambda(t)} \tag{19}$$

where $\lambda(t) = e(t)/\mu(t)$ is the normalized tracking error.

### 3.3 Prescribed Performance Controller

To stabilize the transformed error $z_1$ and to satisfy the prescribed performance of the tracking error $e$, we deduce the derivative of $z_1$ as

$$\dot{z}_1 = \frac{\partial S^{-1}}{\partial \lambda}\dot{\lambda} = \frac{1}{2\mu}\left[\frac{1}{\lambda + \underline{\delta}} - \frac{1}{\lambda - \overline{\delta}}\right]\left(\dot{e} - \frac{e\dot{\mu}}{\mu}\right) \tag{20}$$

define $r = (1/2\mu)[1/(\lambda + \underline{\delta}) - 1/(\lambda - \overline{\delta})]$ and fulfills $0 < r \leq r_M$. The derivative of $r$ with respect to time is deduced as

$$\dot{r} = -\frac{\dot{\mu}}{2\mu^2}\left[\frac{1}{\lambda + \underline{\delta}} - \frac{1}{\lambda - \overline{\delta}}\right] - \frac{\dot{e}\mu - e\dot{\mu}}{2\mu^3}\left[\frac{1}{(\lambda + \underline{\delta})^2} - \frac{1}{(\lambda - \overline{\delta})^2}\right] \tag{21}$$

Moreover, we can further obtain

$$\ddot{z}_1 = \dot{r}\left(\dot{e} - \frac{e\dot{\mu}}{\mu}\right) + r\left[\ddot{e} - \frac{\dot{e}\dot{\mu}}{\mu} - \frac{e\ddot{\mu}}{\mu} + \frac{e\dot{\mu}^2}{\mu^2}\right] \tag{22}$$

We define a filtered error as

$$s = \Lambda z_1 + \dot{z}_1 \tag{23}$$

where $\Lambda > 0$ is a positive constant, thus the tracking error $z_1$ is bounded as long as $s$ is bounded.

The derivative of $s$ can be obtained that

$$\dot{s} = (\Lambda r + \dot{r})\left(\dot{e} - \frac{e\dot{\mu}}{\mu}\right) + r\left[\ddot{e} - \frac{\dot{e}\dot{\mu}}{\mu} - \frac{e\ddot{\mu}}{\mu} + \frac{e\dot{\mu}^2}{\mu^2}\right] \tag{24}$$

Consequently, we can design the control law as

$$u = M\left(-\hat{F}(x) + \ddot{q}_d + \frac{\dot{e}\dot{\mu}}{\mu} + \frac{e\ddot{\mu}}{u} - \frac{e\dot{\mu}^2}{\mu^2} - \frac{1}{r}\left[(\Lambda r + \dot{r})(\dot{e} - \frac{e\dot{\mu}}{\mu}) + k_1 s\right]\right) \tag{25}$$

where $k_1$ is a positive constant.

*Remark 1* All parameters used in the proposed controller can be taken into two groups: the PPF parameters $\mu_0$, $\mu_\infty$, $\alpha$, $\underline{\delta}$ and $\overline{\delta}$ mainly affect the prescribed performance; the control parameters $k_1$ and $\Lambda$ are selected online based on a trial-and-error method.

### 3.4 Stability Analysis

In this section, we present the stability analysis of the robotic closed-loop system with the proposed control. The following theorem is given as

**Theorem 1** *For the robotic system (1) with the estimator (6) and the proposed control (25), the closed-loop control system is stable. If any initial values fulfill $-\underline{\delta}\mu(t) < e(t) < \overline{\delta}\mu(t)$, $\forall t > 0$, the tracking error $e$ can be guaranteed within the prescribed bound defined in (15).*

*Proof* Consider a Lyapunov function as

$$V = \frac{1}{2}s^T s + \frac{1}{2}z^T z \tag{26}$$

The derivative $\dot{V}$ with respect to $t$ can be obtained as

$$\begin{aligned}
\dot{V} &= s^T\dot{s} + z^T\dot{z} \\
&= s^T(-k_1 s + rz) + z^T\big(-\frac{1}{k}z - \frac{1}{k}\zeta - \dot{\xi} + \dot{F}(x)\big) \\
&\le -k_1 s^2 + r_M\|s\|\|z\| - \frac{1}{k}z^2 + \frac{1}{k}\|z\|\gamma + \|z\|\xi + \|z\|z_\xi \\
&\le -(k_1 - \frac{r_M}{2})s^2 - (\frac{1}{2k} - \frac{3}{2})z^2 + \frac{1}{2k}\gamma^2 + \frac{1}{2}\xi^2 + \frac{1}{2}z_\xi^2
\end{aligned} \tag{27}$$

Then, the following form can be obtained:

$$\dot{V} \le -\varphi V + \vartheta \tag{28}$$

where $\varphi = min\{(k_1 - r_M/2)\,,(1/2k - 3/2)\}$, $\vartheta = \frac{1}{2}(\frac{1}{k}\gamma^2 + \xi^2 + z_\xi^2)$.

Consequently, we can conclude that the errors $s$ and $z$ are bounded. Thus, the transformed error $z_1$, $\dot{z}_1$ and the control law $u$ are also bounded. The tracking error of the system (2) can be retained within the prescribed performance bound by tuning the PPF parameters.

## 4 Simulations

In this part, simulation results will be carried out to demonstrate the validity of the proposed controller.

The model of SCARA robot is described as

$$\begin{bmatrix} M_{11} & M_{12} \\ M_{21} & M_{22} \end{bmatrix}\ddot{q} + \begin{bmatrix} C_{11} & C_{12} \\ C_{21} & C_{22} \end{bmatrix}\dot{q} + \begin{bmatrix} G_1 \\ G_2 \end{bmatrix} = \tau \tag{29}$$

Consider the state-space equation as

$$\dot{x} = \begin{bmatrix} \dot{x}_1 \\ \dot{x}_2 \end{bmatrix} = \begin{bmatrix} x_2 \\ -M^{-1}(Cx_2 + G) + M^{-1}u \end{bmatrix} \tag{30}$$

where $x_1 = [q_1\ q_2]$ and $x_2 = [\dot{q}_1\ \dot{q}_2]$ denote the SCARA robot's two joint position and velocity vectors. $m_1 = 1.975\,\text{kg}$, $m_2 = 0.8\,\text{kg}$, $m_3 = 1\,\text{kg}$ are the mass of each robot arm, $l_1 = l_2 = 0.25\,\text{m}$ are the length of each link. We define auxiliary variables as $c_2 = \cos(q_2)$, $s_2 = \sin(q_2)$, $a = m_1 l_1 l_1$, $b = m_2 l_2 l_2$, $c = m_3 l_1 l_2 c_2$, $d = m_3 l_1 l_2 s_2$. Then, the inertia matrix, the Coriolis/centripetal matrix and the gravity matrix can be reformulated as

$$M = \begin{bmatrix} a + b + 2c & b + c \\ b + c & b \end{bmatrix},\ C = \begin{bmatrix} -2d\dot{q}_2 & -d\dot{q}_2 \\ d\dot{q}_1 & 0 \end{bmatrix},\ \text{and } G = \begin{bmatrix} 0 \\ 0 \end{bmatrix}.$$

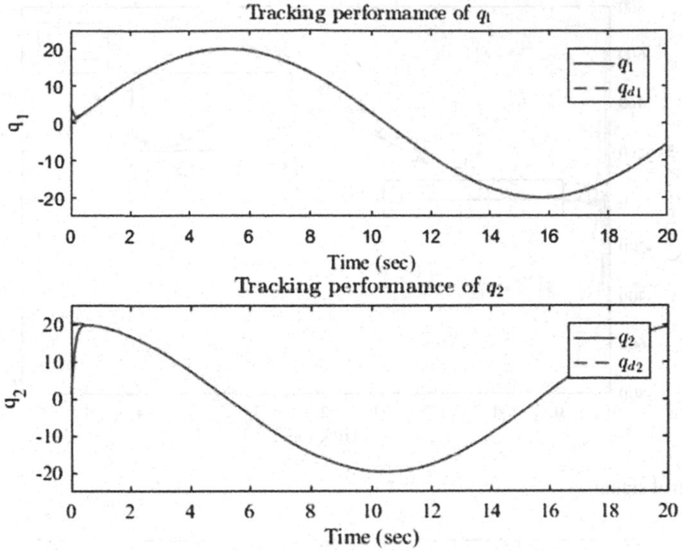


**Fig. 1** Tracking performance of $q$

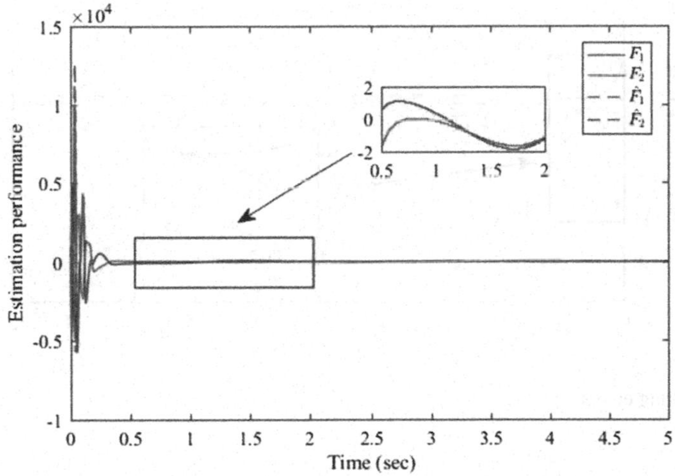


**Fig. 2** Estimation performance

The reference trajectories are chosen as $q_{d1} = 20\sin(0.3t)$, $q_{d2} = 20\cos(0.3t)$. The PPF parameters are selected as $\mu_0 = 22$, $\mu_\infty = 0.5$, $\delta = \delta = 1$, $\alpha = 8$, and the control parameters are chosen as $k_1 = 10$, $\Lambda = 180$.

Figure 1 illustrates the robot's two link tracking profiles. It is shown that the reference signal $q_d$ can be tracked in a short time. The unknown dynamics $F(x)$ can be accurately estimated as shown in Fig. 2. The control signal is given in Fig. 3. As

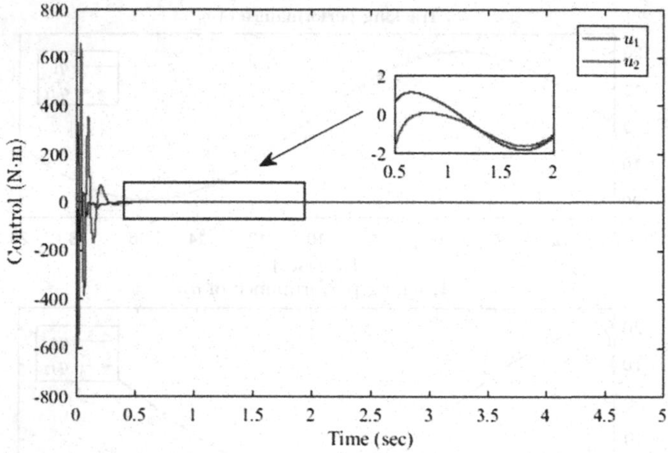


Fig. 3 Control signal

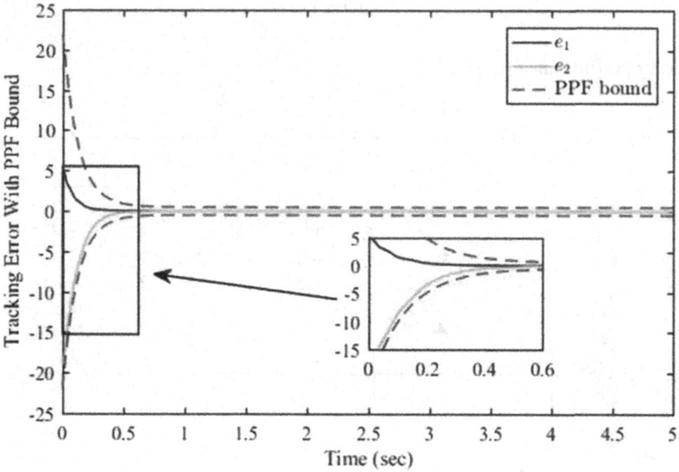


Fig. 4 Tracking errors

shown in Fig. 4, the tracking errors can be prescribed within the PPF bound. One can see that the satisfactory transient and steady-state performance of robotic systems are obtained by using the proposed control scheme.

## 5 Conclusions

In this paper, a prescribed performance control is proposed for a class of robotic systems with unknown dynamics. The unknown dynamics were accurately estimated and then the estimated dynamics are incorporated into the controller. An output error transformation is introduced such that both the transient and steady-state performance (e.g., overshoot, convergence speed, and steady-state error) can be guaranteed by stabilizing the transformed system. The stability of the closed-loop system is proved through Lyapunov theory. Simulation results validate the effectiveness of the proposed control scheme.

**Acknowledgements** This work was supported by the National Natural Science Foundation of China (grant 61573174).

## References

1. F.L. Lewis, D.M. Dawson, C.T. Abdallah, *Robot Manipulator Control: Theory And Practice* (CRC Press, 2003)
2. S. Yuri, E. Christopher, F. Leonid, L. Arie, *Sliding Mode Control and Observation* (Springer, New York, 2014)
3. Y. Jia, Robust control with decoupling performance for steering and traction of 4WS vehicles under velocity-varying motion. IEEE Trans. Control Syst. Technol. **8**(3), 554–569 (2000)
4. Y. Jia, Alternative proofs for improved LMI representations for the analysis and the design of continuous-time systems with polytopic type uncertainty: a predictive approach. IEEE Trans. Autom. Control **48**(8), 1413–1416 (2003)
5. C. Yang, T. Teng, B. Xu, Z. Li, J. Na, C.Y. Su, Global adaptive tracking control of robot manipulators using neural networks with finite-time learning convergence. Int. J. Control, Autom. Syst. **15**, 1916–1924 (2017)
6. J. Han, Extended state observer for a class of uncertain plants in Chinese. Control Decision **10**(1), 85–88 (1995)
7. Q-C. Zhong, A. Kuperman, R.K. Stobart, Design of UDE-based controllers from their two-degree-of-freedom nature. Int. J. Robust Nonlinear Control **21**(17), 1994–2008 (2011)
8. W-H. Chen, D.J. Ballance, P.J. Gawthrop, J. O'Reilly, Nonlinear disturbance observer for robotic manipulators. IEEE Trans. Ind. Electron. **47**(4), 932–938 (2000)
9. W-H. Chen, J. Yang, L. Guo, S. Li, Disturbance-observer-based control and related methods-an overview. IEEE Trans. Ind. Electron. **63**(2), 1083–1095 (2016)
10. L. Wang, Y. Li, J. Na, G. Gao, Q. Chen, Nonlinear servo motion control based on unknown input observer, in *Proceedings of 2016 Chinese Intelligent Systems Conference*, Vol. 405, (Springer, 2016), pp. 541–550
11. C.P. Bechlioulis, G.A. Rovithakis, Robust adaptive control of feedback linearizable MIMO nonlinear systems with prescribed performance. IEEE Trans. Autom. Control **53**(9), 2090–2099 (2008)
12. C.P. Bechlioulis, G.A. Rovithakis, Adaptive control with guaranteed transient and steady state tracking error bounds for strict feedback systems. Automatica **45**(2), 532–538 (2009)
13. J. Na, Q. Chen, X. Ren, Y. Guo, Adaptive prescribed performance motion control of servo mechanisms with friction compensation. IEEE Trans. Ind. Electron. **61**(1), 486–494 (2014)
14. Y. Huang, J. Na, X. Wu, X. Liu, Y. Guo, Adaptive control of nonlinear uncertain active suspension systems with prescribed performance. ISA Trans. **54**, 145–155 (2015)
15. P.A. Ioannou, J. Sun, *Robust Adaptive Control* (Upper Saddle River, PTR Prentice-Hall, NJ, 1996)

# An Extreme Gradient Boosting Algorithm for Short-Term Load Forecasting Using Power Grid Big Data

**Liqiang Ren, Limin Zhang, Haipeng Wang and Qiang Guo**

**Abstract** Directed at the problem of more and increasing data types and volume in power grid, a short-term power load forecasting algorithm based on big data and Extreme Gradient Boosting (XGBoost) is proposed, based on the analysis of power grid load big data low. The algorithm includes the following steps. First, the outlier data and missing data are preprocessed. Then, the K-means algorithm is used to cluster the load big data of the power grid. Finally, The XGBoost algorithm was used to train the load forecasting model, based on the impact of historical load, calendar effect and meteorological factors on the load. Simulation results show that compared with support vector machine, random forest and decision tree, the proposed algorithm has a higher prediction accuracy and smoother prediction error, with smaller mean absolute percentage error, mean absolute error and relative error.

**Keywords** Power grid big data · Power load forecasting
Extreme gradient boosting algorithm · Cluster analysis

## 1 Introduction

Short-term load forecasting (STLF) is an important part of power system load forecasting, and it is also one of the basic links in smart grid construction [1]. The main application of STLF is to provide load forecasting for unit combination and economic

L. Ren · L. Zhang (✉) · H. Wang · Q. Guo
Institute of Information Fusion, Naval Aviation University,
Yantai 264001, Shandong, China
e-mail: iamzlm@163.com

L. Ren
e-mail: 1194153993@qq.com

H. Wang
e-mail: whp5691@163.com

Q. Guo
e-mail: iamgq@163.com



Y. Jia et al. (eds.), *Proceedings of 2018 Chinese Intelligent Systems Conference*,
Lecture Notes in Electrical Engineering 528,
https://doi.org/10.1007/978-981-13-2288-4_46

dispatch. For example, if the load demand is known in advance, the generator can be operated at the lowest possible cost, which can increase the economic efficiency of the power grid and the power plant [2]. The second application of STLF is safety evaluation of power system. The prediction result is of great significance to the safe and stable operation of the power grid. In addition, short-term load forecasting results contribute to not only the smooth development of power management, but also the goal of energy conservation, emission reduction and environmental protection [3]. However, due to the randomness, non-linearity, and instability of the factors affecting power load changes, it is difficult to achieve high-precision predictions.

In the past decades, scholars have put forward many methods for improving the accuracy of STLF, which can be mainly divided into three categories: classical methods, artificial intelligence methods, and machine learning methods. Classical prediction methods are based on mathematical models, including time series method [4], regression analysis method [5], etc. The classical prediction model is a simple linear method. Its advantages are simple model structure and fast prediction speed. Due to the complex nonlinear characteristics of the short-term power load, the prediction accuracy of STLF by classical prediction methods is not sufficient. Artificial intelligence methods mainly include neural networks [6] and extreme learning machines [7]. Neural network algorithm has better self-learning ability and nonlinear fitting ability. Extreme learning machine solves the problem of network structure selection in neural network. Machine learning prediction models mainly include support vector machines [8] and random forests [9]. Support vector machine has a good effect on solving practical problems such as non-linearity and high dimensional. The short-term load forecasting model based on support vector machine has the characteristics of high precision, strong generalization ability and global optimization. Compared with support vector machines, random forests are more robust to outliers, and fewer parameters are needed for model training. However, as the scale of power grids continues to expand, the types of data available in the power grid are becoming increasingly abundant, and the number of data is rapidly increasing. The above methods have limitations when dealing with large-scale, complex, and diverse data.

To solve the above problem, XGBoost [10] was selected as the short-term power load forecasting algorithm. This algorithm has the feature of fast calculation speed, high prediction accuracy and flexible definition of target function. It has excellent performance in many fields, such as biomedical engineering [11], economic and financial [12]. This paper analyzes the law of power load big data, constructs characteristic variables from historical load, calendar effect and meteorological factors, and preprocesses the original data. At the same time, in order to improve the adaptability and accuracy of the XGBoost model, power load big data is clustered before the prediction. The XGBoost algorithm is used to train and test the measured power load data after clustering. The experimental results show that the method has higher prediction accuracy and more stable prediction error.

The rest of this paper is organized as follows. Section 2 presents the change law of load big data of power grid. Section 3 presents the full procedure of short-term power load forecasting, including data preprocessing, load data clustering, and XGBoost algorithm for power load forecasting. Section 4 details the experimental results. The conclusion is described in Sect. 5.

## 2 Analysis of Power Load Variation Based on Big Data

The change law of power load is explored with big data of power grid, and the relationship between power load and its influence factors is emphatically analysed, which lays the foundation for the next step of power load forecasting.

### *2.1 Sources of Short-Term Power Load Influencing Factor Data*

Sensors and monitors, such as power generation, transmission, power use, dispatch, detection and communication, provide a large-scale data set for load forecasting, which have become the main sources of big data in the power grid.

(1) the monitoring data of the power grid operation. It is mainly about the data of power generation, voltage stability and meteorological environment.
(2) the operation data of the electric power enterprise. There are mainly transaction price, electricity sales and customer information, the peak season of industrial electricity customers, the population and geographical location of electricity consumption areas, etc.
(3) power enterprise management data. It is mainly information about Power Grid Corp finance, power dispatching, power network planning, etc. these factors can indirectly reflect users' power consumption.

The above actual data based on grid operation is the basis for analyzing the law of load changes.

### *2.2 Analysis of Change Law of Power Grid Load*

This paper selects hourly electricity load data for a city from 2010 to 2015 and weather data for the corresponding time. Electricity demand is affected by a series of common factors such as meteorological conditions and calendar effects. On this basis, a preliminary exploration of the law of power grid load changes is conducted to provide reference value for the next modelling and forecasting work.

The calendar effect is one of the factors that affect power load changes. Figure 1a shows the hourly load curve in 2013. January, February and July are the peak periods of electricity consumption, and April, May and October are the lows of electricity demand. In other words, the demand for electricity in winter and summer is strong,

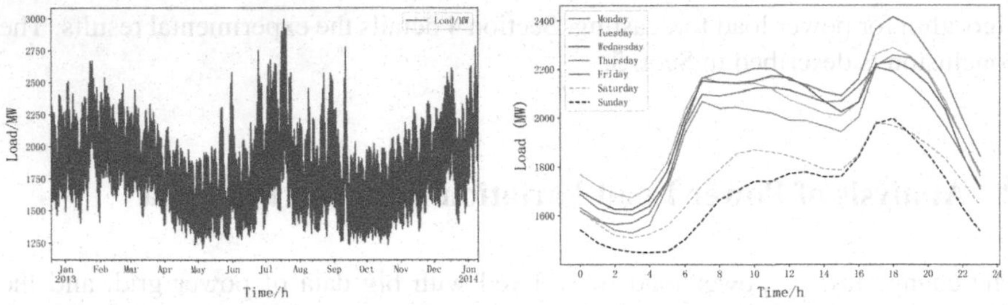


(a) Hourly power load curve of 2013 (b) Daily power load curve for a week

**Fig. 1** Historical load curve

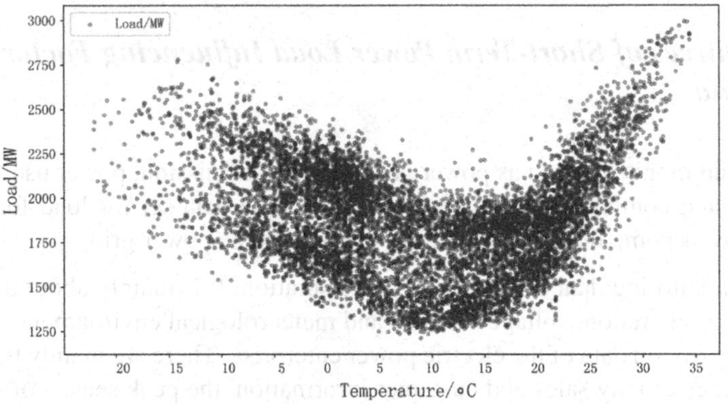


**Fig. 2** The relationship between hourly load and temperature

and the demand for electricity in autumn and spring is reduced. These characteristics indicate that there is a significant time effect on the power load data.

Figure 1b shows the power load curve of 24 h per day in the week of December 2013. It can be seen that from the beginning of 24:00, electricity demand began to decline, and the minimum value was reached between 3:00 and 4:00. After 5 o'clock, it began to rise rapidly. From 18 to 19 o'clock, it reached the highest peak of electricity consumption. At the same time, the load on Saturdays and weekends was generally lower than the load on the working day, but the overall trend of daily load was not changed. Figure 1b shows the time distribution of peak load and low vale load, which indicates that the power load follows the specific pattern day and night, weekly and daily.

Figure 2 shows a significant nonlinear relationship between the load and the current temperature, which indicates that temperature is an important indicator of power load forecasting. When the temperature falls below 10 °C, the power demand increases due to heating requirements. Similarly, when the temperature rises above 23 °C, the power demand also increases due to cooling requirements.

**Table 1** Description of the variables of the dataset used in this study

| Variables | Description | Variables | Description |
|---|---|---|---|
| Load | Present load | Dewpoint | Present dewpoint |
| peak_load | Peak load of the day before | Humidity | Present humidity |
| L_m24 | Load of 24 h before | Visibility | Present visibility |
| L_m48 | Load of 48 h before | Dow | Day of the week, integer, between 0 and 6 |
| Ldif | Difference of L_m24 and L_m48 | Doy | Days of the year, integer, between 0 and 365 |
| Temperature | Present temperature | Day | Days of the month, integer between 1 and 31 |
| Temper_h | 1 h before temperature | Woy | Week of the year, integer between 1 and 52 |
| Humidity | Present humidity | Hour | Hours of the day, integer, between 0 and 23 |
| Windspeed | Wind speed | Month | Months of the year, integer, between 1 and 12 |
| Pressure | Sealevel pressure | Minute | Minutes of the day, integer, between 1 and 720 |

# 3 Algorithm Description

## *3.1 Data Description and Preprocessing*

This paper considers the following input feature variables, as shown in Table 1. In Table 1, the current load factors include historical load, calendar variables and meteorological variables. Due to systematic errors, "bad data", such as missing data and outlier data, is inevitable. These data seriously interfere with the training and prediction process of the model, and affect the accuracy of the prediction results. Therefore, to ensure the smoothness of the load sequence, the consistency of the order of magnitude and the good performance of the algorithm model, it is necessary to preprocess the original data.

(1) Missing data processing

For the missing data, the m data closest to the missing data is selected according to the Euclidean distance and then the weighted average is calculated.

$$l(h) = \omega_1 l(h_1) + \omega_2 l(h_2) + \cdots \omega_m l(h_m) \tag{1}$$

where $l(h)$ is the missing data and does not exist, $[l(h_1), l(h_2), \ldots, l(h_m)]$ is the closest data vector to $l(h)$, and $[\omega_1, \omega_2, \ldots, \omega_m]$ is the weight vector determined by the Euclidean distance. The small Euclidean distance has a high weight coefficient.

(2) Outlier data processing

Due to systematic errors and unexpected events, outlier data may sometimes appear in the load data. These outlier data will disturb the regularity of the entire data sequence and thus affect the prediction accuracy. Therefore, it is necessary to correct outlier data.

The load curve not only has obvious seasonal features, but also presents a specific daily load pattern. In a season, the load data with the same date type are like each other, that is, the data in the same season has a periodicity. Therefore, it is assumed that $l(h, k, n)$ is the load value, h represents the hour of the day, $h = 1, 2, \ldots, 24$; k represents the day of the week, $k = 1, 2, \ldots, 7$; n represents the week of the season, $n = 1, 2, \ldots, N$.

Firstly, we calculate the mean $E(h)$ and the hourly squared error $V(h)$ of the same date type based on formulas (2) and (3).

$$E(h) = \frac{1}{N}\sum_{n=1}^{N} l(h, k, n) \tag{2}$$

$$V(h) = \sigma_h^2 = \frac{1}{N}\sum_{n=1}^{N} [l(h, k, n) - E(h)]^2 \tag{3}$$

Then, the deviation rate is defined as $\rho(h, k, n)$.

$$\rho(h, k, n) = \frac{|l(h, k, n) - E(h)|}{\sigma_h} \tag{4}$$

The process of correcting outliers is as follows:

$$\begin{cases} \rho(h, k, n) \geq \gamma, \begin{cases} 1 < n < N, \hat{l}(h, k, n) = \frac{l(h,k,n-1)+l(h,k,n+1)}{2} \\ n = 1, \hat{l}(h, k, n) = l(h, k, n+1) \\ n = N, \hat{l}(h, k, n) = l(h, k, n-1) \end{cases} \\ \rho(h, k, n) < \gamma, \hat{l}(h, k, n) = l(h, k, n) \end{cases} \tag{5}$$

where $\gamma$ is the expected deviation rate. According to Ref. [13] and several experiments, set $\gamma = 1.25$ in spring and autumn, $\gamma = 1.56$ in summer and winter.

(3) Data normalization

Data normalization is the foundation of data mining. The effect of dimension and range difference between feature variables can be eliminated by normalization. The following formula is used to standardize the original data, so that the average value of converted data is 0 and the standard deviation is 1.

$$x' = \frac{x - \bar{x}}{\sigma} \tag{6}$$

where $\bar{x}$ is the mean of the original data, $\sigma$ is the standard deviation of the original data, and $x'$ is the normalized data.

## 3.2 *Clustering of Power System Load*

The load data of power grid is huge, and there are various kinds of influencing factors. There are significant differences in the consumption of electricity between different meteorological conditions and date types. Therefore, the load data of the power grid are clustered to make the samples with similar characteristics gather in the same group, thereby improving the adaptability and accuracy of the short-term load forecasting method. The k-means algorithm [14] is selected as a clustering method for grid load data, because it is not only simple and fast, but also is relatively scalable and effective for processing big data.

The calculation of distance between samples is the basis of cluster analysis. Considering the significant differences between the dimensions of variables, The Euclidean distance between samples is calculated using the normalized data above.

$$dist(p_\iota, p_j) = \sqrt{\sum_{k=1}^{m} (x_{i,k} - x_{j,k})^2} \tag{7}$$

where $dist(p_i, p_j)$ is the distance between the $i$ sample and the $j$ sample, and $x_{i,k}$, $x_{j,k}$ are the $k$ variable of the $i$ sample and the $j$ sample respectively.

The main steps of K-means algorithm are as follows: (1) Randomly select W objects as the initial cluster center. (2) Assign all unselected samples to the cluster represented by the closest center point according to the distance function. (3) Calculate the mean value of each cluster as a new cluster center. (4) Return to step 2 until the allocation is stable and the algorithm stops.

## 3.3 *XGBoost*

The XGBoost 10 algorithm is a modified algorithm based on Gradient Boosting Decision Tree. In recent years, it has good performance in the scientific research of machine learning and data mining. XGBoost algorithm not only has the advantage of high accuracy of traditional boosting algorithm, but also can handle sparse data efficiently and realize distributed parallel computing flexibly. Therefore, the XGBoost algorithm is applicable to big data. In addition, XGBoost adds regular items to the loss function to control the complexity of the model and reduce the variance of the model. XGBoost can improve the prediction accuracy at a certain speed.

There are n samples and m features in dataset D = $\{(\boldsymbol{x}_i, y_i)\}(\boldsymbol{x}_i \in \boldsymbol{R}^m, y_i \in \boldsymbol{R})$. It is assumed that the XGBoost model has $K$ decision trees. The prediction function is as follows:

$$\hat{y}_i = \sum_{k=1}^{K} f_k(\boldsymbol{x}_i) \tag{8}$$

where $\hat{y}_i$ is the predicted value of the $i$th target variable, $\boldsymbol{x}_i \in R^m$ is the input variable corresponding to $\hat{y}_i$. $f_k$ is the prediction function corresponding to the $k$ th decision tree. It is defined as follows:

$$f_k(\boldsymbol{x}_i) = \boldsymbol{\omega}_{q(\boldsymbol{x}_i)}, q : R^m \rightarrow T, \boldsymbol{\omega} \in R^T \tag{9}$$

where $q(\boldsymbol{x}_i)$ represents the structure function of the $k$ th decision tree that maps $\boldsymbol{x}_i$ to the corresponding leaf node. $\boldsymbol{\omega}$ is the quantization weight vector of the leaf node, and $T$ is the number of leaf nodes in the tree.

The XGBoost algorithm adds regularization terms to the loss function, considering the accuracy and complexity of the model. It learns the prediction function by minimizing the following loss function.

$$L(\emptyset) = \sum_{i=1}^{n} l(\hat{y}_i, y_i) + \sum_{k=1}^{K} \Omega(f_k) \tag{10}$$

where $\sum_{k=1}^{K} \Omega(f_k)$ is a regular term, which helps to prevent model over fitting, $\Omega(f) = \gamma T + \frac{1}{2}\lambda||\omega||^2$ represents the complexity of the model, $\gamma$ is a complex parameter, $\lambda$ is a fixed coefficient, and $T$ represents the number of leaf nodes. In this study, square loss is used as a loss function, $l(\hat{y}_i, y_i) = (\hat{y}_i - y_i)^2$.

### 3.4 The Process of Short-Term Power Load Forecasting Algorithm

The complete process of short-term power load forecasting based on XGBoost is illustrated as follows and presented in Fig. 3.

(1) Data preprocessing and sample clustering. First, the missing values and outliers are corrected. Then the input samples are clustered into three groups using the k-means algorithm.

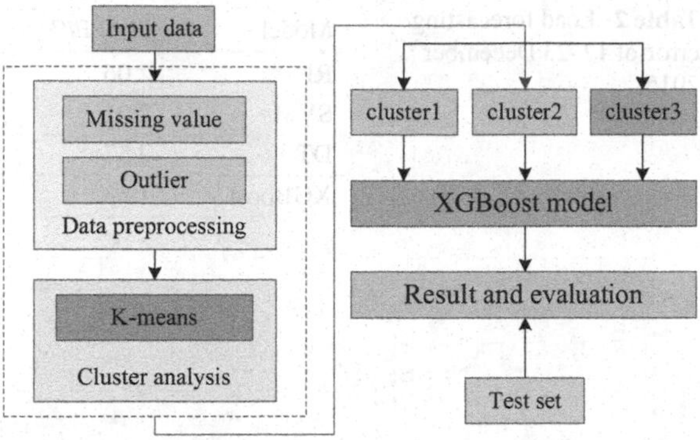


**Fig. 3** The flowchart of short-term power load forecasting algorithm

(2) Model training. Establish an independent XGBoost model for each clustering result and determine the model parameters for each cluster.
(3) Results and analysis, use the above model to predict the daily load in the test set and evaluate the performance of the model.

# 4 Results and Discussion

The experimental training set selects the data from 2010 to 2014, and the test set selects the 2015 data. We use training set to train the above models and verify them on the test set. All algorithms in this study are implemented in python3.6.

Mean absolute percent error (MAPE), mean absolute error (MAE), and relative error (RE) were used as criteria for error evaluation to analyze model prediction performance.

$$MAPE = \frac{1}{n}\sum_{i=1}^{n}\left|\left(Y_i - \hat{Y}_i\right)/Y_i\right| * 100\% \quad (11)$$

$$MAE = \frac{1}{n}\sum_{i=1}^{n}\left|Y_i - \hat{Y}_i\right| \quad (12)$$

$$RE = \frac{Y_i - \hat{Y}_i}{Y_i} * 100\% \quad (13)$$

where $Y_i$ represents the true value of the power load, $\hat{Y}_i$ represents the power load predicted value, and $n$ is the predicted time point. In power load forecasting, the smaller the *MAPE* value, the more accurate the load forecast.

Random forest (RF), support vector machine (SVM), decision trees (DT) and XGBoost models are applied to predict the daily hourly load of December 2015 5–11. The statistical results of prediction error are shown in Table 2.

**Table 2** Load forecasting error of 17–23 December 2015

| Model | MAPE/% | MAE/MW | Max RE/% |
|---|---|---|---|
| RF | 2.06 | 37.57 | 6.84 |
| SVM | 2.01 | 37.44 | 6.28 |
| DT | 2.57 | 46.47 | 11.25 |
| XGBoost | 1.53 | 27.90 | 4.64 |

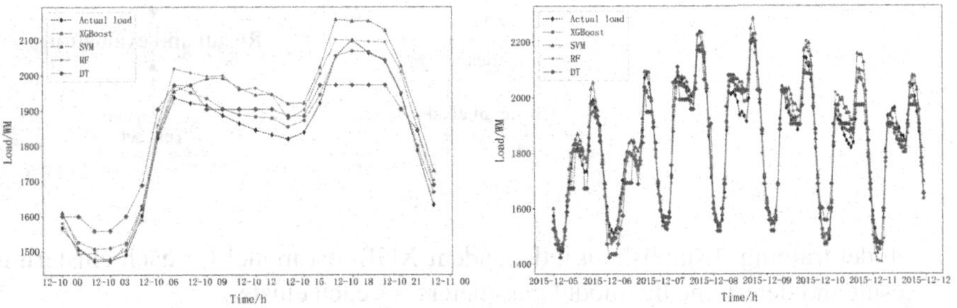


(a) December 10, 2015 load forecast (b) December 5-11, 2015 load forecast

**Fig. 4** Actual and predicted power loads

Figure 4a and b show daily and hourly load forecast results for a day and a week for random forest, support vector machine, decision tree, and XGBoost model, respectively. The prediction curve of the proposed XGBoost model fits the actual load curve better than the prediction results of the other three models, especially near the peak and trough of the load. Although the other three single models can also determine the overall trend of actual load, their overall forecast error is relatively high. Obviously, the proposed XGBoost model can improve the accuracy of load forecasting.

Figure 5 shows the relative error of daily load forecasting for each model over the week of December 5–11. It can be seen from Fig. 5 that each model has a different prediction effect at the same time. The overall fluctuation of the relative error of the proposed XGBoost model is relatively small, with a relative error of less than 3% at a prediction time of about 78.43%. This shows that the proposed XGBoost model has higher prediction accuracy, smoother prediction error, and stronger generalization ability.

## 5 Conclusion

This paper presents a short-term load forecasting algorithm based on load clustering and XGBoost. The proposed algorithm can comprehensively consider the historical load data, weather data and calendar effect data to predict the short-term power load. In addition, the K-means algorithm is used to cluster the load samples before training, so that samples with similar feature are clustered in the same group, which improves

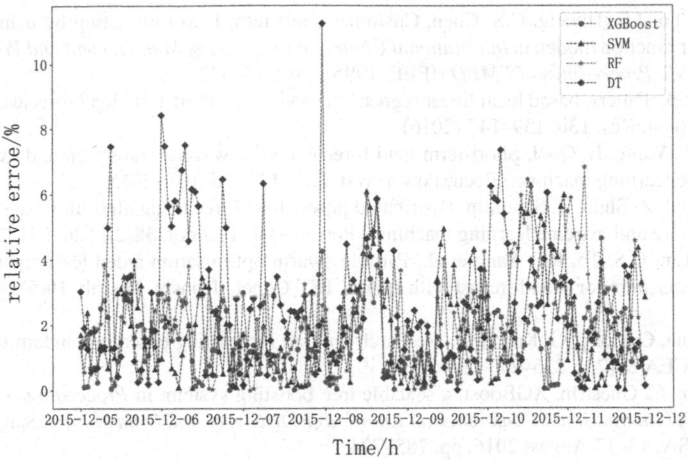


**Fig. 5** Relative error of load forecasting

the adaptability and accuracy of the model. Compared with support vector machine, random forest and decision tree, the proposed algorithm has a higher prediction accuracy and smoother prediction error, with smaller mean absolute percentage error, mean absolute error and relative error. However, the parameters of the XGBoost model in this paper are adjusted by the grid method, which takes more time and cannot achieve global optimization. How to optimize the parameters of the model is a matter for further study.

**Acknowledgements** This research was supported by the National Natural Science Foundation of China (Grant No. 91538201); Taishan scholar project special fund project (Grant No. Ts201511020).

**Author Contributions** Limin Zhang, Qiang Guo design experiments and collect data. Liqiang Ren and Haipeng Wang conducted a case study and analyzed the results. Liqiang Ren wrote this paper. All authors read and approved the final manuscript.

**Conflicts of Interest** The authors declare no conflict of interest.

## References

1. S. Li, L. Goel, P. Wang, An ensemble approach for short-term load forecasting by extreme learning machine. Appl. Energy **170**, 22–29 (2016)
2. H.S. Hippert, C.E. Pedreira, R.C. Souza, Neural networks for short-term load forecasting: a review and evaluation. IEEE Trans. Power Syst. **16**(1), 44–55 (2001)
3. F. Javed, N. Arshad, F. Wallin et al., Forecasting for demand response in smart grids: an analysis on use of anthropologic and structural data and short term multiple loads forecasting. Appl. Energy **96**(8), 150–160 (2012)

4. M.Y. Cho, J.C. Hwang, C.S. Chen, Customer short term load forecasting by using ARIMA transfer function model, in *International Conference on Energy Management and Power Delivery, 1995. Proceedings of EMPD* (IEEE, 1995), pp. 317–322
5. G. Dudek, Pattern-based local linear regression models for short-term load forecasting. Electr. Power Syst. Res. **130**, 139–147 (2016)
6. S. Li, P. Wang, L. Goel, Short-term load forecasting by wavelet transform and evolutionary extreme learning machine. Electr. Power Syst. Res. **122**, 96–103 (2015)
7. W. Baoyi, Z. Shuo, Z. Shaomin, Distributed power load forecasting algorithm based on cloud computing and extreme learning machines. Power Syst. Technol. **38**(2), 526–531 (2014)
8. W.M. Lin, C.S. Tu, R.F. Yang et al., Particle swarm optimisation aided least-square support vector machine for load forecast with spikes. IET Gener. Transm. Distrib. **10**(5), 1145–1153 (2016)
9. L. Yuhua, C. Hong, G. Kun et al., Research on load forecasting based on random forest algorithm. CEA **52**(23), 236–243 (2016)
10. T. Chen, C. Guestrin, XGBoost: a scalable tree boosting system, in *Proceedings of the ACM SIGKDD International Conference on Knowledge Discovery and Data Mining*, San Francisco, CA, USA, 13–17 August 2016, pp. 785–794
11. I. Babajide Mustapha, F. Saeed, Bioactive molecule prediction using extreme gradient boosting. Molecules **21**(8), 983 (2016)
12. Y. Xia, C. Liu, N. Liu, Cost-sensitive boosted tree for loan evaluation in peer-to-peer lending. Electron. Commer. Res. Appl. **24**, 30–49 (2017)
13. M. Liu, *The Power System Load Forecasting Research Based on Wavelet and Neural Network Theory* (Nanjing University of Science and Technology, Nanjing, 2012)
14. R.J. Broderick, J.R. Williams, Clustering methodology for classifying distribution feeders, in *Photovoltaic Specialists Conference* (IEEE, 2013), pp. 1706–1710

# Remote-Monitoring Alarm System from Vehicle Burglar Design—Based on Single Chip Microcomputer

**Haiyan Zhao**

**Abstract** This paper designs a remote-monitoring alarm system from vehicle burglar, based on MCU 52 series as main control chip. Through vibration sensor and pyroelectric infrared sensor, once Single Chip Microcomputer obtains monitoring signals, it can transmit signals to users' mobile terminal by GSM communication chip of SIM900A, so as to real-time monitoring. This design module is stable, cost effective, and is ideal for experiment teaching.

**Keywords** Sensors · AT89C52 · SIM900A

## 1 Introduction

As car holdings constantly increase in China, cases of vehicle burglar get increased, too. The existing car alarms have old-fashioned machinery lock, which is heavy, and customer-unfriendly. The old-fashioncd alarm system has short transmitting distance, and would create lots of noises once it gets shaken. The alarm system from vehicle burglar based on GPS is too costly for accessing the network, and is unsuitable for the configuration of mid-range cars and cheap cars; thus is difficult to be popularized. The system this paper designs for is controlled by MCU, so that it could signal anomaly to distant user by GSM in time.

The full name of GSM [1] is Global System for Mobile Communications, which is the widest used, and the most complete kind of mobile communication system using time division multiple access at this time. GSM has already nationally covered, and is the main kind of public mobile communication network in our country. Among all the services, the service of message texting can transmit and connect by Signaling rather than by dialing. This requires to texting the intended message attached with destination address to the Short Message Service Center (SMSC). After being saved

H. Zhao (✉)
School of Electrical and Control Engineering, North University of China, No. 3 Xueyuan Road, Taiyuan 030051, China
e-mail: zhy19581204@nuc.edu.com



Y. Jia et al. (eds.), *Proceedings of 2018 Chinese Intelligent Systems Conference*, Lecture Notes in Electrical Engineering 528,
https://doi.org/10.1007/978-981-13-2288-4_47

through the SMSC, the message can finally be transmitted to the terminal of mobile phone.

In this modern society, infrared sensing consists of an important part of the advanced technology, and is widely employed in many kinds of scientific field. Since infrared light is invisible, it can be concealed and widely employed. It can especially be employed to the burglar alarm system, in which it can detecting all day long and hardly be disturbed by the outside.

Thus this paper argues that we can use the technology of sensor detecting to design a kind of remote detecting vehicle burglar alarming system with MCU controller based on GSM network message texting, with sound functions and reasonable price.

This paper points to the increasing criminals of vehicle burglary, and aims to improve the technology of burglar alarm system so that it can make alarm more accurate, and can reduce the false alarms and noises.

In this paper, the main control system of AT89C52 single chip microcomputer is discussed, when vehicles are guardless, the proof-burglar configuration would be unlocked, the sensor would enter into standby. When the sensor detects any signal of burglar, it will give an alarm the pre-bundled mobile phone by GSM wireless communication.

This paper would discuss that the detection system of sensor can get detecting signals through vibration sensor and PIR (pyroelectric infrared sensor). The vibration sensor can transfer the mechanical quantity to electric quantity in accordance to the proportional relation between the input vibrating mechanical quantity and the output electric quantity. By Fresnel lens, the PIR in the detecting zone can transfer the change of temperature of moving objects to pyroelectric infrared signal; in this situation, the sensor would output the high level signals. The change of electric quantities of these two sensors would make the external interrupting pin of MCU change, thereby activating the corresponding operations.

## 2 System Description

The Remote-Monitoring Alarm System from Vehicle Burglar this paper design consist of power module, main control module, vibration sensor, infrared sensor and communicational module (see Fig. 1). When vehicles are guardless, the proof-burglar devices would be started. The sensor would enter the standby mode. If the infrared sensor detects anyone approaching the vehicle, or the vibration sensor detects vehicle vibrating, after the two sensors would signal interrupting trigger to the MCU controlling system after they process information analysis. The MCU would trigger signals according to different pulses of pins and input such signals to communication module. The GSM communicating module adopts SIM900A chip, which send message to the mobile phone using "at command language", through GSM mobile logic channels, so as to alarm remotely.

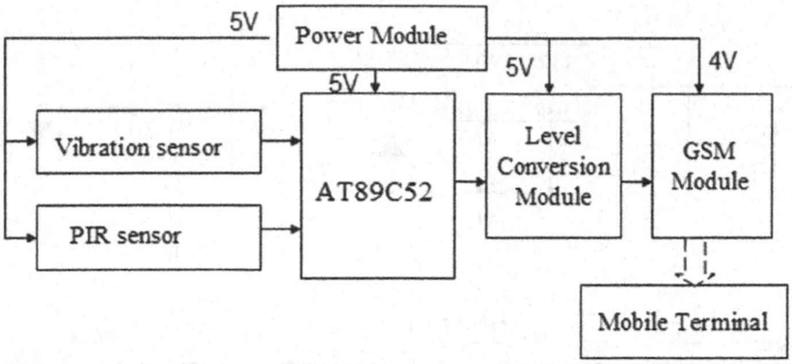


**Fig. 1** Structure chart for remote alarming system

# 3 Mechanism of the Remote Alarm

## *3.1 Power Module*

The power module is the key to the whole system. The working voltage of the chip of main MCU controller AT89C52 is 0–5.0 V. The working voltage of those two sensors is 0–5.0 V. However, the GSM communication chip SIM900A [2] chip supply electricity by single power, whose working voltage is 3.1–4.6 V. Different with the working voltage of MCU, the maximized electric current of GSM module can not exceed 4.6 V, but needs the maximized electric current of 2 A. The power of searching network, receiving and sending message of GSM module is relatively large and needs more electrical energy. To ensure the 2 A electric current, the power supply needs 4 V voltage at least. To get the system work, it is imperative to supply power in according to those two power-supplying situation respectively (see Fig. 2). In Fig. 2, the V12 plug in the outside power supply of 12 V, and was input to LM2576 [3] by capacitor filter, so as to change 12–4 V.

## *3.2 Main Control Module*

The chip of main controlled MCU employs AT89C52. The pin of MCU P3.2 (INT0) is linked to vibration sensor; the pin of MCU P3.3 (INT1) is linked to infrared sensor. But the serial port pin of SIM900A couldn't get linked to the serial port pin of P3.0 (RXD) and P3.1 (TXD) of AT89C52, because the electrical level of serial port communication of SIM900A [4] is 2.8 V, while the electrical level of AT89C52 is TTL. Thus, through MAX232, the electrical level TTL of AT89C52 could be switched into electrical level of SIM900A.

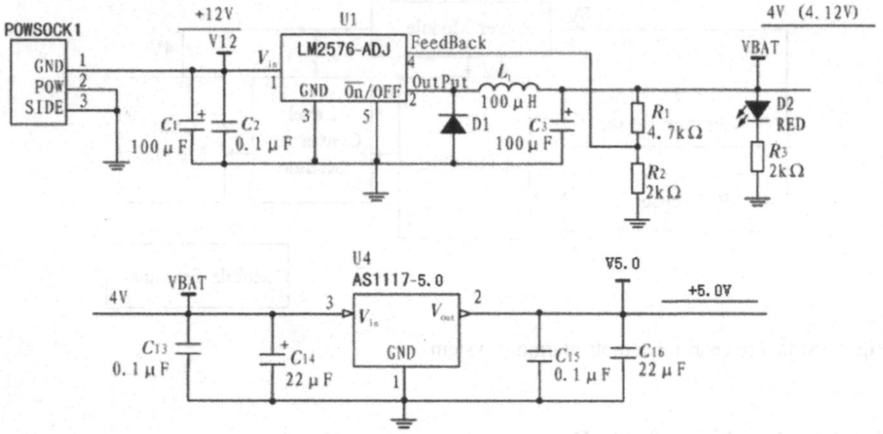


**Fig. 2** Power module

## 3.3 GSM Communication Module

The GSM communication module employs SIM900A chip (see Fig. 3), which is the most important module of the system. There are three considerations. One is for the 4.0 V power supply among 55, 56, 57 three pins of SIM900A module VBAT; Second is for switching connection of electrical level between serial port pins of 9 (TXD), 10 (RXD) of SIM900A and serial port pins of AT89C52. The third is for the design of GND—in case of the decrease of power voltage, it is necessary to drill some more holes on PCB so that the reflux's route between power supply and ground would be short; the impedance would be large, the decrease of voltage would be small; those unused pins would be suspended.

The main functions of SIM900A are: baseband of GSM [5], memorizer, GSM RF, antenna interface, and other interface. By GSM baseband chips, the message would be modulated, and be sent to the mobile phone by GSM RF. Thanks to transmitting power EGSM 900 and DCS 1800 of SIM900A, mobile phones no matter of different frequency bands, or different places of the city, could receive message all the time.

## 3.4 Sensor Module

The sensor modules that this paper designs are the vibration sensor SW1810p and PIR sensor HC-SR501. There are three pins of vibration sensor. Two of them are 5 V voltage and ground voltage, the other is "DO" pin to MCU pin P3.2 (INT0). The PIR sensor also has three pins, among which, two of them are 5 V voltage and ground voltage; the other is "OUT" pin to MCU pin P3.3 (INT1).

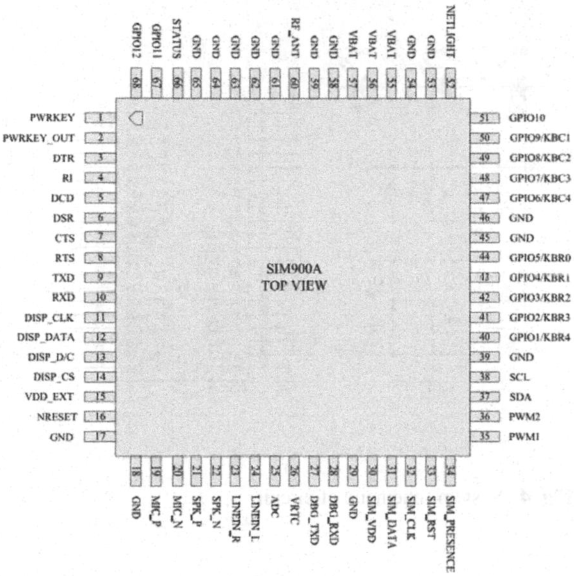


**Fig. 3** Pins of SIM900A

# 4 Hardware Circuit Design Method

## *4.1 Design for the Hardware Circuit*

When the vibration sensor senses any suspected vibrating, the pin of MCU P3.2 (INT0) would follow a triggering signal of falling edge. The MCU [6] controlling system would run interrupt program I, and send corresponding data through serial pin P3.1 (TXD) of MCU. The data would transfer electrical level into the module SIM900A by chip MAX232. Then alarming message would be sent to the mobile phones after SIM900A encodes. In the same way, if the PIR sensor detects anyone approaches, it would send a triggering signal of falling edge to MCU P3.3 (INT1) as the triggering signal of MCU controlling system's external interrupt INT1, which would lead the MCU control system to run interrupt program II, and send alarming messages to the mobile phones. The overall circuit diagram of the proof-burglar vehicle alarming system of remote monitoring could be seen in Fig. 4.

## *4.2 Program Design*

When the system begins to work, it would first initialize serial port and those two external interrupts of MCU. Once detecting any anomalies of the vibration sensor, MCU would be switched to INT0 to go to the entry address of 0003H unit, and then jumps to the interrupt service routine I. Through serial port, the SIM900A would

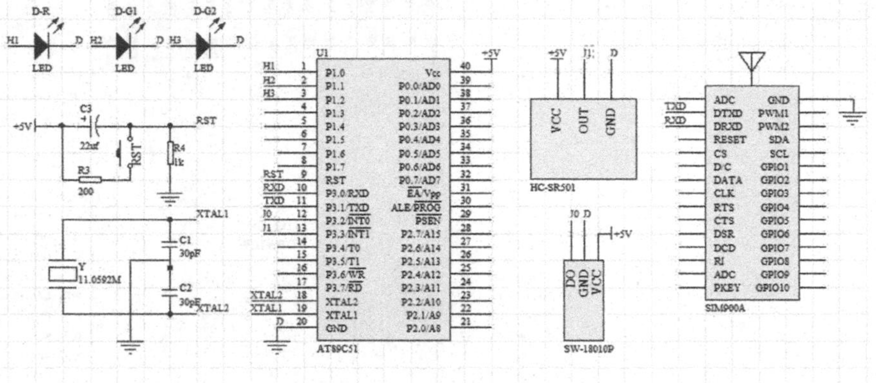


Fig. 4 System schematic of circuit

send a message like “The car is under insecurity” to the binding mobile phones. Once the PIR sensor detecting any anomalies, MCU would be switched to INT1 to enter the address of 0012H unit, and then jumps to the interrupt service routine II. The SIM900A would send a message like “The car is getting approached” to the binding mobile phones.

For serial communication, the Baud rate should be firstly set. The SIM900A defaults to 9600 bit/s, crystal frequency to 11.059 MHz, Timer of T1 to mode 2;Serial port to working mode 1. The overall flow chart could be seen as Fig. 5.

# 5 Prototype and Test Results

## *5.1 Hardware Modulate*

For serial debugging, the first thing to do is to connect SIM900A to the computer. Since serial ports does not support electrical level of RS232, and are only support electrical level of TTL, it is requisite to transfer the electrical level to RS232 after the modules SIM900A connects with CP1020. Then the SIM900A could be connected to the computer. The steps are: 1. Connects the serial port line USB-232 to the computer and install the driver; 2. Configure the COM port that the serial port line allocated, in “my computer—attributes—hardware—device manager—port”; 3. Open the sscom.exe which is the debugging software of SIM900A’s serial port so as to set for serial ports; 4. Insert the SIM card to its seat; 5. Open the GSM antenna.

It needs one second to start and close the SIM900A module.

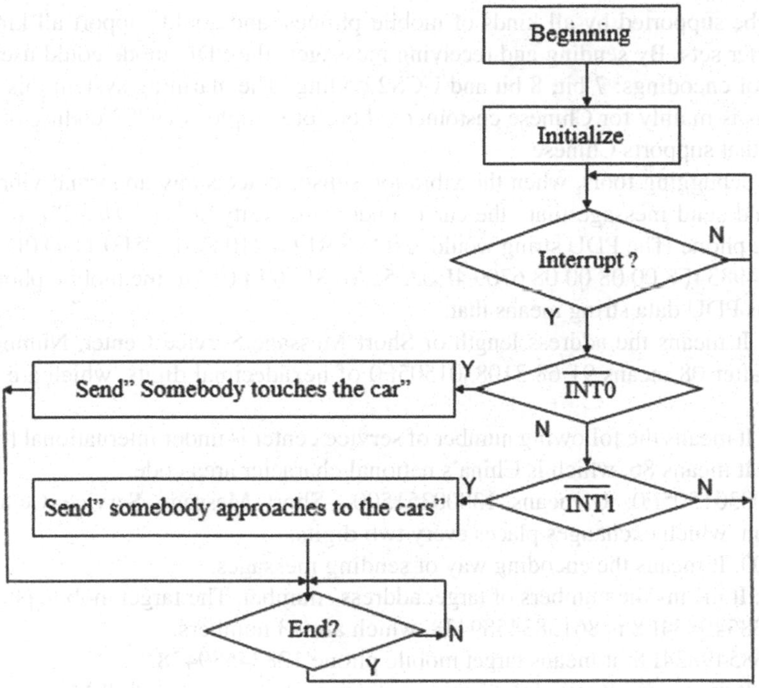


**Fig. 5** The overall flow chart of system's software

## 5.2 *Software Modulate*

The command of SIM900A module is AT [7] instructions, by which the intentional message could be sent to mobile phones. After the module starts up, it will send AT instruction, to synchronize the Baud rates of SIM900A module and the computer. The SIM900A module's Baud rate is default to 9600 bit/s.

Steps for software modulating: 1. Sending "AT+CPIN?" to inquire the module as to whether it could detect commands from SIM card. If the module return to "READY", it means that the module has detected the SIM card. If it return to "ERROR", it means that the module has not detected the SIM card. In this case, it needs to examine whether the SIM card is 2G, since the SIM900A only support 2G both for China Mobile and China Unicode. 2. Sending "AT+COPS?" command to inquire the module as to whether it could register to the network. If the module return to "China Mobile", it means that China Mobile SIM card of 2G has already registered to China Mobile. 3. Sending "AT+CMGF=0" to select PDU mode. 4. Sending "AT+CSCS=UCS2" to select UCS2 code.

There are three main encodings of message: BLOCK mode, TEXT mode, and PDU (Protocol Data Unit) mode. Among them, the BLOCK mode is barely employed now; TEXT mode couldn't receive and send Chinese messages; while the PDU mode could be supported by all kinds of mobile phones, and could support all kinds of character sets. By sending and receiving messages, the PDU mode could use three kinds of encodings: 7 bit, 8 bit and UCS2 coding. The alarming system this paper designs is mainly for Chinese customers, therefore employs UCS2 coding of PDU mode that supports Chinese.

By debugging tools, when the vibration sensor detects any abnormal vibrating, it would send message that “the car is under insecurity” (“有人动车!”) to user’s mobile phone. The PDU string would text “08 91 68 3108301505F0 1100 0D 91 68 3138549834F8 00 08 00 08 6709 4EBA 52A8 8F66 FF01” to the mobile phones.

This PDU data string means that:

08: It means the address length of Short Message Service Center. Numbers of bytes after 08 means 91 68 3108301505F0 of hexadecimal digits, which are just 8 bytes.

91: It means the following number of service center is under international format.

68: It means 86, which is China’s national character area code.

3108301505F0: It means 13800351500—Short Message Service Center of Taiyuan, which exchanges places every two digits.

1100: It means the encoding way of sending messages.

0D: It means the numbers of target address’ number. The target mobile phone of 68 3138549834F8 is 8613834589438, which are 13 numbers.

3138549834F8: it means target mobile phone 13834589438.

00: It means point to point communication, a common type of GSM.

08: It means coding of UCS2 is employed.

00: It means 5 min, which is the effective time of sending message to the target mobile phone.

08: It means Chinese “有人动车!”, (“the car is under insecurity”) which takes 8 bytes.

6709 4EBA 52A8 8F66 FF01: It means the coding of “the car is under insecurity.”

After the debugging, a test is performed. Once the vehicle is vibrating or get approached, the mobile phone would receive corresponding Chinese message that “the car is under insecurity” or “the car is getting approached” after 1 min.

# 6 Conclusion

After the simulation test and hardware welding, debugging, the system is better and accurately completes the remote monitoring vehicle anti-theft alarm function. The system has simple structure, high stability, high cost performance, convenient use and can be widely used in automobile anti-theft field. In the future design also need to resist the interference ability, as well as the system function diversity further research and the consummation, causing it to have the practical application significance.

## References


1. H. Cai, S. Zhai, M. Cai et al., Design of smart home control system based on GSM and STM32. Tech. Autom. Appl. **30**(8), 37–40 (2013)
2. S. Qu, W. Wang, K. Zhang et al., IOT SMS alarm system based on SIM900A. Mod. Electron. Technol. **35**(5):86–89 (2012)
3. LM2576/LM2576HV Series Simple Switcher ® 3A Step-Down Voltage Regulator, http://www.waveshare.net/datasheet/NS_PDF/LM2576.PDF
4. L. Weiping, X. Yang, W. Yuchuan, Research and implementation of the GPS/GSM mobile unit for vehicle monitor system based on single-chip microcomputer. J Wuhan Inst. Sci. Technol. (04), 1–4 (2007)
5. W. Jingwei, The *Research of Preventing Stealing or Robbing and Alarming System of Vehicle Based on GSM*. Master's degree thesis (Northeastern University, Liaoning)
6. M.A. Yongxiang, Design of automobile anti-collision alarming system based on single chip computer. Mod. Electron. Tech. (19), 166–167, 172 (2009)
7. BENQ Corporation. AT Command User Manual, 2006.9, Version7.7

# Optimization Control for Wastewater Treatment Process Based on Data and Knowledge Decision

**Wei Zhang, Ruifei Bai and JiaoLong Zhang**

**Abstract** In this paper, a whole process optimization control (WPOC) method is proposed for the wastewater treatment process (WWTP). The WPOC method is studied under the scheme of hierarchical control. First, the intelligent decision part is designed based on the data and knowledge information of the system. The optimal direction is adjusted according to the preference of decision makers and the current system performance. Then, the weight coefficients of the performance indexes are provided to the optimization layer. The NSGA-II algorithm is adopted for solving the multi-objective optimization problem. The tracking control task is finished using the neural network control method. Simulation results, based on the international benchmark simulation model no. 1 (BSM1), show that WPOC method can achieve the energy saving with meeting effluent discharge, and the comprehensive evaluation of energy consumption and effluent quality is also improved.

**Keywords** Whole process optimization · Hierarchical control · Intelligent decision · Knowledge and data · Wastewater treatment

## 1 Introduction

Wastewater treatment process (WWTP) is a complex and energy-intensive system, whereas its operation must continuously work with effluent requirements. Moreover, stringent discharge standards and regulations have been introduced to protect the environment from the harmful effluent discharge to receiving waters [1, 2]. Therefore, from the points of energy saving and environment protection, optimal control of the WWTP has become potential developing strategy, and has attracted considerable attention recently.

W. Zhang (✉) · R. Bai · J. Zhang
School of Electrical Engineering and Automation,
Henan Polytechnic University, Henan 454000, China
e-mail: zwei1563@hpu.edu.cn



Y. Jia et al. (eds.), *Proceedings of 2018 Chinese Intelligent Systems Conference*,
Lecture Notes in Electrical Engineering 528,
https://doi.org/10.1007/978-981-13-2288-4_48

Some researchers have proposed some optimal control strategies, ranging from fuzzy logic to neural networks and non-linear techniques. Based on the benchmark simulation model 1 (BSM1), Vrecko et al. presented an analysis and optimization of a wastewater treatment benchmark, which provided general research foundation for understanding the optimization control of WWTP [3]. Piotrowski et al. proposed a two-level hierarchical controller for tracking the dissolved oxygen (DO) reference trajectory in activated sludge processes [4]. The upper level control unit generates trajectories of the desired airflows, and the lower-level control-ler forces the aeration system to follow these set-point trajectories. The proposed controller is validated by simulation based on real data records. Chachuat et al. studied the optimization of a small size wastewater treatment plants, where the problem was stated as a hybrid dynamic optimization problem and was solved using a gradient-based method [5]. The results show that the optimized aeration profiles lead to reductions of energy consumption of at least 30%. Based on the relative gain array (RGA) analysis, Machado et al. designed different effluent quality controllers and a cost controller for WWTPs [6]. The cost controller is proved to be a good tool for automating the search of the most profitable set-points of the effluent quality controllers for a given cost set-point. Chen et al. studied optimal design of an activated sludge process (ASP) using multi-objective optimization based on the BSM1 with four indexes of percentage of effluent violation (PEV), overall cost index (OCI), total volume and total suspended solids [7]. The results indicate that multi-objective optimization is a useful method for optimal design ASP. However, the optimal strategy mainly optimizes the design parameters of the system by an offline manner. The optimal control method from the perspective of the whole process optimization is scarce. Actually, it is necessary to carry the whole optimization of WWTP to improve the optimal performance further.

In this paper, the whole process optimization control (WPOC) scheme is proposed for WWTP, which has a three-layer hierarchical structure, including decision management layer, optimal layer and tracking control layer. The main idea of this paper is that the optimal control of WWTP is studied from the perspective of the whole process real-time optimization control, and the knowledge and data information of the system is used in merge way for the decision part. The decision direction is adjusted dynamically based on the decision-maker preference (knowledge information) and the current system performance (data information), and then the weight coefficients of performance indexes are provided to the optimization layer. The NSGA-II algorithm is adopted for solving the multi-objective optimization problem. The tracking control task is finished using the neural network control method. Simulation results show that the WPOC method can achieve energy saving effectively with meeting effluent discharge.

## 2 Scheme of WPOC Method

A Hierarchical control is an appropriate method for the complex process industry system [8, 9]. To improve the optimal performance, it is required to study the whole

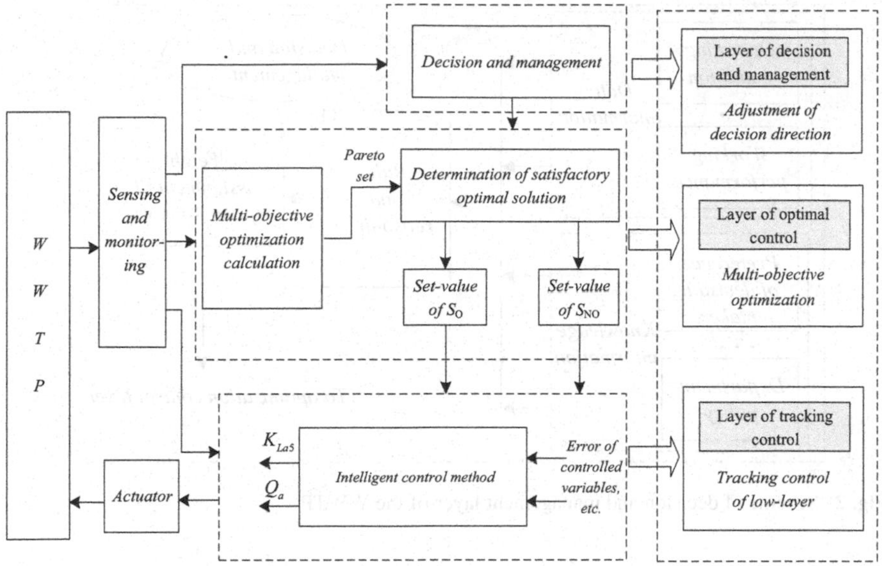


**Fig. 1** Hierarchical control structure of WPOC method for WWTP

process optimization range from the decision management to the tracking control. Moreover, with the development of management level, the role of intelligent decision is becoming more and more prominent and a superior decision management mechanism has an important impact on the information utilization and the guidance of the optimal direction. A hierarchical control structure is adopted as the whole process optimization control scheme in this paper, which is demonstrated in Fig. 1.

The hierarchical control in this paper mainly includes three layers: decision and management layer, optimal control layer and tracking control layer. The upper layer treats the decision information based on various resources. The optimal layer provides the optimal manipulated variables or optimal set-values, where the performance optimization is focused. The lower-layer mainly realizes the tracking control. In this paper, the design of the decision and management layer is paid more attention. The weight factors can be adjusted dynamically based on the data information of the system and the knowledge information from the decision-maker. In Optimization layer, the optimized set-values of the dissolved oxygen and nitrate concentration are obtained using the multi-objective optimization method. And the optimized set-values are tracked using the neural network control method.

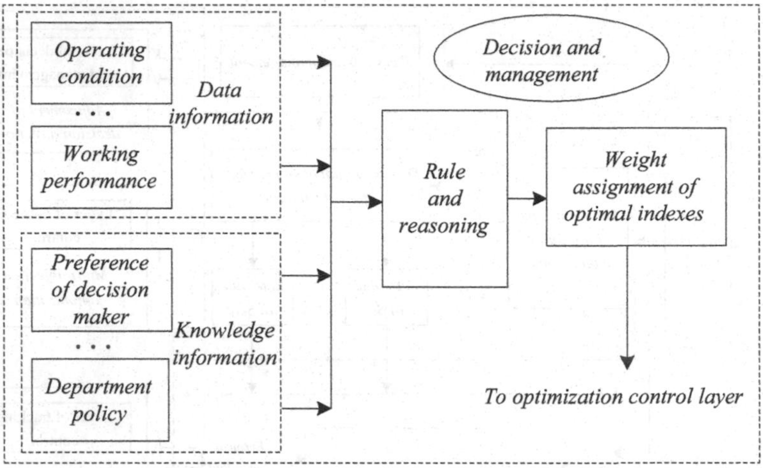


**Fig. 2** Scheme of decision and management layer of the WWTP

## 3 Design of Decision and Management Layer

An advantage of decision and management level is becoming more and more distinct with the system intelligent degree. The common method is employed based on the preference of decision makers, which is too strong subjectivity and the adjustment of external decision information cannot reflect into the system optimization. To reduce the subjectivity when assigning the weight coefficients, an intelligent decision method based on data and knowledge is proposed for the optimization control of WWTP. The direction of optimal decision can be adjusted dynamically according to the system performance, the preference of decision maker and the department policy information. The design scheme of decision and management layer is shown in Fig. 2.

The key issue for the decision layer is that how to adjust the weight coefficients based on the information of effecting the decision direction. In this paper, the information for determining the weight coefficients is divided into two types: one comes from the language information based on the knowledge (such as preference of the decision maker and department policy); another one comes from the process data information of WWTP. In this paper, we focus on the weight values assignment using the knowledge and data information simultaneously.

The indexes of energy consumption (EC) and effluent quality (EQ) are paid more attention in the WWTP. Thus the performance indexes of EC and EQ are studied in this paper. Denote the weight factors of EC and EQ as $\omega_1$ and $\omega_2$. First, the basic weight values ($\omega_{10}$ and $\omega_{20}$) are provided based on the preference of the decision maker and department, and this part is determined as the main body of weight coefficients. Then, the weight coefficients are adjusted dynamically based on the data information of WWTP (related to the real-time indexes of EC and EQ). The inference mechanism of weight adjustment adopts IF-THEN rules.

For the current optimal problem of WWTP, the basic weight values are given $\omega_{10} = 0.6$ and $\omega_{20} = 0.4$ considering the preference of the decision maker and department policy. Since the effluent quality has a much bigger influence in the overall performance index, the general evaluation index is chosen as $\mathrm{COST} = 50 \times \mathrm{EQ} + 25 \times \mathrm{EC}$ [11] that is used extensively in the optimal control of WWTP. After comprehensively analysis, the basis weight values are given $\omega_{10} = 0.5$ and $\omega_{20} = 0.5$ based on the knowledge information.

Denote $\rho_{EC}$ as the change rate of the EC in every optimal cycle; denote $\rho_{EQ}$ as the change rate of EQ in every optimal cycle. According to the system performance, we set the following rules for weight coefficients adjustment.

$$\omega_1 = \begin{cases} \omega_{1,0}, & \rho_{EC} < \alpha \\ 1, & \rho_{EC} \geq \beta \\ \omega_{1,0} + a_0 i, & \rho_{1,i}^l \leq \rho_{EC} < \rho_{1,i}^u \end{cases} \tag{1}$$

$$\omega_2 = \begin{cases} 1 - \omega_{1,0}, & \rho_{EQ} < \alpha \\ 1, & \rho_{EQ} \geq \beta \\ \omega_{2,0} + a_0 i, & \rho_{2,i}^l \leq \rho_{EQ} < \rho_{2,i}^u \end{cases} \tag{2}$$

where $\alpha$ and $\beta$ are the set-values of the performance indexes. $\rho_{1,i}^l$, $\rho_{1,i}^u$ are the lower limit and upper limit values of $\rho_{EC}$ respectively ($i = 1, 2, \ldots, n$). $\rho_{2,i}^l$, $\rho_{2,i}^u$ are the lower limit and upper limit values of $\rho_{EQ}$ respectively. $n$ is related to the fineness of the rules. $a_0$ is the adjustment increment of weight coefficients, and is set as a positive number less than 1. First, the weight of EC ($\omega_1$) is adjusted based on data information of the system. If the EQ index has a stable variation in the optimal cycle ($\rho_{EQ} < \alpha$), the weight of EQ ($\omega_2$) need not be adjusted. If the EQ index has a much bigger variation ($\rho_{EQ} > \alpha$), it means that the effluent equality need be emphasized and the weight $\omega_2$ should be adjusted. When the variation of EC and EQ both have the big values ($\rho_{EC} > \alpha$ and $\rho_{EQ} > \alpha$), the weight with the bigger variation is adjusted preferentially, and $\omega_1 + \omega_2 = 1$ should be satisfied.

## 4 Multi-objective Optimization Method

The optimization problem of WWTP belongs to a multi-objective optimization essentially. In this paper, NSGA-II algorithm is employed for solving the multi-objective problem. The multi-objective optimal model is built using the neural network, where the function relationships of the optimized variables (set-values of DO concentration and nitrate concentration) and optimal performance indexes are established. This method solves the difficulty of no existing optimization model of WWTP.

Denote $x_1(k)$ as the set-value of DO concentration, $x_2(k)$ as the set-value of nitrate concentration, $\mathbf{x}(k) = [x_1(k), x_2(k)]$; $f_{AE}(\mathbf{x})$ represents the function relationship between the optimized variables and aeration energy EC; $f_{PE}(\mathbf{x})$ represents the function relationship between the optimized variables and pump energy EQ; define $f_{EC}(\mathbf{x}) = f_{AE}(\mathbf{x}) + f_{PE}(\mathbf{x})$; $f_{EQ}(\mathbf{x})$ is represents the function relationship between the optimized variables and effluent equality index. And the multi-objective optimization model established can be expressed by (3).

$$\min \quad F(\mathbf{x}) = \{f_{EC}(\mathbf{x}), f_{EQ}(\mathbf{x})\}$$
$$\text{s.t.} \begin{cases} g_1(\mathbf{x}) = g_{NH}(\mathbf{x}) < 4 \\ g_2(\mathbf{x}) = g_{Ntot}(\mathbf{x}) < 18 \\ x_1^l < x_1(k) < x_1^u \\ x_2^l < x_2(k) < x_2^u \end{cases} \tag{3}$$

where $x_1^l$, $x_1^u$ are the lower limit and upper limit of DO concentration respectively; $x_2^l$, $x_2^u$ are the lower limit and upper limit of nitrate concentration respectively; $g_i(\mathbf{x})$ $(i = 1, 2)$ express the function relationships between the optimized variables and two key effluent parameters, and the function mapping relationships are finished by fuzzy neural network.

To handle the effluent constraints, the penalty function method is used. Let the penalty function define as

$$f_{\text{penalty}}(\mathbf{x}) = \max\{g_1(\mathbf{x}) - 4, 0\} + \max\{g_2(\mathbf{x}) - 18, 0\} \tag{4}$$

Then the optimal indexes with penalty item can be expressed by

$$\begin{cases} f_1'(\mathbf{x}) = f_{EC}'(\mathbf{x}) = f_{EC}(\mathbf{x}) + C \cdot f_{\text{penalty}}(\mathbf{x}) \\ f_2'(\mathbf{x}) = f_{EQ}'(\mathbf{x}) = f_{EQ}(\mathbf{x}) + C \cdot f_{\text{penalty}}(\mathbf{x}) \end{cases} \tag{5}$$

where $C$ is the penalty factor. Since a group of equal excellent Pareto solutions are provided by NSGA-II, only one satisfactory optimal solution is required for realizing the closed control. Thus, only one pair of the optimal set-values of DO concentration and nitrate concentration need be provided after multi-objective optimization.

The key step of determining the satisfactory optimal solution from the Pareto set is the weight assignment for optimal performance indexes. In this paper, the satisfactory optimal solution $(\mathbf{x}^K)$ is determined according to the defined utility function value, which is defined as (6).

$$d_{\text{utility}}(\mathbf{x}^p) = \sum_{i=1}^{2} \omega_i f_i' i(\mathbf{x}^p), \quad \sum_{i=1}^{2} \omega_i = 1, p = 1, 2, \ldots, k \tag{6}$$

$$K = \arg \min_{p=1,2,\ldots,k} \{d_{\text{utility}}(\mathbf{x}^p)\} \tag{7}$$

where $k$ is the number of solutions in Pareto set; $\omega_i$ is the weight value of each optimal performance index (that is determined by the part of decision and management layer). $K$ is the label of solutions corresponding to the maximum utility function value.

## 5 Tracking Control of Lower-Layer

The main task of tracking control layer is realizing the high-accuracy tracking and guaranteeing the system stability. To improve the control performance, a directed adaptive neural network control method is proposed in the previous study work [10] and this method is adopted as the lower layer control method in this paper. Further information can be found in the Ref. [12].

Since the DO concentration and nitrate concentration are two key process parameters, they are chosen as the controlled variables just as most of Ref. [13–15]. The manipulated variables are the oxygen transfer coefficient ($K_{La5}$) and the internal recycle flow rate ($Q_a$) for DO concentration and nitrate concentration respectively.

## 6 Simulation Experiment Studies

### *6.1 Experiment Design and Parameter Setting*

All of experiments are operated in the environment of MATLAB 2010b and based on the BSM1. The sample cycle is 15 min and optimal cycle is chosen as 2 h. The main aim of this experiment is to evaluate the optimal performance indexes (EC and EQ) and the overall cost function (COST) under WPOC method. In the part of decision management, the related parameters are set as follows: $\alpha = 0.05$, $\beta = 0.4$, $n = 4$, $a_0 = 0.1$. For the modeling of aeration energy and pump energy, the structure of neural network is chosen as 2-10-1; for the modeling of effluent parameters, the structure of neural network is chosen as 3-20-1 and the inputs of network are DO concentration, nitrate concentration and influent flow rate respectively; the learning rate of neural network is taken as $\eta = 0.01$. In NSGA-II algorithm, the parameters are set as: the variable dimension $D = 2$, the size of population $N = 40$, the maximum evolution epoch $M = 30$, the mutation rate $C_r = 0.9$. These parameters are proved to be appropriate for the current optimal control problem of WWTP.

## 6.2 Experiment Results and Analysis

In this experiment, dry weather data is chosen as the simulation study. The tracking control curves and optimal set-values of DO concentration and nitrate concentration under the proposed WPOC method is shown in Fig. 3.

From the simulation results, it can be seen that the optimized set-values of DO concentration and nitrate concentration are adjusted dynamically with the working condition of WWTP, and the variation basically matches the influent characteristics of WWTP. Simultaneously, the lower-layer controller exhibits an excellent tracking performance within the optimal cycle.

The curve of manipulated variables (oxygen transfer coefficient $K_{La5}$ and internal recycle flow rate $Q_a$) under different control methods are demonstrated in Fig. 4, including PID method, the proposed WPOC method and MOO-PID method. In MOO-PID method, the weight coefficients are fixed and PID method is adopted for low-layer tracking.

From the variation curve of manipulated variables, we can see that there is much more similar alteration for the aeration rate $K_La_5$ under different control methods, especially for the weekend. Variation of $K_La_5$ is much slower under PID that means PID control method has a good control performance within the scope of some certain error limit, but further accuracy control is hard to realize using PID method. The average aeration rate under two optimal control methods is slightly lower than the PID, and thus the aeration consumption (EC) can be decreased accordingly. For $Q_a$, the variation under three control strategies is much more obvious and the general tendency is that the average value of $Q_a$ under two optimal control methods is much higher than PID. Variation of $Q_a$ is rapid and the value is much bigger under the proposed WPOC method compared with PID method, which implies that intensifying the internal recycle rate can improve the removing rate of pollutants in WWTP. Since lower-layer tracking control performance and the guidance of decision direction have an important influence on the whole optimal performance, the proposed strategy can obtain satisfactory optimal performance.

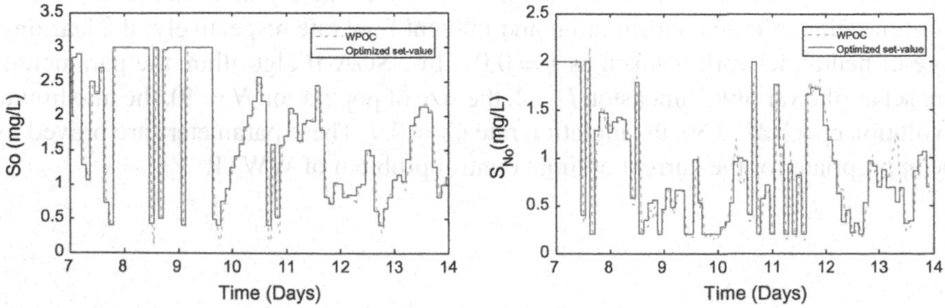


(a)Tracking performance of DO (b) Tracking performance of and nitrate

**Fig. 3** Tracking performance of DO concentration and nitrate concentration under WPOC

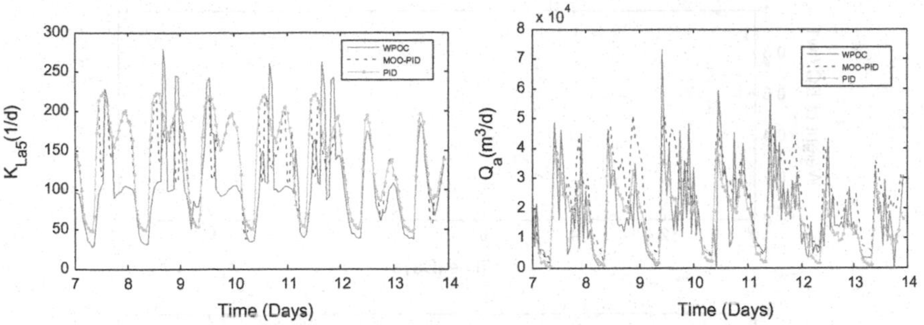


(a) Variation of $K_{La5}$ (b) Variation of $Q_a$

**Fig. 4** Variation of manipulated variables ($K_{La5}$ and $Q_a$)

**Table 1** A comparison of optimal performance indexes under different control methods

| Methods | AE (kWh $d^{-1}$) | PE (kWh $d^{-1}$) | EC (kWh $d^{-1}$) | EQ (kg pollution $d^{-1}$) | COST |
|---|---|---|---|---|---|
| PID | 3675.07 | 231.47 | 3906.54 | 6567.31 | – |
| MOO-PID | 3515.32 | 246.05 | 3761.37 | 6898.17 | 438942.75 |
| WPOC | 3531.39 | 260.20 | 3791.59 | 6733.12 | 431445.75 |

The comparison of optimal performance indexes discussed under different control methods, including AE, PE, EC, EQ and total COST, are given in Table 1.

It can be seen that the values of AE, PE, EC and EQ under WPOC are 3531.39, 260.20, 3791.59 and 6733.12 respectively. Compared with PID method, the AE index is decreased by 3.90%, the PE index is increased by 12.41%, the sum of energy consumption (EC) is decreased by 2.94%, and the energy saving is much more obvious. Compared with MOO-PID (multi-objective optimal method is adopted, PID is used as the tracking control and no intelligent decision design is considered), the similar EC decrement is obtained by the MOO-PID and WPOC method. However, for the effluent EQ index, a much lower EQ value is achieved by WPOC and thus the total COST index is decreased by 1.71% compared with MOO-PID method.

Since the indexes of EC and EQ are a pair of conflicting performance indexes, which means EC is decreased while EQ is increased and vice versa. The multi-objective optimal method can achieve a satisfactory optimal solution based on the Pareto set and an appropriate decision guidance. Besides the knowledge information of the decision maker and policy, the data information (it can reflect the current system performance) is also used for the decision adjustment in our study. Therefore, the optimal weight factors can be achieved dynamically. The adjustment process of weight factors is demonstrated in Fig. 5.

Several key effluent parameters under three different control strategies are demonstrated in Table 2.

It can be drawn from the values in Table 2 that five key effluent parameters can satisfy the effluent discharge standard under all of three methods. The average values

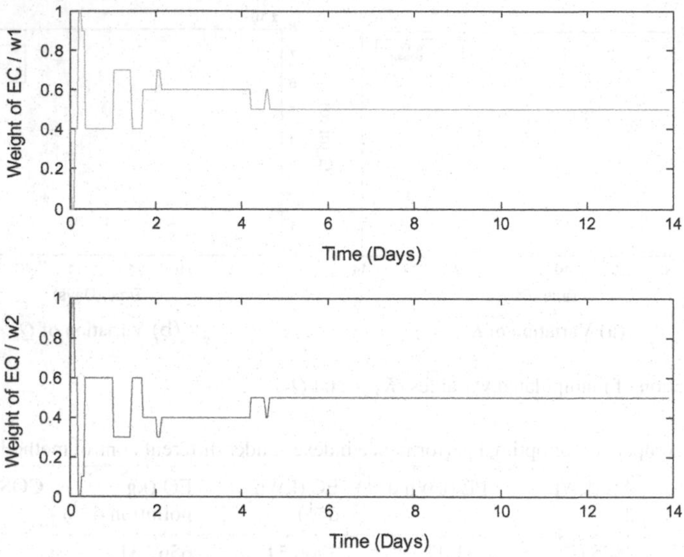


**Fig. 5** Weight factors adjustment of EC and EQ

**Table 2** A comparison of effluent pollutant concentration

| | $BOD_5$ (mg/l) | COD (mg/l) | $N_{tot}$ (mg/l) | $S_{NH}$ (mg/l) | TSS (mg/l) |
|---|---|---|---|---|---|
| Influent | 183.49 | 167.31 | 81.62 | 30.14 | 198.57 |
| PID | 2.67 | 47.48 | 17.24 | 2.31 | 12.60 |
| MOO-PID | 2.90 | 46.36 | 16.34 | 3.46 | 12.71 |
| WPOC | 2.89 | 46.35 | 16.71 | 3.17 | 12.72 |

of effluent parameters (COD, BOD5 and TSS) have a little change, but the indexes of effluent $N_{tot}$ and effluent $S_{NH}$ have a big change under different control strategies. Removing rates of average effluent total nitrogen ($N_{tot}$) under WPOC methods are 79.98%, which are satisfactory. For ammonia-nitrogen ($S_{NH}$) removing rate, this value can reach up to 89.48%.

Analyzing all experiment results above, it can be seen that the multi-objective optimal methods can obtain better optimal performance and can realize energy saving compared with PID. Since WPOC method adopts the intelligent decision and the lower-layer controller with better performance, a superior effluent quality is obtained under a comparable EC index, and the overall evaluation index (COST) is reduced largely.

## 7 Conclusion

In this paper, the WPOC method is proposed for the WWTP, where the hierarchical control structure is adopted. The control system is composed by intelligent decision layer, multi-objective optimal layer and tracking control layer. The decision adjustment is realized based on the knowledge information and the data information, where the weight factors can be adjusted dynamically together with the system working condition. The NSGA-II algorithm is adopted for solving the multi-objective optimization problem. The tracking control task is finished using the neural network control method. The WPOC method provides a significant benefit for the comprehensive evaluation of energy consumption and effluent quality. However, it still is a challenging task for the optimal control of WWTP from the perspective of whole process optimization. Some research work need to be done further.

**Acknowledgements** This work is supported by National Science Foundation of China under Grant 61703145, Doctor Fund Project of Henan Polytechnic University of China under Grant B2017-21.

## References

1. R. Hamilton, B. Braun, R. Dare et al., Control issues and challenges in wastewater treatment plants. IEEE Control Syst. Mag. **26**(4), 63–69 (2006)
2. H.G. Han, H.H. Qian, J.F. Qiao, Nonlinear multi-objective model-predictive control scheme for wastewater treatment process. J. Process Control **24**(3), 47–59 (2014)
3. D. Vrecko, N. Hvala, J. Kocijan et al., System analysis for optimal control of a wastewater treatment benchmark. Water Sci. Technol. **43**(7), 199–206 (2001)
4. R. Piotrowski, M.A. Brdys, K. Konarczak et al., Hierarchical dissolved oxygen control for activated sludge processes. Control Eng. Pract. **16**(1), 114–131 (2008)
5. B. Chachuat, N. Roche, M.A. Latifi, Dynamic optimisation of small size wastewater treatment plants including nitrification and denitrification processes. Comput. Chem. Eng. **25**(4–6), 585–593 (2001)
6. V.C. Machado, D. Gabriel, J. Lafuente et al., Cost and effluent quality controllers design based on the relative gain array for a nutrient removal WWTP. Water Res. **43**(20), 5129–5141 (2009)
7. W.L. Chen, C.H. Yao, X.W. Lu, Optimal design activated sludge process by means of multi-objective optimization: case study in Benchmark Simulation Model 1 (BSM1). Water Sci. Technol. **69**(10), 2052–2058 (2014)
8. I. Santin, C. Pedret, R. Vilanova, Applying variable dissolved oxygen set point in a two level hierarchical control structure to a wastewater treatment process. J. Process Control **28**(1), 40–55 (2015)
9. P. Vega, S. Revollar, M. Francisco et al., Integration of set point optimization techniques into nonlinear MPC for improving the operation of WWTPs. Comput. Chem. Eng. **68**, 78–95 (2014)
10. J. Guerrero, A. Guisasola, J. Comas et al., Multi-criteria selection of optimum WWTP control setpoints based on microbiology-related failures, effluent quality and operating costs. Chem. Eng. J. **188**, 23–29 (2012)
11. B. Beraud, J.P. Steyer, C. Lemoine et al., Towards a global multi objective optimization of wastewater treatment plant based on modeling and genetic algorithms. Water Sci. Technol. **56**(9), 109–116 (2007)
12. W. Zhang, Multi-objective intelligent optimization control study of wastewater treatment process (Beijing University of Technology, 2016)

13. W. Zhang, J.F. Qiao, Direct adaptive neural network control for wastewater treatment process, in *Proceedings of the 11th World Congress on Intelligent Control and Automation, WCICA*, Shenyang, China, 2014
14. J.F. Qiao, G. Han, H.G. Han, Neural network on-line modeling and controlling method for multi-variable control of wastewater treatment processes. Asian J. Control **16**(4), 1213–1223 (2014)
15. N.A. Wahab, R. Katebi, J. Balderud, Multivariable PID control design for activated sludge process with nitrification and denitrification. Biochem. Eng. J. **45**(3), 239–248 (2009)

# Distributed Cooperative Guidance Strategy for Multi-missile Attack the Maneuvering Target

**Xi Zhang, Yonghai Wang, Yongtao Shui, Linlin Wang and Pinghui Jia**

**Abstract** This paper proposed a distributed cooperative guidance strategy in order to achieve multi-missile cooperative attack the maneuvering target. Firstly, $H_\infty$ robust guidance law is adopted to intercept the maneuvering target, which can keep strong robustness and eliminate the prediction of the acceleration of the target. Then the theory of network synchronization is used to design the cooperative guidance components, which is no need to estimate the impact time manually and can make state of all the missiles gradually converge automatically. Finally, De Bruijn network is used to build the local interactions, which has good fault tolerant property, flexibility and high reliability to support large-scale missiles cooperative attacks. The simulation results verifies that the law can drive all the missiles salvo attack the maneuvering target simultaneously.

**Keywords** $H_\infty$ control · De Bruijn network · Network synchronization · Cooperative guidance law

## 1 Introduction

With the continuous enhancement of missile defense system, great challenges have been posed to the capacities of a single missile for defense and attack. As a strategy of collective battling, cooperative operation of multi-missile is an effective measure to improve the damage probability and efficiency by cooperation and collaboration for multi-missile with communication data sharing.

In the previous study of the cooperative guidance, time coordination variable rely on expected designated value [1–3], which is not necessary and generally hard to be obtained in practice. And coordination only needs to meet the demand condition that missiles can reach the destination simultaneously. Distributed network synchronization [4, 5] is applied to construct the cooperative guidance law based

X. Zhang (✉) · Y. Wang · Y. Shui · L. Wang · P. Jia
Beijing Institute of Space Long March Vehicle, Beijing 100076, China
e-mail: calt14_zhangx3@163.com



Y. Jia et al. (eds.), *Proceedings of 2018 Chinese Intelligent Systems Conference*,
Lecture Notes in Electrical Engineering 528,
https://doi.org/10.1007/978-981-13-2288-4_49

on leader-follower strategy [6–8], in which every missile communicates with the adjacent local missiles. Although conventional distributed control only needs a little state information, the effect of distributed control is relatively worse since the information interaction is insufficient among missiles. And the network of conventional distributed control is vulnerable to be disrupted with single missile failed and hard to be extended flexibly. Furthermore, there is less study on cooperative guidance law for maneuvering target interception.

For solving above problems and improving the application value in practical, this paper introduces a new cooperative guidance strategy that has never been proposed previously. Firstly, $H_\infty$ robust guidance law is applied to the interception of the maneuvering target, which is notable for the strong robustness under the external interference. Then, the principle of network synchronization as the guarantee of missiles cooperation is used in designing the cooperative guidance components so that the state of missiles can converge automatically. Finally, On the basis of the undirected De Bruijn network, we design the interactions topology among the missiles, which can share and exchange information better with fault tolerant, flexibility and high reliability.

## 2 Description of $H_\infty$ Robust Guidance Problem

In this paper, suppose that $N(N \geq 2)$ tactical missiles participate in a salvo attack. Consider a two-dimensional planar engagement scene of missile-target illustrated in Fig. 1, where the velocity of all missiles is constant and the target is in maneuvering flight. It is assumed that both missiles and target are viewed as rigid body without dynamic delay, so only kinematics are considered. The location of missiles and target are denoted by $M_{\mathrm{i}}$ and $T$ respectively. Subscript $i$ is the serial number of the missile, and $R_i$, $v_i$, $q_i$, $\theta_i$ represent the relative distance, velocity, line-of-sight (LOS) angle and lead angle of $i$th missile respectively in the geometry. The planar $M_{\mathrm{i}}$-T relative motion equations are described by

$$\dot{R}_i = v_T \cos(q_i - \theta_T) - v_i \cos(q_i - \theta_i) \tag{1}$$

$$R_i \dot{q}_i = -v_T \sin(q_i - \theta_T) + v_i \sin(q_i - \theta_i) \tag{2}$$

Differentiate Eq. (2) with respect to time and plug Eq. (1) into Eq. (2), and we can obtain

$$\ddot{q}_i = -\frac{2\dot{R}_i}{R_i}\dot{q}_i - \frac{1}{R_i}u_{q_i} + f_{q_i} \tag{3}$$

where

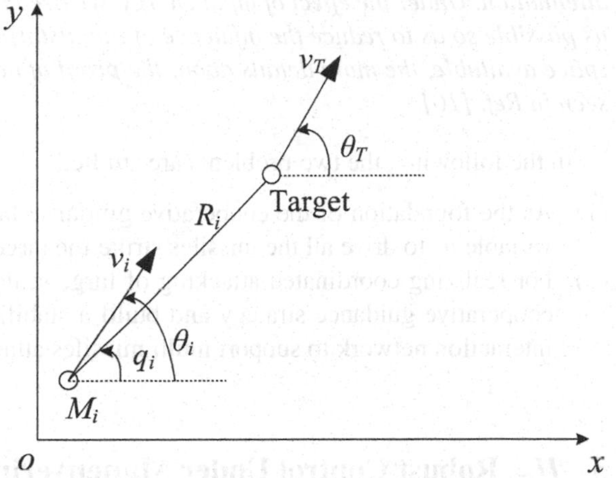

**Fig. 1** Missile-target engagement geometry

$$u_{q_i} = v_i \dot{\theta}_i \cos(q_i - \theta_{M_i}) - \dot{v}_i \sin(q_i - \theta_i)$$
$$f_{q_i} = \frac{1}{R_i}(v_T \dot{\theta}_T \cos(q_i - \theta_T) - \dot{v}_T \sin(q_i - \theta_T))$$

Define the new variables $(u_{q_i}, f_{q_i})$, where $u_q$ denotes the normal overload that is used as the control input to the model and $f_{q_i}$ is assumed to be an unpredictable external disturbance. In practical engineering application, it is difficult to realize the control from the radial velocity of the missile. Thus, the paper just consider the control variable of the normal acceleration to design the guidance law.

For reflecting the guidance performance of the missile, we need to specify the output signal z to be controlled. A good guidance algorithm must ensure the LOS angle rate as small as possible and less affected by the maneuvering target. So we choose $z_{q_i}$ as

$$z_{qi} = k\dot{q}_i \tag{4}$$

where $k(k>0)$ is the weight coefficient about the LOS angle rate.

**Lemma 1** *The problem of the $H_\infty$ robust guidance law design can be described as follows: by giving the weight coefficient k appropriately and choosing the control variable $u_{q_i}$, we must make the gain of the system described [9] by Eq. (3) is lower than or equal to $\gamma$, that is*

$$\frac{\int_0^\infty z_{qi}^T z_{qi} dt}{\int_0^\infty f_{qi}^T f_{qi} dt} \le \gamma^2 \tag{5}$$

*where $\int_0^\infty f_{qi}^T f_{qi} dt$ denotes the input energy of the system, $\int_0^\infty z_{qi}^T z_{qi} dt$ denotes the output energy of the system, $\gamma$ is the inhibitory gain coefficient that reflect disturbance* *attenuation. Under the effect of input energy, we need to keep the value of $\gamma$ as small as possible so as to reduce the influence of the disturbance outside. Due the limited space available, the more details about the proof of nonlinear robust theory can be seen in Ref. [10].*

In the following, the two problems are studied:

(1) As the foundation of the cooperative guidance law, how to design the control variable $u_q$ to drive all the missiles strike the target simultaneously.
(2) For realizing coordinated attacking of large scale missiles, how to design the cooperative guidance strategy and build a stability and highly fault tolerance interaction network to support multi missiles attack.

## 3 $H_\infty$ Robust Control Under Maneuvering Target

In this paper, $H_\infty$ control theory is applied to design the robust guidance law on the basis of the $M_i$-T relative motion model. The normal acceleration is used as the control variable and target maneuvering is viewed as the external interference in the guidance law, then a controlled output signal is chosen to reflect the performance on the missile trajectory. Furthermore, the application of Lyapunov stability analysis method [11, 12] can guarantee the inhibitory effect of the control variable on external interference.

When target maneuvering is not taken into consideration, that is $f_q = 0$, the intercept model of $M_i$-T can be given by

$$\ddot{q}_i = -\frac{2\dot{R}_i}{R_i}\dot{q}_i - \frac{1}{R_i}u_{q_i} \tag{6}$$

Under the guidance, the LOS angle rate will converge to zero for diminishing the miss distance when approaching the target. We can make the LOS angle rate decrease by exponential form if the control variable is designed properly. Then the control variable $u_{q_i}$ is proposed as follows

$$u_{q_i} = -2\dot{R}\dot{q}_i + nR\dot{q}_i \tag{7}$$

where $n > 0$ is the parameter needed to be set.

By substituting Eq. (7) into Eq. (6), the loop system of the missile under the action of control variable is shown as

$$\ddot{q}_i = -n\dot{q}_i \tag{8}$$

The important part of the guidance design is to choose a appropriate value of $n$ so as to restrain the disturbance outside and meet the requirement of Eq. (5). In the following sections, the inference of $n$ will be given.

Defining the positive definite Lyapunov function [13] $V$ as follows

$$V = \frac{1}{2}\dot{q}_i^2 \tag{9}$$

Differentiating Eq. (9) with respect to time and then substituting the control variable $u_{q_i}$, then yields

$$\dot{V} = -n\dot{q}_i^2 + \dot{q}_i w_{q_i} \tag{10}$$

Substituting the inequality: $q_i w_{qi} \leq \frac{\gamma^2}{2} w_{qi}^2 + \frac{1}{2\gamma^2} q_i^2$ into Eq. (10), and using the integral transformation, then induces

$$V(t_f) - V(t_0) \leq \frac{\gamma^2}{2}\int_{t_0}^{t_f} w_{q_i}^2 dt - (n - \frac{1}{2\gamma^2})\int_{t_0}^{t_f} \dot{q}_i^2 dt \tag{11}$$

When the missiles intercept the maneuvering target successfully, the relative distance and the LOS angle rate will gradually converge to zero. Because the relative distance is small relatively, the LOS angle can change sharply. So the relationship between the initial state and the terminal state is: $V(t_f) \geq V(t_0)$. After applying the relationship, rearranging Eq. (11) obtains

$$\frac{2}{h^2}(n - \frac{1}{2\gamma^2})\int_{t_0}^{t_f} z_{q_i}^2 dt \leq \gamma^2 \int_{t_0}^{t_f} w_{q_i}^2 dt \tag{12}$$

When the Eq. (12) meet the condition of robust performance bound Eq. (5), we can compute the range of $n$.

$$n \leq \frac{h^2}{2} + \frac{1}{2\gamma^2} \tag{13}$$

By choosing the proper value of $n$ on the basis of Eq. (13) and substituting it into the control variable $u_{q_i}$, thus the guidance law can limit the LOS angle rate to an expected boundary so that the capability of intercepting process can satisfy the requirement.

It is obvious that the form of $u_{q_i}$ in Eq. (7) consists of two pieces:

(1) The former item is the PNG guidance law at navigation coefficient 2.
(2) The correction item is for reducing the effect of disturbance outside. When the inhibitory gain coefficient $\gamma$ is decreasing or weight coefficient $k$ is increasing, it means that the target maneuvering must have smaller effect on the LOS angle rate. Thus, by adjusting the value of $n$ to get larger, the control variable can meet the requirements of trajectory of missiles.

# 4 Distributed Cooperative Guidance Strategy

## 4.1 Cooperative Guidance Algorithm

Transforming the Eqs. (1)–(2) into the standard state-space form Eq. (14), rewrite the missile-target dynamic model as

$$\begin{bmatrix} \dot{R}_i \\ \dot{q}_i \end{bmatrix} = \begin{bmatrix} v_T \cos(q_i - \theta_T) - v_i \cos(q_i - \theta_i) \\ (-v_T \sin(q_i - \theta_T) + v_i \sin(q_i - \theta_i))/R_i \end{bmatrix} \tag{14}$$

Considering the theory of network synchronization [4] for distributed dynamic network, we start for instance with the synchronized network

$$\dot{x}_i = f(x_i, t) + \sum_{j \in \Omega_i} k_{ji}(x_j - x_i)\ i = 1, 2 \ldots, N \tag{15}$$

Thus the form of distributed cooperative guidance strategy is induced as

$$\begin{bmatrix} \dot{R}_i \\ \dot{q}_i \end{bmatrix} = \begin{bmatrix} v_T \cos(q_i - \theta_T) - v_i \cos(q_i - \theta_i) \\ (-v_T \sin(q_i - \theta_T) + v_i \sin(q_i - \theta_i))/R_i \end{bmatrix} + \begin{bmatrix} u_{ic1} \\ u_{ic2} \end{bmatrix} \tag{16}$$

where $u_{ci} = \begin{bmatrix} u_{ic1} & u_{ic2} \end{bmatrix}^T$ represents the control component of cooperative guidance strategy, and the form of $u_i$ is defined as

$$u_i = \sum_{j \in \Omega_i} k_{ji}(x_j - x_i) \text{ with } k_{ji} = \begin{bmatrix} \phi_{ij} & \delta_{ij} \\ \delta_{ij} & \gamma_{ij} \end{bmatrix}$$

where $k_{ji}$ specified as the positive definite weighted adjacent matrix, of which the elements are positive real number. Denote $\Omega_i$ as the set of local communication network for $M_i$ (a collection of other missiles that have communications with $M_i$).

**Lemma 2** *To achieve cooperative interception of multi missiles, it is necessary to make the state information of $M_i$ transformation:* $\lim_{t\to\infty}|r_i - r_j| \to 0$, $\lim_{t\to\infty}|\eta_i - \eta_j| \to 0$. *Therefore, the study of the cooperative guidance law focus on designing the control input of $u_{ci}$.*

In summary, the additional control component of cooperative guidance is applied to make all missiles attack simultaneously on the basis $H_\infty$ robust guidance law, and the form of control component can be chosen in accordance with the synchronization theory [5]. Thus, the state information of $M_i$ is related to the other missiles which are in the set of communication network $\Omega_i$. The next section will focus on the construction of $\Omega_i$.

## 4.2 De Bruijn Communication Network

In this section, the interaction topology network of information exchange among the missiles can be described by the undirected De Bruijn graph. Let $G = \langle V, E \rangle$ denote the De Bruijn graph [14], where $V$ and $E$ are the set of nodes and edges respectively.

**Definition** Assuming that there are $N$ missiles in total participate in the attack, let the node $i(0 \le i \le N - 1)$, denote the missile $M_{i+1}$ and let the edge be the information link between $M_i$ and $M_j$.

**Definition** Specified $d_i$ as the degree of node $i$, which means the number of edges connected to the node $i$.

**Definition** Denoted $c_{ij}$ be the number of paths which do not have any node in common [16].

The undirected binary for different values of $m$ with radix $r = 2$ $(N = 2^m)$ can be applied to build various interactions, that is, we will represent a node $i$ by a $m$-bit binary number as the form of $(i_{m-1}, i_{m-2}, \ldots, i_0)$. It is noticeable that the node $i$ will have information interaction with node $j$, if meets the condition [15] Eq. (17)

$$i_w = j_{w-1}(1 \le w \le m - 1) \quad \text{or} \quad i_w = j_{w+1}(0 \le w \le m-2) \tag{17}$$

Figure 2a illustrates the system communication topology $n$=8 and $r$=2. It may be noted that in this topology there are always two node disjoint paths between any two nodes in the topology, which can enhance the fault-tolerance of the system by increasing the value of radix $r$.

*Remark 1* It is noticeable that nodes is vulnerable to be attacked in practical application. The degree $d_i$ is viewed as the number of backup in the communication graph, and it can keep communication normally with any node failure. Apparently, the De Bruijn network has good fault tolerant to avoid the interruption of communication.

*Remark 2* Compared with increasing by factorial times in traditional communication graph, the number of the network connections grows linearly, so it will improve the economic feasibility for large-scale system.

*Remark 3* In most cases, the total number $n$ is not just equivalent to $r^m$, then we will choose a new number $n'$ as substitute which is slightly greater than $n$ to meet the demand of the equation $n' = r^m$, and the spare nodes $n' - n$ can be used as a backup of virtual nodes for supplementing the needed nodes timely. Thus the reliability of the network can be greatly strengthened.

A 4-missile communication topology based on binary De Bruijn network ($r$=2, $m$=4) is depicted in Fig. 2b, where the binary code from $M_1$ to $M_3$ is (0, 0), (0, 1), (1, 0) and (1, 1) respectively. Thus in accordance with the condition Eq. (17), we can obtain the connection among the missiles. Thus $i$-th missile can obtain information from which missiles are decided.

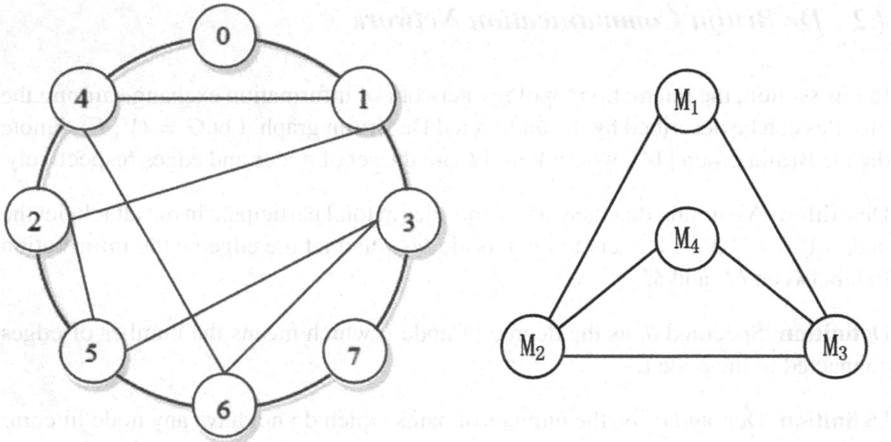


(a) A 8-node interaction topology

(b) A 4-node interaction topology of missiles

**Fig. 2** Interaction topology graph

### 4.3 Analysis of Convergence Conditions of Cooperative Guidance

**Definition** The weighted laplacian matrix of communication topology $G$ between $M_i$ and $M_j$ is defined as, $L_f = (l_{ij})_{N \times N} \in R^{N \times N}$, thus,

$$l_{ij} = \begin{cases} \sum_{j \in \Omega_i} \alpha_{ij} k_{ij} & i = j, i = 1, 2, \ldots, N \\ -\alpha_{ij} k_{ij} & i \neq j, j \in \Omega_i, i = 1, 2, \ldots, N \end{cases}$$

where $\alpha_{ij}$, $k_{ji}$ represent connecting coefficient and weight respectively, $\alpha_{ij}$ is defined as:

$$\alpha_{ij} = \begin{cases} 1 & M_j \text{ is connected to } M_i \\ 0 & \text{others} \end{cases}$$

**Theorem 1** *The states $x_i$ of all the missiles will converge together [4] asymptotically if the coupling strengths are strong enough and Eq. (18) is verified.*

$$\lambda_{min}(L_f) > \max_{i=1}^{N} \lambda_{max}\left(\frac{\partial f}{\partial x}(x_i, t)\right) \tag{18}$$

*where $\lambda$ represents an eigenvalue. Equation (18) is the sufficient condition for convergence of distributed cooperative guidance law. Thus we can make the state infor-*

**Table 1** Initial parameters and simulation results

| Missiles | $R/m$ | $q/\text{deg}$ | $\theta/\text{deg}$ | $t_{png}/\text{s}$ | $t_{co}/\text{s}$ |
|---|---|---|---|---|---|
| $M_1$ | 15,768 | 2.73 | 40 | 88.75 | 19.88 |
| $M_2$ | 925 | 4.09 | 45 | 60.31 | 19.88 |
| $M_3$ | 925 | −4.09 | 30 | 57.56 | 19.88 |
| $M_4$ | 15,768 | −2.73 | 60 | 88.97 | 19.88 |

*mation converge:* $\lim_{t\to\infty}|r_i - r_j| \to 0$, $\lim_{t\to\infty}|\eta_i - \eta_j| \to 0$. *The proof of Theorem 1 is based on the contraction theory [11], the details of which can be found in [4].*

## 5 Simulation Results

Given an engagement scenario that four missiles by applying distributed cooperative guidance law salvo attack a maneuvering target at (10000, 0). The distributed cooperative guidance law proposed consists of two parts: firstly, $H_\infty$ robust control is adopted as the basis guidance law to realize the attack of maneuvering target, then the distributed cooperative guidance strategy is designed so that we can obtain the cooperative guidance components in order to achieve the cooperative strike for multi missiles.

Assuming four missiles located in different initial positions have a constant speed at 400 m/s. And the target implements sine level maneuver flight, of which the normal acceleration is $a_T(t) = 6g\,\sin(\pi t/6)$ and have the speed at 200 m/s. In order to verify the performance of the guidance law, the traditional PNG law and the cooperative guidance law proposed in this paper are compared in the same simulation conditions presented as Table 1, where the PNG law is shown as: $u_{png=} - K\dot{\theta}$ with the navigation coefficient $K$ as 4. In the robust guidance law, the external interference will be restrained if the gain coefficient is set less than 1, so the value of $\gamma$ and $h$ are chosen as 0.5 and 1.5 and $n$ can be obtained by Eq. (13). The interaction communication topology among 4 missiles is shown as Fig. 2.

The trajectory by conventional PNG law is depicted in Fig. 3a, with different attack time $t_{png}$ from maximum 88.97 to 57.56 s. However the trajectory adopting cooperative guidance law is shown in Fig. 3b, where all missiles can reach the target simultaneously at 19.88 s denoted by $t_c$, which has greatly reduced the time difference. The simulation results tells that the cooperative guidance law can drive all the missiles to attack the maneuvering target simultaneously. By applying $H_\infty$ robust guidance law, it is obviously superior to the PNG law that the trajectory is more sensitive to the maneuvering target and straight relatively, which reduces the required overload. In the terminal stage of actual attack, the missile slows down and available overload provided by the rudder based on air lift decreases, thus the law under cooperative guidance is more beneficial for overload allocation in the flight.

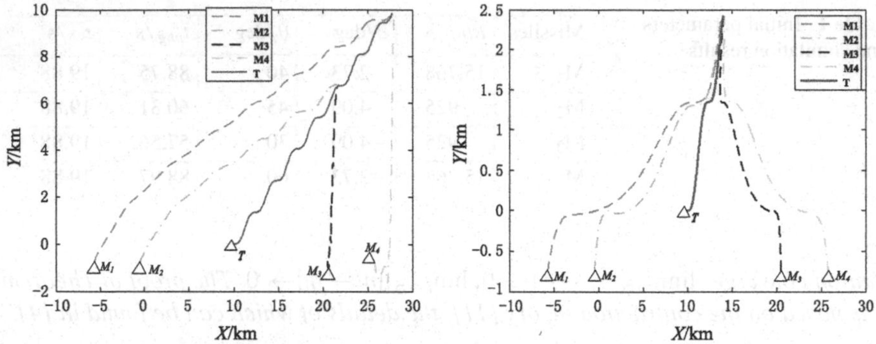


(a) Trajectory by PNG law (b) Trajectory by Cooperative guidance law

**Fig. 3** Trajectory by different guidance law

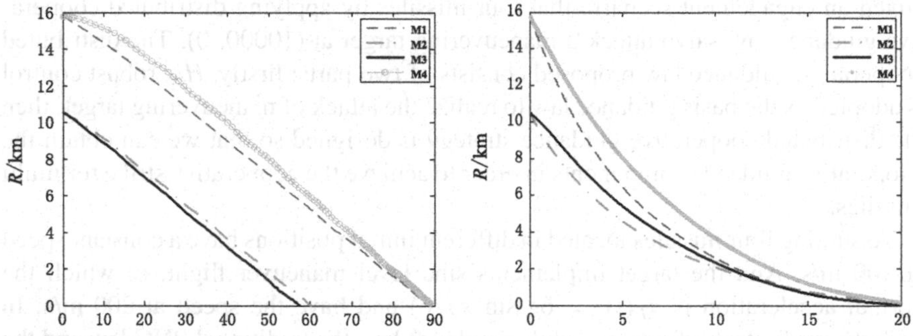


(a) Relative distance by PNG law (b) Relative distance by cooperative guidance law

**Fig. 4** Relative distance by different guidance law

The relative distance by cooperative guidance law is illustrated in Fig. 4b, where the distance gradually converges to zero of all missiles. Compared with PNG law, the distance decrease approximately linearly, which is out of the demand of Lemma 2. The angle $q$ under the normal control depicted in Fig. 5a keeps a small amplitude concussion with the amplitude of less than 1.3°, which is within the available range of the control component. The control components of distributed cooperative guidance law can be seen in Fig. 5b.

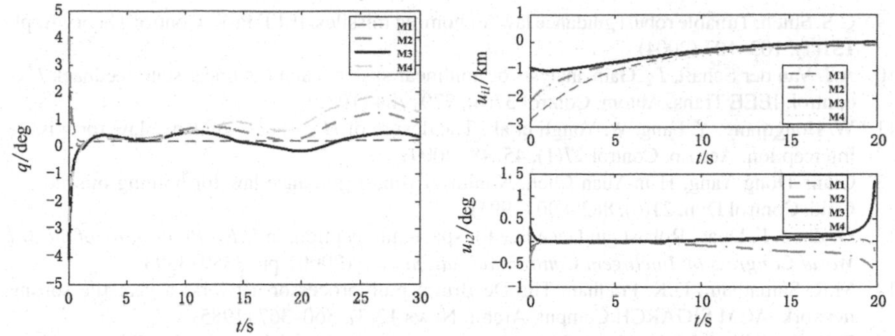


(a) The LOS angle $q$ (b) Control component $u_{i2}$, $u_{i1}$

**Fig. 5** Angle $q$ and control component $u_{i1}$, $u_{i2}$

# 6 Conclusions

This paper introduced a distributed cooperative guidance law to achieve multi-missile cooperative intercept a maneuvering target on the basis of $H_\infty$ robust guidance law, the theory of network synchronization and De Bruijn network, which has strong robustness in attacking the maneuvering target The paper is no need to estimate the impact time and can realize the state of all missiles converge automatically in short time. It is noted that the De Bruijn communication network has high stability and fault tolerance to support large-scale missiles engaged in the combat, so the communication performance will not be influenced by some single failed. The numerical simulations is performed to verify the effectiveness of the guidance law.

# References


1. I.S. Jeon, J.I. Lee, M.J Tahk, Impact-time-control guidance law for anti-ship missiles. IEEE Trans. Control Syst. Technol. **14**(2), 260–266 (2006)
2. S. Zhao, R. Zhou, Cooperative guidance for multi-missile salvo attack. Chin. J. Aeronaut. **21**(6), 533–539 (2008)
3. I.S. Jeon, J.I. Lee, M.J. Tahk, Homing guidance for cooperative attack of multiple missiles. J. Guid. Control Dyn. **33**(1), 275–280 (2010)
4. J.J.E. Slotine, W. Wang, *A Study of Synchronization and Group Cooperation Using Partial Contraction Theory*. Lecture Notes in Control and Information Sciences, vol. 309 (2004), pp. 443–446
5. Wang Wei, J.J.E. Slotine, A theoretical study of different leader roles in networks. IEEE Trans. Autom. Control **51**(7), 1156–1161 (2006)
6. J. Zhou, J. Yang, Cooperative simultaneous attack with multiple missiles under leader-follower communication graphs, in *Control Conference IEEE* (2016), pp. 7763–7768
7. X. Sun, R. Zhou, J. Wu et al., Distributed cooperative guidance and control for multiple missiles. J. Beijing Univ. Aeronaut. Astronaut. **40**(1), 120–124 (2014)
8. L. Zou, F. Kong, R. Zhou et al., Distributed adaptive cooperative guidance for multi-missile salvo attack. J. Beijing Univ. Aeronaut. Astronaut. **38**(1), 128–132 (2012)

9. C.S. Shieh, Tunable robust guidance law for homing missiles. IEEE Proc. Control Theory Appl. **151**(1), 103–107 (2004)
10. A.J. Van der Schaft, $L_2$-Gain analysis of nonlinear systems and nonlinear state feedback $H_\infty$ control. IEEE Trans. Autom. Control **37**(6), 770–784 (1992)
11. W. Hongqiang, Y. Fang, W. Yongli et al., The design of $H_\infty$ robust guidance law for missile interception. Aerosp. Control **27**(1), 45–48 (2009)
12. Ciann-Dong Yang, Hsin-Yuan Chen, Nonlinear robust guidance law for homing missiles. J. Guid. Control Dyn. **21**(6), 882–890 (1998)
13. D. Zhou, F. Duan, Robust guidance law for space interception, in *IEEE Proceeding of the 3rd World Congress on Intelligent Control and Automation* (2000), pp. 3389–3393
14. M.R. Samatham, D.K. Pradhan, The De Bruijn multiprocessor network: a versatile sorting network. ACM SIGARCH Comput. Archit. News **13**(3), 360–367 (1985)
15. D.K. Pradhan, S.M. Reddy, A fault-tolerant communication architecture for distributed systems. IEEE Trans. Comput. C **31**(9), 863–870 (1982)
16. W. Lohmiller, J.J.E. Slotine, On contraction analysis for nonlinear systems. Automatica **34**(6) (1998)

# Control Strategy of PV-Hybrid Energy Storage Device Based on Household Load

**Shu Tian and Shiyuan Feng**

**Abstract** In order to improve the stability and economy of roof photovoltaic system. The frequency distribution of photovoltaic power can be obtained by using Hilbert-Huang transform (HHT), and a boundary frequency algorithm is proposed to determine the frequency of photovoltaic power generation. Then, the roof photovoltaic-hybrid energy storage control model based on household load is constructed, and the model is solved by LINGO software. Data analysis and comparison show that under the condition of full life cycle and real time price of photovoltaic power generation, user economy is superior to that of fixed price condition. The effectiveness of the proposed model coordination control strategy is verified by the user's economy.

**Keywords** Photovoltaic power generation · Maximum power point tracking · Hybrid energy storage · Capacity configuration · Energy management · Control strategy

## 1 Introduction

With the development of photovoltaic technology and the decrease of photovoltaic price, more and more users can use the clean energy from photovoltaic. However, photovoltaic power generation is easy to be affected by environmental factors, so it can not be directly used as power source [1]. As a bridge between photovoltaic cell and load, the energy storage device can solve the problem of low power quality and low efficiency of power generation. There are many forms of energy storage available. Considering the high efficiency and economy of energy storage devices in practical applications, this paper selects batteries and supercapacitors to form a

S. Tian · S. Feng (✉)
School of Electrical Engineering and Automation, Henan Polytechnic University, Henan Jiaozuo 454000, China
e-mail: 211608020039@home.hpu.edu.cn

S. Tian
e-mail: 1755831726@qq.com

 
Y. Jia et al. (eds.), *Proceedings of 2018 Chinese Intelligent Systems Conference*, Lecture Notes in Electrical Engineering 528,
https://doi.org/10.1007/978-981-13-2288-4_50

hybrid energy storage device. The fluctuating power of photovoltaic cell is divided into high frequency component and low frequency component by the algorithm of dividing frequency. The high frequency component is allocated to supercapacitor and the low frequency component is distributed to battery. The maximum power tracking algorithm (MPPT) of Ref. [2] is used to control the output power of photovoltaic cells in this paper. There are many methods of time—frequency analysis [3]. In this paper, empirical mode decomposition (EMD) and Hilbert transform are used to analyze the wave power produced by photovoltaic cells and determine the frequency distribution. Finally, considering the fixed cost, maintenance cost, and hybrid energy storage device in the mode of "spontaneous self-use, surplus Internet access" [4], as well as considering the different modes electricity price of power grid, the price of photovoltaic grid-connected government subsidy and user satisfaction. Under the comprehensive factors, the optimal control mode of user is obtained by using LINGO software to find the optimal solution of the proposed model.

## 2 Hybrid Energy Storage Capacity Allocation Method Based on HHT Transformation

The HHT transform [8] mainly consists of two parts, one is empirical mode decomposition (EMD), the other is Hilbert transform, where EMD is the core. Traditionally, the dividing frequency between high frequency and low frequency is judged by human experience, and its accuracy is low, which leads to the excessive division of high frequency components into storage battery, which is not conducive to battery service life, and increase the user's economic expenditure. Therefore, this paper proposes a method to judge the frequency of the boundary. The specific steps are as follows:

Step 1: Set the dividing frequency $f_g$, step size *Sl* and the Hilbert spectrum of the original signal is obtained by HHT transformation, finally set the iteration stop threshold $\varepsilon$

Step 2: $f_a1$ and$f_a2$ is the first and second order frequencies of each IMF in the Hilbert spectrum, and do the difference with the set frequency $f_g$, as follows:

$$a = f_a1 - f_g \tag{2.1}$$

$$b = f_a2 - f_g \tag{2.2}$$

Step 3: Determine if $a > 0$? the value greater than 0 set 0; determine if $b > 0$? the value is less than 0 set 0. The newly determined value is set to $a'$ and $b'$.

Step 4: To make $a'$ and $b'$ integral in the whole domain … as follows:

$$\alpha = \int_D a' dt \tag{2.3}$$

$$\beta = \int_D b' dt \tag{2.4}$$

Step 5: Set

$$\omega = \alpha + \beta \tag{2.5}$$

Step 6: Judging $|\omega| < \varepsilon$?, if the suspension condition is met, output $f_g$; If it don't satisfy the suspension condition, judging $\omega > 0$?, yes, set $f_g = f_g + Sl$; no, set $f_g = f_g - Sl$; Returns the second step to re-iterate the calculation
After the boundary frequency is obtained, the Butterworth low-pass filter [9] is used to divide the original signal into high-frequency component and low-frequency component.

# 3 Establishment and Solution of PV-Hybrid Energy Storage Mathematical Model

## *3.1 Construction of Objective Functions*

$$\min = \frac{C_0 + C_1 + C_2 + C_3}{I_0 + I_1 + I_2} \tag{3.1}$$

Among them: $I_0+I_1+I_2$ represents the total income of the user; $C_0+C_1+C_2+C_3$ represents the total expenditure of the user. $I_0$ represents the total electricity cost saved by the user through photovoltaic power generation; $I_1$ represents government subsidies to generate electricity from all photovoltaics; $I_2$ represents economic income from grid-connected photovoltaics. $C_0$ represents the cost of hardware investment made up of photovoltaic equipment, rectifier inverter equipment, and hybrid energy storage equipment; $C_1$ represents the maintenance costs of all hardware equipment; $C_2$ represents the total cost that the user has to pay to consume the electrical energy of the power grid; $C_3$ represents the replacement cost of batteries and supercapacitor banks throughout the photovoltaic life cycle.

$$I_0 = \sum_i \sum_j \sum_k p_i(j,k) \cdot w_i'(j,k) \tag{3.2}$$

$$I_1 = \sum_i \sum_j \sum_k p_0' \cdot w_i(j,k) \tag{3.3}$$

$$I_2 = \begin{cases} \sum_i \sum_j \sum_k p_0 \cdot [w_i(j,k) - w_i'(j,k)], & w_i(j,k) > w_i'(j,k) \\ \sum_i \sum_j \sum_k [p_0 - p_i(j,k)] \cdot w_i(j,k), & \frac{p_0}{p_i(j,k)} \geq 0.5 \end{cases} \tag{3.4}$$

$$C_0 = p_{pv} \cdot c_{pv} + p_b \cdot c_b + p_{sc} \cdot c_{sc} \tag{3.5}$$

$$C_1 = \sum_i \sum_j p_{bm} \cdot c_b'(i,j) + p_{scm} \cdot c_{sc}'(i,j) + p_{pvm} \cdot c_{pv}'(i,j) \tag{3.6}$$

$$C_2 = \sum_i \sum_j \sum_k p_i(j,k) \cdot w_i^{grid}(j,k) \tag{3.7}$$

$$C_3 = n_1 \cdot p_b \cdot c_b + n_2 \cdot p_{sc} \cdot c_{sc} \tag{3.8}$$

Among them: $p_i(j,k)$ represents the real time price of the $k$ time slot on the $j$ day of the $i$ year; $w_i'(j,k)$ represents the electrical energy consumed by the user in the $k$ time slot on the $j$ day of the $i$ year; $p_0'$ represents the government subsidies for photovoltaic power; $w_i(j,k)$ represents the electrical energy generated by photovoltaic in the $k$ time slot on the $j$ day of the $i$ year; $p_0$ represents the price of photovoltaic electricity; the $p_{pv}$, $p_b$ and $p_{sc}$ represents the capacity cost of photovoltaic accessory device, battery and supercapacitor bank, respectively; $c_{pv}$, $c_b$ and $c_{sc}$ respectively represents the capacity of PV, battery, and super capacitor banks; $p_{bm}$, $p_{scm}$ and $p_{pvm}$ respectively represent the maintenance costs of batteries, supercapacitors and photovoltaic devices; $c_b'$, $c_{sc}'$, $c_{pv}'$ respectively represent the total storage energy of the battery and the supercapacitor, and the total generating power of photovoltaic in the k time slot on the $j$ day of the $i$ year; $w_i^{grid}(j,k)$ represents the energy consumed by the user under the condition of power supply by grid in the k time slot on the $j$ day of the $i$ year; $n_1$ and $n_2$ respectively indicate the number of battery and super capacitor banks in the whole life cycle of PV.

### 3.2 Objective Function Constraints

$$c_{pv\_\min} \le c_{pv} \le c_{pv\_\max} \tag{3.9}$$

Among them: $c_{pv\_\min}$ and $c_{pv\_\max}$ respectively represent the minimum and maximum PV installed capacity of the user.

$$SOC_{\min} \le SOC \le SOC_{\max} \tag{3.10}$$

Among them: $SOC_{\min}$ and $SOC_{\max}$ respectively represent the lowest and the highest value of the charged state of the energy storage device.

$$SOC(j,k) = SOC(j,k-1) \pm \alpha \cdot P(j,k) \cdot T \tag{3.11}$$

$\alpha$ is a constant, $P(j,k)$ represents the charge/discharge power of the energy storage device, the charge is positive and the discharge is negative.

$$w_b'(j,k) = w_b'(j,k-1) + P_b(j,k) \cdot T \tag{3.12}$$

$$P_b(j,k) = \begin{cases} \frac{1}{SOC(j,k)+\alpha_1} + \beta_1, \text{charge} \\ \ln[SOC(j,k) + \alpha_2] + \beta_2, \text{discharge} \end{cases} \tag{3.13}$$

$\alpha_1, \alpha_2$ and $\beta_1, \beta_2$ are real numbers.

Equation (2.3), it is stipulated that the charging time slot of the energy storage unit can only be carried out in $16 \leq T \leq 31$, and the charge and discharge can't be carried out at the same time.

User satisfaction mainly depends on the ratio of a day's expenditure to income. This paper uses gauss function to simulate user satisfaction:

$$\eta = e^{\frac{-x^2}{0.5^2}} \tag{3.14}$$

Among them:

$$x = \frac{\sum_j \sum_k p(j,k) \cdot w^{grid}(j,k)}{\sum_j \sum_k p_0 \cdot [w(j,k) - w'(j,k)]}, w(j,k) - w'(j,k) > 0. \tag{3.15}$$

when the user's satisfaction is not less than 0.5, the maximum value of the objective function is obtained.

## 3.3 *Model Solving*

This paper uses LINGO software to solve the mathematical model. The biggest feature of LINGO software is that it is convenient to set decision variables as integers, and it can execute quickly and occupy less memory in solving nonlinear programming problems, so as to obtain the optimal solution of the model in this paper.

# 4 Example Analysis

The maximum of PV installed capacity is $c_{pv\_max} = 10$ kW, the minimum of PV installed capacity is $c_{pv\_min} = 1$ kW. The power price of photovoltaic to grid is set at 0.58 yuan/kWh, the government subsidy is 0.37 yuan/kWh, and the initial charge is 0.5. Typical household seasons—energy consumption data are shown in Fig. 1.

The fixed electricity price is 0.58 yuan/kWh. In this paper, the real time price of electricity is given by Ref. [4].

The photovoltaic cell specifications and other parameters are shown in Table 1. The parameters of the mixed energy storage are shown in Table 2.

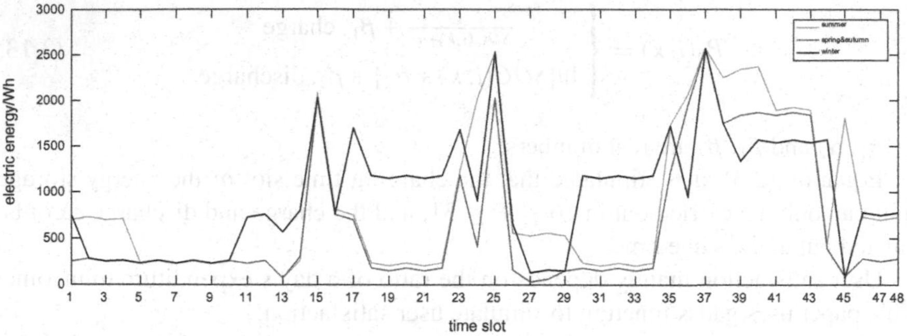


**Fig. 1** Typical daily load changes for each quarter of a typical family

**Table 1** Photovoltaic cell specifications and other parameters

| Product information | Parameters |
|---|---|
| Open circuit voltage $V_{oc}$ | 20 V |
| Short circuit current $I_{sc}$ | 5.4 A |
| Maximum working voltage $V_m$ | 18 V |
| Maximum working current $I_m$ | 5 A |
| Rated output power/$E_0$ | 100 W |
| Compositive output power | 90% $E_0$ |
| Output voltage | 18 V |
| Integrated PV cost | 10 yuan/W |
| Minimum installed capacity | 1 kW |
| Maximum installed capacity | 10 kW |
| Service life of the PV | 20 years |

According to Ref. [7], The random property of irradiance as a random variable can be approximated as following beta distribution. Taking summer as an example, the variation of typical daily irradiance is illustrated in this paper. As shown in Fig. 2.

It is shown that capacity allocation problem of hybrid energy storage device is the only one in the summer, and the calculation method of other seasons is the same. Through the photovoltaic power output characteristic [5, 6] and the summer typical day irradiance, the summer typical day photovoltaic array output power is determined, as shown in Fig. 3.

The photovoltaic array output power by the EMD decomposition, will get 6 IMF components and a residual. The residual component was discarded and HHT transformation was carried out for 6 IMF components. Finally, the output power Hilbert graph of photovoltaic array was obtained, as shown in Fig. 4.

According to the dividing frequency algorithm, the normalized dividing frequencies of summer, winter and spring-autumn are 0.1954 Hz, 0.1891 Hz and 0.1928 Hz,

**Table 2** The parameters of the mixed energy storage

| Hybrid energy storage element | Products indicators | Product parameters |
|---|---|---|
| Storage battery | Capacity cost (yuan/kWh) | 640 |
| | Operation maintenance cost (yuan/kWh) | 0.05 |
| | Charge and discharge efficiency (%) | 80 |
| | Charge state upper limit | 0.9 |
| | Charge state lower limit | 0.2 |
| | Cycle life | 3 years |
| Supercapacitor | Capacity cost (yuan/kWh) | 27,000 |
| | Operation maintenance cost (yuan/kWh) | 0.05 |
| | Charge and discharge efficiency (%) | 98 |
| | Charge state upper limit | 0.9 |
| | Charge state lower limit | 0.2 |
| | Cycle life | 20 years |

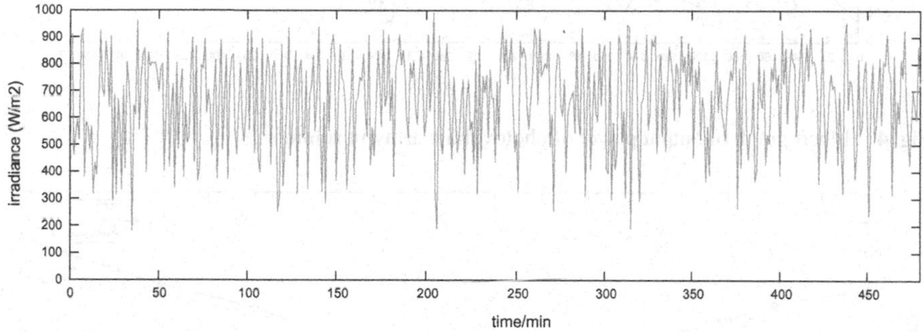


**Fig. 2** Typical daily irradiance in summer

respectively. Then, the output power of the typical daily photovoltaic array in summer is taken as the original signal, and passed into the Butterworth low-pass filter.

Thus, the allocation capacity of storage battery and supercapacitor for each season of each photovoltaic installation capacity can be obtained, as shown in Table 3.

By solving the mathematical model by LINGO software, the optimal value of the objective function under the each installed capacity of photovoltaic in each season is shown in Fig. 5.

The change of the charge state of the battery in each season under the optimal value of the objective function is selected, as shown in Fig. 6.

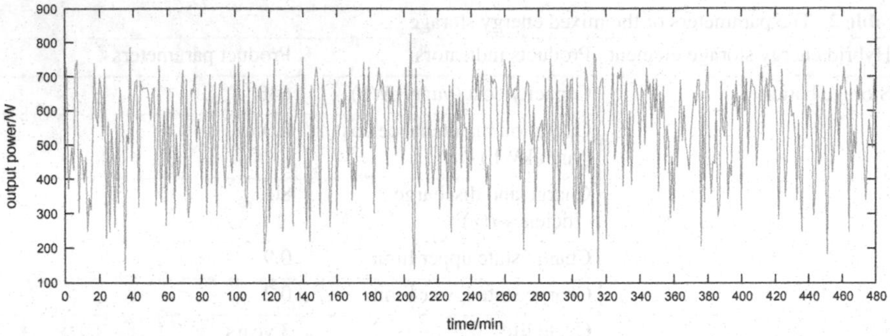


**Fig. 3** Typical photovoltaic array output power in summer

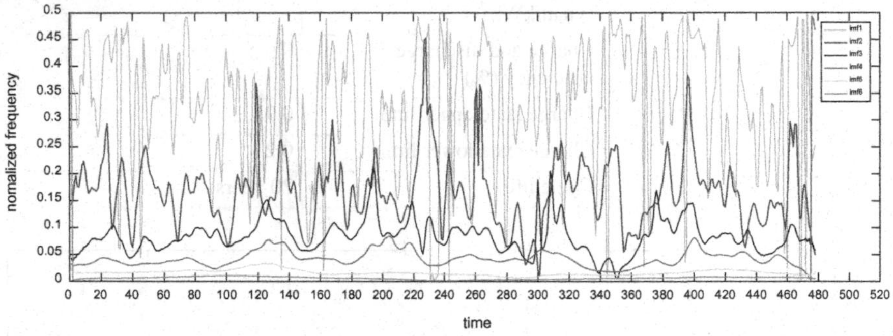


**Fig. 4** Hilbert graph of output power of photovoltaic array in summer

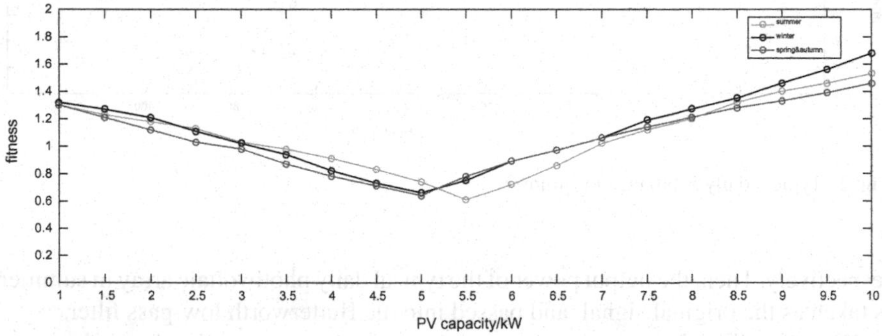


**Fig. 5** Objective function values

As shown in Figs. 5 and 6, it can be seen that the variation of the state of charge and the range of the time gap between charge and discharge meets the constraint conditions.

**Table 3** Distribution capacity of batteries and supercapacitors for each season of each photovoltaic installation capacity

| PV capacity/kW | 1 | 1.5 | 2 | 2.5 | 3 |
|---|---|---|---|---|---|
| Summer/kW | 4.26/0.08 | 6.39/0.13 | 8.52/0.17 | 10.66/0.21 | 12.79/0.25 |
| Winter/kW | 2.33/0.04 | 3.49/0.07 | 4.66/0.09 | 5.82/0.11 | 6.99/0.13 |
| Spring-autumn/kW | 3.23/0.06 | 4.84/0.10 | 6.46/0.13 | 8.07/0.16 | 9.68/0.19 |
| PV capacity/kW | 3.5 | 4 | 4.5 | 5 | 5.5 |
| Summer/kW | 14.92/0.29 | 17.05/0.33 | 19.18/0.38 | 21.31/0.42 | 23.44/0.46 |
| Winter/kW | 8.15/0.15 | 9.31/0.18 | 10.48/0.20 | 11.64/0.22 | 12.81/0.24 |
| Spring-autumn/kW | 11.30/0.23 | 12.91/0.26 | 14.52/0.29 | 16.14/0.32 | 17.75/0.36 |
| PV capacity/kW | 6 | 6.5 | 7 | 7.5 | 8 |
| Summer/kW | 25.57/0.50 | 27.70/0.54 | 29.83/0.58 | 31.97/0.62 | 34.10/0.67 |
| Winter/kW | 13.97/0.26 | 15.14/0.29 | 16.30/0.31 | 17.46/0.33 | 18.63/0.35 |
| Spring-autumn/kW | 19.37/0.39 | 20.98/0.42 | 22.59/0.45 | 24.21/0.48 | 25.82/0.52 |
| PV capacity/kW | 8.5 | 9 | 9.5 | 10 | |
| Summer/kW | 36.23/0.71 | 38.36/0.75 | 40.49/0.79 | 42.62/0.83 | |
| Winter/kW | 19.79/0.37 | 20.95/0.40 | 22.12/0.42 | 23.29/0.44 | |
| Spring-autumn/kW | 27.43/0.55 | 29.05/0.58 | 30.66/0.62 | 32.28/0.65 | |

*Note* Battery capacity/supercapacitor capacity

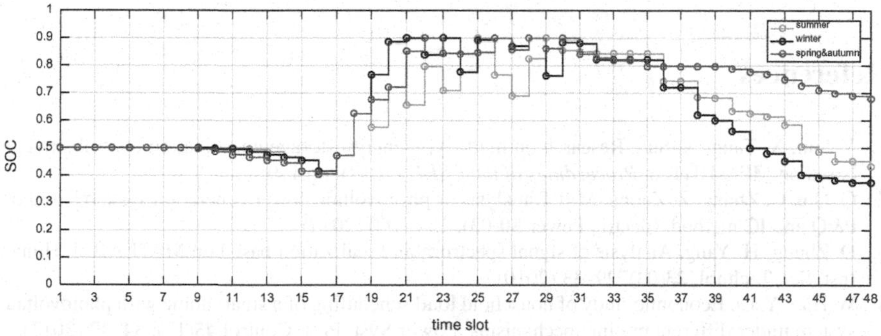


**Fig. 6** Changes in charged state

Lastly, that total income and total expenditure of the different price mode user in the whole life cycle of the photovoltaic cell are compared with the optimal value of the objective function as shown in Table 4.

From Table 4, it can be obtained that the income of rooftop photovoltaic based on real-time electricity price is higher than that of fixed electricity price.

**Table 4** Total income and expenditure of photovoltaic cells throughout their life cycle

| | Summer | Winter | Spring-autumn |
|---|---|---|---|
| *Real-time price* | | | |
| Users's expenditure of PV full life cycle/yuan | 149906.81 | 95651.44 | 115712.14 |
| Users's income of PV full life cycle/yuan | 245748.87 | 144926.42 | 180800.21 |
| *Fixed price* | | | |
| Users's expenditure of PV full life cycle/yuan | 164346.25 | 112674.24 | 133684.61 |
| Users's income of PV full life cycle/yuan | 214691.44 | 129864.71 | 154381.66 |

# 5 Conclusions

In this paper, Hilbert diagram was obtained by Hilbert-Huang transformation of photovoltaic output power, and the normalized dividing frequency was obtained by using the dividing frequency algorithm. Using low pass filter, the original signal is divided into low frequency signal and high frequency signal, and the storage capacity of battery and supercapacitor is determined. Finally, by constructing the mathematical model of rooftop photovoltaic-hybrid energy storage, the effectiveness of optimal control strategy to improve the economic benefits of users is analyzed and verified. At the same time, the effect of real-time price and fixed price on users' income is compared. Finally, it is concluded that the combination of rooftop photovoltaic and real-time price can bring more benefits to users.

# References


1. Y. Sun, X. Tang, X. Sun, Research on multi-type energy storage coordination control strategy based on MPC-HHT, in *Proceedings of the CSEE* (2017), pp.1–9
2. G. Qiu, C. Zhang, Z. Zhong, MPPT analysis of photovoltaic power generation system based on P&Q and IC method. Electric Power **50**(03), 154–160 (2017)
3. D. Zhang, H. Yang, Analysis of signal spectrum and realization based on MATLAB. J. Hunan Inst. Sci. Technol. **23**(03), 29–33 (2010)
4. W. Hao, Y. Li, Economic study of household load scheduling of a smart home with photovoltaic system under different pricing mechanisms. Power Syst. Prot. Control **45**(17), 34–42 (2017)
5. X. He, Y. Yang, R. Feng, Characteristics of photovoltaic cells and maximum power point tracking. Ind. Control Comput. **30**(09), 143–145 (2017)
6. L. Chen, P. Hu, G. Lu, P. Liu, Mechanism analysis and simulation study of MPPT based on improved incremental conductance method. Smart Grid **5**(02), 172–177 (2017)
7. M. Wang, *The Probabilistic Modeling of Distributed Generation and Its Influences on Power Systems* (HeFei University of Technology, 2010)
8. Y. Zhang, HHT analysis of blasting vibration and its application. Proc. CSU-EPSA **25**(05), 60–64 (2013)
9. P. Li, K. Duan, Y. Dong, Energy management strategy of photovoltaic DC microgrid with distributed hybrid energy storage system. Power Syst. Prot. Control **45**(13), 42–48 (2017)

# Structural Controllability of Optimized Networks with Onion-Like Topologies

**Manli Li, Shiwen Sun, Yafang Wu and Chengyi Xia**

**Abstract** Recently, an optimization method has been proposed to increase the ability of complex networks to resist intentional attacks on hub nodes. The finally optimized networks exhibit a novel type of "onion-like" structure. At the same time, structural controllability of complex networks also has been a hot research topic in recent years. Thus, structural controllability of "onion-like" networks deserves sufficient discussion. In this study, we explored the relationship between the attack robustness and structural controllability of scale-free networks before and after optimization. After implementing large quantity of numerical simulations, it has been found that the optimized scale-free networks have both increased robustness and enhanced structural controllability. Current research results can shed some light on the deep understanding of structural complexity and dynamical properties of real-world networked systems.

**Keywords** Structural controllability · Network attack · Scale-free networks
Network optimization · Onion-like topologies

## 1 Introduction

In recent years, complex networks have been regarded as the most important models of many natural and manmade complex systems, which have attracted a lot of research efforts [1–5]. Especially, the problem of how to control complex networks is of great significance from the viewpoints of both science and engineering, including pinning control [6–8], structural controllability [9, 10], exact controllability [11] and so on.

---

M. Li · S. Sun (✉) · Y. Wu · C. Xia
Tianjin Key Laboratory of Intelligence Computing and Novel Software Technology,
Tianjin University of Technology, Tianjin 300384, China
e-mail: sunsw80@126.com

M. Li · S. Sun · Y. Wu · C. Xia
Key Laboratory of Computer Vision and System,
Tianjin University of Technology, Tianjin 300384, China



Y. Jia et al. (eds.), *Proceedings of 2018 Chinese Intelligent Systems Conference*,
Lecture Notes in Electrical Engineering 528,
https://doi.org/10.1007/978-981-13-2288-4_51

The concept of structural controllability of a graph is firstly proposed by Lin [9]. A dynamical network is considered to be controllable if it can be driven from any initial state to any desired final state with external inputs in finite time. Recently, Liu et al. [10] extended this concept of structural controllability to complex networks. They utilized the the maximum matching algorithm [12] to estimate the minimum number of controllers and their locations to control the entire network. Inspired by this work, many researchers have devoted the research efforts on different aspects on controllability, such as control centrality [13, 14], control profile [15], control energy [16], controllability optimization [17, 18], structural role of complex networks on controllability [19–21] and so on.

"Robustness and fragility" is regarded as the most important feature of scale-free networks [22–25]. For a network, after a fraction of nodes or edges have been attacked, if as many as possible elements remain globally connected, the network can be considered as an *robust* network. Moreover, the study of how to generate optimal networks and how to make existing networks more robust resisting all kinds of attacks still remains as a very challenging problem.

An efficient method to construct robust networks against malicious attacks is proposed and developed by Schneider et al. [26]. Their findings revealed that the robustness of a given network can be improved significantly after small changes in the network structure. Moreover, after optimization, all the finally optimized networks exhibit a type of "onion-like" topology, i.e., there exists a core composed of highly connected hub nodes, and nodes with decreasing degree hierarchically surround the core. Since "onion-like" networks have special structural properties as well as enhanced attack robustness [27–30], structural controllability of this kinds of novel networks deserves extensive exploration.

In this study, as preliminary knowledge a brief overview of basic concepts of structural controllability is presented. Nextly, the quantity $R$ used to evaluate network's robustness resisting malicious attacks is discussed. Then, research emphasis is placed on the analysis of relationship between structural controllability and attack robustness of scale-free networks. Additionally, after performing robustness optimization on these networks, the changes of controllability of optimized networks are analyzed. Finally in the last section the whole paper is concluded.

## 2 Preliminaries

### 2.1 *Structural Controllability*

Here we consider a static network, which means that the nodes and edges are fixed without evolving or switching with time. Also, considering a linear time invariant (LTI) system,

$$\dot{x} = A'x(t) + Bu, \quad (1)$$

where $A' \in R^{N\times N}$ is the transpose of the adjacency matrix $A$, and $N$ denotes the number of nodes. For $A = \{a_{ij}\}$, if there is an link from node $i$ to node $j$, $a_{ij} = 1$; otherwise $a_{ij} = 0$. At time $t$, $x(t) = [x_1(t), x_2(t), \ldots, x_N(t)]' \in R^N$ is the state of a system of $N$ nodes. $B \in R^{N\times M}$ denotes the input matrix, and $M$ is the number of controllers on the network. For $\forall b_{ij} \in B$, $b_{ij} \neq 0$ if there is a controller $j$ placed on node $i$, or $b_{ij} = 0$. $u(t) = [u_1(t), u_2(t), \ldots, u_M(t)] \in R^M$ denotes the vector of input signals from the external controllers on the network.

The controllability of network can be measured by the number of driver nodes $N_{\mathrm{D}}$. Based on the maximal matching algorithm [12], Liu et al. [10] proposed a method to calculate the minimum number of driver nodes to control the whole network. If in the network there exists a perfect matching, $N_{\mathrm{D}}$ is equal to 1. Otherwise, it is equal to the number of unmatched nodes to a maximum matching, and these unmatched nodes are called driven nodes. That is, if $N$ is denoted to be the total number of nodes in the network and $|m^*|$ to be the number of matched nodes to a maximum matching,

$$N_{\mathrm{D}} = \max(N - |m^*|, 1). \tag{2}$$

Also, $n_{\mathrm{D}}$ can be denoted to be

$$n_{\mathrm{D}} = N_{\mathrm{D}}/N. \tag{3}$$

Clearly, $0 < n_{\mathrm{D}} \leq 1.0$.

Apparently, for a fixed network, the value of $n_{\mathrm{D}}$ only depends on its topology. Thus, the value of $n_{\mathrm{D}}$ can be used to represent thc impact of a particular topology on the network controllability. The smaller the value $n_{\mathrm{D}}$, the easier the networked system can be controlled, and *vice versa*.

## 2.2 Onion-Like Networks

Recently, node robustness $R$ was proposed by Schneider et al. [26] to evaluate the robustness of networks under attacks.

$$R = \frac{1}{N^2}\sum_{q=1/N}^{1} S(q), \tag{4}$$

where $N$ denotes the total number of nodes and $S(q)$ is the number of nodes in the largest connected cluster after the fraction $q$ of nodes with highest degrees are removed. In general, the larger the value of $R$, the more robust the network resisting attacks on high-degree nodes. Based on $R$, by exchanging only a small number of edges while keeping the degree of every node unchanged, they also developed an edge-swap method to improve the robustness [26]. Here consider a scale-free network with $N = 50$ nodes, average node degree $\langle k\rangle = 4$ and degree distribution $p(k) \sim k^{-3}$. The topologies of the network before and after optimization are shown

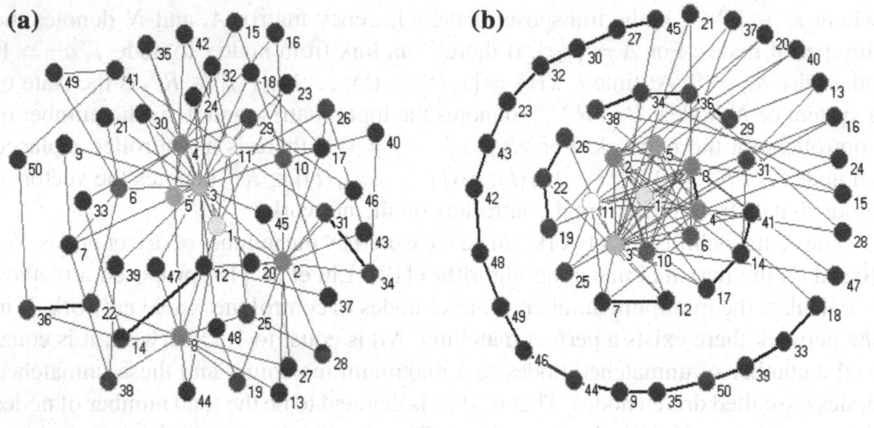


**Fig. 1** Visualization of the topologies of a scale-free network ($N = 50$, $\langle k \rangle = 4$ and $p(k) \sim k^{-3}$) before and after optimization. Nodes with the same degree are shown with the same color. **a** Initial network without any optimization; **b** optimized network with onion-like topology (Color figure online)

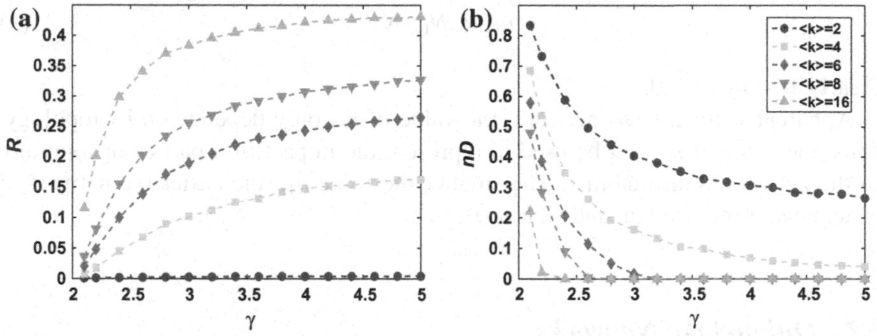


**Fig. 2** $n_D$ and $R$ of scale-free networks

in Fig. 1a and Fig. 1b respectively. In Fig. 1b, optimized network exhibits a type of "onion-like" structure, i.e. there exists a core composed of highly connected hub nodes, and nodes with decreasing degree hierarchically surround the core [30].

# 3 Numerical Simulations and Analysis

## 3.1 The Relationship Between Attack Robustness and Structural Controllability

Firstly, the effects of structural properties on attack robustness and structural controllability of typical scale-free networks is explored. For scale-free networks with degree distributions following a power-law form: $p(k) \propto k^{-\gamma}$, the scaling exponent $\gamma$ and the average node degree $\langle k \rangle$ are two important parameters, which play important roles in both the dynamical and topological properties of networks. In the following numerical simulations (Fig. 2), for different scaling exponents $\gamma$, scale-free networks with $N = 1000$ and $\langle k \rangle = 2, 4, 6, 8, 16$ are investigated. The values of $R$ and $n_D$ of scale-free networks are calculated and shown in Fig. 2.

For scale-free networks with fixed $\langle k \rangle$, $R$ increases as $\gamma$ becomes larger (see Fig. 2a). The results demonstrates that scale-free networks with larger $\gamma$ have improved ability to resist intentional attacks. On the other hand, scale-free networks with fixed $\gamma$ are examined. It can be verified by the numerical results shown in Fig. 2a that $R$ is observed to increase with $\langle k \rangle$ becoming larger, indicating that scale-free networks with larger $\langle k \rangle$ are more robust to resist malicious attacks.

Furthermore, the values of $n_D$ of different scale-free networks are examined (see Fig. 2b). For scale-free networks with fixed $\langle k \rangle$, $n_D$ is observed to decrease with increasing $\gamma$, implying the enhancement of corresponding networks. Additionally, once $\gamma$ is fixed, with the increase of $\langle k \rangle$ $n_D$ becomes smaller which indicates the resultant networks are easier to be controlled.

## 3.2 *Enhanced Controllability*

The impact of a particular topology on structural controllability can be characterized by the following quantities: $n_D$. The change of the controllability of scale-free networks before and after robustness optimization can be investigated by the change of $n_D$. Figure 3a, b present the optimization results of scale-free networks with $N = 1000$, $\langle k \rangle = 4$ and different scaling exponents $\gamma$. At each step $T$ a 20% increase of $R$ is recorded (see Fig. 3a). Figure 3b shows the values $n_D$ of corresponding networks and the behaviors with the optimization step $T$. It can be found that $n_D$ becomes smaller with step $T$, which strongly demonstrates that the optimized networks with improved $R$ have enhanced controllability.

Furthermore, scale-free networks with fixed $N$ and $\gamma$ for different $\langle k \rangle$ are also investigated. The numerical simulation results of scale-free networks with $N = 1000$, $\gamma = 3$ and different $\langle k \rangle$ are shown in Fig. 3c, d. A 10% increase of $R$ is recorded in Fig. 3c, d at each optimization step $T$. The behaviors of $n_D$ which measures the structural controllability of corresponding scale-free networks with the optimization step $T$ are displayed in Fig. 3d. Clearly, with the optimization step $T$, $n_D$ is observed to decease, which implies that structural controllability is highly improved during optimization.

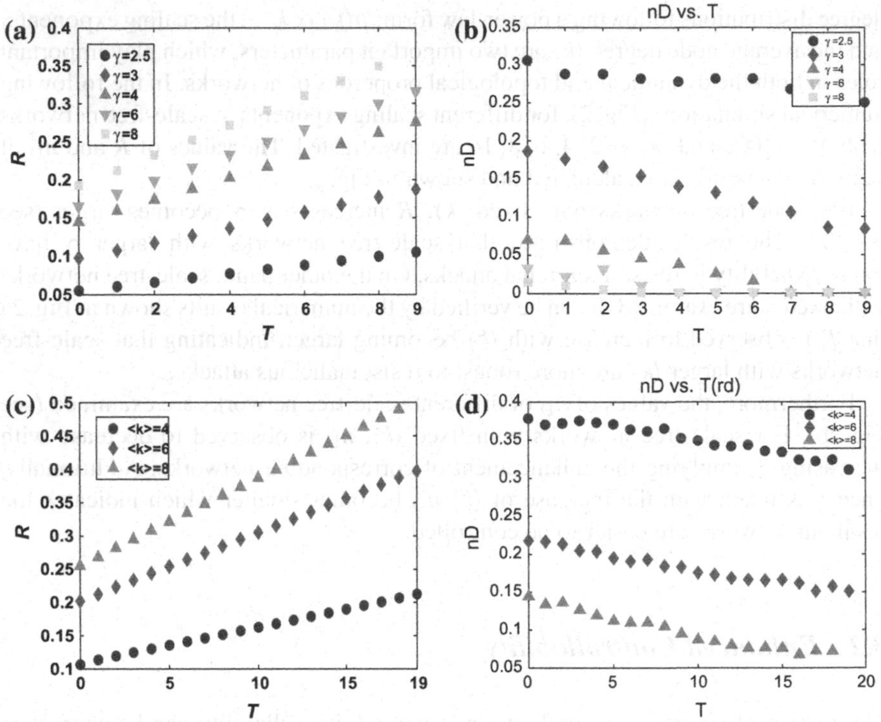


**Fig. 3** The changes of $R$ **(a)**, **(c)** and $n_D$ **(b)**, **(d)** of scale-free networks during optimization

# 4 Conclusions

In this study, through numerical simulations, scale-free networks with different scaling exponent $\gamma$ and average node degree $\langle k \rangle$ are explored to study the relationship between structural controllability and attack robustness. Firstly, it has been found that for scale-free networks, the controllability indicator $n_{\mathrm{D}}$ decreases as $\gamma$ and $\langle k \rangle$ become larger, implying that corresponding scale-free network is more easier to be controlled.

Nextly, the effect of the optimization of attack robustness against attacks on structural controllability is studied. The numerical results demonstrated that the structural controllability of optimized networks also can be enhanced greatly. Thus, it is verified that after the edge-swap method aiming at maximizing $R$ is implemented, the optimized scale-free networks can have both increased ability to resist attacks and enhanced controllability. These findings can help to improve the thorough understanding of both structural complexity and dynamical properties of real-world networks, also, shed some light on the design of resilient complex systems.

**Acknowledgements** SWS and CYX acknowledge the support from Middle-aged and Young Innovative Talents Training Project of the Higher Education Institutions of Tianjin.

## References

1. R. Albert, A.L. Barabási, Statistical mechanics of complex networks. Rev. Modern Phys. **74**, 47–97 (2002)
2. M.E.J. Newman, The structure and function of complex networks. SIAM Rev. **45**(2), 167–256 (2003)
3. S. Boccalettia, V. Latorab, Y. Morenod, Complex networks: structure and dynamics. Phys. Rep. **424**, 175–308 (2006)
4. D.J. Watts, S.H. Strogztz, Collective dynamics of small world networks. Nature **393**, 440–442 (1998)
5. A.L. Barabási, R. Albert, Emergence of scaling in random networks. Science **286**, 509–512 (1999)
6. X. Li, X.F. Wang, G.R. Chen, Pinning a complex dynamical network to its equilibrium. IEEE Trans. Circuit Syst. I: Regul. Pap. **51**(10), 2074–2087 (2004)
7. G.R. Chen, Pinning control and synchronization on complex dynamical networks. Int. J. Control Autom. **12**(2), 221–230 (2014)
8. X.F. Wang, X. Li, G.R. Chen, *Network Science: An Introduction* (Higher Education Press, Beijing, China, 2012)
9. C.T. Lin, Structural controllability. IEEE Trans. Autom. Control **19**(3), 201–208 (1974)
10. Y.Y. Liu, J.J. Slotine, A.L. Barabási, Controllability of complex networks. Nature **473**, 167–173 (2011)
11. Z.Z. Yuan, Z. Chen, Z.R. Di, W.X. Wang, Y.C. Lai, Exact controllability of complex networks. Nat. Commun. **4**, 2447 (2013)
12. J.E. Hopcroft, R.M. Karp, An $n^{5/2}$ algorithm for maximum matchings in bipartite graphs. SIAM J. Comput. **2**(4), 225–231 (1973)
13. Y.Y. Liu, J.J. Slotine, A.L. Barabási, Control centrality and hierarchical structure in complex networks. PLoS ONE **7**, e44459 (2012)
14. Y. Pan, X. Li, Structural controllability and controlling centrality of temporal networks. PLoS ONE **9**(4), e94998 (2014)
15. J. Ruths, D. Ruths, Control profiles of complex networks. Science **343**, 1373 (2014)
16. G. Yan, J. Ren, Y.C. Lai, C.H. Lai, B. Li, Controlling complex networks: how much energy is needed. Phys. Rev. Lett. **108**(21), 218703 (2012)
17. W.X. Wang, N. Xuan, Y.C. Lai, C. Grebogi, Optimizing controllability of complex networks by minimum structural perturbations. Phys. Rev. E **85**, 026115 (2012)
18. Y.D. Xiao, S.Y. Lao, L.L. Hou, L. Bai, Edge orientation for optimizing controllability of complex networks. Phys. Rev. E **90**, 042804 (2014)
19. M. Pósfai, Y.Y. Liu, J.-J. Slotine, A.L. Barabási, Effect of correlations on network controllability. Sci. Rep. **3**, 1067 (2013)
20. G. Menichetti, L. DallÁsta, G. Bianconi, Network controllability is determined by the density of low in-degree and out-degree nodes. Phys. Rev. Lett. **113**, 078701 (2014)
21. S.W. Sun, Y.L. Ma, Y.F. Wu, L. Wang, C.Y. Xia, Towards structural controllability of local-world networks. Phys. Lett. A **380**(22–23), 1912–1917 (2016)
22. R. Albert, H. Jeong, A.L. Barabási, The Internets Achilles Heel: error and attack tolerance of complex networks. Nature **406**, 378–382 (2000)
23. R. Cohen, K. Erez, D. ben-Avraham, S. Havlin, Resilience of the internet to random breakdowns. Phys. Rev. Lett. **85**(21), 4626–4628 (2000)
24. R. Cohen, K. Erez, D. ben-Avraham, S. Havlin, Breakdown of the internet under intentional attack. Phys. Rev. Lett. **86**(16), 3682–3685 (2001)

25. D.S. Callaway, M.E.J. Newmann, S.H. Strogatz, D.J. Watts, Network robustness and fragility: percolation on random graphs. Phys. Rev. Lett. **85**(25), 5468–5471 (2000)
26. C.M. Schneider, A.A. Moreira, J.S. Andrade, S.Havlin, H.J. Herrmann, Mitigation of malicious attacks on networks. Proc. Natl. Acad. Sci. (USA) **108**(10), 3838–3841 (2011)
27. Z.X. Wu, P. Holme, Onion structure and network robustness. Phys. Rev. E **84**, 026106 (2011)
28. T. Tanizawa, S. Havlin, H.E. Stanley, Robustness of onionlike correlated networks against targeted attacks. Phys. Rev. E **85**, 046109 (2012)
29. S.W. Sun, R.Q. Li, L. Wang, C.Y. Xia, Reduced synchronizability of dynamical scale-free networks with onion-like topologies. Appl. Math. Comput. **252**, 249–256 (2015)
30. S.W. Sun, Y.L. Ma, R.Q. Li, L. Wang, C.Y. Xia, Tabu search enhances network robustness under targeted attacks. Phys. A **446**, 82–91 (2016)

# Exponential Synchronization Control of Neural Networks with Time-Delays and Markovian Jumping Parameters

**Yuqing Sun, Yiyuan Zheng, Xiangwu Ding, Yiming Gan, Wuneng Zhou, Xin Zhang and Lifei Yang**

**Abstract** In this paper, the exponential synchronization control is considered for neural networks with time-delays and Markovian jumping parameters. The jumping parameters are modeled as continuous-time finite-state Markov chain. By resorting to the Lyapunov functional method, a linear matrix inequality (LMI) approach is developed to derive the synchronization required. Simulations with Matlab verify the effectiveness of the proposes criteria.

**Keywords** Exponential synchronization · Markovian jumping · Time-delays Linear matrix inequality

## 1 Introduction

The seminal works for the neural networks in the past few decades have witnesses the successful application of them in a variety of areas. Among dynamical behaviors of the neural networks, synchronization is one of the most important ones that has been found considerable and received researchers' attentions due to its great potential applications in secure communication, chaos generators design, optics, social science, harmonic oscillation generation, and power system protection [1–8].

Y. Sun · X. Ding · W. Zhou (✉) · X. Zhang
College of Information Science and Technology, Donghua University,
Shanghai 201620, China
e-mail: zhouwuneng@163.com

Y. Zheng
Wenlan School of Business, Zhongnan University of Economics and Law,
Wuhan 430073, China

Y. Gan
Bros Eastern Stock Co., Ltd, Ningbo 315040, China

L. Yang
Glorious Sun School of Business and Management, Donghua University,
Shanghai 201620, China



Y. Jia et al. (eds.), *Proceedings of 2018 Chinese Intelligent Systems Conference*,
Lecture Notes in Electrical Engineering 528,
https://doi.org/10.1007/978-981-13-2288-4_52

After the pioneering work of Zhou et al. [9] and Wu et al. [10], many different various approaches have been applied theoretically and experimentally to synchronize the chaotic systems, such as complete synchronization, phase synchronization, lag synchronization, adaptive synchronization [11], anti-synchronization, projective synchronization, exponential synchronization.

Recently, systems with Markovian jump parameters have received a great deal of attention [12–16]. A network with such a 'jumping' character may be modeled as a hybrid one; that is, the dynamics of the network is continuous, but the parameter jump among different modes may be seen as discrete events.

As we have stated that there are few studies about exponential synchronization of the networks with time-delays and Markovian jumping parameters, we aim to solve this problem. By utilizing a Lyapunov functional and LMI approach, a sufficient condition for the neural networks with time-delays and Markovian jumping parameters to be exponential synchronization is derived.

**Notions**: Let $(\Omega, F, \{F_t\}_{t\geq 0}, P)$ be a complete probability space with a filtration $\{F_t\}_{t\geq 0}$ satisfying the usual conditions (i.e., the filtration contains all $P$-null sets and is right continuous). Denote by $L^p_{F_0}([-d,0];\mathbb{R}^n)$ the family of all $F_0$-measurable $C([-d,0];\mathbb{R}^n)$-valued random variables $\xi = \{\xi(\theta) : -d \leq \theta \leq 0\}$ such that $\sup_{-d\leq\theta\leq 0} E|\xi(\theta)|^P < \infty$ where $E\{\cdot\}$ stands for the mathematical expectation operator with respect to the given probability measure $P$.

Let $\{r(t), t \geq 0\}$ be a right-continuous Markovian chain on the probability space taking values in a finite state space $S = \{1, 2, \ldots, N\}$ with generator $\Gamma = (\gamma_{ij})(i, j \in S)$ given by $P\{r(t+\Delta) = j | r(t) = i\} = \begin{cases} \gamma_{ij}\Delta + o(\Delta), \textit{if } i \neq j \\ 1 + \gamma_{ij}\Delta + o(\Delta), \textit{if } i = j \end{cases}$ where $\Delta > 0$ and $\lim_{\Delta\to 0} o(\Delta)/\Delta = 0$, $\gamma_{ij} \geq 0$ are the transition rate from $i$ to $j$ if $i \neq j$ and $\gamma_{ii} = -\sum_{i\neq j}\gamma_{ij}$.

## 2 Problem Formulation

Most of the exponential synchronization methods belong to master-slave type.

In this paper, the master system which we consider as a network with mixed time-variant time-delays and Markovian jumping parameters is as follows:

$$dx(t) = \Big[-A(r(t))x(t) + W_0(r(t))l_0(x(t)) + W_1(r(t))l_1(x(t-h(t))) + W_2(r(t))\int_{t-\tau(t)}^{t} l_2(x(s))ds\Big]dt \tag{1}$$

where $x(t) = [x_1(t), x_2(t), \ldots, x_n(t)]^T \in \mathbb{R}^n$ is the state vector of the transformed system. The diagonal matrix $A(r(t)) = A_i = diag\{a_{i1}, a_{i2}, \ldots, a_{in}\}$ has positive entries $a_{ij} > 0$. The matrices $W_0 = (w^0_{ij})_{n\times n}$, $W_1 = (w^1_{ij})_{n\times n}$ and $W_2 = (w^2_{ij})_{n\times n}$

are, respectively, the connection weight matrix, the discretely delayed connection weight matrix, and the distributive delayed connection weight matrix. $l_i(x(t)) = [l_{i1}(x_1), l_{i2}(x_2), \ldots, l_{in}(x_n)]^T\,(i = 0, 1, 2)$ denotes the neuron activation function with $l_i(0) = 0$. The scalar $h(t) > 0$ with $h(t) < \bar{h}$ (a constant) $|\dot{h}(t)| < \hat{h} < 1$ ($\hat{h}$ is a constant), which may be unknown, denotes the discrete time-delay, whereas the scalar $\tau(t) > 0$, with $\tau(t) < \bar{\tau}$, $|\dot{\tau}(t)| < \hat{\tau} < 1$ ($\bar{\tau}, \hat{\tau}$ are positive constants), is the known distributed time-delay. Let $d = \max\left\{\bar{h}, \bar{\tau}\right\}$.

For the master system (1), a response system is constructed as follows:

$$dy(t) = \Big[-A(r(t))y(t) + W_0(r(t))l_0(y(t)) + W_1(r(t))l_1(y(t-h(t)))$$
$$+\ W_2(r(t)) \int_{t-\tau(t)}^{t} l_2(y(s))ds + u(t)\Big]dt + \sigma(t, \beta(t), \beta(t-h(t)))d\omega(t) \quad (2)$$

where $y(t)$ is the state vector of the response system, $u(t)$ is the controller, $\beta(t) = y(t) - x(t)$ is the error vector. $\omega(t) = (\omega_1, \omega_2, \ldots, \omega_n)$ is an n-dimensional Brownian motion defined on a complete probability space $\left(\Omega, F, \{F_t\}_{t\geq 0}, P\right)$. Here the white noise $d\omega_i(t)$ is independent of $d\omega_j(t)$ for the mutually different $i$ and $j$, and $\sigma : \mathbb{R}_+ \times \mathbb{R}^n \to \mathbb{R}^{n\times n}$ is called the noise intensity function matrix.

In order to realize the exponential synchronization of the master and slave system, we define synchronization error system as follows:

$$d\beta(t) = \Bigg[(-A(r(t)))\beta(t) + W_0(r(t))(l_0(y(t)) - l_0(x(t)))$$
$$+\ W_1(r(t))(l_1(y(t-h(t))) - l_1(x(t-h(t))))$$
$$+ W_2(r(t)) \int_{t-\tau(t)}^{t} (l_2(y(s)) - l_2(x(s)))ds + u(t)\Bigg]dt$$
$$+\ \sigma(t, \beta(t), \beta(t-h(t)))d\omega(t) \quad (3)$$

Let $\beta(t;\xi)$ denote the state trajectory of the neural network (3) for the initial data $\beta(\theta) = \xi(\theta) on -d \leq \theta \leq 0$. It can be easily seen that the system (3) admits a trivial solution $\beta(t;0) \equiv 0$ corresponding to the initial data $\xi = 0$.

For a special case that the information on the time-varying delay $\tau(t)$ is available, we consider a delayed feedback controller of the following form:

$$u(t) = G\beta(t), \quad (4)$$

and then we can have the error system can be represented as follows:

$$
\begin{aligned}
d\beta(t) &= dy(t) - dx(t) \\
&= \Big[ (-A(r(t)) + G)\beta(t) + W_0(r(t))(l_0(y(t)) - l_0(x(t))) \\
&\quad + W_1(r(t))(l_1(y(t-h(t))) - l_1(x(t-h(t)))) \\
&\quad + W_2(r(t)) \int_{t-\tau(t)}^{t} (l_2(y(s)) - l_2(x(s)))ds \Big] dt \\
&\quad + \sigma(t, \beta(t), \beta(t-h(t)))d\omega(t)
\end{aligned} \tag{5}
$$

To simplify (5), we have $L_i(\beta(t)) = l_i(y(t)) - l_i(x(t)), i = 0, 1, 2$ and (5) can be expressed as follows:

$$
\begin{aligned}
d\beta(t) = [(-A(r(t)) + G)\beta(t) + W_0(r(t))L_0(\beta(t)) + W_1(r(t))L_1(\beta(t-h(t))) \\
+ W_2(r(t)) \textstyle\int_{t-\tau(t)}^{t} L_2(\beta(s))\, ds]dt + \sigma(t, \beta(t), \beta(t-h(t)))d\omega(t)
\end{aligned} \tag{6}
$$

For the master system and the slave system, we give the following assumptions.

**Assumption 1** The activation function of the neurons $l_i(\beta(t))$ is bounded and satisfies the Lipschitz condition:

$$
|l_i(x) - l_i(y)| \le G_i|x - y|, \forall x, y \in R \tag{7}
$$

where $M$ is a constant matrix.

**Assumption 2** $\sigma : \mathbb{R}_+ \times \mathbb{R}^n \to \mathbb{R}^{n\times n}$ is locally Lipschitz continuous and satisfies the linear growth condition. Moreover, there exists two matrices $H_1$ and $H_2$, such that

$$
\begin{aligned}
&trace\big[\sigma^T(t, \beta(t), \beta(t-h(t)))\sigma(t, \beta(t), \beta(t-h(t)))\big] \\
&\le \|H_1\beta(t)\|^2 + \|H_2\beta(t-h(t))\|^2
\end{aligned} \tag{8}
$$

Now we give the main concept of exponentially synchronization.

**Definition 1** The master system (1) and the slave system (2) is exponentially synchronization in the mean square, if there exist positive constants $\alpha > 0$ and $\mu > 0$ such that

$$
E|\beta(t; \xi(s))|^2 \le \mu e^{-\alpha t} \sup_{-d\le s\le 0} E|\beta(s)|^2, \forall t > 0.
$$

Now, we describe the problem to solve in this paper as follows.

## 3 Main Results and Proofs

For the proof of the main result, the following lemmas are needed.

**Lemma 1** *[14] Let* $x \in \mathbb{R}^n, y \in \mathbb{R}^n$ *and* $\varepsilon > 0$. *Then* $x^T y + y^T x \leq \varepsilon x^T x + \varepsilon^{-1} y^T y$.

**Lemma 2** *[14] Given constant matrices* $\Sigma_1, \Sigma_2, \Sigma_3$, *where* $\Sigma_1 = \Sigma_1^T$ *and* $0 < \Sigma_2 = \Sigma_2^T$, *then* $\Sigma_1 + \Sigma_3^T \Sigma_2^{-1} \Sigma_3 < 0$ *if and only if*

$$\begin{bmatrix} \Sigma_1 & \Sigma_3^T \\ \Sigma_3 & -\Sigma_2 \end{bmatrix} < 0, \text{ or } \begin{bmatrix} -\Sigma_2 & \Sigma_3 \\ \Sigma_3^T & \Sigma_1 \end{bmatrix} < 0$$

**Lemma 3** *[16] For any positive definite matrix* $M > 0$, *scalar* $\gamma > 0$, *vector function* $\omega : [0, \gamma] \to \mathbb{R}^n$ *such that the integrations concerned are well defined, the following inequality holds:*

$$\left( \int_0^\gamma \omega(s) ds \right)^T M \left( \int_0^\gamma \omega(s) ds \right) \leq \gamma \left( \int_0^\gamma \omega^T(s) M \omega(s) ds \right)$$

We write, $W_j(r(t)) = W_{ji}\, j = 0, 1, 2$ for the sake of simplicity. At the same time, we denote

$$\Omega_1 = P_i(-A_i + G) + (-A_i + G)^T P_i + \left(1 + \eta \bar{h}\right) Q_1 + \left(1 - \hat{\tau} - \eta\right)^{-1} \hat{\tau} \bar{\tau} \varepsilon_{3i}^{-1} G_2^T G_2 + \sum_{j=1}^{N} \gamma_{ij} P_j \tag{9}$$

$$\Pi_{11} = \varepsilon_{1i} P_i W_{0i} W_{0i}^T P_i + \varepsilon_{1i}^{-1} G_0^T G_0 + \varepsilon_{2i} P_i W_{1i} W_{1i}^T P_i + \varepsilon_{3i} P_i W_{2i} W_{2i}^T P_i \tag{10}$$

**Theorem 1** *Let* $\eta > 0$ *be arbitrarily given constant satisfying* $\hat{\tau} + \eta < 1$. *If there exist positive scalars* $\varepsilon_i > 0 (i = 1, 2, 3)$ *and positive definite matrices* $P = P^T > 0, Q_1 = Q_1^T > 0$ *following the inequalities:*

$$\Omega_1 + \Pi_{11} + \rho_i H_1^T H_1 < 0 \tag{11}$$

*and*

$$-\left(1 - \hat{h}\right) Q_1 + \varepsilon_{2i}^{-1} G_1^T G_1 + \rho_i H_2^T H_2 < 0 \tag{12}$$

*where* $\Omega_1$ *and* $\Pi_{11}$ *are defined in (9) and (10), then the master system (1) and slave system (2) achieve exponential synchronization.*

*Proof* Define a Lyapunov functional candidate

$$\begin{aligned} V(t) &= V(x(t), r(t) = i) \\ &= \beta^T(t)P_i\beta(t) + \int_{t-h(t)}^{t} \beta^T(s)Q_1\beta(s)ds + \eta \int_{-h}^{0}\int_{t+s}^{t} \beta^T(k)Q_1\beta(k)dkds \\ &\quad + \int_{-\tau(t)}^{0}\int_{t+s}^{t} \beta^T(k)Q_2\beta(k)dkds \end{aligned} \tag{13}$$

where $P_i > 0$, $Q_1 > 0$ and $Q_2 \geq 0$ is given by

$$Q_2 = \left(1 - \hat{\tau} - \eta\right)^{-1}\hat{\tau}\varepsilon_{3i}^{-1}G_2^TG_2 \tag{14}$$

The time derivate of $V(t)$ along the trajectory is calculated as follows:

$$\begin{aligned} \frac{dV(t)}{dt} &= \beta^T(t)[P_i(-A_i + G) + (-A_i + G)^TP_i + (1 + \eta\bar{h})Q_1 + \tau(t)Q_2 + \sum_{j=1}^{N}\gamma_{ij}P_j]\beta(t) \\ &\quad + 2\beta^T(t)P_iW_{0i}L_0(\beta(t)) + 2\beta^T(t)P_iW_{1i}L_1(\beta(t-h(t))) \\ &\quad + 2\beta^T(t)P_iW_{2i}\int_{t-\tau(t)}^{t} L_2(\beta(s))ds - \eta\int_{t-h}^{t}\beta^T(s)Q_1\beta(s)ds \\ &\quad - (1 - \dot{h}(t))\beta^T(t-h(t))Q_1\beta(t-h(t)) + (\dot{\tau}(t) - 1)\int_{t-\tau(t)}^{t}\beta^T(s)Q_2\beta(s)ds \\ &\quad + trace\left[\sigma^T(t, \beta(t), \beta(t-h(t)))P_i\sigma(t, \beta(t), \beta(t-h(t)))\right] \end{aligned} \tag{15}$$

From Lemma 1 and (7), we have

$$\begin{aligned} 2\beta^T(t)P_iW_{0i}L_0(\beta(t)) &\leq \varepsilon_{1i}\beta^T(t)P_iW_{0i}W_{0i}^TP_i\beta(t) + \varepsilon_{1i}^{-1}L_0^T(\beta(t))L_0(\beta(t)) \\ &\leq \varepsilon_{1i}\beta^T(t)P_iW_{0i}W_{0i}^TP_i\beta(t) + \varepsilon_{1i}^{-1}\beta^T(t)G_0^TG_0\beta(t) \end{aligned} \tag{16}$$

$$\begin{aligned} 2\beta^T(t)P_iW_{1i}L_1(\beta(t-h(t))) &\leq \varepsilon_{2i}\beta^T(t)P_iW_{1i}W_{1i}^TP_i\beta(t) + \varepsilon_{2i}^{-1}L_1^T(\beta(t) - h(t))L_1(\beta(t-h(t))) \\ &\leq \varepsilon_{2i}\beta^T(t)P_iW_{1i}W_{1i}^TP_i\beta(t) + \varepsilon_{2i}^{-1}\beta^T(t-h(t))G_1^TG_1\beta(t-h(t)) \end{aligned} \tag{17}$$

$$\begin{aligned} &2\beta^T(t)P_iW_{2i}\int_{t-\tau(t)}^{t} L_2(\beta(s))ds \\ &\leq \varepsilon_{3i}\beta^T(t)P_iW_{2i}W_{2i}^TP_i\beta(t) + \varepsilon_{3i}^{-1}\left(\int_{t-\tau(t)}^{t} L_2(\beta(s))ds\right)^T\left(\int_{t-\tau(t)}^{t} L_2(\beta(s))ds\right) \end{aligned} \tag{18}$$

Moreover, it can be seen from Lemma 3, (7) and (14) that

$$\begin{aligned} &\varepsilon_{3i}^{-1}\left(\int_{t-\tau(t)}^{t} L_2(\beta(s))ds\right)^T\left(\int_{t-\tau(t)}^{t} L_2(\beta(s))ds\right) \leq \varepsilon_{3i}^{-1}\tau(t)\int_{t-\tau(t)}^{t} L_2^T(\beta(s))L_2(\beta(s))ds \\ &\leq \varepsilon_{3i}^{-1}\dot{\tau}(t)\int_{t-\tau(t)}^{t}\beta^T(s)G_2^TG_2\beta(s)ds \leq \varepsilon_{3i}^{-1}\hat{\tau}\int_{t-\tau(t)}^{t}\beta^T(s)G_2^TG_2\beta(s)ds \\ &= \left(1 - \hat{\tau} - \eta\right)\int_{t-\tau(t)}^{t}\beta^T(s)Q_2\beta(s)ds \end{aligned} \tag{20}$$

From Assumption 2, we have

$$
\begin{aligned}
& trace\left[\sigma^T(t, \beta(t), \beta(t-h(t)))P_i\sigma(t, \beta(t), \beta(t-h(t)))\right] \\
& \leq \rho_I trace\left[\sigma^T(t, \beta(t), \beta(t-h(t)))\sigma(t, \beta(t), \beta(t-h(t)))\right] \\
& = \rho_I\left[\beta^T(t)H_1^T H_1\beta(t), \beta^T(t-h(t))H_2^T H_2\beta(t-h(t))\right]
\end{aligned} \tag{21}
$$

Using (16)–(21), we obtain from (15) that

$$
dV(t, \beta(t)) \leq \begin{bmatrix} \beta(t) \\ \beta(t-h(t)) \end{bmatrix}^T \begin{bmatrix} \Omega_1 + \Pi_{11} + \rho_i H_1^T H_1 & 0 \\ 0 & -\left(1-\hat{h}\right)Q_1 + \varepsilon_{2i}^T G_1^T G_1 + \rho_i H_2^T H_2 \end{bmatrix} \begin{bmatrix} \beta(t) \\ \beta(t-h(t)) \end{bmatrix} \tag{22}
$$
$$
-\eta \int_{t-h}^{t} \beta^T(s)Q_1\beta(s)ds - \eta \int_{t-\tau(t)}^{t} \beta^T(s)Q_2\beta(s)ds
$$

where $\Omega_1$ is defined in (9), $\Pi_{11}$ is defined in (10).

Denote

$$
\zeta(t) := \begin{bmatrix} \beta(t) \\ \beta(t-h(t)) \end{bmatrix} \tag{23}
$$

and

$$
\Xi := \left[\begin{bmatrix} \Omega_1 + \Pi_{11} + \rho_i H_1^T H_1 & 0 \\ 0 & -\left(1-\hat{h}\right)Q_1 + \varepsilon_{2i}^T G_1^T G_1 + \rho_i H_2^T H_2 \end{bmatrix}\right] \tag{24}
$$

If

$$
\Xi < 0, \tag{25}
$$

then by $\lambda_{\max}(\Xi) < 0$ *and* $|\beta(t)| \leq |\zeta(t)|$, it can be deduced from (22) that

$$
dV(t, \beta(t)) \leq \lambda_{\max}(\Xi)|\beta(t)|^2 - \eta \int_{t-\bar{h}}^{t} \beta^T(s)Q_1\beta(s)ds - \eta \int_{t-\tau(t)}^{t} \beta^T(s)Q_2\beta(s)ds \tag{26}
$$

To this end, let $\mathbb{V}(t, \beta(t)) = e^{kt}V(t, \beta(t))$, where k is to be determined. It can be checked that

$$
V(t, \beta(t)) \leq \lambda_{\max}(P_i)|\beta(t)|^2 + \left(1 + \eta\bar{h}\right) \int_{t-\bar{h}}^{t} \beta^T(s)Q_1\beta(s)ds + \hat{\tau} \int_{t-\tau(t)}^{t} \beta^T(s)Q_2\beta(s)ds \tag{27}
$$

Therefore,

$$
\begin{aligned}
& d\mathbb{V}(t, \beta(t)) = e^{kt}[kV(t, \beta(t)) + dV(t, \beta(t))] \\
& \leq e^{kt}\left((k\lambda_{\max}(P_i) + \lambda_{\max}(\Xi))|\beta(t)|^2 + \left(k\left(1+\eta\bar{h}\right) - \eta\right) \int_{t-\bar{h}}^{t} \beta^T(s)Q_1\beta(s)ds + \left(k\hat{\tau} - \eta\right)\right. \\
& \int_{t-\tau(t)}^{t} \beta^T(s)Q_2\beta(s)ds
\end{aligned} \tag{28}
$$

Choosing k sufficiently small such that

$$k\lambda_{\max}(P_i) + \lambda_{\max}(\Xi) \le 0, \quad k\big(1 + \eta\bar{h}\big) - \eta < 0, \quad k\hat{\tau} - \eta \le 0$$

and taking the mathematical expectation of both sides of (28), one gets

$$\frac{dE\mathbb{V}(t, \beta(t))}{dt} \le 0 \tag{29}$$

which implies $E\mathbb{V}(t, \beta(t)) \le E\mathbb{V}(0, \beta(0))$. Therefore,

$$\begin{aligned} e^{kt}EV(t, \beta(t)) &\le EV(0, \beta(0)) \\ &\le E\Big[\lambda_{\max}(P_i)|\beta(0)|^2 + \big(1 + \eta\bar{h}\big)\int_{-\bar{h}}^{0} \beta^T(s)Q_1\beta(s)ds + \hat{\tau}\int_{-\tau(t)}^{0} \beta^T(s)Q_2\beta(s)ds\Big] \\ &\le \big(\lambda_{\max}(P_i) + \big(1 + \eta\bar{h}\big)\bar{h}\lambda_{\max}(Q_1) + \hat{\tau}^2\lambda_{\max}(Q_2)\big) \max_{-d\le s\le 0} E|\beta(s)|^2 \end{aligned} \tag{30}$$

Also, it can be seen that

$$EV(t, \beta(t)) \ge \lambda_{\min}(P_i)E|\beta(t)|^2 \tag{31}$$

From (30) and (31), it follows that

$$\begin{aligned} EV(t, \beta(t)) \le \lambda_{\min}^{-1}(P_i)\big(&\lambda_{\max}(P_i) + \big(1 + \eta\bar{h}\big)\bar{h}\lambda_{\max}(Q_1) \\ &+ \hat{\tau}^2\lambda_{\max}(Q_2)\big)e^{-kt} \max_{-d\le s\le 0} E|\beta(s)|^2 \end{aligned} \tag{32}$$

To this end, from Definition 1, we can conclude that the neural networks (1) and (2) are exponentially synchronization which competes the proof of Theorem 1.

## 4 Numerical Example

In this section, we will give a numerical example to illustrate the effective of the main result proposed in this paper.

Consider the master system (1) of a two-neuron delayed neural network with two models. The network parameters are given as follows:

$$A_1 = \begin{bmatrix} 3.1 & 0 \\ 0 & 3.7 \end{bmatrix} A_2 = \begin{bmatrix} 3.2 & 0 \\ 0 & 3.5 \end{bmatrix} W_{01} = \begin{bmatrix} 0.4 & -1.8 \\ -1.1 & 1.6 \end{bmatrix} W_{02} = \begin{bmatrix} 0.3 & -1.7 \\ -1 & 1.5 \end{bmatrix}$$

$$\gamma = \begin{bmatrix} -2 & 2 \\ 1 & -1 \end{bmatrix} W_{11} = \begin{bmatrix} 0.3 & -0.8 \\ -0.1 & 0.6 \end{bmatrix} W_{12} = \begin{bmatrix} 0.3 & -0.9 \\ -0.1 & 0.5 \end{bmatrix}$$

$$W_{21} = \begin{bmatrix} 0.2 & -0.64 \\ -0.08 & 0.48 \end{bmatrix} W_{22} = \begin{bmatrix} 0.2 & -0.6 \\ -0.08 & 0.5 \end{bmatrix}$$

We choose
$h(t) = 0.4 \sin t + 0.4,$ $\tau(t) = 0.1 \sin t + 0.4$
$l_0(\cdot) = 0.6 \tanh(\cdot),$ $l_1(\cdot) = 0.3 \tanh(\cdot)$
$l_2(\cdot) = 0.8 \tanh(\cdot)$, then we can obtain
$\bar{h} = 0.8, \hat{h} = 0.4, \bar{\tau} = 0.5, \hat{\tau} = 0.1$
$G_0 = 0.6I_3, G_1 = 0.3I_3, G_2 = 0.8I_3$ , so we have $H_1 = H_2 = I$
$\sigma(t, \beta(t), \beta(t - h(t)))=\frac{1}{\sqrt{2}}(\beta(t) + \beta(t - h(t)))$

By using LMI toolbox of Matlab, we can obtain the feasible solutions of the parameters in Theorem 1 as follows.

$$P_1 = \begin{bmatrix} 0.4395 \; 0.2306 \\ 0.2306 \; 0.2552 \end{bmatrix} \quad P_2 = \begin{bmatrix} 0.4270 \; 0.2108 \\ 0.2108 \; 0.2528 \end{bmatrix}$$

$$Q_{11} = \begin{bmatrix} 9.4155 & 0 \\ 0 & 9.4155 \end{bmatrix} \quad Q_{12} = \begin{bmatrix} 9.4160 & 0 \\ 0 & 9.4160 \end{bmatrix}$$

$\varepsilon_{32} = 4.9416$ $\varepsilon_{11} = 2.9749$ $\varepsilon_{21} = 3.1942$ $\varepsilon_{31} = 4.9420$ $\varepsilon_{12} = 2.9745$
$\varepsilon_{22} = 3.1950$

$$\rho = 2.4409 \quad G_{U1} = \begin{bmatrix} -11.0265 & 0.7134 \\ 0.7134 & -11.3660 \end{bmatrix} \quad G_{U2} = \begin{bmatrix} -11.0265 & 0.6365 \\ 0.6365 & -11.4248 \end{bmatrix}.$$

So by Theorem 1, the neural networks (1) and (2) are exponentially synchronization. At the same time, in order to make the estimation parameters gradually approach to the real value, we give the response of error system by using Matlab simulation as follows:

From these Figs. 1, 2, 3 and 4, we can see the error system has a good and fast convergence, and the exponential synchronization control of the two systems is verified again.

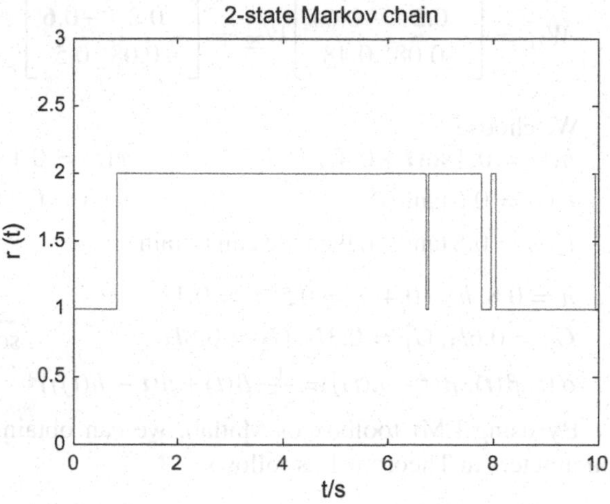


**Fig. 1** 2-state Markov chain

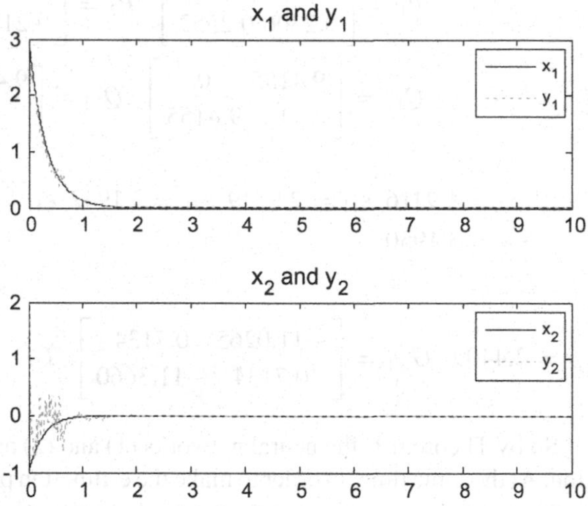


**Fig. 2** The state master and slave systems without control input

# 5 Conclusions

In this paper, we have dealt with the exponential synchronization control of neural networks with time-delays and Markovian jumping parameters. Lyapunov stability theorem and LMI technique have been used to solve the problem. A sufficient condition has been derived to ensure the global stability of the error system, and the exponential synchronization of the master system and salve system is obtained. Finally, we have given a numerical simulation to verify the feasibility and effec-

**Fig. 3** The error state

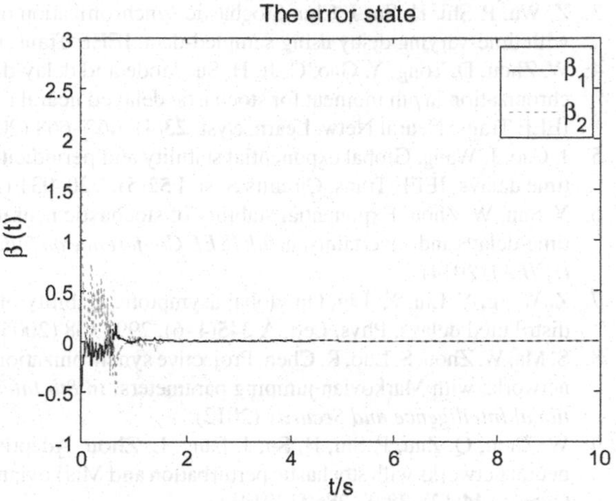


**Fig. 4** The control input

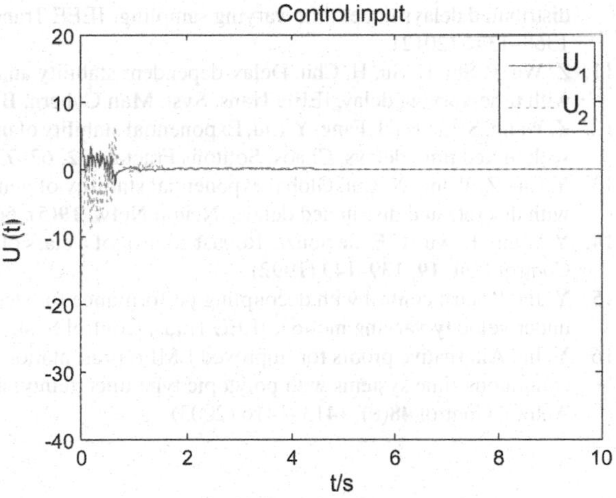


tiveness of the proposed synchronization scheme. This paper is supported by the Specialized Research Fund for the Doctoral Program of Higher Education under grant no. 20120075120009, the Natural Science Foundation of Shanghai under grant no. 12ZR1440200.

## References

1. S. Arik, Stability analysis of delayed neural networks. IEEE Trans Circuits Syst-I **47**, 1089–1092 (2000)
2. Z. Wu, H. Su, J. Chu, W. Zhou, Improved result on stability analysis of discrete stochastic neural networks with time delay. Phys. Lett. A **373**(17), 1546–1552 (2009)

3. Z. Wu, P. Shi, H. Su, J. Chu, Stochastic synchronization of Markovian jump neural networks with time-varying delay using sampled-data. IEEE Trans. Cybern. **43**(6), 1796–1806 (2013)
4. W. Zhou, D. Tong, Y. Gao, C. Ji, H. Su, Mode and delay-dependent adaptive exponential synchronization in *p*th moment for stochastic delayed neural networks with Markovian switching. IEEE Trans. Neural Netw. Learn. Syst. **23**(4), 662–668 (2012)
5. J. Cao, J. Wang, Global exponential stability and periodicity of recurrent neural networks with time delays. IEEE Trans. Circuits Syst. I **52**(5), 920–931 (2005)
6. Y. Sun, W. Zhou, Exponential stability of stochastic neural networks with time-variant mixed time-delays and uncertainty, in *9th IEEE Conference on Industrial Electronics and Applications (ICIEA)* (2014)
7. Z. Wang, Y. Liu, X. Liu, On global asymptotic stability of neural networks with discrete and distributed delays. Phys. Lett. A **345**(4–6), 299–308 (2005)
8. S. Ma, W. Zhou, S. Luo, R. Chen, Projective synchronization control of delayed recurrent neural networks with Markovian jumping parameters, in *8th International Conference on Computational Intelligence and Security* (2012)
9. W. Zhou, Q. Zhu, P. Shi, H. Su, J. Fang, L. Zhou, Adaptive synchronization for neutral-type neural networks with stochastic perturbation and Markovian switching parameters. IEEE Trans. Cybern. **44**(12), 2848–2860 (2014)
10. Z. Wu, P. Shi, H. Su, J. Chu, Exponential synchronization of neural networks with discrete and distributed delays under time-varying sampling. IEEE Trans. Neural Netw. Learn. Syst. **23**(9), 1368–1376 (2012)
11. Z. Wu, P. Shi, H. Su, H. Chu, Delay-dependent stability analysis for switched neural networks with time-varying delay. IEEE Trans. Syst. Man Cybern. B Cybern. **41**(6), 1522–1530 (2011)
12. Z. Wang, S. Lauria, J. Fang, Y. Liu, Exponential stability of uncertain stochastic neural networks with mixed time-delays. Chaos, Solitons Fractals **32**, 62–72 (2007)
13. Y. Liu, Z. Wang, X. Liu, Global exponential stability of generalized recurrent neural networks with discrete and distributed delays. Neural Netw. **19**(5), 667–675 (2006)
14. Y. Wang, L. Xie, C.E. de Souza, Robust control of a class of uncertain nonlinear systems. Syst. Control Lett. **19**, 139–149 (1992)
15. Y. Jia, Robust control with decoupling performance for steering and traction of 4WS vehicles under velocity-varying motion. IEEE Trans. Control Syst. Technol. **8**(3), 554–569 (2000)
16. Y. Jia, Alternative proofs for improved LMI representations for the analysis and the design of continuous-time systems with polytopic type uncertainty: a predictive approach. IEEE Trans. Autom. Control **48**(8), 1413–1416 (2003)

# Research on a Fast Matching Method of K Nearest Neighbor for WiFi Fingerprint Location

**Lijun Hou, Yufeng Luo and Yanhui Liu**

**Abstract** Aiming at the problem of low speed and positioning fluctuations of indoor WiFi fingerprints. Firstly, we use the method of Gauss fitting and averaging to acquire the average value of the received signal. Secondly, we use a distance to be similarity measure to define a threshold to classify the fingerprint database. Finally, By improving the K nearest neighbor algorithm and on the basis of classification, Implement fast matching of K nearest neighbor. The experimental results show that the time efficiency of the classified location system has been greatly improved, with an average decrease of 62.8%; In the positioning accuracy, WiFi fingerprint positioning of the average error from 4.17 m down to 2.12 m.

**Keywords** Gaussian fitting · Multiple measurements for averaging · Database classification · Fast matching of K ncarcst neighbor

## 1 Introduction

The maturity of IEEE802.11 based technology and the popularity of mobile devices and wireless LAN in all parts of the world, meanwhile, WiFi indoor fingerprint positioning has the advantages of low cost, easy deployment and no sight constraint. It has become a hot in current indoor positioning research. The study of WiFi fingerprint localization in the indoor Yang et al. [1] monitor the state of the physical layer channel

L. Hou (✉) · Y. Luo · Y. Liu
School of Electrical Engineering & Automation,
Henan Polytechnic University, Jiaozuo 454003, China
e-mail: 2693267873@qq.com

Y. Luo
e-mail: 375524863@qq.com

Y. Liu
e-mail: 1299280109@qq.com

Y. Luo
Wuxi Jiuyu Architectural Design Institute Co. Ltd., Wuxi 214000, China



Y. Jia et al. (eds.), *Proceedings of 2018 Chinese Intelligent Systems Conference*,
Lecture Notes in Electrical Engineering 528,
https://doi.org/10.1007/978-981-13-2288-4_53

and improve the precision of the WiFi fingerprint location. Laouudias [2] create a systems are developed to reduce the error of WiFi fingerprint positioning to 2–4 m on the basis of fingerprint matching. Fang et al. [3] By extracting the characteristic signal from the measured RSSI signal intensity, it reduces the effect of indoor multipath effect on WiFi signal fluctuation, and improves indoor fingerprint location accuracy by 40%. Zandbergen [4] combined WiFi with A-GPS and honeycomb positioning to further improve the accuracy of indoor positioning. Different network structure, wireless positioning technology and application environment, scholars at home and abroad put forward different algorithms. According to the common wireless location technology, there are [5, 6] such as Bluetooth, infrared, WiFi, Zigbee, ultra wideband and so on.

At present, the location of WiFi is mainly divided into two methods of ranging and non range finding. A class of RSSI range signal attenuation model is used to calculate the distance D, and then the three edge location method is applied to the obtained D value to determine the location of the unknown point. The other is a non range—based—RSSI intensity based fingerprint localization technique, As a whole, the location accuracy of this kind of location is higher than that of the range finding, but the defect of fingerprint localization is the cumbersome process of fingerprint acquisition. The RSSI vector plays an important role in the WiFi fingerprint location. However, the actual RSSI vector has a great fluctuation due to the reflection and diffraction. By Gauss function fitting and obtained the average value to eliminate the environmental factor of RSSI wave, the fingerprint data more accurate and reliable. In the WiFi fingerprint location technology, the second factors that affect the positioning accuracy are the advantages and disadvantages of the WiFi database. No longer just focus on optimizing the fingerprint matching algorithm to improve the accuracy of fingerprint location. Starting with the quality of the fingerprint database, this paper proposes a method of optimizing the database by the method of distance similarity measure. Third factors that affect the positioning accuracy are the online matching algorithm. The traditional K nearest neighbor matching algorithm has many disadvantages, such as large computation, slow speed and poor accuracy. By introducing the number of the same AP as the parameter, the K nearest neighbor matching algorithm is improved, and the fast K nearest neighbor matching algorithm is realized on the basis of the database classification. This algorithm can remove redundant data and effectively improve the accuracy and speed of matching.

## 2 Description of Fingerprint Location Based on WiFi

Environment is very restrictive to the multipath propagation of WiFi signals, and the characteristics of signals are very strong. For every fixed sampling point, the multipath structure of the fixed sampling channel is determined. The feature of such a multipath structure is the "fingerprint" of the sampling point. The location card receives the signal strength from each AP at the unknown location point and matches

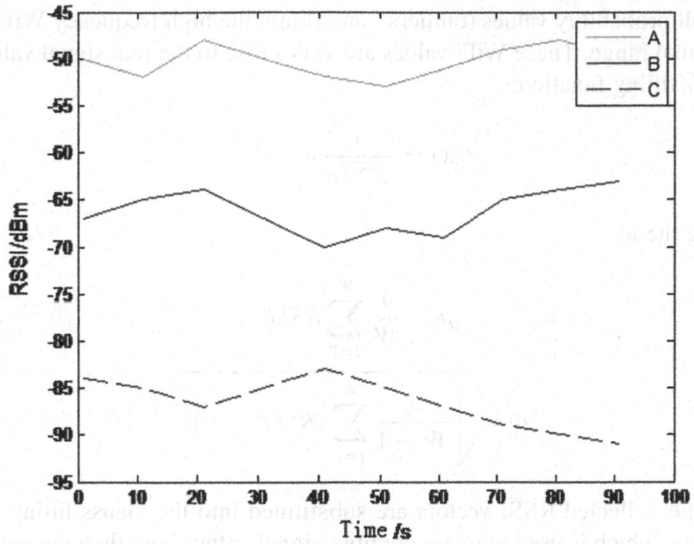


**Fig. 1** The RSSI change trend of three different sampling points

the signal strength stored in the WiFi fingerprint Library in advance, so as to achieve the best estimation of the unknown location point.

## *2.1 RSSI Signal Processing*

RSSI plays an important role in positioning, the greater the value of RSSI, the greater the intensity of the signal. In A, B and C, the distance between the sampling points from three different locations is farther and farther away from a fixed AP, and the trend of RSSI value from three points is collected in 100 s, As shown in Fig. 1.

From Fig. 1, we can see that because the signal's multipath propagation is greatly influenced by the environment, the value of RSSI decreases with the increase of measurement distance, and the RSSI value also fluctuates to with time. In order to ensure the reliability and accuracy of fingerprint database, we first use Gauss function fitting to fit the collected RSSI signal, and quickly select the RSSI of high frequency. For the RSSI fluctuation, by averaging the value method for smoothing the signal sampling point received.

### 2.1.1 Gauss Fitting

Gauss's function fitting is a way to deal with the field values in the data. Applying the Gauss function to the RSSI data received by the sampling points, we can remove some small probability values (outliers), and retain the high frequency WiFi signals in the normal range. These WiFi values are very close to the real signal values.

Gauss's fitting function:

$$f(x) = \frac{1}{\sqrt{2\pi}\sigma} e^{-\frac{(x-\mu)^2}{2\sigma^2}} \tag{1}$$

Among them:

$$\mu = \frac{1}{W} \sum_{i=1}^{W} RSSI^i \tag{2}$$

$$\sigma = \sqrt{\frac{1}{W-1} \sum_{i=1}^{W} (RSSI^i - \mu)^2} \tag{3}$$

First, the collected RSSI vectors are substituted into the Gauss fitting function respectively, which is used to retain credible signal values, and then the mean value of the retained data is obtained. The mean value is the RSSI vector of the sampling points. In order to quickly screen out the desired data, The method of near Gauss fitting is adopted. The formula is as follows:

$$\left| RSSI^i - \mu \right| < k\sigma \tag{4}$$

In the formula, $\mu$ is the mean; $\sigma$ is the standard deviation; K is the number of RSSI values of the screening interval, which can be determined according to the percentage M of the data reservation.

Among them:

$$M = \frac{k}{N} \times 100\% \tag{5}$$

In the type: N is the total number of collections.

## 2.2 *WiFi Fingerprint Library Training Stage*

### 2.2.1 The Setting of the Location Area

First, the positioning of the region according to the designed size, divided into several grids, each grid vertex is a sampling point; secondly, to determine the origin of coordinates and the reference points are numbered according to certain rules, to determine the coordinates of the sampling points, as shown in Fig. 2.

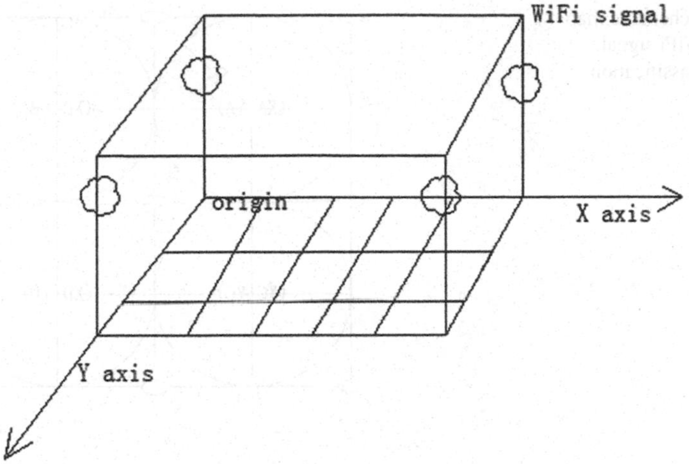


**Fig. 2** Coordinate diagram of the establishment of a location table

**Table 1** WiFi fingerprint database

| Coordinate | The vector of fingerprint points |
|---|---|
| $(X1, Y1)$ | $(RSSI_1^1, RSSI_1^2, \ldots, RSSI_1^W)$ |
| $(X2, Y2)$ | $(RSSI_2^1, RSSI_2^2, \ldots, RSSI_2^W)$ |
| ... | ... |
| $(Xn, Yn)$ | $(RSSI_n^1, RSSI_n^2, \ldots, RSSI_n^W)$ |

### 2.2.2 The Establishment of WiFi Fingerprint Database

Use the location card to collect the RSSI values from different AP. Assuming that there are n sampling points and w AP, each of the sampling points can be collected as a w RSSI as a fingerprint, A total of N fingerprints can be collected. Each fingerprint is represented by $RSSI_i^j$, where I and J represent the I sampling points and AP from the J, respectively, and the N fingerprints are saved to the WiFi fingerprint database. The following is shown in Table 1.

### 2.2.3 Classification of Fingerprint Database

Because of the increasing size of the location area and the more and more data stored in the database, the computational complexity of these matching algorithms becomes very large, It becomes particularly important to find a fast and accurate positioning method. in order to realize the matching of fast positioning, it is necessary to classify the establishment of WiFi fingerprint database, the classification thought is in the positioning area, select multiple appropriate mesh vertices as the center, the radius

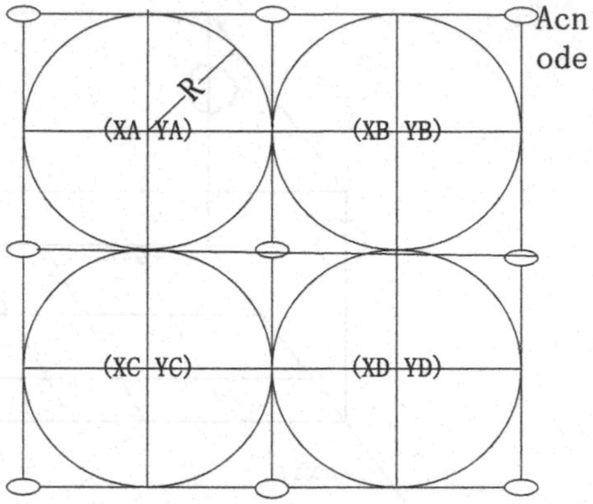


**Fig. 3** A schematic map based on WiFi signal intensity classification

of the R circle, the circle which contains all the sampling points are divided into a class, and will not be included in the circular classification sampling point called acnode and, all in the region as the location of acnode database, as shown in Fig. 3.

The steps of the classification of the fingerprint Library:

1. The distance between a number of sampling points and the center point A (XA, YA) in a good fingerprint library is less than or equal to R as a Class A, as shown by (6).

$$\sqrt{(Xj - XA)^2 - (Yj - YB)^2} \leq R \tag{6}$$

2. According to the classification of the radius of the circle, there will be a number of isolated points can not be classified, all the rest of the database as acnode.

## 2.3 Classification of Unknown Location Points

Assuming the coordinates of an unknown position point (Xt Yt) = [$R_1$, $R_2$, …, $R_j$, …, $R_w$], The thought of classification: The signal intensity of unknown position and type of the center points of the minimum mean square error P as the basis of classification, if the mean square error of more than P of the unknown point to acnode database, the value of P is based on the actual geographical environment and the value of R.

$$\sum_{i=1}^{W} \left(R_i - RSSI_m^i\right)^2 < P \tag{7}$$

where $RSSI_m^w$ is the vector of the center of the circle.

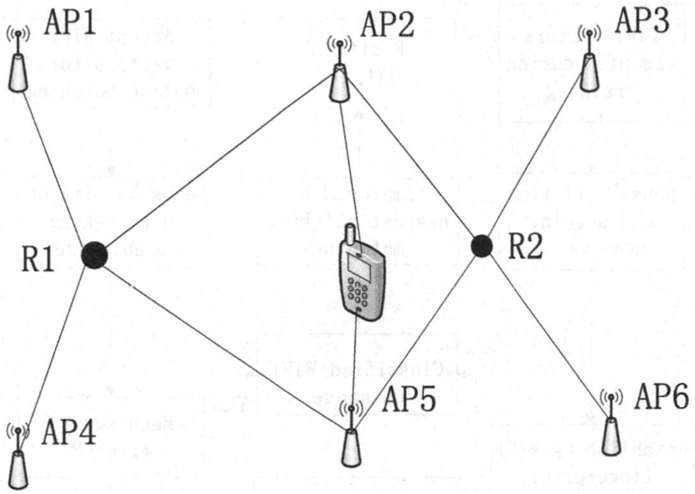


**Fig. 4** A schematic diagram of a WiFi signal

## *2.4 Improved K Nearest Neighbor Matching Algorithm*

Considering the AP transmission distance is 100 m, when the positioning area is too large when the sampling point cannot accept all signals from the AP, the space distance of similar area collected more, the same number of AP on the contrary, the space distance of the region to the same number of less AP acquisition, Therefore, the same number of AP vectors is similar in space distance. By introducing the same number of AP as parameters, the matching degree of WiFi fingerprint can be improved when the similarity of L is calculated.

$$L = \left(1 - \frac{M_{si}}{M_i}\right) \sum_{K=1}^{W} \sqrt{\left(R_K - R_{i,K}\right)^2} \tag{8}$$

In the form, $M_i$ is the AP of the online matching phase location card at the unknown location point and the total number of AP at the i sampling point; $M_{si}$ is the number of AP in the unknown location point of the online matching phase location card and the same number of AP at the I sampling point. Figure 3 is a schematic diagram for receiving WiFi signals, in which R1, R2 are sampling points, and R1 receives signals from AP1, AP2, AP4, AP5, and R2 accepts AP2, AP3, AP5, and signals. Locate the mobile phone to accept the AP2, AP5 signal, in calculating the similarity of R1, M1 = 6, Ms1 = 2 (Fig. 4).

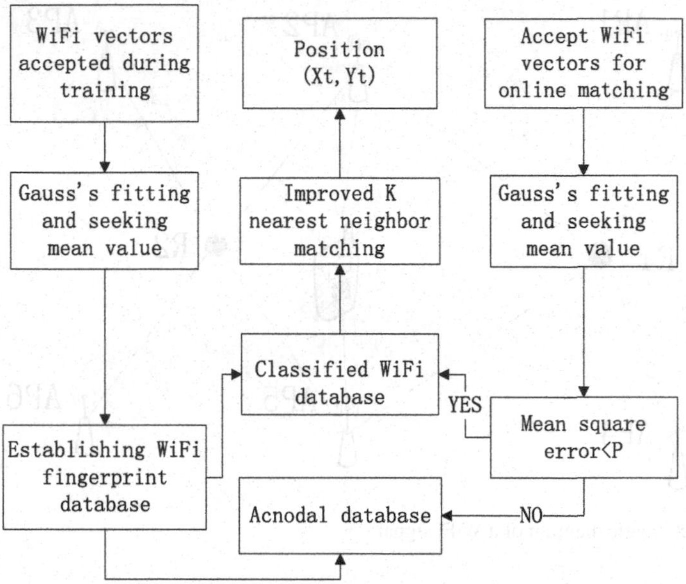


**Fig. 5** Schematic diagram of fast matching algorithm for K nearest neighbor

To sort the L and select the first K small L corresponding to the coordinates of, $\{P_1, P_2, \ldots, P_i, P_k\}(1 << i << K)$, $P_i = (X_i\ Y_i)$, the final positioning result.

$$P = \sum_{i=1}^{K} W_i P_i \tag{9}$$

$W_i$ represents the weight of the unknown location point, and the calculation formula is as follows

$$W_i = \frac{1/L_i}{\sum_{i=1}^{K} 1/L_i} \tag{10}$$

This paper uses improved K nearest neighbor matching positioning principle as shown in Fig. 5, the first Gauss fitting and mean processing on the received WiFi vector in training, the establishment of classification database; secondly, the online matching WiFi vector received are classified; finally, using improved K nearest neighbor matching location.

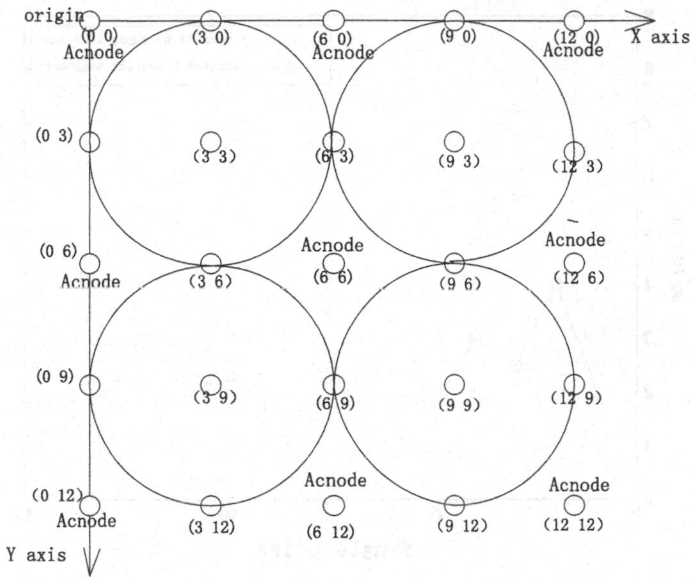


**Fig. 6** Experimental deployment diagram

# 3 Experiment and Error Analysis

## *3.1 Deployment of Experiments*

The experiment was carried out in the electrical building of Henan Polytechnic University, 205. In an experimental area, it is long 12 m, wide 12 m, One sampling point is taken every 3 m, and there are 25 sampling points, and 100 RSSI values are collected at each sampling point, and the mean value is stored in the fingerprint library after Gauss's fitting. To realize fast matching algorithm of K nearest neighbor classification of fingerprint data, radius R classification 3 m, A, B, C, a D and all acnode for a large class of acnode database. As shown in Fig. 6.

In the process of experiment, the WiFi analyzer based on Android platform is used to collect the signal intensity in monitoring area. sing SQL SERVE 2008 to create a fingerprint database, configuring it to a location server and providing data support to the server—side program. The positioning server based on the J2EE platform is responsible for receiving the data uploaded by the client and analyzing, calculating and matching the final implementation.

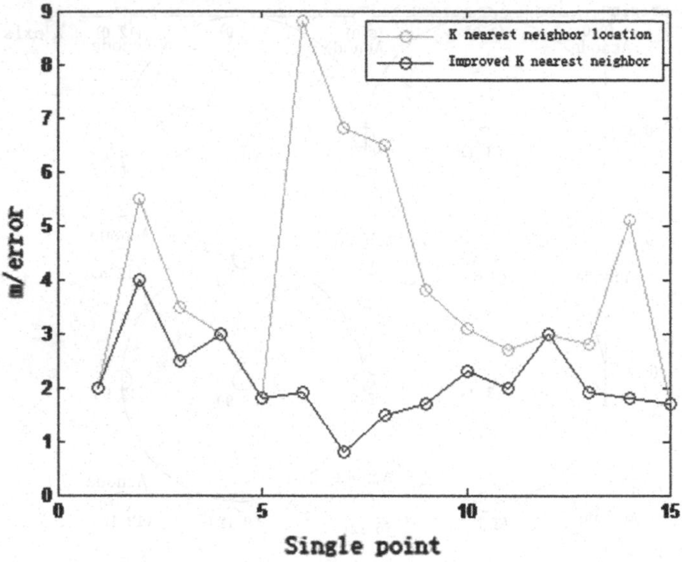


**Fig. 7** Single point positioning error effect

## 3.2 *Analysis of the Results of Experiments*

### 3.2.1 The Accuracy of Positioning

Several localization experiments were carried out in the experimental area, and the results of the positioning were recorded. From the maximum error analysis of location results, the maximum error of location after classification has been greatly improved. Through Fig. 7, we can see that K nearest neighbor matching after data classification is superior to the traditional k nearest neighbor matching in location accuracy. The maximum average error is reduced from 4.17 to 2.12 m, of which the greatest improvement is reduced from 8.8 to 1.9 m. It shows that the improved K nearest neighbor makes the range of the runout range of the positioning result reduced.

### 3.2.2 The Time of Positioning

The time consumption of K nearest neighbor method is mainly spent on the calculation of mean square error. In the experiment, 25 sampling points, divided into a class of A, B, C, D and acnode, circular category there are 5 sampling points, acnode database has 9 sampling points. The traditional K nearest neighbors need to traverse the entire database, calculation of 25 times the mean square error, a maximum of 9 times the mean square error calculation over category. With the increasing number of sampling points in the database, the advantages of the improved K nearest neighbor

**Table 2** Time comparison of the of two algorithms

| Algorithm | Time for average location/ms |
|---|---|
| K nearest neighbor | 52.8 |
| Improved K nearest neighbor | 19.6 |

algorithm are becoming more and more obvious. The average time spent by the two algorithms is as shown in the Table 2.

## 4 Conclusion

In this paper, a fast K nearest neighbor matching location method based on WiFi signal intensity classification is proposed, and the principle and steps of the algorithm are introduced in detail and the algorithm proposed in this paper is measured in the campus. The experimental results show that the proposed K nearest neighbor algorithm reduces the location cost time by 62.8%, and the location accuracy is improved compared with the traditional K nearest neighbor matching. Experimental results show that the proposed method can effectively improve the performance of WiFi fingerprint location system.

## References

1. Z. Yang, Z. Zhou, Y. Liu, From RSSI to CSI: indoor localization via channel response. ACM Comput. Surv. **7**(3), 165–181 (2013)
2. C. Laoudias, Device self-calibration in location systems using signal strength histograms. J. Locat. Based Serv. **7**(3), 165–181 (2013)
3. S.H. Fang, T.N. Lin, K.C. Lee, A novel algorithm for multipath fingerprinting in indoor WLAN environments. IEEE Trans. Wirel. Commun. **7**(9), 3579–3588 (2008)
4. P.A. Zandbergen, Accuracy of iPhone locations: a comparison of assisted GPS, WiFi and cellular positioning. Trans. GIS **13**(s1), 5–25 (2009)
5. K. Pahlavan, X. Li, J.P. Makela, Indoor geolocation science and technology. IEEE Commun. Mag. (IEEE Press) **40**(2), 112–118 (2002)
6. A. Kushki, K.N. Plataniotis, A.N. Venetsanopoulos, Kernel-based positioning in wireless local area networks. IEEE Trans. Mobile Comput. **6**(6), 689–705 (2007)

# The Fuzzy Attitude Control of Visual Servo System

**Yuying Zhang and Yingmin Jia**

**Abstract** The attitude control problem of multi-degree-of-freedom systems has received considerable attention. In this paper, a fuzzy attitude control method is proposed for the multi-degree-of-freedom visual servo system, which is comprised of a target device and a tracking device. The target device has a single freedom, while the track device has four. Firstly, based on characteristics of this system, the kinematic model is established. Next, the excepted attitude is computed from the feature point on the pictures got by the CMOS camera. Then, a fuzzy controller is designed to make the device's attitude meet our requirement. Finally, a simulation platform of the system is developed in MATLAB environment. The result demonstrates that the proposed fuzzy controller can improve the nonlinear coupling system performance.

**Keywords** Visual servo · Fuzzy control · Attitude control · MATLAB

## 1 Introduction

Visual servo systems have gained increasing attention both in robotics and control communities in past decades. The first application of visual information in robot loop control system increased the accuracy of the task by Shirai and Inoue in 1973 [1]. The main goal of visual servo is to make the end-effectors achieve a certain attitude with respect to particular feature in the image. Visual servo control methods can be categorized into the position-based servo control and the image-based servo control. The former is an open-loop visual servo method which only observes the position of the target and relies on precise extrinsic calibration. However, the image-

Y. Zhang · Y. Jia (✉)
The Seventh Research Division and the Center for Information and Control,
School of Automation Science and Electrical Engineering, Beihang University (BUAA),
Beijing 100191, China
e-mail: ymjia@buaa.edu.cn

Y. Zhang
e-mail: carol_zyy@126.com

 
Y. Jia et al. (eds.), *Proceedings of 2018 Chinese Intelligent Systems Conference*,
Lecture Notes in Electrical Engineering 528,
https://doi.org/10.1007/978-981-13-2288-4_54

based servo control is a close-loop visual servo method, which needs sophisticated graphics-processing algorithm [2].

The attitude control of multi-degrees-of-freedom system is one of the most common applications of visual servo control, because of its stability and attitude-maneuverability [3, 4]. Multi-degrees-of-freedom (DOF) system means that it needs two or more generalized coordinates to determine its location completely. Multi-DOF system can be used in varies of areas, such as boats and ships, aircraft, vehicle and so on. Nonetheless, it has some disadvantages like coupled non-linear dynamics, highly susceptible to disturbances, under actuated characteristic, as well as open-loop instability [5–7].

Recently, with the development of intelligent technology, the intelligence control has been used more and more widely. Intelligent control is a sort of control techniques which using various of artificial intelligence computing approaches such as neural network, fuzzy logic, machine learning, evolutionary computation, genetic algorithms and so on [8]. Fuzzy logic was put forward by Lotfi A. Zadeh of the University of California in paper [9]. Fuzzy control, one of the earliest intelligent control methods, is based on applying a set of fuzzy rules to describe a global non-linear system in terms of local linear models which are smoothly connected by fuzzy membership functions [10]. The logic used in fuzzy control is fuzzy logic which input values is logical variables that take on continuous value between 0 and 1, while the digital logic operates on discrete value of true or false. The fuzzy controller can solve the nonlinear and coupling problem without precise model. Compared to other alternative approaches like the neural network and genetic algorithms, only fuzzy logic can use the human experience in the process of designing controller. This makes the mechanize tasks which are already successfully performed by human easier to be finished [11].

The purpose of our work is controlling the track device to the excepted attitude based on fuzzy controller. The excepted attitude is computing by the picture of target device. There are two devices in our platform, track device and target device. The track device has four degrees of freedom, while the target only has one. We established a kinematic model based on the kinematic analysis of the two devices. There are five feature points on the target device to help us confirm the target attitude and calculate the excepted attitude of track device. The whole principle of the system is based on position. The system is an open-loop, nonlinear, strong coupling system, so we choose fuzzy controller which is appropriate for the nonlinear system in our system.

In this paper, we first introduce the whole system, and then present the process of building kinematics model. After that the design process of controller will be written. At the last is the result and discussion part.

## 2 System Configuration

The whole system is designed as a track device and a target device. The track device has four degrees of freedoms include lifting, precession, pitch, rolling. Each of them is driven by a stepping motor. To calculate the excepted attitude, the track device has

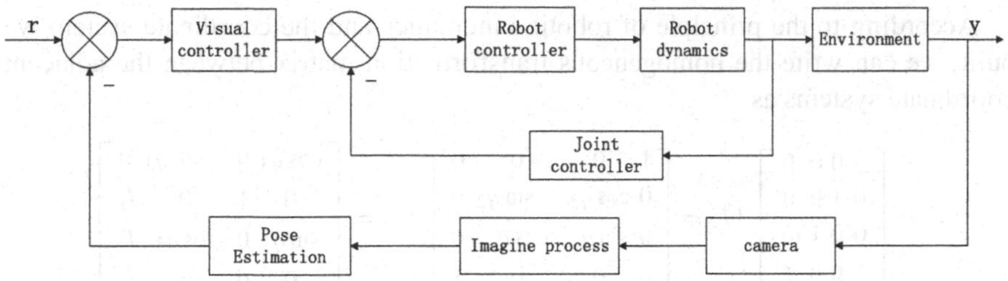


**Fig. 1** Position-based method structure

a CMOS camera to obtain the picture of target device. In front of its lens, we put a filter to remove the noise. The target device which only has one single freedom: yaw, is also controlled by a stepping motor. On the target device, there is a feature board having five infrared ray light-emitting diodes (LED) as feature points.

The design principle of this system is based on the basic visual servo control method called the position-based servo control. The structure of position-based servo control is shown in Fig. 1.

The camera we installed on the track device is the COMS camera, it means that the core imaging component of the camera is complementary metal-oxide-semiconductor transistor. The image processing algorithm we used is very simple, just grey processing and binarization processing.

The combination of position and orientation is defined as the pose of an object. The pose estimation is a typical task that identifies specific objects and each object's position and orientation with imagine. The algorithm we used is Perspective N Points (called as PnP) which is defined as the method of calculating the transformation matrix between the target coordinate system and camera coordinate system based on a series of point coordinates and its projection in the imagine plane, on the condition that the camera intrinsic parameters are known. The transformation matrix contains 3 rotation parameters and 3 translation parameters. We used five LED as feature point on the board which is installed on the target device.

The all constituent serve the track device to the excepted attitude through the picture of target device.

## 3 Kinematics Analysis

To make an analysis, we established six coordinate systems as Fig. 2 shown. There into, the $o_0o_1o_2o_3o_4o_t$ are on the track device, while the other origins is on the target device. The $o_1x_1y_1z_1$ is coincident with $o_2x_2y_2z_2$, and $o_5x_5y_5z_5$ is coincident with $o_6x_6y_6z_6$. At the same time, the environment makes sure that the origin of No. 6 coordinate system $o_6$ is in the plane $o_0y_0z_0$ of No. 0 coordinate system.

According to the principle of robotic kinematics and the coordinate system we built, we can write the homogeneous transformation matrix between the adjacent coordinate systems as:

$$
{}^{0}T_1 = \begin{bmatrix} 1 & 0 & 0 & 0 \\ 0 & 1 & 0 & 0 \\ 0 & 0 & 1 & q_1 \\ 0 & 0 & 0 & 1 \end{bmatrix} \quad
{}^{1}T_2 = \begin{bmatrix} 1 & 0 & 0 & 0 \\ 0 & \cos q_2 & -\sin q_2 & 0 \\ 0 & \sin q_2 & \cos q_2 & 0 \\ 0 & 0 & 0 & 1 \end{bmatrix} \quad
{}^{2}T_3 = \begin{bmatrix} \cos q_3 & 0 & -\sin q_3 & 0 \\ 0 & 1 & 0 & l_1 \\ \sin q_3 & 0 & \cos q_3 & l_2 \\ 0 & 0 & 0 & 1 \end{bmatrix}
$$
$$
{}^{3}T_4 = \begin{bmatrix} 1 & 0 & 0 & 0 \\ 0 & 1 & 0 & q_4 \\ 0 & 0 & 1 & 0 \\ 0 & 0 & 0 & 1 \end{bmatrix} \quad
{}^{4}T_5 = {}^{c}T_t \quad
{}^{5}T_6 = \begin{bmatrix} \cos q_5 & \sin q_5 & 0 & 0 \\ -\sin q_5 & \cos q_5 & 0 & 0 \\ 0 & 0 & 1 & 0 \\ 0 & 0 & 0 & 1 \end{bmatrix} \tag{1}
$$

$$
{}^{c}T_t = \begin{bmatrix} \cos\theta_z\cos\theta_y & -\sin\theta_z\cos\theta_x + \cos\theta_z\sin\theta_y\sin\theta_x & \sin\theta_z\sin\theta_x + \cos\theta_z\sin\theta_y\cos\theta_x & p_x \\ \sin\theta_z\cos\theta_y & \cos\theta_z\cos\theta_x + \sin\theta_z\sin\theta_y\sin\theta_x & -\cos\theta_z\sin\theta_x + \sin\theta_z\sin\theta_y\cos\theta_x & p_y \\ -\sin\theta_y & \cos\theta_y\sin\theta_x & \cos\theta_y\cos\theta_x & p_z \\ 0 & 0 & 0 & 1 \end{bmatrix} \tag{2}
$$

$$
p_x = p_z \sin(q_3)/\cos(q_3) \tag{3}
$$

In the equations, the $q_1, q_2, q_3, q_4, q_5$ is the variable of the joints, the $\theta_x, \theta_y, \theta_z$ is the rotation parameters and the $p_x, p_y, p_z$ is the translation parameters computed from the PnP algorithm. The track device attitude is defined as a state variable

$$
P(t) = [q_1(t), q_2(t), q_3(t), q_4(t), q_5(t)] \tag{4}
$$

so the next moment state can be described as:

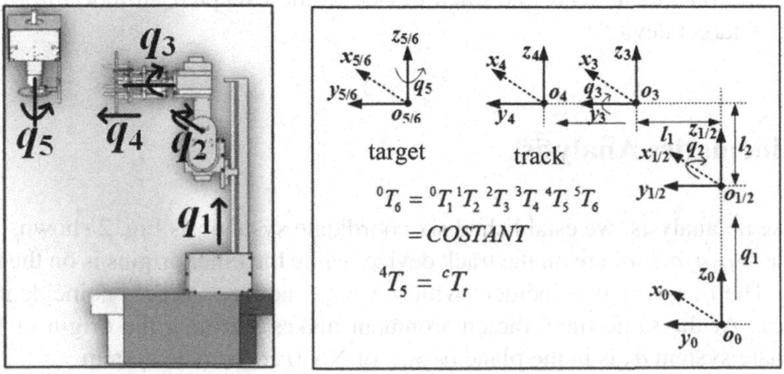


**Fig. 2** The coordinate systems

$$P(t+1) = [q_1(t+1), q_2(t+1), q_3(t+1), q_4(t+1), q_5(t+1)] \tag{5}$$

Because the result of the PnP is the relative attitude, so in the ideal state all the parameters should be 0. Because the No. 0 system and No. 6 system is relative static, so we can get the equation:

$$^0T_6(t) = {}^0T_6(t+1) \tag{6}$$

From the equation, we can get the next moment excepted state of the track device and send to the joint controller. Then calculate the new state according to the picture that the camera get, and solve the new excepted state of the joint until achieve matching state. The matching state means that all the rotation parameters of relative attitude between the track device and target device become zero and the transformation matrix is as following:

$$^cT_t = {}^cT_t^{\mathrm{d}} = \begin{bmatrix} 1\,0\,0\,0 \\ 0\,1\,0\,p_y^d \\ 0\,0\,1\,0 \\ 0\,0\,0\,1 \end{bmatrix} \tag{7}$$

## 4 Controller Design

In our work, in order to control the track device achieve the excepted attitude, we designed a visual controller and a joint controller.

The input of visual controller is the relative attitude between the track device and target device in this moment, while the output is the next moment parameters which can be used to calculate the next moment joint variables.

The visual controller is designed as anti-saturation ratio controller as follows:

$$\begin{cases} \theta_x(t+1) = \theta_x(t) - sat_{L_{\theta x}}\left(k_{\theta_x}(\theta_x(t) - 0)\right) \\ \theta_y(t+1) = \theta_y(t) - sat_{L_{\theta y}}\left(k_{\theta_y}(\theta_y(t) - 0)\right) \\ \theta_z(t+1) = \theta_z(t) - sat_{L_{\theta z}}\left(k_{\theta_z}(\theta_z(t) - 0)\right) \\ p_z(t+1) = p_z(t) - sat_{L_{pz}}\left(k_{p_z}(p_z(t) - 0)\right) \\ p_y(t+1) = p_y(t) - sat_{L_{py}}\left(k_{p_y}(p_y(t) - p_y^d)\right) \end{cases} \tag{8}$$

There into, $k_{\theta_x}, k_{\theta_y}, k_{\theta_z}, k_{p_y}, k_{p_z}$ is the proportional coefficient, and the $L_{\theta_x}, L_{\theta_y}, L_{\theta_z}, L_{p_y}, L_{p_z}$ is the maximum amplitude coefficient to make sure the output of controller meet the mechanical characteristic. The function $sat_L(x)$ is defined as:

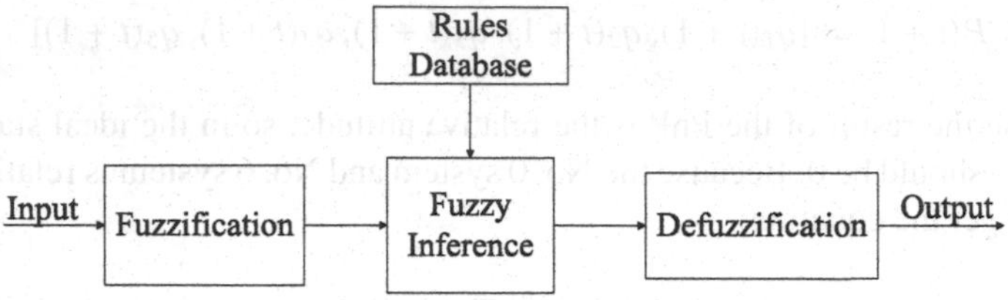


**Fig. 3** The fuzzy controller structure

**Table 1** The fuzzy rules

| Rate\error | NB | NS | ZO | PS | PB |
|---|---|---|---|---|---|
| NB | PB | PS | PS | ZO | ZO |
| NS | PB | PS | PS | ZO | NS |
| ZO | PS | PS | ZO | NS | NS |
| PS | PS | ZO | NS | NS | NB |
| PB | ZO | ZO | NS | NB | NB |

$$sat_L(x) = \begin{cases} x & abs(x) \leq L \\ L & x > L \\ -L & else \end{cases} \tag{9}$$

The relative attitude has six parameters. Five of them can be calculated from the visual controller as we can see in the equations. The last parameter of the attitude can be computed according to the equations.

According to the conversion of coordinates and the result of visual controller, we can get the next moment track device attitude. We used fuzzy controller to control the velocity of motors. Fuzzy controller can improve the response performance without accurate model. The structure of fuzzy controller is in Fig. 3

First, the input is fuzzified into degree of membership corresponding to the fuzzy membership functions. Taking the roll angle as an example, the input is error and error rate. The range of both of them is [−3, 3]. Triangular membership functions are used to fuzzify the input. There are five degrees, "PB", "NB", "NS", "ZO", "PS". They represents the linguistic variable "Negative Big", "Negative Small", "Zero", "Positive Small", and "Positive Big".

The design of fuzzy rules is based on the experience of human expert. For example, if error is negative big but the error rate is zero or positive small, the motor should take a large speed to improve the rapidity of the system. The all fuzzy rules for output are shown in Table 1.

The accurate output is calculated by membership function. The range of output is [−1, 1]. The symbol of the output represents the direction of the motor, and the value is the duty ratio to control the speed of the motor.

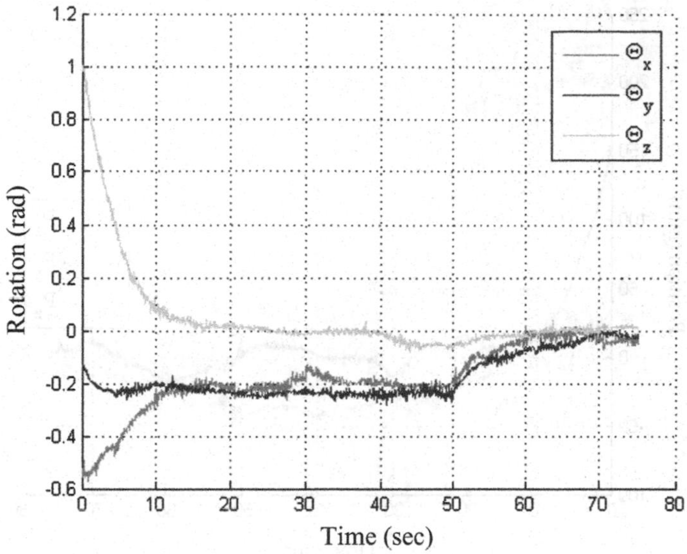


**Fig. 4** The result of rotation parameters

## 5 Result and Discussion

We use the tool box of MATLAB to implement the simulations, the result of which is shown in the figure. The parameters of relative attitude between the camera and target device is shown in the Figs. 4 and 5. Figure 4 shows the rotation parameters, while the Fig. 5 is the position parameters. We can see that the three rotation parameters converge to 0, and the position parameters converge to 0, 0 and $p_y^d$. In the other word, the track device has already shown the excepted attitude got from the target device. After 60 s, the errors of the parameters are in the tolerance interval. The whole system meets our requirement.

## 6 Conclusion

In this paper, we designed a visual servo system with a track device and a target device. The track device has four degrees of freedom, while the target device has one. We designed a fuzzy controller for the joint controlling and an anti-saturation ratio controller for the visual servo. The method was examined in MATLAB successfully. As shown in the simulation result, the two controller can meet the requirement of the visual servo system. It also insures the precision and the rapidity.

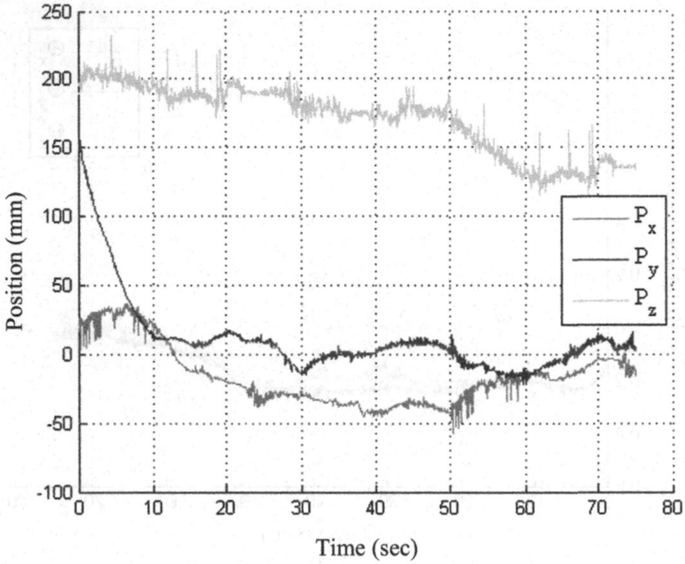


**Fig. 5** The result of position parameters

# References

1. Y. Shirai, H. Inoue, Guiding a robot by visual feedback in assembling tasks. Pattern Recogn. **2**(5), 99–106 (1973)
2. D.-M. Chuang, S.-C. Wu, M. Hor, Adaptive fuzzy visual servoing in robot control, in *IEEE International Conference on Robotics and Automation* (1997), pp. 811–816
3. M. Sidi, *Spacecraft Dynamics and Control, A Practical Engineering Approach* (Cambridge University Press, 1997)
4. M. Navabi, M. Rajab Ali Fardi, Based fuzzy gain scheduled PD law for spacecraft attitude control, in *Congress on Fuzzy and Intelligent System* (2018), pp. 149–151
5. J. Ghommam, N. Fethalla, M. Saad, Quad rotor circumnavigation of an unknown moving target using camera vision-based measurements. IET Control Theory Appl. **10**(15), 1874–1887 (2016)
6. F. Muñoz, I. González-Hernández, S. Salazar et al., Second order sliding mode controllers for altitude control of a quad rotor UAS: real-time implementation in outdoor environments. Neurocomputing **233**, 61–71 (2017)
7. Y. Zou, Nonlinear robust adaptive hierarchical sliding mode control approach for quad rotors. Int. J. Robust Nonlinear Control **27**, 925–941 (2017)
8. R.F. Stengel, Intelligent failure tolerant control. IEEE Control Syst. Mag. **11**(4), 14–23 (1991)
9. L.A. Zadeh, Fuzzy sets. Inf. Control **8**(3), 338–353
10. L. Mazmanyan, M.A. Ayoubi, Fuzzy attitude control of spacecraft with fuel sloshing via linear matrix inequalities. IEEE Trans. Aerosp. Electron. Syst. (2018)
11. W. Pedrycz, *Fuzzy Control and Fuzzy System*, 2nd edn. (Research Studies Press Ltd.)

# Grey Markov Model Prediction Method for Regular Pedestrian Movement Trend

**Xiaoyu Fang, Xiaobin Li, Tianyang Yu, Zhen Guo and Tao Ma**

**Abstract** This paper focuses on the problem that mobile vehicles easily collide with regular pedestrians in dangerous area, and the gray prediction algorithm is applied to establish the markov model of regular pedestrian data. Predict their walking trajectory according to the regular pedestrian movement trend, and provide active and safe predictive control for vehicle braking in the region. Taking the coke oven coal transportation area as an example, a set of regular pedestrian trajectory data is selected to verify the model and prediction method. The experimental results show that this method can predict this type of pedestrian trajectory. When it is compared with the results of the traditional gray model prediction, the error is smaller and the accuracy is higher.

**Keywords** Regular pedestrian · Trend of movement · Grey Markov model
Trajectory prediction

## 1 Introduction

In order to ensure the safety of pedestrians in the region, many dangerous places only allow those people who have access to enter the area. Pedestrians of this type carry out their work in this area. Their walking trajectories have certain rules and are called regular pedestrians. For the sake of avoiding collision accidents between vehicles and regular pedestrians in the working area, it is necessary to predict the position coordinates and trends of regular pedestrians at a certain time in the future, and then carry out the operation of deceleration or brake.

For the problem of pedestrian walking trajectory prediction, Hoogendom et al. [1] proposed a method for predicting pedestrian movement based on Kalman filter; Morzy [2] uses matching functions to refine the best rules, mine frequent trajecto-

X. Fang · X. Li (✉) · T. Yu · Z. Guo · T. Ma
School of Electrical and Electronic Engineering, Shanghai Institute of Technology, Shanghai 201418, China
e-mail: 1486126523@qq.com



Y. Jia et al. (eds.), *Proceedings of 2018 Chinese Intelligent Systems Conference*, Lecture Notes in Electrical Engineering 528,
https://doi.org/10.1007/978-981-13-2288-4_55

ries, and predict the position of moving objects; Burbey [3] matches the recent and historical activity of the observation target and proposes the PPM-C algorithm to predict the future position and arrival time of the moving target; Wang Hongmei [4] uses the background difference method to coarsely segment the moving target object, and combines the feature matching of the GM(1, 1) target to predict the position of the moving target; According to historical trajectory, Li et al. [5] sets up Markov model for moving targets, and puts forward location prediction algorithm according to the movement trend of observation targets. However, most of the above methods are applied to regional venues with more pedestrian obstacles such as streets and crossroads, and many methods need to combine video images in order to predict pedestrian walking trajectories, while the number of regular pedestrians in the study area is relatively small, and the type of exercise is special. Therefore, it is necessary to predict the coordinates of the walking position by studying the regular pedestrian movements and historical movement trends. A grey markov model prediction method based on the regular pedestrian movement trend is proposed to provide an anti-collision active safety control method for vehicles in the area that does not affect vehicles' normal work.

Take the coke oven as an example. By collecting the data of the pedestrian walking position of the sweeping furnace cover and combining the gray forecasting method with the characteristics of predicting the small sample data, a markov model of pedestrian walking trajectory in the coal carriage region is established based on the gray prediction method. It can predict the movement positions of the workers in the furnaces of the coke ovens accurately and identify the characteristics of the trajectories clearly when they walk, dynamically establish the predicted trajectories. It can also improve the prediction accuracy and solve the problem of regular pedestrian walking trajectory prediction.

## 2 Grey Markov Prediction Model

### 2.1 *Gray Prediction [6, 7]*

The gray prediction method can reflect the changes in the development of objects within the system. The prediction steps are as follows:

(1) Mark $x^{(0)}(k)$, $k = 1, 2, \ldots, n$, as the original sequence, $z^{(1)}(k)$, $k = 2, 3, 4, \ldots, n$, as new sequence after accumulating raw data at a time:

$$z^{(1)}(k) = 0.5x^{(1)}(k) + 0.5x^{(1)}(k-1), k = 2, 3, \ldots, n \quad (1)$$

$$x^{(1)}(k) = \sum_{i=1}^{k} x^{(0)}(i), k = 1, 2, \ldots, n \quad (2)$$

The $x^{(0)}(k) + az^{(1)}(k) = u$ is a gray model. Estimating the parameter column, and the least squares solution is used to obtain:

$$C = (B^T B)^{-1} B^T Y \tag{3}$$

In this formula, $Y = \begin{bmatrix} x^{(0)}(2) \\ x^{(0)}(3) \\ \cdots \\ x^{(0)}(n) \end{bmatrix}$, $B = \begin{bmatrix} -z^{(1)}(2) & 1 \\ -z^{(1)}(3) & 1 \\ \cdots & \cdots \\ -z^{(1)}(n) & 1 \end{bmatrix}$, $\frac{dx^{(1)}(t)}{dt} + ax^{(1)}(t) = u$,

for the differential equation: $x^{(0)}(k) + az^{(1)}(k) = u$

(2) The solution of differential equation is available:

$$x^{(0)}(k) = (\beta - \alpha x^{(0)}(1)) \times e^{(-a(k-2))} \tag{4}$$

(3) The cumulative number of calculated values of the data are calculated:

$$f(k) = (x^{(0)}(1) - \frac{u}{a}) \times e^{(-a(k-1))} + \frac{u}{a}, k = 1, 2, \ldots, n \tag{5}$$

(4) Through these reductions restore $x^{(0)}(k)$:

$$X(k+1) = f(k+1) - f(k), k - 1, 2, \ldots, n \tag{6}$$

$$X(k+1) = (x^{(0)}(1) - \frac{u}{a}) \times e^{(-a(k-1))} \times (1 - e^{a}), k = 1, 2, \ldots, n \tag{7}$$

## 2.2 The Markov Model

The markov model squares up a stochastic system, which uses the transitional probability matrix of the system's different states to predict the future state of the target. It is suitable for predicting randomness. Due to the randomness of the regular pedestrian movement trajectory, the grey markov model can handle the problem of modeling large random fluctuations in this data, obtain accurate prediction values, and have good stability. Therefore, a grey markov model is established to predict the trajectory of regular pedestrians. Firstly, calculating the gray prediction value, then dividing the state range of the prediction value and calculating the markov transition probability matrix. Finally, taking the product of the median of the state interval and the prediction value as the correction value [8–10].

#### 2.2.1 Dividing State Intervals

The state division in this study is divided according to the relative value sequence, i.e. the relative value between the original sequence and the grey prediction value is calculated: $Q = x(t)/X(t)$. According to the size of the relative value, divide the state interval: $S_i = [Q_{i1}, Q_{i2}], i = 1, 2, \ldots, k$. In the formula, $Q_{i1}, Q_{i2}$ are the lower and upper limits of the relative value respectively.

#### 2.2.2 Calculate State Transition Probability Matrix

The gray forecast should be divided into different states, formula: $A_{ij} = n_{ij}(k)/n_i$. It indicates that the number of occurrences of the system from state $S_i$ through $k$ to state $S_j$ is $n_{ij}(k)$, and the occurrence of $S_i$ is $n_i$. Get the state transition probability matrix, such as shown below:

$$A_k = \begin{bmatrix} A_{11k} & A_{12k} & \cdots & A_{1mk} \\ A_{21k} & A_{22k} & \cdots & A_{2mk} \\ \cdots & \cdots & \cdots & \cdots \\ A_{m1k} & A_{m2k} & \cdots & A_{mmk} \end{bmatrix} \quad (8)$$

#### 2.2.3 Correction of Gray Prediction Value

Through the state transition probability matrix, the relative value between the original sequence and the gray prediction value is obtained, and the state interval is divided into $[Q_{i1}, Q_{i2}]$. Take the product of the median of the interval and the original data as the final prediction result, that is $Y(t)$, the correction value of the prediction value. The formula is:

$$Y(t) = 0.5 \times (Q_{i1} + Q_{i2}) \times X(t) \quad (9)$$

## 3 Experiments

### *3.1 Regular Pedestrian Trajectory Prediction Simulation Experiment*

For the purpose of verifying the feasibility and superiority of grey markov model in predicting the regular pedestrian trajectory, the simulation experiment was carried out with MATLAB R2014a as experimental platform. Taking the coke oven coal car running area as an example, the motion trajectory data of a group of regular pedestrians (sweeping furnace cover workers) was selected to simulate and apply

**Table 1** The original data table of the trajectory position of regular pedestrians (sweeping furnace cover personnel)

| Walk step | Y axis coordinate position |
|---|---|
| 1 | 17.22 |
| 2 | 17.35 |
| 3 | 17.8 |
| 4 | 18.2 |
| 5 | 18.56 |
| 6 | 19.2 |
| 7 | 19.71 |
| 8 | 20.08 |

experiments, and the prediction of the trajectory of the pedestrians under the gray model and grey markov model was compared.

The statistics of the walking trajectory data of the regular pedestrians is used to take the 7-step position data to predict the position of the 8th step (Table 1). Compare and analyze the prediction results of the two models, and verify the superiority of the grey markov forecast model.

Since the predicted target is walking on the track in a straight line, combined with the need to study the problem, only the ordinate of its' moving position is predicted in the next step. Combined with Eqs. (1)–(7), a = −0.0254, u = 16.6651, can be obtained through matlab programming. Therefore, the trend curve function of the position of the walking trajectory of the sweeping cover worker is $x(k+1) = 17.3215 \times e^{0.0254k}$. The y axis position of the 7 steps of the regular pedestrians is predicted to the ordinate of the 8th step position. The experimental results are shown in Fig. 1, the error comparison is shown in Fig. 2:

In Fig. 1, the blue line indicates the actual value, the red line represents the gray prediction value, the green line indicates the prediction value of grey markov model. The green line is more close to the blue line which represents actual value. In Fig. 2, the blue dotted line is the relative error of the forecast results of the grey markov model. Obviously, the predicted value obtained by the grey markov model is closer to the actual value. Therefore, this system is suitable for predicting the walking trajectory of regular pedestrians.

## 3.2 *Regular Pedestrian Motion Track Prediction Application Experiment*

In order to verify the effectiveness of the grey markov model in predicting regular pedestrian trajectories, collecting trajectory data of regular pedestrians on site and predicting. And then comparing the predicted results and the actual trajectory.

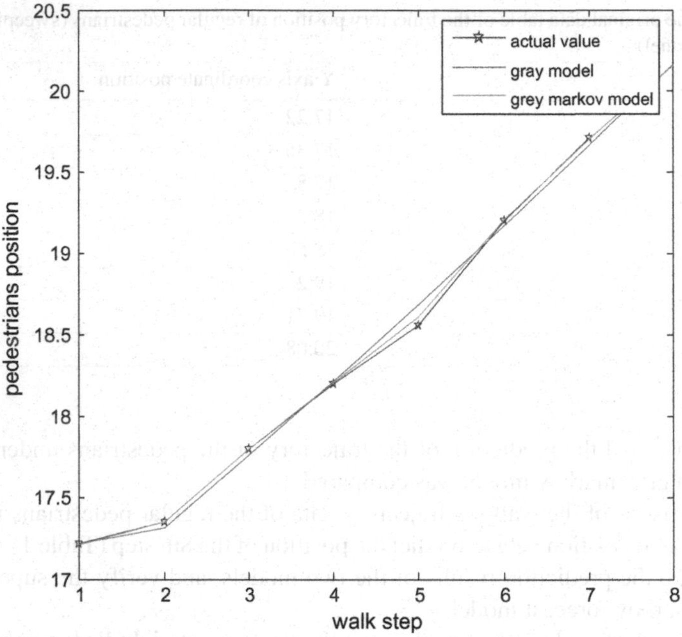


**Fig. 1** Simulation of regular pedestrian position (Color figure online)

**Table 2** Predicted values and relative values tables 1–8

| Step | Predictive value | Relative value |
|---|---|---|
| 1 | 17.22 | 1.00 |
| 2 | 17.3178 | 1.0019 |
| 3 | 17.7627 | 1.0021 |
| 4 | 18.2190 | 0.9989 |
| 5 | 18.6870 | 0.9932 |
| 6 | 19.1671 | 1.0017 |
| 7 | 19.6595 | 1.0026 |
| 8 | 20.1645 | |

#### 3.2.1 Dividing the Trajectory Status Range

According to the obtained trend curve function of the position of the cleaning furnace cover worker's walking trajectory: $x(k+1) = 17.3215 \times e^{0.0254k}$, the predicted value and the relative value of the 8th step can be calculated, as shown in Table 2, and the motion states of Step 1 to Step 7 are divided into three states in Table 3.

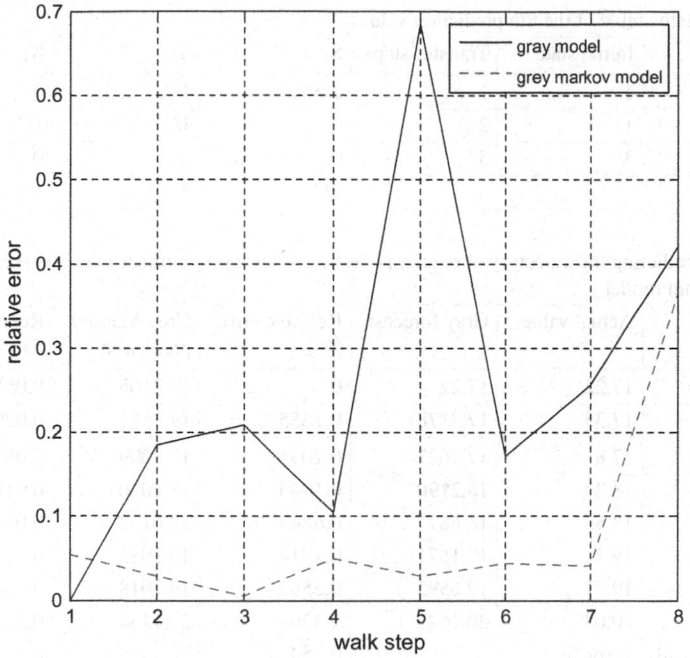


**Fig. 2** Relative error comparison chart (Color figure online)

**Table 3** Status of relative values and state of each step

| State | Meaning | Interval range | Step |
|---|---|---|---|
| $S_1$ | Overestimate | [1.0017, 1.0026] | 2, 3, 6, 7 |
| $S_2$ | Relatively accurate | [0.9989, 1.00] | 1, 4 |
| $S_3$ | Underestimate | [0.9932, 0.9989] | 5 |

### 3.2.2 Calculating the Transition Probability Matrix of Trajectories

Establishing a corresponding state transition probability matrix by formula (8).

$$A(1) = \begin{bmatrix} \frac{2}{3} & \frac{1}{3} & 0 \\ \frac{1}{2} & 0 & \frac{1}{2} \\ 1 & 0 & 0 \end{bmatrix} \quad A(2) = \begin{bmatrix} 0 & \frac{1}{2} & \frac{1}{2} \\ 1 & 0 & 0 \\ 1 & 0 & 0 \end{bmatrix} \quad A(3) = \begin{bmatrix} \frac{1}{2} & 0 & \frac{1}{2} \\ \frac{1}{2} & \frac{1}{2} & 0 \\ 0 & 0 & 0 \end{bmatrix}$$

### 3.2.3 Correction of Gray Prediction Values of Walking Trajectories

Using the gray prediction method to deal with such random data as the regular pedestrian movement trajectory, the error between the obtained result and the actual

**Table 4** Status table of the 8th prediction value

| Step | Initial state | Transfer steps | $S_1$ | $S_2$ | $S_3$ |
|---|---|---|---|---|---|
| 7 | 1 | 1 | 2/3 | 1/3 | 0 |
| 6 | 1 | 2 | 0 | 1/2 | 1/2 |
| 5 | 3 | 3 | 0 | 0 | 0 |
| Total | | | 2/3 | 5/6 | 1/2 |

**Table 5** Predicting the trajectory of regular pedestrians (sweeping furnace cover workers) by grey markov chain model

| Step | Actual value | Gray forecast | Relative error (%) | Grey Markov prediction | Relative error (%) |
|---|---|---|---|---|---|
| 1 | 17.22 | 17.22 | 0 | 17.2105 | 0.0551 |
| 2 | 17.35 | 17.3178 | 0.1855 | 17.3551 | 0.0293 |
| 3 | 17.8 | 17.7627 | 0.2095 | 17.8009 | $5.06 \times 10^{-3}$ |
| 4 | 18.2 | 18.2190 | 0.1043 | 18.2090 | 0.0494 |
| 5 | 18.56 | 18.687 | 0.6843 | 18.6132 | 0.0287 |
| 6 | 19.2 | 19.1671 | 0.1714 | 19.2083 | 0.0432 |
| 7 | 19.71 | 19.6595 | 0.2562 | 19.7018 | 0.0416 |
| 8 | 20.08 | 20.1645 | 0.4208 | 20.1534 | 0.3655 |
| Average relative error % | | | 0.254 | | 0.0772 |

value is large. Therefore, it is essential to revise the gray prediction value through the markov model, and then obtaining an accurate trajectory of regular pedestrian movement prediction.

The 3-step transition probability matrix is used to calculate the state of the trajectory data for the 8th step of the predicted target. Select the three steps that are closest to the number of steps for the prediction: In the seventh, sixth, and fifth steps, choose the number of steps to move in according to the distance to predict the number of steps: one, two, three, and in the corresponding transition matrix, selecting the row vector corresponding to the initial status to constitute a new state probability matrix. Then sum the column vectors of the matrix to determine the future transition state of the system, where the largest sum is the motion state of the 8th step of the prediction target. See Table 4.

As can be seen from Table 4, the probability of $S_2$ is the largest in the total line, so it can be predicted that the motion trajectory of the predicted target in the 8th step is in the $S_2$ state, i.e., within [0.9989, 1.00]. According to the correction formula of formula (9), the predicted value of the obtained grey markov model is: $Y(t) = 0.5 \times (0.9989 + 1) \times 20.1645 = 20.1534$. The grey markov model predicts the predicted and relative error data for the target 1–8 trajectories, as shown in Table 5.

As shown in the Table 5, the average fractional error among gray prediction and actual value is 0.254%, while the average fractional error among forecasted data of grey markov model and actual values is 0.0772%. Combining with Figs. 1 and 2, it can be concluded that the grey markov prediction model is used to predict the walking trajectory of regular pedestrians with higher accuracy and smaller error.

## 4 Conclusions

This article addresses the issue of collisions between vehicles that are likely to run in hazardous areas and regular pedestrians. Considering the special characteristics of the regular pedestrian movement, the existing road pedestrian trajectory prediction method can not guarantee the condition of normal safety production. A grey markov model prediction method based on regular pedestrian movement trend is proposed. Using gray forecasting method to perform pedestrian trajectory prediction processing, dividing the relative value state of gray prediction values, and calculating the corresponding state transition probability matrix. The markov model is used to correct the predicted value, which compensates for the large error of the common gray model for dealing with random data. Taking the coke oven coal car running area as an example, the feasibility and superiority of the grey markov model in predicting the regular pedestrian trajectory are verified through the trajectory predicting imitation and application experiments of regular pedestrians in the area. Experiments show that the grey markov model can accurately predict the trajectory of regular pedestrians. Compared with the traditional gray model predicted results, the model predicted data proposed in this paper are closer to the general trend of the actual walking trajectory, and the relative error among the predicted data and the actual value is smaller, meanwhile the accuracy is higher.

## References


1. S.P. Hoogendoom, W. Daamen, P.H.L. Bovy, Extracting microscopic pedestrian characteristics from video data, in *Transportation Research Board Annual Meeting* (National Academic Press, Washington, DC, 2003)
2. M. Morzy, Mining frequent trajectories of moving objects for location prediction, in *Machine Learning and data Mining in Pattern Recognition* (2007), pp. 667–680
3. I.E. Burbey, *Predicting Future Locations and Arrival Times of Individuals* (Virginia Polytechnic Institute and State University, Virginia, 2011)
4. W. Hongmei, Research on human detection and counting based on video. Sichuan Normal University (2011)
5. W. Li, S. Xia, F. Liu, L. Zhang, G. Yuan, Position prediction of moving objects based on movement trend. J. Commun. **35**(02), 46–53 (2014)
6. C. Yang, M. Yang, Q. Wang, Design and implementation of grey Markov chain prediction system. Surv. Sci. **34**(06), 182–183 (2009)
7. Q. Fan, W. Li, X. Wang, M. Fan, X. Yang, A memory cutting algorithm of shearer based on grey Markov combination model. J. Central South Univ. (Sci. Technol.) **42**(10), 3054–3058 (2011)
8. J. Yang, Z. Hou, A real-time prediction model of large passenger flow based on gray Markov. J. Beijing Jiaotong Univ. **37**(02), 119–123 (2013)

9. W. Chen, A. Xu, Prediction of gas flow from boreholes based on grey Markov model. China Saf. Sci. J. **22**(03), 79–85 (2012)
10. Z. Xiong, X. Wang, Dynamic unbiased gray Markov prediction of relative gas emission from mine. J. Saf. Environ. **15**(03), 15–18 (2015)

# An Improved Algorithm for Maximum Power Point Tracking of Photovoltaic Cells Based on Newton Interpolation Method

**Yuanyuan Li, Sumin Han and Fuzhong Wang**

**Abstract** Photovoltaic (PV) arrays are power generation equipment in PV systems. Maximum power point Tracking (MPPT) scheme in the PV array affects the power generation efficiency of the PV system. In this paper, based on the deficiencies of existing MPPT methods, an algorithm by combining the increment conductance method with variable step size and Newton interpolation method is proposed, which can automatically adjust the step size according to changes in the external environment to avoid power loss and improve the photovoltaic power generation efficiency. The results show the improved MPPT algorithm can efficiently control the vibration amplitude of the power waveform output compared with the traditional conductance increment method. The problem studied in this paper is somewhat interesting. I have the following comments. Meanwhile, it presents a faster tracking speed and a good adaptability for the environment.

**Keywords** Maximum power point tracking · Newton interpolation
Variable step length · Conductance increment method · Photovoltaic array

## 1 Introduction

As a new type of renewable energy source, solar energy has the advantages of being renewable, non-polluting, and widely sourced, which are incomparable with traditional energy sources [1]. At present, photovoltaic power generation technology has become a new energy technology developed globally. As the photovoltaic cell,

Y. Li (✉) · S. Han · F. Wang
School of Electrical Engineering and Automation,
Henan Polytechnic University, Jiaozuo 454000, China
e-mail: 1591456477@qq.com

S. Han
e-mail: hansumin@hpu.edu.cn

F. Wang
e-mail: wangfzh@hpu.edu.cn



Y. Jia et al. (eds.), *Proceedings of 2018 Chinese Intelligent Systems Conference*,
Lecture Notes in Electrical Engineering 528,
https://doi.org/10.1007/978-981-13-2288-4_56

power generation output presents a very distinctive nonlinearity. If the environment in which the PV array operates changes, then the maximum power point of the PV array will also change. Tracking the maximum power point of photovoltaic power generation, the efficiency of photovoltaic power generation can be greatly improved. In recent years, scholars at home and abroad have put forward a variety of different MPPT algorithms, such as constant voltage method, perturbation observation method, admittance increment method etc. [2]. Literature [3] applies the Newton interpolation method to the maximum power tracking of photovoltaic power generation, the Simulink simulation shows that the algorithm can quickly and accurately track the maximum power and improve the utilization of photovoltaic energy. Literature [4] was aimed to explore the performance of a maximum power point tracking system, which implements Incremental Conductance method, from the simulation, the method shows a better performance and has a lower oscillation. Literature [5] proposes maximum power tracking based on perturbation observation method, simulation and experimental results are compared with other methods and the effectiveness of the proposed method is evaluated. The constant voltage method has poor adaptability to the change of environmental conditions, and perturbation observation method oscillates at the maximum power point voltage. The control method of the admittance increment method is more complicated. Newton interpolation method can adapt to changes in the external environment conditions, avoid voltage oscillations and improve the photoelectric conversion efficiency, but the selection of interpolation points is more difficult to control.

Therefore, this paper combining the advantages of the Newton interpolation method and the increment conductance method with variable step size, this paper proposes an algorithm combining the increment conductance method with variable step size and Newton interpolation method. Section 2 discusses PV cell output characteristics. Section 3 proposes an improved maximum power point tracking algorithm based on Newton interpolation. In Sect. 4, the improvement and implementation of MPPT algorithm is discussed. Section 5 uses Simulink to build a simulation model and performs simulation analysis. Finally, this paper is concluded in Sect. 6.

## 2 Photovoltaic Cell Output Characteristics

The photovoltaic cell mathematical model is shown in Fig. 1. The photovoltaic cell is regarded as the nonlinear element of the current source and produces current that is positively correlated with the light intensity, the resistors $R_S$ and $R_{Sh}$ representing the equivalent resistance of the photovoltaic cell power consumption.

The P-U and I-U curve characteristics of the resulting photovoltaic cell are nonlinear and generally changeable with temperature and light intensity. Formula Eq. (1) gives a mathematical model with the working principle of PV cells:

$$I = I_{ph} - \mathrm{I}_s[e^{\frac{q(U + IR_S)}{KAT}} - 1] - \frac{U + IR_S}{R_{sh}} \tag{1}$$

where q is the amount of charge; *A* is the ideal diode factor, the Boltzmann constant is denoted by k, the output current of the battery is represented by I, photocurrent is represented by $I_{ph}$, $I_s$ represents the reverse saturation current of the diode, $R_s$ and $R_{sh}$ are the series and parallel resistances of the PV cells. In addition, T represents PV cell temperature; U represents the PV cell output voltage.

The output characteristics of PV cells are affected by external factors. In different light and temperature, the output curves of photovoltaic cells are different, and the corresponding MPP is different. The article simulates the model through the Simulink platform. The curves of I-U and P-U under different sunlight irradiation intensity and temperature conditions can be obtained as shown in Figs. 2 and 3.

From Figs. 2 and 3, the photovoltaic system has a single MPP at a certain temperature and light intensity. When the temperature is constant, the maximum short-circuit current increases with the increase of the light intensity, and the relative change of the maximum open-circuit voltage is small, at the same time, the MPPT also increases. When the light intensity is constant, the maximum open circuit voltage decreases with the increase of temperature, and the relative change of the maximum short-circuit current is small, at the same time, the MPPT also decreases.

# 3 An Improved Algorithm for MPPT Based on Newton Interpolation

## *3.1 Increment Conductance Method with Variable Step Size*

For the conventional fixed-step conductance increment method, a smaller disturbance step length can improve the tracking accuracy under steady state. If the step length is too small, the expected result of the photovoltaic array can easily stay in the low power output area, and will slow down the pace of tracking. Selecting a larger disturbance step length can increase the tracking speed, and a larger disturbance step can increase the tracking speed, if the step length is too large, the detected result is likely to produce a large vibration, and the power loss in the steady state is large. In order to solve the

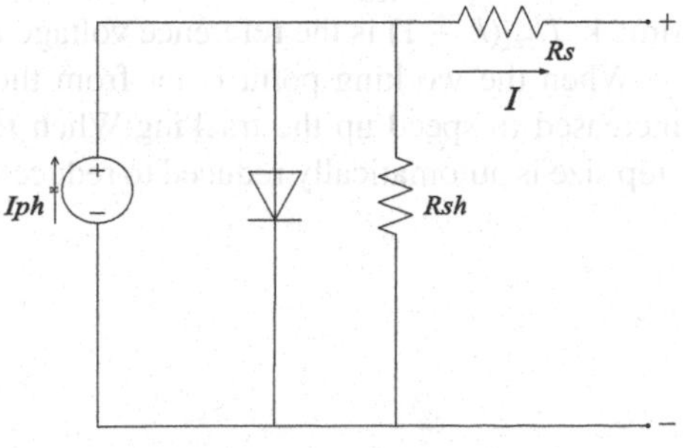


**Fig. 1** Solar cell equivalent circuit

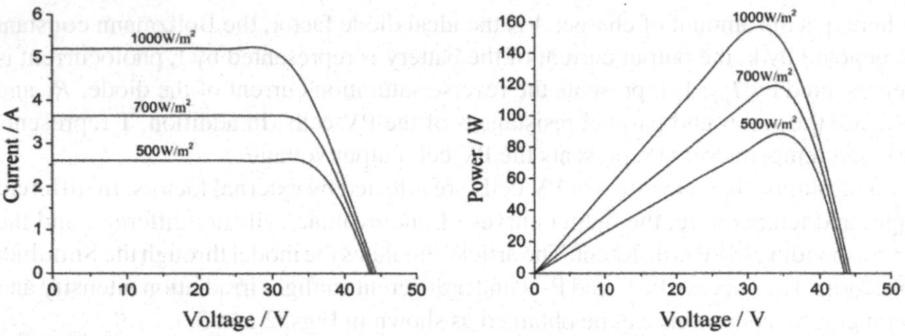


Fig. 2 I-U and P-U characteristic curves at different light levels

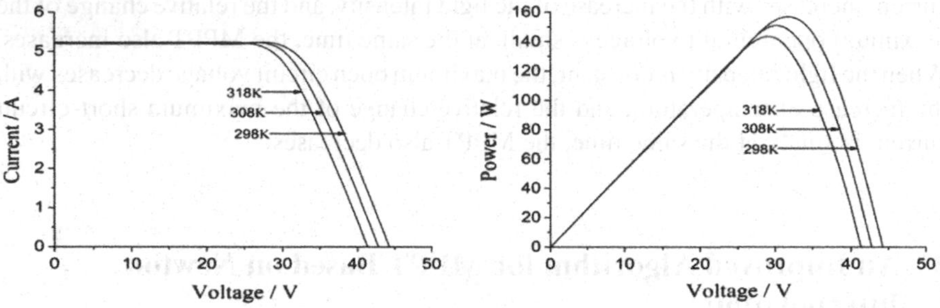


Fig. 3 I-U and P-U characteristic curves at different temperatures

problem of the dynamic speed and steady-state accuracy of the photovoltaic system caused by the selection of step length, this paper chooses increment conductance method with variable step size. The step length is $h = a \times \Delta U$. Among them, $a = \left|\frac{dP}{dU}\right| = \left|I + U\frac{dI}{dU}\right|$ is the step length control factor to adjust the size of each step; $\Delta U$ is a fixed constant.

The step update rule for variable step increment method is:

$$U_{\text{ref}}(\text{k}) = U_{\text{ref}}(\text{k} - 1) \pm \text{h} \quad (2)$$

In Eq. (2), $\Delta U_{\text{ref}}$ is the disturbance step size; $U_{\text{ref}}(k)$ is the reference voltage at time k; $U_{\text{ref}}(k-1)$ is the reference voltage at time k – 1.

When the working point is far from the MPP, the step length is automatically increased to speed up the tracking;When the operating point is near the MPP, the step size is automatically reduced to reduce the power loss caused by the fluctuation.

**Table 1** The second order difference quotient table

| U | P(U) | First-order difference quotient | Second-order difference quotient |
|---|---|---|---|
| $U_0$ | $P(U_0)$ | | |
| $U_1$ | $P(U_1)$ | $P[U_0, U_1]$ | |
| $U_2$ | $P(U_2)$ | $P[U_1, U_2]$ | $P[U_0, U_1, U_2]$ |

## 3.2 *Variable Step Size Conductance Increment Method Based on Newton Interpolation Method*

(1) The basic principle of Newton interpolation method

The basic principle of the Newton interpolation method is based on the output characteristics of photovoltaic cells, if the system works in the constant voltage region, it can be considered that voltage U and MPP voltage $U_m$ are equal. According to the polynomial approximation approach, polynomial approximation of several working voltages is used to obtain the output power near the maximum power point position. It is based on the numerical calculation that leads to an improved MPPT algorithm. At present, Lagrange interpolation and Newton interpolation are commonly used [6, 7].

Based on the idea of quadratic interpolation [8, 9], the Newton interpolation method uses the speed of the second approximation to quickly track to the MPP. Compared with other algorithms, the Newton interpolation method has significant advantages in accuracy and speed. The principle of Newton interpolation MPPT algorithm is: collect three operating points ($U_0$, $P(U_0)$), ($U_1$, $P(U_1)$), ($U_2$, $P(U_2)$) around the MPP. Photovoltaic cell P-U curves were constructed using Newton interpolation. The Newton interpolation second-order difference quotient table is shown in Table 1.

From the Table 1:

$$P[U_0, U_1] = \frac{P(U_1) - P(U_0)}{U_1 - U_0} \tag{3}$$

$$P[U_1, U_2] = \frac{P(U_2) - P(U_1)}{U_2 - U_1} \tag{4}$$

$$P[U_0, U_1, U_2] = \frac{P(U_1, U_2) - P(U_0, U_1)}{U_2 - U_0} \tag{5}$$

Newton's quadratic interpolation equation is:

$$N(U) = P(U_0) + P(U_0, U_1)(U - U_0) + P(U_0, U_1, U_2)(U - U_0)(U - U_1) \tag{6}$$

The Eq. (6) is derived and its derivative is 0, and the voltage $U_m$ at MPP can be obtained.

$$U_m = \frac{1}{2}[(U_0 + U_1) - \frac{P[U_0, U_1]}{P[U_0, U_1, U_2]}] \tag{7}$$

(2) The combination of increment conductance method with variable step and Newton interpolation method

Because the Newton interpolation method has various advantages such as high tracking accuracy, rapidity, and easy operation, it can reduce the need for system handler performance. However, this algorithm only exists in the assumption that the PV array has always been working in the constant voltage region. Based on the above, this paper proposes an improved MPPT control algorithm that combines incremental variable conductance conductivity method with Newton interpolation method.

Assumptions:

$$\left|\frac{dP(k)}{dU(k)}\right| = S_1 \tag{8}$$

$$\left|\frac{dP(k-1)}{dU(k-1)}\right| = S_2 \tag{9}$$

Due to the larger rate of change of the power of PV cells in the initial stage of MPPT, generally $a = |\frac{dP}{dU}| > 5.0$. In addition, when near the MPP, generally a $\leq$ 1.0. Therefore, it is divided into four situations for discussion:

(1) If $1.0 < S_1 < S_2$, it shows that it has not reached the point near MPP at this time, and increase the step length continuous tracking. Step length $l = S_1 \times S_2 \times \Delta U$, $\Delta U$ is a fixed constant.
(2) If $S_1 < 1.0 < S_2$, it shows that it has reached near MPP, reducing the step length continuous tracking. Step length $l = \Delta U_{\min}$, $\Delta U_{\min} = 0.1$.
(3) If $S_1 < S_2 < 1.0$, it shows that at this time MPP is between three sampling points, The Newton interpolation output voltage equation is as follows:

$$U_{out}(n-1) = \frac{[dP(k-1) \times dU(k-2) \times dU(k-1) + (U(k-1) + U(k-2)) \times m]}{(2 \times m)} \tag{10}$$

Let $m = dP(k-1) \times dU(k) - dP(k) \times dU(k-1)$, $U_{out}(n-1)$ as a reference point, reduce the step size to the previous one-half, then keep tracking to get two other points, repeat this process, the interpolation measurement is continued until the current measured voltage value and the last measured voltage value satisfy:
$|U_{out}(n) - U_{out}(n-1)| < \varepsilon$, $\varepsilon$ is a sufficiently small quantity, then the estimated point is the MPP.

(4) If $S1 \times S2 = 0$, it means that the MPP is reached and the voltage at the MPP is the same as the voltage at this time.

The MPPT flow chart of variable step conductance increment method based on Newton interpolation is shown in Fig. 4.

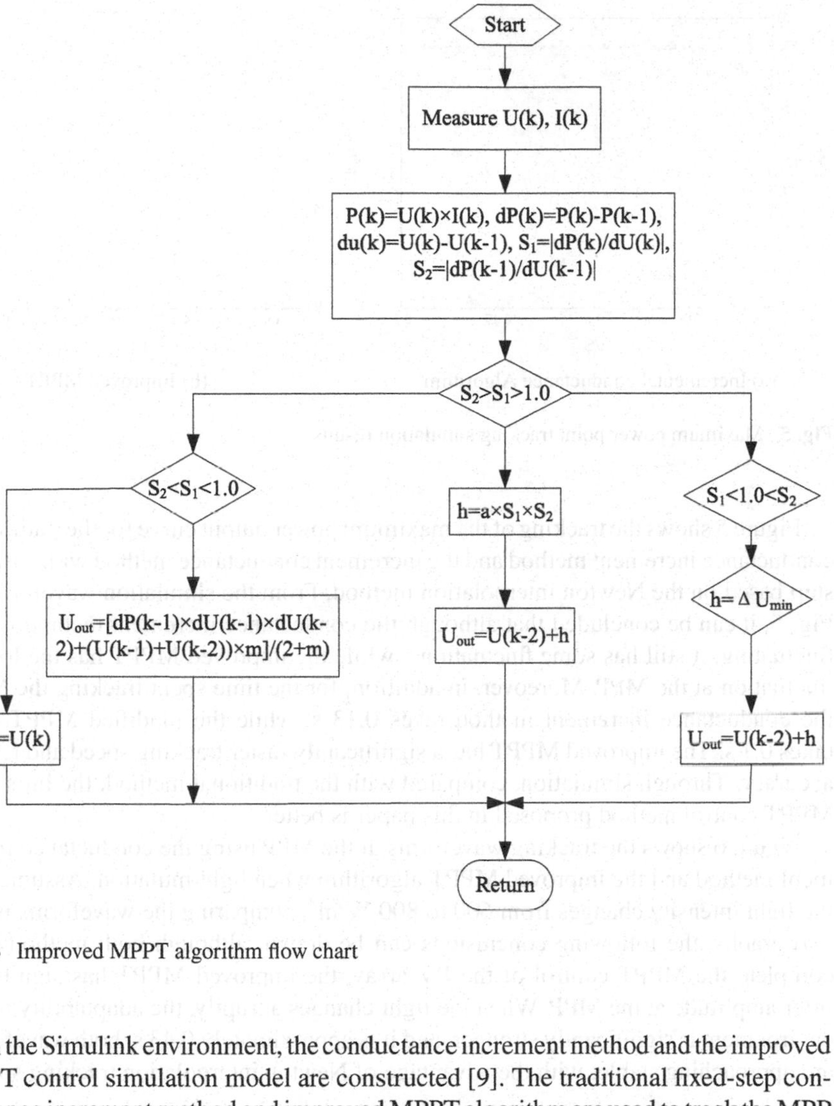


**Fig. 4** Improved MPPT algorithm flow chart

In the Simulink environment, the conductance increment method and the improved MPPT control simulation model are constructed [9]. The traditional fixed-step conductance increment method and improved MPPT algorithm are used to track the MPP of the PV system. Duty cycle adjust PWM output according to MPPT algorithm [10]. Then control the switch transistor turned on and off, and until the completion of the MPPT. The system forms the PWM waveform switching frequency f = 20 kHz. Simulation conditions are as follows: short circuit current is 5.2 A, open circuit voltage is 44.3 V, maximum power point voltage is 35.1 V, maximum power point current is 4.96 A.

The MPPT simulation model is established using the conductance increment method and the improved MPPT algorithm respectively. The simulation results are shown in Fig. 5.

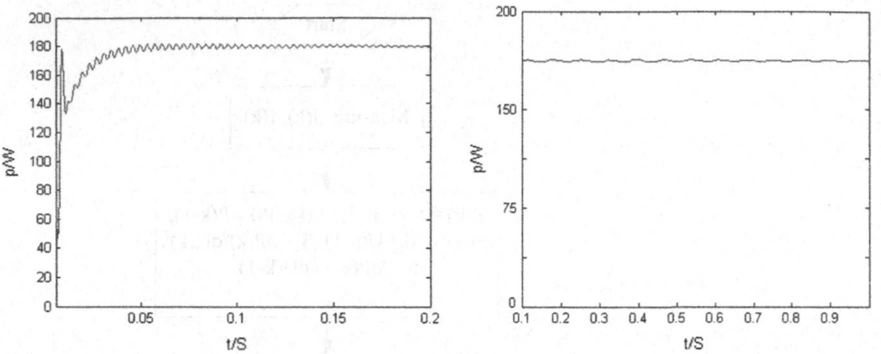


(a) Incremental conductance Algorithm (b) Improved MPPT

**Fig. 5** Maximum power point tracking simulation results

Figure 5 shows the tracking of the maximum power output curve for the traditional conductance increment method and the increment conductance method with variable step based on the Newton interpolation method. From the simulation waveforms in Fig. 5, it can be concluded that although the conductance increment method is less fluctuating, it still has some fluctuations, while the improved MPPT has the lowest fluctuation at the MPP. Moreover, in addition, for the time spent tracking the MPP, the conductance increment method takes 0.13 s, while the modified MPPT only takes 0.1 s. The improved MPPT has a significantly faster tracking speed and higher accuracy. Through simulation, compared with the traditional method, the improved MPPT control method proposed in this paper is better.

Figure 6 shows the tracking waveforms at the MPP using the conductance increment method and the improved MPPT algorithm when light mutation. Assume that the light intensity changes from 600 to 800 W/m$^2$, comparing the waveforms in the two graphs, the following conclusions can be drawn: although both methods can complete the MPPT control of the PV array, the improved MPPT has significant jitter amplitude at the MPP. When the light changes abruptly, the adaptability to the environment is significantly stronger, and it is approximately 0.43 s, both waveforms are approaching stable, with the advantage of Newton interpolation tracking speed.

## 4 Conclusions

The MPPT algorithm combined with the variable-step conductance increment method and the Newton interpolation proposed in this paper not only can quickly and accurately achieve the MPPT of the PV array, but also can effectively suppress the oscillation of the PV system at the MPP, making the PV array operates smoothly at the MPP. In addition, after a variety of experimental conditions, such as the sudden change in sunlight, this algorithm still has a very strong adaptability, which can be

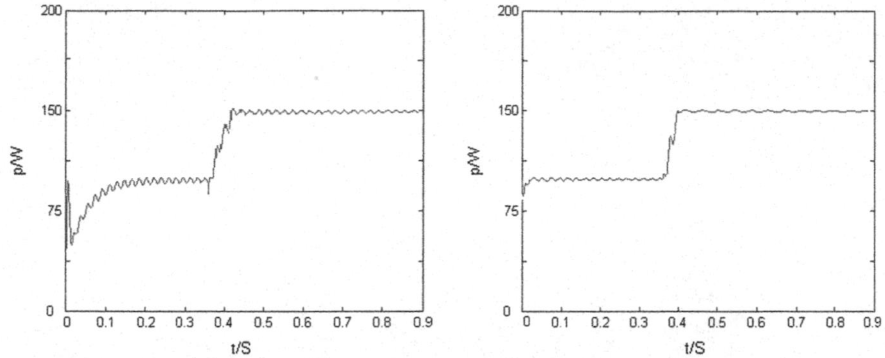


(a) Incremental conductance Algorithm (b) Improved MPPT

**Fig. 6** Maximum power point tracking waveform when the light mutation

adjusted very quickly and tracked to the MPP of the PV array, it is proved that the MPPT algorithm given in this paper has good steady-state and dynamic performance.

## References

1. B. Sun, J. Mei, J. Zheng, An improved algorithm for maximum power point tracking under local shadow conditions. Electr. Power Autom. Equip. **34**(01), 115–119+127 (2014)
2. L. Zhou, J. Wu, Q. Liu, et al., Survey of PV array maximum power point tracking control method. High Volt. Eng. (06), 1145–1154 (2008)
3. G. Wu, X. Li, Application of Newton interpolation method in maximum power tracking of photovoltaic power generation. Power Technol. **39**(7), 1432–1434 (2015)
4. R.I. Putri, S. Wibowo, M. Rifa'i, Maximum power point tracking for photovoltaic using incremental conductance method. Energy Procedia **68**, 22–30 (2015)
5. M.H. Moradi, A.R. Reisi, A hybrid maximum power point tracking method for photovoltaic systems. Sol. Energy **85**(11), 2965–2976 (2011)
6. Y. Jia, Robust control with decoupling performance for steering and traction of 4WS vehicles under velocity-varying motion. IEEE Trans. Control Syst. Technol. **8**(3), 554–569 (2000)
7. J. Hu, J. Zhang, Research on MPPT control algorithm for photovoltaic power generation system based on numerical method. Power Sci. Eng. **25**(07), 1–6 (2009)
8. Y. Liu, K. Ying, H. Xin, et al., Control strategy of photovoltaic power generation system based on quadratic interpolation method. Autom. Electr. Power Syst. **36**(21), 29–35 (2012)
9. Y. Jia, Alternative proofs for improved LMI representations for the analysis and the design of continuous-time systems with polytopic type uncertainty: a predictive approach. IEEE Trans. Autom. Control **48**(8), 1413–1416 (2003)
10. X. Wen, *Application Design of MATLAB Neural Network* (Science Press, Beijing, 2001), pp. 35–72
11. J. Jiang, Design and transient simulation of grid-connected photovoltaic system controlled by improved MPPT (Yanshan University, 2013)

# Research on Multi-view Based Embedded Software Safety Mechanism

**Shidong Luo, Xiaohong Bao, Wang Wang and Tingdi Zhao**

**Abstract** As embedded software is widely used in a variety of safety-critical control systems, the scale of the software and the complexity are constantly increasing, and the system safety problems caused by software have become more serious. Related fields of software for the current accident mechanism described angle is not comprehensive, so this research focuses on the safety mechanism of the embedded software. Through analysis and empirical data collection, the accident model of embedded system software is given. According to the accident model, the control strategies based on development view, structure view, logic view and environment view are proposed. Using this control strategy, the safety design, analysis, and software safety related development and management of embedded system software can be performed more comprehensively to provide effective protection for system safety.

**Keyword** Embedded system software · Software safety mechanism
Software safety control

## 1 Introduction

The mechanism of software safety is the analysis of the correlation and interaction of the related factors in the process of the formation, development and evolution of the accident. Measures are taken in the corresponding stage to eliminate, block, detect and suppress the possible dangers in order to improve the safety of the software system. The mechanism of software safety mainly includes two aspects: software accident mechanism and accident control mechanism. At present, there are few researches on accident models for software systems. In the field of industrial safety, the theory of accident causation has been iterated and updated several times, and gradually scientific and systematic. At present, the accident model can be classified into traditional accident model and modern accident model according to the time and

S. Luo (✉) · X. Bao · W. Wang · T. Zhao
School of Reliability and Systems Engineering, Beihang University, Beijing 100191, China
e-mail: sdluo@buaa.edu.cn

 
Y. Jia et al. (eds.), *Proceedings of 2018 Chinese Intelligent Systems Conference*,
Lecture Notes in Electrical Engineering 528,
https://doi.org/10.1007/978-981-13-2288-4_57

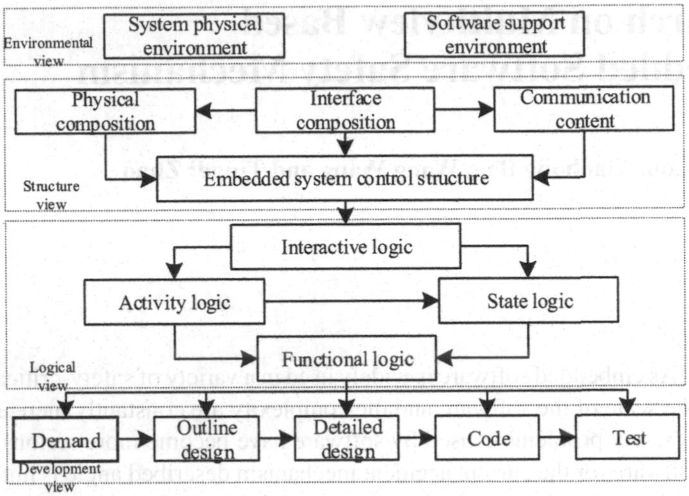


**Fig. 1** Multi view model composition

characteristics of the model generation [1]. Most of these models are not applicable to software intensive systems. At present, Nancy discusses the causes of software accidents on the basis of the system level [2, 3]. At same time, some domestic scholars have discussed the accident mechanism of the software system based on the mechanism of accident occurrence and the formation mechanism of the deep defect [4]. But the description of the cause of the accident of the software system is not deep enough, and the control idea of the system is not given to the problem.

Based on the above situation, this paper gives a control strategy based on the development view, structure view, logical view and environmental view on the basis of the process of embedded system software accident, which provides effective guarantee for the safety of embedded system software.

## 2 Multi View Model of Embedded System Software

Referring to the "4+1" model in software architecture design [5], this paper establishes a multi view model based on the software development process and the running characteristics of the software, which can be used to classify the reasons for the subsequent software problems and lay the foundation for the system. The multi view description model mainly includes four view angles: development view, structural view, logical view and environment view [6].

Each view represents different viewing angles, but these views are not completely independent, as shown in Fig. 1.

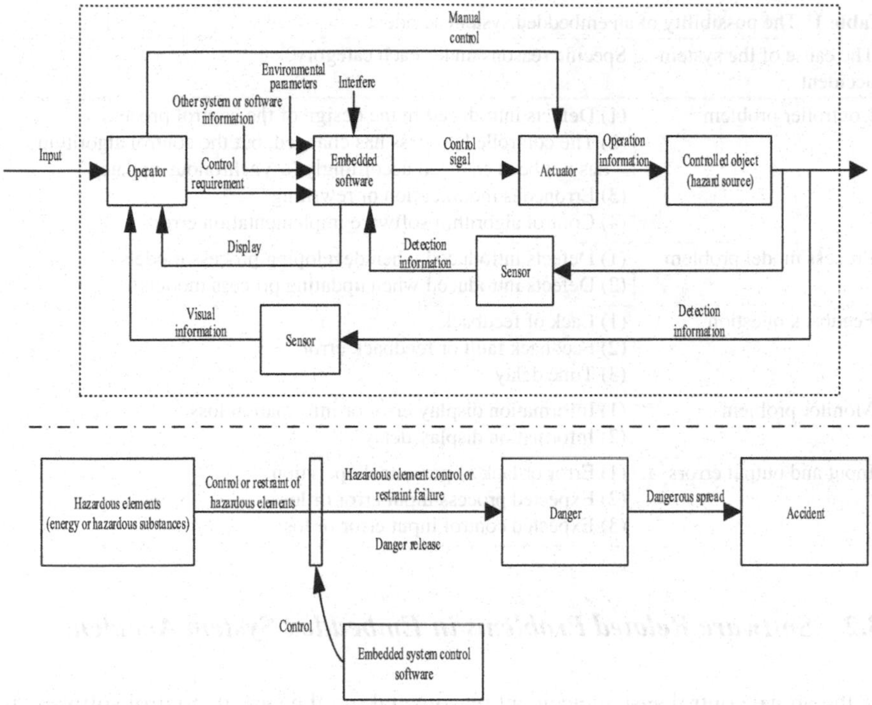


**Fig. 2** Embedded system and accident process

# 3 Accident Mechanism of Embedded System Software

## *3.1 The Accident Process of Embedded System*

In the principle of safety, the accident is composed of three parts: dangerous element, trigger mechanism, target and threat [7]. The components related to dangerous elements are divided into two categories, one is the storage, transmission and use of dangerous elements, and the other is the components related to monitoring, controlling and executing mechanism of dangerous elements. The safety critical system belongs to the latter, as shown in Fig. 2.

In the safety key system, basing on system level and other mature existing research [8, 9], combined with the system characteristics discussed in this paper, the possible causes of the accident in the embedded control system are shown in Table 1.

**Table 1** The possibility of an embedded system accident

| The cause of the system accident | Specific reasons under each category |
|---|---|
| Controller problem | (1) Defects introduced in the design of the control process<br>(2) The controlled process has changed, but the control algorithm has not been changed accordingly (asynchronous update)<br>(3) Erroneous modification or rewriting<br>(4) Control algorithm software implementation error |
| Process model problem | (1) Defects introduced when developing process models<br>(2) Defects introduced when updating process models |
| Feedback question | (1) Lack of feedback<br>(2) Feedback fault or feedback error<br>(3) Time delay |
| Monitor problem | (1) Information display error or information loss<br>(2) Information display delay |
| Input and output errors | (1) Error or lack of personnel operation<br>(2) Expected process input error or loss<br>(3) Expected control input error or loss |

## 3.2 *Software Related Problems in Embedded System Accident*

In the above control system accident reasons, taking the system control software as an example, the reasons related to the control software include the following two categories: (1) the control software itself fails, that is, the control system itself is designed correctly, but the software itself, such as a defect leads to the occurrence of the accident. (2) There is an interaction exception between the software, the design of the system itself is mistaken, and the individual software can realize the control requirements at their own level, but when the system is running at the system level, there will be an interaction anomaly, thus causing an accident.

(1) Software failure

The failure of embedded system software mainly means that the output of the software does not meet the requirements of the system. Software failure is an abnormal result of software. There are three main reasons for software failure in embedded system. Figure 3 shows.

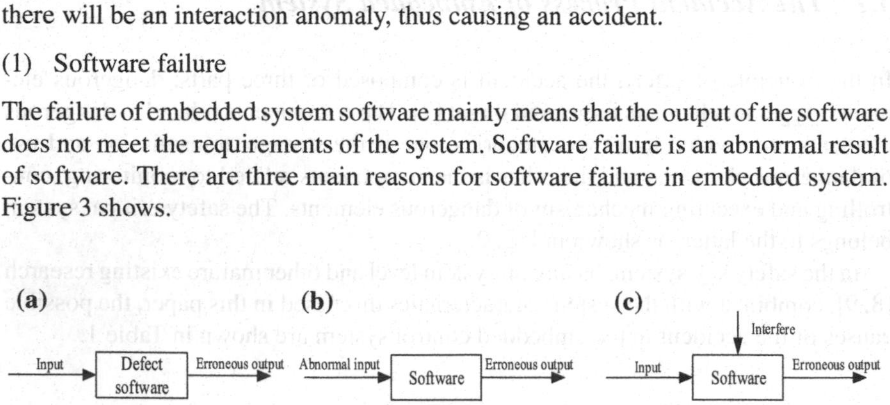


**Fig. 3** The three reasons for the failure of software

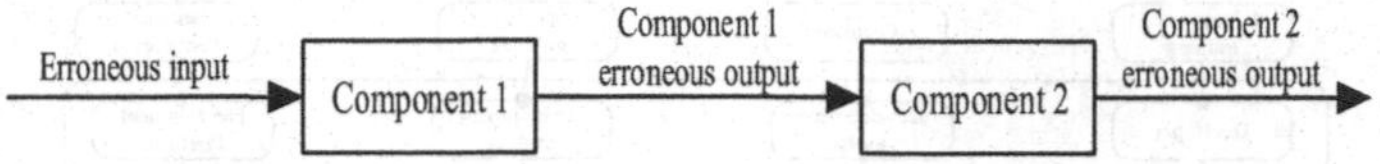


**Fig. 4** Linear interaction anomaly

**Fig. 5** Parallelism interaction exception

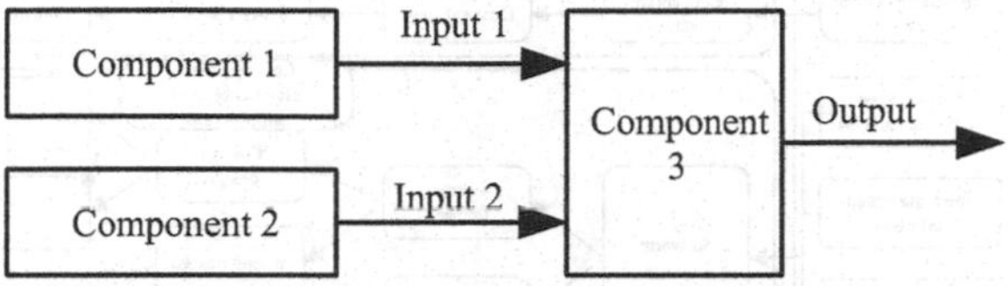


① Software failure caused by software defects

According to the theory of software reliability, the root cause of software failure is internal defects in software products. If the related defects in the software are executed, it may lead to software fault or even failure. As shown in Fig. 3a.

② Software failure caused by abnormal input

Even if reliable software developed according to system and software requirements, when the input of the software is abnormal, it can also cause software failure, as shown in Fig. 3b.

③ Software failure caused by abnormal operating conditions

Embedded system software needs to ensure a certain computing speed, read and write speed and data storage space. These rates and spaces usually require embedded system hardware and operating systems, middleware, and driver support. When the aging of the hardware is damaged or other software resources fail, it will also affect the operation ability of the embedded system software itself, as shown in Fig. 3c.

(2) Software related interactive abnormal behavior

① Linear interaction anomal

Linear interaction means that certain interactions in the system have certain inheritance in time. Common linear interaction exceptions such as inconsistent data transmission between systems and so on (Fig. 4).

② Parallel interaction exception

Parallelism interaction refers to processing multiple signals within the same specified period of time. The common parallel interaction exception coordination control occurs in the control conflict or lack of boundary control (Fig. 5).

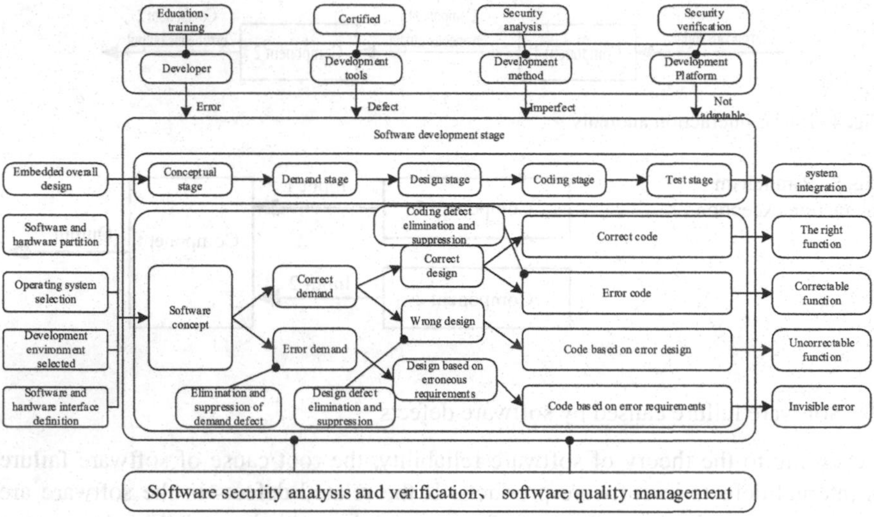


**Fig. 6** Software safety control based on development view

# 4 Safety Control of Embedded System Software Based on Multiple Views

Based on the above two points, a multi-view software model is given. For each view, the failure reason is analyzed and a corresponding safety control scheme is given.

(1) Software failure reason and safety control based on development view

The development view focuses on the introduction and transfer of defects in the embedded system software development process. The defects of the embedded system software come from the software developers and development conditions (including development tools, development methods, and development platforms, etc.).

The errors of the above developers and development conditions are introduced into software at various stages of software development to form software defects. At the same time, if the defects of the previous development stage are not corrected, they will be transferred to the next development stage. In order to reduce the defects in the embedded system software, we need to control the development personnel, development conditions, development process and development management. The control methods mainly include:

- Training for developers.
- Safety analysis and certification of development tools, methods, platforms.
- Analyze software defects at various stages and take measures to eliminate or suppress software defects.
- Software quality management (Fig. 6).

**Table 2** Based on the type of structure view defect types

| | Type of problem | Control measures description |
|---|---|---|
| Composition | Missing necessary hardware devices | The design software detects the composition of the necessary hardware devices |
| | Low CPU speed | Design degradation mode |
| | Memory read and write speed is too low | Designed to detect read and write status, alert when write/read writes are slow |
| | No software partition | Make a clear division between software and operating chips |
| Interface | Operating interface is not conducive to personnel operation | Optimizing man-machine interface and reducing human operation |
| | Interface inconsistency | A strict definition of each interface |
| Communication | Data not defined or incorrect in format or inaccuracy | Strictly define the format and accuracy of data |
| | Data overflow | Design data spillover detection and alarm |

(2) Software safety control based on Structure View

Through the analysis of the problem types in the structural view and the selection of relevant safety strategies, the results of the final safety design measures are shown in Table 2.

(3) Software safety control based on logical view

By analyzing the types of problems in the logical view and selecting the relevant safety strategies, the results of the final safety design measures are shown in Table 3.

(4) Software safety control based on Environmental View

By analyzing the types of problems in the environment view and selecting the relevant safety strategies, the results of the final safety design measures are shown in Table 4.

## 5 Conclusion

The main research content of this paper is the safety mechanism of embedded system software, and the safety mechanism is divided into two parts: accident mechanism and accident control mechanism. According to the accident mechanism, on the basis of the investigation of the accident model, the causes of the possible accident in the embedded system are analyzed and summarized, and the reasons for the control software in the system are mainly composed of two points: The failure of the software itself and the abnormal interaction of the software. Aiming at the causes of the above two kinds of accidents, a multi-view model was constructed to control the safety of

**Table 3** Safety design control for logic view defect types

| | Type of problem | Control measures description |
|---|---|---|
| Interaction problem | Object deletion or object error | Design object loss alarm |
| | Unable to transmit information | Design information transfer failure detection and alarm |
| | Unacceptable information | Design interface selfcheck program |
| | Repeatedly sending information | Send confirmation process for safety critical information design |
| Activity | Erroneous action | Redundancy mechanism is used in design |
| | Action sequence error | Redundancy mechanism is used in design |
| | Boundary value is not defined | The strict definition of boundary value condition of software operation |
| | The opposite of judgment condition | Redundancy mechanism is used in design |
| State | State error | Strictly define the content of the operation in each state |
| | Event and state erroneous mapping | Design to request confirmation from operator when switching state. |
| Function | Incompleteness of lower function | Redesign the function |
| | Module is not independent | Redesign the function |
| | Functional module duplication | Redesign the function |

**Table 4** Safety design control for defect types in environment view

| | Type of problem | Control measures description |
|---|---|---|
| Software support environment | Incompatibility between software and operating system | Design incompatible alarm |
| | Runtime CPU load | Design abnormal alarm |
| | The use and availability of memory are not reasonable | Design abnormal alarm |
| | Low reading and writing rate of memory | Design abnormal alarm |
| System external environment | Environmental parameter anomaly | Design check mask to receive erroneous messages |
| | Abnormal radiation condition interference software operation | Design check mask to receive erroneous messages |

the embedded system software from the four aspects: structure view, composition view, logical view and environment view. At the end of the article, the model of the embedded system software safety mechanism is given. By using this model, the safety design can be better performed, and the safety analysis can be carried out effectively. According to the above safety control method, provide an effective protection for the operation of embedded systems.

**Acknowledgements** This research has been supported by grants from the Major State Basic Research Development Program of China (973 Program) (No. 2014CB744904), and grants from a project of Ministry of Industry and Information Technology of China (No. JSZL2015601C008), and Civil Aviation Joint Funds established by National Nature Science Foundation of China and Civil Aviation Administration of China (No. U1533201).

## References

1. X. Sun, X. Zhou, J. Lin, et al., Accident model theory development and applied research. Qual. Reliab. (2), 19–23 (2014)
2. N. Leveson, A new accident model for engineering safer systems. Saf. Sci. **42**(4), 237–270 (2004)
3. N. Leveson, J.P. Thomas, *STPA Handbook* (2018)
4. X. Zhou, X. Sun, Space software system accident mechanism and model research. Qual. Reliab. (4), 1–5 (2014)
5. P. Kruchten, Architecture blueprints—the "4+1" view model of software architecture (ACM, 1995), pp. 540–555
6. W. Wang, X. Bao, T. Zhao, A research for embedded system software accident mechanism, in *International Conference on System Reliability and Safety* (2017), pp. 460–464
7. P. Sui, B. Chen, *Safety Principle and Accident Prediction* (Metallurgical Industry Press, 1988)
8. Y. Jia, Robust control with decoupling performance for steering and traction of 4WS vehicles under velocity-varying motion. IEEE Trans. Control Syst. Technol. **8**(3), 554–569 (2000)
9. Y. Jia, Alternative proofs for improved LMI representations for the analysis and the design of continuous-time systems with polytopic type uncertainty: a predictive approach. IEEE Trans. Autom. Control **48**(8), 1413–1416 (2003)

# Containment Control of Second-Order Multi-agent Systems with Mismatched Disturbances

**Yuling Li, Hongyong Yang, Fan Liu, Yize Yang and Yuanshan Liu**

**Abstract** For the cluster movement problem of multi-agent systems, this paper studies containment control of second-order multi-agent systems with mismatched disturbances. State observers and disturbance observers are designed to estimate the unknown states and disturbances of the systems, and a control protocol based on the active anti-disturbance observers is constructed. By applying matrix theory and modern control theory, the dynamic characteristics of second-order multi-agent systems based on disturbance observers are studied. In numerical simulations, the observers based containment control algorithm is applied to make the motion states of the systems eventually converge to the target area formed by multiple leaders, which verifies the validity of the conclusions in this paper.

**Keywords** Containment control · Mismatched disturbances · Multi-agent systems
Active anti-disturbance control

Y. Li · H. Yang (✉) · F. Liu · Y. Liu
School of Information and Electrical Engineering, Ludong University, Yantai 264025, China
e-mail: hyyang@yeah.net

Y. Li
e-mail: liyuling822@163.com

F. Liu
e-mail: jsgyliufan@163.com

Y. Liu
e-mail: yuanshanliu@163.com

Y. Yang
School of Electrical Engineering and Telecommunications,
The University of New South Wales, Sydney, Australia
e-mail: yangyz1994@126.com

 
Y. Jia et al. (eds.), *Proceedings of 2018 Chinese Intelligent Systems Conference*,
Lecture Notes in Electrical Engineering 528,
https://doi.org/10.1007/978-981-13-2288-4_58

## 1 Introduction

Recently, multi-agent system control is a hot research topic in the complex systems. Distributed coordination which is motivated by biology promotes the rapid development of multi-agent systems research, due to its broad applications in various areas such as mobile robots cooperation control, military reconnaissance and formation control of drones.

In the multi-agent systems, the followers dynamically follow the motion states of leaders, which can be called the cooperative control of cluster systems [1–3]. As a kind of cooperative control problems of multi-agent systems with multiple leaders, containment control [4, 5] regulates followers eventually converge to a target area by designing a control protocol. Wang et al. [6] discussed distributed containment control problems for multi-agent systems with nonlinear dynamics.

In reality, the multi-agent systems are usually affected by all kinds of disturbances. The existence of disturbances affects the motion states of systems, so it is significant to study the effect of disturbances on the systems. In most literatures, the multi-agent systems are affected by matched disturbances [7, 8], that is, the disturbances are in the same channels as the control inputs. In addition, in the multi-agent systems, if the disturbances enter systems through different channels from control inputs, then they are called mismatched disturbances. Wang et al. [9] studied consensus problem for higher-order multi-agent systems with mismatched disturbances. The mismatched disturbances make impact on the system performance in different ways from matched disturbances. The existing control methods which deal with matched disturbances can't effectively handle the effect of the mismatched disturbances. Therefore, it is of great significance to study the anti-disturbance control for the containment control problem of multi-agent systems.

In this paper, containment control algorithms of second-order multi-agent systems are studied. The innovation of this paper is to use the observers to estimate the unknown states and the disturbances affected in the systems. Then a nonlinear containment control algorithm is proposed to research the cluster movements of multi-agent systems with mismatched disturbances.

## 2 Preliminaries

Assume that $n$ agents constitute a network topology graph $G = (V, E, A)$, in which $V = \{v_1, v_2, \ldots, v_n\}$ represents a set of n nodes, and its edges set is $E \subseteq V \times V$. $I = \{1, 2, \ldots, n\}$ is the node subscripts set, $A = [a_{ij}] \in R^{n \times n}$ is an adjacency matrix with elements $a_{ij} \geq 0$. An edge of the graph $G$ is denoted by $e_{ij} = (v_i, v_j) \in E$. Let the adjacency element $a_{ij} > 0$ when $e_{ij} \in E$, otherwise $a_{ij} = 0$. The neighbors' set of node $i$ is denoted by $N_i = \{v_j \in V \mid (v_i, v_j) \in E\}$.

Let graph $G$ be a weighted graph without self-loops, i.e. $a_{ii} = 0$, and let matrix $D = \mathrm{diag}\{d_1, d_2, \ldots, d_n\} \in R^{n \times n}$ be the diagonal matrix with the diagonal elements

$d_i = \sum_{j=1}^{n} a_{ij}$. $L = D - A \in R^{n\times n}$ is the Laplacian matrix of the weighted graph $G$.

**Definition 1** [10] Assume that the finite set $X = \{x_1, x_2, \ldots, x_m\}$ is a subset of real vector space $R^n$, then the convex hull of set $X$ is defined as $CO(X) = \{\sum_{i=1}^{m} \alpha_i x_i | x_i \in X, \alpha_i \geq 0, \sum_{i=1}^{m} \alpha_i = 1\}$.

**Lemma 1** *[11] Suppose $f(x, u, t)$ is continuously differentiable and globally Lipschitz in $(x, u)$, uniformly in $t$. If the unforced system $\dot{x} = f(x, 0, t)$ is globally uniformly exponentially stable at the origin, when $\lim_{t\to\infty} u(t) = 0$, then system $\dot{x} = f(x, u, t)$ is asymptotically approach zero, i.e., $\lim_{t\to\infty} x(t) = 0$.*

**Lemma 2** *[12] Assume that polynomial $q(s) = \rho_0 + \rho_1 s + \cdots + \rho_n s^n$, and let $s = j\omega$ denotes*

$$q(j\omega) = m(\omega) + jn(\omega)$$

*And then $q(s)$ is Hurwitz stable if and only if the following conditions are satisfied*

*(1) The root of $m(\omega) = 0$, $m_1 < m_2 < \cdots$ and root of $n(\omega) = 0$, $n_1 < n_2 < \cdots$ meet*

$$m_1 < n_1 < m_2 < n_2 < \cdots \text{ or } n_1 < m_1 < n_2 < m_2 < \cdots$$

*(2) $m(0)n'(0) - m'(0)n(0) > 0$.*

## 3 Containment Control of Multi-agent Systems with Mismatched Disturbances

Suppose that the second-order multi-agent systems with mismatched disturbances are consisted of $n$ followers and $m$ leaders, and the dynamical equations of multi-agent systems are defined as

$$\begin{cases} \dot{\bar{q}}_i(t) = \bar{p}_i(t) + d_{i2}(t) \\ \dot{\bar{p}}_i(t) = u_i(t) + d_{i1}(t) \\ y_{i1}(t) = \bar{q}_i(t) \\ y_{i2}(t) = \bar{p}_i(t) + d_{i2}(t) \end{cases} \quad i = 1, \ldots, n \tag{1a}$$

$$\begin{cases} \dot{\bar{q}}_i(t) = \bar{p}_i(t) \\ \dot{\bar{p}}_i(t) = 0 \end{cases} \quad i = n+1, \ldots, n+m \tag{1b}$$

where $\bar{q}_i(t)$ and $\bar{p}_i(t)$ are the agent position and velocity, $u_i(t)$ is the control input, $d_{i1}(t)$ and $d_{i2}(t)$ are matched disturbances and mismatched disturbances respectively.

$F = \{1, 2, \ldots, n\}$ and $K = \{n+1, n+2, \ldots, n+m\}$ are the sets of $n$ follower agents and $m$ leader agents.

**Assumption 1** In the system (1a), the disturbances $d_{i1}(t)$ and $d_{i2}(t)$ are first-order differentiable and second-order differentiable respectively, and they are both Lipschitz stable.

**Definition 2** $L = \begin{bmatrix} L_F & L_{FK} \\ 0_{m\times n} & 0_{m\times n} \end{bmatrix}$ is the Laplacian matrix of graph $G$, where $L_F \in R^{n\times n}$ is the Laplacian matrix of followers, $L_{FK} \in R^{n\times m}$.

Let $q_i(t) = \bar{q}_i(t)$, $p_i(t) = \bar{p}_i(t) + d_{i2}(t)$, $d_i(t) = \dot{d}_{i2}(t) + d_{i1}(t)$, then system (1a, 1b) can be rewritten as

$$\begin{cases} \dot{q}_i(t) = p_i(t) \\ \dot{p}_i(t) = u_i(t) + d_i(t) \\ y_{i1}(t) = q_i(t) \\ y_{i2}(t) = p_i(t) \end{cases} \quad i = 1, \ldots, n \tag{2a}$$

$$\begin{cases} \dot{q}_i(t) = p_i(t) \\ \dot{p}_i(t)=0 \end{cases} \quad i = n+1, \ldots, n+m \tag{2b}$$

**Lemma 3** *[13] Assume that the communication topology of n followers in the multi-agent system (2a, 2b) with multiple leaders is undirected, then matrix $L_F$ is position definite, and $-L_F^{-1}L_{FK}$ is a non-negative matrix whose elements sum in each row is 1.*

**Lemma 4** *Let $X_F = [X_1, \ldots, X_n]^T$, $X_k = [X_{n+1}, \ldots, X_{n+m}]^T$, the networked systems can achieve containment control if the condition $X_F \to -L_F^{-1}L_{FK}X_K$ is satisfied.*

*Proof* From Lemma 3, $-L_F^{-1}L_{FK}$ is a non-negative matrix whose elements sum in each row is 1, which is satisfied with the condition in Definition 1. Let $X_k = [X_{n+1}, \ldots, X_{n+m}]^T$, then $-L_F^{-1}L_{FK}X_K$ is in the convex hull formed by the leaders. Therefore, the networked systems can achieve containment control if the condition $X_F \to -L_F^{-1}L_{FK}X_K$ is satisfied.

Let $q_F(t) = [q_1(t), q_2(t), \ldots, q_n(t)]^T$, $p_F(t) = [p_1(t), p_2(t), \ldots, p_n(t)]^T$, $x_F(t) = [q_F(t), p_F(t)]^T$, $q_K(t) = [q_{n+1}(t), q_{n+2}(t), \ldots, q_{n+m}(t)]^T$, $p_K(t) = [p_{n+1}(t), p_{n+2}(t), \ldots, p_{n+m}(t)]^T$, $x_K(t) = [q_K(t), p_K(t)]^T$, $y(t) = [y_1(t), y_2(t)]^T$, $y_1(t) = [y_{11}(t), y_{21}(t), \ldots, y_{n1}(t)]^T$, $y_2(t) = [y_{12}(t), y_{22}(t), \ldots, y_{n2}(t)]^T$, $d(t) = [d_1(t), d_2(t), \ldots, d_n(t)]^T$, $g(t) = [g_1(t), g_2(t), \ldots, g_n(t)]^T$, then the system (2a, 2b) can be rewritten as

$$\begin{cases} \dot{x}_F(t) = Ax_F(t) + Du(t) + Dd(t) \\ y(t) = Gx_F(t) \end{cases} \tag{3a}$$

$$\dot{x}_K(t) = Ax_K(t) \tag{3b}$$

where $A = \begin{bmatrix} 0 & I \\ 0 & 0 \end{bmatrix}, D = \begin{bmatrix} 0 \\ I \end{bmatrix}, G = \begin{bmatrix} I & 0 \\ 0 & I \end{bmatrix}$.

To establish a state observer for system (3a)

$$\begin{cases} \dot{\hat{x}}_F(t) = A\hat{x}_F(t) + Du(t) + D\hat{d}(t) + H(y(t) - \hat{y}(t)) \\ \hat{y}(t) = G\hat{x}_F(t) \end{cases} \tag{4}$$

where $\hat{x}_F(t), \hat{d}(t)$ and $\hat{y}(t)$ are the estimated values of $x_F(t), d(t)$ and $y(t)$ respectively. $H = \begin{bmatrix} H_1 & \\ & H_2 \end{bmatrix}$ is an undetermined matrix, where $H_1 = diag\{h_{11}, \ldots, h_{1n}\}$, $H_2 = diag\{h_{21}, \ldots, h_{2n}\}$.

The disturbances in the system (3a, 3b) can be estimated by the disturbance observer as follows

$$\begin{cases} \dot{g}(t) = -MD(g(t) + Mx_F(t)) - M(Ax_F(t) + Du(t)) \\ \hat{d}(t) = g(t) + Mx_F(t) \end{cases} \tag{5}$$

where $g(t)$ is an auxiliary intermediate matrix, $M$ is an undetermined gain matrix of the disturbance observer.

The control protocol for i-th agent is designed as

$$u_i(t) = c\left(\begin{array}{c} \sum_{j=1}^{n} a_{ij}\left(\hat{q}_j(t) - \hat{q}_i(t)\right) + \sum_{j=1}^{n} a_{ij}\left(\hat{p}_j(t) - \hat{p}_i(t)\right) \\ - \sum_{j=n+1}^{n+m} a_{ij}\left(\hat{q}_i(t) - q_j(t)\right) - \sum_{j=n+1}^{n+m} a_{ij}\left(\hat{p}_i(t) - p_j(t)\right) \end{array}\right) - \hat{d}_i(t) \tag{6}$$

where $c > 0$ is an undetermined control gain.

Let $\theta(t) = x_F + I_2 \otimes \left(L_F^{-1}L_{FK}\right)x_K = \begin{bmatrix} q_F(t) + L_F^{-1}L_{FK}q_K(t) \\ p_F(t) + L_F^{-1}L_{FK}p_K(t) \end{bmatrix}$, $e_x(t) = x_F(t) - \hat{x}_F(t) = \begin{bmatrix} e_q(t) \\ e_p(t) \end{bmatrix} = \begin{bmatrix} q_F(t) - \hat{q}_F(t) \\ p_F(t) - \hat{p}_F(t) \end{bmatrix}$, $e_d(t) = d(t) - \hat{d}(t)$, where $I_2 = \begin{bmatrix} 1 & 0 \\ 0 & 1 \end{bmatrix}$.

Then, combining with Eqs. (3a, 3b)–(8), the following equations can be got:

$$\dot{\theta}(t) = (A - N)\theta(t) + Ne_x(t) + De_d(t) \tag{7a}$$

$$\dot{e}_x(t) = (A - HG)e_x(t) + De_d(t) \tag{7b}$$

$$\dot{e}_d(t) = \dot{d}(t) - MDe_d(t) \tag{7c}$$

where $N = \begin{bmatrix} cDL_F & cDL_F \end{bmatrix}$.

**Assumption 2** In the system (3a), $\lim_{t\to\infty} \dot{d}(t) = 0$.

*Remark 1* Assumption 2 is used to prove that the disturbance estimation error system (7c) asymptotically converges. However, the disturbance observer (5) can also estimate fast time-varying disturbances as long as the dynamic of the observer is fast enough and the appropriate observation gain is selected.

**Theorem 1** *Suppose the second-order multi-agent systems (1a, 1b) with mismatched disturbances consisting of n followers and m leaders are satisfied with Assumption 1. When the connection topology of the systems is undirected and at least one follower can sense the information of leaders. Based on the state observer (4) and disturbance observer (5), we can design a control protocol to achieve containment control of the system (1a, 1b) if the following conditions are satisfied.*

*(1)* $H_1 = diag\{h_{11}, \ldots, h_{1n}\}$, $H_2 = diag\{h_{21}, \ldots, h_{2n}\}$, *where* $h_{1i}$ *and* $h_{2i}$ *are positive real numbers.*
*(2) The control gain* $c > max\left\{\frac{|Im(\mu_i)|^2}{Re(\mu_i)|\mu_i|^2}\right\}$, *where* $\mu_i$ *is the eigenvalue for the matrix* $\bar{L} = cL_F$, *and* $\mu_i = Re(\mu_i) + jIm(\mu_i)$.
*(3) The matrix* $-MD$ *is Hurwitz.*

*Proof* Let $Q(t) = [\theta(t), e_x(t), e_d(t)]^T$, according to (7a–7c), the following equation can be got

$$\dot{Q}(t) = \psi Q(t) + T\dot{d}(t) \tag{8}$$

$$\text{where } \psi = \begin{bmatrix} A-N & N & D \\ 0 & A-HG & D \\ 0 & 0 & -MD \end{bmatrix}, T = \begin{bmatrix} 0 \, 0 \, I \end{bmatrix}^T.$$

By applying Lemma 1, $\dot{d}(t)$ is treated as the input of the system (8). According to Assumption 2, assume that $\lim_{t\to\infty} \dot{d}(t) = 0$, then the system (8) is rewritten as

$$\dot{Q}(t) = \psi Q(t) \tag{9}$$

To obtain the eigenvalues of the matrix $\psi$, let $|\lambda E - \psi| = 0$, the following equations can be got

$$|\lambda E - (A - N)| = 0 \tag{10a}$$

$$|\lambda E - (A - HG)| = 0 \tag{10b}$$

$$|\lambda E + MD| = 0 \tag{10c}$$

(i) For the Eq. (10a), $A - N = \begin{bmatrix} 0 & I \\ -c\bar{L} & -c\bar{L} \end{bmatrix}$, where $\bar{L} = cL_F$. Then (10a) can be rewritten as

$$\left|\lambda^2 E + \lambda \bar{L} + c\bar{L}\right| = 0 \tag{11}$$

Assume $\mu_i$ is the eigenvalue of matrix $\bar{L}$, let $\mu_i = a + jb$, where $a > 0$. Therefore, according to Lemma 3, the Eq. (11) can be simplified to the following equation

$$\lambda^2 + \lambda c\mu_i + c\mu_i = 0 \tag{12}$$

Let $r(\lambda) = \lambda^2 + \lambda c(a + jb) + c(a + jb)$, $\lambda = j\omega$, then (12) becomes

$$r(j\omega) = m(\omega) + jn(\omega) \tag{13}$$

where $m(\omega) = -\omega^2 - bc\omega + ac$, $n(\omega) = \omega ac + bc$. Due to $m(0) = ac$, $m'(0) = -bc$, $n(0) = bc$ and $n'(0) = ac$, there is $m(0)n'(0) - m'(0)n(0) = c^2|\mu_i|^2 > 0$. The roots of equations $m(\omega) = 0$ and $n(\omega) = 0$ are

$$m_{1,2} = \frac{-bc \pm \sqrt{b^2c^2 + 4ac}}{2}, n = \frac{-b}{a} \tag{14}$$

Let

$$\frac{-bc - \sqrt{b^2c^2 + 4ac}}{2} < \frac{-b}{a} < \frac{-bc + \sqrt{b^2c^2 + 4ac}}{2} \tag{15}$$

there is

$$c > \frac{b^2}{a|\mu_i|^2} \tag{16}$$

Therefore, according to Lemma 2, when the control gain $c$ meets the condition (2) of Theorem, the eigenvalues of (10a) are all in the left half-plane of the complex space. $r(\lambda)$ is Hurwitz stable, then the eigenvalues of the characteristic Eq. (12) are all negative.

(ii) For the Eq. (10b), we can obtain $A - HG = \begin{bmatrix} -H_1 & I \\ 0 & -H_2 \end{bmatrix}$. Due to $H_1$ and $H_2$ are both the diagonal matrixes with positive real numbers, the matrix $A - HG$ is full rank and its eigenvalues have negative real part. Therefore, the eigenvalues of the characteristic Eq. (10b) are all negative.

(iii) For the Eq. (10c), the eigenvalues of the characteristic Eq. (10c) are all negative due to the matrix $-MD$ is Hurwitz.

In conclusion, the system (9) is stable. According to ISS Theorem, the system (8) is stable, so $\lim_{t\to\infty} \theta(t) = 0$, i.e., $\lim_{t\to\infty} \left(x_F(t) + I_2 \otimes \left(L_F^{-1} L_{FK}\right) x_K(t)\right) = 0$. That is, $q_F(t) \to -L_F^{-1} L_{FK} q_K(t)$, $p_F(t) \to -L_F^{-1} L_{FK} p_K(t)$. According to Lemma 5, the system (1a, 1b) achieves the containment control, Theorem is proved. Therefore, by applying the disturbance observer (5) based control protocol (6), the multi-agent system (1a, 1b) can achieve containment control.

## 4 Numerical Simulations

In order to verify the effectiveness of the control protocol (6), consider the communication topology with four followers and three leaders (illustrated as 5, 6, 7) shown in Fig. 1, where the connection weights of each edge is 1.

Assume initial positions and velocities of followers are respectively taken as $q_1(0) = [2\ 0]^T$, $q_2(0) = \left[4\ 0\right]^T$, $q_3(0) = \left[0\ 2\right]^T$, $q_4(0) = \left[0\ 4\right]^T$, $p_1(0) = [2\ 6]^T$, $p_2(0) = \left[3\ 8\right]^T$, $p_3(0) = \left[4\ 6\right]^T$, $p_4(0) = \left[5\ 5\right]^T$. The initial positions and velocities of dynamic leaders are respectively taken as $q_5(0) = \left[6\ 8\right]^T$, $q_6(0) = \left[8\ 8\right]^T$, $q_7(0) = \left[8\ 6\right]^T$, $p_5(0) = \left[1\ 1\right]^T$, $p_6(0) = \left[1\ 1\right]^T$, $p_7(0) = \left[1\ 1\right]^T$. The parameter of the control protocol (6) is taken $c = 20$. The parameter of the disturbance observer (5) is taken $M = \left[10\ 10\right]^T$, and the parameter of the state observer (4) is taken $H = \begin{bmatrix} 50 & \\ & 50 \end{bmatrix}^T$. The matched disturbances and mismatched disturbances of the four followers are respectively taken $d_{11} = 1$, $d_{12} = 2$, $d_{21} = 0.5$, $d_{22} = 1$, $d_{31} = 0.5 \sin 2t$, $d_{32} = 2 \cos 2t$, $d_{41} = 0.5 \cos 2t$, $d_{42} = 2 \sin 2t$. Then we can obtain the final disturbances of the four followers by the equation $d_i(t) = \dot{d}_{i2}(t) + d_{i1}(t)$, i.e., $d_1 = 2$, $d_2 = 1$, $d_3 = 3 \cos 2t$, $d_4 = \sin 2t$.

Figure 2 shows the motion tracks of the second-order system with four followers and three leaders. From Fig. 2, it is clearly that the four followers eventually converge to the convex hull formed by the three dynamic leaders. Figure 3 shows that the estimated values of disturbances $d$ are eventually close to the actual values.

## 5 Conclusions

This paper studies containment control of multi-agent systems with multiple dynamic leaders and mismatched disturbances. The observers are designed to estimate the unknown states and disturbances of the systems, and a control protocol based on the active anti-disturbance observers is constructed. The numerical simulations results

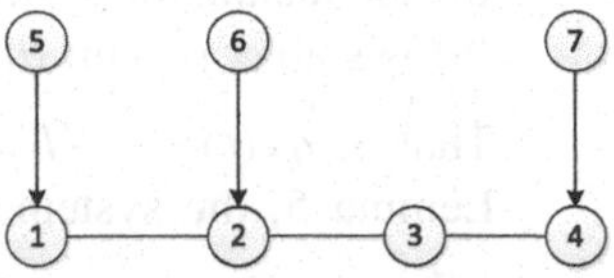


**Fig. 1** The communication topology of multi-agent systems

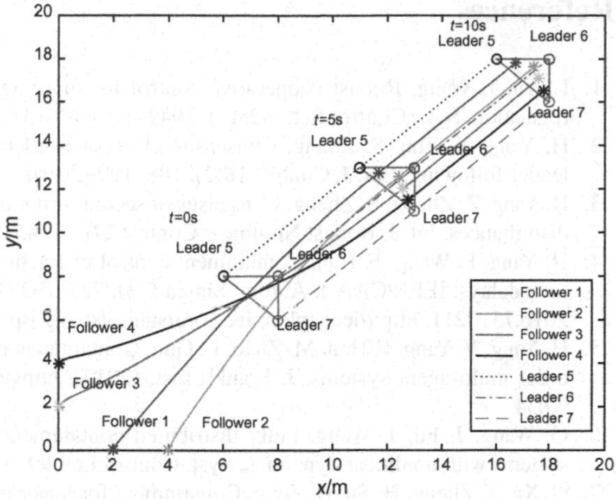


**Fig. 2** The motion tracks of the second-order system

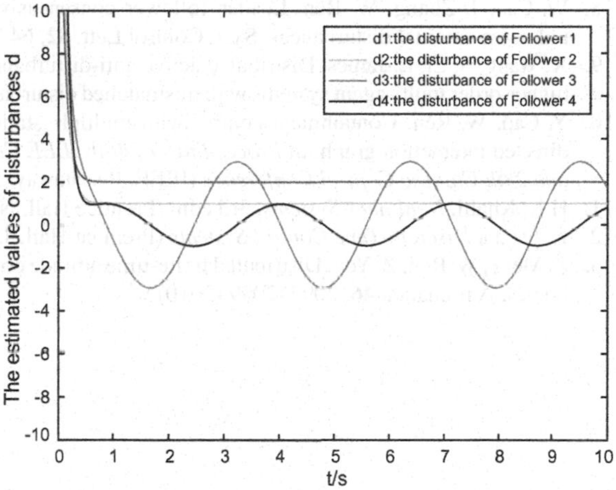


**Fig. 3** The estimated values of disturbances $d$

show that the designed observers based control algorithm can achieve containment control of the second-order multi-agent systems.

**Acknowledgements** This research is supported in part by the National Natural Science Foundation of China (61673200, 61472172, 61471185), the Natural Science Foundation of Shandong Province of China (ZR2017PF010, ZR2018ZC0438), the Key R&D Program of Yantai City of China (2016ZH061).

## References

1. L. Yu, J. Wang, Robust cooperative control for multi-agent systems via distributed output regulation. Syst. Control Lett. **62**(11), 1049–1056 (2013)
2. H. Yang, X. Zhu, S. Zhang, Consensus of second-order delayed multi-agent systems with leader-following. Eur. J. Control **16**(2), 188–199 (2010)
3. H. Yang, Z. Zhang, S. Zhang, Consensus of second-order multi-agent systems with exogenous disturbances. Int. J. Robust Nonlinear Control **21**(9), 945–956 (2011)
4. H. Yang, F. Wang, F. Han, Containment control of fractional order multi-agent systems with time delays. IEEE/CAA J. Autom. Sinica **5**(3), 727–732 (2018). https://doi.org/10.1109/JAS.2016.7510211, http://ieeexplore.ieee.org/stamp/stamp.jsp?tp=&arnumber=7783963
5. H. Yang, Y. Yang, F. Han, M. Zhao, L. Guo, Containment control of heterogeneous fractional-order multi-agent systems. J. Frankl. Inst. (2017), https://doi.org/10.1016/j.jfranklin.2017.09.034
6. Q. Wang, J. Fu, J. Wang, Fully distributed containment control of high-order multi-agent systems with nonlinear dynamics. Syst. Control Lett **99**, 33–39 (2017)
7. C. Xu, Y. Zheng, H. Su, H. Zeng, Containment for linear multi-agent systems with exogenous disturbances. Neurocomputing **160**, 206–212 (2015)
8. W. Cao, J. Zhang, W. Ren, Leader–follower consensus of linear multi-agent systems with unknown external disturbances. Syst. Control Lett. **82**, 64–70 (2015)
9. X. Wang, S. Li, L. James, Distributed active anti-disturbance output consensus algorithms for higher-order multi-agent systems with mismatched disturbances. Automatica **74**, 30–37 (2016)
10. Y. Cao, W. Ren, Containment control with multiple stationary or dynamic leaders under a directed interaction graph, in *Proceedings of 48th IEEE Conference on Decision and Control and 28th Chinese Control Conference* (IEEE, Piscataway, 2009), pp. 3014–3019
11. H.K. Khalil, *Nonlinear Systems*, 3rd edn. (Prentice Hall, New York, 2002)
12. K. Ogata, *Discrete-Time Control Systems* (Prentice-Hall, 1995)
13. Z. Meng, W. Ren, Z. You, Distributed finite-time attitude containment control for multiple rigid bodies. Automatica **46**, 2092–2099 (2010)

# Automated Guided Vehicle Indoor Positioning Method Based Cellular Automata

**Jian Sun, Yongling Fu, Shengguang Li and Yujie Su**

**Abstract** Automated guided vehicle (AGV) can greatly improve warehousing operation efficiency and reduce labor costs. As a key technology, the positioning method is crucial for the path planning and cruising of AGVs. In view of existing technology, the method with high positioning precision is too expensive, and the low cost method performance is too poor. In this paper, a low cost and high precision positioning method based on cellular automata is proposed. This method utilizes the wireless communication system that AGV has equipped to complete the positioning through the continuous iteration of simple cell evolution rules in the cell space mapped by the positioning space. For the problem of positioning errors caused by environmental factors changing and equipment aging, this method integrates spatiotemporal correlation and differential calculations as constraints in the evolution rules to achieve high-precision positioning. Through simulation experiments, the feasibility and effectiveness of the method are verified and analyzed. It has a good application prospect.

**Keywords** Automated guided vehicle (AGV) · Indoor positioning · Cellular automata · Intelligent warehouse · Wireless communication

---

J. Sun · Y. Fu
School of Mechanical Engineering & Automation, Beihang University, Beijing 100191, China
e-mail: shijsunbj@139.com

Y. Fu
e-mail: fuyongling@126.com

J. Sun · S. Li (✉)
The F. R. I. of Ministry of Public Security, Beijing 100048, China
e-mail: lishengg@163.com

Y. Su
North China University of Water Resources and Electric Power, Zhengzhou 450045, China
e-mail: winnersu@163.com



Y. Jia et al. (eds.), *Proceedings of 2018 Chinese Intelligent Systems Conference*,
Lecture Notes in Electrical Engineering 528,
https://doi.org/10.1007/978-981-13-2288-4_59

## 1 Introduction

AGV is a kind of power device equipped with wireless communication and various sensing components that can travel according to pre-specified guidance routes, with operation, stop, safety protection and various transfer functions [1]. It can significantly improve logistics efficiency, reduce operating costs and save labor force [2]. With the rapid development of modern storage industry, the demand for smart AGV is urgent. As one of the key technologies of AGV, positioning method solves the core technical problem of how to perceive its own position and provide accurate and reliable localization for path planning and cruising [3]. Widely used AGV positioning technology can be divided into four categories [4]: (1) AGV uses a magnetic sensor to detect the laying of the positioning magnetic strip, which is laid according to a preset route. Some scholars have upgraded magnetic strips to magnetic nails or used light reflection beacons to achieve positioning [5]. Such methods appeared earlier and are simple to implement [6]. However, this method has a completely fixed guided path and is difficult to update later. Although the cost is lower but the effect of positioning and guidance is limited [7]. (2) Based on the SLAM (Simultaneous Localization and Mapping) device to achieve the positioning guidance of AGV, this method has high positioning accuracy and can realize navigation of free path. However, due to using the laser radar and other high-precision distance measuring devices, the manufacturing cost of AGV is greatly increased [8]. (3) The positioning method based on visual analysis is to use the camera to identify image markers containing location information deployed on the ground or on the shelves. However, the two-dimensional codes number is large, error-prone, and the cost of image recognition devices is high, also regular cleaning and maintenance are required [9]. (4) Based on the positioning method of the wireless communication system, the AGV is installed with a wireless communication module for data exchange with the control center, the distance from the base station is obtained from the AGV wireless signal feature value, and the AGV coordinates can be calculated according to the geometric relationship. This method utilizes existing wireless communication networks and equipment, low cost and positioning and guidance route is relatively flexible [10]. However, the multi-path and non-line-of-sight effects of wireless signals seriously affect positioning accuracy and positioning stability [11].

High-precision positioning method has high cost, and low-cost method positioning accuracy is difficult to meet the demand, to address this problem, the paper, based on the complexity of wireless positioning system, the complex influence of AGV acceleration and braking control on location is fully considered [12], and a AGV location method based on cellular automata is proposed. Establish the mapping relationship between positioning space and cell space, design positioning evolution rules, and get AGV localization through continuous iteration of simple rules. In addition, the spatial-temporal correlation and signal eigenvalue difference calculations are integrated into the rules of cell evolution to constrain the positioning error and effectively improve the accuracy. The purpose method has the characteristics of low cost, simple implementation and high positioning accuracy, and it has broad application prospects.

In the second part, the method of indoor positioning and cellular automata is introduced. The third part describes the principle and implementation process of indoor positioning method based on cellular automata. The fourth part carries on the simulation verification and analysis to the method proposed in this article, the fifth part summarizes the full text.

# 2 Related Work

## *2.1 Wireless Indoor Positioning*

Most intelligent warehouses are built in buildings. The locating of AGV belongs to the research field of indoor positioning (IP). The wireless indoor positioning relies on the existing wireless communication equipment and network architecture to achieve the collection and transmission of wireless location ranging values. Common wireless positioning systems include WiFi, ZigBee, BT, RFID, UWB, etc. Most of them are based on the ISM band.

Position Measurement Value (PMV) is the data foundation of wireless indoor positioning system, Commonly used wireless location ranging values include: Received Signal Strength (RSS), Time of Arrived (TOA), Angle of Arrived (AOA), and Time Difference of Arrived (TOA). Based on the transmission characteristics of wireless signals, IP system using RSS as PMV calculates the distance of the point from a known location node to the unknown node, and finally uses the geometric relationship or scene characteristics to obtain the position.

IP methods can be divided into three categories: (1) approximate positioning method, (2) geometric positioning method, (3) scene analysis and positioning method. The approximate method is to use the nearest known location node to find the unknown node. The geometric positioning method calculates the coordinates according to the geometric relationship between the distance between the unknown node and base stations. This system is simple to implement, but the positioning complexity is low. The scene analysis method needs to collect the positioning distance value, and it establishes the prior relationship between the PMV and the localization offline.

Wireless indoor positioning system mainly consists of three parts: wireless communication network, positioning server and positioning display, as shown in Fig. 1.

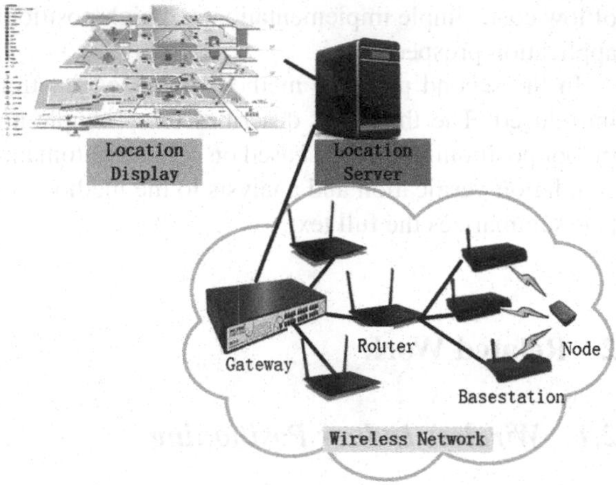


**Fig. 1** Architecture of wireless positioning system

## 2.2 *Cellular Automata*

Cellular automata (CA) was proposed by the famous computer scientist Von Neumann in the 1950s and is one of the important methods used to solve complexity problems. The mechanism of CA is iteratively evolved dynamic systems in discrete time dimensions according to the same local evolution rules [13], which is defined in a cellular space composed of discrete and finite state cells. CA replaced strict mathematical functions with simple rules that collectively obey interactions, which portrays and describes objective things and phenomena. In both time and space, both bounded prediction and unbounded simulation can be achieved, so it can defined as a model, it can also be understood as a modeling method.

The basic components of cellular automata include five parts: cell, neighbor type, cell state, cell space, and cell evolution rules. CA can be described as in mathematical form. A stands for CA, stands for cell space and d is spatial dimension, N represents the combination of all the cells in a neighboring area which include Center Cell (CC) and Neighbor Cell (NC), Local state transition function. The composition of the cellular automata is shown in Fig. 2.

The remarkable features of cellular automata are spatio-temporal discrete, state discrete and parallel computing. It has been widely and maturely applied in biological evolution, traffic flow simulation, urban system simulation and path planning [14].

# 3 CA Positioning Method

The proposed method is based on the existing wireless communication network and equipment, and AGV positioning and data exchange share the same wireless communication systems, and uses RSS as PMV to achieve high-precision positioning. It

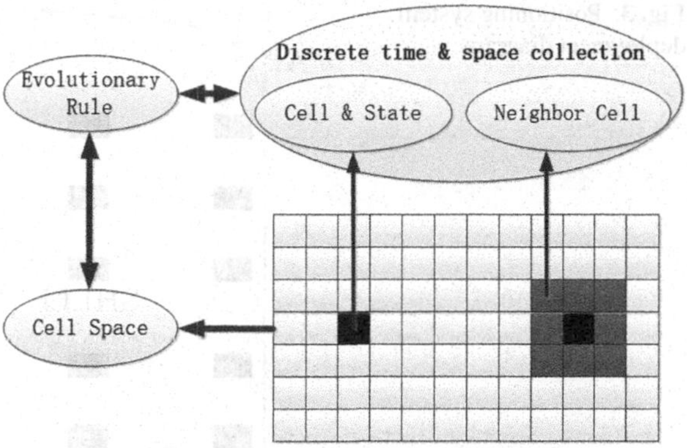


**Fig. 2** Composition of cellular automata

is simple to implement and addresses the positioning error caused by factors such as environmental changes, long-term aging of equipment, and obstruction of obstacles. This paper is inspired by the solution of LMI in control problems [15]. It proposes a cell evolution rule with time-space correlation and differential calculation of location ranging values to constrain the ranging error and improve positioning. Accuracy.

## 3.1 *System Setup*

This paper is based on the WiFi wireless network which is widely deployed in smart warehouses. It chooses the 2.4 GHz band and complies with the 802.11.n protocol for star network topology. No less than 4 WiFi base stations are deployed in the intelligent warehouse and installed in the four corners of the warehouse respectively. The AGV performs positioning and ranging with the base station according to a fixed period, and the positioning ranging values in the same time are packed and uploaded to the positioning server for calculation. Taking four base stations as an example, the positioning system deployment diagram is shown in Fig. 3.

## 3.2 *Methodology*

The core idea of the method is the use of simple logic to determine the iterative realization of the rules of the indoor positioning process simulation. According to the radio free space propagation mechanism, the rules of cell evolution are designed to achieve the goal of positioning.

**Cell**

The cell is a basic unit in CA. Cell size and state type are set according to the positioning requirements. The state type is determined by the state parameters

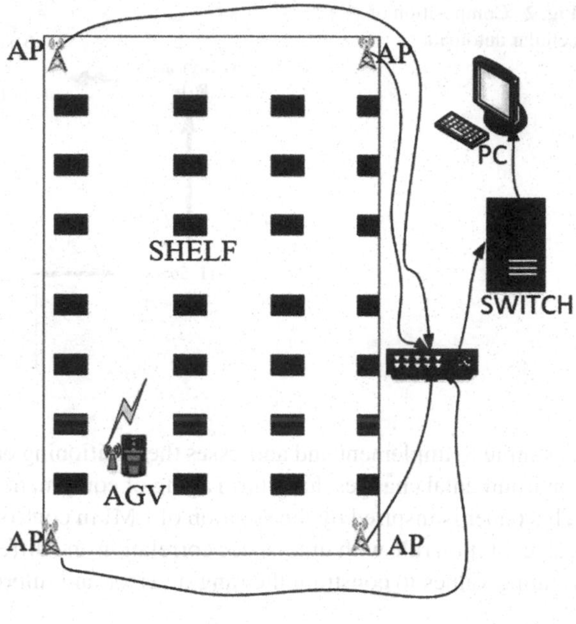


**Fig. 3** Positioning system deployment diagram

which is determined by the signal strength of the four base stations. The amount of change at the time is composed of discrete components. This can be expressed as $(\omega_A, \omega_B, \omega_C, \omega_D)$, the correspondence between cell state and state parameters is shown in Table 1.

**Cell Neighbor**
Cell neighbor consist of the CC and its surrounding NCs. The state of the cell neighbor determines the state of the central cell. The common cell neighbor types are Moore type and von Neumann type.

**Cell Space**
The indoor space has a mapping relationship with the cell space and is composed of all the cells.

**Evolutionary Rules**
Cell evolution rules are the soul of CA. Based on the accurate description of the positioning mechanism, it directly relates to the feasibility, effectiveness, and positioning performance of the indoor positioning method. The previous moment states of CC and NC determined the current moment state of CC, but has nothing to do with the cell state at other times and other locations.

**Differential Calculation**
Radio propagation characteristics in the free space are the theoretical basis of the wireless indoor positioning method, as shown in Formula (1). When the environmental factors such as temperature and humidity change, the media attenuation coefficient $\beta$ in the formula changes, resulting in RSS changes. Due to aging of transceiver com-

**Table 1** State truth table

| No. | $S^t_{(i,j)}(\omega_A, \omega_B, \omega_C, \omega_D)$ | | | | State |
|---|---|---|---|---|---|
| 1 | 0 | 0 | 0 | 0 | Hold |
| 2 | 1 | 1 | −1 | −1 | Up |
| 3 | 1 | 1 | 0 | 0 | Up |
| 4 | 1 | 1 | −1 | 0 | Up |
| 5 | 1 | 1 | 0 | −1 | Up |
| 6 | 0 | 0 | −1 | −1 | Up |
| 7 | 1 | 0 | −1 | −1 | Up |
| 8 | 0 | 1 | −1 | −1 | Up |
| 9 | −1 | −1 | 1 | 1 | Down |
| 10 | 0 | 0 | 1 | 1 | Down |
| 11 | −1 | 0 | 1 | 1 | Down |
| 12 | 0 | −1 | 1 | 1 | Down |
| 13 | −1 | −1 | 0 | 0 | Down |
| 14 | −1 | −1 | 0 | 1 | Down |
| 15 | −1 | −1 | 1 | 0 | Down |
| 16 | 1 | −1 | −1 | 1 | Left |
| 17 | 1 | 0 | 0 | 1 | Left |
| 18 | 1 | −1 | 0 | 1 | Left |
| 19 | 1 | 0 | −1 | 1 | Left |
| 20 | 0 | −1 | −1 | 0 | Left |
| 21 | 1 | −1 | −1 | 0 | Left |
| 22 | 0 | −1 | −1 | 1 | Left |
| 23 | −1 | 1 | 1 | −1 | Right |
| 24 | 0 | 1 | 1 | 0 | Right |
| 25 | −1 | 1 | 1 | 0 | Right |
| 26 | 0 | 1 | 1 | −1 | Right |
| 27 | −1 | 0 | 0 | −1 | Right |
| 28 | −1 | 1 | 0 | −1 | Right |
| 29 | −1 | 0 | 1 | −1 | Right |

ponents, the standard signal strength coefficient $P_o$ in the formula will change, which will also cause RSS fluctuations.

$$P_i = P_o - 10\beta \log_{10}\left(\frac{d_i}{d_o}\right) \tag{1}$$

wherein, $P_i$ represents the RSS (dB); $P_0$ represents the standard signal strength coefficient, i.e. the received signal power at the distance $d_0$, usually $d_0 = 1\text{m}$; $d_i$ represents the Euclidean distance (m) from the positioning of $i$ base station to the positioning terminal; $\beta$ represents the channel attenuation factor. The difference calculation is integrated into the cell evolution rule, and the variation value of RSS in the adjacent positioning period is used to calculate, which can effectively suppress the ranging error caused by the change of environmental factors, and eliminate the common mode error caused by the aging of the equipment.

**Time Correlation**

Time correlation means that the position of the CC at the current time is determined by the combination of CC & NC RSS changes at the previous time and the coordinates of the CC position at the previous time. It can be expressed as:

$$f : S_i^{t+1} = f(S_i^t, S_N^t) \tag{2}$$

In this equation, $f$ represents the cell evolution rule; $S_i^t$ represents the state of the central cell i at time t; $S_N^t$ represents the state of the neighbor cell at time t and $S_i^{t+1}$ represents the state of the central cell at time $t+1$.

**Spatial Correlation**

Once the type of the cell neighbor has been selected, the spatial relative relationship between the CC and the NC has a known correlation. Spatial correlation means that the calculation of the RSS in the cell evolution rule fuses the fixed spatial relationship between the CC and the NC, and reflects it in the form of weights, the smaller the far distance is, the greater the distance is.

The weight of the neighbor cell $i$ is $w_i = \frac{S_i}{S}$. The sum of the PMV in the neighborhood of the CC is represented as $\sum RSS_m$, Can be obtained by formula (3).

$$\sum RSS_m = \sum \left(RSS_{m,i} \times w_i\right) + RSS_{m,c}, i \in (1, \ldots, 8) \tag{3}$$

$RSS_{m,i}$ represents the amount of change in signal strength of the $NC_i$ relative to the positioning base station; $w_i$ represents the weight of the $NC_i$ relative to the positioning base station; $RSS_{m,c}$ represents the amount of change in signal strength of the central cell relative to the base station m.

**Specific Process**

Since most intelligent warehouses use standardized shelves and shelf layouts at specific distances, the shelf size and shelf separation are usually integer multiples of the positioning step size, as shown in Fig. 4. Therefore, this method sets the distance which the AGV moves in a single step to be one cell, just as the magnetic nail and the two-dimensional code visual servo method adopt the same setting which the single-step moves fixed length [16]. The positioning period is 1 s.

The process of the method is shown in Fig. 5 and is described as follows:

Firstly, the signal strength variation $\Delta RSS_m (m \in A, B, C, D)$ between the CC and the NC relative to the four positioning base stations is collected from $t$ to $t+1$. Then, using GRss for discretization, obtain a combination of $(\omega_A, \omega_B, \omega_C, \omega_D)$, and the direction of movement of CC is determined according to different combinations,

**Fig. 4** Standardized shelves

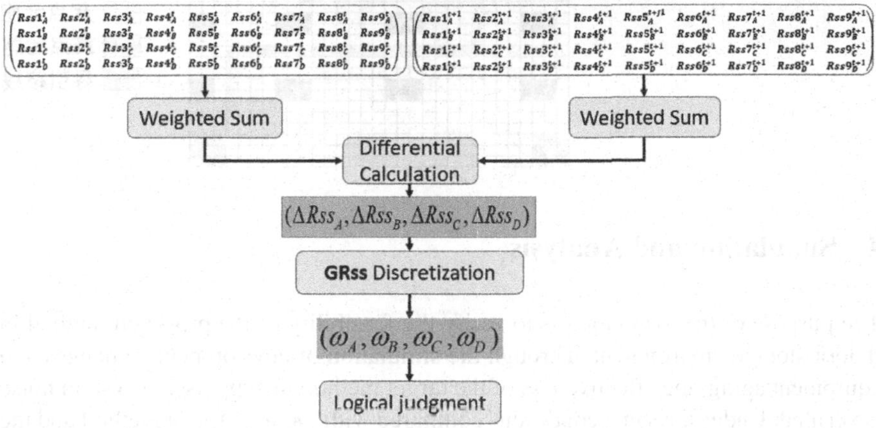


**Fig. 5** Method flow

as shown in Table 1, realizing cellular positioning. The discrete method of $\omega_m$ is shown in Eq. (4):

$$\omega_m = \begin{cases} -1 & \Delta RSS_m < -GRss \\ 0 & -GRss \le \Delta RSS_m \le GRss \\ 1 & \Delta RSS_m > GRss \end{cases} \quad (4)$$

A standard assumption in the classical control theory is that the data transmission required by the control or state estimation algorithm can be performed with infinite precision. However, due to the growth in communication technology, it is becoming more common to employ digital limited capacity communication networks for exchange of information between system components.

**Fig. 6** Cellular space initial state diagram

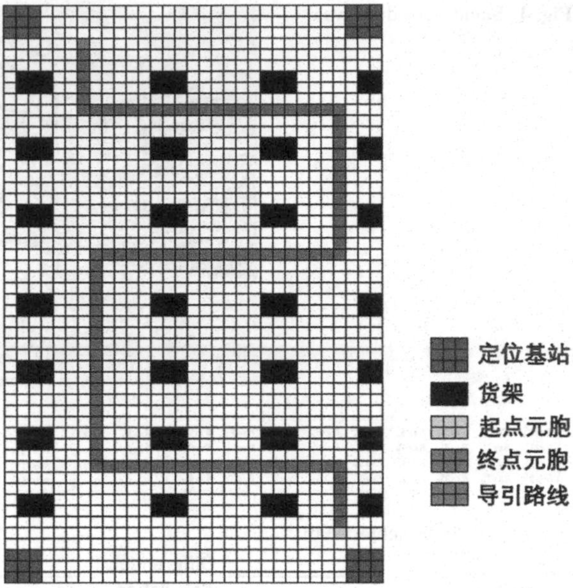


# 4 Simulation and Analysis

The purpose of the simulation is to verify the feasibility of the proposed method in indoor storage environment. Through the simulation of environmental changes and equipment aging, the effectiveness of the target method in suppressing system noise is verified. Under the same conditions compared with the fingerprint method and the trilateration method.

## *4.1 Parameter Settings*

(1) Set cell size to $r_b = r_l$=1 m;
(2) Set five kinds of cell states, HOLD, FORWARD, BACKWARD, LEFT, AND RIGHT;
(3) Definition of signal intensity variation discrete threshold GRss = 3 dB;
(4) Set the initial state of cell evolution as shown in Fig. 6, in which the black cell represents the shelf and the red cell represents the array antenna of the positioning base station;
(5) The mobile route is arbitrarily set in advance and is composed of 106 cells, as shown by blue cells in Fig. 6.

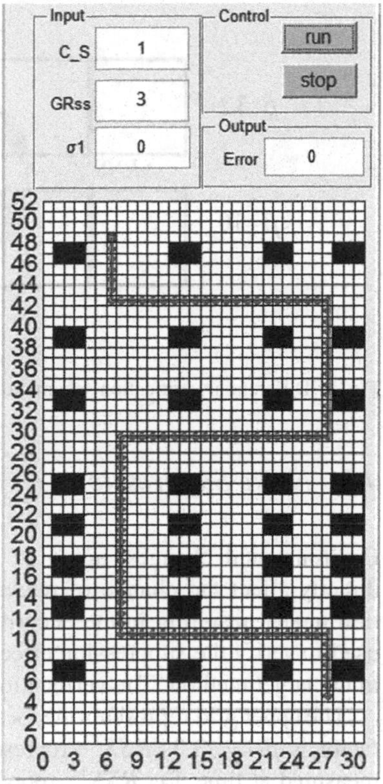


**Fig. 7** Noise-free position trajectory

## 4.2 *Feasibility*

The wireless ranging value used in the simulation is generated according to the wireless signal propagation empirical formula (1). According to the warehouse environment, set the parameters as $P_0 = 16.7$ (dBm), $\beta = 3.2$, $d_0 = 1$ m, $\sigma_n = 0$, Calculate the variation of the four base station signal strengths received at 106 preset cells and their NC positions. Running the CALoc method to obtain its simulation position trajectory is shown in Fig. 7.

The simulation trajectory (red star line) exactly matches the preset route (blue solid line), which verifies the validity of the CALoc algorithm for indoor positioning and the rationality of $\Delta RSS$ as the cell state parameter. According the formula (1), the change of the AGV position directly leads to the $\Delta RSS$ change. The CALoc method incorporates the CA evolution rule and obtains the $t + 1$ position coordinates. Then the CA continues to iterate to achieve the real-time tracking of the target trajectory.

This reveals the essence of the evolutionary rules—changes in RSS resulting from changes in the current position, and the combination of GRss discretizations determines the evolutionary outcome of the next moment. The above accurate positioning of dynamic objects is reasonable and effective.

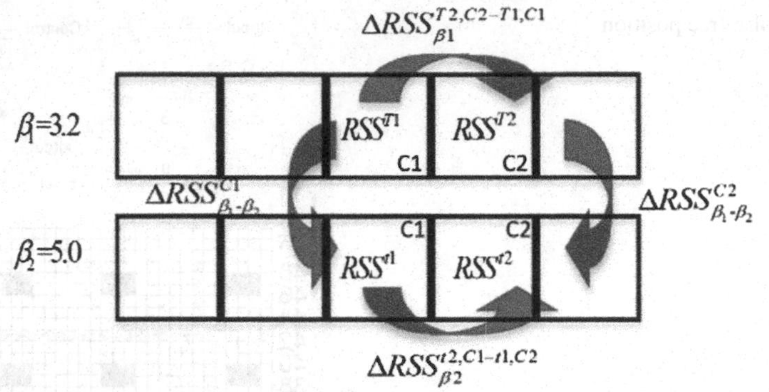


**Fig. 8** Differential verification schematic

## *4.3 Noise Suppression*

Environmental changes, aging of equipment, and obstruction of obstacles will directly affect the change of relevant parameters in the wireless signal propagation model, resulting in the occurrence of positioning errors. The simulation was used to verify the suppression effect of differential calculation on the system error in the target method. The verification scheme is shown in Fig. 8.

The upper grid represents the local cells intercepted when the channel attenuation factor is $\beta_1 = 3.2$, $RSS^{T1}$ represents the signal strength of the central cell C1 measured at time T1, $RSS^{T2}$ represents the signal strength of the central cell C2 measured at time T2. The amount of change in signal intensity from cell C1 to C2 from time T2 to time T2 is represented $\Delta RSS_{\beta 1}^{T2,C2-T1,C1}$. The same as the upper grid, lower grid represents the channel attenuation factor $\beta_1 = 5.0$, the corresponding parameter represents the same meaning as above. Also, $\Delta RSS_{\beta_1-\beta_2}^{C1}$ represents the amount of $\Delta RSS$ change under different environmental factors of the same C1 cell position, and $\Delta RSS_{\beta_1-\beta_2}^{C2}$ represents the signal intensity difference of the same position of the C2 cell position under different environmental factors. Set the cell evolution cycle of $T_2 - T_1 = t_2 - t_1 = 1$ to move the distance of 1 cell in the same direction.

The upper grid represents the local cells intercepted when the channel attenuation factor is $\beta_1$=3.2, $RSS^{T1}$ represents the signal strength of the central cell C1 measured at time T1, $RSS^{T2}$ represents the signal strength of the central cell C2 measured at time T2. The amount of change in signal intensity from cell C1 to C2 from time T2 to time T2 is represented $\Delta RSS_{\beta 1}^{T2,C2-T1,C1}$. The same as the upper grid, lower grid represents the channel attenuation factor $\beta_1$= 5.0, the corresponding parameter represents the same meaning as above. Also, $\Delta RSS_{\beta_1-\beta_2}^{C1}$ represents the amount of $\Delta RSS$ change under different environmental factors of the same C1 cell position, and $\Delta RSS_{\beta_1-\beta_2}^{C2}$ represents the signal intensity difference of the same position of the C2 cell position under different environmental factors. Set the cell evolution cycle of $T_2 - T_1 = t_2 - t_1 = 1$ to move the distance of 1 cell in the same direction.

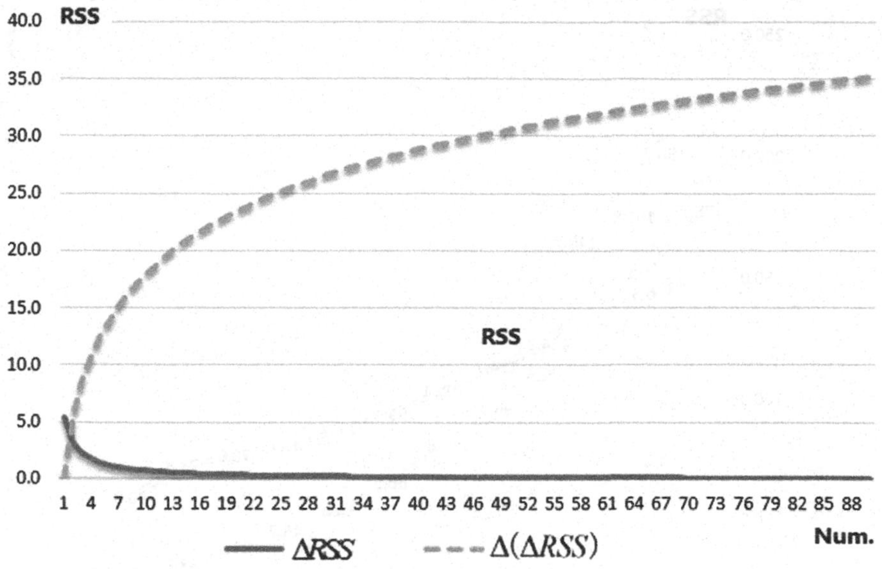


**Fig. 9** Comparison of RSS and $\Delta RSS$ under different $\beta$

The solid line in Fig. 9 indicates the difference between the signal intensity variations $\Delta RSS_{\beta 1}^{T2,C2-T1,C1}$ and $\Delta RSS_{\beta 2}^{t2,C1-t1,C2}$ of adjacent cells under the same conditions, and it is seen that it is much smaller than the difference between $\Delta RSS_{\beta_1-\beta_2}^{C1}$ and $\Delta RSS_{\beta_1-\beta_2}^{C2}$. Therefore, instead of directly using the wireless ranging value RSS to perform the positioning calculation using the variation of the signal strength, i.e., the difference calculation result, the positioning error caused by the environmental factors can be effectively suppressed.

Figure 10 the wireless signal propagation curve under the same environmental conditions before and after the aging of the equipment. Only the standard signal strength coefficient $P_0$ of the two curves is changed. However, since the $\Delta RSS$ attenuation trend is exactly the same, the change trend remains unchanged. Therefore, the RSS difference calculation is used. As a result, positioning calculations instead of RSS can remove the errors of the common mode caused by the device's own factors.

## 5 Conclusion

In this paper, an AGV wireless location method based on cellular automata is proposed to solve the problem of positioning error, which is due to factors such as changing of temperature or humidity and equipment aging. First, the cellular automata are introduced into the AGV positioning method due to the complexity of the wireless indoor location problem, then CA is introduced into the indoor positioning from

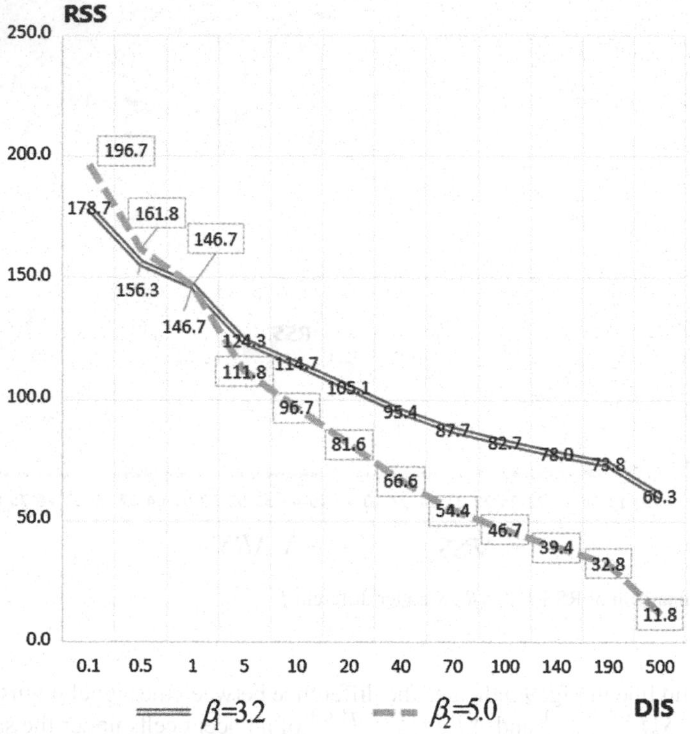


**Fig. 10** Comparison of propagation characteristics

the perspective of solving complexity problem. Then the mapping of indoor positioning space to cell space is established, and the spatiotemporal correlation and difference calculations are integrated into the cell evolution rules. Positioning error is constrained in the process of cell evolution. Finally, the target method is modeled and simulated to verify and analyze the feasibility and the effectiveness of mitigating errors caused by environmental noise. It is proved that the method can achieve high precision positioning at low cost and has a good application prospect. Further extensive warehouse verification will become the focus of the next stage of work.

A standard assumption in the classical control theory is that the data transmission required by the control or state estimation algorithm can be performed with infinite precision. However, due to the growth in communication technology, it is becoming more common to employ digital limited capacity communication networks for exchange of information between system components.

## References

1. I.F.A. Vis, Survey of research in the design and control of automated guided vehicle systems. Eur. J. Oper. Res. **170**(3), 677–709 (2006)
2. H. Zhu, D. Chen, Study and implementation of infrared digital guiding on AGV, in *Proceedings of the, World Congress on Intelligent Control and Automation, 2002*, vol. 4 (IEEE, 2002), pp. 3224–3227
3. S. Berman, Y. Edan, M. Jamshidi, Navigation of decentralized autonomous automatic guided vehicles in material handling. IEEE Trans. Robot. Autom. **19**(4), 743–749 (2003)
4. H. Martínez-Barber, D. Herrero-Perez, et al., Autonomous navigation of an automated guided vehicle in industrial environments. Robot. Comput. Integr. Manuf. **26**(4), 296–311 (2010)
5. X. Tan, G. Dai, S.O. Business, et al., Research on magnetic navigation sensor applied in logistics AGV[J]. Logistics Sci-Tech (2017)
6. Z. Song, X. Wu, T. Xu, et al., A new method of AGV navigation based on Kalman filter and a magnetic nail localization, in *IEEE International Conference on Robotics and Biomimetics* (IEEE, 2017), pp. 952–957
7. J. Borenstein, The OmniMate: a guidewire- and beacon-free AGV for highly reconfigurable applications. Int. J. Prod. Res. **38**(9), 1993–2010 (2000)
8. M. Jung, J.B. Song, Graph SLAM for AGV using geometrical arrangement based on lamp and SURF features in a factory environment, in *International Conference on Control, Automation and Systems* (IEEE, 2017), pp. 844–848
9. Y. Kizu, S. Kuchii, Research and development of the trackless type AGV by the image analysis (2015). https://doi.org/10.1299/jsmermd.2015_2A1-V01_1
10. S.J. Guo, T. Luo, Q. Tai-Ping, Positioning of AGV based on wireless sensor network. Logist. Technol. (2010)
11. I.S. Lee, J.Y. Kim, J.H. Lee, et al., Kalman filter-based sensor fusion for improving localization of AGV. Adv. Mater. Res. **488–489**, 1818–1822 (2012)
12. Y. Jia, Robust control with decoupling performance for steering and traction of 4WS vehicles under velocity-varying motion. IEEE Trans. Control Syst. Technol. **8**(3), 554–569 (2000)
13. J. Tian, B. Jia, S. Ma, et al., Cellular automaton model with dynamical 2D speed-gap relation. Trans. Sci. 2017(3) (2017)
14. V.J. Blue, J.L. Adler, Cellular automata microsimulation for modeling bi-directional pedestrian walkways. Trans. Res. Part B Methodol. **35**(3), 293–312 (2001)
15. Y. Jia, Alternative proofs for improved LMI representations for the analysis and the design of continuous-time systems with polytopic type uncertainty: a predictive approach. IEEE Trans. Autom. Control **48**(8), 1413–1416 (2003)
16. Z.H. Chang, W. Wang, Based on BD and QR for research and design of AGV. J. Qingdao Univ. (2014)

# Optimal Sensor Configuration for Three-Dimensional Range-Only Target Localization

**Yueqian Liang**

**Abstract** Optimal sensor configuration for range-only target localization in three-dimensional (3D) space is investigated in this paper. Based on the fact that to achieve more accurate localization, larger amount of information should be gathered, the maximization of the determinant of Fisher information matrix (FIM) is chosen as the optimality criterion. And by regarding the determinants as continuous polynomial functions of multiple formal variables, the optimal geometric configuration is systematically discussed.

**Keywords** Optimal sensor configuration · Sensor network · Target localization Fisher information · Cramér-Rao lower bound (CRLB)

## 1 Introduction

Target localization, determining the position (or, a generalized state) of a friendly or hostile target, covers a broad set of applications in both military and civilian areas [1, 2]. An inherent requirement of target localization is the localization accuracy. External sensors, which can provide a variety of measurements, such as range [3, 4], angle-of-arrival (AOA) [3, 5], time-of-arrival (TOA) [3, 6], received signal strength (RSS) [7, 8], etc., are usually employed to achieve target localization.

Three main strategies can usually be used to improve the localization performance. The first one is to let the sensor move along a designed path [9–11], which can be well guaranteed by some advanced control strategies (See Refs. [12–15] for some examples). The second one is to exploit multiple sensors [11, 16–18]. And the last one is the combination of the former two strategies, i.e., making multiple sensors moving along their respective designed paths [11, 17, 18]. The essence of these strategies is to acquire more information for localization usage.

Y. Liang (✉)
China Academy of Electronics and Information Technology,
Beijing 100041, China
e-mail: liangyueqian3002@126.com



Y. Jia et al. (eds.), *Proceedings of 2018 Chinese Intelligent Systems Conference*,
Lecture Notes in Electrical Engineering 528,
https://doi.org/10.1007/978-981-13-2288-4_60

When multiple sensors are available, it turns out that their geometric configuration with respect to the target significantly influences the localization performance [3, 8, 19–21]. In [3, 8], the maximization of the determinant of the Fisher information matrix (FIM), which is equivalent to minimize the lower bound of any unbiased estimator, i.e., the well-known Cramér-Rao lower bound (CRLB), is chosen by Bishop et al. as the optimality criterion to systematically identify the optimal configuration of multiple homogeneous sensors of the above-mentioned types for two-dimensional (2D) target localization. Some interesting results were derived, equal angular spacing configuration is usually a special optimal configuration for homogeneous sensors, and there are infinite number of optimal configurations when the sensor number exceeds a certain value. Martínez et al. gave out the optimal configuration result for range-only sensors in both 2D and three-dimensional (3D) spaces, and also discussed a technique to maintain the obtained configuration in Ref. [19]. In [20, 21], the optimal configuration of multiple heterogeneous sensors was studied.

Motivated by the current situation that the existing results are mainly on the 2D space, and few are based on 3D space, we consider the problem of 3D optimal target localization using multiple range-only sensors in this paper. The specific problem is formulated in the second section. And the optimal configuration of the range-only sensors for 3D target localization is discussed in detail in the third section, followed by a short conclusion in the last section.

## 2 Problem Formulation

### 2.1 Measurement Model

An appropriate inertial frame XYZ with origin **O** is firstly chosen. The inertial location of the target is assumed to be $\mathbf{p}_t = (x_t, y_t, z_t)^T$. $N \geq 2$ range-only sensors are employed, with their inertial locations $\mathbf{p}_i = (x_i, y_i, z_i)^T$ for $i = 1, 2, \ldots, N$.

The range between sensor $i$ and the target is denoted as $r_i = \|\mathbf{p}_{it}\| = \|\mathbf{p}_i - \mathbf{p}_t\|$, where $\|\cdot\|$ is the 2-norm operator and $\mathbf{p}_{it}$ is the line-of-sight from the target to sensor $i$. We use $d_i = \sqrt{(y_i - y_t)^2 + (x_i - x_t)^2}$ to denote the horizontal range from sensor $i$ to the target, i.e., the length of the projection of the line-of-sight $\mathbf{p}_{it}$ onto the XY plane. The relative azimuth angle and elevation angle of sensor $i$ with respect to the target are indicated as $\phi_i = \arctan2(y_i - y_t, x_i - x_t) \in [0, 2\pi)$ and $\theta_i = \arctan2(z_i - z_t, d_i) \in [-\pi/2, \pi/2]$ respectively. Here $\arctan2(\cdot)$ is the four-quadrant inverse tangent function. It can be seen that the three-tuple $\{r_i, \phi_i, \theta_i\}$ constitutes the relative spherical coordinates of sensor $i$ with respect to the target. Additionally $\{r_i, \phi_i, \theta_i\}_{i=1}^N$ can be regarded as the sensor configuration with respect to the target.

The noisy range measurements for sensor $i$ with $i$ ranging from 1 to $N$ are modeled as follows, and as usual the noises between different sensors are assumed to be independent identically distributed.

$$\hat{r}_i(\mathbf{p}_t) = r_i(\mathbf{p}_t) + \varepsilon_{r,i} \tag{1}$$

Here $\varepsilon_{r,i} \sim \mathcal{N}(0, \sigma_r^2)$ denotes the additive Gaussian distributed noise of the measured range with zero mean and standard deviation $\sigma_r$.

## 2.2 *Optimality Metric*

The CRLB is known to represent the best estimation of an unbiased estimator. That is, for a general vector $\mathbf{x}$ and its any unbiased estimation $\hat{\mathbf{x}}$ from the observable noisy measurement vector $\hat{\mathbf{z}} = \mathbf{z}(\mathbf{x}) + \varepsilon$, we have

$$\mathbb{E}\left[(\hat{\mathbf{x}} - \mathbf{x})(\hat{\mathbf{x}} - \mathbf{x})^T\right] \geq \mathcal{C}(\mathbf{x}) \tag{2}$$

where $\mathbb{E}$ is the expectation operator, and $\mathcal{C}(\mathbf{x})$ is the CRLB of $\mathbf{x}$, which can be determined from the FIM $\mathcal{I}(\mathbf{x})$ as $\mathcal{C}(\mathbf{x}) = \mathcal{I}(\mathbf{x})^{-1}$ if $\mathcal{I}(\mathbf{x})$ is non-singular. If $\mathcal{I}(\mathbf{x})$ is singular, then no unbiased estimator for $\mathbf{x}$ exists with a finite variance. The $(i, j)$-th element of $\mathcal{I}(\mathbf{x})$ can be calculated as [3]

$$(\mathcal{I}(\mathbf{x}))_{i,j} = \mathbb{E}\left[\frac{\partial}{\partial x_i} \log\left(f_{\hat{\mathbf{z}}}(\hat{\mathbf{z}}; \mathbf{x})\right) \frac{\partial}{\partial x_j} \log\left(f_{\hat{\mathbf{z}}}(\hat{\mathbf{z}}; \mathbf{x})\right)\right] \tag{3}$$

where $f_{\hat{\mathbf{z}}}(\hat{\mathbf{z}}; \mathbf{x})$ is the likelihood function of $\mathbf{x}$ evaluating at the measurement $\hat{\mathbf{z}}$. Under the assumptions that (i) the noise $\varepsilon$ is Gaussian distributed, (ii) the noise covariance is independent of the unobservable vector $\mathbf{x}$, and (iii) the noises of the $N$ sensors are independently distributed, we have [3]

$$\mathcal{I}(\mathbf{x}) = \sum_{i=1}^{N} (\nabla_{\mathbf{x}}\, \mathbf{z}_i(\mathbf{x}))^T\ \mathbf{R}_{\varepsilon,i}^{-1}\ (\nabla_{\mathbf{x}}\, \mathbf{z}_i(\mathbf{x})) \tag{4}$$

where $\nabla_{\mathbf{x}}\, \mathbf{z}_i(\mathbf{x})$ is the Jacobian matrix of the measurement function of sensor $i$ with respect to its corresponding variable $\mathbf{x}$, and $\mathbf{R}_{\varepsilon,i}$ is the variance or covariance matrix of the measurement noise of sensor $i$.

In general the "larger" the FIM $\mathcal{I}(\mathbf{x})$ is, the larger amount of information about the unobservable vector $\mathbf{x}$ the measurement $\hat{\mathbf{z}}$ carries, and this further implies more accurate target localization result theoretically. The magnitude of $\mathcal{I}(\mathbf{x})$ can usually be measured by its determinant, $\det(\mathcal{I}(\mathbf{x}))$, which is also adopted herein.

Then the problem of seeking optimal sensor configuration for 3D target localization can be summarized as follows: To find a certain sensor configuration $\{r_i, \phi_i, \theta_i\}_{i=1}^N$ for the sensors to maximize $\det(\mathcal{I}(\mathbf{x}))$, i.e.,

$$\max_{\{r_i, \phi_i, \theta_i\}_{i=1}^N} \det(\mathcal{I}(\mathbf{x})) \tag{5}$$

subject to the following constraints

$$r_i > 0,\ \phi_i \in [0, 2\pi),\ \text{and}\ \theta_i \in [-\pi/2, \pi/2] \tag{6}$$

## 3 Optimal Sensor Configuration

In this section we discuss the optimal configuration for the rang-only sensor target localization. Hereinafter if it is not explicitly specified, $\sum$ is used to represent the summation from $i = 1$ to $i = N$.

The main result is concluded in the following theorem.

**Theorem 1** *Consider that N identical range-only sensors are used to localize the target. The maximum amount of information (representing by the determinant of the FIM), which is $N^3/(27\sigma_r^6)$, can be achieved if the following condition holds*

$$X_1 = N/3,\ X_2 = X_3 = X_4 = X_5 = 0 \tag{7}$$

*where the formal variables are given by*

$$X_1 = \sum \cos 2\theta_i \tag{8}$$

$$X_2 = \sum \cos 2\phi_i (1 + \cos 2\theta_i) \tag{9}$$

$$X_3 = \sum \sin 2\phi_i (1 + \cos 2\theta_i) \tag{10}$$

$$X_4 = \sum \cos \phi_i \sin 2\theta_i \tag{11}$$

$$X_5 = \sum \sin \phi_i \sin 2\theta_i \tag{12}$$

*Proof* From the range measurement Eq. (1) we have

$$\nabla_{\mathbf{x}} \mathbf{z}_i(\mathbf{x}) = \begin{bmatrix} \cos\phi_i \cos\theta_i & \sin\phi_i \cos\theta_i & \sin\theta_i \end{bmatrix} \tag{13}$$

$$\mathbf{R}_{\varepsilon,i} = \sigma_r^2 \tag{14}$$

Substituting (13) and (14) into (4) and simplifying it yield

$$\mathcal{I}(\mathbf{x}) = \frac{1}{4\sigma_r^2} \sum \begin{bmatrix} (1+\cos 2\phi_i)(1+\cos 2\theta_i) & \sin 2\phi_i (1+\cos 2\theta_i) & 2\cos\phi_i \sin 2\theta_i \\ \sin 2\phi_i (1+\cos 2\theta_i) & (1-\cos 2\phi_i)(1+\cos 2\theta_i) & 2\sin\phi_i \sin 2\theta_i \\ 2\cos\phi_i \sin 2\theta_i & 2\sin\phi_i \sin 2\theta_i & 2(1-\cos 2\theta_i) \end{bmatrix} \tag{15}$$

Taking the determinant of (15) and simplifying it, we obtain

$$\det(\mathcal{I}(\mathbf{x})) = \frac{1}{32\sigma_r^6}\{N^3 + N^2X_1 - N\left[X_1^2 + 2X_4^2 + 2X_5^2 + X_2^2 + X_3^2\right] \\ -X_1\left[X_1^2 + 2X_4^2 + 2X_5^2 - X_2^2 - X_3^2\right] + 2X_2\left(X_4^2 - X_5^2\right) + 4X_3X_4X_5\} \tag{16}$$

For a fixed $N$, it can be seen that $X_1, X_4$ and $X_5$ are in $[-N, N]$, while $X_2$ and $X_3$ are in $[-2N, 2N]$. Hence $\det(\mathcal{I}(\mathbf{x}))$ can be regarded as a 5-variable continuous function (actually a multivariate polynomial function) with respect to $\{X_1, X_2, X_3, X_4, X_5\}$ defining at the closed region $[-N, N] \times [-2N, 2N] \times [-2N, 2N] \times [-N, N] \times [-N, N]$. According to the extrema theory of multivariate continuous function, one necessary condition for $\max\{\det(\mathcal{I}(\mathbf{x}))\}$ is that the partial derivatives of the function with respect to all the variables vanish, which leads to

$$\begin{cases} N^2 - 2NX_1 - (3X_1^2 + 2X_4^2 + 2X_5^2 - X_2^2 - X_3^2) = 0 \\ -NX_2 + X_1X_2 + X_4^2 - X_5^2 = 0 \\ -NX_3 + X_1X_3 + 2X_4X_5 = 0 \\ -NX_4 - X_1X_4 + X_2X_4 + X_3X_5 = 0 \\ -NX_5 - X_1X_5 - X_2X_5 + X_3X_4 = 0 \end{cases} \tag{17}$$

Solving (17) yields nine groups of extreme points which are listed in Table 1.

Compared the values of (16) at these extreme points with those at the vertices of the closed region, we find that the fourth extreme point is the maxima, i.e., when $X_1 = N/3$ and $X_2 = X_3 = X_4 = X_5 = 0$, $\det(\mathcal{I}(\mathbf{x}))$ is maximized, and its maximum value is $N^3/(27\sigma_r^6)$. ■

To localize the target, at least 3 (stationary) range-only sensors should be used. Letting $N = 3$, consider an approximate solution for (7), $\phi_1 = 2.8278$, $\phi_2 = 0.2433$, $\phi_3 = 1.5279$, $\theta_1 = -0.7356$, $\theta_2 = -0.7605$ and $\theta_3 = 0.2869$. Making four of these angles fixed and the other two vary, we plot the contours of the (normalized) determinant with respect to the two varying angles. The results are illustrated in Fig. 1, from which we can see that except the given optimal solution, a second solution exists for each considered case.

Considering $N = 4$ and an approximate solution for (7), $\phi_1 = 5.6359$, $\phi_2 = 3.9845$, $\phi_3 = 5.3571$, $\phi_4 = 0.6051$, $\theta_1 = -0.7467$, $\theta_2 = 0.3772$, $\theta_3 = 0.4581$ and $\theta_4 = 0.8313$, varying $\{\phi_4, \theta_4\}$ and $\{\phi_2, \phi_3\}$, we have the contour plots given in Fig. 2. Similar phenomenon as for the case $N = 3$ can be observed.

**Corollary 1** *Given a certain optimal range-only sensor configuration $\{r_i, \phi_i, \theta_i\}_{i=1}^N$, the following actions do not change its optimality: (i) Changing the range $r_i$ along the line-of-sight $\mathbf{p}_{it}$ for some or all $i \in \{1, 2, \ldots, N\}$; (ii) Changing the sign of the elevation angle $\theta_i$ (i.e., $\theta_i$ to $-\theta_i$) and simultaneously the azimuth angle $\phi_i$ to $\pi + \phi_i$ for some or all $i \in \{1, 2, \ldots, N\}$; and (iii) Changing the signs of all the elevation angles.*

**Table 1** Extreme points

| No. | $X_1$ | $X_2$ | $X_3$ | $X_4$ | $X_5$ |
|---|---|---|---|---|---|
| 1 | $-N$ | 0 | 0 | 0 | 0 |
| 2 | $\neq \pm N$ | $-N - X_1$ | 0 | 0 | $\pm\sqrt{N^2 - X_1^2}$ |
| 3 | $\in [-N, N]$ | $\neq 0$ | $\pm\sqrt{(N + X_1)^2 - X_2^2}$ | $\pm\frac{\sqrt{(2N+X_2)^2-(2X_1+X_2)^2}}{2}$ | $\frac{X_3(N-X_1)}{X_4}$ |
| 4 | $\frac{N}{3}$ | 0 | 0 | 0 | 0 |
| 5 | $\frac{N}{3}$ | 0 | $\pm\frac{4N}{3}$ | $\pm\frac{2N}{3}$ | $\frac{2N}{3}$ |
| 6 | $N$ | 0 | $\pm 2N$ | 0 | 0 |
| 7 | $N$ | $\pm 2N$ | 0 | 0 | 0 |
| 8 | $N$ | $\neq 0$ | $\pm\sqrt{4N^2 - X_2^2}$ | 0 | 0 |
| 9 | $\neq \frac{N}{3}$ | 0 | $\pm\sqrt{(N + X_1)^2}$ | $\pm\sqrt{\frac{N^2-X_1^2}{2}}$ | $\frac{(N-X_1)X_3}{2X_4}$ |

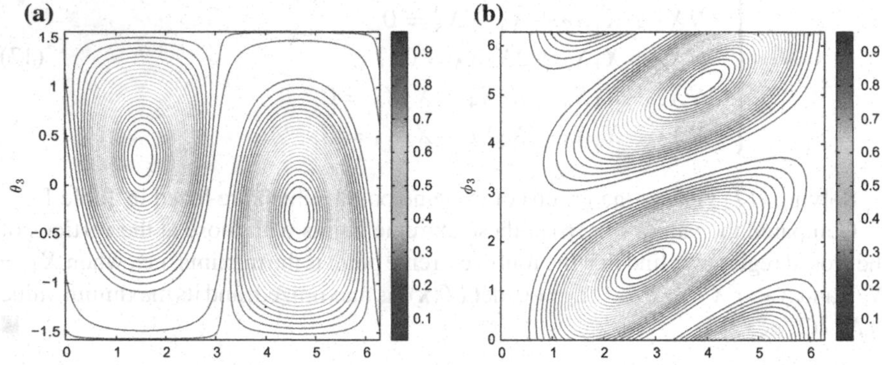


**Fig. 1** Contour plots of the (normalized) determinant with 3 range-only sensors with respect to: **a** $\phi_3$ and $\theta_3$; **b** $\phi_1$ and $\phi_3$

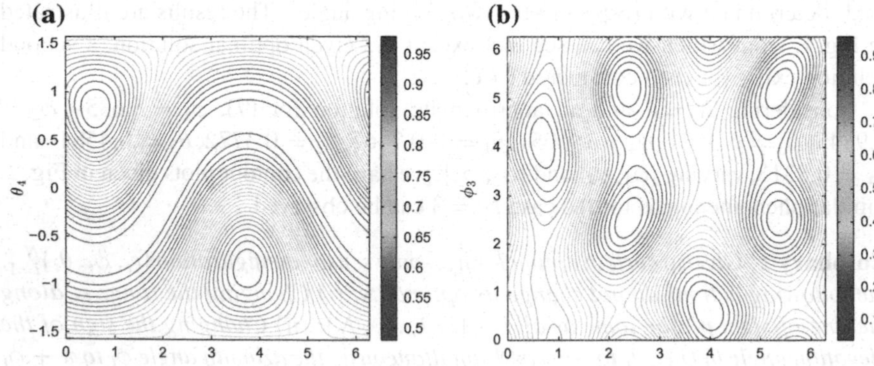


**Fig. 2** Contour plots of the (normalized) determinant with 4 range-only sensors with respect to: **a** $\phi_4$ and $\theta_4$; **b** $\phi_2$ and $\phi_3$

*Proof* Since condition (7) is independent of $r_i$, its change does not affect the original optimality.

Replacing $\theta_i$ and $\phi_i$ by $-\theta_i$ and $\pi + \phi_i$ respectively in (8)–(12), it can be found that they remain unchanged. Therefore, (ii) does not change the original optimality either.

Noticing that $\cos(-2\theta_i) = \cos 2\theta_i$ and $\sin(-2\theta_i) = -\sin 2\theta_i$, (iii) can be easily concluded. ∎

It can be seen that the additional solutions shown in Figs. 1a and 2a are just the case described in (ii) of Corollary 1.

**Corollary 2** *One particular optimal range-only sensor configuration* $\{r_i, \phi_i, \theta_i\}_{i=1}^N$ *is such that (i) all the sensors share the same elevation angle, and in specific* $\theta_i = \pm\frac{1}{2}\arccos\frac{1}{3}$ *for all* $i = 1, 2, \ldots, N$, *and (ii) the angle subtended at the target by each two azimuth-angularly adjacent sensors is* $2\pi/N$, *i.e., with no loss of generality,* $\phi_{j+1} = \phi_1 + 2\pi j/N$ *for* $j = 1, 2, \ldots, N-1$.

*Proof* Consider the case when all the sensors have the same elevation angles with respect to the target, i.e., let $\theta_i = \theta$ for all $i = 1, 2, \ldots, N$. Then from $X_1 = N/3$, we can easily have $\theta = \pm\frac{1}{2}\arccos\frac{1}{3}$.

In this case we can see from the remaining equalities in (7) that

$$\sum \cos 2\phi_i = \sum \sin 2\phi_i = \sum \cos \phi_i = \sum \sin \phi_i = 0 \qquad (18)$$

Then from the following facts

$$\sum \cos\left[2\left(\phi_0 + \frac{2(i-1)\pi}{N}\right)\right] = \sum \sin\left[2\left(\phi_0 + \frac{2(i-1)\pi}{N}\right)\right] = 0 \qquad (19)$$

$$\sum \cos\left[2\left(\phi_0 + \frac{(i-1)\pi}{N}\right)\right] = \sum \sin\left[2\left(\phi_0 + \frac{(i-1)\pi}{N}\right)\right] = 0 \qquad (20)$$

(ii) can be easily obtained. ∎

Figure 3 shows us this particular optimal configuration when 3 or 4 range-only sensors are used. It can be seen that both of the two contours are symmetric about the northeast direction, since the two varying azimuth angles play the same role in the determinant of the FIM.

## 4 Conclusions

The 3D optimal geometric configuration for range-only sensors has been studied systematically in this paper. The determinant of the FIM is treated as a continuous multivariate polynomial function, and the optimal configuration conditions are explicitly given out by solving the maximization problem. It is shown that the optimal

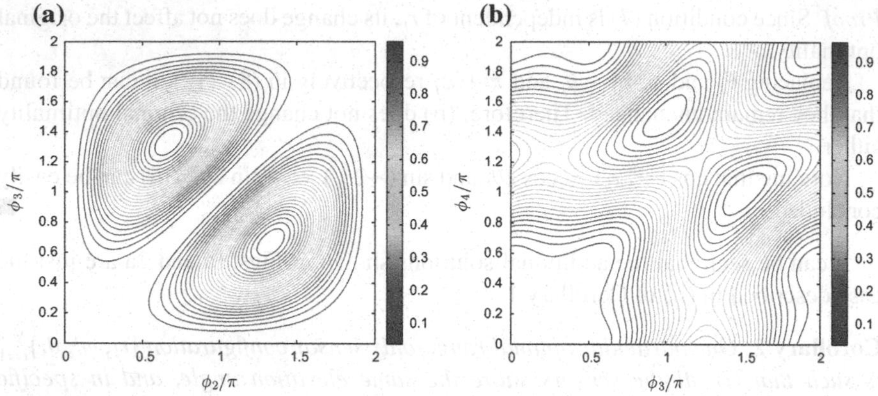


**Fig. 3** Contour plots of the (normalized) determinant of the particular optimal configuration in Corollary 2 with: **a** $N = 3$, and $\phi_1 = 0$; **b** $N = 4$, $\phi_1 = 0$ and $\phi_2 = \pi/2$

configuration is independent of the sensor-target ranges, and some particular optimal configurations have also been discussed.

## References


1. J.R. Lowell, Military applications of localization, tracking, and targeting. IEEE Wirel. Commun. **18**(2), 60–65 (2011)
2. J. Rantakokko, P. Händel, M. Fredholm, F. Marsten-Eklöf, User requirements for localization and tracking technology: a survey of mission-specific needs and constraints, in *Proceedings of International Conference on Indoor Positioning and Indoor Navigation (IPIN)*, Zürich, Switzerland, September 2010, pp. 1–9
3. A.N. Bishop, B. Fidan, B.D. Anderson, K. Doğançay, P.N. Pathirana, Optimality analysis of sensor-target localization geometries. Automatica **46**(3), 479–492 (2010)
4. G. Marani, S.K. Choi, Underwater target localization: autonomous intervention with the DIDSON sonar in SAUVIM. IEEE Robot. Autom. Mag. **17**(1), 64–70 (2010)
5. J. Zhong, Z. Lin, Z. Chen, W. Xu, Cooperative localization using angle-of-arrival information, in *Proceedings of 2014 11th IEEE International Conference on Control & Automation (ICCA)*, Taichung, Taiwan, June 2014, pp. 19–24
6. E. Xu, Z. Ding, S. Dasgupta, Source localization in wireless sensor networks from signal time-of-arrival measurements. IEEE Trans. Signal Process. **59**(6), 2887–2897 (2011)
7. X. Li, Collaborative localization with received-signal strength in wireless sensor networks. IEEE Trans. Veh. Technol. **56**(6), 3807–3817 (2007)
8. A.N. Bishop, P. Jensfelt, An optimality analysis of sensor-target geometries for signal strength based localization, in *Proceedings of the 3rd International Conference on Intelligent Sensors, Sensor Networks, and Information Processing*, Melbourne, Australia, December 2009, pp. 127–132
9. Y. Oshman, P. Davidson, Optimization of observer trajectories for bearings-only target localization. IEEE Trans. Aerosp. Electron. Syst. **35**(3), 892–902 (1999)
10. M.L. Hernandez, Optimal sensor trajectories in bearings-only tracking, in *Proceedings of the 7th International Conference on Information Fusion*, Stockholm, Sweden, 28 June–1 July 2004, pp. 893–900

11. K. Doğançay, Single- and multi-platform constrained sensor path optimization for angle-of-arrival target tracking, in *Proceedings of the 18th European Signal Processing Conference (EUSIPCO-2010)*, Aalborg, Denmark, August 2010, pp. 835–839
12. Y. Jia, Robust control with decoupling performance for steering and traction of 4WS vehicles under velocity-varying motion. IEEE Trans. Control Syst. Technol. **8**(3), 554–569 (2000)
13. Y. Jia, Alternative proofs for improved LMI representations for the analysis and the design of continuous-time systems with polytopic type uncertainty: a predictive approach. IEEE Trans. Autom. Control **48**(8), 1413–1416 (2003)
14. A.P. Aguiar, J.P. Hespanha, Trajectory-tracking and path-following of underactuated autonomous vehicles with parametric modeling uncertainty. IEEE Trans. Autom. Control **52**(8), 1362–1379 (2007)
15. P.B. Sujit, S. Saripalli, J.B. Sousa, Unmanned aerial vehicle path following: a survey and analysis of algorithms for fixed-wing unmanned aerial vehicles. IEEE Control Syst. Mag. **42**(1), 42–59 (2014)
16. S.R. Sukumar, H. Bozdogan, D.L. Page, A.F. Koschan, M.A. Abidi, Uncertainty minimization in multi-sensor localization systems using model selection theory, in *Proceedings of the 19th International Conference on Pattern Recognition*, Tampa, FL, USA, December 2008, pp. 1–4
17. S.R. Semper, J.L. Crassidis, Decentralized geolocation and optimal path planning using limited UAVs, in *Proceedings of the 12th International Conference on Information Fusion*, Seattle, WA, USA, July 2009, pp. 355–362
18. K. Doğançay, H. Hmam, S.P. Drake, A. Finn, Centralized path planning for unmanned aerial vehicles with a heterogeneous mix of sensors, in *Proceedings of the 3rd International Conference on Intelligent Sensors, Sensor Networks, and Information Processing*, Melbourne, Australia, December 2009, pp. 91–96
19. S. Martínez, F. Bullo, Optimal sensor placement and motion coordination for target tracking. Automatica **42**(4), 661–668 (2006)
20. A.N. Bishop, B. Fidan, K. Doğançay, B.D. Anderson, P.N. Pathirana, Exploiting geometry for improved hybrid AOA/TDOA based localization. Signal Process. **88**(4), 1775–1791 (2008)
21. K. Doğançay, UAV path planning for passive emitter localization. IEEE Trans. Aerosp. Electron. Syst. **48**(2), 1150–1166 (2012)

# Optimal Placement of Wireless Sensor Networks for 2-Dimensional Source Localization

**Yueqian Liang**

**Abstract** Optimal sensor network configuration for 2-Dimensional (2D) source localization is investigated systematically in this paper. The maximization of the determinant of Fisher information matrix (FIM) is chosen as the optimality criterion. Homogeneous range, received signal strength (RSS), time-of-arrival (TOA) and angle-of-arrival (AOA) sensor networks with different measurement noises are considered. The optimal configuration conditions for these four types of sensors are given out. Discussions based on these conditions are done to derive the optimal sensor configurations.

**Keywords** Optimal sensor configuration · Sensor network · Source localization
Fisher information

## 1 Introduction

Source localization, determining the position (or, a generalized state) of an emitting or a reflective source, covers a broad set of applications in both military and civilian areas [1, 2]. An inherent requirement of source localization is the localization accuracy. External sensors, which can provide a variety of measurements, such as range [3, 4], angle-of-arrival (AOA) [3, 5], time-of-arrival (TOA) (or equivalently, time-difference-of-arrival (TDOA)) [3, 6], received signal strength (RSS) [7, 8], etc., are usually employed to achieve accurate source localization.

Three main strategies can usually be used to improve the localization performance. The first one is to let the sensor move along a designed path [9, 10], which can be well guaranteed by some advanced control strategies (See Refs. [11–14] for some examples). The second one is to exploit multiple sensors to constitute a wireless sensor network (WSN) [10, 15, 16]. And the last one is the combination of the former two strategies, i.e., making multiple sensors moving along their respective designed

Y. Liang (✉)
China Academy of Electronics and Information Technology, Beijing 100041, China
e-mail: liangyueqian3002@126.com



Y. Jia et al. (eds.), *Proceedings of 2018 Chinese Intelligent Systems Conference*,
Lecture Notes in Electrical Engineering 528,
https://doi.org/10.1007/978-981-13-2288-4_61

paths [10, 16]. The essence of these strategies is to collect as much information as possible for localization usage.

When multiple sensors are available, it turns out that the sensor network configuration with respect to the source location can significantly influence the localization performance. Martínez et al. gave out the optimal configuration result for range-only sensors in both 2-dimensional (2D) and 3-dimensional (3D) spaces, and also presented a technique to maintain the obtained configuration in [17]. In [3, 8, 18, 19], the maximization of the determinant of the Fisher information matrix (FIM) (det(FIM)), which is equivalent to minimize the lower bound of any unbiased estimator, i.e., the well-known Cramér-Rao lower bound (CRLB), is chosen by the authors as the optimality criterion to systematically identify the optimal configuration of multiple homogeneous sensors of the above-mentioned types for 2D source localization. Some interesting results were derived therein. Recently Zhao et al. formulated the 2D and 3D optimal homogeneous sensor (including range-only, AOA-only, and RSS-only ones) placement problems as a unified parameter optimization problem and then solved it using the frame theory [20]. They concluded that det(FIM) is maximized when the unit line-of-sights from the source location to the sensor locations form a tight frame.

In this paper, we systematically consider the optimal sensor network configuration problem for 2D source localization using homogeneous range, RSS, TOA and AOA sensor networks with different measurement noise variances. Multiple formal variables are firstly chosen, which allows us to model the optimal configuration problem as the extreme problem of multivariate continuous functions [21]. Compared with the mostly-used traditional algebraic method, the proposed method need few complex algebraic simplifications. And compared with the newly-developed frame theory method, the proposed method can address the homogeneous TOA sensor deployment problem.

## 2 Problem Formulation

The source is assumed to locate at $\mathbf{p}_t = (x_t, y_t)^T$ in inertial XY frame. The sensor locations are denoted as $\mathbf{p}_i = (x_i, y_i)^T$ for $i = 1, 2, \ldots, N$. The distance between sensor $i$ and the source is denoted as $r_i = \|\mathbf{p}_{it}\| = \|\mathbf{p}_i - \mathbf{p}_t\| = \sqrt{(y_i - y_t)^2 + (x_i - x_t)^2}$, where $\|\cdot\|$ is the 2-norm operator and $\mathbf{p}_{it}$ is the line-of-sight from the source to sensor $i$. The relative azimuth angle of sensor $i$ with respect to the source is indicated as $\phi_i = \arctan2(y_i - y_t, x_i - x_t) \in [-\pi, \pi)$. Here $\arctan2(\cdot)$ is the four-quadrant inverse tangent function. Denote $\mathbf{g}_i = (g_{xi}, g_{yi})^T = (\cos\phi_i, \sin\phi_i)^T$ as the unit line-of-sight.

The noisy range, RSS, TOA, AOA measurements for sensor $i$ (with $i$ ranging from 1 to $N$) are modeled as follows, and as usual the noises between different sensors are assumed to be independently distributed. But different measurement noises for different sensors are allowed.

- Range measurement.

$$\hat{r}_i(\mathbf{p}_t) = r_i(\mathbf{p}_t) + \varepsilon_{r,i} \tag{1}$$

Here $\varepsilon_{r,i} \sim \mathcal{N}(0, \sigma_{r,i}^2)$ denotes the additive Gaussian distributed noise of the measured range with zero mean and standard deviation $\sigma_{r,i}$.

- RSS measurement.

$$\hat{\mathcal{R}}_i(\mathbf{p}_t) = \mathcal{R}_i(\mathbf{p}_t) + \varepsilon_{RSS,\,i} = \mathcal{R}_0 - 10\alpha_i \log_{10} r_i + \varepsilon_{RSS,\,i} \tag{2}$$

Here $\mathcal{R}_0$ is the RSS calculated at one unit reference distance using the free-space Friis model, and $\alpha_i$ is the assumed known relevant path loss exponent [7, 8]. And $\varepsilon_{RSS,\,i} \sim \mathcal{N}(0, \sigma_{RSS,i}^2)$ is the RSS measurement noise.

- TOA measurement.

$$\hat{\mathcal{T}}_i(\mathbf{x}_t) = \mathcal{T}_i(\mathbf{x}_t) + \varepsilon_{TOA,\,i} = \tau + \frac{r_i}{c_0} + \varepsilon_{TOA,\,i} \tag{3}$$

where $\mathbf{x}_t = (\mathbf{p}_t^T, \tau)^T$ denotes the source state, and $\tau$ is the signal emitting/reflecting time which is usually unknown to us. $c_0$ is the signal propagation speed. $\varepsilon_{TOA,\,i} \sim \mathcal{N}(0, \sigma_{TOA,\,i}^2)$ denotes the Gaussian noise involving in the TOA measurement.

- AOA measurement.

$$\hat{\mathcal{A}}_i(\mathbf{p}_t) = \mathcal{A}_i(\mathbf{p}_t) + \varepsilon_{AOA,\,i} = \phi_i + \varepsilon_{\phi,i} \tag{4}$$

where $\varepsilon_{\phi,i} \sim \mathcal{N}(0, \sigma_{\phi,i}^2)$ denotes the azimuth angle measurement noise. And $\varepsilon_{AOA,\,i} \sim \mathcal{N}(0, \sigma_{AOA,i}^2)$ denotes the Gaussian noise involving in the AOA measurement.

As mostly done in the literature, we choose the maximization of the determinant of the FIM as the optimality criterion.

**Table 1** Computation parameters of the four types of sensors

| Type | $c_i$ | $Y_1$ | $Y_2$ | $Y_3$ | $Y_4$ |
|---|---|---|---|---|---|
| Range | $1/\sigma_{r,i}^2$ | $\sum c_i \cos 2\phi_i$ | $\sum c_i \sin 2\phi_i$ | – | – |
| RSS | $(10\alpha_i/(r_i\sigma_{RSS,i} \log 10))^2$ | $\sum c_i \cos 2\phi_i$ | $\sum c_i \sin 2\phi_i$ | – | – |
| AOA | $1/(r_i\sigma_{\phi,i})^2$ | $\sum c_i \cos 2\phi_i$ | $\sum c_i \sin 2\phi_i$ | – | – |
| TOA | $1/\sigma_{TOA,i}^2$ | $\sum c_i \cos 2\phi_i$ | $\sum c_i \sin 2\phi_i$ | $\sum c_i \cos \phi_i$ | $\sum c_i \sin \phi_i$ |

## 3 Optimal Configuration Conditions

In this paper if it is not explicitly specified, we use $\sum$ to represent the summation from $i = 1$ to $i = N$.

We firstly compute the sensor characteristic coefficients (SCCs) ($c_i$'s) and define the formal variables ($Y_i$'s) as in Table 1. $\{c_i\}_{i=1}^N$ is rearranged to be non-increasing. We say that $\{c_i\}_{i=1}^N$ is regular if

$$c_1 \leq \sum_{i=2}^{N} c_i \tag{5}$$

and otherwise, we say that $\{c_i\}_{i=1}^N$ is irregular (with irregularity 1). We denote $C = \sum c_i$.

### 3.1 Range-Only, RSS-Only and AOA-Only Sensors

When homogeneous range, RSS and AOA sensors are exploited to achieve the source localization, their unified optimal configuration condition is given in the following theorem.

**Theorem 1** *For homogeneous range, RSS and AOA sensors, the maximum amount of information, $C^2/4$, can be achieved if and only if the following condition satisfies*

$$Y_1 = Y_2 = 0 \tag{6}$$

*which is equivalent to*

$$\sum c_i \mathbf{g}_i \mathbf{g}_i^T = \frac{C}{2} \mathbf{I}_2 \tag{7}$$

*where $\mathbf{I}_2$ is the 2D identity matrix.*

*Proof* For homogeneous range, RSS and AOA sensor networks, it can be easily found through similar computations that

$$\det(\mathcal{I}(\mathbf{x})) = \frac{1}{4}\left[C^2 - Y_1^2 - Y_2^2\right] \tag{8}$$

From this we can easily obtain the optimal sensor configuration condition (6) and the maximum amount of information $C^2/4$. The conclusions can then be easily drawn. ■

### 3.2 TOA-Only Sensor

The optimal configuration condition of homogeneous TOA sensor networks is illustrated in the following theorem.

**Theorem 2** *For homogeneous TOA sensors, the maximum amount of information,* $C^3/(8c_0^4)$, *can be achieved if and only if the following condition satisfies*

$$Y_1 = Y_2 = Y_3 = Y_4 = 0 \tag{9}$$

*which is equivalent to*

$$\sum c_i \mathbf{g}_i \mathbf{g}_i^T = \frac{C}{2} \mathbf{I}_2 \tag{10a}$$

$$\sum c_i \mathbf{g}_i = 0 \tag{10b}$$

*Proof* For TOA sensor networks, after some symbolic computations, it can be found that

$$\det(\mathcal{I}(\mathbf{x})) = \frac{1}{8c_0^4}\left[C^3 - (Y_1^2 + Y_2^2 + Y_3^2 + Y_4^2)C + Y_1(Y_3^2 - Y_4^2) + 2Y_2Y_3Y_4\right] \tag{11}$$

For a fixed $C$, it can be seen that $Y_1$, $Y_2$, $Y_3$ and $Y_4$ are in $[-C, C]$. Hence $\det(\mathcal{I}(\mathbf{x}))$ can be regarded as a 4-variable continuous function (actually a 4-variable polynomial function) with respect to $\{Y_1, Y_2, Y_3, Y_4\}$ defined at the closed region $[-C, C] \times [-C, C] \times [-C, C] \times [-C, C]$. According to the extrema theory of multivariate continuous function [21], one necessary condition for the function maximization is that the partial derivatives with respect to all the formal variables are zero. This leads to

$$\begin{cases} -2CY_1 + Y_3^2 - Y_4^2 = 0 \\ -2CY_2 + 2Y_3Y_4 = 0 \\ -2CY_3 + 2Y_1Y_3 + 2Y_2Y_4 = 0 \\ -2CY_4 + 2Y_2Y_3 - 2Y_1Y_4 = 0 \end{cases} \tag{12}$$

Solving (12) yields 3 groups of extreme points which are listed in Table 2. Comparing the values of (11) at these extreme points with those at the vertices of the closed region and further considering the dependency of $Y_1$ with $Y_2$ and $Y_3$ with $Y_4$ (e.g., when $Y_1 = \pm C$, $Y_2$ can only be 0.), we find that the 3rd extreme point is the maxima, i.e., if and only if (9) holds, det(FIM) is maximized, and its maximum value is $C^3/(8c_0^4)$. As discussed in the former theorem, $Y_1 = Y_2 = 0$ is equivalent to (10a). And from the definition of the unit line-of-sight $\mathbf{g}_i = (\cos\phi_i, \sin\phi_i)^T$, we can find that $Y_3 = Y_4 = 0$ is equivalent to (10b). ■

# 4 Optimal Configuration Conclusions

In this section we aim to derive the optimal sensor configuration conclusions from the conditions obtained in the former section. We assume that all the sensor-source distances are fixed, i.e., we regard $\{\phi_i\}_{i=1}^N$ (or, $\{\mathbf{g}_i\}_{i=1}^N$) as a certain sensor configuration.

## 4.1 *Range-Only, RSS-Only and AOA-Only Sensors*

For the homogeneous range, RSS and AOA sensors, if we denote $\mathbf{F}_i = c_i (\cos 2\phi_i, \sin 2\phi_i)^T$ as a force vector in 2D space, then the optimal condition $Y_1 = Y_2 = 0$ becomes

$$\sum \mathbf{F}_i = 0 \tag{13}$$

which is to find an optimal azimuth angular sequence $\{\phi_i\}_{i=1}^N$ making the resultant force vanish. We have the following conclusion from this.

**Theorem 3** *For the 2D homogeneous range, RSS and AOA sensor networks,*

- *when* $\{c_i\}_{i=1}^N$ *is regular, the optimal condition* (6) *is solvable, and the maximum amount of information,* $C^2/4$*, can be achieved. A particular solution can be achieved using* ***Algorithm 1*** *(See also* ***Algorithm 1*** *in [20]).*
- *when* $\{c_i\}_{i=1}^N$ *is irregular, the optimal condition* (6) *is unsolvable, and the achievable maximum amount of information is* $c_1(C - c_1)$ *if and only if* $\phi_i =< \phi_1 \pm \pi/2 >$ *for* $i = 2, 3, \ldots, N$*. Here* $< \cdot >$ *is used to wrap the angle to the range* $[-\pi, \pi)$.

*Proof* When $\{c_i\}_{i=1}^N$ is regular, from the definition of regularity, we can find an integer $m$ such that

$$\sum_{i=1}^{m} c_i \le \frac{C}{2}, \sum_{i=1}^{m+1} c_i \ge \frac{C}{2} \tag{14}$$

Then we let

**Table 2** Extreme points

| No. | $Y_1$ | $Y_2$ | $Y_3$ | $Y_4$ |
|---|---|---|---|---|
| 1 | 0 | $\pm C$ | $\pm C$ | $C$ |
| 2 | $\neq -C$ | $\pm\sqrt{C^2 - Y_1^2}$ | $\pm\sqrt{C^2 + CY_1}$ | $Y_2Y_3/(C + Y_1)$ |
| 3 | 0 | 0 | 0 | 0 |

$$l_1 = \sum_{i=1}^{m} c_i,\ l_2 = c_{m+1},\ l_3 = \sum_{i=m+2}^{N} c_i \tag{15}$$

It can be found from (14) that $l_1 < l_2 + l_3$, $l_2 < l_1 + l_3$ and $l_3 < l_1 + l_2$ hold. This means that $l_1$, $l_2$ and $l_3$ can form a triangle, or equivalently, there exist $\varphi_1$, $\varphi_2$ and $\varphi_3$ such that

$$l_1 \begin{bmatrix} \cos 2\varphi_1 \\ \sin 2\varphi_1 \end{bmatrix} + l_2 \begin{bmatrix} \cos 2\varphi_2 \\ \sin 2\varphi_2 \end{bmatrix} + l_3 \begin{bmatrix} \cos 2\varphi_3 \\ \sin 2\varphi_3 \end{bmatrix} = 0 \tag{16}$$

A feasible solution of (16) is

$$\varphi_2 = < \varphi_1 + \frac{\gamma_1 + \gamma_2}{2} >,\ \varphi_3 = < \varphi_1 + \frac{\pi + \gamma_1}{2} > \tag{17}$$

where $\varphi_1 \in [-\pi, \pi)$ can be arbitrarily chosen, and

$$\gamma_1 = \arccos\left(\frac{l_1^2 + l_3^2 - l_2^2}{2l_1 l_3}\right),\ \gamma_2 = \arccos\left(\frac{l_2^2 + l_3^2 - l_1^2}{2l_2 l_3}\right) \tag{18}$$

Then (6) is solved by letting

$$\phi_1 = \phi_2 = \cdots = \phi_m = \varphi_1,\ \phi_{m+1} = \varphi_2,\ \phi_{m+2} = \phi_{m+3} = \cdots = \phi_N = \varphi_3 \tag{19}$$

And the maximum det(FIM), $C^2/4$, is achieved.

When $\{c_i\}_{i=1}^N$ is irregular, it can be observed that if all of the remainder force vectors ($\mathbf{F}_i$'s) are in the opposite direction of the first force vector ($\mathbf{F}_1$), i.e., $2\phi_i = 2\phi_1 \pm \pi$ for $i = 2, 3, \cdots, N$, the magnitude of the resultant force (or equivalently, $Y_1^2 + Y_2^2$) is minimized, and therefore (8) is maximized. Substituting this solution into (8), we can get the achievable maximum amount of information, which is $c_1 \sum_{i=2}^N c_i = c_1(C - c_1)$. ■

---

**Algorithm 1.** A particular optimal configuration solution for 2D homogeneous range, RSS and AOA sensor networks when the SCC sequence $\{c_i\}_{i=1}^N$ is regular.

---

1. Find $m$ satisfying (14).
2. Compute $l_1$, $l_2$ and $l_3$ according to (15) and further $\gamma_1$ and $\gamma_2$ according to (18).
3. Choose azimuth angles for the sensors according to (17) and (19).

---

**Theorem 4** (See also Corollary 1 in [3].) *Consider the special situation that all the sensors share the same SCC, i.e., $c_i = c$ for $i = 1, 2, \ldots, N$ with $N \geq 3$. Two special optimal configurations are given below.*

- $\phi_1 \in [-\pi, \pi)$, $\phi_i = < \phi_{i-1} + \frac{2\pi}{N} >$ *for* $i = 2, 3, \ldots, N$.

- $\phi_1 \in [-\pi, \pi)$, $\phi_i =< \phi_{i-1} + \frac{\pi}{N} >$ *for* $i = 2, 3, \ldots, N$.

*Proof* In this case, the optimal configuration condition becomes $\sum \cos 2\phi_i = \sum \sin 2\phi_i = 0$. And the conclusions can be easily drawn using the following facts that, for $N \geq 3$,

$$\sum_{i=1}^{N} \cos\left[2\left(\phi_1 + \frac{2(i-1)\pi}{N}\right)\right] = \sum_{i=1}^{N} \sin\left[2\left(\phi_1 + \frac{2(i-1)\pi}{N}\right)\right] = 0 \tag{20}$$

$$\sum_{i=1}^{N} \cos\left[2\left(\phi_1 + \frac{(i-1)\pi}{N}\right)\right] = \sum_{i=1}^{N} \sin\left[2\left(\phi_1 + \frac{(i-1)\pi}{N}\right)\right] = 0 \tag{21}$$

■

**Theorem 5** *If* $\{\phi_i^*\}_{i=1}^N$ *(or equivalently,* $\{\mathbf{g}_i^*\}_{i=1}^N$*) corresponds to an optimal configuration, then the following three actions do not change its optimality.*

- *Add all the azimuth angles (*$\phi_i^*$*'s) with an arbitrary* $\phi_0$*. Or equivalently, execute an orthogonal transformation* $\mathbf{U}$ *to all the unit line-of-sights (*$\mathbf{g}_i^*$*'s). Here* $\mathbf{U}\mathbf{U}^T = \mathbf{I}_2$*.*
- *Change some or all* $\phi_i^*$*'s to* $< \phi_i^* + \pi >$*, i.e., reflect some or all the sensors about the source.*
- *Change the signs of all* $\phi_i^*$*'s.*

*Proof* The conclusions can be easily drawn from Theorem 1. ■

## 4.2 TOA-Only Sensor

For the homogeneous TOA sensor networks, if we adopt the indication $\mathbf{F}_i$ and further denote $\mathbf{G}_i = c_i\,(\cos\phi_i, \sin\phi_i)^T$, then the optimal condition becomes

$$\sum \mathbf{F}_i = 0,\ \sum \mathbf{G}_i = 0 \tag{22}$$

It is not easy for the solutions given in Theorem 3 satisfying $\sum \mathbf{F}_i = 0$ to solve $\sum \mathbf{G}_i = 0$ as well.

**Theorem 6** (See also Proposition 3 in [3].) *Consider the special situation that all the sensors share the same SCC, i.e.,* $c_i = c$ *for* $i = 1, 2, \ldots, N$ *with* $N \geq 3$*. One special optimal configuration is given by letting* $\phi_1 \in [-\pi, \pi)$ *and* $\phi_i =< \phi_{i-1} + 2\pi/N >$ *for* $i = 2, 3, \ldots, N$.

*Proof* In this case, the optimal configuration condition becomes $\sum \cos 2\phi_i = \sum \sin 2\phi_i = \sum \cos\phi_i = \sum \sin\phi_i = 0$. And again using (20) and (21), we get this special optimal configuration. ■

**Theorem 7** *If $\{\phi_i^*\}_{i=1}^N$ (or equivalently, $\{\mathbf{g}_i^*\}_{i=1}^N$) corresponds to an optimal configuration, then the following three actions do not change its optimality.*

- *Add all the azimuth angles ($\phi_i^*$'s) with an arbitrary $\phi_0$. Or equivalently, execute an orthogonal transformation $\boldsymbol{U}$ to all the unit line-of-sights ($\mathbf{g}_i^*$'s).*
- *Change the signs of all $\phi_i^*$'s.*
- *Change some or all the sensor-source distances.*

*Proof* Proofs of the first two conclusions are similar to those of Theorem 5. And since the SCCs of TOA sensors are independent of the sensor-source distances, the third action does not change the original optimality either. ■

## 5 Conclusions

The 2D general optimal sensor network configuration problems for homogeneous sensors, including range, RSS, TOA and AOA ones, have been studied systematically in this paper. The determinants of the corresponding FIMs are chosen as the optimality criterion, and the optimal configuration conditions are then given out. The solvability of the conditions and the optimal configuration constructions are discussed in detail.

## References


1. J.R. Lowell, Military applications of localization, tracking, and targeting. IEEE Wirel. Commun. **18**(2), 60–65 (2011)
2. J. Rantakokko, P. Händel, M. Fredholm, F. Marsten-Eklöf, User requirements for localization and tracking technology: a survey of mission-specific needs and constraints, in *Proceedings of International Conference on Indoor Positioning and Indoor Navigation (IPIN)* (Zürich, Switzerland, 2010), pp. 1–9
3. A.N. Bishop, B. Fidan, B.D. Anderson, K. Doğançay, P.N. Pathirana, Optimality analysis of sensor-target localization geometries. Automatica **46**(3), 479–492 (2010)
4. G. Marani, S.K. Choi, Underwater target localization: autonomous intervention with the DIDSON sonar in SAUVIM. IEEE Rob. Autom. Mag. **17**(1), 64–70 (2010)
5. J. Zhong, Z. Lin, Z. Chen, W. Xu, Cooperative localization using angle-of-arrival information, in *Proceedings of 2014 11th IEEE International Conference on Control & Automation (ICCA)* (Taichung, Taiwan, 2014), pp. 19–24
6. E. Xu, Z. Ding, S. Dasgupta, Source localization in wireless sensor networks from signal time-of-arrival measurements. IEEE Trans. Signal Process. **59**(6), 2887–2897 (2011)
7. X. Li, Collaborative localization with received-signal strength in wireless sensor networks. IEEE Trans. Veh. Technol. **56**(6), 3807–3817 (2007)
8. A.N. Bishop, P. Jensfelt, An optimality analysis of sensor-target geometries for signal strength based localization, in *Proceedings of the 3rd International Conference on Intelligent Sensors, Sensor Networks, and Information Processing* (Melbourne, Australia, Dec 2009), pp. 127–132
9. M.L. Hernandez, Optimal sensor trajectories in bearings-only tracking, in *Proceedings of the 7th International Conference on Information Fusion* (Stockholm, Sweden, 28-July 1 2004), pp. 893–900

10. K. Doğançay, Single- and multi-platform constrained sensor path optimization for angle-of-arrival target tracking, in *Proceedings of the 18th European Signal Processing Conference (EUSIPCO-2010)* (Aalborg, Denmark, Aug 2010), pp. 835–839
11. Y. Jia, Robust control with decoupling performance for steering and traction of 4WS vehicles under velocity-varying motion. IEEE Trans. Control Syst. Technol. **8**(3), 554–569 (2000)
12. Y. Jia, Alternative proofs for improved LMI representations for the analysis and the design of continuous-time systems with polytopic type uncertainty: a predictive approach. IEEE Trans. Autom. Control **48**(8), 1413–1416 (2003)
13. A.P. Aguiar, J.P. Hespanha, Trajectory-tracking and path-following of underactuated autonomous vehicles with parametric modeling uncertainty. IEEE Trans. Autom. Control **52**(8), 1362–1379 (2007)
14. P.B. Sujit, S. Saripalli, J.B. Sousa, Unmanned aerial vehicle path following: a survey and analysis of algorithms for fixed-wing unmanned aerial vehicles. IEEE Control Syst. Mag. **42**(1), 42–59 (2014)
15. S.R. Sukumar, H. Bozdogan, D.L. Page, A.F. Koschan, M.A. Abidi, Uncertainty minimization in multi-sensor localization systems using model selection theory, in *Proceedings of the 19th International Conference on Pattern Recognition* (Tampa, FL, USA, 2008), pp. 1–4
16. S.R. Semper, J.L. Crassidis, Decentralized geolocation and optimal path planning using limited UAVs, in *Proceedings of the 12th International Conference on Information Fusion* (Seattle, WA, USA, July 2009), pp. 355–362
17. S. Martínez, F. Bullo, Optimal sensor placement and motion coordination for target tracking. Automatica **42**(4), 661–668 (2006)
18. K. Doğançay, H. Hmam, Optimal angular sensor separation for AOA localization. Signal Process. **88**(5), 1248–1260 (2008)
19. K. Doğançay, H. Hmam, On optimal sensor placement for time-difference-of-arrival localization utilizing uncertainty minimization, in *Proceedings of the 17th European Signal Processing Conference (EUSIPCO 2009)* (Glasgow, Scotland, UK, 24–28 Aug 2009), pp. 1136–1140
20. S. Zhao, B.M. Chen, T.H. Lee, Optimal sensor placement for target localisation and tracking in 2D and 3D. Int. J. Control **86**(10), 1687–1704 (2013)
21. R. Larson, B.H. Edwards, *Calculus*, 9th edn. (Brooks/Cole, Belmont, CA, USA, 2010)

# Non-rigid 3D Shape Classification Based on Low-Level Features

**Yujuan Wu, Haisheng Li, Yujia Du and Qiang Cai**

**Abstract** Non-rigid 3D shape classification is an important issue in digital geometry processing. In this paper, we propose a novel non-rigid 3D shape classification method using Convolutional Neural Networks (CNNs) based on the scale-invariant heat kernel signature (SIHKS). Firstly, SIHKS feature is extracted and we can get a matrix for every 3D shape. Then CNNs is employed to shape classification. The matrix of 3D shapes can be the input of CNNs. Finally, we can obtain the category probability of 3D shapes. Experimental results demonstrate the proposed method can get better results compared with SVM.

**Keywords** Non-rigid 3D shape classification · Low-level feature · Scale-invariant heat kernel signature (SIHKS) · Convolutional neural networks

## 1 Introduction

Three-dimensional shapes have been extensively applied in the domains of virtual reality, amusement, multimedia, graphics, design, and manufacturing [1] due to the abundant information preserving the surface, color, and texture of real objects. With the improvement of computer hardware performance and the rapid development of the three-dimensional modeling technology, more and more 3D shapes can be obtained, which require efficient classification methods for effective application and management [2]. 3D shape can be divided into rigid 3D shape and non-rigid 3D shape according to its structure. Compared with rigid 3D shape, non-rigid 3D shapes have an articulated structure, which can produce a variety of posture deformation. The task of classifying non-rigid 3D shapes is more challenging [3]. In recent years,

Y. Wu · H. Li (✉) · Y. Du · Q. Cai
School of Computer and Information Engineering, Beijing Technology and Business University, Beijing 100048, China
e-mail: lihsh@th.btbu.edu.cn

Y. Wu · H. Li · Y. Du · Q. Cai
Beijing Key Laboratory of Big Data Technology for Food Safety, Beijing 100048, China



Y. Jia et al. (eds.), *Proceedings of 2018 Chinese Intelligent Systems Conference*, Lecture Notes in Electrical Engineering 528,
https://doi.org/10.1007/978-981-13-2288-4_62

CNNs have been extensively used in image feature extraction and have shown better performance than traditional manually-crafted feature descriptors. However, the 3D shape structure is distinct from the 2D image. 2D images can be classified based on pixels directly to extract features, but analogy to 3D shapes, it's not easy to categorize by voxel information [4].

The feature descriptors based on spectral analysis can express the intrinsic properties of 3D shapes and have been widely utilized in the task such as non-rigid 3D shape analysis [5]. Local feature descriptors are also called point features. Eigenvalues and eigenvectors are used to describe the regional structural characteristics around each point of the shape surface more particularly [6]. In a small region on the meshed surface, local features are built which can capture the intrinsic geometric structure of the shape [7]. 3D shape classification is based on features to determine the category. Due to the special nature of non-rigid 3D shape structure, many researches often describe it using spectral features. Rustamov [8] proposed the Global Point Signature (GPS). This method associates each point on the surface of the 3D shape with the sequence formed by the eigenvalues and eigenvectors of the Laplace-Beltrami (LB) operator. Rustamov [9] further presented template-based feature descriptors. Some 3D models are selected as shape templates randomly, and a mapping of models to these shape templates is established. By calculating GPS features in the shape template and connecting descriptors from different templates, the shape descriptor can be obtained for 3D shape retrieval. Sun et al. [10] proposed Heat Kernel Signature (HKS) built on analysis of the heat diffusion process. Bronstein and Kokkinos [11] presented Scale-invariant Heat Kernel Signature (SIHKS) based on HKS. Time-shift invariance of the amplitude in Fourier transform is used to make HKS feature scale-invariant. This method offers an opportunity to deal with global and local scaling transformations. Aubry et al. [12] proposed the Wave Kernel Signature (WKS) by analyzing the shape using the wave equation in the framework of quantum mechanics. Li et al. [6] presented Spectral Graph Wavelet Signature (SGWS) which is a multi-resolution descriptor based on cubic spline generation kernel.

Li et al. [13] used manifold learning method to reduce feature dimensions of shape and the distinguishing feature of 3D shape can be obtained. Then the classifier is trained with this feature which used for classification. Qin et al. [14] applied deep learning algorithm to the classification problem of 3D shapes. Features of the input 3D shape can be extracted using a deep neural network. Bu et al. [15] presented a multi-level 3D shape feature extraction framework based on deep learning algorithm. Low-level feature descriptors are encoded into geometric bag-of-features, and intermediate features from them can be obtained. Then high-level features are learned through the deep belief network which is used for 3D shape classification. Leng et al. [16] used Deep Boltzmann Machine (DBM) to get the distribution of approximate input data, and highly abstract feature representation is obtained. Then semi-supervised learning algorithm is utilized to train and learn the extracted features. Dai et al. [7] proposed a learning framework to extract concise data-driven model feature descriptors. Most of above methods use different algorithms to extract features first, and then traditional classifiers are employed to classify 3D shapes. This makes it hard to solve the relatively difficult classification problems.

In this paper, we develop a novel network structure based on SIHKS feature, for non-rigid 3D shape classification. Low-level feature SIHKS of the shape is extracted firstly, which is regarded as the input of the proposed network. By performing multiple convolutional and pooling operations on SIHKS, the high-level feature is generated to describe the shape. Finally, we can classify shapes through fully connected layers. Experimental results indicate the proposed method can obtain better results compared with SVM.

# 2 Non-rigid 3D Shape Classification via CNNs

The proposed algorithm is mainly divided into two parts: the extraction of low-level features and the classification of 3D shapes based on low-level features. The overall algorithm framework is shown in Fig. 1.

## *2.1 Low-Level Feature Extraction*

In this paper, the SIHKS feature is employed as low-level feature.
**Heat Kernel Signature (HKS)**. The heat equation is as follows,

$$\frac{\partial k_t}{\partial t} + \Delta k_t = 0 \tag{1}$$

where $k_t$ is the heat kernel, $\Delta$ is the LB operator.

HKS at vertex $x$ on the 3D shape can be obtained:

$$h(x,t) = k(x,x) = \sum_{i=0}^{\infty} e^{-\lambda_i t} \phi_i^2(x) \tag{2}$$

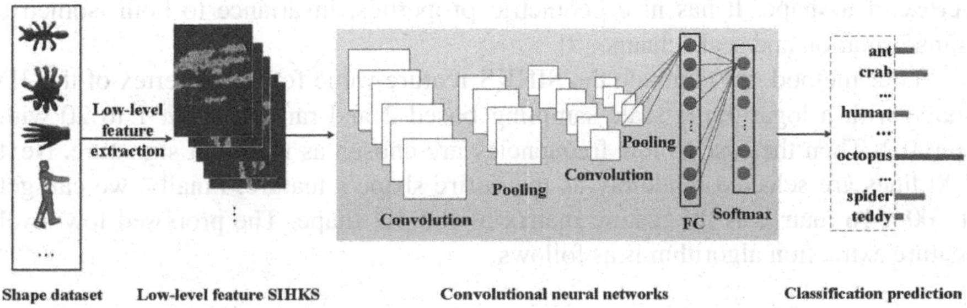


**Fig. 1** The framework of proposed algorithm

where $\lambda_i$ is the $i$th eigenvalue of the LB operator $\Delta$, $\lambda_i \geq 0$, $\phi_i$ is the $i$th eigenfunction of the LB operator $\Delta$, and $\Delta\phi_i = \lambda_i\phi_i$. HKS $h(x, t)$ shows the amount of heat remaining at vertex $x$ after a period of time $t$.

Given a 3D shape $S$, the scaled shape is $S' = \beta S$, where $\beta$ is the scale factor. The scaled HKS is as follows,

$$h'(x,t) = \beta^2 h(x, \beta^2 t) \tag{3}$$

**Scale-invariant Heat Kernel Signature** (**SIHKS**). In Bronstein's work [11], the author takes three steps to remove the scale factor in Eq. (3). At first, sample the heat kernel signature logarithmically to remove the scale change and get the following equation,

$$h'_\tau = h(x, \alpha^\tau)' = \beta^2 h(x, \beta^2\alpha^\tau) = \beta^2 h(x, \alpha^s\alpha^\tau) = \beta^2 h_{\tau+s} \tag{4}$$

with $\alpha^s = \beta^2$, $s = 2\log_\alpha \beta$.

Then take the logarithm and derivative on both sides to remove the scale change in amplitude and get the following equation,

$$\dot{f}'_\tau = \dot{f}'_{\tau+s} \tag{5}$$

where $f'_\tau = \log h'_\tau$, $\dot{f}'_\tau$ and $\dot{f}'_{\tau+s}$ denote the partial derivatives of $f'_\tau$ and $f'_{\tau+s}$, respectively.

Finally, take the Fourier transform and modulus so that the phase difference generated in the first step can be removed,

$$F'(\omega) = F(\omega)e^{2\pi\omega s}, \omega \in [0, 2\pi] \tag{6}$$

$$\left|F'(\omega)\right| = |F(\omega)| \tag{7}$$

where $F$ and $F'$ denote the Fourier transforms of $\dot{f}$ and $\dot{f}'$, respectively. $|F|$ and $\left|F'\right|$ denote the moduli of $F$ and $F'$

**The proposed low-level feature**. SIHKS is a local feature used to describe each vertex of a shape. It has nice geometric properties, invariance to both isometric transformation and scale change [7].

In our method, we compute the SIHKS feature value for every vertex of the 3D shape with a logarithmic scale sampling based 2 and ranging from 1 to 20 with step 0.2. Then the first 16 low frequencies are chosen as the point signature. Next 100 lines are selected randomly as the entire shape's feature. Finally, we can get a $100 \times 16$ matrix as the feature matrix of the 3D shape. The proposed low-level feature extraction algorithm is as follows.

**Algorithm1: Low-level Feature Extraction**

Input: non-rigid 3D shape $G = (V, E, F)$;
Output: low-level feature $100 \times 16$ matrix ;
Procedure:
1. Compute shapes' eigenvalues and eigenvectors;
2. Time scales T for SIHKS: range from 1 to 20 with step 0.2;
3. Frequencies $\omega$ for SIHKS: range from 2 to 17 with step 1;
4. for each $x \in V$ do
    compute the feature SIHKS;
    store SIHKS in a feature file by line;
   end
5. Sample 100 lines randomly from the feature file;
6. Save selected point features as the final feature.

## 2.2 The Proposed Network Structure

In our paper, we developed a 4-layer network structure based on the CNNs. Feature matrixes of 3D shapes can be input to the proposed network, while output is the category probability of 3D shapes. The network structure is as follows:

$$U_1 \rightarrow U_2 \rightarrow FC(256) \rightarrow \text{softmax}$$

where $U_i$ = Convolution + BN + Sigmoid + Maxpooling, and i = 1, 2. Here 256 is equal to the number of neurons in fully connected layer (FC).
**Input data**. The SIHKS feature matrixes of 3D shapes are input of this network structure. And we make each 3D shape correspond to a $100 \times 16$ matrix.
**Convolution**. Convolution operations are performed on input data, which a $100 \times 16$ matrix. The convolution calculation process is as follows:

$$x_j^l = f\left(b_j^l + \sum_i^m x_i^{l-1} * k_{ij}^l\right) \tag{8}$$

where $x_j^l$ is output of the $j$th channel of convolutional layer $l$, $k_{ij}^l$ is the convolutional kernel, $b_j^l$ is the bias of convolutional feature map, $m$ is the collection of feature maps of the previous layer, $*$ is the convolution symbol, and $f(\cdot)$ is activation function. We use the sigmoid function to activate the output, and the formula of nonlinear operating function is as follows:

$$g(z) = \frac{1}{1 + e^{-z}} \tag{9}$$

In addition, batch normalization (BN) is employed to reduce the internal-covariate-shift.

**Max-pooling**. The main function of pooling is to decrease the spatial size of the feature representation. The input of pooling comes from convolution layer of the preceding layer, and output results are regarded as the input of the next convolution layer. In our paper, max pooling is used to reduce the dimensionality of features.

After two rounds of convolution and pooling operation, a fully connected layer is added. Finally 3D shapes can be classified by the softmax layer.

# 3 Experiments

The experiments were implemented on the workstation with an Intel(R) Xeon(R) CPU E5-2630 v3 (2.40 GHz) and a Tesla K40 m GPU (11.17GiB memory).

## *3.1 Experimental Dataset*

SHREC 2010 retrieval dataset [17] is employed for training and testing, which contains 200 non-rigid watertight 3D triangular meshes, subdivided into 10 categories, each of which contains 20 objects with distinct postures. We annotated the dataset with multi-class labels according to the given classification file.

We choose 150 3D shapes from the dataset, with each of category contains 15 objects. The remaining dataset is selected for testing which contains 50 3D shapes, with each of category contains 5 objects.

## *3.2 Experimental Results*

For the first convolutional layer, we set the size of the kernel to 11 and the number to 64. In addition, for the subsequent max-pooling layer, we set the kernel size 4 with a stride of 4. For the second convolutional layer, we set the size of the kernel to 5 and the number to 64. And for the max-pooling layer, we set the kernel size 4 with a stride of 4. For the fully connected layer, we define the number of neurons is 256. And for softmax layer, the number of neurons is the number of 3D shape categories 10. The proposed network is optimized by stochastic gradient descent (SGD) with a weight decay of $1e-4$, and a batch size of 30. The learning rate is 0.001 and epochs are 200. Parameters' set is fixed as shows in Fig. 2.

In Fig. 2, Convi is the $i$th convolutional operation, the corresponding $m \times m$ ($11 \times 11$, $5 \times 5$) is the kernel size, $n$ (64) is the number of the kernel. Pooli is the $i$th pooling operation, the corresponding $k \times k$ ($4 \times 4$) is the kernel size, $s$ (4) is the

**Table 1** Classification results on SHREC 2010 retrieval dataset

| Method | Average accuracy (%) |
|---|---|
| Ours | 90 |
| SVM | 68 |

number of the kernel, $i = 0, 1$. CP is the category probability of the input non-rigid 3D shape.

We compare our method with SVM. Table 1 gives average classification accuracy for this dataset. Table 2 lists classification accuracies on each type of the shape. The results of the proposed algorithm and SVM are presented in Fig. 3.

## 4 Conclusions

In this paper, we proposed a novel method for non-rigid 3D shape classification. We have given a low-level feature processing method which can extract the SIHKS feature descriptor and obtain a matrix for every 3D shape. And a novel network structure is presented for non-rigid 3D shape classification. The matrix obtained at the data processing stage can be used as the input of the network. Experiments conducted on SHREC 2010 retrieval dataset show the proposed algorithm obtains better performance compared with SVM. In the future, we would like to improve our network structure to make it more robust to different types and sizes of low-level features.

**Acknowledgements** This work was partially supported by Beijing Natural Science Foundation (4162019).

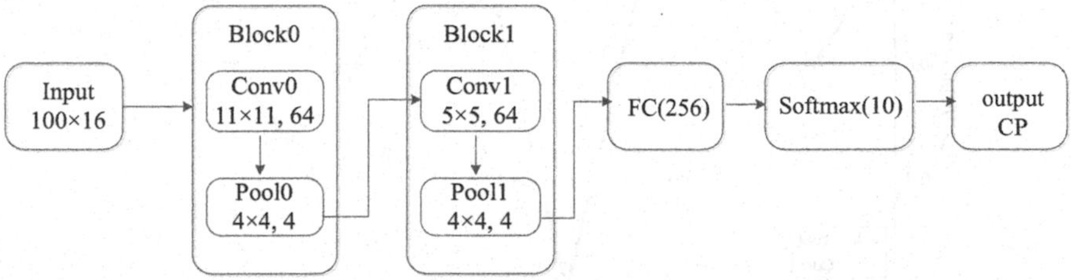


**Fig. 2** Parameters set in the proposed network structure

**Table 2** Classification results on each type of the 3D shape

| 3D shapes | SVM | Ours |
|---|---|---|
|  | **0.80** | **0.80** |
|  | 0.60 | **1.00** |
| 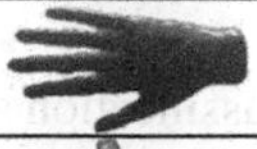 | 0.80 | **1.00** |
|  | 0.80 | **1.00** |
|  | 0.20 | **0.80** |
| 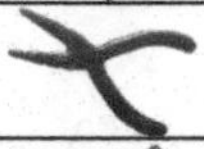 | 0.80 | **1.00** |
|  | 0.80 | **1.00** |
|  | 0.40 | **0.80** |
|  | **0.60** | **0.60** |
|  | **1.00** | **1.00** |

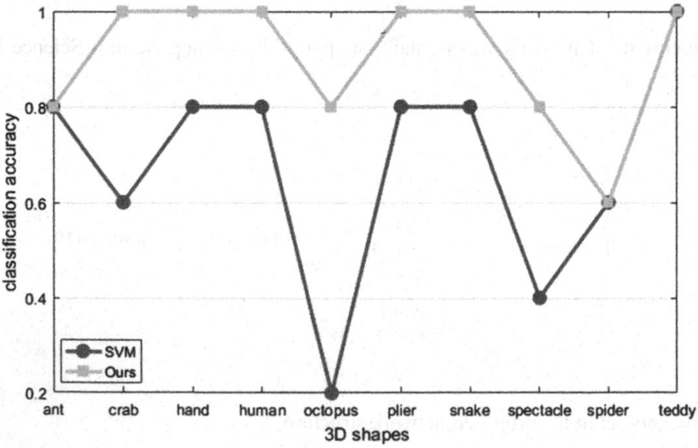


**Fig. 3** Classification results compared with SVM

## References


1. J.W.H. Tangelder, R.C. Veltkamp, A survey of content based 3D shape retrieval methods, in *Proceedings of International Conference on Shape Modeling Applications* (2004), pp. 145–156
2. A.D. Bimbo, P. Pala, Content-based retrieval of three-dimensional models. ACM Trans. Multimedia Comput. Commun. Appl. (TOMM) **2**(1), 20–43 (2006)
3. R. Jain, J. Tyagi, S.K. Singh et al., Hybrid context aware recommender systems, in *Advancement in Mathematical Sciences: Proceedings of the, International Conference on Recent Advances in Mathematical Sciences and ITS Applications*, pp. 020–028 (2017)
4. P.S. Wang, Y. Liu, Y.X. Guo et al., O-CNN: octree-based convolutional neural networks for 3D shape analysis. ACM Trans. Graph. **36**(4), 72 (2017)
5. Y. Yubin, L. Hui, Z. Qing, Content-based 3D model retrieval: a survey. Chin. J. Comput. **27**(10), 1297–1310 (2004)
6. C. Li, A.B. Hamza, A multiresolution descriptor for deformable 3D shape retrieval. Vis. Comput. **29**(6–8), 513–524 (2013)
7. G. Dai, J. Xie, F. Zhu et al., Learning a discriminative deformation-invariant 3D shape descriptor via many-to-one encoder. Pattern Recogn. Lett. **83**, 330–338 (2016)
8. R.M. Rustamov, Laplace-Beltrami eigenfunctions for deformation invariant shape representation, in *Proceedings of the fifth Eurographics symposium on Geometry processing* (Eurographics Association, 2007), pp. 225–233
9. R.M. Rustamov, Template based shape descriptor, in *Proceedings of the 2nd Eurographics conference on 3D Object Retrieval* (Eurographics Association, 2009), pp. 1–7
10. J. Sun, M. Ovsjanikov, L. Guibas, A concise and provably informative multi-scale signature based on heat diffusion. Comput. Graph. Forum **28**(5), 1383–1392 (2010)
11. M.M. Bronstein, I. Kokkinos, Scale-invariant heat kernel signatures for non-rigid shape recognition. Comput. Vis. Pattern Recog (IEEE), 1704–1711 (2010)
12. M. Aubry, U. Schlickewei, D. Cremers, The wave kernel signature: a quantum mechanical approach to shape analysis, in *IEEE International Conference on Computer Vision Work-shops* (IEEE, 2011), pp. 1626–1633
13. Z. Li, D. Wang, L. Boyang et al., 3D model classification using salient features for content representation, in *International Conference on Natural Computation* (IEEE, 2010), pp. 3541–3545
14. F.W. Qin, L.I. Lu-Ye, S.M. Gao et al., A deep learning approach to the classification of 3D CAD models. Front. Inf. Technol. Electr. Eng. **15**(2), 91–106 (2014)
15. S. Bu, Z. Liu, J. Han et al., Learning high-level feature by deep belief networks for 3-D model retrieval and recognition. IEEE Trans. Multimed. **16**(8), 2154–2167 (2014)
16. B. Leng, X. Zhang, M. Yao et al., A 3D model recognition mechanism based on deep Boltzmann machines. Neurocomputing **151**(151), 593–602 (2015)
17. Z. Lian, A. Godil, T. Fabry et al., SHREC'10 track: non-rigid 3D shape retrieval, in *Eurographics Workshop on 3D Object Retrieval*, Norrköping, Sweden, 2 May 2010, pp. 101–108

# The Development of a Charge Movement Model for Blast Furnace Based on Extended Kalman Filter

**Jingchu Duan and Weicun Zhang**

**Abstract** In this paper, by analyzing the movement of blast furnace charge in the void area after the chute, the system's state equations and measurement equations are established, and the extended Kalman filter is used to predict and track the data. Extended Kalman filter is a method that can expand nonlinear equations into linear equations through Taylor expansion to make better observations. Then three-dimensional equations of state are established. The position, velocity, and acceleration of the charge are used to fit. Finally, the blast furnace is verified through experiments state of charge equation and measurement equation of charge and validity of three-dimensional extended Kalman filter.

**Keywords** Blast furnace · Extended Kalman filter · State equations · Measurement equations

## 1 Introduction

The material level of blast furnace is an important control parameter in iron and steel smelting, the accuracy of measurement data and continuity is not only to enhance the production efficiency and ensure the smooth operation of blast furnace and the necessary conditions, but also to make the shape distribution and the accuracy improved [1]. Therefore, real-time and accurate information acquisition of blast furnace material level is of great significance for improving blast furnace distribution, optimizing the distribution of gas flow in the furnace, improving the quality of the blast furnace, reducing energy consumption and ensuring the smooth operation.

At present, the mechanical probe and radar detection are the usual method of measuring the position of blast furnace [2]. Mechanical probe detection is the use

J. Duan (✉) · W. Zhang
College of Automation, University of Science and Technology Beijing, Beijing, China
e-mail: 18611533898@163.com

W. Zhang
e-mail: weicunzhang@263.net



Y. Jia et al. (eds.), *Proceedings of 2018 Chinese Intelligent Systems Conference*,
Lecture Notes in Electrical Engineering 528,
https://doi.org/10.1007/978-981-13-2288-4_63

of mechanical probe in blast furnace burden of direct contact measurement, high precision and stability; but there will be some problems, such as the measurement cycle is too long, not for continuous measurement of the material level, and the blast furnace feeding does not work and other defects. Radar detection is a non-contact measurement of the surface using electromagnetic wave ranging principle, which is not affected by the feeding process constraints. It can realize continuous measurement of material level, so the existing technology is mostly used in radar data fusion of material analysis, but radar waves are easily disturbed by factors such as dust and air flow in the furnace [3]. The measured data are the average of the microwave coverage area, which has the disadvantage of low detection precision and poor stability.

In addition, the blast furnace level detection methods include laser detection, infrared imaging, video monitoring, etc. These methods are completed by direct measurement or indirect calculation. However, due to the harsh environment of the blast furnace, the test results are also extremely affected by high temperature, high dust, and gas. Due to the influence of flow and other factors, it is difficult for these methods to obtain high-precision material surface detection information continuously, and can only be used as an auxiliary method for the open furnace test or for the qualitative analysis of the material surface [4].

To sum up, the current detection methods and modeling methods are difficult to achieve continuous and high-precision detection of blast furnace material level at the same time, which makes the blast furnace operation has a certain degree of blindness, and affects the rationality and safety. Therefore, it is necessary to better predict the real-time change of material level, so as to make the error smaller. Kalman filtering is an algorithm that uses the state equation of a linear system to input and output observation data through the system to make an optimal estimation of the system state. Since the observation data includes the effects of noise and interference in the system, the optimal estimation can also be seen as a filtering process. Kalman filtering is currently the most widely used filtering method, and it has been widely applied in many fields such as communication, navigation, guidance and control.

## 2 Blast Furnace Mathematical Model

After leaving the chute, the blast furnace charge fell into the empty area and was subjected to rising gas resistance in addition to continued gravity. Assume that the rising gas resistance is $P$; The gravity of the burden material is $Qg$; According to the principle of fluid mechanics, it can be written as:

$$P = ks\frac{\gamma v^2}{2g} \tag{1}$$

In the expression: $k$ is the resistance coefficient; $\gamma$ is the gas density,the unit of $\gamma$ is kg/m$^3$; $s$ is the maximum cross-sectional area of the largest charge, the unit of $s$ is m$^2$; $g$ is the acceleration of gravity, the unit of $g$ is m/s$^2$ (Fig. 1).

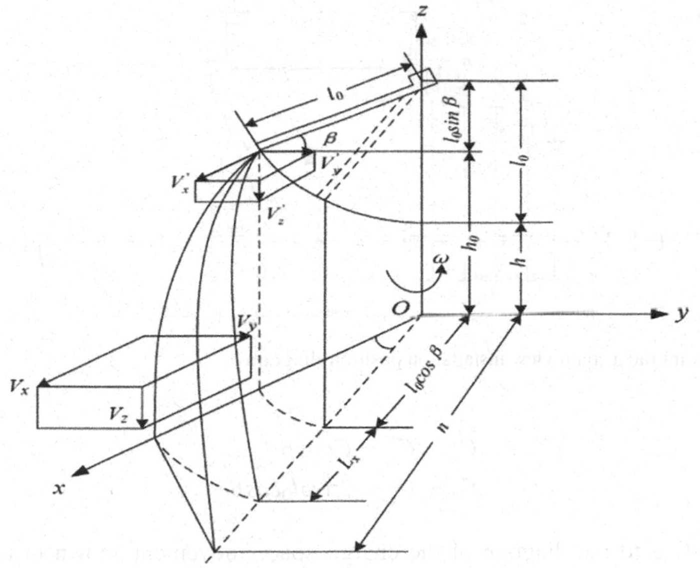


**Fig. 1** Blast furnace charge state model

Assume that the time the charge leaves the chute to the material surface (the movement time in the empty area) is $t$; Take the No. 1 radar as an example, establish a space rectangular coordinate system $O_{xyz}$, The initial velocity of the charge just leaving the end of the chute is $C$, Then decompose the speed to get $C_x'$, $C_y'$, $C_z'$, the chute rotates at the speed ω, and the length of the chute is $l_0$:

$$\begin{cases} C_x' = Ccos\beta \\ C_y' = 2\pi\omega l_0\ cos\beta \\ C_z' = Csin\beta \end{cases} \tag{2}$$

After the charge leaves the chute, it enters the empty space of the blast furnace and is subjected to downward gravity $Qg$ and upward gas resistance $P$ in the direction. According to Newton's second law and kinematics formula, the speed in the $z$ direction when the charge falls to the bottom of the blast furnace and the distance from the charge to the zero material surface can be obtained and denoted by $h_0$.

$$C_z = C_z' + \frac{Qg - P}{Q}t = Csin\beta + \frac{Qg - P}{Q}t$$

$$h_0 = C_z' t + \frac{Qg - P}{2Q}t^2 = Csin\beta t + \frac{Qg - P}{2Q}t^2 \tag{3}$$

Since the charge is not affected by force in the $x$, $y$ directions, so its expression is:

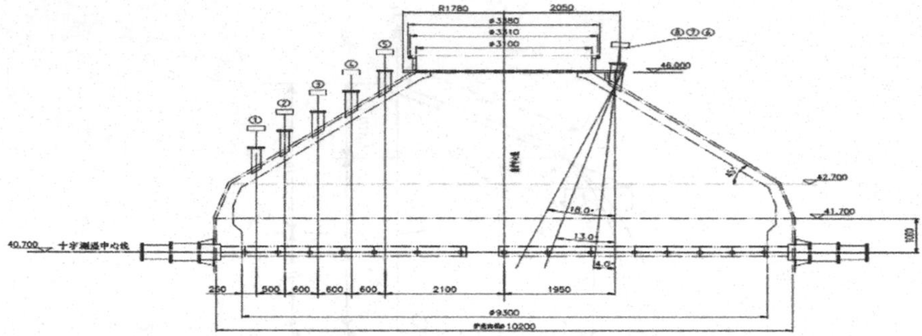


**Fig. 2** 8 point radar main view installation position diagram

$$C_x = C_x' = C\cos\beta$$
$$C_y = C_y' = 2\pi\omega l_0 \cos\beta \tag{4}$$

According to the diagram of the charge space movement shown in the figure above:

$$h_0 + l_0 \sin\beta = l_0 + h \tag{5}$$

Finishing the expression, the solution is (Fig. 2):

$$h_0 = h + l_0(1 - \sin\beta) \tag{6}$$

Assume that the charge falls from the end of the chute to the material level, moving distance in the $c$ direction in the empty area is $L_x$. According to the same principle of object motion time, there are formulas:

$$t = \frac{L_x}{C\cos\beta} \tag{7}$$

The above equations are used to determine the relationship between the depth of the feed line and the horizontal position of the charge stack tip:

$$h = L_x \tan\beta + \frac{Qg - P}{2QC^2\cos^2\beta} L_x^2 - l_0(1 - \sin\beta) \tag{8}$$

Transform the above equation to get:

$$L_x = \frac{QC^2\cos^2\beta}{Qg - P}\left\{\sqrt{\tan^2\beta + \frac{2(Qg - P)}{QC^2\cos^2\beta}[l_0(1 - \sin\beta) + h]} - \tan\beta\right\} \tag{9}$$

Similarly, we can get the horizontal position in the $y$ direction $L_y$:

$$L_y = C_y t = 2\pi \omega l_0 cos\beta t = 2\pi \omega l_0 cos\beta \frac{L_x}{Ccos\beta} \tag{10}$$

Finishing the expression, the solution is:

$$L_y = 2\pi \omega l_0 \frac{QCcos^2\beta}{Qg - P} \times \left\{ \sqrt{tan^2\beta + \frac{2(Qg - P)}{QC^2cos^2\beta}[l_0(1 - sin\beta) + h]} - tan\beta \right\} \tag{11}$$

Similarly, we can get $L_z$

$$L_z = h_0 = Csin\beta t + \frac{Qg - P}{2Q} t^2 \tag{12}$$

## 3 Establish Kalman Filter Time Update Equation

Kalman filtering is a real-time recursive algorithm. It uses the system's material height measurement value h($k$) as the input of the filter. Under the condition of known material height measurement h($k$), the minimum linear variance estimation $\hat{x}_k$ of the system state $x(k)$ is obtained and finally the material level of the blast furnace is realized. The optimal estimate of the prediction bias. The input and output of the Kalman filter are linked by a time update and measurement update algorithm. They mainly include three models: the system state transfer model, the height measurement model and the filter model of the blast furnace, plus the corresponding System dynamic noise and radar measurement noise. The Kalman filter tracking algorithm is based on the state equation of the system and the height measurement equation of the blast furnace material level. We first establish the system state equation and the height measurement equation of the blast furnace material level.

Kalman filtering is usually used to estimate the state variables of a discrete process $x \in R^n$, which can be described by the following linear stochastic difference equation:

$$x_{k+1} = Ax_k + Bu_k + C\omega_k \tag{13}$$

The observed variable is $z_k$, And the measurement equation is:

$$z_k = Hx_k + \vartheta_k \tag{14}$$

Among them, $A$ is n × n dimensional state transition matrix of the dimension, which converts the state of $k$ at the previous moment to the current moment $k$ + 1, $B$ is $n \times l$ dimensional state control input matrix of the dimension, $C$. is the $n \times l$ dimensional state noise driving matrix of the dimension, $H$ is m × n dimensional state observation matrix of the dimension, which reflects the state variable $x_k$ pair Observed variable $z_k$ gain, $x_k$ is the system's dimensional of n state vector, $z_k$ is the dimensional observation of m state in the system, $u_k$ is the dimensional control input vector, $\omega_k$ is the process noise, and $\vartheta_k$ is the observed noise. Both $\omega_k$ and $\vartheta_k$ are Gaussian white noise sequences that are independent and have a normal distribution.

In the process of fabric distribution, the radar is fixed on the workbench, the measurement points are fixed, and the radar is taken as the reference system. The detected blast furnace charge is regarded as the particle, and the space rectangular coordinate system is established. Assume that at time k, the radar detects that the coordinate value of the blast furnace material level is $P = \left(r_x(k), r_y(k), r_z(k)\right)$, and at time k+1, the material level of the blast furnace becomes $P = \left(r_x(k+1), r_y(k+1), r_z(k+1)\right)$, and the sampling time is t. Assume at time k that the velocity of the blast furnace charge in the $X$, $Y$, $Z$ planes is $v_x(k)$, $v_y(k)$, $v_z(k)$. The accelerations are $a_x(k)$, $a_y(k)$, $a_z(k)$ respectively. The influence of noise interference $\omega_x(k)$, $\omega_y(k)$, $\omega_z(k)$ is received during the blast furnace charge movement. So set the charge material point $Q$ of motion in the three-dimensional plane, and set $\Delta t = t(k+1) - t(k)$, then at time k its state vector $x_k$ is expressed as:

$$x_k = \left[r_x(k)r_y(k)r_z(k)v_x(k)v_y(k)v_z(k)a_x(k)a_y(k)a_z(k)\right] \tag{15}$$

$$\begin{cases} r_x(k+1) = r_x(k) + \Delta t v_x(k) + \frac{1}{2}\Delta t^2 a_x(k) \\ r_y(k+1) = r_y(k) + \Delta t v_y(k) + \frac{1}{2}\Delta t^2 a_y(k) \\ r_z(k+1) = r_z(k) + \Delta t v_z(k) + \frac{1}{2}\Delta t^2 a_z(k) \end{cases} \tag{16}$$

$$\begin{cases} v_x(k+1) = v_x(k) + a_x(k)\Delta t \\ \quad = Ccos\beta + a_x(k)\Delta t \\ v_y(k+1) = v_y(k) + a_y(k)\Delta t \\ \quad = 2\pi\omega l_0 cos\beta + a_y(k)\Delta t \\ v_z(k+1) = v_z(k) + a_z(k)\Delta t \\ \quad = Csin\beta + \frac{Qg-P}{Q}\Delta t + a_z(k)\Delta t \end{cases} \tag{17}$$

$$\begin{cases} a_x(k+1) = a_x(k) + \omega_x(k) = \omega_x(k) \\ a_y(k+1) = a_y(k) + \omega_y(k) = \omega_y(k) \\ a_z(k+1) = a_z(k) + \omega_z(k) = \frac{Qg-P}{Q} + \omega_z(k) \end{cases} \tag{18}$$

In summary, the state equation of the material of the furnace can be expressed as $\mathbf{x_{k+1} = Ax_k + Bu_k + C\omega_k}$由, By the above three equations, the state transition matrix is $A$ $(9 \times 9)$, the control input matrix is $B$ $(9 \times 3)$, and the noise drive matrix $C(9 \times 1)$.

$$A = \begin{bmatrix} I_3 & \Delta t I_3 & \frac{1}{\lambda^2}\left(e^{-\lambda\Delta t} + \lambda\Delta t - 1\right)I_3 \\ O_3 & I_3 & \frac{1}{\lambda}\left(1 - e^{-\lambda\Delta t}\right)I_3 \\ O_3 & O_3 & e^{-\lambda\Delta t}I_3 \end{bmatrix} \tag{19}$$

$$B = \begin{bmatrix} O_2 & 0 \\ 0 & \frac{1}{2}\Delta t^2 \\ O_2 & 0 \\ 0 & \Delta t \\ O_2 & 0 \\ 0 & 1 \end{bmatrix} \tag{20}$$

$$C = \left[ 0\,0\,0\,0\,0\,0\,\omega_x(k)\omega_y(k)\omega_z(k) \right]^T \tag{21}$$

Among them:

$$\mathrm{E}[c\omega_k] = q_1 = 0_{9\times1}$$

$$\mathrm{E}\left[c\omega_k c\omega_k^T\right] = Q_1 = \begin{bmatrix} 0_6 & 0_{6\times3} \\ 0_{3\times6} & \sigma^2 I_3 \end{bmatrix}$$

The radar uses azimuth observations on the charge material point, observed as pitch and horizontal declination, In actual measurement, radar has additive measurement noise $\vartheta_k$, and the measurement equation is:

$$z_k = Hx_k + \vartheta_k \tag{22}$$

$$Hx_k = \left[ arctan\frac{r_y(k)}{\sqrt{r_x(k)^2 + r_y(k)^2}}, arctan\frac{-r_x(k)}{r_z(k)} \right] \tag{23}$$

$\vartheta_k$ is the measurement noise and it is a white Gaussian noise random sequence white noise.

$$\mathrm{E}[\vartheta_k] = r_1 = 0_{2\times1}, \mathrm{E}\left[\vartheta_k\vartheta_k^T\right] = R_1$$

$$R_1(\mathrm{k}) = D^{-1}(k)xD^{-T}(k), x = 0.1I_2 \tag{24}$$

$$D(k) = \begin{bmatrix} \sqrt{r_x(k)^2 + r_y(k)^2 + r_z(k)^2} & 0 \\ 0 & \sqrt{r_x(k)^2 + r_y(k)^2 + r_z(k)^2} \end{bmatrix} \tag{25}$$

## 4 Blast Furnace Charge Tracking Algorithm Steps

The Kalman filter estimates the state of the process in a feedback-controlled manner: the state of the filter is estimated at some point in the process, and feedback is obtained

Fig. 3 Kalman filtering process cycle diagram

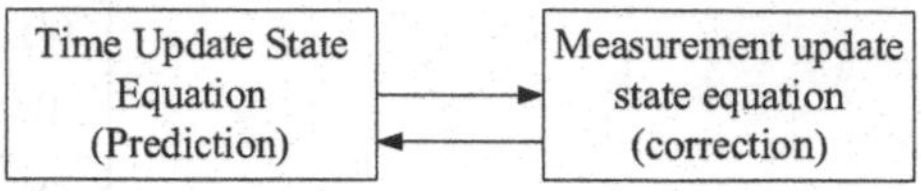


in the form of observation variables (include noise). Therefore, the Kalman filter equation can be divided into two parts: the time update equation and the measurement update equation.

1. The time update equation forwards the state variable and the error covariance estimate at the current moment to obtain the a priori estimate at the next moment.
2. The measurement update equation can be viewed as a correction equation. The final estimation algorithm is called a prediction-correction algorithm with numerical solutions, as shown in the schematic figure (Fig. 3).

The collection of blast furnace level data will be affected by radar measurement noise. In order to eliminate the influence of noise, the Kalman filter is used to carry out an optimal estimation of the unknown state of the blast furnace's material level detected by the radar, and according to the principle of minimum mean square error between the measured signal and the estimated signal, The recursive formula of the Kalman filter for the material level deviation can be deduced. Each iteration of the loop is mainly based on the prediction based on the previous estimated value and the information obtained by the measurement data to correct the prediction, and the best estimate $x(k|k)$ is obtained.

1. State prediction and observation prediction

$$\begin{aligned} \hat{x}(k+1|k) &= A(k+1|k)\hat{x}(k|k) + Bu_k + C\omega_k \\ z_k &= Hx_k + \vartheta_k \end{aligned} \tag{26}$$

At the time $k$, the system state estimate $\hat{x}(k|k)$ is known. The state transition matrix $A(k+1|k)$ is multiplied by the state estimation value to obtain the prediction of the blast furnace material level from the time $k$ to the state of the $k+1$ moment.

2. Update covariance of $x(k|k-1)$

$$P(k+1|k) = A(k+1|k)P(k+1|k)A^T(k+1|k) + Q \tag{27}$$

Since the state vector contains the material level deviation information, the prediction of the material level deviation at the time of $k+1$ is realized, but the predicted value at this time has already brought the system process noise ω(k), obeys the normal distribution $N(0, Q)$.

3. Nonlinear time update equations for prediction

The time update equation includes the equations of motion and measurement equations of the system. The motion equation of the system is linear, and the measurement matrix is nonlinear. So:

$$H_{[i,j]} = \frac{\partial h_{[i]}}{\partial x_{[j]}}\Big|_{x(k)=\hat{x}(k+1|k)}$$

$$H_{[i,j]} = \begin{pmatrix} \frac{-r_x(k)r_y(k)}{\sqrt{r_x^2(k)+r_y^2(k)+r_z^2(k)}} & \frac{\sqrt{r_x(k)^2+r_z(k)^2}}{\sqrt{r_x^2(k)+r_y^2(k)+r_z^2(k)}} \\ \frac{r_x(k)}{r_x^2(k)+r_z^2(k)} & 0 \end{pmatrix}$$

$$\begin{pmatrix} \frac{-r_y(k)r_z(k)}{r_x^2(k)+r_y^2(k)+r_z^2(k)} & 0\,0\,0\,0\,0\,0 \\ \frac{-r_x(k)}{r_x^2(k)+r_z^2(k)} & 0\,0\,0\,0\,0\,0 \end{pmatrix} \tag{28}$$

$$z_k \approx \tilde{z}_k + H_{[i,j]}(x_k - \tilde{x}_k) + V\vartheta_k \tag{29}$$

4. Calculate filter gain $K_k$

$$K_k = P(k|k-1)H^T\left(HP(k|k-1)H^T + R\right)^{-1} \tag{30}$$

When the measurement noise covariance $R$ is smaller, the residual gain $K_k$ is larger. In particular, when $R$ approaches zero, there is

$$\lim_{R_k \to 0} K_k = H^- \tag{31}$$

On the other hand, the smaller the a priori estimated error covariance $P(k|k-1)$ is, the smaller the residual gain $K$ is. In particular, when $P(k|k-1)$ approaches zero, there are:

$$\lim_{P(k|k-1)\to 0} K_k = 0 \tag{32}$$

5. Forecast Correction, Status Vector Update:

$$\hat{z}(k+1) = z(k+1) - H\hat{x}(k+1|k) \tag{33}$$

Using the measurement matrix $H$ multiplied by the state prediction value to obtain the predicted value of the blast furnace level measurement at $k+1$.

$$\hat{x}(k+1|k+1) = \hat{x}(k+1|k) + K_{k+1}\hat{z}(k+1) \tag{34}$$

Then use the filter gain $K_{k+1}$ multiplied by the innovation value to get the filter correction at time $k+1$.

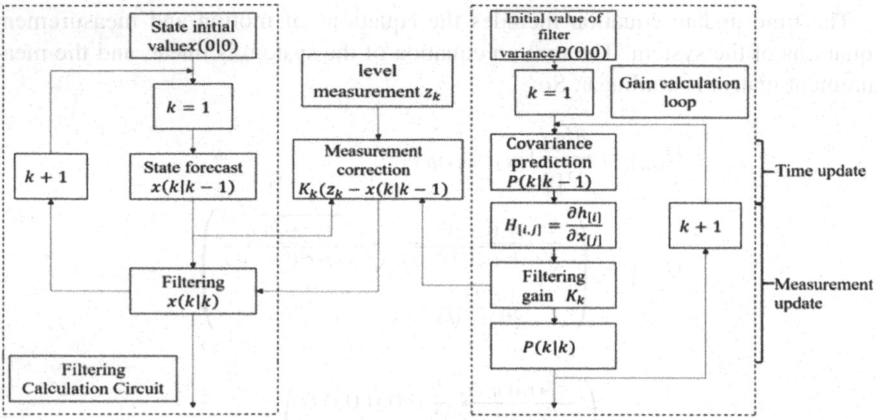


**Fig. 4** Extended Kalman filter based algorithm for blast furnace charge status

6. Covariance update of $x(k|k)$

$$P(k+1|k+1) = [I - K_k H]P(K+1|K) \quad (35)$$

The state prediction value $\hat{x}(k+1|k)$ is added to the correction amount to obtain the filter value at the time of the state vector $k+1$ to eliminate the influence of noise, and at this time the filter error variance is updated. Given the initial value $\hat{x}(0|0)$, $P(0|0)$, the above ground-pull formula is unbiased (Fig. 4).

# 5 Extended Kalman Filter Simulation Experiment

The Extended Kalman Filter (EKF) algorithm is described in detail based on the actual conditions of the blast furnace (using No. 1 radar as an example), perform initialization settings, set sampling time t and simulation time t, and maneuver frequency λ. Assume that the initialization sampling time is $\Delta t = 1$ s, the simulation time is 500 s, and $\lambda = 10{,}000$ HZ

The radar installation data, as shown in Table 1, B is the radial distance between the radar installation position and the furnace center. $\alpha$ is the angle between the position of 0°, and the $\beta$ is the angle of radar incidence. The installation coordinates of each point under the overlook of the blast furnace is $(x_i, y_i)$.

The initial value of the equation of state:

$$x(0) = \left[r_x(0) r_y(0) r_z(0) v_x(0) v_y(0) v_z(0) a_x(0) a_y(0) a_z(0)\right]$$

**Table 1** 8 point radar unknown parameters

| Radar | B (m) | α | β | $(x_i, y_i)$ |
|---|---|---|---|---|
| #1 | 4.4 | 342° | 4° | (−1.30, 3.99) |
| #2 | 3.9 | 285° | 3° | (−3.80, 1.02) |
| #3 | 3.4 | 105° | 3° | (3.08, −1.44) |
| #4 | 2.4 | 15° | 5° | (0.62, 2.32) |
| #5 | 2.2 | 145° | 15° | (1.26, 1.80) |
| #6 | 2.0 | 27° | 4° | (0.91, 1.78) |
| #7 | 2.0 | 0° | 3° | (0.00, 2.00) |
| #8 | 2.2 | 215° | 12° | (−1.26, −1.8) |

$$\begin{cases} r_x(0) = 1.3\,\text{m} \\ r_y(0) = 3.99\,\text{m} \\ r_z(0) = 2.10\,\text{m} \end{cases}$$

$$\begin{cases} v_x(0) = C_x' = Ccos\beta \\ v_y(0) = C_y' = 2\pi\omega l_0\ cos \\ v_z(0) = C_z' = Csin\beta \end{cases}$$

Set the chute length $l_0 = 2.6\,\text{m}$, chute angle $\beta \approx 60°$, chute speed $\omega \approx 0.16$ laps/s, according to experience to take charge at the end of the chute speed $c = 4.2\,\text{m/s}$.

$$\begin{cases} v_x(0) = 2.1\,\text{m/s} \\ v_y(0) = 1.31\,\text{m/s} \\ v_z(0) = 3.64\,\text{m/s} \end{cases} \quad \begin{cases} a_x(0) = 0 \\ a_y(0) = 0 \\ a_z(0) = 4.5\,\text{m/s}^2 \end{cases}$$

Set the process noise variance:

$$\sigma^2 = 0.1\,Q = \left[0_{6\times6},\ \ 0_{3\times6},\ \ 0_{6\times3},\ \ \sigma^2 I_{3\times3}\right]$$

Initialize the state covariance matrix estimated by the EKF filter:

$$P_0 = \left[10^4 I_6, 0_{6\times3}, 0_{3\times6}, 10^2 I_3\right]$$

In summary, write the program through matlab and get the simulation results (Fig. 5).

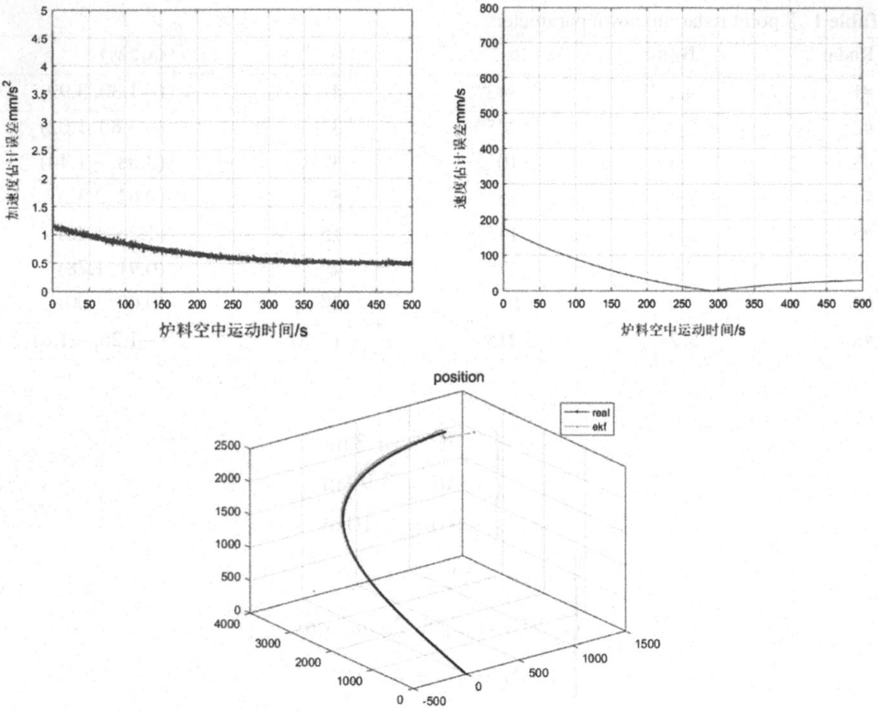


**Fig. 5** Matlab simulation image

# 6 Conclusions

In this paper, a blast furnace charge model was established based on the Tangshan Iron and Steel Plant. According to the actual charge material state, the state equations and measurement equations of the charge materials are established, and a three-dimensional motion model is established. Since the measurement equations are nonlinear equations, the Kalman filter is used to predict the state equations, and finally it is obtained through simulation. The predicted tracking effect is good, acceleration error, speed error, and distance error all decrease with time.

# References

1. Y. Liu, *The Regulation on burden Distribution* (Metallurgical Industry Press, Beijing, 2005)
2. W. Huang, Y. Xu, J. Wang, Algorithm of data fusion based on mode recognition and regression analysis. Control Decis. **25**(1), 110–114 (2010)
3. D. Liu, X. Li, D. Ding et al., Multi-model control of blast furnace burden surface based on observed data of radars. Control Theor. Appl. **29**(10), 1277–1283 (2012)
4. L. He, PLC optimized application on blast furnace stock rod control. Autom. Instrum. **29**(6), 74–74 (2009)

# An Overview of SLAM

**Sufang Wang, Zheng Wu and Weicun Zhang**

**Abstract** Simultaneous Localization and Mapping (SLAM) based on LIDAR and Visual SLAM (VSLAM) are key technologies for mobile robot navigation. In this paper, the SLAM algorithm based on these two types of sensors is described, and their advantages and disadvantages are comprehensively analyzed and compared. In order to better achieve active navigation and positioning, path planning and obstacle avoidance, the advantages of both should be brought into full play. In the end, the future development direction of mobile robot is discussed.

**Keywords** SLAM · Mobile robot · Navigation · Sensor

## 1 Introduction

For autonomous mobile robots, there are three aspects involved in navigation and positioning: (1) where am I? (2) where am I going? (3) How should I get there? [1]. These three issues correspond to three aspects respectively: (1) positioning, get the robot's current position. (2) Perception, get target location. (3) Path planning. SLAM stated that in an unknown environment, the mobile robot can perceive the surrounding environment according to its own sensors during the movement so as to perform autonomous positioning and incrementally construct an environmental map [2, 3].

S. Wang · W. Zhang (✉)
School of Automation and Electrical Engineering, University of Science and Technology Beijing, Beijing 100083, China
e-mail: weicunzhang@263.net

S. Wang
e-mail: s20170610@xs.ustb.edu.cn

Z. Wu
Youer (Beijing) Robot Technology Ltd., Beijing 100081, China
e-mail: wuzheng@youerobot.com



Y. Jia et al. (eds.), *Proceedings of 2018 Chinese Intelligent Systems Conference*, Lecture Notes in Electrical Engineering 528,
https://doi.org/10.1007/978-981-13-2288-4_64

Although the research on autonomous mobile robots has become a hot topic in the field of high-tech, we have seen and applied very few in our lives. Most of them are based on theoretical research. According to published papers at home and abroad, most of them are based on a single sensor, such as LIDAR SLAM and visual SLAM. SLAM using LIDAR has mature algorithms and solutions. With the follow-up of hardware devices, visual SLAM has also developed. In order to meet the requirements as much as possible, different products use different sensors.

The SLAM technology of mobile robot has very important theoretical significance and application value. For driverless cars, SLAM can build 3D environment models and position navigation by LIDAR. In terms of military, SLAM can allow mobile robots to reach many harsh environments that humans cannot reach. It can help realize intelligent reconnaissance and combat of robots. It can also be used to search for and remove hazardous explosives [4]. In life, SLAM can achieve household robots to walk autonomously and successfully avoid obstacles to complete high-quality tasks.

## 2 SLAM Classification

Intelligent robots gradually enter our lives. It makes our lives easier and more convenient. However, simple robots cannot walk autonomously. At this time, the auxiliary role of the LIDAR is needed to realize the robot's intelligence. It can acquire the information of the robot's environment in real time. However, robot cannot understand the information scanned by LIDAR. The powerful SLAM navigation algorithm is depended to achieve the intelligent walking of the robot. Common LIDAR, such as SICK, Velodyne and rplidar, can be used for SLAM.

VSLAM is mainly implemented with cameras. According to different working methods, VSLAM is classified into Monocular, Stereo Vision and RGB-D.

Using only one camera for SLAM is called monocular SLAM. Using multiple cameras as sensors is called stereo vision SLAM, the most widely used is the stereo camera. The combination of a monocular camera and an infrared sensor to form a sensor is called RGB-D SLAM.

## 3 LIDAR-SLAM

The key issue in SLAM is positioning [5]. The methods to solve positioning problems are divided into probabilistic and non-probabilistic method. The method based on probability is the mainstream method, and the method based on Bayesian estimation is the basis of probability estimation method. There are mainly Kalman filter (KF) method and particle filter (PF) method.

(1) Kalman Filter Method

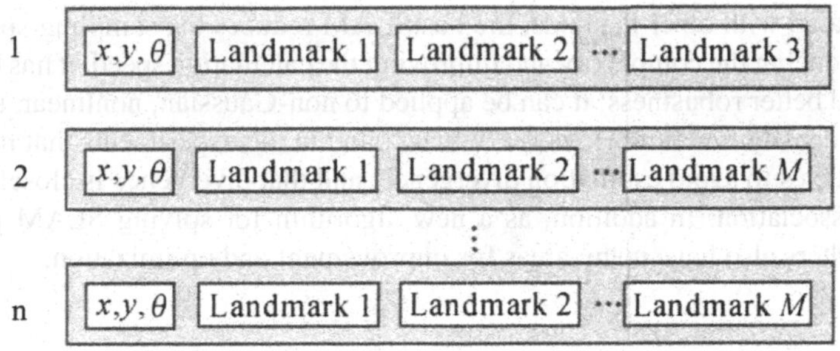


**Fig. 1** RBPF-SLAM algorithm

In the case that the state noise and observation noise are gaussian distributions, data is observed by system input and output. Then KF makes the optimal estimate of the system state [6–9]. The premise of using KF is that the system is a linear system. However, the actual system is often a nonlinear system. Extended Kalman Filter (EKF) linearizes the nonlinear system through first-order Taylor expansion [10]. Unscented Kalman Filter(UKF) is a method of approximating nonlinear distribution using a sampling strategy [11]. It does not need to calculate the Jacobian, and has a higher linearization accuracy, and its performance is better than that of EKF.

KF-SLAM occupies a dominant position in many solutions, because it has greater advantages in convergence and implementation complexity. But it has the problem of lack of self-closed loop capability and associated fragility. When an error occurs in the data association, it will eventually be brought into the entire SLAM state estimation, sometimes even causing the entire prediction process to diverge. Therefore, a robust data association method is very important.

(2) Particle Filter Method

As a new type of filter, the particle filter can get rid of the linear assumption of the system and the constraints of the Gaussian noise assumption of the sensor. It can approximate any probability distribution and the calculation is simple and convenient. It can effectively solve the robot positioning problem [12]. Murphy [13], Murphy and Russell [14] found that if a robot's motion trajectory is known, then the probability between landmark positions is conditionally independent. Therefore, Rao-Blackwellised decomposition is proposed and implemented, which provides a theoretical basis for the particle filter to solve the SLAM problem. Based on this, Montemerlo et al. [15] demonstrated the feasibility of using Rao-Blackwellised Particle Filter (RBPF) to solve the SLAM problem, and proposed the FastSLAM algorithm.

The FastSLAM uses a improved particle filter to estimate the posterior distribution of the robot path. In Fig. 1 [16], each particle maintains a state estimate and a set of individual feature location information. Each particle represents a path traveled by a robot. Each feature is individually estimated using an EKF. Each particle maintains M EKFs and there are a total of N particles to predict the robot state [17].

Compared with other methods, the FastSLAM reduces the sampling space, thus greatly reducing the complexity and improving the calculation speed; it has high precision and better robustness; it can be applied to non-Gaussian, nonlinear, unknown posterior density function. However, it was found in the experiments that it requires more particles to avoid estimation divergence, and that divergence is closely related to data association. In addition, as a new algorithm for solving SLAM problem, particle filters also have many areas for improvement and optimization.

# 4 Visual SLAM Algorithm

Visual SLAM is mainly divided into visual front-end and optimized back-end. The front-end is also called visual odometry (VO). It estimates rough camera movement based on the information of adjacent images and provides a good initial value for the back-end. The implementation method of VO is divided into feature point method and direct method according to whether or not features need to be extracted. Feature-based method is stable in operation and insensitive to light and dynamic objects. So it is considered as the mainstream method of VO [18].

(1) Feature-based method

Davision et al. [19] first proposed a monocular visual SLAM (MonoSLAM) system that uses EKF as the back-end to track sparse feature points on the front-end. After the keyframe-based monocular visual SLAM gradually developed [20–23]. The most representative of these is parallel tracking and mapping (PTAM). Klein and Murray [21] proposed a simple and effective method for extracting keyframes, parallelizing the tracking and mapping. And for the first time PTAM used nonlinear optimization as the back-end. Mur-Artal et al. [23] inherited and improved PTAM and innovatively proposed the three thread to implement the monocular visual SLAM system based on PTAM of the dual thread. The entire system is implemented around the ORB feature, namely ORB-SLAM. Subsequent studies have shown that the ORB-SLAM system is equally applicable to monocular, stereo, and RGB-D models, and has good general-purpose use.

Because a monocular camera captures a two-dimensional projection of a three-dimensional space object, which is a single image, and there is uncertainty in the motion and trajectory estimated by moving the camera, the true depth of the object cannot be determined. Therefore, there are enormous difficulties in the three-dimensional reconstruction work. At this time, stereo cameras and RGB-D cameras appear, but stereo cameras have a major problem in terms of calculations under the existing conditions. However, SLAM based on RGB-D data simplifies the complexity of 3D reconstruction. Henry et al. [24] first proposed method of 3D reconstruction of indoor environment using RGB-D camera, extracting the SIFT features in the color image and finding the corresponding depth information on the depth image. Then Random Sample Consensus (RANSAC) method is used to match the 3D feature points and the corresponding rigid motion transformation is calculated. RANSAC is applied to many cases with incorrect data, which can handle data with error matching and use it as the initial value of the iterative closest point (ICP) to find more accurate position and pose. RGB-D SLAM usually uses ICP algorithm to estimate pose and optimize the camera's motion transfer matrix. The basic process of ICP algorithm in reference [25] is as follows:

Step 1: Read the point sets P1, P2.
Step 2: Select the pair of points. Search P2 for the closest point to P1 to form a point pair. Find out all the pairs of points in the two point sets.
Step 3: Calculate two barycentric coordinates based on the pairs of points in the two point sets.
Step 4: From the new point set, calculate the rotation matrix R and the translation matrix t.
Step 5: From the obtained R and t, calculate the new point set $P2'$ after the rigid transformation of the point set P2.
Step 6: If the absolute value of the difference between the sum of squares of the distances from P2 to $P2'$ is less than the threshold value for two consecutive times, then it is converged and the iteration is stopped. Otherwise, the steps 1–6 are repeated until convergence.

Among many matching algorithms for depth images, such as Generic Algorithm, ANSAC, and ICP. ICP is the most widely used one. It does three-dimensional data processing directly to depth images, and does not need to assume and divide of object features. After selecting the initial value, the algorithm has good convergence, thus getting the global optimal value and obtaining more accurate matching results. So it quickly becomes a mainstream algorithm for depth image matching [26].

Although feature-based SLAM method has many advantages, the extraction of key points and the calculation of feature points are very time-consuming, and some useful image information may be discarded because only feature points are used. Such as a white wall, where there are few feature points. We will not be able to accurately calculate the motion of the camera.

(2) Direct method

According to the classification of the number of pixels used, direct method can be divided into sparse, dense and semi-dense. Assume that P is a spatial point of a known location. When P is derived from a sparse key point, it is called a sparse direct method. When P comes from a portion of pixels, it is called semi-dense direct method. If all pixels are used, it is called dense direct method, which can build a complete map.

Many people have been devoted to the study of direct method [27–33]. Irani and Anandan [27] gave a detailed and in-depth description of direct method. Silveira et al. [28] applied direct method to the visual SLAM and described main advantages and limitations. Subsequently, a sparse-based semi-direct monocular visual odometry (SVO) was proposed [29]. This algorithm has high accuracy, good robustness, and is faster than the most advanced methods currently available. This sparse method eliminates the needs of feature extraction and robust matching techniques for motion estimation, and is suitable for estimating the state of microcars in GPS denied environments. Next, Engel et al. [30] proposed a large-scale direct monocular SLAM (LSD-SLAM). Compared to other direct methods, it reconstructs the keyframe's pose map and the environment's semi-dense and highly accurate three-dimensional map in real time. Usenko et al. [33] proposed a novel direct visual-inertial odometry method for stereo cameras that use the complementarity of visual and inertial data to improve the accuracy of 3D reconstruction maps.

Newcombe et al. [34] integrate all the deep data and image information from Kinect into the observation scene, and reconstruct the 3D model, so as to get the global map. Particularly, it allows reconstruction of dense maps in real time, making a big step towards augmented reality (AR). Henry et al. [35] use a joint optimization algorithm to apply RGB-D cameras to the robot field in indoor environments. Kerl et al. [36] proposed a visual SLAM method based on a direct dense RGB-D camera. The error terms of this method are the photometric and the depth error. The optimal camera pose is solved using the g2o optimization library. And a keyframe selection and loop closure detection method entropy-based is proposed. Thus, the path error is greatly reduced.

Compared to feature method, direct method does not need to extract image features. It has a fast speed of execution, high robustness to the photometric error of the image, but a high requirement for the camera internal reference. When there is geometric noise, the algorithm performance decreases quickly. In the event of image motion blur, camera position can still be achieved. But direct method has poor robustness to large baseline motion.

## 5 Scheme Comparison

The advantages of LIDAR is that it has a wide range of visibility, and can detect the angle and distance of the obstacle points with high accuracy to achieve obstacle avoidance. But it's expensive. However, the camera has the advantages of small size, light weight, easy installation, abundant information extraction, convenient and flexible, and low price. Therefore, VSLAM has become a hot topic in the research of SLAM algorithm in recent years [37].

The performance of visual SLAM depends on the environment in which it is operating. Ideal conditions are as follows:

(1) Light is sufficient, without large changes in lighting. The camera must be able to identify features in the scene. Generally, a more complex scene, with lots of objects or geometry is best. Blank walls, floors, or ceilings are the worst cases. Many reflective surfaces, such as glass or mirrors, can cause problems. Also, direct sunlight can interfere with the depth cameras, which affects the accuracy of occupancy mapping.

(2) When the scene is mostly motionless, Visual SLAM works best. If people or objects are moving, performance will be affected more or less. If the entire scene is moving, such as in an elevator, SLAM will not work at all.
(3) When the camera motion is primarily translation, not a rotation. Visual SLAM works best. When rotation is necessary, it is best to rotate slowly.
(4) When SLAM begins, the camera must be stationary, and there must be sufficient visual features. If the camera is aimed at a blank wall, floor, or ceiling that is enough to block the camera's view, SLAM may not initialize properly.

LIDAR and camera have their own advantages and disadvantages. It is very easy for the camera to identify the same object, but it is difficult for the LIDAR. if the camera tells the LIDAR that the two frames are the same object, then it is possible to know what the speed and displacement of this object are between the two frames by the LIDAR. It can be seen that recognition and tracking are easy to achieve, resulting in more accurate maps.

## 6 Conclusions

After more than thirty years of research and efforts by predecessors, SLAM has made great progress. Because of the limitations of the indoor operating environment, GPS cannot be used to restrict positioning errors, and SLAM opens the door for the development of the indoor robot field. SLAM based on LIDAR has already a relatively mature scheme, but the high cost is still the primary problem. Therefore, the low-cost visual SLAM has become a research hotspot in recent years. However, no matter which sensor is used alone, there are some defects. The multi-sensor fusion technology based on LIDAR, vision sensor and inertial measurement unit [38, 39] not only can realize the cooperative operation among sensors, but also greatly enhance the robustness. It is believed that the research and application of multi-sensor fusion technology will bring wider space to driverless, robotics, augmented reality and virtual reality. In addition, SLAM is combined with deep learning to perform image processing [40], to generate semantic maps of the environment and improve the human-computer interaction techniques, so that intelligence can be better realized.

**Acknowledgements** This work was supported by National Natural Science Foundation of China (No. 61520106010; 61741302).

## References

1. J.J. Leonard, H.F. Durrant-Whyte, I.J. Cox, Dynamic map building for an autonomous mobile robot, vol. 11, no. 4 (Sage Publications Inc., 1992), pp. 286–298
2. W. Chen, F. Zhang, Review on the achievements in simultaneous localization and map building for mobile robot. Control Theory Appl. **22**(3), 455–457 (2005)

3. M. Csorba, *Simultaneous Localization and Mapping* (University of Oxford, 1997), pp. 56–89
4. M. Montemerlo, S. Thrun, D. Koller, Fast SLAM (Simultaneous Localization And Mapping), in *Proceedings of the AAAI National Conference on Artificial Intelligence* (Menlo Park, CA, USA, 2002), pp. 593–598
5. H. Durrant-Whyte, Where am I? A tutorial on mobile vehicle localization. Ind. Robot **21**(2), 11–16 (1994)
6. S. Yavuz, Z. Kurt, M.S. Bicer, Simultaneous localization and mapping using extended Kalman filter. in *2009 IEEE 17th Signal Processing and Communications Applications Conference* (IEEE, Antalya, 2009), pp. 700–703
7. Y. Wei, Z. Zuo, Improvement of the simultaneous localization and map building algorithm applying scaled unscented transformation, in *International Conference on Industrial Mechatronics and Automation, 2009, ICIMA 2009* (IEEE, Chengdu, 2009), pp. 371–374
8. J.G. Kang, W.S. Choi, S.Y. An et al., Augmented EKF based SLAM method for improving the accuracy of the feature map, in *2010 IEEE/RSJ International Conference on Intelligent Robots and Systems (IROS)* (IEEE, Taipei, 2010), pp. 3725–3731
9. D. Wang, H. Liang, T. Mei, Lidar scan matching EKF-SLAM using the differential model of vehicle motion, in *Intelligent Vehicles Symposium (IV), 2013 IEEE*, vol. 36, no. 1 (IEEE, Gold Coast, QLD, 2013), pp. 908–912
10. S.J. Julier, J.K. Uhlmann, A counter example to the theory of simultaneous localization and map building. in *IEEE International Conference on Robotics and Automation* (Seoul, Korea: [s. n.], 2001), pp. 4238–4243
11. S.J. Julier, The spherical simplex unscented transformation, in *Proceedings of the American Control Conference* (Denver: [s. n.], 2003), pp. 2430–2434
12. M.S. Arulampalam, S. Maskell et al., A tutorial on particle filters for online nonlinear/non-Gaussian Bayesian tracking. IEEE Trans. Signal Process. **50**(2), 174–188 (2002)
13. K. Murphy, Bayesian map learning in dynamic environments. Adv. Neural Inform. Process. Syst. **12**, 1015–1021 (2008)
14. K. Murphy, S. Russell, Rao-Blackwellised particle filtering for dynamic bayesian networks, vol. 43, no. 2 (Springer New York, 2001), pp. 499–515
15. M. Montemerlo, S. Thrun, D. Koller, Fast SLAM (Simultaneous Localization And Mapping), in *Proceedings of the AAAI National Conference on Artificial Intelligence* (AAAI, Menlo Park, CA, USA, 2002), pp. 593–598
16. E. Wu, X. Zhiyu, M. Shen, J. Liu, Robot SLAM algorithm based on laser range finder for large scale environment. J. Zhejiang Univ. **41**(12), 1982–1986 (2007)
17. S. Thrun, D. Fox, W. Burgard, F. Dallaert, Robust Monte Carlo localization for mobile robots. Artif. Intell. **128**(1–2), 99–141 (2001)
18. X. Gao, T. Zhang, Y. Liu, Q. Yan, *Visual SLAM Fourteen Lectures: From Theory to Practice* (Publishing House of Electronics Industry, Beijing, 2017), pp. 132–204
19. A.J. Davison, I.D. Reid, N.D. Molton, O. Stasse, Monoslam: real-time single camera SLAM. IEEE Trans. Pattern Anal. Mach. Intell. **29**(6), 1052–1067 (2007)
20. M. Quan, S. Piao, L. Guo, An overview of visual SLAM. CAAI Trans. Intell. Syst. **11**(6), 768–776 (2016)
21. G. Klein, D. Murray, Parallel tracking and mapping for small AR workspaces, in *IEEE and ACM Intemational Symposium on Mixed and Augmented Reality* (Nara, Japan, 2007), pp. 225–234
22. R. Mur-Artal, J.D. Tardós, Fast relocalisation and loop closing in keyframe-based SLAM, in *IEEE International Conference on Robotics and Automation* (New Orleans, LA, 2014), pp. 846–853
23. R. Mur-Artal, J.M.M. Montiel, J.D. Tardós, ORB-SLAM: a versatile and accurate monocular SLAM system. IEEE Trans. Rob. **31**(5), 1147–1163 (2015)
24. P. Henry, M. Krainin, E. Herbst et al., RGB-D mapping: using depth cameras for dense 3D modeling of indoor environments, vol. 31, no. 5 (Springer Berlin Heidelberg, 2014), pp. 647–663
25. K. Zhu, H. Liu, Q. Xia, Survey on monocular visual SLAM algorithms. Appl. Res. Comput. **35**(1), 1–6 (2018)

26. L. Shifei, P. Wang, S. Zhenkang, A survey of iterative closest point algorithm. Sig. Process. **25**(10), 1582–1588 (2009)
27. M. Irani, P. Anandan, About direct methods, in *International Workshop on Vision Algorithms: Theory & Practice* (1999), pp. 267–277
28. G. Silveira, E. Malis, P. Rives, An efficient direct approach to visual slam. IEEE Trans. Rob. (IEEE Press) **24**(5), 969–979 (2008)
29. C. Forster, M. Pizzoli, D. Scaramuzza, SVO: fast semi-direct monocular visual odometry. in *IEEE International Conference on Robotics and Automation (ICRA)* (2014), pp. 15–22
30. J. Engel, T. Schöps, D. Cremers, LSD-SLAM: large-scale direct monocular SLAM, in *Computer Vision-ECCV 2014* (Springer International Publishing, 2014), pp. 834–849
31. J. Engel, J. Sturm, D. Cremers, Semi-dense visual odometry for a monocular camera, in *Proceedings of the IEEE International Conference on Computer Vision* (2013), pp. 1449–1456
32. J. Engel, V. Koltun, D. Cremers, Direct sparse odometry. IEEE Trans. Pattern Anal. Mach. Intell. **PP**(99), 1 (2017)
33. V. Usenko, J. Engel, J. Stückler, D. Cremers, Direct visual-inertial odometry with stereo cameras, in *IEEE International Conference on Robotics and Automation (ICRA)* (2016), pp. 1885–1892
34. R.A. Newcombe, S. Izadi, O. Hilliges et al., KinectFusion: real-time dense surface mapping and tracking, in *IEEE International Symposium on Mixed and Augmented Reality* (2011), pp. 127–136
35. P. Henry, M. Krainin, E. Herbst et al., RGB-D mapping: using kinect-style depth cameras for dense 3D modeling of indoor environments. Int. J. Rob. Res. **31**(5), 647–663 (2012)
36. C. Kerl, J. Sturm, D. Cremers, Dense visual SLAM for RGB-D cameras, in *IEEE/RSJ International Conference on Intelligent Robots and Systems*, vol. 8215, no. 2 (2014), pp. 2100–2106
37. M. Wu, J. Sun, Extended Kalman filter based moving object tracking by mobile robot in unknown environment. ROBOT **32**(3), 334–343 (2010)
38. W. Zhang, *Research on Autonomous Navigation Method for Indoor Robots Based on Multi-sensor Fusion* (University of Science and Technology of China, 2017)
39. Q. Lai, S. Yu, S. Ding, Mobile robot SLAM and path planning system. Comput. Knowl. Technol. **33**(13), 24–37 (2017)
40. Y. Zhao, G. Liu, G. Tian et al., A survey of visual SLAM based on deep learning. ROBOT **39**(6), 889–896 (2017)

# The Hardware System Based on Wi-Fi Failure Prevention System for Photovoltaic Arrays

**Jili Ren, Fuzhong Wang and Sumin Han**

**Abstract** Real-time online monitoring the working status of the photovoltaic array and warning potential faults will play a crucial role in the safe operation of the photovoltaic power generation system. This paper presents a hardware system which is based on Wi-Fi failure prevention system of photovoltaic arrays. The collected data from each collecting node are transmitted to the STM32 chip to process and analysis. The system achieves the data transmission between the upper computer and the lower computer through the Wi-Fi technology and accomplish the real-time monitoring the working status of the photovoltaic array by the upper computer. The test results show that the system is not only stable but also reliable and it can monitor the working environment and status of the photovoltaic array in real time accurately.

**Keywords** Wi-Fi · Hardware design · STM32 control chip · Data acquisition · Fault early warning

## 1 Introduction

Photovoltaic arrays are key equipment of photovoltaic power generation systems [1, 2], the working environment and working status of the photovoltaic array are accurately monitored, potential faults of photovoltaic arrays are discovered in advance and take appropriate measures, which is a great significance to reduce the incidence of photovoltaic array faults and improve the safe operation of photovoltaic power generation systems. In order to improve the efficient and safe operation of photo-

J. Ren (✉) · F. Wang · S. Han
School of Electrical Engineering and Automation,
Henan Polytechnic University, Jiaozuo 454000, China
e-mail: 374039706@qq.com

F. Wang
e-mail: wangfzh@hpu.edu.cn

S. Han
e-mail: hansumin@hpu.edu.cn

 
Y. Jia et al. (eds.), *Proceedings of 2018 Chinese Intelligent Systems Conference*,
Lecture Notes in Electrical Engineering 528,
https://doi.org/10.1007/978-981-13-2288-4_65

voltaic systems, domestic and foreign scholars have also conducted research and design of photovoltaic system hardware equipment. In the literature [3], a low-cost, high-reliability and highly-flexible hardware test circuit for photovoltaic cell test system was designed. The design scheme of the hardware circuit, the module division, and the specific implementation of the corresponding circuit were discussed, photovoltaic cell test systems have been developed to meet actual needs. Literature [4] proposed the use of zigbee's wireless transmission technology in the monitoring system for the problems of remote monitoring and centralized control in photovoltaic power generation systems, and designed a communication protocol based on self-development, which is a great reference for fault warning. In the literature [5], the data collected by each collection node are processed by the STM32F103 single-chip microcomputer. Then the temperature and humidity parameters of each collection node are sent to the remote server through the Wi-Fi module and the wireless router. In this way, remote monitoring of temperature and humidity is realized and developed.

Based on the study of the above ideas and related aspects, this paper designs the hardware part of the fault diagnosis and early warning device of the photovoltaic array, and Wi-Fi technology is applied to the PV array early warning system, the remote data transmission of the system is realized, and the working environment and status of the photovoltaic array are monitored in real time through the PC and the mobile client.

The arrangement of this paper is as follows: Sect. 2 presents an overview of some related works, and the overall design is also performed, in Sect. 3, the main circuit design of the system is carried out based on the analysis of relevant hardware modules to complete the collection and transmission of the data of each acquisition module and the function realization of the terminal module. Section 4 has been selected for the acquisition module hardware and circuit design. Section 5 test the hardware system. Finally, this article's concludes are in Sect. 6.

## 2 Overall System Design

The hardware system that based on Wi-Fi failure prevention system for photovoltaic arrays is composed of an upper computer and a lower computer. The upper computer includes a PC terminal and a mobile phone client. It can monitor the photovoltaic array fault status online through Wi-Fi data transmission technology and send control commands to achieve control of the lower machine. The lower unit consists of STM32 control chip, voltage detection module, current detection module, light intensity detection module and temperature detection module. Its overall design is shown in Fig. 1.

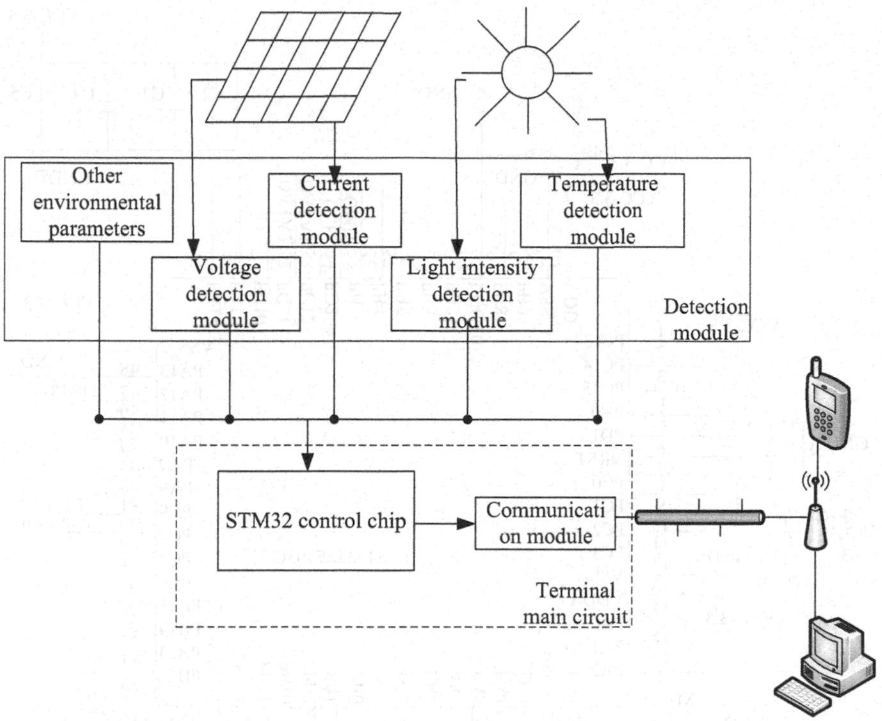


**Fig. 1** Overall design of hardware system based on Wi-Fi photovoltaic array early warning diagnosis device

# 3 The Design of Terminal Main Circuit

## *3.1 The Design of STM32 Processor Circuit*

The acquisition circuit of the WI-FI based on photovoltaic array fault early warning device adopts the STM32F103C as the processor to take the collected information as input, and its output is connected with several pins of the Wi-Fi module to control the Wi-Fi module to connect with the external wireless router so as to achieve the purpose of data transmission. The circuit diagram of the processor is shown in Fig. 2. The processor circuit is powered by 3.3 V. Its reset circuit uses the property that the capacitor voltage cannot be changed. When the power is turned on, the capacitor voltage is zero to reset the chip, then the power supply charges capacitor C8 through resistor R10 until the capacitor voltage rises high level, the reset pin is the NRST pin as in the figure.

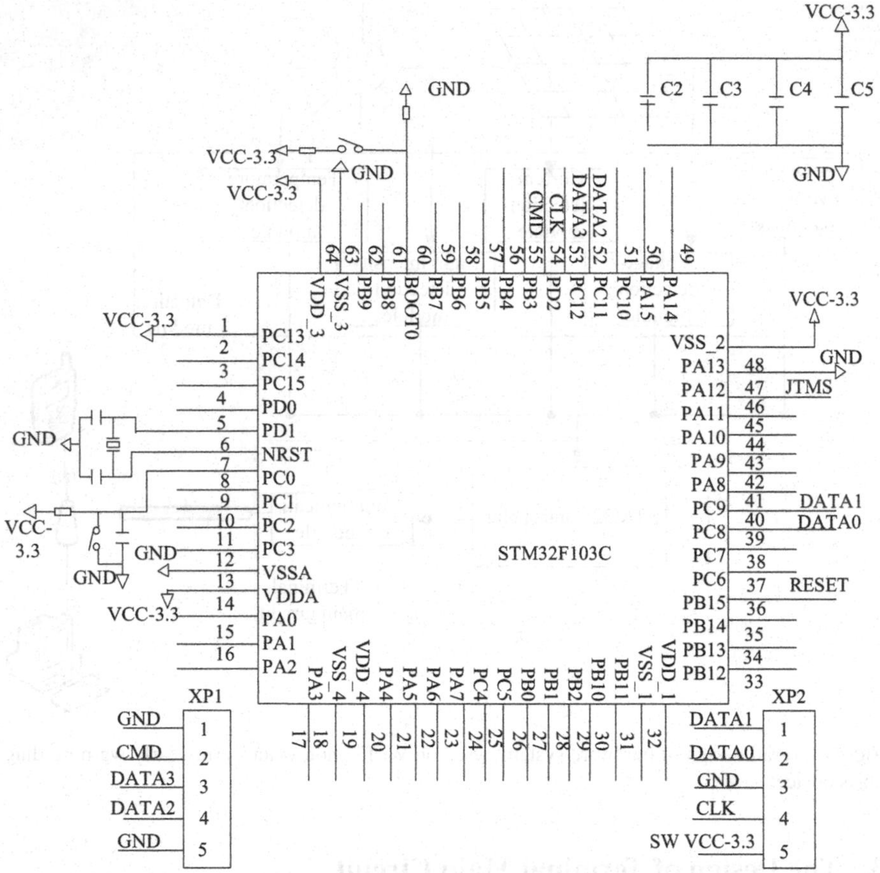


**Fig. 2** Processor circuit schematic

## 3.2 *The Design of Power Circuit*

In order to ensure that the WIFI-based Photovoltaic array failure warning system can work stably, the power supply circuit plays an important role. The power module uses Freescale's special nickel-cadmium battery for power. Figure 3 shows the power module circuit diagram of the termination circuit. The 5 V output voltage of the circuit is 3.3 V after being processed by the ASM1117 chip module to power the STM32 processor. The ASM1117 module is a regulator that internally integrates both thermal protection and current limit protection. Taking into account that the power supply output voltage waveform will often contain certain fluctuation components, in order to make the main circuit to get a more smooth DC voltage needing to use components with reactive properties (such as capacitors) to form a filter circuit to filter out the fluctuation components in the output voltage, the filtering part of the power circuit of the hardware terminal designed in this paper mainly uses the capacitor filter circuit

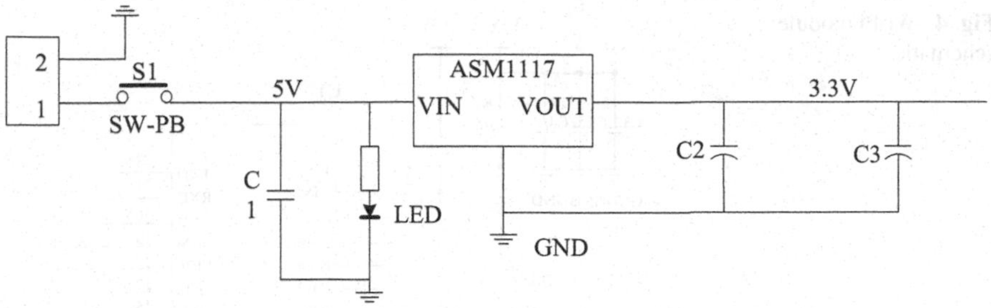


**Fig. 3** Power supply circuit schematic

(such as the capacitor C2, C3 in the circuit diagram) to ensure the stability of the output voltage.

## 3.3 *The Design of Wi-Fi Communication Module Circuit*

Wi-Fi [6, 7] is a technology and specification for wireless LAN data transmission. It conforms to the communication standard of IEEE. 802.11 and the rate of transmission can reach 11 Mbit/s. Wi-Fi can adjust its transmission rate by itself to protect the reliability and stability of the network. This system uses ESP8266-12F wireless communication module. The wireless communication module has a complete self-contained internet of things solution that can not only operate independently but also the auxiliary machine is equipped with other microcontroller units. And in terms of software, SDK is provided to researchers for secondary development, support for custom servers, and mobile phone software can be used to monitor failure data of PV arrays.

The Wi-Fi channel is effectively connected to the host computer by means of a routing system. This system uses the TL-WR703N router to brush into the OpenWrt system. The advantages of this router are its small size, low power consumption, and high security. Wi-Fi module interface schematic has been shown in Fig. 4.

# 4 The Design of Detection Modules

## 4.1 *The Design of Voltage Detection Module*

WI-FI which is based on photovoltaic array fault early warning system circuit voltage measurement circuit design uses an isolated DC voltage detection module, isolated voltage detection module schematic diagram shown in Fig. 5. The voltage detection module can introduce 5 V power from the TTL interface pin and collect the DC voltage data of −1 to 150 V of the photovoltaic array. It also can be sent to other

**Fig. 4** Wi-Fi module schematic

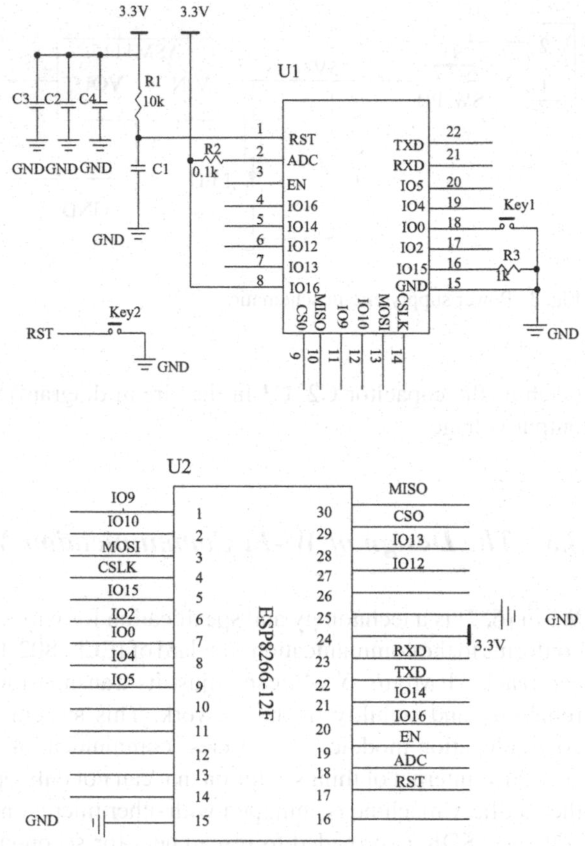


devices with TTL serial port via TTL serial port (such as Bluetooth module, GSM module, etc.), and the serial port baud rate is 9600 or 19,200.

The isolated voltage detection module uses the hall sensor of the BJHVS-AS5 series, which can collect DC voltage under the condition of electrical isolation according to the hall effect principle. It has the advantages of fast response time, stable level switching, and convenient pressure limiting setting.

## 4.2 The Design of Current Detection Module

Photovoltaic array fault early warning device based on Wi-Fi current measurement circuit design uses a model of WCS270 current detection sensor, with adjustable 2A short circuit/over current protection module, and it has the advantages of fast response time, good sensitivity, precise current setting, etc., WCS270 current detection sensor schematic has been shown in Fig. 6.

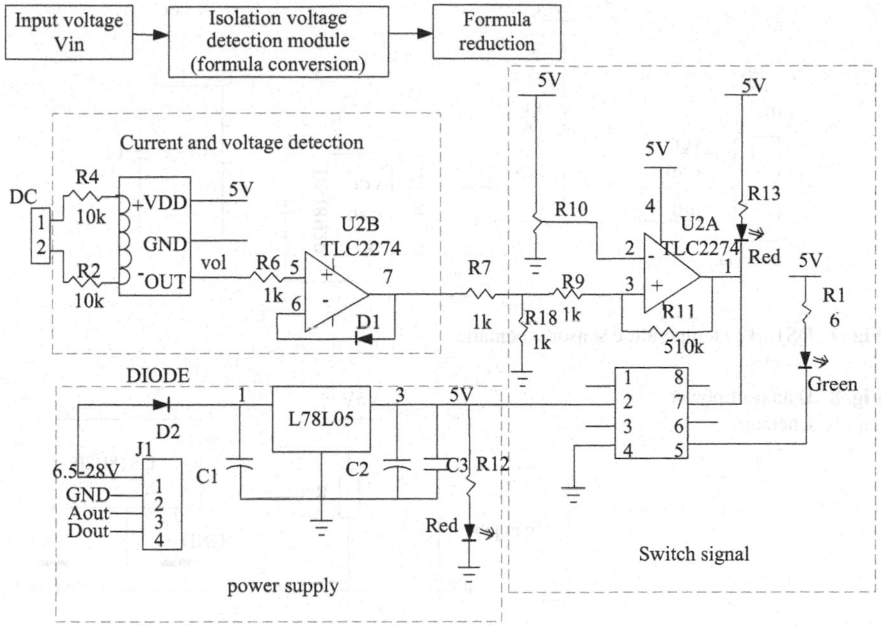


Fig. 5 Schematic diagram of isolated voltage detection module

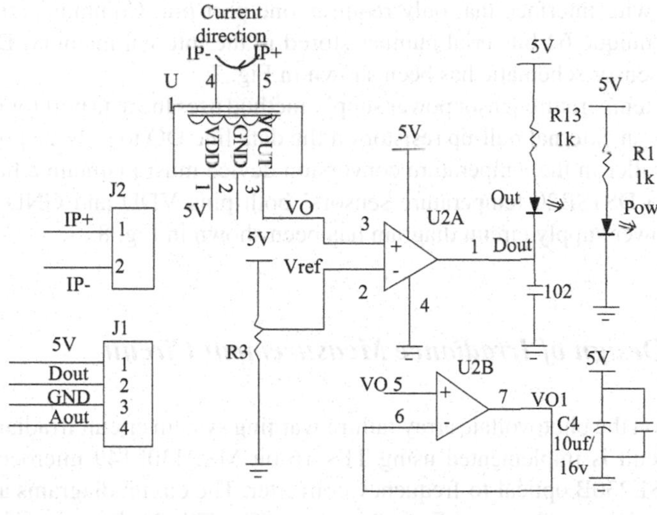


Fig. 6 Schematic of the WCS270 current sensing sensor

## 4.3 The Design of Temperature Measurement Circuit

The Wi-Fi based on photovoltaic array fault early warning system circuit temperature measurement circuit is designed by using a DS18B20 temperature sensor. The sensor

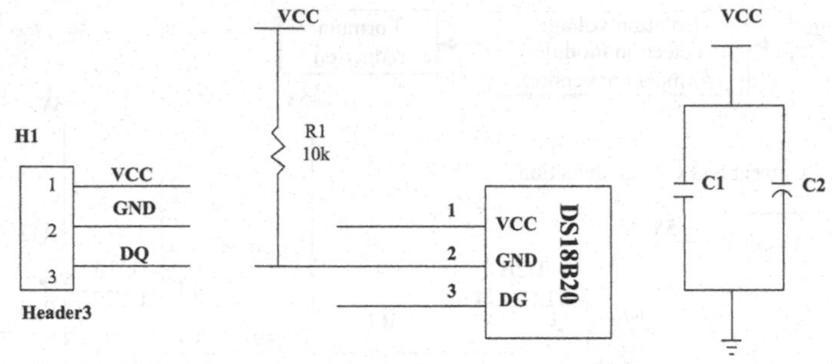


**Fig. 7** DS18B20 temperature sensor schematic

**Fig. 8** Data port power supply schematic

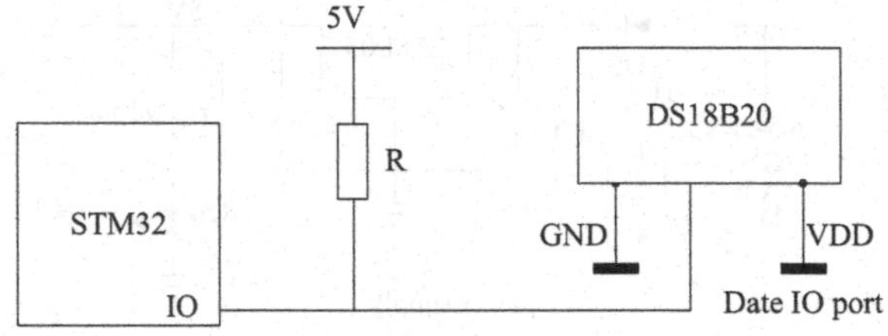


is a "single-wire communication" interface temperature sensor. Its characteristic is the single-wire interface that only requires one port pin. Communication, each device has a unique 64-bit serial number stored in the internal memory, DS18B20 temperature sensor schematic has been shown in Fig. 7.

DS18B20 temperature sensor power supply method uses the data port for its power supply [8, 9], an external pull-up resistor on the data line DQ to provide power, and the bus controller in the temperature conversion device must maintain a high state, the other two DS18B20 temperature sensors' both pins VDD and GND must be grounded. Power supply circuit diagram has been shown in Fig. 8.

## *4.4 The Design of Irradiance Measurement Circuit*

Wi-Fi based on the photovoltaic array failure warning system circuit irradiance measurement circuit is implemented using TI's 16-bit MSP430F149 microcontroller-controlled TSL230B optical-to-frequency converter. The circuit diagrams and functional schematics are shown in Figs. 9 and 10. The TSL230B optical frequency converter includes not only poly-silicon photodiodes but also CMOS current frequency converters, the photodiode converts the light intensity into a current signal and then converts it into a frequency square through the CMOS. The signal is sent to the SCM to capture and analyze the frequency, and the processing operation results in the irradiance.

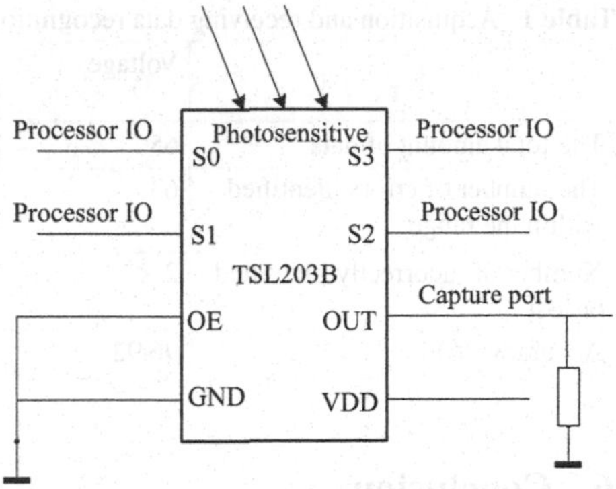


**Fig. 9** TSL230B circuit diagram

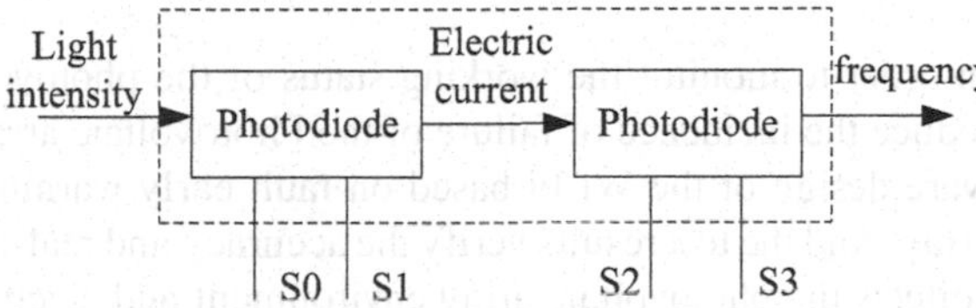


**Fig. 10** Schematic diagram of the TSL230B

## 5 System Test

In order to verify the performance of the Wi-Fi based on solar array warning device hardware system, the system conducts data acquisition experiments, and conducts wireless transmission debugging based on the design of the ESP8266-12F wireless communication module. The program is designed on the KEIL MDK platform, and the ST-LINK emulator is used to simulate the Microcontroller Unit of the control chip through the SWD to USB interface. The designed program is downloaded from the computer and input to the STM32 chip [11]. The sensors, detection circuits, and other hardware devices measure the voltage, current, temperature, and irradiance parameters of the photovoltaic array. This data is transferred to Lab-VIEW in the PC on the host PC by using the serial port assistant. Then the host computer on the PC side is connected to a router, and the mobile terminal is connected to the shared variable network which is issued by Lab-VIEW on the PC through the same router. In the data dashboard software, click the variable configuration button below the control, enter the IP address of the corresponding PC, and then find the group name of the corresponding variable, and finally complete the connection and run the program. As shown in Table 1, the results of acquisition, reception and recognition of photovoltaic array voltage, current, temperature and irradiance are presented. The results show that the system has high accuracy, low packet loss rate, good real-time performance, and can accurately and reliably transmit the collected data to a PC client or a mobile client.

**Table 1** Acquisition and receiving data recognition results

| | Voltage | Electric current | Temperature | Radioactivity |
|---|---|---|---|---|
| The total amount of data | 65 | 65 | 70 | 70 |
| The number of errors identified within the range | 63 | 62 | 63 | 66 |
| Number of incorrectly identified or lost | 2 | 3 | 7 | 4 |
| Accuracy (%) | 96.92 | 95.4 | 90.00 | 94.3 |

# 6 Conclusion

In order to monitor the working status of the photovoltaic system in real time and reduce the incidence of failure of the Photovoltaic array, this paper presents a hardware design of the Wi-Fi based on fault early warning device for the photovoltaic array. And the test results verify the accuracy and real-time performance. The system collects the photovoltaic array environment and electrical data, sends signals to the PC or mobile client in a timely and accurate manner, and completes real-time and accurate monitoring of the photovoltaic array.

# References

1. W. Chen, X. Ai, T. Wu et al., A review of the influence of photovoltaic grid-connected power generation system on the power grid. Elect. Power Autom. Equip. **33**(02), 26–32 (2013)
2. R. Wu, Z. Li, Hardware circuit design of solar photovoltaic cell testing system. Autom. Instrum. **03**, 156–157 (2010)
3. G. Yang, Z. Yang, Q. Peng et al., Hardware design of concentrator for power information acquisition system. Autom. Electr. Power Syst. **34**(09), 106–108 (2010)
4. F. Shariff, N.A. Rahim, W.P. Hew, Zigbee-based data acquisition system for online monitoring of grid-connected photovoltaic system. Expert Syst. Appl. **42**(3) (2015)
5. J. Liu, Application of WiFi technology in remote temperature and humidity monitoring system. Autom. Instrum. **39**(06), 79–82 (2014)
6. Z. Zhou, C. Wu, Z. Yang et al., Sensorless sensing with WiFi. Tsinghua Sci. Technol. **20**(01), 1–6 (2015)
7. L. Zeng, H. Zhang, H. Weiyan, Design and implementation of wireless measurement and control system based on WiFi. Electr. Measur. Instrum. **48**(07), 81–83 (2011)
8. Y. Jia, Robust control with decoupling performance for steering and traction of 4WS vehicles under velocity-varying motion. IEEE Trans. Control Syst. Technol. **8**(3), 554–569 (2000)
9. Y. Wang, Wireless data acquisition system based on STM32. Instrum. Technol. Sens. **03**, 64–66 (2018)
10. Y. Jia, Alternative proofs for improved LMI representations for the analysis and the design of continuous-time systems with polytopic type uncertainty: a predictive approach. IEEE Trans. Autom. Control **48**(8), 1413–1416 (2003)
11. A. Vijayakumari, A.T. Devarajan, N. Devarajan, Design and development of a model-based hardware simulator for photovoltaic array. Int. J. Electr. Power Energy Syst. **43**(1) (2012)

# Performance Analysis of Flux-Switching Stator Permanent Magnet Motor Based on Linear Active Disturbance Rejection Control

**Kelei Wang, Zengqiang Chen, Mingwei Sun and Qinglin Sun**

**Abstract** Flux-switching permanent magnet motor (FSPMM) is a new stator permanent magnet brushless motor. It overcomes many shortcomings of the conventional permanent magnet motor having magnets in the rotor and has a well application prospect. The three-phase 12-slots/10-poles FSPMM is used as the control object. On the basis of the working principles, the mathematical models have deduced and the mechanical properties are calculated. The characteristics of the electromagnetic are analysed by setting up the steady and dynamic-state models of the FSPMM. Linear Active Disturbance Rejection Control (LADRC) is designed in the speed loop of the FSPMM to realize the linear control of the nonlinear system. By using the Linear Extended State Observer (LESO), the total disturbances can be estimated and compensated in real time. The performance robustness is verified by the Monte Carlo experiments of the two control strategies, including the LADRC algorithm, and the traditional PI control strategy. The results show that LADRC strategy has a greater capability of disturbance rejecting and stronger performance robustness.

**Keywords** Flux-switching permanent magnet motor · Speed control algorithm
Linear active disturbance rejection control · Linear extented state observer
robustness · Monte Carlo

## 1 Introduction

Flux-Switching Permanent Magnet Motor (FSPMM) is the latest stator permanent magnet brushless motor, it placed the permanent magnets in the stator teeth and the rotor is very simple and robust. The FSPMM overcomes many shortcomings of the traditional permanent magnet machines having magnets in the rotor, including the complexity of the motor structure, the high cost of manufacturing and so on. FSPMM

K. Wang · Z. Chen (✉) · M. Sun · Q. Sun
College of Computer and Control Engineering, Nankai University, Tianjin 300071, China

Z. Chen
Key Lab of Intelligent Robotics of Tianjin, Tianjin 300071, China
e-mail: chenzq@nankai.edu.cn



Y. Jia et al. (eds.), *Proceedings of 2018 Chinese Intelligent Systems Conference*,
Lecture Notes in Electrical Engineering 528,
https://doi.org/10.1007/978-981-13-2288-4_66

can be operated at a very high speed, so it is considered to be the most popular of the stator permanent magnet brushless motor and it has great application value [1–5]. Owing to the torque generation mechanism and the structure of the stator permanent magnet motor have a significant difference with the traditional rotor permanent stator magnet motor, the traditional rotor permanent magnet motor analytical methods and design theories are difficult to apply to the stator permanent magnet motor directly. The stator permanent magnet brushless motor also has some unique electromagnetic phenomena, including the DC bias magnetic field, the stator leakage magnetic and the end magnetic flux leakage [6], so that its analysis and calculation are more difficult. To improve fault tolerance or availability of electric drive for the FSPMM in case of an open-circuit fault, VSI topology is applied to allow flowing zero sequence current. The space-vector modulation and space-phasor modulation have been analysed in this type of machine for control scheme design under three-phase operation [7]. As for control requirements of different control applications, the well-developed voltage space-vector PWM control strategy and field weakening control strategy have widely been applied to the FSPMM control system [8, 9]. In recent years, some advanced control methods have also been adopted in the speed control system of motors [10–12], such as the stator-flux-oriented control method, maximum efficiency tracking control method and position-sensorless control method. A novel control strategy of stator-flux orientation for FSPMM based on voltage space-vector is proposed in [13], the fundamental relationship and equalization between the two magnetic fields where the magnets are in the stator and rotor, respectively, are analyzed. A new self-tuning fuzzy PI controller with conditional integral offers better adaptability than the normal linear PI control and that the developed motor drive offers better steady-state and dynamic performances [14].

After J. Han developed the three major tools: the Tracking Differentiator (TD), the Nonlinear State Error Feedback (NLSEF) and the Extended State Observer (ESO), he proposed the Active Disturbance Rejection Control (ADRC) theory in 1998 [15, 16]. The ADRC is a kind of original control method which does not depend on the controlled process model [17], the core is putting the external disturbance and the internal unmodeled dynamic together as the 'total disturbance', the system state and the total disturbance are estimated online through the ESO, in addition to the ESO can timely compensate the total disturbance to the feedback control. Linear Active Disturbance Rejection Control (LADRC) [18] is proposed by Gao and other scholars,they linearized the main sections of the ADRC. LADRC has the characteristics of simple structure, easy to performance analysis and parameter settings, so it greatly promoted the development of ADRC theory and its application in engineering [19–21].

Up to now, the investigation on the control strategies and the operation performance of the FSPMM drive lack theoretical and practical analysis, and the existing control strategies of the FSPMM have poor anti-interference ability. Hence, in this paper a novel control scheme of LADRC for FSPMM based on PWM vector control with current hysteresis comparison is proposed. This paper chooses the three-phase 12/10-poles FSPMM as the control object, its stator windings are centrally distributed and the rotor structure is similar to the switched reluctance motor. The FSPMM's topology and operating principles were analysed firstly. On the basis of the theoretical analysis,the mathematical models in the stator coordinate system and rotor coordinate system are deduced, respectively. The electromagnetic characteristics of the four different current control methods are analysed and calculated. The steady-state and dynamic-state simulation models of FSPMM are established, and the robustness of traditional PI control and LADRC is verified by Monte Carlo experiments. The numerical simulation results show that the proposed control scheme can effectively improve the robustness and the anti-interference ability of the FSPMM.

## 2 Topology and Operating Principles

A Three-phase 12/10-poles FSPMM topology is illustrated in Fig. 1. The stator contains twelve U-shaped laminated segments, PMs. It placed two concentrated winding coils side by side in each U-shaped magnet core. The twelve armature coils are divided into three groups, every four series become a phase armature winding. For example, A1–A4 in the figure are four coils of phase A. The windings of each coil with a PM embedded therein spans the two stator teeth, and the PMs are alternately magnetized in the tangential direction. The rotor has 10 teeth, called the 10 pole, the poles number of the rotor determines the rotor cycle, that is, the mechanical angle of the no-load back electromotive force (EMF) in a cycle. Based on the 'reluctance minimum principle', the flux linkages are always closed by the path with the smallest reluctance.

The operating principles of the FSPMM are illustrated in Fig. 2. When the rotor teeth are aligned with the stator teeth belonging to the two U-shaped units under the same phase coil respectively, the polarity of the flux linkages in the winding will change owning to this unique design. That is, with the cyclical changes in the area between the rotor poles and the stator teeth, the air gap flux linkages in the main magnetic circuit also changes, resulting in a change in the size and direction of the EMF with the rotor position changing, which achieves the so-called 'flux-switching'.

Fig. 1 Sectional view of three-phase 12/10 pole FSPMM

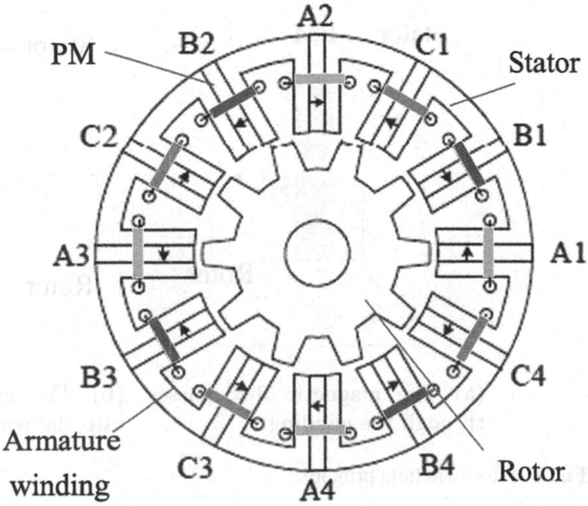

# 3 Mathematical Model

Assuming the same structure and symmetry of each phase winding and neglecting the influence of core loss and saturation, the voltage balance equation of the three-phase FSPMM in the stator coordinate system is as follows

$$\begin{bmatrix} u_a \\ u_b \\ u_c \end{bmatrix} = R_{ph} \begin{bmatrix} i_a \\ i_b \\ i_c \end{bmatrix} + L_1 \frac{d}{dt} \begin{bmatrix} i_a \\ i_b \\ i_c \end{bmatrix} + \frac{d}{dt} \begin{bmatrix} \psi_a \\ \psi_b \\ \psi_c \end{bmatrix} + \begin{bmatrix} e_{ma} \\ e_{mb} \\ e_{mc} \end{bmatrix} \tag{1}$$

Three-phase back electromotive force equation of the FSPMM can be described as follows

$$\begin{cases} e_{ma} = E_m \sin(P_r\theta_r) \\ e_{mb} = E_m \sin(P_r\theta_r - 120^\circ) \\ e_{mc} = E_m \sin(P_r\theta_r + 120^\circ) \end{cases} \tag{2}$$

where $u_a, u_b, u_c$ are the three-phase voltage, $i_a, i_b, i_c$ are the three-phase aramture current, $R_{ph}$ is the resistance of each phase winding, $L_1 = L_0 - M_0$, $L_0$, $M_0$ are the self-inductances and mutual-inductances of the fundamental DC component, respectively, $\psi_a$, $\psi_b$, $\psi_c$ are the flux linkage vector associated with the change of self-inductance of the three-phase synthetic flux linkage in the stator side, $e_{ma}, e_{mb}, e_{mc}$ are the three-phase back electromotive force, $E_m = -P_r\psi_m\omega_r$ is the amplitude of the EMF of the stator side winding, $\psi_m$ are the permanent magnetic flux linkages, $P_r$ is the pairs of the rotor poles, $\theta_r$ is the rotor position in mechanical degrees.

As a result of the permanent magnetic air gap flux density of the FSPMM is so high, the motor cogging torque is far greater than the traditional rotor permanent

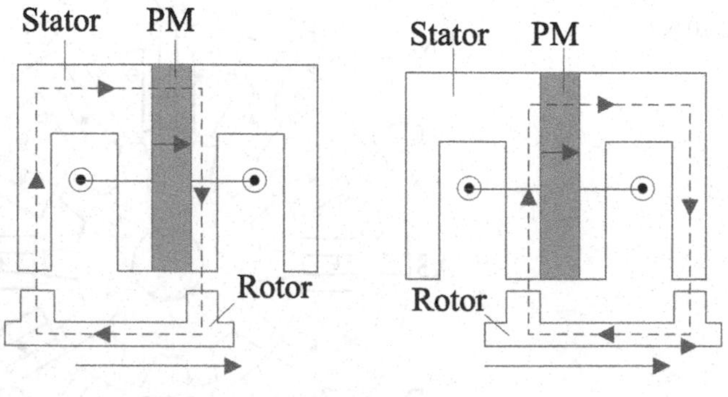


(a) The magnetic flux passes through the winding. (b) The magnetic flux enters into the winding.

**Fig. 2** Flux switching principle

magnet motor. The cogging torque will be the most important factor in the torque ripple when the disturbance caused by the commutation of the winding current is not taken into account.

$$T_{em} = \frac{P_{em}}{\omega_r} = T_{pm} + T_r + T_{cog} \tag{3}$$

The mechanical equation of motion of the FSPMM can be expressed as

$$T_{em} = J\frac{d\omega_r}{dt} + T_L + B_v\omega_r \tag{4}$$

$$\omega_r = \frac{d\theta_r}{dt} \tag{5}$$

where $P_{em}$ is the electromagnetic power, $T_{pm}$ is the torque generated by PM flux linkage and armature current, $T_r$ is the reluctance torque, $T_{cog}$ is the cogging torque, $\omega_r$ is the mechanical angular velocity of the motor, $T_L$ is the load torque, $J$ is the moment of inertia, $B_v$ is the coefficient of friction of the drive system.

Ignore the motor core saturation, eddy current loss and hysteresis loss, assuming that the motor current is symmetrical three-phase sinusoidal current. When the flux linkage in the three-phase stator windings are transformed into the rotating rotor coordinate system, only a constant flux linkage in the $d$-axis winding, that is, the $d$-axis and the $q$-axis permanent magnet flux linkages of the three-phase FSPMM can be expressed as

$$\begin{cases} \psi_{md} = \psi_m \\ \psi_{mq} = 0 \end{cases} \tag{6}$$

The flux linkage equations of the FSPMM can be expressed as

$$\begin{cases} \psi_d = \psi_{md} + L_d i_d \\ \psi_q = L_q i_q \end{cases} \tag{7}$$

The voltage equations of the FSPMM can be described as

$$\begin{cases} u_d = \frac{d\psi_d}{dt} + R_{ph} i_d - \omega_e \psi_q \\ u_q = \frac{d\psi_q}{dt} + R_{ph} i_q + \omega_e \psi_d \end{cases} \tag{8}$$

The electromagnetic torque of the FSPMM is calculate by

$$
\begin{aligned}
T_{em} &= \frac{3}{2}\frac{P_{em}}{\omega_r} = \frac{3}{2}P_r\left[\psi_m i_q + \left(L_d - L_q\right) i_d i_q\right] + T_{cog} \\
&= T_{pm} + T_r + T_{cog}
\end{aligned}
\tag{9}
$$

The following parameters are defined as

$$
\begin{cases}
T_{pm} = \frac{3}{2}P_r\psi_m i_q \\
T_r = \frac{3}{2}P_r\left(L_d - L_q\right) i_d i_q \\
T_{cog} = 1.11\sin\left(2mP_r\theta_r + 0.24\right) + 0.59\sin\left(4mP_r\theta_r + 0.48\right)
\end{cases}
\tag{10}
$$

where $u_d, u_q$ are the $d$-axis and $q$-axis components of the armature voltage, respectively, $i_d, i_q$ are the $d$-axis and $q$-axis components of the armature current, respectively, $L_d, L_q$ are the $d$-axis and $q$-axis components of the inductance, respectively, $\psi_d, \psi_q$ are the $d$-axis and $q$-axis components of the total flux linkage, respectively, $\omega_e = P_r\omega_r$ is the electrical angular frequency, $m$ is the number of phases.

## 4 Control Strategy for FSPMM

### *4.1 Design of Current Loops with Vector Control*

Although FSPMM is a new type of stator permanent magnet brushless motor, but the more mature vector control algorithm can also be used in the control algorithm, which is very favorable for FSPMM to applicating in the industry. The purpose of vector control is to control the torque by controlling the stator current's phase and amplitude. Therefore, the vector control strategy is adopted in the current loop. The speed control system is divided into below the rated speed constant torque area and higher than the rated speed constant power area. In the constant torque region, it is mainly to study how to allocate the $d$-axis current $i_d^*$ for regulating the magnetic field and the $q$-axis current $i_q^*$ for adjusting the magnitude of the electromagnetic torque by the given value of torque.

In the constant torque region often used in the current vector control method has $i_d = 0$ control, the maximum torque current ratio control, constant flux linkage control and full power factor control ($\cos\phi = 1$).

The $i_d = 0$ control is simple and has no demagnetizing effect, and the output torque is proportional to the input current. When the $i_d = 0$ control is adopted, the reluctance torque component is zero, so the electromagnetic torque equation can be simplified as

$$
T_{em} = T_{pm} + T_{cog} = \frac{3}{2}P_r\psi_m i_q + T_{cog}
\tag{11}
$$

The maximum speed when the limit voltage of the inverter is satisfied as

$$\omega_{e\,\max} = U_{\max} / \sqrt{\psi_m^2 + \left(\frac{2T_{em}L_q}{3P_r\psi_m}\right)^2} \tag{12}$$

Due to the inverter capacity limit, there is a maximum limited current $I_{max}$, so that the motor at run time always satisfy the following relations

$$i_d^2 + i_q^2 \le I_{\max}^2 \tag{13}$$

If the maximum output voltage vector amplitude of inverter is $U_{max}$, then the $d$-axis and $q$-axis voltage should meet

$$u_d^2 + u_q^2 \le U_{\max}^2 = \left(\frac{U_{dc}}{\sqrt{3}}\right)^2 \tag{14}$$

When the motor speed reaches a certain value, considering the steady state and neglecting the resistance terms, (8) is expressed as

$$\begin{cases} u_d = -\omega_e L_q i_q \\ u_q - \omega_e \psi_m + \omega_e L_d i_d \end{cases} \tag{15}$$

Substituting (15) into (14), the voltage constraints are given as

$$\left(-\omega_e L_q i_q\right)^2 + (\omega_e \psi_m + \omega_e L_d i_d)^2 \le U_{\max}^2 \tag{16}$$

Equation (16) can be rewritten as

$$(\psi_m + L_d i_d)^2 + \left(L_q i_q\right)^2 \le \left(\frac{U_{\max}}{\omega_e}\right)^2 \tag{17}$$

As obviously seen in Eq. (17), the voltage limit is an ellipse of center $(-\psi_m/L_d, 0)$. When the speed $\omega_e$ increases, the voltage limit $U_{max}$ does not change, so the operating area of $i_d$, $i_q$ are reduced with the increase in speed. So the relationship between the speed is $\omega_{e1} < \omega_{e2} < \omega_{e3} < \omega_{e4}$, as shown in Fig. 3.

When the speed reaches the rated speed, the operating point is located at the intersection of the voltage limit ellipse and the current limit ellipse. If the speed continues to rise, the induced EMF of the motor will exceed the maximum voltage limit of the inverter. The actual current will no longer follow the command current value, resulting in the power and output torque will drop together. Therefore, in the constant

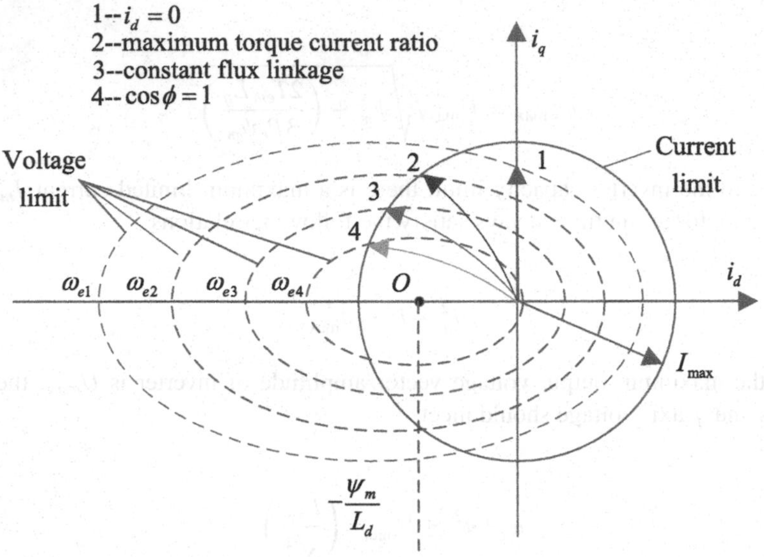


**Fig. 3** Voltage and current limits

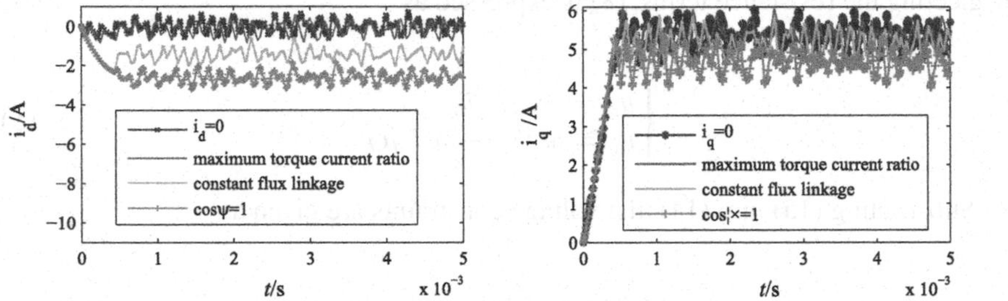

(a) *d*-axis current of four current control algorithms.

(b) *q*-axis current of four current control algorithms.

**Fig. 4** *d*-axis and *q*-axis current of four current control algorithms

power area, the flux weakening control is adopted to produce a polarized reactivity flux in the opposite direction by applying a weak magnetic current component $i_d$ opposite to the direction of the PM flux-linkages in order to reduce or offset the PM flux-linkages, which can reduce the combined electromotive force, to ensure that the armature winding voltage balance and improve the system speed range.

The *d*-axis and *q*-axis current curves of the four current control algorithms, as shown in Fig. 4. Owing to the $L_d$ and $L_q$ is not same in the FSPMM, there is only a small *d*-axis current when the maximum torque current control algorithm is adopted, this current control algorithm produces a reluctance torque in the same direction as the permanent magnet torque, and the offset current is almost equal to the control method $i_d = 0$.

**Table 1** FSPMM mechanical properties of the theoretical calculation results

| Current control method | $T_{em}\ (N \cdot m)$ | $i_d\ (A)$ | $i_q\ (A)$ |
|---|---|---|---|
| $i_d = 0$ | 13.34 | 0 | 5.37 |
| Maximum torque current ratio | 13.35 | −0.25 | 5.36 |
| Constant flux linkage | 13.06 | −1.55 | 5.22 |
| $\cos\phi = 1$ | 12.16 | −2.89 | 4.73 |

The calculation results of the mechanical properties of the three-phase FSPMM using four current control algorithms are given by Table 1.

## 4.2 Design of Speed Loop with LADRC

LADRC does not depend on the exact mathematical model. Although it doesn't have integral link, it still can achieve no static error. The total disturbance of the system is estimated by LESO and compensated timely, the system is linearized into the form of series integrator and the parameter setting is simple. The controller of the speed loop adopts a first-order LADRC in this paper.

The speed equation of the FSPMM is expressed as

$$\begin{aligned} \dot{\omega}_r &= \frac{1}{J}\left(T_{em} - T_L - B_v \omega_r\right) \\ &= a\left(x\right) + bu \end{aligned} \tag{18}$$

where $u = T_{em}$ is the input of the speed loop, $b = \frac{1}{J}$, the friction coefficient is generally taken as $B_v = 0$. Setting $f = a\left(x\right) + \left(b - b_0\right)u$, $f$ is referred to as the generalized disturbance of the speed loop, $b_0$ is the estimation of $b$.

Then the system (18) can be rewritten as the following standard form

$$\begin{cases} \dot{\omega}_r = f + b_0 u \\ y = \omega_r \end{cases} \tag{19}$$

A second-order Linear Extented State Observer (LESO) is constructed as follows

$$\begin{cases} e = z_1 - y \\ \dot{z}_1 = z_2 - \beta_1\left(z_1 - \omega_r\right) + b_0 u \\ \dot{z}_2 = -\beta_2\left(z_1 - \omega_r\right) \end{cases} \tag{20}$$

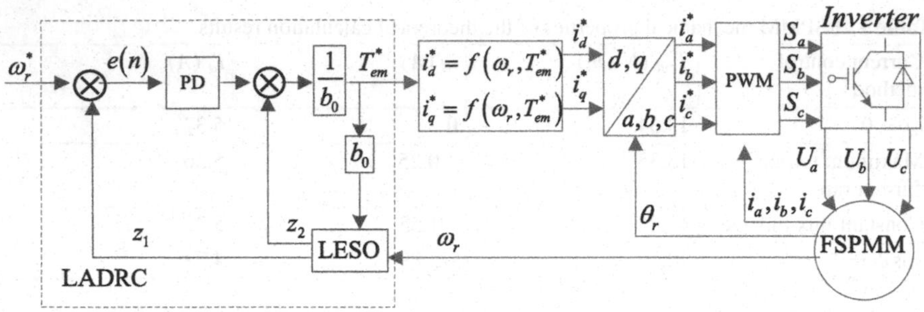


**Fig. 5** Vector control diagram of FSPMM

where $\beta_1$ and $\beta_2$ are the gains of the LESO, $z_1$ and $z_2$ are the estimations of the state variables $x_1$ and $x_2$. In order to simplify the parameter adjustment process, the observer gains can be obtained using any known method such as the pole placement technique. Therefore, the observer gain satisfies the following relation: $s^2 + \beta_1 s + \beta_2 = (s + \omega_o)^2$, then $\beta_1 = 2\omega_o$, $\beta_2 = \omega_o^2$, where $\omega_o$ is the bandwidth of the LESO. In general, the larger the $\omega_o$, the higher the accuracy of the observer estimation, but the larger $\omega_o$ increases the noise sensitivity (introducing high frequency noise). So reasonable choice of the observer gains is critical to the accuracy of the estimation, it is need to consider the performance and noise tolerance.

Here, $u_0$ chooses the P controller, which can be written as the following form

$$u_0 = k_p (v - z_1) \tag{21}$$

where $k_p$ are the controller parameters. The control law is expressed as

$$u = \frac{(u_0 - z_2)}{b_0} = \frac{k_p (\omega_r^* - \omega_r) - z_2}{b_0} \tag{22}$$

Then the Eq. (18) is equivalent to the following linear system:

$$\dot{\omega}_r = u_0 \tag{23}$$

For the first order system of the speed loop, the system stability can meet the performance requirement by adjusting the observer bandwidth $\omega_o$, the controller bandwidth, the control gain $b_0$.

The current hysteresis comparison PWM vector control strategy is adopted in this paper, the $i_d = 0$ control strategy is used in the current loop and the LADRC is adopted in the speed loop of FSPMM. The vector control diagram of the whole closed-loop system is shown in Fig. 5.

# 5 Simulation Results and Analysis

In order to verify the correctness of the mathematical models of the FSPMM. Firstly the steady state and dynamic state simulation models are established to verify the steady and dynamic characteristics of the FSPMM, respectively.

The steady state model is a speed open-loop system, which consists of the main body module of the motor (excluding the equation of motion), the power converter and the current controller. Under the conditions of specified speed and torque, the mechanical properties of the motor are studied in the steady state model. The dynamic state model is used to study the dynamic response characteristics of the motor in the event of a sudden change in speed or load torque. Therefore, based on the steady state model, the dynamic state model also includes the speed and rotor position feedback modules, the speed controller and the modules of the mechanical and kinematic equations. The parameters of the FSPMM under investigation are listed in Table 2.

Here the first-order LADRC and PI control algorithms are adopted of the speed loop, respectively. During the dynamic simulation of the system, the parameters of the PI are selected as $K_P = 1.105$, $K_I = 0.01$, LADRC parameters are setting as $b_0 = 6.25$, $\omega_o = 5.25$, $k_p = 3.87$.

## 5.1 *Steady-State Simulation Results*

The three-phase FSPMM steady-state simulation results are shown in Fig. 6. The curves of the three-phase current in steady state can be seen in Fig. 6a; The curves of the electromagnetic torque and its components can be seen from the Fig. 6b, the average value of the $T_{em}$ is around 13 N m that is corresponds to the theoretical value. $T_r$ is almost zero, it is consistents with the $i_d$=0 current vector control method. It can be seen from Fig. 6c that the FSPMM has a high cogging torque and the cogging torque is the main factor causing the electromagnetic motor to pulsate, that is consistents with the theoretical analysis.

**Table 2** Key parameters of the FSPMM

| Parameters | Value | Parameters | Value |
|---|---|---|---|
| $n_N$, r/min | 600 | $R_{ph}$, $\Omega$ | 1.436 |
| $T_N$, N · m | 13 | $L_d$, mH | 14.308 |
| $I_N$, A | 3.8 | $L_q$, mH | 15.533 |
| $\psi_m$, Wb | 0.166 | $U_{dc}$, V | 440 |
| $P_r$ | 10 | $J$, kg m$^2$ | $8 \times 10^{-4}$ |

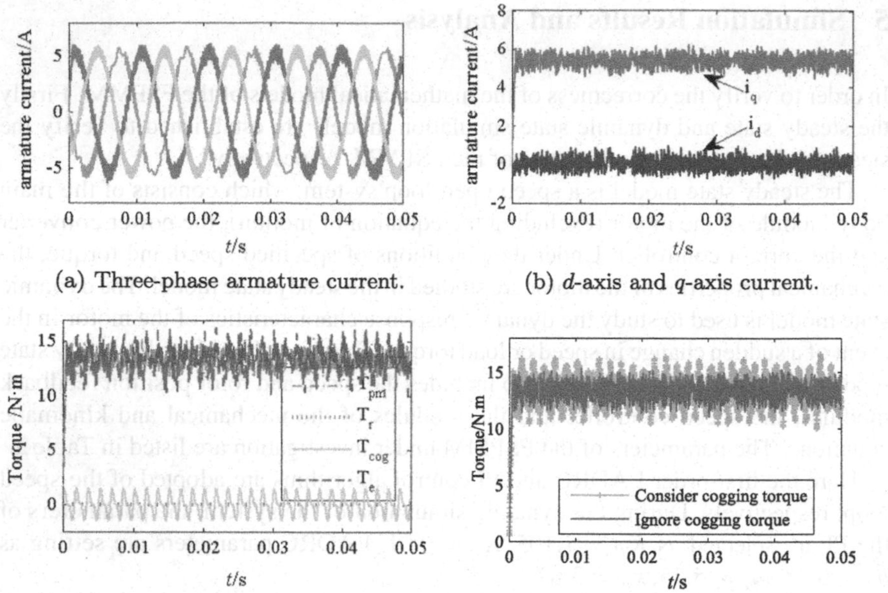


(a) Three-phase armature current. (b) *d*-axis and *q*-axis current.

(c) Torque components. (d) Influence of cogging torque on electromagnetic torque.

**Fig. 6** The steady-state results of FSPMM

## 5.2 Dynamic Simulation Results

Dynamic simulation is a powerful tool to analyze the dynamic response performance of the motor. This section focuses on the dynamic performance after introducting the first-order LADRC into the speed loop of the FSPMM. The novel LADRC control scheme is compared with the traditional PI control strategy.

The simulation with a given speed of 600 r/min, the load changes from 0 to 8 N · m at 0.05 s, and then back to 0 N · m at 0.08 s, the dynamic simulation results is shown in Fig. 7.

As the load increases, the three-phase armature current increases while the speed decreases but it returns rapidly to a given speed after adjusting. We can see from the Fig. 7a, LADRC has a smaller overshoot than PI for the speed response. With the load increasing, the FSPMM need to increase the $q$-axis current in order to increase the electromagnetic torque, the $q$-axis current fluctuates around 3.21 A, which is consistents with the theoretical analysis. We can learn from the Eq. (18) that when the electromagnetic torque is balanced with the load torque, the speed change is almost zero, which means that the system is in steady state, therefore the electromagnetic torque fluctuates around 8 N · m in Fig. 7d.

The $d$-axis current is always around zero because of the current vector control strategy of $i_d = 0$ is employed. The simulation results show that, with the FSPMM

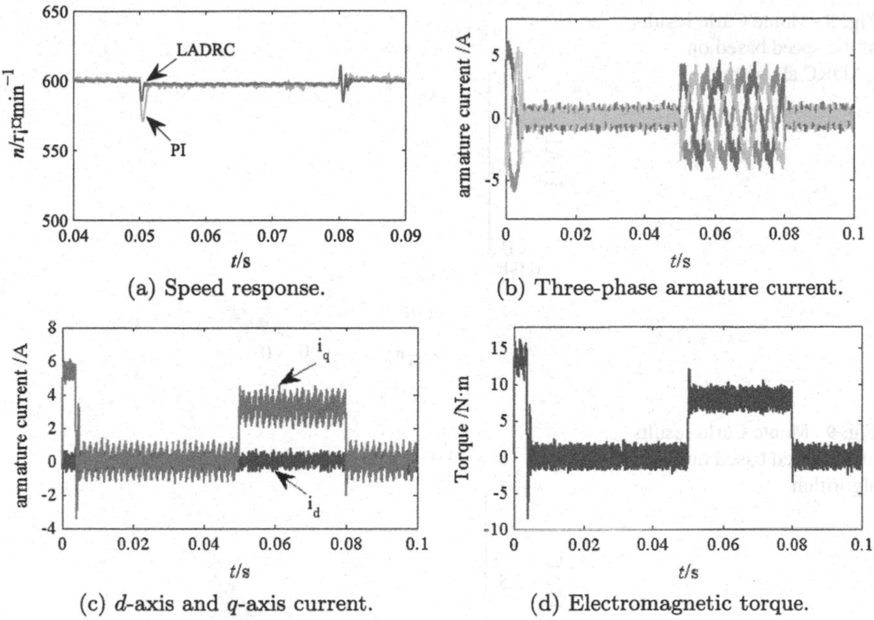


**Fig. 7** The dynamic-state results of FSPMM

load disturbance, LADRC can be stabilized with shorter settling time and smaller overshoot, so LADRC has stronger anti-interference performance than PI.

## 5.3 Monte Carlo Experiment

Moreover, in order to further analyse the performance robustness of PI control and LADRC method,the Monte Carlo approach is adopted in this paper. The Monte Carlo, also known as random sampling, is a probabilistic and statistical based approach. The principle of the Monte Carlo method is to sample variables that affect system performance, the quantitative performance evaluation of the system design is provided by repeating the assigned experiments within a reasonable range of system parameters. A performance evaluation index called ITAE is introduced to test the dynamic performance of the system

$$ITAE = \int_0^T t\,|e(t)|dt \tag{24}$$

where $e(t)$ is the error between setting value and the practical output, $T$ is the simulation time, $t_s$ is the settling time, $\sigma\%$ is the overshoot.

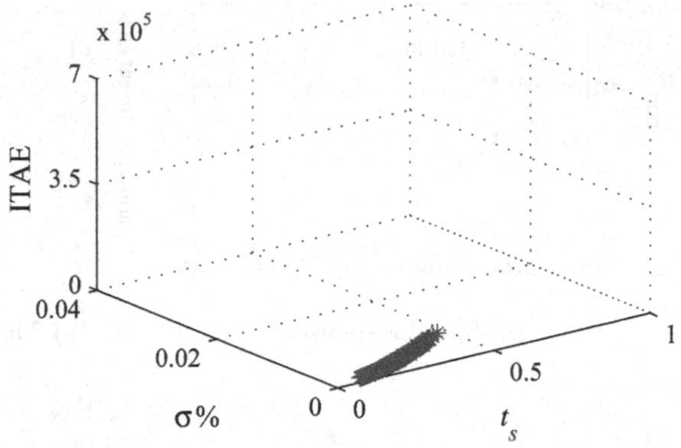


**Fig. 8** Monte Carlo results of the speed based on LADRC algorithm

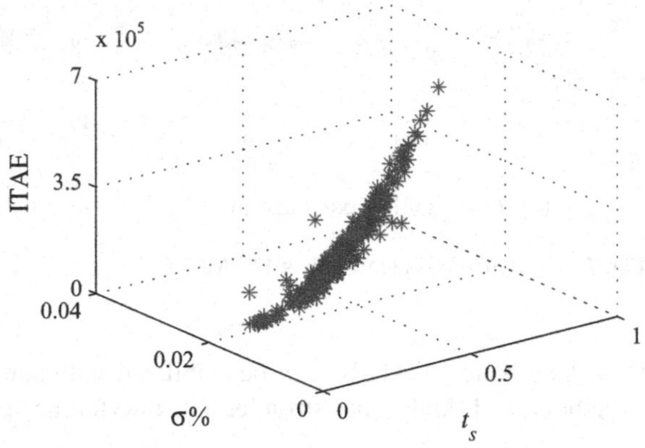


**Fig. 9** Monte Carlo results of the speed based on PI algorithm

As the operating temperature of the motor increases, some parameters will perturbed. In the case of keeping the LADRC parameters unchanged, the FSPMM's per-winding resistance and the $d$-axis and $q$-axis inductors are subject to a random perturbation of $\pm 20\%$, and then perform 500 times Monte Carlo experiments. Putting the $t_s$, $\sigma\%$ and ITAE as the $x$-axis, $y$-axis, $z$-axis, respectively. The Monte Carlo experiment results of the speed loop are shown in Figs. 8 and 9 as well as Table 3.

As we all known that the more concentrated the experimental results of Monte Carlo, the stronger the performance robustness. And the smaller the value of ITAE, the better the system performance. It can be seen clearly from Figs. 8 and 9, the LADRC algorithm has superior dynamic performance and stronger performance robutess, comparing with the PI control method.

In theory, the output of the PI controller is basically derived from its integral term and feedforward. In the face of disturbance, the feedforward is incapable of doing nothing, and the integral term is lagging behind. However, LADRC can take the initiative from the controlled object input and output signals to extract the disturbance

**Table 3** Monte Carlo results of speed based on PI and LADRC algorithms

| Control algorithm | PI | LADRC |
|---|---|---|
| Settling time/s | 0.2–0.7 | 0.08–0.48 |
| Overshoot | 0.019–0.030 | 0.00035–0.00043 |
| ITAE | $(1-3.5)\times 10^5$ | $(0.05-8.5)\times 10^4$ |

information, and then as soon as possible with the control signal to eliminate the error, thus it can greatly reduce the impact of the disturbance on the controlled amount.

The numerical simulation results show that LADRC algorithm has better anti-interference ability and stronger robustness than PI control in the case of load mutation and parameter perturbation, which is consistent with theoretical analysis.

## 6 Conclusion

In order to realize the high performance control of the new FSPMM, a novel control strategy of LADRC has been proposed in this paper, and the simulation results have compared with the traditional PI control. The topology and operating principles of FSPMM are first analysed. On the basis of the $i_d = 0$ current vector control method, the mathematical model has been deduced in $d$-$q$ axis coordination system, and the mechanical properties were calculated of the four current vector control strategies, including the $i_d = 0$, the maximum torque current ratio control, constant flux linkage control, and full power factor control. The steady and dynamic simulation models of the FSPMM provide a theoretical analysis tool for systematically analyzing the electromagnetic characteristics and the dynamic performances. The numerical simulation results show that LADRC has better anti-interference ability with the load torque mutation and the well speed tracking capability after parameter tuning. The Monte Carlo results show that LADRC has stronger performance robustness than PI in terms of parameter perturbation. Moreover, the cogging torque is the main reason for the torque ripple of the FSPMM verified by simulation. So it is helpful for further study the problem of reducing torque ripple.

**Acknowledgements** This work was supported by National Natural Science Foundation of China (Grant No. 61573199 and 61573197) and the Tianjin Municipal Natural Science Foundation (Grant No. 14JCYBJC18700).

## References

1. W. Hua, M. Cheng, Static characteristics of doubly-salient brushless machines having magnets in the stator considering end-effect. Electr. Mach. Power Syst. **36**(7), 754–770 (2008)

2. Z.Q. Zhu, Y. Pang, D. Howe, S. Iwasaki, R. Deodhar, Analysis of electromagnetic performance of flux-switching permanent-magnet machines by nonlinear adaptive lumped parameter magnetic circuit model. IEEE Trans. Magn. **41**(11), 4277–4287 (2005)
3. Z.Q. Zhu, J.T. Chen, Advanced flux-switching permanent magnet brushless machines. IEEE Trans. Magn. **46**(6), 1447–1453 (2010)
4. W. Zhao, M. Cheng, W. Hua, H. Jia, Ruiwu Cao, Back-emf harmonic analysis and fault-tolerant control of flux-switching permanent-magnet machine with redundancy. IEEE Trans. Ind. Electron. **58**(5), 1926–1935 (2011)
5. Y. Wang, M.J. Jin, M.J. Fei, J.X. Shen, Cogging torque reduction in permanent magnet flux-switching machines by rotor teeth axial pairing. IET Electr. Power Appl. **4**(7), 500–506 (2010)
6. M. Cheng, W. Hua, J. Zhang, W. Zhao, Overview of stator—permanent magnet brushless machines. IEEE Trans. Ind. Electron. **58**(11), 5087–5101 (2011)
7. L. Wang, S. Aleksandrov, Y. Tang, J.J.H. Paulides, E.A. Lomonova, Fault-tolerant electric drive and space-phasor modulation of flux-switching permanent magnet machine for aerospace application. IET Electr. Power Appl. **11**(8), 1416–1423 (2017)
8. W. Hua, M. Cheng, A new model of vector-controlled doubly-salient permanent magnet motor with skewed rotor, in *Proceedings of the International Conference on Electrical Machines and Systems* (Wuhan, China, 2009), pp. 3026–3031
9. H. Jia, M. Cheng, W. Hua, W. Lu, X. Fu, Investigation and implementation of control strategies for flux-switching permanent magnet motor drives, in *Proceedings of the IEEE Industry Applications Society Meeting* (Edmonton, AB, Canada, 2008), pp. 1–6
10. H.H. Choi, J.W. Jung, R.Y. Kim, Fuzzy adaptive speed control of a permanent magnet synchronous motor. Int. J. Electron. **99**(5), 657–672 (2012)
11. S. Ma, W. Peijun, J. Ji, X. Li, Sensorless control of salient pmsm with adaptive integrator and resistance online identification using strong tracking filter. Int. J. Electron. **103**(2), 217–231 (2016)
12. C. Zhong, Y. Lin, Model reference adaptive control (MRAC)-based parameter identification applied to surface-mounted permanent magnet synchronous motor. Int. J. Electron. **104**(11), 1854–1873 (2017)
13. H. Jia, M. Cheng, W. Hua, W. Lu, A new stator-flux orientation strategy for flux-switching permanent motor drive based on voltage space-vector, in *Proceedings of the International Conference on Electrical Machines and Systems* (Wuhan, China, 2009), pp. 3032–3036
14. M. Cheng, Q. Sun, E. Zhou, New self-tuning fuzzy PI control of a novel doubly salient permanent magnet motor drive. IEEE Trans. Ind. Electron. **53**(3), 814–821 (2006)
15. J. Han, From pid to active disturbance rejection control. IEEE Trans. Ind. Electron. **56**(3), 900–906 (2009)
16. J. Han, *Active Disturbance Rejection Control Technique-the Technique for Estimating and Compensating the Uncertainties* (National Defense Industry Press, Beijing, 2008)
17. J. Li, Y. Xia, X. Qi, Z. Gao, On the necessity, scheme and basis of the linear-nonlinear switching in active disturbance rejection control. IEEE Trans. Ind. Electron. **64**(2), 1425–1435 (2016)
18. W. Tan, F. Caifen, Linear active disturbance-rejection control: analysis and tuning via imc. IEEE Trans. Ind. Electron. **63**(4), 2350–2359 (2016)
19. G. Wang, B. Wang, C. Li, X. Dianguo, Weight-transducerless control strategy based on active disturbance rejection theory for gearless elevator drives. IET Electr. Power Appl. **11**(2), 289–299 (2017)
20. Y. Jiang, Q. Sun, X. Zhang, Z. Chen, Pressure regulation for oxygen mask based on active disturbance rejection control. IEEE Trans. Ind. Electron. **64**(8), 6402–6411 (2017)
21. Q. Zheng, L. Dong, D.H. Lee Zhang, Z. Gao, Active disturbance rejection control for mems gyroscopes. IEEE Trans. Control Syst. Technol. **17**(6), 1432–1438 (2009)

# Flocking of Heterogeneous Multi-agent with Time Delay and Nonlinear Inner-Coupling Functions

**Jianli Ding, Yang Li and Zhengquan Yang**

**Abstract** In this paper, flocking of heterogeneous multi-agent is researched. The multi-agent consists of two types of agents with different dynamics. Based on some the certain assumptions, a controller with time delay and nonlinear inner-coupling function is given. It can be proved that each agent's velocity tracks the leader's by using the Lyapunov stability theory. Particularly, Collisions can be avoided between agents. Finally, a simulation is proposed to test and verify the availability of the controller.

**Keywords** Multi-agent · Flocking · Nonlinear dynamic · Time delay

## 1 Introduction

Flocking is a manifestation of individual collective behavior with universal group goals, and it can be coordinated collective movement by limited environmental information and simple rules. In recent years, researchers in various disciplines have done a lot of research on the flocking of multi-agent [1].

In 1986, Reynolds [2] first used computers to simulate the flocking problem and gave three rules to achieve rendezvous. Since then, Vicsek et al. [3] gave a simple flocking model that all multi-agents can move with at same velocity. In 2003, Jadbabaie et al. [4] gave the first accurate proof of the stability of Vicseks model. Afterwards, Olfati-Saber [5] studied the flocking of multi-agent having switching topologies and time-delay. Tanner et al. [6] proposed a simple control law that allows multi-agent to attain flocking in fixed and switching networks. Recently, researchers studied flocking of multi-agent having more complex dynamics. Su et al. [7] designed

J. Ding · Y. Li
College of Computer Science and Technology, Civil Aviation University of China, Tianjin 300300, China

Z. Yang (✉)
College of Science, Civil Aviation University of China, Tianjin 300300, China
e-mail: zquanyang@163.com



Y. Jia et al. (eds.), *Proceedings of 2018 Chinese Intelligent Systems Conference*, Lecture Notes in Electrical Engineering 528,
https://doi.org/10.1007/978-981-13-2288-4_67

an adaptive control rule to solve flocking of multi-agent having local Lipschitz nonlinear dynamics. Ghapani et al. [8] investigated the flocking problem with uncertain lagrange systems.

In reality and engineering applications, the dynamics of the coupling between agents are always dissimilar due to the external environmental influence or restriction of information transmission. Therefore, it is very important to study some behaviors of heterogeneous multi-agent. With the deep study of this problem, some new models and protocols were proposed for heterogeneous multi-agent [9–15]. Kim et al. [9] researched the output consensus problem for a class of heterogeneous multi-agent that had uncertain linear dynamics. Hong et al. [10] studied a leader-following problem that the followers had first-order integrator and the leader had second-order integrator. Song and Gu [11] researched the quasi-average consensus in undirected networks of multi-agent with mixed order integrators. Zheng and Wang [12] studied the finite-time consensus of heterogeneous multi-agent systems. In recent years, some researchers devoted to studying flocking of heterogeneous multi-agent. Wang et al. [13] investigated the flocking of heterogeneous multi-agent having nonlinear dynamics. Zhang et al. [14] investigated flocking for heterogeneous multi-agent with different dynamics and designed an adaptive controller to realize flocking. Over all flocking heterogeneous multi-agents is an interesting topic to be further researched.

In this paper, we propose an adaptive controller with time delay and nonlinear inner-coupling function to solve the flocking of heterogeneous multi-agent. There are two reasons for us to study this topic. Firstly, as everyone knows that time delay exists widely in reality owing to switching speed, memory effect and limited signal transmission efficiency. Whether the time delay is negligible or not depends on the properties of the event. Note, the time delay cannot be ignored in long-distance communication and traffic congestion.Secondly, in effect, the inner-coupling of multi-agents are much more complex than expected. In order to reflect the real flocking problem as possible, inner-couplings should be nonlinear. Accordingly, the main contributions of this article are summed as follows: First of all, the researched multi-agent system is heterogeneous. The second, the designed controller has time delay and nonlinear inner-coupling function.

The remaining parts of the paper are arranged as follows: In Sect. 2, we give the definition of flocking of heterogeneous multi-agent and present the known results that are used in this paper. In Sect. 3 we study flocking of heterogeneous multi-agent. We design an controller with time delay and nonlinear inner-coupling function such that all agent velocities become asymptotically the same and avoidance of collisions between the agents is ensured. In Sect. 4, a simulation is given to test and verify the availability of the proposed algorithm. In Sect. 5, we conclude this article.

## 2 Problem Formulations

We consider multi-agent system having $N(N \geq 2)$ agents in this paper. There are two kinds of agents in the multi-agent system. The former $l(l < N)$ agents are assumed having dynamic function $f_1$. While the last $(N - l)$ agents have dynamic function $f_2$. The heterogeneous multi-agent system we considered as following:

$$\begin{cases} \dot{x}_i = v_i \\ \dot{v}_i = f_1(v_i) + u_i \end{cases} \qquad i = 1, 2, \ldots, l \tag{1}$$

$$\begin{cases} \dot{x}_i = v_i, \\ \dot{v}_i = f_2(v_i) + u_i \end{cases} \qquad i = l + 1, 2, \ldots, N \tag{2}$$

where $x_i, v_i \in R^n$ and $u_i \in R^n$ represent the position, the velocity nd the control input vector of the $i$ agent, correspondingly. $f_1, f_2 : R^n \to R^n$ describe the intrinsic dynamic of the two kinds of agents respectively. The virtual leader about the system (1) and (2) is given by

$$\begin{cases} \dot{x}_0 = v_0 \\ \dot{v}_0 = f_1(v_0) \end{cases} \tag{3}$$

where $x_0 \in R^n$ and $v_0 \in R^n$ are the virtual leader's position vector and velocity vector, respectively.

Our purpose is to devise $u_i$ for (1) and (2) using the local interaction from its neighbors such that all agents track the leader while maintaining connectivity and avoid collisions.

Let $G(t) = (V, E(t), A(t))$ is an undirected time-varying graph. that is if $(i, j) \in E(t)$ and only if $(j, i) \in E(t)$. And then, any points $i$ and $j$ in the graph $G(t)$ are joined by a connection $(i, j) \in E(t)$, which means the agent $i$ and $j$ are neighbors at time $t$. $A(t) = [a_{ij}(t)] \in R^{N \times N}$ is the weighted adjacency matrix for the information graph of $N$ agents, where $a_{ij}(t) = a_{ji}(t) > 0$ if $(i, j) \in E(t)$, $a_{ij}(t) = a_{ji}(t) = 0$ otherwise. The neighbors of vertex $i$ are given by $N_i(t) = \{j \in \mathcal{V} : \|x_i - x_j\| \leq R\}$, where $R > 0$ is a constant and can be viewed as the sensing radius of the sensors. An undirected network $G(t)$ is connected if there is a path between any pair of distinct nodes in $G(t)$ .

*Remark 1* In this paper, the considered topology is time-varying. In the remain parts of this paper, for simplicity, we rewrite $a_{ij}(t)$ and $N_i(t)$ as $a_{ij}$ and $N_i$ respectively.

In this paper we will need to use the following assumption.

**Assumption 1** The activation function $f_1(\cdot)$ and $f_2(\cdot)$ satisfy the Lipschitz condition, that is, for any $x, y \in R^n$, there exists a non-negative constant $\alpha$ satisfying

$$\begin{aligned} \|f_1(x) - f_1(y)\| \leq \alpha\|x - y\| \\ \|f_2(x) - f_2(y)\| \leq \alpha\|x - y\| \end{aligned} \tag{4}$$

## 3 Flocking with Time Delay

In this subsection, we propose $u_i$ for (1) and (2) such that all followers follow the virtual leader with time delay and nonlinear inner-coupling function. To deal with this problem, we offer the following control law:

$$\begin{cases} u_i = -\sum_{j=1}^{N} a_{ij}\nabla_{x_i} V_{ij} - \sum_{j=1}^{N} a_{ij}(v_i - v_j(t-\tau)) - d_i e_i, \quad i = 1, 2, \ldots, l \\ u_i = -\sum_{j=1}^{N} a_{ij}\nabla_{x_i} V_{ij} - \sum_{j=1}^{N} a_{ij}\phi[v_i - v_j(t-\tau)] - f_2(v_0) + f_1(v_0) - d_i e_i, \\ i = l+1, 2, \ldots, N \end{cases} \tag{5}$$

$$\dot{d}_i = k_i e_i^T e_i \tag{6}$$

where $\phi : R^n \to R^n$ is a continuous nonlinear function, $e_i = v_i - v_0, i = 1, 2, \ldots, N$ is velocity error, $d_i$ is updating laws, $k_i$ is the positive constant, and $V_{ij}$ is an artificial potential function of agents $i$ and $j$ to be definite.

*Remark 2* The inner-coupling function of the former $l$ controller is linear and the inner-coupling function $\phi$ of the $(N-l)$ controller is nonlinear.

*Remark 3* The researched multi-agent system is heterogeneous. Firstly, the dynamics of the multi-agent system are different. Secondly, the inner-coupling functions of the controller are also different.

**Definition 1** [15] The potential function $V_{ij}$ is a differentiable nonnegative function of $\|x_i - x_j\|$ which satisfies the following conditions
(1) $V_{ij} = V_{ji}$ has a unique minimum in $\|x_i - x_j\| = d_{ij}$, where $d_{ij}$ is a desired distance between agents $i$ and $j$ and $R > \max_{i,j} d_{ij}$.
(2) $V_{ij} \to \infty$ if $\|x_i - x_j\| \to 0$.
(3) $\begin{cases} \frac{\partial V_{ij}}{\partial(\|x_i - x_j\|)} = 0, & \|x_i(0) - x_j(0)\| \geq R, \|x_i - x_j\| \geq R \\ \frac{\partial V_{ij}}{\partial(\|x_i - x_j\|)} \to \infty, & \|x_i(0) - x_j(0)\| < R, \|x_i - x_j\| \to R \end{cases}$

A potential function partial derivatives is given in Eqs. (36) and (37) in [15], where $d_{ij} = 0.5, \forall i, j$. We will also use that potential function in this paper. About the nonlinear function $\phi(\cdot)$, we have the following assumption.

**Assumption 2** For any $x, y \in R^n$, there exists a non-negative constant $\gamma$ satisfying

$$\|\phi(x-y)\| \leq \gamma(\|x\| + \|y\|).$$

**Theorem 1** *Assume that the Assumption 1 and Assumption 2 are satisfied and the initial graph G(0) is connected. Then, under the control laws (5), the complicated multi-gent systems (1) and (2), all agents in the system can keep up the virtual leader's velocity while preserving connectivity and avoiding collisions.*

*Proof* From the heterogeneous multi-agent system (1) and (2), the leader system (3) and the control laws (5), we will get the following error equation

$$\begin{cases} \dot{e}_i = f_1(v_i) - f_1(v_0) - \sum\limits_{j=1}^{N} a_{ij}\nabla_{x_i} V_{ij} - \sum\limits_{j=1}^{N} a_{ij}(e_i - e_j(t-\tau)) - d_i e_i, \\ \quad i = 1, 2, \ldots, l \\ \dot{e}_i = f_2(v_i) - f_2(v_0) - \sum\limits_{j=1}^{N} a_{ij}\nabla_{x_i} V_{ij} - \sum\limits_{j=1}^{N} a_{ij}\phi[e_i - e_j(t-\tau)] - d_i e_i, \\ \quad i = l+1, 2, \ldots, N \end{cases} \tag{7}$$

For proof the theorem, let us consider the Lyapunov functional candidate:

$$W = \sum_{i=1}^{N} {e_i}^T e_i + \sum_{i=1}^{N} \int_{t-\tau}^{t} e_i^T(\eta) e_i(\eta) d\eta + \sum_{i=1}^{N} \frac{(d_i - \hat{d})^2}{2k_i} + \sum_{i=1}^{N}\sum_{j=1}^{N} a_{ij} V_{ij}$$

Taking time derivative of $W$, we can get

$$\begin{aligned} \dot{W} &= 2\sum_{i=1}^{N} e_i^T \dot{e}_i + \sum_{i=1}^{N} [e_i^T e_i - e_i^T(t-\tau)e_i(t-\tau)] + \sum_{i=1}^{N} \frac{(d_i - \hat{d})}{k_i}\dot{d}_i + \sum_{i=1}^{N}\sum_{j=1}^{N} a_{ij}\dot{V}_{ij} \\ &= 2\sum_{i=1}^{l} {e_i}^T \left\{ f_1(v_i) - f_1(v_0) - \sum_{j=1}^{N} a_{ij}[e_i - e_j(t-\tau)] - \sum_{j=1}^{N} a_{ij}\nabla_{x_i} V_{ij} - d_i e_i \right\} \\ &\quad +2\sum_{i=l+1}^{N} {e_i}^T \left\{ f_2(v_i) - f_2(v_0) - \sum_{j=1}^{N} a_{ij}\phi(e_i - e_j(t-\tau)) - \sum_{j=1}^{N} a_{ij}\nabla_{x_i} V_{ij} - d_i e_i \right\} \\ &\quad + \sum_{i=1}^{N} [e_i^T e_i - e_i^T(t-\tau)e_i(t-\tau)] + \sum_{i=1}^{N} (d_i - \hat{d}) e_i^T e_i + \sum_{i=1}^{N}\sum_{j=1}^{N} a_{ij}\dot{V}_{ij} \\ &= -2\sum_{i=1}^{N}\sum_{j=1}^{N} a_{ij} e_i^T \nabla_{x_i} V_{ij} + 2\sum_{i=1}^{l} e_i^T[f_1(v_i) - f_1(v_0)] + 2\sum_{i=l+1}^{N} e_i^T[f_2(v_i) - f_2(v_0)] \\ &\quad -2\sum_{i=1}^{l} e_i^T \sum_{j=1}^{N} a_{ij}[e_i - e_j(t-\tau)] - 2\sum_{i=l+1}^{N} e_i^T \sum_{j=1}^{N} a_{ij}\phi(e_i - e_j(t-\tau)) \\ &\quad + \sum_{i=1}^{N} [e_i^T e_i - e_i^T(t-\tau)e_i(t-\tau)] - \hat{d}\sum_{i=1}^{N} e_i^T e_i^v + \sum_{i=1}^{N}\sum_{j=1}^{N} a_{ij}\dot{V}_{ij} \end{aligned} \tag{8}$$

Note However, we can get the following equation from the symmetry of $V_{ij}$.

$$\sum_{i=1}^{N}\sum_{j=1}^{N} a_{ij}\dot{V}_{ij} = 2\sum_{i=1}^{N}\sum_{j=1}^{N} a_{ij}e_i^T\nabla_{x_i}V_{ij}$$

Thus, (8) can be simplified to

$$\begin{aligned}
\dot{W}_{\mathcal{G}} &= 2\sum_{i=1}^{l} e_i^T[f_1(v_i) - f_1(v_0)] + 2\sum_{i=l+1}^{N} e_i^T[f_2(v_i) - f_2(v_0)] \\
&\quad -2\sum_{i=1}^{l}\sum_{j=1}^{N} a_{ij}e_i^T[e_i - e_j(t-\tau)] - 2\sum_{i=l+1}^{N}\sum_{j=1}^{N} a_{ij}e_i^T\phi(e_i - e_j(t-\tau)) \\
&\quad +\sum_{i=1}^{N}[e_i^T e_i - e_i^T(t-\tau)e_i(t-\tau)] - \hat{d}\sum_{i=1}^{N} e_i^T e_i \\
&\le \sum_{i=1}^{N}(\alpha + 1 - \hat{d})e_i^T e_i - 2\sum_{i=1}^{l}\sum_{j=1}^{N} a_{ij}\|e_i^T\|\|e_i\| + 2\sum_{i=1}^{l}\sum_{j=1}^{N} a_{ij}\|e_i^T\|\|e_j(t-\tau))\| \\
&\quad +2\sum_{i=l+1}^{N}\sum_{j=1}^{N} a_{ij}\gamma\|e_i^T\|[\|e_i\| + \|e_j(t-\tau))\|] - \sum_{i=1}^{N} e_i^T(t-\tau)e_i(t-\tau) \\
&= \sum_{i=1}^{N}(\alpha + 1 - \hat{d})e_i^T e_i - 2\sum_{i=1}^{l} k_i e_i^T e_i + 2\sum_{i=l+1}^{N} \gamma k_i e_i^T e_i - \sum_{i=1}^{N} e_i^T(t-\tau)e_i(t-\tau) \\
&\quad +2\sum_{i=1}^{l}\sum_{j=1}^{N} a_{ij}\|e_i^T\|\|e_j(t-\tau))\| + 2\sum_{i=l+1}^{N}\sum_{j=1}^{N} \gamma a_{ij}\|e_i^T\|\|e_j(t-\tau))\| \qquad (9)
\end{aligned}$$

where $k_i = \sum_{j=1}^{N} a_{ij}$.

If we let

$$\Lambda = diag\{-k_1, \ldots, -k_l, \gamma k_{l+1}, \ldots, \gamma k_N\},$$

$$e = \left(\|e_1\|, \|e_2\|, \ldots, \|e_N\|\right)^T, B = \begin{pmatrix} E_{l\times N} \\ 0 \end{pmatrix}, C = \begin{pmatrix} 0 \\ E_{(N-l)\times N} \end{pmatrix},$$

Then, the above equation can rewrite as

$$\begin{aligned}
\dot{W}_{\mathcal{G}} &= e^T[(\alpha + 1 - \hat{d})I]e + 2e^T\Lambda e - e^T(t-\tau)e(t-\tau) \\
&\quad +2e^T ABe(t-\tau) + 2\gamma e^T ACe(t-\tau) \\
&= e^T[(\alpha + 1 - \hat{d})I + 2\Lambda]e - e^T(t-\tau)e(t-\tau) + e^T A(B + \gamma C)e(t-\tau) \\
&= e^T[(\alpha + 1 - \hat{d})I + 2\Lambda]e - e^T(t-\tau)e(t-\tau) + 2e^T\tilde{A}e(t-\tau) \\
&\le e^T[(\alpha + 1 - \hat{d})I + 2\Lambda]e + e^T\tilde{A}^T\tilde{A}e \\
&= e^T[(\alpha + 1)I + 2\Lambda + \tilde{A}^T\tilde{A} - \hat{d}I]e \qquad (10)
\end{aligned}$$

where

$$\tilde{A} = A(B + \gamma C)$$

As $\alpha, \gamma$ are non-negative, and $(\alpha + 1)I + 2\Lambda + \tilde{A}^T\tilde{A}$ is a symmetric matrix, We are able to choose proper positive constant $\hat{d}$ to make $(\alpha + 1)I + 2\Lambda + \tilde{A}^T\tilde{A} - \hat{d}I \leq 0$. Thus, one obtains $\dot{W} \leq 0$. So, we have $W \geq 0$ and $\dot{W} \leq 0$, which implies that $W(t) \leq W(0) < \infty$. Furthermore, because $V_{ij}$ is bounded, by Definition 1, it is guaranteed that collisions can be avoided between the agents and the connectivity is maintained. Meantime, by LaSalle's invariance principle, we can get $e_1 = \cdots = e_N = 0$. Moreover, it is natural to see $v_i = v_0, i = 1, 2, \ldots, N$. Therefore all agent velocities can track the leaders velocity in the steady state, that is $v_1 = v_2 = \cdots = v_N = v_0$.

## 4 Simulation

In this section, we give a simulation to test and verify the availability of the controller for the heterogeneous multi-agent (1) and (2). And then, in the simulation, we use the Lorenz system and the Chen system as the two dynamics functions. The dynamics of the former $l$ agents is Lorenz system

$$\dot{v}_i = \begin{cases} 10(v_{i2} - v_{i1}) \\ v_{i2} - v_{i1}v_{i3} + 28v_{i1} \\ v_{i1}v_{i2} - \frac{8}{3}v_{i3} \end{cases} \tag{11}$$

where $1 \leq i \leq l$.

On the contrary, the $(N - l)$ agents with Chen system as follows:

$$\dot{v}_i = \begin{cases} 35(v_{i2} - v_{i1}) \\ -7v_{i1} - v_{i1}v_{i3} + 28v_{i2} \\ v_{i1}v_{i2} - 3v_{i3} \end{cases} \tag{12}$$

Especially, we chose $N = 30$ agents to initialize to a line and the distance between adjacent agents is 0.25, initialize velocity selected randomly in the unit square (Fig. 1a). Meanwhile, setting link range $R = 5$, it means that when the distance between agents is smaller than that of $R$, they are neighbors. Agents are represented by points, and neighbor-ship (links) between agents are represented by hard lines. The length of the solid line attached to each agent represents the velocity of the agents, and the arrow corresponds to the direction of the agent's movement. In the simulation, we choose $k_i = 1, i = 1, 2, \ldots, N, \tau = 0.6$. Figure 1b plots the ending stable network and the ending velocity of the heterogeneous multi-agent. We can see from it that the connectivity of the multi-agent for is guaranteed and asymptotic flocking of multi-agent is realized.

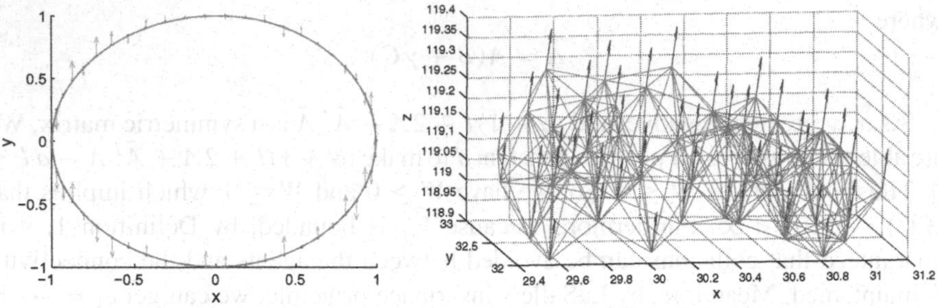


(a) Initialize configuration (b) End configuration

**Fig. 1** Flocking of multi-agent

**Fig. 2** Velocity error

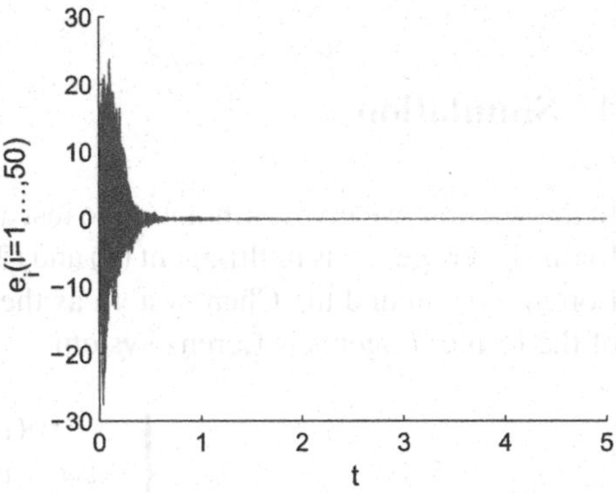


Figure 2 shows the velocity error between the agents and the leader. From that, we can see that the velocities of all agents can track the velocity of the leader in deed.

## 5 Conclusions

In this paper, we researched flocking of heterogeneous multi-agent with time delay. The multi-gent we considered are consisted of two kinds of agents with different dynamics. Based on the some assumptions, a controller with time delay and nonlinear inner-coupling function is proposed. We proved that each agent's velocity tracks the leader's velocity by using the Lyapunov stability theory. Particularly, collisions between agents can be avoided. Finally, we gave a simulation to test and verify the availability of the proposed algorithm.

**Acknowledgements** This work is supported by the National Natural Science Foundation of China (61573199). Civil Aviation Science and Technology Innovation Directs Fund Major Special Project (MHRD20150107, MHRD20160109). Basic Research Projects of High Education (3122015C025).

## References


1. E. Shaw, Fish in schools. Nat. Hist. **84**(8), 40–45 (1975)
2. C.W. Reynolds, Flocks, herds, and schools: a distributed behavioral model, vol. 21, no. 4, in *Computer Graphics, ACM SIGGRAPH, 87 Conference Proceedings*, pp. 25–34 (1987)
3. T. Vicsek, A. Czirok, E. Ben-Jacob, Novel type of phase transition in a system of self-driven particles. Phys. Rev. Lett. **75**(6), 1226–1229 (1995)
4. A. Jadbabaie, J. Lin, A.S. Morse, Coordination of groups of mobile autonomous agents using nearest neighbor rules. IEEE Trans. Autom. Control **48**(6), 988–1001 (2003)
5. R. Olfati-Saber, Flocking for multi-agent dynamic systems: algorithms and theory. IEEE Trans. Automat. Control **51**(3), 401–420 (2006)
6. H.G. Tanner, A. Jadbabaie, G.J. Pappas, Flocking in fixed and switching networks. IEEE Trans. Automat. Control **52**(5), 863–868 (2007)
7. H. Su, N. Zhang, M.Z.Q. Chen, H. Wang, X. Wang, Adaptive flocking with a virtual leader of multiple agents governed by locally Lipschitz nonlinearity. Nonlinear Anal. Real World Appl. **14**, 798–806 (2013)
8. S. Ghapani, J. Mei, W. Ren, Flocking with a moving leader for multiple uncertain Lagrange systems. Am. Control Conf. 3189–3194 (2014)
9. H. Kim, H. Shim, J.H. Seo, Output consensus of heterogeneous uncertain linear multiagent systems. IEEE Trans. Autom. Control **56**(1), 200–206 (2011)
10. Y. Hong, J. Hu, L. Gao, Tracking control for multi-agent consensus with an active leader and variable topology. Automatica **42**, 177–1182 (2000)
11. Y. Song, M. Gu, Quasi-average consensus in undirected networks of multi-agents with mixed order integrators. Control Eng. China **16**(2), 220–222 (2009)
12. Y. Zheng, L. Wang, Finite-time consensus of heterogeneous multi-agent systems with and without velocity measurements. Syst. Control Lett. **61**(8), 871–878 (2012)
13. M. Wang, H. Su, M. Zhao, M.Z.Q. Chen, H. Wang, Flocking of multiple autonomous agents with preserved network connectivity and heterogeneous nonlinear dynamics. Neurocomputing **115**, 169–177 (2013)
14. Q. Zhang, Y. Hao, Z. Yang, Z. Chen, Adaptive flocking of heterogeneous multi-agents systems with nonlinear dynamics. Neurocomputing **216**, 72–77 (2016)
15. Y. Cao, W. Ren, Distributed coordinated tracking with reduced interaction via a variable structure approach. IEEE Trans. Autom. Control **57**(1), 33–48 (2012)

# Research on an Advanced Cooperative Censored Positioning Algorithm in Wireless Sensor Networks

**Guolong Zhang, Lan Zhang, Jinpu Li and Bing Wang**

**Abstract** In cooperative positioning, the additional information from agent-to-agent links helps more agents complete their position estimates, but at the same time, leads to increased network traffic and even degraded positioning accuracy. Transmit and receive censoring can significantly reduce complexity and network traffic without hampering the positioning performance. In this paper, an advanced combined censored positioning algorithm is proposed to further reduce the use of invalid information links. Specifically, a method in the algorithm with variable step size has a positive impact on accuracy. The simulation results manifest that the proposed algorithm is efficient to reduce computational complexity and improve positioning performance.

**Keyword** Cooperative positioning · Cramer-Rao bound · Transmit censoring
Receive censoring · Variable step size

## 1 Introduction

Network location awareness (NLA) describes the location of a node in a network. It is a fundamental requirement for mobile systems such as wireless sensor networks (WSN) because location information facilitates navigation, path planning, military, environmental monitoring and office automation [1, 2].

A location aware wireless network generally includes two types of nodes: Anchors whose locations have been known as a priori information and agents whose locations to be determined [3]. In addition, both nodes can send or receive ranging information. There are two types of positioning schemes in WSNs: non-cooperative positioning (nCP) and cooperative positioning (CP) [4].

---

G. Zhang · L. Zhang (✉) · J. Li
School of Automation Electronic Engineering, University of Science & Technology Beijing, Beijing 100083, China
e-mail: zhanglan2013@ustb.edu.cn

B. Wang
Tianjin Zhongwei Aerospace Data System Technology Co., Ltd., Tianjin 300301, China



Y. Jia et al. (eds.), *Proceedings of 2018 Chinese Intelligent Systems Conference*,
Lecture Notes in Electrical Engineering 528,
https://doi.org/10.1007/978-981-13-2288-4_68

In the non-cooperative scheme, the agents' locations are deduced from just anchor-to-agent ranging information [5]. In the cooperative scheme, the additional information gained from agent-to-agent ranging information helps more agents to complete positioning [6, 7]. However, redundant measurement information can trigger computation complexity or even decrease positioning accuracy [8, 9]. How to reduce the complexity of calculation and network traffic becomes the key to cooperative positioning research [10].

As clearly reported in the related papers [11, 13], several censored methods can reduce redundant measurement information. A quintessential example should be cited in reference [11] proposes a censored technique based on the Cramer-Rao Bound (CRB) that takes into account both the uncertainties of the neighbors and the link quality, but the use of prior location information results in the high computation complexity and makes it challenging for practical implementation [12]. Furthermore, reference [13] proposes a method that uses CRB as the censored parameter, not only to remove bad links after receiving information from neighbors (receive censoring), but also to block the broadcast of unreliable nodes (transmit censoring). By implementing the method in [13], the network can reduce the computational complexity while ensuring the positioning accuracy. In addition, reference [10] proposes a new CP algorithm based on the AOA (Angle-of-Arrival) analyzing the CRB for cooperative AOA positioning. Reference [14] proposes a cluster nodes selection strategy based on the CRB for CP algorithm in WSN defining clusters for every specific agent by setting an RSS threshold which greatly saving the cost of energy.

In this paper, we use CRB as the censored parameter in the CP network, including two processes: the initial positioning estimation of agents and positions iterative update [15]. Precision censoring in process 1 and the method with variable step size in process 2 are proposed to further reduce redundant links and improve positioning speed while achieving better positioning accuracy.

The rest of the paper is organized as follows: Sect. 2 present principle description of cooperative positioning and introduce some criteria for evaluating the positioning performance. Section 3 introduces an advanced cooperative censored positioning algorithm. After that, Sect. 4 presents simulation results and the corresponding analysis. Finally, Sect. 5 concludes the paper.

## 2 Principle Description

### *2.1 Non-cooperative and Cooperative Networks*

Figure 1 illustrated a group of specific non-cooperative and cooperative networks. We consider 13 anchors and 100 agents, with a communication range of 20 m. The solid lines are anchors-to-agents measurement links, and the dotted lines are agents-to-agents measurement links. Anchors are aware of their locations, and agents determine their locations using internode ranging information. Note the large number of usable

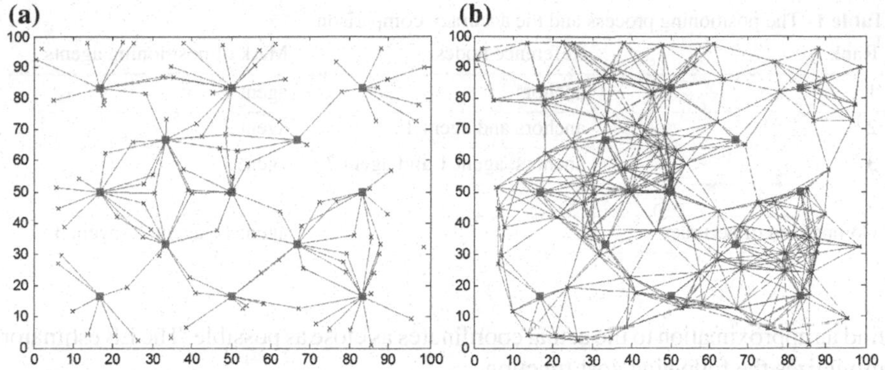


**Fig. 1** The links between different nodes: **a** non-cooperative network. **b** Cooperative network

links in the cooperative network due to huge amount of incoming information from the neighboring nodes, making cooperative positioning promising, but challenging to implement.

## 2.2 *Cooperative Positioning*

Cooperative positioning includes two processes: the initial positioning estimation and position iterative update. They are discussed in detail are as follows:

### 2.2.1 The Initial Positioning Estimation

In the CP network, both the anchors and the agents that have completed positioning can be used for positioning other nodes. We call these nodes as reference nodes.

For example, as can be seen in Table 1: Firstly, the anchors are used for positioning the agent 1. Secondly, all anchors and agent 1 as the reference nodes are used for positioning agent 2. Then, agent 2 will be seen as a reference node. With the increase of reference nodes, more ranging information can be applied to positioning more agents.

The position estimation of the latter agents has a lower accuracy.

### 2.2.2 Position Iterative Update

In the iterative process, agents update their position estimation at the same time based on position information of neighboring nodes. The iterative LS algorithm attempts to

**Table 1** The positioning process and the accuracy comparison

| Rank | Reference nodes | Mark of positioning agents |
|---|---|---|
| 1 | anchors | agent 1 |
| 2 | anchors and agent 1 | agent 2 |
| 3 | anchors, agent 1 and agent 2 | agent 3 |
| … | … | … |
| Comparison of accuracy | | agent 1 > agent 2 > agent 3 > … |

find an approximation to the actual coordinates as close as possible. The LS estimator minimizes the following cost function.

$$C_{LS}(x) = \sum_{i=1}^{N} \sum_{j \in S_{\rightarrow i}} \|\hat{d}_{ij} - \|\hat{x}_i - \hat{x}_j\|\|^2 \quad (1)$$

We denote that $\hat{x}_i$ is the estimated position coordinate of the agent $i$, $S_{\rightarrow i}$ is the set of nodes from which node $i$ can receive signals. $\hat{d}_{ij}$ is obtained by measuring the distance between agent $i$ and agent $j$, which is actually determined by the actual distance difference between two nodes ($d_{ij}$) and noise deviation ($e_{noise}$). In the simulation process, $e_{noise}$ obeys a Gaussian distribution N (0, $\delta^2$). $\delta^2$ is the noise variance, which is set as 0.16 in the later simulation. The $\hat{d}_{ij}$ is composed according to the follow equation:

$$\hat{d}_{ij} = d_{ij} + e_{noise} \quad (2)$$

The update phase of the iterative LS algorithm becomes:

$$\hat{x}_i^{(l+1)} = \hat{x}_i^{(l)} + k^{(l)} \sum_{j \in S_{\rightarrow i}} (\hat{d}_{ij}^{(l)} - ||\hat{x}_i^{(l)} - x_j^{(l)}||) \frac{(\hat{x}_i^{(l)} - \hat{x}_j^{(l)})}{||\hat{x}_i^{(l)} - \hat{x}_j^{(l)}||} \quad (3)$$

where $l$ is the number of iterations, $0<k<1$ is the step size. We stop updating the positional information of MS when $l$ reaches the stop number of iterations.

## 2.3 *Criteria to Evaluate the Positioning Performance*

### 2.3.1 Cumulative Distribution Function (CDF)

CDF is used to describe the positioning accuracy of a nCP/CP network. When positioning is completed, calculate the positioning error of each agent firstly and then get the CDF curve based on the positioning error. The CDF curve is always a monotonically increasing curve.

#### 2.3.2 Average Number of Links Per Agent

The links include anchors-to-agents measurement links and agents-to-agents measurement links. Each agent needs at least 3 links (TOA) for positioning. In the nCP network, average number of links per agent is usually less than 3 so there must be some agents that cannot complete the positioning. However, this number is always much larger than 3 in the CP network. Hence, some methods of links selection can be applied as information censoring to reduce complexity of CP network.

This number can also be seen as average number of reference nodes per agent. The large amount of network traffic has always been a drawback of CP network. Therefore, the average number of links per agent should be a criterion to present the ability of a CP algorithm reducing the network traffic without degrading positioning accuracy.

## 3 Advanced Cooperative Positioning

### *3.1 Cramer-Rao Bound*

The Cramer-Rao bound is a lower bound on the performance of any unbiased estimator. It is calculated by taking the inverse of the Fisher Information matrix (FIM). The FIM is given by:

$$\mathrm{F}(x_i) = -E\left\{\frac{\partial^2 \Delta(x_i)}{\partial x_i^2}\right\} \tag{4}$$

Then the FIM of $x_i$ will be of the form:

$$\mathrm{F}(x_i) = \sum_{j \in S_{\rightarrow i}} \frac{1}{\sigma_{j\rightarrow i}^2} \frac{x_i - x_j}{\| x_i - x_j \|} \left(\frac{x_i - x_j}{\| x_i - x_j \|}\right)^T \tag{5}$$

$\sigma_{j\rightarrow i}^2$ is the measured standard deviation of 10 measurements of $\hat{d}_{ij}$. Finally, the CRB can be calculated as:

$$\mathrm{CRB}(x_i) = trace([F(x_i)]^{-1}) \tag{6}$$

### *3.2 Advanced Cooperative Positioning Process*

In order to reduce complexity and traffic without decreasing positioning accuracy in the CP network, this paper proposes an advanced algorithm based on the combined censoring method. There are two main improvements in the algorithm compared to

**Table 2** The comparison of traditional and advanced algorithms

| Traditional CRB censored algorithm | Advanced combined censored algorithm |
|---|---|
| (1) Censoring by CRB:<br>Set $CRB(\hat{x}_i) = 0$ for anchors, if<br>$CRB(\hat{x}_i) = trace([F(\hat{x}_i)]^{-1}) < \gamma_{TX}$,<br>broadcast position information of $\hat{x}_i$;<br>if $CRB(\hat{x}_i) = trace([\tilde{F}(\hat{x}_i)]^{-1}) < \gamma_{RX}$,<br>discard the information of the $k$th receiving position.<br>(2) Iterative by a changeless step, $k^{(l)}$ is a changeless step in formula (3) | (1) Receive censoring by the accuracy of the neighboring nodes<br>Remove receive information come from lower precision agents<br>(2) Censoring by CRB as previous iterative process<br>(3) Iterative LS algorithm by a variable step, $k^{(l)}$ is a variable step in formula (3): ($l$ is the number of iterations)<br>$k^{(l)} = 0.8/b, b = (l/5) + 5$ |

reference [8]. Firstly, in the case of adequate positioning information, we remove receive information coming from lower precision agents before censoring by CRB, and we call this method as precision censoring. We call the algorithm with precision censoring as advanced cooperative censored algorithm. It can reduce the computational complexity of CRB in the process of censoring and improve the positioning speed. Secondly, an iterative method of variable step size is proposed to take into account both the number of iterations and the positioning accuracy. With the same number of iterations, it can achieve better positioning accuracy. Moreover, we call the algorithm with the above two methods as advanced combined censored algorithm. The comparison of traditional CRB censored algorithm and advanced algorithm is shown in Table 2.

# 4 Simulation Results and Analysis

In this section, we conduct several simulations regarding advanced positioning algorithms. The following simulations are based on two dimensions. Firstly, simulation setup and some performance criteria are shown below:

1. We consider a $100 \times 100\ m^2$ area, 13 anchor nodes are placed regularly while 100 agents are placed randomly.
2. Noise in simulations is a line of sight (LOS) error. It is white Gaussian noise with zero mean and variance of 0.16.
3. We set the communication range within 20 m.
4. Performance criteria include CDF and average number of links per agent. The paper had discussed both of them in detail in Sect. 2.3.

**Table 3** The comparison of average links used for update phase

| Cooperative positioning algorithm | Average links used for update phase |
|---|---|
| Advanced combined censored | 3.7580 |
| Advanced CRB censored | 5.5180 |
| Traditional CRB censored | 9.5800 |
| Non censored | 12.6200 |

## 4.1 Comparison of Average Links Used for Update Phase

As can be seen in Table 3, the advanced combined censored algorithm can effectively reduce the number of links used for updating locations. Compared with traditional CRB censored algorithm, the average links of it reduce from 9.5800 to 3.7580 for per agent. The result manifests that the advanced combined censored algorithm works out the reducing of redundant links, the computational complexity and the network traffic problem.

## 4.2 The Influence of Different Step Size on Positioning Performance

Different curves shown in the Fig. 2 display the CDF of positioning error in 70 iterations in the condition of different step size between 0.01, 0.05 and 0.1 respectively. From the figure, we can see clearly that the three curves usually have different derivatives at the same abscissa. Obviously, the algorithm with smaller step size can achieve better performance when the number of iterations is sufficient to reach convergence. In addition, the improvement of performance slows down when the step size less than 0.05 (the blue curve).

## 4.3 The Influence of Different Number of Iterations on Positioning Performance

Refer to Fig. 3 which demonstrates three different growth curves of different number of iterations. Each of them represents an algorithm with 10, 30 or 70 iterations. At the same time, for equality, both of the three algorithms have the same step size of 0.05 and other parameter setting. The comparing of the three algorithms manifests that the algorithm with larger number of iterations can achieve better performance. In addition, the improvement of performance slows down when the number of iterations larger than 70.

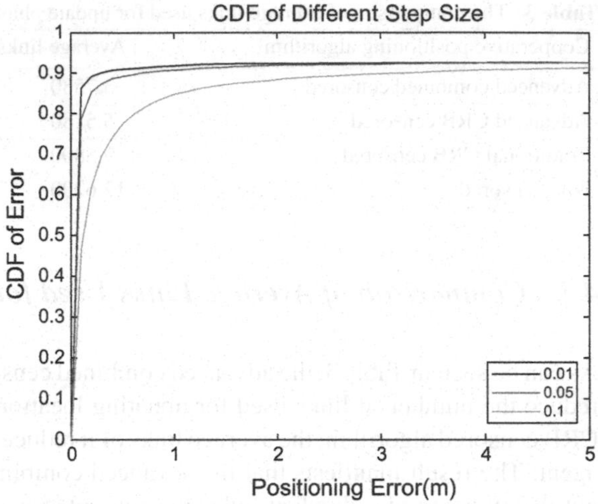


**Fig. 2** The CDF curves of different step size with the same number of iterations

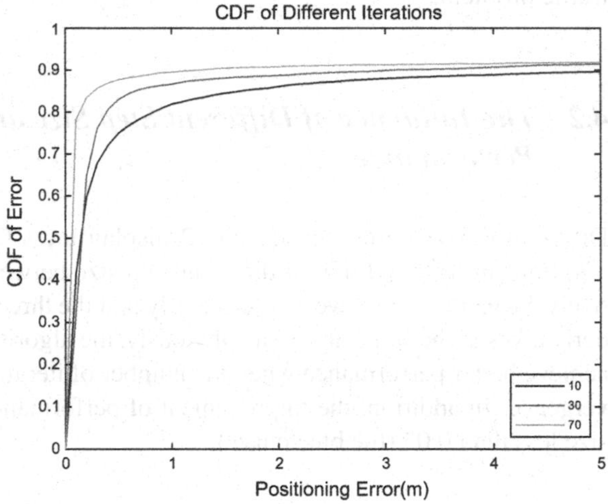


**Fig. 3** The CDF curves of different iterations with same step size

## *4.4 The Influence of Variable Step Size on Single Node Positioning Performance*

Figure 4 illustrates the relationship between the number of iterations and the root mean square error (RMSE) in the condition of step size between 0.1, 0.005 and the variable step size respectively. It can be seen that the RMSE decreases as the number of iterations increasing. Obviously, compared with the step size of 0.1, the method with the variable step size k can achieve higher accuracy. Simultaneously, it can achieve the same accuracy but use fewer iterations compared with the step size of 0.005. The result manifests that a suitable variable step size can bring about faster convergence and higher accuracy.

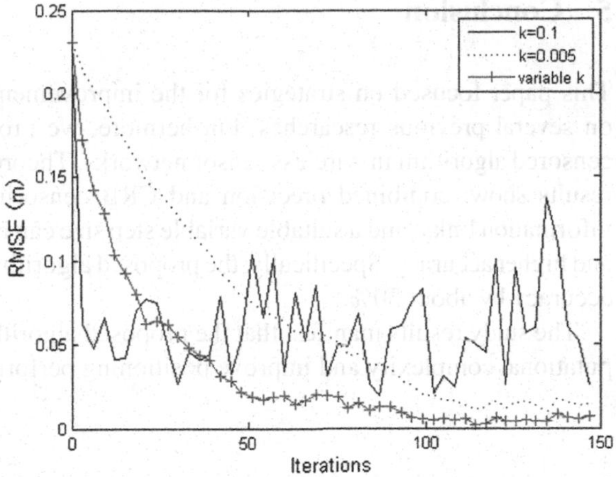


**Fig. 4** The relationship between RMSE and iterations in different step size

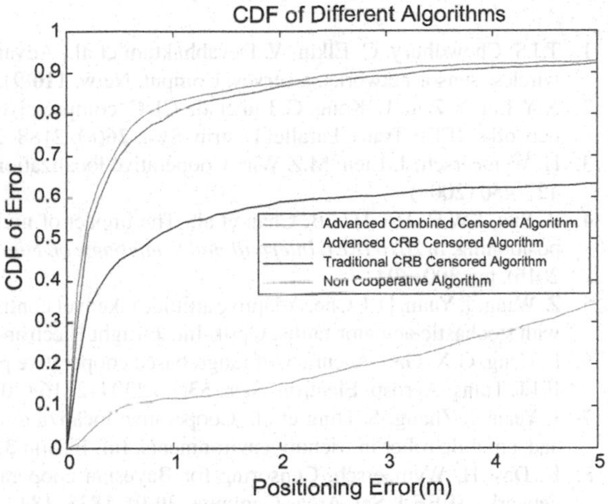


**Fig. 5** The CDF curves of four algorithms

## 4.5 *Comparison of Performance Between Several Positioning Algorithms*

As demonstrated in Fig. 5, four increasing curves display four different algorithms. By referring to the CDF, in the worst case, nCP algorithm can just guarantee 5 m accuracy for 35% of agents. Besides, the others algorithms can achieve it for over than 63%. Especially, concerning the advanced combined censored algorithm and advanced CRB censored algorithm, over 92% of agents' accuracy is less than 5 m. In addition, compared with the others algorithms, the advanced combined censored algorithm has the best performance under the same positioning error. The results manifest that the methods we have proposed including precision censoring and variable step size have a positive impact on positioning performance of algorithm.

## 5 Conclusion

This paper focused on strategies for the improvement of censoring method based on several previous researches. Furthermore, we propose an advanced combined censored algorithm in wireless sensor networks. Theoretical analyses and simulation results show: combined precision and CRB censoring can reduce the redundant information links; and a suitable variable step size can bring about faster convergence and higher accuracy. Specifically, the proposed algorithm can improve the positioning accuracy by about 30%.

The study results manifest that the proposed algorithm is efficient to reduce computational complexity and improve positioning performance.

## References


1. T.J.S. Chowdhury, C. Elkin, V. Devabhaktuni et al., Advances on localization techniques for wireless sensor networks: a survey. Comput. Netw. **110**(9), 284–305 (2016)
2. X.Y. Liu, Y. Zhu, L. Kong, C. Liu et al., CDC: compressive data collection for wireless sensor networks. IEEE Trans. Parallel Distrib. Syst. **26**(8), 2188–2197 (2015)
3. H. Wymeersch, J. Lien, M.Z Win, Cooperative localization in wireless networks. Proc. IEEE 427–450 (2009)
4. Z. Situ, I.W.H. Ho, H.C.B. Chan et al., The impact of node reliability on indoor cooperative positioning, in *2016 IEEE International Conference on Digital Signal Processing (DSP)* (IEEE, 2016), pp. 300–304
5. Z. Wang, J. Yuan, D.J. Che, Adaptive attitude takeover control for space non-cooperative targets with stochastic actuator faults. Optik-Int. J. Light Electron Optics **137**, 279–290 (2017)
6. L. Heng, G.X. Gao, Accuracy of range-based cooperative positioning: a lower bound analysis. IEEE Trans. Aerosp. Electron. Syst. **53**(5), 2304–2316 (2017)
7. J. Yuan, J. Zhang, S. Ding et al., Cooperative localization for disconnected sensor networks and a mobile robot in friendly environments. Inf. Fusion **37**, 22–36 (2017)
8. K. Das, H. Wymeersch, Censoring for Bayesian cooperative positioning in dense wireless networks. IEEE J. Sel. Areas Commun. **30**(9), 1835–1842 (2012)
9. R. Raulefs, S. Zhang, C. Mensing, Bound-based spectrum allocation for cooperative positioning. Trans. Emerg. Telecommun. Technol. **24**(1), 69–83 (2013)
10. N. Patwari, J.N. Ash, S. Kyperountas et al., Locating the nodes: cooperative localization in wireless sensor networks. IEEE Signal Process. Mag. **22**(4), 54–69 (2005)
11. S. Van de Velde, G.T.F. de Abreu, H. Steendam, Improved censoring and NLOS avoidance for wireless localization in dense networks. IEEE J. Sel. Areas Commun. **33**(11), 2302–2312 (2015)
12. A.T. Ihler, J.W. Fisher, R.L. Moses et al., Nonparametric belief propagation for self-localization of sensor networks. IEEE J. Sel. Areas Commun. **23**(4), 809–819 (2005)
13. K. Das, H. Wymeersch, Censored cooperative positioning for dense wireless networks, in *2010 IEEE 21st International Symposium on Personal, Indoor and Mobile Radio Communications Workshops (PIMRC Workshops)* (IEEE, 2010), pp. 262–266
14. J. Xu, M. Ma, C.L. Law, AOA cooperative position localization, in *IEEE Global Telecommunications Conference* (New Orleans, USA, 2008), pp. 1–5
15. T. Van Nguyen, Y. Jeong, H. Shin et al., Least square cooperative localization. IEEE Trans. Veh. Technol. **64**(4), 1318–1330 (2015)

# Event-Triggering Consensus of Second-Order Multi-agent Systems

**Yuanshan Liu, Hongyong Yang, Yize Yang, Yuling Li and Fan Liu**

**Abstract** The consensus problem of second-order multi-agent system with event-trigger control is studied in this paper. In order to reduce the waste of hardware resources, improve the communication efficiency and increase the convergence speed of multi-agent systems, distributed event-trigger control algorithm is presented by means of the information of the neighborhood nodes. A closed-loop system model of second-order multi-agent under event-trigger strategy is established. By applying the of analytical tools such as matrix theory and modern control theory, the convergence condition of multi-agent systems with event-trigger control is obtained. Finally, the simulation results show the effectiveness of the algorithm.

**Keywords** Multi-agent systems · Consensus · Event-triggering · Undirected graph

## 1 Introduction

In recent years, with the development of science technology and network communication technology, the multi-agent systems have been widely used in robot coop-

Y. Liu · H. Yang (✉) · Y. Li · F. Liu
School of Information and Electrical Engineering, Ludong University, Yantai 264025, China
e-mail: hyyang@yeah.net

Y. Liu
e-mail: yuanshanliu@163.com

Y. Li
e-mail: liyuling822@163.com

F. Liu
e-mail: jsgyliufan@163.com

Y. Yang
School of Electrical Engineering and Telecommunications, The University of New South Wales, Sydney, Australia
e-mail: yangyz1994@126.com

 
Y. Jia et al. (eds.), *Proceedings of 2018 Chinese Intelligent Systems Conference*, Lecture Notes in Electrical Engineering 528,
https://doi.org/10.1007/978-981-13-2288-4_69

erative control, traffic management, network distribution control and other fields. The consensus problem of multi-agent systems is an important research direction of distributed cooperative control. Now it has become a hot research topic in the field of artificial intelligence. Ren et al. [1] studied the consensus algorithm of the coordinated control of multi-agent systems. Olfati-Saber et al. [2] have studied the consensus problem of dynamic multi-agent network under fixed and switching topology. Ma et al. [3] have studied the consensus problem of multi-agent systems with parameter uncertainties in directed communication topology. Yang et al. [4] have studied on the consensus of fractional order multi-agent systems with time delays. Li et al. [5] have studied on the consensus of multi-agent with disturbances.

In practice, if we reduce the number of communication between agents, the amount of data transmission in the network can be decreased to improve the convergence speed of the system. Zhang and Liu [6], Tang et al. [7] studied the problem of distributed consensus of multi-agent systems in sampled data. Sampling control is easy to implement in practice application, but the difference between sampled data is small or when the system is stable, the system still sampling and there will be a waste of resources. Guo et al. [8] proposed a distributed event-triggered sampling data transmission strategy, which improves the performance of the system. Dimarogonas et al. [9] proposed event-triggering functions that only depend on the state of the agent itself and its neighboring agents. Huang et al. [10] studied average consensus of second-order multi-agent systems based on event-triggering. Tabuadad [11] designed a simple event-triggered scheduler, enabling it to schedule stable control tasks on embedded processors.

The research of event-triggering strategy is mostly focused on the first-order multi-agent systems. This paper studies the consensus problem of second-order multi-agent based on event-triggering strategy, and constructs the unified control protocol and event-trigger function of the distributed second-order intelligence systems.

## 2 Preliminaries and Problem Description

### 2.1 Preliminaries Knowledge

The communication topology between agents of multi-agent systems is usually described by structural diagrams. Suppose $G = \{V, \varepsilon, A\}$ is a communication structure diagram for a system with $n$ agents, therein $V = \{1, 2, \ldots, n\}$ representing a set of nodes, $\varepsilon \subseteq V \times V$ is set of edges, $A$ is denoted adjacency matrix. An edge $e_{ij} \in \varepsilon$ in a directed graph represents an ordered node, where $i$ is the starting point, $j$ is the target node, indicating that node $i$ can receive the information of node $j$. In an undirected graph, the edges of any node are undirected, and if there is an edge connection between the node $i$ and the node $j$, the two nodes can convey information to each other, in other words $(i, j) \in \varepsilon \Leftrightarrow (j, i) \in \varepsilon$. The relation between the element and the edge in the adjacency matrix is $a_{ij} = 1 \Leftrightarrow e_{ij} \in \varepsilon$, assume $a_{ii} = 0$. Let

$D = diag\{d_{in}(i)\} \in R^{n\times n}$ represent degree matrix, Then the Laplace matrix of graph $G$ can be expressed as $L = D - A$.

**Lemma 1** *If $G$ is an undirected connected graph, Laplacian matrix $L$ has a unique zero eigenvalue, and other eigenvalues are positive real numbers; if $L = \begin{bmatrix} l_{11} & * \\ * & \hat{L} \end{bmatrix}$, among $L \in R^{n\times n}$, $\hat{L} \in R^{(n-1)\times(n-1)}$, therefore $\hat{L}$ is a positive definite matrix.*

**Lemma 2** [10] (Schur Complement) *For given symmetric matrix $S$ with the from*

$$S = \begin{bmatrix} S_{11} & S_{12} \\ S_{21} & S_{22} \end{bmatrix},$$

*then $S > 0$, if and only if:*

$$S_{11} > 0, S_{22} - S_{12}^T S_{11}^{-1} S_{12} > 0$$

*or*

$$S_{22} > 0, S_{11} - S_{12}^T S_{22}^{-1} S_{12} > 0$$

# 3 Event-Triggering Consensus of Second-Order Multi-agent Systems

## 3.1 Problem Description

A distributed second-order multi-agent network, with $n$ agents communicating with each other is considered. The dynamic equations of each agent $i$ can be described as:

$$\begin{cases} \dot{x}_i(t) = v_i(t) \\ \dot{v}_i(t) = u_i(t) \end{cases} \tag{1}$$

For arbitrary $i \in V$, if $x_i$, $v_i \in R$ represents respectively the location and speed of the $i$ agents, $u_i(t) \in R$ is the control input for the $i$ agent. We call the consensus of multi-systems has been achieved, if and only if $\lim_{t\to\infty} |x_i(t) - x_j(t)| = 0$, $\lim_{t\to\infty} |v_i(t) - v_j(t)| = 0$, where $i, j \in N$.

### 3.2 The Design of the Controller

In the engineering application, event-triggered strategy can reduce the execution times of the intelligent control protocol and save the computing resources of each agent. It is assumed that the event-triggered control protocol of multi-agent systems is:

$$u_i(t) = -\alpha \sum_{j\in N_i} a_{ij}\left[x_i(t_k) - x_j(t_k)\right] - \beta \sum_{j\in N_i} a_{ij}\left[v_i(t_k) - v_j(t_k)\right] \tag{2}$$

where $\alpha \geq 0$, $\beta \geq 0$, $t \in [t_k, t_{k+1})$, $t_k$ is the trigger time for the $k$ event of the controller.

The definition error function is:

$$\begin{cases} e_x^i(t) = x_i(t_k) - x_i(t) \\ e_v^i(t) = v_i(t_k) - v_i(t) \end{cases}$$

then:

$$\begin{aligned} u_i(t) = & -\alpha \sum_{j\in N_i} a_{ij}\left[e_x^i(t) + x_i(t) - e_x^j(t) - x_j(t)\right] \\ & -\beta \sum_{j\in N_i} a_{ij}\left[e_v^i(t) + v_i(t) - e_v^j(t) - v_j(t)\right] \end{aligned} \quad i = 1, 2, \ldots, n \tag{3}$$

The dynamic equations of the system are as follows:

$$\begin{cases} \dot{x}_i(t) = v_i(t) \\ \dot{v}_i(t) = -\alpha \sum_{j\in N_i} a_{ij}\left[(x_i(t) - x_j(t)) + (e_x^i(t) - e_x^j(t))\right] \quad i = 1, 2, \ldots, n \\ - \beta \sum_{j\in N_i} a_{ij}\left[(v_i(t) - v_j(t)) + (e_v^i(t) - e_v^j(t))\right] \end{cases} \tag{4}$$

Let $x(t) = [x_1(t), \ldots, x_n(t)]^T v(t) = [v_1(t), \ldots, v_n(t)]^T e_x(t) = \left[e_x^1(t), \ldots, e_x^n(t)\right]^T$, $e_v(t) = \left[e_v^1(t), \ldots, e_v^n(t)\right]^T$. Then the system dynamical Eq. (4) is rewritten as: $\begin{bmatrix} \dot{x}(t) \\ \dot{v}(t) \end{bmatrix} = A \begin{bmatrix} x(t) \\ v(t) \end{bmatrix} + B \begin{bmatrix} e_x(t) \\ e_v(t) \end{bmatrix}$ where $A = \begin{bmatrix} 0 & I \\ -\alpha L & -\beta L \end{bmatrix}$, $B = \begin{bmatrix} 0 & 0 \\ -\alpha L & -\beta L \end{bmatrix}$. Let $\xi(t) = \left[x(t)^T, v(t)^T\right]^T$, $e(t) = \left[e_x(t)^T, e_v(t)^T\right]^T$, then the dynamical equations of the system are described as follows:

$$\dot{\xi}(t) = A\xi(t) + Be(t) \tag{5}$$

### 3.3 Event-Triggering Consensus of Second-Order Multi-agent Systems

The consensus of second-order multi-agent systems based on event-triggered strategy will be analyzed.

Let

$$\begin{cases} \delta_{x_i}(t) = x_i(t) - x_1(t) \\ \delta_{v_i}(t) = v_i(t) - v_1(t) \end{cases}, \quad i = 2, 3, \ldots, n$$

then

$$\begin{cases} \dot{\delta}_{x_i}(t) = \delta_{v_i}(t) \\ \dot{\delta}_{v_i}(t) = -\alpha \sum_{j \in N_i} a_{ij}\left[(\delta_{x_i}(t) - \delta_{x_j}(t)) + (e_x^i(t) - e_x^j(t))\right] \end{cases} \quad i = 2, 3, \ldots, n$$

$$- \beta \sum_{j \in N_i} a_{ij}\left[(\delta_{v_i}(t) - \delta_{v_j}(t)) + (e_v^i(t) - e_v^j(t))\right] \tag{6}$$

The above formula can be written as $\dot{\delta}(t) = \hat{A}\delta(t) + \hat{B}\hat{e}(t)$, where $\hat{A} = \begin{bmatrix} 0 & I_{n-1} \\ -\alpha\hat{L} & -\beta\hat{L} \end{bmatrix}$, $\hat{B} = \begin{bmatrix} 0 & 0 \\ -\alpha\hat{L} & -\beta\hat{L} \end{bmatrix}$, it is known by Lemma 1 $\hat{L} \in R^{(n-1)\times(n-1)}$, $\delta(t) = \left[\delta_{x_2}(t), \ldots, \delta_{x_n}(t), \delta_{v_2}(t), \ldots, \delta_{v_n}(t)\right]^T \in R^{2(n-1)}$, $\hat{e}(t) = \left[e_x^2(t), \ldots, e_x^n(t), e_v^2(t), \ldots, e_v^n(t)\right]^T \in R^{2(n-1)}$.

**Theorem 1** *In the distributed second-order multi-agent systems (1). Assume that the system graph G has a spanning tree, with the control protocol (2), the trigger rule (7) is designed as follows:*

$$f(\hat{e}(t), \delta(t)) = \left\|\hat{e}(t)\right\| - \theta \frac{\lambda_{min}(Q)}{2(\alpha + \beta)\lambda_{max}(\hat{L})\|P\|}\|\delta(t)\| \tag{7}$$

*therein* $\alpha > 1$, $\beta > \frac{1}{\hat{\mu}} + \frac{\alpha}{4\hat{\mu}^2}$, *When the trigger is* $f(\hat{e}(t), \delta(t)) \geq 0$, *the controller can adjust the state of the system to achieve asymptotic consensus. Where* $\hat{B} = \begin{bmatrix} 0 & 0 \\ -\alpha\hat{L} & -\beta\hat{L} \end{bmatrix}$, $Q = \begin{bmatrix} 2\alpha\hat{L} & -\alpha I \\ -\alpha I & 2(\beta\hat{L} - I) \end{bmatrix}$, $P = \begin{bmatrix} (\alpha + \beta)\hat{L} + \alpha I & I \\ I & I \end{bmatrix}$, *I is a unit matrix of* $n - 1$ *order,* $\hat{\mu}$ *is the minimum eigenvalue of* $\hat{L}$.

*Proof* Assumption of Lyapunov function $V(t) = \delta(t)^T P\delta(t)$

Where $P = \begin{bmatrix} (\alpha+\beta)\hat{L}+\alpha I & I \\ I & I \end{bmatrix}$, when the parameter $\alpha$ is selected, $\beta$ makes $(\alpha+\beta)\hat{L}+(\alpha-1)I > 0$ (Positive definite matrix), Lemma 1 and Lemma 2 show that $P$ is a positive definite symmetric matrix. The derivation of $V(t)$ can be obtained:

$$\begin{aligned}\dot{V}(t) &= (\hat{A}\delta(t)+\hat{B}\hat{e}(t))^T P\delta(t) + \delta(t)^T P\hat{A}\delta(t) + \delta(t)^T P\hat{B}\hat{e}(t) \\ &\leq -\delta(t)^T \begin{bmatrix} 2\alpha\hat{L} & -\alpha I \\ -\alpha I & 2(\beta\hat{L}-I) \end{bmatrix} \delta(t) + 2(\alpha+\beta)\lambda_{\max}(\hat{L})\left\|\hat{e}(t)\right\| \|P\| \|\delta(t)\|\end{aligned}$$

Let $Q = \begin{bmatrix} 2\alpha\hat{L} & -\alpha I \\ -\alpha I & 2(\beta\hat{L}-I) \end{bmatrix}$, the following proves that the matrix $Q$ is positive definite

$$\begin{aligned}|\lambda E - Q| &= \begin{vmatrix} \lambda - 2\alpha\hat{L} & \alpha I \\ \alpha I & \lambda - 2(\beta\hat{L}-I) \end{vmatrix} \\ &= \left|(\lambda - 2\alpha\hat{L})(\lambda - 2(\beta\hat{L}-I)) - \alpha^2 I\right| \\ &= \prod_{i=1}^{n-1} \left|\lambda^2 - (2\alpha\cdot\mu_i + 2(\beta\cdot\mu_i - 1))\lambda + 4\alpha\cdot\mu_i(\beta\cdot\mu_i - 1) - \alpha^2\right|\end{aligned}$$

Let $g(\lambda) = \lambda^2 - (2\alpha\cdot\mu_i + 2(\beta\cdot\mu_i - 1))\lambda + 4\alpha\cdot\mu_i(\beta\cdot\mu_i - 1) - \alpha^2$.
Suppose that the roots of $g(\lambda)$ as $\lambda_1$, $\lambda_2$, they satisfy
$\begin{cases} \lambda_1\cdot\lambda_2 = 4\alpha\cdot\mu_i(\beta\cdot\mu_i - 1) - \alpha^2 > 0 \\ \lambda_1+\lambda_2 = 2\alpha\cdot\mu_i + 2(\beta\cdot\mu_i - 1) > 0 \end{cases}$ we have $\alpha > 1$, $\beta > \frac{1}{\mu_i}+\frac{\alpha}{4\mu_i^2}$
Since $\mu_i$ is an arbitrary eigenvalue of positive definite symmetric matrix $\hat{L}$, the above inequality can be established.
Let $\hat{\mu} = \lambda_{\min}(\hat{L})$, thus $\beta > \frac{1}{\hat{\mu}}+\frac{\alpha}{4\hat{\mu}^2}$, $\alpha > 1$. The $Q$ can be guaranteed to be a positive definite matrix, in that way,

$$\dot{V}(t) \leq -\lambda_{\min}(Q)\|\delta(t)\|^2 + 2(\alpha+\beta)\lambda_{\max}(\hat{L})\left\|\hat{e}(t)\right\| \|P\| \|\delta(t)\|$$

If $0 < \theta < 1$, then:

$$\left\|\hat{e}(t)\right\| \leq \theta \frac{\lambda_{\min}(Q)}{2(\alpha+\beta)\lambda_{\max}(\hat{L})\|P\|} \|\delta(t)\|$$

Since $\dot{V}(t) \leq -(1-\theta)\|\delta(t)\| < 0$, for any $x_i(t) \in R$, $v_i(t) \in R$, $i = 2, 3, \ldots, n$, we obtain from *Lyapunov* theory of stability $\lim_{t\to\infty} |x_i(t) - x_1(t)| = 0$, $\lim_{t\to\infty} |v_i(t) - v_1(t)| = 0$, so the design event-triggers function (7), When the trigger is $f(\hat{e}(t), \delta(t)) \geq 0$, it can make the second-order system (1) achieve *Lyapunov* consensus stability.

In order to show that the system will not be triggered many times in the whole operation process, the following theorem is proved for obtaining the minimum interval of any two event-triggering.

**Theorem 2** *Consider a second-order multi-agent systems (1) with the control protocol (2) and trigger function (7). Assume the graph G has a spanning tree. Trigger times are* $\tau_0, \ldots, \tau_k, \tau_{k+1}$, *the minimum trigger interval* $\tau_{min} = \min_s\{\tau_{s+1}, \tau_s\}$ *for all* $s = 0, 1, \ldots, k$ *is not less than:* $\frac{-\theta\lambda_{min}(Q) - 4(\alpha+\beta)\lambda_{max}(\hat{L})\|P\|}{\mu\cdot(\theta\lambda_{min}(Q) - 2(\alpha+\beta)\lambda_{max}(\hat{L})\|P\|)}$, *where* $0 < \theta < 1$, $\mu = max\{\left\|\hat{A}\right\|, \left\|\hat{B}\right\|\}$.

*Proof* The dynamic equations for constructing $\frac{\|\hat{e}(t)\|}{\|\delta(t)\|}$ are as follows:

$$\frac{d}{dt}\left(\frac{\|\hat{e}(t)\|}{\|\delta(t)\|}\right) = \frac{\hat{e}(t)^T\dot{\hat{e}}(t)}{\|\delta(t)\|\|\hat{e}(t)\|} - \frac{\|\hat{e}(t)\|\delta(t)^T\dot{\delta}(t)}{\|\delta(t)\|^3}$$

Let $t \in [t_k, t_{k+1})$, $\hat{e}(t) = \delta(t_k) + \delta(t)$, therefore $\dot{\hat{e}}(t) = \dot{\delta}(t_k) - \dot{\delta}(t) = 0 - \dot{\delta}(t)$. So $\frac{d}{dt}\left(\frac{\|\hat{e}(t)\|}{\|\delta(t)\|}\right) \le \frac{\|\dot{\delta}(t)\|}{\|\delta(t)\|}\left(1 + \frac{\|\hat{e}(t)\|}{\|\delta(t)\|}\right)$, Also because of $\dot{\delta}(t) = \hat{A}\delta(t) + \hat{B}\hat{e}(t)$

$$\begin{aligned}\frac{d}{dt}\left(\frac{\|\hat{e}(t)\|}{\|\delta(t)\|}\right) &\le \frac{\left\|\hat{A}\delta(t) + \hat{B}\hat{e}(t)\right\|}{\|\delta(t)\|}\left(1 + \frac{\|\hat{e}(t)\|}{\|\delta(t)\|}\right) \\ &\le \max\{\left\|\hat{A}\right\|, \left\|\hat{B}\right\|\}\left(1 + \frac{\|\hat{e}(t)\|}{\|\delta(t)\|}\right)^2\end{aligned}$$

Let $\psi(t) = \frac{\|\hat{e}(t)\|}{\|\delta(t)\|}$, $\mu = \max\{\left\|\hat{A}\right\|, \left\|\hat{B}\right\|\}$, the solution of the above differential equation can be obtained as follows: $\psi(t) \le \frac{-\mu\cdot t - C - 1}{\mu\cdot t + 1}$. The shortest event-trigger interval is $\tau_{\min}$, the expression is as follows:

$$\tau_{\min} = \frac{-\theta\lambda_{\min}(Q) - 4(\alpha+\beta)\lambda_{\max}(\hat{L})\|P\|}{\mu\cdot(\theta\lambda_{\min}(Q) - 2(\alpha+\beta)\lambda_{\max}(\hat{L})\|P\|)}$$

## 4 Simulation Experiment

Consider the specific 6 node multi-agent systems, we can get the graph theory model of multi-agent systems (Fig. 1), and the corresponding Laplace matrix of the system is:

$$L = \begin{bmatrix} 3 & -1 & 0 & 0 & -1 & -1 \\ -1 & 2 & -1 & 0 & 0 & 0 \\ 0 & -1 & 3 & -1 & -1 & 0 \\ 0 & 0 & -1 & 2 & -1 & 0 \\ -1 & 0 & -1 & -1 & 4 & -1 \\ -1 & 0 & 0 & 0 & -1 & 2 \end{bmatrix}$$

Under the topology shown in Fig. 1, the initial state of the system is respectively $x(0) = [-4\ 6\ 1\ 16.2\ 2.4\ 8]$, $v(0) = [-0.5\ 1.3\ \ -3.2\ 8.4\ 3\ 2.2]$. In this initial state, the system can get the following position and speed state diagram (Figs. 2 and 3) according to the given controller, and all the positions and speed states of all the agents in the two graphs are finally converged. Figure 4 shows the error of the trigger when the trigger is triggered. In the initial stage, the trigger interval is small and the value of the error norm is quite large. When the whole systems is gradually stable, the trigger interval of the trigger will be reduced greatly. Figure 5 shows the change of control input.

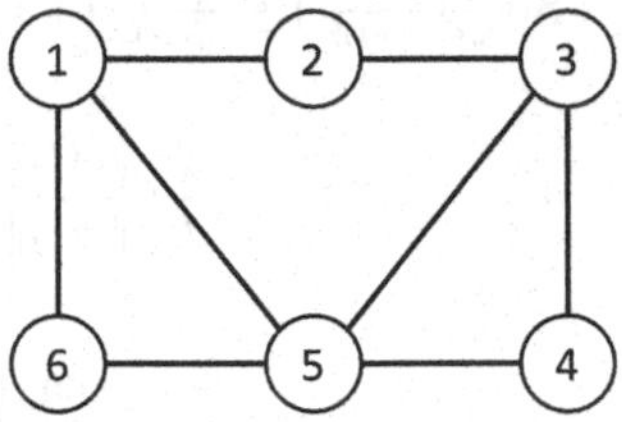


**Fig. 1** Second-order multi-agent systems topology diagram

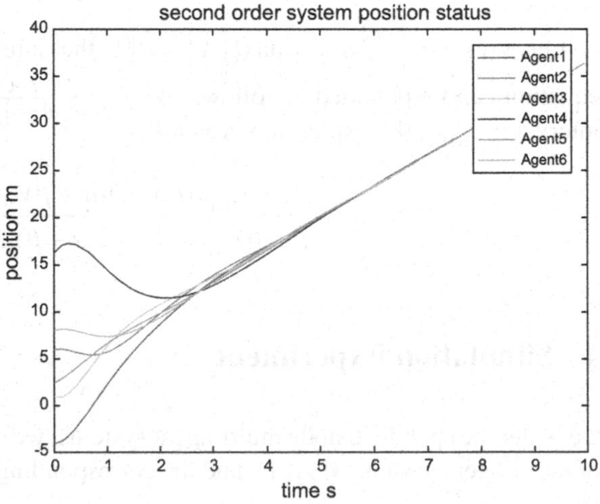


**Fig. 2** The position state of the second-order systems

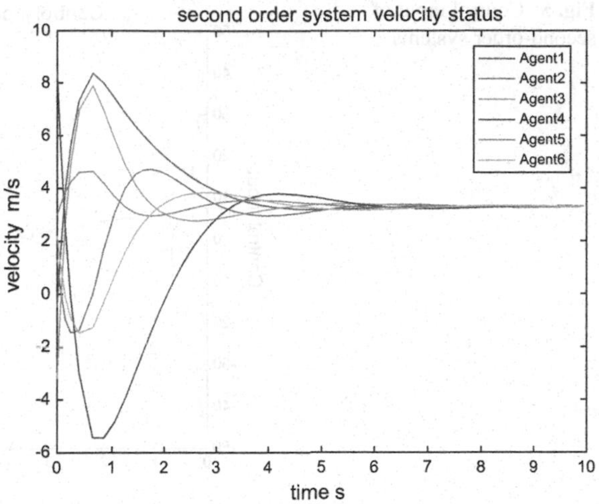


**Fig. 3** The velocity state of the second-order systems

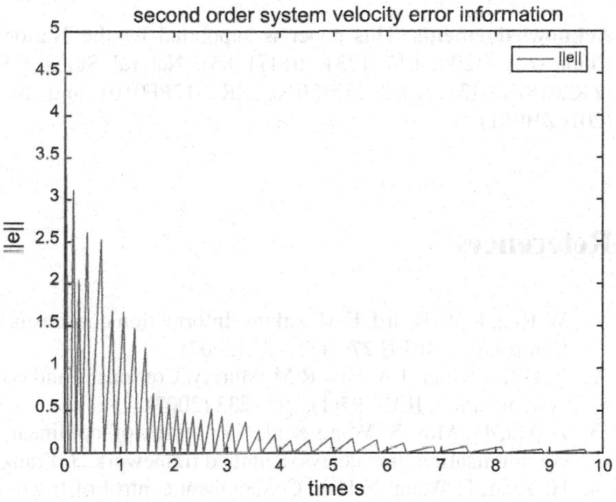


**Fig. 4** The error function of flip-flops

# 5 Conclusion

In this paper, we study the consensus of distributed second-order multi-agent systems, and design a control algorithm based on event-triggered strategy. By introducing the deviation vector of each agent and one agent, the consensus of second-order differential systems is transformed into the stability problem of the second-order closed loop system. Based on the theory of algebraic graph, Lyapunov stability theory, the sufficient conditions for the convergence of the multi-agent systems based on the event-trigger strategy are obtained. Finally, the validity of the theoretical solution is verified by Matlab simulation.

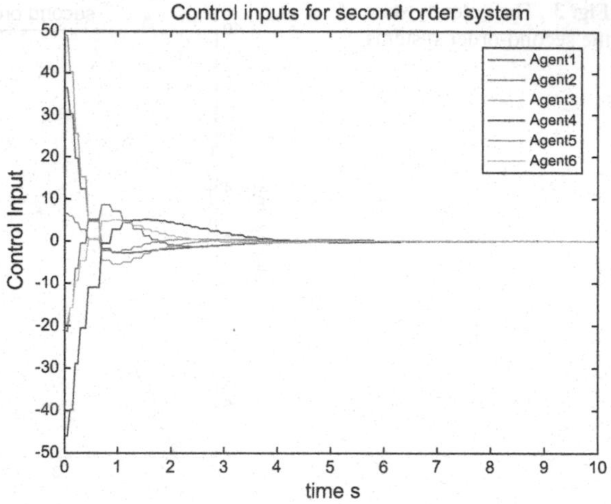


**Fig. 5** Control input of second-order systems

**Acknowledgements** This paper is supported by the National Natural Science Foundation of China (61673200, 61771231, 61471185), Natural Science Foundation of Shandong Province (ZR2018ZC0438, ZR2017MF010, ZR2017PF010) and the Key R&D Program of Yantai (2016ZH061)

# References

1. W. Ren, R.W. Beard, E.M. Atkins, Information consensus in multivehicle cooperative control. Control Syst. IEEE **27**(2), 71–82 (2007)
2. R. Olfati-Saber, J.A. Fax, R.M. Murray, Consensus and cooperation in networked multi-agent systems. Proc. IEEE **95**(1), 215–233 (2007)
3. L. Ma, H. Min, S. Wang et al., Consensus of nonlinear multi-agent systems with self and communication time delays: a unified framework. J. Frankl. Inst. **352**(3), 745–760 (2015)
4. H. Yang, F. Wang, F. Han, Containment control of fractional order multi-agent systems with time delays. IEEE/CAA J. Autom. Sinica **5**(3), 727–732 (2018). https://doi.org/10.1109/JAS.2016.7510211, http://ieeexplore.ieee.org/stamp/stamp.jsp?tp=&arnumber=7783963
5. Y. Li, H. Yang, F. Liu, Y. Yang, Finite-time containment control of second-order multi-agent systems with unmatched disturbances. Acta Automat. Sinica (2018). https://doi.org/10.16383/j.aas.2018.c170571
6. W. Zhang, Y. Liu, Distributed consensus for sampled-data control multi-agent systems with missing control inputs. Appl. Math. Comput. **240**(4), 348–357 (2014)
7. Z.J. Tang, T.Z. Huang, J.L. Shao et al., Brief paper: leader-following consensus for multi-agent systems via sampled-data control. IET Control Theory Appl. **5**(14), 1658–1665 (2011)
8. G. Guo, L. Ding, Q.L. Han, A distributed event-triggered transmission strategy for sampled-data consensus of multi-agent systems. Automatica **50**(5), 1489–1496 (2014)
9. D.V. Dimarogonas, E. Frazzoli, K.H. Johansson, Distributed event-triggered control for multi-agent systems. IEEE Trans. Autom. Control **57**(5), 1291–1297 (2012)
10. H. Huang, T. Huang, S. Wu, Event-triggered average consensus of second-order multi-agent systems. Inf. Control **45**(6), 729–734 (2016)
11. P. Tabuada, Event-triggered real-time scheduling of stabilizing control tasks. IEEE Trans. Autom. Control **52**(9), 1680–1685 (2007)

# Research on Loop Closing for SLAM Based on RGB-D Images

**Hongwei Mo, Kai Wang, Haoran Wang and Weihao Ding**

**Abstract** This paper mainly studies a loop closing detection method based on visual SLAM. We used RGB-D image as data source. The main idea is to construct a word bag based on DBoW3. Using rBRIEF makes it possible to perform feature extraction after the image is rotated. And added the elimination of mis-match links to improve the accuracy of detection. In order to ensure the reliability of the loop closing test results, the matching image is also verified. RGB-D image is rich in information and can synchronously extract the depth and color information of the main objects in the scene. The depth information directly reflects the distance information of each object in the scene.

**Keywords** SLAM · Loop closing · RGB-D · DBoW3

## 1 Introduction

SLAM is the abbreviation of Simultaneous Localization and Mapping. It refers to the subject carrying a specific sensor. In the absence of environmental prior information, it establishes an environmental model during the exercise process and estimates its own movement. Because of different sensors, there are laser-based SLAM and vision-based SLAM. This paper is focus on visual SLAM. The advantage of visual SLAM is that the camera generates rich texture information, which has great advantages in repositioning and scene classification. Especially in recent years, rapid development of science and technology has brought a new vision SLAM.

Visual SLAM system gave up the expensive laser and inertial measurement units, replaced it with a cheaper camera. The visual SLAM algorithm can build a 3D map of the world in real time and track the position and orientation of the camera. So, SLAM involved positioning technology, tracking technology and path planning technology. Loop closing detection is an important part of the entire SLAM system. The close loop

H. Mo (✉) · K. Wang · H. Wang · W. Ding
School of Automation, Harbin Engineering University, Harbin 150000, China
e-mail: mhonwei@163.com

 
Y. Jia et al. (eds.), *Proceedings of 2018 Chinese Intelligent Systems Conference*,
Lecture Notes in Electrical Engineering 528,
https://doi.org/10.1007/978-981-13-2288-4_70

detection module can provide some long-distance constraints apart from adjacent frames. The key of loop closing detection is how to effectively detect whether the camera has passed the same place. If it can be successfully detected, it can make the back end get more effective data, and finally get a globally consistent estimate. Therefore, in order to eliminate the error, loop closing detection is necessary in visual SLAM. It is also the key to ensure the quality of the construction and repositioning after losing position information, and the loop closing detection is more accurate constraint method than BA (Bundle Adjustment).

## 2 Related Work

There has been no major breakthrough in the loop closing detection method in recent years. Sivic and Zisserman [1] and others proposed in 2003 that the bag of words in text information retrieval converts the continuously changing features into discretized "words", and then use the statistical histogram of words to describe the scene. Newman and Ho [2] proposed to extract the feature descriptors from the pictures in 2005 and store them in a database. When matching, the images in the database are used for comparison and a value is returned. This value is used to detect whether a closed-loop event has occurred. This method relies on accurate position estimation and accurate positioning. Niser and Stewenius [3] proposed in 2006 that the local feature descriptors were hierarchically quantified into a vocabulary tree. Using the tree structure to increase the clustering speed on the basis of BoW, it was better at that time. Angeli and Filliat [4] proposed an online method for visual recognition of the effect of closed-loop detection on the size of the dictionary, using local shape and color information to monitor previous scenes, extending the BoW method, and constructing A quantitative dictionary that uses the classified image and Bayesian formula to estimate the likelihood of a loop closing. Cadena and Galvez-Lopez [5] used the complementary features of binocular cameras in 2010 to test loop closing detection in conjunction with the pouch method, ignoring geometrically discontinuous pictures, and experimentally verifying that CR-matching performed on the appearance of the scene. Liu [6] proposed a method for describing children using a compact feature in 2012. This method uses the low-dimensional descriptors available in a single image to perform image matching and PCA (Principal Component Analysis) for dimensionality reduction. At the same time, the computing efficiency of the computer can also be improved, and the correlation between different pictures can be found by using a particle filter method. Cumminsk and Newman [7, 8] proposed in 2007 the use of a probabilistic framework model to calculate two similar obstacles and whether they are the starting position. This method assumes that the map is known to achieve positioning, and in 2011 A SLAM construction method in a large scene is proposed [8], which is called a sparse approximation FAB-MAP model. The minimum spanning tree is used to describe the relationship between words, and the context information is used to further reduce the perceived ambiguity, achieving an ideal effect and becoming a test. The benchmark of the visual closed-loop detection method. The improved closed-loop detection method is compared with the FAB-MAP method to verify the real-time, high-efficiency and robustness of the method. Tully and Kantor [9] proposed a hybrid map representation method in 2012. They use topological maps globally to optimize maps, locally use maps that are more suitable for human observation (such as grid maps and point clouds), and then use recursive Bayesian leaves to solve the closed-loop detection problem.

## 3 System Description

The hardware platform uses TurtleBot2 mobile robot, its mainly includes Kobuki mobile base, Kinect vision sensor, 2200 mAh battery. It need install ROS (robot operating system) on Ubuntu 14.04 as a development platform. And take a notebook to run the SLAM algorithm. The Kinect is used to capture RGB-D images. TurtleBot2 is an open source hardware platform and mobile base station. When using ROS software, TurtleBot2 can handle vision, positioning, communication and mobility, It can autonomously move to designated place.

## 4 RGB-D Feature

The RGB-D camera can observe the three-dimensional position of the landmark points each time. It can easily track of image feature pairs with more information. As show follows.

We need to get each pixel depth value in depth images, and the pixel should be marked by the index. If a pixel is marked, we should find the corresponding coordinate in color images.

## 5 DBoW3 Algorithm

BoW (bag of words) was originally used in text categorization and later used in image feature extraction and target detection. DBoW3 can sort image features and convert images into visual words bag. It uses the similar image features of hierarchical tree structures to gather together on physical storage to create a visual dictionary. It will generate an image database with sequential and reverse indexing. As shown in Fig. 1, the word in the dictionary is the leaf node in the tree. The weight of the word is stored in the inverted index. The contents of the direct index are mainly the characteristics of the image and the nodes in the dictionary tree related to the features.

In this paper, the rBRIEF used to make the binary descriptor space discretized, and generates a more concise dictionary, ranks the word bag by rank, and the whole dictionary is a above tree structure. Here rBRIEF is used instead of BRIEF in order to let the descriptor have rotation invariance. And find the direction of FAST key point as the direction of BRIEF. So, we will get a coordinate matrix S:

Fig. 1 RGB-D feature

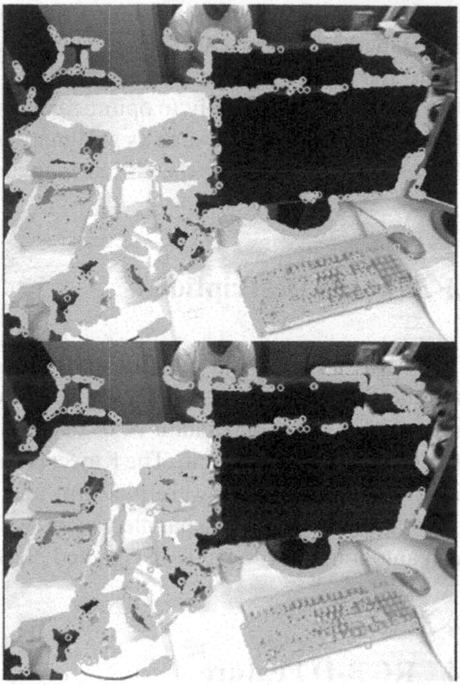

$$S = \begin{pmatrix} x_1, x_2, \ldots, x_n \\ y_1, y_2, \ldots, y_n \end{pmatrix} \tag{1}$$

In this way can easily get the coordinates of the picture after the rotation transformation.

$$S_\theta = R_\theta S \tag{2}$$

In order to generate a dictionary tree, a large number of features are extracted from the training images. Their corresponding descriptors are discrete into $K_w$ binary clusters according to the K-means++ algorithm. These clusters are the first level nodes of the dictionary tree, and the K-means++ algorithm for each node is carried out again. In order to generate second level nodes, $L_w$ (L indicates the number of pairs of feature points) steps are followed according to this step, and eventually a W word dictionary tree is generated.

The K-means++ algorithm steps are as follows (Table 1).

As show in Fig. 2, a bag of word dictionary is established (Fig. 3).

The establishment process is as follows (Fig. 4).

According to the relevance of the word in the training center, each word is assigned a weight, those which appear frequently and do not have much effect on different images, assign a small weight, and have a significant allocation right for the words that distinguish the significant image.

**Table 1** K-means++ algorithm

| K-means++ algorithm step |
|---|
| Step 1: We select a sample point randomly in the data set as the first initialized cluster center |
| Step 2: We calculate the distance between each sample point in the sample and the cluster center that has been initialized, and select the shortest distance between them |
| Step 3: We select the sample with the largest distance as the new cluster center and repeat the above steps until the *K*th cluster centers have been find |
| Step 4: For the k initial cluster centers, we calculate the final cluster center using the K-means algorithm |

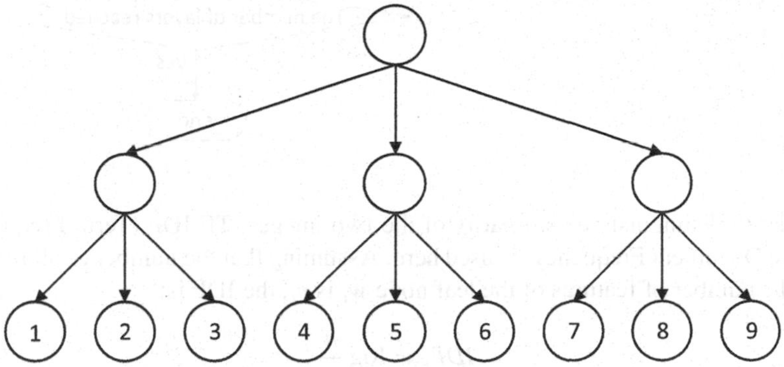


**Fig. 2** Vocabulary tree

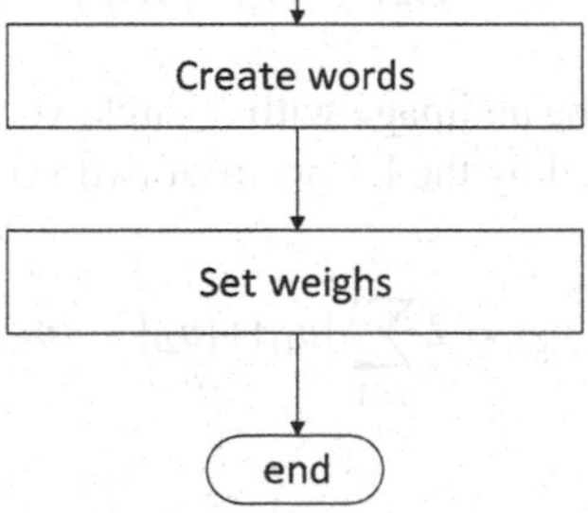

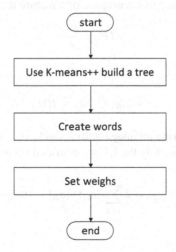


**Fig. 3** The algorithm flow to create a dictionary

Given any eigenvalue $f_i$, as long as search in the dictionary layer-by-layer, and finally can find the corresponding word $w_j$, so, $f_i$ and $w_j$ come from same object.

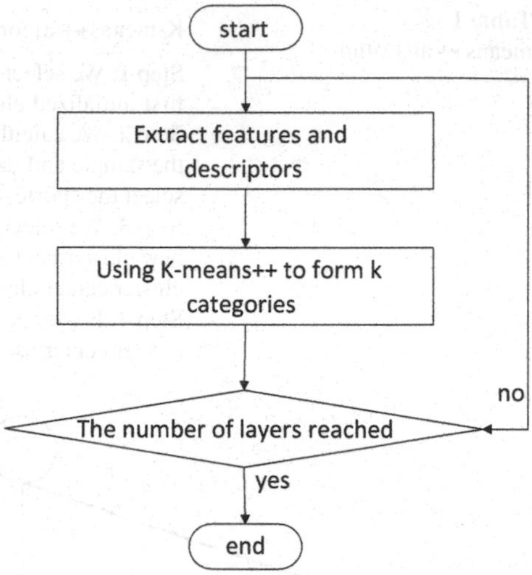


**Fig. 4** The algorithm flow to create a tree

In order to distinguish the similarity of the two images, TF-IDF (Term Frequency-Inverse Document Frequency) is used here. Assuming that the number of all features is n, the number of features of the leaf node $w_i$ is $n_i$, the IDF is:

$$IDF_i = \log \frac{n}{n_i} \tag{3}$$

TF refers to the frequency of occurrence of a feature in a single image. So, TF is:

$$TF_i = \frac{n_i}{n} \tag{4}$$

And the weight is:

$$weight_i = TF_i \times IDF_i \tag{5}$$

In this way, it can describe an image with a single vector. The similarity between two images can be calculated by the L1 norm of two vectors:

$$s(v_1 - v_2) = 2\sum_{i=1}^{N} |v_{1i}| + |v_{2i}| - |v_{1i} - v_{2i}| \tag{6}$$

## 6 RANSAC Algorithm

In this paper, the most important step is use RANSAC (Random Sample Consensus) algorithm to eliminate mis-match after extract features. It is based on a set of sample

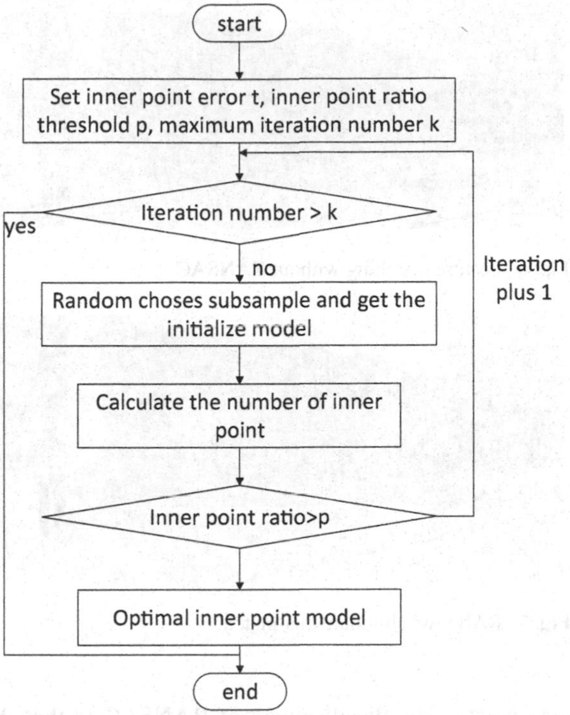


**Fig. 5** The RANSAC algorithm flow

data sets containing abnormal data, calculating the mathematical model parameters of the data, and obtaining effective sample data algorithm. It first proposed in 1981 by Newman and Ho [2]. RANSAC algorithm is often used in computer vision. For example, in the field of stereo vision, the matching points of a pair of cameras and the calculation of the fundamental matrix are simultaneously solved.

The main step as follows (Fig. 5).

The input data of RANSAC algorithm is a set of observation data, a parameterized model that can be interpreted or adapted to observation data and some credible parameters. The model corresponds to the rotation and translation of one point cloud data in another space to another point cloud data. The first step is to get the point pair in a point cloud, and to use its invariant feature (distance of two points, normal vector angle of two point) as the index value of the hash table to search for a pair of corresponding points in the other point cloud, and it will calculate the parameter values of the rotation and translation. Then the transformation is applied to find other local points. And the algorithm need to recalculate the rotation and translation to the next state after finding the inner point. Then it iterates the above process to find the final location.

Apply RANSAC algorithm can remove points outside of consistency. The advantage of RANSAC is its robust estimation of model parameters. For example, it can estimate high-precision parameters from data sets containing a large number of out-

Fig. 6 Feature matching without RANSAC

Fig. 7 RANSAC eliminate mis-match

side points. The disadvantage of RANSAC is that there is no upper limit on the number of iterations that it calculates, and if the upper limit of the number of iterations is set, the result may not be the best result or even the wrong result. RANSAC has only a certain probability to get a credible model, and the probability is proportional to the number of iterations. Another drawback of RANSAC is that it requires setting the threshold associated with the problem.

In the feature matching, the following situation occurs (Fig. 6).

As is show above, although most of the matches are correct, there are some matching errors, which constitute "contaminated observation data" and also the application conditions of RANSAC (Fig. 7).

Table 2 shows the similarity matrix obtained after entering the above algorithm with ten pictures. It can be seen that after setting a certain threshold, the algorithm cam be more accurately judged whether it is a loop closing.

## 7 Conclusions

In visual SLAM system, loop closing detection is an extension of data association. It is a process from point to point to face to face. In this paper, the image database structure used in the closed loop detection algorithm has a direct index in addition to the rank word bag and the inverted index. This structure makes the efficiency of

**Table 2** Similar matrix

| | $I_1$ | $I_2$ | $I_3$ | $I_4$ | $I_5$ | $I_6$ | $I_7$ | $I_8$ | $I_9$ | $I_{10}$ |
|---|---|---|---|---|---|---|---|---|---|---|
| $I_1$ | 1 | 0.251869 | 0.264077 | 0.258784 | 0.250996 | 0.268458 | 0.269992 | 0.257847 | 0.28638 | 0.241537 |
| $I_2$ | | 1 | 0.236969 | 0.264267 | 0.252454 | 0.245609 | 0.251969 | 0.2404556 | 0.254931 | 0.23764 |
| $I_3$ | | | 1 | 0.227906 | 0.23721 | 0.238696 | 0.261195 | 0.21486 | 0.23448 | 0.235304 |
| $I_4$ | | | | 1 | 0.281962 | 0.242712 | 0.266108 | 0.225938 | 0.241647 | 0.238004 |
| $I_5$ | | | | | 1 | 0.255106 | 0.248995 | 0.236341 | 0.213317 | 0.240724 |
| $I_6$ | | | | | | 1 | 0.239303 | 0.231276 | 0.2399 | 0.228895 |
| $I_7$ | | | | | | | 1 | 0.246679 | 0.25424 | 0.2588312 |
| $I_8$ | | | | | | | | 1 | 0.217322 | 0.216933 |
| $I_9$ | | | | | | | | | 1 | 0.228173 |
| $I_{10}$ | | | | | | | | | | 1 |

the geometric verification better. Using of binary descriptor rBRIEF makes it faster to extract image features and to calculate the distance between descriptors. When reference the scale, illumination and camera rotate, the BRIEF descriptor is unstable. So, using rBRIEF can improve the SLAM system performance.

Geometric verification is also needed for candidate loop closing images. And using RANSAC algorithm to find correspondence of local features between two images. The matching images are validated for determining the similarity between the two images when use the geometric verification, and the correct data association is also needed after the verification.

In an unknown environment, the mobile robot's awareness of the surrounding environment is a very basic function. Therefore, SLAM is an important direction to solve this problem. Loop closing detection can improve the accuracy of building a map and make the robot's autonomous positioning more accurate. Using RGB-D information can improve the efficiency of loop closing detection.

## References


1. J. Sivic, A. Zisserman, Video google: a text retrieval approach to object matching in videos, in *2003 IEEE International Conference on Computer Vision*, vol. 2 (Nice, France, 2003), p. 1470
2. P. Newman, K. Ho, SLAM-loop closing with visually salient features, in *Proceedings of the 2005 IEEE International Conference on Robotics and Automation* (Washington, America, 2005), pp. 635–642
3. D. Nister, H. Stewenius, Scalable recognition with a vocabulary tree, in *IEEE Computer Society Conference on Computer Vision and Pattern Recognition* (New York, America, 2006), pp. 2161–2168
4. A. Angeli, D. Filliat, Fast and incremental method for loop-closure detection using bags of visual words. IEEE Trans. Rob. **24**(5), 1027–1037 (2008)
5. C. Cadena, D. Galvez-Lopez, Robust place recognition with stereo cameras, in *IEEE/RSJ International Conference on Intelligent Robots and Systems*, vol. 25, no. 1 (Taipei, Taiwan 2010), pp. 5182–5189
6. Y. Liu, H. Zhang, Visual loop closure detection with a compact image descriptor, in *2012 IEEE/RSJ International Conference on Intelligent Robots & Systems*, vol. 57, no. 1 (Vilamoura, Algarve, 2012), pp. 1051–1056
7. M. Cummins, P. Newman, Probabilistic appearance based navigation and loop closing, in *IEEE International Conference on Robots & Automation* (Roma, Italy, 2007), pp. 2042–2048
8. M. Cummins, P. Newman, Appearance-only SLAM at large scale with FAB-MAP 2.0. Int. J. Robot. Res. **30**(9), 1100–1123 (2011)
9. S. Tully, G. Kantor, A unified bayesian framework for global localization and SLAM in hybrid metric topological maps. Int. J. Robot. Res. **31**(31), 271–288 (2012)
10. F.M. Campos, L. Correia, Mobile robot global localization with non-quantized SIFT features, in *IEEE 15th International Conference on Advanced Robotics: New Boundaries for Robotics* (Tallin, 2011), pp. 582–587
11. R. Mur-Artal, J.D. Tardos, ORB-SLAM2: an open-source system for monocular, stereo, and RGB-D cameras. IEEE Trans. Rob. **PP**(99), 1–8 (2016)
12. C. Kerl, J. Sturm, Dense visual SLAM for RGB-D cameras, in *2014 IEEE/RSJ International Conference on Intelligent Robots & Systems*, vol. 8215, no. 2 (Chicago, America, 2014), pp. 2100–2106
13. Y. Latif, C. Cadena, Robust loop closing over time for pose graph SLAM. Int. J. Rob. Res. **32**(14), 1611–1626 (2013)

# Autonomous Localization and Mapping for Mobile Robot Based on ORB-SLAM

**Hongwei Mo, Xiaosen Chen, Kai Wang and Haoran Wang**

**Abstract** We aim to realize autonomous localization and mapping for mobile robot while no prior knowledge of its environment provided, based on one of the state-of-the-art SLAM algorithm called ORB-SLAM. A local 3D point cloud map is constructed, through the depth information acquired from RGB-D sensor and corresponding camera poses estimated from ORB-SLAM, which is then transformed to a 2D occupancy grid map using octree. Based on the 2D map, an information-theoretic exploration algorithm is used to travel through all the environment. Finally, experiments are carried out in a mobile robot.

**Keywords** Autonomous exploration · Mobile robot · ORB-SLAM

## 1 Introduction

In the field of mobile robot, SLAM (Simultaneous Localization And Mapping) has become an increasingly popular research topic in the past 30 years [1]. A large number of SLAM systems have been proposed by researchers, using different sensors and optimization techniques. VSLAM (Visual SLAM) system uses cameras as its main sensors to estimate poses of the camera and map of the environment, which is cheaper and containing more information than the system using laser radar.

At the beginning, filtering is the main optimization methodology for VSLAM, like EKF-SLAM [2]. PTAM [3] firstly used nonlinear optimization approach to estimate motion and structure, which achieves robustness and high precision though dividing tracking and mapping into two parallel threads in monocular SLAM. Then many direct-based approaches are employed to VSLAM, including SVO [4], LSD-SLAM [5], etc. ORB-SLAM [6] is one of the most noticeable approaches based on sparse feature, achieves unprecedented performance with its fittest strategy and compact structure, and constructs a sparse 3D map while tracking.

---

H. Mo (✉) · X. Chen · K. Wang · H. Wang
School of Automation, Harbin Engineering University, Harbin 150000, China
e-mail: mhonwei@163.com

 
Y. Jia et al. (eds.), *Proceedings of 2018 Chinese Intelligent Systems Conference*,
Lecture Notes in Electrical Engineering 528,
https://doi.org/10.1007/978-981-13-2288-4_71

For a mobile robot, there are some requirements for map in the applications of exploration and obstacle avoidance. Robot must know where is free or occupied in the space of the environment, which asks for a dense map, and requires dynamically local map when encountering changeable obstacles. Besides, with no prior knowledge of the contents of the environment, mobile robot must make sequential decisions about where to travel next in the autonomous exploration.

## 2 Related Work

Methods that simultaneously recover camera poses and scene structure from images can be divided into two classes, feature-based methods and direct methods. Direct methods optimize the geometry directly on the image intensities instead of the set of features, which enable using most of information in the image and are more likely to construct dense map. Newcombe firstly proposed a monocular SLAM system that achieves real-time camera tracking and reconstruction which relies not on feature extraction but every pixel in the image [7], but it needs powerful GPU to accelerate. To reduce computation, the semi-dense approach that only extracts corners and edges in images was gradually proposed, and LSD-SLAM [5] is a direct VSLAM which allows to build semi-dense map of large-scale environment in real time. Besides, RGB-D sensors, through physical method instead of geometric computing, provide high-precision depth information, which can reduce resource consumption and improve the real-time performance of the system. Based on RGB-D sensors, Newcombe et al. [8] proposed a system named Kinect Fusion which is a real-time object surface reconstruction method, however, it still needs powerful GPU and can't reduce cumulative error. Then Felix Endres proposed RGBD-SLAM [9] which extracts visual keypoints from the color images and localizes them in 3D using the depth images, reducing cumulative error with loop closing.

The information-theoretic exploration problem is solved by maximizing the mutual information (MI) function associated with entropy between sensor observations and an occupancy grid map [10]. It is acknowledged that the evaluation of MI is expensive, so S. Bai. et al. applied supervised learning to the prediction of information actions without evaluating the expected MI exhaustively for every possible action [11], however, the approach can be ineffective when its prediction of information gain is provided insufficient support from its training data. Bayesian optimization has been successfully used in a variety of robotics applications, including robot gait optimization, environmental monitoring and rough-terrain navigation. Therefore, the writers of [12] used Bayesian optimization for selecting the candidate actions in information-theoretic exploration, and only the MI of these candidate actions will be explicitly evaluated, which is computational efficiency and robustness in comparison with competing approaches.

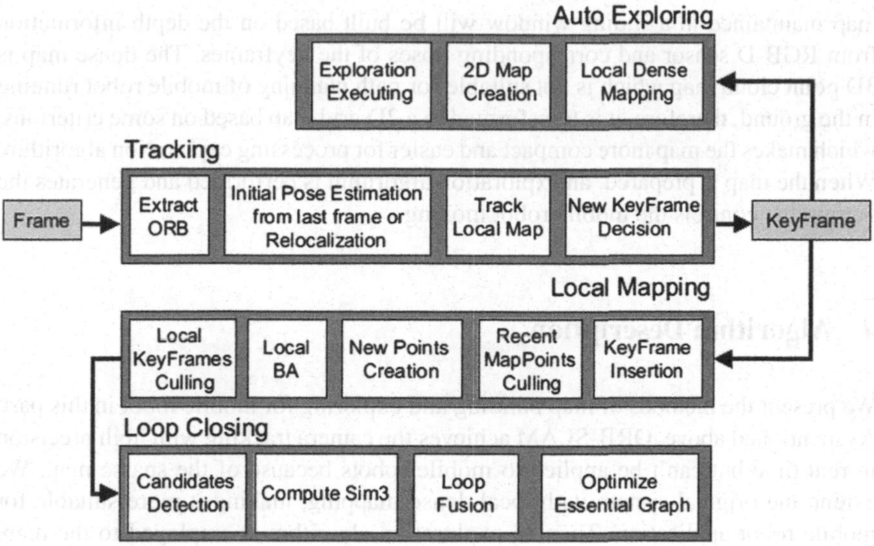


**Fig. 1** System overview

## 3 System Overview

Our system consists of the original system (ORB-SLAM) and the extended part which contains local dense mapping and exploration processing for mobile robot. The system, see an overview in Fig. 1, incorporates four threads that run in parallel: tracking, local mapping, loop closing and auto exploring.

### *3.1 Original System*

The original system contains three threads, tracking, local mapping and loop closing. The tracking is in charge of localizing the camera with every frame, deciding whether a frame is a new keyframe and inserting the keyframe. And the local mapping processes new keyframes and performs local BA to achieve an optimal reconstruction in the surroundings of the camera pose. Besides, the thread of loop closing tries to detect and close loop for every new keyframe.

### *3.2 Auto Exploring*

Auto exploring thread aims at constructing local dense map and exploring the environment. Once acquiring the new keyframe in the tracking thread, the local dense map maintained in a sliding window will be built based on the depth information from RGB-D sensor and corresponding poses of the keyframes. The dense map is 3D point cloud map which is not suitable for path planning of mobile robot running in the ground, therefore, it is transformed to a 2D grid map based on some criterions, which makes the map more compact and easier for processing exploration algorithm. When the map is prepared, an exploration algorithm is performed and generates the output that controls the mobile robot moving.

## 4 Algorithm Description

We present the methods of map building and exploring for mobile robot in this part. As mentioned above, ORB-SLAM achieves the camera tracking with high precision in real time but can't be applied to mobile robots because of the sparse map. We extend the original system with local dense mapping, making it more suitable for mobile robot application. Then an exploration algorithm is employed to the map, achieving autonomous localization and mapping for mobile robot.

### 4.1 Map Building

The environment can be represented as a series of points: $X = \{x_1, x_2, \ldots, x_n\}$, where $x_i = [x, y, z]$, denoting the location of every point. For RGB-D sensor, the color image can be used for estimating the motion of camera, while the depth image contains the absolute distance information of the environment.

According to pinhole camera model, the relationship between position in the world coordinate and pixel coordinate is expressed as

$$s \cdot \begin{bmatrix} u \\ v \\ 1 \end{bmatrix} = C \cdot \left( R \cdot \begin{bmatrix} x \\ y \\ z \end{bmatrix} + t \right) \tag{1}$$

where, $C$ is the camera internal matrix, $R$ and t represent the camera rotation and translation respectively, and s is the depth of scale factor. In order to create dense map, we use the open-source PCL (Point Cloud Library) to process the point cloud acquired from RGB-D sensor. The type of the point cloud we use don't contain color information, since it is useless and wastes computing resource in the robotic applications. Therefore, the i element in the point cloud map $p_i$ consists of $p_{i,x}$, $p_{i,y}$ and $p_{i,z}$ of the position, which can be computed from the depth image:

$$\begin{bmatrix} p_{i,z} \\ p_{i,x} \\ p_{i,y} \end{bmatrix} = \begin{bmatrix} d \\ (n - c_x) * p_{i,z}/f_x \\ (m - c_y) * p_{i,z}/f_y \end{bmatrix} \tag{2}$$

where d represents the depth value associated with the pixel in m rows and n columns of the depth image, $f_x, f_y$, $c_x$, and $c_y$ are the camera internal parameters.

To create local dense map, we select a certain amount of keyframes to construct the local map. Keyframe is selected based on the ratio of common points between the last keyframe and current keyframe. Therefore, we use a criterion that the horizontal angle of rotation between the first keyframe and last keyframe is over a certain threshold, because the motion model of mobile robot is 2D and mainly cause the change of yaw. Then the local point cloud map is created using the depth images and corresponding poses of keyframes, but the map contains many outliners. A statistical filtering is employed to remove outliners of the map, meanwhile not destroying the structure of point cloud. For a mobile robot running on the ground, 3D map seems to be redundant and not practical, so we ignore the point cloud beyond the top of mobile robot, which is actually not the part of obstacles for mobile robot, and project a certain range of point cloud to the horizontal plane that we define in the world coordinate. However, the projected map is still not suitable for robot applications since the point cloud map is not intuitional for robot and contains too much points to operate exploring algorithm in real time. For this sake, we use octree to create the occupancy grid map based on the project map, and the resolution of it is adjustable according to the requirement of applications. Once the map is created, it is used to the exploration, and the set of keyframes is soon updated as the mobile robot travels around. The pipeline of map building is in Fig. 2.

## *4.2 Information-Theoretic Exploration with Bayesian Optimization*

The algorithm of exploration we use is mainly from [12], which is designed for the range sensor providing a 360° field of view (FOV). In order to make it useful with the RGB-D sensor whose FOV is narrow, we control the mobile robot to spin round, imitating the range sensor. Using Bayesian optimization to predict the next sensing action of mobile robot, the algorithm can guide mobile robot to traveling all the environment with real-time performance, and it achieves admirable performance in the complex environment. We present its contents as follows.

We define the space of mobile robot sensing actions to be the configuration space $\mathcal{L} \in R^2$, a subset of 2D Euclidean space. As we use occupancy grid map, the robot is assumed capable of traveling from any grid cell in the map to any other cell, in the absence of obstacles. We represent Shannon's entropy over the grid map m as

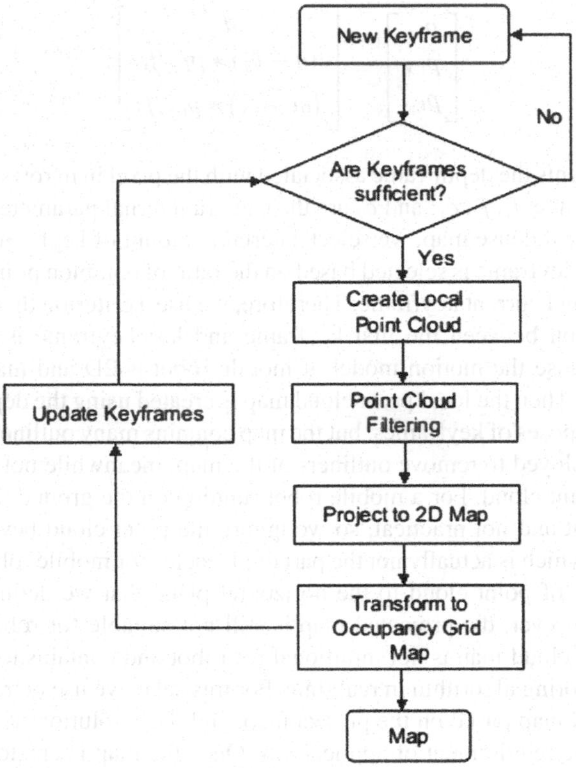


**Fig. 2** Pipeline of map building

$$\mathrm{H(m)} = -\sum_i \sum_j p(m_{i,j}) \log p(m_{i,j}) \tag{3}$$

where index i is the individual grid cells of the map and index j refers to the possible output of the Bernoulli random variable representing each grid cell, which is either free or occupied.

Mutual information $\mathrm{I}(\mathrm{m}, x_i)$ is used to evaluate the expected information gain associated with a specific configuration $x_i$, defined as follows:

$$\mathrm{I}(\mathrm{m}, x_i) = \mathrm{H(m)} - H(m|x_i) \tag{4}$$

where H(m) is the current entropy of the map, and $H(m|x_i)$ refers to the expected entropy of the map provided with a new sensor observation at configuration $x_i$. The optimal configuration $x^*$ can be calculated from (5), showing that the value of $\mathrm{I}(\mathrm{m}, x_i)$ is large where the information gain is taking high through sensing action.

$$x^* = argmax_{x_i \in \mathscr{L}_{action}} \mathrm{I}(\mathrm{m}, x_i) \tag{5}$$

Based on Bayesian optimization, the candidate sensing action is selected using an acquisition function, which has a high value where Gaussian process (GP) regression predicts high value of the MI objective function. The algorithm adopts the Gaussian process upper confidence bound algorithm to select the test configurations $x_t$ for the evaluation of MI function, as follows:

$$x_t = argmax_{x_i \in \mathcal{L}_{action}} \mu(x) + \beta\sigma(x) \tag{6}$$

where $\beta$ is an adjustable parameter, $\mu(x)$ and $\sigma(x)$ are the predicted mean and variance derived from Gaussian process regression.

We have acquired the set of test configuration $x_t$ above, then assuming a set of training data x that represents the candidate sensing configurations $x_i$ for which I(m, $x_i$) has been calculated.

$$y_t = \frac{k(x_t, x)}{\left[k(x, x) + \sigma_n^2 I\right]} y \tag{7}$$

In the above equations, $y_t$ are the estimated values I(m, $x_{it}$) for the test data $x_t$, $\sigma_n^2$ is a vector of Gaussian noise variances associated with the observed outputs y, and $k(x_t, x)$ is the kernel function, which is the Matérn kernel function.

## 5 Experimental Results

To evaluate the performance of the proposed system, we apply our system to a mobile robot called Turtlebot 2.0. Besides, we use the ASUS laptop with i5 CPU, 4G RAM and 256G SDD and Kinect 1.0 to acquire the depth images and color images for experiments. And the assembled hardware system is presented in Fig. 3.

We tested our designed system in our laboratory room (see Fig. 4) which is almost 30 $m^2$. Putting the Turlebot 2.0 on one side of the room, repeatedly performing the experiment several times, it can always arrive the other side of the room. In practice, we stop the robot when it returns to the started place after a certain time and record the time of the process, and the average time of the period is almost eight minutes. Besides, we save the keyframe trajectory (see Fig. 5), global point cloud map (see Fig. 7) and its Octree map (see Fig. 8) in the experimental process.

### *5.1 Results of Exploration*

We sign five points representing different locations in the environment in Fig. 5, and for intuitional cognition, the associated points in Fig. 5 are highlighted in Fig. 6 which is the global octree map in the view from the top. Among these points, the first point is the starting point while the fifth point is the farthest point that robot

**Fig. 3** Assemble system

can travel to. Besides, the deep violet areas are the free spaces that mobile robot can move around in Fig. 6, and comparing Figs. 5, 6, it shows that the mobile robot, in unknown environment, can produce reliable trajectory that avoids obstacles and spreads all over the space that it can pass through.

## 5.2 *Results of Mapping*

Our system is proposed to explore the unknown environment as well as construct the map of the environment. Therefore, we save the global point cloud map, whose resolution is as high as 0.01 m, as the consequence of map construction, and for a better visual experience, the map is transformed to octree map offline, whose resolution is lower but structure is more compact. From Figs. 7 and 8, we can see that the whole structure of map is relatively accurate, but it contains many outliners. Besides, some structures of the objects are distorted in the map, caused by the shake of mobile robot while moving and the error of state estimation of our system, for example, the rectangular desks' edges are not so straight in the Fig. 8.

Fig. 4 Laboratory room

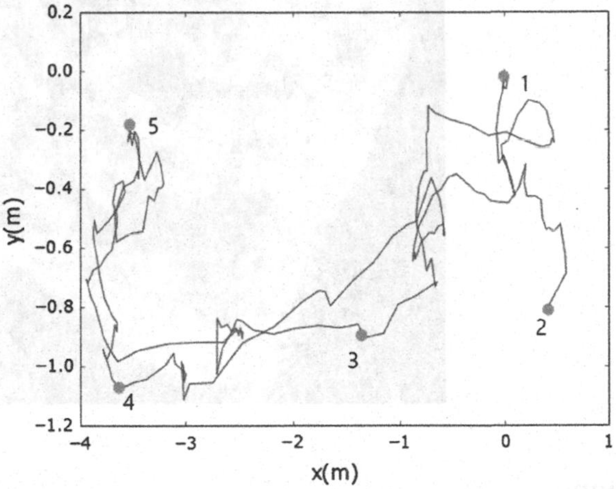


Fig. 5 Keyframe trajectory

Fig. 6 Octree map (top view)

Fig. 7 Point cloud map

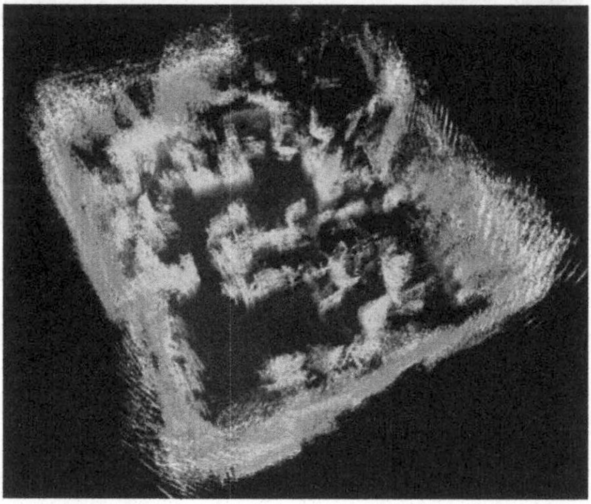

## 6 Conclusion

In this paper, we improve the system of ORB-SLAM, building the dense map for the applications of mobile robot, and use an information-theoretic exploration algorithm based on Bayesian optimization to achieve the autonomous exploration of mobile robot, while the robot can localize its position and construct the map of environment, which can be seen by the camera in mobile robot. In order to verify the performance

**Fig. 8** Octree map

of the designed system, we test it in a mobile robot, Turtlebot 2.0, testifying that the system can achieve autonomous exploration while capable of localization and mapping, with real-time performance. For the future improvement, the system can be tested in other kinds of environments, and the map resolution can be automatically adjusted, according to the complexity of environment, to accelerate the speed of the algorithm.

## References


1. S. Thrun, W. Burgard, D. Fox, *Probabilistic Robotics* (MIT Press, 2005)
2. S. Thrun, Y. Liu, D. Koller et al., Simultaneous localization and mapping with sparse extended information filters. Int. J. Rob. Res. **23**(7–8), 693–716 (2004)
3. G. Klein, D. Murray, Parallel tracking and mapping for small AR workspaces, in *IEEE and ACM International Symposium on Mixed and Augmented Reality* (IEEE Computer Society, 2007), pp. 1–10
4. C. Forster, M. Pizzoli, D. Scaramuzza, SVO: fast semi-direct monocular visual odometry, in *IEEE International Conference on Robotics and Automation* (IEEE, 2014), pp. 15–22
5. J. Engel, T. Schöps, D. Cremers, LSD-SLAM: large-scale direct monocular slam, in *Computer Vision—ECCV 2014* (Springer International Publishing, 2014), pp. 834–849
6. R. Mur-Artal, J.M.M. Montiel et al., ORB-SLAM: a versatile and accurate monocular SLAM system. IEEE Trans. Rob. **31**(5), 1147–1163 (2015)
7. R.A. Newcombe, S.J. Lovegrove, A.J. Davison, DTAM: dense tracking and mapping in real-time, in *IEEE International Conference on Computer Vision* (IEEE, 2011), pp. 2320–2327
8. R.A. Newcombe et al., KinectFusion: real-time dense surface mapping and tracking, in *IEEE International Symposium on Mixed and Augmented Reality* (IEEE, 2012), pp. 127–136

9. F. Endres, J. Hess, J. Sturm et al., 3-D mapping with an RGB-D camera. IEEE Trans. Rob. **30**(1), 177–187 (2017)
10. A. Elfes, Robot navigation: integrating perception, environmental constraints and task execution within a probabilistic framework. Lect. Notes Comput. Sci. **1093**, 91–130 (1996)
11. S. Bai, J. Wang, K. Doherty et al., Inference-enabled information-theoretic exploration of continuous action spaces. Rob. Res (2018)
12. S. Bai, J. Wang, F. Chen et al., Information-theoretic exploration with Bayesian optimization, in *IEEE/RSJ International Conference on Intelligent Robots and Systems* (IEEE, 2016), pp. 1816–1822

# User Behavior Prediction with SVM for Garment Ordering System

**Yimin Zhou and Xiaohai Chen**

**Abstract** In this paper, a iPAD based garment ordering system is first developed via support vector machine (SVM) learning algorithm. The garment ordering system is introduced with its development history and current situation. SVM algorithm has the advantages of pattern recognition which is used to deal with the binary issues for the user selection. From the perspective of requirement of the ordering meeting, the ordering system is designed and accomplished with module decomposition. The data are collected from the central unit of actual ordering meetings and uploaded in the system based on the potential selection so as to reduce the loading pressure and waiting time of the end users. Experiments are performed to testify the efficacy of the proposed algorithm for users behaviour prediction with higher system performance and user satisfaction.

**Keywords** Support vector machine · User behaviour prediction
Garment ordering system · iPAD

## 1 Introduction

Recently, along with the vigorous popularization of mobile tablets such as iPAD and the rapid development of mobile communication technology, the garment ordering system has become full visualization and mobility, and some of the less confidential ordering can be controlled remotely. This project is carried out to study the iPAD-based clothing ordering system design, implementation and user behavior prediction.

The iPAD was released in China at 2010 and began popular since 2012, while some companies starts to transplant the ordering meetings on iPAD from 2013. Since China is the largest garment manufacturers and sellers around the world, the

Y. Zhou (✉) · X. Chen
Shenzhen Institutes of Advanced Technology, Chinese Academy of Sciences,
Shenzhen 518055, China
e-mail: ym.zhou@siat.ac.cn

 
Y. Jia et al. (eds.), *Proceedings of 2018 Chinese Intelligent Systems Conference*,
Lecture Notes in Electrical Engineering 528,
https://doi.org/10.1007/978-981-13-2288-4_72

developed iPAD-based footwear ordering system in China maintains the leading position.

The Shanghai Guanxin company was founded in 2011, whose main business is to provide the iPAD intelligent ordering meetings or conferences and its program takes high market share in China. Guangzhou LENX Company was founded in 2006, and they provide the whole solution with wireless ordering for footwear industry. Another Xiaobang Technology company was also founded in 2006, their main business is the handhold PDA Scanner and the ordering supporting system, which occupied half share of the shoes and apparel industry at their development peak period. Due to the rise of the iPAD-based ordering, they have been attracted as well.

Compared with the manually or PDA ordering, there are a lot of advantages for iPAD based garment ordering system. It has high efficiency with quick cloth browsing, retrieval and ordering, which can reduce the ordering cycle, along with catering, accommodation, venues and other operating costs reduction. The clothes can be displayed with vivid pictures intuitively. Besides, the ordering can be placed with simple operation and instant response from the system. The real-time data analysis can be notified to the users, i.e., list of the hot sales, real-time interaction or multi-dimensional data comparison.

However, there could be some drawbacks during iPAD-based ordering, such as system break down due to overload or too much waiting time due to too many users with limited network bandwidth. Through the user behavior prediction, the most possible next selection of the pictures from the users can be predicted so that the system would load these potential set in advance to prevent the big data concurrency under limited bandwidth [1, 2]. Then the slow loading of large pictures can be improved and the block or delayed waiting at the user-end can be optimized, thus lowering the investment cost on the server and network construction.

There are a lot of methods have been developed for the analysis of the user behaviour preference along with the computer technology and big data analyzation. The advantage of data mining becomes gradually obvious, neural network, decision tree, Bayes methods are widely applied. Among these methods, support vector machine (SVM) is the most applied method. SVM is one type of learning machine algorithms proposed by Vapnik et al. in 1990s [3, 4]. It is more suitable for solving the pattern recognition problems with high dimension and nonlinearities. Compared with other learning machines, i.e., neural network [5] or genetic algorithm [6], it has the generalization ability and nonlinear processing capability, especially in dealing with data with high dimension. It has been widely applied in facial detection [7], web page classification [8], data fusion [9] and function estimation [10] etc. To deal with regression and classification issues, several extended SVM algorithms have been developed, such as V-SVM [11], Generalized SVM [12], and least squared SVM [13].

In this paper, a iPAD-based garment ordering system is developed for commercial use. The data loading process is optimized via SVM to predict the users behaviour at the mobile end to avoid abnormalities [1], such as break down or long time waiting under limited network resources.

The remainder of the paper is organized as follows. Section 2 explains the binary classification via SVM. The iPAD ordering system is introduced in Sect. 3, while the user behaviour prediction by SVM is also explained. The experiments are performed to testify the effectiveness of the proposed optimization program of the iPAD ordering system in Sect. 4. Conclusion is given in Sect. 5.

## 2 The Binary Classification of SVM

The SVM is mainly suitable for the classification in small sample space, which is applicable in linear and nonlinear situations via the SVM kernel function transformed from the nonlinear function into linear function.

Suppose there are $n$ samples with $d$ dimension in the sample space, i.e., $(X_i, Y_i)$, $X_i \in R^d$, $y \in (-1, +1)$, $i = 1, \ldots, n$, and the space consists of two categories. If $X_i$ belongs to the first category, it is labeled as positive ($y_i = +1$), otherwise as negative ($y_i = -1$), where a hyperplane function is used to separate the space, written as,

$$w \cdot x + b = 0 \tag{1}$$

where $w \in R^n$, $x \in R^n$, and $b$ is a scalar. There are more than one such hyperplanes, where the one with the most maximum distance from the two samples categories is the optimal hyperplane and these sample points are the support vectors. Through the normalization of Eq. (1), the optimization problem is expressed as,

$$y_i(w \cdot x_i + b) \geq 1, i \in (1, 2, \ldots, n) \tag{2}$$

The interval of the classification is $2/\|w\|$ ($\| \cdot \|$ is the norm operation). In order to maximize the interval, it is equivalent to min $\|w\|$ and the question of the optimal separation hyperplane solution can be written as,

$$\begin{aligned} &\min_{w,b,\xi} \frac{1}{2} \parallel w \parallel^2 + C \sum_{i=1}^{n} \xi_i \\ &s.t. \quad y_i \left( (w \cdot x_i) + b \right) \leq 1 - \xi_i, \xi_i \leq 0, i = 1, 2, \ldots, n \end{aligned} \tag{3}$$

where $C$ is the penalty coefficient, the larger the higher penalty degree. $C$ is a super parameter, which is determined based on the human experience.

For the purpose of alleviating the over fitting problem or due to the noise in the data, the hyperplane would not be expected to separate the samples perfectly, i.e., allowing mistakes in the samples separation. Therefore, the concept of the ‘soft interval’ is introduced, so that some samples do not satisfy the constraints and a relaxation function $\xi$ is used to describe the violation degree.

In order to easily solve problem, the original problem can be transformed to the dual problem with the introduction of the Lagrange function for the constrained optimization,

$$L(w, b, a) = \frac{1}{2}\|w\|^2 - \sum_{i=1}^{m} a_i(y_i((w \cdot x_i) + b) - 1) \quad (4)$$

where $a = (a_1, \ldots, a_m)' \in R_+^m$ is the Lagrange multiplier. Since the derivative of $L$ to $w$ and $b$ is 0, the partial derivative of the Lagrange function to the $w$ and $b$ can be derived from the extremum condition based on the dual principle,

$$\nabla_b L(w, b, a) = 0, \quad \nabla_a L(w, b, a) = 0 \quad (5)$$

Then it can be derived as,

$$\sum_{i=1}^{m} a_i y_i = 0, \quad w = \sum_{i=1}^{m} a_i y_i x_i \quad (6)$$

Put Eqs. (3) and (6) into Eq.(4), the dual problem of the original optimization problem is calculated as,

$$\max_a -\frac{1}{2}\sum_{i=1}^{m}\sum_{j=1}^{m} a_i a_j y_i y_j (x_i \cdot y_j) + \sum_{j=1}^{m} a_j \quad (7)$$

$$s.t. \sum_{i=1}^{m} a_i y_j = 0 \quad (8)$$

where $a_i \leq 0, i = 1, 2, \ldots, m$, the input $x_i$ of the corresponding training set $T$ is called support vectors. Suppose $a^* = (a_1^*, \ldots, a_m^*)$ is the any solution of the dual problem, then the solution of the original problem can be calculated as,

$$w^* = \sum_{i=1}^{m} a_i^* y_i x_i, \quad b^* = y_j - \sum_{i=1}^{m} a_i^* y_i (x_i \cdot x_j) \quad (9)$$

where $j \in \{j | a_i^* > 0\}$, $w^*$ and $b^*$ are the optimal solution. Thus the linear support classification machine algorithm can be summarized as:

Suppose the known training set $T = \{(x_1 \cdot y_1), \ldots, (x_i \cdot y_i), \ldots, (x_m \cdot y_m)\}(\in (X \times Y)^m)$, $x_i \in X = R_n$, $y_i \in Y = -1, 1$, to configure and solve the optimal problem described in Eq.(7) as the optimized solution $a_i \leq 0, i = 1, 2, \ldots, m$. The $w^* = \sum_{i=1}^{m} a_i^* y_i x_i$ is calculated and an element $a_j^*$ is selected to calculate $b^* = y_j - \sum_{i=1}^{m} a_i^* y_i (x_i \cdot x_j)$; Then put $w^* = \sum_{i=1}^{m} a_i^* y_i x_i$ into the hyperplane function to obtain the decision function as,

$$f(x) = sgn\left(\sum_{i=1}^{m} a_i^* y_i (x \cdot x_i) + b^*\right) \quad (10)$$

For more details, the readers can refer [4].

# 3 The Users Behaviour Prediction for the Ordering Meeting

## *3.1 The iPAD based Ordering System*

The design of the ordering system is composed of several modules, shown in Fig. 1.

The hardware requirement is: MAC computer, iPAD, system ver10.0 above. The ordering system was developed on the latest mac10.13 operating system and development tool is xcode9.1. Due to the space limits, the design of the ordering system will not be explained and only the data processing will be discussed in details.

During the ordering meeting, the corresponding products have not been manufactured in mass production yet and the images of the products can not be leaked out, especially to the competitors. Since the requirements for secrecy are extremely

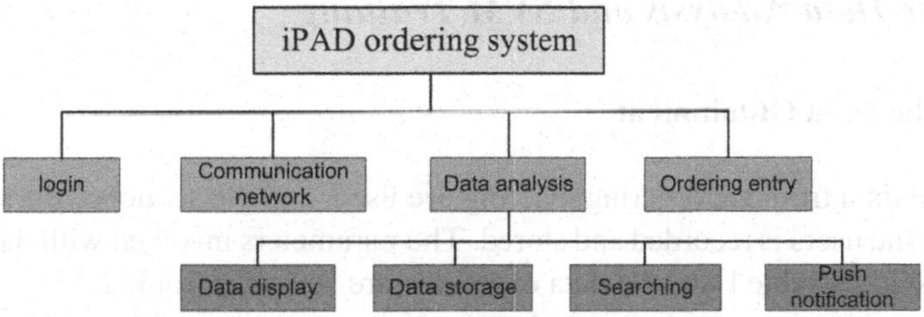


**Fig. 1** The System design of the iPAD based ordering system

**Fig. 2** The display of the ordering system

high, the organizer of the order meeting requests that the mobile terminal can not have picture data stored in the hard disk. On the other hand, the organizer only has constrained budget with limited number of the servers and bandwidth. When more HD (high-definition) images are acquired in real-time, the data coming from the APP end will be obviously loaded in longer waiting time along with the increase of the users number. Therefore, the user behaviour are studied for their selection prediction to solve the above problems.

Considering limited bandwidth, the most possible next selection of the images order from the users can be estimated via the user behaviour prediction. At most ordering meetings, 10–15 pictures will be loaded on the large screen, then the problems can be summarized as: to predict the highest probability of the first 5 pictures and load them in advance to avoid system bottleneck caused by too much data loading. The sample pictures displayed on the screen is shown in Fig. 2.

## *3.2 The Data Analysis and SVM Training*

### 3.2.1 The Data Obtainment

The actual data from the ordering meeting are used, and the sequence during download from the users is recorded and stored. The parameters involved with the pictures are set, listed in Table 1 and the data examples are shown in Table 2.

### 3.2.2 Data Preprocessing

The data has to be processed before entering the procedure to remove abnormal data or noise and the obtained raw data are summarized in Table 3.

Whether the click sequence belongs to the first five clicks is used as the condition. If there is no click at all, the question is simplified as a binary classification problem. In addition, the user data with less total clicks will be discarded to avoid interference, for they are not from the actual ordering staff, and here the click rate is set 30%. Therefore, the processed user number is 621 after the noise data removal and the average click rate is around 89%. The data from the central screen is used. The reason why the click rate is less than 100% is that not all the data have been counted due to other entries.

### 3.2.3 Model Selection

In the paper, the experiments are performed based on the libSVM, so the data has to be transferred into the appropriate format. LibSVM was developed by the researchers from Taiwan University, which is quite simple and efficient in pattern recognition and regression. It is applicable for Windows operating system and Linux operating system, and the program can be altered to realize the transplantation. The libSVM is quite powerful, which can solve the C-SVM classification, V-SVM and e-SVM regression such questions. Here, C-SVM is selected for classification. During the configured decision function, the kernel function and $C$ in the model C-SVM have to be determined.

### 3.2.4 The Selection of the Kernel Function

The selection of parameters in the learning algorithm is quite important to directly determine the accuracy of the prediction results. Currently, there are four commonly used kernel functions, i.e., linear kernel, polynomial kernel, radial basis function (RBF) and sigmoid function. RBF has less parameters selection and easy calculation compared to polynomial kernel. Besides, it has higher precision and general applicability compared with Sigmoid function. RBF is then adopted here,

$$k(x, x_i) = exp(-\gamma \|x - x_i\|^2), \quad \gamma > 0 \tag{11}$$

In this experiment, the parameters in two parts are selected, model parameter $C$ and the parameters involved in the kernel function, $Y$, which are explained in details.

The model parameter $C$ is used as a penalty parameter, i.e., the larger the value of the parameter, the higher the penalty, shown on the punishment of the wrong

**Table 1** The parameters of the pictures

| | |
|---|---|
| No. | Label of the picture; data type: integer |
| Push | Main push; data type: Boolean |
| Guanggao | Advertisement type; data type: Boolean |
| Chuxiao | Sales type; data type: Boolean |
| MustOrder | Bound type; data type: Boolean |
| Score | Number of stars; data type: integer within the range [0–5] |
| Tag | Identifier; data type: integer |
| OrderQty | Order quantity; data type: integer |
| Price | Price of the product; data type: integer (original float, for less calculation consideration) |
| X | Horizontal position of the picture; data type: integer within the range [0–4] |
| Y | Vertical position of the picture; data type: integer within the range [1–2] |
| Page | The page of the picture |
| Class | To classify whether it is the first five clicks; data type: boolean (1: Yes; 0: No) |

**Table 2** The data samples

| No. | 0 | 1 | 2 | 3 |
|---|---|---|---|---|
| Push | 0 | 1 | 0 | 0 |
| Guanggao | 0 | 1 | 0 | 0 |
| Chuxiao | 0 | 0 | 1 | 0 |
| MustOrder | 1 | 0 | 1 | 0 |
| Score | 4 | 1 | 5 | 3 |
| Tag | 3 | 1 | 4 | 0 |
| OrderQty | 50 | 10 | 0 | 0 |
| Price | 399 | 319 | 399 | 559 |
| X | 1 | 2 | 3 | 4 |
| Y | 1 | 1 | 1 | 1 |
| Page | 1 | 1 | 1 | 1 |
| Class | 1 | 1 | 1 | 0 |

**Table 3** The raw data

| Project | Number | Note |
|---|---|---|
| User number | 653 | The total login users |
| Display number | 82 | Small-scale spring ordering meeting |
| Effective click rate (%) | 85 | Total clicks/total users number |
| The percentage of the advertisement (%) | 32 | Advertising fees/total amount |
| The percentage of the bounded amount (%) | 55 | Bounded amount/total amount |
| The percentage of the sales amount (%) | 31 | Sales amount/total amount |
| The range of the wholesale price | 111–1099 | From the lowest to the highest |
| Number of the click rate ≤ 30% | 32 | Personnel or random users |

classification. Based on the previous analysis, the model parameter in SVM is also called adjustment factor, when the ratio of the model structure complexity of the objective function and the empirical risk changes but not as required, the parameter $C$ will be adjusted to normalize the ratio. The structural risk of the minimum model can be represented by a mathematical model, i.e., the sum of the weights between the complexity of the model and the training error. Then the value of $C$ is regarded as a percentage of empirical risk, and the greater the $C$, the higher the empirical risk, and vice versa. When $C$ becomes larger, the empirical risk in the actual risk increases and the actual risk is entirely empirical risk if $C$ is infinitive.

The parameter $Y$ in the kernel function directly indicates the distribution of the input value, called the width distribution in the RBF, which has a heavy effect on the final result of the fitting. Through comparison, the width coefficient $Y$ in SVM can be regarded as the sensory interval of neurons in the neural network and the support vector can be regarded as the center of the neuron kernel function.

$Y$ reflects the closeness between the neuron and the peripheral neurons, i.e., the smaller the tighter of the connections between neurons and more easily affected them. The support vector would not be adjusted by the punishment factor. It can be seen that the selection of the width coefficient is quite important, not too larger or too small. When large width coefficients are selected by all the neurons, the curvature of the model curve would be large which would result in overfitting. On the other hand, if the width coefficients of the neurons are too small, the model curve will be relatively smooth and the penalty factor would not be effect any more.

### 3.2.5 The Parameter Selection and the Prediction Accuracy

During the procedure, the final objective is to optimize the sample prediction with the premise of the optimized parameters. Here, an interactive verification technique is proposed to deal with the data samples and the asymptotic convergence of the model parameter to achieve the optimal result of the final prediction. The cross correlation is also known as the cross ratio, and $n$ data will be divided equally while one subset is used as the test data and the other $n-1$ subsets are used as the training data. Then another subsect is selected as the test set to proceed the above process, so each subset is trained and predicted. It can be seen that the accuracy of the method is the average of the $n$ operation. Through cross testing, the precision of the classifier can be verified with less learning data and multiple learning angles for local extrema avoidance, but the parameters of SVM and kernel function can be obtained to avoid overfitting.

The cross validation is summarized as: Suppose there are $n$ times cross testing, and the training data are equally divided into $n$ portions, $I_1, I_2, \ldots, I_n$, the model parameter is $C$ and the kernel parameter is $Y$, and the execution steps are as follows:

1. During the first operation, the subset $I_1$ in the $n$ equally divisions of the data is used as the test set and the rest $n-1$ subsets are the training data set, and the current prediction accuracy is $a_1$;
2. During the second operation, the subset $I_2$ is used as the test set and the rest $n-1$ subsets are the training data set, and the current prediction accuracy is $a_2$;
3. During the next operation, the subset $I_i$ is used as the test set and the rest $n-1$ subsets are the training data set, and the current prediction accuracy is $a_i$;

Then the final total prediction accuracy is calculated as,

$$P_{accu} = \frac{1}{n}\sum_{i=1}^{n} a_i \qquad (12)$$

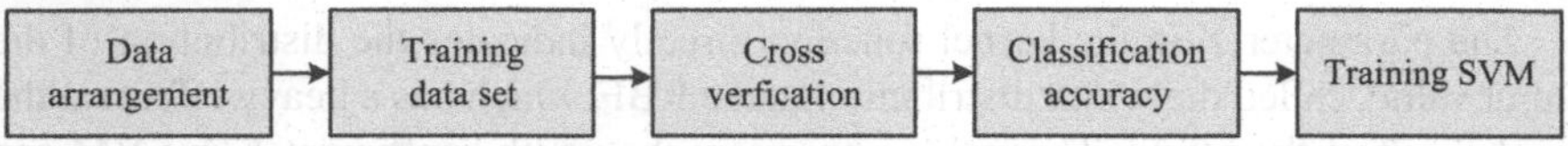


**Fig. 3** The diagram of the SVM algorithm

Through the implementation of these steps, the optimal selection of the parameters of the model parameter and the kernel function can be determined from the experiments. It is also noted that the classification results can be analyzed with the cross testing, which can deal with the large amount of data but also with generalization.

### 3.2.6 SVM Training

First, the training data set and the test data set are determined so as to extract the data for preprocessing. Then the cross verification is used to select the optimal parameters $C$ and $Y$ for SVM training. The obtained model is used for the test data prediction and the algorithm procedure is demonstrated in Fig. 3.

The previously collected data after normalization can be used directly and they are labelled and classified. The first 5 clicked pictures in the same page is ascribed as the first category, i.e., labelled as 1 and the rest are ascribed as the second category, i.e., labelled as 0. Then 50,922 effective data from 621 users are equally divided into two sets, one set is used for the training set and the other is used for testing set.

RBF kernel function is adopted and the optimized parameters involved are determined as described in the paper via the function 'SVMcgForClass' in the Matlab,

$$Output = SVMcg\ For\ Class\ (train_{labels}, train, -10, 10, -10, 10)$$

where the output is a vector with three elements, [bestacc, bestc, bestg], representing the best cross validation accuracy and optimal $C$ and $Y$. After running the program, the obtained results is printed on the screen as,

$$\text{Best Cross Validation Accuracy} = 85.4\% \ \text{Best c} = 2 \ \text{Best g} = 0.1$$

where $g$ in the Matlab is the parameter $Y$ of the kernel function. So the model parameter $C = 2$, $Y = 0.1$ and cross validation accuracy is 85.4%. The obtained parameters are used for SVM training and classification prediction via the 'svmtrain' and 'svmpredict' two functions. The final classification accuracy for the training data set is obtained as: 88.1% and the prediction accuracy of the test set is 85.4%.

The kernel code of the proposed potential pictures selection to realize the cell generation is shown in (Tables 4 and 5).

# 4 Experiments and Analysis

Matlab is used as a tool for the programming operation.

## *4.1 Comparison Experiments*

Different kernel functions are used in SVM for the prediction classification accuracy comparison with the obtained parameters, $C = 2,\ g = 0.1$, and comparison results are shown in Table 6.

It can be seen that the highest classification rate is obtained when the RBF is used as the kernel function with the same model parameters settings.

The next group experiments are performed to compare the classification rate with various parameters. Here, two functions $\text{rand}_c$ and $\text{rand}_g$ are programmed to generate the parameters *C* and *g* randomly, and they are used to train the SVM. Then the obtained SVM is applied for prediction to acquire the classification accuracy rate (CAR), and some results are exemplified in Table 7.

**Table 4** The kernel code for the cell generation

| |
|---|
| (UITableViewCell∗)tableView : (UITableView∗)tableView cellForRowAtIndexPath :{ |
| (NSIndexPath∗)indexPathGoodsTableViewCell ∗ viewCell1 = |
| [[GoodsTableViewCellalloc]initWithFrame : rect1]]; |

**Table 5** The kernel code for the cell generation

| |
|---|
| $\text{dispatch}_\text{a}\text{sync}(\text{dispatch}_\text{g}\text{et}_\text{g}\text{lobal}_\text{q}\text{ueue}(0, 0),$ |
| NSData ∗ data = [NSDatadataWithContentsOfURL : url]; |
| image = [UIImageimageWithData : newData]; |
| $\text{dispatch}_\text{a}\text{sync}(\text{dispatch}_\text{g}\text{et}_\text{m}\text{ain}_\text{q}\text{ueue}(), \text{imageView.image} = \text{image})\};$ |

**Table 6** The prediction classification accuracy comparison results for different kernel functions

| Kernel function | RBF | Linear | Sigmoid | Polynomial |
|---|---|---|---|---|
| CAR (%) | 85.4 | 78.1 | 82.7 | 81.6 |

**Table 7** The classification accuracy comparison with different model parameters

| No. | 1 | 2 | 3 | 4 |
|---|---|---|---|---|
| *C* | 5.7765 | 29.0935 | 87.9639 | 44.7941 |
| *g* | 35.5479 | 21.4365 | 3.4572 | 48.9834 |
| CAR (%) | 55.8722 | 47.3333 | 82.6666 | 51.3458 |

It is shown that the CAR can not be guaranteed via the random model parameters. Sometimes human experience could be assisted for more appropriate parameters selection, though, which can not always guarantee the optimal parameters, even if with highest classification accuracy. However, with the proposed algorithm, the obtained model parameters can be ensured with certain optimization and higher classification accuracy rate.

## 4.2 *The Efficiency of the Users's Behaviour Prediction*

**(1) The average waiting time from clicking to full picture loading**

The code has been embedded in the iPAD system and the finger touching screen event is triggered by the 'tableView' function. 'didSelectRowAtIndexPath' is used to record the starting time and the system completion time till the full data loading on the big screen, where the duration is called the effective time, $T_{eff}$. Considering the same process of each page, the loading time of the first page is used as a case study, i.e., the average $T_{eff}$ of the whole pictures fully loaded in the first page with the optimized system.

Before the system optimization, the loading time of each picture should be taken with the same time, excluding the abnormalities. During the ordering meeting, the data will be sampled with 10 min intervals, totally 10 times for each user and the average value will be recorded for the experiments. The results are shown in Table 8.

With the proposed SVM-based user behaviours prediction algorithm, the hit rate of the first 5 pictures is 85.4%, and the last 5 pictures would not be loaded in advance. Then the load waiting time at the user terminal is reduced greatly with 47.4% ($P_{eff} = \frac{1.48}{3.12}$) efficiency improvement. The calculated theoretical efficiency is basically in accordance with the actual efficiency.

**(2) The peak pressure comparison**

Since it is impossible to obtain two copies of the data at the same ordering meeting, the peak pressure is compared between different ordering meetings. The data from the optimized system and the non-optimized ordering meeting is used for comparison, and the peak amount is the highest amount during the recorded continuously 24 hours. The data unit is GB (Table 9).

**Table 8** The indexes involved in pictures loading

| | $T_b$ | $T_a$ | $T_s$ | $P_{eff}$ (%) |
|---|---|---|---|---|
| The average loading time of the homepage | 3.12 | 1.64 | 1.48 | 47.4 |
| The average loading time of the single graph | 0.312 | 0.164 | 0.148 | |

$T_b$: the duration before the optimization, unit: s
$T_a$: the duration after the optimization, unit: s
$T_s$: the duration reduction, unit: s

**Table 9** The result of the c

| | $D_p$(GB) | $N_t$ | $N_h$ | $Ds$(GB) | $P_{eff}$ (%) |
|---|---|---|---|---|---|
| Without optimization | 84.5 | 348 | 281 | 0.301 | 9.63 |
| With optimization | 158.3 | 653 | 582 | 0.272 | |

$N_t$: The total user number; $N_h$: The most online user number
$D_p$: The peak data flow; $Ds$: The data exchange per second

There are many devices to exchange the data with the data control center simultaneously, and it is impossible to extract only two groups of data under the same condition. Hence, the latest collected peak flow data and the highest online data is used for comparison.

Although there are certain data errors, the peak data flow after system optimization can reduce 9.63%($P_{eff} = \frac{0.301-0.272}{0.301}$). The abnormal phenomena such as user break off or long time waiting status have been improved significantly during the data peak time.

## 5 Conclusion

In this paper, an optimization algorithm is proposed to improve the iPad based garment ordering system performance. Based on SVM, the developed algorithm can predict the users behaviour with satisfied accuracy to load the most possible selected pictures in advance before the click/selection of the users. So the system would not incur bottleneck to result in longer waiting time under limited network resources and it can significantly increase the system efficiency.

Although there are great improvement for the ordering system, the developed algorithm is focused on the iPAD, and the ordering system in PC has certain difference, which should be further investigated. Different machine learning algorithms could be applied for the performance comparison with SVM.

**Acknowledgements** This work is supported under the Shenzhen Science and Technology Innovation Commission Project Grant Ref. JCYJ20160510154736343 and JCYJ201703071654 42023, and Guangdong Provincial Engineering Technology Research Center of Intelligent Unmanned System and Autonomous Environmental Perception.

## References

1. Z. Dong, The user behaviour preference of mobile internet based support vector machine. Beijing University of Posts and Telecommunications. Ph.D. Thesis, in Chinese, 2014
2. V. Geetha, K.A. Rangarajan, A conceptual Framework for Perceived Rish in Consumer Online Shopping. Glob. Manag. Rev. (2015)

3. V. Vapnik, E. Levin, Y. Cun, Measuring the VC-dimension of a learning machine. Neural Comput. **6**(5), 851 (1994)
4. B. Schölkopf, C. Burges, V. Vapnik, Incorporating invariances in support vector learning machines, in *International Conference on Artificial Neural Networks* (1996), pp. 47–52
5. J. Peng, Y. Guo, Multi-parameter prediction of drivers' lane-changing behaviour with neural network model. Appl. Ergon. **50**, 207–217 (2015)
6. J. Hauschild, A. Kazeminia, A. Braasch, Reliability prediction for automotive components using Real-Parameter Genetic Algorithm, in *Proceedings of the Joint ESREL and SRA-Europe Conference*, vol. 3 (2009), pp. 2245–2249
7. M. Nguyen, J. Perez, F. De La Torre, Facial feature detection with optimal pixel reduction SVM, in *The 8th IEEE International Conference on Automatic Face and Gesture Recognition*, Article number: 4813372 (2008)
8. C. Chen, H. Lee, M. Kao, Multi-class SVM with negative data selection for web page classification, in *IEEE International Conference on Neural Networks* vol. 3 (2004), pp. 2047–2052
9. A. Braun, U. Weidner, S. Hinz, Classifying roof materials using data fusion through kernel composition—Comparing v-SVM and one-class SVM, in*Proceedings of the Joint Urban Remote Sensing Event* (2011), pp. 377–380
10. S. Watanabe, Y. Kimura, A methodology using EMO for parameter estimation of SVM kernel function, in *Proceedings of the IEEE Conference on Soft Computing on Industrial Applications* (2008), pp. 211–216
11. P. Chen, C. Lin, B. Schölkopf, A tutorial on v-support vector machines. Appl. Stoch. Models Bus. Ind. **21**(2), 111–136 (2005)
12. F. Cai, V. Cherkassky, Generalized SMO algorithm for SVM-based multitask learning. IEEE Trans. Neural Netw. Learn. Syst. **23**(6), 997–1003 (2012)
13. N. Zhang, C. Williams, P. Behera, Water quantity prediction using least squares support vector machines (LS-SVM) method, in *Proceedings of the 18th World Multi-Conference on Systemics, Cybernetics and Informatics*, (2014), pp. 251–256

# Flocking Motion of Second-Order Multi-agent Systems with Mismatched Disturbances

**Fan Liu, Hongyong Yang, Yize Yang, Yuling Li and Yuanshan Liu**

**Abstract** Based on the problem of flocking for second-order multi-agent systems with mismatched disturbances, a distributed control algorithm with individual local information is investigated. For each agent, a disturbance observer is designed. And then based on disturbance observer, a distributed control protocol with feed-forward compensation term is proposed. By using Lyapunov stability and input to state stability theory, it proves that the distributed control law enables to make all agents eventually converge to the leader's velocity when at least one agent can receive the leader's information. Finally, a numerical simulation example illustrates the effectiveness of the conclusion.

**Keywords** Distributed control · Flocking motion · Mismatched disturbances Disturbance observer · Weak connectivity

F. Liu · H. Yang (✉) · Y. Li · Y. Liu
School of Information and Electrical Engineering,
Ludong University, Yantai 264025, Shandong, China
e-mail: hyyang@yeah.net

F. Liu
e-mail: jsgyliufan@163.com

Y. Li
e-mail: liyuling822@163.com

Y. Liu
e-mail: yuanshanliu@163.com

Y. Yang
School of Electrical Engineering and Telecommunications,
University of New South Wales, Sydney, Australia
e-mail: yangyz1994@126.com



Y. Jia et al. (eds.), *Proceedings of 2018 Chinese Intelligent Systems Conference*,
Lecture Notes in Electrical Engineering 528,
https://doi.org/10.1007/978-981-13-2288-4_73

## 1 Introduction

In recent years, the study of flocking behaviors has been increasingly focused by the scholars from diverse fields such as biology, physics and engineering. Due to the wide control applications of flocking in engineering, there has been a surge of interest among control theorists in flocking problems.

In 1987, Reynolds [1] proposed the a classical flocking model which is based on three heuristic rules: (I) Separation; (II) Alignment; (III) Cohesion. Subsequently, many improvements and additional new rules have been suggested, such as rendezvous, goal seeking, and formation control. Olfati-Saber proposed two control algorithms [2, 3]: the former algorithms reflected the Reynolds's three rules and had the ability of obstacle avoidance, and the later algorithm realized goal tracking by adding navigational feedback term when all agents could receive the information of leader. However, this assumption is not correspond to the examples in nature and is hard to realize in engineering. On this basis, Su [4] investigated the case of part of agent receiving the leader's information by using the results of algebraic graph theory. Then, Su [5] proposed the control approach based only on position measurements about the case of multiple leaders. Yang [6] studied the containment control of fractional order multi-agent systems with time delays. Yang et al. [7] also studied distributed containment control of heterogeneous fractional-order multi-agent systems. Then Yang et al. [8] studied the case of communication delays. From the above results, we can see that the containment control can also be considered as flocking motion.

In practical applications, multi-agent systems usually suffer from the disturbances. Yang et al. [9] designed a disturbance observer for the second-order systems with exogenous disturbances and proposed a consensus control algorithm based on disturbance observer. However, the above disturbances belong to matched disturbances, i.e., the uncertainties or disturbances are in the same channels as the control inputs. In multi-agent systems, there are many disturbances entering systems through different channels from the control inputs, which called mismatched disturbances [10, 11]. Since mismatched disturbances cannot be eliminated directly through the feedback controller, the cooperative control for multi-agent systems with mismatched disturbances is more challenging, and the study of flocking motion of multi-agent systems with mismatched disturbances is rarely reported.

In this paper, a disturbance observer is designed to estimate the agents' disturbances for the multi-agent systems with mismatched disturbances. Based on artificial potential function gradient and the approach of disturbance observer, a distributed control protocol with feed-forward compensation term is proposed. Then, the multiple agents asymptotically track the leader and attain flocking motion under the situation of disturbances.

## 2 Preliminaries and Problem Description

### 2.1 Graph Theory Notions

The communication topology of n agents is denoted by a connected graph $\boldsymbol{G} = (V, E, \boldsymbol{A})$, where $V = \{v_1, v_2, \cdots, v_n\}$ is the set of $n$ nodes, $N = \{1, 2, \cdots, n\}$ is the set of subscript, $E \subseteq V \times V$ is the set of edges and $\boldsymbol{A} = [a_{ij}] \in \boldsymbol{R}^{n\times n}$ is the weighted adjacency matrix where the matrix element $a_{ij}$ denotes the connected weight between node $v_i$ and node $v_j$. The set of neighbors of node $v_i$ is denoted as $N_i = \{v_j \in V|(v_i, v_j) \in E\}$. If $v_j \in N_i$, then $a_{ij} > 0$ otherwise $a_{ij} = 0$. We assume that each node of graph $\boldsymbol{G}$ doesn't have self-loop that $a_{ii} = 0, \forall i \in N$. The degree diagonal matrix of graph $\boldsymbol{G}$ is denoted as $\boldsymbol{D} = diag(d_1, d_2, \cdots, d_n)$, where the degree of the node $v_i$ is $d_i = \sum_{j=1}^{n} a_{ij}$. The Laplacian matrix of graph $\boldsymbol{G}$ is $\boldsymbol{L} = \boldsymbol{D} - \boldsymbol{A}$.

### 2.2 Problem Description

We consider $n$ agents moving in the 2 dimensional Euclidean space. The dynamic model of $i$ agent with disturbances is described by two integrators as

$$\begin{cases} \dot{\boldsymbol{x}}_i(t) = \boldsymbol{v}_i(t) + \boldsymbol{d}_{i1}(t) \\ \dot{\boldsymbol{v}}_i(t) = \boldsymbol{u}_i(t) + \boldsymbol{d}_{i2}(t) \end{cases} \tag{1}$$

where $i \in N$, $\boldsymbol{x}_i(t), \boldsymbol{v}_i(t) \in \boldsymbol{R}^2$ are respectively the position and velocity vectors of agent $i$, $\boldsymbol{u}_i(t) \in \boldsymbol{R}^2$ is the (acceleration) control input acting on agent $i$, $\boldsymbol{d}_{i1}(t)$ and $\boldsymbol{d}_{i2}(t)$ are respectively the mismatched disturbances and matched disturbances.

And we assume the model of motion of the virtual leader is $\dot{\boldsymbol{x}}_r(t) = \boldsymbol{v}_r(t)$, $\dot{\boldsymbol{v}}_r(t) = 0$,where $\boldsymbol{x}_r(t), \boldsymbol{v}_r(t) \in \boldsymbol{R}^2$ are respectively the position and velocity vectors of leader.

**Definition 1** If every agent (asymptotically) attains the same velocity, collision between agents can be avoided, and the final distance between agents keeps stable, this agent group (asymptotically) attains flocking motion.

The main goal is to design a control protocol to eliminate the effect of disturbances and attain the flocking motion under the meaning of Definition 1.

**Definition 2** Potential function $U_{ij}$ is a differentiable, non-negative and unbounded function about the relative distance $\|\boldsymbol{x}_i - \boldsymbol{x}_j\|$ between agent $i$ and $j$, and it satisfies that when $\|\boldsymbol{x}_i - \boldsymbol{x}_j\| \to 0$, $U_{ij}(\|\boldsymbol{x}_i - \boldsymbol{x}_j\|) \to \infty$; and $U_{ij}$ can attain the only minimum value in an appropriate distance.

We assume potential function as

$$U_{ij}(\|\boldsymbol{x}_i - \boldsymbol{x}_j\|) = c\left(\frac{\varepsilon^2}{\|\boldsymbol{x}_i - \boldsymbol{x}_j\|^2} + \ln\left(\|\boldsymbol{x}_i - \boldsymbol{x}_j\|^2 + a^2\right)\right) \tag{2}$$

where $\varepsilon > 0$, $a > 0$, $c > 0$. It is thus clear that $U_{ij}$ can attain the only minimum value when $\|\boldsymbol{x}_i - \boldsymbol{x}_j\| = \sqrt{\frac{\varepsilon^2+\sqrt{\varepsilon^4+4\varepsilon^2 a^2}}{2}}$. Separation and cohesion in the group can be steered by the potential function defined by Definition 2. The total potential energy of agents can be denoted as $U_i = \sum_{j\in N_i} \Phi(\|\boldsymbol{x}_i - \boldsymbol{x}_j\|)$.

# 3 Analysis of Flocking Motion for Multi-agent Systems with Mismatched Disturbances

**Lemma 1** *([12]) (Input-to-State Stability Theorem, ISS Theorem) Consider the nonlinear system $\dot{x} = f(x, u, t)$, if the system $\dot{x} = f(x, 0, t)$ is globally uniformly asymptotically stable, and $\lim_{t\to\infty} u(t) = 0$, then the states of system $\dot{x} = f(x, u, t)$ are asymptotically convergent to zero, i.e. $\lim_{t\to\infty} x(t) = 0$.*

## *3.1 Disturbance Observer Design*

**Assumption 1** The disturbances $\boldsymbol{d}_{ik}(t)$ and $\dot{\boldsymbol{d}}_{ik}(t)$, $i \in N, k = 1, 2$ are bounded.

**Assumption 2** $\lim_{t\to\infty} \dot{\boldsymbol{d}}_{ik}(t) = 0$, $i \in N, k = 1, 2$.

The disturbance observer is designed as follow by referring [9, 10]

$$\begin{cases} \hat{\boldsymbol{d}}_{i1} = \boldsymbol{\theta}_1(\boldsymbol{x}_i - \boldsymbol{p}_{i1}) \\ \dot{\boldsymbol{p}}_{i1} = \boldsymbol{v}_i + \hat{\boldsymbol{d}}_{i1} \\ \hat{\boldsymbol{d}}_{i2} = \boldsymbol{\theta}_2(\boldsymbol{v}_i - \boldsymbol{p}_{i2}) \\ \dot{\boldsymbol{p}}_{i2} = \boldsymbol{u}_i + \hat{\boldsymbol{d}}_{i2} \end{cases} \tag{3}$$

where $i \in \{1, 2, \cdots, n\}$, $\hat{\boldsymbol{d}}_{i1}$ and $\hat{\boldsymbol{d}}_{i2}$ are respectively estimated disturbance vectors, $\boldsymbol{p}_{i1}$ and $\boldsymbol{p}_{i2}$ are auxiliary vectors, and $\boldsymbol{\theta}_1$, $\boldsymbol{\theta}_2 > 0$ are observation gain matrices.

The error systems can be got from systems (1) and (3) that

$$\begin{cases} \dot{\boldsymbol{e}}_{d_{i1}}(t) = -\boldsymbol{\theta}_1 \boldsymbol{e}_{d_{i1}}(t) + \dot{\boldsymbol{d}}_{i1}(t) \\ \dot{\boldsymbol{e}}_{d_{i2}}(t) = -\boldsymbol{\theta}_2 \boldsymbol{e}_{d_{i2}}(t) + \dot{\boldsymbol{d}}_{i2}(t) \end{cases}, i \in N \tag{4}$$

**Theorem 1** *If Assumptions 1 and 2 hold, the estimates $\hat{d}_{ik}$, $i \in N$, $k = 1, 2$, can asymptotically track the disturbances $d_{ik}$ of system (1).*

*Proof* From Eq. (4), we can see that $e^{\theta_k t}\dot{e}_{d_{ik}}(t)+e^{\theta_k t}\theta_k e_{d_{ik}}(t)=e^{\theta_k t}\dot{d}_{ik}(t)$. So It can be got by integral that

$$e_{d_{ik}}(t)=e^{-\theta_k t}e_{d_{ik}}(0)+\int_0^t e^{\theta_k(s-t)}\dot{d}_{ik}(s)ds \tag{5}$$

We can see that $\lim_{t\to\infty}e^{\theta_k(s-t)}<1$, where $\theta_k>0, t>s$. When $\lim_{t\to\infty}\dot{d}_i(t)=0$, we can get the result. And from Eq. (5) we can also obtain that greater the observation gain $\theta_k$, faster the response speed of observer. This completes the proof.

## 3.2 Distributed Control Protocol Design

The distributed controller is designed as follow under the assumption that the communication or interaction between agents is time-invariant.

$$\begin{aligned}\boldsymbol{u}_i=&-\sum_{j\in N_i}a_{ij}\varphi\left(\|\boldsymbol{x}_i-\boldsymbol{x}_j\|\right)\vec{\boldsymbol{n}}_{ij}-k\sum_{j=1}^{n}a_{ij}\left(\left(\boldsymbol{v}_i+\hat{\boldsymbol{d}}_{i1}\right)-\left(\boldsymbol{v}_j+\hat{\boldsymbol{d}}_{j1}\right)\right)\\&-h_i c_1\left(\boldsymbol{v}_i+\hat{\boldsymbol{d}}_{i1}-\boldsymbol{v}_r\right)-\hat{\boldsymbol{d}}_{i2}\end{aligned} \tag{6}$$

where $k,\ c_1>0, \vec{\boldsymbol{n}}_{ij}=\frac{\boldsymbol{x}_i-\boldsymbol{x}_j}{\|\boldsymbol{x}_i-\boldsymbol{x}_j\|}$ is unit vector, and $\hat{\boldsymbol{d}}_{i1},\ \hat{\boldsymbol{d}}_{i2}$ are respectively estimated disturbance vectors.

**Theorem 2** *Consider a network system consisting of n agents and a leader, which dynamic model can be denoted as Eq. (1) with the effect of mismatched disturbances. If at least one agent can receive the leader's information, the distributed control law (6) enables to make all agents eventually converge to the leader's velocity and attain flocking motion.*

*Proof* Denote tracking error as $\bar{\boldsymbol{x}}_i=\boldsymbol{x}_i-\boldsymbol{x}_r, \bar{\boldsymbol{v}}_i=\boldsymbol{v}_i+\hat{\boldsymbol{d}}_{i1}-\boldsymbol{v}_r$, so the error system is that $\dot{\bar{\boldsymbol{x}}}_i=\bar{\boldsymbol{v}}_i+\boldsymbol{e}_{d_{i1}}, \dot{\bar{\boldsymbol{v}}}_i=\boldsymbol{u}_i+\boldsymbol{d}_{i2}+\boldsymbol{\theta}_1\boldsymbol{e}_{d_{i1}}$

Combine tracking error system and disturbance error system (4), then

$$\begin{cases}\dot{\bar{\boldsymbol{x}}}_i=\bar{\boldsymbol{v}}_i+\boldsymbol{e}_{d_{i1}}\\ \dot{\bar{\boldsymbol{v}}}_i=\boldsymbol{u}_i+\boldsymbol{d}_{i2}+\boldsymbol{\theta}_1\boldsymbol{e}_{d_{i1}}\\ \dot{\boldsymbol{e}}_{d_{i1}}=-\boldsymbol{\theta}_1\boldsymbol{e}_{d_{i1}}+\dot{\boldsymbol{d}}_{i1}\\ \dot{\boldsymbol{e}}_{d_{i2}}=-\boldsymbol{\theta}_2\boldsymbol{e}_{d_{i2}}+\dot{\boldsymbol{d}}_{i2}\end{cases} \tag{7}$$

Denote $\boldsymbol{\zeta}=\left[\bar{\boldsymbol{x}},\bar{\boldsymbol{v}},\boldsymbol{e}_{d_1},\boldsymbol{e}_{d_2}\right]^{\mathrm{T}}$, $\dot{\boldsymbol{d}}=\left[\dot{\boldsymbol{d}}_1,\dot{\boldsymbol{d}}_2\right]^{\mathrm{T}}$, where $\bar{\boldsymbol{x}}=[\bar{\boldsymbol{x}}_1,\bar{\boldsymbol{x}}_2,\ldots,\bar{\boldsymbol{x}}_n]^{\mathrm{T}}$, $\bar{\boldsymbol{v}}=[\bar{\boldsymbol{v}}_1,\bar{\boldsymbol{v}}_2,\ldots,\bar{\boldsymbol{v}}_n]^{\mathrm{T}}$, $\boldsymbol{e}_{d_1}=\left[\boldsymbol{e}_{d_{11}},\boldsymbol{e}_{d_{21}},\ldots,\boldsymbol{e}_{d_{n1}}\right]^{\mathrm{T}}$, $\boldsymbol{e}_{d_2}=\left[\boldsymbol{e}_{d_{12}},\boldsymbol{e}_{d_{22}},\ldots,\boldsymbol{e}_{d_{n2}}\right]^{\mathrm{T}}$,

$\dot{\boldsymbol{d}}_1 = \left[\dot{\boldsymbol{d}}_{11}, \dot{\boldsymbol{d}}_{21}, \ldots, \dot{\boldsymbol{d}}_{n1}\right]^{\mathrm{T}}$, $\dot{\boldsymbol{d}}_2 = \left[\dot{\boldsymbol{d}}_{12}, \dot{\boldsymbol{d}}_{22}, \ldots, \dot{\boldsymbol{d}}_{n2}\right]^{\mathrm{T}}$, then system (7) is turned into

$$\dot{\boldsymbol{\zeta}} = \boldsymbol{A}\boldsymbol{\zeta} + \nabla U + \boldsymbol{D}\dot{\boldsymbol{d}} \tag{8}$$

where $\boldsymbol{A} = \begin{bmatrix} \boldsymbol{0} & \boldsymbol{I}_{2n} & \boldsymbol{I}_{2n} & \boldsymbol{0} \\ \boldsymbol{0} & -(k\boldsymbol{L} + c_1\boldsymbol{H}) \otimes \boldsymbol{I}_2 & \boldsymbol{I}_n \otimes \boldsymbol{\theta}_1 & \boldsymbol{I}_{2n} \\ \boldsymbol{0} & \boldsymbol{0} & -\boldsymbol{I}_n \otimes \boldsymbol{\theta}_1 & \boldsymbol{0} \\ \boldsymbol{0} & \boldsymbol{0} & \boldsymbol{0} & -\boldsymbol{I}_n \otimes \boldsymbol{\theta}_2 \end{bmatrix}$, $\boldsymbol{D} = \begin{bmatrix} \boldsymbol{0} & \boldsymbol{0} \\ \boldsymbol{0} & \boldsymbol{0} \\ \boldsymbol{I}_{2n} & \boldsymbol{0} \\ \boldsymbol{0} & \boldsymbol{I}_{2n} \end{bmatrix}$, $\boldsymbol{L}$ is the Laplacian matrix of connected graph $\boldsymbol{G}$, $\boldsymbol{H} = diag\{h_1, h_2, \ldots, h_n\}$ is the weighted adjacency matrix of leader, and $\nabla U$ is a function about $\left(\bar{\boldsymbol{x}}, \bar{\boldsymbol{v}}, \boldsymbol{e}_{d_1}, \boldsymbol{e}_{d_2}\right)$, which denotes the gradient of potential energy. By using ISS theory, $\dot{\boldsymbol{d}}$ is considered as the input of system (7). So we can firstly assume $\dot{\boldsymbol{d}} = \boldsymbol{0}$, and then analyze the stability of system $\dot{\boldsymbol{\zeta}} = \boldsymbol{A}\boldsymbol{\zeta} + \nabla U$. So system (8) is turned to

$$\begin{cases} \dot{\bar{\boldsymbol{x}}}_i = \bar{\boldsymbol{v}}_i + \boldsymbol{e}_{d_{i1}} \\ \dot{\bar{\boldsymbol{v}}}_i = \boldsymbol{u}_i + \boldsymbol{d}_{i2} + \boldsymbol{\theta}_1\boldsymbol{e}_{d_{i1}} \\ \dot{\boldsymbol{e}}_{d_{i1}} = -\boldsymbol{\theta}_1\boldsymbol{e}_{d_{i1}} \\ \dot{\boldsymbol{e}}_{d_{i2}} = -\boldsymbol{\theta}_2\boldsymbol{e}_{d_{i2}} \end{cases} \tag{9}$$

For system (9), the Lyapunov function is chosen as

$$V = \sum_{i=1}^{n}\sum_{j\in N_i}\Phi\left(\|\bar{\boldsymbol{x}}_i - \bar{\boldsymbol{x}}_j\|\right) + \frac{1}{2}\sum_{i=1}^{n}\left(\bar{\boldsymbol{v}}_i + \boldsymbol{e}_{d_{i1}}\right)^{\mathrm{T}}\left(\bar{\boldsymbol{v}}_i + \boldsymbol{e}_{d_{i1}}\right) + \frac{1}{2}\sum_{i=1}^{n}\boldsymbol{e}_{d_{i1}}^{\mathrm{T}}\boldsymbol{e}_{d_{i1}} + \frac{1}{2}\sum_{i=1}^{n}\boldsymbol{e}_{d_{i2}}^{\mathrm{T}}\boldsymbol{e}_{d_{i2}} \tag{10}$$

The derivative of $V$ function is that

$$\dot{V} = \sum_{i=1}^{n}\left(\bar{\boldsymbol{v}}_i + \boldsymbol{e}_{d_{i1}}\right)^{\mathrm{T}}\left(-k\sum_{j=1}^{n}a_{ij}\left(\bar{\boldsymbol{v}}_i - \bar{\boldsymbol{v}}_j\right) - h_i c_1\bar{\boldsymbol{v}}_i + \boldsymbol{e}_{d_{i2}}\right) - \sum_{i=1}^{n}\boldsymbol{e}_{d_{i1}}^{\mathrm{T}}\boldsymbol{\theta}_1\boldsymbol{e}_{d_{i1}} - \sum_{i=1}^{n}\boldsymbol{e}_{d_{i2}}^{\mathrm{T}}\boldsymbol{\theta}_2\boldsymbol{e}_{d_{i2}}$$

$$= \sum_{i=1}^{n}\bar{\boldsymbol{v}}_i^{\mathrm{T}}\left(-k\sum_{j=1}^{n}a_{ij}\left(\bar{\boldsymbol{v}}_i - \bar{\boldsymbol{v}}_j\right) - h_i c_1\bar{\boldsymbol{v}}_i\right) + \sum_{i=1}^{n}\bar{\boldsymbol{v}}_i^{\mathrm{T}}\boldsymbol{e}_{d_{i2}} +$$

$$\sum_{i=1}^{n}\boldsymbol{e}_{d_{i1}}^{\mathrm{T}}\left(-k\sum_{j=1}^{n}a_{ij}\left(\bar{\boldsymbol{v}}_i - \bar{\boldsymbol{v}}_j\right) - h_i c_1\bar{\boldsymbol{v}}_i\right) + \sum_{i=1}^{n}\boldsymbol{e}_{d_{i1}}^{\mathrm{T}}\boldsymbol{e}_{d_{i2}} - \sum_{i=1}^{n}\boldsymbol{e}_{d_{i1}}^{\mathrm{T}}\boldsymbol{\theta}_1\boldsymbol{e}_{d_{i1}} - \sum_{i=1}^{n}\boldsymbol{e}_{d_{i2}}^{\mathrm{T}}\boldsymbol{\theta}_2\boldsymbol{e}_{d_{i2}}$$

$$= \bar{\boldsymbol{v}}^{\mathrm{T}}(-(k\boldsymbol{L} + c_1\boldsymbol{H}) \otimes \boldsymbol{I}_2)\bar{\boldsymbol{v}} + \bar{\boldsymbol{v}}^{\mathrm{T}}\boldsymbol{I}_{2n}\boldsymbol{e}_{d_2} + \boldsymbol{e}_{d_1}^{\mathrm{T}}(-(k\boldsymbol{L} + c_1\boldsymbol{H}) \otimes \boldsymbol{I}_2)\bar{\boldsymbol{v}} + \boldsymbol{e}_{d_1}^{\mathrm{T}}\boldsymbol{I}_{2n}\boldsymbol{e}_{d_2} -$$

$$\boldsymbol{e}_{d_1}^{\mathrm{T}}(\boldsymbol{I}_n \otimes \boldsymbol{\theta}_1)\boldsymbol{e}_{d_1} - \boldsymbol{e}_{d_2}^{\mathrm{T}}(\boldsymbol{I}_n \otimes \boldsymbol{\theta}_2)\boldsymbol{e}_{d_2}$$

$$= \begin{bmatrix} \bar{\boldsymbol{v}} \\ \boldsymbol{e}_{d_1} \\ \boldsymbol{e}_{d_2} \end{bmatrix}^{\mathrm{T}}\begin{bmatrix} -(k\boldsymbol{L} + c_1\boldsymbol{H}) \otimes \boldsymbol{I}_2 & \boldsymbol{0} & \boldsymbol{I}_{2n} \\ -(k\boldsymbol{L} + c_1\boldsymbol{H}) \otimes \boldsymbol{I}_2 & -\boldsymbol{I}_n \otimes \boldsymbol{\theta}_1 & \boldsymbol{I}_{2n} \\ \boldsymbol{0} & \boldsymbol{0} & -\boldsymbol{I}_n \otimes \boldsymbol{\theta}_2 \end{bmatrix}\begin{bmatrix} \bar{\boldsymbol{v}} \\ \boldsymbol{e}_{d_1} \\ \boldsymbol{e}_{d_2} \end{bmatrix} = \begin{bmatrix} \bar{\boldsymbol{v}} \\ \boldsymbol{e}_{d_1} \\ \boldsymbol{e}_{d_2} \end{bmatrix}^{\mathrm{T}}\boldsymbol{\Omega}\begin{bmatrix} \bar{\boldsymbol{v}} \\ \boldsymbol{e}_{d_1} \\ \boldsymbol{e}_{d_2} \end{bmatrix}$$

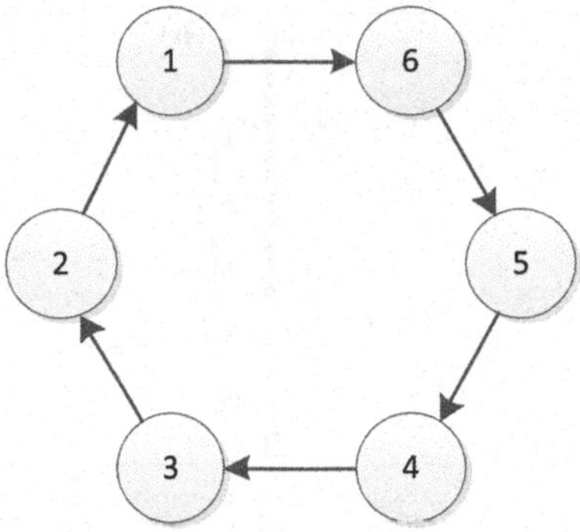


**Fig. 1** Directed topology construct of 6 agents

Since both $\boldsymbol{L}$ and $\boldsymbol{H}$ are positive semi-definite matrices, $k\boldsymbol{L} + c_1\boldsymbol{H} > 0$ is positive definite matrix. From the design rules of disturbance observer, we can see that observation gain matrices $\boldsymbol{\theta}_1,\ \boldsymbol{\theta}_2 > 0$ are positive definite matrices. Thus, the matrix $\boldsymbol{\Omega}$ is negative definite. So $\dot{V} < 0$. According to Lyapunov stability theory, system (9) is asymptotically stable.

According to Assumption 2, the disturbances satisfy $\lim_{t\to\infty} \dot{d}_{ik}(t) = 0,\ i \in N, k = 1, 2$. By using ISS theory, the error system (8) is asymptotically stable. Thus, all agents eventually converge to the leader's velocity and attain flocking motion. This completes the proof.

## 4 Numerical Simulations

The connected graph shown in Fig. 1 consists of 6 agents with the same motion model. Its adjacent matrix is $\boldsymbol{A} = \begin{bmatrix} 0_{5\times1} & E_5 \\ 1 & 0_{1\times5} \end{bmatrix}$ and potential energy function is chosen as (2), where $c = 2$, $\varepsilon = 5$, $a = 1$. The control gain factors are $k = 20$ and $c_1 = 10$. The observer gains are set as $\theta_1 = 10$ and $\theta_2 = 10$. The initial position is randomly chosen in the area of [0 10; 0 10] and the initial velocity with uncertain direction is set in [0 5]. We assume that only agent 1 can receive the leader's information, and the leader's motion is uniform with the initial velocity $v_{rx} = 1.2$, $v_{ry} = 1.2$. We assume that all agents suffer from the disturbances. The results are shown in Figs. 2, 3 and 4.

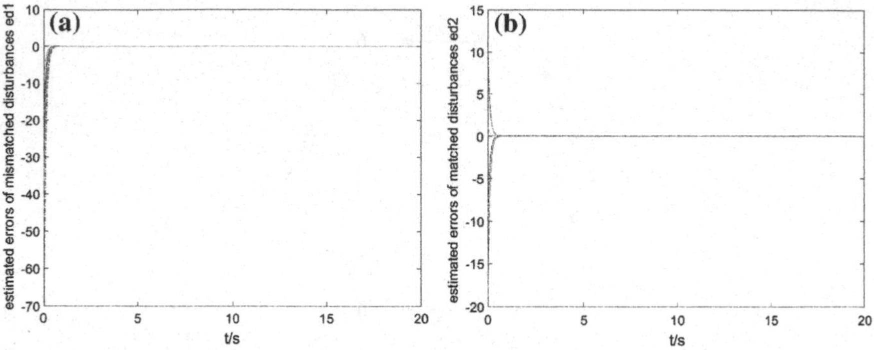


**Fig. 2** Estimated errors of DO (3): **a** the estimated errors of mismatched disturbances; **b** the estimated errors of matched disturbances

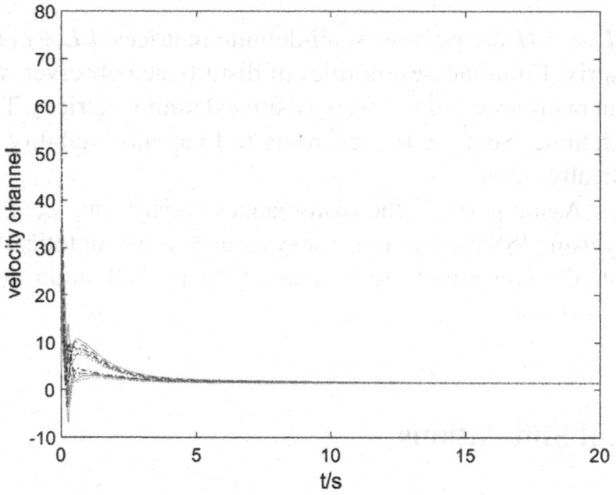


**Fig. 3** The velocity of each agent

The estimations and estimated errors of mismatched and matched disturbances for each agent of DO (3) are shown in Fig. 2. It presents better observation performance. The changing situation of velocity for each agent are illustrated in Fig. 3, i.e. the value of $v_i + \hat{d}_{i1}$, where $i = 1, 2, \cdots, 6$. This is owing to the mismatched disturbances in the velocity channels. The disturbance compensation is required to overcome the effect of disturbances in velocity channel. Thus, the state in velocity channel is $\tilde{v}_i = v_i + \hat{d}_{i1}$. According to Fig. 3, the velocities of 6 agents reach an agreement with leader after 5 s. Figure 4 is the trajectories of relative position in two coordinate axes, where asterisk denotes followers. We can see from Fig. 4 that the trajectories of agents and leader can reach parallel and avoid collision after a long time.

**Fig. 4** The relative position of agents

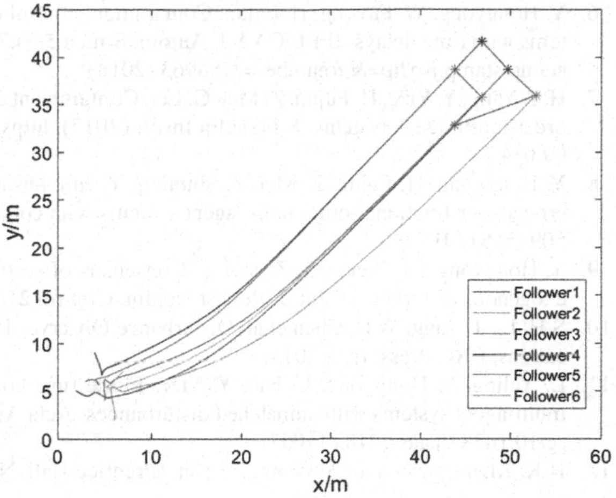


# 5 Conclusions

In this paper, the effect of external disturbances of multiple agents, especially the case of existing mismatched disturbances, is considered. Based on the control approach of disturbance observer, a distributed control protocol is proposed to make agents attain flocking motion. The feed-forward compensation term in the controller will effectively eliminate the disturbances. Furthermore, by utilizing ISS theory and Lyapunov stability theory, the stability of the dynamic behaviors of multiple agents is analyzed. Finally, the numerical simulations illustrate the effectiveness of the proposed control algorithms.

**Acknowledgements** This paper is supported by the National Natural Science Foundation of China (61673200, 61771231, 61471185), Natural Science Foundation of Shandong Province (ZR2018ZC0438, ZR2017MF010, ZR2017PF010) and the Key R&D Program of Yantai City (2016ZH061)

# References

1. C.W. Reynolds, Flocks, herds and schools: a distributed behavioral model. ACM SIGGRAPH Comput. Graph. **21**(4), 25–34 (1987)
2. R. Olfati-Saber, R.M. Murray, Consensus problems in networks of agents with switching topology and time-delays. IEEE Trans. Autom. Control **49**(9), 1520–1533 (2004)
3. R. Olfati-Saber, Flocking for multi-agent dynamic systems: algorithms and theory. IEEE Trans. Autom. Control **51**(3), 401–420 (2006)
4. H. Su, X. Wang, Z. Lin, Flocking of multi-agents with a virtual leader. IEEE Trans. Autom. Control **54**(2), 293–307 (2009)
5. H. Su, Flocking in multi-agent systems with multiple virtual leaders based only on position measurements. Commun. Theor. Phys. **57**(5), 801–807 (2012)

6. Y. Hongyong, W. Fuyong, H. Fujun, Containment control of fractional order multi-agent systems with time delays. IEEE/CAA J. Autom. Sinica **5**(3):727–732. http://ieeexplore.ieee.org/stamp/stamp.jsp?tp=&arnumber=7783963 (2018)
7. H.Y. Yang, Y. Yize, H. Fujun, Z. Mei, G. Lei, Containment control of heterogeneous fractional-order multi-agent systems. J. Franklin Instit. (2017). https://doi.org/10.1016/j.jfranklin.2017.09.034
8. Y. Hongyong, H. Fujun, Z. Mei, Z. Shuning, Y. Jun, Distributed containment control of heterogeneous fractional-order multi-agent systems with communication delays. Open Phys. **15**, 509–516 (2017)
9. Y. Hongyong, Z. Zhenxing, Z. Siying, Consensus of second-order multi-agent systems with exogenous disturbances. Int. J. Robust Nonlin. Control **21**(9), 945–956 (2011)
10. S.H. Li, J. Yang, W.H. Chen et al. Disturbance Observer-Based Control: Methods and Applications. CRC Press, Inc. (2014)
11. L. Yuling, Y. Hongyong, L. Fan, Y. Yize, Finite-time containment control of second-order multi-agent systems with unmatched disturbances. Acta Autom. SINICA (2018). https://doi.org/10.16383/j.aas.2018.c170571
12. H.K. Khalil, *Nonlinear Systems*, 3rd edn. (Prentice Hall, New York, 2002)

# Multi-objective Robust Output Feedback Control for Receiver Station-Keeping in Boom and Receptacle Refueling

**Liang Chang and Yingmin Jia**

**Abstract** This paper considers the multi-objective reliable robust output feedback control problem for receiver station-keeping in boom and receptacle refueling (BRR), which considers the main features of BRR, i.e., mass and inertia variation of receiver aircraft, sensors failure, input constraints and disturbance attenuation. A new receiver aircraft model is firstly established in terms of those main features of BRR. Then, a new multiple objectives robust output feedback controller is designed for this control problem. The controller's existence is derived by using the Lyapunov method and linear matrix inequalities (LMIs) technique; and then the desired controller can be achieved by the LMI tools. A practice example is presented to demonstrate that the proposed controller design method can successfully solve the multi-objective control problem.

**Keywords** Robust output feedback control · Mass and inertia variation
Sensors failure

## 1 Introduction

This paper considers the receiver station-keeping problem in BRR. There have been plenty of work on this problem during the last decades [1–11], however, challenges still remain.

One of the challenges ignored by most of the previous work in BRR is sensors failure. The relative position and orientation between aerial refueling aircrafts are measured by machine vision-based techniques [7, 12, 13], these techniques often

---

L. Chang · Y. Jia (✉)
School of Automation Science and Electrical Engineering,
The Seventh Research Division and the Center for Information and Control,
Beihang University (BUAA), Beijing 100191, China
e-mail: ymjia@buaa.edu.cn

L. Chang
e-mail: chlchang01@163.com

 
Y. Jia et al. (eds.), *Proceedings of 2018 Chinese Intelligent Systems Conference*,
Lecture Notes in Electrical Engineering 528,
https://doi.org/10.1007/978-981-13-2288-4_74

need some assumptions which can not be satisfied in practise [12, 14]. Thus, it is necessary to design a controller that can tolerate sensors failure. Furthermore, many machine vision-based techniques can not measure the relative velocity between aerial refueling aircrafts, therefore a output feedback scheme is also essential.

Besides sensors failure, input constraints is also a challenge for this control problem. In BRR, it may cause plane collision or refueling equipment damage if the control input (such as thrust) is beyond the limit or there exists external disturbance (turbulence). Thus, control input constraints and disturbance attenuation ought to be taken into account in BRR. So far, some researchers have studied the control input constraints problem in flight control system [15, 16].

We concern the multiple objectives robust output feedback control problem for receiver station-keeping in this paper. The outstanding features of this paper are: (1) Compared with other flight control literatures, this paper considers the receiver's mass variation; (2) Present a new multi-objective control problem for receiver station-keeping in BRR; this control problem has more engineering signification than existing control problems because the control objects, i.e. mass and inertia variation, sensors failure, input constraints and external disturbance, are all the actual requirements of BRR.

This paper is organized as follows: Sect. 2 formulates the multi-objective control problem studied in this paper; Sect. 3 solves this control problem and presents the multiple objectives robust output feedback controller; a illustrative instance is presented in Sect. 4 and Sect. 5 has the conclusions.

## 2 Problem Formulation

The center of gravity of the receiver is assumed to be invariant. The perturbed velocity, flight path angle of attack, pitch angular velocity, flight path angle, slant range and altitude are defined as $\delta V_K$, $\delta\alpha_K$, $\delta q_K$, $\delta\gamma$, $\delta R$, $\delta H$, respectively; the perturbed elevator deflection and throttle position are defined as $\delta_\eta$ and $\delta_F$. Setting $\boldsymbol{x} = [\delta q_K, \delta\alpha_K, \delta V_K, \delta\gamma, \delta H, \delta R]^T$ as state, $\boldsymbol{u} = [\delta_F, \delta_\eta]^T$ as control input, $\boldsymbol{w} = [w_1, w_2, w_3]^T$ as external disturbance and $z = [\delta\gamma, \delta H, \delta R]^T$ as output. Then in the body coordinate, the dynamic equations of receiver is written as [12]

$$\begin{cases} \dot{\boldsymbol{x}} = A\boldsymbol{x} + B\boldsymbol{u} + B_w\boldsymbol{w} \\ z = C_0\boldsymbol{x}, \end{cases} \tag{1}$$

where

$$A = \begin{bmatrix} M_q & M_\alpha & M_u & 0 & 0 & 0 \\ 1 & Z_\alpha & Z_u & 0 & 0 & 0 \\ 0 & X_\alpha - g & X_u & -g & 0 & 0 \\ 0 & -Z_\alpha & -Z_u & 0 & 0 & 0 \\ 0 & 0 & 0 & V_0 & 0 & 0 \\ 0 & 0 & -1 & 0 & 0 & 0 \end{bmatrix}, B = \begin{bmatrix} M_f & M_\eta \\ Z_f & Z_\eta \\ X_f & X_\eta \\ -Z_f & -Z_\eta \\ 0 & 0 \\ 0 & 0 \end{bmatrix}, B_w = |X_u| \begin{bmatrix} 1 & 0 & 0 \\ 0 & 1 & 0 \\ 0 & 0 & 1 \\ 0 & -1 & 0 \\ 0 & 0 & 0 \\ 0 & 0 & 0 \end{bmatrix},$$

$C_0 = [0, I]_{6\times 3}$.

The following features of BRR should be considered in our control problem.

(1) Mass and Inertia Variation: As the analysis of [17], the coefficient matrices of (1) have the following decompositions:

$$A = A_0 + \Delta_A = A_0 + M_A F_A N_A, \tag{2}$$

$$B = B_0 + \Delta_B = B_0 + M_B F_B N_B, \tag{3}$$

where the expressions of $A_0$, $M_A$, $F_A$, $N_A$, $B_0$, $M_B$, $F_B$, $N_B$ can be found in [17].

(2) Sensors Failure: Define measured output vector $\boldsymbol{y} = [\delta\gamma', \delta H', \delta R']^T$. Due to the measuring error of sensors, system (1) can be rewritten as

$$\begin{cases} \dot{\boldsymbol{x}} = A\boldsymbol{x} + B\boldsymbol{u} + B_w\boldsymbol{w}, \\ \boldsymbol{y} = C\boldsymbol{x}, \\ \boldsymbol{z} = C_0\boldsymbol{x}, \end{cases} \tag{4}$$

where $C = C_0 + \Delta_C = M_C F_C N_C, M_C = I, F_C = \frac{1}{\eta_C} diag\{\delta_{C_1}, \delta_{C_2}, \delta_{C_3}\}, N_C = \eta_C C_0$, $\delta_{C_i} \le \eta_C$, for $1 \le i \le 3$.

(3) Output Feedback: Adopt a output feedback controller with the following structure

$$\begin{cases} \dot{\boldsymbol{x}}_c = A_c\boldsymbol{x}_c + B_c\boldsymbol{y}, \\ \boldsymbol{u} = C_c\boldsymbol{x}_c, \end{cases} \tag{5}$$

where $\boldsymbol{x}_c = [x_{c_1}, \ldots, x_{c_6}]^T$ is the state vector of controller (5), $A_c$, $B_c$ and $C_c$ will be presented later.

(4) Input Constraint: The control input $\boldsymbol{u}$ has the following constraints

$$\begin{cases} |u_1| \le u_{max_1}, \\ |u_2| \le u_{max_2}, \end{cases} \tag{6}$$

where $u_1 = \delta_F$, $u_2 = \delta_\eta$, $u_{max_1} > 0$, $u_{max_2} > 0$.

From the discussion above, we can construct the close-loop system of receiver. Define $\bar{\boldsymbol{x}} = [\boldsymbol{x}^T, \boldsymbol{x}_c^T]^T$ as the state vector of the closed-loop system. Apply controller (5) to system (4) and consider Eqs. (2) and (3), the closed-loop system is:

$$\begin{cases} \dot{\bar{x}} = \bar{A}\bar{x} + \bar{B}_w w \\ z = \bar{C}\bar{x}, \end{cases} \tag{7}$$

where $\bar{A} = \bar{A}_0 + \bar{\Delta}_A = \bar{A}_0 + \bar{\Delta}_{A_1} + \bar{\Delta}_{A_2} + \bar{\Delta}_{A_3}$,

$$\bar{A}_0 = \begin{bmatrix} A_0 & B_0 C_c \\ B_c C_0 & A_c \end{bmatrix}, \ \bar{\Delta}_A = \begin{bmatrix} \Delta_A & \Delta_B C_c \\ B_c \Delta_C & 0 \end{bmatrix}, \ \bar{B}_w = \begin{bmatrix} B_w \\ 0 \end{bmatrix}, \ \bar{C}^T = \begin{bmatrix} C_0 \\ 0 \end{bmatrix},$$

$\bar{\Delta}_{A_1} = \bar{M}_A F_A \bar{N}_A, \bar{\Delta}_{A_2} = \bar{M}_B F_B \bar{N}_B, \bar{\Delta}_{A_3} = \bar{M}_C F_C \bar{N}_C, \bar{M}_A^T = [M_A \ 0], \bar{N}_A = [N_A \ 0]$, $\bar{M}_B^T = [M_B \ 0], \bar{N}_B = [0 \ N_B C_c], \bar{M}_C^T = [0 \ B_c M_C], \bar{N}_C = [N_C \ 0]$.

Besides, the control input vector $u$ in controller (5) can be rewritten as

$$u = K\bar{x} = \begin{bmatrix} 0 & C_c \end{bmatrix} \bar{x}, \tag{8}$$

Then the proposed multiple objectives control problem is as follows: find a output feedback controller of the form (5) such that the closed-loop system (7) achieves the following objectives:

(i) It is asymptotically stable with sensors failure and mass variation, the system transfer function $T_{wz}$ satisfies

$$\|T_{wz}\|_\infty < \gamma_0, \quad \gamma_0 > 0 \tag{9}$$

(ii) The control input constraints (6) hold.

## 3 Multi-objective Reliable Robust Output Feedback Control

This section will address the multi-objective control problem mentioned above for receiver aircraft in BRR. Lemma 1 gives the conditions under which the close-loop system (7) achieves objective (i) mentioned above.

**Lemma 1** *Given scalar $\gamma_0 > 0$, the closed-loop system (7) achieves objective (i), if there exist $\hat{A}$, $\hat{B}$, $\hat{C}$, symmetric matrices $Y$, $S$ and constants $\gamma_A > 0$, $\gamma_B > 0$, $\gamma_C > 0$ such that inequalities*

$$\begin{bmatrix} \Xi_{Y_{11}} & \Xi_{Y_{12}} & B_w & SC_0^T & M_A & SN_A^T & M_B & \hat{C}^T N_B^T & 0 & SN_C^T \\ * & \Xi_{Y_{22}} & YB_w & C_0^T & YM_A & N_A^T & YM_B & 0 & \hat{B} & N_C^T \\ * & * & -\gamma_0 I & 0 & 0 & 0 & 0 & 0 & 0 & 0 \\ * & * & * & -\gamma_0 I & 0 & 0 & 0 & 0 & 0 & 0 \\ * & * & * & * & -\frac{1}{\gamma_A} I & 0 & 0 & 0 & 0 & 0 \\ * & * & * & * & * & -\gamma_A I & 0 & 0 & 0 & 0 \\ * & * & * & * & * & * & -\frac{1}{\gamma_B} I & 0 & 0 & 0 \\ * & * & * & * & * & * & * & -\gamma_B I & 0 & 0 \\ * & * & * & * & * & * & * & * & -\frac{1}{\gamma_C} I & 0 \\ * & * & * & * & * & * & * & * & * & -\gamma_C I \end{bmatrix} < 0 \quad (10)$$

$$G_0 = \begin{bmatrix} -S & -I \\ -I & -Y \end{bmatrix} < 0 \quad (11)$$

*hold, where* $\Xi_{Y_{11}} = A_0 S + B_0 \hat{C} + SA_0^T + \hat{C}^T B_0^T$, $\Xi_{Y_{12}} = A_0 + \hat{A}^T$, $\Xi_{Y_{22}} = YA_0 + \hat{B}C_0 + A_0^T Y + C_0^T \hat{B}^T$.

*Proof* From Jia [14] and Boyd et al. [18], the close-loop system (7) achieves objective (i) if there has a symmetric matrix $P > 0$ such that

$$\Xi_P + \widetilde{M}_A F_A \widetilde{N}_A + \widetilde{N}_A^T F_A^T \widetilde{M}_A^T + \widetilde{M}_B F_B \widetilde{N}_B + \widetilde{N}_B^T F_B^T \widetilde{M}_B^T + \widetilde{M}_C F_C \widetilde{N}_C + \widetilde{N}_C^T F_C^T \widetilde{M}_C^T < 0 \quad (12)$$

holds, where

$$\Xi_P = \begin{bmatrix} \bar{A}_0^T P + P\bar{A}_0 & P\bar{B}_w & \bar{C}^T \\ * & -\gamma_0 I & 0 \\ * & * & -\gamma_0 I \end{bmatrix}, \ \widetilde{M}_A^T = \begin{bmatrix} \bar{M}_A^T P \ 0 \ 0 \end{bmatrix}, \widetilde{N}_A = \begin{bmatrix} \bar{N}_A \ 0 \ 0 \end{bmatrix},$$

$$\widetilde{M}_B^T = \begin{bmatrix} \bar{M}_B^T P \ 0 \ 0 \end{bmatrix}, \widetilde{N}_B = \begin{bmatrix} \bar{N}_B \ 0 \ 0 \end{bmatrix}, \ \widetilde{M}_C^T = \begin{bmatrix} \bar{M}_C^T P \ 0 \ 0 \end{bmatrix}, \widetilde{N}_C = \begin{bmatrix} \bar{N}_C \ 0 \ 0 \end{bmatrix}.$$

For any scalars $\gamma_A > 0, \gamma_B > 0, \gamma_C > 0$, (12) is ensured if inequality

$$\begin{bmatrix} \Xi_{P1} & P\bar{B}_w & \bar{C}^T \\ * & -\gamma_0 I & 0 \\ * & * & -\gamma_0 I \end{bmatrix} < 0 \quad (13)$$

holds, where $\Xi_{P1} = \bar{A}_0^T P + P\bar{A}_0 + \gamma_A P\bar{M}_A\bar{M}_A^T P + \gamma_A^{-1}\bar{N}_A^T\bar{N}_A + \gamma_B P\bar{M}_B\bar{M}_B^T P + \gamma_B^{-1}\bar{N}_B^T\bar{N}_B + \gamma_C P\bar{M}_C\bar{M}_C^T P + \gamma_C^{-1}\bar{N}_C^T\bar{N}_C$. By the Schur complement, (13) can be obtained if

$$\begin{bmatrix} \bar{A}_0^T P + P\bar{A}_0 & P\bar{B}_w & \bar{C}^T & P\bar{M}_A & \bar{N}_A^T & P\bar{M}_B & \bar{N}_B^T & P\bar{M}_C & \bar{N}_C^T \\ * & -\gamma_0 I & 0 & 0 & 0 & 0 & 0 & 0 & 0 \\ * & * & -\gamma_0 I & 0 & 0 & 0 & 0 & 0 & 0 \\ * & * & * & -\frac{1}{\gamma_A} I & 0 & 0 & 0 & 0 & 0 \\ * & * & * & * & -\gamma_A I & 0 & 0 & 0 & 0 \\ * & * & * & * & * & -\frac{1}{\gamma_B} I & 0 & 0 & 0 \\ * & * & * & * & * & * & -\gamma_B I & 0 & 0 \\ * & * & * & * & * & * & * & -\frac{1}{\gamma_C} I & 0 \\ * & * & * & * & * & * & * & * & -\gamma_C I \end{bmatrix} < 0 \quad (14)$$

holds. Next we define $P$, $P^{-1}$ and a transform matrix $G_1$ as

$$P = \begin{bmatrix} Y & Z \\ Z^T & W \end{bmatrix}, \; P^{-1} = \begin{bmatrix} S & T \\ T^T & U \end{bmatrix}, \; G_1 = \begin{bmatrix} S & I \\ T^T & 0 \end{bmatrix}, \quad (15)$$

and define new matrix variables $\hat{A}$, $\hat{B}$ and $\hat{C}$ as

$$\begin{aligned} \hat{A} &= YA_0S + ZB_cC_0S + YB_0C_cT^T + ZA_cT^T, \\ \hat{B} &= ZB_c, \quad \hat{C} = C_cT^T, \end{aligned} \quad (16)$$

then use $diag\{G_1^T, I, I, I, I, I, I, I, I\}$ and its transpose premultiply and postmultiply (14), thus we have inequality (14) is feasible if inequality (10) is feasible.

Notice that the above prove requires $P > 0$. According to (11) and

$$P = -(G_1^T)^{-1} G_0 G_1^{-1}, \quad (17)$$

we can meet this requirement. This completes the proof.

Lemma 2 considers the input constraint problem of the close-loop system (7).

**Lemma 2** *Consider the closed-loop system (7). Given the initial state* $\bar{\boldsymbol{x}}(0) = \left[\boldsymbol{x}(0)^T \;\; 0\right]^T$, *positive scalars* $u_{max_1}$, $u_{max_2}$, *the symmetric matrices* $Y$, $S$ *and matrix* $\hat{C}$ *proposed in Lemma 1. Then, the closed-loop system (7) achieves objective (ii), if there exists a constant* $\rho > 0$ *such that inequalities*

$$\begin{bmatrix} -u_{max_j}^2 & -\hat{C}_j & 0 \\ * & -\frac{1}{\rho} S & -\frac{1}{\rho} I \\ * & * & -\frac{1}{\rho} Y \end{bmatrix} \le 0, \quad j = 1, 2; \quad (18)$$

$$\begin{bmatrix} -\rho & -\boldsymbol{x}(0)^T & -\boldsymbol{x}(0)^T Y \\ * & -S & -I \\ * & * & -Y \end{bmatrix} \le 0 \quad (19)$$

*hold, where* $\hat{C}_1 = [\,1\; 0\,]\hat{C}$, $\hat{C}_2 = [\,0\; 1\,]\hat{C}$.

*Proof* The closed-loop system (7) achieves objective (ii), if the following inequalities

$$\bar{x}^T K_j^T K_j \bar{x} \le u_{max_j}^2, \quad j = 1, 2; \tag{20}$$

hold, where $K_1 = [\,1\;0\,]K$, $K_2 = [\,0\;1\,]K$. Equation (20) can be ensured by

$$\frac{\bar{x}^T K_j^T K_j \bar{x}}{u_{max_j}^2} \le \frac{\bar{x}^T P \bar{x}}{\rho}, \quad j = 1, 2; \tag{21}$$

$$\frac{\bar{x}^T P \bar{x}}{\rho} \le 1, \tag{22}$$

where $P$ can be found in (15), $\rho$ is a positive constant.

It can be seen that (21) can be guaranteed if the following inequalities

$$\begin{bmatrix} -u_{max_j}^2 & -K_j \\ * & -\frac{1}{\rho}P \end{bmatrix} \le 0, \quad j = 1, 2; \tag{23}$$

hold. Use $diag\{I, G_1^T\}$ and its transpose premultiply and postmultiply (23), then we have (23) hold if (18) hold.

From [19], inequality (22) can be ensured by the following inequality,

$$\begin{bmatrix} -\rho I & -\bar{x}^T(0) \\ * & -P^{-1} \end{bmatrix} \le 0. \tag{24}$$

Define

$$G_2 = \begin{bmatrix} I & Y \\ 0 & Z^T \end{bmatrix}, \tag{25}$$

Use $diag\{I, G_2^T\}$ and its transpose premultiply and postmultiply (24), then we have (24) is ensured by (19). This completes the proof.

Based on lemmata 1 and 2, the following theorem solves the multi-objective control problem mentioned in the last section, and presents the multiple objectives robust output feedback controller.

**Theorem 1** *Consider refueling aircraft system (4), output feedback controller (5) and their closed-loop system (7). Given the initial state of the closed-loop system (7)* $\bar{x}(0) = \left[x(0)^T\; 0\right]^T$, *positive scalars* $\gamma_0$, $u_{max_1}$, $u_{max_2}$. *Then system (7) is asymptotically stable with mass variation, sensors failure and input constraints (6), the closed-loop transfer function satisfies* $\|T_{wz}\|_\infty < \gamma_0$, *if there exist* $\hat{A}$, $\hat{B}$, $\hat{C}$, *symmetric matrices* $Y$, $S$ *and constants* $\gamma_A > 0$, $\gamma_B > 0$, $\gamma_C > 0$, $\rho > 0$ *such that (10–11)(18–19) are feasible. Furthermore, the parameter matrices of controller (5)* $A_c$, $B_c$, $C_c$ *are given by*

$$A_c = Z^{-1}(\hat{A} - YA_0S - ZB_cC_0S - YB_0C_cT^T)(T^T)^{-1},$$
$$B_c = Z^{-1}\hat{B}, \quad C_c = \hat{C}(T^T)^{-1}. \tag{26}$$

From lemmata 1 and 2, it is obvious that system (7) meets the multi-objectives if inequalities (10–11) (18–19) hold. And from Eq. (16), we can obtain those parameter matrices $A_c$, $B_c$, $C_c$ by Eq. (26). This completes the proof.

## 4 Illustrative Example

To illustrate how to solve the receiver aircraft control problem by the proposed method, we present the following example. Consider a A300 aircraft as the receiver aircraft and make some assumptions as follows. The linear models of the receiver aircraft is proposed in [12, 17]. The receiver has a level and straight flight at 3000 m altitude, its velocity is 131.5 m/s. At initial time, the receiver locates 3 m below, 0.5 m behind the contact location. Then, the receiver flies to the contact location and begins refueling. It is supposed that the receiver's mass rises to 140 tones from 120 tones during aerial refueling. The following assumptions are also necessary:

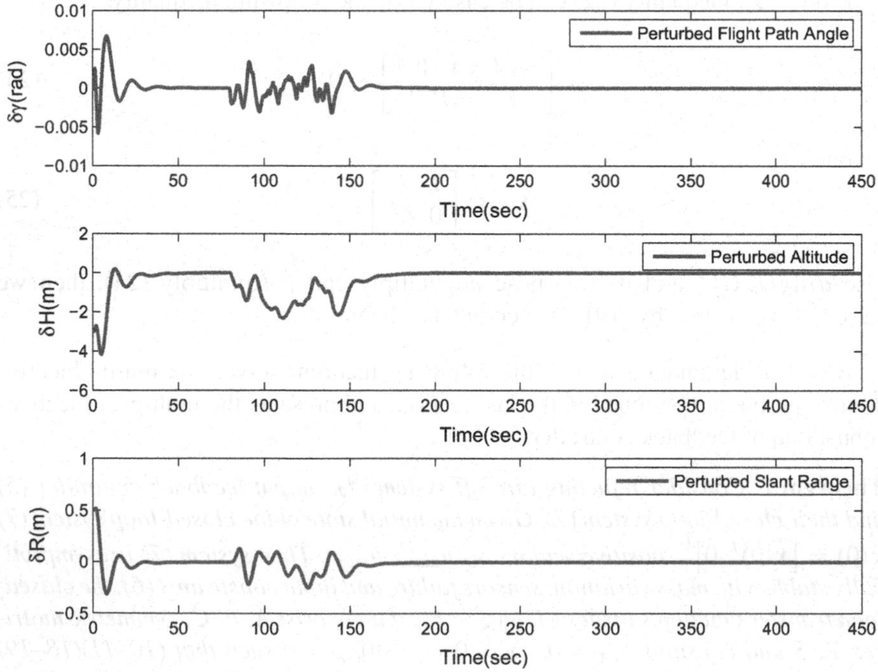


**Fig. 1** Output of the longitudinal closed-loop system

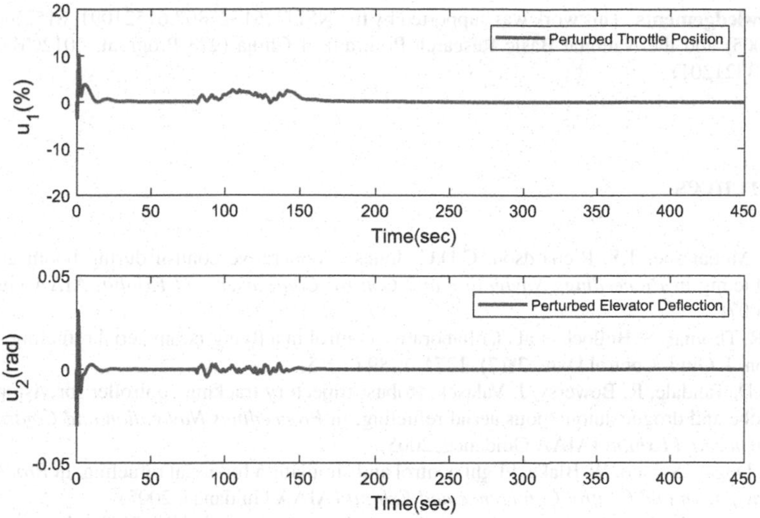


**Fig. 2** Longitudinal control input

the longitudinal uncertainty parameter is $\eta_m = 0.1$ and $\eta_C = 0.05$; the longitudinal control input need satisfy $|\boldsymbol{u}_1| \leq 0.14$, $|\boldsymbol{u}_2| \leq 0.17$.

The simulation results are shown in Figs. 1 and 2. Figure 1 illustrate the perturbed slant range and altitude of the receiver longitudinal system. As shown in figure, the longitudinal system is asymptotically stable with sensors failure and mass variation by the presented controller, and the system transfer function has $\|T_{wz}\|_{\infty} < 2.78$.

Figure 2 depicts the receiver longitudinal input. It is obvious that the input of the receiver longitudinal system satisfies the input restrictions, which means that the proposed controller make the closed-loop system achieves objective (ii).

## 5 Conclusions

This paper has addressed the reliable robust output feedback control problem of receiver in boom and receptacle refueling. A new receiver aircraft system has been formed firstly in terms of the multi-objects, i.e., mass and inertia variation of receiver aircraft, sensors failure, input constraints and disturbance attenuation. Then, by using the Lyapunov method, a reliable robust output feedback controller has been designed for the proposed control problem; the LMI approach has been used to convert the controller design into convex optimization problem, and then the desired controller has been obtained by the LMI tools. Finally, An illustrative example has been presented to show the availability of the proposed controller design method.

**Acknowledgements** This work was supported by the NSFC (61327807,61521091, 61520106010, 61134005) and the National Basic Research Program of China (973 Program: 2012CB821200, 2012CB821201).

## References

1. C. McFarlane, T.S. Richardson, C.D.C. Jones, Cooperative control during boom air-to-air refueling in *Proceedings Navigation and Control Conference and Exhibit*, AIAA Guidance (2007)
2. P.R. Thomas, S. Bullock et al., Collaborative control in a flying-boom aerial refueling simulation. J. Guid. Control Dyn., **38**(7), 1274–1289 (2015)
3. M.D. Tandale, R. Bowersy, J. Valasek, Robust trajectory tracking controller for vision based probe and drogue autonomous aerial refueling, in *Proceedings Navigation and Control Conference and Exhibit* (AIAA Guidance, 2005)
4. A. Dogan, S. Sato, W. Blake, Flight control and simulation for aerial refueling, in *Proceedings Navigation and Control Conference and Exhibit* (AIAA Guidance, 2005)
5. C.M. Elliott, A. Dogan, Improving receiver station-keeping in aerial refueling by formulating tanker motion as disturbance, in *Proceedings AIAA Atmospheric Flight Mechanics Conference* (2009)
6. J. Valasek, K. Gunnam et al., Vision-based sensor and navigation system for autonomous air refueling. J. Guid. Control Dyn. **28**(5), 979–989 (2005)
7. J. Doebbler, T. Spaeth et al., Boom and receptacle autonomous air refueling using visual snake optical sensor. J. Guid. Control Dyn. **30**(6), 1753–1769 (2007)
8. C.M. Elliott, A. Dogan, Investigating nonlinear control architecture options for aerial refueling, in *Proceedings AIAA Atmospheric Flight Mechanics Conference* (2010)
9. J. Wang, N. Hovakimyan, C. Cao, Verifiable adaptive flight control: unmanned combat aerial vehicle and aerial refueling. J. Guid. Control Dyn. **33**(1), 75–87 (2010)
10. S. Venkataramanan, A. Dogan, Dynamic effects of trailing vortex with turbulence and time-varying inertia in aerial refueling, in *Proceedings AIAA Atmospheric Flight Mechanics Conference and Exhibit* (2004)
11. M. Pachter, C.H. Houpis, D.W. Trosen, Design of an air-to-air automatic refueling flight control system using quantitative feedback theory. Int. J. Robust Nonlin. **7**(6), 561–580
12. R. Brockhaus, W. Alles, R. Luckner, Flugregelung (Springer, Germany, 2011)
13. M.V. Basin, A.E. Rodkina, On delay-dependent stability for a class of nonlinear stochastic delay-difference equations.Dyn. Continuous, Discr. Impul. Syst. **12**(5), 663–675 (2005)
14. Y. Jia, *Robust* $H_\infty$ *control* (Science Press, Beijing, 2007)
15. S. Patra, S. Sen, G. Ray, Local stabilisation of uncertain linear time-invariant plant with bounded control inputs: parametric $H_\infty$ loop-shaping approach. IET Contr. Theory Appl. **6**(11), 1567–1576 (2012)
16. J. Fan, Y. Zhang, Z. Zheng, Robust fault-tolerant control against time-varying actuator faults and saturation. IET Contr. Theory Appl. **6**(14), 2198–2208 (2012)
17. L. Chang, Y. Jia, Robust $H_\infty$ control for tanker station-keeping with mass and inertia variation, in *Proceedings 13th IEEE Conference on Automation Science and Engineering* (2017)
18. S. Boyd, L.E. Ghaoui et al., *Linear matrix inequalities in systems and control theory* (SIAM, Philadelphia, PA, 1994)
19. H. Gao, X. Yang, P. Shi, Multi-objective robust $H_\infty$ control of spacecraft rendezvous. IEEE Trans. Control Syst. Technol. **17**(4), 794–802 (2009)

# Distributed Fixed-Time Consensus Algorithm for Multiple Nonholonomic Chained-Form Systems

**Yutao Jiang, Zhongxin Liu and Zengqiang Chen**

**Abstract** In this paper, the fixed-time control algorithm is used to address the consensus problem for multiple nonholonomic chained-form systems. For the sake of analysis, a switching control strategy is introduced to solve the fixed-time consensus problem. Compared with the finite-time control algorithm, the convergence time of the fixed-time consensus protocol, can be guaranteed regardless of the initial conditions. Rigorous proof using Lyapunov theory shows that the sates of multiple nonholonomic chained-form systems can reach a consensus in a fixed time. To further illustrate the effectiveness of the control algorithm, a numerical simulation is given.

**Keywords** Distributed control · Nonholonomic chained-form systems
Lyapunov theory · Finite-time consensus · Fixed-time consensus

## 1 Introduction

In recent years, with the development of swarm intelligence technology, the distributed consensus control problems of multi-agent systems have attracted considerable attention due to its wide applications, such as unmanned vehicles formation control [1], network synchronization [2, 3], rendezvous problem [4], attitude alignment [5], flocking control [6] and so forth. The mean of consensus control is to design an appropriate algorithm based on local information exchange such that groups of agents agree upon certain quantities of common interest. Due to the different types of systems, multi-agent consensus problem includes linear systems, nonlinear systems, continuous systems and discrete systems. The consensus problems in the process of

Y. Jiang · Z. Liu (✉) · Z. Chen
College of Computer and Control Engineering, Nankai University, Tianjin 300353, China
e-mail: lzhx@nankai.edu.cn

Y. Jiang · Z. Liu · Z. Chen
Tianjin Key Laboratory of Intelligent Robotics, Nankai University, Tianjin 300353, China



Y. Jia et al. (eds.), *Proceedings of 2018 Chinese Intelligent Systems Conference*,
Lecture Notes in Electrical Engineering 528,
https://doi.org/10.1007/978-981-13-2288-4_75

network communication with switching topology and time-delays have been studied in [7] and [8].

In the practical applications, the convergence rate for many closed-loop systems is always regarded as an important index to evaluate the consensus behavior of some control algorithm. However, so far, the majority of consensus control algorithms just can keep the multi-agent systems asymptotically stable or exponential stable in infinite time. Therefore, in order to improve the convergence rate, the finite-time control algorithm was proposed in [9]. Besides, [10] demonstrates that the finite-time control algorithm also has a better disturbances rejection properties for the closed-loop systems. Due to the advantages mentioned above, the finite-time control algorithm was used to solve the consensus problem of multi-agent systems with single-integrator and double-integrator dynamics in [11] and [12], respectively. In [13], the numerical simulation shows that the formation control with finite-time control algorithm has a better accuracy of tracking then the traditional one. Up to now, the finite-time control algorithm has been applied in many fields, however, the convergence time or the settling time is upper bounded depends on the initial conditions. In some leader-follower consensus problems, the followers that cannot directly access the leader information usually need observers to estimate leader's states, then the settling time is always regarded as the moment for observers to track certain state. Such as, the finite-time observer, which tracks the states of virtual leader was proposed in [14, 15] and the finite-time observers will track the objective states completely during a finite time. However, in most of literatures, the settling time got from the figures of simulations is not a precise way, and when the initial states of agents are some big values, the boundary of the convergence time is also a value, which is much bigger than the actual settling time. Therefore, how to get more precise time is an important problem.

Fixed-time control algorithm, proposed in [16], makes the boundary of settling time independent from initial conditions. Therefore, the fixed-time consensus control can provide a more precise estimation of the settling time than the finite-time control. In the latest literatures, the fixed-time control has been applied to the single-integrator multi-agent networks in [17], and Liu et al. [18] used the fixed-time control algorithm to solve the event-triggered consensus control with nonlinear uncertainties. In reality, many mathematical models of multi-agent systems are more complicated than the single or double integrator models. Such as the nonholonomic mobile robots and unmaned vehicles are all the nonholonomic dynamic systems. Because, many nonholonomic dynamic systems can be transformed into the nonholonomic chained-form system, it's significant to solve the consensus problem of multiple nonholonomic chained-form system.

In this paper, the fixed-time control algorithm will be applied in the consensus problem for the multiple nonholonomic chained-form systems. The model of each agent is the nonholonomic chained-form dynamics. Inspired by [1], the main system can be divided into two subsystems, single-integrator case and a double-integrator case. For the double-integrator subsystems, a virtual state is introduced to design the fixed-time control algorithm. When the states of the the double-integrator subsystem achieve consensus within a fixed time, another control algorithm is proposed for the single-integrator subsystems achieve consensus. Finally, all the states of the nonholonomic chained-form systems achieve consensus in a fixed time.

The remainder of this paper is organised as follows. The preliminaries and problem formulation is given in Sect. 2. The fixed-time control for multiple nonholonomic chained-form system is proposed in Sect. 3. In Sect. 4, the numerical simulations are provided to verify the efficiency of the proposed method. Conclusions and future work are presented in Sect. 5.

## 2 Preliminaries and Problem Formulation

### 2.1 Graph Theory

Assuming that $G = (V, E, A)$ is an undirected weighted graph with $n$ nodes, where $V = \{\nu_1, \nu_2, \ldots, \nu_n\}$ represents the set of nodes, and $E \subseteq V \times V$ represents the set of edges. $A = (a_{ij})_{n\times n}$ is the weighted adjacency matrix of $G$. $a_{ij} > 0 \Longleftrightarrow (\nu_j, \nu_i) \in \varepsilon$ which means that node $j$ is connected with node $i$ and can obtain information from $i$. Obviously, we also have $(\nu_j, \nu_i) \notin E \Longleftrightarrow a_{ij} = 0$. It is assumed that node cannot be available from itself, i.e., $a_{ii} = 0$ for all $i \in V$. The neighbor set of the node $i$ is denoted by $N_i = \{j \in V : (\nu_j, \nu_i) \in E\}$. The Laplacian matrix of $G$ is defined by $L = D - A$, where $D = diag\{d_1, d_2, \ldots, d_n\}$ and $d_i = \sum_{j\in N_i} a_{ij}$. In the graph $G$, if any two distinct nodes are connected via a path, then $G$ is called connected. $R_+ = \{x > 0 : x \in R\}$, $N_+ = N \setminus \{0\}$. $\mathbf{1}$ is a column vector and its entries are all number 1, $\mathbf{0}$ is a zero column vector.

### 2.2 Some Lemmas and Definitions

**Lemma 1** *[7] If an undirected graph G is connected, these following properties for the Laplacian matrix L are satisfied.*
*(1)* $\forall x \in R^n$, $x^T Lx = \frac{1}{2}\sum_{i=1}^{n}\sum_{j=1}^{n} a_{i,j}\left(x_i - x_j\right)^2$, *which implies L is semidefinite. Therefore, all the eigenvalues of L are real and not less than zero.* $\mathbf{1}$ *is an eigenvector of L, which is associated with 0 eigenvalue.*
*(2)* $\lambda_2$ *denotes the second smallest eigenvalue of L. Furthermore,*

$$\lambda_2 = \min_{x\neq 0,\ \mathbf{1}^T x=0}\left(\frac{x^T Lx}{x^T x}\right) > 0.$$

*If* $\mathbf{1}^T x = 0$, *then* $x^T Lx \geq \lambda_2(L)\, x^T x$.

*Consider the nonlinear system:*

$$\dot{x} = f(x),\ x(0) = x_0, \tag{1}$$

*where $x \in R^n$ and $f : R_+ \times R^n \rightarrow R^n$ is a continuous nonlinear function.*

**Definition 1** [16] $H$ denotes some set. If the state of system (1) reaches the set $H$ in a finite time $T(x_0)$, $H$ is called the globally finite-time attractive. $T : R^n \rightarrow R_+ \cup \{0\}$ is a function with respect to the initial sate $x_0$.

**Definition 2** [16] $H$ denotes some set. For system (1), if $H$ is the globally finite-time attractive and the function $T(x_0)$ mentioned in Definition 1 is bounded by some value, $H$ is called fixed-time attractive.

According to the Lemma 2 of [16], Lemma 2 in this paper can be derived as follows: (The proof of Lemma 2 below has been mentioned in [19]).

**Lemma 2** *$V : R^n \rightarrow R_+ \cup 0$ is a continuous radially unbounded function and such that,*
*(1) $V(x) = 0 \Leftrightarrow x \in H$, $H$ is mentioned in Definition 2.*
*(2) For system (1), $V(x(t))$ satisfies the inequality*

$$DV(x(t)) \leq -\alpha V^p(x(t)) - \beta V^q(x(t)),$$

*where $\alpha, \beta > 0$, $p = 1 - \frac{1}{2\gamma}$, $q = 1 + \frac{1}{2\gamma}$ and $\gamma > 1$; then, the set $H \subset R^n$ is globally fixed-time attractive for (1) and the convergence time $T \leq \frac{\pi\gamma}{\sqrt{\alpha\beta}}$. When the condition (1) is changed to $V(x) = 0 \Leftrightarrow \dot{x} = 0$, the set $H$ is invariant. Further, $H = 0$, Lemma 3 can be used to analyze fixed-time stability of the origin.*

**Lemma 3** *[20] $\forall \varepsilon_j \in R$, $j = 1, 2, \ldots, S$, $S \in N_+$, $\kappa_1 \in (0, 1]$, $\kappa_2 \in (1, \infty)$, then*

$$\sum_{i=1}^{S} |\varepsilon_i|^{\kappa_1} \geq \left( \sum_{i=1}^{S} |\varepsilon_i| \right)^{\kappa_1},$$

$$\sum_{i=1}^{S} |\varepsilon_i|^{\kappa_2} \geq S^{1-\kappa_2} \left( \sum_{i=1}^{S} |\varepsilon_i| \right)^{\kappa_2}.$$

**Lemma 4** *[21] $\forall z \in R, \rho > 0$, the equation $\frac{d}{dz}|z|^{\rho+1} = (\rho + 1)sig(z)^\rho$ is valid. Let $sign()$ denotes the standard symbolic function, $sig(z)^\rho = sign(z)|z|^\rho$, then $\frac{d}{dz}sig(z)^{\rho+1} = (\rho + 1)|z|^\rho$.*

### 2.3 *Problem Formulation*

Let $\Gamma = \{1, 2, \ldots, n\}$, denotes the set of n agents. The model of each agent is described as follows

$$\begin{cases} \dot{q}_{1i} = u_{1i}, \\ \dot{q}_{2i} = q_{3i} u_{1i}, \\ \dot{q}_{3i} = u_{2i}, \qquad i \in \Gamma, \end{cases} \tag{2}$$

where $q_{1i}$, $q_{2i}$, $q_{3i}$ are the system states of the $i$ th agent, and $u_{1i}$, $u_{2i}$ are control inputs, $i \in \Gamma$. Throughout the paper, the following states vector will be used. $q_1 = [q_{11}, q_{12}, \ldots, q_{1n}]^T, q_2 = [q_{21}, q_{22}, \ldots, q_{2n}]^T, q_3 = [q_{31}, q_{32}, \ldots, q_{3n}]^T, u_1 = [u_{11}, u_{12}, \ldots, u_{1n}]^T$, $u_2 = [u_{21}, u_{22}, \ldots, u_{2n}]^T$, $\mathbf{1} = [1, 1, \ldots, 1]^T \in R^n$, $\mathbf{0} = [0, 0, \ldots, 0]^T \in R^n$.

**Assumption 1** In this paper, the communication topology among these agents is described by an undirected connected graph.

Under Assumption 1, this paper provide a consensus control algorithm. With this control algorithm, the fixed-time consensus problem of nonholonomic chained-form systems can be solved.

## 3 Main Results

According to the switching strategy [1], system (2) is divided into two subsystems i.e. the single-integrator subsystem (3) and the double-integrator subsystem (4) as follows:

$$\dot{q}_{1i} = u_{1i}, \quad i \in \Gamma \tag{3}$$

$$\begin{cases} \dot{q}_{2i} = q_{3i} u_{1i}, \\ \dot{q}_{3i} = u_{2i}, \quad i \in \Gamma. \end{cases} \tag{4}$$

Firstly, a fixed-time consensus algorithm is designed to make states of (4) reach consensus in a fixed time and states of (3) are bounded. Then, Theorem 1 is proposed.

**Theorem 1** *Under Assumption 1, $u_{1i}$ and $u_{2i}$ are designed as*

$$u_{1i} = k, \quad i \in \Gamma, \tag{5}$$

$$\begin{aligned} u_{2i} = & -c_1\alpha_1 \Big|\sum_{j\in N_i} a_{ij}(q_{2i} - q_{2j})\Big|^{\alpha_1 - 1} \sum_{j\in N_i} a_{ij}(q_{3i} - q_{3j}) \\ & - c_2\beta_1 \Big|\sum_{j\in N_i} a_{ij}(q_{2i} - q_{2j})\Big|^{\beta_1 - 1} \sum_{j\in N_i} a_{ij}(q_{3i} - q_{3j}) \\ & - c_3 sig(e_i)^p - c_4 sig(e_i)^q, \qquad i \in \Gamma. \end{aligned} \tag{6}$$

*Then the states of subsystem (4) achieve consensus in a fixed time, and the states of subsystem (3) are bounded. Here, $k \in R_+$, $e_i = q_{3i} - q_{3i}^*$, $i \in \Gamma$. $q_{3i}^*$ is a virtual state,* *which is defined as*

$$q_{3i}^* = -c_1 sig(\sum_{j\in N_i} a_{ij}(q_{2i}-q_{2j}))^{\alpha_1} - c_2 sig(\sum_{j\in N_i} a_{ij}(q_{2i}-q_{2j}))^{\beta_1}, \quad i \in \Gamma,$$

*$c_1$, $c_2$, $c_3$, $c_4$ are positive constants; $\alpha_1$, $p \in (0,1)$, $\beta_1 = 2-\alpha_1$, $q = 2-p$.*

*Proof* For a better illustration, the proof is processed by two steps.
**Step 1**: According to Lemma 4, take the derivative of $e_i$ with (5), (6), $i \in \Gamma$.

$$\begin{aligned}\dot{e}_i &= \dot{q}_{3i} - \dot{q}_{3i}^* \\ &= u_{2i} + c_1\alpha_1 |\sum_{j\in N_i} a_{ij}(q_{2i}-q_{2j})|^{\alpha_1-1} \sum_{j\in N_i} a_{ij}(q_{3i}-q_{3j}) \\ &\quad + c_2\beta_1 |\sum_{j\in N_i} a_{ij}(q_{2i}-q_{2j})|^{\beta_1-1} \sum_{j\in N_i} a_{ij}(q_{3i}-q_{3j}) \\ &= -c_3 sig(e_i)^p - c_4 sig(e_i)^q, \qquad i \in \Gamma.\end{aligned} \tag{7}$$

For the subsystem (7), choose the Lyapunov candidate function

$$V_1(e(t)) = \frac{1}{2}\sum_{i=1}^n e_i^2, \tag{8}$$

where $e(t) = [e_1, e_2, \ldots, e_n]^T$. Take the derivative of $V_1(e(t))$ by Lemma 3

$$\begin{aligned}\dot{V}_1(e(t)) &= \sum_{i=1}^n e_i\dot{e}_i = -c_3\sum_{i=1}^n |e_i|^{p+1} - c_4\sum_{i=1}^n |e_i|^{q+1} \\ &\le -c_3(\sum_{i=1}^n e_i^2)^{\frac{p+1}{2}} - c_4 n^{\frac{1-q}{2}}(\sum_{i=1}^n e_i^2)^{\frac{q+1}{2}} \\ &= -c_3 2^{\frac{p+1}{2}} V_1(e(t))^{\frac{p+1}{2}} - c_4 n^{\frac{1-q}{2}} 2^{\frac{q+1}{2}} V_1(e(t))^{\frac{q+1}{2}}.\end{aligned} \tag{9}$$

Due to $p \in (0,1)$, $q > 1$, then $\frac{p+1}{2} \in (0,1)$, $\frac{q+1}{2} > 1$. According to Lemma 2, $e_i$ converges to 0 in a fixed time $T_1$, i.e. $q_{3i}$ converges to $q_{3i}^*$ in a fixed time, $\forall i \in \Gamma$. $T_1$ satisfies

$$T_1 \le T_1^* := \frac{\pi\gamma_1}{\sqrt{c_3c_4 n^{\frac{1-q}{2}} 2^{\frac{p+q+2}{2}}}} = \frac{\pi\gamma_1}{2\sqrt{c_3c_4 n^{\frac{1-q}{2}}}}, \tag{10}$$

where $\gamma_1 > 0$ satisfies $p = 1 - \frac{1}{2\gamma_1}$, $q = 1 + \frac{1}{2\gamma_1}$.
**Step 2**: When $t > T_1^*$, $e_i = 0$, i.e. $q_{3i} = q_{3i}^* = -c_1 sign(\sum_{j\in N_i} a_{ij}(q_{2i}-q_{2j})) - c_2\beta_1 sig(\sum_{j\in N_i} a_{ij}(q_{2i}-q_{2j}))^{\beta_1}$, $i \in \Gamma$. Then, for the subsystem (4), consider the continuous radially unbounded function

$$V_2(q_2) = \frac{1}{2}q_2^T L q_2 = \frac{1}{4}\sum_{i=1}^{n}\sum_{j\in N_i} a_{ij}(q_{2i} - q_{2j})^2. \tag{11}$$

Let $s_i = \sum_{j\in N_i} a_{ij}(q_{2i} - q_{2j})$, $i \in \Gamma$. According to Lemma 3, take the derivative of $V_2(q_2)$ with (5), (6)

$$\begin{aligned}\dot{V}_2(q_2) &= q_2^T L q_2 = \sum_{i=1}^{n}\sum_{j\in N_i} a_{ij}(q_{2i} - q_{2j})\dot{q}_{2i} \\ &= -c_1 k \sum_{i=1}^{n}(s_i^2)^{\frac{1}{2}} - kc_2 \sum_{i=1}^{n}(s_i^2)^{\frac{1+\beta_1}{2}} \\ &\le -c_1 k\Big(\sum_{i=1}^{n} s_i^2\Big)^{\frac{1}{2}} - kc_2 n^{\frac{1-\beta_1}{2}}\Big(\sum_{i=1}^{n} s_i^2\Big)^{\frac{1+\beta_1}{2}}.\end{aligned} \tag{12}$$

Since $Lq_2 = [s_1, s_2, \dots, s_n]^T$, i.e. $\sum_{i=1}^{n} s_i^2 = (Lq_2)^T Lq_2 = q_2^T L^2 q_2$. Let $L^{\frac{1}{2}}\mathbf{1} = h = [h_1, h_2, \dots, h_n]^T$. Then, $h^T h = (L^{\frac{1}{2}}\mathbf{1})^T L^{\frac{1}{2}}\mathbf{1} = \mathbf{1}^T L \mathbf{1} = 0$, which implies $h = \mathbf{0}$, therefore, $\mathbf{1}^T L^{\frac{1}{2}} q_2 = h^T q_2 = 0$. By Lemma 1, then

$$\sum_{i=1}^{n} s_i^2 = (L^{\frac{1}{2}}q_2)^T L(L^{\frac{1}{2}}q_2) \ge \lambda_2 q_2^T L q_2 = 2\lambda_2 V_2(q_2). \tag{13}$$

Therefore, by (13), (14),

$$\dot{V}_2(q_2) \le -c_1 k(2\lambda_2)^{\frac{1}{2}} V_2(q_2)^{\frac{1}{2}} - c_2 k n^{\frac{1-\beta_1}{2}}(2\lambda_2)^{\frac{1+\beta_1}{2}} V_2(q_2)^{\frac{1+\beta_1}{2}} \tag{14}$$

Due to $\beta_1 > 1$, then $\frac{\beta_1+1}{2} > 1$. When $V_2(q_2) = 0$, $q_2 \in H$ mentioned in Lemma 2, and due to Lemma 1, $H$ is a consensus attractive. According to Lemma 2, the $q_2$ achieves consensus in a fixed time noted by $T_2$ and $T_2$ satisfies

$$T_2 \le T_2^* := \frac{2}{c_1 k\sqrt{2\lambda_2}} + \frac{2}{c_2 k n^{\frac{1-\beta_1}{2}}(2\lambda_2)^{\frac{1+\beta_1}{2}}(\beta_1 - 1)}. \tag{15}$$

When $t > (T_1^* + T_2^*)$, $q_{3i} = q_{3i}^*$, $i \in \Gamma$ and $q_2$ achieves consensus. Due to $q_{3i}^* = -c_1 sign(\sum_{j\in N_i} a_{ij}(q_{2i} - q_{2j})) - c_2\beta_1 sig(\sum_{j\in N_i} a_{ij}(q_{2i} - q_{2j}))^{\beta_1}$, $q_3$ achieves consensus. Moreover, $q_3 = \mathbf{0}$. Due to (3), when $t \in (0, T_1^* + T_2^*]$, $q_1$ is bounded.

To sum up, the control algorithms (5), (6) of Theorem 1 solve the fixed-time consensus problem of subsystem (4), and keep the state of subsystem (3) is bounded.

For the nonholonomic chained-form system (1), on the basis of Theorem 1, Theorem 2 is proposed as follows.

**Theorem 2** *Under Assumption 1, let $T_{switch} := T_1^* + T_2^*$, where $T_1^*$, $T_2^*$ are given in the proof of Theorem 1. When $t \le T_{switch}$, $u_1$ and $u_2$ are designed as (5), (6). When*

*$t > T_{switch}$, $u_2$ is the same as (6), $u_1$ are designed as follows*

$$u_{1i} = -c_5 sig(\sum_{j \in N_i} a_{ij}(q_{1i} - q_{1j}))^{\alpha_2} - c_6 sig(\sum_{j \in N_i} a_{ij}(q_{1i} - q_{1j}))^{\beta_2}, i \in \Gamma. \quad (16)$$

*Proof* According to Theorem 1, when $t \geq T_{switch}$, the control algorithms (5), (6) make the states $q_2$, $q_3$ converge in a fixed time, respectively, and $q_3 = \mathbf{0}$. From (4), $\dot{q}_2 = \mathbf{0}$, i.e., the change of $u_1$ has no effect on the states $q_2$, $q_3$. For the subsystem (3), choose the continuous radially unbounded function

$$V_4(q_1) = \frac{1}{2} q_1^T L q_1 = \frac{1}{4} \sum_{i=1}^{n} \sum_{j \in N_i} a_{ij}(q_{1i} - q_{1j})^2. \quad (17)$$

Let $g_i = \sum_{j \in N_i} a_{ij}(q_{1i} - q_{1j})$. Take the derivative of $V_4(q_1)$ with Lemma 3

$$\begin{aligned} \dot{V}_4(q_1) = q_1^T L q_1 &= \sum_{i=1}^{n} \sum_{j \in N_i} a_{ij}(q_{1i} - q_{1j}) \dot{q}_{1i} \\ &= -c_5 \sum_{i=1}^{n} (g_i^2)^{\frac{1+\alpha_2}{2}} - c_6 \sum_{i=1}^{n} (g_i^2)^{\frac{1+\beta_2}{2}} \\ &\leq -c_5 \Big(\sum_{i=1}^{n} g_i^2\Big)^{\frac{1+\alpha_2}{2}} - c_6 n^{\frac{1-\beta_2}{2}} \Big(\sum_{i=1}^{n} g_i^2\Big)^{\frac{1+\beta_2}{2}}. \end{aligned} \quad (18)$$

The remained proof is similar to the Step 2 in Theorem 1. Then we can obtain

$$\dot{V}_4(q_1) \leq -c_5 (2\lambda_2)^{\frac{1+\alpha_2}{2}} V_4(q_1)^{\frac{1+\alpha_2}{2}} - c_6 n^{\frac{1-\beta_2}{2}} (2\lambda_2)^{\frac{1+\beta_2}{2}} V_4(q_1)^{\frac{1+\beta_2}{2}} \quad (19)$$

Therefore, $q_1$ achieves consensus in a fixed time $T_4$, and $T_4$ satisfies

$$T_4 \leq T_4^* := \frac{\pi \gamma_3}{2\lambda_2 \sqrt{c_5 c_6 n^{\frac{1-\beta_2}{2}}}}, \quad (20)$$

where $\gamma_3 > 0$ satisfies $\alpha_2 = 1 - \frac{1}{2\gamma_3}$, $\beta_2 = 1 + \frac{1}{2\gamma_3}$.

In conclusion, Theorem 2 solves the fixed-time consensus problem of nonholonomic chained-form system (2), and makes all the states of system (2) achieve consensus in a fixed time notated by $T' := T + T_4^*$.

*Remark* The fixed-time consensus algorithm can get a more precise estimate of settling time than the finite-time control. Therefore, the fixed time control algorithm can be better used in many fields, such as fixed time observers, fixed time switch control strategies. When $c_2 = c_4 = c_6 = 0$, the control algorithm of Theorem 2 can solve the finite time consensus problem for the the multiple nonholonomic chained-form systems. Therefore the fixed time control algorithm is a more general result.

## 4 Numerical Simulation

In this section, the validity of the proposed control algorithm is verified by a numerical simulation.

Consider a multi-agents system with 5 agents. Its undirected commutation topology is shown in Fig. 1, where the weights of the edges are: $a_{12} = a_{21} = 0.5$, $a_{23} = a_{32} = 0.2$, $a_{25} = a_{52} = 0.3$, $a_{34} = a_{43} = 0.5$.
From the graph, the Laplacian matrix can be given:

$$L = \begin{bmatrix} 0.5 & 0 & -0.5 & 0 & 0 \\ -0.5 & 1 & -0.2 & 0 & -0.3 \\ 0 & -0.2 & 0.7 & -0.5 & 0 \\ 0 & 0 & -0.5 & 0.5 & 0 \\ 0 & -0.3 & 0 & 0 & 0.3 \end{bmatrix} \tag{21}$$

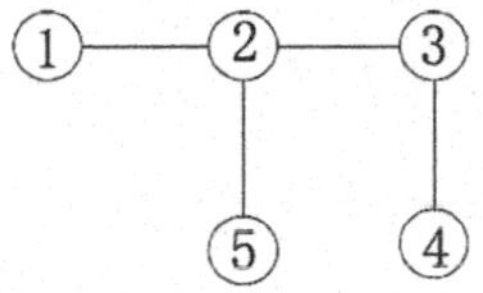


**Fig. 1** Figure of system communication topology

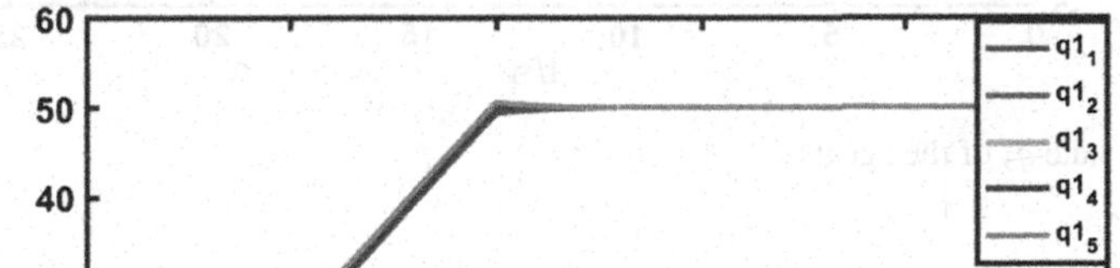

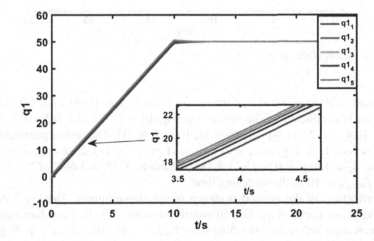


**Fig. 2** The state $q_1$ of the agents

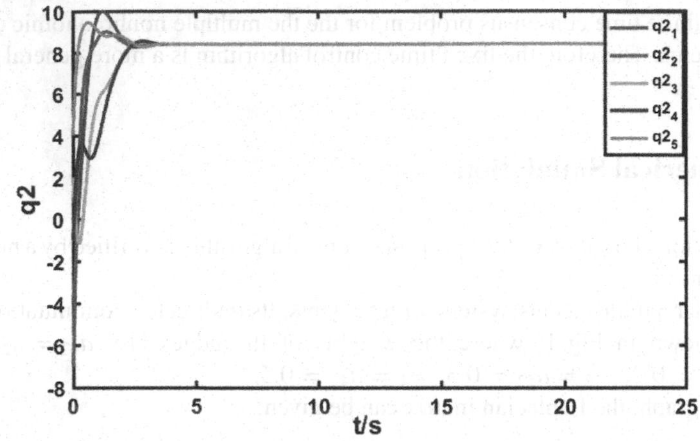


**Fig. 3** The state $q_2$ of the agents

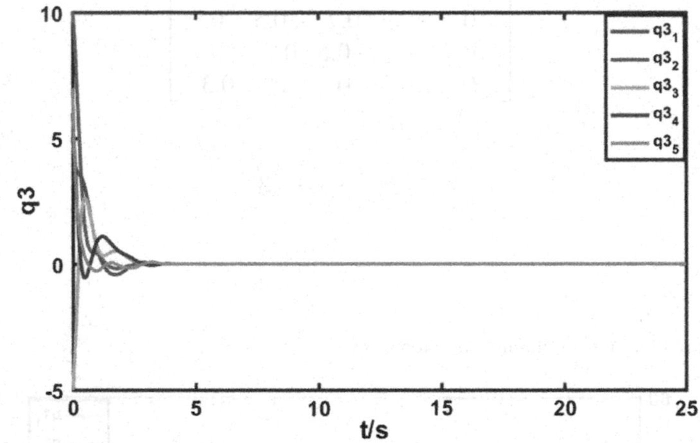


**Fig. 4** The state $q_3$ of the agents

The second smallest eigenvalue of the matrix $L$ can be calculated $\lambda_2 = 0.1328$. The simulation is conducted by the initial states $q_1(0) = [-0.7, 0.1, 0.3, -0.2, 0.5]^T$, $q_2(0) = [-4, -7, 2, -1, 4]^T, q_3(0) = [5, 10, -5, 6, 7]^T$. The control parameters are chosen as $c_1 = 1, c_2 = c_3 = c_4 = c_5 = c_6 = 1, \gamma_1 = \gamma_2 = 1.25, k = 5, \alpha_1 = \alpha_2 = 0.6, \beta_1 = \beta_2 = 1.4, p = 0.6, q = 1.4$. Due to Lemma 3, $T_1^* = 3.6902, T_2^* = 5.5572$, choose $T_{switch} = 10$ as the switching time.

The effectives of the method is shown in the above figures. The Fig. 2 depicts the convergence curve of $q_1$. Due to control algorithm (5) is a constant value, $q_1$ keeps increasing before the switching time $T_{switch} = 10$, after $T_{switch}$, $q_1$ begins to converge by (17). The Figs. 3 and 4 show that the states $q_2$ and $q_3$ achieve the fixed-time consensus, and $q_3$ converges to $\mathbf{0}$.

## 5 Conclusions

A fixed-time consensus problem for the multiple nonholonomic chained-form system is considered in this paper. Due to the complexity of nonholonomic systems, systems are divided into two subsystem, and a switching control strategy is introduced in the control algorithm. The switching time can be acquired by Lemma 2. Comparing with [1], the switching time is more reasonable. Finally, the numerical simulation is used to show that the effectiveness of the fixed-time consensus algorithm. In the future, the fixed-time consensus algorithm will be used to address the formation control problem for the nonholonomic mobile robots.

**Acknowledgements** This work is supported by the National Natural Science Foundation of China (Grant Nos. 61573200, 61573199).

## References

1. D. HaiBo, G.H. Wen, Y.Y. Cheng, Y.G. He, R. Jia, Distributed finite-time cooperative control of multiple high-order nonholonomic mobile robots. IEEE Trans. Neural Netw. Learn. Syst. **28**(12), 2998–3006 (2017)
2. Y. WenWu, G.R. Cheng, J.H. Lü, On pinning synchronization of complex dynamical networks. Automatica **45**(2), 429–435 (2009)
3. Z. Li, Z.S. Duan, L. Huang, Consensus of multiagent systems and synchronization of complex networks: a unified viewpoint. IEEE Trans. Circuits Syst. I **57**(1), 213–224 (2010)
4. A.S. Jun Lin, B.D.O. Morse Anderson, The multi-agent rendezvous problem. I: the synchronous case. SIAM J. Control Opt. **46**(6), 1508–1513 (2008)
5. W. Ren, Distributed attitude alignment in spacecraft formation flying. Int. J. Adap. Control Signal Process. **21**(2), 95–113 (2010)
6. R. Olfati-Saber, Flocking for multi-agent dynamic systems: algorithm and theory. IEEE Trans. Auto. Control **49**(4), 622–629 (2004)
7. R. Olfati-Saber, R.M. Murray, Consensus problems in network of agents with switching topology and time-delays. IEEE Trans. Auto. Control **49**(9), 1520–1533 (2004)
8. W.Y. Hou, F. MinYue, H.S. Zhang, Consensusability of linear multi-agent systems with time delay. Int. J. Robust Nonlinear Control **26**(12), 2529–2541 (2016)
9. P. Sanjay, Bhat, Dennis S. Bernstein, Finite-time stability of continuous autonomous systems. SIAM J. Control Opt. **38**(3), 751–766 (2000)
10. S.H. Li, S.H. Ding, Q. Li, Global set stabilisation of the spacecraft attitude using finite-time control technique. Int. J. Control **82**(5), 822–836 (2009)
11. J. Corts, Finite-time convergent gradient flows with applications to network consensus. Int. J. Control **42**(11), 1993–2000 (2006)
12. S.H. Li, D. HaiBo, XianZe Li, Finite-time consensus algorithm for multi-agent systems with double-integrator dynamics. Automatica **42**(11), 1993–2000 (2006)
13. F. Xiao, L. Wang, J. Chen, Y.P. Gao, Finite-time formation control for multi-agent systems. Automatica **45**(11), 2605–2611 (2009)
14. H.B. Du, G.H. Wen, X.H. Yu, S.H. Li, M.Z.Q. Che, Finite-time consensus of multiple nonholonomic chained-form systems based on recursive distributed observer. Automatica **62**(C), 236–242 (2015)
15. O. MeiYing, D. HaiBo, S.H. Li, Finite-time formation control of multiple nonholonomic mobile robots. Int. J. Robust Nonlinear Control **24**(1), 140–165 (2014)

16. A. Polyakov, Nonlinear feedback design for fixed-Time stabilization of linear control systems. IEEE Trans. Auto. Control **57**(8), 2106–2110 (2012)
17. Z.Y. Zuo, L. Tie, A new class of finite-time nonlinear consensus protocols for multi-agent systems. Int. J. Control **87**(2), 363–370 (2014)
18. J. Liu, Y. Yao, Q. Wang, C. Sun, Fixed-time event-triggered consensus control for multi-agent systems with nonlinear uncertaintie. Neurocomputing **260**(18), 497–504 (2017)
19. S. Parsegov, A. Polyakov, P. Shcherbakov, Nonlinear fixed-time control protocol for uniform allocation of agents on a segment, in *51st Conference on Decision and Control*, pp. 7732–7737 (2012)
20. Z.Y. Zuo, L. Tie, Distributed robust finite-time nonlinear consensus protocols for multi-agent systems. Int. J. Syst. Sci. **47**(6), 1366–1375 (2016)
21. V.T. Haimo, Finite time controllers. SIAM J. Control Opt. **24**(4), 760–770 (1986)

# Consensus of Linear Multi-agent Systems with a Smart Leader

**Yangbo Li, Hui Liu, Zhongxin Liu and Zengqiang Chen**

**Abstract** This paper addresses the consensus problem with linear systems via a smart leader under directed topology. The smart leader tracks a given reference model and a control function is constructed to control the smart leader. The smart leader can utilize its neighboring followers' feedback information when the control function meet certain condition, which can effectively reduce the leader's controller cost and the tracking error among the leader and followers. By utilizing the relative output message of neighboring agents, a reduced-order observer is adopted under the assumption that the directed topology have a directed spanning tree. A sufficient condition is given to guarantee that the leader-following system can achieve consensus. Finally, simulation examples are given to demonstrate the effectiveness of the obtained results.

**Keywords** Smart leader · Linear system · Leader-following consensus
Reduced-order observer

## 1 Introduction

In recent years, an increasing number of researchers focus on the distributed coordination of groups of agents due to their broad applications in many fields. The consensus problem is a fundamental problem in cooperative control, whose intention is to design efficient control protocols such that every agent can reach agreement of their common states in finite time or asymptotically. Much effort has been taken to solve all sorts of consensus problem in the early literature,such as formation control

Y. Li · H. Liu · Z. Liu (✉) · Z. Chen
College of Computer and Control Engineering, Nankai University,
Tianjin, China
e-mail: lzhx@nankai.edu.cn

Y. Li · H. Liu · Z. Liu · Z. Chen
Tianjin Key Laboratory of Intelligent Robotics, Nankai University,
Tianjin, China



Y. Jia et al. (eds.), *Proceedings of 2018 Chinese Intelligent Systems Conference*,
Lecture Notes in Electrical Engineering 528,
https://doi.org/10.1007/978-981-13-2288-4_76

[1, 2], tracking control [3], disturbed sensor network [4], state constraint consensus problem [5] and so on.

In the existing studies, stochastic matrix and Lyapunov-based methods were frequently used in solving first-order [6, 7] and second-order consensus problems [8–10]. Besides, the consensus problems with linear multi-agent system [11–13] had also been investigated. The literatures mentioned above all assume that agents could pick up neighbours state information through communication channel unerringly. However, for many practical systems, the agent can only measure output information of its neighbors, so full state information is not available. To solve such consensus problems, a common method is to propose control protocols comprising of observers to evaluate those unmeasured variables. In [8], a neighbor-based estimation method was presented for the first-order follower to estimate the leader's state whose velocity is unmeasured. Abdessameud and Tayebi [9] offered an observer-based consensus protocol for second-order system lacking of velocity measurement. In [14], the researchers probed the consensus problem with general linear dynamics by adopting a new reduced-order observer. Then, on the basis of [14], an observer-based protocol was presented in [15] for investigating the general consensus problem and model-reference consensus problem via the reduced-order observer.

So far, most of existing works on leader-following consensus largely assumed the leader is individual. Followers can track the leader but the leader is not affected by its neighboring follower. However, there are that can better describe the leader's character. In [16], a modified leader-following architectures was proposed for improving the robustness of the system and a optimal control strategy is designed to minimize individual cost functions. Ma et al. [10] investigated the leader-following consensus problem of the first-order system with a smart leader which can gain advantage in fault-tolerance and robustness with the tranditional leader.

In many practical systems, the communication distance of agents is limited and the followers may have a loss of communication once some faults happened. In addition, the controller cost of the leader is also considered in some situation. Based on the above factors, this paper considers the consensus problem of linear leader-following system with a smart leader. An objective function is constructed and the leader can use the feedback information from its neighbors when the objective function is positive. By utilizing the relative outputs of neighboring agents, a reduced-order observer is then applied to the system. The main contribution of this paper can be concluded into two points: (i) The consensus problem of linear multi-agent systems with a smart leader was investigated. The leader can utilize the feedbacks from neighboring agents when necessary. A reduced-order observer only relied on the relative output information is then adopted to modify the linear system. (ii) By comprehensive consideration of leader energy consumption and Euclidean distance between leader and followers, an objection function is designed to control the smart leader.

The remainder of the paper is designed as follows. Section 2 contains the problem formulation and preliminaries, some lemmas and assumptions are also provided here, Sect. 3 gives the main results. Numerical simulation results are presented in Sect. 4 and the conclusion of this article is given in Sect. 5.

Throughout this paper, some notations used in the following part are provided here. Let $\mathbb{R}^{n\times m}$ be the set of all $n \times m$ real matrices. $A^T$ and $A^H$ denote the transpose and conjugate transpose of matrix $A$, respectively. $I_N$ is the $N \times N$ identity matrix. For symmetric matrix $P$, when $P > 0(\geq 0, < 0, \leq 0)$, we say it is positive definite (positive semi-definite, negative definite, negative semi-definite). $1_N = [1, 1, \ldots, 1]^T$. $\|\cdot\|$ denotes the Euclidean norm. $|\cdot|$ denotes the module of a complex number or the absolute value of a real number. $\otimes$ denotes the Kronecker product.

# 2 Preliminaries and Problem Formulation

## 2.1 Graph Theory and Lemmas

With the graph theory, the interaction relationship among $N$ agents can be described by a directed graph $\mathcal{G} = (\mathcal{V}, \mathcal{E}, \mathcal{A})$. $\mathcal{V} = \{\nu_1, \nu_2, \ldots, \nu_N\}$ denotes the set of nodes and $\mathcal{E} \subseteq \mathcal{V} \times \mathcal{V}$ is the edges set. The weighted adjacency matrix $\mathcal{A} = \left[a_{ij}\right] \in \mathbb{R}^{n\times n}$ is defined such that $a_{ii} = 0$, $a_{ij} > 0$ if $(i, j) \in \mathcal{E}$ and $a_{ij} = 0$ otherwise. The Laplacian matrix $\mathcal{L}$ is defined as $l_{ii} = \sum_{j=1, j\neq i}^{N} a_{ij}$ and $l_{ij} = -a_{ij}, i \neq j$. A directed spanning tree of a directed graph is a directed tree formed by graph edges that connect from the root node to every other node in the graph. According to the definition, we can obtain that $\mathcal{L}$ satisfies $\mathcal{L}1_N = 0$. Now some basic lemmas are introduced in the following for further study.

**Lemma 1** *([7]) The Laplacian matrix $\mathcal{L}$ is positive semi-definite and satisfies $\mathcal{L}1_N = 0$. If weighted diagraph $\mathcal{G}$ contains a directed spanning tree, then the corresponding Laplacian matrix $\mathcal{L}$ has exactly one zero eigenvalue.*

**Lemma 2** *([17]) A symmetric matrix $S$ can be partitioned into block form $S = \begin{bmatrix} S_{11} & S_{12} \\ S_{12}^T & S_{22} \end{bmatrix}$, when $S_{11}$ and $S_{22}$ are symmetric matrix. Then $S < 0$ if and only if*

$$S_{11} < 0, S_{22} - S_{12}^T S_{11}^{-1} S_{12} < 0 \quad or \quad S_{22} < 0, S_{11} - S_{12} S_{22}^{-1} S_{12}^T < 0$$

**Lemma 3** *([18]) For a switching system, if a common Lyapunov function can be established for all subsystems, then the switching system can guarantee stability when switching arbitrarily.*

**Lemma 4** *([19]) If $(A, B)$ is stabilizable and $Q$ is symmetric positive definite matrix, there exist a unique positive-definite solution $P$ such that the Riccati equation (1) holds*

$$A^T P + PA - PBB^T P + Q = 0 \tag{1}$$

### 2.2 Model Formation

Consider a leader-following system consisting of $N$ followers and one leader. Let $\mathcal{G}'$ be a directed graph with these $N+1$ nodes, and use $\mathcal{G}'$ to model the communication topology of the leader-following multi-agent systems. $\mathcal{G}'$ contains a graph $\mathcal{G}$ of N nodes and $\nu_0$(denoting the leader) with the directed edges from some following agents to the leader. If at least one agent of $\mathcal{G}$ is connected to leader by a direct edge, the graph $\mathcal{G}'$ is said to be connected.

The dynamics of $i$-th follower is modeled by the following linear system

$$\dot{x_i} = Ax_i + Bu_i, y_i = Cx_i, i \in S_r \tag{2}$$

where $S_r = \{1, 2, \ldots, N\}$, $x_i \in \mathbb{R}^n$, $u_i \in \mathbb{R}^p$ and $y_i \in \mathbb{R}^q$ are the state, control input and measured output of agent $i$, respectively. $A \in \mathbb{R}^{n\times n}$, $B \in \mathbb{R}^{n\times p}$ and $C \in \mathbb{R}^{q\times n}$ are constant matrices.

The leader's dynamics described is by

$$\dot{x_0} = Ax_0 + Bu_0, y_0 = Cx_0 \tag{3}$$

where $x_0 \in \mathbb{R}^n$ is the leader's state, $y_i \in \mathbb{R}^q$ is the output of leader and $u_i \in \mathbb{R}^p$ is the input of leader.

Then some necessary assumptions are listed here for our study.

**Assumption 1** The interaction topology $\mathcal{G}'$ contains a directed spanning tree.

**Assumption 2** The matrix pair $(A, B)$ is stabilizable.

**Assumption 3** $(A, C)$ is detectable. $C$ is a full row rank matrix.

Define a target model that is represented as $\dot{x_d} = Ax_d$. Then the leader-following systems (2) and (3) are deemed to achieve consensus, if

$$\lim_{k\to\infty} \|x_i - x_d\| = 0, \forall i = 0, 1, ..., N \tag{4}$$

hold.

## 3 Main Results

### 3.1 Objective Function Design

For a smart leader, there are two situations in the control protocol, that is, using $(u_{0s})$ or without using $(u_{0l})$ the feedback information from its neighbors. So an objective function will be designed to determine whether the smart leader should use the neighbors' information. By comprehensive consideration of leader energy consumption and distance between leader and followers, the objective function can be designed as,

$$\begin{cases} J_1 = a\int_{t_k}^{t_{k+1}} u_{0s}^T u_{0s} dt + bd \\ J_2 = a\int_{t_k}^{t_{k+1}} u_{0l}^T u_{0l} dt + b\frac{1}{m}\sum\limits_{i \in N_0} \|a_{0i}(x_0 - x_i)\| \end{cases} \tag{5}$$

where $a_{0i}$ is the connection weight between the leader with $i$-th agent, $d$ is a positive constant reflecting the critical range of the smart leader, a and b are positive weight parameters. Let $\Delta t = t_{k+1} - t_k$, the first term of the formula indicates the energy consumption of in $\Delta t$ and the second term shows the leader's care for the followers. Our purpose is to choose the smaller one at every $\Delta t$. Then we define the symbolic function $\mathrm{sgm}(x) = \begin{cases} 0, & x \le 0 \\ 1, & x > 0 \end{cases}$ and $J = J_1 - J_2$, which will be used in the following sections.

### 3.2 Consensus Protocol

In this part, our goal is to design the control protocols for the followers and smart leader. For the $i$-th follower, the control protocol is designed as,

$$u_i(t) = -K_1[\sum_{j \in N_i} a_{ij}(x_i(t) - x_j(t)) + b_i(x_i(t) - x_0(t))] \tag{6}$$

with the matrix $K_1$ is the gain matrix and $b_i$ denotes the relationship between the $i$-th follower with the leader. Then the control law of the smart leader is designed as,

$$u_0(t) = -\gamma K_1[x_0(t) - x_d(t)] - K_1[\sum_{j \in N_0} a_{0j}(x_0(t) - x_j(t))]sgm(J) \tag{7}$$

where $a_{0j}$ is the connection weight between the leader with $j$-th agent of graph $\mathcal{G}'$ and $\gamma > 0$.

Let $\bar{x}_i(t) = x_i(t) - x_d(t)$, $\bar{x}(t) = \left[\bar{x}_0^T(t), \bar{x}_1^T(t), \ldots, \bar{x}_N^T(t)\right]^T$, then we can easily obtain $\dot{\bar{x}}_i(t) = A\bar{x}_i(t) + Bu_i(t)$. Now we consider the two cases of the leader's protocol:

Case 1: $sgm(J) = 0$, the system's consensus error can be rewritten in the compact form, one as

$$\dot{\bar{x}}(t) = (I \otimes A)\bar{x}(t) + (M_1 \otimes BK)\bar{x}(t) \tag{8}$$

with the matrix $M_1 = \begin{bmatrix} \gamma & 0_N^T \\ -\Omega & \mathcal{L}+D \end{bmatrix}$, $D = diag\{b_1, \ldots, b_N\}$ and $\Omega = [b_1, \ldots, b_N]^T$.

Case 2: $sgm(J) = 1$, the system's consensus error can be rewritten as

$$\begin{aligned} \dot{\bar{x}}(t) &= (I \otimes A)\bar{x}(t) + (\bar{\mathcal{L}} + \gamma\Gamma) \otimes BK\bar{x}(t) \\ &= (I \otimes A)\bar{x}(t) + M_2 \otimes BK\bar{x}(t) \end{aligned} \tag{9}$$

where $\bar{\mathcal{L}}$ denotes the Laplacian matrix of the graph $\mathcal{G}'$ when the smart leader can utilize the feedbacks and $M_2 = \bar{\mathcal{L}} + \gamma\Gamma$, $\Gamma = diag\{1, 0, \ldots\} \in \mathbb{R}^{(N+1)\times(N+1)}$.

**Theorem 1** *For the leader-following system (2) and (3) with graph $\mathcal{G}'$, suppose Assumptions 1 and 2 hold, then the consensus problem with control protocols (6) and (7) can be solved under the following conditions:*
*$K = \alpha B^T P$, where $P$ is the unique positive-definite solution of Riccati equation (1) and the gain $\alpha$ satisfies $\alpha \geq \frac{1}{\lambda_{MIN}}$, $\lambda_{MIN} = \min\{\lambda_{M_2+M_2{}^T}, \lambda_{M_1+M_1{}^T}\}$*

*Proof* For case 1, construct the Lyapunove function as,

$$V_1(t) = \bar{x}^T(t)(I_{N+1} \otimes P)\bar{x}(t) \tag{10}$$

Then we can get the time derivative of $V_1(t)$

$$\dot{V}_1(t) = \bar{x}^T(t)(I \otimes A^T P + I \otimes PA - \alpha(M_1 + M_1^T) \otimes PBB^T P)\bar{x}(t) \tag{11}$$

For case 2, construct the Lyapunove function as,

$$V_2(t) = \bar{x}^T(t)(I_{N+1} \otimes P)\bar{x}(t) \tag{12}$$

Similarly, we can get

$$\dot{V}_2(t) = \bar{x}^T(t)(I \otimes A^T P + I \otimes PA - \alpha(M_2 + M_2{}^T) \otimes PBB^T P)\bar{x}(t) \tag{13}$$

From the conditions, it's easily to get $V_1(t) > 0$, $\dot{V}_1(t) < 0$, $V_2(t) > 0$, $\dot{V}_2(t) < 0$, then according to Lemma 3, the systems (2) and (3) are stable and the consensus problem is solved. The proof is completed.

*Remark 1* In the leader's control protocol (7), it can be seen that only in the case of $sgm(J) = 1$, the smart leader use the information from its neighbors. By forecasting the controller energy consumption in the next $\Delta t$ and considering whether the followers need to be taken care of, the leader will alway choose the strategy that minimized the objective function.

### 3.3 Consensus with Reduced-Order Observer

In this part, a distributed reduced-order observer is adopted for the following agents and we construct a distributed control protocol only utilized the relative output information for leader-following system. Considering the consensus definition (4) a new state vector can be designed as follows,

$$\psi_i = \sum_{j=0}^{N} l'_{ij} x_j, i = 1, 2, \ldots, N \tag{14a}$$

$$\psi_0 = sgm\,(J) \sum_{j=1}^{N} a_{0j}(x_0 - x_j) + \gamma(x_0 - x_d) \tag{14b}$$

where $\mathcal{L}' = [l'_{ij}] \in \mathbb{R}^{(N+1)\times(N+1)}$ denotes the Laplacian matrix of the graph $\mathcal{G}'$, $a_{0j}$ is the connection weight between the leader with $j$-th agent of graph $\mathcal{G}'$ and $\gamma > 0$. Then we get the derivate of $\psi_i$ as

$$\dot{\psi}_i = \sum_{j=1}^{N} l'_{ij}\dot{x}_j = \sum_{j=1}^{N} l'_{ij}\left(Ax_j + Bu_j\right), i = 1, 2, \ldots, N \tag{15a}$$

$$\begin{aligned}\dot{\psi}_0 &= sgm\,(J)\,(\sum_{j=1}^{N} a_{0j}(\dot{x}_0 - \dot{x}_j)) + \gamma(\dot{x}_0 - \dot{x}_d)\\ &= sgm\,(J)\,(\sum_{j=1}^{N} a_{0j}\left(A(x_0 - x_j) + B(u_0 - u_j)\right)) + \gamma(A(x_0 - x_d) + Bu_0)\end{aligned} \tag{15b}$$

Denote $\psi = \left[\psi_0^T, \psi_1^T, \ldots, \psi_N^T\right]^T$ and $u = [u_0^T, u_1^T \ldots, u_N^T]^T$, the vector can be rewritten in the compact form as

$$\dot{\psi} = ((I_{N+1} \otimes A)\psi + ((\widehat{\mathcal{L}} + \gamma\Gamma) \otimes B)u) \tag{16}$$

with the matrix $\widehat{\mathcal{L}}$ is the Laplacian matrix of the graph $\mathcal{G}'$ that changes with the smart leader and $\Gamma = diag\,\{1, 0, \ldots\} \in \mathbb{R}^{(N+1)\times(N+1)}$. Then set the relative output vector

$$\xi = (I_N \otimes C)\psi \tag{17}$$

Now we can design the control protocol

$$u_i = -cKS_1\xi_i - cKS_2\varepsilon_i \tag{18}$$

with a reduced-order observer for agent $i$ to estimate the state of $\psi_i$ as follow,

$$\begin{cases} \dot{z}_i = W(Nz_i + K\xi_i + B\sum_{j=0}^{N}\widehat{l}_{ij}u_j) \quad i \neq 0 \\ \dot{z}_0 = W(Nz_i + K\xi_i + B((\sum_{j=0}^{N}\widehat{l}_{ij}u_j) + \gamma u_0) \end{cases} \tag{19a}$$

$$\varepsilon_i = z_i + E\xi_i \tag{19b}$$

where $c$ is a positive gain to be determined and $K \in \mathbb{R}^{q\times n}$ is given gain matrix. Then according to [14], the coefficient matrices $W, N, K$ are designed as $W = T - EC, N = AS_2, K = AS_2E + AS_1$ with $\begin{bmatrix} C \\ T \end{bmatrix}^{-1} = \begin{bmatrix} S_1 & S_2 \end{bmatrix}$, $S_1 \in \mathbb{R}^{m\times q}$, $S_2 \in \mathbb{R}^{m\times\beta}$. Then an algorithm is presented for constructing the matrices as follow,

**Algorithm 1** *Under the Assumptions 1–3, the conditions below are satisfied,*

*(i) There exist positive definite matrix $P_1$, $Q_1 \in \mathbb{R}^{n\times n}$ satisfying the Riccati equations*

$$A^T P_1 + P_1 A - P_1 B B^T P_1 + Q_1 = 0 \tag{20}$$

*(ii) Let $\lambda_i$ be the i-th eigenvalue of $\widehat{L} + \gamma\Gamma$, select c satisfying $c \geq \frac{1}{2\min_{\lambda_i\neq 0}\mathrm{Re}(\lambda_i)}$*

*(iii) There exist positive define matrices $P_2$, $Q_2 \in \mathbb{R}^{\beta\times\beta}$ and a gain matrix E satisfying the following condition,*

$$(TAS_2 - ECAS_2)^T P_2 + P_2(TAS_2 - ECAS_2) = -Q_2 \tag{21}$$

**Theorem 2** *Suppose that Assumptions 1–3 hold, consider the leader-following systems (2) and (3). Under the control protocol (18) and the reduced-order observer (19), the systems can achieve consensus if Algorithm 1 hold. The gain matrix K can be constructed by*

$$K = B^T P_1 \tag{22}$$

*Proof* A new error vector can be designed as

$$e' = (I_{N+1} \otimes T)\psi - \varepsilon \tag{23}$$

where $e' = [e_0'^T, e_1'^T, \ldots, e_N'^T]^T$. Then the dynamics of $e'(k)$ is obtained as,

$$\begin{aligned}
\dot{e}' &= (I_{N+1} \otimes T)\dot{\psi} - \dot{\varepsilon} \\
&= (I_{N+1} \otimes T)\dot{\psi} - \dot{z} - (I_{N+1} \otimes EC)\dot{\psi} \\
&= (I_{N+1} \otimes TA)\psi + ((\widehat{\mathcal{L}} + \gamma\Gamma) \otimes TB)u) - (I_{N+1} \otimes (T - EC))[(I_{N+1} \otimes N)z \\
&\quad + (I_{N+1} \otimes KC)\psi + ((\widehat{\mathcal{L}} + \gamma\Gamma) \otimes B)u)] - (I_N \otimes EC)[(I_{N+1} \otimes A)\psi \\
&\quad + ((\widehat{\mathcal{L}} + \gamma\Gamma) \otimes B)u] \\
&= [I_{N+1} \otimes (TA - ECA - WNT + WNEC - WKC))]\psi \\
&\quad + [((\widehat{\mathcal{L}} + \gamma\Gamma) \otimes (TB - ECB - WB)]u + (I_{N+1} \otimes (WN))e' \\
&= (I_{N+1} \otimes (TAS_2 - ECAS_2))e'
\end{aligned} \tag{24}$$

Due to $\begin{bmatrix} C \\ T \end{bmatrix}^{-1} = \begin{bmatrix} S_1 & S_2 \end{bmatrix}$ we can get $S_2T+S_1C=I_n$, combining with (16) we have

$$\dot{\psi} = [(I_{N+1} \otimes A) - (\widehat{L} + \gamma\Gamma) \otimes (cBK)]\psi + [(\widehat{L} + \gamma\Gamma) \otimes cBKS_2]e' \tag{25}$$

Let $\eta = [\psi^T, e'^T]^T$ ,according to (24) and (25) we can obtain

$$\begin{aligned}
\dot{\eta} &= \begin{bmatrix} I_{N+1} \otimes A - (\widehat{L} + \gamma\Gamma) \otimes (cBK) & (\widehat{L} + \gamma\Gamma) \otimes (cBKS_2) \\ 0 & I_{N+1} \otimes (TAS_2 - ECAS_2) \end{bmatrix} \eta \\
&= F\eta
\end{aligned} \tag{26}$$

Now construct a common Lyapunov function for dynamic system (26)

$$V = \eta^T \tilde{P} \eta \tag{27}$$

where $\tilde{P} = \begin{bmatrix} I_{N+1} \otimes P_1 & 0 \\ 0 & \omega I_{N+1} \otimes P_2 \end{bmatrix}$ is the parameter-dependent matrix and $\omega$ is a given positive parameter. From (27) we can obtain

$$\dot{V} = \eta^T (F^T \tilde{P} - \tilde{P} F)\eta \tag{28}$$

According to Lemma 4 and Algorithm 1, $P_1$ is the unique positive-definite solution of (20). Then, we have

$$\begin{aligned}
&(A - c\lambda_i BK)^H P_1 + P_1(A - c\lambda_i BK) \\
&= A^T P_1 - P_1 A - 2c\mathrm{Re}\,(\lambda_i)\, P_1 BB^T P_1 \\
&\leq A^T P_1 - P_1 A - P_1 BB^T P_1 < 0
\end{aligned} \tag{29}$$

Then by condition 3 of Algorithm 1, there exist a positive define matrix $P_2$ satisfying

$$(TAS_2 - ECAS_2)^T P_2 - P_2(TAS_2 - ECAS_2) = -Q_2 \tag{30}$$

with the matrices $(TAS_2 - ECAS_2)$ and $(A - c\lambda_i BK)$ are Hurwitz, it's easily to get $\dot{V} < 0$ by choosing appropriate $\omega$. Thus, the system (26) is asymptotically stable and the leader-following consensus with a smart leader is achieved by the control protocol (18) and the reduced-order observer (19), that is, $\lim_{k\to\infty} \|x_i - x_d\| = 0, \forall i \in [0, 1, \ldots, N]$. The proof is thus complete.

*Remark 2* In the control protocol (18) and estimation law (19), the agent $i$'s state has not been used directly. Since the relative state information $\psi_i$ converge to zero, the observer state $z_i$ will also converge to zero. Meanwhile, the observer states will not be affected by the agents's common consensus states.

## 4 Simulation Results

In this part, several numerical simulations is provided to verify the effectiveness of the theoretical results.Consider a group of agents consisting of four followers and one leader. The coefficient matrices of both leader and follower dynamics are given as

$$A = \begin{bmatrix} -1 & 0 & -1 \\ 0 & 1 & -1 \\ 1 & 4 & -1 \end{bmatrix}, B = \begin{bmatrix} 1 & 0 \\ 1 & 1 \\ 1 & 1 \end{bmatrix}, C = \begin{bmatrix} 1 & 0 & 0 \\ 0 & 1 & 0 \end{bmatrix}$$

Consider the following Laplacian matrix $\mathcal{L}'$ for graph $\mathcal{G}'$ and the Laplacian matrix $\widehat{\mathcal{L}}$ when the smart leader utilizes the feedbacks

$$\mathcal{L}' = \begin{bmatrix} 0 & 0 & 0 & 0 & 0 \\ 0 & 3 & -1 & 0 & -2 \\ 0 & -1.5 & 3.5 & -2 & 0 \\ -1 & 0 & -1 & 2 & 0 \\ -1 & -1 & 0 & 0 & 2 \end{bmatrix}, \widehat{\mathcal{L}} = \begin{bmatrix} 2 & 0 & 0 & -1 & -1 \\ 0 & 3 & -1 & 0 & -2 \\ 0 & -1.5 & 3.5 & -2 & 0 \\ -1 & 0 & -1 & 2 & 0 \\ -1 & -1 & 0 & 0 & 2 \end{bmatrix}$$

with $\widehat{\mathcal{L}} = \mathcal{L}'$ when the smart leader don't utilizes the feedbacks. Let the coefficient matrices $T = \begin{bmatrix} 0 & 0 & 1 \end{bmatrix}$, $S_1 = \begin{bmatrix} 1 & 0 & 0 \\ 0 & 1 & 0 \end{bmatrix}'$, $S_2 = \begin{bmatrix} 0 & 0 & 1 \end{bmatrix}'$, $E = \begin{bmatrix} 0.08 & 0.02 \end{bmatrix}$. Set $Q = diag([3, 4, 5])$, then we can get

$$P = \begin{bmatrix} 1.2309 & -0.5837 & 0.1388 \\ -0.5837 & 2.3326 & -0.3314 \\ 0.1388 & -0.3314 & 1.3978 \end{bmatrix}, K = \begin{bmatrix} 0.7860 & 1.4174 & 1.2052 \\ -0.4449 & 2.0012 & 1.0664 \end{bmatrix}$$

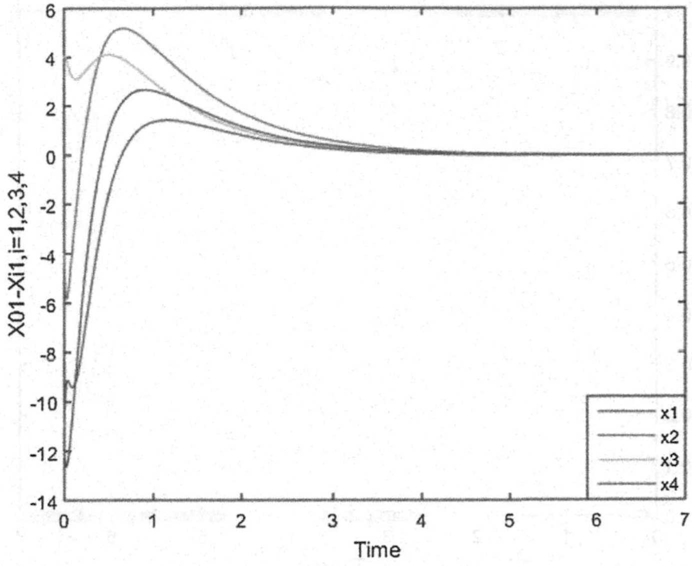


**Fig. 1** Position error with tranditional leader

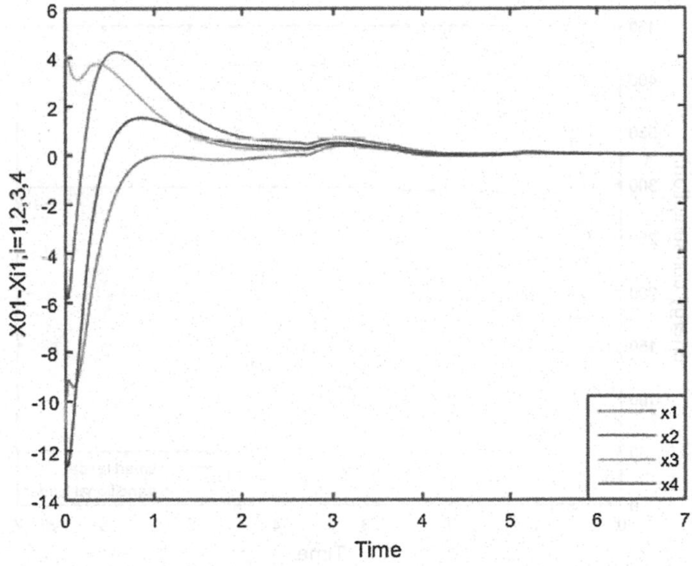


**Fig. 2** Position error with smart leader

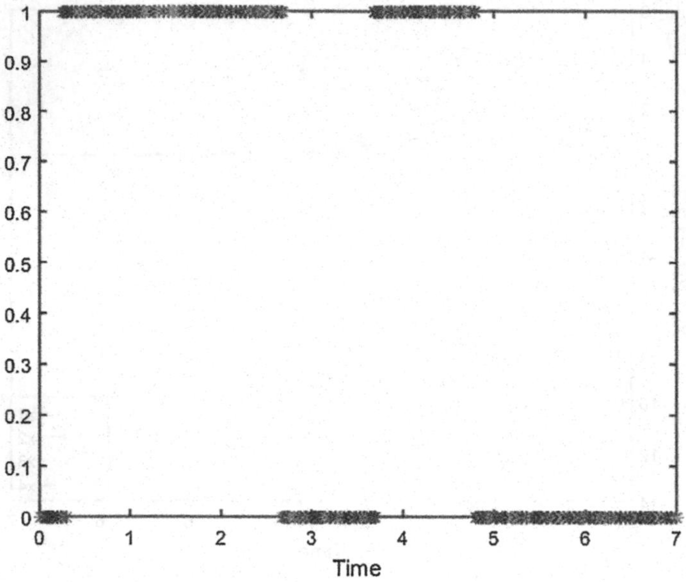


**Fig. 3** The trajectories of sgm($J$)

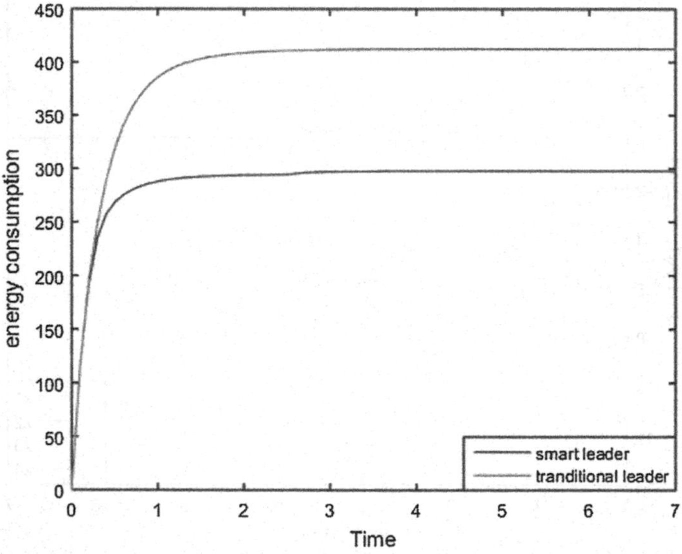


**Fig. 4** The trajectories of leader's cost

In this simulation, we choose the parameters $a = 1, b = 100, c = 2, \gamma = 0.9$, the critical range $d = 2$ and $\Delta t = 0.1$. the position error of leader and followers with a traditional/smart leader are showed in Figs. 1 and 2. Figure 3 shows the trajectories of $\text{sgn}(J)$ which indicate the leader can arbitrary switching in two cases. It shows that the tracking error $x_i - x_0 (i = 1, 2, 3, 4)$ with a smart leader is reduced compared with the tranditional case. Let $\int_0^{\infty} u_0^T u_0 dt$ represents the energy consumption of the leader, Fig. 4 shows that the energy consumption of a smart leader is significantly lower than the traditional one.

## 5 Conclusions

In this paper, the leader-following consensus of linear multi-agent systems with a smart leader is investigated under directed topology. A reduced-order observer is then proposed to modified the protocol. An objective function is used in the control strategy of the smart leader. With the smart leader, the leader's energy consumption and the tracking error among the leader and follower can be reduced effectively.

**Acknowledgements** This work is supported by the National Natural Science Foundation of China (Grant Nos. 61573200, 61573199).

## References

1. P.P. Menon, C. Edwards, N.M. Gomes, Paulino. Observer-based controller design with disturbance feedforward framework for formation control of satellites. IET Control Theor. Appl. **9**(8), 1285–1293 (2015)
2. W. Qin, Z. Liu, Z. Chen, Observer-based controller design with disturbance feedforward framework for formation control of satellites. IEEE/CAA J. Automat. Sinica **1**(2), 171–179 (2014)
3. W. Wang, C. Wen, J. Huang, Distributed adaptive asymptotically consensus tracking control of nonlinear multi-agent systems with unknown parameters and uncertain disturbances. Automatica **77**, 133–142 (2017)
4. H. Ji, F.L. Lewis, Z. Hou, et al. Distributed information-weighted Kalman consensus filter for sensor networks, Automatica **77**, 18–30 (2017)
5. Z. Liu, Z. Chen, Discarded consensus of network of agents with state constraint. IEEE Trans. Automat. Control **57**(11), 2869–2874 (2012)
6. Z.H. Guan, Y. Wu, G. Feng, Consensus analysis based on impulsive systems in multiagent networks. IEEE Trans. Circ. Syst. I Regular Papers **59**(1), 170–178 (2012)
7. W. Ren, R.W. Beard, Consensus seeking in multiagent systems under dynamically changing interaction topologies. IEEE Trans. Automat. Control **50**(5), 655–661 (2005)
8. Y. Hong, J. Hu, L. Gao, Tracking control for multi-agent consensus with an active leader and variable topology. Automatica **42**(7), 1177–1182 (2006)
9. A. Abdessameud, A. Tayebi, On consensus algorithms for double-integrator dynamics without velocity measurements and with input constraints. Syst. Control Lett. **59**(12), 812–821 (2010)
10. Z. Ma, Z. Liu, Z. Chen, Leader-following consensus of multi-agent system with a smart leader. Neurocomputing **214**, 401–408 (2016)
11. Z. Li, Z. Duan, G. Chen, On dynamic consensus of linear multi-agent systems. IET Control Theor. Appl. **5**(1), 19–28 (2011)

12. L. Gao, J. Li, X. Zhu, Leader-following consensus of linear multi-agent systems with state-observer under switching topologies. Math. Prob. Eng. **2013**(3), 572–577 (2013)
13. F. Xiao, L. Wang, Consensus problems for high-dimensional multi-agent systems. IET Control Theor. Appl. **1**(3), 830–837 (2007)
14. L. Gao, B. Xu, J. Li, Distributed reduced-order observer-based approach to consensus problems for linear multi-agent systems. IET Control Theor. Appl. **9**(5), 784–792 (2015)
15. T. Yang, P. Zhang, S. Yu, Consensus of linear multi-agent systems via reduced-order observer. Neurocomputing **240**, 200–208 (2017)
16. E. Semsar-Kazerooni, K. Khorasani, Optimal consensus algorithms for cooperative team of agents subject to partial information. Automatica **44**(11), 2766–2777 (2008)
17. R. Horn, C. Johnson, Matrix Analysis. Cambridge University Press (1985)
18. E. Semsar-Kazerooni, K. Khorasani, Switching control of a modified leader-follower team of agents under the leader and network topological changes. IET Control Theor. Appl. **5**(12), 1369–1377 (2011)
19. W. Wonham, *Linear Multivariable Control* (Springer, New York, 1985)

# Attitude Control of Oblique Cross Quad-Rotors UAV

**Hui Ji, Qing Li, Jiarui Cui and Wenhao Wang**

**Abstract** Compared with cross quad-rotors, oblique cross quad-rotors are more flexible, stable, and suitable for expansion, so that a growing number of attentions are paid on them. However, some drawbacks such as high model dependency, poor anti-interference ability and robustness are appeared when traditional controllers are used for the attitude control of oblique cross quad-rotors, which is because of strong coupling among the various channels. These problems can be solved by designing a controller based on Linear Active Disturbance Rejection Control (LADRC) algorithm. Firstly, the unmanned aerial vehicle (UAV) modeled is established by the Newton-Euler formula. Secondly, a series of simulation including tracking, anti-interference and robustness experiments are carried out. Finally, the comparison with classic proportional-integral-derivative (PID) controller are analyzed. The results show that LADRC controller has higher performance, such as better tracking capacity, stronger anti-interference ability and robustness.

**Keywords** Oblique cross quad-rotors · LADRC · Anti-interference · Robustness

## 1 Introduction

The quad-rotor vehicle is a kind of unmanned aircraft with axially symmetric rotors, good structure, and wide range of applications. In recent years, it has been widely used in many fields because of its unique advantages. According to the rotor structure, it can be divided into cross type and oblique cross type.

The structure of the cross UAV is the most classic one, and its manipulation is also simple. At present, many scholars have studied it deeply. At the beginning of this study, PID [1], sliding mode [2] and back-stepping controllers [3] were applied

H. Ji · Q. Li (✉) · J. Cui · W. Wang
School of Automation and Electrical Engineering, University of Science and Technology Beijing, Beijing, China
e-mail: liqing@ies.ustb.edu.cn



Y. Jia et al. (eds.), *Proceedings of 2018 Chinese Intelligent Systems Conference*, Lecture Notes in Electrical Engineering 528,
https://doi.org/10.1007/978-981-13-2288-4_77

in the design of flight control system. With the development of research, improved linear quadratic regulator (LQR) [4], ADRC [5] and optimal generalized predictive control (OGPC) [6] have been applied gradually.

In contrast, the coupling relationship among channels of oblique cross UAV is more complicated. Therefore, reliable results of the attitude control is difficult to achieve in terms of stability control as well as resistance to unknown disturbances. However, oblique cross UAV is more scalable (such as equipped with aerial platform, etc.), more sensitive and stable. In addition, its research results can also be applied directly to the aerial hybrid tilting aircraft, which has a higher research value. The most communal problem is that the research for the oblique cross UAV is relatively rare.

The control of this type of UAV is dominated by some common algorithms, such as PID [7], back-stepping [8] controller and so on. Moreover, multi-channel dual-loop vector PD control system [9] and attitude control of UAV with a tilting rotor structure [10] are also mentioned.

All the above algorithms have gained advantageous effects in application, but this often depends on the exact model of the controlled object. Firstly, the fling environment is usually complex and changeable, and it is very difficult to establish an accurate model. Secondly, UAV is small in size, light in weight and sensitive to external disturbances, which leads to higher requirements for algorithm.

Therefore, this paper chooses LADRC technology to design the attitude controller to solve above problems. To verify the effectiveness of LADRC controller better, all the simulation results are compared with classical PID controller.

## 2 Dynamic Model of Quad-Rotor Aircraft

A typical structural diagram of the oblique cross UAV wing (the rotor is simplified to Brushless direct current (DC) motor) is shown in Fig. 1. Compared with the cross UAV, the differences mainly include establishment of coordinate system and flight control strategy. To build the mechanism model, first of all, two coordinate systems are needed, which are the Body coordinate system (B system) in Fig. 1a and the Earth coordinate system (E system) in Fig. 1b. The normal direction of $X_b$ is consistent with the UAV flight direction, and $Z_b$ is perpendicular to the plane of the body. The positive axis of $X_e$ is consistent with $X_b$, and $Z_e$ is perpendicular to the ground.

Because of the limited influence of air resistance and elastic deformation of the body on the model accuracy, some simple assumptions are put forward to simplify its process.

**Fig. 1** Coordinate system

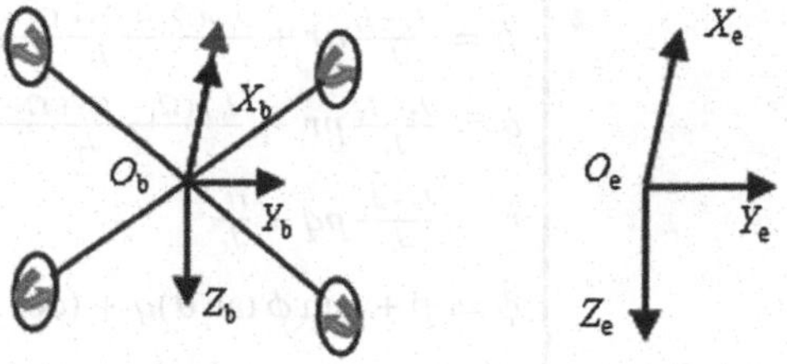


(a) Body coordinate system (b) Earth coordinate system

The Newton Euler equation is a set of equations based on Newton's law to describe the motion and angular motion of the UAV. It is the theoretical basis for the establishment of the whole model, and its matrix form is shown in formula (1).

$$\begin{bmatrix} m\boldsymbol{I}_{3*3} & \boldsymbol{0} \\ \boldsymbol{0} & \boldsymbol{J} \end{bmatrix} \bullet \begin{bmatrix} \dot{\boldsymbol{v}}_e \\ \dot{\boldsymbol{\omega}}_b \end{bmatrix} + \begin{bmatrix} \boldsymbol{0} \\ \boldsymbol{\omega}_b \times \boldsymbol{J}\boldsymbol{\omega}_b \end{bmatrix} = \begin{bmatrix} \boldsymbol{F}_e \\ \boldsymbol{T}_b \end{bmatrix} \tag{1}$$

where $m$ is the mass of the UAV; $J = [\,J_x\ J_y\ J_z\,]^{\mathrm{T}}$ is rotary inertia of the body; $v_e = [\,u\ v\ w\,]^{\mathrm{T}}$ is line speed of the aircraft in E coordinate system; $\omega_b = [\,p\ q\ r\,]^{\mathrm{T}}$ is angular velocity in B coordinate system; $F_e = [\,F_x\ F_y\ F_z\,]^{\mathrm{T}}$ is external force in the E coordinate system; $T_b = [\,\tau_x\ \tau_y\ \tau_z\,]^{\mathrm{T}}$is external torque for aircraft in the B coordinate system.

Usually, the external torque can be expressed as the vector sum of the gyroscopic moment and the aerodynamic torque. Unlike the cross UAV, the expression of the aerodynamic torque of the oblique cross UAV is shown in formula (2).

$$T_T = \begin{bmatrix} U_2 \\ U_3 \\ U_4 \end{bmatrix} = \begin{bmatrix} -\frac{\sqrt{2}}{2}l(T_1 - T_2 - T_3 + T_4) \\ -\frac{\sqrt{2}}{2}l(-T_1 - T_2 + T_3 + T_4) \\ d\left(\Omega_1^2 - \Omega_2^2 + \Omega_3^2 - \Omega_4^2\right) \end{bmatrix} \tag{2}$$

In this equation, $U_2$, $U_3$ and $U_4$ are rolling, pitching and yawing moments respectively; $T_i, i = 1, 2, 3, 4$ is the lift of the ith rotor; $\Omega_i, i = 1, 2, 3, 4$ is the speed of the ith rotor; $l$ is the distance from rotor to center of mass; $d$ is the reverse torque coefficient of rotor.

Then, the whole dynamic model of the oblique cross UAV is shown as formula (3).

$$\begin{cases} \dot{p} = \frac{J_y - J_z}{J_x} qr + \frac{J_r q(\Omega_1 - \Omega_2 + \Omega_3 - \Omega_4)}{J_x} - \frac{U_2}{J_x} \\ \dot{q} = \frac{J_z - J_x}{J_y} pr - \frac{J_r p(\Omega_1 - \Omega_2 + \Omega_3 - \Omega_4)}{J_y} - \frac{U_3}{J_y} \\ \dot{r} = \frac{J_x - J_y}{J_z} pq + \frac{U_4}{J_z} \\ \dot{\phi} = p + (\sin\phi \tan\theta) q + (\cos\phi \tan\theta) r \\ \dot{\theta} = q\cos\phi - r\sin\phi \\ \dot{\psi} = (\sin\phi / \cos\theta) q + (\cos\phi / \cos\theta) r \\ \ddot{x} = -U_1(\cos\psi \sin\theta \cos\phi + \sin\psi \sin\phi) \\ \ddot{y} = -U_1(sin\psi \sin\theta \cos\phi - \cos\psi \sin\phi) \\ \ddot{z} = g - U_1(\cos\phi \cos\theta) \end{cases} \tag{3}$$

where $J_r$ is the rotational inertia of the rotor; $U_1$ is the torque for lift, and $U_1 = \sum_{i=1}^{4} T_i/m$.

## 3 Control System Design

### 3.1 Design Scheme of Control System

It can be known that each channel of UAV is coupled and interacted with each other. An extended state observer (ESO) can take the coupling effect among different channels as unknown disturbance. By estimating and compensating it, effective decoupling control is finally achieved. Therefore, in the process of designing the controller, there is no need to consider decoupling link, and the controller can be designed independently for each channel. The attitude controller is divided into four channels: height $Z$, roll $\varphi$, pitch $\theta$ and yaw $\psi$. The final control system is shown in Fig. 2.

Figure 2 still adopts the strategy of PID controller, and the error signal is used as the input of each channel. The control volume is transformed into pulse width modulation (PWM) signal by the tension distribution link to realize effective control

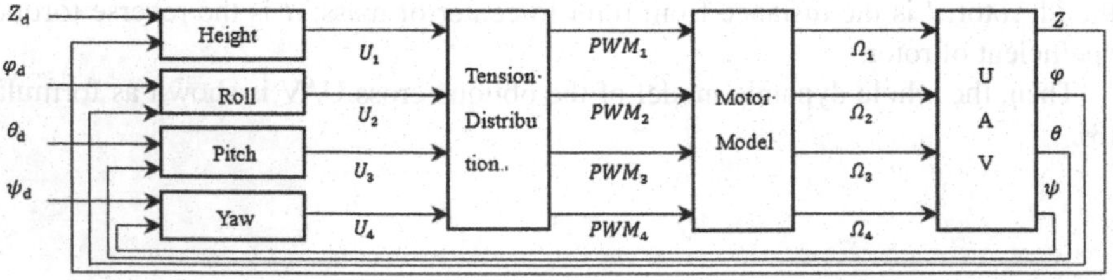


**Fig. 2** Structure of the control system

of the motor. The UAV generally utilizes brushless DC motor to provide power, and the motor model can be simplified to a first order inertia link.

## 3.2 Design of Linear Auto Disturbance Rejection Controller

LADRC is a simplified controller based on ADRC, which preserves the transition process of ADRC. The linear function is adopted in both linear ESO (LESO) and linear state error feedback (LSEF) parts, which can greatly reduce the difficulty of parameter tuning and guarantee the superiority of the algorithm at the same time [11]. According to the separation principle, the controller can be divided into three parts: tracking differentiator (TD), LESO and LSEF. Taking the $\psi$ channel as an example, the design process is introduced as follows:

(1) TD: the main function of this part is to arrange a reasonable transition process for a given input signal to solve the conflict between speed and overshoot. If an input signal $\psi_d(k)$ is given to the TD, then two output $V_1(k)$ and $V_2(k)$ can be obtained, where $V_1(k)$ tracks input signal and $V_2(k)$ is the differential of $V_1(k)$ which can be regarded as the approximate differential of input signal. The second order discrete tracking differentiator is shown as formula (4).

$$\begin{cases} v_1(k+1) = v_1(k) + hv_2(k) \\ v_2(k+1) = v_2(k) + hfhan(v_1(k) - \psi_d(k), v_2(k), r, h_0) \end{cases} \tag{4}$$

The $fhan$ function is the steepest control synthesis function, which is a useful tool for the ADRC controller to arrange the transition process. The definition of this function is shown as formula (5).

$$\begin{cases} d = rh^2 \; a_0 = hx_2 \; y = x_1 + a_0 \\ a_1 = \sqrt{d^2 + 8d|y|} \\ a_2 = a_0 + sign(y)(a_1 - d)/2 \\ s_a = (\sin(a+d) - sin(a-d))/2 \\ fhan = -r(a/d - \sin(a))s_a - r\sin(a) \end{cases} \tag{5}$$

(2) LESO: the most important function of the ESO is to expand the unknown external disturbance and uncertain model into a new state $Z_3$ of the system, and to estimate and compensate it. The input of LESO is the real output of the system $\psi_m(k)$, and state parameters $Z_1$ as well as $Z_2$ are estimated values of the given value $\psi_m(k)$ and yaw rate $r$ respectively. Symbol $\varepsilon$ is an error having $\varepsilon = z_1 - \psi_m(k)$. The discrete form of the third order extended state observer is shown as formula (6).

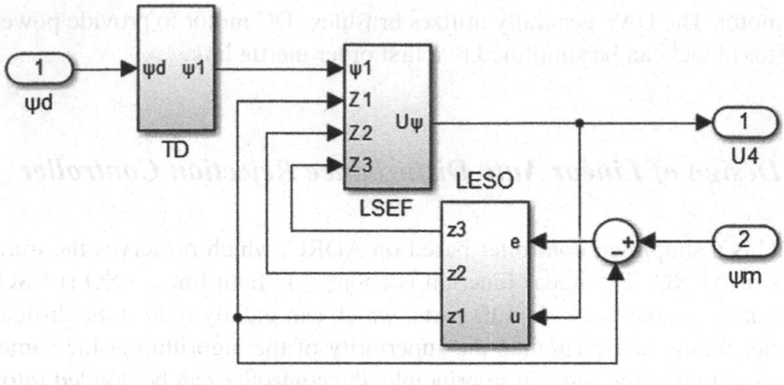


**Fig. 3** Internal structure of the LADRC controller

$$\begin{cases} z_1(k+1) = z_1(k) + h(z_2(k) - \beta_1 \varepsilon) \\ z_2(k+1) = z_2(k) + h(z_3(k) - \beta_2 \varepsilon + b_0 u) \\ z_3(k+1) = z_3(k) + h(-\beta_3 \varepsilon) \end{cases} \tag{6}$$

(3) LSEF: The control quantity we need is formed in this module. Results are shown in formula (7). We can see that the control law is only related to the error $v_1 - z_1$ and $\varepsilon$, like PID, thus, LADRC does not depend on the model. Principle of parameter tuning can refer to PD controller. The controller's structure diagram is shown in Fig. 3.

$$\begin{cases} u_0 = k_p(v_1 - z_1) - k_d z_2 \\ u = (u_0 - z_3)/b_0 \end{cases} \tag{7}$$

## 4 Simulation Experiment and Result Analysis

In order to simulate real flight of the UAV more accurately, it is necessary to consider the different situations, such as the instantaneous change of the control signal, the disturbance in the environment, and even parameter changes of the model. Aiming at these situations, tracking control, anti-interference and robustness experiments are designed respectively. In addition, this part also shows the simulation results of classical PID controller which adopts a dual loop structure to enhance its ability to resist disturbance. Parameters of PID controller are listed in Table 1.

## 4.1 Tracking Experiment

In tracking experiments, the expected inputs of all channels are step signals to simulate the change of control commands in actual flight. The amplitude of the three expected values of the attitude angle is set as 0.5 rad, and the expected value of the height channel is set as 1 m, assuming that the initial values of each channel are zero. The response curves of the PID controller and the LADRC are shown in Fig. 4, and the performance indicators of each channel are listed in Table 2.

From the above results, we can see that although PID controller can track the desired curve quickly, there is a certain overshoot in each channel. This is mainly caused by the coupling among the various channels of the system. Besides, the addition of integral links will lead to poor dynamic performance. Compared with the PID controller, the LADRC controller reduces the overshoot when the adjustment time does not increase. This is because the latter can arrange a reasonable transition process for a given value which can slow down the change of input values and avoid overshooting. At the same time, LSEF can compensate the error in real time, eliminating the necessity of integrating the links, so its rapidity can also be achieved.

## 4.2 Anti-interference Experiment

The actual flight conditions of UAVs are complex and usually affected by gust. In order to simulate this situation, we use rectangular wave signal to simulate gust to conduct anti- interference experiments. At different times, the rectangular wave interference signal, whose duration is 2 s and amplitude is 20% of the steady-state

**Table 1** Parameters of PID controller

| | Outer loop | | | Inner loop | |
|---|---|---|---|---|---|
| | P | I | D | P | D |
| $\varphi$ | 13 | 0.2 | 1.2 | 13 | 1.7 |
| $\theta$ | 16 | 0.4 | 1.5 | 15 | 2.3 |
| $\psi$ | 20 | 0.1 | 1 | 15 | 7 |
| Z | 23 | 10 | 4 | 40 | 8 |

**Table 2** Performance index of tracking experimental

| | Overshoot $\sigma$/% | | Adjusting time $t_s$/s | |
|---|---|---|---|---|
| | PID | LADRC | PID | LADRC |
| $\varphi$ | 10 | 0 | 1.25 | 1.4 |
| $\theta$ | 5 | 0 | 1.1 | 1.2 |
| $\psi$ | 6 | 1 | 2.1 | 1.6 |
| Z | 21 | 1.1 | 3.8 | 1.3 |

**Fig. 4** Response curve of attitude tracking experiment

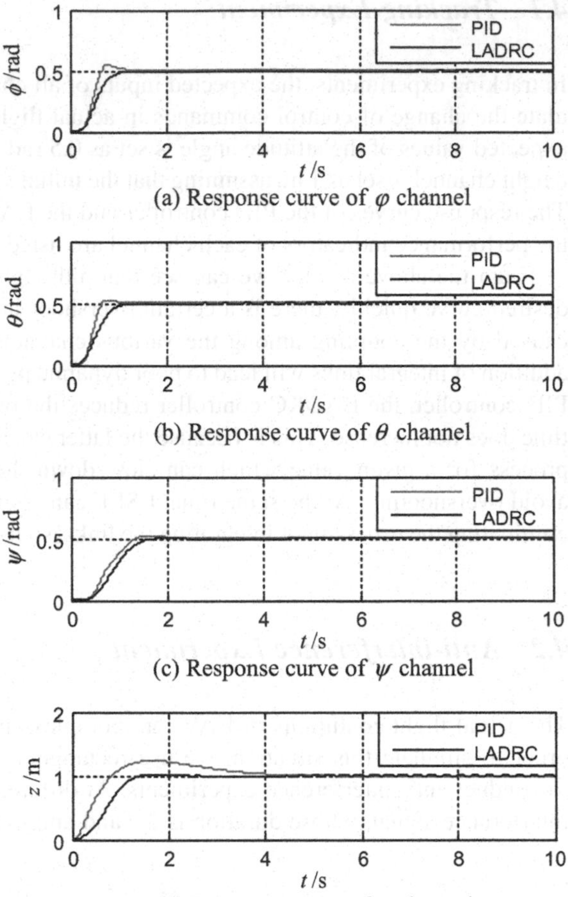


(a) Response curve of $\varphi$ channel

(b) Response curve of $\theta$ channel

(c) Response curve of $\psi$ channel

(d) Response curve of $z$ channel

**Table 3** Performance index of Fig. 5

| | Maximum deviation/rad | | Recovery time $t_s$/s | |
|---|---|---|---|---|
| | PID | LADRC | PID | LADRC |
| $\varphi$ | 0.07 | 0.02 | 2.4 | 2.25 |
| $\theta$ | 0.06 | 0.02 | 2.45 | 2.6 |
| $\psi$ | 0.05 | 0.06 | 2.7 | 2.9 |

value, are added to the $\varphi$, $\theta$ and $\psi$ channels. The results of the experiment are shown in Fig. 5. Performance indicators are listed in Table 3.

From the above comparison, when the UAV is affected by small disturbances, the two controllers of PID and LADRC can quickly restore and achieve stable, and their maximum deviations are similar. In this case, LADRC performs slightly better than PID.

**Table 4** Performance index of Fig. 6

| | Maximum deviation/rad | | Recovery time $t_s$/s | |
|---|---|---|---|---|
| | PID | LADRC | PID | LADRC |
| $\varphi$ | 0.67 | 0.1 | 3.6 | 4.5 |
| $\theta$ | 0.61 | 0.11 | 3.45 | 4.3 |
| $\psi$ | 1.95 | 0.27 | 65.2 | 11.2 |

In order to verify the anti-interference capability of LADRC, we set the amplitude of the interference from 20 to 100%. The results are shown in Fig. 6, and the performance indexes are listed in Table 4.

In Table 4, in the maximum deviation aspect, we can see that LADRC performs much better than PID. Meanwhile, the recovery time of $\psi$ channel obtained by LADRC is much shorter than that obtained by PID, although LADRC executes longer than PID in this performance index.

To sum up, although the PID controller has a certain anti-interference ability, it is obviously impossible to obtain better control result when the system appears frequent and large disturbance. In contrast, LADRC controller can estimate and compensate the interference in real time. It can not only avoid the poor stability and even oscillation caused by the excessive integral constant, but also can overcome

**Fig. 5** Interferences curves with 20% amplitude of each steady-state value

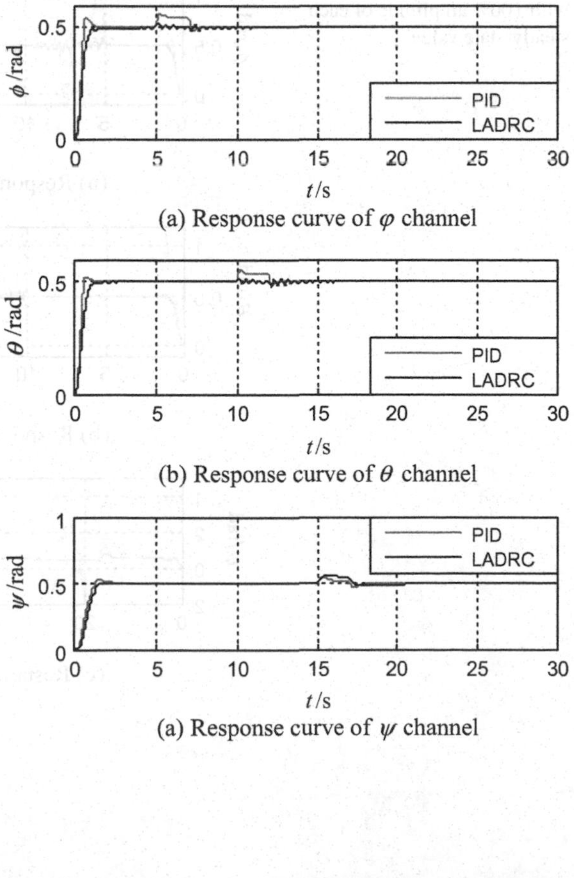

the overreaction of differential module to abrupt interference. In short, the anti-interference capability of LADRC controller is better than that of PID controller.

### 4.3 Robustness Experiment

After several flights, the body may have some loss and slight deformation, which will result in the changes in parameters of UAV mass and the moment of inertia. Therefore, the controller is required to have good robustness. In order to compare the robustness of the two controllers, we keep the parameters of the controller the same as that of the tracking experiment and increase the quality $m$ and the values of $J_x$, $J_y$ and $J_z$ to 120% at the same time. The experimental results are shown in Fig. 7 and the performance indexes are listed in Table 5.

Comparing the above experimental results with that in Sect. 1.4.1, we can see that the results of the two controllers become slightly worse. Therefore, it is known that the two controllers are all robust, and results are similar. In order to better verify the robust characteristics of LADRC controller, the above parameters are set to be doubled. The experimental results are shown in Fig. 8 and performance indexes are listed in Table 6.

**Fig. 6** Interferences curves with 100% amplitude of each steady-state value

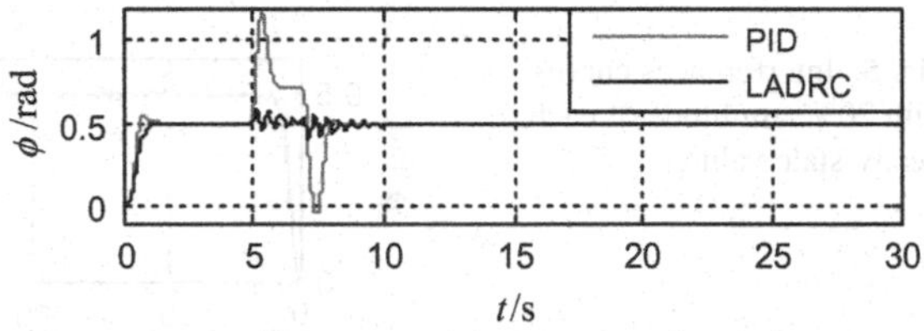


(a) Response curve of $\varphi$ channel

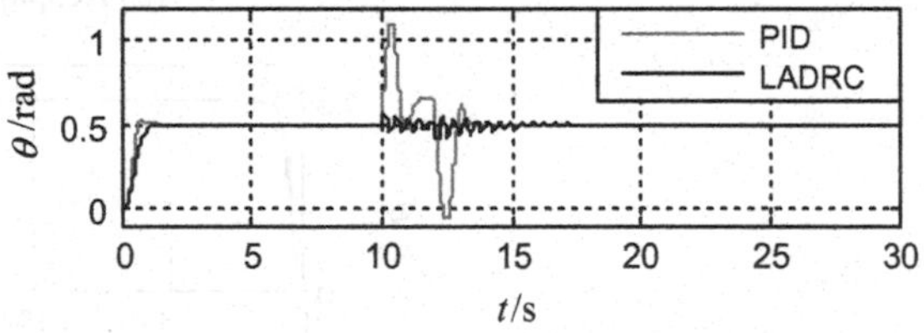


(b) Response curve of $\theta$ channel

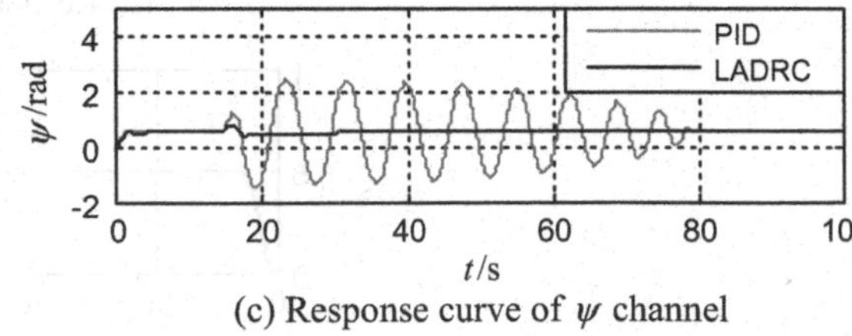


(c) Response curve of $\psi$ channel

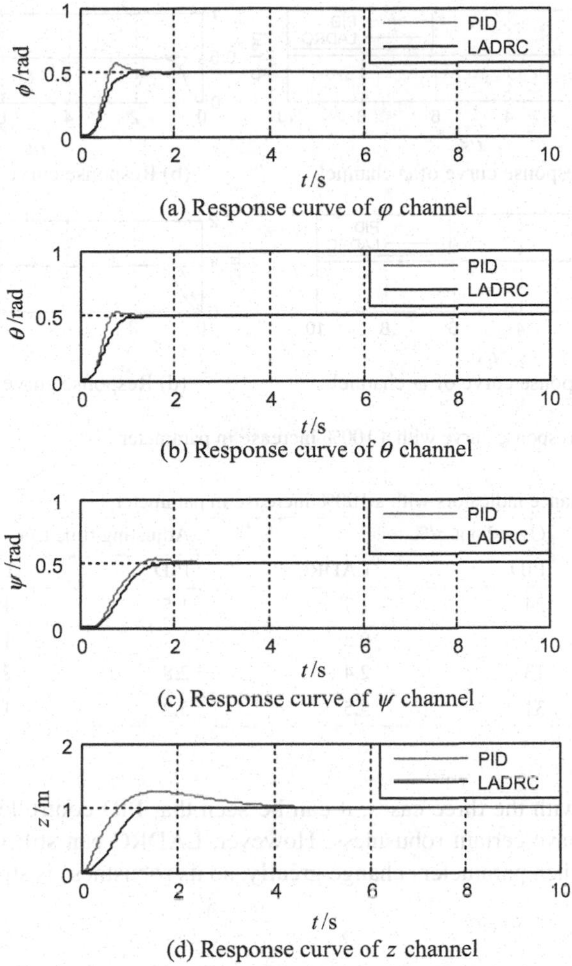


(a) Response curve of $\varphi$ channel

(b) Response curve of $\theta$ channel

(c) Response curve of $\psi$ channel

(d) Response curve of $z$ channel

**Fig. 7** Tracking response curve with a 20% increase in parameter

**Table 5** Performance indicators with a 20% increase in parameter

| | Overshoot σ/% | | Adjusting time $t_s$/s | |
|---|---|---|---|---|
| | PID | LADRC | PID | LADRC |
| $\varphi$ | 13 | 0.7 | 1.4 | 1.47 |
| $\theta$ | 5.4 | 0.2 | 1.2 | 1.1 |
| $\psi$ | 7 | 1.2 | 2.28 | 1.67 |
| Z | 26 | 1.4 | 5.5 | 1.31 |

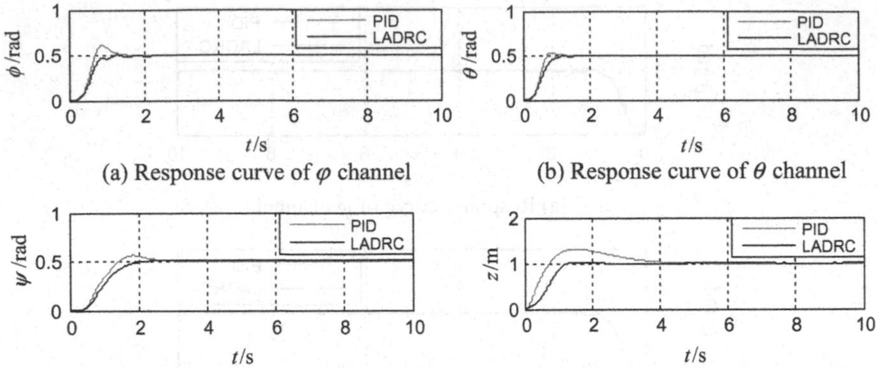


(a) Response curve of $\varphi$ channel (b) Response curve of $\theta$ channel

(c) Response curve of $\psi$ channel (d) Response curve of $z$ channel

**Fig. 8** Tracking response curve with a 100% increase in parameter

**Table 6** Performance indicators with a 100% increase in parameter

| | Overshoot σ/% | | Adjusting time $t_s$/s | |
|---|---|---|---|---|
| | PID | LADRC | PID | LADRC |
| $\varphi$ | 24 | 1.7 | 1.5 | 1.6 |
| $\theta$ | 6 | 0.2 | 1.2 | 1.1 |
| $\psi$ | 13 | 2.4 | 2.8 | 2 |
| Z | 31 | 5.5 | 5.2 | 1.33 |

Compared with the three cases, it can be seen that PID controller and LADRC controller do have certain robustness. However, LADRC can still maintain better performance when parameters change greatly, so its robustness is stronger.

## 5 Conclusion

This paper focuses on the simulation of attitude control for an oblique cross UAV. Some conclusions can be drawn from the results. Firstly, LADRC controller has good tracking performance. It can still track the given value quickly and accurately when the PID controller is unable to coordinate the contradiction between speediness and overshoot. Secondly, LADRC controller has good anti-interference ability and can effectively resist strong gust effect. Thirdly, when the parameters of the controlled object change significantly, LADRC controller can still maintain good control performance and has better robustness.

To sum up, LADRC controller has good tracking performance, anti-interference ability and strong robustness in the attitude control of the oblique cross UAV.

However, the parameter tuning of LADRC controller is still a complicated process. In the future work, attentions will be paid to the study of this problem in order to realize the self-tuning of the controller parameters.

## References


1. S.F. Ahmed, K. Kushsaiiry, M.I.A. Bakar, et al. Attitude stabilization of quad-rotor (UAV) system using Fuzzy PID controller (an experimental test). In *Second International Conference on Computing Technology and Information Management*. IEEE (2015)
2. H.J. Jayakrishnan, Position and Attitude control of a Quad-rotor UAV using super twisting sliding mode. IFAC-Papers Line **49**(1), 284–289 (2016)
3. X. Teng, H.Y. Wu, Y. Chen et al. Trajectory tracking of Quad-Rotor aircrafts base on back-stepping. Comput. Simulat. **33**(5), 78–83 (2016)
4. D. Tang, X. Tang, *Design of UAV attitude controller based on improved robust LQR control*. Youth Academic Conference of Chinese Association of Automation (2017)
5. M. Zheng, M.J. Song, Research on the attitude control of Quad-rotor UAV based on active disturbance rejection control. In *IEEE International Conference on Control Science and Systems Engineering*. IEEE (2017)
6. L.Z. Zhang, Q.Y. Wang, M.M. Wang, et al. Design of UAV attitude control law based on the OGPC. Electron. Opt. Control (2017)
7. H. Wang, X. Fang, Z. CAI, et al. Modelling and control of oblique cross quad-rotors vehicle. In *International Conference on Artificial Intelligence, Management Science and Electronic Commerce*. IEEE (2011)
8. S.Y. Shen, *Modelling based on experiment and control technology of an oblique cross quad-rotor* (Nanjing University of Aeronautics and Astronautics, Nanjing, 2013)
9. S.B. Ding, C.S. Xiao, J.G. Liu, et al. Modeling and quaternion control of X type Quad-rotor. J. Syst. Simulat. **27**(12), 3057–3062 (2015)
10. K. Zhou, *Design and performance analysis of four-rotor aircraft* (Shenyang Aerospace University, Shenyang, 2017)
11. Z. Gao, Scaling and bandwidth-parameterization based controller tuning. In American Control Conference. IEEE, pp. 4989–4996 (2003)

# Fast Convergence for Flocking Motion of Discrete-Time Multi-agent System with Disturbance

**Yize Yang, Hongyong Yang, Fan Liu, Yuling Li and Yuanshan Liu**

**Abstract** For discrete-time flocking problems of networked systems with multiple leaders, containment control algorithms converged in finite time is presented. Based on modern control theory, algebraic graph theory and linear matrix inequality method, the proposed control algorithm is analyzed theoretically. The convergence condition is obtained to ensure the flocking motion in the finite time for discrete-time multi-agent systems with disturbance. Finally, the system simulation results are given to illustrate the correctness of the conclusion.

**Keywords** Multi-agent systems · Discrete-time · Flocking motion · Finite time
Disturbances

## 1 Introduction

As a main branch of distributed systems, multi-agent systems have been widely concerned by many researchers in the fields of UAV formation, robot control and

Y. Yang · H. Yang (✉) · F. Liu · Y. Li · Y. Liu
School of Information and Electrical Engineering,
Ludong University, Yantai 264025, Shandong, China
e-mail: hyyang@yeah.net

Y. Yang
e-mail: yangyz1994@126.com

F. Liu
e-mail: jsgyliufan@163.com

Y. Li
e-mail: liyuling822@163.com

Y. Liu
e-mail: yuanshanliu@163.com

Y. Yang
School of Electrical Engineering and Telecommunications,
University of New South Wales, Sydney, Australia

 
Y. Jia et al. (eds.), *Proceedings of 2018 Chinese Intelligent Systems Conference*,
Lecture Notes in Electrical Engineering 528,
https://doi.org/10.1007/978-981-13-2288-4_78

distributed sensors network. Multi-agent systems flocking movement has become a very important research topic in distributed systems. In order to study the consistency problem of multi-agent continuum system, the literature [1] studies the consistency of delay multi-agent systems with dynamic leaders. The following consistency control problem is discussed in [2, 3] respectively on the second-order multi-agent systems based on uniform sampling control and the generalized linear multi-agent systems based on events.

The research results of the previous literature mainly focus on the infinite time asymptotic convergence of multi-agent systems. In practical engineering application, systems need to achieve the expected goal in finite time, which is to achieve stability in finite time. For the finite time convergence problem of multi-agent systems, the finite time control problem of multi-agent systems with general linear dynamics is studied in [4]. Based on the principle of graph theory, matrix theory, homogeneity expansion and LaSalle invariance, the literature [5] designs multi-agent control protocols and analyzes the finite time consistency of leader-following systems in detail.

In practical engineering applications, the performance of systems is affected by external disturbances. For the study of the consistency problem of discrete-time systems with disturbances, the paper [6], the finite time control problem of discrete-time systems with modelling uncertainties and perturbation terms is studied, and sufficient conditions for finite time-boundedness of the system are given. The extended LMI features of robust finite time control for discrete-time uncertain linear systems are presented in [7]. More results about consensus problem of multi-agent with disturbances were obtained by many outstanding researchers [8, 9].

In this paper, using modern control theory and linear matrix inequalities, the problem of flocking motion of multi-agent systems with disturbances in discrete time is studied. The state feedback control protocol of first-order networked system is designed, and the finite time convergence condition of the system can be realized under the two-way communication network structure. Finally, the numerical simulation of the system shows that the multi-agent systems can be stable in a finite time.

## 2 Preliminaries and Problem Description

Let the graph $\zeta = (V, \omega, A)$ be a weighted bidirected graph of nodes $n$, $V = \{1, 2, \ldots, n\}$ is a collection of vertices (or nodes). $\omega \subseteq V \times V$ is a collection of edges. $A = [a_{ij}] \in R^{n\times n}$ is the node adjacency matrix ($\forall i \in V$, $a_{ii} = 0$). Considering $\forall i, j \in V$, $i \neq j$, if $(i, j) \in \omega$ , so $a_{ij} > 0$, otherwise $a_{ij} = 0$. The Neighbor collection of node $i$ is defined as $N_i = \{j \in V | (i, j) \in \omega\}$. Let the matrix $D = diag\{d_1, d_2, \ldots, d_n\} \in R^{n\times n}$ be a measure matrix of the graph $\zeta$ , where $d_i = \sum_{j\in N_i} a_{ij}$ , $i = 1, 2, \ldots, n$. The analogous Laplacian matrix of topology graph $\zeta$ is defined as $L = D - A \in R^{n\times n}$ .

**Definition 1** Assuming the set $X = \{x_1, x_2, \ldots, x_n\}$ is the subset of Real vector space $R^n$ , the definition of convex hull $X$ is

$$CO(X) = \left\{ \sum_{i=1}^{n} \alpha_i x_i | x_i \in X, \alpha_i \geq 0, \sum_{i=1}^{n} \alpha_i = 1 \right\}.$$

**Lemma 1 [6]** (Finite time boundedness) *Considering discrete-time linear systems*

$$\begin{aligned} x(k+1) &= Ax(k) + Gw(k) \\ w(k+1) &= Fw(k) \end{aligned} \tag{1}$$

*is related to* $(\delta_x, \delta_w, \varepsilon, R, N)$ *in finite time bounded. If* $\forall k \in \{1, \ldots, N\}$ ,

$$\begin{cases} x^T(0)Rx(0) \leq \delta_x^2 \\ w^T(0)w(0) \leq \delta_w^2 \end{cases} \Rightarrow x^T(k)Rx(k) < \varepsilon^2,$$

*where* R *is Positive definite matrices,* $0 \leq \delta_x < \varepsilon$ , $\delta_w \geq 0$.

# 3 Flocking Motion of Discrete Time Multi-agent System with Leaders

## *3.1 A Model of Containment Control Algorithm for Discrete-Time Networked Systems*

Considering a networked system composed of $n$ followers and $m$ leaders, the follower set and the leader set are $\Gamma = \{1, 2, \ldots, n\}$ and $\Upsilon = \{n+1, n+2, \ldots, n+m\}$ respectively.

This paper assumes that the state of the system is in real space $R$, But the research conclusion can be extended to $R^n$ by applying Kronecker operator.

Assuming leaders are static, the speed is zero, then the discrete time multi-agent system can be described as:

$$x_i(k+1) = x_i(k) + Tu_i(k) + Gw_i(k) \tag{2}$$

where $x_i(k) \in R$ denotes the position state of the agent $i$, $u_i(k) \in R$ denotes the control input of the agent $i$, $w_i(k)$ denotes the disturbance of the agent $i$. $T$ denotes the sampling period, $G$ denotes external disturbance matrix. Assuming external disturbances are satisfied $w_i(k+1) = Fw_i(k)$, $F$ denotes the discrete time gain. Let the control protocol for multi-agent discrete-time networked systems,

$$\begin{cases} u_i = \sum\limits_{j \in N_i} a_{ij}(x_j - x_i), & i \in \Gamma \\ u_i = 0, & i \in \Upsilon \end{cases} \tag{3}$$

This means that,

$$u(k) = -Lx(k) \tag{4}$$

where $x(k) = [x_F(k), x_{F\Upsilon}(k)]^{\mathrm{T}}$, $x_F(k) = [x_1(k), \ldots, x_n(k)]^{\mathrm{T}}$, $x_{F\Upsilon}(k) = [x_{n+1}(k), \ldots, x_{n+m}(k)]^{\mathrm{T}}$. $L$ denotes the Laplacian matrix of graph $\zeta$, and $L = \begin{bmatrix} L_F & L_{F\Upsilon} \\ 0_{m\times n} & 0_{m\times m} \end{bmatrix}$, where $L_F$ is n-order phalanx, $L_{F\Upsilon}$ is the $n \times m$ order matrix.

*Remark 1* If a network of $n$ followers is a bidirectional connected graph, at least one follower is connected to the leader, $L_F$ is the positive definite matrix.

*Remark 2* If a network of $n$ followers is a unidirectional connected graph, at least one follower is connected to the leader, $L_F$ is the positive semidefinite matrix.

This paper mainly considers the networked system with two-way connectivity, but it can still achieve systems stability without losing generality.

**Definition 2** Let $X_F = [x_1, \ldots, x_n]^T$, $X_k = [x_{n+1}, \ldots, x_{n+m}]^T$, such that $X_F \to -L_F^{-1} L_{F\Upsilon} X_k$, the networked system (2) realizes containment control.

*Remark 3* If leaders are dynamic, it is necessary to estimate the motion trajectory of the multi-agent by using the state observer. Based on the speed estimation combined with definition 2, the networked system containment control of dynamic leaders can be realized. In this paper, it is assumed that leaders in multi-agent networked systems are static, and the state observer should not be considered.

## 3.2 Analysis of Containment Control Algorithm for First-Order Discrete-Time Networked Systems

**Theorem 1** *Considering a networked system composed of n followers and m leaders. Assuming that a network topology composed of n followers is bidirectional connected, at least one follower can receive information from the leader. The discrete-time multi-agent networked system with leaders (2) based on control protocol (3) can realize finite time bounded, if there are positive definite matrices $P_1$ and $P_2$ satisfied*

$$\begin{bmatrix} (I - TL_F)^T P_1 (I - TL_F) - \gamma P_1 & (I - TL_F)^T P_1 G \\ G^T P_1 (I - TL_F) & G^T P_1 G + F^T P_2 F - \gamma P_2 \end{bmatrix} < 0,$$

$$\frac{1}{\lambda_{min}(\tilde{P}_1)}\gamma^N\left[\lambda max(\tilde{P}_1)\delta_y^2 + \lambda max(P_2)\delta_w^2\right] \leq \varepsilon^2.$$

*where* $\gamma > 1$, $\tilde{P}_1 = R^{-1/2}P_1R^{-1/2}$ .

*Proof* Assuming that leaders are unaffected by the disturbance, discrete-time linear systems based on the control protocol (3) are

$$\begin{cases} x_F(k+1) = x_F(k) - T[L_F x_F(k) + L_{F\Upsilon} x_{F\Upsilon}(k)] + Gw(k) \\ x_{F\Upsilon}(k+1) = x_{F\Upsilon}(k) \end{cases} \tag{5}$$

where $w(k+1) = Fw(k)$ denotes disturbance in this system, $T$ denotes the sampling period.

Let $y(k) = x_F(k) + L_F^{-1}L_{F\Upsilon}x_{F\Upsilon}(k)$, then we have

$$\begin{aligned} y(k+1) &= x_F(k+1) + L_F^{-1}L_{F\Upsilon}x_{F\Upsilon}(k+1) \\ &= x_F(k) - TL_F\left[x_F(k) + L_F^{-1}L_{F\Upsilon}x_{F\Upsilon}(k)\right] + Gw(k) + L_F^{-1}L_{F\Upsilon}x_{F\Upsilon}(k) \\ &= x_F(k) + L_F^{-1}L_{F\Upsilon}x_{F\Upsilon}(k) - TL_F y(k) + Gw(k) \\ &= y(k) - TL_F y(k) + Gw(k) \end{aligned}$$

The system equations change to

$$\begin{cases} y(k+1) = y(k) - TL_F y(k) + Gw(k) \\ x_{F\Upsilon}(k+1) = x_{F\Upsilon}(k) \\ w(k+1) = Fw(k) \end{cases} \tag{6}$$

From Lemma 1, assume that $\forall k \in \{1, \ldots, N\}$

$$\begin{cases} y^T(0)Ry(0) \leq \delta_y^2 \\ w^T(0)w(0) \leq \delta_w^2 \end{cases}, \tag{7}$$

where $R$ is the positive definite matrix, $0 \leq \delta_y < \varepsilon$ , $\delta_w \geq 0$.

The Lyapunov function is constructed as,

$$V(k) = y^{\mathrm{T}}(k)P_1 y(k) + w^{\mathrm{T}}(k)P_2 w(k) \tag{8}$$

where $P_1 \in R^{n\times n}$ and $P_2 \in R^{n\times n}$ are positive definite matrices. Thus, it is easy to get

$$\begin{aligned}\Delta &= V(K+1) - \gamma V(K) \\ &= \left[y^{\mathrm{T}}(k+1)P_1 y(k+1) + w^{\mathrm{T}}(k+1)P_2 w(k+1)\right] \\ &\quad - \left[y^{\mathrm{T}}(k)P_1 y(k) + w^{\mathrm{T}}(k)P_2 w(k)\right] \\ &= \left[y^T(k)\ w^T(k)\right]\begin{bmatrix}(I-TL_F)^T P_1(I-TL_F) - \gamma P_1 & (I-TL_F)^T P_1 G \\ G^T P_1(I-TL_F) & G^T P_1 G + F^T P_2 F - \gamma P_2\end{bmatrix}\begin{bmatrix}y(k) \\ w(k)\end{bmatrix}\end{aligned} \tag{9}$$

where $\gamma > 1$. According to the condition of Theorem 1,

$$\begin{bmatrix}(I-TL_F)^T P_1(I-TL_F) - \gamma P_1 & (I-TL_F)^T P_1 G \\ G^T P_1(I-TL_F) & G^T P_1 G + F^T P_2 F - \gamma P_2\end{bmatrix}$$

is a negative definite, that is $\Delta = V(k+1) - \gamma V(k) < 0$.

For discrete-time linear systems (6), let $\tilde{P}_1 == R^{-1/2}P_1R^{-1/2}$ , then

$$y^T(k)\lambda_{\min}(\tilde{P}_1)Ry(k) \le y^T(k)P_1 y(k) \le y^T(k)\lambda_{\max}(\tilde{P}_1)Ry(k) \tag{10}$$

Thus, according to $V(k) = y^T(k)P_1y(k) + w^T(k)P_2w(k)$, we have

$$\begin{aligned}V(k) &\ge y^T(k)P_1 y(k) \\ &\ge \lambda_{\min}(\tilde{P}_1)y^T(k)Ry(k)\end{aligned} \tag{11}$$

Since $\Delta < 0$, Multiple iterations of $V$ function,it follows that

$$V(k) < \gamma^k V(0). \tag{12}$$

Since

$$\begin{aligned}V(0) &= \left[y^{\mathrm{T}}(0)P_1 y(0) + w^{\mathrm{T}}(0)P_2 w(0)\right] \\ &\le \left[\lambda_{\max}(\tilde{P}_1)y^{\mathrm{T}}(0)Ry(0) + \lambda_{\max}(\tilde{P}_1)w^{\mathrm{T}}(0)w(0)\right]\end{aligned} \tag{13}$$

It follows that

$$\gamma^k V(0) \le \gamma^k\left[\lambda_{\max}(\tilde{P}_1)y^T(0)Ry(0) + \lambda_{\max}(P_2)w^T(0)w(0)\right] \tag{14}$$

It then follows from $y^{\mathrm{T}}(0)Ry(0) \le \delta_y^2$ , $w^{\mathrm{T}}(0)w(0) \le \delta_w^2$ , $\forall k \in \{1, 2, \ldots, N\}$ we have

$$\gamma^k V(0) \le \gamma^N\left[\lambda_{\max}(\tilde{P}_1)\delta_y^2 + \lambda_{\max}(P_2)\delta_w^2\right] \tag{15}$$

where $\gamma > 1$. According to the condition of theorem 1 and Eq. (11), we can obtain

$$y^T(k)Ry(k) < \frac{1}{\lambda_{\min}(\tilde{P}_1)}\gamma^N\left[\lambda\max(\tilde{P}_1)\delta_y^2 + \lambda\max(P_2)\delta_w^2\right] \le \varepsilon^2 \tag{16}$$

Therefore, by the assumption $\begin{cases} y^{\mathrm{T}}(0)Ry(0) \le \delta_y^2 \\ w^{\mathrm{T}}(0)w(0) \le \delta_w^2 \end{cases}$, it follows that

$$y^{\mathrm{T}}(k)Ry(k) < \varepsilon^2. \tag{17}$$

From Lemma 1, It can be concluded that the discrete-time multi-agent systems realizes finite time bounded on $(\delta_y, \delta_w, \varepsilon, R, N)$.

**Lemma 2 [7]** *Considering a linear stationary discrete system is*

$$\begin{cases} x(k+1) = Gx(k) \\ x_e = 0 \end{cases} \tag{18}$$

*where $x$ is the $n$ dimension vector, $x_e$ is the equilibrium point, $G$ the $n \times n$ dimensional constant coefficient nonsingular matrix. The necessary and sufficient conditions for the asymptotic stability of systems at the equilibrium point are: For any given positive definite matrix $Q$, there is a positive definite real symmetric matrix $P$ and satisfies the following matrix equation,*

$$G^{\mathrm{T}}PG - P = -Q, \tag{19}$$

*and $v[x(k)] = x^{\mathrm{T}}(k)Px(k)$ is the Lyapunov function of this system.*

**Corollary 1** *Considering a networked system composed of n followers and m leaders. Assuming that a network topology composed of n followers is two-way connected, at least one follower can receive information from the leader. The discrete-time multi-agent networked system with leaders based on control protocol (3) can realize finite time bounded, if there are positive definite matrices $P_1$ and $P_2$ satisfied*

$$\begin{bmatrix} (I - TL_F)^T P_1 (I - TL_F) - P_1 & (I - TL_F)^T P_1 G \\ G^T P_1 (I - TL_F) & G^T P_1 G + F^T P_2 F - P_2 \end{bmatrix} < 0,$$

$$\frac{1}{\lambda_{min}(\tilde{P}_1)}\left[\lambda max(\tilde{P}_1)\delta_y^2 + \lambda max(P_2)\delta_w^2\right] \le \varepsilon^2.$$

*Proof* The Lyapunov function can be described as $V(k) = y^{\mathrm{T}}(k)P_1y(k) + w^{\mathrm{T}}(k)P_2w(k)$. According to the conditions in Theorem 1, let $\gamma = 1$, then

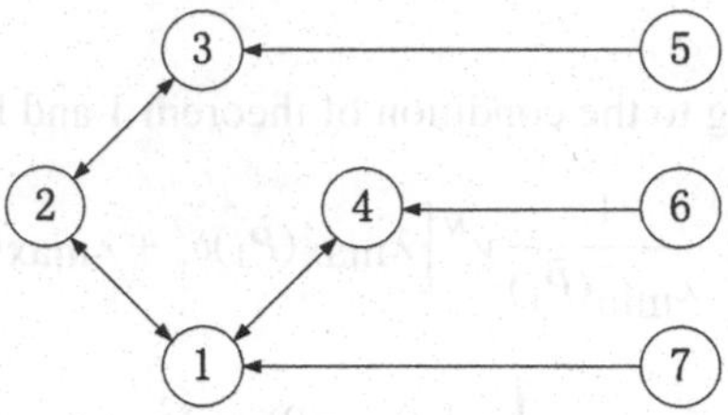


**Fig. 1** Multi-agent connected graphs

$$\begin{bmatrix} (I - TL_F)^T P_1 (I - TL_F) - P_1 & (I - TL_F)^T P_1 G \\ G^T P_1 (I - TL_F) & G^T P_1 G + F^T P_2 F - P_2 \end{bmatrix}$$

is a negative definite matrix, which meet the condition of Lemma 2.$y(k)$ is Asymptotic stability $\lim_{k\to\infty} y(k) = 0$. That is $\lim_{k\to\infty} \left[x_F(k) + L_F^{-1} L_{F\Upsilon} x_{F\Upsilon}(k)\right] = 0$, then $x_F(k) \to -L_F^{-1} L_{F\Upsilon} x_{F\Upsilon}(k)$. This means that the system achieves containment control in finite time.

From Lemma 1, it can be concluded that the discrete-time multi-agent systems realizes finite time asymptotic stability on $(\delta_y, \delta_w, \varepsilon, R, N)$, that is containment control in finite time.

## 4 Numerical Simulations

Consider a topological structure consisting of 3 leaders and 4 followers, as shown in Fig. 1, where agent 5, 6, 7 are leaders, the rest of agents are followers. Assuming that the weights of all the lines in the topology diagram are 1, the parameters in the control protocol are set to $G = 1$, $F = 0.6$, $\gamma = 1$. Let $R = \begin{bmatrix} 1 & \\ & 1 \end{bmatrix}$, $\delta_y = 1.3$, $\delta_w = 0.01$, then we can get $P_1 = 0.14$, $P_2 = 1.66$, $T = 0.4$, $\varepsilon = 1.6$ by using computer simulation.

Assuming that the initial state of followers is $x_1 = (0, 4)$, $x_2 = (4, 0)$, $x_3 = (0, 2)$, $x_4 = (2, 0)$. The initial state of leaders is $x_5 = (6, 8)$, $x_6 = (8, 8)$, $x_7 = (8, 6)$. It is shown from Fig. 2 that the position state of followers eventually converges to the planar triangular region surrounded by the position of three leaders, and the dynamic multi-agent system realizes the containment control.

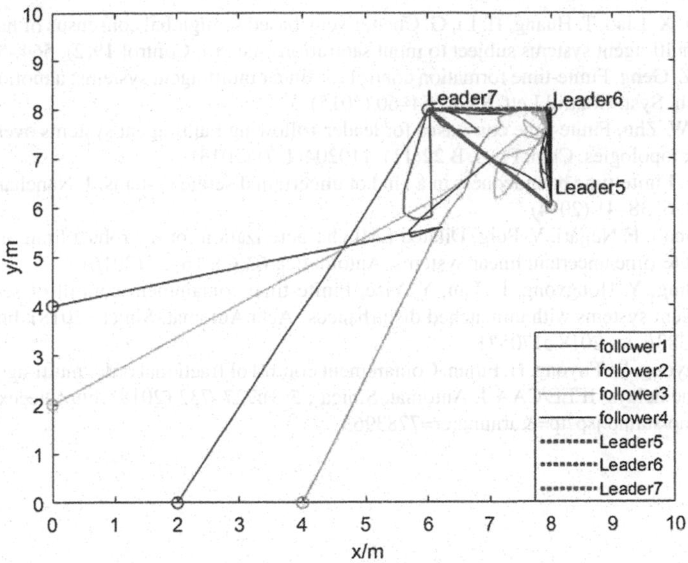


**Fig. 2** Motion trajectory of discrete-time multi-agent systems without disturbances

# 5 Conclusions

In this paper, the problem of containment control for the discrete time of multi-agent systems is studied, and the finite time control algorithm of multi-agent with external disturbances is considered. The coordinated control protocol of the first-order networked system is designed, and the flocking motion of the discrete-time multi-agent systems with disturbances is studied by means of Lyapunov function and algebraic graph theory. Finally, the simulation of the algorithm is carried out by computer simulation, and the multi-agent systems can converge quickly to the area surrounded by multi-agent.

**Acknowledgements** This paper is supported by the National Natural Science Foundation of China (61673200, 61771231, 61471185), Natural Science Foundation of Shandong Province (ZR2018ZC0438, ZR2017MF010, ZR2017PF010) and the Key R&D Program of Yantai City (2016ZH061).

# References

1. H.Y. Yang, S.W. Tian, S.Y. Zhang, Consensus of multi-agent systems with heterogeneous delays and leader-following. Acta Electron. Sinica **39**(4), 872–876 (2011)
2. B. Zhou, X. Liao, Leader-following second-order consensus in multi-agent systems with sampled data via pinning control. Nonlinear Dyn. **78**(1), 555–569 (2014)

3. B. Zhou, X. Liao, T. Huang, H. Li, G. Chen, Event-based semiglobal consensus of homogenous linear multi-agent systems subject to input saturation. Asian J. Control **19**(2), 564–574 (2016)
4. Y. Liu, Z. Geng, Finite-time formation control for linear multi-agent systems: a motion planning approach. Syst. Control Lett. **85**(9), 54–60 (2015)
5. F. Sun, W. Zhu, Finite-time consensus for leader-following multi-agent systems over switching network topologies. Chin. Phys. B **22**(11), 110204(1-7) (2013)
6. Zhu Lin, Finite-time boundedness in a kind of uncertain discrete systems. J. Nanchang Institute Technol. **6**, 38–41 (2014)
7. D. Rotondo, F. Nejjari, V. Puig, Dilated LMI characterization for the robust finite time control of discrete-time uncertain linear systems. Automatica **63**(C), 16–20 (2016)
8. L.I. Yuling, Y. Hongyong, L. Fan, Y. Yize, Finite-time containment control of second-order multi-agent systems with unmatched disturbances. Acta Automat. Sinica (2018). https://doi.org/10.16383/j.aas.2018.c170571
9. Y. Hongyong, W. Fuyong, H. Fujun. Containment control of fractional order multi-agent systems with time delays. IEEE/CAA J. Automat. Sinica , **5**(3):727–732 (2018). http://ieeexplore.ieee.org/stamp/stamp.jsp?tp=&arnumber=7783963

# Fault-Tolerant Time-Varying Formation Tracking for Second-Order Multi-agent Systems Subjected to Directed Topologies and Actuator Failures with Application to Cruise Missiles

**Xingguang Xu, Zhenyan Wei, Zhang Ren and Shusheng Li**

**Abstract** Fault-tolerant time-varying formation tracking problems for second-order multi-agent systems with actuator failures and directed topologies are investigated. Firstly, a distributed formation tracking control protocol is constructed using the adaptive law. In the case where the information of actuator failures remain unknown and only the local information of neighboring agents is available. Then the formation tracking condition is provided, and it is proven that by designing the formation tracking protocol using the proposed approaches, time-varying formation tracking can be achieved by the multi-agent system in the presence of actuator failures. The obtained results are applied to solve the formation tracking problem of a multi cruise missile system through acceleration tracking. Finally, numerical simulations are presented to demonstrate the effectiveness of the theoretical results.

**Keywords** Fault-tolerant control · Formation tracking · Multi-agent system
Cruise missile

## 1 Introduction

Recently formation control of multi-agent system has been paid considerable attention from different research areas, such as unmanned aerial vehicles (UAVs) formation [1–3], mobile robots formation [4], spacecraft formation [5], autonomous underwater vehicles [6] etc. The object of formation control is to make a cluster of agents form a specific shape in the state space. How to design distributed formation control protocols using neighboring agents' local information has long been the common concern.

X. Xu · Z. Wei (✉) · Z. Ren · S. Li
School of Automation Science and Electrical Engineering,
Beihang University, Beijing, China
e-mail: weizhenyan@buaa.edu.cn



Y. Jia et al. (eds.), *Proceedings of 2018 Chinese Intelligent Systems Conference*,
Lecture Notes in Electrical Engineering 528,
https://doi.org/10.1007/978-981-13-2288-4_79

There exist several classic formation control approaches, including leader-follower, virtual structure, and behavior based ones, but these three strategies have their own weaknesses [7].

In the past decade, consensus-based formation control strategies have been extensively studied, and the above-mentioned strategies can be converted into the consensus based ones. Because of the universality of the second-order system in describing the agent's dynamics, distributed formation control protocols for second-order multi-agent systems under directed topologies are proposed in [8–15].

Nowadays, formation tracking problems, where followers are supposed to maintain a time-varying formation while achieving target enclosing, have gained more attention [7, 16–21]. In [7], a formation tracking protocol was presented for second-order multi-agent systems under switching interaction topologies, and the obtained results were applied to the target enclosing problem of a multi quadrotor system. Time-varying formation tracking analysis and design problems using only neighboring relative information for second-order multi-agent systems with one leader were investigated in [16], and an engineering application to multiple vehicles was addressed. In [17], time-varying formation tracking strategies were applied for swarm systems with multiple leaders. Two formation tracking protocols were investigated for second-order multi-agent systems with time-varying delays using neighboring position and full state information respectively in [18, 19].

It should be stressed that, each agent of the multi-agent system may encounter actuator malfunction as a result of the complexity and uncertainty of the agents' dynamics, and the faults may spread across neighboring agents through the interaction topology, thus cause destructive malfunction in the whole system. However, the invalidation problem of actuators was not taken into account in [7–21].

Fault-tolerant control (FTC) has long been considered as one of the most effective control approaches for maintaining good performance for each agent under circumstance of unexpected faults, so it is significant to go into the fault-tolerant control problem of the multi-agent system with actuator malfunction. The FTC problem for the multi-agent system in the presence of actuator failures was addressed in [22–28]. To the best of our knowledge, fault-tolerant time-varying formation tracking problems for multi-agent systems subjected to directed topologies and actuator failures are still open.

In the current paper, fault-tolerant time-varying formation tracking control problems for the second-order multi-agent system with directed topologies and actuator failures are dealt with. Firstly, a distributed formation tracking control protocol is constructed using the adaptive law, where the information of actuator failures remain unknown and only the local information of neighboring agents is available. Then the formation tracking condition is addressed, and it is verified that by designing the formation tracking protocol using the proposed approaches, time-varying formation tracking can be achieved by the multi-agent system in the presence of actuator failures. The obtained results are applied to investigate the formation tracking problem of a multi cruise missile system through acceleration tracking. Finally, practical applications of the theoretical results to the formation tracking control of the multi-agent system are presented.

Compared with the existing results, the main features of the current paper are threefold. Firstly, the current paper studies time-varying formation tracking control approaches where there are one leader and multiple followers, in contrast to [7–21]. In [22–28], fault-tolerant control protocols are also proposed for the multi-agent system with actuator failures. However, only formation problem without target enclosing was considered in the fault-tolerant control protocol design. Secondly, both the unknown gain fault and bias fault are addressed in the time-varying formation tracking control approaches in the current paper. In [7, 16–21], the invalidation problem of actuators was not considered when designing the time-varying formation tracking control protocols and the multi-system could not maintain desired performance applying these proposed protocols directly in the presence of actuator failures. Thirdly, the engineering application of consensus-based formation control strategies for aircrafts has not been studied fully yet, the obtained results can be extended to a class of cruise missiles powered by turbojet engine in this current paper. The formation tracking problems of multi UAV systems is investigated through an outer-loop control and an inner-loop control, of which the former is derived from the obtained protocols and used as the control objects of the latter. In [1–3, 7, 21], the proposed approaches were applied to UAVs, quadrotor aircrafts and ballistic missiles respectively.

The remainder of this paper is organized as follows. Basic concept on graph theory and the problem description are presented in Sect. 2. Main results on the formation tracking FTC problem are derived in Sect. 3. Engineering applications to formation tracking of cruise missiles are given in Sect. 4. The numerical simulation is conducted in Sect. 5. Finally, Sect. 6 concludes this paper.

## 2 Preliminaries and Problem Description

In this section, some basic concepts on graph theory are introduced and the problem description is presented.

### 2.1 *Basic Concepts on Graph Theory*

The interaction topology of a multi-agent system can be represented by a directed graph $G$ denoted by $\{V, \varepsilon, W\}$, where $V = \{v_1, v_2, \ldots, v_N\}$, $\varepsilon \subseteq \{(v_i, v_j) : v_i, v_j \in V\}$, and $W = [w_{ij}] \in \mathbb{R}^{N\times N}$, $w_{ij} \geq 0$ represent the node set, the edge set and the weighted adjacency matrix associated with $G$, respectively. The edge is denoted by $e_{ij} = (v_i, v_j)$ in $G$. The neighbor set of node $v_i$ is represented by $N_i = \{v_j \in V : (v_j, v_i) \in \varepsilon\}$. The in-degree of node $i$ is defined as $\deg_{in}(v_i) = \sum_{j=1}^{N} w_{ij}$, let $D = diag\{\deg_{in}(v_i), i = 1, 2, \ldots, N\}$ be the degree matrix of $G$. The Laplacian matrix of $G$ is defined as $L = D - W$. A directed path from node $v_i$ to node $v_j$ is a sequence of ordered edges with the form of $(v_{i_k}, v_{i_{k+1}})$, where $v_{i_k} \in V (k = 1, 2, \ldots, l-1)$. A directed graph is said to have a spanning tree if there exists at least one node having a directed path to all the other nodes. The interaction topology among the $N$ agents can be denoted by the graph $G$ with each agent being a node in $G$ and the interaction from agent $i(i \in \{1, 2, \ldots, N\})$ to agent $j(j \in F)$ is represented by the edge $e_{ij}$. An agent is called a leader if it has no neighbors and is called a follower if it has at least one neighbor.

**Lemma 1** *[12, 23] If the directed graph $G$ is connected, then*

*(1) 0 is a simple eigenvalue of $L$ and all the other $N-1$ eigenvalues have positive real parts.*

*(2) There exists a positive vector $\varphi = [\varphi_1, \varphi_2, \ldots, \varphi_N]^T$ with $\sum_{i=1}^N \varphi_i = 1$ such that $\varphi^T L = 0$.*

*(3) Let $\psi = diag\{\varphi_1, \varphi_2, \ldots, \varphi_N\}$, then $\hat{L} = \psi L + L^T \psi$ is a symmetric Laplacian matrix associated with an undirected connected graph. Moreover, let $\varsigma \in \mathbb{R}^{N\times 1}$ stand for any positive column vector and $\lambda_2(\hat{L})$ denote the minimum nonzero eigenvalue of $\hat{L}$. For $\phi(t) \in \mathbb{R}^{N\times 1}$, it holds that $\frac{\lambda_2(\hat{L})}{N}\phi^T(t)\phi(t) < \min_{\phi^T(t)\varsigma=0} \{\phi^T(t)\hat{L}\phi(t)\}$.*

## 2.2 Problem Description

Consider a second-order multi-agent system with $N$ agents. Assume that there exist one leader and $N-1$ followers and the leader and followers suffer from both loss of effectiveness and bias faults.

The control objective is to let all the $N-1$ followers form a predefined time-varying formation while tracking the trajectory of the leader in the presence of actuators malfunction. Consider the following dynamics for the leader

$$\begin{cases} \dot{x}_1(t) = v_1(t) \\ \dot{v}_1(t) = \rho_1(t)u_1(t) + b_1(t) \end{cases} \tag{1}$$

where $x_1(t) \in \mathbb{R}^n$ and $v_1(t) \in \mathbb{R}^n$ are the position and velocity vectors of the leader, respectively, with $n \geq 1$ being the dimension of the space, $\rho_1(t) = diag\{\rho_{11}(t), \rho_{12}(t), \ldots, \rho_{1n}(t)\}$ and $b_1(t) = [b_{11}(t), b_{12}(t), \ldots, b_{1n}(t)]^T$, where $0 < \rho_{1j}(t) \leq 1$ denotes the unknown gain fault and $b_{1j}(t)$ is the unknown bias fault of actuator channel $j(j = 1, 2, \ldots, n)$ for the leader.

The dynamics of the $N-1$ followers are described by

$$\begin{cases} \dot{x}_i(t) = v_i(t), \\ \dot{v}_i(t) = \rho_i(t)u_i(t) + b_i(t), \end{cases} \tag{2}$$

where $n \geq 1$ is the dimension of the space, $x_i(t) \in \mathbb{R}^n$, $v_i(t) \in \mathbb{R}^n$ and $u_i(t) \in \mathbb{R}^n$ denote the position, velocity and control input vectors of agent $i$ respectively.

**Table 1** Actuator failure modes of the multi-agent system

| Actuator failure modes | Gain fault parameter | Bias fault parameter |
|---|---|---|
| Fault-free | $\rho_i(t) = 1$ | $b_i(t) = 0$ |
| Gain fault | $0 < \rho_i(t) < 1$ | $b_i(t) = 0$ |
| Bias fault | $\rho_i(t) = 1$ | $b_i(t) \neq 0$ |
| Both gain and bias fault | $0 < \rho_i(t) < 1$ | $b_i(t) \neq 0$ |

$\rho_i(t) = diag\{\rho_{i1}(t), \rho_{i2}(t), \ldots, \rho_{in}(t)\}$ and $b_i(t) = [b_{i1}(t), b_{i2}(t), \ldots, b_{in}(t)]^T$, where $0 < \rho_{ij}(t) \leq 1$ denotes the unknown gain fault and $b_{ij}(t)$ is the unknown bias fault of actuator channel $j(j = 1, 2, \ldots, n)$ for the follower $i(i = 2, 3, \ldots, N)$.

In the following, for simplicity of notation, let $n = 1$ if not otherwise specified. However, all the results hereafter can be directly extended to the higher dimensional case by using the Kronecker product.

The actuator failure modes of the multi-agent system in the current paper are presented in Table 1.

**Assumption 1** The time-varying gain fault and bias fault parameter $\rho_i(t)$, $b_i(t)$ are assumed to be bounded and unknown, and there exist some positive constants $\underline{\rho}_i$, $\bar{b}_i$ such that $0 < \underline{\rho}_i \leq \rho_i(t) \leq 1$, $\|b_i(t)\| \leq \bar{b}_i$, $i = 1, 2, \ldots, N$.

**Assumption 2** The directed graph $G$ of this multi-agent system is connected.

The time-varying formation for the followers to achieve is described by $h_F(t) = [h_2^T(t), h_3^T(t), \ldots, h_N^T(t)]^T$ with $h_i(t) = [h_{ix}(t), h_{iv}(t)]^T (i \in F)$ piecewise continuously differentiable. Let $\xi_k(t) = [x_k(t), v_k(t)]^T (k = 1, 2, \ldots, N)$.

**Definition 1** For any given bounded initial states, multi-agent system (1), (2) is said to achieve time-varying formation if

$$\lim_{t\to\infty} ((\xi_i(t) - h_i(t)) - (\xi_j(t) - h_j(t))) = 0_{2\times1}(i, j \in F) \tag{3}$$

**Definition 2** For any given bounded initial states, multi-agent system (1), (2) is said to achieve target enclosing if

$$\lim_{t\to\infty} \left( \sum_{i=2}^{N} \xi_i(t)/(N-1) \right) - \xi_1(t)) = 0_{2\times1}(i \in F) \tag{4}$$

**Definition 3** For any given bounded initial states, multi-agent system (1), (2) is said to achieve time-varying formation tracking if Eqs. (3) and (4) hold simultaneously.

Let $h_1(t) = \sum_{i=2}^{N} h_i(t)/N - 1$, therefore $h_1(t) = 0$ by choosing $\lim_{t\to\infty} \sum_{i=2}^{N} h_i(t) = 0_{2\times1}$. Then the time-varying formation tracking can be specified by the time-varying vector $h(t) = [h_1^T(t), h_2^T(t), \ldots, h_N^T(t)]^T$. Consider the following time-varying formation tracking protocols for multi-agent system (1), (2)

$$\begin{cases} u_i(t) = -\hat{c}_i(t)\vartheta_i(t) - \hat{c}_i(t)\left(1 + \left\|\dot{h}_{iv}(t)\right\|\right)\mathrm{sgn}(\vartheta_i(t)) \\ \dot{\hat{c}}_i(t) = \gamma_i\|\vartheta_i(t)\|^2 + \gamma_i\left(1 + \left\|\dot{h}_{iv}(t)\right\|\right)\|\vartheta_i(t)\| \end{cases}, i \in 1, 2, \ldots, N \quad (5)$$

Let $\vartheta_i(t) = v_i(t) - h_{iv}(t) + \sum_{j=1}^{N} w_{ij}((x_i(t) - h_{ix}(t)) - (x_j(t) - h_{jx}(t))), i = 1, 2, \ldots N$, $\hat{c}_i(t)$ is an time-varying adaptive parameter, $\mathrm{sgn}(\vartheta_i(t))$ represents the sign function of $\vartheta_i(t)$ and $\gamma_i$ is a positive constant to be designed.

*Remark 1* Time-varying formation tracking problems for second-order multi-agent systems are dealt with in [7], but protocol (3) in [7] is not applicable for the reason that actuator failures are not considered therein, and the derivative of the time-varying formation cannot be counteracted by the control input directly. A fault-tolerant control protocol (5) is presented for the multi-agent system (1), (2) using the adaptive law to accommodate the actuator bias fault and gain fault, while compensating the derivative of the time-varying formation. Protocol (5) is a passive FTC approach which is designed for both the normal case and the faulty case. It should be noted that when choosing $i = 1$, protocol (5) regresses to a fault-tolerant control scheme designed for a single agent. Moreover, FTC problem for a single isolated agent is not a research priority of this current paper.

The current paper mainly focuses on the following two problems for multi-agent system (1), (2) under protocols (5) under directed topologies in the presence of actuator failures:

(1) under what conditions the time-varying formation tracking can be achieved;
(2) how to design the distributed fault-tolerant control protocols (5).

## 3 Main Results

In this section, for system (1), (2) under protocols (5), sufficient conditions to achieve the time-varying formation tracking are first presented, an algorithm is investigated to determine the parameters in the distributed fault-tolerant control protocols.

**Theorem 3.1** *Multi-agent system (1), (2) under protocols (5) with actuator failures achieves time-varying tracking under the following condition*

*Proof* For all $i \in F$, $\lim_{t\to\infty}(h_{iv}(t) - \dot{h}_{ix}(t)) = 0$.

Let $z_i(t) = x_i(t) - h_{ix}(t)$ and $r_i(t) = v_i(t) - h_{iv}(t)$, $i = 1, 2, \ldots N$. Then multi-agent system (1), (2) can be transformed into the following form.

$$\begin{cases} \dot{z}_i(t) = r_i(t) + h_{iv}(t) - \dot{h}_{ix}(t) \\ \dot{r}_i(t) = \rho_i(t)u_i(t) + b_i(t) - \dot{h}_{iv}(t) \end{cases} \quad (6)$$

Let

$$\tilde{z}_i(t) = z_i(t) - \sum_{j=1}^{N} \varphi_j z_j(t) \quad \tilde{z}(t) = \left[\tilde{z}_1^T(t), \tilde{z}_2^T(t), \ldots, \tilde{z}_N^T(t)\right]^T \quad \vartheta(t) = \left[\vartheta_1^T(t), \vartheta_2^T(t), \ldots, \vartheta_N^T(t)\right]^T$$

$$\rho(t) = \left[\rho_1^T(t), \rho_2^T(t), \ldots, \rho_N^T(t)\right]^T, b(t) = \left[b_1^T(t), b_2^T(t), \ldots, b_N^T(t)\right]^T, \dot{h}_v(t) = \left[\dot{h}_{1v}^T(t), \dot{h}_{2v}^T(t), \ldots, \dot{h}_{Nv}^T(t)\right]^T$$

where $\varphi_j$ is denoted in Lemma 1.

Then multi-agent system (6) can be rewritten as follows.

$$\begin{cases} \dot{\tilde{z}}(t) = (I_{N-1} - 1_{N-1}\varphi^T)\vartheta(t) - L\tilde{z}(t) \\ \dot{\vartheta}(t) = -\rho(t)\hat{c}(t)\vartheta(t) - \rho(t)\hat{\lambda}(t)\text{sgn}(\vartheta(t)) + b(t) - \dot{h}_v(t) + L\vartheta(t) - L^2\tilde{z}(t) \end{cases} \tag{7}$$

where

$$\begin{aligned} \hat{c}(t) &= diag\{\hat{c}_1(t), \hat{c}_2(t), \ldots, \hat{c}_N(t)\}, \hat{\lambda}(t) \\ &= diag\{\hat{\lambda}_1(t), \hat{\lambda}_2(t), \ldots, \hat{\lambda}_N(t)\}, \hat{\lambda}_i(t) \\ &= \hat{c}_i(t)\left(1 + \left\|\dot{h}_{iv}(t)\right\|\right), i \\ &= 1, 2, \ldots, N \end{aligned}$$

Choose a Lyapunov functional candidate as follows.

$$V(t) = \tilde{z}^T(t)\psi\tilde{z}(t) + \frac{1}{2}\vartheta^T(t)\vartheta(t) + \sum_{i=1}^{N} \frac{\underline{\rho}_i \tilde{c}_i^2(t)}{2\gamma_i} \tag{8}$$

where $\psi = diag\{\varphi_1, \varphi_2, \ldots, \varphi_N\}$ and $\tilde{c}_i(t) = \hat{c}_i(t) - \bar{c}$. $\bar{c}$ is a positive constant to be designed later.

Taking the time derivative of $V(t)$ along the trajectory of (7)

$$\begin{aligned} \dot{V}(t) = {} & 2\tilde{z}^T(t)(\psi - \varphi\varphi^T)\vartheta(t) - \tilde{z}^T(t)\hat{L}\tilde{z}(t) + \vartheta^T(t)L\vartheta(t) - \vartheta^T(t)L^2\tilde{z}(t) - \vartheta^T(t)\rho(t)\hat{c}\vartheta(t) \\ & + \vartheta^T(t)b(t) - \vartheta^T(t)\rho(t)\hat{\lambda}\text{sgn}(\vartheta(t)) - \vartheta^T(t)\dot{h}_v(t) \\ & + \sum_{i=1}^{N} \underline{\rho}_i(t)\tilde{c}_i(t)\left(\vartheta_i^T(t)\vartheta_i(t) + (1 + \left\|\dot{h}_{iv}(t)\right\|)\|\vartheta_i(t)\|\right) \end{aligned} \tag{9}$$

where $\hat{L} = \psi L + L^T\psi$.

Under Assumption 1, one can obtain that

$$-\vartheta^T(t)\rho(t)\hat{c}(t)\vartheta(t) = -\sum_{i=1}^{N} \rho_i(t)\hat{c}_i(t)\vartheta_i^T(t)\vartheta_i(t) \le -\sum_{i=1}^{N} \underline{\rho}_i \hat{c}_i(t)\vartheta_i^T(t)\vartheta_i(t) \tag{10}$$

$$\vartheta^T(t)b(t) \le \sum_{i=1}^{N} \|b_i(t)\| \|\vartheta_i(t)\| \le \sum_{i=1}^{N} \bar{b}_i \|\vartheta_i(t)\| \tag{11}$$

Substituting (10) and (11) into (9) yields

$$\dot{V}(t) \leq 2\tilde{z}^T(t)(\psi - \varphi\varphi^T)\vartheta(t) - \tilde{z}^T(t)\hat{L}\tilde{z}(t) + \vartheta^T(t)L\vartheta(t) - \vartheta^T(t)L^2\tilde{z}(t) - \sum_{i=1}^{N}\underline{\rho}_i\hat{c}_i(t)\vartheta_i^T(t)\vartheta_i(t) + \sum_{i=2}^{N}\bar{b}_i\|\vartheta_i(t)\| - \vartheta^T(t)\rho(t)\hat{\lambda}(t)\mathrm{sgn}(\vartheta(t)) + \sum_{i=1}^{N}\left\|\dot{h}_{iv}(t)\right\|\|\vartheta_i(t)\| + \sum_{i=1}^{N}\underline{\rho}_i(t)\tilde{c}_i(t)(\vartheta_i^T(t)\vartheta_i(t) + (1 + \left\|\dot{h}_{iv}(t)\right\|)\|\vartheta_i(t)\|) \tag{12}$$

Since

$$-\vartheta^T(t)\rho(t)\hat{\lambda}(t)\mathrm{sgn}(\vartheta(t)) \leq -\sum_{i=1}^{N}\underline{\rho}_i\hat{c}_i(t)(1 + \left\|\dot{h}_{iv}(t)\right\|)\|\vartheta_i(t)\| \tag{13}$$

$$\vartheta^T(t)L\vartheta(t) \leq \|\vartheta(t)\|\,\|L\vartheta(t)\| \leq \sigma_{\max}(L)\|\vartheta(t)\|^2 \tag{14}$$

$$-\vartheta^T(t)L^2\tilde{z}(t) \leq \sigma^2_{\max}(L)\|\vartheta(t)\|\,\|\tilde{z}(t)\| \tag{15}$$

$$2\tilde{z}^T(t)(\psi - \varphi\varphi^T)\vartheta(t) \leq 2\|\vartheta(t)\|\,\|\tilde{z}(t)\| \tag{16}$$

Using (13), (14), (15) and (16), it can be shown that

$$\dot{V}(t) \leq (\sigma_{\max}(L) - \underline{\rho}\bar{c})\|\vartheta(t)\|^2 + \sigma^2_{\max}(L)\|\vartheta(t)\|\,\|\tilde{z}(t)\| + 2\|\vartheta(t)\|\,\|\tilde{z}(t)\| - \tilde{z}^T(t)\hat{L}\tilde{z}(t) - \sum_{i=1}^{N}(\underline{\rho}_i\bar{c} - \bar{b}_i)\|\vartheta_i(t)\| - \sum_{i=1}^{N}(\underline{\rho}_i\bar{c} - 1)\left\|\dot{h}_{iv}(t)\right\|\|\vartheta_i(t)\| \tag{17}$$

For the reason that $\tilde{z}^T\varphi = 0$ and $\varphi > 0$, based upon Lemma 1 and Assumption 2, one has

$$\tilde{z}^T(t)\hat{L}\tilde{z}(t) > \frac{\lambda_2(\hat{L})}{N}\|\tilde{z}\|^2 \tag{18}$$

Moreover,

$$\sigma^2_{\max}(L)\|\vartheta(t)\|\,\|\tilde{z}(t)\| \leq \frac{\lambda_2(\hat{L})}{4N}\|\tilde{z}(t)\|^2 + \frac{N\sigma^4_{\max}(L)}{\lambda_2(\hat{L})}\|\vartheta(t)\|^2 \tag{19}$$

$$2\|\vartheta(t)\|\,\|\tilde{z}(t)\| \leq \frac{\lambda_2(\hat{L})}{4N}\|\tilde{z}(t)\|^2 + \frac{4N}{\lambda_2(\hat{L})}\|\vartheta(t)\|^2 \tag{20}$$

Substituting (18), (19), (20) and into (17) yields

$$\dot{V} \leq -\frac{\lambda_2(\hat{L})}{2N}\|\tilde{z}(t)\|^2 - \left(-\sigma_{\max}(L) - \frac{N(\sigma_{\max}^4(L)+4)}{\lambda_2(\hat{L})} + \underline{\rho}\bar{c}\right)\|\vartheta(t)\|^2$$
$$-\sum_{i=1}^{N}(\underline{\rho}_i\bar{c} - \bar{b}_i)\|\vartheta_i(t)\| - \sum_{i=1}^{N}(\underline{\rho}_i\bar{c} - 1)\left\|\dot{h}_{iv}(t)\right\|\|\vartheta_i(t)\| \tag{21}$$

Let

$$\bar{c} > \max\left\{\frac{1}{\underline{\rho}}\left(\frac{N(\sigma_{\max}^4(L)+4)}{\lambda_2(\hat{L})} + \sigma_{\max}(L)\right), \frac{1}{\underline{\rho}}, \frac{\bar{b}_i}{\underline{\rho}_i}\right\}, i = 1, 2, \ldots, N \tag{22}$$

Hence, one has $\dot{V}(t) \leq 0$. Since $\dot{V}(t) \equiv 0$ indicates that $\vartheta(t) = 0$ and $\tilde{z}(t) = 0$, one can obtain that $\lim_{t\to\infty}(z_i(t) - z_j(t)) = 0$ and $\lim_{t\to\infty}(r_i(t) - r_j(t)) = 0$, $i, j = 1, 2, \ldots, N$. The time-varying formation tracking errors $\tilde{z}(t)$, $\vartheta(t)$ and the adaptive control gain $\hat{c}_i(t)$ are uniformly ultimately bounded, multi-agent system (1), (2) under protocols (5) subjected to actuator failures can achieve time-varying tracking, and (22) provides us with an approach to determine the positive constant $\bar{c}$ of protocol (5). Theorem 3.1 is completed.

# 4 Application to Formation Tracking of Cruise Missiles

In this section, the theoretical results are applied to treat with the formation tracking of a multi cruise missile system in the presence of actuator failures.

The multi cruise missile system is required to carry out the formation tracking task in the horizontal *X–O–Y* plane. The dynamics of the cruise missile consist of the attitude dynamics and the trajectory dynamics, of which the time constants for the attitude dynamics are much smaller than the ones for the trajectory dynamics. Thus, the following assumption can be made throughout this section.

**Assumption 3** The cruise missile flies at a constant altitude. Short period process and the effect of rotation can be neglected in the trajectory realization.

Consider the horizontal mathematical mode of the Tomahawk cruise missile, which was a single engine subsonic cruise missile designed and manufactured by the Convair Division of General Dynamics Corporation [29–31]. The model parameters can be found in Table 2.

Based upon Assumption 3, the horizontal mathematical mode of the cruise missile is presented as

$$\begin{cases} \dot{v}_x = (P\cos\alpha - C_x qS)/m \\ \dot{v}_z = g\tan\gamma_v \\ (P\sin\alpha + C_y qS)\cos\gamma_v = mg \end{cases} \tag{23}$$

**Table 2** Model parameters of the Tomahawk cruise missile

| Parameter | Value |
|---|---|
| Mass | 1315 kg |
| Fuselage length | 5560 mm |
| Wingspan | 2650 mm |
| Engine trust | 267 kgf |
| Flight speed | 0.72 Ma |
| Flight height | 100 m |

where $v_x, v_z, m, \alpha, g, \gamma_v, q, S$ represent acceleration in X-direction, acceleration in Y-direction, engine trust, mass, angle of attack, gravitational acceleration, tilt angle, dynamic pressure, and reference area, respectively. $C_x$, $C_y$, $P$ denote drag coefficient, lift coefficient, turbojet engine trust and satisfy

$$C_x = c_{x0} + c_x^\alpha \alpha^2,\ C_y = c_y^\alpha \alpha,\ P = P(\phi) \tag{24}$$

where $P$ is related to the throttle setting of engine $\phi$.

From (23), it can be known that acceleration tracking in X-direction and Y-direction can be achieved by adjusting the setting of the engine throttle and the speed tilt angle.

The model of the turbojet engine can be approximated by a first-order inertial element, that is

$$G_f(s) = \frac{1}{\tau s + 1},\quad 0.2 < \tau < 0.4 \tag{25}$$

The formation tracking of the cruise missiles can be achieved by acceleration tracking for every agent. The acceleration tracking process consist of an outer-loop control and an inner-loop control, of which the former is derived from the fault-tolerant control protocols (5) and used as the control objects of the latter. The acceleration tracking process can be seen in Fig. 1.

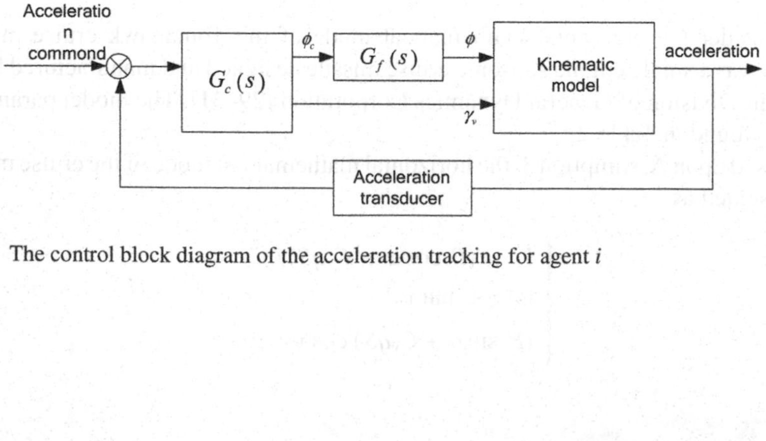


**Fig. 1** The control block diagram of the acceleration tracking for agent $i$

It is essential that the dynamic lag, which can be described by the low-pass filter, should be taken into account in the controller design of the cruise missile slant angle motion. Then

$$\gamma_v = \left( \tan^{-1}(a_{zc}/g) - k_{Pz}(a_z - a_{zc}) - k_{Iz} \int (a_z - a_{zc})dt - k_{Dz}(\dot{a}_z - \dot{a}_{zc}) \right) \frac{1}{0.8s + 1} \tag{26}$$

The engine throttle controller is designed as follows

$$\phi_c = -k_{px}(a_x - a_{xc}) - k_{ix} \int (a_x - a_{xc})dt \tag{27}$$

where $a_{xc}$, $a_{zc}$ denote the control commands of acceleration in X-direction and Y-direction, $a_x$, $a_z$ represent the acceleration measurement in X-direction and Y-direction, $k_{Px}$, $k_{Ix}$, $k_{Pz}$, $k_{Iz}$ and $k_{Dz}$ are proportional-integral (PI) and proportional-integral-derivative (PID) control parameters to be further chosen.

## 5 Numerical Simulations

In this section, a numerical example is given to illustrate the effectiveness of theoretical results obtained in the previous sections. The interaction topology of the multi-agent system is demonstrated in Fig. 2. Without loss of generality, it is assumed that the interaction topology has 0–1 weights.

Consider a second-order multi-agent system with one leader and six followers. The dynamics of these agents are described in (1), (2) with $x_i(t) = [x_{Xi}(t), x_{Yi}(t)]^T$, $v_i(t) = [v_{Xi}(t), v_{Yi}(t)]^T$ and $u_i(t) = [u_{Xi}(t), u_{Yi}(t)]^T$, $i = 1, 2, \ldots, 7$.

To verify the effectiveness of the proposed fault-tolerant formation tracking control scheme, actuator failure modes are presented as follows. There exist bias fault in agent 3 and gain fault in agent 7, agent 1, 2, 4, 5 are considered to suffer from

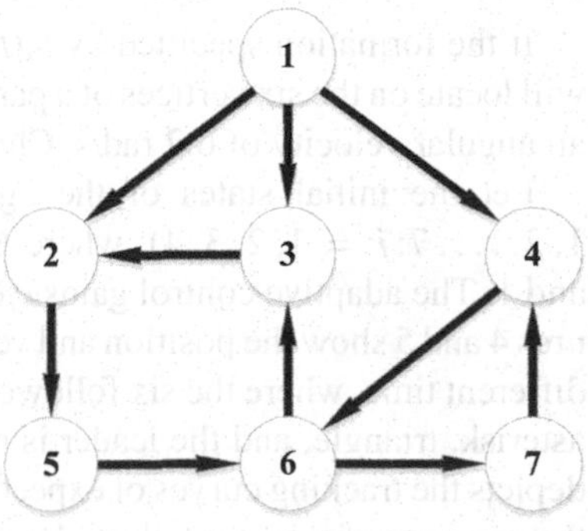


**Fig. 2** Directed interaction topology *G*

**Table 3** Actuator failure modes of agent 1–7

| Agents | Gain fault | Bias fault |
|---|---|---|
| Agent 1 | $\begin{bmatrix} 0.8+0.1\cos(t) & 0 \\ 0 & 0.7-0.1\sin(t) \end{bmatrix}$ | $[0.15\sin(2t), -0.2\cos(2t)]^T$ |
| Agent 2 | $\begin{bmatrix} 0.6+0.2\sin(2t) & 0 \\ 0 & 0.5+0.1\sin(t) \end{bmatrix}$ | $[0.1\cos(t), -0.1\sin(t)]^T$ |
| Agent 3 | $\begin{bmatrix} 1 & 0 \\ 0 & 1 \end{bmatrix}$ | $[0.15\cos(3t), -0.2e^{-t}]^T$ |
| Agent 4 | $\begin{bmatrix} 0.6+0.15\cos(3t) & 0 \\ 0 & 0.8-0.1\sin(2t) \end{bmatrix}$ | $[0.2\cos(t), -0.3\sin(t)]^T$ |
| Agent 5 | $\begin{bmatrix} 0.7+0.1e^{-2t} & 0 \\ 0 & 0.6+0.2\cos(t) \end{bmatrix}$ | $[0.3\cos(t), -0.2\sin(t)]^T$ |
| Agent 6 | $\begin{bmatrix} 1 & 0 \\ 0 & 1 \end{bmatrix}$ | $[0, 0]^T$ |
| Agent 7 | $\begin{bmatrix} 0.5+0.1\cos(t) & 0 \\ 0 & 0.6-0.1\cos(t) \end{bmatrix}$ | $[0, 0]^T$ |

both gain and bias faults, while agent 6 is free from actuator faults, as are shown in Table 3.

The followers are required to preserve a periodic time-varying parallel hexagon formation described as $h_i(t)$ and keep rotation around the leader at the same time.

$$h_i(t) = \begin{bmatrix} 12\sin(0.2t + (i-2)\pi/3) \\ 2.4\cos(0.2t + (i-2)\pi/3) \\ 36\cos(0.2t + (i-2)\pi/3) \\ -7.2\sin(0.2t + (i-2)\pi/3) \end{bmatrix} (i = 2, 3, \ldots, 7) \tag{28}$$

If the formation specified by $h_i(t)(i = 2, 3, \ldots, 7)$ is achieved, the six followers will locate on the six vertices of a parallel hexagon and revolve around the leader with an angular velocity of 0.2 rad/s. Choose $\gamma_{iX} = 1$, $\gamma_{iY} = 1$, $\hat{c}_{iX}(0) = 0$, $\hat{c}_{iY}(0) = 0$.

Let the initial states of the agents be given by $\xi_{ij}(0) = 8(\Theta - 0.5)(i = 1, 2, \ldots, 7; j = 1, 2, 3, 4)$, where $\Theta$ represents a pseudorandom value between 0 and 1. The adaptive control gains $\hat{c}_i(t)(i = 1, 2, \ldots, N)$ are displayed in Fig. 3. Figures 4 and 5 show the position and velocity trajectory snapshots of the seven agents at different time, where the six followers are denoted by circle, diamond, point, square, asterisk, triangle, and the leader is represented by pentagram respectively. Figure 6 depicts the tracking curves of expected acceleration for a group of winged missiles in the presence of actuator failures based upon (5). It can be seen from Figs. 3, 4, 5 and 6

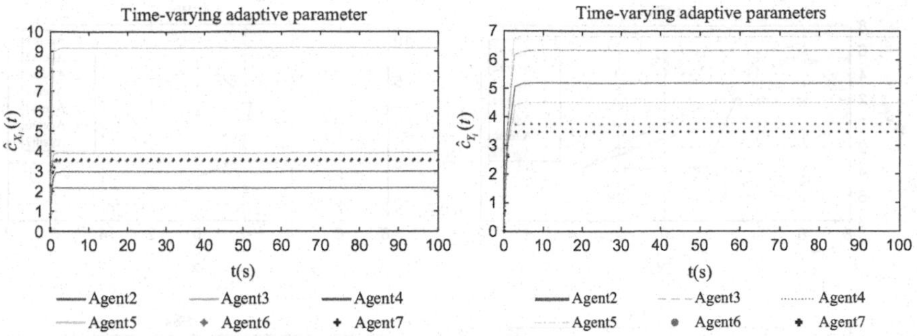


**Fig. 3** Adaptive control gains $\hat{c}_i(t)(i = 1, 2, \ldots, N)$

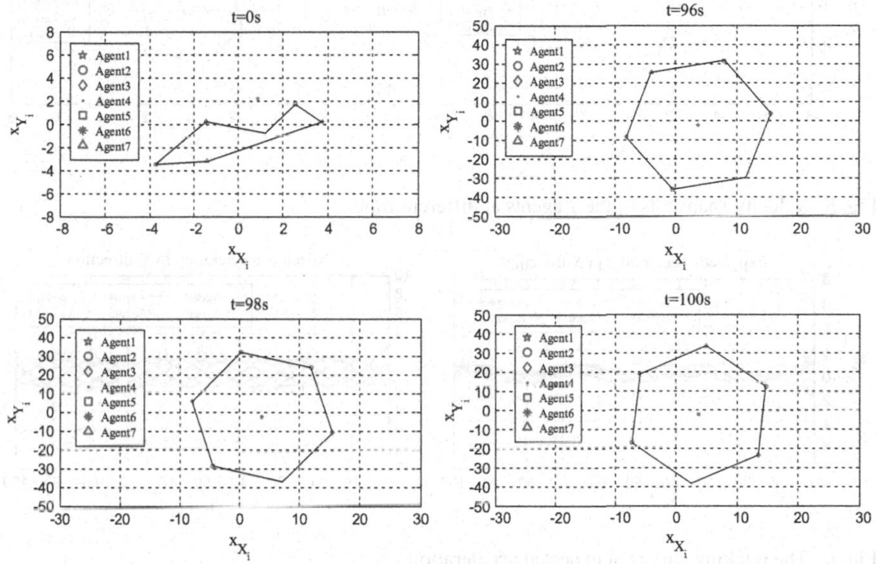


**Fig. 4** Position snapshots of the 7 agents at different time

that the both the position and velocity of these six followers achieve the time-varying parallel hexagons while keep rotating around the leader in spite of actuator failures, which demonstrates the effectiveness of the proposed FTC time-varying formation tracking protocols.

The application of the proposed approach to cruise missiles is shown in Figs. 7, 8 and 9. The speed tilt angle motion and engine throttle setting are controlled by controller (26) and (27) respectively so as to track the reference signals set by the expected acceleration for every missile. Assume that each agent in this simulation stands for a cruise missile. Take agent 4 for example, the speed tilt angle is controlled by the PID controller with 0.5, 1.2 and 0.2 as the coefficients for the proportional, integral, and derivative terms, respectively. The engine throttle setting is controlled

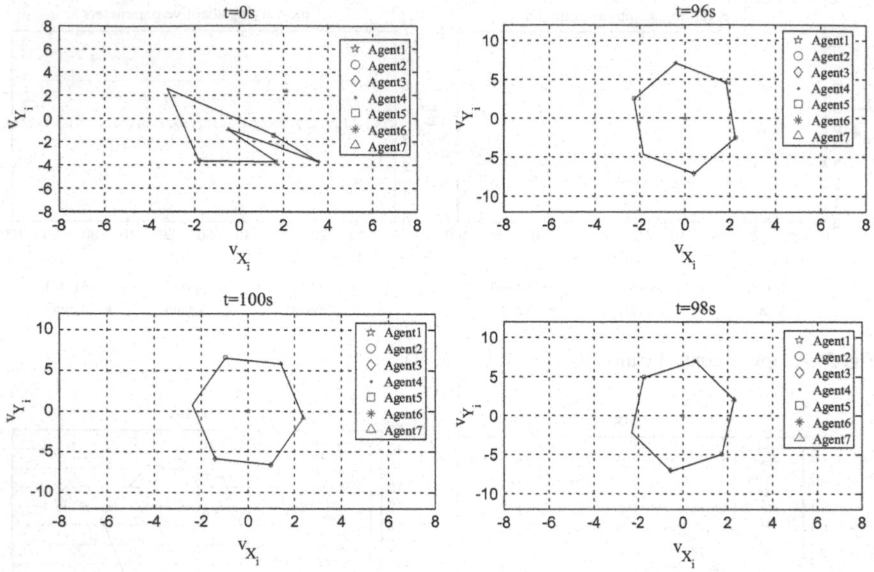


**Fig. 5** Velocity snapshots of the 7 agents at different time

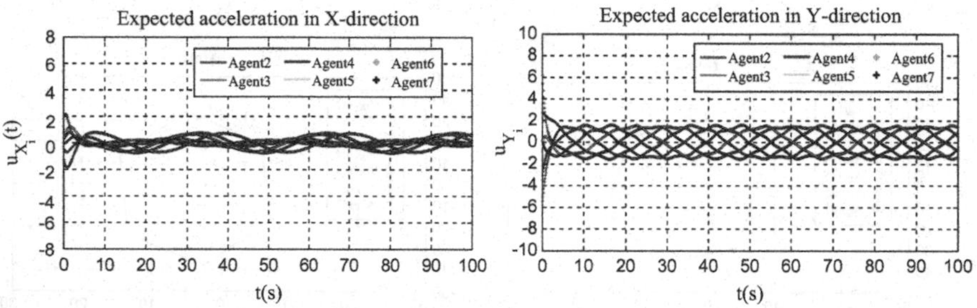


**Fig. 6** The tracking curves of expected acceleration

by the PI controller with 12 and 10 as the coefficients for the proportional and integral terms, respectively. Figure 7 illustrates that the output acceleration can effectively keep up with the command acceleration under the proposed control scheme. From Figs. 8 and 9, one sees the amplitude and transition of the controller output is rather smooth and reasonable small, which shows the feasibility of the proposed approach for application in aeronautical engineering.

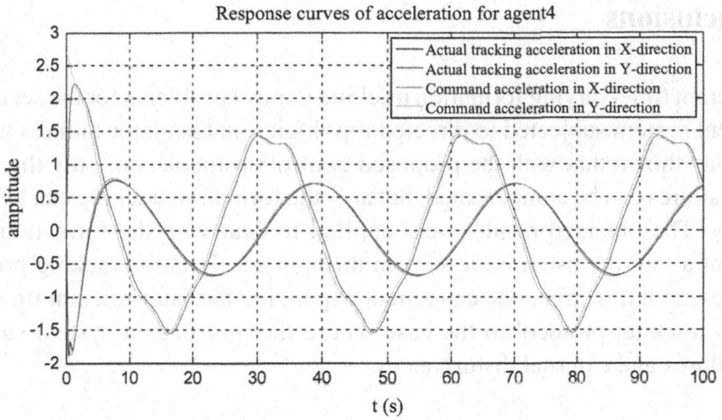


**Fig. 7** Response curves of acceleration for agent 4

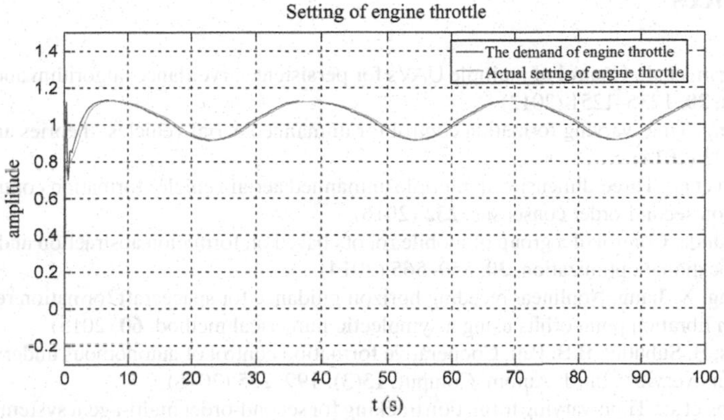


**Fig. 8** Throttle setting of turbojet engine

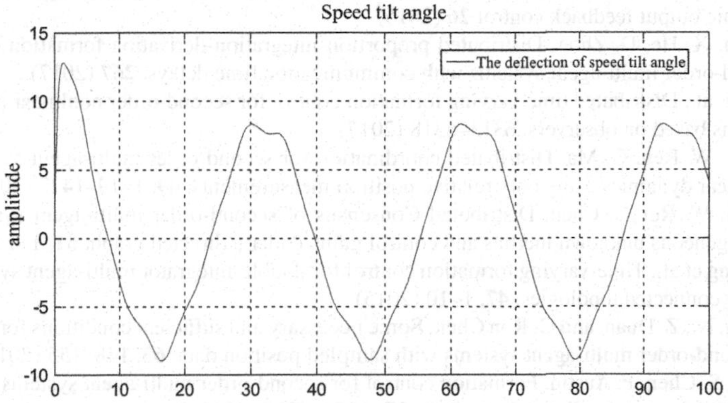


**Fig. 9** The deflection of speed tilt angle

## 6 Conclusions

Fault-tolerant time-varying formation tracking control problems for the second-order multi-agent system subjected to directed topologies and actuator failures were dealt with. It was shown that with the proposed control protocols, not only the unknown actuator failures can be compensated, but also the formation tracking can be achieved eventually. The obtained results were applied to deal with the formation tracking problem of a multi cruise missile system through acceleration tracking process. As a future research direction, one can aim at expanding the fault-tolerant time-varying formation tracking protocol to the case where the multi-agent system suffer from sensor failures and external disturbances.

## References

1. N. Nigam et al., Control of multiple UAVs for persistent surveillance: algorithm and flight test results. **20**, 1236–1251 (2012)
2. X. Dong, Time-varying formation control for unmanned aerial vehicles: theories and applications. **23** (2014)
3. X. Zhu et al., Three-dimensional multiple unmanned aerial vehicles formation control strategy based on second-order consensus. **232** (2016)
4. K. Yoshida, Control of a group of mobile robots based on formation abstraction and decentralized locational optimization. **30**, 550–565 (2014)
5. H. Peng, X. Jiang, Nonlinear receding horizon guidance for spacecraft formation reconfiguration on libration point orbits using a symplectic numerical method. **60** (2015)
6. B. Das, B. Subudhi, B.B. Pati, Cooperative formation control of autonomous underwater vehicles: an overview. Int. J. Autom. Comput. **13**(3), 199–225 (2016)
7. X. Dong et al., Time-varying formation tracking for second-order multi-agent systems subjected to switching topologies with application to Quadrotor formation flying. **64** (2016)
8. H. Du et al., Second-order consensus for nonlinear leader-following multi-agent systems via dynamic output feedback control **26** (2015)
9. L. Qin, X. He, D. Zhou, Distributed proportion-integration-derivation formation control for second-order multi-agent systems with communication time delays. **267** (2017)
10. J. Yu et al., Distributed time-varying formation control for second-order nonlinear multi-agent systems based on observers. 6313–6318 (2017)
11. J. Mei, W. Ren, G. Ma, Distributed coordination for second-order multi-agent systems with nonlinear dynamics using only relative position measurements. **49**, 1419–1427 (2013)
12. J. Mei, W. Ren, J. Chen, Distributed Consensus of second-order multi-agent systems with heterogeneous unknown inertias and control gains under a directed graph. **61**, 1–1 (2015)
13. X. Dong et al., Time-varying formation control for double-integrator multi-agent systems with jointly connected topologies. **47**, 1–10 (2015)
14. Huang, N., Z. Duan, and G. Ron Chen, Some necessary and sufficient conditions for consensus of second-order multi-agent systems with sampled position data. **63**, 148–155 (2016)
15. X. Lu, S. Chen, F. Austin, Formation control for second-order multi-agent systems with time-varying delays under directed topology. **17**, 1304–1308 (2011)
16. X. Dong et al., Time-varying formation tracking for second-order multi-agent systems with one leader. 1046–1051 (2015)
17. X. Dong, G. Hu, Time-varying formation tracking for linear multiagent systems with multiple leaders. 1–1 (2017)

18. L. Han et al., Formation tracking control for second-order multi-agent systems with time-varying delays. 7902–7907 (2016)
19. L. Han et al., Formation tracking control for time-delayed multi-agent systems with second-order dynamics. Chin. J. Aeronaut. **30**(1), 348–357 (2017)
20. M. Liu et al., Formation tracking control for multi-agent systems with nonlinear dynamics via impulsive control. 3669–3674 (2014)
21. Q. Zhao et al., Distributed group cooperative guidance for multiple missiles with switching directed communication topologies. 5741–5746 (2017)
22. Wang, Y., Y.-D. Song, F.L. Lewis, Robust adaptive fault-tolerant control of multiagent systems with uncertain nonidentical dynamics and undetectable actuation failures. **62**, 1–1 (2015)
23. Y. Hua et al. Fault-tolerant time-varying formation control for second-order multi-agent systems with directed topologies. 467–472 (2017)
24. Y. Wang et al., Fault-tolerant finite time consensus for multiple uncertain nonlinear mechanical systems under single-way directed communication interactions and actuation failures. **63**, 374–383 (2016)
25. Salimifard, M. and H. Ali Talebi, Robust output feedback fault-tolerant control of non-linear multi-agent systems based on wavelet neural networks. **11** 3004–3015 (2017)
26. M. Tri Nguyen, et al., Fault tolerant predictive control for multi-agent dynamical: formation reconfiguration using set-theoretic approach (2014)
27. M. Khalili, M., et al., Distributed Adaptive fault-tolerant consensus control of multi-agent systems with actuator faults (2015)
28. J. Guo, et al., Distributed fault-tolerant topology control in cooperative wireless ad hoc networks. **26**, 2699–2710 (2015)
29. E. Craig, R.R.J. Reich, Flight test aerodynamic drag characteristics development and assessment of in-flight propulsion analysis methods for the AGM-109 cruise missile. **1** (1981)
30. B. McGrath, Subsonic aerodynamic fin-folding moments for the tactical tomahawk missile configuration, in *22nd Applied Aerodynamics Conference and Exhibit 2004*, American Institute of Aeronautics and Astronautics (2004)
31. M. Newman, A., et al., Optimizing assignment of Tomahawk cruise missile missions to firing units. **58**, 281–294 (2011)

# Position/Force Control of the Mobile Manipulator with Rheonomic Constraints

**Baigeng Wang and Shurong Li**

**Abstract** In the view of the rheonomic constraints problem of the mobile manipulator, the corrected orthogonalization method is adopted to solve the problem that velocity and force are not orthogonal under rheonomic constraints. By transforming the system with the designed transformation matrix, the velocity and the force are mapped to the orthogonal spaces. Then, a position/force strategy is designed, which drives the position and force converge to zero. Furthermore, we consider the situation where there is interference in the motor and propose the desired torque control strategy and the desired motor control strategy. The robustness of the system is enhanced. By selecting the proper Lyapunov function, the effectiveness of the proposed strategy is proved. Through simulation, the validity of the above conclusions are verified.

**Keywords** Mobile manipulator · Rheonomic constraints · Position/force control · Lyapunov function

## 1 Introduction

The position and force control for the mobile manipulator has been widely concerned by academic communities. Compare with the fixed manipulator, this system combines the advantages of the mobile platform and the fixed manipulator. Because of these advantages, such systems could be applied in construction, planetary exploration, and military [1–3].

Because the mobile base is subject to the nonholonomic constraints, it cannot follow an arbitrary path in the configuration space. Therefore the controller design and

B. Wang
College of Information and Control Engineering, China University of Petroleum, Qingdao 266580, China
e-mail: upc_wbg@163.com

S. Li (✉)
Automation School, Beijing University of Posts and Telecommunications, Beijing 100876, China
e-mail: lishurong@bupt.edu.cn



Y. Jia et al. (eds.), *Proceedings of 2018 Chinese Intelligent Systems Conference*, Lecture Notes in Electrical Engineering 528,
https://doi.org/10.1007/978-981-13-2288-4_80

system analysis of this system face much more problems. To solve these problems, many researchers have done a lot of effort. Wu et al. [4] proposed a position/force hybrid control law to decompose the task space into two orthogonal subspaces. In [5] a robust adaptive position/force control with hybrid variable signals is proposed to compensate for parametric uncertainties and suppress bounded disturbances. In [6], a nonlinear input–output decoupling controller is designed to satisfy the velocity subsystem which controlled by the longitudinal acceleration/braking force. In [7], the author adopted a sliding-mode law to control the mobile manipulator system. In [8], several new linear matrix inequality conditions have been obtained to reduce the conservativeness for the design of continuous-time systems.

However, the literatures mentioned above do not pay attention to rheonomic constraints problem of the manipulator. Rheonomic constraints will affect the stability of the system and increase the control difficulty of the terminal manipulator. In [9], a new method of hybrid controller is established based on a modified joint-space othorgonalization scheme to solve the problem of rheonomic constraints.

In this paper, the dynamic model of mobile manipulator under rheonomic constraints is established. Moreover, a motor model is used in this system. And then, a sliding-mode control strategy is proposed to solve the problems of controller design. Through simulation the effectiveness knowledge of the control strategy is vividly shown.

## 2 Dynamics of the System

Figure 1 shows the practical model of the manipulator system in this paper. It consists of a mobile platform and a manipulator. Consider a n-DOF mobile manipulator system under rheonomic constraints. The dynamic model can be expressed by the following formula:

$$D(q)\ddot{q} + C(q, \dot{q})\dot{q} + G(q) = B(q)\tau + f \tag{1}$$

where $q = \left(q_v^T \ q_a^T\right)^T \in R^n$ denotes the generalized state vector of the system, in which $q_v \in R^{n_v}$ represents the state of the mobile platform and $q_a \in R^{n_a}$ repre-

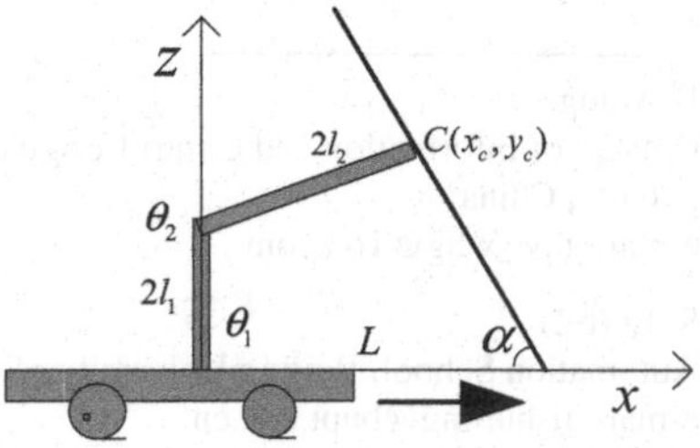


**Fig. 1** The mobile manipulator system with rheonomic constraints

sents the state of the manipulator. $D(q) \in R^{n \times n}$ is the symmetric bounded positive definite inertia matrix. $C(q, \dot{q}) \in R^{n \times n}$ denote the Centripetal and Coriolis torques. $G(q) \in R^n$ is the gravitational torque vector. $B(q) \in R^{n \times (n-l)}$ is a full rank input transformation matrix. $\tau = (\tau_v^T \ \ \tau_a^T)^T \in R^{n-l}$ is the control input of the system, in which $\tau_v \in R^{n_v - l}$ denotes the control input of the mobile platform and $\tau_a \in R^{n_a}$ denotes the control input of the manipulator. $f \in R^n$ is the rheonomic constraint forces. Using Lagrange multiplier $\lambda$, we can obtain $f = J^T \lambda$. And $J$ is the Jacobian matrix of the system.

Each part of the dynamic model is given in detail as below.

$$D(q) = \begin{bmatrix} D_v & D_{va} \\ D_{av} & D_a \end{bmatrix}, \quad C(q, \dot{q}) = \begin{bmatrix} C_v & C_{va} \\ C_{av} & C_a \end{bmatrix}, \quad G(q) = \begin{bmatrix} G_v \\ G_a \end{bmatrix},$$

$$B(q) = \begin{bmatrix} B_v & 0 \\ 0 & B_a \end{bmatrix}, \tau = \begin{bmatrix} \tau_v \\ \tau_a \end{bmatrix}, \quad J = \begin{bmatrix} A & 0 \\ J_v & J_a \end{bmatrix}, \lambda = \left(\lambda_n^T, \lambda_h^T\right)^T$$

In this paper, the mobile platform is assumed to be completely nonholonomic. The effect of the constraints can be viewed as a restriction of the dynamics on the manifold $\Omega_n$ as:

$$\Omega_n = \{(q_v, \dot{q}_v) | A(q_v)\dot{q}_v = 0\} \tag{2}$$

where $A(q_v) = \left[A_1^T(q_v), \ldots, A_l^T(q_v)\right]^T \in R^{l \times n_v}$ is the kinematic-constraint matrix.

The generalized constraint forces for the nonholonomic constraints can be given by:

$$f_n = A^T(q_v)\lambda_n \tag{3}$$

Assume $H(q_v) \in R^{n_v \times (n_v - l)}$ is orthogonal to $A(q_v)$, and $H^{\mathrm{T}}(q_v)A^{\mathrm{T}}(q_v) = 0$. Then there exists a vector $\eta \in R^{n_v - l}$ satisfying the following relation:

$$\begin{aligned} \dot{q}_v &= H(q_v)\dot{\eta}, \\ \ddot{q}_v &= \dot{H}(q_v)\dot{\eta} + H(q_v)\ddot{\eta}. \end{aligned} \tag{4}$$

Let $\zeta = [\eta^{\mathrm{T}}, q_a^{\mathrm{T}}]^{\mathrm{T}} \in R^{n-l}$ be the state vector under nonholonomic constraints, and the dynamic model is obtained as:

$$\bar{D}(\zeta)\ddot{\zeta} + \bar{C}(\zeta, \dot{\zeta})\dot{\zeta} + \bar{G}(\zeta) = \bar{B}(\zeta)\tau + \bar{J}^T\lambda_h \tag{5}$$

where

$$\bar{D}(\zeta) = \begin{bmatrix} H^{\mathrm{T}} D_v H & H^{\mathrm{T}} D_{va} \\ D_{av} H & D_a \end{bmatrix}, \quad \bar{C}(\zeta, \dot{\zeta}) = \begin{bmatrix} H^{\mathrm{T}} D_v \dot{H} + H^{\mathrm{T}} C_v H & H^{\mathrm{T}} C_{va} \\ D_{av}\dot{H} + C_{av} H & C_a \end{bmatrix},$$

$$\bar{G}(\zeta) = \begin{bmatrix} H^{\mathrm{T}} G_v \\ G_a \end{bmatrix}, \quad \bar{B}(\zeta) = \begin{bmatrix} H^{\mathrm{T}} B_v & 0 \\ 0 & B_a \end{bmatrix}, \quad \bar{J} = \begin{bmatrix} J_v & J_a \end{bmatrix}.$$

Consider the rheonomic constraints at the end of the manipulator as follows:

$$\phi(\zeta, t) = 0 \tag{6}$$

Take the derivative of both sides of the Eq. (7), and the relation can be obtained as:

$$\dot{\phi}(\zeta, t) = \bar{J}(\zeta, t)\dot{\zeta} + \varphi = 0 \tag{7}$$

where $\bar{J}(\zeta, t) = \frac{\partial \phi(\zeta,t)}{\partial \zeta}$, and $\varphi = \frac{\partial \phi(\zeta,t)}{\partial t}$ is the velocity of rheonomic constraints.

In view of the existence of rheonomic constraints, the velocity and the constraint force do not satisfy the condition of orthogonal decomposition. We need to make proper modifications to orthogonalize to meet the design requirements.

Define matrix $P(x, t)$ and $Q(x, t)$:

$$P(\zeta, t) = \bar{J}^{+}(\zeta, t)\bar{J}(\zeta, t) \tag{8}$$

$$Q(\zeta, t) = I - P(\zeta, t) = I - \bar{J}^{+}(\zeta, t)\bar{J}(\zeta, t) \tag{9}$$

where $\bar{J}(\zeta, t)$ is full rank.

The velocity $\dot{\zeta}$ is decomposed as follows:

$$\dot{\zeta} = Q(\zeta, t)\dot{\zeta} + (I - Q(\zeta, t))\dot{\zeta} = Q(\zeta, t)\dot{\zeta} + \bar{J}^{+}(\zeta, t)\bar{J}(\zeta, t)\dot{\zeta} \tag{10}$$

According to the formula (7), we can obtain:

$$\varphi = -\bar{J}(\zeta, t)\dot{\zeta} \tag{11}$$

Substituting (11) into (10) yields:

$$\dot{\zeta} = Q(\zeta, t)\dot{\zeta} - \bar{J}^{+}(\zeta, t)\varphi \tag{12}$$

**Property 1** *$D, C, G, B, \bar{D}, \bar{C}, \bar{G}, \bar{B}$ are uniformly continuous and bounded if $q, \zeta$ are uniformly continuous and bounded.*

**Property 2** *The inertia matrix $\bar{D}$ after dimensionality reduction is positive definite and satisfies*

$$\lambda_m\{\bar{D}(\zeta)\}I \leq \bar{D}(\zeta) \leq \lambda_M\{\bar{D}(\zeta)\}I, \quad \forall \zeta \in R^{n-l}$$

*where $\lambda_m\{\cdot\}(\lambda_M\{\cdot\})$ donate eigenvalue of the minimum (maximum) matrix.*

**Property 3** *The matrix* $\dot{\bar{D}} - 2\bar{C}$ *is skew symmetric.*

**Lemma 1** *If* $x : [0, \infty) \to R$ *is square-integrable, that is* $\lim_{t\to\infty} \int_0^t x^2(\tau)d\tau < \infty$. *If* $\dot{x}(t), t \in [0, \infty)$ *exists and is bounded, then* $\lim_{t\to\infty} x(t) = 0$.

## 3 Controller Design

Define:

$$e = \zeta - \zeta_d, \quad e_\lambda = \lambda_h - \lambda_{hd} \tag{13}$$

where $e$ is the position error. $\zeta_d$ is the desired position. $e_\lambda$ is the force error.

Define an auxiliary variable $\zeta_r$:

$$\dot{\zeta}_r = Q(\zeta, t)(\dot{\zeta}_d - \Lambda_1 e) - \bar{J}^+(\zeta, t)(\varphi - \Lambda_2 e_\lambda) \tag{14}$$

The sliding variable are also defined as:

$$\begin{aligned} s = \dot{\zeta} - \dot{\zeta}_r &= Q(\zeta, t)(\dot{e} + \Lambda_1 e) - \bar{J}^+(\zeta, t)\Lambda_2 e_\lambda + (I - Q(\zeta, t))\dot{\zeta} + \bar{J}^+(\zeta, t)\varphi \\ &= Q(\zeta, t)(\dot{e} + \Lambda_1 e) - \bar{J}^+(\zeta, t)\Lambda_2 e_\lambda + \bar{J}^+(\zeta, t)\bar{J}(\zeta, t)\dot{\zeta} - \bar{J}^+(\zeta, t)\bar{J}(\zeta, t)\dot{\zeta} \\ &= Q(\zeta, t)(\dot{e} + \Lambda_1 e) - \bar{J}^+(\zeta, t)\Lambda_2 e_\lambda = s_1 + s_2 \end{aligned} \tag{15}$$

where $s_1$ is the error of position and velocity, and $s_2$ is the error of force. $\Lambda_1$ is the positive definite diagonal matrix, and $\Lambda_2$ is a positive constant.

From formula [8, 9], it is known that $s_1$ and $s_2$ are orthogonal, therefore we can design the control law as follow:

$$\tau = \bar{B}^+(-Ks + \bar{D}\ddot{\zeta}_r + \bar{C}\dot{\zeta}_r + \bar{G}) - \bar{B}^+\bar{J}^T\lambda_h \tag{16}$$

Consider the following Lyapunov function:

$$V_1 = \frac{1}{2}s^{\mathrm{T}}\bar{D}s \tag{17}$$

Take the derivative of both sides of Eq. (17):

$$\dot{V}_1 = s^{\mathrm{T}}\bar{D}\dot{s} + \frac{1}{2}s^{\mathrm{T}}\dot{\bar{D}}s \tag{18}$$

Since the control law (16):

$$\bar{D}\dot{s} = \bar{D}(\ddot{\zeta} - \ddot{\zeta}_r) = -\bar{C}\dot{\zeta} - \bar{G} + \bar{B}\tau + \bar{J}^T\lambda_h - \bar{D}\ddot{\zeta}_r = -Ks - \bar{C}s \tag{19}$$

According to (19), (18) can be expressed as:

$$\dot{V}_1 = s^{\mathrm{T}}(-Ks - \bar{C}s) + \frac{1}{2}s^{\mathrm{T}}\dot{\bar{D}}s = -s^{\mathrm{T}}Ks + \frac{1}{2}s^{\mathrm{T}}(\dot{\bar{D}} - 2\bar{C})s = -s^{\mathrm{T}}Ks \le 0 \quad (20)$$

Integrate the two side of formula (20):

$$K\int_0^{\infty} s^2 dt \le V_1(0) - V_1(\infty) < \infty \quad (21)$$

Based on Lemma 1, $s \to 0$ when $t \to \infty$. Owing to $s = s_1 + s_2$, we can obtain $s_1 \to 0, s_2 \to 0$ when $t \to \infty$. And then $e \to 0, \dot{e} \to 0, e_\lambda \to 0$. In practice, the external electromagnetic interference often causes the perturbation of the motor parameters, which makes the actual output torque deviate from the expected. In this paper, the influence of the motor model on the control system is considered:

$$U_i = \bar{L}_i\dot{\tau}_i + \bar{R}_i\tau_i + \bar{K}_i\dot{p}_i + W_i(t) \quad (22)$$

where $U_i$ is motor armature voltage. $\tau_i$ is the motor output torque. $p_i$ is the motor rotation angle. $W_i(t)$ is the uncertain parameter. $\bar{L}_i = \frac{L}{nk_t}$, $\bar{R}_i = \frac{R}{nk_t}$, $\bar{K}_i = nk_e$, $n$ is velocity reduction ratio of motor gear. $k_t, k_e$ are motor torque constants. $L$ and $R$ are motor armature inductance and motor armature resistance.

The dynamic model of $n - m$ motor system can be expressed:

$$U = \bar{L}\dot{\tau} + \bar{R}\tau + \bar{K}\dot{p} + W(t) \quad (23)$$

where $W(t) = \left[W_1(t), \ldots, W_{n-m}(t)\right]^{\mathrm{T}}, \quad \|W(t)\| \le \rho_m$.

In this paper, select a mobile manipulator with a 2-DOF manipulator as the research object. $p = [p_1, \ldots, p_{n-m}]^{\mathrm{T}}$ can be described as $p = [\theta_l, \theta_r, \theta_1, \theta_2]^{\mathrm{T}}$. According to $\zeta = [\theta_l, \theta_r, \theta_1, \theta_2]^{\mathrm{T}}$, $p = \zeta$.

The dynamic formula (5) can be changed into the following form:

$$\bar{D}(p)\ddot{p} + \bar{C}(p, \dot{p})\dot{p} + \bar{G}(p) = \bar{B}(p)\tau + \bar{J}^T\lambda_h \quad (24)$$

Based on the previous proof, the desired torque can be obtained:

$$\tau_d = \bar{B}^{+}(-Ks + \bar{D}\ddot{p}_r + \bar{C}\dot{p}_r + \bar{G}) - \bar{B}^{+}\bar{J}^T\lambda_h \quad (25)$$

Define

$$e_\tau = \tau - \tau_d \quad (26)$$

Design the motor control law:

$$U = \bar{L}\left[\dot{\tau}_d + e_\tau + \rho_m \mathrm{sgn}(e_\tau)\left\|\bar{L}^{+}\right\|\right] + \bar{R}\tau + \bar{K}\dot{p} \quad (27)$$

Consider the following Lyapunov function:

$$V_2 = V_1 + \frac{1}{2} e_\tau^{\mathrm{T}} e_\tau \tag{28}$$

Take the derivative of both sides of Eq. (28):

$$\begin{aligned} \dot{V}_2 &= \dot{V}_1 + e_\tau^{\mathrm{T}} \dot{e}_\tau = -s^{\mathrm{T}} K s + e_\tau^{\mathrm{T}} (\dot{\tau}_d - \dot{\tau}) \\ &= -s^{\mathrm{T}} K s + e_\tau^{\mathrm{T}} [\dot{\tau}_d - \bar{L}^{+} (U - \bar{R} \tau - \bar{K} \dot{p} - W)] \end{aligned} \tag{29}$$

Substituting (27) into (29):

$$\begin{aligned} \dot{V}_2 &= -s^{\mathrm{T}} K s + e_\tau^{\mathrm{T}} \left[ \dot{\tau}_d - \bar{L}^{+} (\bar{L} (\dot{\tau}_d + e_\tau + \rho_m \mathrm{sgn}(e_\tau) \| \bar{L}^{+} \|) - W) \right] \\ &= -s^{\mathrm{T}} K s - e_\tau^{\mathrm{T}} e_\tau + e_\tau^{\mathrm{T}} (\bar{L}^{+} W - \rho_m \mathrm{sgn}(e_\tau) \| \bar{L}^{+} \|) \end{aligned} \tag{30}$$

Since $\|W(t)\| \le \rho_m$,

$$\dot{V}_2 \le -\sigma_{\min}(K) \|s\|^2 - \|e_\tau\|^2 + \left( \|e_\tau^{\mathrm{T}}\| \|\bar{L}^{+}\| \|W\| - \rho_m \|e_\tau^{\mathrm{T}}\| \|\bar{L}^{+}\| \right) \le 0 \tag{31}$$

Because of $V_2 > 0$, $\dot{V}_2 \le 0$, it is known that the system is Lyapunov stable. Moreover $\dot{V}_2$ is not equal to 0 constantly when $s \ne 0, e_\tau \ne 0$. And then we can obtain $s \to 0, e_\tau \to 0$ when $t \to \infty$. Therefore $e \to 0$, $\dot{e} \to 0$, $e_\lambda \to 0$. The asymptotic convergence of the position and force error can be guaranteed.

## 4 Simulation Analysis

A motion control system with two two-link mobile manipulators is considered to verify the validity of the proposed controller, the dynamic equation of the system can be shown as follows (Table 1):

$$D_{v11} = D_{v22} = m_{p12}, D_{v12} = D_{v21} = 0, D_{v13} = D_{v31} = m_2 l_2 \sin\theta_2 \sin(\theta + \theta_1),$$
$$D_{v33} = I_{zp12} + m_2 \sin\theta_2 l_2^2 \sin\theta_2, D_{a11} = I_{z12}, D_{a22} = I_{y2} + m_2 l_2^2, D_{a12} = D_{a21} = 0$$

**Table 1** Parameter values of the model

| $m_1$ | $m_2$ | $m_p$ | $l_1$ | $l_2$ | $l$ | $r$ | $I_{z1}$ | $I_{z2}$ | $I_{zp}$ |
|---|---|---|---|---|---|---|---|---|---|
| 1 kg | 1 kg | 6 kg | 0.5 m | 0.3 m | 0.5 m | 0.5 m | 1 kg m$^2$ | 1 kg m$^2$ | 19 kg m$^2$ |

$$D_{va} = D_{av}^T = \begin{bmatrix} m_2 l_2 \sin\theta_2 \sin(\theta+\theta_1) & -m_2 l_2 \cos\theta_2 \cos(\theta+\theta_1) \\ -m_2 l_2 \sin\theta_2 \cos(\theta+\theta_1) & -m_2 l_2 \cos\theta_2 \sin(\theta+\theta_1) \\ I_{z12} + m_2 \sin\theta_2 l_2^2 \sin\theta_2 & 0 \end{bmatrix},$$

$$C_{v11} = C_{v22} = C_{v12} = C_{v21} = 0, \quad C_{v13} = m_2 l_2 \sin\theta_2 \cos(\theta+\theta_1)(\dot{\theta}+\dot{\theta}_1),$$

$$C_{v23} = m_2 l_2 \sin\theta_2 \sin(\theta+\theta_1)(\dot{\theta}+\dot{\theta}_1), C_{v33} = 0,$$

$$C_{v31} = m_2 l_2 (\cos\theta_2 \sin(\theta+\theta_1)\dot{\theta}_2 + \sin\theta_2 \sin(\theta+\theta_1)(\dot{\theta}+\dot{\theta}_1)),$$

$$C_{v32} = -m_2 l_2 (\cos\theta_2 \cos(\theta+\theta_1)\dot{\theta}_2 - \sin\theta_2 \sin(\theta+\theta_1)(\dot{\theta}+\dot{\theta}_1)),$$

$$C_{va11} = m_2 l_2 \sin\theta_2 \cos(\theta+\theta_1)(\dot{\theta}+\dot{\theta}_1), C_{va21} = m_2 l_2 \sin\theta_2 \sin(\theta+\theta_1)(\dot{\theta}+\dot{\theta}_1),$$

$$C_{va12} = m_2 l_2 (\sin\theta_2 \cos(\theta+\theta_1)\dot{\theta}_2 + 2\cos\theta_2 \sin(\theta+\theta_1)(\dot{\theta}+\dot{\theta}_1)),$$

$$C_{va22} = m_2 l_2 (\sin\theta_2 \sin(\theta+\theta_1)\dot{\theta}_2 - 2\cos\theta_2 \cos(\theta+\theta_1)(\dot{\theta}+\dot{\theta}_1)),$$

$$C_{va32} = 2m_2 \cos\theta_2 (l_2^2 \sin\theta_2 - dl\ \cos\theta_1)(\dot{\theta}+\dot{\theta}_1), C_{va31} = 0,$$

$$C_{av11} = m_2 l_2 (\sin\theta_2 \cos(\theta+\theta_1)(\dot{\theta}+\dot{\theta}_1) + \cos\theta_2 \sin(\theta+\theta_1)\dot{\theta}_2)$$

$$C_{av12} = m_2 l_2 (\sin\theta_2 \sin(\theta+\theta_1)(\dot{\theta}+\dot{\theta}_1) - \cos\theta_2 \cos(\theta+\theta_1)\dot{\theta}_2)$$

$$C_{av21} = m_2 l_2 (\sin\theta_2 \cos(\theta+\theta_1)\dot{\theta}_2 + \cos\theta_2 \sin(\theta+\theta_1)(\dot{\theta}+\dot{\theta}_1))$$

$$C_{av13} = 2m_2 \cos\theta_2 l_2^2 \sin\theta_2 \dot{\theta}_2, \quad C_{a11} = 2m_2 \sin\theta_2 l_2^2 \cos\theta_2 \dot{\theta}_2,$$

$$G_v = [0, 0, 0]^{\mathrm{T}}, G_a = [0, -m_2 g l_2 \sin\theta_2]^{\mathrm{T}}, J_v = [0\ 0\ 0], J_a = [0\ l_2 \cos\theta_2]$$

$$B_{v11} = B_{v12} = \cos\theta / r, B_{v21} = B_{v22} = \sin\theta / r, B_{v31} = -B_{v32} = l/r$$

$$B_{a11} = B_{a22} = 1, A = [-\sin\theta\ \cos\theta\ 0]$$

The angle between the constraint surface and the X axis is given as:

$$\alpha(t) = 0.2\pi + 0.03t, \quad t \in [0, 31]$$

Dynamic constraint equation can be expressed as: $(L - x_c)\tan\alpha - y_c = 0$
Assume the expected joint trajectory:

$$x_{1d} = x_{2d} = \left(L - \frac{[2l_1 + 2l_2 \cos(\sin(t))]}{\tan\alpha} + 2l_2 \sin(\sin(t))\right)/r, x_{3d} = 0, x_{4d} = \sin t$$

Suppose the initial value of the system: $x(0) = [0, 1, 0.5, 0.28]^{\mathrm{T}} m$, $L = 2m$
The parameter of (30) is: $\bar{L} = \mathrm{diag}[4]$, $\bar{R} = \mathrm{diag}[0.8]$, $\bar{K} = \mathrm{diag}[0.06]$, $\rho_m = 1$
In [1], the dynamic model is expressed by the following formula: $K = \mathrm{diag}[40]$, $\Lambda_1 = \mathrm{diag}[1.5]$, $\Lambda_2 = 1$
The simulation results are shown in Figs. 2, 3, 4, 5 and 6. It can be seen that the trajectory tracking error and force error are asymptotically convergent.

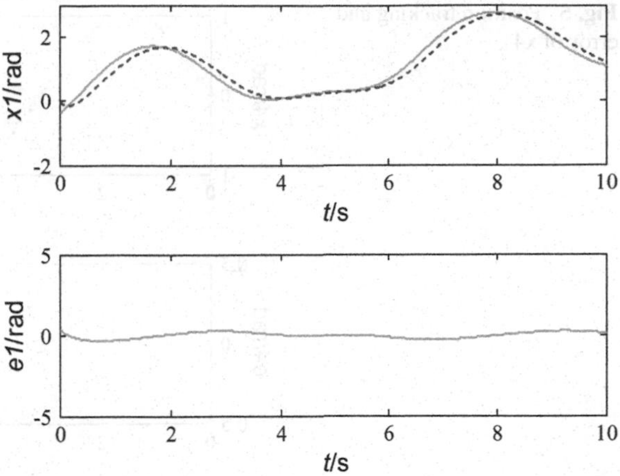


**Fig. 2** Position tracking and error of x1

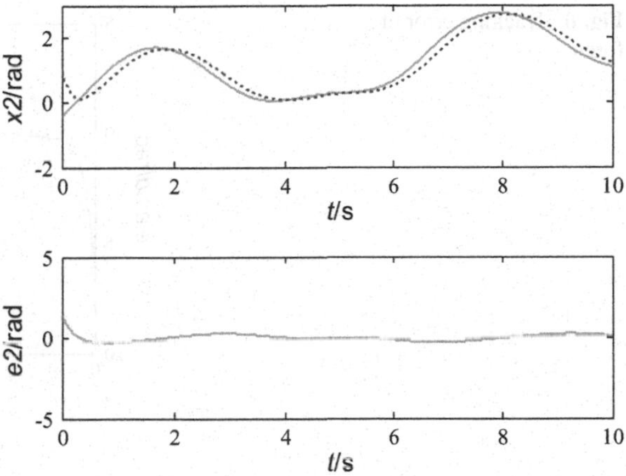


**Fig. 3** Position tracking and error of x2

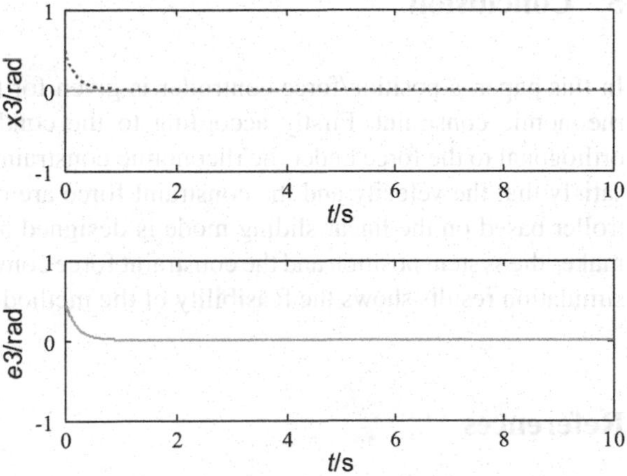


**Fig. 4** Position tracking and error of x3

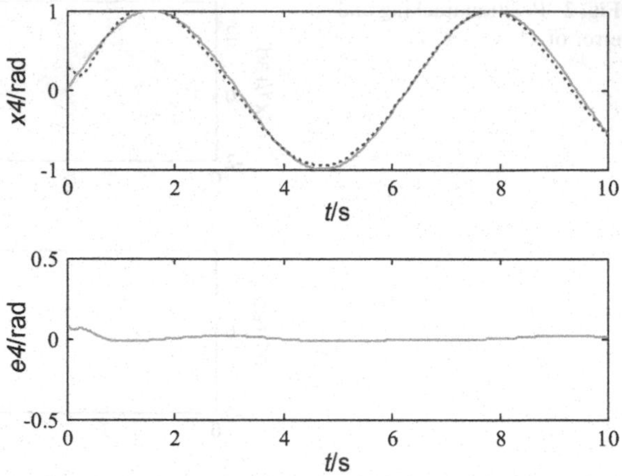


**Fig. 5** Position tracking and error of x4

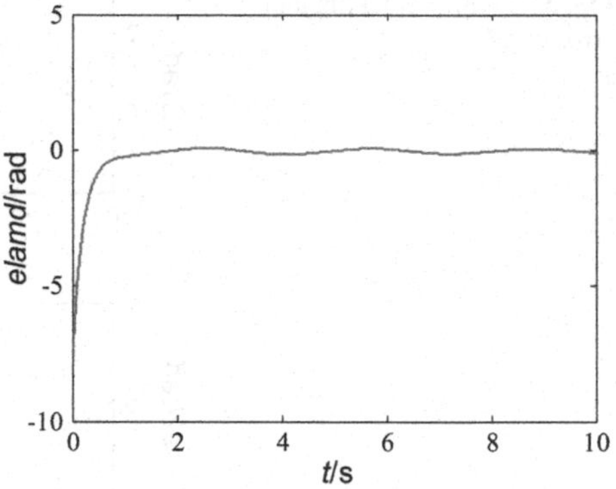


**Fig. 6** Tracking error of force

## 5 Conclusion

In this paper, a position/force controller is given for the mobile manipulator under rheonomic constraint. Firstly according to the condition that the velocity is not orthogonal to the force under the rheonomic constraint, the system is transformed to satisfy that the velocity and the constraint force are orthogonal. And then the controller based on the linear sliding mode is designed according to the model, which makes the system position and the constraint force converge to zero gradually. Finally simulation results shows the feasibility of the method vividly.

## References

1. J. Chung, H. Velinsky et al., Modeling and control of a mobile manipulator. Robotica **16**(16), 607–613 (1998)

2. K. Watanabe, K. Sato, K. Izumi et al., Analysis and control for an omnidirectional mobile manipulator. J. Intell. Rob. Syst. **27**(1–2), 3–20 (2000)
3. M. Galicki, Control of mobile manipulators in a task space. IEEE Trans. Autom. Control **57**(11), 2962–2967 (2012)
4. M.H. Wu, S. Ogawa, A. Konno, Symmetry position/force hybrid control for cooperative object transportation using multiple humanoid robots. Adv. Robot. **30**(2), 131–149 (2016)
5. Z. Li, S.S. Ge, M. Adams et al., Robust adaptive control of uncertain force/motion constrained nonholonomic mobile manipulators. Automatica **44**(3), 776–784 (2008)
6. Y. Jia, Robust control with decoupling performance for steering and traction of 4WS vehicles under velocity-varying motion. IEEE Trans. Control Syst. Technol. **8**(3), 554–569 (2000)
7. N. Chen, F. Song, G. Li et al., An adaptive sliding mode backstepping control for the mobile manipulator with nonholonomic constraints. Commun. Nonlinear Sci. Numer. Simul. **18**(10), 2885–2899 (2013)
8. Y. Jia, Alternative proofs for improved LMI representations for the analysis and the design of continuous-time systems with polytopic type uncertainty: a predictive approach. IEEE Trans. Automat. Control **48**(8), 1413–1416 (2003)
9. Y.H. Liu, S. Arimoto, *Adaptive and Nonadaptive Hybrid Controllers for Rheonomically Constrained Manipulators* (Pergamon Press, Inc., 1998)

# A Ship Tracking Algorithm of Harbor Channel Based on Orthogonal Particles Filter

**Lei Xiao, Hui-Gang Wang, Mian-Lu Zou and Zhong-Yi Hu**

**Abstract** This paper, employing Bayes state estimation, proposes a ship tracking algorithm of harbor channel based on orthogonal particle filter. (1) The dynamic model fully takes speed of state change into consideration during the movement of target ship, to improve the problem that the existing correlation algorithms have poor adaptability to the target ship tracking of the complex mode. (2) The proposed algorithm reorganizes and estimates particles by using orthogonal particles arrays, which can avoid particles degradation problems caused by resampling. Experimental results demonstrate that our algorithm outperforms other algorithms.

**Keywords** Ship tracking · Bayes state estimation · Orthogonal particles filter

## 1 Introduction

In the sense of harbor channel, the relative position of the target ship with the CCTV camera usually changes during the movement. Therefore, robust harbor channel ship tracking system should be able to adapt to ship motion changes. However, from the papers [1–4], we can know that the existing ship tracking algorithms fail to implement perfect prediction of ship scale. To deal with this problem, this paper, employing Bayes state estimation, proposes a ship tracking algorithm of harbor channel based on orthogonal particle filter. Firstly, the dynamic model fully takes speed of state change into account in the movement process of target ship, and can effectively improve the problem that the existing correlation algorithms have poor adaptability to the target ship tracking of the complex motion mode. In order to avoid particles

L. Xiao (✉) · H.-G. Wang
School of Marine Science and Technology, Northwestern Polytechnical University, Xi'an 710072, Shanxi, China
e-mail: xiaolei@wzu.edu.cn

M.-L. Zou · Z.-Y. Hu
Intelligent Information Systems Institute, Wenzhou University, Wenzhou 325035, Zhejiang, China



Y. Jia et al. (eds.), *Proceedings of 2018 Chinese Intelligent Systems Conference*, Lecture Notes in Electrical Engineering 528,
https://doi.org/10.1007/978-981-13-2288-4_81

degradation problems caused by resampling, the proposed algorithm reorganizes and estimates particles by using orthogonal particles array. Considering the problem of low tracking precision about existing algorithms in target ship change, the observation model adopts the strategy merging ship vessel contour with color feature to realize robust representation of target ship, making the algorithm more stable [5–9].

## 2 Bayes State Estimation

In order to maintain the integrity of content, simultaneously expediently describe our proposed algorithm, this paper firstly elaborates and deduces basic theory of Bayes state estimation. Then it analyses the key problem of Bayes state estimation.

We assume that mathematical model system is described with formulas (1) and (2), when target ship is moving.

$$s_t = f_t(s_{t-1}, v_{t-1}) \tag{1}$$

$$z_t = h_t(s_t, n_t) \tag{2}$$

In the formulas (1) and (2), $s_t$ describes the state of target ship at time t, $z_t$ describes the observation of target ship at time t. $f_t$ and $h_t$ are defined as the linear or non-linear functions of state variables, $v_t$ and $n_t$ respectively represent process noise and observation noise which are independent of system state variables.

The aim of Bayes state estimation is to recursively express the posterior probability density function $p(s_t|z_{1:t})$, i.e., to express the distribution of target state $s_t$ at time $t$ by getting target observations $z_1, z_2, \ldots, z_t$ at time $1, 2, \ldots, t$. Bayes formula is as follow (3):

$$p(s_t|z_{1:t}) = \frac{p(z_{1:t}|s_t)p(s_t)}{p(z_{1:t})} = \frac{p(z_t, z_{1:t-1}|s_t)p(s_t)}{p(z_t, z_{1:t-1})} \tag{3}$$

## 3 Orthogonal Experiment Setup

In the initial state, the loss of generality can be ignored, $a_j$ denotes the j$^{\text{th}}$ column of orthogonal array $[a_{i,j}]_{M\times N}$. If $j$=$1, 2, (Q^2-1)/(Q-1)+1, (Q^2-1)/(Q-1)+1, \ldots, (Q^J-1)/(Q-1)+1$, $a_j$ represents standard columns; else, $a_j$ represents non-standard columns. The pseudo-code of constructing the orthogonal arrays is as shown in Table 1.

**Table 1** The pseudo-code for constructing the orthogonal array

| |
|---|
| Step 1: construct the standard columns: |
| for $k = 1$ to $J$ |
| $\quad j = \frac{Q^{k-1} - 1}{Q - 1} + 1$; |
| $\quad$ for $i = 1$ to $Q^J$ |
| $\qquad a_{i,j} = \left\lfloor \frac{i-1}{Q^{J-k}} \right\rfloor mod\,Q$; |
| $\quad$ end |
| end |
| Step 2: construct the non-standard columns: |
| for $k = 2$ to $J$ |
| $\quad j = \frac{Q^{k-1} - 1}{Q - 1} + 1$ |
| $\quad$ for $s = 1$ to $j$-1 |
| $\qquad$ for $t = 1$ to $Q$-1 |
| $\qquad\quad a_{j+(s-1)(Q-1)+t} = (a_s * t + a_j)\, mod\,Q$ |
| $\qquad$ end |
| $\quad$ end |
| end |
| For all, $1 \le i \le M, 1 \le j \le N$, $a_{i,j} = a_{i,j} + 1$ |

## 4 Experiment

In this paper, we compare our algorithm with other tracking algorithms like CT, TLD, MTT, IVT and etc. [10–15], which are usually used in visual tracking. The experimental environment configuration: CPU Intel Core(TM) i7-4500U@1.80 GHz, 8G memory. CT, TLD, MTT, IVT and Ours are all implemented in Matlab. In our algorithm, the state of ship motion $s_t = \{x_t, y_t, w_t, h_t\}$, parameter of motion model $A = I$, the initial standard deviations of the state are set as $\sigma_{x_t} = \sigma_{y_t} = 2.5$, $\sigma_{w_t} = \sigma_{h_t} = 0.1$.

## 5 Qualitative Experimental Result and Analysis

In Fig. 1a, b, there is no interference in all image sequences, therefore all tracking algorithm perform robust tracking for target ship. In Fig. 1a CCTV_3747, influenced by foggy weather, and great background noise, MTT gradually tracks drift during the movement. Influenced by ripple and illumination, CT appears serious drift under the background blur (as shown in Fig. 1c CCTV_9352 #000100–#000210). While our

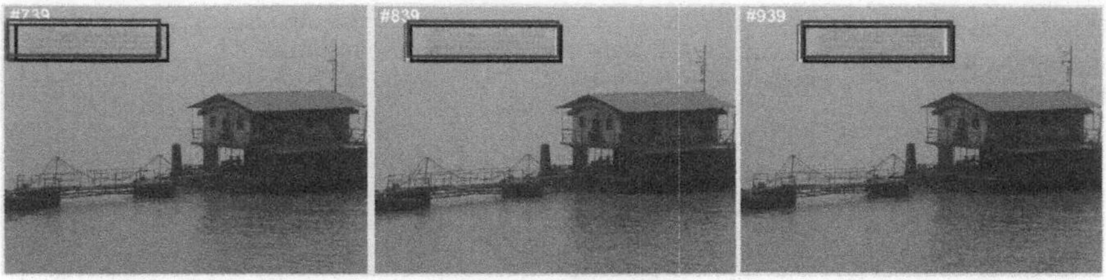


(a) CCTV_3747

(b) CCTV_9240

(c) CCTV_9352

OURS IVT MTT TLD CT

**Fig. 1** Representative track results of various algorithms

algorithm can effectively adapt background blur and interference in the process of ship motion, which can achieve observation of distance gain coefficient by utilizing differences of the target ship state in adjacent time.

## 6 Quantitative Experimental Result and Analysis

From the experimental success curve of TRE, we can observe that IVT algorithm performs slightly high compared with our proposed algorithm under the background blur and illumination variation (as shown in Fig. 2b); while our proposed algorithm performs better than IVT in success under the low quality images. The results further indicate that the discriminant performance can almost match the performance of several strong detectors. In addition, the proposed algorithm (LRCF) better rank in the experiments of OPE and SRE (as shown in Fig. 2a, c). The results demonstrate that LRCT has certain resistance to the noise of tracking initialization process.

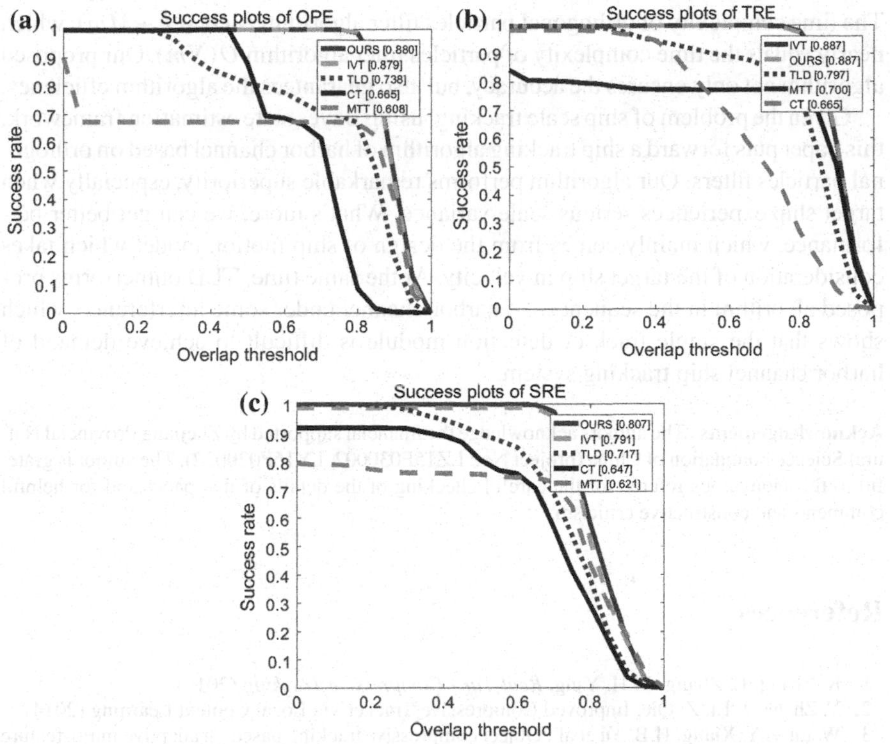


**Fig. 2** **a** Denotes experimental success curve of OPE. **b** Denotes experimental success curve of TRE. **c** Denotes experimental success curve of SRE

## 7 Algorithm Analysis and Conclusion

In this paper, the proposed algorithm based on orthogonal particles filter and current ship tracking algorithms [16, 17] based on vision, are both implemented by utilizing Bayes state estimation framework, i.e., translate ship tracking task into recursive reasoning of ship motion state. Therefore, position of target in current frame can be determined by employing particles with high weight, to avoid tracking drift and failure. Furthermore, Bayes state estimation framework can also directly implement real-time prediction and assessment of the target multiple mode states (such as ship's position, scale, rotation angle, etc.), simultaneously, which is simple and efficient compared with algorithms of other type like visual tracking algorithms based on discriminant model. However, the proposed algorithm has two significant differences compared with [16, 17] the existing models. On one hand, dynamic model makes full use of ship movement information in the image sequences. We can achieve prediction of distance gain coefficient by utilizing differences of the target ship state in adjacent time, contributing to target ship adapt displacement and velocity changes during the movement. On the other hand, we can effectively avoid degradation of representativeness and diversity caused by resampling in traditional particles filter, by using orthogonal experiment of orthogonal array to reorganize and evaluate particles.

The time complexity of orthogonal particles filter algorithm is $O(MF + Mm)$, which nearly equals the time complexity of particles filter algorithm $O(Nm)$. Our proposed algorithm not only ensures the accuracy, but also guarantees the algorithm efficiency.

Given the problem of ship scale tracking, using Bayes state estimation framework, this paper puts forward a ship tracking algorithm of harbor channel based on orthogonal particles filters. Our algorithm performs remarkable superiority, especially when target ship experiences serious scale variance. What's more, we can get better performance, which mainly comes from the design of ship motion model which takes consideration of the target ship in velocity. At the same time, TLD outperforms proposed algorithm in the sequences of harbor channel under some interference, which shows that the single track or detection module is difficult to achieve demand of harbor channel ship tracking system.

**Acknowledgements** The authors acknowledge the financial supported by Zhejiang Provincial Natural Science Foundation of China (project No.: LZ15F030002, LY16F020022). The author is grateful to the anonymous referee for the careful checking of the details of this paper and for helpful comments and constructive criticism.

## References

1. K. Zhang, L. Zhang, M.H. Yang, *Real-Time Compressive Tracking* (2012)
2. Y. Zhang, J. Li, Z. Qie, Improved Compressive Tracker via Local Context Learning (2014)
3. W. Lu, Z.Y. Xiang, H.B. Yu et al., Object compressive tracking based on adaptive multi-feature appearance model. J. Zhejiang Univ. Eng. Sci. **48**(12), 2132–2138 (2014)
4. Y. Wu, N. Jia, J. Sun, Real-time multi-scale tracking based on compressive sensing. Visual Comput. **31**(4), 471–484 (2015)
5. K. Zhang, L. Zhang, M.H. Yang, Fast compressive tracking. IEEE Trans. Pattern Anal. Mach. Intell. **36**(10), 2002–2015 (2014)
6. L. Qing-wu, Z. Guo-qing, Z. Yan et al., Object compressive tracking via online feature selection. Acta Automatica Sinica **41**(11), 1961–1970 (2015)
7. H.L. Luo, B.K. Zhong, F.S. Kong, Object tracking algorithm by combining the predicted target position with compressive tracking. J. Image Graph. **19**(6), 875–885 (2014)
8. Z. Lu-ping, H. Jian-tao, L. Biao et al., The scale adaptive feature compressed tracking. J. Natilnal Univ. Defense Technol. **35**(5), 146–151 (2013)
9. Z. Qiu-ping, Y. Jia, H. Zhang et al., Real-time tracking using multiple features based on compressive sensing. Opt. Precision Eng. **21**(2), 437–444 (2013)
10. D.C. Montgomery, Design and analysis of experiments. J. Am. Statistic. Assoc. **81**(16), 308 (2005)
11. W.H. Carter, Fundamental Concepts in the Design of Experiments (Oxford University Press, 1974), pp. 652–653
12. Y.W. Leung, Y. Wang, An orthogonal genetic algorithm with quantization for global numerical optimization. IEEE Trans. Evol. Comput. **5**(1), 41–53 (2001)
13. Z. Kalal, K. Mikolajczyk, J. Matas, Tracking-learning-detection. IEEE Trans. Pattern Anal. Mach. Intell. **34**(7), 1409–1422 (2011)
14. B. Babenko, M.H. Yang, S. Belongie, *Visual Tracking with Online Multiple Instance Learning* (2009)
15. D. Comaniciu, V. Ramesh, P. Meer, *Real-Time Tracking of Non-Rigid Objects Using Mean Shift* (2000)
16. L. Jing, L. Yi-an, Y. Xin-gang, Shipping trace technology based on Kalman filter. Microcomput. Informat. **23**(31), 187–189 (2007)
17. S.L. Wang, Research and Implement on Video Tracking Technology Based on Kalman Filter (Herbei University of Technology, 2010)

# Study on the Changes of Macular Retinal Thickness in Myopia

**Hejun Tong and Dongmei Fu**

**Abstract** With the increasing use of eyes at close range, myopia has become a public health problem in many areas. Fundus changes caused by high myopia have become one of the main causes of vision loss. In order to have a deeper understanding of myopia, we need to clarify the characteristics of myopia. In this paper, the retinal tomographic images obtained by optical coherence tomography (OCT) were used to analyze the changes of retinal thickness and retinal boundary morphology of myopia with the myopic diopter. The results showed that the retinal thickness and diopter have the highest correlation in the 2–2.5 mm region of the macula from the nasal side. The retinal thickness in this region was reduced by approximately 3.189 μm per unit of myopic diopter degrees. The subretinal boundary can be modeled by a quadratic function, and the quadratic coefficient decreases with increasing diopter.

**Keywords** Myopia · Diopter · Retinal macula · OCT · Morphologic modeling

## 1 Introduction

With the increasing need for close-up eyes, myopia has become a very common phenomenon. According to relevant reports, the number of myopia has reached nearly 5 billion in China [1]. Although myopia can be corrected by proper optical correction, the retina changes in myopia [2], especially in high myopia, cannot be corrected by optical correction, which will eventually lead to permanent loss of vision. Therefore, it is particularly important to understand the characteristics of myopic fundus changes so as to gain a better understanding of myopia.

Optical coherence tomography (OCT) is a technique that can obtain high-resolution tomographic images of living body tissues without damage and fast [3]. The most significant medical contribution of OCT is in the ophthalmic field,

H. Tong · D. Fu (✉)
School of Automation and Electrical Engineering, University of Science and Technology Beijing, Beijing 100083, China
e-mail: fdm_ustb@ustb.edu.cn



Y. Jia et al. (eds.), *Proceedings of 2018 Chinese Intelligent Systems Conference*, Lecture Notes in Electrical Engineering 528,
https://doi.org/10.1007/978-981-13-2288-4_82

which can provide images of the structure and function of the retina that other non-destructive diagnostic techniques cannot provide. At present, OCT has been widely used in the detection of retinal diseases, especially for the diagnosis of macular degeneration [4, 5].

In this paper, OCT will be used to model and analyze the changes of retinal thickness and morphological changes in the macular area in myopic patients with different diopter degrees.

## 2 Data Source Description

The OCT image data used in this article comes from the Department of Ophthalmology of the 306 Hospital of the Chinese People's Liberation Army. The image capture instrument is Optovue Avanti RTVue XR (Optovue Corp., Fremont, CA). The shooting mode is Enhanced HD Line. This mode can obtain a cross sectional view of the retina with a 12 mm width between the fovea and the center of the optic disc, as shown in Fig. 1. This study selected 64 OCT images of emmetropic and myopic patients, all aged between 18 and 29 years old. The diopter of myopic patients ranges from −1.00 to −9.00 D.

The retina resides in the inner layer of the human eye and is a transparent film. The retina is formed by the evolution of the optic cup. The outer layer of the optic cup evolves into pigment epithelium layer. The inner layer is highly differentiated to form

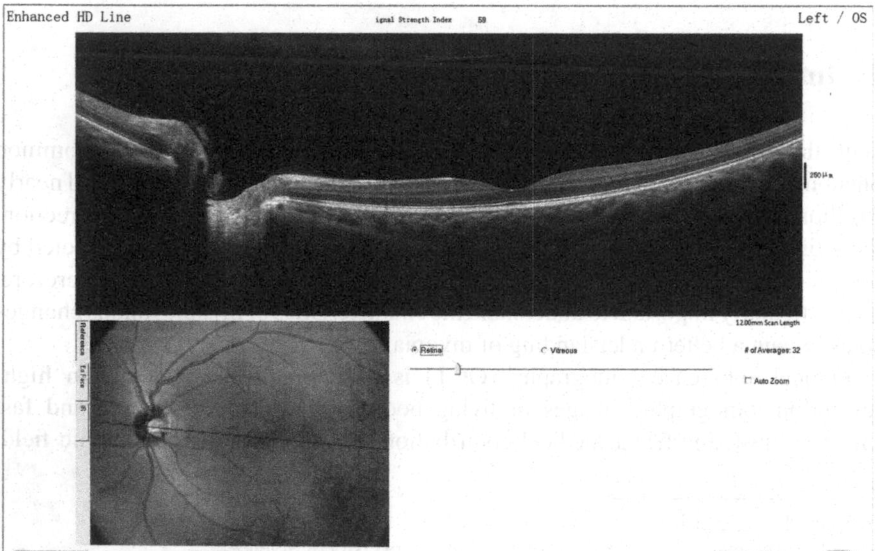


**Fig. 1** Cross-sectional view of the retina and fundus images obtained by OCT

the retinal neuroepithelium layer (including cone, rod cell layer, outer membrane, outer granular layer, outer plexiform layer, inner granular layer, inner plexiform layer, ganglion cell layer, nerve fiber layer, inner limiting membrane). The pigment epithelial layer and the retina sensory layer are two components of the retina. In the pathological condition, the two layers can be separated and called retinal detachment. Among them, the pigment epithelium consists of pigment epithelial cells and is tightly connected with the choroid. It has support and nutrition effects on photoreceptor cells, and also has the functions of shading, heat dissipation, regeneration and repair. In the posterior pole of the retina, there is a shallow funnel-shaped depression, which is the optical center of the human eye called the macular area. The retinal macular area can be subdivided into three areas from the depodization: Fovea, the center of the macular area, is a concave surface, a diameter of about 1.5 mm, which is a diameter of the optic disc. Its center is called foveola and is about 0.35 mm in diameter. Parafovea, an annular zone approximately 0.5 mm in width outside the macula, which contains ganglion cells, inner nuclear layers, and outer plexiform layers called Henle fibers. Perifovea, an annular zone about 1.5 mm wide outside the macular periarea. The fovea is the most sensitive part of the retina, and in the macular area, its resolving power is also the strongest. The macular retinal morphology appears as follows: The other retinal layers below the fovea disappear and gradually appear and thicken toward the macula. Therefore, in the macular retinal imaging of the OCT scan, depressions in the central region appear, gradually becoming thicker and thinner on both sides.

## 3 Image Processing

During the imaging process of the OCT instrument, the focus will be automatically adjusted according to the shape of the shooting target, and the axial resolution of the image will be changed, and the retina in the image will have different degrees of tilt. Therefore, the obtained OCT image should be standardized first. The goal of image normalization is to: (1) make the image uniform in terms of x-axis size and y-axis resolution; (2) focus on the macular center area; (3) position the entire retina horizontally to ensure consistency of subsequent feature extraction. The standardization results are shown in Fig. 2b.

The retinal OCT image has the characteristics of unclear edges, low target blurring and signal-to-noise ratio, and therefore needs denoising during preprocessing. This article uses the BM3D method [6] to perform de-noising of OCT images. This method not only has significant denoising effect, but also preserves the contrast between the boundary information and the layers of the retina, as shown in Fig. 2c.

When lesions occur in the macular area of the retina, morphological variations often occur in the ILM and RPE of the pigmented epithelial layer, which is an important indicator of disease condition. The ILM is the inner boundary of the retina, and the lower boundary of the RPE is the outer boundary membrane of the retina. The retina is defined medically between these two boundary membranes, and some OCT instruments also use the probe data between these two boundary membranes as the thickness data of the retina. Therefore, the ILM and RPE boundaries are extracted on the basis of image preprocessing, as shown by the red line in Fig. 2d.

## 4 Changes in Retinal Thickness with Diopter in Different Areas

The literature on OCT scanning of retinal thickness in myopia confirms that there is no significant difference in the retinal thickness of normal macular areas between male and female, and there is no correlation with age [7]. Therefore, gender and age differences are not analyzed in this paper. In order to more accurately represent the rules of change, the right eye OCT image in the experimental example is flipped, unified to the left eye perspective, and the left side of the image is closer to the optic disc. On the basis of obtaining the boundary of the retina, combined with the axial resolution provided by the OCT instrument, the thickness information of each position in the horizontal direction of the retina can be obtained, so as to analyze the change of retinal thickness with diopter. In order to reduce the error, the boundary and thickness data with the same diopter were averaged, and finally 15 sets of statistics with diopter from 0 to −9.00 D were obtained.

The correlation between diopter and retinal thickness at different distances from the macular center fovea is shown in Fig. 3, where the blue line indicates the correlation coefficient and the red line is the significance test of the correlation coefficient (P-value). It can be seen from the figure that there is a highly linear correlation between retinal thickness and diopter outside the central concave area 1.5 mm, and there is a statistically significant difference ($P<0.05$). The correlation between the

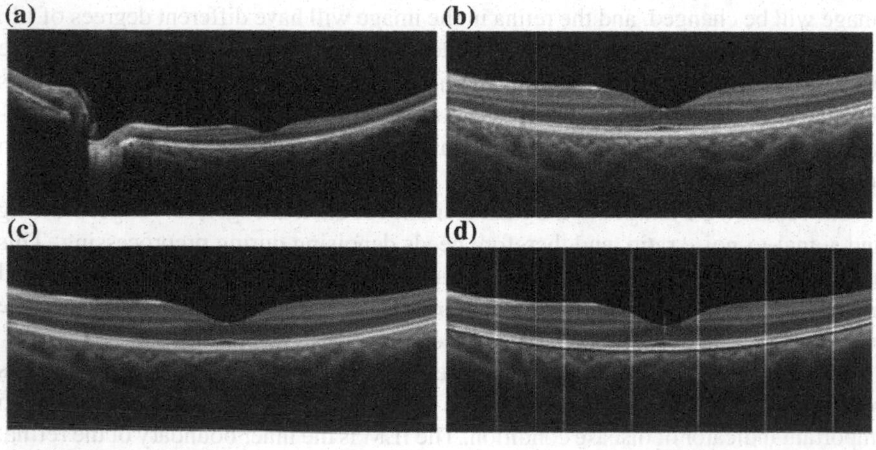


**Fig. 2** Image processing process. **a** Original OCT image; **b** image normalization; **c** image denoising; **d** retina boundary extraction

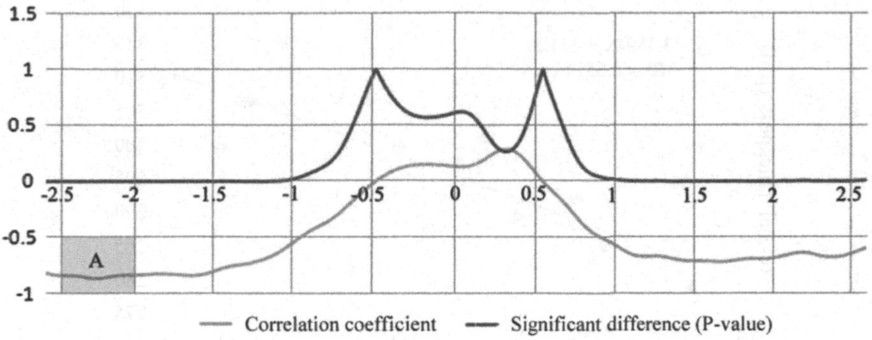


**Fig. 3** Correlation analysis of retinal thickness and diopter in different locations

**Table 1** Retinal thickness (μm) at different diopter levels

| Diopter | 0 D | −1.0 D | −1.50 D | −2.00 D | −2.50 D | −3.00 D | −3.5 D | −4.0 D |
|---|---|---|---|---|---|---|---|---|
| Thickness | 309.4 | 299.8 | 306.6 | 315.0 | 295.3 | 307.2 | 306.6 | 299.5 |
| Diopter | −4.50 D | −5.00 D | −5.50 D | −6.00 D | −7.00 D | −7.50 D | −9.00 D | |
| Thickness | 287.5 | 298.0 | 301.2 | 289.2 | 291.9 | 285.1 | 287.7 | |

area of 1.5 and 1 mm from the foveola has a decreasing trend. There was a low degree of linear correlation in the foveal area (−1 to 1 mm), and the difference was not statistically significant ($P > 0.05$). The results showed that with the increase of myopia, the macular area's retinal thickness gradually decreased in the peripheral area (1.5–2.5 mm from the foveola), while there was no significant change in the fovea area (1.5 mm from the center pit). This is in accordance with the results of literature [8–11] and other studies.

From Fig. 3, it can be seen that the retinal thickness is best correlated with diopter in the nasal region from the 2–2.5 mm area of the central fovea (yellow region A in Fig. 3). Therefore, the mean retinal thickness data of this region is used to analyze the relationship of diopters. Table 1 shows the retinal thickness of area A under different diopter. Figure 4 shows a linear regression analysis between retinal thickness and diopter. From the statistics it can be seen that for every unit of myopic diopter increase, the retinal thickness in this region is reduced by approximately 3.189 μm.

# 5 Myopia Retinal Morphology Analysis

Studies have shown that as the degree of myopia increased, the axial length gradually increased [12], and with the increase of the axial length of the myopia, the thickness of the parafoveal retina thins, and the outer ring is thinner than the inner ring, except for the thickness of the macular retina [8]. Studies have shown that the growth of the

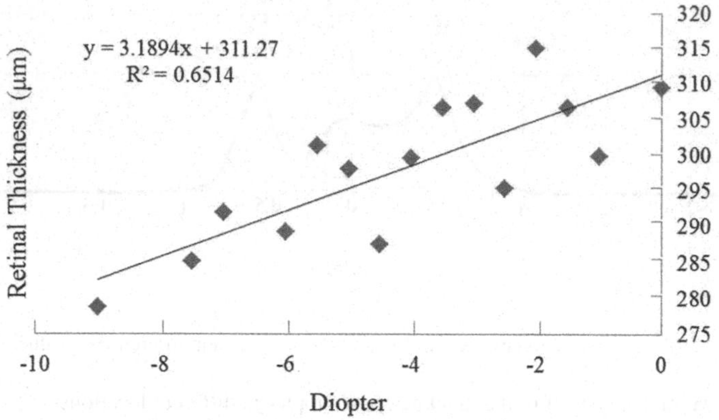


**Fig. 4** Relationship between retinal thickness and diopter in the periphery of the nasal fovea

axial axis leads to a significant thinning of the thickness of the posterior pole retina and a myopic diopter [12].

The increase in axial length will bring about changes in the morphology of the retina and the change in the curvature of the RPE border is more obvious [13]. In this paper, the quadratic function model $y = ax^2 + bx + \mathbf{c}$ ($a, b, c$ is constants, $a \neq 0$) is used to fit the RPE boundary, and the obtained parameters are analyzed. Among them, the quadratic coefficient $a$ determines the direction and magnitude of the opening of the parabola. The coefficient of the first order term $b$ and the quadratic coefficient $a$ determine the position of the symmetry axis. The constant term $c$ determines the intersection of the parabola and the Y axis. In this paper, $a > 0$ in the quadratic function model fitted according to the RPE boundary. When the left and right eyes are not distinguished, the average retinal morphology shows symmetry, and $b = 0$. As the left eye is unified during preprocessing, the axis of symmetry of the fitted line is to the left. In the pretreatment process, the position of the central foveal was unified. In addition, because the macular foveal retinal thickness has no correlation with the refraction, the retinal thickness in the central fovea is relatively stable, and the constant term $c$ can be set as a fixed value in this article. The comparison of the retinal morphology between normal eyes and high myopia and the quadratic function fitted line of the RPE boundary are shown in Fig. 5.

The quadratic coefficient a plays a decisive role in the change of the RPE boundary shape, so this paper mainly discusses the relationship between the coefficient $a$ and the diopter. Fifteen groups of RPE statistics with different diopters were fitted with quadratic functions, and three regression coefficients were obtained. The negative correlation between coefficient $a$ and diopter is shown in Fig. 6. By constructing the regression relationship between the two-parameter coefficient $a$ and the diopter, the relationship between RPE boundary morphology and diopter can be further obtained.

# 6 Conclusion

In this paper, the retinal tomographic images obtained by optical coherence tomography (OCT) were used to analyze the changes of retinal thickness and retinal boundary morphology of myopia with myopia diopter. The image processing method was used to obtain the upper and lower boundaries of the retina, combined with axial resolution to obtain the thickness of the retina. The relationship between retinal thickness and diopter in different locations was analyzed to find the best relative position, and a regression model for retinal thickness and diopter was constructed. The subretinal boundary is modeled by quadratic function and the law of the quadratic coefficient with diopter is analyzed. The results showed that the correlation between retinal thickness and diopter were highest in the 2–2.5 mm region of the macula from the nasal side. The retinal thickness in this region was reduced by 3.189 μm per unit of myopic diopter. The binomial coefficients of the quadratic function modeling of the subretinal boundary decrease with increasing diopter.

# References


1. L. Xu, Y. Wang, S. Wang et al., High myopia and glaucoma susceptibility the Beijing Eye study. Ophthalmology **114**, 216 (2007)
2. Z.F. Ismael, E.F. El-Shazly, Y.A. Farweez et al., Relationship between functional and structural retinal changes in myopic eyes. Clinic. Experiment. Optom. **100**(6), 695–703 (2017)
3. D. Huang, E.A. Swanson, C.P. Lin, J.S. Schuman, W.G. Stinson, W. Chang, M.R. Hee, T. Flotte, K. Gregory, C.A. Puliafito, Optical coherence tomography. Science **254**(5035), 1178–1181 (1991)
4. D.S. Ng, C.Y. Cheung, F.O. Luk et al., Advances of optical coherence tomography in myopia and pathologic myopia. Eye **30**(7), 901–916 (2016)


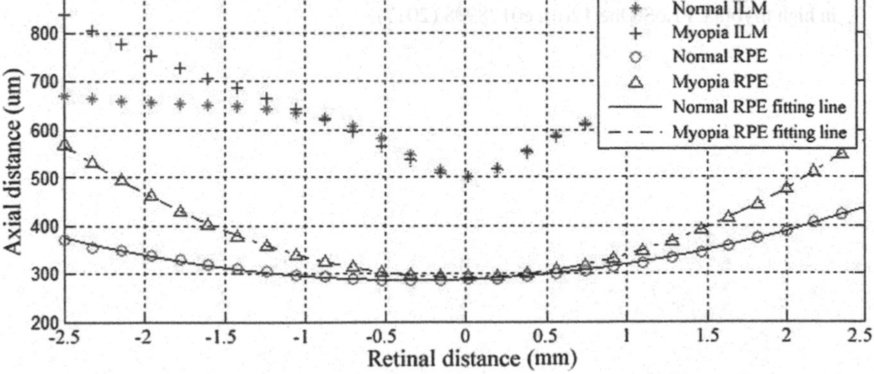


**Fig. 5** Retinal morphology difference between normal and high myopia eyes and quadratic function fitting lines of RPE

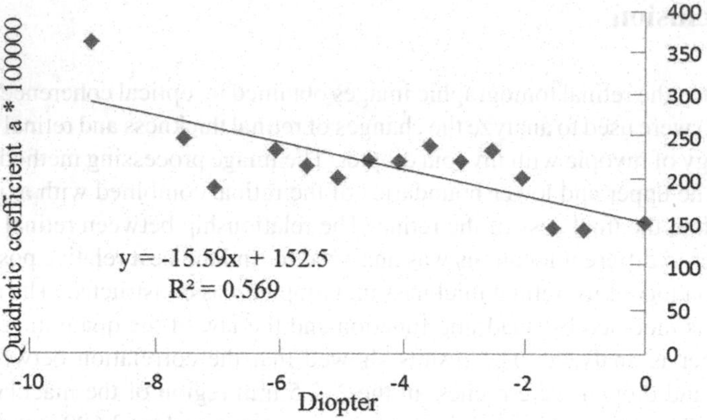


**Fig. 6** The relationship between quadratic coefficient *a* and diopter


5. H.L. Rao, A.U. Kumar, S.R. Bonala et al., Repeatability of spectral domain optical coherence tomography measurements in high myopia. J. Glaucoma **25**(5), E526–E530 (2016)
6. Y.J. Li, J.W. Zhang, M.N. Wang, Improved BM3D denoising method. IET Imag. Proc. **11**(12), 1197 (2017)
7. W. Wang, N. Wang, H. Wang, Measurement of normal macular thickness with optical coherence tomography in healthy people. Chinese J. Rehab. Theor. Pract. **14**(2), 1072–1074 (2008)
8. N. Li, H. Liao, Y. Fangzhi, L. Wang, Study on retina macular thickenss in myopia. Chinese Ophthal. Res. **26**(6), 436–438 (2008)
9. R. Gao, Y. Wei, Z. Li, Macular area observation in myopic patients with different diopter by OCT. Recent Advanc. Ophthalmol. **32**(1), 71–74 (2012)
10. Y. Zhang, W. Zhang, Investigation on the variation in macular retinal thickness of high myopia. Int. Eye Sci. **13**(5), 950–952 (2013)
11. K. Teberik, M. Kaya, Retinal and choroidal thickness in patients with high myopia without maculopathy. Pakistan J. Med. Sci. **33**(6), 1438–1443 (2017)
12. Z. Deng, S. Liu, J. Tan et al., The correlation between retinal thickness at posterior pole in myopia and axis length of eyeball. Int. J. Ophthalmol. **4**(4), 654–656 (2004)
13. Y. Fang, J.B. Jonas, T. Yokoi et al., Macular Bruch's membrane defect and dome-shaped macula in high myopia. PLoS One **12**(6), e0178998 (2017)

# Backstepping Based Neuroadaptive Control for Uncertain Robot Systems

**Fanfeng Meng, Lin Zhao and Jinpeng Yu**

**Abstract** This paper gives a command filter backstepping method to design an adaptive controller to achieve position tracking for robot systems with uncertain parameters. Command filter is used to deal with computing complex problem of classical backstepping strategy. The neutral network is used to approximate uncertain dynamics. The error compensation signal is used to eliminate the error caused by the filtering. An example is applied to demonstrate effectiveness of control method.

**Keywords** Robot systems · Backstepping · Command filtering

## 1 Introduction

At present, the scope of application of robots has been extended to all aspects of human production and life, which has greatly influenced and promoted the development of industry, national defense, and science and technology [1–3]. Industrial robots appear precisely to meet the higher requirements of production automation and market adaptability, and advances in production and technology also place new demands on the control of industrial robots [4–7].

The backstepping method is a common tool for designing nonlinear system controllers, but there are two main difficulties: one is the computation complex problem since each step needs the derivation of virtual control signal; the other is the practical system usually with uncertain parameters, but the traditional backstepping requires the function must be linear [8]. In order to solve the above drawbacks, [9, 10] proposed a command filtering based backstepping method, which can overcome the first difficulty. In the design of each step, the output of the command filter is used to approximate the differential of virtual control signal. The error caused by the command filter is eliminated with the error compensation function. Then, in [11, 12], it was further extended to an adaptive command filtering backstepping method for

F. Meng · L. Zhao (✉) · J. Yu
School of Automation, Qingdao University, Qingdao 266071, China
e-mail: zhaolin1585@163.com



Y. Jia et al. (eds.), *Proceedings of 2018 Chinese Intelligent Systems Conference*,
Lecture Notes in Electrical Engineering 528,
https://doi.org/10.1007/978-981-13-2288-4_83

strict-feedback SISO nonlinear systems with uncertain parameters. However, to the best of our knowledges, no work has been given to study the command filtering based backstepping control for Industrial robots with uncertain parameters.

According to the above discussion, the adaptive command filtering backstepping method is applied to the industrial robot system with uncertain parameters to achieve desired tracking performance. In the proposed control scheme, the command filter is used to deal with computing complex problem of classical backstepping strategy, the neutral network is used to approximate uncertain dynamics and the error compensation signal is used to eliminate the error caused by the filtering.

## 2 Robot System Description

Considering the robot dynamics equation as:

$$M(q)q + C(q, \dot{q})\dot{q} + G(q) = \tau \tag{1}$$

where $q \subset R^n$ is the joint configuration variable, $M(q) \in R^{n\times n}$ is the symmetric inertia matrix, $C(q, \dot{q}) \in R^{n\times n}$ is the centripetal and coriolis torques matrix, and $G(q) \in R^n$ is the gravity term, and $\tau$ is the driving force of the robot actuator on each rod. In this paper, we consider $M(q)$, $C(q, \dot{q})$ and $G(q)$ are all with uncertain parameters.

Let $q = x_1$ $\dot{q} = x_2$, we get the following equation:

$$\begin{cases} \dot{x}_1 = \dot{x}_2 \\ \dot{x}_2 = -M(q)^{-1}C(q, \dot{q})\dot{q} - M(q)^{-1}G(q) + M(q)^{-1}\tau \end{cases} \tag{2}$$

Moreover, (2) can be rewritten as

$$\begin{aligned} \dot{x}_1 &= x_2 \\ \dot{x}_2 &= f + u \end{aligned} \tag{3}$$

where

$$\begin{aligned} u &= M(x_1)^{-1}\tau \\ f &= -M(x_1)^{-1}C(x_1, x_2)x_2 - M(x_1)^{-1}G(x_1) \end{aligned}$$

## 3 Control Law Design

Firstly, define the command filter as:

$$\begin{aligned}\dot{\varphi}_1 &= \iota_1 \\ \iota_1 &= -r_1|\varphi_1-\alpha_1|^{\frac{1}{2}}\mathrm{sign}(\varphi_1-\alpha_1)+\varphi_2 \\ \dot{\varphi}_2 &= -r_2\mathrm{sign}(\varphi_2-\iota_1)\end{aligned} \tag{4}$$

For the filter in (4), from [11], we know that if input signal $\alpha_1$ is not affected by noise, (4) is finite-time stability, and if input signal is affected by noise and satisfies $|\alpha_1-\alpha_{10}| \le \rho_1$, we have $|\varphi_1-\alpha_{10}| \le \mu_1\rho_1=\varpi_1$, $|\iota_1-\dot{\alpha}_{10}| \le \mu_1\rho_1^{\frac{1}{2}}-\varpi_2$.

The tracking errors are defined as:

$$\tilde{x}_1 = x_1 - x_d \tag{5}$$

$$\tilde{x}_2 = x_2 - x_{2,c} \tag{6}$$

where $x_d$ is a given reference signal and it with its first-order parameters are smooth, well-known and bounded. $x_{2,c}$ is the output of the command filter and the input is the virtual controller $\alpha_1$. The applied command filter can produce errors that affect control strategies, in order to deal with this problem, an error compensation mechanism is introduced to eliminate the error $(x_{2,c}-\alpha_1)$ generated by the command filtering process.

The error compensation mechanism $\varepsilon_i (i = 1, 2)$ is proposed as follows:

$$\dot{\varepsilon}_1 = -\kappa_1\varepsilon_1 + x_{2,c} - \alpha_1 + \varepsilon_2 \tag{7}$$

$$\dot{\varepsilon}_2 = -\kappa_2\varepsilon_2 - \varepsilon_1 \tag{8}$$

where $\kappa_i(i = 1, 2) > 0$, $\varepsilon(0) = 0$, $\alpha_1$ is virtual control function.

For (7) and (8), we define $U = \frac{\varepsilon_1^T\varepsilon_1+\varepsilon_2^T\varepsilon_2}{2}$, then we have

$$\begin{aligned}\dot{U} &= \varepsilon_1^T\dot{\varepsilon}_1 + \varepsilon_2^T\dot{\varepsilon}_2 = -\kappa_1\varepsilon_1^T\varepsilon_1 - \kappa_2\varepsilon_2^T\varepsilon_2 + \varepsilon_1^T(x_{2,c}-\alpha_1) \\ &\le -\left(\kappa_1-\frac{1}{2}\right)\varepsilon_1^T\varepsilon_1 - \kappa_2\varepsilon_2^T\varepsilon_2 + \frac{\sqrt{3}}{2}\varpi_2 \\ &\le -\kappa_0 U + \frac{\sqrt{3}}{2}\varpi_2\end{aligned} \tag{9}$$

where $\kappa_0 = \min(2\kappa_1 - 1, \kappa_2)$. Then, we have for $t$ tend to $\infty$, we can obtain $\lim_{t\to\infty}\|\varepsilon_i\| \le \sqrt{\frac{\sqrt{3}\varpi_2}{\kappa_0}}$. Define the compensated tracking error signal:

$$v_2 = \tilde{x}_2 - \varepsilon_2 \tag{10}$$

The virtual control function $\alpha_i(i = 1, 2)$ is defined as:

$$\alpha_1 = -\kappa_1 \tilde{x}_1 + \dot{x}_d \tag{11}$$

$$u = \alpha_2 = -\kappa_2 \tilde{x}_2 - \frac{1}{2} v_2 + \dot{x}_{2,c} - \tilde{x}_1 - \frac{1}{2l^2} \begin{pmatrix} v_{21} \hat{\theta}_1 S_1^T S_1 \\ v_{22} \hat{\theta}_2 S_2^T S_2 \\ \vdots \\ v_{2n} \hat{\theta}_n S_n^T S_n \end{pmatrix} \tag{12}$$

The recursion of the design process is divided into the following two steps:
Step 1: Set the Lyapunov equation as:

$$V_1 = \frac{1}{2} v_1^T v_1 \tag{13}$$

Then, differentiating $V_1$, we have

$$\begin{aligned} \dot{V}_1 = v_1^T \dot{v}_1 &= v_1^T \left( \dot{\tilde{x}}_1 - \dot{\varepsilon}_1 \right) \\ &= v_1^T (\dot{x}_1 - \dot{x}_d - \dot{\varepsilon}_1) \\ &= v_1^T (x_2 - \dot{x}_d - \dot{\varepsilon}_1) \\ &= v_1^T \left( \alpha_1 + \left( x_{2,c} - \alpha_1 \right) + \left( x_2 - x_{2,c} \right) - \dot{x}_d - \dot{\varepsilon}_1 \right) \end{aligned} \tag{14}$$

Substituting (5), (7) into (14) yields

$$\begin{aligned} \dot{V}_1 &= v_1^T (-\kappa_1 \tilde{x}_1 + \kappa_1 \varepsilon_1 + (\tilde{x}_2 - \varepsilon_2)) \\ &= v_1^T (-\kappa_1 v_1 + v_2) \\ &= -\kappa_1 v_1^T v_1 + v_1^T v_2 \end{aligned} \tag{15}$$

Step 2: Define another Lyapunov equation as:

$$V_2 = V_1 + \frac{1}{2} v_2^T v_2 \tag{16}$$

Then, we have

$$\begin{aligned} \dot{V}_2 &= \dot{V}_1 + v_2^T \dot{v}_2 \\ &= \dot{V}_1 + v_2^T \left( \dot{\tilde{x}}_2 - \dot{\varepsilon}_2 \right) \\ &= \dot{V}_1 + v_2^T \left( \dot{x}_2 - \dot{x}_{2,c} - \dot{\varepsilon}_2 \right) \\ &= \dot{V}_1 + v_2^T \left( f + u - \dot{x}_{2,c} - \dot{\varepsilon}_2 \right) \end{aligned} \tag{17}$$

Since the function $f = [f_1, \cdots, f_n]^T$ contains uncertainties, it cannot be directly applied for controller construction, so the neural network is used to approximate it. As in [11], $f_i, i = 1, \cdots, n$ can be approximated by

$$f_i = W_i^T S_i + \delta_i \tag{18}$$

where $W_i$ is ideal weight matrix, $S_i$ is vector of basis function and $\|\varepsilon_i\| \le \varepsilon, \varepsilon > 0$ is approximation error. By Young's inequality, we have

$$v_2^T f \le \sum_{i=1}^{n} \left( \frac{1}{2l^2} v_{2i}^2 \| W_i \|^2 S_i^T S_i + \frac{1}{2} l^2 + \frac{1}{2} v_{2i}^2 + \frac{1}{2} \varepsilon^2 \right) \tag{19}$$

where $l > 0$ is a constant.

By using (8) and (12), (19) can be given as:

$$\dot{V}_2 \le - \sum_{i=1}^{2} \kappa_i v_i^T v_i + \sum_{i=1}^{n} \left( \frac{1}{2l^2} v_{2i}^2 \left( \| W_i \|^2 - \hat{\theta}_i \right) S_i^T S_i + \frac{1}{2} l^2 + \frac{1}{2} \varepsilon^2 \right) \tag{20}$$

Define $\theta_i = \| W_i \|^2$, then the updating law for estimating $\theta_i$ is given as following:

$$\dot{\hat{\theta}}_i = -r_i \rho_i \hat{\theta}_i + \frac{1}{2l^2} r_i v_{2i}^2 S_i^T S_i \tag{21}$$

where $r_i > 0$ and $\rho_i > 0$ are constants.

Further define $\tilde{\theta}_i = \theta_i - \hat{\theta}_i$, we choose the Lyapunov equation as

$$V = V_2 + \sum_{i=1}^{n} \frac{1}{2r_i} \tilde{\theta}_i^2 \tag{22}$$

Differentiating $V$, we can obtain

$$\dot{V} = \dot{V}_2 + \sum_{i=1}^{n} \frac{1}{r_i} \tilde{\theta}_i \dot{\hat{\theta}}_i \tag{23}$$

Substituting (20) and (21) into (23), we get

$$\dot{V} \le - \sum_{i=1}^{2} \kappa_i v_i^T v_i + \sum_{i=1}^{n} \rho_i \tilde{\theta}_i \hat{\theta}_i + \frac{1}{2} n \left( l^2 + \varepsilon^2 \right) \tag{24}$$

By Young's inequality, we have $\tilde{\theta}_i^T \hat{\theta}_i \le -\frac{1}{2} \tilde{\theta}_i^2 + \frac{1}{2} \theta_i^2$, substituting it into (24) yields

$$\begin{aligned} \dot{V} &\le - \sum_{i=1}^{2} \kappa_i v_i^T v_i - \sum_{i=1}^{n} \frac{\rho_i}{2} \tilde{\theta}_i^2 + \sum_{i=1}^{n} \frac{\rho_i}{2} \theta_i^2 + \frac{1}{2} n \left( l^2 + \varepsilon^2 \right) \\ &\le -a + b \end{aligned} \tag{25}$$

where $a = (2\kappa_i, \rho_i r_i), b = \sum_{i=1}^{n} \frac{\rho_i}{2}\theta_i^2 + \frac{1}{2}n(l^2 + \varepsilon^2)$.

From (25), we can get

$$V(t) \le \left(V(t_0) - \frac{b}{a}\right)e^{-a(t-t_0)} + \frac{b}{a} \le V(t_0) + \frac{b}{a}\ \forall t \ge t_0 \tag{26}$$

which means that $v_i$,$\varepsilon_i$ are bounded. Since $\theta_i$ is constant, so $\hat{\theta}_i$ is also bounded. We also obtain $\tilde{x}_i$ is bounded. At any time, all control signals are bounded. Then, we can obtain

$$\lim_{x\to\infty} \tilde{x}_1 \le \sqrt{\frac{2b}{a}} + \sqrt{\frac{\sqrt{3}\varpi_2}{\kappa_0}} \tag{27}$$

**Theorem 1** *For the robot system (1), the designed error compensation mechanism (7) and (8), and control signals (11) and (12) with adaptive law (21) can guarantee the position tracking error $\tilde{x}_1$ converges to a desired neighborhood of origin and all signals in closed-loop system are bounded.*

## 4 Simulation

In this section, a simulation example is given to testify the advantage of method. The matrix $M(q)$=$[M_{mn}] \in R^{2\times2}$ and centripetal matrix $C(q, \dot{q})$ = $[C_{mn}] \in R^{2\times2}$ are defined as

$$M_{11} = a_1 + 2a_2\cos(q_2), M_{12} = M_{21} = a_3 + a_2\cos(q_2), M_{22} = a_3,$$
$$C_{11} = -a_2\sin(q_2)\dot{q}_2, C_{12} = -a_2\sin(q_2)(\dot{q}_1 + \dot{q}_2), C_{21} = a_2\sin(q_2)\dot{q}_1,$$
$$C_{22} = 0, a_1 = I_1 + m_1 l_{c1}^2 + m_2 l_1^2 + I_2 + m_2 l_{c2}^2, a_2 = m_2 l_1 l_{c2}, a_3 = I_2 + m_2 l_{c2}^2$$

with $m_1$ and $m_2$ being the masses of the links, $I_1$ and $I_2$ being the moments of inertia, $l_{c1}$ and $l_{c2}$ being the mass centers of the links. The gravitational torque $G(q)$ are assumed to be zero. The parameters are chosen as

$$I_1 = 0.52, I_2 = 0.41, m_1 = 1.6, m_2 = 1.7, L_1 = 2, L_2 = 1.8, L_{c1} = 1.2, L_{c2} = 1, 2$$

The control parameters are chose $\kappa_1$=30, $\kappa_2$=30, $r_1$ = 40, $r_2$ = 40.Figure 1 shows the trajectories of $q$ and $x_d$ under the proposed control scheme.

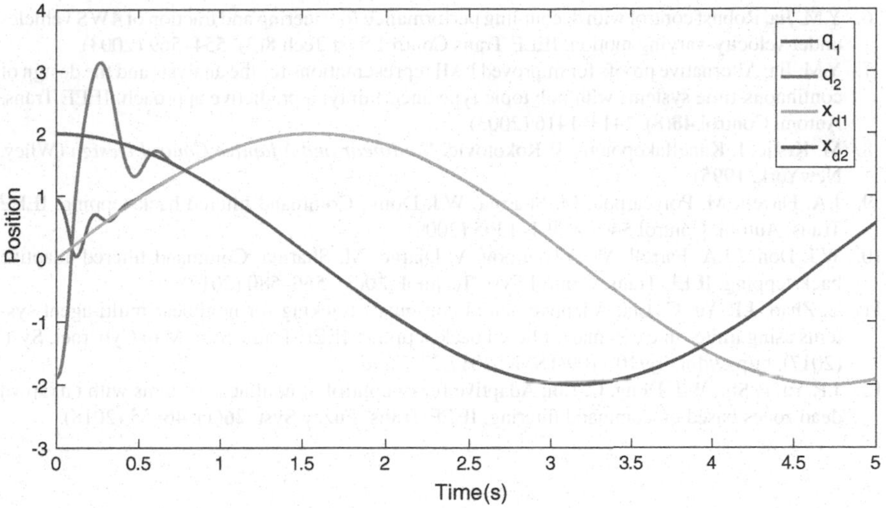


**Fig. 1** The trajectories of $q$ and $x_d$

# 5 Conclusion

The paper mainly uses the command filter backstepping method to control the industrial robot system. The command filter is used to deal with the computing complex problems of classical backstepping method, and the error compensation mechanism is designed to overcome errors between filter and virtual signal. It was proved that the position tracking error converges to a sufficiently small neighborhood of the origin and all signals in closed-loop systems are bounded.

**Acknowledgements** This work was supported by the NSFC (61603204), the Shandong Province Outstanding Youth Fund (ZR2018JL020), the Natural Science Foundation of Shandong Province (ZR2016FP03) and the Qingdao Application Basic Research Project (16-5-1-22-jch).

# References

1. W. Huo, *Robotic Dynamics Control*. Higher Education Press (2004)
2. A.R. Angeles, H. Nijmeijer, Mutual synchronization of robots via estimated state feedback: a cooperative approach. IEEE Trans. Control Syst. Technol. **12**(4), 542–554 (2004)
3. E. Nuno, R. Ortega, L. Basanez, D. Hill, Synchronization of networks of nonidentical Euler-CLagrange systems with uncertain parameters and communication delays. IEEE Trans. Autom. Control **56**(4), 935–941 (2011)
4. H. Wang, Consensus of networked mechanical systems with communication delays: a unified framework. IEEE Trans. Autom. Control **59**(6), 1571–1576 (2014)
5. L. Zhao, J.P. Yu, H.S. Yu, Adaptive finite-time attitude tracking control for spacecraft with disturbances. IEEE Trans Aero Elec Sys **54**(3), 1297–1305 (2018)

6. Y.M. Jia, Robust control with decoupling performance for steering and traction of 4WS vehicles under velocity-varying motion. IEEE Trans Control Syst Tech **8**(3), 554–569 (2003)
7. Y.M. Jia, Alternative proofs for improved LMI representations for the analysis and the design of continuous-time systems with polytopic type uncertainty: a predictive approach. IEEE Trans. Autom. Control **48**(8), 1413–1416 (2003)
8. M. Krstic, I. Kanellakopoulos, P. Kokotovic, *Nonlinear and Adaptive Control Design* (Wiley, NewYork, 1995)
9. J.A. Farrell, M. Polycarpou, M. Sharma, W.J. Dong, Command filtered backstepping. IEEE Trans. Autom. Control **54**(6), 1391–1395 (2009)
10. W.J. Dong, J.A. Farrell, M. Polycarpou, V. Djapic, M. Sharma, Command filtered adaptive backstepping. IEEE Trans. Control Syst. Technol. **20**(3), 566–580 (2012)
11. L. Zhao, J.P. Yu, C. Lin, Adaptive neural consensus tracking for nonlinear multi-agent systems using finite-time command filtered backstepping. IEEE Trans. Syst. Man Cybernet. Syst. (2017). https://doi.org/10.1109/TSMC.2017.2743696
12. J.P. Yu, P. Shi, W.J. Dong, C. Lin, Adaptive fuzzy control of nonlinear systems with unknown dead zones based on command filtering. IEEE Trans. Fuzzy Syst. **26**(1), 46–55 (2018)

# Prediction Model of Steel Mechanical Properties Based on Integrated KPLS

**Ling Wang, Hui Zhu and Ruixia Huang**

**Abstract** In this paper, an integrated KPLS (Kernel Partial Least Square) prediction model for steel mechanical property is proposed. To eliminate the heterogeneity among variables in the hot rolling process, the KFA (Kernel Factor Analysis) is used to obtain the latent factor load vectors. Then the variables with large factor load were clustered into subsets, and the KPLS components are extracted respectively for each subset variable and target variable. Finally, the KPLS results of all subsets were integrated as input, and an integral KPLS prediction model is constructed with the target variables. An application study was carried out on the real production data of a steel-making plant. The experimental result shows that the precision of the presented method is greatly improved.

**Keywords** Kernel factor analysis · Kernel partial least square
Steel mechanical property

## 1 Introduction

The mechanical property of steel materials [1] is an important index to evaluate the steel quantity of steel products, which is influenced by the microstructure, chemical content and all kind of factors in manufacturing. In order to optimize the hot rolling process, it is of great significance to build an exact model to predict the mechanical property. So far, a series of prediction models for the mechanical properties have been developed such as mechanism-based model [1], Multivariate statistical regression model [2], Principal Component Analysis and Artificial Neural Network [3] and Support Vector Regression Model [4]. However, the correlation between process variables and mechanical property is often inconsistent with diverse data subset, operating on very different time scales. For example, the process variable tempera-

L. Wang (✉) · H. Zhu · R. Huang
School of Automation & Electrical Engineering, University of Science and Technology, Beijing, China
e-mail: lingwang@ustb.edu.cn



Y. Jia et al. (eds.), *Proceedings of 2018 Chinese Intelligent Systems Conference*,
Lecture Notes in Electrical Engineering 528,
https://doi.org/10.1007/978-981-13-2288-4_84

ture may be strongly correlated with mechanical property in data subset D1, while weakly correlated with mechanical property in data subset D2. In consequence, the correlation patterns in all kind of factors during the hot rolling process are unfixed on every time scale. To implement more accurate predictions, the paper has presented a new prediction strategy which could extract the key features of hot rolling process. In order to solve the aforementioned problems, the kernel techniques are introduced into the formulation of Factor Analysis (FA) in this paper, thereby achieving a nonlinear generalization called kernel Factor Analysis (KFA). The developed KFA is a nonlinear model which could capture higher order statistics and could deal with non-Gaussian distribution in feature space. And then the feature selection is investigated according to the heterogeneity of feature variables.

In [5, 6], the nonlinear kernel-based partial least squares (KPLS) methodology and support vector regression were presented, which was urged by the results of kernel-based learning. Generally speaking, partial least square (PLS) method [7, 8] creates score vectors (components, latent vectors) by using the existing covariances between input and output variables while keeping most of the variance of both data sets.

The innovation point of this paper is its combination of KFA clustering with KPLS regression for mechanical properties forecasting. In the first stage, the KFA is used to get the latent factor load vectors, and the variables with large factor load are clustered into sub-clusters to eliminate heterogeneity; In the second stage, components variable of each sub-cluster and object variable are extracted with the KPLS; In the third stage, the KPLS components of all sub-clusters are integrated as independent variables, combining with the object variables, the KPLS regression model is built. We applied this method to predict the mechanical properties of steel process, and simulation results verified the superiority of this method.

The remaining structure of the paper is as follows. In Sect. 2, the variable-clustering method based on KFA is described. Section 3 presents the integrated KPLS prediction model. We analyze the hot rolling process and compare our algorithm with others in Sect. 4. The last part is conclusion.

## 2 Variable Clustering Method Based on KFA

Factor analysis (FA) [9] is a statistical technique, which attempts to explain covariation among a set of observed variables by introducing unobserved variables that are presumed to be causes of the observed variables. However, FA is a linear model based on the second-order statistics. In other words, the processed data needs to satisfy the Gaussian distributions. To address the fore-mentioned issues, the kernel factor analysis method is adopted.

Compared with the factor analysis method, the basic idea of KFA [10] is to map input $x$ into high-dimensional feature space $F$ by the non-linear function $\phi(x)$. Then, the linear factor analysis algorithm is adopted in the feature space. The non-linear mapping is implemented on the original space by using the kernel function inner product operation without focusing on the specific mapping form.

Let us first begin with $x_i(x_i \in R^d, i = 1, 2, \ldots, n)$ in the input space. For the kernel version of FA, a nonlinear mapping of the centered input data in the feature space is first defined as $\phi : R^d \rightarrow F$. Using the covariance matrix,

$$C = \frac{1}{n}\sum_{i=1}^{n}\phi(x_i)\phi(x_j)^T \tag{1}$$

In feature space, the KFA transformation can be achieved by the eigenvalue decomposition of the covariance matrix:

$$CV = \lambda V \tag{2}$$

where, $V$ is the eigenvector of covariance matrix $C$ and $\lambda$ is the corresponding eigenvalue.

Also, the coefficients $\alpha_1, \alpha_2, \ldots, \alpha_n$ can be related to eigenvector $V$, which can be represented linearly as

$$V = \sum_{i=1}^{n}\alpha_i\phi(x_i) \tag{3}$$

Now we take the inner product of the both sides of Eq. (2) with $\phi(x_j)$, the equivalent relation can be written as

$$(\phi(x_j), CV) = \lambda(\phi(x_j), V) j = 1, 2, \ldots, n \tag{4}$$

Combination of Eqs. (1), (3) and (4) yields

$$n\lambda\alpha = K\alpha \tag{5}$$

where $K = K(x_i, x_j) = \phi(x_i)\phi(x_j)^T, i, j = 1, 2, \ldots, n.$

The eigenvalue $\lambda$ and correlation coefficient $\alpha$ of $C$ can be obtained by solving the Eq. (5).

To extract the $k$th ($k = 1, 2, \ldots, d$) nonlinear principal component, the projection of $\phi(x)$ on the $k$th feature vector $V^k$ need to be calculated. In order to satisfy $(V^k, V^k) = 1$, the appropriate $\alpha^k$ is adopted to meet the formula $n\lambda(\alpha^k, \alpha^k) = 1$. And the projection of $\phi(x)$ on $V^k$ can be expressed as

$$g_k(x) = (V^k, \phi(x)) = \sum_{i=1}^{n}\alpha^k K(x_i, x) \tag{6}$$

where, the vectors $(g_1(x), g_2(x), \ldots, g_k(x))$ denote the nonlinear principal components of sample $x$.

According to the accumulative variance contributing rate $\sum_{i=1}^{m} \lambda_i \Big/ \sum_{i=1}^{d} \lambda_i \geq 80\%$, the first $m$ principal components are generally extracted as the common factors. Moreover, the loading matrix is obtained.

$$P = (\sqrt{\lambda_1} V^1, \sqrt{\lambda_2} V^2, \ldots, \sqrt{\lambda_m} V^m) \tag{7}$$

Based on the theoretic explanation above, the variable with larger loading on common factor can be clustered as a subset, each variable only belongs to a subset.

## 3 Mechanical Performance Prediction Model Based on Integrated Kernel Partial Least Squares

In this section, a new integrated KPLS model is proposed, which can capture the heterogeneity relations between diverse input variables and output variables. First, based on the KFA clustering, the variable set are divided into several subsets, which avoid the heterogeneity of the variables and assure the correlation between the process variable and the target variable in each subset is consistent. Then, the KPLS model is adopted to each variable subset to extract the partial least squares components, which are integrated as the input of the whole KPLS to establish the prediction model.

The prediction model based on the integrated KPLS method first uses the kernel factor analysis clustering method to divide the variable set into several subsets. After the feature extraction process of each subset using the KPLS method, all PLS components are integrated as the input of the KPLS to establish the prediction model. The model not only has higher information synthesis and feature extraction capability, but also can reduce the difficulty of storage and calculation of the kernel function matrix and reduce the memory space occupied. The structure of the model is shown in Fig. 1.

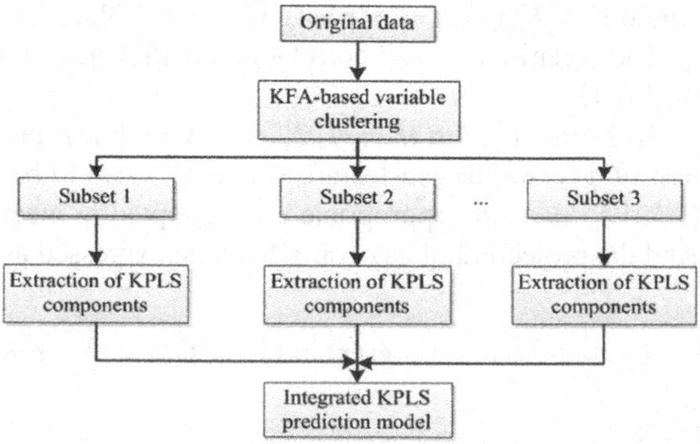


**Fig. 1** Integrated KPLS prediction model

Assume that all input variables affecting the mechanical properties of steel are divided into $p$ subsets $X = (X_1, X_2, \ldots, X_p)$ based on the kernel factor clustering method.

Where, $X_i(i = 1, 2, \ldots, p)$ is the input sample set of size $n \times p_i$, $p_i$ denotes the number of input variables. For each subset, a KPLS regression model of output $Y$ and input $X_i$ is established. Assuming that the principal components extracted from $X_i$ and $Y$ are expressed as $t$ and $u$, respectively.

The KPLS algorithm is as follows:

Step 1: $K^0 = \phi(X_i)\phi(X_i)^T$, after processing, we obtained

$$K = \left(I - \frac{1}{n}\mathbf{1}\mathbf{1}^T\right)K^0\left(I - \frac{1}{n}\mathbf{1}\mathbf{1}^T\right) \tag{8}$$

where,$I$ denotes the Unit Matrix and 1 represents the all-one vector.

Step 2: $t = KY, t \leftarrow t/\|t\|$;

Step 3: $u = YY^T t, u \leftarrow u/\|u\|$;

Step 4: $K \leftarrow K - tt^T K, Y \leftarrow Y - tt^T Y$;

Step 5: Repeat Steps 2–4 until convergence is obtained.

Assuming the corresponding KPLS components extracted from each sub-block are $t_{i1}, t_{i2}, \ldots, t_{im_i}(i = 1, 2, \ldots, p)$, where $m_i$ denotes the number of KPLS components extracted from $X_i$. $t_{i1}, t_{i2}, \ldots, t_{im_i}$ and $Y$ are performed KPLS regression, where $t_{i1}, t_{i2}, \ldots, t_{im_i}$ denotes the KPLS components extracted from $X_i(i = 1, 2, \ldots, p)$.

Based on the cross-validity, final extracted pivot components $t$, i.e. $t_1, t_2, \ldots, t_m$, is obtained. After combining $t$ with the output principal component $u$ into the principal element matrix $T$ and $U$, we can get a simplified KPLS regression model in matrix form which can be expressed as

$$\hat{Y} = \phi B \tag{9}$$

where, the regression coefficient matrix behaves as

$$B = \phi^T U(T^T KU)^{-1} T^T Y \tag{10}$$

For the training dataset, KPLS regression model can be rewritten as

$$\hat{Y} = \phi B = KU(T^T KU)^{-1} T^T Y = TT^t Y \tag{11}$$

When predicted with (11), we have

$$\hat{Y}^{test} = \phi B = K^{test} U(T^T K^{train} U)^{-1} T^T Y \tag{12}$$

Since the output in this article is one-dimensional, the KPLS regression can be expressed as

$$f(x) = \sum_{i=1}^{n} d_i K(x, x_i) \tag{13}$$

where $d_i$ is the $i$th principal element of $d = U(T^T K^{train} U)^{-1} T^T Y$.

# 4 Prediction Model of Mechanical Properties of Steel

The process of hot rolling includes the following four areas: heating area, rough rolling area, finishing rolling area and coiling area [11], and seven procedures: slab preparation, slab heating, roughing, finishing, cooling after rolling, coiling and finishing operation. Each procedure has an impact on the final performance of hot-rolled products.

Due to the irregular data distribution throughout the hot rolling process, it leads to diverse correlation between target variables and process variables for diverse data subsets on different time scale. In order to capture the characteristics of hot rolling data for making a good prediction, we developed a new forecasting strategy using clustering algorithms and kernel regressors. Although traditional clustering algorithm is widely adopted to partition variable, it cannot be applied to partition variables with different correlations for different data subsets. Therefore, in this paper, based on the KFA, the original process variables are transformed into the potential common factors to reflect the heterogeneity of the process variables. Then the clustering analysis is used to obtain variable subsets with a consistent correlation to construct the KPLS model.

## *4.1 Pre-processing*

In this study, we selected 1000 data samples for different batches of steels, in which the first 700 samples were used to create a training data set and the remaining 300 samples were used as a test set. There are 29 input process variables, which mainly contain original chemical components, some alloying elements, and gas composition. The process parameters are consisted of the beginning and ending temperature of finishing rolling, the thickness of steel coil, the coiling average temperature, finish rolling temperature and reduction rate. Furthermore, only the elongation is chosen as the output variable. Taking into account the huge differences in magnitude of different variables, we normalize the original input and output to the [0, 1] by using the z-score methods.

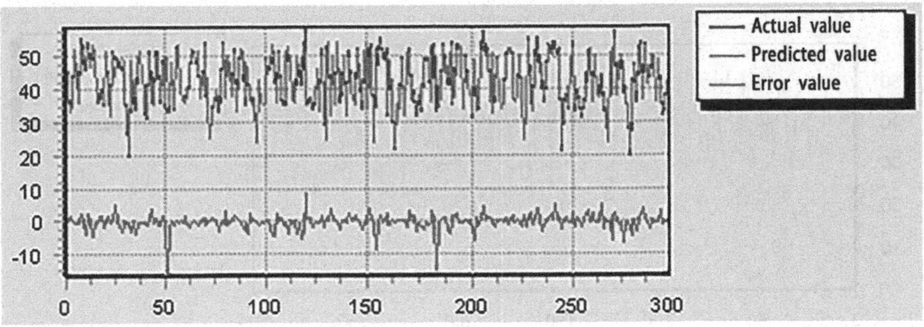


**Fig. 2** Prediction result with KPLS prediction model

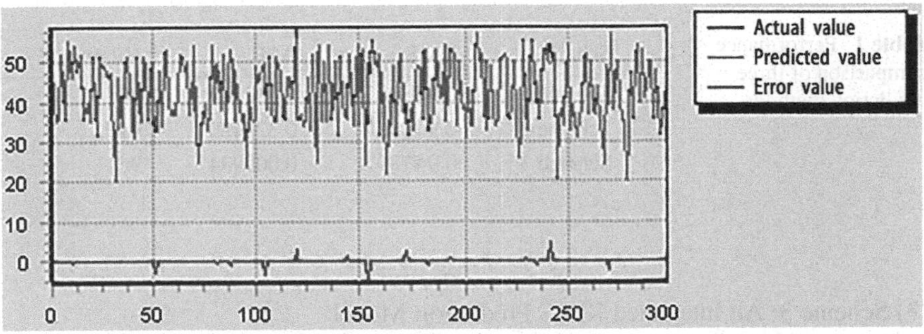


**Fig. 3** Prediction results based on variable clustering and KPLS

## 4.2 *Experimental Results*

In order to test the efficiency and effectiveness of the proposed algorithm, we compared the experimental results with three prediction schemes.

(1) Scheme 1: Kernel Partial Least Squares Regression Model
All input variables are applied as input to the model, and only kernel partial least-squares regression is adopted to predict. The parameters of KPLS are set as follows: the number of principal components is 7, and Gaussian kernel function is selected with $\sigma = 1$. The prediction result of the model is shown in Fig. 2.

(2) Scheme 2: Prediction Model Based on Variable Clustering and KPLS
Hierarchical Clustering Method [12] is used to analyze the correlation between the original input variables, and Euclidean distance is applied to evaluate the validity of clustering results and all variables are divided into 4 categories. Then kernel partial least squares component analysis is applied on each subclass, and all the kernel partial least square components are integrated as input to the kernel partial least squares regression model for training. During the test, the samples are directly classified, and kernel partial least square components (contribution ratio is higher than 80%) are extracted for test sample. The predicted result is shown in Fig. 3.

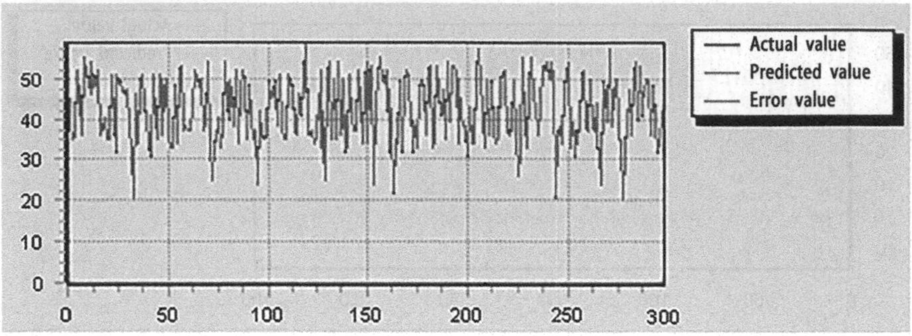


**Fig. 4** Prediction result of integrated KPLS

**Table 1** Performance comparison of three prediction methods

| Scheme | RMSE | ASE | Hit rate (%) |
|---|---|---|---|
| Scheme 1 | 0.988083 | 4.832063 | 80 |
| Scheme 2 | 0.978997 | 0.325480 | 91 |
| Scheme 3 | 0.947892 | 0.009984 | 96 |

(3) Scheme 3: An Integrated KPLS Prediction Model
Based on the analysis of the correlation between the 30 original variables in Scheme 2, for the target variable elongation, among the 29 input variables, there are different correlations in different subsets. Therefore, it is more suitable to adopt the KFA method to reflect the heterogeneity between variables. According to the nonlinear covariance matrix of the sample, the principal component method is used to extract the initial common factor in the feature space to obtain the initial factor load matrix, and the maximum variance orthogonal rotation method is adopted to rotate the initial factor load matrix to get the rotated factor load matrix. In this paper, we chose Gaussian kernel as the kernel function. After orthogonal rotation of factors, the variables with large load factors of each factor are clustered into a group, and the original input variables can be divided into four groups. The variance and contribution rate of the first partial least square component of each group variables has reached up to 80%, therefore, the first partial least squares component can be extracted. Then, the kernel partial least squares components of each group are integrated as input to establish the prediction model. The prediction result is shown in Fig. 4.

From Figs. 2, 3 and 4, it can be seen that our proposed model in this paper has the best prediction result, and the trend fits the real situation well. Furthermore, the measures for generalization performance evaluation are listed in Table 1, which are defined as follows:

(1) Root Mean Square Error (RMSE)

$$RMSE = \sqrt{\frac{1}{n}\sum_{i=1}^{n}(y_i - \hat{y}_i)^2} \tag{14}$$

(2) Asymptote Standard Error (ASE)

$$ASE = \sum_{i=1}^{n}(y_i - \hat{y}_i)^2 \bigg/ n \tag{15}$$

where $n$ represents the number of test samples, $\hat{y}_i$ and $y_i$ denote the prediction value and real value, respectively.

(3) Hit rate. The deviation between the prediction value and the real value is less than 2, which can be looked as the prediction hit. Hit rate refers to the ratio of the number of hits to the total number of samples.

As shown in Table 1, the RMSE and ASE of our proposed model are smaller than that of the other methods. In addition, the hit rate reaches up to 96%, which shows the prediction accuracy is obviously improved.

## 5 Conclusion

To solve the problems of strongly nonlinear, multivariate correlation and heterogeneity in hot rolling process, an integrated KPLS modeling method for the prediction of steel mechanical property is presented. Based on the variable clustering method with KFA, the problem of heterogeneity between process variables and target variables is effectively solved, which makes the correlation between independent variable and the target variable in each variable group consistent. On this basis, computational complexity of kernel matrix is further considered, and an integrated KPLS regression model is built to reduce the computational complexity of kernel matrix. In this paper, we first extract the KPLS components for each variable group, and then KPLS prediction model is constructed by comprehensively using all KPLS components and the target variables. Compared with various methods, it shows that our proposed model can effectively improve the prediction effect of the mechanical properties of steel.

**Acknowledgements** This research work was supported by the National Natural Science Foundation of China (Grant No. 61572073), National Key R&D Program of China (NO. 2017YFB0306403) and the Fundamental Research Funds for the China Central Universities of USTB (FRF-BD-17-002A).

## References

1. D. Hodouin, Methods for automatic control, observation, and optimization in mineral processing plants. J. Proc. Control **21**(2), 211–225 (2011)
2. Y. Zongshen, Y. Zexi, L. Shiqi, et al. *Quantitative relationship of composition, residual elements and Properties of Steel* (Metallurgical Industry Press, 2001)
3. Y. Sun, Z. Weidong, Z. Yongqing et al., Modeling the correlation of composition-processing-property for TC11 titanium alloy based on principal component analysis and artificial neural network. J. Mater. Eng. Perform. **21**(11), 2231–2237 (2012)
4. T. Jiali, C. Qiuru, L. Yijun, Prediction of Material Mechanical Properties with Support Vector Machine, in *International Conference on Machine Vision and Human-Machine Interface* (IEEE, 2010), pp. 592–595
5. R. Rosipal, Kernel partial least squares for nonlinear regression and discrimination. Neural Netw. World **13**(3), 291–300 (2002)
6. R. Rosipal, L.J. Trejo, *Kernel partial least squares regression in reproducing kernel hilbert space*. JMLR.org (2002)
7. R. Rosipal, N. Krämer, Overview and recent advances in partial least squares. Subspace Latent Struct. Feature Selection Techniques **3940**, 34–51 (2006)
8. G. Mateos-Aparicio, Partial least squares (PLS) methods: origins, evolution, and application to social sciences. Commun. Stat. **40**(13), 2305–2317 (2011)
9. K.M. Hayden, R.N. Jones, C. Zimmer et al., Factor structure of the national Alzheimer's coordinating Centers uniform dataset neuropsychological battery: an evaluation of invariance between and within groups over time. Alzheimer Dis. Assoc. Disord. **25**(2), 128–137 (2011)
10. X. Guoen, S. Peiji, Factor analysis algorithm with Mercer Kernel, in *International Symposium on Intelligent Information Technology and Security Informatics* (IEEE, 2009), pp. 202–205
11. W. Ling, F. Dongmei, L. Qing, Samples selection based on SVR for prediction of steel mechanical property, in *International Conference on Intelligent System Design and Engineering Application* (IEEE Computer Society, 2012), pp. 909–912
12. I. Baruchi, D. Grossman, V. Volman et al., Functional holography analysis: Simplifying the complexity of dynamical networks. Chaos Interdiscip. J. Nonlinear Sci. **16**(1), 86–92 (2006)

Printed in the United States
By Bookmasters